POWER PLANT ENGINEERING

FOR

[BE (Mech.), B. Tech. (Mech.); UPSC (Engineering Services), AMIE etc.]

(In SI Units)

By

Dr. D.K. Chavan
Professor,
Maharashtra Institute of Technology
M.I.T., Pune – 38

Prof. G.K. Pathak
Sr. Faculty Member,
Mechanical Engineering Department,
Maharashtra Institute of Technology
M.I.T., Pune – 38

STANDARD BOOK HOUSE

Unit of: **RAJSONS PUBLICATIONS PVT. LTD.**

1705-A, Nai Sarak, PB. No. 1074, Delhi-110006 Ph.: +91-(011)-23265506
Show Room: 4262/3, First Lane, G-Floor, Gali Punjabian, Ansari Road, Darya Ganj,
New Delhi-110002 Ph.: +91-(011) 43751128 Tele Fax : +91-(011)43551185,
Fax: +91-(011)-23250212

E-mail: sbh10@hotmail.com, www.standardbookhouse.in

Published by:
RAJINDER KUMAR JAIN
Standard Book House
Unit of: Rajsons Publications Pvt. Ltd.
1705-A, Nai Sarak, Delhi - 110006
Post Box: 1074
Ph.: +91-(011)-23265506 Fax: +91-(011)-23250212

Showroom:
4262/3. First Lane, G-Floor, Gali Punjabian,
Ansari Road, Darya Ganj,
New Delhi-110002
Ph.: +91-(011)-43751128, +91-(011)-43551185
E-mail: sbhl0@ hotmail.com
Web: www.standardbookhouse.in

First Published : 2012
Second Edition : 2017

Price: **₹ 640.00**

ISBN: 978-81-89401-42-9

Typeset by:
C.S.M.S. Computers, Delhi.

Printed by:
R.K. Print Media Company, New Delhi

DEDICATED
To

Late Shri Kishanrao L Chavan
Late Smt. Umabai Kishanrao Chavan
Late Shri Ramkrishna I Zad

Preface

It gives us great pleasure to present a book on "Power Plant Engineering", for the Final year Mechanical Engineering students from various Universities.

In this book we will study Thermal Power Plant, Hydro-electric Power Plant, Nuclear Power Plant, Non-conventional Power Generation, Power Plant economics in detail.

Every care has been taken to present the matter in precise and simple language. Self explanatory figures are included to enable the students to draw the same in the exams very easily.

From a long teaching experience, it has been observed that students commit mistakes in the units, as such, for every step in the solution of problems, units have been made clear.

To develop more confidence in students appropriate questions from various universities are given at the end of each chapter. In the entire book SI system of units is used.

All necessary care has been taken to avoid mistakes/misprints in the book. However it is quite likely that some mistakes/misprints might have been left unnoticed. Such mistakes and misprints of the book if brought to the notice will be gratefully acknowledged. Also any suggestions to improve the usefulness of the book will be gratefully considered and appreciate.

We express our sincere thanks to,

(*i*) Prof. Dr. Mangesh Karad, Executive Director, MIT, Pune.

(*ii*) Prof. Dr. S.P. Kallurkar, Principal, MMCOE, Pune.

(*iii*) Prof. Dr. G.S. Tasgamkar Zeal, COE, Pune.

(*iv*) Prof. Raghubir Singh, Principal, Wingss Aeronautical College, Pune.

(*v*) Prof. Molkeri. M., HOD Mech. Engg. Deptt. REC Bhalki, Karnataka.

(*vi*) Prof. N.K. Joshi, HOD Mech. Engg. Deptt. Tolain Maritime Institute, Pune.

(*vii*) Prof. A.D. Desai, HOD; Mech. Engg. Deptt. PES, Modern COE, Pune.

(*viii*) Prof. S.M. Nakate, HOD; Mech. Engg. Deptt. MMIT, Pune.

We also express our sincere thanks to entire staff of Standard Books House New Delhi, those have taken sincere and commendable efforts to bring out this edition in very short span of time.

Jan. 2011

Prof. D.K. Chavan
Prof. G.K. Pathak

Contents

PART 1

PART 2

PART 3

PART 4

PART 5

PART 6

SYMBOLS USED IN THIS BOOK

A = Area
a = Acceleration
C = Specific heat, velocity
^{o}C = Degree centigrade
C_p = Specific heat at constant pressure
C_v = Specific heat at constant volume
F = Force
g = Acceleration due to gravity
g_c = Constant of proportionally in Newton's law
h = Specific enthalpy
H = Enthalpy
J = Mechanical equivalent of heat or joules equivalent
K = Degree Kelvin
kg = Kilograms
KE = Kinetic Energy
L = Length
m = Metre
M = Molecular weight
n = Number of moles, Polytropic index
N = Revolutions per minute (RPM)
P = Pressure
PE = Potential Energy
q, Q = Heat transfer
R = Specific gas constant
$\overline{R}$ = Universal gas constant
sS = Entropy
t = Temperature celsius
T = Absolute temperature
u, U = Internal energy
v, V = Volume
w, W = Work
x = Dryness fraction
Z = Elevation above datum

Greek Symbols

γ = Specific heat ratio
Δ = Change in a property
δ = Small change in a property
η = Efficiency
θ = Temperature
ρ = Density

SI Units and Conversion Factors

I. Base Units

Quantity	*Name*	*Symbol used*
Length	Metre	m
Mass	Kilogram	kg
Time	Second	s
Amount of substance	Mole	Mol

II. Derived Units

Quantity	*Name*	*Symbol used*
Area	Square metre	m^2
Volume	Cubic metre	m^3
Speed or velocity	metre/second	m/s
Acceleration per second	metre per second	m/s^2
Density	Kilogram per cubic metre	kg/m^3
Specific volume	Cubic metre per kilogram	m^3/kg
Force	Newton	N
Pressure	Force per unit area, Pascal	N/m^2
Work	Joule	J = N – m

III. CONVERSION FACTORS

1. **Force**

$$1 \text{ Newton} = \frac{\text{kg-m}}{\text{sec}^2} = 0.012 \text{ kg}_f$$

$1 \text{ kg}_f = 9.81 \text{ N}$

2. **Pressure**

1 bar = 750.06 mm Hg

= 0.9869 atm.

$= 10^5$ N/m^2

1 N/m^2 = 1 Pascal

$= 10^{-5}$ bar

1 atm = 760 mm Hg

= 1.01325 bar

$= 1.01325 \times 10^5$ N/m^2

3. Work, Energy or Heat

1 Joule = 1 Newton-metre

= 1 Watt-sec

$= 2.7778 \times 10^{-7}$ kWh

= 0.239 cal

1 cal = 4.184 Joule

1 kWh = 3600 Joule

= 860.42 Kcal

$$1\ kg_f - m = \frac{1}{427}\ Kcal$$

= 9.81 Joules

4. Power

1 Watt = 1 Joules/sec = 0.86 Kcal/h

1 h.p. = 75 kg_f.m/sec = 735.3 Watt

1 KW = 1000 Watts = 860 Kcal/h

5. Specific Heat

1 Kcal/kg–K = 0.4184 Joules/kg–K

= 4.184 KJ/kg–K

6. Thermal Conductivity

1 Watt/m–K = 0.8598 Kcal/h–m–°C

1 Kcal/h–m.°C = 1.16123 Joules/s-m-K

7. Heat Transfer Coefficient

1 Watt/m^2–K = 0.86 Kcal/m^2–h–°C

1 Kcal/m^2–h–°C = 1.163 Watt/m^2–K

8. Other Constants

(a) Volume of g_o = 9.81 kg–m/kg_f–sec^2

= 1 kg–m/N–sec^2

(b) Universal gas constant $\bar{R}$

$\bar{R}$ = 848 kg_f–m/kg.mole–°K

= 8314 J/kg–mol–°K

(c) Specific gas constant (R)

R = 287 Joules/kg–K

(d) Specific heats (for air)

C_v = 0.17 Kcal/kg–°K

= 0.718 KJ/kg–°K

C_p = 1.005 KJ/kg–°K

PART 1

Power Plant Engineering

Chapter 1 Thermal Power Plants

1.1 INTRODUCTION

If coal is used to generate electric energy as a fuel, then the steam power plant is known as **Thermal Power Plant.**

The economic development of any country is measured by the per capita energy consumption/generation.

It is to be noted that developing countries like India are required to generate more power, as there is large gap between demand and supply.

Generally energy is available in various forms such as—

(i) Mechanical Energy

(ii) Thermal Energy

(iii) Electrical Energy etc.

Out of these electrical energy is preferred mostly, because of the following advantages :

(i) Less transmission and distribution losses

(ii) Electrical energy can be easily converted into other forms, etc.

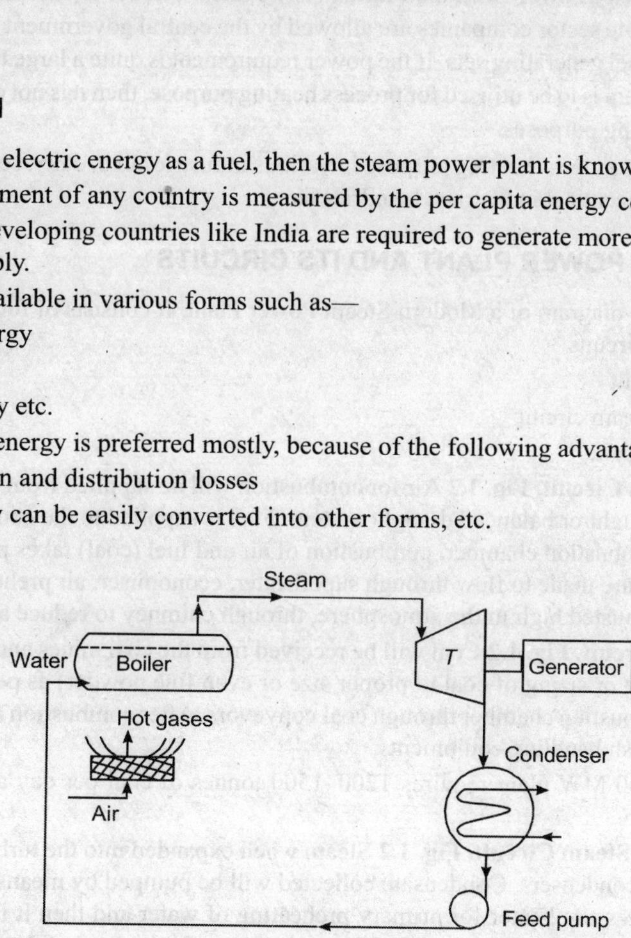

Fig. 1.1

Figure 1.1 shows the simple steam power plant, it mainly consists of (i) Boiler (ii) Turbine, (iii) Condenser, (iv) Feed pump etc. In this fossil fuels such as gas, oil, coal are used. Combustion of air and fuel takes place to generate hot gases. Heat of hot gases is used to heat the water and convert the water into steam. This steam after it is being superheated is expanded in the turbines for generating power. The steam afterwards is exhausted either into the atmosphere or in the condenser. If it is exhausted in the condenser, gets condensed and if the condensate is pure, it is pumped back to the boiler by means of feed pump.

1.2 CLASSIFICATION OF STEAM POWER PLANTS

Steam power plants are broadly classified as following :

(i) Central (or General sale) power plants

(ii) Captive (or Self use) power plants.

(i) Central (or General sale) Power Stations Plants. These are large power plants, generating large amount of power. The power generated is supplied to the grid, from where it is distributed to consumers like domestic and industrial usage.

(ii) Captive (or Self use) Power Stations Plants. As we know there is always more demand of power than supply. So, some of the private sector companies are allowed by the central government to generate power for their own utilisation by using diesel generating sets. If the power requirement is quite a large then it is advisable to go for steam power plant. If the steam is to be utilised for process heating purpose, then it is not condensed in the condenser but is used for process heating purposes.

Thus the steam is used for process heating purpose in various industries like, dairy, sugar industries, paper mills, chemical industries, pharmaceutical industries, laundries etc.

1.3 MODERN STEAM POWER PLANT AND ITS CIRCUITS

Figure 1.2 shows schematic diagram of a Modern Steam Power Plant. It consists of four circuits.

(a) Air and fuel gas circuits

(b) Coal and ash circuit

(c) Feed water and steam circuit

(d) Cooling water circuit.

(a) Air and Fuel Gas Circuit. Fig. 1.2 Air for combustion will be supplied from the atmosphere, because of forced draught/induced draught/or balanced draught system. Fuel is supplied to the combustion chamber from the fuel feed system. In the combustion chamber, combustion of air and fuel (coal) takes place and hot fuel gases are generated. These hot gases are made to flow through superheater, economiser, air preheater. After recovering heat from the flue gases are exhausted high in the atmosphere, through chimney to reduce air pollution.

(b) Coal and Ash Circuit. Fig. 1.2 Coal will be received from the coal mines and is stored in the coal yards. Coal is pulverised (grinding or sizing of coal to proper size or even fine powder) as per requirement of the plant. The coal is supplied to combustion chamber through coal conveyors. After combustion ash produced is collected in ash storage yards through ash handling equipments.

Note that normally a 100 MW plant requires 1200–1500 tonnes of coal per day and a reserve of 15 days is generally maintained.

(c) Feed Water and Steam Circuit. Fig. 1.2 Steam when expanded into the turbines it produces power and then it is condensed in the condensers. Condensate collected will be pumped by means of feed pump through low pressure heater and high pressure heater for primary preheating of water and then it is made to pass through the economiser, where heat of hot gases will be needed for preheating of water and preheated water is supplied to the boiler drum.

In the boiler, water is heated and steam is generated. After it is being superheated, it is expanded in the turbine.

In passing through all these equipments, there is some loss of water and this is compensated by adding make up water. The make up water is treated so as to avoid scale formation on the inner surface of tubes and boiler drum.

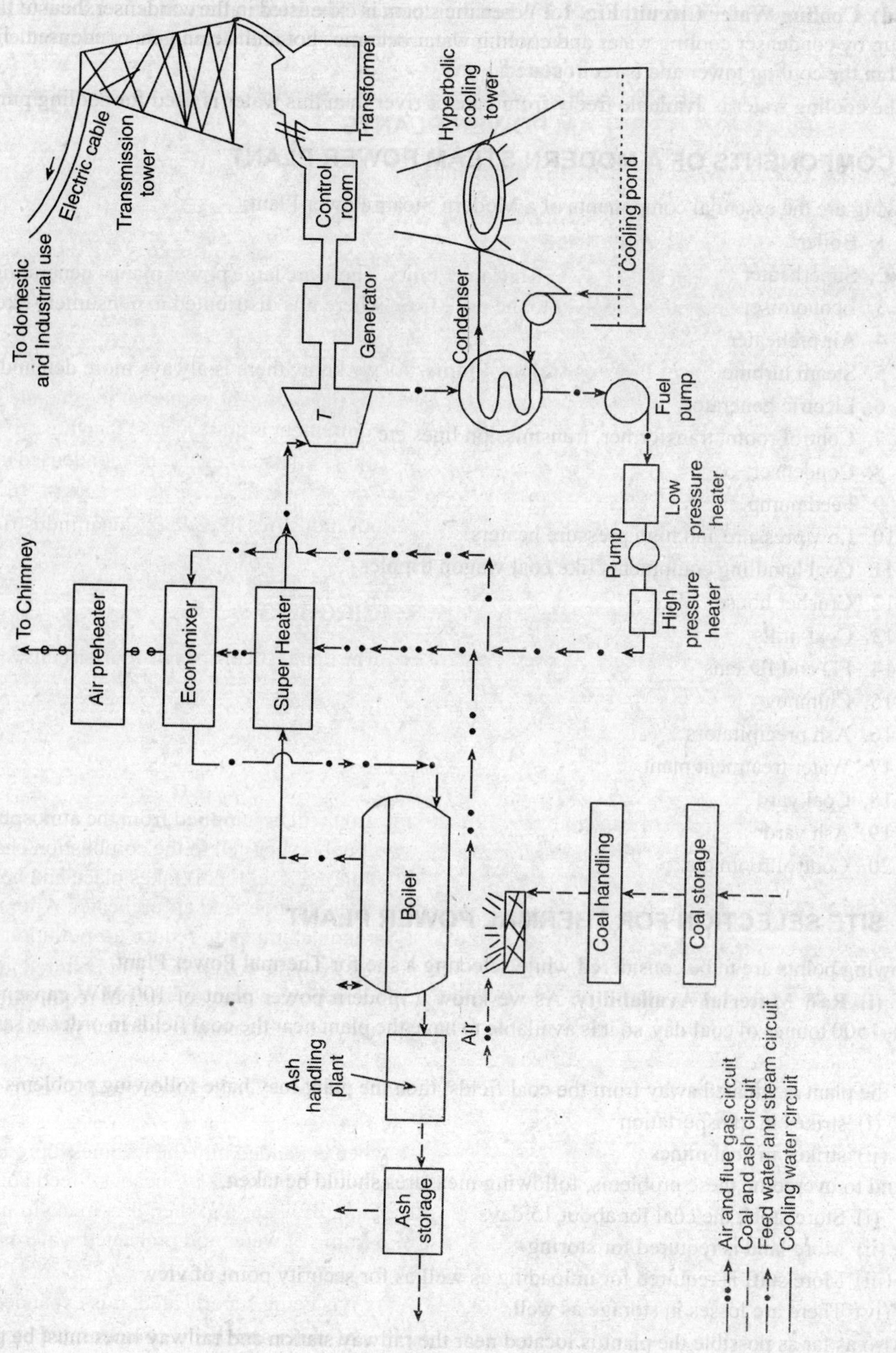

Fig. 1.2 *Modern steam power plant and its circuits*

(d) Cooling Water Circuit. Fig. 1.2 When the steam is exhausted in the condenser, heat of the steam will be taken up by condenser cooling water and cooling water becomes hot and steam gets condensed. This hot water is cooled in the cooling tower and is recirculated.

If the cooling water is available freely from lake or river then this water is used for cooling purpose.

1.4 COMPONENTS OF A MODERN STEAM POWER PLANT

Following are the essential components of a Modern Steam Power Plant:

1. Boiler
2. Superheater
3. Economiser
4. Air preheater
5. Steam turbine
6. Electric generator
7. Control room, transformer, transmission lines etc.
8. Condenser
9. Feed pump
10. Low pressure and high pressure heaters
11. Coal handling equipments like coal wagon trippler
12. Crusher house
13. Coal mills
14. FD and ID fans
15. Chimney
16. Ash precipitators
17. Water treatment plant
18. Coal yard
19. Ash yard
20. Control room etc.

1.5 SITE SELECTION FOR THERMAL POWER PLANT

Following points are to be considered while selecting a site for Thermal Power Plant.

(i) Raw Material Availability. As we know a modern power plant of 100 MW capacity requires about 1200–1500 tonnes of coal/day, so it is available to have the plant near the coal fields in order to save transportation cost.

If the plant is situated away from the coal fields, then the plant may have following problems due to,

(i) strikes in transportation
(ii) strikes in coal mines

And to overcome these problems, following measures should be taken,

(i) Store/stock the coal for about 15 days
(ii) More land is required for storing
(iii) More staff is required for unloading as well as for security point of view
(iv) There are losses in storage as well.

Also as far as possible the plant is located near the railway station and railway lines must be provided upto the plant for the transportation of coal upto the plant.

(ii) Availability of Water. In thermal power plants, large amount of water is converted into steam for power generation. This steam is again condensed in the condensers. For condensing the steam, large amount of cooling water is required. Hence the plant must be located near the water sources such as river, lake or pond. Also we have to add about 2–5% of make up water due to evaporation losses.

(iii) Land. Considerably large area of land is required for the power plant. The site selected should have good load bearing capacity, since it has to withstand dead load of plant and vibrations generated due to rotary machines like pumps, turbines etc.

(iv) Cost of Land. Cost of land should be less as large amount of land is required for the power plant. Hence land away from city places is to be selected as the cost is less, also from pollution point of view.

(v) Ease of Transport. Site selected for the plant should be such that, there is no any difficulty in the transportation of raw material and machinery for the plant by road and as well as by rail.

(vi) Availability of Labour. For the construction, erection, commissioning of plant, large amount of man power is required and it should be available locally and at the reasonable wages.

(vii) Ash Disposable Facilities. While selecting the site, ash disposal facility is to be considered, since ash which is of fine size and when hot it is highly corrosive.

(viii) Load Centre. Power station must be located as far as possible near load to which the power is supplied. However power station cannot be located near all the loads. Since the centre of gravity of the load is determined with respect to the reference axis. This centre of gravity of load centre and power station should be as far as possible near the load centre to reduce transmission cost and losses. It is to be taken into consideration the future development during selection of load centre.

(ix) Air Fields. Considering the aerial warfare, the power station should be located away from cities and industrial areas.

(x) Town Planning Department. Concent must be taken fromTown Planning Department.

(xi) Future Extension. As the demand for electricity is increasing day by day. It is advisable to make provision for extension of existing plant. It is always preferable to do the extension of existing plant, instead of going for totally new plant.

A site selected may not necessarily satisfy all the factors mentioned above but however a suitable compromise must be made between the major factors.

(xii) Plant Capacity. Capacity of a plant can be determined by studying the load duration curve and the anticipated future demands. As a minimum requirement, the plant capacity must be equal to atleast the peak load. In case of small loads, it may prove to be economical to instal two units of equipment, each being capable of supplying the maximum demand independently. In the event of failure of one unit or during maintenance etc. atleast one unit can be used to maintain uninterrupted supply of energy.

In case of medium power plants, usually the number of units is more than two with the total installed capacity equal to the maximum demand plus the capacity of two large units.

Large units are generally conservately rated. In the case of steam turbines, there is an overload capacity of 10 to 15% of the rated capacity. With a number of units, peak load can be easily adjusted by overloading some units temporarily.

The load on a profit can be reduced by dropping the supply voltage. Thus a 5% reduction in supply voltage results in similar reduction in the load. As per Indian Electricity Act (see Appendix), an electric supply undertaking has to maintain the voltage within 10% of the declared pressure. So during peak hours the voltage can be reduced within the allowable limits in order to meet the demand without use of additional units. This technique results in saving of the capital cost also.

When it is decided to add a unit to the existing station, the size of the unit is decided on the following considerations:

(a) the expected rate of increase of the maximum demand over the next few years,

(b) suitability of the generator to the existing system regarding pressure, temperature etc.,

(c) the room available for the additional unit,

(d) effect of additional unit on the thermal efficiency of the plant.

(xiii) Fixation of Steam Conditions. The main features to be considered in fixing the initial steam conditions are:

(a) the capital cost of the plant

(b) thermal efficiency obtainable

(c) price of coal

(d) load factor on the station

(e) time available for erection

Modern trend is towards higher pressures and temperatures. The problem of deciding upon the steam pressure and temperature to be adopted for a plant is purely economic one. In considering the merits of increase in temperature and pressure on overall efficiency it will be seen that whilst the former obeys the straight line law, the latter obeys the law of diminishing returns indicating the desirability of adopting the highest possible steam temperature.

Boilers have been built to generate steam at supercritical pressures where the densities of steam and water are equal. As far as the temperature is concerned, it is restricted by the metallurgical considerations of the material. In considering the economy which theoretically may result from the use of higher steam pressures, a reference to the temperature entropy diagram is useful. It will be observed that with increased pressure, heat available for work increases whereas heat lost in condenser remains almost the same. Consumption of steam per kilowatt-hour also decreases with the increased pressure. It would appear that present practice favours the use of steam pressures in the neighbourhood of 60 km/cm^2 for entirely new stations, but that there is a profitable field for higher pressures of the order of 100 kg/cm^2, when the problem is that of increasing thermal efficiency of existing medium pressure

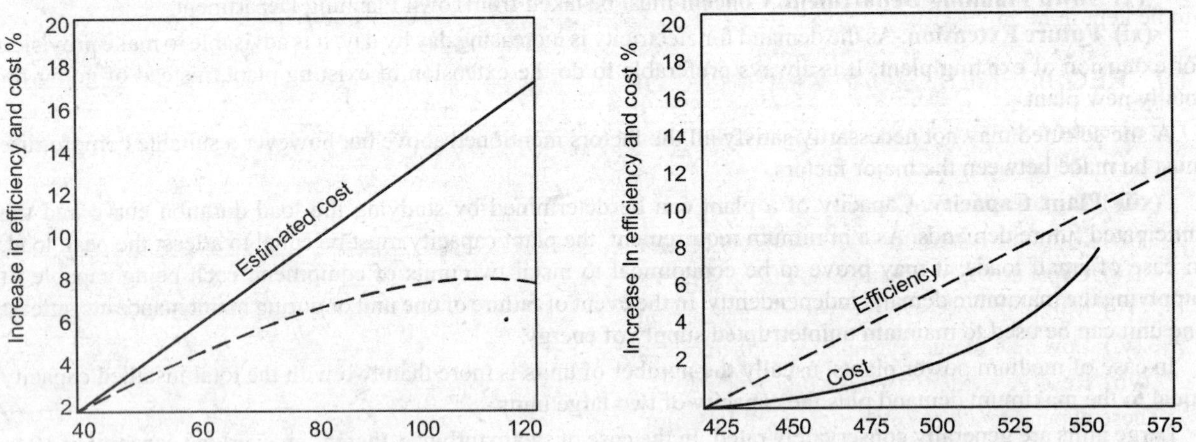

Fig. 1.3

units. Investigations have shown that a 100 MW turboalternator under certain conditions of fuel prices provide adequate return on the capital and running charges when operating at 80 kg/cm^2, 510°C as compared with 40 kg/cm^2, 450°C.

(xiv) Rating of Units. It is usual practice to classify the output of units under either of the following heads:

(a) Economical rating.

(b) Maximum continuous rating.

It is not very necessary that a generator may operate most economically at full load. Present trend is towards economical running at 75–85% of the full load, for the most economical operation.

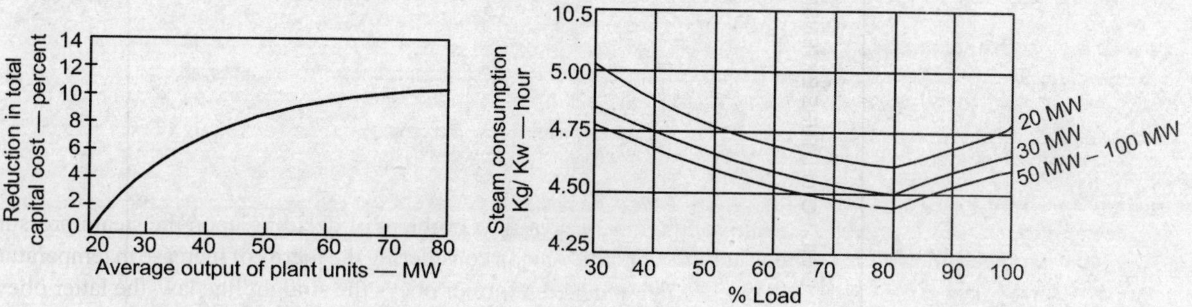

Fig. 1.4

Maximum continuous rating of a generating unit is the maximum load at which it can be run continuously for severel hours. It is normally 10 to 15% less than the maximum capacity of the unit.

1.6 LAYOUT OF THERMAL POWER PLANT

Figure 1.5 shows (given on page 10) the general layout of a Thermal Power Plant. In actual practice there is no fixed layout as such. But the layout should be designed in such a way that there should be less movements of men, machinery and there should be easy access to each equipment for maintenance. Also scope for future expansion is to be kept in mind.

1.7 REQUIREMENTS OR PRINCIPLES OF STEAM POWER STATION DESIGN

The essential requirements of steam power station design are :

(i) Reliability
(ii) Minimum capital cost
(iii) Minimum operating and maintenance cost
(iv) Capacity to meet peak load effectively
(v) Minimum losses of energy in transmission
(vi) Low cost of energy supplied to the consumers
(vii) Reserve capacity to meet future demands.

These depend to a large extent on the following :

1. Simplicity of Design. Simplicity in design and layout in all sections of the station are the desirable features. This involves minimum number of auxiliaries, spacious but economical layout and minimum number of sections. A simple design generally requires reduced capital cost and is easy for maintenance work.

2. Subdivision of Plant and Apparatus. A continuous supply of energy can be assured only by subdividing the plant into smaller units. Total capacity of a plant may be subdivided into four or five units so that during reduced load periods, some of the units may be stopped. This helps in maintenance of the units and also reduces the operation cost. It is not necessary to subdivide the plant into smaller units of equal capacity but the capacity of individual units may be divided on the basis of load curve and economic considerations, if any. The present trend is towards one boiler per turbine and one alternator for each section of feeders with arrangements for interconnection under maintenance and emergency.

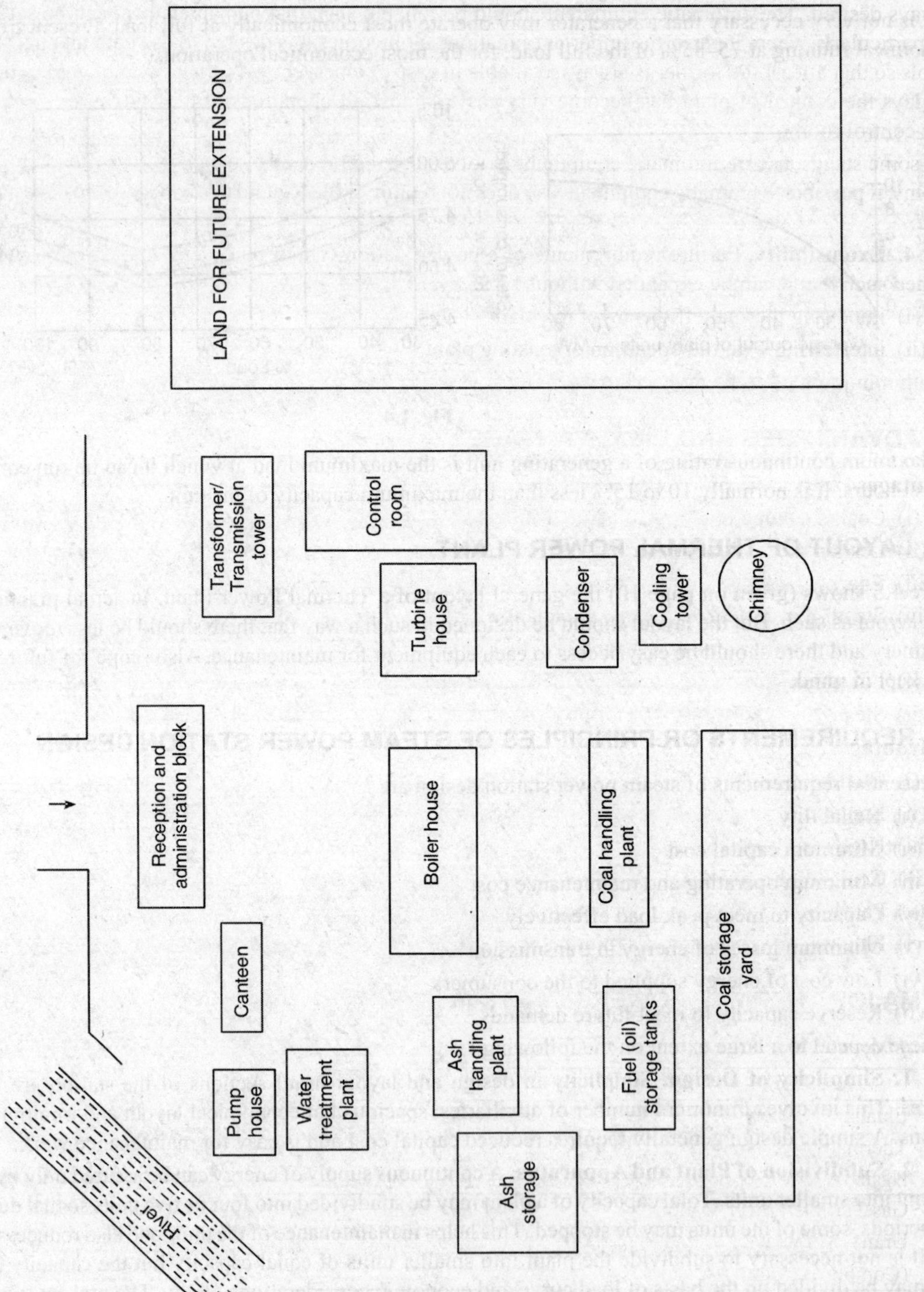

Fig. 1.5 *Layout of thermal power plant.*

3. Use of Automatic Equipment. In order to achieve reliability of service, the use of automatic equipment is always desired. The automatic equipment should be reliable such that human element can be safely dispensed with, particularly where it performs important functions. Sometimes it is desired to use both manual and automatic controls so that atleast the former is always available in case of emergency. With the help of automatic equipments nowadays the control of plant has become very easy. Almost all operations of a plant can be controlled from a single control room.

At some stages, use of automatic equipment involves considerable cost, which may be avoided for the sake of economy, if possible. Automatic equipment also does not require skilled attendants which would have been otherwise a *must*.

4. Extensibility. Furture requirements of a power station cannot be exactly known but a plant should be designed such that it can be expanded without :

(i) impairing thermal efficiency of the plant,

(ii) interferring with the operation of existing plant,

but with minimum of civil engineering work.

1.8 ADVANTAGES AND DISADVANTAGES OF THERMAL POWER PLANT

Advantages

(i) Coal is available in huge quantities and is considerably cheap compared to other fuels.

(ii) Less space is required per MW of power generated.

(iii) Energy transmission costs are less as the plant may be located near the load centre.

(iv) Steam power plant takes up load change if any (i.e., upto 15–25%).

(v) Steam turbines are rotary machines, hence balancing problems are less.

(vi) Power generation cost is less in thermal power plants.

(vii) Steam also can be utilised for process heating purpose in addition to power generation.

Disadvantages

(i) Quality of water required in steam power plants is considerably large.

(ii) Erection, commissioning time required is more compared with diesel and gas turbine power plants.

(iii) High maintenance cost.

(iv) Coal storage and ash disposal requires very large land area.

(v) Part load efficiency is low, above 65–75% load.

(vi) Flue gas or exhaust gases from the plant cause large amount of air pollution.

1.9 MAJOR THERMAL POWER PLANTS IN INDIA

1. (*Coal based Power Plants*)

S.No.	*Name of Plant*	*State*	*Capacity*
1.	Eklahare Nasik Thermal Power Station Nasik	Maharashtra	. 2 × 140 = 280 MW . 3 × 210 = 630 MW 910 MW
2.	Gas Turbine Power Station Uran – Mumbai (Combined cycle)	Maharashtra	

Contd.

Contd.

S.No.	*Name of Plant*	*State*	*Capacity*
	(Waste heat recovery plant)		
	Stage III A		. 2 × 120 = 240 MW
	Stage III B		. 1 × 120 = 120 MW
3.	Parali Extension	Maharashtra	250 MW
4.	Paras Extension	Maharashtra	250 MW
5.	Royalseema	Andhra Pradesh	2 × 420 = 840 MW
6.	Vermagiri	Andhra Pradesh	137 MW
7.	Gowthami	Andhra Pradesh	464 MW
8.	Koonaseema	Andhra Pradesh	445 MW
9.	Kahalgaon	Bihar	2 × 500 = 1000 MW
10.	Korba Extension	Chhattisgarh	500 MW
11.	Sipat Stage II	Chhattisgarh	1000 MW
12.	Raigarh (Private)	Chhattisgarh	1000 MW
13.	KLTPS Extension	Gujarat	75 MW
14.	Dhuvaran Extension	Gujarat	40 MW
15.	Akhakh	Gujarat	365 MW
16.	Bellary	Karnataka	500 MW
17.	Vindhyachal	Madhya Pradesh	1000 MW
18.	Birsingpur Extension	Madhya Pradesh	500 MW
19.	Amarkantak	Madhya Pradesh	210 MW
20.	Deemapur	Nagaland	23 MW
21.	GHTP – Stage II	Punjab	500 MW
22.	Giral	Rajasthan	125 MW
23.	Dholpur	Rajasthan	330 MW
24.	Unchahar Stage III	Uttar Pradesh	210 MW
25.	Parichha Extension	Uttar Pradesh	210 MW
26.	Mejia Stage V	West Bengal	250 MW
27.	Mejia Stage VI	West Bengal	250 MW
28.	Sagardighi Unit I and II	West Bengal	600 × 2 = 1200 MW
29.	Santaldih	West Bengal	250 MW
30.	Durgapur	West Bengal	300 MW
31.	Bakreshwar	West Bengal	210 MW
32.	Singrauli	Uttar Pradesh	2000 MW
33.	Ramagundam	Andhra Pradesh	2600 MW
34.	Farakka	West Bengal	1600 MW
35.	Talcher Thermal	Orissa	460 MW
36.	Badarpur	Delhi	705 MW

THEORY QUESTIONS

1.1. Explain with a neat schematic diagram modern steam power plant showing its circuits.
1.2. What are the different components of modern steam power plant?
1.3. What are the different factors which are to be considered while selecting the site for thermal power plant? Explain each in brief.
1.4. Draw a layout of thermal power plant and explain in brief.
1.5. What are the requirements or principles and steam power station design?
1.6. What are the advantages and disadvantages of thermal power plants?
1.7. Give the list of thermal power plants in India.

Chapter 2 Fuels for Thermal Power Plants and Combustion

More than half the disposable coal output of the country is used for steam raising. In last few years, sudden shift from oil to coal—where practical—has triggered the need for a searching look into the modern technology of handling and burning coal.

The basic requirements of fuels for power generation are :

1. It should have sufficient energy content.
2. It should be easy to handle.
3. It should be economically available in large quantities.
4. It should bs free from fire hazards, as far as possible.
5. It should not chemically react with the parts of equipment where it is burnt.
6. It should be easy to store.
7. It should easily mix with air for combustion.
8. It should have requisite combustion qualities.
9. Products of combustion should not be corrosive and should be in gaseous form, as far as possible.

2.1 TYPES OF FUELS

There are three general types of fuels :

1. Solid 2. Liquid 3. Gaes

Solid, including coal, coke, peat, briquets, wood, charcoal and waste products.

Liquid, including petroleum and its derivatives, synthetic liquid fuels manufactured from natural gas and coal, shale oil, coal by-products (including tar and light oil) and alcohols.

Gases, including natural gas, manufactured and industrial by-product gases and the propane and butane or liquefied petroleum (LP) gases that are stored and delivered as liquids under pressure but used in gaseous state.

2.2 EFFECT OF FUEL PROPERTIES ON DESIGN

While considering furnace and firing equipment, it will be found that each element of design must be considered in relation to burning rates encountered and nature of load. For example, minimum ash fusibility (fusion) temperature for clinker free operation on an underfed stoker depends largely on combustion rate. Likewisely nature of load affects importance of caking qualities of coal ; a fluctuating load may bar a strongly caking coal from use on a stoker with inadequate agitation of the fuel bed, or from use on a hand fired grate.

The plant engineer must remember that one property of a coal may modify another; thus, the presence of excessive fines may aggrevate clinkering tendencies, producing trouble with a coal having apparently satisfactory ash fusibility characteristics.

While analysing fuel for a plant, engineer must distinguish between limitations and preferences. For example, if grindability drops below a certain critical point in a plant with barely enough pulverizer capacity, the load cannot be carried, and coal with lower grindabilities is worthless. If pulverizer capacity is ample, however, the engineer may still prefer high grindability coal, because grinding costs will be less. Savings in power consumption and in cost of replacing worn out parts will indicate how much can be paid for higher grindability.

Some properties of coal have importance only as limits, never as preferences. For example, once the minimum ash fusibility temperature for trouble free operation has been determined, there is no value in any further increase.

The completed plant analysis is primarily a set of limitations, expressed (i) in coal properties that can be measured (ash content, fusibility temperature, sulphur content, etc.), and (ii) in properties that still defy exact expression (caking quality, weathering properties but stated as clearly as possible. Armed with the analysis, the power engineer can tackle the list of coal available, which is another challenge.

The effect of change in fuel properties on the coal handling and burning equipment can be visualised from the following practical case:

A power plant was specified to operate with coal having following analysis:

Moisture	12.43%
Volatile matter	35.17%
Fixed carbon	43.45%
Ash	8.95%
Heating value	26700 kJ/kg

Between the time the equipment was purchased and actual operation, however, the coal supply situation changed drastically, leading to the put chase of coal with following analysis:

Moisture	11.44%
Volatile matter	29.42%
Fixed carbon	39.14%
Ash	20.02%
Heating value	22500 kJ/kg

Burning of this fuel resulted in severe overtaxing of both the coal handling and ash handling equipment. Not only does the purchased coal contain better than twice the ash content of the coal for which the plant was designed, it also requires burning of a significantly greater quantity to meet the load demand. Bunkers designed for a six hour capacity only give four hours of operation with the lower grade coal. As a result of this decreased reserve, less time is available for repair of critical items of the coal handling system which could force a plant shutdown.

The original system consisted of one 25 cm sluice line running from the plant to a sluice pond. This line was soon found to be completely inadequate. Since initial operation, it has been necessary to add this equipment; a second ash sluice line ; a third ash sluice pump; and a fly ash storage silo, which acts as a surge bin and also provides for separate unloading facilities. The plant was able to remain fully operational at rated capacity only because of this added equipment. Also failure to provide adequate ash removal facilities could cause damage to the electrostatic precipitator in addition to its reduced operating efficiency.

2.3 SELECTION OF COAL

Because coal is a complex substance, selection cannot be reduced to a simple formula. The right coal for any plant is the one that will carry the load, meet all design operating requirements and produce lowest overall steam cost. Although any coal can be buried in any plant, after a fashion, every practical engineer knows that many plants have design and equipment limitations that rule out certain coal either because they will not carry the load, because reliability will be seriously impaired, or because intolerable operating difficulties will be encountered.

A thorough going fuel analysis should precede everything but the decision to build or modify the plant since choice of coal affects, practically everything in the boiler house. Each design decision made before the analysis may prove a later limitation in buying coal for the plant. The purpose of coal analysis is therefore threefold :

(i) To find out what fuels are available to the plant, especially what grades of coal ;
(ii) To establish the relative values of these fuels ; and
(iii) To attempt to predict or atleast allow for, changes in relative value in future years.

This is a complex undertaking these days to be sure, but there is no part of the design that pays bigger dividends over the life of the plant.

Limiting factors on coal usage will take various shapes in different plants. For example, inadequate ash handling equipment may require pumping a limit on the ash content of the fuels burned. Or a low boiler setting at high burning rate may lead to smoke problems if the amount of volatile matter in the coal is not limited.

2.4 SAMPLING OF COAL

As a result of its mode of formation, coal is a very complex substance. Any two lumps from the same bed are never absolutely the same in every respect. It is fundamental in coal sampling and analysis, therefore, that although general characteristics may be determined for a given quantity of coal, not every small piece in that quantity of coal will have the same characteristics as the average characteristics for the lot as determined by the analysis. Because of this variation on the coal substance, it is extremely important to follow proper sampling procedures. The most accurate analysis possible means little if the sample is not representative of the lot of coal being sampled.

2.4.1 Methods of Sampling Coal and Coke

Mine Samples. The mine sample represents the quality of coal that can be obtained when the impurities are readily separated from the coal by washing or other coal preparation methods and when extreme care is taken in the mine and in the preparation to obtain the cleanest possible coal. However, mine samples may not be representative of commercial coal in ash content and heating value and caution must be exercised in estimating the grade of commercial coal on the basis of mine samples.

Tipple, Breaker, Truck, Railroad, Car and Ship Samples of Coal. Most coal samples are taken after the coal has been removed from the mine—at mine tipples ; at breaker and preparation plants ; from trucks, railroad cars, and ships ; or upon delivery at the ultimate destination 'increments', each consisting of the coal obtained by a single motion of the sampling instrument through or into the coal being sampled, are taken in accordance with standardised procedures. The sample is crushed, mixed and reduced by riffling until a sample of not less than 1 kg is obtained. This sample is then sealed in a special jar and sent to laboratory. Special procedures are required in sampling for total moisture content.

2.5 ANALYSIS OF COAL

Various tests and methods of analysis express coal qualities in figures instead of words. Principal characteristics are expressed in what is known as proximate analysis, as distinguished from an ultimate analysis, which shows the exact chemical composition of a fuel without reference to the physical form in which the compounds appear. The ultimate analysis provides data needed for combustion calculations.

The proximate analysis gives the power engineer a good picture of coal's behaviour in a furnace. It is a relatively simple procedure, which involves determining the percentage of (1) moisture, (2) ash, and (3) volatile matter, in accordance with prescribed ASTM test methods, and calculating the percentage of fixed carbon by difference. Since the percentages of these four variables add up to 100, fixed carbon can be found arithmatically once the other three are known. It is also customary to determine separately the total amount of sulphur contained in the coal, the ash-fusibility (fusion) temperature and the fuel's heating value.

Coal analyses can be made on several bases. For power plant work, as-received, air-dried or moisture-free analyses generally are used. Moisture and ash-free and moisture and mineral-free analyses usually are employed in classifying coal.

As the name implies, the as-received analysis reports the condition of coal as delivered to the laboratory. This comes closest to the condition as-shipped or as-fired, the values desired in practical work. Loss or gain of moisture between the times of sampling and of analysis depends on the type of coal, its size, the weather conditions and the method of handling the sample.

2.5.1 Moisture

All coals contain some natural moisture which may vary from 10% to as high as 45%. This moisture lies in the pores and forms a true part of the coal, being retained when the coal is air-dried. Surface moisture, on the other hand, depends on conditions in the mine, and the weather during transit.

2.5.2 Ash

Ash is the incombustible mineral matter left behind when coal burns completely. Like moisture, ash is an impurity that increases shipping and handling costs. It must be removed from the furnace and the plant, thus requiring additional equipment and expenses knowing the amount and character of ash in the coal, fuel bed problems such as clinkering and slagging can be avoided. An increase in ash content generally means an increase in carbon carried to the ash pit.

2.5.3 Volatile Matter

Volatile matter is that portion of the coal which is driven off in gaseous form when the fuel is subjected to a standardized temperature test. It consists of combustible gases, such bs methane and other hydrocarbons, hydrogen and carbon-monoxide and non-combustible gases. Since the quantity of volatile matter indicates the amount of gaseous fuel present, it affects firing mechanics. It also influences furnace volume and the arrangement of heating surfaces.

2.5.4 Fixed Carbon

Fixed carbon is the combustible residue left after the volatile matter distills off. It consists mainly of carbon, but contains some hydrogen, oxygen, sulphur, and nitrogen not driven off with the gases. The form and hardness of this residue are an indication of the caking properties of a fuel and are, therefore, a guide in the selection of combustion equipment.

2.5.5 Sulphur

Sulphur content in coal may vary from traces to as high as 8%. When coal is burned, sulphur oxides are omitted from the chimney, thereby creating problem of air pollution. In addition to this sulphur plays a part in corrosion of air heaters, economizers and stacks.

2.6 CLASSIACATION OF COAL

Three methods of classifying coal have been adopted. These classifications are :

(1) by rank (degree of metamorphism, or progressive alteration, in the natural series from lignite to anthracites);

(2) by grade (quality determined by designation, calorific value, ash, ash softening temperature and sulphur); and

(3) by type or variety (determined by the nature of the original plant material and subsequent alteration thereof).

Other methods of coal classification are by use or suitability for specific purpose or types of combustion equipment, and by various trade systems set up to meet particular conditions in a given area or time.

2.6.1 Rank

This is the widely adopted method of classifying coal. In this the coal is arranged according to fixed carbon content and heating value calculating on mineral-matter-free basis. The higher rank coal is classified according to fixed carbon on dry basis, the lower rank coals, according to heating value on a moist basis (containing natural bed moisture).

2.6.2 Grade

Classification of coal by grade is indicated by a symbol designation system indicating size, heating value in Kcal/kg, ash, ash softening temperature and sulphur content of the coal. Thus a 10 ×5 cm, 115—A_8–F_{13}–$S_{1.6}$ indicates a coal 10 cm by 5 cm in size (through a 10 cm and a 5 cm round hole screen); with approximately 11,200 Kcal per kg of heating value, an ash content of 6.1% to 8%. inclusive, ash softening temperature of 1300°C to 1400°C, inclusive, and sulphur content of 1.4% to 1.6% inclusive. Analysis are expressed on the basis of the coal as sampled.

TABLE 2.1 *Classification of coal on tbe Basis of Grade*

Ash		*Softening Temperature of ash*		*Sulphur*	
Symbol	*% inclusive*	*Symbol*	*°C inclusive*	*Symbol*	*% inclusive*
A_4	0.0–4.0	F_{15}	1500 and higher	$S_{0.7}$	00–0.7
A_6	4.1–6.0	F_{14}	1400–1500	$S_{1.0}$	0.8–1.0
A_8	6.1–8.0	F_{13}	1300–1400	$S_{1.3}$	1.1–1.3
A_{10}	8.1–10.0	F_{12}	1200–1300	$S_{1.6}$	1.4–1.6
A_{12}	10.1–12.0	F_{11}	1100–1200	$S_{2.0}$	1.7–2.0
A_{14}	12.1–14.0	F_{10}	<1100°C	$S_{3.0}$	2.1–3.00

2.6.3 Type

In classifying coals by type or variety, consideration is given to the nature of the original plant ingredients and the bio-chemical alteration of these ingredients during the various steps in transforming living plant tissues into coal. Commercial varieties of bituminous and sub-bituminous coal have been accepted as standard.

Common-bonded Coal is the common variety of bituminous and sub-bituminous coal. It consists of a sequence of irregularly alternating layers or lenses of (1) homogeneous black material with a brilliant vitreous lustre, (2) greyish black, less brilliant, striated material, usually of silky lustre, and (3) generally thinner bands or lenses of soft, powdery, fibrous particles of mineral charcoal. The difference in lustre of the bands is greater in bituminous than in sub bituminous coal.

Splint Cool is a variety of bituminous and sub-bituminous coal, commonly having a dull lustre and greyish-black colour, of compact structure, often containing a few thin irregular bands with vitreous lustre. It is hard and resonant when struck.

Cannel Coal is a variety of bituminous or sub-bituminous coal of uniform and compact fine grained texture, with a general absence of banded structure. It is dark grey to black in colour and has a greasy lustre.

Boghead Coal is a variety of bituminous or sub-bituminous coal resembling cannel coal in appearance and behaviour during combustion. It is characterised by a high percentage of algel remains and volatile matter. Upon distillation, it gives exceptionally high yields of tar and oil.

2.7 DIFFERENCE BETWEEN CAKING AND COKING

Confusion often exists with respect to the proper use of these two terms. When coal is heated in the absence of air or in an atmosphere very deficient in oxygen, volatile matter is driven off, leaving behind a residue of carbon. This is *coke.* It may take the form of small powdery particles, or it may fuse into lumps of varying size and strength. And swelling may occur. In commercial coke making, the term 'Coke' refers to lumps of marketable size and quality ; coking coals are used to produce them in a coke oven.

Coke formation, in one shape or another, represents an intermediate combustion stage tin any fuel bed. In a boiler furnace, for example, some coals become plastic and form lumps or masses of coke. (This type of coke usually is not of metallurgical quality, and is often referred to as semi-coke). These are called caking coals. Those coals that show little or no fusing action are called free burning. Caking properties of a coal and the nature of the coke masses formed (size, strength, etc.) are valuable indicators of a fuel performance in the furnace.

Although both caking and free burning coals can be burned without difficulty, and equally well, in boilers fired with pulverized coal, this is not the case with all types of stoker firing. In general, caking coals are burned on underfed stokers, which have moving rams or other means of breaking the masses of semi-coke formed in the fuel bed. Free burning coals usually are burned on travelling grate or spreader stokers because there is no agitation of the fuel bed. Agitation should be avoided to prevent small pieces of coal from shifting through the grate and into the ash pit, where they might continue to burn. Also, realize that any unburned carbon lost to the ash-pit reduces combustion efficiency.

2.7.1 Indian Coals

The most important coal fields in India are Haniganj in West Bengal and Jharia and Bokaro in Bihar. The West Bengal and Bihar fields contribute major share of the total output. The best coking coal comes from Giridih, Jharia and Raniganj. Other states in which coal is found are Madhya Pradesh, Orissa, Andhra Pradesh, Chennai, Maharashtra, Assam, Rajasthan and Kashmir.

Indian coals fall mainly into two classes namely high grade steam and low grade steam coals. Heating values of some of the varieties are given below:

TABLE 2.2

(A) Lew Moisture coal		*(B) High Moisture coal*	
Grade	*Useful Heat value kJ/kg*	*Grade*	*Useful Heat value kJ/kg*
Sel.A	Exceeds 27720	Sel. A	Exceeds 25690
Sel. B	27720–26590	Sel. A and Sel. B	25690–24830
Gr. 1	26590–24870	Gr. 1 and Assam coal	24830–21650
Gr. 2	24870–22360	Gr. 2	21650–19140
Gr.3 A	22360–19720	Gr. 3	19140–15970
Gr.3 B	19720–15910	Ungradable	15790–10470
Ungradable	15910–10470		

The coal reserves in India have been estimated at 329.6 million tons in respect of selected grade, 395.7 million tons of grade I quality and 49.1 million tons of grade II quality.

Because of paucity of coal deposits in south India, high priority has been given to the development of the multipurpose "South Arcot Lignite Project" at Neyveli. This project is planned to increase the mine output to 5.5 million tons per annum. Lignite also occurs at Palana and Khari in Bikaner district, Rajasthan, Nichahom in J & K, Kutch and Broach districts in Gujarat.

2.7.2 Liquid Fuels

Petroleum consists of a very complex mixture of hydrocarbons. These chemical compounds of carbon and hydrogen fall naturally into several homologous series, which are classified according to the chemical construction or structure of their molecules. Each series exhibits a steady gradation of properties, both physical and chemical, as the number of carbon with associated hydrogen atoms is increased, ranging from gaseous hydrocarbons to solid waxes and asphalt. Fractional distillation serves as the simplest commercial method of separating crude petroleum into usable products, by which process the various hydrocarbons are grouped, together roughly according to their boiling points as shown below:

CRUDE PETROLEUM

			Undistilled residue
Gasoline 30–200°C	Kerosene 150–300°C	Gas oil and light diesel fuels (200–370°C and upwards)	For further processing or use as residual oil

2.8 PROBLEMS CONNECTED WITH THE USE OF HEAVY FUELS

Common to all types of equipment using heavy fuels are problems of distribution, storage and handling. The heavier grades of the marine fuel oil type are extremely viscous at atmospheric temperatures and sometimes have rather high pour points. When burning heavy residual fuels in diesel engines, it is most essential to maintain the fuel injection system in proper order, so that fouling and increased wear of vital parts will be obviated. If there is an insufficient degree-of atomisation coupled with improper mixing and over-penetration, combustion will be impaired, incompletely burnt fuel particles will pass the piston, causing excessive sooting of the lubricating oil and possibly ring sticking.

Sulphur, which is partly responsible for increased cylinder wear is usually associated with petroleum in chemical combination with the hydrocarbons, the proportion depending upon the origin of fuel. Fortunately, in the large engines which would normally use these heavier fuels containing greater proportions of sulphur, due to the lower gas movement and the greater temperature difference over the thicker cylinder walls, the internal skin temperatures are maintained well above the critical value under most operating conditions.

In the gas turbine, formation of deposits, particularly from the ash content of the fuel, has been considered to be a major problem connected with the use of heavy fuels. Most crudes naturally contain a small proportion of foreign matter which, on ignition, leaves the ash. The distillate portions prepared from the crude are, of course, freed from such impurities in the distillation process, but the impurities remain in the residual fractions. Some ashes pass harmlessly through the unit and give rise to no trouble, but others show some tendency to adhere to the turbine blades etc. Much depends upon the nature of the ash, and it would seem that ways and means of obviating the trouble are likely to be found.

In oil-fired steam raising plant provided the equipment is designed to handle heavy fuels and appropriate preheating is provided, no really serious problems are encountered, and although in exceptional cases we sometimes hear of corrosion troubles from the sulphur in the fuel when the flue gases are cooled beyond the dew-point, it is normally possible to find a remedy by careful study of combustion conditions and the nature of the surfaces affected.

2.8.1 Combustion of Fuels

Furnaces for boilers may use solid, liquid or gaseous fuels. Heating of all the furnaces is carried out, as a result of heat energy liberated due to chemical reaction occurring in it. In order to have complete knowledge of the reactions, it is necessary to know :

(i) the energy released by the reaction when they proceed to completion,

(ii) the condition of reaction under finite known heat loss to surroundings,

(iii) with the limited time available for the fuel to burn, the degree of completion of reaction.

Most of the chemical reactions occurring inside a furnace are exothermic and certain amount of energy is liberated.

The requirements of boiler furnaces are :

(i) to make available required quantity of energy in specified lime,

(ii) self-sustained combustion of fuel,

(iii) required temperature,

(iv) high efficiency,

(v) ability to operate efficiently under variable load conditions.

For self-sustained combustion of fuel, it is necessary to control the heat loss from reaction zone to the surroundings. If the heat loss is high, the temperature of the reaction zone will fall, consequently it will loose its capacity of self sustained combustion. Green fuel entering the furnace may not be ignited completely due to lower temperature in the furnace. For this the knowledge of the factors governing the rate of chemical reaction between fuel and air or oxygen is necessary.

2.8.2 Reaction Energies

In boiler furnace firing, most of the reactions, chemical reactions, take place at atmospheric pressure. Let us consider the case of combustion of carbon. The reaction is :

$$C + O_2 \rightarrow CO_2 + E$$

where E is the energy liberated. Since the reaction occurs at constant pressure, the energy liberated at 25°C, when 12 kg of carbon react with 32 kg of ogygen, is 393500 kJ. Corresponding weight of carbon dioxide formed is 44 kg. In case of oxygen 32 kg of it will occupy 22.35 m^3 when air is used for combustion, the volume of air required for 22.35 m^3 of oxygen, is 106.6 m^3. So the complete reaction may be written as

$$\begin{array}{ccccccc} C & + & O_2 & \rightarrow & CO_2 & + 393500\,\text{kJ} \\ 12\text{ kg} & + & 32\text{ kg} & & 44\text{ kg} & \end{array}$$

In a furnace using coke, it is always desired that the reaction should proceed to completion as rapidly as possible in order to liberate all the chemical energy of the carbon in the useful furnace space.

Reaction of the formation of carbon monoxide may be written as

$$\begin{array}{cccccc} C & + & \frac{1}{2}O_2 & \rightarrow & CO & + 110540\text{ kJ} \\ 12\text{ kg} & + & 16\text{ kg} & \rightarrow & 28\text{ kg} & \end{array}$$

In this case energy liberated is less than that in previous case where the reaction was complete with the formation of carbon-di-oxide. However, carbon-mono-oxide may have further reaction with oxygen to form carbon di-oxide

$$CO + \tfrac{1}{2}O_2 \rightarrow CO_2 + 283000\text{ kJ}$$

In case carbon-dioxide is reduced to carbon-mono-oxide reaction may be written as

$$\begin{array}{cccccc} CO & + & C & \rightarrow & 2CO & + 172925\text{ kJ} \\ 44\text{ kg} & + & 16\text{ kg} & & 56\text{ kg} & \end{array}$$

The reaction of hydrogen may be expressed as

$$\begin{array}{lllll} H_2 & + & \frac{1}{2}O_2 & \rightarrow H_2O & + 285555 \text{ kJ} \\ 2 \text{ kg} & + & 16 \text{ kg} & & \text{(Heat liberated, if the water is condensed} \end{array}$$

$$\begin{array}{lllll} H_2 & + & \frac{1}{2}O_2 & \rightarrow H_2O & + 242000 \text{ kJ} \\ 2 \text{ kg} & + & 16 \text{ kg} & 18\text{kg} & \text{(Heat liberated, when the water remains in vapour conditions)} \end{array}$$

At any temperature and pressure a reaction will not proceed completely in either direction, but will result in certain mixture of compounds on either side. The rate of reaction is low at low temperatures. As the temperature is increased the reaction is accelerated but at high temperatures again dissociation of carbon into carbon monoxide and oxygen starts. This reaction is endothemic. Initially the reaction starts in forward direction at a higher rate. With the time, as the products are formed, the rate of reaction in the forward direction is reduced. As the products begin to dissociate a sort of equilibrium is reached between the two directions.

Let us consider the ignition of particles on the actual fuel bed. As the air is supplied the reaction starts rapidly and almost entire oxygen of air is consumed at a short distance from the inlet edge. At this, the percentage of carbon dioxide is maximum. Thereafter carbon dioxide dissociates and due to lack of oxygen, carbon monoxide is formed. Typical analysis of gases in fuel bed with the distance measured along the path of gases is shown in Fig. 2.1. The

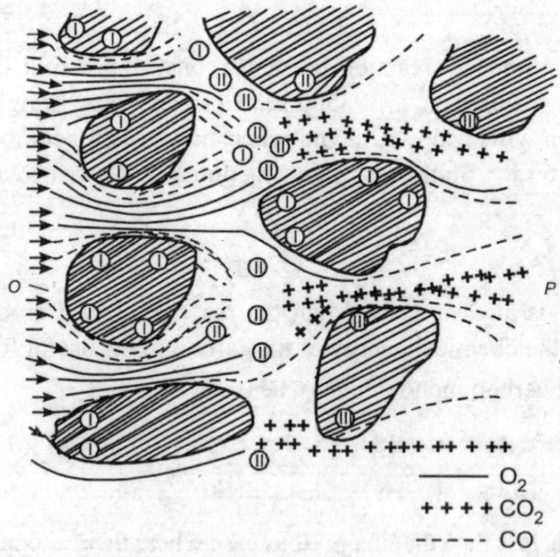

Fig. 2.1 *Combustion on actual fuel bed.*

rate of reaction is high in case of fine grades of coal as the reaction proceeds to completion immediately. In case of big lumps of coal the rate of reaction is low and sometimes incomplete reaction results. The performance of a boiler improves and also its thermal efficiency increases as the size of coal is reduced. See Fig, 2.2.

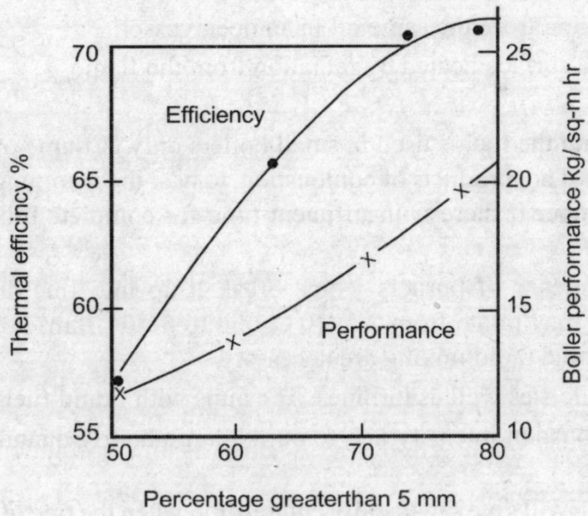

Fig. 2.2

2.8.3 Pulverised fuels

A pulverised fuel consists of fine particles of coal. A moving dust cloud is formed by blowing finally powdered coal into a hot combustion chamber on a stream of air so that a three-dimensional stationary flame is formed inside the chamber. Such a flame has something in common both with solid and combustion, as the fuel "flows" in the atmosphere of air.

The existence of a P.F. flame depends upon two complimentary factors

(1) small particle size,

(2) a large flame or a supporting ignition source.

Since a P.F. flame is composed of multitudes of particles, the radiation intensity from the flame, and hence the radiative heat loss, is high. Obviously, if this is too high compared to the rate of energy generation by the reaction, the flame will be extinguished. Therefore, to maintain the flame, the ratio of heat generated to heat lost must exceed some critical value. This may be achieved by :

(a) fine grinding of the dust to create a large surface area and hence increase the rate of heat generation per unit volume, or

(b) increasing the total size of the flame so that the surface to volume ratio is reduced and hence the average heat loss per unit volume of flame is reduced.

Air is supplied in stages as primary air and secondary air and sometimes tertiary air also. Besides large capacity high pressure boilers, super-critical pressure plants, pulverised fuel may also be used for small boilers, gas turbines etc.

2.8.4 Oil Firing

Oil is fired for heating purposes mainly in two ways :

(1) It is substantially evaporated before it ignites so that it burns as a vapour.

(2) It can be broken up into fine droplets which are heated by radiation or turbulents mixing with combustion products as they mix with air, so that the droplets are evaporating while combustion is proceeding.

The first method is essentially the same as the method of combustion of gases, with the only difference in the method of evaporating the fuel. This may be done by

(1) heating of a tube in the flame through which the oil passes.

(2) radiation and conduction from the same oil in an open vessel.

(3) spraying of oil on to a surface heated by radiation from the flame.

(4) by the use of wick.

Vaporising method of burning the fuel is used in small boilers only. A flame of this type is usually stabilised at high velocities by recirculation of hot products of combustion, to heat the incoming stream. Incomplete combustion may occur in combustion chamber if there is insufficient time for complete mixing and chemical reaction after mixing is not complete.

The combustion of oil by means of burners which break it up into fine droplets is usually referred to as 'atomisation'. Size of droplets may range from 2.5×10^{-4} mm to 5×10^{-3} mm. For high intensity combustion fine atomisation, high recirculation and rapid mixing are necessary.

Liquid fuels are used for boilers as well as turbines. The aims with liquid fuels are :

(1) to obtain a high combustion intensity i.e., to burn a large hourly quantity of fuel completely in a given volume,

(2) to have a system which will give satisfactory combustion when the rate of firing is varied due to fluctuating load,

(3) to obtain high combustion efficiency (i.e., the fraction of calorific value of fuel which is liberated by combustion within the flame) of the order of 95% to 98%,

(4) to avoid damage to refractory and heat receiving equipment,

(5) to avoid deposition of carbon or soot on any region of the chamber.

2.10 GASES

Gases are burned for heating purposes by mixing with air at atmospheric pressure or at higher pressures. Air for combustion is supplied in parts, that is, as primary air, secondary air and sometimes tertiary air also. There are two types of flames common with gaseous flames e.g., premixed and diffusion flame. In premixed flame, a mixture of gas and air passes through the burner and is ignited at the burner mouth. In diffusion flame, only gas is passed through the burner and the air diffuses through the sides. Because of difficulty of obtaining complete combustion in a diffusion flame, it is generally necessary to operate such flames with appreciably more excess air than would be necessary with premixed flame, otherwise a mixture of oxygen and unbumt combustibles will be found in the flue gases.

2.11 MASS BALANCE

Fuel and air enter the furnace of a boiler and flue gases and ash leave it. The atmospheric air may be regarded s mostly dry air with a relatively small amount of humidity. The wet flue gas contains dry flue gas

Fig. 2.3. *Mass balance for Combustion Process WFG—Wet FluelGas, DFG—Dry Rue Gas*

a and water vapour. This water vapour comes from the combustion of hydrogen and it may also come from moisture in the fuel and a slight amount may come from the humidity in the air supplied. Ash may also contain some unburnt carbon.

THEORY QUESTIONS

2.1. What is the difference between caking and coking ?

2.2. How is coal classified according to Rank ?

2.3. Discuss briefly the possibility of using coal for power generation in India.

2.4. Discuss the factors to be considered while selecting coal for a power plant.

2.5. How is coal analysed ?

2.6. What is the difference between 'proximate analysis' and 'ultimate analysis' ?

2.7. What is the effect of following on the preformance of coal in a boiler furnace:

(a) Moisture. (b) Volatile matter,

(c) Ash, (d) Sulphur ?

2.8. What is the impact of recent oil crises on the power plants burning liquid fuels ?

2.9. As a Power Plant Engineer how you will select and analyse coal for your plant ?

2.10. What is the effect of change of properties of coal on the performance of a power plant ?

2.11. With present 'oil crises', what is the future of coal fired thermal power plants in India ?

2.12. What changes you consider are necessary for converting a oil fired boiler into a coal fired boiler.

2.13. What information you can get from ultimate analysis of coal?

2.14. How is moisture content of coal restricted in power plants during rainy season ?

2.15. A coal contains 3.5% sulphur by weight, as fired, and all of the sulphur may be assumed to go to formation of sulphur dioxide. How many kilograms of SO_2 are produced per day if the firing rate is 35 tonnes of coal per hour? **[Ans.** 58,800 kg per day]

Chapter 3 Coal Handling, Preparation and Feeding

The sudden shift of both industry and electric utilities from oil to coal—where practical—has triggered the need for a searching look into the modern technology of handling and burning coal. To some, use of this energy source will come as a new technology, since many engineers today have never been exposed to the realities of coal combustion and allied needs.

Fuel handling plant needs extra attention, while designing a thermal power station, as almost 50% to 60% of the total operating costs consist of fuel purchasing and handling. Fuel system is design ed in accordance with the type and nature of fuel. Plants may use coal, oil or gas as the fuel. Plants using coal as fuel will be considered first.

3.1 COAL HANDLING SYSTEMS

Mechanical handling of coal is preferred over manual handling due to the following reasons:

(1) It is economical for medium and large capacity plants.

(2) It is much more reliable than manual handling.

(3) It requires comparatively less number of labourers.

(4) Operation is comparatively easy and smooth. Also the supervision requirements are limited in mechanical handling.

(5) Minimum labour is put to unhealthy working conditions.

(6) With reduced labour, management and control of the plant becomes easy and smooth.

(7) In large capacity plants it is almost impossible to use manual means.

(8) Losses in transport are minimised by mechanical handling.

(9) Mechanical handling system can be easily started and can be economically adjusted according to the need.

Disadvantages

(1) In mechanical handling some power generated is usually consumed, resulting in less net power available for supply to consumers.

(2) Mechanical handling needs continuous maintenance and repairs.

(3) Plants depending totally on mechanical means, sometimes, in the event of breakdown, face serious problems as they cannot be immediately substituted by manual labour.

(4) Capital cost of the plant is increased.

As has already been stated, the cost of coal in a plant may be of the order of 50% to 60% of the total station operating cost, as such handling of coal should be carried out as expeditiously and economically as possible. Requirements of a good coal handling plant are:

(1) It should be reliable.
(2) It should be sound, simple and require a minimum of operatives.
(3) There should be minimum wear in running the equipment due to abrasive action of the coal particles.
(4) It should need minimum maintenance.
(5) It should be able to deliver requisite quantity of coal at the destination during peak periods.

In essence, the job of a handling system is to move coal from a receiving point to the firing equipment. The simplicity or complexity of the plant depends upon the way in which the coal is received, orientation of the plant, desired capacity and flexibility of the arrangement. In order to satisfy a variety of conditions to meet several requirements, we have an extensive array of mechanized handling devices, which may be combined in almost innumerable ways.

No coal handling system can be termed as typical. Figure 3.1 shows the various steps in coal handling. It is not very necessary that the flow chart may be followed as such in all the plants. Depending upon the type of plants, load factor, plant capacity, some of the intermediate steps may be eliminated or rearranged.

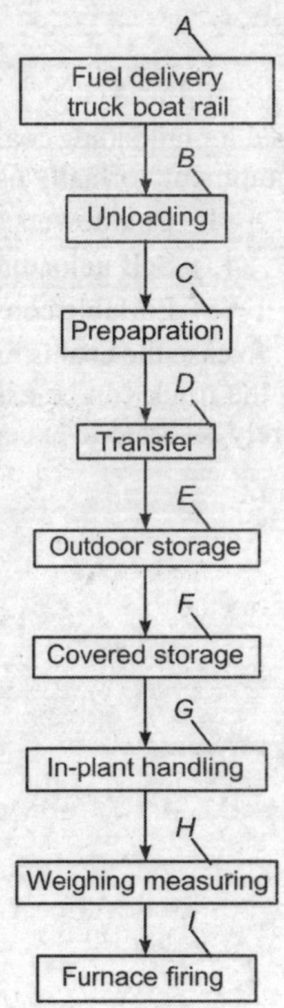

Fig. 3.1

3.1.1 Delivery

The method of transporting coal to a power station depends on the location of the plant, but may be one or more of the following: rail, road, river or sea. Plants situated near river or sea may make use of the navigation facilities Stations which cannot possibly make use of these facilities may be supplied coal either by trucks or by rail. Transportation of coal by trucks is usually used in case the mines are not far off or when the necessary railway facilities are not available. In case rail transport is to be adopted, the necessary siding for receiving the coal should be brought as near the station as possible. It is always desired to have separate in-coming and out-going tracks to the main lines thereby enabling the wagons to be emptied and returned back from the site with minimum of shunting operations.

3.1.2 Unloading

Just what kind of equipment will do the best job for unloading depends first of all on how the coal is received. If the coal is delivered in dump trucks and if the plant site is favourable, we may not need additional unloading equipment. The trucks may dump the coal to small outdoor storage or into a basement bin. In case the coal is received by rail in hopper cars, again the coal may be unloaded directly into a underground bin. For this it is necessary to have some arrangements for bringing the cars to the right spot for quick and clean unloading. In case the above facilities for unloading cannot be used as such, suitable unloading devices have got to be provided. These include the following for unloading from railway wagons :

(a) Car shakers.
(b) Car throwing equipment.
(c) Car dumpers (rotary).
(d) Coal accelerators.

For bringing the cars to the right spot, car pullers are used which may be manually or mechanically operated depending upon the requirements. In large operations mini-locomotives may be used. Equipment mentioned above is used for speedy unloading. 10 to 20 tonne capacity wagons can be unloaded within 15 minutes. Car shakers are used for unloading the wagons having bottom hoppers. In case the wagons with hopper are not available, car throwers may be used with advantage. Rotary car dumpers give fast high capacity loading in case of closed wagons.

Coal accelerators are normally aaed for the unloading of wet coal (due to its compacting property). For high capacity plants, specially designed coal towers (Fig. 3.2a) or unloading bridges may be used. These may also be

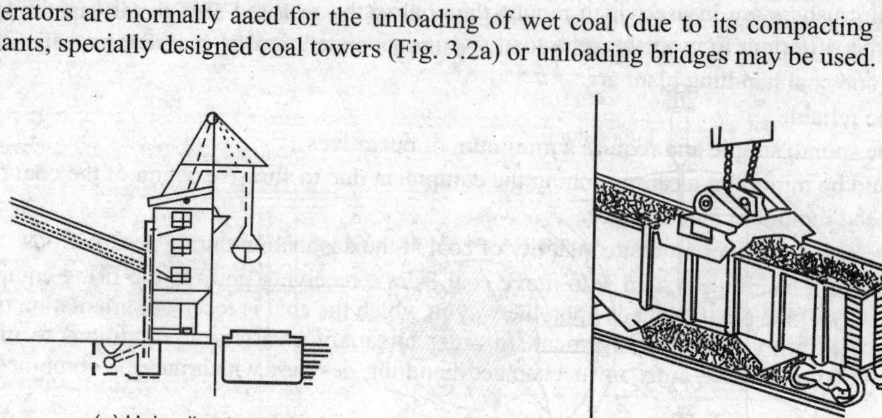

(a) Unloading tower

(b) Car Shakers

Fig. 3.2

used for unloading coal when the same is supplied by river or sea. When the coal is transported by sea, the unloading equipment normally used, includes :

(1) Coal towers.
(2) Unloading bridges.
(3) Self unloading boats.
(4) Coal accelerators.
(5) Portable conveyors.

In case the coal is brought to the site by road, use of lift trucks with scoops is always preferred as the entire load of the truck can be easily unloaded without additional equipment. In large stations, transport of coal by road is rarely resorted to except in case of emergency.

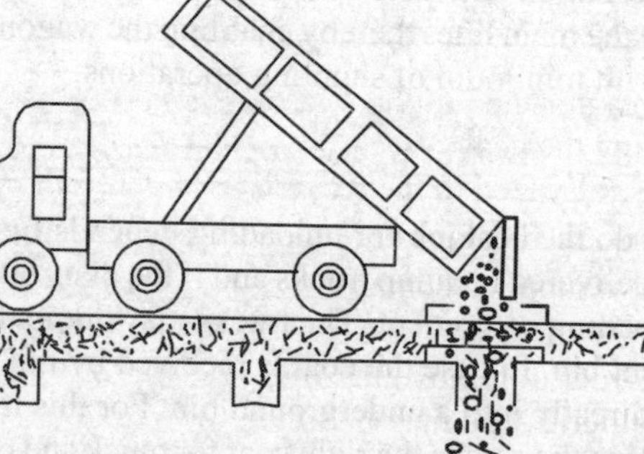

Fig. 3.3 *Lift-truck with scoop.*

3.1.3 Preparation

If the coal is brought to the site unsized and sizing is desirable for storage or firing purposes, preparation plants may be used either near the coal receiving point or at the point of actual use. A coal preparation plant may include one of the following :

(a) Crushers
(b) Breakers
(c) Sizers
(d) Dryers
(e) Magnetic separators.

Many types of coal crushers are in use which reduce the coal to the required size. Total capacity or a coal crushing plant should be sufficient to meet the peak load requirements. In case the coal obtained from the mine

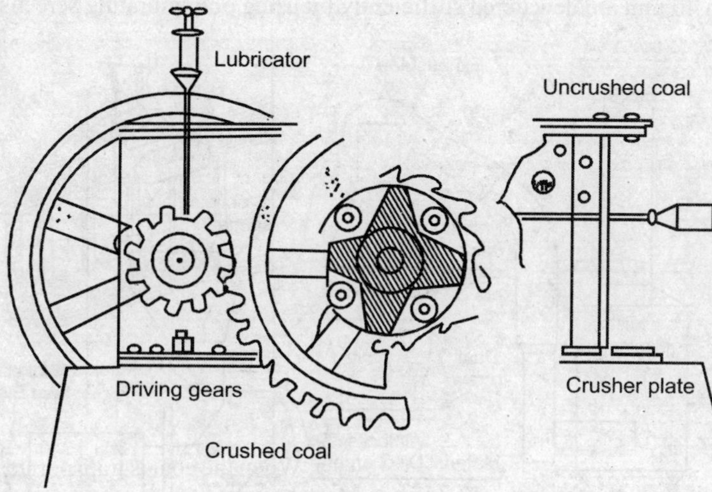

Fig. 3.4 *Coal crushers.*

does not require sizing, crushing plant may be by-passed. Regular use of crushers in a plant may prove to be expensive due to :

(a) higher price paid for large-size coal,

(b) cost of crushing,

(c) combustion in the boiler may be impaired due to higher proportion of duff unavoidably produced and carried into the boiler furnace.

Coal breakers are used in large central stations and have got the capacity of breaking 100 to 600 tonnes per hour. Coal is crushed, sized and cleaned in one opreation. In this case coal enters slowly in resolving drum at one end and is caught up by special lifting shelves attached to the screen plates. As each shelf approaches the top, its load is automatically dumped, and, on striking the heavy steel screen plates, coal is broken down. Remaining coal as well as over-size coal is recirculated until it is sized to the requirements.

Sizers are used along with crushers and breakers or independently for the separation of the coal of required size.

Coal dryers are used in case the coal supplied contains excessive free moisture or is wetted during transport. Hot flue gases are passed through the coal storage in closed spaces for removing the moisture.

Magnetic separators are used in order to remove the iron particles accompanying the coal. Iron particles may find their way into the coal at the collieries and may be in the form of bolts, nails, rivets, nuts etc. Entire mass of coal is passed over the separator in small layers. A strong magnet arrests the iron particles. Presence of iron particles may choke the burners and may increase the wear of the handling equipment.

3.2 DEWATERING AND DRYING

Water remaining on coal is a contaminant as serious as the undesirable ash. It may cause problems in handling and reduces the heating value of fuel. The problem is common in power stations particularly during rainy season. As the coal is generally stored in open, rain causes moisture contamination of coal.

There are two general processes that can be used to eliminate undesirable water. It may be removed mechanically or thermally; and sometimes both methods may be used successively on the same product.

Since water is a surface contaminant, the difficulty of dewatering coal will increase with increases in the surface area of the material. The finer the coal, the greater the surface area available for the adherence of water and hence the problems in moisture removal.

There are many mechanical dewatering devices most of which are designed to work most efficiently with coal particles within a certain size range.

Coal sizes larger than 40 mm are dewatered sufficiently by using only vibrating screens.

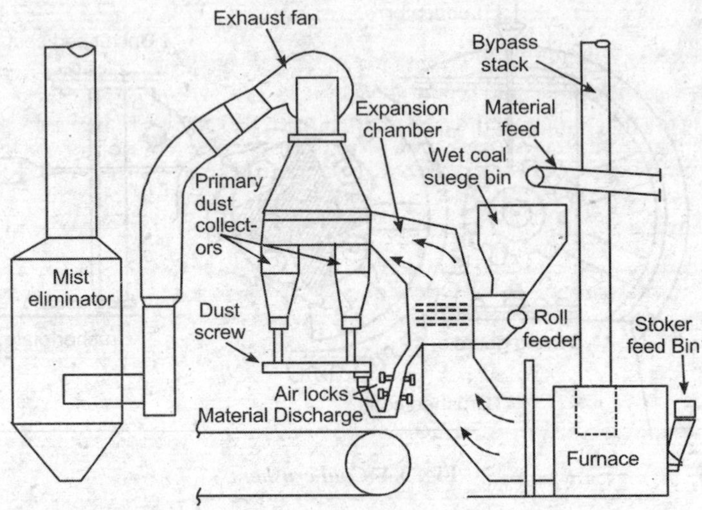

Fig. 3.5 *Coal drier.*

For smaller sizes vibrating-screen type centrifuges are used. Coal driers, as shown in Fig 3.5 are also sometimes used to control the moisture in coal.

Transfer. 'Transfer' means the handling of coal between the unloading point and the final storage point from where it is discharged to the firing equipment. It may involve only one piece of equipment, or several, depending on several factors including the local conditions. Equipment used for the transfer of coal may be any one of the following or a suitable combination thereof :

(1) Scraper or flight conveyor,
(2) Vee bucket elevator and conveyor,
(3) Pivoted bucket conveyor,
(4) Belt conveyor,
(5) Skip hoist,
(6) Screw conveyor,
(7) Grab bucket conveyor,
(8) Mass flow conveyor,
(9) Chutes.

Transfer operations usually include lifting as well as movement of coal. Most conveying equipment can do both the operations but some designs serve best for vertical or nearly vertical lifts and others for horizontal runs or slight inclines.

3.2.1 Scraper or Flight Conveyor

This type of conveyor is used extensively for conveying coal horizontally and for inclinations upto 35°. It consists of one or two strands of chain, to which steel scraper, or flights, are attached, which scrap the coal through a trough having identical shape. The coal is discharged in the bottom of the trough through gate controlled openings. Capacity of a conveyor of this type may range from 10 to 100 tonnes per hour.

Following are the advantages and disadvantages scraper, or flight conveyor are :

Advantages

(1) It has got a rugged construction and requires little operational care.

(2) It needs small headroom.
(3) It can be used for transfer of coal as well as ash.
(4) Speed of conveyor can be easily regulated to suit the requirements.

Disadvantages

(1) Due to scrapping or dragging action, there is excessive wear.
(2) It requires high maintenance and repair charges.
(3) Owing to the abrasive nature of the material handled, its speeds are low (about 10 to 30 metres/minute).
(4) Power consumption is high due to dragging action.

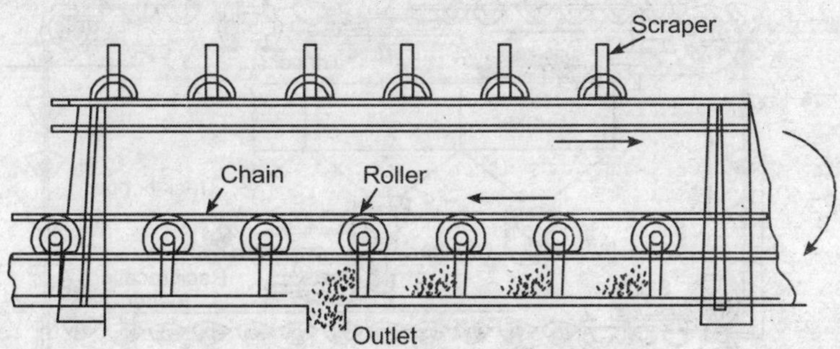

Fig. 3.6 *Scraper or flight conveyors.*

3.2.2 Vee Bucket Elevator

A vee elevator consists of steel V shaped buckets rigidly fastened to an endless chain going round sprockets. The buckets are equally spaced on the chain, and receive their load by dipping into a coal pocket at the lower end of the system. The material elevated in Vee buckets is discharged either by centrifugal force at the top of the elevator or by drawing back the buckets on the discharge side. On horizontal runs the buckets may act as a drag conveyor till they discharge coal through, suitable openings.

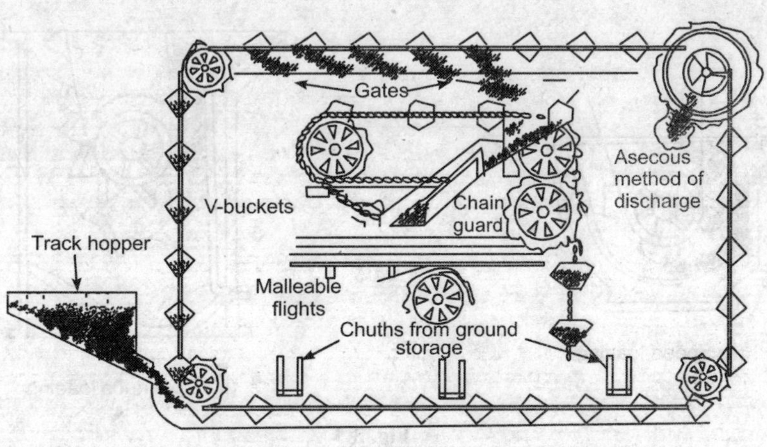

Fig. 3.7 *Vee bucket elevator.*

Following are the advantages and disadvantages of vee bucket elevator

Advantages

(1) Since the coal is carried and not dragged, power requirement for opening the equipment is less.

(2) Less floor area is required.

(3) Coal can be discharged at elevated places.

Disadvantage

It suffers from the disadvantages that its capacity is limited, hence for large capacity stations it may not suit.

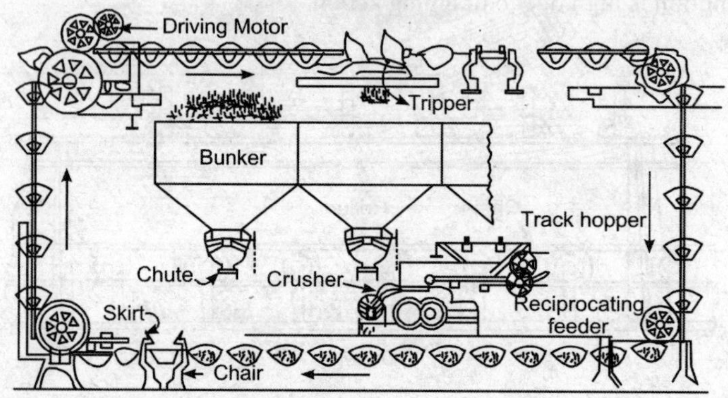

Fig. 3.8 *Bucket elevator.*

3.2.3 Pivoted Bucket Conveyor

The conveyor consists of malleable iron buckets suspended by pivots midway between the joints of two endless chains, which aire driven by a motor located at some convenient points, usually at the top of a vertical rise. While travelling horizontally, buckets maintain their position due to gravity and support the joints.

The conveyor is loaded by passing below a crusher. The coal is discharged into the bunker by a tripping device which engages the cam located on the side of the buckets and tilts the buckets sufficiently to discharge the coal.

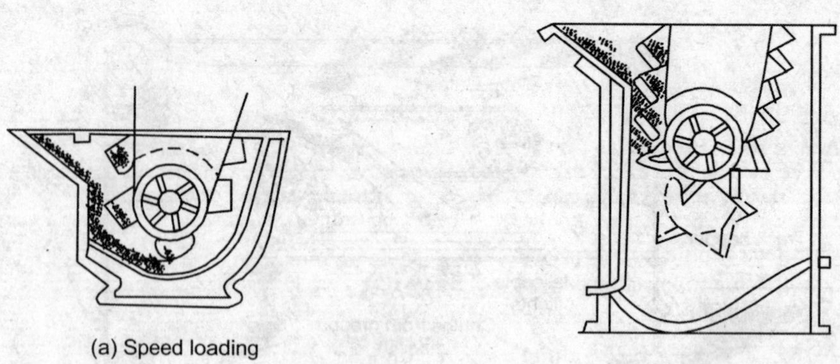

(a) Speed loading

(b) Coatinuos loading

Fig. 3.9

Advantages of this method include low operational cost, high capacity (300 to 400 tonnes per hour) and less floor area requirements. Initial cost of the equipment is high.

3.2.4 Belt Conveyor

The belt conveyor consists of an endless belt which carries coal and runs on 'carrying' pulleys. Carrying pulleys are arranged in such a way so as to give the shape of air trough as shown in Fig. 3.10. Return side of the belt is supported on flat idler pulleys. The driving mechanism of the belt is simple and may be located at any point along the carrying length of the belt.

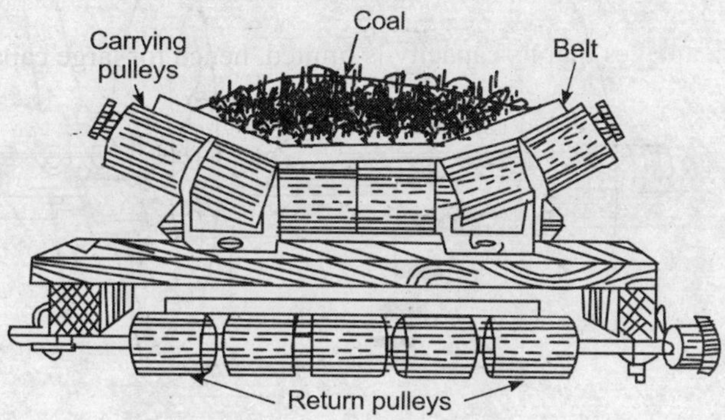

Fig. 3.10 *Belt conveyor*

3.2.5 Belt

The belt is generally 1.5 to 3 mm thick, with nylon or rayon cords or steel wires embedded longitudinally in order to increase the carrying capacity over long distances. In selecting a belt for conveyor, following points should be considered :

(a) Durability
(b) Strength
(c) Toughness
(d) Effect of atmospheric variations e.g., steam, fume, moisture, sun, etc.
(e) Freedom from excessive stretch
(f) Capacity to absorb shocks from fluctuating loads
(g) Lightness
(h) Pliability
(i) Ability to elongate and return to original length without undue heating
(j) Length, speed and type of drive
(k) Cost and expected life

3.2.6 Pulleys

The conveyor 'carrying'pulleys should be of heavy duty cast iron with machined crowned faces. The driving pulley should be faced with some friction material in order to minimise slip. On the carrying side there may be either three or five pulleys depending upon load to be carried, speed and size of coal.

Tripping. Belt conveyors are fitted with travelling trippers (throw off carriages) provided with chutes to discharge the coal into bunkers. The trippers may be of automatic friction driven reversing type or manually operated power propelled gear.

Drive. Conveyors are driven by electric motors through totally enclosed worm reduction gears, preferably mounted on common bed plate. A braking arrangement is also provided to prevent run back when the conveyos is stopped.

Magnetic Pulley. A magnetic pulley is provided to arrest iron particles accompanying the coal. A powerful magnet attracts the iron particles and they are prevented from being discharged along with the coal. Thereafter the cleaning gear is used for removing those iron particles.

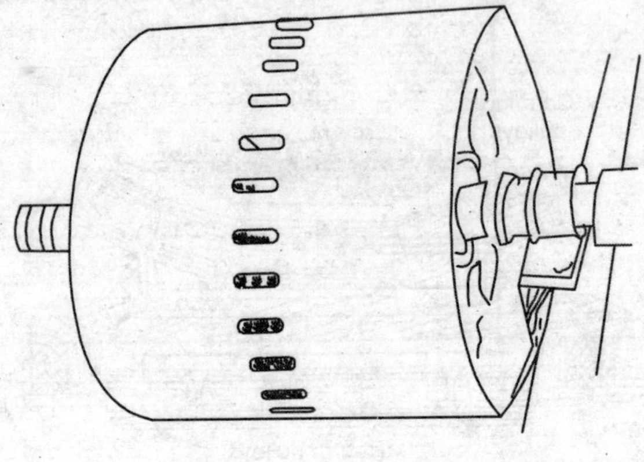

Fig. 3.11 *Magnetic pulley.*

Belt Cleaning. Even after discharging coal, damp coal dust remains on the belt. For this, a belt cleaning arrangement is provided, which uses a rubber scraper or soft wire brush. Cleaners always remain in contact with the belt. The efficient use of a belt cleaner depends on :

(a) Size of coal
(b) Stickiness of coal
(c) Belt material.

Conveyor tracks are usually fitted with sheet metal decking plates to protect the return side of the belt from falling coal. Belt tension is maintained automatically by gravity take-ups. The life of belt is also dependent upon the speed at which the load is carried, excess speed causing accelerated deterioration.

Belt conveyor speeds may range from 80 to 150 metres per second. Distances over which coal can be easily transferred may be as long as 400 metres. Carrying capacity of a conveyor depends upon the size of the belt and working speed and may vary from 50 tonnes per hour to 500 tonnes per hour. Coal can also be transported at inclinations upto 30°.

Advantages

(1) It is the most economical method of coal transfer used in medium as well as large capacity power plants.
(2) The rate of coal transfer can be varied by varying the speed of the conveyor.
(3) It requires minimum maintenance.
(4) Coal being transferred is protected from wind, rain etc. by providing overhead covers.
(5) As coal is carried and not dragged,, so it does not require much power for operation.

Disadvantages

(1) It is not suitable for short distances at inclination.
(2) As the maximum inclination at which coal can be transferred by this arrangement is limited, in order to transfer coal at considerable heights as involved in modern stations, the length of the conveyor becomes excessive.

3.3 SKIP HOIST

The skip hoist is the oldest and simplest means of elevating coal or ash and is favourite of engineers particularly in ash handling. It consists of a vertical or inclined hoistway, a bucket or a car guided by the frame, and a cable for hoisting the bucket. The bucket is so pivoted, with reference to its centre of gravity, as to be held in

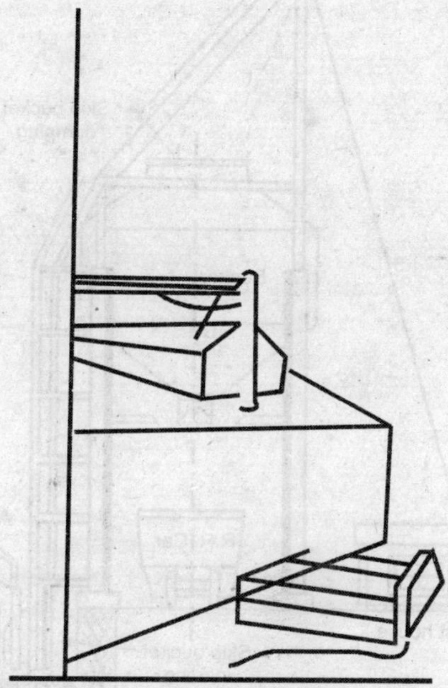

Fig. 3.12 *Skip hoist*

upright position by its own weight and the weight of its contents. A curve guide is located at the dumping point near the top of the hoistway, and engages a roller on each side of the bucket, pulling it into dumping position.

There are three basic types of skip hoists :

(a) Uncounter-weighted (b) Counter-weighted, (c) Balanced.

An uncounter-weighted type skip hoist sends the buckets up fully loaded and, after dumping, lets it return by gravity. This means braking is needed.

In counter-weighted type skip hoist, an auxiliary cable wound in reverse direction on the drum carries a counter-weight equal to weight of the bucket plus 50% of its load. In either direction, the load on motor is half the useful load.

Balanced skip hoists are provided with two buckets operating from a common drum and returning in opposite directions. This arrangement also doubles the capacity at the cost of a similar increase in motor but not in the size of the hoist.

Advantages and disadvantages of skip hoist are as follows :

Advantages

(1) It requires very low maintenance.

(2) Power requirement is also low as the same is required only when the material is being hoisted.

(3) There is no problem in handling large clinkers.

(4) It can be used for handling ash as well as coal.

(5) It needs minimum of floor area.

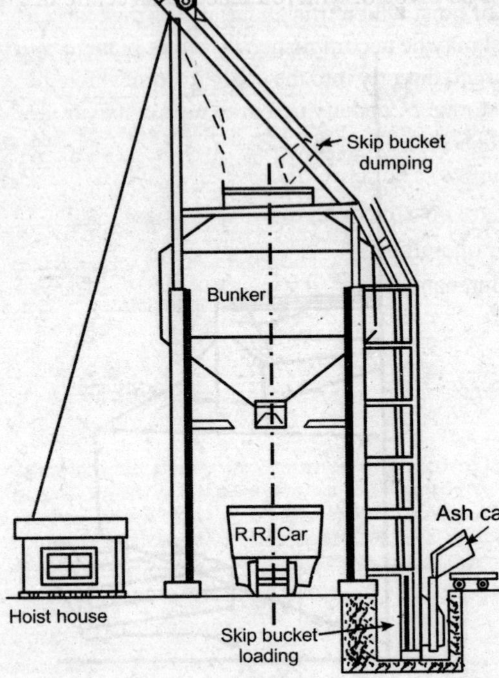

Fig. 3.13 *Skip hoist with bunker and hoist house.*

Disadvantages

(1) Initial cost of the hoist is high. (2) This arrangement gives only intermittent delivery of the material.

(3) There is excessive wear of skips and ropes which need frequent replacement.

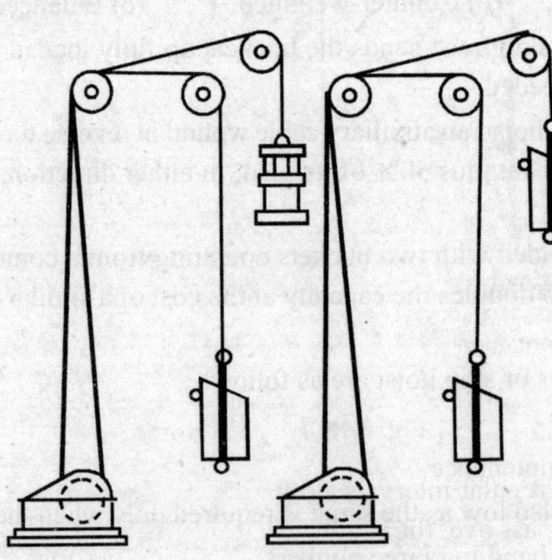

Fig. 3.14

3.4 GRAB BUCKET CONVEYOR

A grab bucket conveyor is a form of hoist which lifts and transfers the load on a single rail or track from one point to another. Both hoisting and travel may be accomplished by either manual operation or power. If the coal car is of dump type, the contents are discharged directly into the coal pit from which the coal is removed by grab bucket and transferred to overhead bunker or storage. Capacity of a grab bucket may be about 50 tonnes per hour. Grab bucket conveyor has the following advantages.

Advantages

(1) Both lifting and transfer are possible with this arrangement.

(2) It requires less power for operation.

(3) It requires minimum maintenance.

3.5 BUCKET ELEVATORS

These are used extensively for vertical lifts, though their possibility for horizontal runs is not ruled out. These elevators consist of relatively small size buckets closely spaced on an endless chain. Three types of elevators e.g., centrifugal, perfect and continuous are shown in Fig. 3.15. In centrifugal discharge type elevator malleable iron

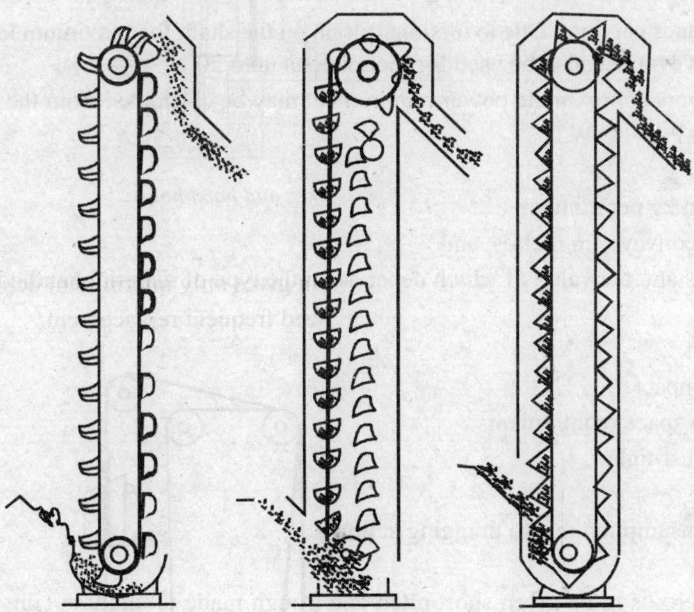

(a) Centrifugal (b) Perfect (c) Continuous

Fig. 3.15 *Bucket conveyor.*

or steel plate buckets are bolted at regular intervals along a chain. The coal is loaded in the buckets in the loading zone at the bottom. As the buckets pass over the headwheel, centrifugal force unloads them. Naturally, high speeds are required for efficient operation. Backets on the chain may be spaced close or apart, depending upon the requirement.

3.6 SCREW CONVEYOR

A screw conveyor consists of a rolled sheet steel spiral secured to a shaft or hollow pipe so as to form a helicoid screw which revolves in an enclosure. Driving mechanism is connected to one side of the shaft and the other end of the shaft is supported in an enclosed ball bearing. Diameter of the screw may vary from 15 cm to 50 cm. The screw is rotated at a slow speed varying from 70 to 125 r.p.m. Capacity of a conveyor of this

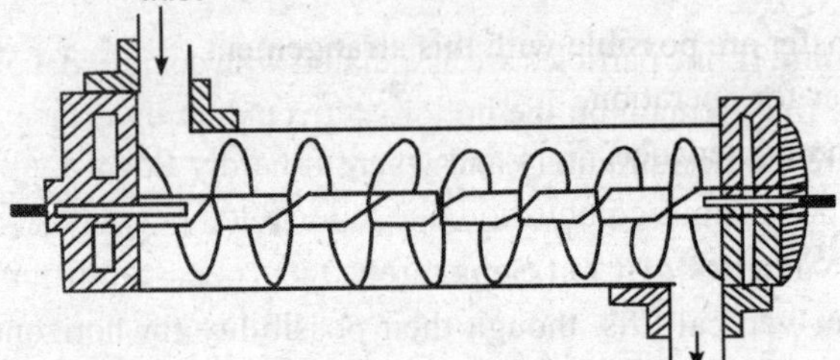

Fig. 3.16 *Screw conveyor.*

type may be upto 125 tonnes per hour. Due to torsional strain on the shaft, the maximum length seldom exceeds 30 metres. This type of conveyor can also be used for inclinations upto 20°.

For horizontal conveyors approximate power requirement may be calculated from the following relation :

$$\text{H.P.} = \text{C.W.L.}$$

where

W = capacity in kg per minute,

L = length of conveyor in metres, and

C = it is a constant, the value of which depends on the type of material handled.

Advantages

(1) Low initial cost

(2) Simple and compact

(3) Adoptability to space requirements

(4) Can be made dust-tight.

Disadvantages

(1) High power consumption due to dragging action

(2) Excessive wear

(3) For inclines, helix is to be given short pitch and trough made tubular, so considerable capacity loss is there.

3.6.1 Storage

Whether the storage is large or small, it needs protection against losses by weathering and by spontaneous combustion. With proper methods adopted even large outdoor storage can remain safe.

All coals tend to combine with oxygen of the air. Sometimes this is a slow process called weathering. By this some heating value of the coal is definitely lost. If the coal has got greater affinity for oxygen under existing conditions, in that case reaction is fast and if the heat of reaction is not carried away, it may result in spontaneous combustion. Following are the methods for avoiding this situation :

(1) By forming the compact layers of coal so that air is not allowed to move through the pile so that the possibilities of such reaction are minimised.

(2) By allowing air to move through the layers evenly so that the heat of reaction is carried away and the temperature of the coal is not allowed to raise and hence the chances of spontaneous combustion are minimised.

(3) **Piling Sized Coal.** With sized coal which leaves more or less uniform passage between the layers, air can move freely and uniformly. In this case the heat of reaction is lost to air and the chances of spontaneous combustion are reduced.

(4) **Avoiding Segregation.** If fine particles are associated with coarse coal pieces, the problem of segregation is created. In normal storage, fines remain on the bottom surrounding by coarse lumps. Such segregation in coal pile gives areas or lanes where air flows radially and where it hardly flows at all. In such case, flow of air is not uniform and the chances of spontaneous combustion due to local temperature rise are there.

(5) **Storing Different Varieties of Coal Separately.** Coal received from different sources or in separate batches is usually stored in separate piles so that even if some variety is liable to spontaneous combustion, rest of the piles are at least saved.

(6) **Sealing the Piles.** Sealing the exterior surface of the piles is probably the best method. Asphalt and other coating materials are generally used for this purpose.

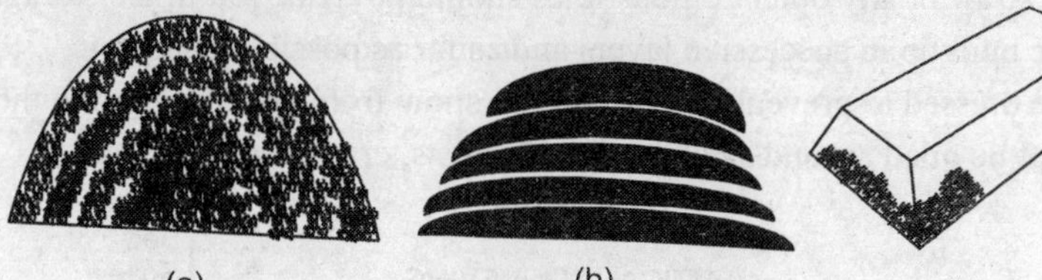

Fig. 3.17 *Coal storage.*

3.6.2 Stock Piling

As far as outdoor stock piling is concerned, a number of means are available. They may be broadly classified as :

(a) fixed (b) mobile

Fixed equipment includes drag scrapers, different types of conveyors etc. Mobile equipment includes lift trucks, bulldozers, scraper with rubber-tired prime-mover, self-loading type twin-engined scrapers, tractors etc. Type of equipment for any given installation will depend on a number of factors, one being initial cost and the operating cost including maintenance.

3.6.3 Storage Procedure

When outdoor storage serves as a reserve, usual practice is to unload the incoming consignment of coal directly into storage within the plant. The proportion of outdoor storage to indoor storage is usually decided on the basis of local factors.

For normal piles the drag scraper arrangement is used. For larger piles and longer hauls, radial set up is preferred. Following points should be considered in the storage of coal:

(1) Storage area should be free from standing water. If necessary, site may be raised.

(2) In case a well-drained area is not available, drainage ditches should be installed.

(3) Storage ground should be solid and not loose or porous. Sometimes the storage areas are top dressed or hard surfaced with paving material.

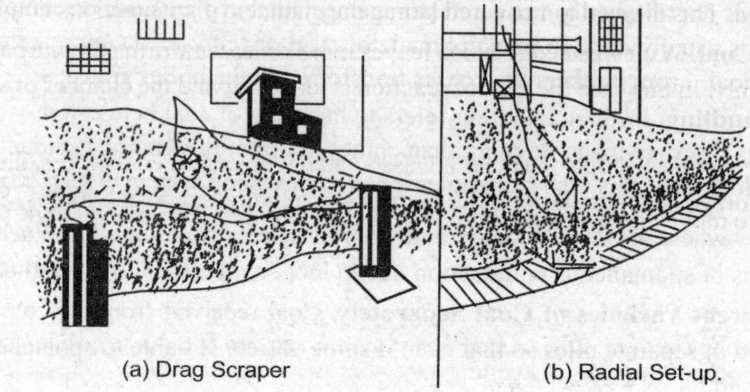

Fig. 3.18

(4) Refuse, wood, straw or any other combustibles should never be left in the storage area.

(5) Piles should be built up in successive layers and as far as possible compact.

(6) Piles should be dressed to prevent rain or melting snow from penetrating into the pile.

(7) Coal should not be piled around steel or wooden posts, crane supports etc.

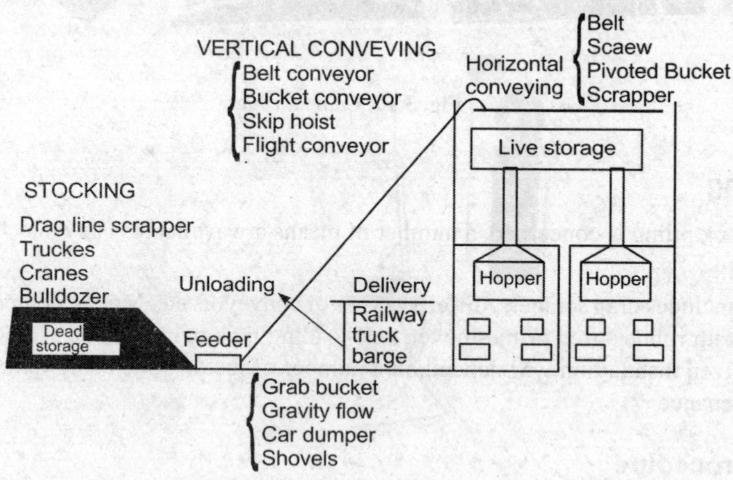

Fig. 3.19 *Coal handling system.*

(8) A pile should not be allowed to be ventilated with pipes or wooden ducts to avoid air and gas movement.

(9) Coal should not be stored near steam pipes, boiler blow-off lines or any other heated elements.

(10) Unsized coal should be invariably compact. For sized coal it is not very necessary.

(11) Conical piling should be avoided. Also sides of the pile should not be steep.

(12) Periodical temperature checks of the area concerned should be carried out.

(13) In order to protect against wind erosion, piles should be sealed.

(14) Storage should be done in such a way that the handling cost is minimum.

(15) Storage area should be minimum and slightly away from the main station.

(16) Fire-fighting equipment should be easily available at the storage site.

1. Live Storage. This is usually a covered storage provided in plants, sufficient to meet a day's requirement of the boiler. Storage is usually done in bunkers made of steel or reinforced concrete, having enough capacity to store the requisite of coal. From bunkers the coal is transferred to the boiler grates.

2. In-plant Handling. In-plant handling refers to handling of coal between the final storage and the firing equipment. Sometimes this may mean no more than chutes to direct flow into individual firing units and grates or valves to control the flow. This may also sometimes mean a conveying system to feed coal from any bunker section to any firing unit and to move coal from one bunker section to another. This may include the equipment used for the transfer of coal.

3.7 WEIGHING

A frequent part of in-plant handling is keeping tabs on quantity and quality of coal fired. Weighing of feed to individual boilers or of total receipts is outlined below with sketches of typical equipment used for the purpose.

3.7.1 Weigh Bridge

With fuel as a major running cost item, keeping watch on its quality and quantity is essential. Total coal received may be weighed at the unloading point in order to have an idea of the total quantity of coal delivered at the site. Sometimes quantity of coal going to individual boiler may also be weighed at the final point. It often pays to weigh the coal at both the points. Weigh bridges (Fig. 3.20) can directly accommodate truck loads on a platform and the measurement is indicated directly by a needle on a graduated scale.

Fig. 3.20 *Weigh bridge*

3.7.2 Belt Scale

Coal is weighed in transit also. In the belt scale shown in Fig. 3.21, the weigh beam connects to a suspended section of the carrying run. An integrator relates running weights and belt travel to record the amount of coal being handled.

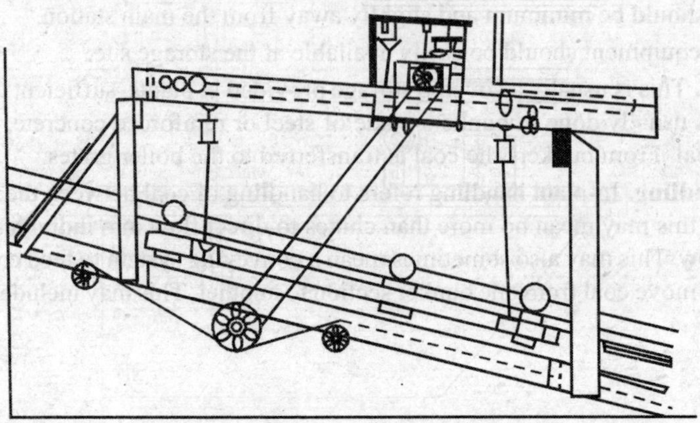

Fig. 3.21 *Belt scale.*

3.7.3 Weigh Lorry

In case it is desired to weigh the quantity of coal delivered to each boiler, weigh lorries, Fig. 3.22, are used. The arrangement may be automatic or manual.

3.7.4 Automatic Scale

It is a device for automatically recording quantity of coal delivered. Accuracy may be upto 1%.

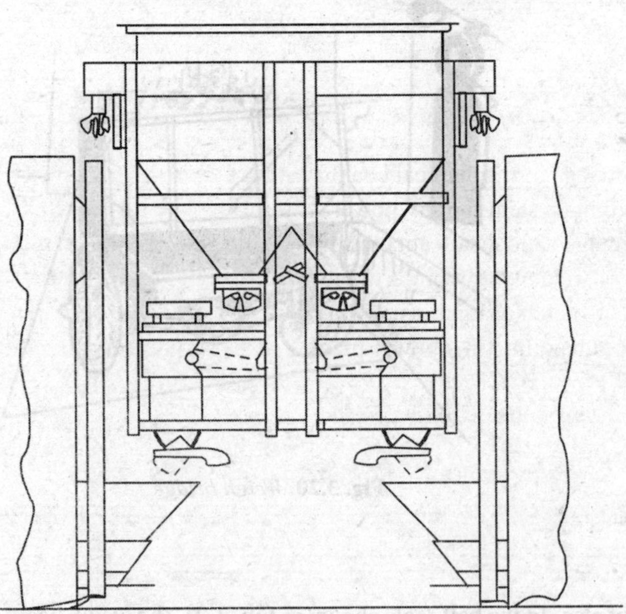

Fig. 3.22 *Weigh lorry.*

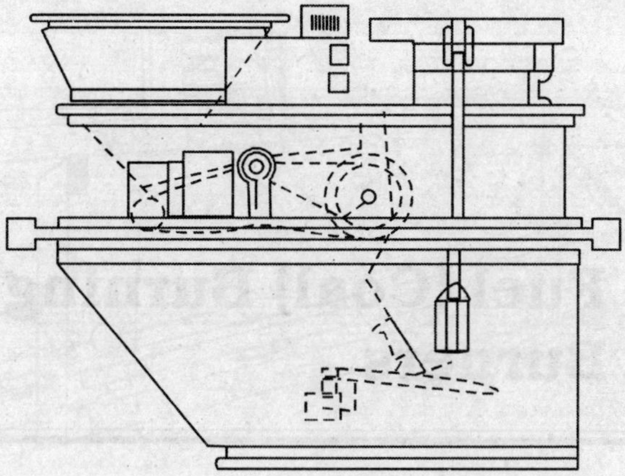

Fig. 3.22 *Automatic scale.*

THEORY QUESTIONS

3.1. Suggest the coal handling system for a big size thermal power station. The site location is within the radius of 30 km of coal mines.

3.2. Describe a modern method of handling coal in power-house of 300 MV capacity.

3.3. What is meant by spinning reserve?

3.4. Describe with a neat sketch a drag scraper.

3.5. Write short notes on :

(a) Skip hoist

(b) Screw conveyor

(c) Belt conveyor.

3.6. What are the advantages of mechanical coal handling system over manual operation?

3.7. Describe coal handling system for any power station which you have visited.

3.8. What are the advantages and disadvantages of scraper or flight conveyor in coal handling?

3.9. Why belt conveyors are generally preferred in power plants over other methods?

3.10. Describe with a neat sketch a skip hoist. What are its advantages over other systems?

3.11. If the choice for coal handling in a power plant is restricted between skip hoist and screw conveyor, which one would you prefer? Give reasons.

3.12. What are the important considerations in the storage of coal at the plant site? What is meant by 'Live Storage'?

Chapter 4 Fuel (Coal) Burning and Burners

A. FUEL (COAL) BURNING

4.1 INTRODUCTION

Fuels burn in a confined space called furnace. The principal requisites for proper combustion of fuel in a furnace are :

(1) Supply of requisite quantity of air to the furnace.
(2) Efficient mixing of air and fuel.
(3) High flame temperature to maintain ignition.
(4) Enough time to complete the burning process within the furnace enclosure.

The process of steam generation can be divided into two parts:

(a) furnace, where heat of the fuel is liberated,
(b) heat transfer surface for the absorption of heat so released.

The way in which fuel is supplied and burned in a furnace has got the maximum influence on the quantity of heat liberated from the fuel. Different methods of firing fuel depend primarily on the physical state of the fuel, e.g., solid, liquid or gas. In this chapter, the methods used for the solid fuels will be discussed.

Based on the type of combustion equipment used, boilers may be classified as:

(1) Solid fuels fired—
  (a) Hand fired,
  (b) Stoker fired,
    (i) Overfeed stokers (chain grate),
    (ii) Underfeed stokers (Single or multiple retort type).
  (c) Pulverised fuel fired,
    (i) Unit system,
    (ii) Central system,
    (iii) Combination of (i) and (ii).
(2) Liquid fuel fired—
  (a) Injection system,

(b) Evaporation system,

(c) Combination of (a) and (b).

(3) Gaseous fuel fired—

(a) Atmospheric pressure system,

(b) High pressure system.

Following considerations should be taken into account while selecting a combustion equipment for solid fuels :

(1) Initial cost.

(2) Adequate combustion space and its ability to withstand high flame temperature.

(3) Grate area over which fuel burns.

(4) In case it is decided to supply air in parts (e.g., primary, secondary, tertiary air etc.), then the ratio of these portions.

(5) Arrangements for thorough mixing of air with fuel for efficient combustion.

(6) Flexibility of operation.

(7) Operating cost of the equipment.

(8) Minimum nuisance due to smoke.

4.2 HAND FIRING

Hand firing is the simplest method of fuel firing and can be used for small capacity units only. Skill on the part of the operator is required in spreading the coal uniformly on the grate. Hand firing is usually done by any one of the methods :

(1) Placing coal on the front edge of the grate, usually known as coking firing.

(2) Spreading the coal uniformly on the grate, known as spread firing.

(3) Placing of fuel on the different parts of the grate alternately known as alternate firing.

(4) In placing the fuel on the grate so as to form channels with ridges and valleys, known as channel firing.

Grates for hand fired furnaces are usually made of cast iron sections in a variety of shapes as shown in Fig. 4.1. In this case primary air is usually supplied from underneath the furnace and the secondary air is supplied at suitable

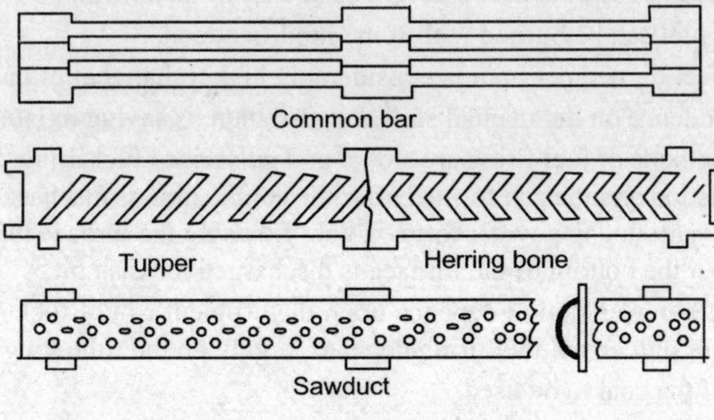

Fig. 4.1 *Different types of grates.*

point, above the grate. Draft is usually controlled by manually operated dampers provided in the flue path. Unless the air spaces are kept free from clinkers and ashes, combustion is hindered and the fire is rendered sluggish. Frequent cleaning of grate is necessary, which is possible only at the cost of furnace efficiency as excess air enters whenever the door is opened.

Stationary grates have been replaced by shaking grates which have the advantage of stoking without opening the fire door and require comparatively less manual labour.

In order to reduce smoke, steam jets are used. A steam jet mixes the air and gases and results in intimate mixing of the combustibles. When steam comes in contact with incandescent carbon it combines with carbon forming carbon monoxide and hydrogen burns in air forming carbon dioxide and water respectively. Clinker formation is reduced and complete combustion is promoted by using steam jet in case of certain types of coals. Of course, there is loss of heat in using steam in jet as steam is usually superheated which takes some heat from the furnace but this loss is offset by the efficient use of fuel due to prevention of clinker formation etc.

Advantages and disadvantages of hand firing method are:

Advantages

(1) It is the simplest method of firing involving practically no capital investment.

(2) Coals of different varieties can be burned without any necessity for adjustment.

(3) Low grade fuels can be economically burned with this arrangement.

Disadvantages

(1) Lower combustion efficiency.

(2) This method does not respond to fluctuations of load immediately.

(3) In case of fixed grate, repeated opening of the fire door for charging and ash door for the removal of ash results in loss of heat as excess air enters the furnace and carries with it the heat of the furnace.

(4) Draft cannot be controlled easily.

4.2.1 Mechanical Methods

Mechanical methods of fuel firing include an arrangement for "carrying" fuel into the furnace enclosure where it is burned and the ash so formed is collected at the bottom.

Advantages of mechanical method of fuel firing are :

(1) There is always uniform layer of fuel on the fuel bed.

(2) Combustion conditions inside the furnace can be easily controlled, hence fluctuating load demands can be easily met with.

(3) The problem of draft control is not so severe as in case of hand firing.

(4) Poorer grades of fuel can be burned with some modifications.

(5) The quantity of fuel burned per hour is considerably higher than that of hand firing.

(6) There is no dependence on the manual skill and also there is saving in labour cost.

Principal mechanical methods of fuel firing are stoker and pulverised fuel firing. In stoker firing coal is carried into the furnace for combustion and the ash formed after the combustion is discharged at the appropriate point. In pulverised fuel firing, fuel is ground to powder form, which 'flows' to the furnace through burners with the help of primary air. Ash collected at the bottom of the furnace is discharged into ash pit.

Economical working of a power station depends upon the efficient use of fuel. Whether this can be obtained more efficiently by stoker or pulverised fuel firing depends largely on the following factors :

(a) Characteristics of the coal to be used.

(b) Capacity of the boiler units.

(c) Load fluctuations.

(d) Station load factor.

(e) Capital cost and maintenance cost of the equipment selected.

Merits and demerits of the two methods e.g., stoker and pulverised fuel firing are given below :

4.3 STOKER FIRING

Advantages of stoker firing are

(1) The coal is burned as obtained from the mines, so there is no necessity of coal preparation plant. Sometimes it may be only necessary to 'size' the coal in order to suit the furnace conditions.

(2) This method can be economically used for medium as well as large capacity boilers.

(3) This method of firing is practically free from the dangers of explosions.

(4) Usually, less building space is required.

(5) Capital investment as compared to pulverised fuel system is less.

(6) It involves less maintenance and operating cost.

(7) Substantial reduction in auxiliary plant.

(8) The method is more reliable.

(9) Some reserve is gained by the large amount of coal stored on the grate in the event of coal handling plant failure.

Disadvantages of stoker firing are as following

(1) Furnaces generally need fire arches or other modifications which increase the cost and are liable to give trouble.

(2) Structural arrangements are not so simple and surrounding floors hare to be designed for heavy loadings.

(3) There is always some loss of coal through riddlings etc.

(4) There is excessive wear of moving parts due to abrasive action of coal.

(5) There is multiplicity of moving parts always under mechanical stresses and high temperature.

(6) Sudden variations in load cannot be met to the same degree of efficiency as in case of pulverised fuel firing.

(7) Troubles due to slagging and clinkering of combustion chamber walls are experienced.

(8) Banking and stand-by losses are always present.

4.4 PULVERISED FUEL FIRING

Advantages

(1) The heating surface area is increased considerably so that high rates of combustion are possible.

(2) Much smaller quantity of air is required as compared to that of stoker firing.

(3) Smaller quantity of excess air and more intense mixing of the fuel and air, produces a high furnace temperature with little smoke.

(4) Fuels of high ash content may be burnt, provided the fusing point of the ash is not too low.

(5) Increased rate of evaporation and increased boiler efficiency can be achieved.

(6) Easier steaming is possible and there is greater capacity of the plant to meet peak loads.

(7) Stand-by losses are minimised.

(8) Boiler can be started from cold conditions rapidly and efficiently. This is highly important in times of emergency.

(9) Practically free from clinker troubles.

(10) Preheated air temperatures of the order of 700°F are possible, thus promoting more rapid flame propagation.

(11) Gives freedom from corrosion of external heating surfaces.

Disadvantages

(1) A separate coal preparation plant is necessary.

(2) The installation is expensive in initial cost and repairs, and unless the calorific value of the fuel is less than 5000 Kcal per kg, or the fuel is available in powdered form, it usually pays to instal mechanical stokers in preference to pulverised fuel firing.

(3) High furnace temperature causes rapid deterioration of the refractory surfaces of the furnace.

(4) Due to high temperature, ash is in the molten form which creates additional problems.

(5) It is difficult to remove economically the fine dust which is suspended in the flue gas.

(6) Fine regular grinding of fuel and proper distribution to burners is usually difficult to achieve.

(7) Large building space is normally required especially with the central system.

(8) Skilled operators are necessary as there is always danger of explosion.

(9) Special starting up equipment is necessary.

(10) Nuisance is caused by the emission of very fine particles of grit and dust.

4.4.1 Stoker firing

Modern stoker-fired boilers exhibit the following features:

(a) The liberty of design enabling the specified continuous maximum duty to be met without excessively high rates of combustion on the grate surface or of heat transmission through the boiler heating surface.

(b) A furnace volume of very ample proportions with a correspondingly large area of radiant heat absorbing surface, whereby the temperature of the gases entering the convection zone of the boiler is kept well below the ash fusion temperature of the fuel or mixture of fuels which may be encountered.

(c) Admission of secondary air in such quantity, pressure and direction as to ensure thorough mixing of the gaseous products of combustion. The furnace design generally will aim at ensuring that all the products pass through a zone of luminous flame, the value of which features in inhibiting harmful deposits has been established by the researchers.

(d) The boiler surface proper will be disposed in well-defined banks and in lateral spacing to establish a regular temperature gradient between gases and metal surface of the tubes, and to afford accessibility both for *on* load and *off* load cleaning.

(e) Accumulation in contact with heating surfaces of dust or grit carried forward by the gases will be avoided by the provision of suitable pockets at appropriate points in the gas passes, whence on load dust extraction can be readily affected.

(f) On load cleaning, in fact, will have received consideration as part of the fundamental design, an aspect facilitated by the general adoption of remote controlled sequence operated soot blowers, the location which can be determined without regard to the limits formerly imposed by manual operation of individual blowers.

4.5 PRINCIPLE OF OVERFEED STOKER

In overfeed stokers, coal is supplied on the top of the grate. As the coal burns, ash and clinkers are formed which collect at the bottom, then the next charges of coal are supplied and the process continues. Primary air is supplied under pressure from the bottom of the grate. As the air flows through the grate openings and the layer of ash, it becomes heated up and as it passes through the layer of incandescent coal, combustion is accelerated in that region. By the time it reaches the region of green coal, entire oxygen of the air is consumed. Combustion of green coal is promoted by supplying secondary air from the side. Gases leaving incandescent region consist of carbon-di-oxide, carbon monoxide, nitrogen, hydrogen and steam with oxygen usually missing. Fresh fuel undergoing distillation of its volatile matter forms the topmost layer of the fuel bed. Heat for distillation and eventual ignition comes from the following sources :

(1) by conduction, from the incandescent coal region.

(2) from the gases at high temperature diffusing through the upper layer.

(3) radiation from flames,

(4) hot gases present in the furnace,

(5) from the hot furnace walls,

(6) secondary air, if pre-heated.

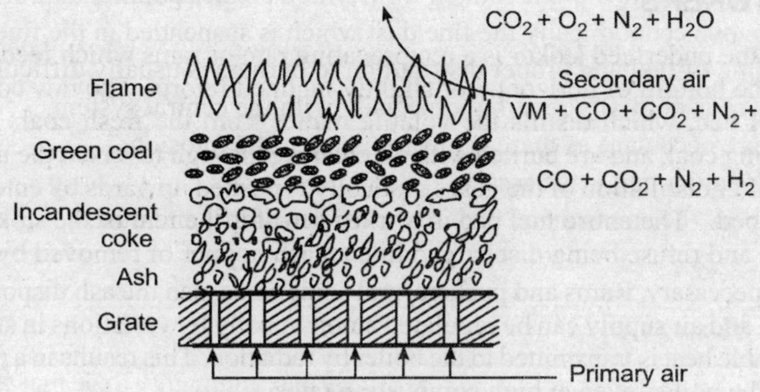

(a) Principle of overfeed stoker

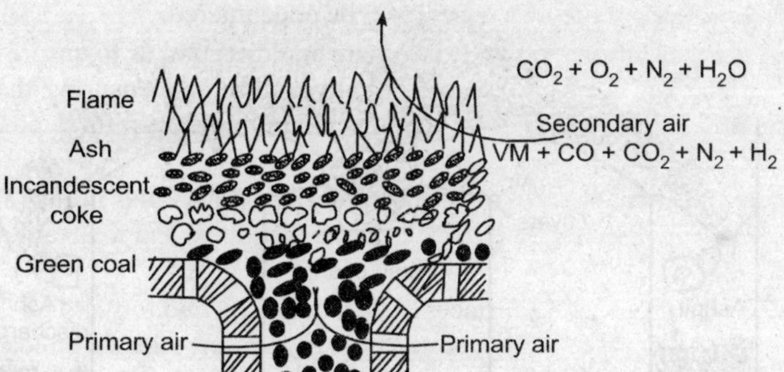

(b) *Principle of underfeed stoker*

Fig. 4.2

For complete combustion in the furnace :

(i) Secondary air is introduced at high velocity so as to create turbulence.

(ii) A fire arch is usually provided for thorough mixing.

4.6 PRINCIPLE OF UNDERFEED STOKER

Underfeed stokers utilise the gas producer principle. Green coal is fed to the lower layer of the fuel bed and is gradually pushed up and coked, giving up its volatile constituents and becoming incandescent by the time it reaches the top layer. Clinkers or ash which is formed on the upper layer is forced to the side or back of the fuel bed, where it is removed by suitable means. Air under pressure is supplied through small openings in the grate. First of all air diffuses through the pores of the raw fuel, thereby taking volatile matter along with it. Then it passes through the ignition zone and enters the region of incandescent coke. Usually the reaction is not complete in this region, so secondary air is required for the same. Underfeed stokers are best suited for semi-bituminous and bituminous coals high in volatile matter content and with caking tendency. The tendency of caking coals to swell augments, the action of the stoker in producing a fuel bed of unusual thickness and the pushing action of the feeding mechanism keeps the bed broken up and porous. All underfeed stokers are essentially forced draft stokers, since they operate with restricted air openings only. There are two general classes of underfeed stokers, e.g.,

(i) Single retort type. (ii) Multiple retort type.

4.7 UNDERFEED STOKERS

The essential principle of the underfeed stoker is a reciprocating ram or rams which feed coal from hoppers at the front of the furnace into the bottom of horizontal or slightly inclined retorts. The raw coal is underneath burning coal at the top of the fuel bed, which distills the volatile matter from the fresh coal. The liberated gases pass upwards through the burning coal, and are burned with air entering through tuyers at the upper edges of the retorts. The coke which remains after distillation of the gases gradually is pushed upwards by entering fresh fuel and burns on the furnace of the fuel bed. The entire fuel bed is worked towards the rear of the stoker or on to dead plates at the sides of the retort, ash and refuse being discharged into an ash hopper or removed by hand.

Forced draft is always necessary. Rams and pushers, and sometimes also the ash disposal equipment, are driven by a motor or engine. Fuel and air supply can be regulated automatically by variations in steam pressure. Arches are unnecessary, and considerable heat is transmitted to the boiler by radiation. This results in a relatively low temperature of gases passing through the boiler, even at high combustion rates.

Single Retort Stokers (Fig. 4.3) use a steam-driven ram or a screw feed, together with supplementary adjustable stroke pushers to distribute coal properly in the retort. From the surface of the fuel bed, refuse is deposited on dead

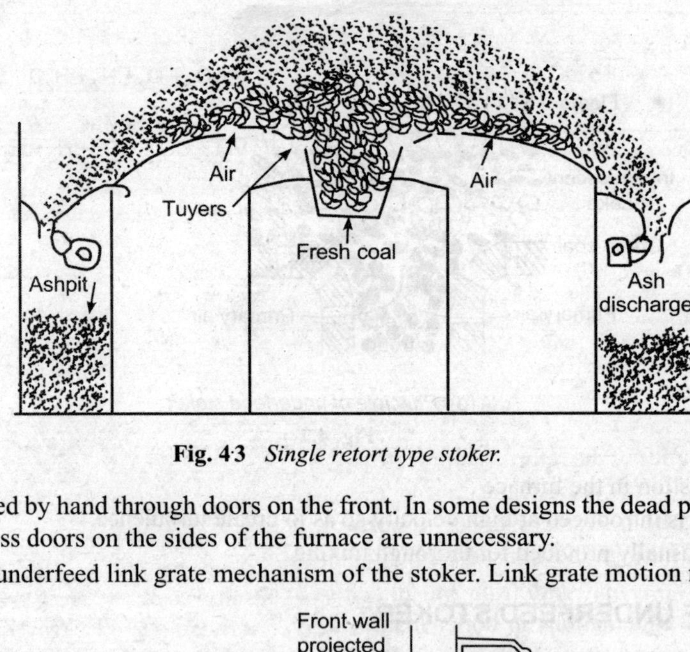

Fig. 4·3 *Single retort type stoker.*

plates where it is removed by hand through doors on the front. In some designs the dead plates may be dropped to dump to the ashpit. Access doors on the sides of the furnace are unnecessary.

Figure 4.4 illustrates underfeed link grate mechanism of the stoker. Link grate motion reduces resistance to air

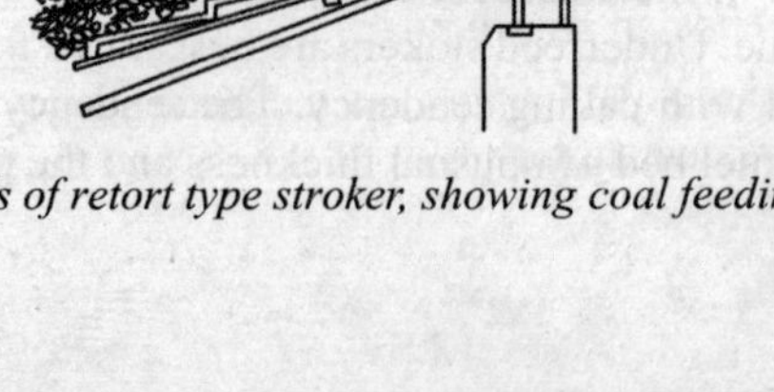

Fig. 4.4 *Details of retort type stroker, showing coal feeding ssytem.*

flow by keeping the fuel bed porous. Air flowing through the grates is not smothered, but is allowed to support combustion by thoroughly permeating all parts of the fuel bed.

At moderate combustion rates, even with high volatile coal, combustion is complete within a short distance of the surface of the fuel bed. The capacity of stroker ranges from 100 to 2000 kg of coal burned per hour.

Multiple Retort underfeed stokers occupy the fuel width of the furnace. The fuel bed consists of alternate retorts and tuyers for the supply of air. The furnace bed may be inclined at about 20°. Below the retorts, the grate surface is continuous so as to burn fuel completely before it is pushed off the ash supporting plates. The distinctive feature

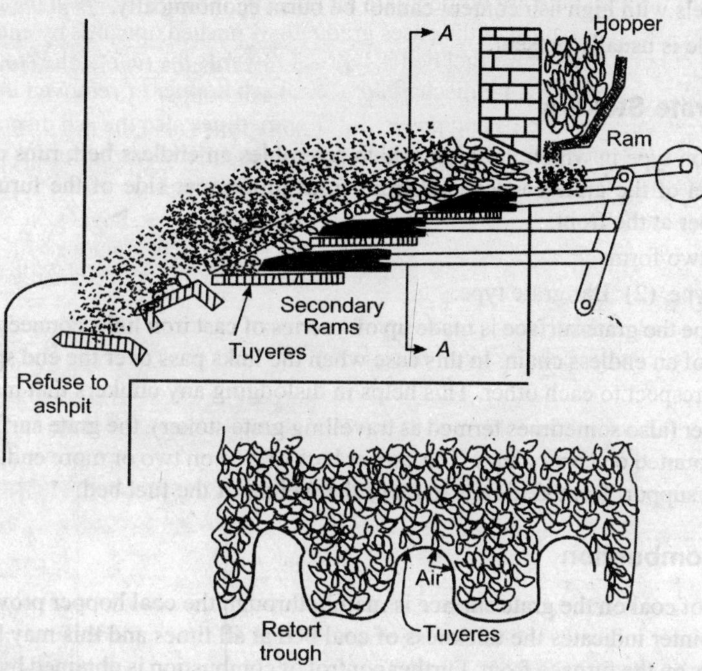

Fig. 4.5 *Multiple retort type stroker.*

of this stoker is that the sides of the retorts reciprocate relative to the bottoms. This provides a means of moving the fuel uniformly along out of the retort. It also provides a moving grate surface on to which the fuel is passed as it leaves the retort. The same movement serves to push the refuse cross the rocker ash dumping plates where it continuously discharges through the adjustable opening next to the bridge wall.

The number of retort units may vary from 2 to 20 with coal burning capacity ranging from 300 kg to 2000 kg of coal per hour per retort. Fluctuations in boiler load can be effectively met in this case. Part load efficiency of multiple retort type stokers is usually high.

Advantages of Retort Stokers

(1) Higher thermal efficiency is possible as compared with chain grate stoker.

(2) Part load efficiency is high particularly with multiple retort type.

(3) Much higher steaming rates are possible with this type of stoker.

(4) Higher temperatures of preheated air are possible.

(5) Substantial amount of coal always remains on the grate so that the boiler may remain in service in the event of temporary breakdown of the coal supply system.

(6) Overload capacity of the boiler is high as large amount of coal is carried on the grate.

(7) Ignition arches, if required, are simple in construction.

(8) With the use of clinker grinder, more heat can be liberated out of the fuel.

(9) The grate is self-cleaning.

(10) Tuyers, grate bars etc., are always in contact with fresh coal and are not subject to high temperature.

Disadvantages

(1) Initial cost is high.

(2) Large building space is required.

(3) Low grade fuels with high ash content cannot be burnt economically.

(4) Clinker trouble is usually present.

4.7.1 Travelling Grate Stokers

This is the most common type in which the grate which resembles an endless belt, runs over two sets of sprocket wheels. The upper level of the grate carries coal from front to the rear side of the furnace. The flow of fuel is regulated through hopper at the front.

These stokers have two forms :

(1) Chain grate type (2) Bar grate type.

In the chain grate type the grate surface is made up of a series of cast iron links connected by suitable bar or pins thereby forming a sort of an endless chain. In this case when the links pass over the end sprockets, they move with a scissor like action in respect to each other. This helps in dislodging any clinkers that may adhere to the grate.

In the bar-grate stoker (also sometimes termed as travelling grate stoker), the grate surface is made up of a series of cast-iron sections mounted on carrier bars. The carrier bars of ride on two or more endless type drive chains and may receive additional support from skid rails to carry the weight of the fuel bed.

4.7.2 Control of Combustion

The thickness of layer of coal on the grate surface is mainly through the coal hopper provided at the furnace front. An index plate with pointer indicates the thickness of coal bed at all times and this may be regulated by adjusting the opening of the doors on the furnace front. Further control of combustion is obtained by speed control (PIV type) of the stoker motor. Also in certain cases the grate can be lowered or raised as desired.

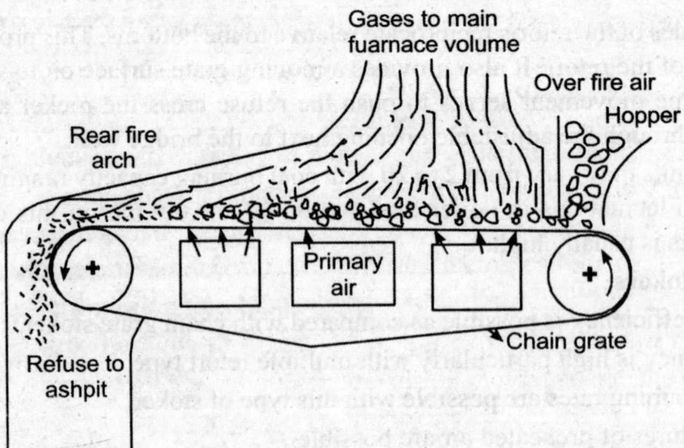

Fig. 4.6 *Chain grate stoker.*

1. Air Supply. The air supply above and below the grate is of prime importance and it is usual to divide the underside of the grate into several compartments, each of which are connected to the air duct or wind box through adjustable dampers, (Fig. 4 5). The air pressure is kept high in the first compartment and it is successively reduced in the remainder. This is done according to the requirements for the combustion of coal, which is maximum at the front and goes on reducing as the grate progresses further inside the furnace. The air may be preheated, the temperature however depends on the type of coal burned and may be as high as 225°C. High air temperatures accelerate ignition, giving more flexibility with changing loads, and improve overall boiler efficiency. However high air temperatures result in warping and overheating of grates.

Secondary air enters the furnace through nozzles from ducts in the furnace walls not shown in the figure.

The provision of simultaneous control for the grate speed, coal layer thickness and flow of air helps in controlling combustion inside the furnace in such a way so that only nothing but ash remains on the furnace by the time it reaches the furnace rear. The ashes fall into the ash pit as the grate turns on the rear sprocket to make the return trip on the lower level to the furnace front.

2. Fire Arch. As the raw or green coal on the grate enters the furnace, the surface coal ignites from heat of the furnace flame and also from heat reflected by the fire arch. Thus coal undergoes the process of coking and it is here that the volatile hydrocarbons are driven off and combustion of the resulting gas can be effected by deflecting them downwards on the incandescent coal further back. For this purpose an ignition arch as shown in Fig. 4.4 is provided. Thus the functions of ignition arch are :

(a) to ignite the coal,

(b) to direct the flow of gases to the heating surface,

(c) to ensure thorough mixing of the furnace gases.

Travelling grate stokers can burn coals at the rates varying from 150 to 250 kg per sq. m per hour of grate surface. As the percentage of fires in the coal drops the rate of heat release drops sharply. In terms of heat release the capacity of these stokers ranges from 1.6×10^6 to 2.5×10^6 Kcal/sq. m/hr depending upon the type of coal burned.

For speed control of grate either variable speed electric motors are used or motor with P.I.V. or gearbox are used for changing the grate speeds.

Secondary air is usually introduced through this arch at the neck of the furnace entrance thereby helping in the mixing of gases and supplying oxygen for completing the combustion. The ignition arches are provided on the front as well as rear of the furnace and their shape varies from manufacturer to manufacturer.

Advantages

(1) Simple in construction.

(2) Lower initial cost.

(3) Maintenance charges are low.

(4) Reliable in service.

(5) Self cleaning.

(6) Higher rates of heat release per cubic volume are possible.

(7) Ash crushers are not required.

(8) Speed of chain can be controlled to suit the furnace requirements.

Disadvantages

(1) The amount of coal carried on the grate is usually small as increase in size of grate creates additional problems.

(2) Preheated air temperatures are limited to 180°C maximum.

(3) There is always some loss of coal in the form of fine particles through riddlings.

(4) Ignition arches to suit the specific furnace conditions are required.

(5) Clinker troubles are present.

4.8 SPREADER STOKERS

These stokers use the overfeed principle of fuel burning. In this case the grate is stationary. The coal is handled by the revolving blades of feeder units of drum type and flung out into the furnace on the top of fire. The spreader rotor revolves at high speed and the paddles of the rapidly revolving spreader hit the particles of the steady coal stream falling in front of them and hurl them into the furnace. The angularity of the pad-dles gives uniform distribution of the coal along the width of the furnace. However, distinction of coal along the length of the furnace is controlled by the rotor speed.

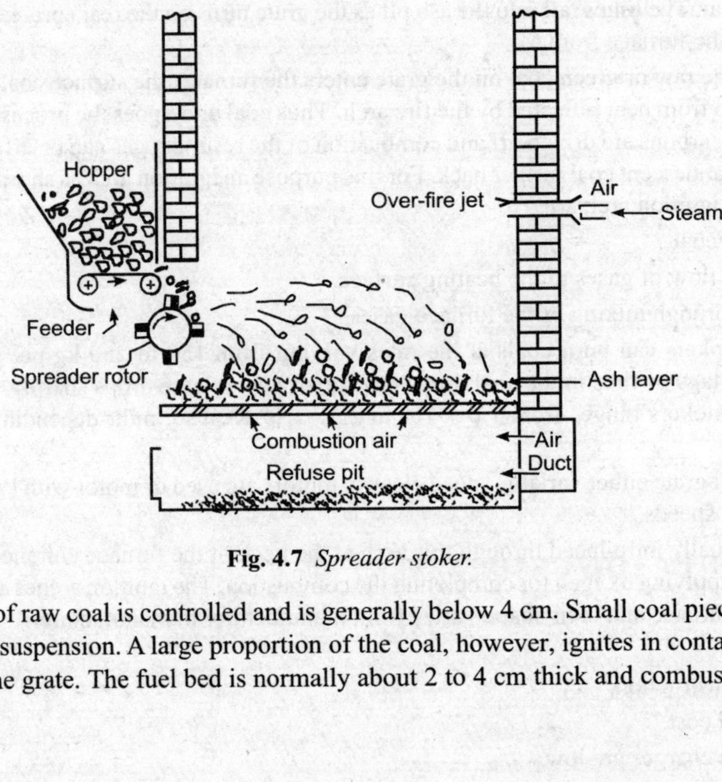

Fig. 4.7 *Spreader stoker.*

In this case the size of raw coal is controlled and is generally below 4 cm. Small coal pieces while flying inside the furnace burn out in suspension. A large proportion of the coal, however, ignites in contact with the active fuel bed and burns out on the grate. The fuel bed is normally about 2 to 4 cm thick and combustion takes place at the surface of the bed.

4.8.1 Grate

The grate in this type of stoker usually consists of perforated cast iron bars about 1 metre long and 15 cm wide with a depth of about 15 cm. The bars are connected in such a way that the operation of a lever back and for makes the bars rock about their pivots. This makes the ash on the bottom of the fire fall through the ashpit underneath the grate. Till the lever is operated, the ash remains below the fuel and both grate surface and ash are maintained at the temperature of the incoming air. This tends to eliminate clinkering even if the fusion temperature of the ash is as low as 1100°C. and ash on the grate tends to protect the bars.

4.8.2 COMBUSTION CONTROL

In this type of stoker there is very little active fuel in the furnace. In such case the fire may go out within a few minutes if the fuel supply is stopped. This feature makes it very responsive to changes in fuel and air supply and an increase or decrease of fuel with constant rate of air flow or vice versa (i.e., constant rate of fuel supply and varying rate of air flow) will produce a marked change of carbon-di-oxide percentage in the flue gases in a matter of seconds. For efficient operation, therefore, constant and simultaneous variation of fuel and supply is necessary.

Secondary air is supplied through air jets about 1 metre above the grate for thorough mixing of furnace gases and air. Some designs may use steam jets for the purpose and force all combustion air through the fuel beds.

Because of the thin fuel bed and suspension burning spreader stokers tend to suffer a fuel loss in cinders up the stacks. This may run from about 1 to 4 percent. Arrangements are usually made to return some of the cinders, trapped in the boiler passes, to the furnace for complete burning. Steam jets help to reduce the cinder loss and smoke.

Large steam generators may have two or more spreader stokers side by side in the furnace front wall. Stationary grates usually serve boilers upto a maximum capacity of about 40,000 kg/hr capacity. For large capacities upto about 200,000 kg/hr, a travelling grate is used.

Advantages

(1) This stoker is simple to operate, easy to light up and bring into commission.

(2) There is flexibility in operation and it is responsive to load changes and is capable of being banked and brought up again with low fuel consumption.

(3) As coal particles are thrown inside the furnace they burn in suspension.

(4) Volatile matter is easily burned.

(5) Fire arches etc. are generally not required with this type of stokers.

(6) High temperatures of preheated air are possible.

(7) As the depth of coal bed on the grate is usually limited to 10 to 15 cm only, fluctuating loads can be easily met with.

Disadvantages

(1) It is difficult to operate spreader with varying sizes of coal.

(2) There is no remedy for clinker troubles.

(3) Because of the thin fuel bed and suspension burning some fuel loss in the cinders up the stack is possible.

4.9 COMBUSTION OF LOW GRADE FUELS ON STOKERS

Low grade fuels are characterised by one or more of the following:

(1) High ash content (18 – 20% and upwards).

(2) Excessive moisture content (such as, for example, to make the total inerts upto about 30% and more).

(3) Low calorific value–say 21000 kJ/kg.

(4) High fines content (say 50% and more of fines smaller than 3 mm).

In addition such fuels will be included which for some other reason (e.g., the chemical or refractory qualities of the mineral matter or very low volatile matter content) cause difficulties in combustion.

With low grade fuels the difficulties caused by fouling of the external heating surfaces of boiler plants are greatly increased, despite improvement in soot blowing techniques. Another natural consequence of high ash and fines content of low grade fuels is the problem of grit emission, a concomitant of which is increased loss of carbon in grit. Designers of grit arresting equipment have, however, successfully applied themselves available which is capable of minimizing grit emission satisfactorily from all sizes of boilers. More recently, attention has been directed to the recovery of the combustibles in the grits, and means have been devised, particularly in connection with the developments of the spreader stokers of returning the large grits to the boiler furnace for reheating.

The major problem in burning low grade fuels on stokers not designed for the purpose, is that of obtaining the rated capacity of the boiler, since it is impossible to obtain the equivalent heat release per square metre of the grate area when using fuels so much inferior to those for which the grates were designed. Attempts to succeed by the usual methods of increasing draught and stoker speed do not generally achieve the desired result and, in the case of chain and travelling grates, give rise to "horse-shoeing" or late ignition.

Once the ignition line has receded beyond about 50 cm from the coal shutter door, the ignition arch becomes cooled below the temperature at which its radiated heat can ignite the incoming coal. The grate speed has to be reduced considerably in order to maintain stable ignition. As a result of this, and also of the lower furnace temperature, the boiler output is seriously reduced.

The difficulty is particularly acute when low volatile coal is used on stoker settings designed for high volatile coal, and, therefore, fitted with comparatively short ignition arches. Late ignition is also accentuated by high ash content and high moisture content.

Although moisture causes difficulties and lowers effective calorific value, the burning of dry smalls and fines is generally assisted by the controlled addition of water at the point of consumption in amounts upto 12% to 15%. Fines in a mixed fuel retard the burning rate, and require a larger grate area for a given boiler duty, this is particularly the case with fines such as slurry, the burning of which is complicated by delay in ignition and consequent risk of loss of combustible in the ash.

The following methods have been found particularly useful in solving the above problems :

(1) Blending
(2) Sandwich system
(3) Conditioning of coal with water and steam
(4) Oil assisted firing
(5) Steam jet ignition system.

1. Blending. The burning properties of a low grade fuel can be improved by premising with better quality coal. Strictly, a low grade solid fuel cannot be upgraded. In a mixture of a low volatile coal or of coke breeze with a higher volatile coal, their respective characteristics are retained ; but when the higher volatile coal burns, it undoubtedly assists the combustion of the other fuel. The essential point is that the mixture does not necessarily behave exactly as would a single solid fuel of the same analytical composition, but the mass of burning volatile matter provides it effect of an ignition arch. That is why retort stokers do not have ignition arches ; they do not require them. The flame arch constituted by the burning volatile matter replaces the ignition arch.

2. Sandwich System. In this system, the two fuels are fed by independent conveyors to the main overhead hoppers, and then by separate chutes to independent hoppers on the stoker front. The coal hopper, which has the normal guillotine door, is nearer to the combustion chamber. The coke hopper is outside it and is fitted with an independent fuel levelling plate. An average fuel bed of about 13 cm thickness is maintained, made up of one layer of coke breeze with the coal on the top. The thickness of each fuel layer is determined by the proportion by weight of the fuel in use.

3. Coal Conditioning. Conditioning of coal is an aid to the efficient combustion of small coals containing high proportions of fines. In this case, moisture is added to incandescent coal on the fuel bed, thereby improving combustion and reducing the "fuel bed resistance". In addition to increasing the boiler output, owing to the lower fuel bed resistance and even burning of the fuel, adequate conditioning of the fuel produces other incidental advantages. The losses due to riddlings and grit carry over are reduced and the "cauliflowering" experienced with dry fuels having a high swelling index is appreciably overcome.

4. Oil Assisted Ignition. In this system, using a quality of oil not exceeding 5% by weight of the total fuel required, the boilers have operated at full rate capacity on poorer grades of coal which it was not possible to use alone. The performance compares favourably with that obtained with good quality coals. The oil supply is maintained through a burner connected to a overhead constant level tank. A feature of the system is its simplicity and the case with which it can be installed on existing plant.

5. Steam Jet Ignition System. This is another system with the merit, of simplicity of design, cases of installation on existing plant and low operating cost, which was designed to facilitate the burning of low grade fuels on travelling grate stokers.

The equipment comprises a number of steam jet nozzles arranged in a horizontal plane, attached to a steam supply bus line which is mounted on the lower end of the coal feed regulating shutter in such a manner that the steam jets constantly retain their position relative to the incoming fuel bed, irrespective of the fuel bed thickness—provision is made for control of the angle of steam injection into the combustion chamber, Fig. 4.8 (a) and (b).

The steam jets have a considerable initial velocity which dimini-shes rapidly with increasing distance from the nozzles, until ultimately the jets of steam lose their identity and become merged with the furnace gases. The operation of the jets causes a greatly reduced furnace pressure in the ignition zone of the combustion chamber. This causes the hot gases and flame to blow back towards the ignition zone. They do so on well-defined path, those below the jets making direct contact with the incoming fuel and giving immediate ignition, whilst those above the jets tend to raise the temperature of the ignition arch thus further promoting ignition. The extent to which the gases are drawn back is controlled by means of the steam pressure to the jets, which may vary from 1.05 to 3.5 kg. per sq. cm gauge. This permits of a very fine degree of control and the maintenance of straight line ignition across the width of the grate.

Fig. 4.8

Untreated bituminous slacks having high ash and fines content and have been successfully used with the help of this system at several places, on both large and small stoker fired boilers.

B. BURNERS

4.10 DEFINITION

Solid fuels are supplied to the furnace by means of stokers. For liquid, gaseous as well as pulverised fuels, burners are used. A burner in its simplest form incorporates an arrangement for supplying the correct amount of fuel and air to the furnace. Depending upon the type of fuel handled the burners are known as liquid fuel burners, gas fuel burners, pulverised fuel burners, etc.

4.10.1 Liquid Fuels for Oil Burners

Most liquid fuels are extremely difficult to burn from the liquid state. In fact, a stream of many liquid fuels will extinguish an ignition source and make the combustion of such liquid fuels impossible. Liquid fuel burners may be classified as :

(a) *Evaporation type.* In this case, fuel is evaporated before ignition.

(b) *Atomisers.* In this case, fuel is atomised or split into very fine particles, before combustion.

4.10.2 Evaporation Type Burners

The requirements of an evaporation type burner are to

(1) vaporise the fuel before ignition.

(2) mix the vaporised fuel thoroughly with the air.

(3) minimise the soot formation. For this the fuel should be evaporated or dispersed in a fine mist before oil begins to undergo cracking due to high temperature of the flame and combustion chamber.

(4) to give high heat release by burning large quantity of oil per hour. For this usually the velocity of mixture of fuel vapour and air through the burner is increased.

(5) to allow for efficient combustion of fuel at part load operation. This is usually done by controlling the quantity of fuel and air supply.

Most of the viscous liquid fuels are difficult to evaporate at atmospheric pressure. An attempt to vaporise these fuels by subjecting them to high temperatures often results in thermal cracking of the oil. In order to promote rapid evaporation of heavy fuel oils, burners are designed to break the fuel up into minute droplets and project these into the combustion chamber in the form of a fine mist. This process is different from atomising. Although in this case ignition of the oil occurs before all of the fuel is evaporated, it may be assumed that fuel entirely surrounds the independent droplets. As the vapour mixes with the air and burns, the radiant heat from the flame promotes evaporation of the remainder of the droplets and the entire combustion process probably occurs between vapour and air.

Vaporising burners may be classified as :

(1) Atmospheric pressure atomising burner.

(2) Rotating cup burner.

(3) Recirculation burner.

(4) Wick type burner.

(1) Atmospheric Pressure Vaporising Burner. This type of burner is used for highly volatile liquid fuels such as neptha, volatile gasoline etc. In this case, fuel at low pressure is passed through a tube adjacent to the

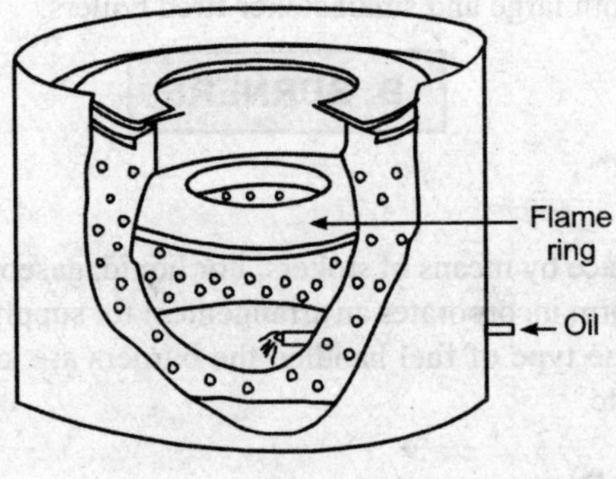

Fig. 4.9 *Vaporising burner.*

flame before being released through an orifice. While passing through the hot tube, most of the fuel is vaporised so that the fluid ejected from the orifice is more or less a vapour. Necessary quantity of primary air is supplied to burn the vapour steam in a cylindrical tube.

(2) Rotating Cup Burner. In this case fuel oil flows through a tube in the hollow shaft of the burner and into the cup at the furnace end. The shaft and cup are rotated at high speeds ranging from 3,000 to 10,000 RPM by an electric motor or an air turbine. Splitting of fuel into small droplets occurs as a result of centrifugal force, acting on the fuel projecting out the rapidly rotating cup. About 10% to 15% of air is supplied as primary air. This air is

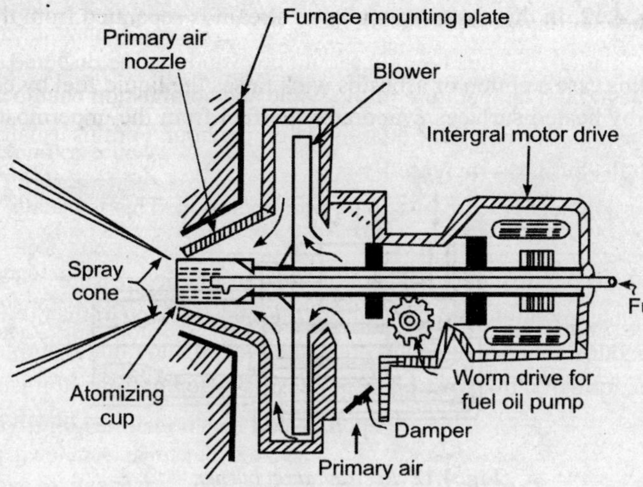

Fig. 4.10 *Rotating cup type burner.*

supplied from a blower surrounding the cup. The angle at which the ait hits the fine oil mist may be adjusted by regulating the relative position of the cup and the air cone. The shape of the flame is governed by the shape of the cup and the position of air nozzle. These burners are used on low as well as medium capacity boilers.

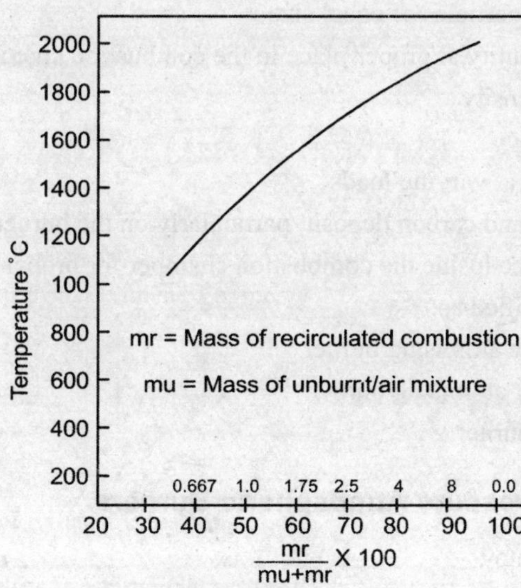

Fig. 4.11 *Effect of recirculation.*

(3) Recirculation Burners. In order to heat up the incoming steam of fuel and air, part of the products of combustion may be recirculated. The effect of recirculation on the temperature of the incoming stream is shown in Fig. 4.11. Low ratio of the mass of recirculated combustion products to the mass of unburnt fuel-air mixture results in less temperature rise of the mixture whereas high ratio may extinguish the flame due to increased proportion of recirculated products. An optimum ratio may be determined for different fuels experimentally. A. burner utilising this principle is shown in Fig. 4.12. In this case, recirculation stream is separated from the combustion zone by a solid wall.

(4) Wick Burners. In this case a cotton or asbestos wick raises the liquid fuel by capiliaty action. Radiant heat from the flame and nearby heated surfaces evaporates the fuel from the uppermost part of the wick. Air is

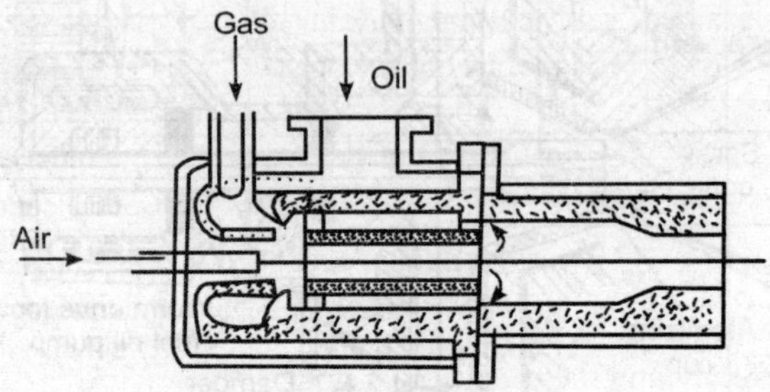

Fig. 4.12 *Recirculation burner.*

admitted through holes in the surrounding walls. This type of burner is suitable where low heating rates are desired e.g., for models or domestic appliances.

4.10.3 Atomising Fuel Burners

Requirements of an atomising fuel burner are to :

(a) atomise the fuel into fine particles of equal size
(b) supply air in required quantity at proper place in the combustion chamber
(c) give high combustion intensity
(d) give high thermal efficiency
(e) operate without difficulty at varying loads
(f) minimise soot formation and carbon deposit, particularly on the burner nozzle
(g) create necessary turbulence inside the combustion chamber for proper combustion of fuel.

Atomising burners may be classified as:

(a) Mechanical or oil pressure atomising burner.
(b) Steam or high pressure air atomising burner.
(c) Low pressure atomising burner.

4.10.4 Mechanical or Oil Pressure Atmospheric Burners

These may be further subdivided into :

(i) Swirl pressure jet burner
(ii) Spill control burner
(iii) Spring loaded piston type burner.

In a swirl pressure jet burner fuel is introduced tangentially through two ports into a swirl chamber of 6 to 9 times the diameter of the port. By the conservatoin of angular momentum fuel spins faster and faster as it moves towards the axis. It leaves through a circular hole in one side of the axis and provided the viscosity of the oil is not high, there will be a hollow air core down the centre where all the inlet static pressure is converted into tangential velocity. This air core will be smaller in size than exit nozzle and the fuel will flow out between the air core and the nozzle. It leaves in the form of a conical thin sheet which is kept thin by the fact that the fuel leaving from the air

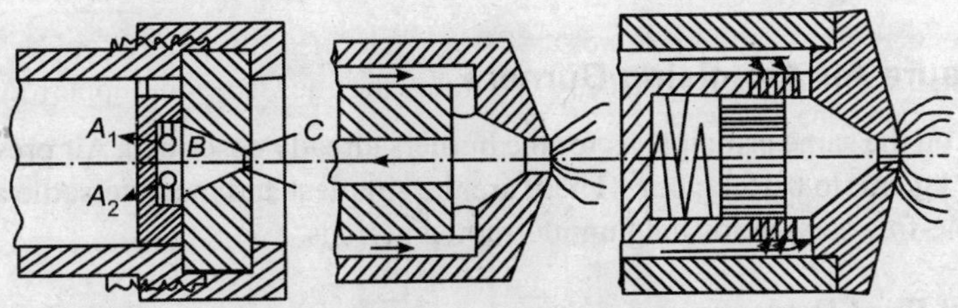

(a) Swirl pressure (b) Spill control (c) Spring loaded pistion type burner

Fig. 4.13

core has greatest tangential velocity and the least axial velocity. This jet is broken up into fine droplets by friction with the atmosphere. The size of the droplets will depend on the ambient pressures. This type of burner is not capable of giving satisfactory results at varying rates of fuel supply.

In spill control burner fuel is pumped into the burner at the constant rate and when less fuel is required to be burnt, excess fuel is returned through overflow pipe in the centre as shown in Fig. 4.13. By this method the size of the droplet can be controlled at varying rates of burning.

In case of a spring loaded piston, the size of the droplets is kept constant over a wide range of fuel input by providing aa air operated piston so that the effective area of ports leading to the swirl chamber rises above a certain value, piston uncovers greater and greater area of ports thereby restricting the fuel supply.

4.10.5 Steam or High Pressure Air Atomising Burners

In this case steam eaters on one side and oil from other side. la case of outside mixing type burners, oil is ejected through oneside of holes and is blasted by a high velocity jet of steam issuing from other holes. Mixing, however, occurs outside the burner. In case of inside mixing type burners steam and oil are mixed inside the burner before the

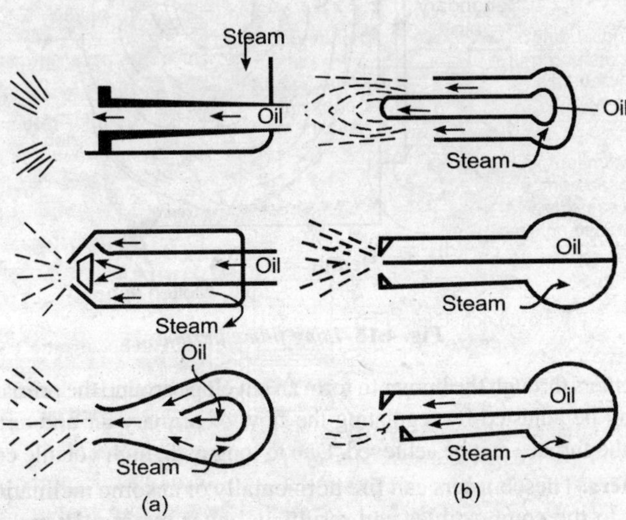

Fig. 4.14 *Steam jet (a) inside mixing (b) outside mixing.*

mixture is projected in the furnace in either a flat spray or in a hollow cone. High pressure air can also be used instead of steam in this type of burner. These burners provide high efficiency at the high firingrates aad flexible flame shape.

4.10.6 Low Pressure Air Atomising Burners

These burners operate on the same principle as for the burners already described. Air pressure required in this case may range from 0.015 kg/cm^2 to 0.15 kg/cm^2. These are the simplest and most versatile atomising type of burners and usually give trouble-free service for long uninterrupted periods.

4.10.7 Pulverised Fuel Burners

Pulverised fuel burners may be classified as:

(1) Long flame burners (2)Turbulent burners (3) Tangential burners (4) Cyclone burners.

(1) Long-Flame Burners. In this case coal is floated on a portion of air supply (primary air) and supplied to burner in one stream. Secondary and tertiary air supplies are maintained as shown in Fig. 4.15. The length of the flame is increased in the combustion chamber by downward initial flow of the flame. The flame produced is stable, long and intense. By admitting much secondary air, the flame produced can be made short and intense. A long, diffusion type of flame may be created by admitting secondary air progressively down the entire height of the

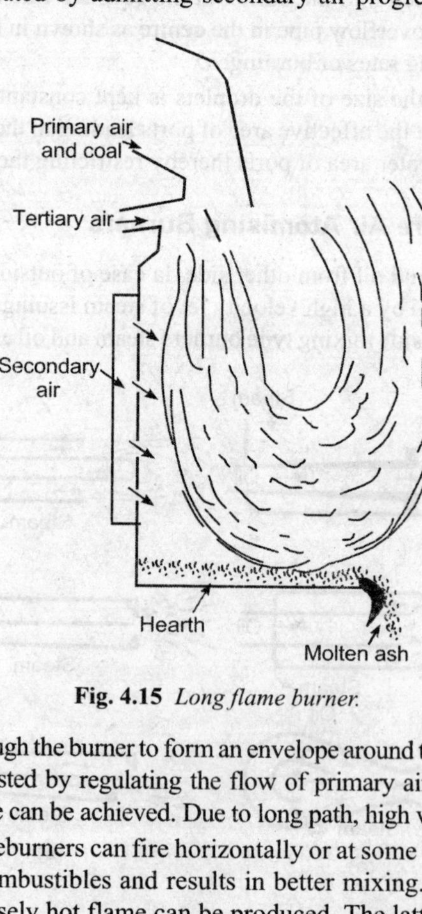

Fig. 4.15 *Long flame burner.*

furnace. Tertiary air which enters through the burner to form an envelope around the primary air and fuel to provide better mixing. Depth of flame may be adjusted by regulating the flow of primary air and tertiary air. In this case, uniform temperature distribution in the furnace can be achieved. Due to long path, high volatile coal can be easily burnt.

(2) Turbulent Burners. Theseburners can fire horizontally or at some inclinations by adjustment. Air ducts arc provided to give swirl to the combustibles and results in better mixing. By proper adjustment, either a long penetrating flame or short and intensely hot flame can be produced. The latter yields higher rates of heat release. Turbulence can be further increased by placing two burners on the opposite sides of the furnace.

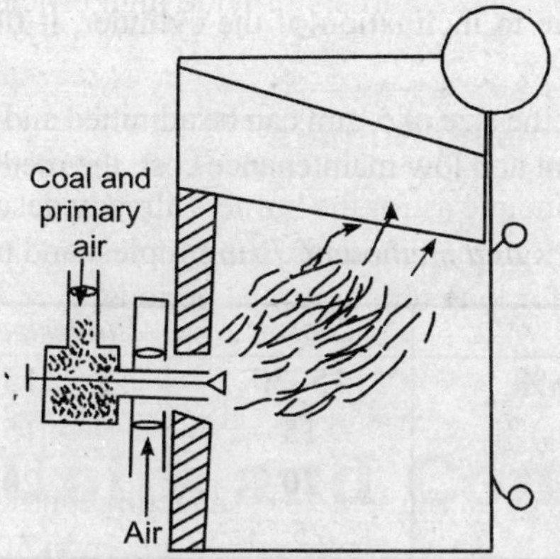

Fig. 4.16 *Turbulent burner.*

(3) Tangential Burners. In this case four burners are located in the four corners of the furnace and are fired in such a way that the four flames are tangential to a imaginary circle formed at the centre. An intensive, turbulent rotative motion is developed where the flames impinge, which promotes rapid and complete combustion. High rates of heat release are possible with this method of firing.

Plan of furnace tangential firing

Fig. 4.17

(4) Cyclone Burners. This is a new method of burning coal practice in suspension which overcomes the disadvantages of other burners i.e., the cost of pulvarisation and the problem of fly-ash. In this case pulverised coal and primary air enters a cylindrical cyclone cylinder. Secondary air results in thorough mixing of fuel and air, and creates a strong vertex, highly turbulent in nature.

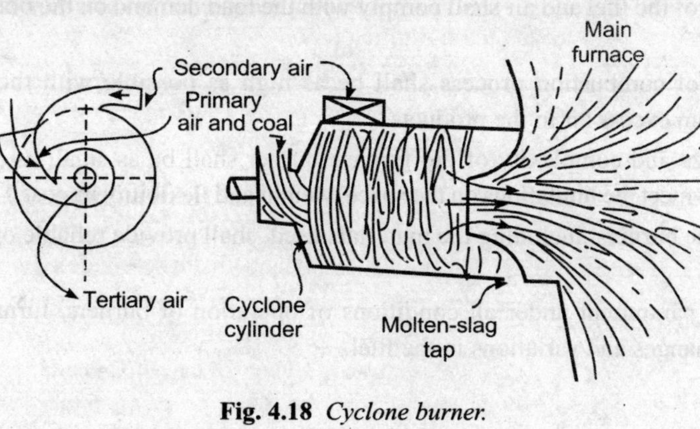

Fig. 4.18 *Cyclone burner.*

Extremely high heat liberation and the use of preheated air cause high temperature in the cyclone. The ash formed is in the molten form and, due to inclination of the cylinder, it flows down and is discharged through a suitable point.

In cyclone firing coal particles upto the size of 5 mm can be admitted and high heat release rates are possible. Due to simplified dust collector arrangement and low maintenance cost, the method has proved to be economical too.

TABLE 4.1 *Coal Characteristics and method of firing*

Characteristic	*Stoker*	*Pulverized coal*	*Cyclone-furnace*
Maximum total moisture (as fired) %	15–20	15	20
Minimum volatile matter (dry basis) %	15	15	15
Maximum total ash (dry basis) %	20	20	25
Maximum Sulphur (as fired) %	5	—	—

(5) Gas Burners. Almost all gas burners work on the same principle i.e., a part or all the air for combustion must be mixed with the gas before ignition. Gas burners may be classified as atmospheric or high pressure type, depending upon whether the gas is admitted at atmospheric pressure or high pressure. Atmospheric pressure gas burners use gas supplied at pressure of about 5 cm to 30 cm water gauge. For high heat release, gas pressures of the order of 2 to 3 kg/cm^2 are commonly used.

Gas burners aim to mix gas and air intimately and evenly so that the gas will be completely burnt before striking the cold surfaces of boiler or furnace.

In ring type burner, gas issues from a number of small holes drilled around the inner periphery of the burner ring, jetting into the stream of air which usually passes through the centre of the ring. Spin is given to the air, for creating turbulence, by means of damper vanes. This type of burner suffers from a disadvantage that with increased amount of air, the heat loss increases and flame may extinguish. Gas is usually supplied at the pressures of the order of 0.025 kg/cm^2. In centrally fired type burners, gas issues semiradially in the centre of the burner through 2 or 3 ports. Air is supplied from sides through dampers for spin. In this case 2 or 3 flames issuing from different parts mix up and give the appearance of a single furnace. Excess air has little effect on the stability of the burners.

4.11 EFFECTIVENESS OF BURNERS

The burners introduce the fuel and air into the surface to sustain the exothermic chemical reactions, for the most effective release of heat. The effectiveness is judged by the following factors:

(1) The rate of feed of the fuel and air shall comply with the load demand on the boiler over a predetermined operating range.

(2) The efficiency of combustion process shall be as high as possible with the minimum of unbumed combustibles and minimum excess air in the products.

(3) The physical size and complexity of the furnace burner shall be as small as possible to minimise the required investment and to meet the limitations on the space, weight and flexibility imposed by the service conditions.

(4) The design of the burners, including the materials used, shall provide reliable operation, under specified service conditions.

(5) Safety shall be paramount under all conditions of operation of burners, furnace and boiler including starting, stopping, load changes and variations in the fuel.

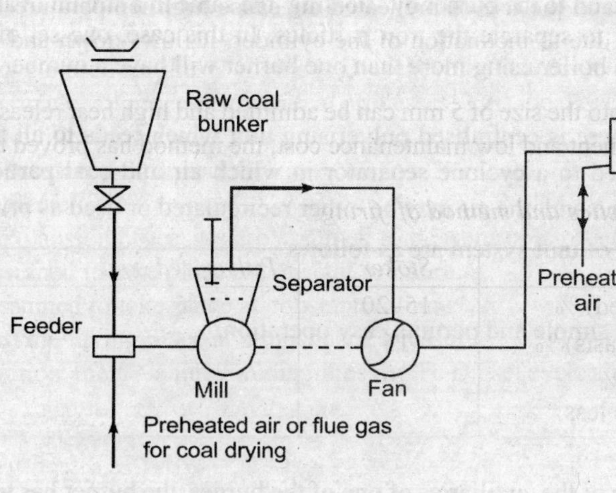

Fig. 4.19 *Unit system.*

4.12 PULVERISED FUEL HANDLING

Basically, pulverised fuel plant may be divided into two systems viz., unit system and central system. In unit system, coal from raw coal bunker is passed on to a feeder. Here preheated air or hot flue gases may be circulated

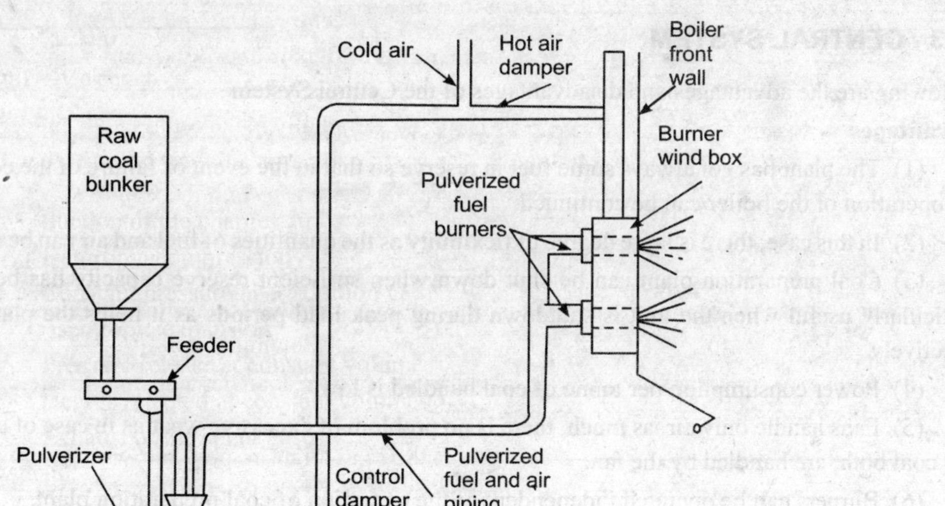

Fig. 4.20 *Pulverised fuel firing—general arrangement.*

through the mass of coal for drying. Dried coal is transferred to the mill for pulverisation. Pulverised coal, which is in the form of fine powder is passed on to the burner by 'floating' the same in a stream of air. A magnetic separation may also be incorporated in order to separate the iron particles. In this case, one set of feeder mill fan is used exclusively for one burner. Thus, a boiler using more than one burner will have a number of such circuits equal to the number of burners.

In the case of central system there is centralised pulverising unit which feeds to all the burners. In this case, pulverised coal from mill is carried to a cyclone separator in which air and coal particles are separated. Coal collects in the bottom of the cyclone and the air may be either recirculated or used as primary air in the furnace.

Advantages and disadvantages of unit system are as follows :

Advantages

(1) In this case, the layout is simple and permits easy operation.

(2) Less spaces are required.

(3) Maintenance charges are less.

Disadvantages

(1) In the event of the failure of the auxiliaries of one of the burner, the burner has to be put off as there is no reserve capacity maintained in this case.

(2) The plant has to operate under fluctuating load conditions.

(3) The degree of flexibility is less than that of central system.

(4) Fans used in this system handle coal particles as well as air, causing excessive wear of the blades.

(5) Stricter maintenance is required as the operation of the plant directly depends on the operation of the coal preparation plant.

4.13 CENTRAL SYSTEM

Following are the advantages and disadvantages of the Central System

Advantages

(1) The plant has got always some fuel in reserve so that in the event of failure of the coal preparation plant, the operation of the boiler can be continued.

(2) In this case, there is large degree of flexibility as the quantities of fuel and air can be separately controlled.

(3) Coal preparation plant can be shut down when sufficient reserve capacity has been achieved. This is particularly useful when the unit is shutdown during peak load periods as it helps the plant to meet peak load effectively.

(4) Power consumption per tonne of coal handled is low.

(5) Fans handle only air, as much, there is no problom of excessive wear as in case of unit system, where air and coal both, are handled by the fan.

(6) Burners can be operated independent of the operation of coal preparation plant.

(7) Less labour is required.

Disadvantages

(1) The initial cost of the plant is high.

(2) Large space is required in this case.

(3) Power consumption of the auxiliaries is high. This some-times results in high overall power consumption per tonne of coal handled as compared to the unit system.

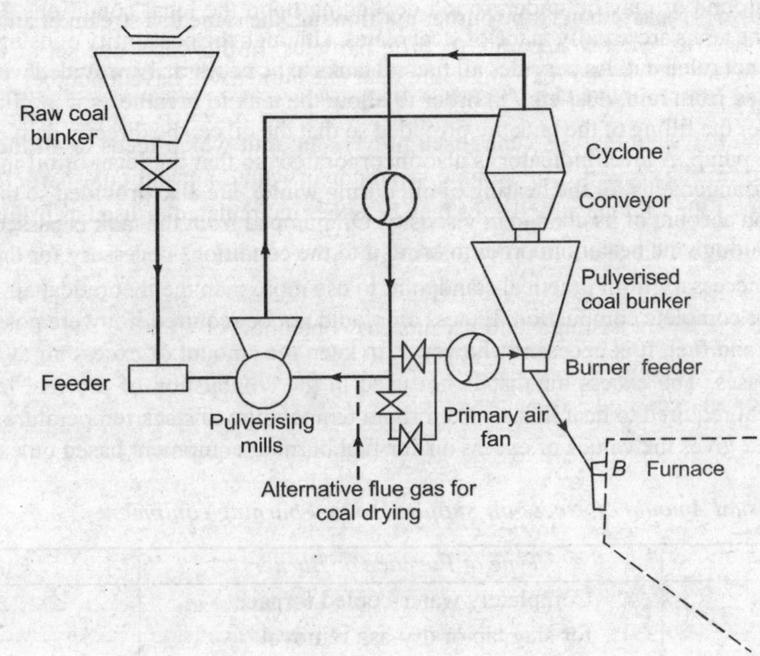

Fig. 4.21 *Central system.*

4.14 LIQUID FUEL HANDLING

The quantity of oil stored in a plant, depends upon the rate of consumption and the regularity of supply. Also oil in large quantities may be purchased in case the price of oil falls down seasonally. In that case also, storage problem of oil is to be solved

Oil is delivered at the site either by trucks, special railway wagons, tankers or in some cases by specially laid down pipe line. Oil so delivered at the site is directly transferred to the storage tanks. The storage tanks may be

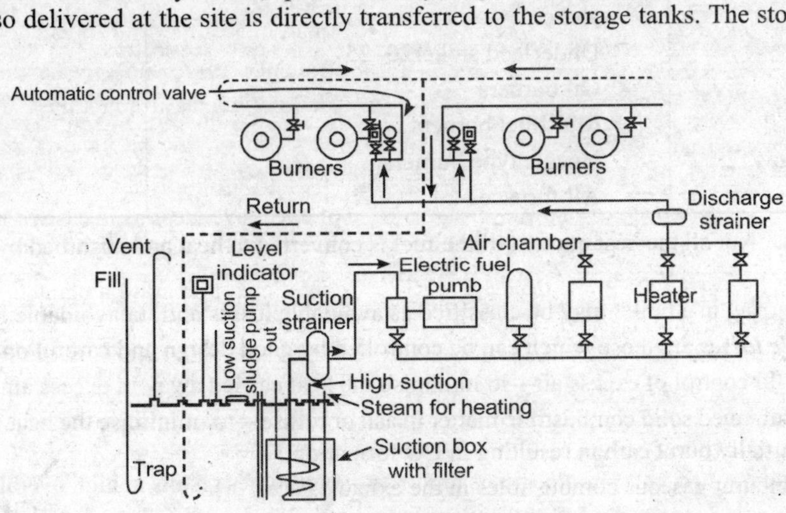

Fig. 4.22 *Liquid fuel handling.*

located above the ground or may be underground depending upon the local conditions. Underground tanks are usually preferred. The tanks are usually made of steel plates, although the possibility of using other materials for the tank construction is not ruled out. As per rules all fuel oil tanks to be necessarily provided with vent pipe open to the atmosphere, protected from rain, dust etc., in order to allow the tank to breathe as it is filled or emptied. Besides this, arrangements for the filling of the tank are provided so that the oil can be directly delivered to the tank by easy connection with the pump. A level indicator is also incorporated so that the level of oil in the tank can be easily noted. Sometimes arrangements for the heating of oil, during winter, are also provided so that there is no difficulty in pumping the oil on account of its change in viscosity. Oil pumped from the tank is passed through the strainers. oil may be passed through the beaters in order to bring it to the conditions necessary for the burners.

Excess Air. It is necessary from practical standpoint to use more than the theoretical air requirements to assure sufficient oxygen for complete combustion. Excess air would not be required if it were possible tp have an ideally perfect union of air and fuel. It is necessary, however, to keep the amount of excess air at a minimum in order to hold down stack losses. The excess air that is not used in the combustion of the fuel leaves the unit at stack temperature. The heat required to heat this air from room temperature to stack temperature serves no purpose and is lost heat. Table 4.2 gives the values of excess air for fuel burning equipment based on experience.

TABLE 4.2 *Usual Amount of excess air supplied to fuel burning equipment*

Fuel	*Type of Furnace or Burners*	*Excess Air % by weight*
Pulverised coal	Completely water-cooled furnace for slag tap or dry-ash removal	15–20
	Partially water-cooled furnace for dry-ash removal	15–40
Crushed coal	Cyclone furnace pressure or sunction	10–15
Coal	Spreader stoker	30–60
	Water-cooled vibrating-grate stoker	30–60
	Chain grate and travelling grate stoker	15–50
	Underfeed stoker20–25	
Fuel oil	Oil burners	5–10
	Multifuel burners	10–20
Blast Furnace gas	Nozzle type burners	15–18
By gases	All furnaces	25–35

Energy Losses. Not all the heat content of the fuel is converted to heat and absorbed by the steam generation equipment.

The losses occurring in a boiler may be classified as avoidable losses and 'unavoidable losses'.

(a) *Avoidable losses* are those which can be controlled by good design and careful operation. These are:

(1) Careful control of excess air—to minimise the heat carried away in excess air.

(2) No unbumed solid combustible matter in ash or refuse—to -minimise the heat lost in unburnt carbon or partially burnt carbon resulting in CO formation.

(3) No unburnt gaseous combustibles in the exhaust gases.

(4) Reducing radiation losses—by a well insulated setting for the steam generating unit,

(b) *Unavoidable losses* are inherent losses over which there is no control: These are :

(1) The discharge of the products of combustion at a tempe-rature higher than ambient temperature.
(2) The moisture content of the fuel.
(3) Combination of some of the hydrogen with the oxygen in the fuel.

THEORY QUESTIONS

A. (Fule (Coal) Burning)

4.1. Describe with the help of line sketch the constructional details and function of any one of the following :
(a) Spreader stoker.
(b) Underfeed stoker (Multiple retort type).
State the type of coal for which they are suitable.

4.2. Draw a neat sketch of spreader stoker and show how the fuel is fed into the furnace and brought back as ash. What will be the size of fuel for this type of boiler?

4.3. What are the advantages of pulverised fuel firing over stoker firing?

4.4. Explain with a neat sketch the principle of a underfeed stoker.

4.5. How can long grade fuels be burnt on stokers, efficiently?

4.6. How will you classify boilers, based on the combustion equipment used?

4.7. What are the factors to be taken into account while deciding whether stoker firing or pulverised fuel firing be adopted?

4.8. What are the advantages of stoker firing over pulverised fuel firing?

4.9. Explain with a sketch the principle of overfeed stoker firing.

4.10. What is meant by a Retort type stoker? Explain with a sketch the coal feeding system for a Retort type stoker. What are the advantages of a Retort type stoker?

4.11. Explain with a sketch the principle of underfeed stoker firing.

4.12. Write short notes on :
(a) Steam Jet Ignition System
(b) Oil Assisted Ignition
(c) Blending of Coal
(d) Grate
(e) Air Supply for Combustion.

B. (Burners)

4.13. What are the essential requirements of Atomising Fuel Type Burners?

4.14. How are Pulverised Fuel Burners classified?

4.15. Describe with a neat sketch a Long Flame Pulverised Fuel Burner.

4.16. Describe with a neat sketch a Cyclone Burner for Pulverised Fuel Firing.

4.17. What are the advantages and disadvantages of a Unit System for Pulverised Fuels.

4.18. What are the advantages and disadvantages of a Central System for Pulverised Fuel firing.

4.19. How are liquid fuels stored and utilised in Power Plants?

4.20. Write short notes on :
(a) Tangential Pulverised Fuel Burner.
(b) Turbulent Burners for Pulverised Fuels.
(c) Atomisers.

Chapter 5 Boiler House, Chimney and Water Treatment Plants

A. BOILER HOUSE

5.1 BOILER FURNACES

A boiler furnace is a confined space in which a fuel is burnt to liberate heat energy. A furnace therefore must have a suitable enclosure for burning the fuel, an arrangement for regulating the flow of gases, heat receiving surfaces for the absorption of heat for useful purposes and an arrangement for the disposal of residue, if any. The design of a boiler furnace depends on the following :

(1) Pressure and temperature desired.

(2) Evaporative capacity.

(3) Type and grade of the fuel used.

(4) Furnace operating temperature.

(5) Characteristics of the ash produced, if any.

(6) Method of fuel firing.

(7) Nature of load demand.

For any fuel and a given set of operating conditions, there is a method which will give the best results but due to several variables involved each installation is considered on its own merits. For example, temperature of combustion should be maximum for highest thermal efficiency but the limitation is imposed by the capacity of unit, heat receiving equipment and furnace walls to withstand such conditions.

Selection of a furnace for steam generation is made on the basis of the following considerations :

(1) Suitability of furnace to burn the type of fuel available.

(2) Initial cost.

(3) Elate of heat release.

(4) Cost of repairs and maintenance.

(5) Thermal efficiency expected.

(6) Capacity to meet fluctuating load demands effectively.

5.2 METHODS OF WATER CIRCULATION

In water tube boilers, water and steam flow in a relatively large number of tubular paths that are heated externally. The basic idea can be simplified by visualising a single circuit like that Fig. 5.1 (a). A drum with steam outlet and water inlet (not shown) has connected it to a loop of tubing, one leg which is heated and the other unheated. Steam bubbles form in the heated leg, generally known as a riser. The steam-water mixture so formed is displaced by the relatively heavy water in the unheated leg or the downcomer, and the circulation is maintained. Under operating conditions there is a continuous flow of water in the loop.

An actual boiler may consist of several such downcomers and riser paths. Circulation maintained in this way is known as the natural circulation as no external means are used for the purpose. The factors influencing circulation are:

(a) difference in density between water and the steam-water mixture,

(b) the frictional losses opposing circulation.

In modern high capacity units, and also in case of modern furnaces when combustion intensity inside the furnace is too high, natural circulation method is not so reliable and safe. Hence in such cases an alternative i.e., forced circulation is used. In this case, water is pumped to overcome the resistance to flow, as well as to maintain

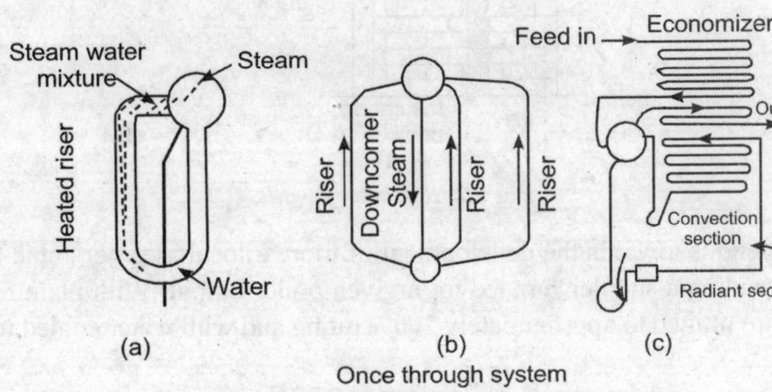

Fig. 5.1 *Methods of water circulation.*

high circulation velocity, Fig. 5 .1 (c). In yet another design known as the 'once through' design there is no recirculation of water. In this case feedwater leaves the tubes as steam. A separator may be used to separate the moisture.

5.3 FURNACE WALLS

There are three types of furnace walls used in furnace construction :

(1) Refractory walls.

(2) Hollow refractory walls with an arrangement for air coolirg.

(3) Water walls.

Boilers having smaller evaporative capacity use refractory walls. Refractories are structural material which can withstand high temperatures. Materials commonly used for refractories include fireclay, silicon carbide, magnesite, magnesia, etc. A good refractory should have the following properties :

(1) High degree of heat resistance.

(2) High resistance to abrasion.

(3) Low thermal expansion.

(4) Should not chemically react with the enclosed combustibles at elevated temperatures.

(5) Its fusion temperature should be higher than temperature inside the furnace.

(6) There should be no breaking or cracking of refractory material while in service.

Refractory may be used in several forms: brick or cast or monolithic covering. As has been already mentioned, for highest thermal effciency of a furnace, the temperature of combustion should be the maximum that can be mentioned but the brick work employed should also be able to whithstand such high temperatures, otherwise cost of repairs will increase substantially. For this purpose either some efficiency must be sacrificed or artificial cooling of refractories should be done. So, in case of furnaces working on high temperatures and having high heat release, refractory walls are made hollow and air is circulated through this space in order to keep the temperature of the refractory low.

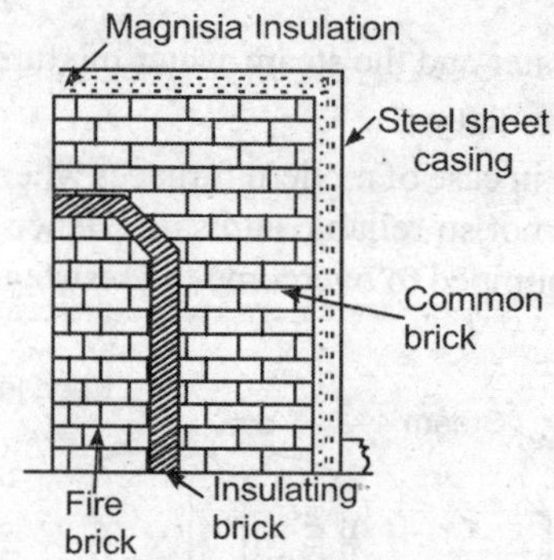

Fig 5.2 *Refractory walls.*

In furnace design the trend is towards the development of more kilocalories per cubic metre of furnace volume which is equivalent to providing a smaller furnace for a given boiler output. With plain refractory walls boiler for capacities for most coals are limited to approximately 200% rating and with water cooled refractories to 300–450%.

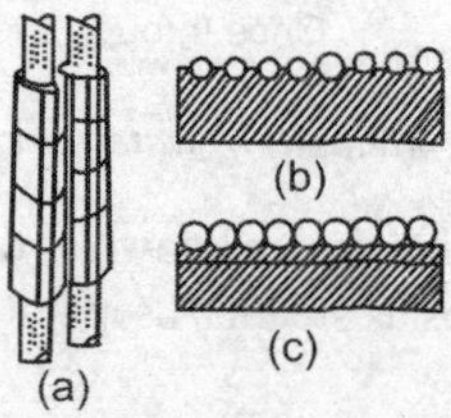

Fig. 5.3 *Examples of water-walls.*

With the use of recently developed waterwalls it has increased to 1200% rating. Waterwalls are composed of a number of elements consisting of either plain tubes, finned tubes, or metal protected tubes, arranged side by side and connected at the ends to upper and lower headers of the boiler water circulation system. Water-walls are intended to absorb the radiant heat which would otherwise heat up the furnace walls. Furnaces may be totally or partially water cooled. Very high heat transfer rates have been achieved in this case of the order of 200,000 kcal per sq m of the tube surface per hour.

Furnace Bottom and Refuse Removal. Design of furnace bottom depends on the following factors :

(1) Characteristics of the fuel burnt.

(2) Rate of burning.

(3) Method of firing.

(4) Characteristics of the ash.

Combustion Equipment for Boilers

- Lump coal (stokers)
  - Underfeed
    - Horizontal retort
    - Sloping retort–Multiple retort
      - Single
      - Twin
        - Ram feed
        - Screw feed
          - Grates stationary
          - Grates agitated
  - Overfeed
    - Conveyor storker
      - Travelling grate
      - Chain grate
        - Natural draft
        - Forced draft
    - Spreader stoker
      - Mechanical throw
        - Over throw
        - Under throw
          - Dump grate
          - Travelling
      - Jet throw
        - Air jet
        - Steam jet
  - Crushed and finely sized
    - Cyclone furnace
    - Pulverized coal
      - Short flame
      - Long flame
      - Tangential
- Fluid fuels (Burners)
  - Coal
  - Oil
    - Vaporizing
      - Wick
      - Hot plate
    - Steam atomizing
    - Rotary cup
      - Outside mix
      - Inside mix
    - Pressure atomizing
      - Multiple burner
      - Replaceable trip
      - Wide range trip
  - Gas
    - Multiple jet
    - Fan mix
    - Pre-mix

(5) Initial cost and maintenance.

A stoker fired furnace has the stoker as the major part of its bottom. Remaining space is occupied by ash pit. In the case of under-feed stoker; dump plates are provided to receive ash from the stoker extension grate. In case of pulverised fuels, due to high heat release rate and high temperature of furnace, special care is to be taken for

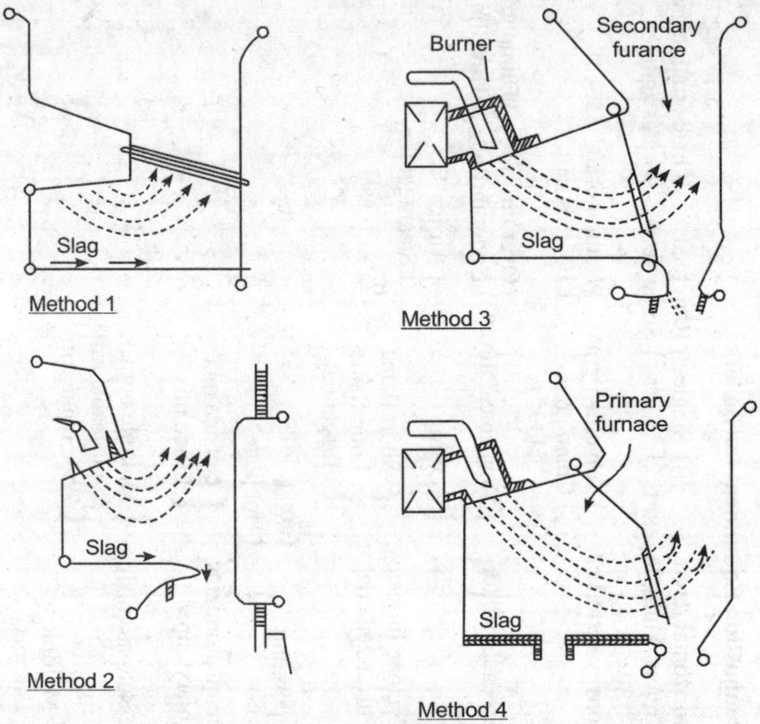

Fig. 5.4 *Typical furnace bottoms.*

designing the furnace bottom. Due to high temperature, ash is usually in the molten form. Water jet may be used to reduce the products to easily sluiced granular particles. Different shapes of furnaced bottoms are shown in Fig. 5.4.

5.4 INSULATION

The purpose of providing insulation on a furnace is to minimise the loss of heat through the structure of the furnace to the surroundings. Requirements of a good insulating material are:

(1) It should have low thermal conductivity, at the temperatures encountered.

(2) It should have low coefficient of expansion.

(3) It should not be affected by the atmospheric conditions.

(4) It should be light in weight.

(5) It should be cheap and easily available.

Principal insulating materials used in furnace design are : asbestos, kaoline, glasswool, alumina, etc.

5.5 SUPERHEATERS AND REHEATERS

Superheated steam provides a thermodynamic gain and ensures minimum wetness in the turbine. These facts have encouraged efforts to obtain the highest temperatures permitted by metallurgical considerations. Superheaters have an arduous duty, working under higher metal temperature conditions than are found elsewhere in a boiler.

Figure 5.5 shows different types of superheaters incorporated in a single boiler. These are:

(1) Radiant Superheater.
(2) Platen Superheater.
(3) Convection Superheater (Vertical).
(4) Horizontal Super heater.

The sequence in which they are connected is shown in Fig. 5.6.

5.5.1 Radiant Superheater

These superheaters are generally installed on high capacity boilers. These superheaters are generally located in the upper furnace walls away from the most intense radiant heat. They are fed with saturated steam, the flow being upward in one section and downward in other section. Radiant heat is applied to one side of the tubes only, so that some temperature gradient exists round the tube circumference. Owing to high rate of heat transfer there is only limited tolerance for changes in operating conditions.

5.5.2 Platen Superheater

This type of superheater is used on boilers having capacity of 200 MW or more. It consists of closed pitch tube loops welded together to form a flat section or "platen". The platens are suspended, as pendants at the top of the furnaces in line with the gas flow and pitched at distances of 60 to 75 cm. It is impossible for deposits to bridge over such a wide gap.

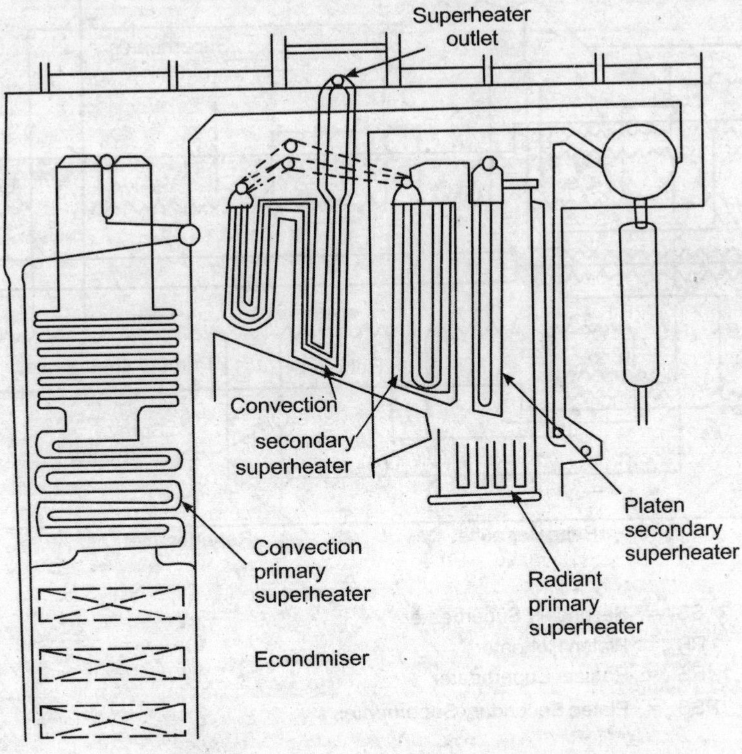

Fig. 5.5 *Arrangement of superheater and reheater of a 200 MW boiler.*

The principle of this type of surface is to utilise the radiant heat component of the gases before they eater the convection surfaces. About one third of the heat available in the gases at this stage is in radiant form. The proportion of convection heat absorbed by the platen is relatively small, because there is no gap between successive tubes to promote turbulent gas flow.

5.5.3 Convection Superheaters

Convection superheaters are situated in gas passes and use mainly convection heat. They are usually divided into two sections, the primary and the secondary. This helps in

(1) Simplification in construction.

(2) Accessibility easy.

(3) Facilitates in steam temperature control.

Convection superheaters are protected from radiant heat or high temperature gases, by the boiler tube banks or screen tubes. This protects the tubes and also prevents fusion of deposits. Also low gas velocities and relatively wide spacing of tubes discourages fouling by reducing the impingement of ash particles, permitting penetration of soot wowing steam and prevents deposits from bridging over gas lanes.

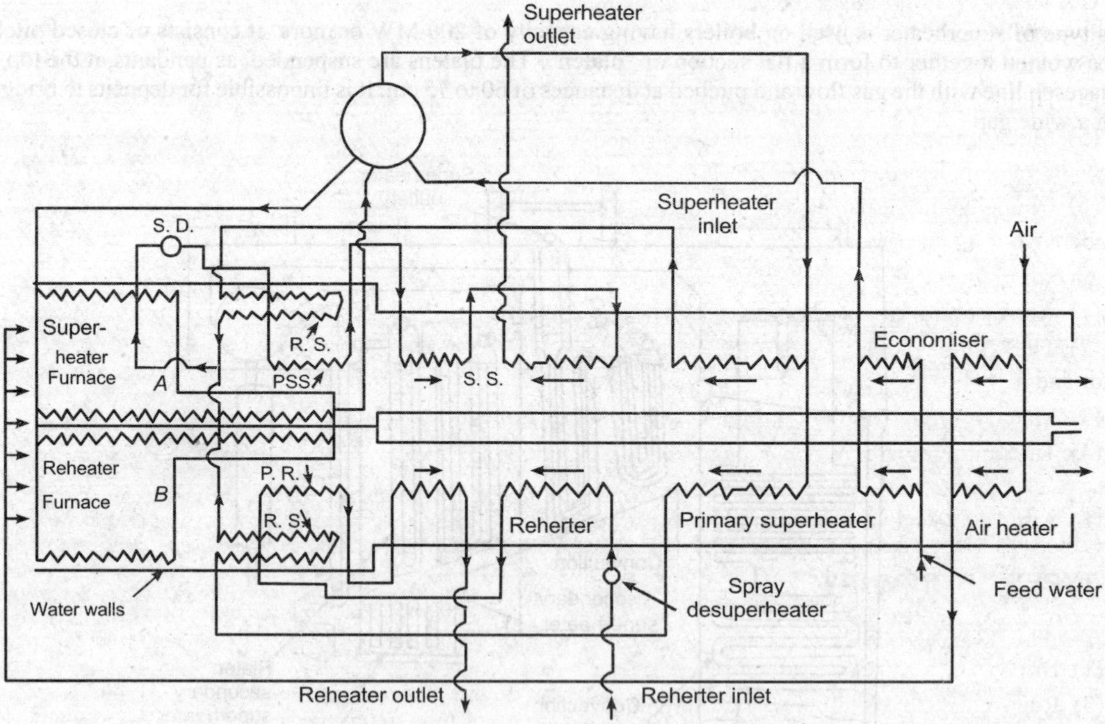

SS = Secondary Superheater

PR = Platen Reheater

RS = Radiant Superheater

PSS = Platen Secondary Superheater.

Fig, 5.6 *Flow sequence diagram for twin furnace unit boiler.*

Reheaters. In order to take full advantage of the improved cycle efficiency offered by the high steam pressure and yet avoid excessive wetness at the low pressure end of the turbine, with the adoption of reheated steam, the steam is withdrawn from the turbine before it becomes saturated during expansion and is reheated before passing back to the turbine to complete its expansion. Advances in steam pressure to 100 kg/cm^2 pressure and above have made the use of reheat desirable to avoid excessive wetness at the low pressure end of the turbine.

In initial designs, reheating was carried out in various ways, amongst them being the use of a separate reheat boiler unit, the use of live steam reheating in the turbine room, or the use of a high boiling point fluid. Nowadays it is invariably carried out in a single boiler unit in which the reheat surface is located in the convection zone of the superheater.

The types of reheaters installed are very similar to those of the superheater, the only difference being in the use, generally, of a slightly larger tube size. As the pressure at which reheater operates is about 25 percent less than the boiler pressure, therefore higher specific volume of steam requires larger tube diameter.

B. CHIMNEY

All boilers are necessarily provided with chimney, the purpose of which may be :

(1) To create natural draft necessary for the efficient working of furnace.

(2) To deliver the products of combustion at a sufficient height above the ground in order to minimise nuisance to public.

(3) Any solid matter accompanying flue gases should be dispersed over a wide area.

(4) Higher rates of combustion are possible with mechanical draft and chimney.

(5) Mechanical draft is more readily controlled to meet varying load conditions and also mechanical draft is independent of atmospheric conditions.

(6) Chimney once erected costs nothing for operation, and has a very low depreciation rate.

Boilers working on induced or forced draft are invariably provided with a chimney.

Choice of the material of which a chimney may be made depends on the following factors :

(1) Abrasive action of the flue gases necessitate a hard and rough material.

(2) The chimney should be made of a material which must be non-hydroscopic so that corrosion due to sulphur and moisture invariably present in flue gases is minimised.

(3) Material of the chimney should be able to withstand high temperature of the flue gases.

(4) The lining provided (if any) should be sufficiently flexible to allow for chimney movement in very high winds.

(5) Atmospheric conditions should not have effect on the material of the chimney.

5.6 TYPES OF CHIMNEY

The types in common use are :

(1) Brick (2)Steel Plate

(3) Reinforced Concrete (4) Aluminium.

Most of the power plants have been provided with masonry chimney because the material for construction is usually available locally at cheaper rates and the masons are available almost everywhere.

Steel chimney has got certain advantages over masonry chimney :

(1) Easy and rapidity of construction.

(2) Less weight for a given internal diameter and height.

(3) Lower overall cost.

(4) Less surface exposed to the wind.

(5) Efficiency slightly high.

(6) Requires smaller space.

Nowadays R.C.C. chimneys are also becoming popular with new plants. These chimneys have got many advantages over other types of chimneys.

5.6.1 Height and Diameter

The height of chimney is generally 2 to 3 times the height of boiler-house. Other conditions, however, may also influence the height, such as proximity to high ground when the station is located in a deep river valley or near to a residential area.

The internal diameter of the chimney must permit the unimpeded emission of flue gases.

5.6.2 Stability

As the height of chimney is considerable, the effect of wind pressure, on its surface, cannot be ignored. A chimney must be stable under high wind pressures causing tension on the windward side and compression on the leeward side. For stability purposes chimney is considered as a hollow cantilever beam. The common flexure for formula applies in this case and for stability the resultant of wind and dead weight forces must lie within the base ring of the chimney.

5.6.3 Foundation

A chimney foundation is generally of monolithic concrete and it should spread the load sufficiently to prevent over-stressing the soil or completely neutralising the windward compression. Typical shapes are shown in Fig. 5.7.

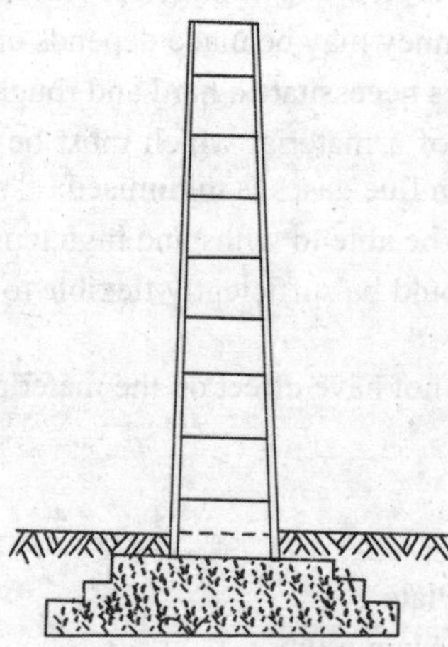

Fig. 5.7 *Chimney.*

Foundations for steel or round masonry chimney are generally octagonal in shape, being about 30 to 50 cm larger than the chimney at the top base. From this they should increase to a lower base sufficiently large to hold the maximum soil pressure to design value, but not to have a width less than one tenth the chimney height plus chimney diameter at its top. The depth of foundation is usually 3 to 4% of the height of chimney, for masonry work and generally not less than 1.25 metre deep.

Due to its height, lightning protection is always provided on chimney. For this a copper conductor runs from the top and terminates at underground buried plate.

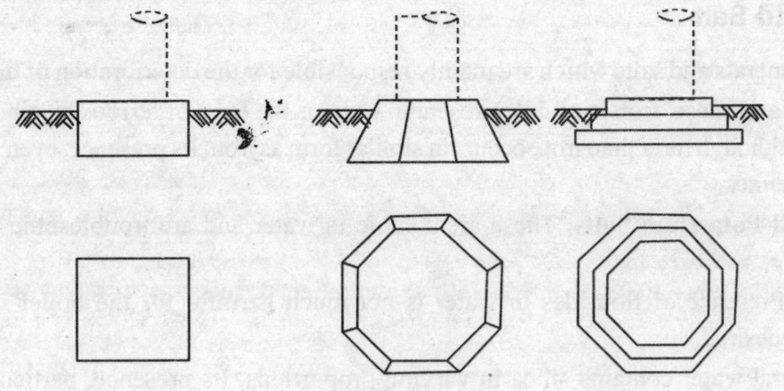

Fig. 5.8 *Chimney foudations.*

C. WATER TREATMENT PLANTS

Water is one of the most important raw materials for steam power plants. In most of the cases, water used for steam power plants contains impurities, which must be treated before use. All natural waters—even rain, snow, hail, treated municipal supplies contain impurities in one form or the other. These impurities may be classified as:

5.6.4 Visible impurities

These may be further classified as:

(1) Microbiological growth. Various growths occur in surface watei particularly in case of stationary water. Micro-organisms may include algae, bacterial slimes, diatoms, sulphur and sulphate, reducing bacteriae etc. Presence of these bacteria is always undesirable as they may produce clogging troubles.

(2) Turbidity and sediments. Turbidity is the suspended insoluble matter whereas sediments are the coarse particles which normally settle down in stationary water. These ate objectionable for almost all the use of water.

5.6.5 Dissolved gases

Water may contain gases dissolved in it such as :

(1) Carbon-di-oxide. Presence of carbon-di-oxide in well water is well-known. Besides, surface water may also contain carbon-di-oxide dissolved in it, percentage of which may vary from sample to sample. Presence of carbon di-oxide helps in corrosive action.

(2) Oxygen. Oxygen is responsible for the corrosive action of water. Oxygen is found in almost all the samples of water and is also responsible for the presence of ozone at low tempera-tures.

(3) Nitrogen. Nitrogen is an inert gas and finds its way in water due to its presence in air.

(4) Methane. Methane is generally found in deep well waters. It can be easily identified by its inflammability and is responsible for fire and explosions.

(5) Hydrogen sulphide. Presence of hydrogen sulphide in water can be identified by its peculiar smell. Its presence results in certain chemical reactions which lead to troubles particularly in case of boiler waters.

5.6.6 Minerals and Salt

Water may contain minerals and salts which are mainly responsible for the deterioration of the boiler metal surface.

(1) Iron and Manganese. Iron is usually present in water is the form of ferrous bicaibonate, which is soluble in water. Manganese, although rarer than iron, occurs in similar form as iron. Its presence, even in smaller proportions, forms deposits in pipelines.

(2) Sodium and Potassium Salts. These are soluble in water and are troublesome particularly for boiler water.

(3) Flourides. Presence of flourides in water is not much harmful for the boiler water but however its elimination is always desired.

(4) Silica. Natural water contains silica in varying proportions. Its presence, particularly in high pressure boilers, is undesirable. It is responsible for the hard scale formation on the boiler surface and the same may pass along with steam, to the turbine resulting in the formation of a glossy surface on the turbine blades.

5.6.7 Mineral Acids

Presence of mineral acids in water is always undesirable as it may result in the chemical reaction with the boiler material.

5.6.8 Hardness

Calcium and Magnesium salts as bicarbonates, chlorides, sulphates etc. are mainly responsible for the formation of a very hard surface which resists heat transfer and clogs the passage in pipes. Presence of these salts is known as hardness.

Presence of above impurities may result in one or more of the following:

(1) Scale formation

(2) Corrosion

(3) Carry-over

(4) Embrittlement.

In order to safeguard the equipment against the above effects, it is always desired to treat the water chemically in such a way that the presence of impurities is either eliminated or minimised to such an extent that the safe working of the equipment is possible.

5.7 ANALYSIS OF WATER

Analysis of water consists in finding out the quantity of impurities and other chemical substances in a sample of water. A typical analysis report is given below :

pH value of Water. pH value of water is the logarithm of the reciprocal of hydrogen ion concentration. It is numbered from zero to 14 with 7 indicating neutral water. A sample of water having pH value less than 7 indicates acidity of water and the samples having value more than 7 are alkaline. As pH is a logarithmic function, solutions having a pH value as 6.0, 5.0 or 4.0 are respectively 10, 100 and 1000 times more acidic than one with pH value as 7·0. In addition to molecules of H_2O, pure water contains molecules known as hydrogen ions (H^+) and hydroxyl ions (OH^-). The extent to which this dissociation is present in a given sample depends on the temperature and the concentration of these ions goes on increasing with the temperature and the change in concentration of the ions of hydrogen is counter-balanced by the equivalent amount of the concentration of the hydroxyl ions. At 68°F the

Water Analysis Report*

S. No.

Date :

Quantity of Sample:

Source of Sample:

Date of Collection of Sample :

Date of Analysis :

Name of Party:

S. No.	*Particulars*	*Parts per million*
1.	Suspended solids	4 ppm
2.	Colour of water	Clear
3.	Turbidity	3 ppm
4.	Organic matter	Nil
5.	Chloroform, oil etc.	Nil
6.	pH value	7.7
7.	Alkalinity as $CaCO_3$	
	(a) Phenolphthalein	Nil
	(b) Methyl Organge	Nil
8.	Hydroxide alkalinity as $CaCO_3$	Nil
9.	Hardness as $CaCO_3$	222 ppm
10.	Silica as SiO_2	3 ppm
11.	Iron as Fe_2O_3	2 ppm
12.	Calcium	98 ppm
13.	Magnesium	22 ppm
14.	Sodium	21 ppm
15.	Bicarbonates as HCO_3	190 ppm
16.	Carbonates as CO_3	Nil
17.	Hydroxides as OH	Nil
18.	Chlorides as Cl	17 ppm
19.	Sulphates as SO_4	54 ppm
20.	Nitrates as NO_3	Nil
21.	Carbon dioxide	8 ppm

*Sometimes these reports are prepared for the data which is actually useful for the purpose. However in case of a new plant, data as shown above is essentially obtained.

TABLE 5.1 *pH Numbers*

Hydrogen-ion concentration in grams per litre (1)	*Column (1) expressed as a fraction* (2)	*Reciprocal of column (2)* (3)	*Logarithm of column (3)* (4)	*pH value* (5)	*Acidity or alkalinity of solution relative to a neutral solution* (6)	
1.0	1	1	0	0	10,000,000	↑
0.1	$1/10^1$	10^1	1	1	1,000,000	
0.01	$1/10^2$	10^2	2	2	100,000	
0.001	$1/10^3$	10^3	3	3	10,000	Acidity
0.0001	$1/10^4$	10^4	4	4	1,000	
0.00001	$1/10^5$	10^5	5	5	100	
0.000001	$1/10^6$	10^6	6	6	10	
0.0000001	$1/10^7$	10^7	7	7	1	← Neutral (pure water)
0.00000001	$1/10^8$	10^8	8	8	10	
0.000000001	$1/10^9$	10^9	9	9	100	
0.0000000001	$1/10^{10}$	10^{10}	10	10	1,000	
0.00000000001	$1/10^{11}$	10^{11}	11	11	10,000	Alkalinity
0.000000000001	$1/10^{12}$	10^{12}	12	12	100,000	
0.0000000000001	$1/10^{13}$	10^{13}	13	13	1,000,000	
0.00000000000001	$1/10^{14}$	10^{14}	14	14	10,000,000	↓

concentration of these ions is of the order of 1×10^{-14}. Due to this, the pH scale has been spread upto 14. Acidic or alkaline nature of a sample depends upon whether the hydrogen or hydroxyl ions predominate. In a mixture of bases and water, hydroxyl ions results. Hydrogen ions make a solution acidic whereas hydroxyl ions make it

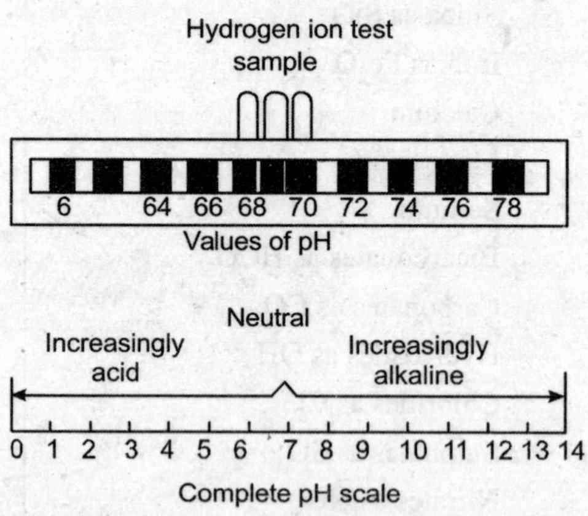

Fig. 5.9

alkaline. When a sample contains one (OH) ion for every (H) ion, acid effect of one balances the alkaline effect of the other. Result is a neutral solution with a pH value of 7.0.

pH value of a sample of water can be easily determined by a pH meter. A pH meter is essentially a comparator which compares sample colour with that of many standards to determine the value.

5.7.1 Removal of Turbidity and Sediments

The process of removal of turbidity and sediments is known as clarification. This is usually the first step in the treatment of water. Coarse particles can be easily removed passing the water through a basin with sudden change in the direction of flow. However, if the water sample is not clear and contains fine particles, then some chemicals may be added to water in order to allow those impurities to settle down under chemical action. Chemicals normally used for this purpose are :

Coagulants: sodium aluminate, ferrous sulphate, ferric chloride etc. These chemicals produce a jelly like spongy mass called floc.

For this purpose several filters are also used working on more or less the same principles. Some of the filters generally used in power plants are given below :

5.8 PRESSURE FILTERS

These filters can be easily installed on the pipe line as much lesser space is required. These filters can handle cold as well as hot water for the treatment purpose. In this case water passes through the layer of graded gravel, coarse sand and fine sand-water at outlet and can be recirculated, if necessary.

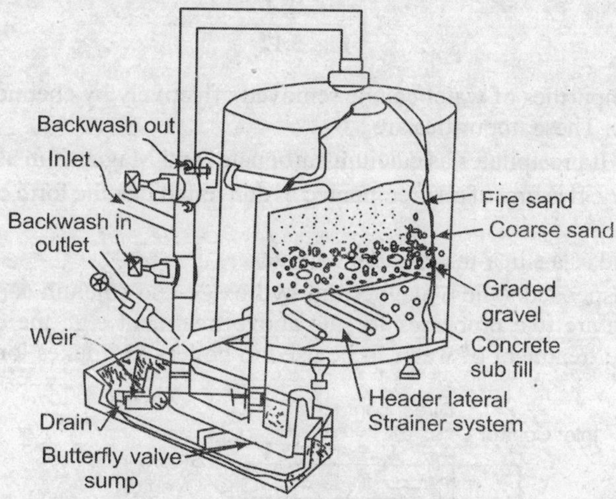

Fig. 5.10 *Pressure filters.*

5.9 GRAVITY FILTERS

These filters also make use of gravel and sand. These filters can handle large quantities of water economically. In this case, also, an arrangement for the back-washing is provided.

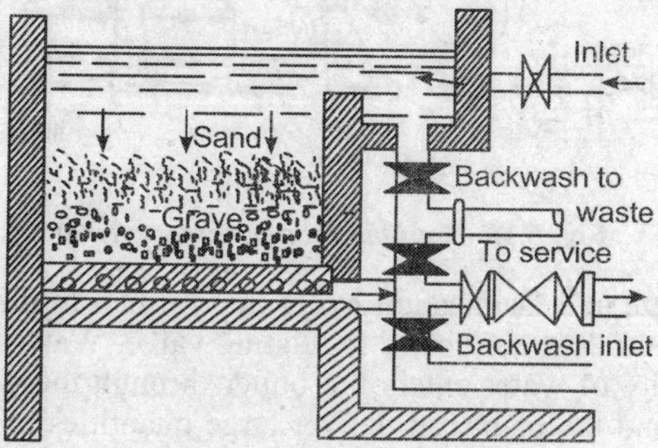

Fig. 5.11 *Gravity filters.*

5.10 CARTRIDGE FILTER

In this case fine solids ranging in size from 1 to 75 microns can be separated. These filters use disposable type elements to remove fine solids from water. These filters can be directly installed in the pipeline.

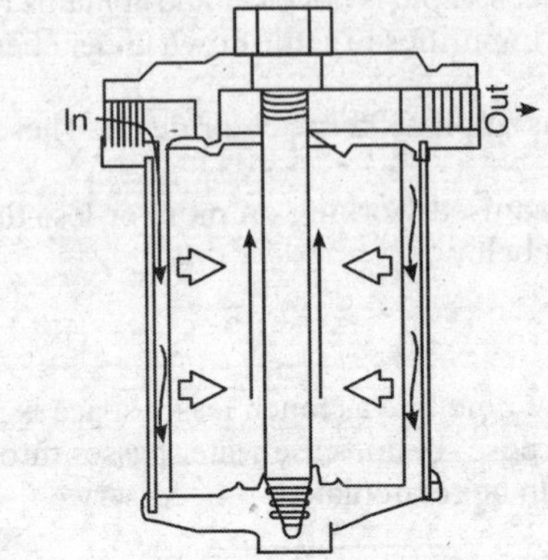

Fig. 5.12

Four most troublesome impurities of water can be removed effectively by chemical reaction with combination of lime, soda ash and caustic. These impurities are :

(1) *Calcium hardness.* It precipitates as calcium carbonate (and Magnesium as Magnesium hydroxide).

(2) *Bicarbonate alkinity*. It is first of all precipitated as converted into the form of carbonate and then removed as calcium carbonate.

(3) *Silica.* It is removed as a silica-magnesium complex.

(4) *Turbidity.* From suspended solids, magnesium hydroxide and calcium carbonate sludge is removed by settling and filtration. There are two processes for the above treatment e.g., the cold and hot processes. Cold process is rarely used for the treatment of water to be used in boilers as it takes long time for the completion of

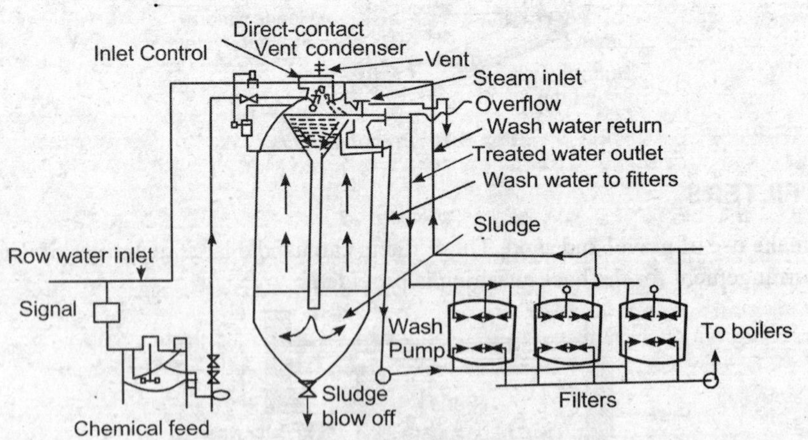

Fig. 5.13 *Hot lime soda process softener.*

process and both calcium carbonate and magnesium hydroxide formed in this case are soluble in water. In hot process, cold water enters through a float controlled regulating valve. Water is heated by means of steam taken from the boilers. Requisite quantity of water enters the boiler through the chemical feed line. In this case the reaction is almost instantaneous and the plant can deliver large quantities of treated water. Sludge collected at bottom is discharged periodically.

5.11 REMOVAL OF DISSOLVED GASES

Dissolved gases in water which do not chemically react, become less and less soluble as the temperature of water is raised. So these gases can be easily removed in deaerators where the temperature

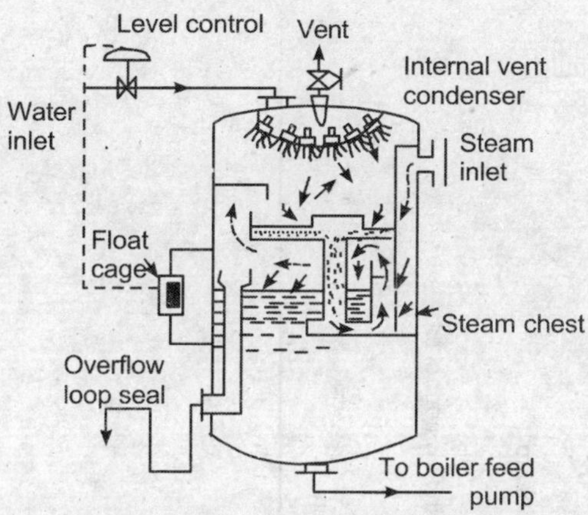

Fig. 5.14 *Spray or pressure deaerator.*

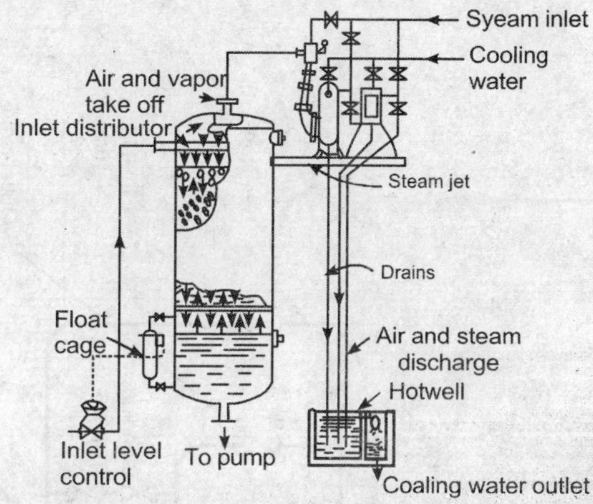

Fig. 5.15 *Vacuum deaerator.*

of water is increased by using the available steam. Deaerators are of two types e.g., vacuum and pressure type. Vacuum type deaerators use steam jet ejectors or vacuum pumps in order to produce required vacuum. As the pressure inside the tank falls, air and other dissolved gases begin to separate out from water and find their way out through suitable passage provided for the purpose. In pressure type deaerators, water is sprayed through suitable nozzles and the temperature of water is raised by means of suitable steam jacketing.

5.12 ION EXCHANGE

Most of the impurities in water have got the tendency of getting dissociated into positively and negatively charged particles known as ions. Positive ions are called cations and the negative ions are called the anions. Generally, all waters for boilers contain electrolytes in varying concentrations. Ion material has been developed which has got the ability to exchange one ion for another hold it temporarily in chemical combination and give it up to a strong regenerating solution. A typical ion exchanger in service has been shown in Fig. 5.16. In this case the ion exchanger, after some time of use, can be regenerated by the help of salt solution.

The methods described so far were meant for the external treatment of water. In this case, water, before being fed to the boiler, is treated chemically, the choice of process however depending on the analysis of water and the nature and concentration of impurities. In spite of the best treatment, some quantity of dissolved and suspended impurities enter the boiler. The concentration of these impurities goes on increasing as steam is raised. Ultimately a point is reached when the concentration of these impurities render the operation completely unsatisfactory. Recommended boiler water concentrations at various operating pressures are given below.

TABLE 5.2

Boiler pressure range	*Total solids solids*	*Alkalinity (total)*	*Suspended solids*	*Silica*
	ppm	ppm	ppm	ppm
0 to 25 kg/cm^2	3000	650	300	100
25 to 50 kg/cm^2	2500	500	150	65
50 to 100 kg/cm^2	1000	200	40	3.0
above 100 kg/cm^2	400	75	4	6.5

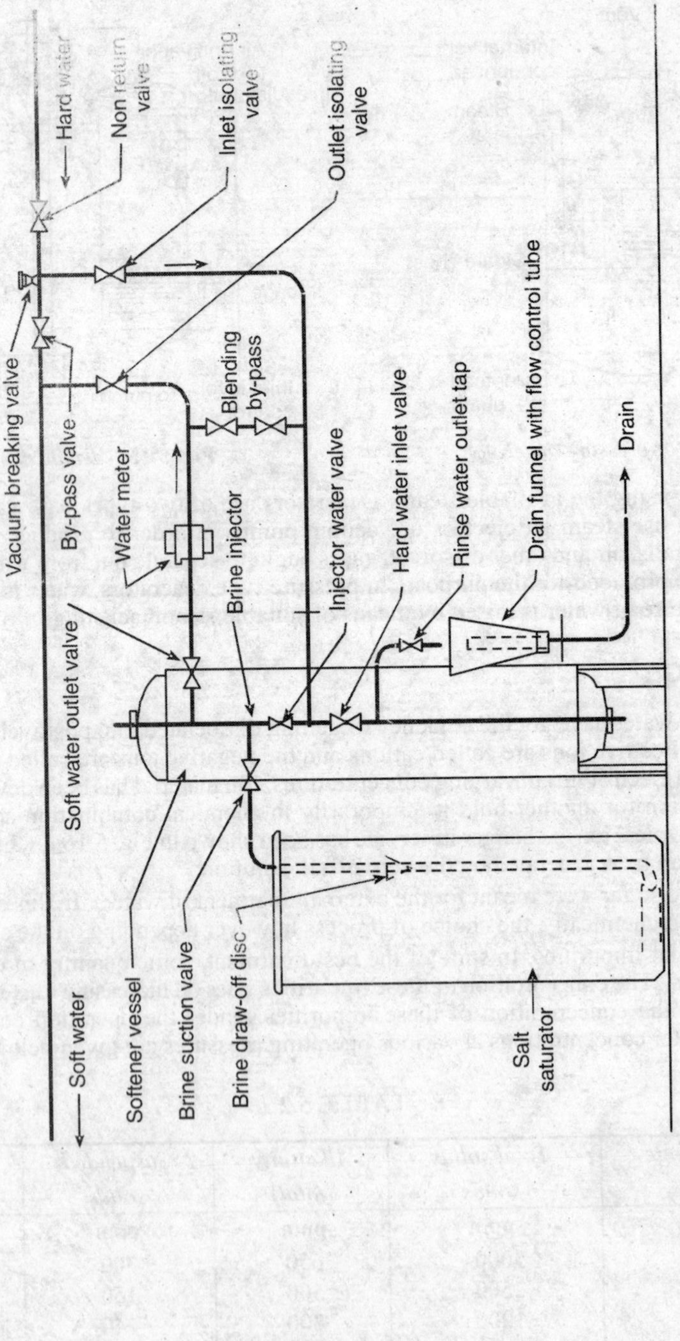

Fig. 5.16 *Ion exchanger*

As is clear from the Table 5.2, the allowable limit for the concentration of the solids goes on decreasing as the working pressure of the boiler goes on increasing. In boiler of any working pressure the concentration of these solids can be reduced by blowing down some quantity of water from the bottom of the drum. This blow-down mainly contains the undesirable impurities which concentrate at the bottom of the drum. As a result of blow-down the concentration of these impurities inside the boiler drum can be temporarily reduced. Thus a boiler may have

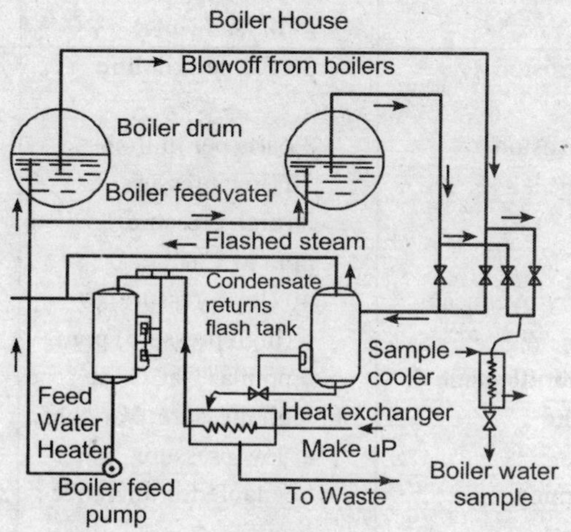

Fig. 5.17

periodical blow-down in order to keep the concentration of the impurities within the permissible limits. However, large capacity units can be designed to have continuous blowdown for the purpose. Naturally the water, when it is blown off, carries away along with it some heat which it had received in the drum. However, this loss of heat can be minimised by using suitable heat exchanger to make use of this heat.

The treatment of water inside the boiler is known as the internal treatment. Some form of internal treatment is always given to a boiler in order to prevent the scale formation, minimise corrosion, system contamination and caustic embrittlement. Every internal treatment is designed to precipitate the salts of calcium and magnesium. This removes the hardness of water but produces an equivalent amount of suspended solids which are responsible for the formation of sludge inside the boiler. This sludge is troublesome to handle and can be only controlled by blow-down organic materials such as tannins, lignins, starches etc. There are several different ways in which these organics hinder the scale formation inside the boiler. Sometimes calcium carbonate crystal particles are coated

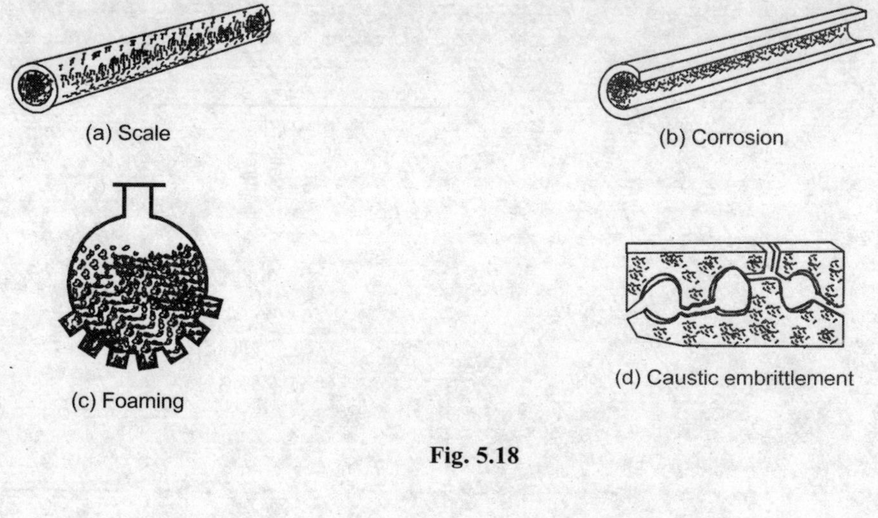

Fig. 5.18

along with a protective layer of suitable colloidal material. By this precipitated particles are prevented from sticking to the boiler dram surface. For low hardness feed waters phosphate treatment is preferred.

TABLE 5.3 *Summary of water treatment*

Impurity	*Trouble*	*Limit of tolerance in feedwater*	*Kind of treatment*
Oxygen	Corrosion	30 CC per kilolitre	Chemical deoxidation; Thermal deaeration
Carbon dioxide	Corrosion	3 parts per million	Thermal deaeration
Calcium, magnesium salts	Scale	20 (low pressure) to 2 (high pressure) ppm as $CaCO_3$	External softening Internal softening
SiO_2	Carry over scale	40 (low pressure) to 5 (high pressure) ppm	Deionization
Na alkalinity	Embrittlement foam	5 ppm as $CaCO_3$ at high pressure; Variable at low pressure	Acid neutralizer
Total dissolved solids	Priming foam	See table for tolerance limits	Deconcentration
Turbidity	Sludge sediment	5 ppm or less	Coagulation sedimentation filtration
Oil	foam	5 ppm or less	Coagulation filteration surface blow-off

Chapter 6 Ash Handling and Dust Collectors

A. ASH HANDLING

6.1 INTRODUCTION

UsualIy 10% to 20% of the coal fired in a station is released in the form of ash. Thus a 200 MW plant burning 400,000 tonnes of coal per annum will release something like 50,000 tonnes of ash. Naturally, mechanical means are to be used for the disposal of such a huge quantity of ash. Equipment used for the handling of coal cannot be necessarily used for the handling of ash. Exclusive problems in the handling of ash are :

(a) Hot materials
(b) Abrasive properties of ash
(c) Dust nuisance
(d) Poisonous gases
(e) Corrosive action.

It is sometimes desired to quench the ash before handling. Following advantages are derived due to this treatment:

(1) Temperature is reduced.
(2) Large clinkers disintegrate due to quenching.
(3) Dust nuisance is minimised.
(4) Quenched ash has got less corrosive action.

Power stations located aear sea or river have got the advantage in disposing of ash. In land stations ash is usually disposed of by auction to the contractors for use in building construction, road making etc.

Principal requirements of an ash handling plant are :

(1) The plant should be able to handle requisite quantity of ash daily.
(2) The plant should be able to deal with hot as well as wet ash.
(3) The plant should be able to handle large clinkers, soot, dust etc., without difficulty.
(4) The ever present feature of abrasion should be avoided by the use of suitable equipment in such a way that long, uninterrupted periods of operation can be obtained.
(5) The operating and maintenance charges on the plant should be minimum.
(6) Capital investment involved should be minimum.

(7) The system should be able to operate under all the variable conditions of operation.

(8) In case of additional units being installed, it should need minimum changes and investment.

(9) The operation should be noiseless.

(10) In case of variation in the characteristics of the ash, due to use of fuels of different compositions, the plant should continue to render efficient service.

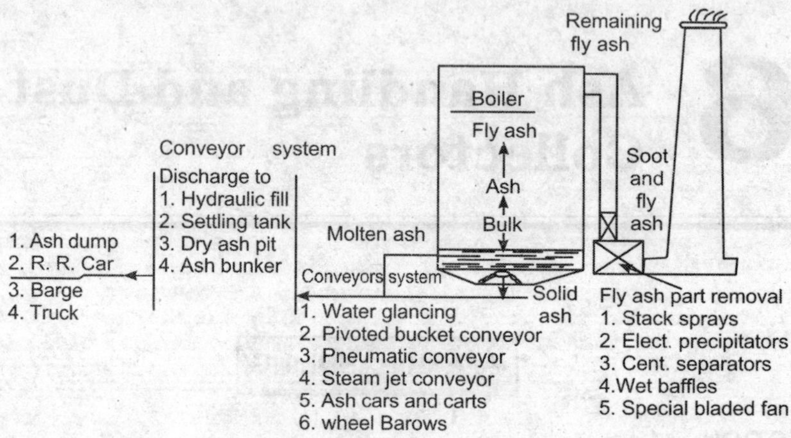

Fig. 6.1 *Ash handling systems.*

Various methods of ash handling are :

(A) Manual handling—

(i) Wheel barrows.

(B) Mechanical handling—

(i) Vee-bucket system.

(ii) Skip hoist.

(iii) Drag chain or scraper conveyors.

(iv) Pivoted or overlapping conveyors,

(v) Belt conveyors.

(C) Steam or vacuum conveyors.

(D) Hydraulic or gravity systems.

(E) Pneumatic suction system.

6.2 WHEEL BARROWS

These are the containers which directly collect ash from the boiler through ash outlet doors, usually operated manually. The barrows are moved on rails either manually or mechanically to the dumping site. This method has got the advantage of its simplicity and minimum capital cost. The method is suitable for small stations and the labour charges are high in this case.

6.3 MECHANICAL EQUIPMENT

Mechanical equipment used in ash handling is more or less the same as used in the handling of coal, with some minor modifications to suit the conditions. For example, in case of belt conveyor, ash must be quenched before being loaded on the conveyor. As compared to coal handling plant, the capacity of an ash handling plant is always much less.

6.4 STEAM CONVEYORS

The basic principle of this system lies in the fact that when steam is passed through a pipe at sufficiently high velocity, it is capable of carrying dry solid materials of considerable size along with it. The high velocity may be established by forcing steam through the pipe under pressure greater than of atmospheric or by creating partial vacuum in the pipe. Usually, a pipe runs from the ash outlets to a bunker outside the boiler house. The ash particles remain in suspension and flow like a fluid through the pipe. Valves are provided in the line wherever necessary.

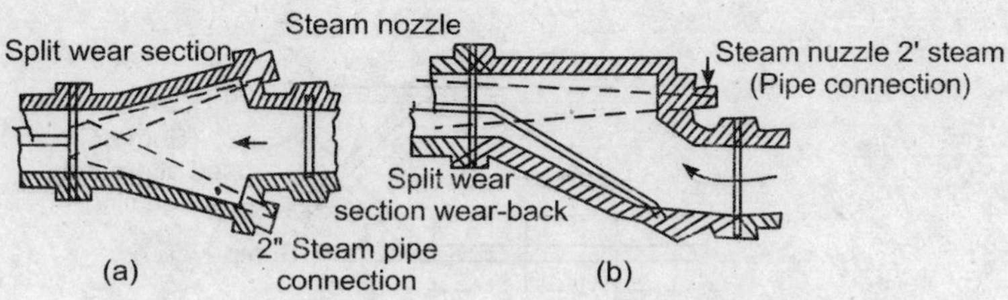

Fig. 6.2 *Use of steam jets in ash handling.*

Sharp bends in this case should be avoided in order to reduce the wear.

Advantages

(1) The method does not require any sort of auxiliary drive as it directly uses the steam generated by the boiler.

(2) Ash can be moved economically by steam jet conveyors through a horizontal distance of about 200 metres and through a vertical distance of about 30 metres.

(3) Initial cost of steam jet conveyors is lower than that of mechanical conveying systems of equivalent capacity.

(4) It requires less space.

(5) The equipment can be installed in awkward positions too.

(6) This method results in cleaner basement and ordinarily requires little attention.

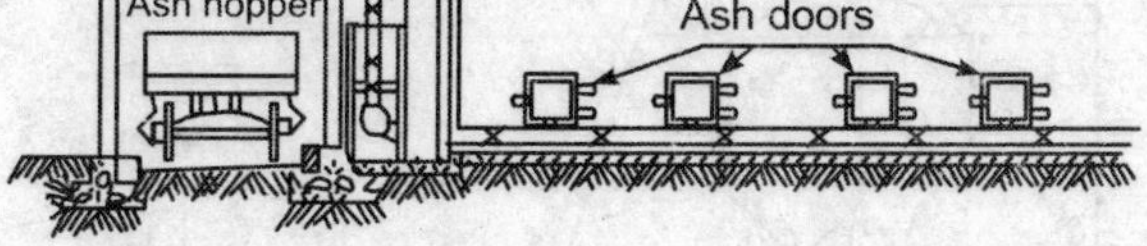

Fig. 6.3 *Layout of steam conveyor system.*

Disadvantages

(1) Large amount of wear occurs in pipe work specially at bends.

(2) Maximum permissible diameter of the pipe is 25 cm and the capacity of such a pipe is only 10 to 15 tonnes per hour. Thus the capacity of the system being less, it needs continuous operation. (Part of the

steam which could have been otherwise used for the purpose of power generation is used up by this system).

(3) Operation is noisy.

6.4 HYDRAULIC OR GRAVITY SYSTEM

In this system a stream of high velocity water flows in open conduit underneath the ashpit and carries the ash to a sump. In case of chain grate stokers the ash is discharged continuously into the running stream of water. In case of

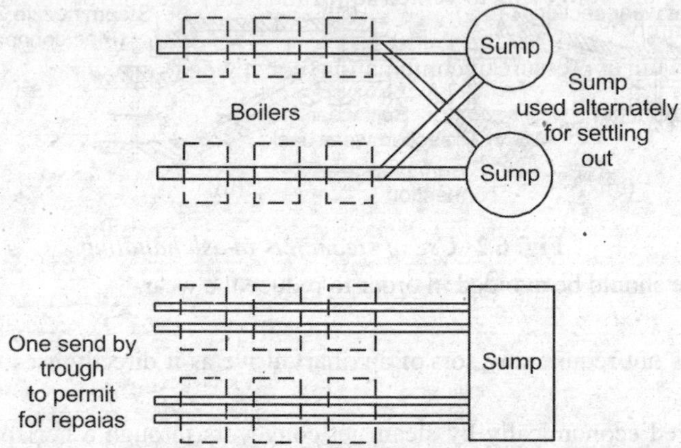

Fig. 6.4 *Hydraulic system*

stokers discharging intermittently, refuse is dumped on to a grizzly or massive bars on which the clinkers may be broken. In this case water is made to flow only when the ash is discharged, for sufficient time to carry the entire mass of refuse. Usually high pressure system is used for intermittent ash disposal system and low pressure system for continuous system. Water supply is usually taken from the condenser. Two alternative arrangements of the trough are shown in Fig. 6.3. The troughs and sumps used in the conveyance of ash are required to be of special design which is corrosive and wear resistant. Normally, these are made of concrete with "wetted" portion usually covered with the lining of vitrified earth and drain tiles.

6.5 PNEUMATIC SUCTION SYSTEM

In this case a suction pipe runs from the boiler ash outlets to a crusher, then to an ash bunker outside the boiler house and continued upto exhauster through a filter. With increased air velocity, even large particles are carried in suspension.

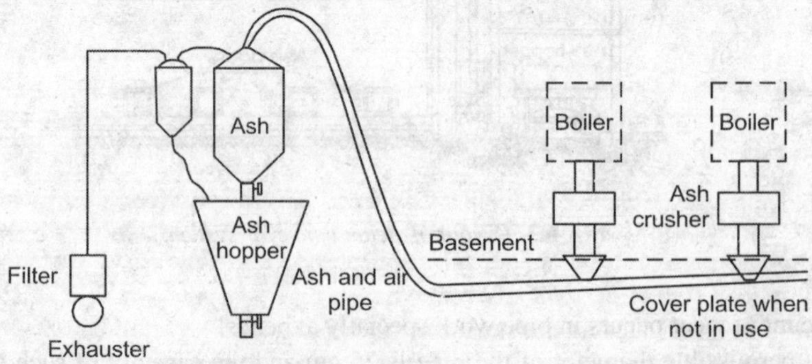

Fig. 6.5 *Pneumatic suction system.*

The ash is collected at the boiler outlet, crushed and then passed on to the pipeline. In this arrangement sharp bends are avoided for smooth flow of ash. In order to protect the blades of the exhauster, a filter is provided to filter the dust-laden air. The system suffers from the disadvantages that the wear of pipe work is excessive and maintenance charges are high.

6.6 FLY-ASH HANDLING

(a) Pneumatic (vacuum) conveying to de-aerating tank and gravity hydraulic conveying to fill area.

(b) Pneumatic (vacuum) conveying to vented sump and hydraulic conveying by either jet or pump to fill area with bottom ash or separately.

(c) Pneumatic (vacuum or pressure of combination) conveying to dry storage silo provided for truck, railroad car or bagging equipment removal.

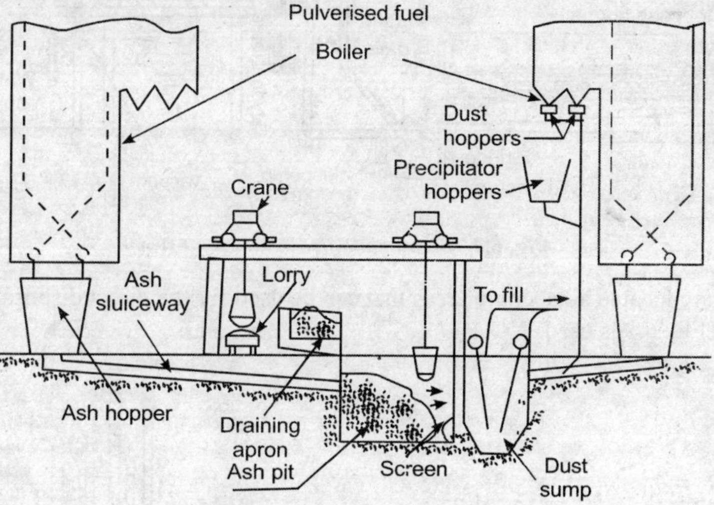

Fig. 6.6 *Fly-ash handling.*

(d) Pneumatic (vacuum and pressure or pressure) conveying for furnace re-injection.

(e) Pneumatic gravity conveying by air slides.

Figure 6.7 illustrates both bottom and fly-ash systems applicable for small or large steam generators capable of handling the products of combustion produced by poorer grades of coal.

The greater quantity of ash ranging from 65% to 85% in a pulverized coal-fired boiler is fly-ash. It is handled through dust valves in a piping system to separating equipment from which it is discharged into a storage silo equipped with a vent filter. The conveying air is discharged with the water by the hydraulic vacuum producer to vented waste line or sump. Below the silo is an unloader known as a hydromixer, which combines a small amount of water with the ash to prevent dusting during both loading of the track or car and transportation to the disposal site.

The illustration indicates a stepped feed plate, end outlet bottom ash hopper in which the heavier ash from the furnace is collected. This hopper contains a pool of water which will disintegrate any sticky slag that may be discharged from the furnace. Were the hopper maintained dry, such slag would agglomerate and make removal difficult, if not impossible. Ash is periodically removed from the hopper by a system of feed nozzles through a suitable hydraulically operated gate and housing at the end of the hopper. A clinker grinder reduces the size of the large pieces of clinker and slag to permit transportation by a jet pump through a pipeline to a de-watering bin where the ash and the water are separated.

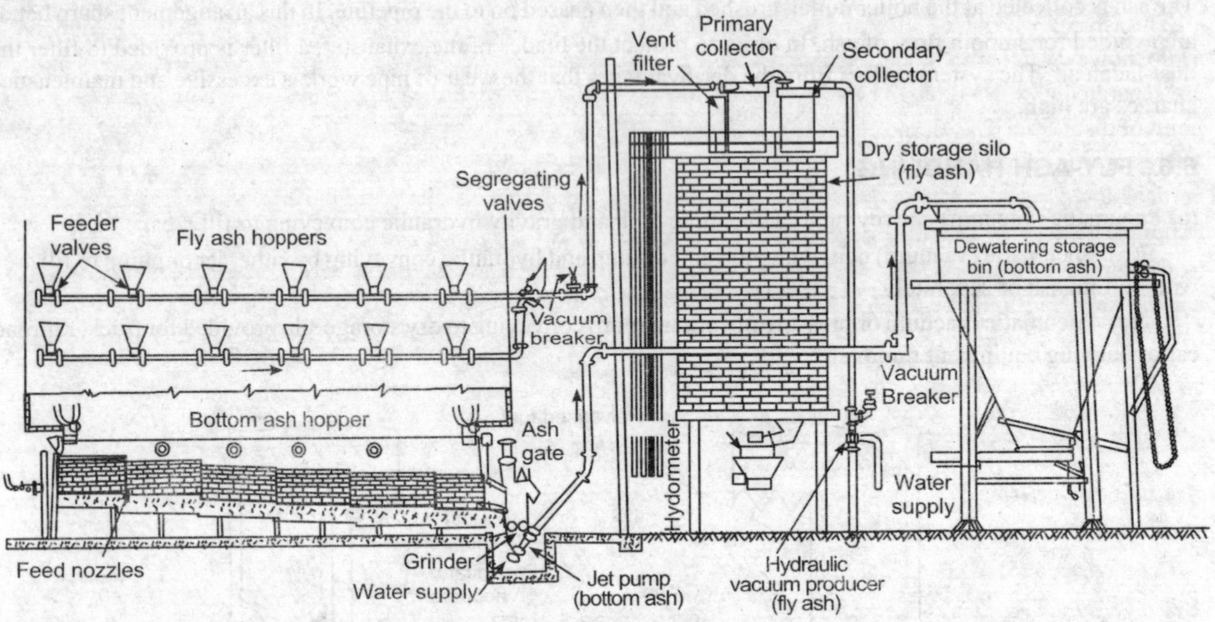

Fig. 6.7 *Bottom and flyash handling system.*

Where the plants are located adjacent to areas that can be used for the direct disposal of ash and dust, a system such as illustrated in Fig. 6.8 is used.

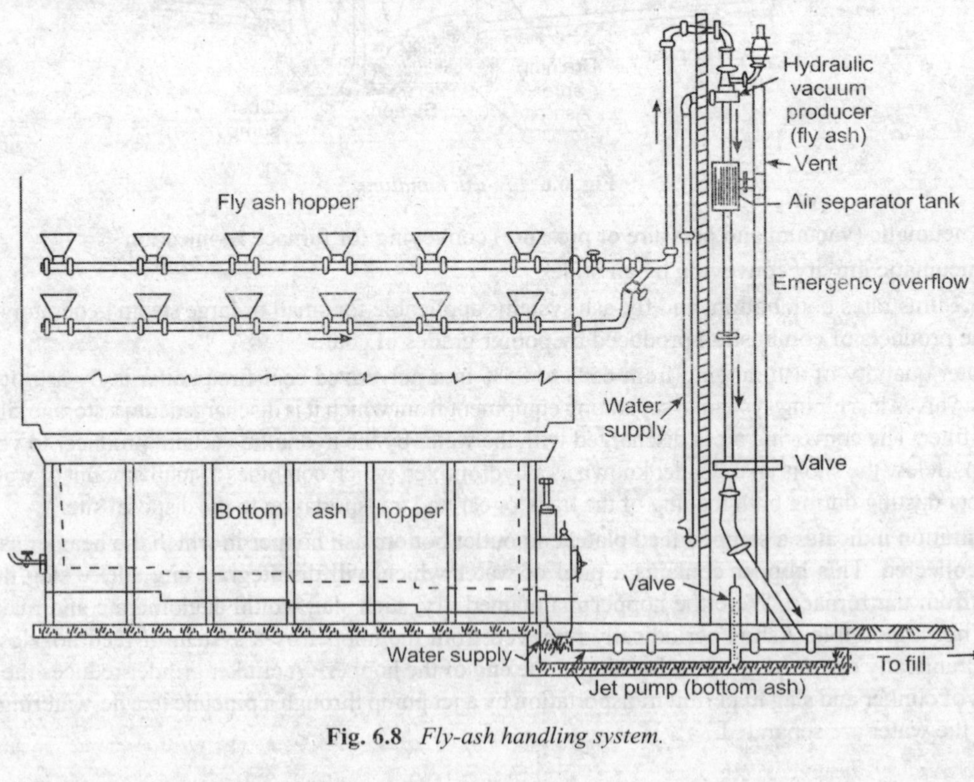

Fig. 6.8 *Fly-ash handling system.*

The ash and dust handling equipment is similar to that indicated in the diagram. Instead of directing the dust into a silo, however, it is run directly through an hydraulic vacuum producer into a de-aerating tank. The slurry and air are separated and the sludge is run by gravity through a downcomer into the bottom ash line and, thence, to the point of discharge. The air is vented to atmosphere. The bottom ash is handled by a jet pump as in above diagram. Valves are provided back of the lateral into which the fly-ash slurry is discharged and above the lateral in the vertical downcomer to prevent backflow when handling, either type of ash through the conveying piping common to both. This type of system is exceedingly simple and highly economical where applicable.

Contaminated water (waste), such as overflow and drain from ash-hoppers, hydrobins, fill areas, vacuum producers, and air washers should be discharged to suitable clarifying basins in order to avoid stream pollution.

B. DUST COLLECTORS

6.7 CLASSIFICATION

Gas-borne particles greater than 0.001 mm size are classified as dust. This may be further classified as follows:

(1) Particles greater than 0.1 mm in size are termed as 'cinders'.

(2) Particles less than 0.1 mm in size termed as fly-ash.

(3) Incomplete combustion of volatile matter contained in the solid fuel forms smoke. These are usually less than 0.01 mm in size.

(4) These smoke particles form what is known as 'soot'.

As per Indian Boilers Act there is limit imposed on the emission of dust through the stacks. As such, the plants have got to make ate of dust collection equipment. The quantity of dust emitted in a power plant depends on the quantity of gas used. Use of high volatile coals or pulverised fuels calls for special attention towards this aspect.

6.8 PRINCIPLE OF DUST COLLECTION

The heavier particles carried along with the flue gases can be arrested by :

(a) Suddden decrease in velocity by enlarging the section of flow. This results in declaration of the particles as compared to the main flue gas.

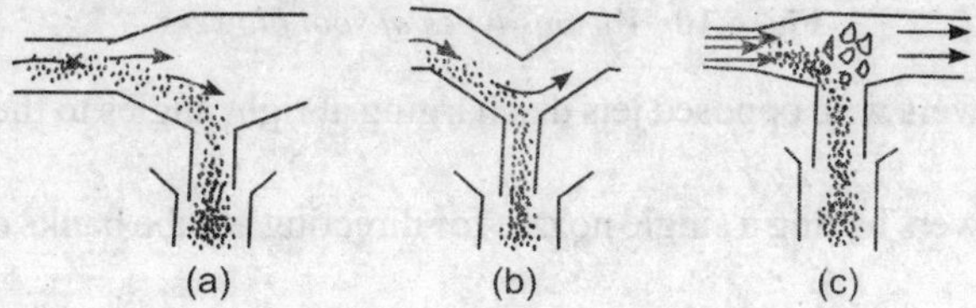

Fig. 6.9 *Principle of dust collectors.*

(b) Abrupt change in the direction of flow. This results in the separation of the path of the heavier particles with respect to the main flue path;

(c) By interposing baffles in the main direction of flow. Various dust collectors used in power plants make use of these principles in one way or the other. Flue dust so collected may be used for fill when mixed with heavier components. Some commercial uses of this have also been established.

6.9 SOOT BLOWERS

It is necessary to maintain gas side services of all sections of boilers in clean condition to ensure good availability and efficiency in heat transfer. Modem boilers are therefore fitted with soot blowers which are remotely operated and sometimes automatically sequence controlled. Steam at high temperature is normally used to serve the blowers. Figure 6.10 shows various types of soot blowers used on boilers :

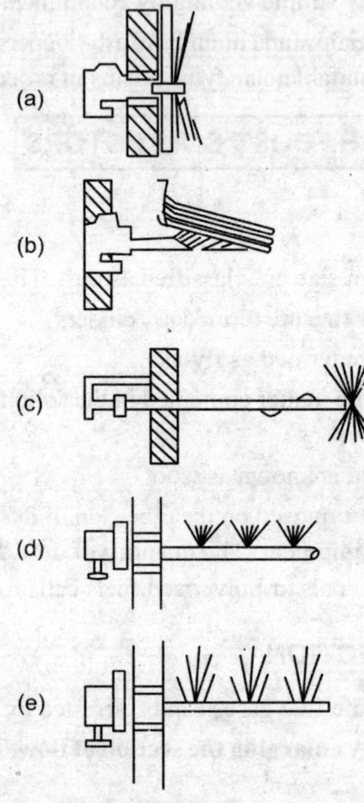

Fig. 6.10 *Various types of soot blowers.*

(a) Retractable gun blowers with opposed jets discharging at right angles to the blower axis to sweep furnace walls. (Fig. 6.10 a).

(b) Retractable gun blowers having a single nozzle for directing at tube banks of boiler and superheater. (Fig. 6.10 b).

(c) Long, retractable lance blowers which traverse between banks of tubes; the nozzle, which has opposed jets to balance the thrust, rotates. This is the only really effective blower for superheaters in modern boilers, for it makes it possible to reach right across wide passes with equal effect. (Fig. 6.10 c).

(d) Multi-jet tube blowers, for use in lower temperature zones such as economisers and air-heater. This type of blower may rotate and/or traverse. (Fig. 6.10 d).

(e) Multi-jet blower used in fixed position to sweep the lanes between tubes. These blowers are again only suitable for low temperature zones and relatively light deposits owing to the reduced power of a large number of small jets. (Fig. 6.10 e).

THEORY QUESTIONS

6.1. Give a rough idea of the quantity of ash to be handled in a power plant having 200 MW capacity.

6.2. Why equipment used for handling coal cannot be used for handling ash in a power plant ?

6.3. What are the principal requirements of an Ash Handling Plant?

6.4. Describe various methods available for handling ash in a power plant using stokers.

6.5. Describe with sketch a steam conveyor system for disposal of ash. What are its advantages over other systems?

6.6. Describe with a neat sketch Pneumatic Ash Handling System for thermal power plants.

6.7. What is 'Flyash' ? What are the problems in handling Flyash in a power plant ?

Chapter 7 Steam Piping and Testing of Boilers

A. STEAM PIPING

7.1 INTRODUCTION

The main object in a steam power station is to use the minimum length of pipes, consistent with the requirements of the plant for safe and smooth operation. Following points must be considered in laying a piping system :

(1) It should ensure maximum reliability of the sections concerned. It is only for this reason that some sections need duplicate line even.

(2) The path chosen should be such that the minimum length of pipe work is required.

(3) The system chosen should be such that inspection and maintenance could be carried without interfering with the rest of the arrangements.

(4) Minimum number of joints, bends should be required.

(5) Pipes carrying fluids at high temperature should allow for expansion and contraction during off periods.

(6) Pipes carrying steam should be lagged so as to minimise the loss of heat.

7.2 MATERIALS FOR PIPES AND FITTINGS

Materials for pipes and fitting are mentioned as following :

Wrought Steel. The greater pipe work of average steam power plant is made of Bessemer or open hearth steel. Wrought steel pipes are cheap and fulfil practically the requirements of general service.

Wrought Iron. These pipes are used for boiler tubes and to a limited extent for water, gas and steam pipes as they have low tensile strength. For high pressure steam service these are never used.

Cast Iron. Cast iron is extensively used for underground water and gas service and replaces wrought steel and wrought iron pipes where the problem of corrosion is excessive. Flanges, valve bodies etc. are also made of cast iron upto steam pressures of 15 bar. For high pressure steam, it is not used because of its density, and low tensile strength.

Steel. All fittings, bodies of valves, for high steam pressures are invariably made of cast or forged steel which may be alloyed, if necessary.

Malleable Iron. Midway between steel and cast iron is the malleable iron which has got twice the tensile strength of cast iron and is not brittle.

Copper. Copper pipes are used on locomotives and marines. Main advantage of a copper pipe is its flexibility. Due to high cost it is being replaced by other materials.

Besides this, brass and galvanised iron fittings are used for low pressure service of water, gas, steam etc.

7.2.1 Size and Strength of Pipes

Wrought iron and M.S. pipes are marketed in standard sizes. These sizes are designated as follows:

1. *Full weight pipe*. It is specified by the normal inside diameter upto 300 mm size. Pipes above 300 mm are specified by outside diameter. Full weight pipes may vary in weight by ± 5%.

2. (a) *Extra heavy or extra strong.*

(b) *Double extra heavy or double extra strong.* These pipes have the same external diameter at the standard pipes in order to accommodate pipe threads of standard size. Thickness of an extra heavy pipe is usually 40% greater than the standard pipe. Double extra heavy pipes are usually 2½ to 3 times thicker than standard pipes. These pipes may be 'reamed' to give smooth inner surface.

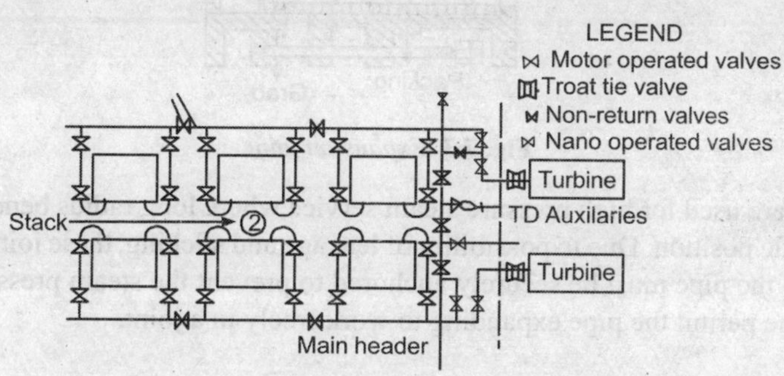

Fig. 7.1 *Pipe work for a plant using two turbines.*

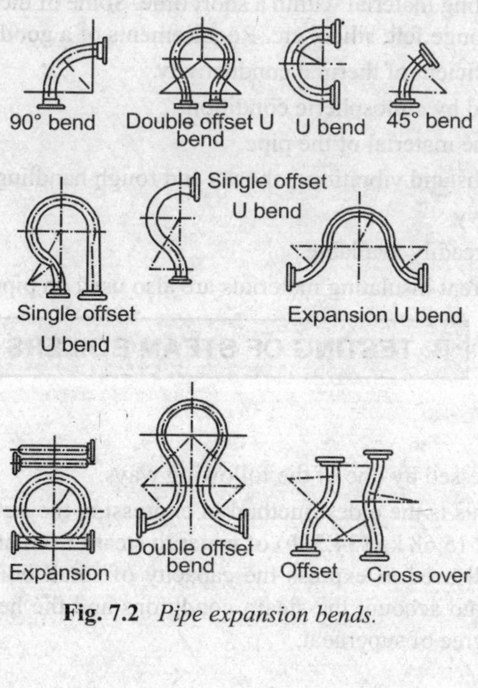

Fig. 7.2 *Pipe expansion bends.*

7.2.2 Expansion Bends

As the forces produced by expansion are practically irresistible, the pipe is invariably allowed to expand and its movement is prevented from unduly stressing the fittings, and connections by providing

(i) suitable bends, or

(ii) expansion joints.

Standard expansion bends are shown in Fig. 7.2. The use of pipe bends reduces internal friction by revolving easy turns and joints.

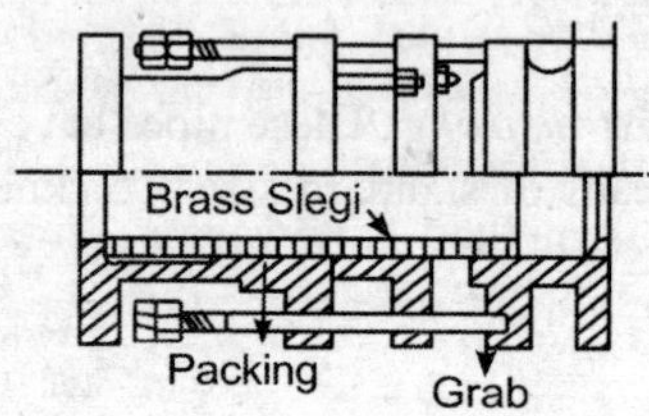

Fig. 7.3 *Expansion joint.*

Slip expansion joints are used for high pressure steam service where long radius bends cannot be provided due to limited space or specific position. Due to possibility of leakage and sticking, these joints are not much preferred. When using these joints, the pipe must be securely anchored to prevent the steam pressure from forcing the joint apart and at the same time permit the pipe expanding to work freely in a joint.

7.2.3 Insulation

Heat loss from pipes carrying fluids at high temperatures is always desired to be minimum. By properly applying a good insulating material covering on the pipe, the heat loss can be reduced by 75 to 90%. Saving in heat loss usually pays for the cost of insulating material within a short time. Some of the good insulating materials used are: 85% magnesia, asbestos fibre, sponge felt, silicol etc. Requirements of a good insulating material are :

(1) It should have low coefficient of thermal conductivity.

(2) It should not be affected by atmospheric conditions.

(3) It should not damage the material of the pipe.

(4) It should be able to withstand vibrations, shocks and rough handling.

(5) It should not be too heavy.

(6) It should be cheap and readily available.

Sometimes two layers of different insulating materials are also used on pipes for the sake of economy.

B. TESTING OF STEAM BOILERS

7.3 RATING OF BOILERS

Capacity of a boiler may be expressed by one of the following ways:

(a) *Boiler Horse Power.* This is the oldest method of expressing the output of a boiler. According to this, a boiler having ability to evaporate 15.68 kg (34.5 lb) of water to steam from and at 212°F (100° C) is rated at one horse power. This method is still used to express the capacity of small boilers. This method suffers from the drawback that it does not take into account the steam conditions and the heating rate varies considerably with change in steam pressure and degree of superheat.

(b) *Evaporation Capacity.* According to the method the capacity of a boiler is expressed in terms of the quantity of steam raised per hour. In this case the condition of steam is also expressed along with the rate of evaporation. In this case capacity of boilers is sometimes expressed in two parts e.g., the maximum continuous rating generally for 24 hours and the two-hour peak output.

(c) *kW or mW Rating.* Recent trend is to express the output of boiler in terms of the output of the connected turbine generator unit. Thus a boiler capable of running a turbo-alternator set of 800 Kw will be rated at S00 Kw. In this case the steam conditions need not be specified and the rating includes the work done by reheater also.

(d) *Heating Surface.* Manufacturers of small capacity fire tube boilers express the capacity of boiler in terms of the area of the heating surface. This method is particularly adopted where operating conditions e.g., draft, type of fuel and firing conditions cannot be ascertained. Formerly, boilers were assumed to evaporate about 16.8 kg of water per sq. m. This rating was found to be ultra conservative and units operating at 200% to 300% of the full load capacity were found to be common. This method is nowadays seldom used to indicate capacity of boiler but however total heating surface is always mentioned in the specification of the boiler in order to have a fair idea about the performance of the boiler.

The heating surface for a steam generator unit consists of the portion of the heat transfer apparatus exposed, on one side to the gas or refractory being cooled, and on the other to the fluid being heated, measured on the side receiving heat. Total heating surface may be divided into boiler, water walls, water screen and water floor, superheating surface, economizer surface, air heater surface and the reheater surface. Heating suiface of a return tubular boiler equals one half the external area of the cylindrical shell plus the inside area of all the tubes plus two-third of the area of the rear head. All areas are expressed in sq m. An approximate rule for estimating the boiler heating surface area is to divide the area of the tubes by 0.80.

Sometimes the rating of the boiler is expressed in terms of the area of the grate surface. One square metre of the grate surface is approximately equal to 30 horse power. However, this may vary substantially with the operating conditions. The ratio of heating surface to grate surface varies from 30 to 40 for water tube boiler and from 45 to 55 for tube boilers.

7.4 BOILER PERFORMANCE

The performance of a steam boiler may be expressed in terms of:

(1) Efficiency.

(2) Rate of combustion in kJ per square metre of the grate area per hour for solid fuels, as fired.

(3) Rate of combustion expressed in terms of kg of fuel fired per hour per cubic metre of the furnace volume.

(4) Heat transferred per square metre of the heating surface area per hour.

(5) Heat release per cubic metre of the furnace volume.

Efficiencies obtained under test conditions are generally higher than the all-year round efficiencies. The best efficiency is usually obtained when operating at above the normal rating.

7.4.1 Testing of Boilers

Periodical testing of the boiler units is necessary as it may serve one of the following purposes:

(a) To compare different conditions and methods of operation.

(b) To determine the cause of inferior or superior performance.

(c) To compare different kinds of fuel.

(d) To determine the effect of changes in design or proportions of the boiler or heating surface upon the capacity and efficiency.

(e) To determine the capacity and efficiency of the boiler for comparison with specified or guaranteed results.

Tests of boilers are generally conducted for such period in which all the observations can be recorded with air accuracy and repeatability. The results of the test are usually incorporated in a report which should present all the leading facts bearing on the design, condition, etc.

The results of the boiler tests are usually recorded as under :

TABLE 7.1 *Data and results of boiler test (Example)*

Manufacturer : M/s Babcock and Wilcock.

Type of boiler : Water tube.

Kind of fuel: Semi-bituminous coal.

Method of firing : Hand firing.

Purpose of test: To conduct trial and to determine boiler thermal efficiency

Date

S.No.	*Name of the Item with*	*Units*	*Observations*
1.	Duration of test	hrs.	12
2.	Grate surface	sq. m.	4.20
3.	Water heating surface	—	190
4.	Air preheater heating surfaces	—	478
5.	Water walls-heating surface	—	–
	Average Pressure		
6.	Steam pressure bar	km/cm^2	10
7.	Barometer	cm-Hg	74.4
8.	Draft between damper and boiler	cm-water	–
9.	Forced draft under grate	cm-water	–
	Steam Conditions		
10.	Degree of superheat	°C	–
11.	Dryness fraction of steam	x	0.98
	Temperatures		
12.	Temperature of feed-water entering boiler	°C	29.7°C
13.	Temperature of gases leaving boiler	°C	276°C
	Fuel		
14.	Proximate analysis (as fired)		
	(percentage by weight)		
	Moisture	—	1.21
	Volatile matter	—	21.53
	Fixed carbon	—	71.80
	Ash	—	5.38
15.	Ultimate analysis (as fixed)		
	(Percentage by weight)		
	Carbon	—	81.50
	Oxygen	—	3.78
	Sulphur	—	0.45

Contd.

Table 7.1 Contd.

S.No.	*Name of the Item with*	*Units*	*Observations*
	Hydrogen	—	4.66
	Nitrogen	—	1.65
	Ash	—	8.06
16.	Heating value of coal	kJ/kg	25000 kJ/kg
17.	Weight of coal fired	kg	6100
18.	Weight of dry coal	kg	6015
19.	Weight of ash and refuse	kg	572
20.	Weight of dry coal consumed per hour	kg	502
21.	Dry coal per sq m of grate area per hour	kg	119.5
22.	Analysis of flue gases (percentage by volume)		
	Carbon-dioxide–	13.77	
	Carbon-monoxide	—	0.00
	Oxygen	—	4.70
	Nitrogen (by difference)	—	81.53
	Water		
23.	Total weight of water fed to the boiler	kg	59,200
24.	Total weight of water evaporated, corrected for moisture	kg	58,000
25.	Water evaporated per hour (corrected for moisture)	kg	4,833
26.	Equivalent evaporation per hour into dry steam	kg	5,517
27.	Power developed	kW	X
28.	Evaporation per sq m of heating surface	kg	29.11
	Results		
29.	Water evaporated per kg of coal	kg/hr	0.807
30.	Efficiency of unit (ηth)	%	71.73

7.5 SUMMARY OF CAUSES OF HEAT LOSS IN BOILERS

1. (A) Loss due to incomplete combustion

(i) Insufficient Air Supply
(ii) Undercooling of furnace at low ratings
(iii) Improper setting of boilers
(iv) Fuel bed in poor condition

(B) Loss due to combustible in ash-pit

(i) Too high rate of combustion attempted
(ii) Grates or stoker not proportioned to the kind of fuel used
(iii) Grates dumped or fuel bed sliced too frequently
(iv) Furnace temperature above fusion temperature of ash
(v) Excess coal dust in fuel.

2. Loss due to radiation and convection heat transfer from boiler and setting

(i) Boiler drum unisulated
(ii) Wall of setting too thin or poor quality
(iii) Furnace refractories in need of repair or renewal
(iv) Higher calorific value fuel used in boiler.

3. Loss due to heat carried away by dry fuel gases

(i) High excess air (revealed in low CO_2 content of flue gas)
(ii) High flue gas temperature
(a) Poor water circulation
(b) Poor head transfer due to scale formation
(c) Dead Gas pockets
(d) Dirty heating surfaces
(e) Gas velocity too high.

4. Loss due to Moisture

(A) Moisture in Coal

(i) Excessive wetting down of coal before firing
(ii) High moisture absorption by coal in yard storage.

(B) Moisture in Air

(i) Moisture-laden air as from steam jet
(ii) High humidity in air.

(C) Loss due to moisture formed by combustion of Hydrogen. Irreducible for any specific fuel. This loss is larger for oil and gas fuels than for coal.

THEORY QUESTIONS

7.1. What are factors to be taken into account while designing a boiler furnace?
7.2. What are the different systems of water circulation in Boilers?
7.3. What are the desirable features of refractories used in boilers?
7.4. What are 'Water walls'? How is heat transfer rate increased through these walls?
7.5. What are the factors on which the design of furnace bottom depends? Give sketches of furnace bottoms commonly used in boilers.
7.6. What are the requirements of a good insulating material for boilers?
7.7. What are the advantages of steel chimney over masonry chimney?
7.8. What are the factors to be taken into account while selecting material for the chimney?
7.9. How is water purified for boilers? Describe with a sketch pressure filter system.
7.10. What is the significance of pH value of water for boilers?
7.11. What is the difference between External and Internal treatment of water for boilers?
7.12. Write short notes on :

(a) Scale formation in boilers
(b) Corrosion
(c) Caustic embrittlement
(d) Hardness of water
(e) Water Analysis Reports
(f) Expansion Bends.

7.13. What is the function of a Dust Collector? How does it work?

7.14. How is the rating of boilers for power plants usually specified? What is meant by 60 MW boiler?

7.15. How is the performance of boiler expressed?

7.16. What are the meanings of the terms "Cations" and "Anions"?

7.17. Explain whether silica removal is or is not affected by precipitation.

7.18. How may a water supply become contaminated by oil? How may the oil be removed?

7.19. What are the essential factors required for corrosion?

7.20. The concentration limit for the boiler saline of a certain 35 kg/cm^2 generator is 2500 ppm. Make up is 30% and the total solids content in the make up is 200 ppm. Determine the amount of continuous blow-down required if the steam purity is 1 ppm. [**Ans.** 2.36% of steaming rate]

7.21. The raw water fed to evaporator has a total solids content of 1000 ppm. If the concentration limit for the evaporation is 5000 ppm and the vapour purity averages 5 ppm, determine the percentage of blow-down required. [**Ans.** 19.9% of feed]

7.22. Controlled circulation boilers are built with smaller tubes and lower water storage capacities than natural circulation boilers of the same output. How does this affect the blow-down requirements of controlled circulation unit in comparison with requirements of natural circulation units?

7.23. Five steam generators are being operated manually in parallel, all supplying steam to a common heater. Each unit has a variable-speed motor-driven stoker, but all receive forced draft through individual dampers from a common plenum chamber. All receive feed water from a common header through individual hand-operated valves. All discharge flue gases to a common breeching through individual uptake dampers.

One unit suddenly suffers a burst boiler tube and must be taken out of service. Outline the series of steps that the operators must take to remove the damaged unit from service and to redistribute the load evenly among the units remaining in service.

7.24. All but one of the following possible steam generator test difficulties would cause the apparent performance to be poorer than it actually is. One item would make the performance look better than it really is. Identify the one item and explain how this error would affect the results :

(a) Steam flow meter reads low

(b) Steam temperature gauge reads too high

(c) HHV of fuel is determined incorrectly being too high

(d) Coal scales are inaccurate, reading high.

Chapter 8 Steam Generators and Draught Systems

8.1 INTRODUCTION

Amongst a number of power plants fulfilling the power requirements of the world, major share is contributed by thermal power plants. As this type of power plants use steam as the working fluid, the generator of the steam can be said to be heart of it.

The steam generator is also known as 'Boiler'. The steam produced in the boiler is used in the industries for different purposes viz:

1. For generation of electrical energy as in power plants.
2. For generating mechanical energy in steam turbines.
3. For process heating purpose as heating of ground nuts in Oil Mills.
4. For sizing and bleaching in textile industries.

Normally a boiler produces steam at high pressure. This steam is first used to drive turbines for generating power. Then the steam exhausted from turbines, which is at lower pressure is used for process heating purpose.

A typical boiler has two different parts known as pressure part and combustion chamber. Pressure part is closed vessel, strongly fabricated out of steel, this contains water in it. The combustion chamber is constructed around pressure part with the help of fire bricks. The fuel undergoes combustion in this combustion chamber producing high temperature gases. These gases are also known as flue gases, hot furnace gases, or products of combustion. When flue gases come in contact with the water container, they give their heat to water, causing generation of steam. Afterwards flue gases are exhausted high into the atmosphere through the chimney.

8.2 HISTORICAL DEVELOPMENT OF THE BOILER

As we know boiler is steam generator, or it is a closed vessel in which water is heated and converted into steam. The boiler was initially developed in the very first century by Mr. Hero of Alexandria. However it was too small and it was used as toy only.

Consider a tea kettle on stove or burner, which has water or tea boiling in it, we can see that steam spreads as it leaves the spout. If the spout is closed by means of a cork, after some time the cork will be blown off.

This shows that the steam generated in a closed vessel keeps on the developing pressure. Such steam at high pressure is utilised to do work in expanding in a steam engine.

At the beginning a small closed vessel was used to generate steam by the application of heat. At the beginning the quantity of steam generated and subsequently used in an engine was not sufficient. Therefore larger vessel called **Boilers** were developed for giving a sufficient quantity of steam at the required pressure. The pressure was maintained constant in the boiler by opening the outlet valve such that the steam generated is the same as the steam supplied to the engine.

At the initial stages boilers used were of the fire tube type or tank boilers.

In our everyday life we used to have hot water for bathing purpose a device similar to a tank boiler but on a very smaller scale as shown Fig. 8.1.

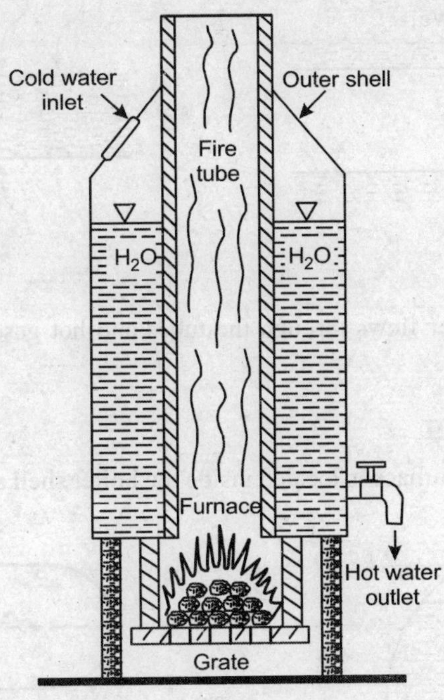

Fig. 8.1. *Small tank boiler used for heating water*

Taking into consideration the safety, the development of simple boiler took place over the years. Till the 17th century, there was no much development in the boiler design and manufacturing. Mr. Denis Papin of France designed and developed the first boiler with safety valve in 1679 A.D.

The major developments took place after 1769 A.D., when James Watt of Scotland developed the practical steam engine.

Afterwards taking into consideration the quantity of steam generated with respect to time, water tube boilers were developed. Subsequently high pressure boilers and supercritical boilers for power generation were developed.

The steam generated in the boilers is utilised for 2-purposes viz:

(i) Power Generation. The boilers generating steam for power generation purpose are bigger in size, generate steam at high pressure and this high pressure steam is first expanded in the turbines for generating power. Then the exhausted steam is utlised for process heating purpose.

(ii) Process Heating Purpose. For example, heating of the oil seeds in the oil mills before they are being crushed.

8.3 CLASSIFICATION

Boilers are mainly classified according to the following factors.

8.3.1 By the Relative Position of Flue Gases and Water (or Tube Contents)

(*a*) *Fire tube.* In this case flue gasses flow through the tubes, which are surrounded by the water which is to be evaporated. Example: Lancashire, Cornish, Cochran, Locomotive, Simple vertical, Scotch marine boiler, etc.

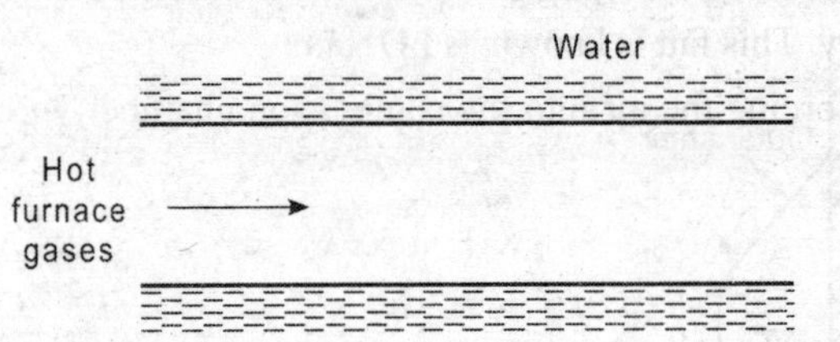

Fig. 8.2

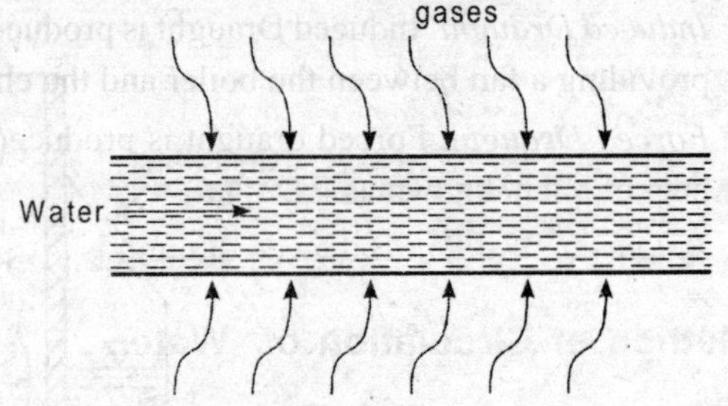

Fig. 8.3

(*b*) *Water tube.* In this case water flows through the tubes and hot gases heat the tubes from out side e.g., Babcock Wilcox, Stirling Boiler, etc.

8.3.2 By the Method of Firing

(*a*) *Internally fired.* In this case the furnace is located inside the boiler shell as in case of Lancashire boiler.

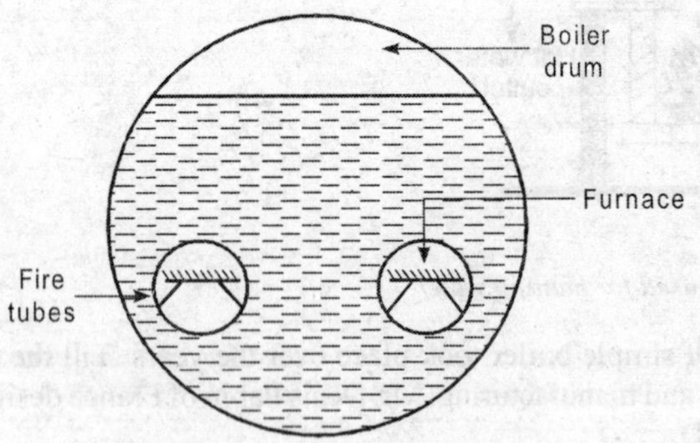

Fig. 8.4. *Internal fired*

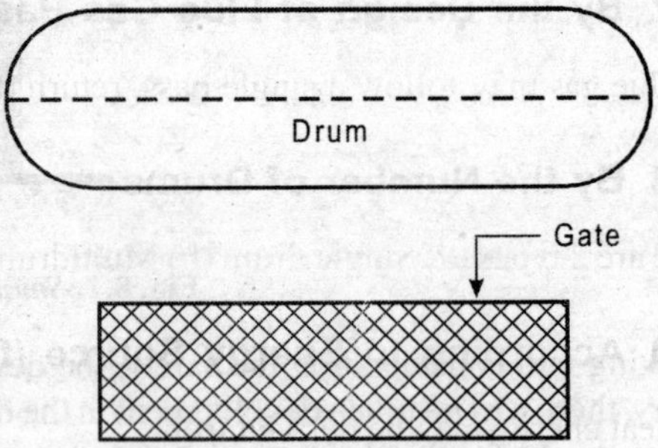

Fig. 8.5. *External fired*

(*b*) *Externally fired.* The furnace is located outside the boiler shell as in case of Babcock Wilcox boiler Fig. 8.5.

8.3.3 According to Pressure of Steam

(*a*) *Low pressure boiler.* Boilers which can develop pressures below 80 bar are known as low pressure boilers e.g., Lancashire, Cochran, Cornish etc.

(*b*) *High pressure boilers.* Boilers which can develop pressures above 80 bar are known as high pressure boilers e.g., Babcock, Lamount, Velox, Benson etc.

8.3.4 Nature of Draught

Draught is the pressure difference, which is necessary to draw the sufficient quantity of it for combustion. It is of following types:

(*a*) *Natural draught.* In this case the draught is produced by mean of chimney only. The amount of draught directly depends upon the height of chimney.

(*b*) *Artificial draught.* Artificial draught is produced by means of fans.

(i) *Induced Draught.* Induced Draught is produced by sucking the flue gases from chamber. It is obtained by providing a fan between the boiler and the chimney. This fan is known as I.D. fan.

(ii) *Forced Draught.* Forced draught is produced by forcing the air into the combustion chamber with the help of a fan, known as F.D. fan.

(iii) *Balanced Draught.* Balanced draught is produced by means of I.D. fan and F.D. fan.

8.3.5 Method of Circulation of Water

(*a*) *Natural circulation.* In which the circulation of water is due to gravity.

(*b*) *Forced circulation.* In which the circulation of water is obtained by a centrifugal pump.

8.3.6 By the Use

(*a*) *Land type.* Boilers used with stationary plants.

(*b*) Marine type

(*c*) Locomotive boilers.

8.3.7 By the Design of Flue Gas Passages

The flue gas may follow a single pass, return pass or multipass.

8.3.8 By the Number of Drums

There are 2 types (a) Single drum (b) Multidrum boilers.

8.3.9 According to Energy Source (fuel) Used

The heat energy may be derived from,

(a) Combustion of fossile fuels such as coal, wood, oil or natural gas etc.

(b) Electric or Nuclear energy.

(c) Hot waste gases of other chemical reactions.

8.3.10 According to the Material of Construction of the Boiler Shell

(*a*) *Steel Boilers*. Power boilers are generally fabricated out of steel plates.

(*b*) *C.I. Boilers*. Generally low pressure boilers are, sometimes built out of castings.

8.4 LANCASHIRE BOILER

It is fire tube, stationary, horizontal straight tube, internally fired, natural circulation boiler. Normal working pressure of this boiler is about 15 bar and steam generation capacity is upto 8*T* /hr.

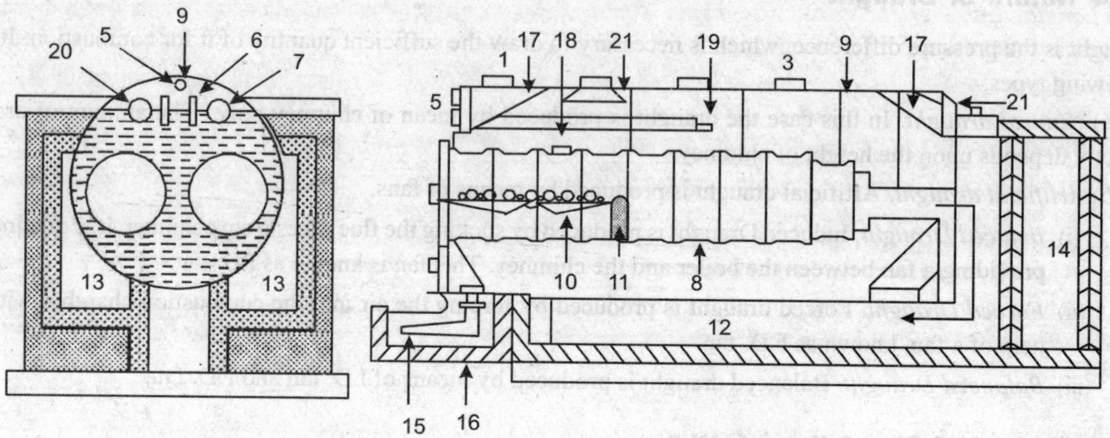

1. Dead weight safety valve.
2. High steam low water alarm.
3. Manhole.
4. Steam stop valve and Antipriming Device.
5. Pressure gauge.
6. Water level indicator.
7. Feed check valve and perforated feed pipe.
8. Fire tubes.
9. Boiler shell.
10. Fire bars.
11. Brick work Bridge.
12. Bottom flue.
13. Side flues.
14. Main flue.
15. Blow-off cock.
16. Blow-off pit.
17. Gusset stays.
18. Fusible plug.
18. Fusible plug.
19. Scum tray.
20. Scum valve.
21. End plate.

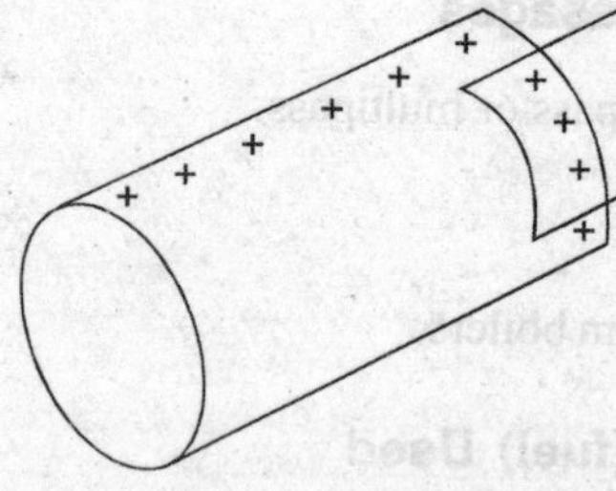

Fig. 8.6 Lancashire boiler

8.4.1 Constructional Details

The main parts of the boiler are as follows.

(1) Boiler Drum. This boiler consists of a very large boiler drum. The size of the boiler drum is approximately 7–9 metres in length and 2–3 metres in diameter. This is a very old type of boiler, in this boiler, the meeting edges of the plate are connected by means of rivetted joints. Steel plates are rolled to form shell and the meetings edges of the shell are connected by means of rivetted joints. The alternate shells are made smaller in a diameter, so that the longer shells of required length can be obtained by inserting one shell into the other and rivetting over the circumference.

The end plates of the boiler are flat. These flat end plates, when they are subjected to steam pressure from inside, they will have a tendency to bulge out. In order to prevent them from bulging, they are strengthened by means of stays. There are 2 types of stays.

(i) *Longitudinal stay*. It consist of round bar which extends from one end plate to the other and are connected to the end plates by means of lock nuts.

(ii) *Diagonal gusset stay*. Consists of a triangular plate, which is rivetted to the shell plate and end plate as shown in Fig. 8.6(a).

(2) Fire Tubes or Flue Tubes. Through the boiler drum passes 2 fire tubes.

Part of each of fire tubes is corrugated, so as to take up expansion and contraction when the boiler gets hot and cold. Also by providing corrugations surface area increases. These tubes are also made of number of shells. The ends of each of the shell are flanged outwards and the flanges of the two adjacent shells are connected by means of rivets, so as to get shells of required length.

It is to be noted that, the centre line of fire tubes is slightly below that of the centre line of boiler shell. Because fire tubes are provided approximately in the centre of water, so that heat can be transferred uniformly in all directions.

(3) Grate, Brickwork Bridge etc. Combustion of fuel takes place over the Grate, Top surface of the grate is made up of fire-bars. Below the grate Ash-pit is provided.

When the combustion of fuel takes place, hot gases are generated and the Brickwork bridge deflects the gases to move upwards.

(4) Manhole. It gives access to the inside of the boiler drum. It will be sufficiently large enough for a man to pass through it, for cleaning and inspection purpose.

(5) Mountings. The boiler is fitted with various mountings. Mountings are necessary for the safe and efficient operation of the boiler.

The term mountings refers to the items such as,

(*i*) *Safety valves*. (a) Dead weight safety valve. (b) Spring loaded safety valve. (c) High steam low water alarm. When the pressure of steam in the boiler drum increases beyond safe value, then these valves will open and allow the surplus steam to escape to the atmosphere, till the pressure of steam in the boiler drum comes to normal working pressure.

(*ii*) *Pressure gauge*. It indicates pressure inside the boiler drum.

(*iii*) *Feed check valve*. Water enters the boiler drum through feed check valve. On the inner side of the feed check valve, perforated feed pipe is fitted for feeding the water uniformly in the boiler drum. Below the feed pipe scum tray is fitted which separates scum and is removed through the scum valve.

(*iv*) *Water level indicator*. It indicates the water level in the drum.

(*v*) *Blow-off valve*. It blows out impurities collected due to gravity.

(*vi*) *Steam stop valve*. Through this steam will be taken out for use.

(*vii*) *Antipriming device*. It is always desirable that the steam should leave the boiler as dry as possible, for that purpose an antipriming device or steam collecting pipe is provided in the steam space. It consists of a pipe and rectangular openings are provided in its bottom surface. It is fitted by means of baffles from inside. When the wet steam (steam and water particles) enters the pipe, the water particles being heavier than steam, after striking the baffles fall back to water space and steam almost dry will be taken out through the steam stop valve.

(6) Accessories. Accessories are used for improving the efficiency of boiler plant. Generally Super heater and Economiser are used as accessories.

(7) Path of Flue Gases. When the combustion of fuel takes place, hot gases are generated and the brick work bridge deflects these gases to move upwards.

During their first pass, these gases flow from the front end of the boiler to the rear end through central tubes. During their I-pass, they pass through the superheater, then they enter the common bottom flue at the rear end. Now their II-pass starts, during this gases flow from rear end to front end through the common bottom flue. At the end of II-pass, the gases are bifurcated into two side flues, when the flue gases enter side flues, their III-pass starts end during their III-pass, they flow from the front end to rear end through side flues and during this pass they heat the water from sides.

After their third pass they meet in main common flue, from where they pass through the economiser and then they are exhausted through the chimney.

8.5 CORNISH BOILER

The only difference between Cornish boiler and Lancashire boiler is that, in case of Cornish boiler there is only one fire tube, whereas in case of Lanchashire boiler there are 2 fire tubes, rest all the arrangement is same.

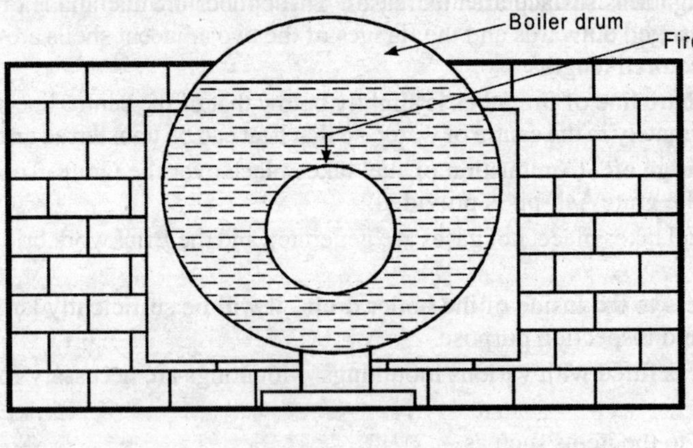

Fig. 8.7. *Cornish boiler*

8.6 LOCOMOTIVE BOILER

It is a Fire tube, Horizontal, Multitubular, Internally fired, Natural circulation boiler.

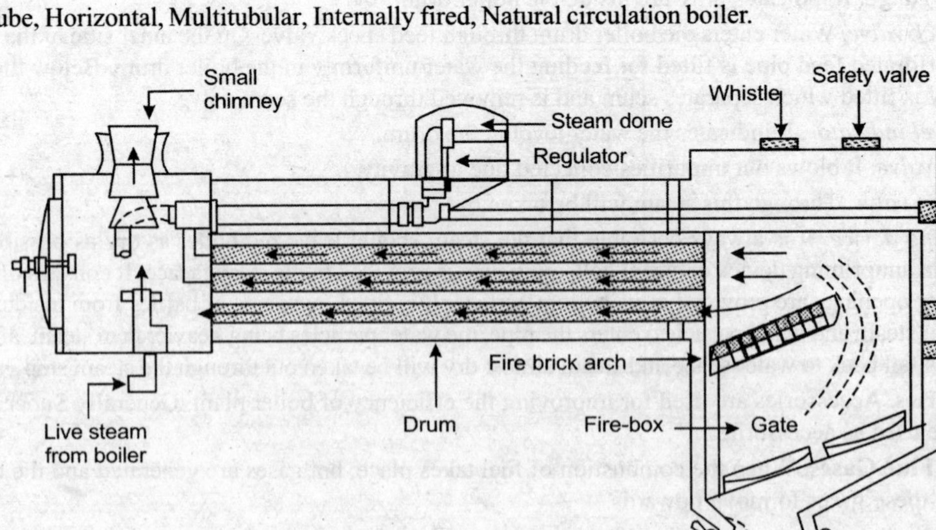

Fig. 8.8. *Locomotive boiler*

As shown in Fig. 8.8 it consist of a large cylindrical shell. This shell is fitted to a rectangular fire box at one end and smoke box at other end. In between the smoke box and fire box a number of tubes are fitted.

When the combustion of fuel (coal) takes place over the grate, hot gases are generated. These hot gases rising from the grate are deflected by the Fire brick arch, so that they come in contact with the complete heating area of the fire box. The dampers as shown, control the flow of air for combustion. These hot gases then flow through the tubes, heat will be transferred to water and water will be converted into steam and accumulates in the steam dome and gases afterward are exhausted to the atmosphere through the small chimney.

The regulator can be operated from the cabin by turning the lever. In order to have superheated steam, the steam is passed through the superheater tubes. These superheater tubes are of smaller diameter and are laid in the large diameter fire tubes as shown. Then this superheated steam is supplied to the engine. This boiler is fitted with various mountings as shown in Fig. 8.8.

8.7 COCHRAN BOILER

It is a Fire tube, Stationary, Multitubular, Internally fired, Vertical boiler.

Cochran boiler is made in many sizes capable of producing steam upto 3 *T*/hour and working pressures upto 17 bar. This boiler gives Thermal efficiency (η_{th}) of 70% with coal firing and about 75% with oil firing.

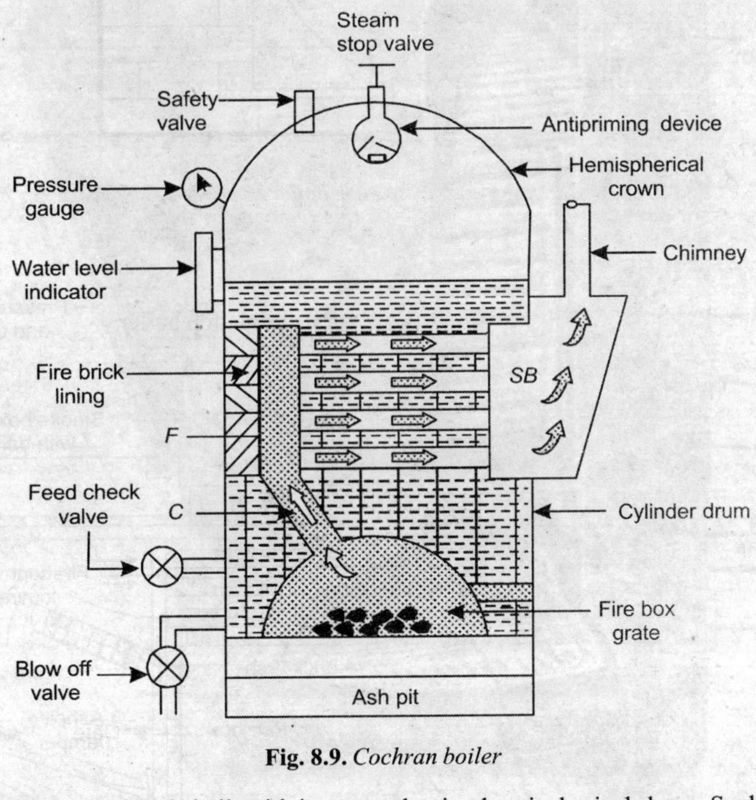

Fig. 8.9. *Cochran boiler*

The boiler consists of a cylindrical shell, with its crown having hemispherical shape. Such a shape of the crown plate, gives sufficient strength to withstand the bulging effect.

The furnace or fire box is also crown shaped or dome shaped, which deflects the unburnt fuel back to the grate. Combustion of fuel takes place over the grate. Hot gases generated in the fire box, enter into the chamber *C*, though a short flue pipe *F* and strikes the boiler shell plate or Back plate which is lined by fire bricks. The back plate directs

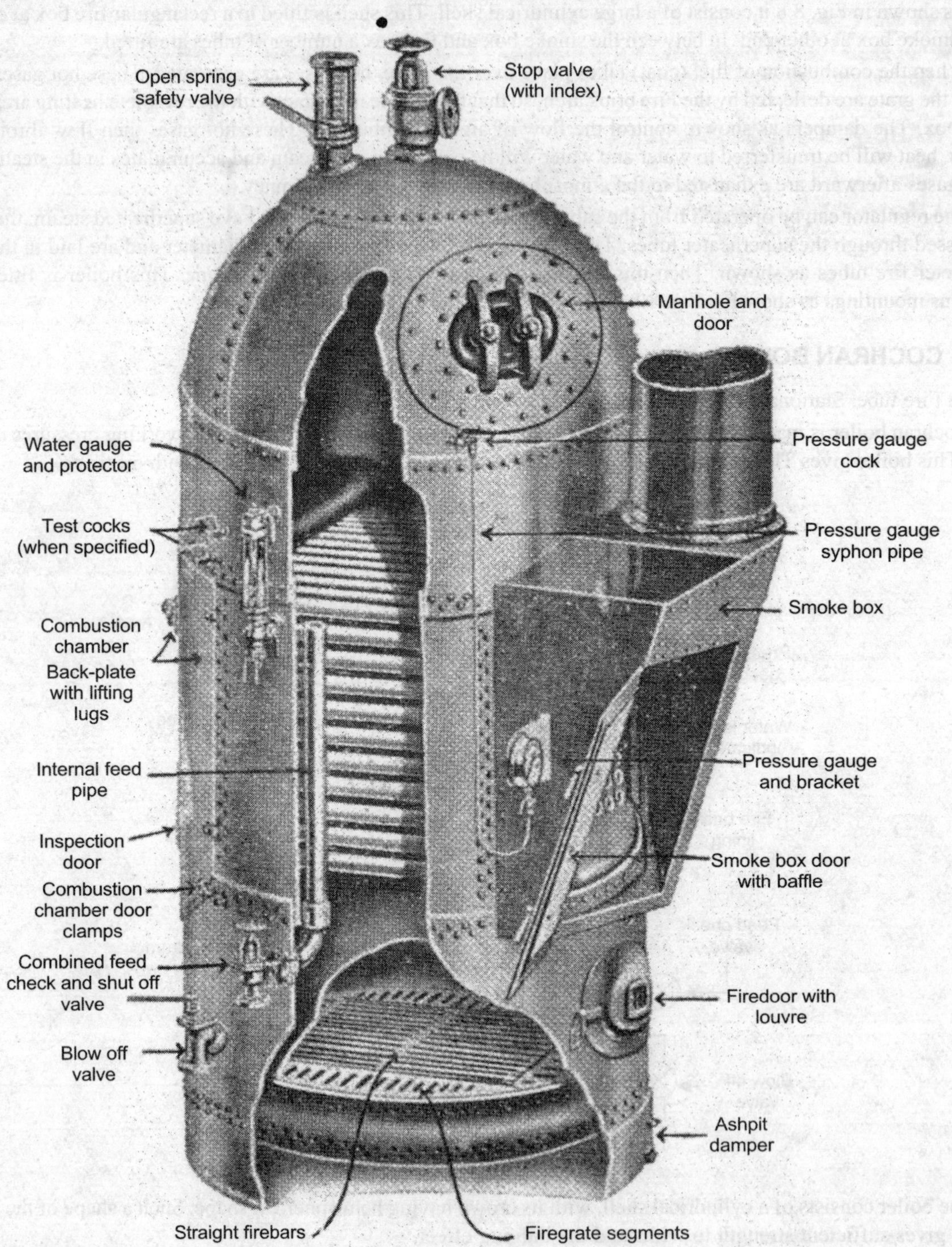

Fig. 8.9 *Cochran boiler*

the gases into the smoke or fire tubes. The gases after passing through the horizontal tubes enter the smoke box *SB* and from where they are exhausted through the chimney.

The hot gases passing through the horizontal tubes, give their heat to the water and in doing so, convert water into steam, which gets accumulated in the steam space.

8.8 BABCOCK WILCOX BOILER

It is a water tube boiler. It mainly consists of 3 parts.

(i) A Horizontal Steam and Water Drum. This is the main part of the Babcock Wilcox boiler. The size of the boiler drum is very much smaller, when compared with boiler drums of fire tube boiler of same capacity. Hence these are simpler from transportation point of view.

(ii) A Bundle of Steel Tubes. It is provided in the combustion chamber.

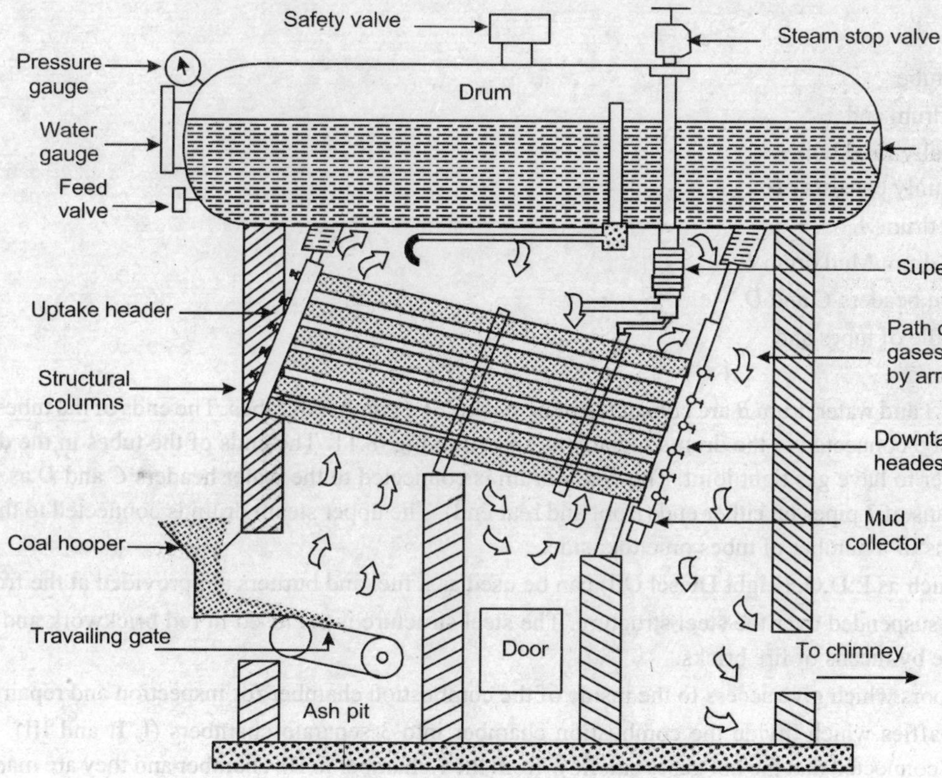

Fig. 8.10. *Babcock wilcox boiler.*

(iii) Combustion Chamber. It is the space above the grate and below the drum where the combustion of fuel takes place. This complete chamber is divided into three separate chambers. These chambers are so arranged, that, the gases can flow from one chamber to the other and give out heat in each chamber gradually. Hence the first chamber is hottest and the last chamber is at lowest temperature.

The front end of the boiler is connected to the uptake header by a short tube and the rear end is connected to the downtake header by a long tube. In between the headers a number of tubes are fitted. The ends of the tubes in the headers are either expanded or welded in order to have gas tight joint. These tubes are arranged in zig-zag manner, so that more surface area of the tubes is exposed to hot gases. These tubes are sloping, downwards towards the rear end. The inclination of the tubes with the horizontal, will be in between 5–15°. A mud collector is provided at the bottom most position of down take header for collecting and disposing the mud or any other suspended impurities. The entire boiler except the furnace in suspended on steel structure, so that the boiler can expand or contract freely without producing the cracks into the brick work, which encloses the boiler and furnace. This brickwork is lined from inside by means of fire bricks. Doors provided give access for cleaning inspection and repair purpose. Dampers provided are used for regulating draught.

This boiler is fitted with various mountings viz., safety valve, steam stop valve, pressure gauge, water level indicator, fusible plug, feed check valve, blow-off valve etc. This boiler is also fitted with superheater as shown in Fig. 8.10.

8.9 STERLING BENT TUBE BOILER

It is a,

(a) water tube,

(b) multi drum and

(c) externally fired boiler.

The boiler mainly consists of,

(i) Steam drum *A*.

(ii) Water drum/Mud drum *B*.

(iii) Bottom beaders *C* and *D*.

(iv) A bundle of tubes and

(iv) Burners.

The steam drum *A* and water drum *B* are connected of by means of a number of tubes. The ends of the tubes are bent so that they can be connected to the drums radially as shown in Fig. 8.11. The ends of the tubes in the drums are expanded in order to have gas tight joint. The water drum is connected to the water headers *C* and *D* as shown in Fig. 8.11 by means of 2 pipes on either end (front and rear end). The upper steam drum is connected to the bottom headers by means of a number of tubes on either side.

Liquid fuel such as L.D.O. (Light Diesel Oil) can be used as a fuel and burners are provided at the front end.

The boiler is suspended over the steel structure. The steel structure is enclosed in red brickwork and which is lined from inside by means of fire bricks.

E—are the doors which give access to the inside of the combustion chamber for inspection and repair purpose.

F—are the baffles which divide the combustion chamber into 3 separate chambers (I, II and III). These 3-chambers are so connected that the hot gases can flow from the I-chamber to III-chamber and they are made to pass through economiser and finally they are exhausted through chimney. The water in the tubes will absorb heat in each chamber, so the temperature of flue gases will go on decreasing towards the rear end.

The boiler is fitted with various mountings viz. :

(a) Spring loaded safety valves

(b) Pressure gauges

(c) Thermometers

(d) Water level indicators

(e) Feed check valve.

(f) Steam stop valve.

(g) Blow off valve.

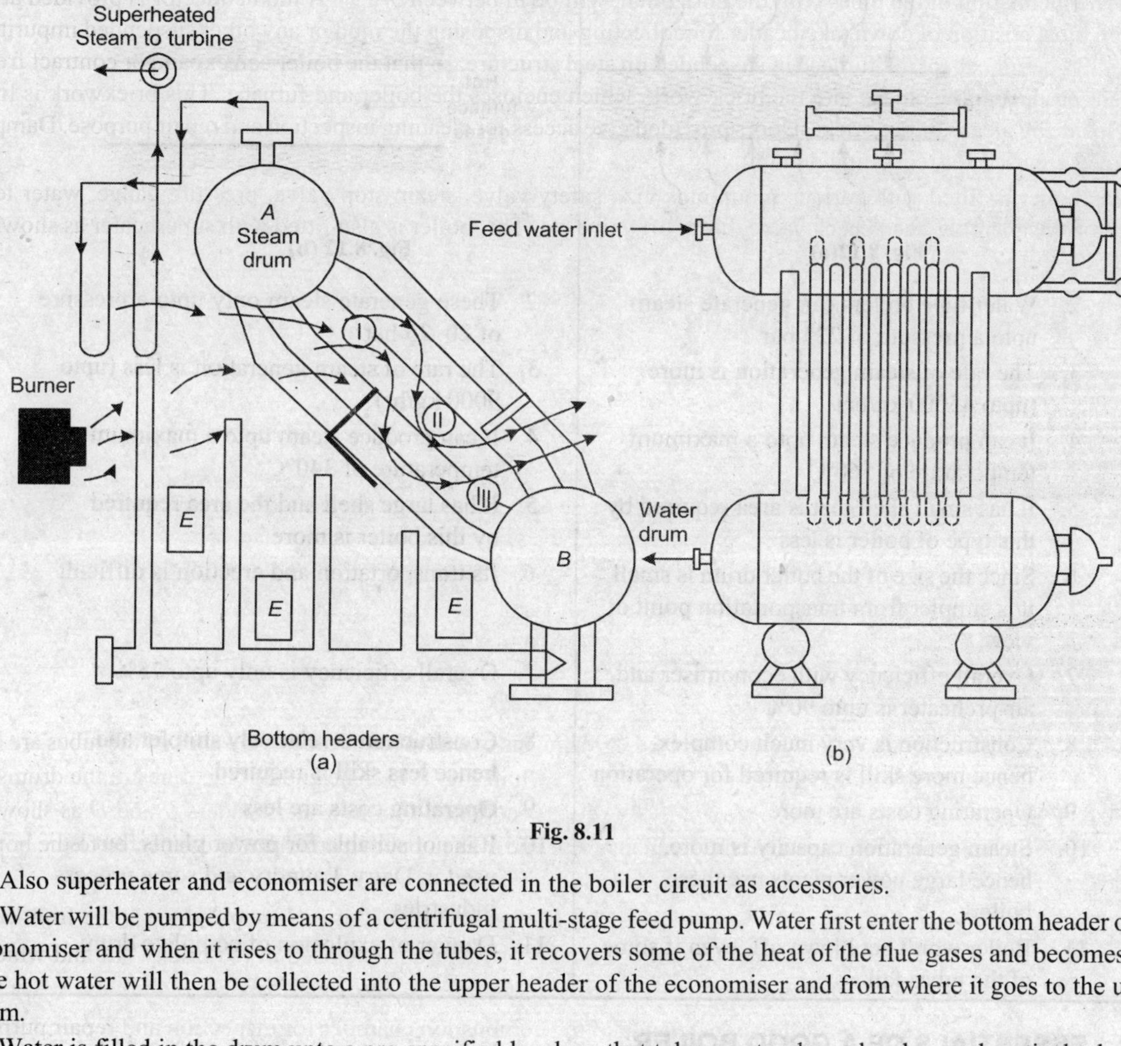

Fig. 8.11

Also superheater and economiser are connected in the boiler circuit as accessories.

Water will be pumped by means of a centrifugal multi-stage feed pump. Water first enter the bottom header of the economiser and when it rises to through the tubes, it recovers some of the heat of the flue gases and becomes hot. The hot water will then be collected into the upper header of the economiser and from where it goes to the upper drum.

Water is filled in the drum upto a pre-specified level, so that tubes, water drum, headers and vertical tubes are flooded by means of water.

We know that the I-chamber is at the hottest temperature, II-chamber is at the medium hot temperature and III-chamber is at the lowest temperature. So, due to this difference, in temperature, difference in densities of water is created and water will rise from water drum to steam drum. As the water rises, some of the water gets converted into steam. Water and steam then enters the steam drum *A*. In this drum steam gets accumulated in the steam space and the water which has not converted into steam; flows down to bottom headers through vertical tubes, and from bottom headers water goes to the water drum and again the cycle repeats.

The steam so generated will be wet, in order to have superheated steam, it is made to pass through the superheater and then it is utilised for power generation purpose.

8.10 COMPARISON BETWEEN WATER TUBE AND FIRE TUBE BOILERS

Water Tube	Fire Tube
1. 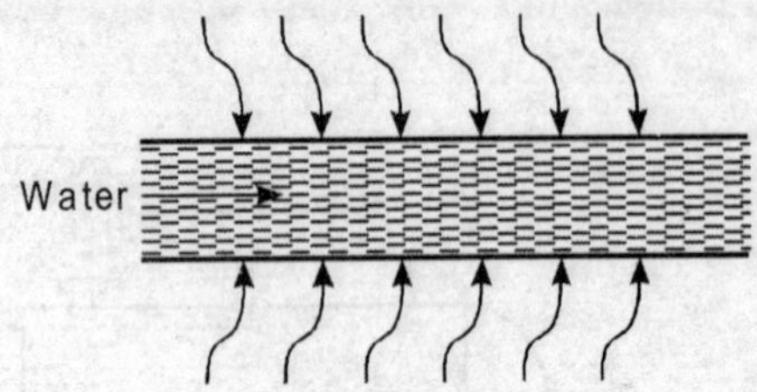 Fig. 8.12(a)	1. 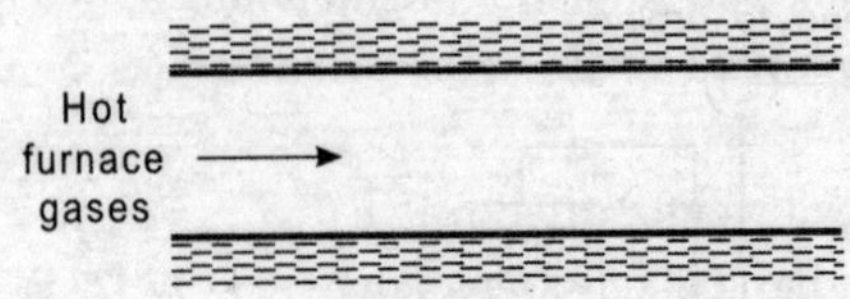 Fig. 8.12 (b)
2. Water tube boilers can generate steam upto a pressure of 225 bar	2. These generate steam only upto a pressure of 20–25 bar
3. The rate of steam generation is more (upto 45000 kg/hr)	3. The rate of steam generation is less (upto 9000 kg/hr)
4. It can produce steam upto a maximum temperature of 560ºC	4. It can produce steam upto a maximum temperature of 340°C
5. It has small shell that is area required by this type of boiler is less	5. It has large shell and the area required by this boiler is more
6. Since the size of the boiler drum is small it is simpler from transportation point of view	6. Its transportation and erection is difficult
7. Overall efficiency with economiser and air-preheater is upto 90%	7. Overall efficiency is only upto 75%
8. Construction is very much complex, hence more skill is required for operation	8. Construction is relatively simpler and hence less skill is required
9. Operating costs are more	9. Operating costs are less
10. Steam generation capacity is more, hence large power plants use these boilers	10. It is not suitable for power plants, but is used in Dairy, Laundry and some process industries
11. Boiler won't get blown off, even if some of the tubes fail	11. Danger of explosion of complete drum

8.11 ESSENTIALS OF A GOOD BOILER

1. With minimum fuel consumption, the boiler should be capable of generating steam at the required quality.
2. It should be light and occupy less space.
3. It should be capable of quick starting.
4. The initial cost, installation cost and maintenance cost of the boiler should be minimum.
5. The different parts of the boiler should be easily approachable for repairs.
6. The boiler should have minimum number of joints and those too should be away from direct flames.
7. There should be no deposition of mud and other deposits on the inner surface of the boiler drum.
8. Boiler should conform to safety regulations, laid down in Boiler act [Indian Boiler Regulations (IBR) Act].

8.12 BOILER MOUNTINGS AND ACCESSORIES

Mountings. Mountings are the fittings mounted on the pressure part of the boiler. Mountings are necessary for the safe and efficient operation of the boiler. Without mountings boiler cannot work safely. Special provisions are made on the pressure part to mount the mountings.

The term mountings refers to the items such as,

(1) Safety valves
   (a) Dead weight safety valve
   (b) Spring loaded safety valve
   (c) High steam and low water alarm.
(2) Water level indicators
(3) Fusible plug
(4) Pressure gauges
(5) Steam stop valve
(6) Feed check valve
(7) Blow-off valve etc.

The above mountings are usually installed in accordance with IBR.

Accessories. Accessories are used to improve the efficiency of the boiler plant. Without accessories boiler can work safely. These are not mounted on the boiler, but these are connected in the boiler circuit.

Usually following accessories are provided with the boiler.

1. Super heater
2. Economiser
3. Air-preheater
4. Steam injector
5. Feed pump
6. Steam trap
7. Induced draught fan
8. Forced draught fan etc.

8.12.1 Dead Weight Safety Valve

As the name implies, this valve is used to assure safe working of the boiler. When the pressure of steam inside the boiler drum increases beyond the safe value, it may cause damage to the pressure part, or even a severe accident. It can be prevented by allowing the surplus steam to escape to the atmosphere, by means of the safety valve.

This valve was first introduced by M/s. J.J. Hopkinson and Co. Ltd. It consists of a vertical pipe *A*, bottom end of which is flanged and through this flanged end it is mounted on the mounting blocks of the boiler. *B* is the Gun metal valve, resting over the valve seat *C*. Valve seat *C* is fixed on the top of the pipe *A* by means of a securing ring and screw. *F* is the feather which guides the valve *B*. Valve *B* is also guided by a ring in steam discharge casing. The weight carrier is suspended from the top of the valve *B*. Dead weights are placed on the weight carrier and these are covered by the Cast Iron cover.

Weight of the valve *B*, spindle, carrier, dead weights and cover will be acting downwards. During normal working conditions, this downward acting load will be balanced by the steam pressure from below. When the steam pressure exceeds pre specified or safe value, then the valve will be lifted up and the surplus steam will be exhausted to the atmosphere.

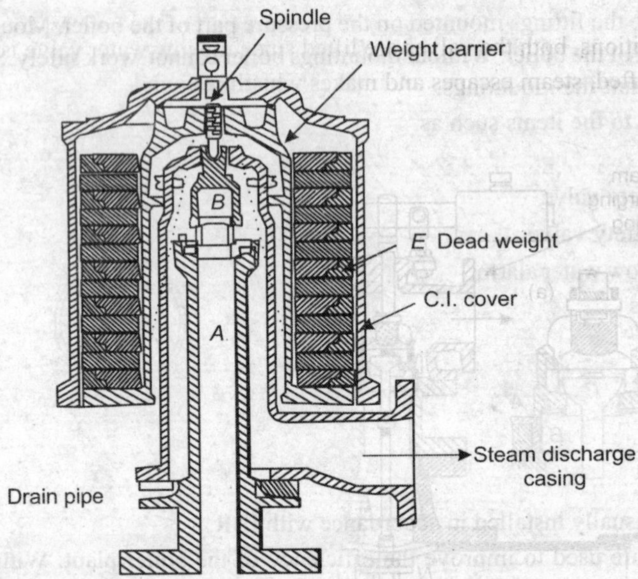

Fig. 8.13. *Dead-weight safety valve*

Example, Dead weight safety valve used on pressure cookers.

8.12.2 High Steam Low Water Alarm

This valve was also introduced by M/s Hopkinson and Co. Ltd. This is a two-in-one valve, i.e., it consists of 2 valves.

(i) *High steam valve.* During high steam conditions this valve opens and allows surplus steam to escape to the atmosphere.

(ii) *Low water valve.* When the water level in the boiler drum falls below the safe level, then this valve opens and allows the steam to escape. When the steam is allowed to escape, it makes whistling sound i.e., it gives warning to the boiler attendant. The whistling sound for high steam conditions will be louder than low water conditions, so that one can distinguish between the two. Since it makes whistling sound it is known as alarm.

As shown in Fig. 8.13 *A* is the high steam valve, resting over the valve seat *B*. Valve seat *B* in secured in position by means of a ring and feather *S* is cost on it. This feather guides the valve *A*. *C* is the low water valve, which is resting over the inner edge of the valve *A* as shown. *R* is the rib which guides the valve *C*. Low water valve. *C* will be held in position by means of the Dead weight, *D* which is connected to the rod *E*, which position by means of the weight *D* and partly by means of weight and counter weight outside the steam discharge casing.

In the steam space lever *FK* is provided. A large fire brick float *L* is suspended from the end *K*.

During normal working conditions, the float will be submerged in water and the weight of this float in partly balanced by means of counter weight *M* and partly by means of buoyant force due to water, which is acting in the upward direction. But when the water level goes below the safe level, the float gets uncovered by means of water, then there won't be buoyant force and counter, weight. *M* alone will not be sufficient to balance the float. So, the lever *FK* tilts in the clockwise direction. Then the knife edge projection of the lever *FK* comes in contact with the collar N and in turn lifts the hemispherical valve C. When the valve C is lifted, steam escapes and makes whistling sound.

During high steam conditions, both the valves are lifted since the low water valve is resting over inner edge of the high steam valve C is lifted, steam escapes and makes whistling sound.

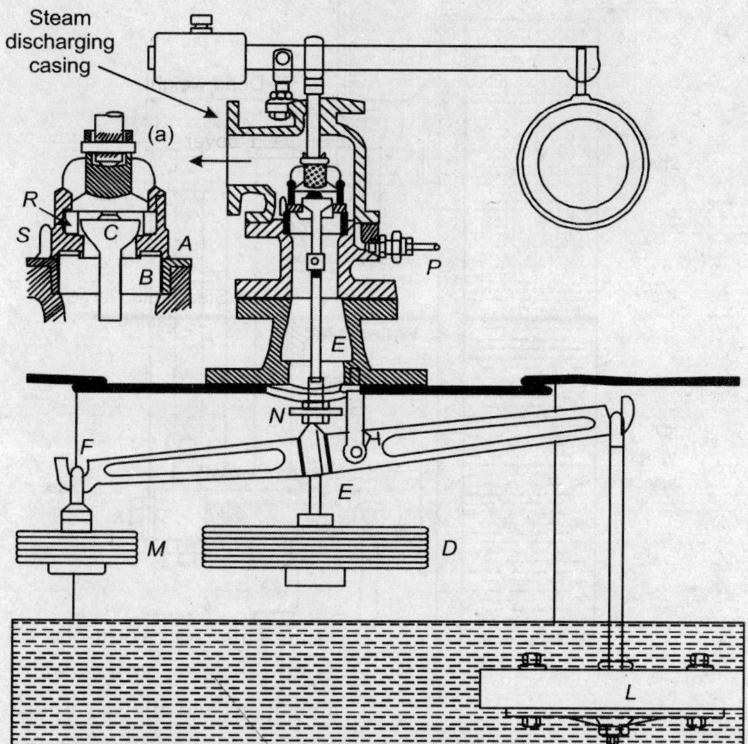

Fig. 8.14. *High steam low water alarm*

During high steam conditions, both the valves are lifted since the low water valve is resting over inner edge of the high steam valve, and surplus steam will be exhausted.

8.12.3 Water Level Indicator

It indicates the water level inside the boiler drum. This water level can be made visible from the boiler room floor by using mirrors. This unit was also introduced by M/s Hopkinson and Co. Ltd.

The unit consist of two hollow gun-metal stuffing boxes. One end of each of the stuffing box will be flanged and through this flanged end it is connected to the front end plate of the boiler. Upper stuffing box is connected to the steam space and lower stuffing box is connected to the water space. In between the two stuffing boxes a strong glass tube is fitted. (In between the glass and metal, rubber packing is provided in order to have gas tight joint).

Thus two valves A and B which control the passage of steam and water between the boiler and the glass tube. When these valves are open, the handles are vertical and the water level in the glass tube will be the water level in the boiler drum. A third valve *C* called as Blow-off valve is generally closed. When this valve is closed, its handle is vertical and to blow out sediments and impurities handle is to be made horizontal.

An arrangement is also provided, in order to shut off automatically the supply of steam and water to the glass tube, if the glass tube breaks due to any reason. This arrangement consists of a hollow gun-metal column *D* and two steel balls as shown.

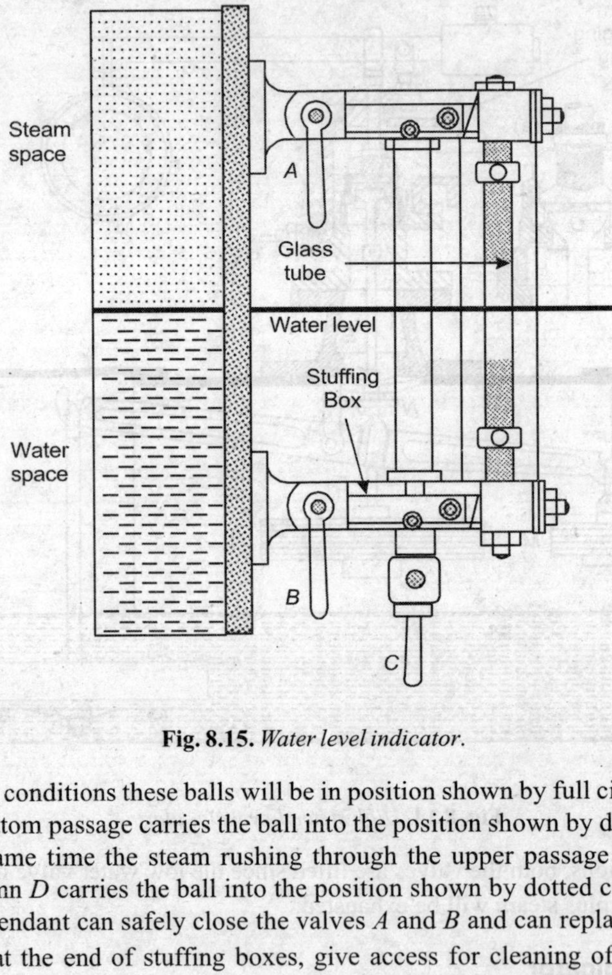

Fig. 8.15. *Water level indicator.*

During normal working conditions these balls will be in position shown by full circles if the glass tube breaks the rush of water in the bottom passage carries the ball into the position shown by dotted circle and shuts off the supply of water. At the same time the steam rushing through the upper passage aided by the water rushing upwards through the column *D* carries the ball into the position shown by dotted circle and shuts off the steam supply. Then the boiler attendant can safely close the valves *A* and *B* and can replace the tube.

Screw plugs provided at the end of stuffing boxes, give access for cleaning of passages. The glass tube is generally covered from the front and from sides by Guard glass in order to protect the boiler attendant from the flying fragments of glass tube, if it breaks.

8.12.4 Fusible Plug

It actuates itself as the last resort, when the water level in the boiler drum falls below the minimum permissible water level. As a result of which, fusible metal melts by the heat of steam, the plug drops out and the steam rushes into the combustion chamber and extinguishes the fire. Thus it safeguards the combustion chamber crown from burning and also warns the boiler attendant about the fall of water level in the drum below minimum permissible level.

Plate 1 *Sterling bent tube boiler.*

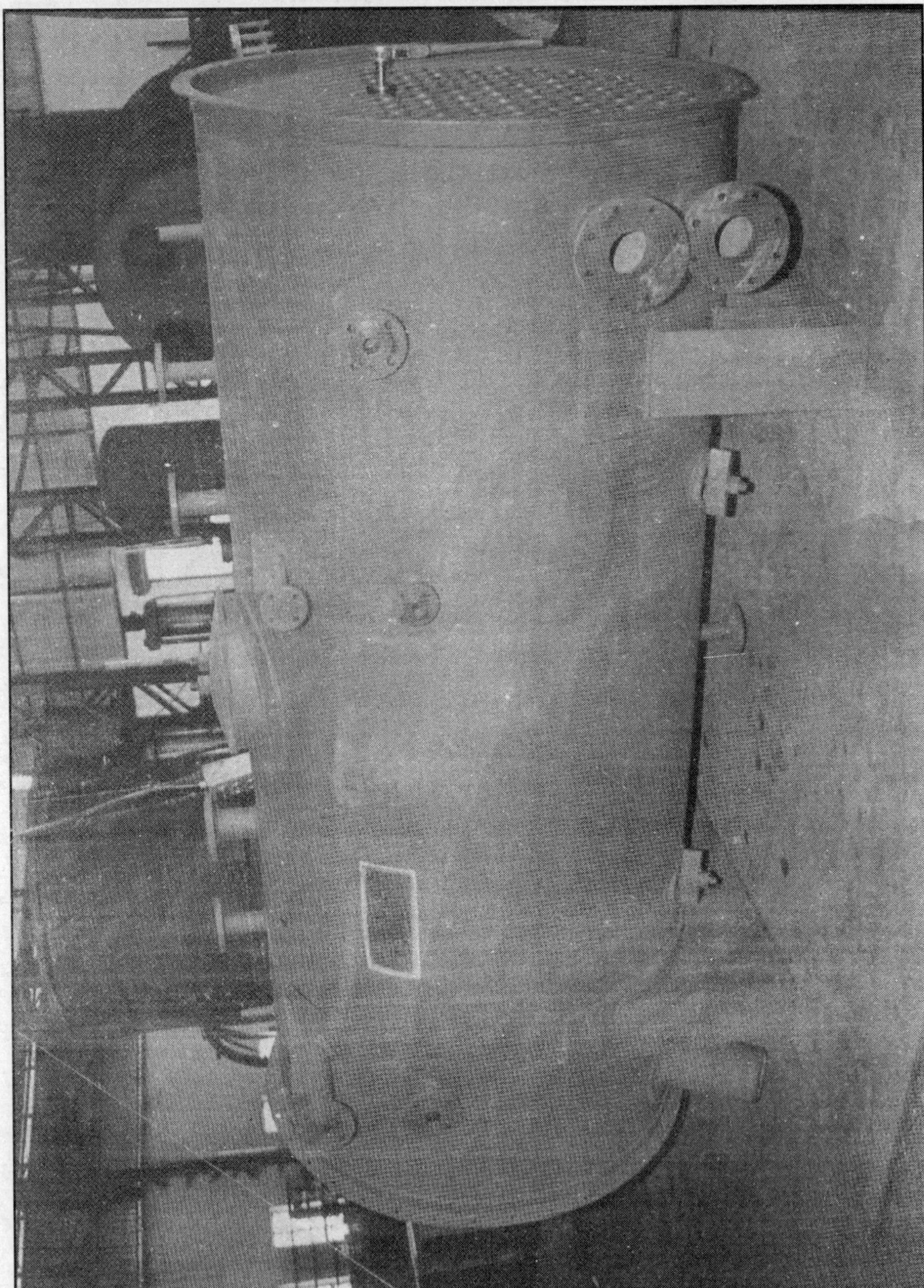

Plate 2 *Fire Tube, 6TPH, LIPI Make Boiler*

Plate 3 *Water Tube, 90TPH, LIPI Make Boiler*

Plate 4 *Sterling Bent tupe boiler*
Courtesy : LIPI BOILER LTD. PUNE-1

Function:

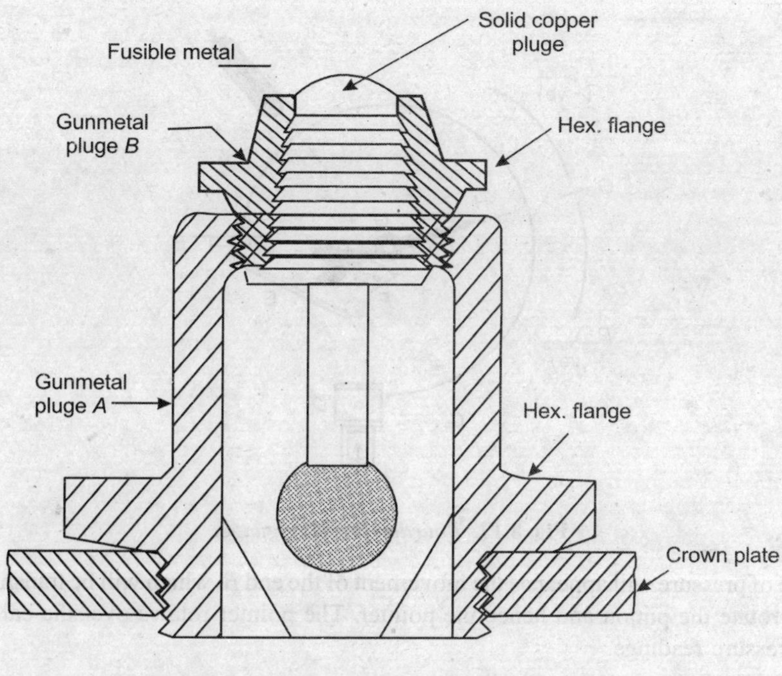

Fig. 8.16. *Fusible plug.*

Fusible plug is fitted on the combustion chamber crown, but in the drum. As shown in Fig. 8.16, *A* is the hollow gun metal plug screwed into the crown plate. *B* is fhe second hollow gun metal screwed into the plug *A* and *C* is the third solid copper conical plug. The inner surface of *B* and *C* for removal and refitting purpose by using spanners.

Under normal working conditions, fusible plug will be covered by means of water in the drum. This water keeps the temperature of the fusible metal below its melting point. But when due to any reason, water level in the drum falls below the minimum permissible water level, the plug gets uncovered from water and is exposed to steam. Due to heat of steam, fusible metal melts and copper plug drops and the steam rushes through the access into the combustion chamber. Before referring the boiler, the attendant removes and refits the plug after interposing the fusible metal between the plugs *B* and *C*.

8.12.5 Bourdon Pressure Gauge

This is the most common type of pressure gauge. This is used to measure pressure inside the containers and pipe lines etc.

The basic element of this gauge is the tube *A*. It is elliptical in cross section and is bent into and arc of circle as shown in Fig. 8.17.

End *B* of this tube is sealed, while open end *C* is connected to the connecting union *D*. Through the connecting union the gauge will be mounted over the containers or pipes of whose pressures are to be measured.

When it is mounted on the systems and if the pressure of the system is more than atmosphere (i.e., + ve pressure) then the tube will tend to curl out. Conversely if the pressure of the system is less than the atmosphere, (i.e., –ve pressure) the tube will tend to curl in. Thus the change of pressure will therefore appear as a movement of the end

B. This end *B* is connected by means of a link *E* to a quadrant gear *F*. The quadrant gear engages with a small gear *G* on to which a point *H* is attached.

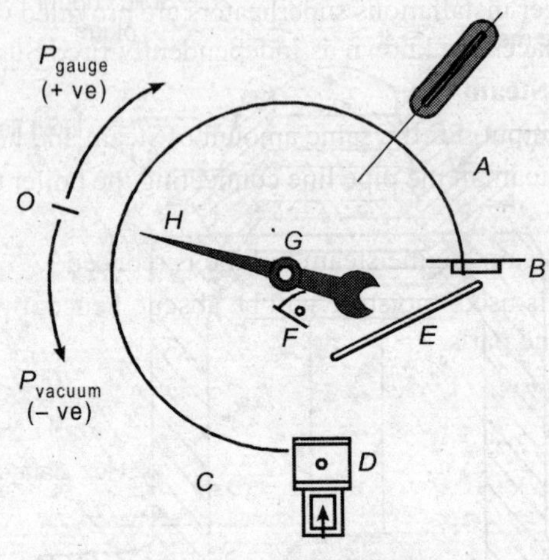

Fig. 8.17. *Bourdon pressure guage.*

Thus then change of pressure will appear as the movement of the end *B*, which will be transmitted to the quadrant gear, which in turn rotate the pinion and hence the pointer. The pointer rotates over the calibrated scale, which directly gives the pressure readings.

8.12.6 Steam Stop Valve

Through the steam stop valve, steam will be taken out for power generation purpose.

As shown in Fig. 8.18 valve body is made up of C.I. Through its bottom flange *A* it is mounted in the boiler drum, and its side flange *F* is connected to the pipe through which steam goes to the steam engine or steam turbine. It is clear from full figure that when the hand wheel is rotated in clockwise or anticlockwise direction, the valve will move up and down. And the required quantity of steam can be taken out for power generation.

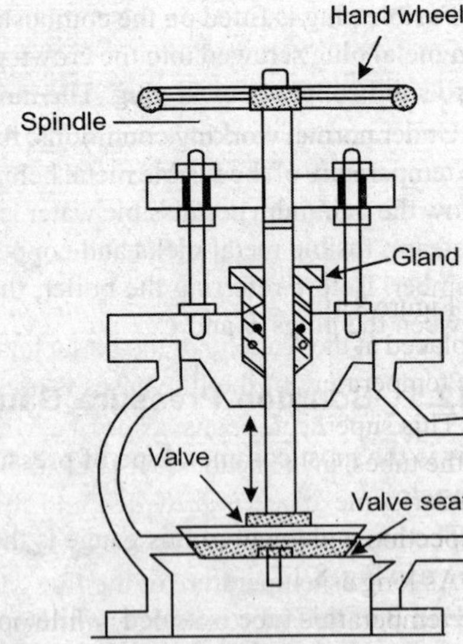

Fig. 8.18. *Steam stop valve.*

8.12.7 Feed Check Valve

Through the feed check valve, water enters the boiler drum. It allows the water to flow into the drum and it will not allow the water to flow out of the drum, so it is called as check valve.

8.12.8 Blow Off Valve

Mud or any other settled impurities are blown out through this valve.

8.12.9 Superheater

The steam generated from a simple low pressure boiler is generally wet. So, for superheating the steam, superheaters are used. In the superheater, wet steam is first dried at the same temperature and pressure and then it is superheated at constant pressure.

Generally, same heat of the flue gases is used for superheating purpose and hence superheaters are placed in the path of flue gases. However in bigger installations superheaters are provided with independent furnaces. And such superheaters with independent furnaces are known as Independently fired Superheaters.

Advantages of Superheating the Steam

(1) Increase in amount of work output, for the same amount of steam and hence increase in cycle efficiency.

(2) Loss due to condensation of steam in the pipe line connecting the boiler to the steam engine or steam turbine is reduced.

(3) Loss due to condensation of steam engine/steam turbine is reduced.

(4) When the superheated steam is used, moisture will be absent, hence it will reduce corrosion and erosion of steam engine and steam turbine parts.

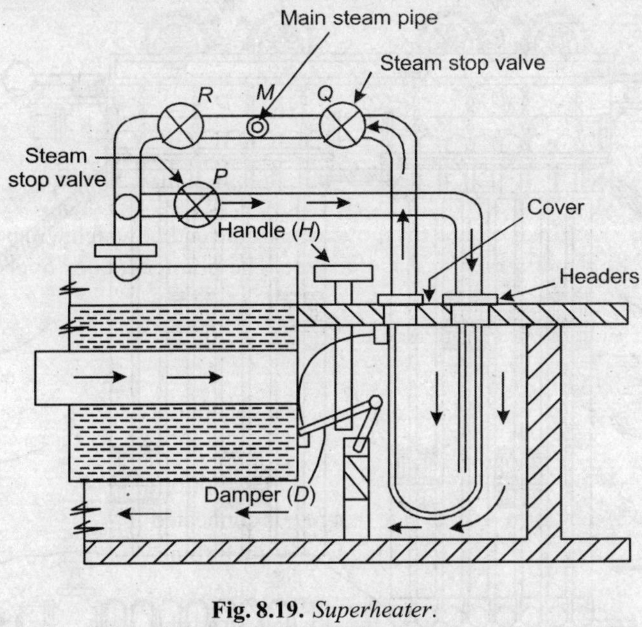

Fig. 8.19. *Superheater.*

Figure 8.19 shows *Sudgen's Hair pin type of superheater* arranged with Lancashire/Cornish boiler. The superheater is placed at the rear end of the boiler and the flue gases after their first pass, they pass through the superheater. Here the temperature of the flue gases is not less than 550°C. This superheater gives a superheat of 40 – 95°C.

This superheater consists of two Mild Steel boxes or headers and a number of tubes bent to U-shape. The ends of the tubes, in the headers will be either expanded or welded in order to have gas tight joint. Between front end rear header some space is provided and this space is covered by means of a cover. This gives access for cleaning, inspection and repair purpose.

As long as temperature of the flue gases is less than 550°C then there is no danger of burning of tubes. But when the temperature increases beyond this and also when the delivery of steam from the boiler to the steam turbine through the superheater is suspended for along time, then the following arrangements are to be made so as to protect the tubes from burning out.

(1) Flooding the Superheater Tubes. This water is to be drained before the supply of steam through the superheater again starts.

(2) Diverting the Flow of Flue Gases. This is done by Damper *D* which is operated by means of handle *H*.

An arrangement is also provided so as to pass the steam through superheater or directly to the main steam pipe as may be required *M* is the main steam pipe, *PQ* and *R* are stop valves. When the superheater is in action valves *P* and *Q* are open *R* is closed. When steam is to be taken out directly from the boiler to the steam pipe, then valves *P* and *Q* are closed and *R* is opened.

8.12.10 Economiser

One of the major heat losses in the boiler plant is the heat carried away by the exhaust gases. An economiser is a device or appliance, which recovers some of the heat of the exhaust gases. This recovered heat in utilised for preheating the water. This preheated water when it is supplied to the boiler drum, it requires less heat for its conversion into steam. This reduces the amount of fuel required and thus the efficiency of boiler plant increases.

Figure 8.20 shows the *Green's economiser*. It is used with stationary low pressure boilers.

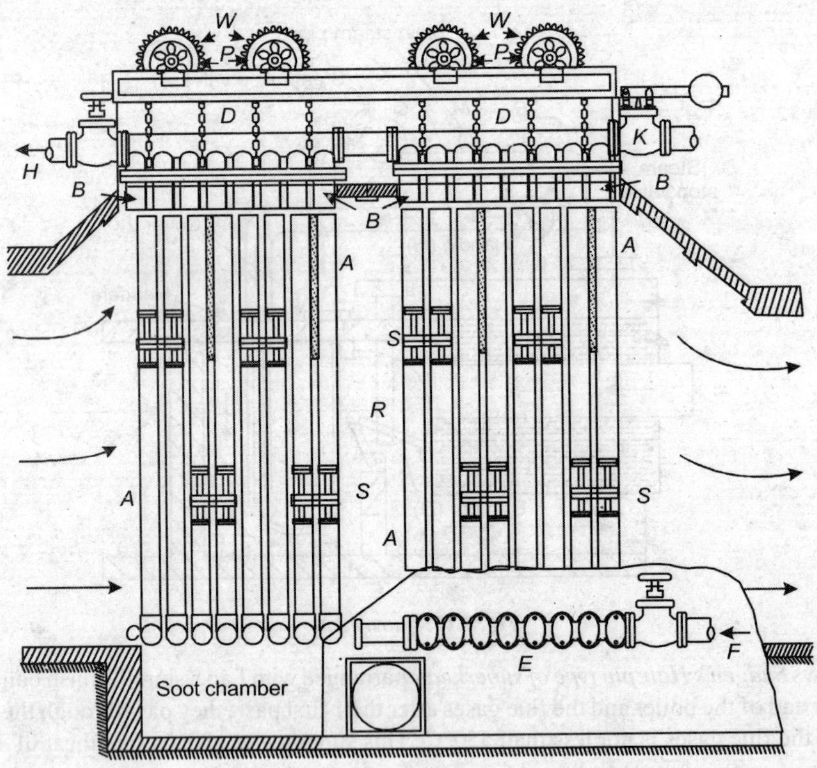

Fig. 8.20. *Green's economiser.*

It consists of a number of vertical pipes *A*. These pipes are arranged in groups of 4, 8, 12,16 etc. Upper ends of the pipes are connected to the upper boxes *B*. These boxes *B* are inturn connected to the common header *D*. Lower ends of the pipes are connected to lower boxes. These lower boxes are connected to the horizontal pipe *C* through the cross pipes. All the tubes are enclosed in the brick work of the economiser except header *D* and pipe *C*.

Water will be pumped by means of a feed pump, water enters the economiser through the stop valve at the inlet *F* and passes into the pipe *C*, from where it flows through the cross pipe and then rises through the vertical tubes *A*. Since the economiser is provided in the path of flue gases the water rising through the tubes recovers heat and becomes hot. Hot water will be then collected into the top common header.

This hot water from the top header is taken out through the stop valve H, from where it goes to the boiler drum. K is the safety valve for the economiser.

Since the economiser is provided in the path of flue gases, soot (or ash) is likely to be collected over the tubes. It retards the rate of heat transfer from the flue gases to the water inside the tubes. In order to avoid this, soot is scrapped by means of scrapers. Scrapers of the two-adjacent tubes are coupled to form one pair and two such pairs of scrapers are connected by means of a chain, which passes over pulley P. Pulley P is connected to worm shaft, in such a way that when one pair of scrappers comes to top most position, another pair will be at the bottom most position, then it automatically reverses the direction of motion. The scrapers are kept in motion as long as economiser is in action.

By Pass Arrangement. By Pass arrangements are provided for the flue gases and for feed water, so that the economiser may be put out of action when not required. Figure 8.21 shows such an arrangement for two Lancashire boilers fitted with an economiser. When the economiser is in service, then dampers L and M are open and the damper N is closed. For isolating the economiser damper N is opened and dampers L and M are closed.

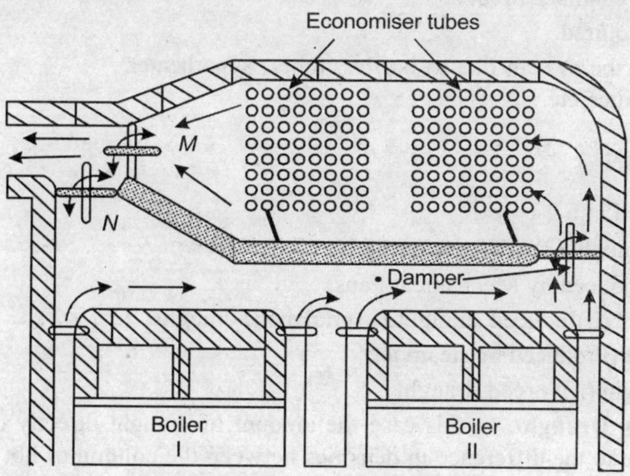

Fig. 8.21. *By pass arrangement.*

8.12.11 Air Preheater

The air is heated, by using the heat of flue gases in a device called preheater, which otherwise would have been lost to the atmosphere. By using air preheater the overall efficiency will increase by about 10%.

8.12.12 Feed Pump

It is an appliance used to pump the feed water into the boiler drum. The feed pump may be either rotary or reciprocating type.

8.12.13 Steam Injector

When the feed pump fails due to any reasons, then steam injector, injects or pumps or delivers feed water to the boiler drum, by the use of steam.

8.12.14 Steam Trap

Steam trap automatically collects and returns to the boiler, the water resulting from partial condensation of steam, without allowing any steam to escape.

8.13 BOILER DRAUGHT

Draught is the pressure difference which is necessary to draw the required quantity of air for combustion and to remove the flue gases out of the system.

Thus the object of producing draught in a boiler is,

(i) To provide sufficient quantity of air for combustion

(ii) To make the resulting hot gases, to flow through the system.

(iii) To discharge these gases to the atmosphere through the chimney.

Usually this drought (pressure difference) in boiler is of small magnitude and is measured in mm of water column by means of draught gauge/manometer.

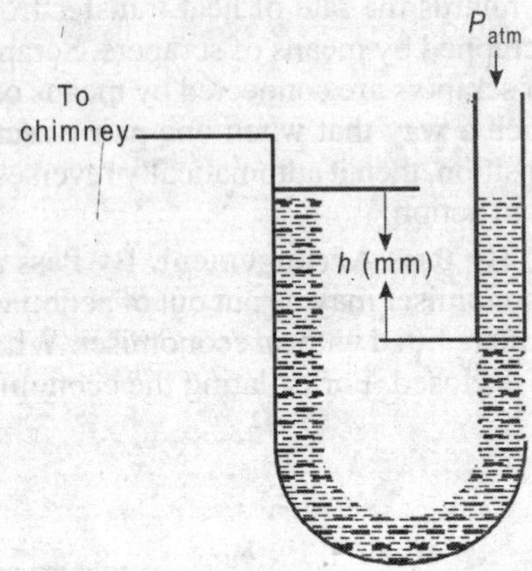

Fig. 8.22

Note. The amount of draught depends upon,

(i) Nature and depth of fuel on the grate.

(ii) Design of combustion chamber/firebox.

(iii) Rate of combustion required.

(iv) Resistance offered in the system due to baffles, tubes, superheater, economiser, air pre-heater, etc.

8.14 CLASSIFICATION

Draught is broadly classified into 2-types,

(1) Natural or chimney draught (2)Artificial draught

(a) Fan Drought (Produced by Mechanical Fans)

(i) Forced draught (ii)Induced draught (iii) Balanced draught

(b) Steam jet draught (Produced by steam jet)

(i) Induced draught (ii)Forced draught

1. Natural or Chimney Draught. In this case the amount of draught directly depends upon the height of chimney. It is produced due to the difference in densities between the column of hot gases in the chimney and a similar column of cold air out side the chimney.

Let us first consider the case when fires are not lighted.

Let, the atmospheric pressure at grate level be P_1 and P_2 be the atmospheric pressure at an altitude H. The pressure P_2 is lower than the pressure P_1 because with the altitude pressure goes on decreasing.

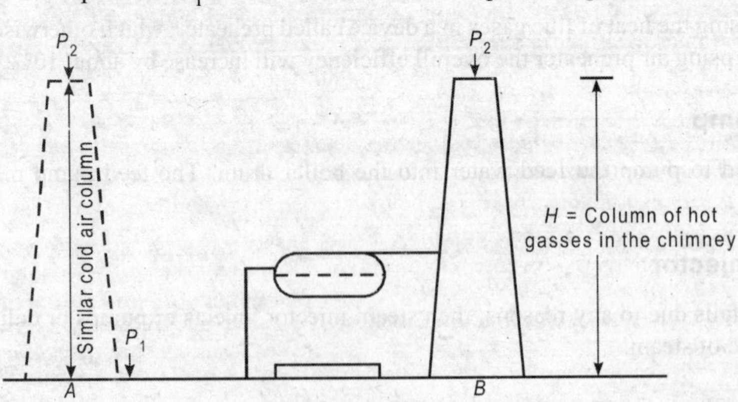

Fig. 8.23

Now let us consider the case when fires are lighted and the chimney is full of hot gases. Under these circumstances, the pressure at the base of the chimney is the sum of pressure P_2 at the top and the pressure due to hot gas column H. But pressure P_1 at grate is the sum of pressure P_2 and the pressure due to similar cold column of air H.

Since, ρ cold air > ρ hot gases

i.e., $$P_A > P_B$$

$\therefore$ P_2 + Pressure due to cold column $H > P_2$ + Pressure due to hot column H.

$\therefore$ Pressure at grate due to cold column > Pressure at the chimney base due to hot column H.

This difference is called **static draught** and because of the pressure difference, (draught) air will rush to the combustion chamber, where combustion of air and fuel takes place and hot gases are generated. Then these hot gases because of draught, flow through the system and finally they are exhausted to the atmosphere through the chimney.

Advantages of Natural Draught

1. Easy to construct.
2. No power is required for producing the draught.
3. Long life of chimney.
4. No maintenance is required.

Disadvantages

1. Tall chimney is required.
2. Poor efficiency.
3. Decreases with increase in outside temperature.
4. No flexibility it create more draught to take peak loads.

2. Artificial Draught. In bigger power plants, the draught of the order of 25–350 mm of H_2O column is required. Far producing this much draught, the chimney height has to be increased considerably. which is neither convenient nor economical. Also, since the draught depends upon the climatic conditions, some mechanical equipments are used for producing the required draught and the draught so produced is called as the Artificial draught.

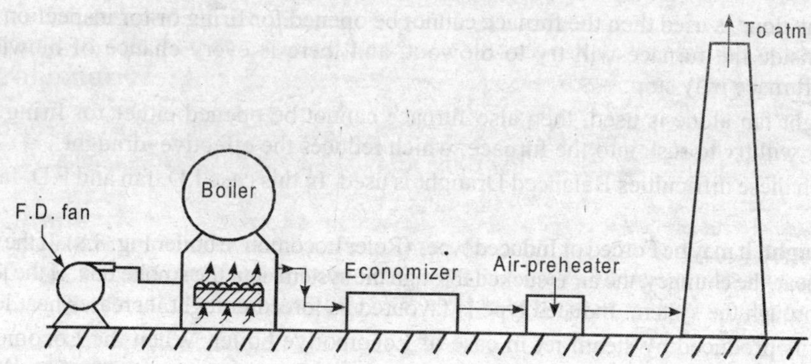

Fig. 8.24

3. Forced Draught. In a Forced draught system, a Fan or Blower is provided is shown in figure which forces the air in the combustion chamber. In the combustion chamber combustion of air and fuel takes place and hot gases generated. These gases are forced to pass through the flues, economiser, air pre-heater and then they are exhausted after recovering heat of flue gases. This draught system is known positive draught system, since the pressure of gases throughout the system is above atmospheric pressure.

It is to be noted that, the function of chimney use is to discharge the gases high in the atmosphere to reduce air pollution and it is not much significant for producing draught.

4. Induced Draught. In this system, the Blower or Induced Draught fan is located near the base of chimney. The air is sucked in the system, by reducing the pressure through the system below atmosphere. The flue gases, generated after combustion are drawn through the system and after recovering heat in the economiser, air-preheater, they are exhausted through the chimney to the atmosphere.

Here it is to be noted that the draught produced is independent of the temperature of hot gases, so the gases may be discharged as cold as possible after recovering as much heat as possible.

Advantage of Forced Draught (F.D.) over Induced Draught (I.D.)

1. The size and power required by I.D. fan is more because this fan handles more gases.
2. Since the I.D. fan handles hot gases, water cooled or air cooled bearings are to be used.
3. F.D. fan consumes less power and normal bearing can be used.

5. Balanced Draught. It is always preferable to use combinations of I.D. and F.D. instead of Forced or Induced draught alone.

Fig. 8.25

If Forced Draught alone is used then the furnace cannot be opened for firing or for inspection. Because the high pressure air/gases inside the furnace will try to blowout, and there is every chance of blowing out of the fire completely and the furnace may stop.

If Induced Draught fan alone is used, then also furnace cannot be opened either for firing or for inspection. Because the cold air will try to rush into the furnace, which reduces the effective draught.

To overcome both these difficulties Balanced Draught is used. In this case I.D. fan and F.D. fans are provided as shown.

6. Steam Jet Draught. It may be Forced or Induced type, (Refer Locomotive boiler Fig. 8.8). If the steam jet is directed into the smoke box, near, the chimney, the air is sucked through the system, into the smoke box. If the jet is Located before grate, air is forced through the system. Induced type is favoured as forced draught increases heat losses.

Induced draught is produced by steam jet in case of Locomotive boiler. When the Locomotive in stationary, steam from the boiler may be supplied to the smoke box through the nozzle to create draught. While locomotive is running, due to motion, the air flows to the furnace.

8.15 CALCULATION OF CHIMNEY HEIGHT

We know that,

$$C + O_2 = CO_2 \quad \text{...(1)}$$

$$1\,\text{vol} + 1\,\text{vol} = 1\,\text{vol}$$

and $$2H_2 + O_2 = 2H_2O$$

$$2\text{ vol} + 2\text{ vol} = 2\text{ vol} \qquad \ldots(2)$$

Therefore, for Eqs (1) and (2), one *volume* of O_2 from air is used for the combustion of carbon in the fuel, produces one volume of gaseous products of combustion CO_2. But one vol. of O_2 from air produces 2 volumes of steam (H_2O) in the gaseous products and since the hydrogen content in solid or liquid fuel is quite small, so steam formed is negligible compared to other constituents. N_2 and other constituents does not flake part in combustion.

So, the assumption that the volume of air required for combustion is approximately the same as the volume of the products of combustion is fairly accurate.

Let m = Mass of air used in kg/kg of fuel.

T_f = Mean absolute temperature of chimney on flue gases in K.

T_a = Absolute temperature of outside air K

∴ Mass of chimney gases = $(m + 1)$ kg/kg of fuel

Volume per kg of flue gases = 273 K or 0°C

= Volume per kg of air at 273 K

From Perfect gas equation, $PV = RT$.

Volume at standard conditions,

$$V_N = \frac{RT}{P} = \frac{287 \times 273}{101325}$$

$$= 0.7734 \text{ m}^3/\text{kg}$$

The pressure difference being very small, of the order of 25 mm of H_2O column, hence the pressure can be considered as constant at furnace and chimney base for the purpose of calculation of volumes at high temperatures.

∴ Volume per kg of air at $T_a K$,

Since we know that

$$\frac{P_N V_N}{273} = \frac{P_a V_a}{T_a}$$

$$V_a = V_N \times \frac{T_a}{273} \times \frac{P_N}{P_a}$$

$$\therefore \quad V_a = 0.7734 \times \frac{T_a}{273} \times \frac{P_N}{P_a}$$

∴ Volume of m kg of air at

$$T_a K = 0.7734 \times m \times \frac{T_a}{273} \times \frac{P_N}{P_a} \text{ and}$$

Volume of $(m + 1)$ kg of chimney gases at $T_f K$,

$$= \frac{0.7734 \times m \times T_f}{273} \times \frac{P_N}{P_a} \text{ and}$$

∴ Density of air at

$$T_a K = \frac{\text{Mass}}{\text{Volume}} = \frac{m \times 273}{0.7734 \times m \times T_a} \times \frac{P_a}{P_N}$$

$$= \frac{273}{0.7734 \times T_a} \times \frac{P_a}{P_s}$$

$$\rho_a = 1.293 \left[\frac{273}{T_a}\right] N \frac{\text{kg}}{\text{m}^3} \times \frac{P_a}{P_N}$$

And density of flue gases at T_f at K,

$$= \frac{273}{0.7734 \times T_a} \times \frac{P_a}{P_s}$$

$$\rho_a = 1.293 \frac{273}{T_a} \frac{\text{kg}}{\text{m}^3} \times \frac{P_a}{P_N}$$

And density of flue gases at $T_f K$

$$= \frac{(m+1) \times 273}{0.7734 \times m \times T_f} \times \frac{P_a}{P_N}$$

$$\rho_g = 1.293 \left[\frac{m+1}{m}\right] \times \frac{273}{T_f} \times \frac{P_a}{P_N}$$

Let H = chimney height in meter from grate level.

Then pressure exerted by the column of hot gases.

$$H = \rho_g \cdot g \cdot H$$

$$P_f = 1.293 \left[\frac{m+1}{m}\right] \times \frac{273}{T_1} \times g \times H \text{ N/m}^2 \qquad \ldots(1)$$

Pressure exerted due to cold air column,

$$\rho_a \times g \times H = 1.293 \times \frac{273}{T_a} \times g \times H \text{ N/m}^2 \qquad \ldots(2)$$

Draught pressure, in terms of static pressure difference between hot gas column in chimney and a similar cold column.

$$P_d = \text{Eq. (2)} - \text{Eq. (1)}$$

$$= g \times H[P_a - P_f]$$

$$= 1.293 \times 273 \times g \times H \left[\frac{1}{T_a} - \left[\frac{m+1}{m}\right] \times \frac{1}{T_f}\right] \text{N/m}^2$$

$$P_d = 353 \times g \times H\left[\frac{1}{T_a} - \left[\frac{m+1}{m}\right] \times \frac{1}{T_f}\right] \text{N/m}^2$$

If P_a and P_N are different, then

$$P_d = 353 \times g \times H \times \frac{P_a}{P_N}\left[\frac{1}{T_a} - \left[\frac{m+1}{m}\right] \times \frac{1}{T_f}\right] \text{N/m}^2$$

To express in mm of H_2O column

$$P_d = \rho_w . g . h_w \frac{\text{N}}{\text{m}^2}$$

where ρ_w = Density of water in kg/m^3

g = Acceleration due to gravity in m/sec^2

h_w = Water column in mm

Then, P_d = Pressure difference wil be in N/m^2

i.e.,
$$P_d = \frac{\text{kg}}{\text{m}^3} \times \frac{\text{m}}{\text{sec}^2} \times m$$

$$= \left[\text{kg} \times \frac{\text{m}}{\text{sec}^2}\right] \times \frac{1}{m^2} = \text{Force} \times \frac{1}{m^2}$$

$$\therefore \quad = \frac{\text{N}}{\text{m}^2} \text{ resultant unit.}$$

$$= h_a = \frac{P_d}{P_w . g}$$

$$= \frac{353 \times g \times H \times \dfrac{P_a}{P_N}\left[\dfrac{1}{T_a} - \left[\dfrac{m+1}{m}\right] \times \dfrac{1}{T_f}\right]}{1000 \times g} \text{metres of } H_2O$$

$$\therefore \quad h_w = 353 \times H \times \frac{P_a}{P_w}\left[\frac{1}{T_a} - \left[\frac{m+1}{m}\right] \times \frac{1}{T_f}\right] \text{mm of water}$$

If $P_a = P_N$ then,

$$h_w = 353 \times H\left[\frac{1}{T_a} - \left[\frac{m+1}{m}\right] \times \frac{1}{T_f}\right] \text{mm of } H_2O$$

To express, P_d in meters of hot flue gas

$$\therefore \quad P_d = \rho_f \cdot g \cdot h_f$$

where r_f = Density of hot flue gas

h_f = height of flue gas column

$$\therefore \quad P_d = r_f \cdot g \cdot H$$

$$H = \frac{P_d}{\rho_f \cdot g}$$

$$= \frac{353 \times g \times H \times \dfrac{P_a}{P_N}\left[\dfrac{1}{T_a}\left[\dfrac{m+1}{m}\right] \times \dfrac{1}{T_f}\right]}{g \times 353 \times \left[\dfrac{m+1}{m}\right] \times \dfrac{1}{T_f} \times \dfrac{P_d}{P_N}}$$

On simplification we get,

$$H' = H\left[\frac{m}{m+1} \times \frac{T_f}{T_a} - 1\right] \text{ metre of hot gas column}$$

Condition for maximum discharge through the chimney.

Mass of the flue gases discharged through the chimney is proportional to the density, cross sectional areas and velocity.

i.e., M α density × cross section area × velocity

Note. Cross section areas is constant because of construction,

And we know that,

$$\rho_f = \frac{1}{T_f}$$

$$M = \text{Constant} \times \frac{1}{T_f} \times \text{Velocity}$$

$$= k \times \frac{1}{T_f} \times \sqrt{2gH}$$

$$= \text{Constant} \times \frac{1}{T_f} \times \sqrt{H}$$

$$\text{Mass of gas} = \frac{1}{T_f} \times \left[H \times \left[\frac{m}{m+1} \times \frac{T_f}{T_a} - 1\right]\right]^{1/2} \times \text{Constant}$$

$$= \frac{1}{T_f} \times \left[\frac{m}{m+1} \times \frac{T_f}{T_a} - 1\right]^{1/2} \times \text{Constant (Since } H = \text{Constant)}$$

$$M = \left[\frac{m}{m+1} \times \frac{1}{T_f . T_a} - \frac{1}{T^2{}_f}\right]^{1/2} \times \text{Constant}$$

For maximum discharge,

$$\frac{dM}{dT_f} = 0$$

$$\frac{dM}{dT_f} = \frac{1}{2}\left[\frac{m}{m+1} \times \frac{1}{T_f . T_a} - \frac{1}{T_f^2}\right]^{1/2} \times \left[\frac{m}{m+1} \times \frac{1}{T_a} \times \frac{1}{T_f^2} - \frac{1}{T_f^3}\right] = 0$$

$$\therefore \quad \left[-\frac{m}{m+1} \times \frac{1}{T_a . T_f{}^3} \right] = 0$$

$$\frac{m}{m+1} \times \frac{1}{T_a \cdot T_f^2} = \frac{2}{T_f^3}$$

$$T_f = 2\left[\frac{m}{m+1}\right] \times T_a$$

We know that, $$H = \left[\frac{m}{m+1} \times \frac{T_f}{T_a} - 1\right] - H$$

Substituting T_f in above equation we get,

$$H' = H[2-1] = H$$

$$\therefore \quad H' = H$$

8.16 DRAUGHT LOSSES

The actual or available draught will be less than the theoretical Draught obtained by calculations, because of the draught losses viz.

(1) Head loss due to fuelled resistance h_{fb}: Fuel bed resistance depends upon the size of the fuel, depth of fuel on grate and combustion rate.

(2) Head loss due to frictional resistance to the flow of flue gases because of the equipments like superheater, economiser, air preheater, etc. (h_e)

(3) Head loss due to friction in bends and curves in the flue gas passages (h_b).

Thus, total draught loss = $h_{\text{Total}} = h_{fb} + h_e + h_b$

8.17 INTRODUCTION TO IBR OR REGULATIONS FOR CHIMNEY (WITH REFERENCE TO IBR/MSNC BOMBAY)

To keep control on boiler accidents, Indian Government after independence made following rules and regulations, which the boiler should conform. They are as under:

1. A plan of the boiler house and chimney is to be approved by the chief inspector of Indian Boiler Regulations (IBR)/Maharashtra Chief Inspector of Nuisance Commission (MSNC)—Bombay.
2. Chimney fabrication drawings are to be approved before fabrication.
3. A plan of 1m =1cm in blue print showing the exact position of furnace, flues, chimney, duly certified by the chartered engineer/owner should be submitted.
4. Substantial structure certificate for the chimney stating that the chimney in self supported/supported by wire ropes and with stands a wind load of 207 kg/m² is to be submitted (Note this should be certified by the chartered engineer).
5. Chimney height minimum required as per IBR/MSNC is 100 feet that i.e., 30.48 m from the firing floor of the boiler.
6. An undertaking for the height of chimney is to be submitted by the owner to MSNC.
7. An undertaking for the use of specific fuel is to be sumitted by the owner.
8. No objection certificate (NOC) is to be obtained from Air Force Deptt.
9. Smoke generated should conform to Bombay Smoke Nuisance Act 1912.

10. Chimney should be day painted. (i.e., Red and white strips of equal width in total 7 strips are to be painted)
11. Anti collision lights are to be fitted, for sighting in the night and poor visibility.
12. Following details are to be submitted,
    (a) Height of own factory building/boiler house.
    (b) Height of tallest building within a radius of 300 m around ground level.
    (c) Capacity of boiler.
    (d) Working pressure
    (e) Type of fuel
    (f) Quantity of fuel/hour
    (g) Type of chimney self-supported/supported by wire ropes
    (h) Material of construction.

8.18 BOILER CALCULATIONS

In case of boilers it is very important to find,

(a) *Heat transfer required to form steam.* We have studied steam generation process,

Let h_2 = Specific enthalpy of steam formed in kJ/kg

h_1 = Specific enthalpy of feed water in kJ/kg

Then, since the steam is generated ($P = C$) at constant pressure,

And for a constant pressure process $\delta Q = dH$

i.e., Heat supplied = Change of enthalpy,

∴ Heat transfer required to form l kg of steam in the boiler,

$$= (h_2 - h_1) \text{ kJ/kg}$$

m_s is the mass of steam formed in kg, then heat transfer required to form total steam.

$$= m_s.(h_2 - h_1)\left[\text{kg} \times \frac{\text{kJ}}{\text{kg}} = \text{kJ}\right] \qquad ...(1)$$

(b) *Energy received from fuel.* This is obtained from the mass of fuel burnt and its calorific value of fuel,

If mass of fuel used = m_f kg

And calorific value of fuel = C.V. kJ/kg

Then,

$$\text{Energy received} = m_f \times \text{C.V.}\left[\text{kg} \times \frac{\text{kJ}}{\text{kg}} = \text{kJ}\right] \qquad ...(2)$$

(c) *Boiler thermal efficiency.* It is the ratio of the energy received by the steam to the energy supplied by the fuel to produce steam.

Thus, Boiler Thermal Efficiency;

$$\eta_{th} = \frac{\text{Energy to steam}}{\text{Energy from fuel}}$$

$$\eta_{th} = \frac{\text{Energy received by the steam}}{\text{Energy supplied by the fuel}}$$

From Eqs (1) and (2)

$$\textbf{Boiler Thermal Efficiency } \mathbf{h_{th}} = \frac{\mathbf{m_s(h_2 - h_1)}}{\mathbf{m_f \times C.V.}} \times \mathbf{100\%}$$

(*d*) *Boiler performance.* As we have studied, the main function of the boiler is to produce steam from feed water, with the use of heat liberated by the combustion of fuel. It is to be noted the number of kg of water evaporated per hour is not an exact measure of the performance of the boiler. Since different boilers may be operating at different at different conditions, so their comparison may be made on the basis of some reference condition known as Equivalent of Evaporation.

8.18.1 Equivalent of Evaporation from and at 100° C

"The evaporation which would be obtained if the feed water were supplied at 100°C and converted into dry and saturated steam at 100°C, at standard atmospheric pressure, is known as equivalent of evaporation from and at 100°C".

We know that from Eq. (1) energy received by the steam

$$= m_s (h_2 - h_1) \text{ kJ}$$

It is to be noted that, at 100°C, it is the enthalpy of evaporation which is supplied and the specific enthalpy of evaporation at 100°C, = 2256.9 kJ/kg. (From steam tables h_{fg} at 100° C and 1 atm.)

The equivalent of evaporation of a boiler from and at 100°C,

$$= \frac{\mathbf{m_s (h_2 - h_1)}}{\mathbf{2256.9}} \textbf{ kg/kg of coal.}$$

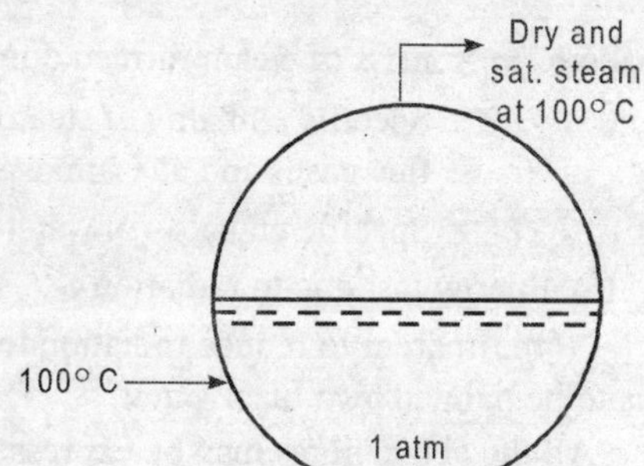

Fig. 8.27

8.18.2 Evaporative Capacity

"It is defined as the amount of water and water evaporated into steam per hour". The evaporative capacity may be expressed in kg/hr. or kg/kg, of fuel or kg/hr-m^2 of heating surface area.

Actual Evaporation

It is defined as the amount of water evaporated into steam, at actual working conditions, per kg of fuel burnt.

8.19 ENERGY BALANCE

Energy released due to the combustion of fuel will be utilised for various purposes and is tabulated in heat balance sheet.

Heat balance sheet gives the account of heat energy released and heat energy utilised

(1) Energy released due to the combustion of fuel

$$= m_f \times \text{C.V.} = \text{kg} \times \frac{\text{kJ}}{\text{kg}} = \text{kJ}$$

(2) This energy released will be utilised for various purposes, viz.

(i) Heat energy utlised to form the steam

$$= m_s (h_2 - h_1)$$

$$\text{Units} = \text{kg} \times \frac{\text{kJ}}{\text{kg}} = \text{kJ}$$

(ii) Heat energy lost to the flue gases

$$= \text{Mass of flue gases} \times C_p \text{ of flue gases } (T_f - T_a)$$

$$= m_{fg} \times C_{pg} \times (T_f - T_a)$$

where T_f = temperature of flue gases

and T_a = atmospheric temperature

$$\left[\text{units} = \frac{\text{kg}}{\text{kg of coal}} \times \frac{\text{kJ}}{\text{kg.K}} \times \text{K} = \frac{\text{kJ}}{\text{kg of coal}}\right]$$

(3) Energy lost due to the unburnt fuel falling through the grate bars. When solid fuels are used, some of the fuel falls through the grate bars and is lost with ash and this is

$$= m_{uf} \times C.V_{uf}$$

(4) Energy lost due to the water vapour formed due to the combustion of H_2 in the flue gases

$$= m_1 (h'_2 - h'_1)$$

where m_1 = mass of steam formed due to the combustion of H_2 in kg/kg of coal burnt.

h'_2 = Specific enthalpy of steam formed due to the combustion of H_2 in the flue gases at the temperature of flue gases and at 1 atmospheric pressure or, at 1 bar pressure or at the given partial pressure.

h'_1 = Specific enthalpy of water at the boiler room temperature

(5) Energy lost due to radiation

Note. In order to reduce radiation losses, Asbestos lagging is to be provided on the exposed boiler surface and is to be painted by a black paint.

All the above items may be expressed in the balance sheet as follows.

Energy supplied	*kJ per minutes or per hr*	*%*	*Energy utilised*	*kJ*	
	say X	100	(i) To form steam	x_1	.....
			(ii) Energy lost to, flue gases	x_2	.....
			(iii) Energy lost unburnt fue!	x_3	.....
			(iv) Energy lost to water vapour	x_4	.....
			(v) Every unaccounted by difference	x_5	.....
Total	X	100	Total	X	100

LIST OF FORMULAE

1. Draught produced in mm of H_2O

$$h_W = 353 \times H\left\{\frac{1}{T_a} - \left[\frac{m+1}{m}\right] \times \frac{1}{T_f}\right\} \frac{P_a}{P_N}$$

where H = Height of chimney in metre

P_a = Actual pressure at boiler location.

Contd.

Contd.

$$P_N = \text{Standard atmospheric pressure}$$
$$T_a = \text{Actual atmospheric temperature K}$$
$$T_f = \text{Temperature of flue gases in K}$$
$$m = \text{Mass of air in kg to burn 1 kg of fuel}$$
$$\therefore \quad m+1 = \text{Mass of flue gases in kg/kg of fuel.}$$

(i.e., when 1 kg of fuel is burnt with m kg of air, then total flue gases produced = $m + 1$)

2. Density of air (ρ_a)

$$\rho_a = 1.293\left[\frac{273}{T_a}\right]\frac{P_a}{P_N} \text{ in } \frac{\text{kg}}{\text{m}^3}$$

3. Density of flue gases

$$\rho_f = 1.293\left[\frac{m+1}{m}\right].\frac{273}{T_f}.\frac{P_a}{P_N}$$

or

$$\rho_f = \frac{m+1}{m}\times\frac{353}{T_f}\times\frac{P_a}{P_N} \text{ kg/m}^3$$

4. Draught pressure in terms of static pressure difference between hot gas column and a parallel cold air column.

$$P_d = g\times H\times[P_\alpha - P_f]\frac{\text{N}}{\text{m}^2}$$

and in terms of mm of water column

$$h_w = H[P_\alpha - P_f] \text{ mm of } H_2O$$

5. Velocity of flue gases = $C = \sqrt{2gH'}$

6. Equivalent height of hot gas column H_{gases} or H'

$$H' = H\left[\frac{m}{m+1}\times\frac{T_f}{T_a} - 1\right] \text{ m of hot gas column.}$$

7. Condition for maximum discharge through chimney $H' = H$ and during maximum discharge conditions, Maximum temperature of flue gases,

$$T_f = 2\left[\frac{m+1}{m}\right]\times T_a$$

SOLVED EXAMPLES
(Problems on Boiler Draught)

Example 8.1 *Find the minimum height of chimney required to produce a draught of 16 mm of H_2O, if 19 kg of air is required per kg of fuel burnt on the grate. The mean temperature of flue gases inside the chimney is 330°C.*

Data: $H = ?$ $h_w = 16 \text{ mm of } H_2O$

$m = 19$ kg/kg of fuel $T_f = 330 + 273 = 603\ K$

$T_a = 30 + 273 = 303\ K$

Solution

We know that, the drought produced in mm of H_2O column is given by,

$$h_w = 353 \times H\left\{\frac{1}{T_a} - \frac{m+1}{m} \times \frac{1}{T_f}\right\}\frac{P_a}{P_N} \quad \text{...(1)}$$

Since in the problem they have not mentioned anything about P_a and P_N

∴ Assuming $P_a = P_N$

Now substituting the values in the Eq. (1) we get,

$$h_w = 353 \times H\left\{\frac{1}{303} - \frac{20}{19} \times \frac{1}{603}\right\} \text{mm of } H_2O$$

$$16 = 353 \times H\{0.00330033 - 0.0017456577\}$$

$$\mathbf{H = 29.155\ m}$$

Example 8.2 *Readings from a boiler trial,*

(i) Height of chimney : 100 m

(ii) Ambient temperature : 32 °C

(iii) Mass of air : 18 kg/kg of fuel

(iv) Flue gas temperature : 300°C

Find natural draught in mm of H_2O

Data: $H = 100$ m $T_a = 32 + 273 = 305\ K$

$m = 18$ kg/kg of fuel $T_f = 300 + 273 = 573\ K$

$h_w = ?$

Solution

We know that, $$h_w = 353 \times H\left\{\frac{1}{T_a} - \left[\frac{m+1}{m}\right] \times \frac{1}{T_a}\right\} \text{ mm of } H_2O$$

Assuming $P_a = P_N$

Now $$h_w = 353 \times 100\left\{\frac{1}{305} - \frac{19}{18} \times \frac{1}{573}\right\}$$

$$h_w = 35300 \times \{0.0014365322\}$$

$$\mathbf{h_w = 50.7096\ m}$$

Example 8.3 *A boiler is equipped with a chimney of 30.48 m height. Atmospheric conditions are 755 mm of Hg and 35°C. Average temperature of flue gases in the Chimney is 300°C. If the boiler is supplied with 20 kg of air kg of fuel.*

Calculate theoretical draught created in mm of H_2O. Take standard atmospheric pressure as 760 mm of Hg.

Solution

We know that $$h_w = 353 \times H \times \frac{P_a}{P_N}\left\{\frac{1}{T_a} - \left[\frac{m-1}{m}\right] \times \frac{1}{T_f}\right\}$$

$$= 353 \times 30.48 \times \frac{755}{760}\left\{\frac{1}{308} - \frac{21}{20} \times \frac{1}{573}\right\}$$

$$\mathbf{h_w = 15.1168 \text{ mm of } H_2O}$$

Example 8.4 *Find the amount of air used in a boiler/kg of fuel, when chimney height is 50 m. Draught produced is 25 mm of H_2O. The temperature of flue gases is 50 m. Draught produced is 25 mm of H_2O. The temperature of flue gases is 353°C and that of room is 35°C.*

Also determine the draught produced in metres of hot gas column and flue gas temperature for maximum discharge conditions.

Data: $m = ?$ $\quad H = 50\ m$

$h_w = 25\ mm\ of\ H_2O$ $\quad T_f = 353 + 273 = 626\ K$

$T_a = 35 + 273 = 308\ K$ $\quad H = ?$

$T_f = ?$ *for max. discharge conditions*

Solution

We know that, $$h_w = 353 \times H\left\{\frac{1}{T_a} - \left[\frac{m+1}{m}\right] \times \frac{1}{T_f}\right\} \text{mm of } H_2O$$

$$25 = 353 \times 50\left\{\frac{1}{308} - \left[\frac{m+1}{m}\right] \times \frac{1}{626}\right\} \text{mm of } H_2O$$

$\therefore$ $$\mathbf{m = 6.8595 \text{ kg} = 6.86 \text{ kg}}$$

(ii) And we know that equivalent height of hot gas column,

$$H' = H\left\{\frac{m}{m+1} \times \frac{T_f}{T_a} - 1\right\} m \text{ of hot gas column}$$

$$= 50\left\{\frac{6.86}{7.86} \times \frac{626}{308} - 1\right\}$$

$$\mathbf{H' = 38.6942 \text{ m of hot gas column}}$$

(iii) We know that for max. discharge conditions

T_f is given by, $$T_f = 2 \times \frac{(m+1)}{m} \times T_a = 2 \times \frac{7.86}{6.86} \times 308$$

$$\mathbf{T_f = 705.7959 \text{ K}}$$

Example 8.5 *Find the minimum temperature of the flue gases required to produce a draught of 1.5 cm of water by a chimney of 30 m height, when air-fuel ratio required in the combustion is 20. Atmospheric air temperature is 27ºC.*

Data: $T_f = ?$ $\quad h_w = 15\ mm$

$H = 30\ m$ $\quad \frac{Air}{Fuel} = \frac{20}{1}$

$\therefore$ *Mass of air,* $m = 20$ *kg*

and mass of flue gases $m + 1 = 21$ *kg/ kg of fuel*

$$T_a = 27 + 273 = 300 \text{ K}$$

Solution

We know that, $$h_w = 353 \times H \times \left[\frac{1}{T_a} - \frac{(m+1)}{m} \times \frac{1}{T_f}\right] \text{ mm of H}_2\text{O}$$

$$15 = 353 \times 30 \times \left[\frac{1}{300} - \frac{21}{20} \times \frac{1}{T_f}\right]$$

$$\therefore \quad \mathbf{T_f = 547.75 \text{ K}}$$

Example 8.6 *How much air is used/kg of coal burnt in a boiler having a chimney of 35 m height to create a draught of 20 mm of water, when the temperature of flue gas in the chimney is 370° C and the boiler house temperature is 34°C. Does this chimney satisfy the condition of max. discharge ?*

Data: $m = ?$ $\qquad H = 35$ *m*

$h_w = 20$ *mm of* H_2O $\qquad T_f = 643$ *K*

$T_a = 307$ *K*

Solution

We know that, $$h_w = 353 \times H \times \left\{\frac{1}{T_a} - \frac{m+1}{m} \times \frac{1}{T_f}\right\} \text{ mm of H}_2\text{O}$$

$$20 = 353 \times 35 \left\{\frac{1}{307} - \frac{m+1}{m} \times \frac{1}{643}\right\}$$

$$\therefore \quad \mathbf{m = 18.66 \text{ kg/kg of coal}}$$

For max. discharge, $$T_f = 2 \times \left[\frac{m+1}{m}\right] \times T_a$$

$$= 2 \times \frac{19.66}{18.66} \times 307$$

$$\mathbf{T_f = 646.905 \text{ K}}$$

Whereas actual T_f given is 643 K which is less than T_f max.

$\therefore$ Condition is satisfied.

Example 8.7 *The draught produced in a certain boiler is 18.5 mm of water when the temperature of the flue gases in the chimney is 370°C and the temperature of the boiler house is 30°C. Estimate the air fuel ratio. Also state whether chimney operates under the maximum discharge condition or not. The height of the chimney is 32 m.*

Data: $h_w = 18.5$ *mm of* H_2O $\qquad T_f = 370 + 273 = 643$ *K*

$T_a = 30 + 273 = 303$ *K* $\qquad m = ?$ *in* $\dfrac{\text{kg of air}}{\text{kg of fuel}}$

$H = 32$ *m*

Solution

Check T_f for maximum discharge conditions.

We know that, the draught produced in mm of H_2O column is given by,

$$h_w = 353 \times H \left\{ \frac{1}{T_a} - \left[\frac{m+1}{m} \right] \times \frac{1}{T_f} \right\}$$

$$18.5 = 353 \times 32 \left\{ \frac{1}{303} - \left[\frac{m+1}{m} \right] \times \frac{1}{643} \right\}$$

$$\mathbf{m = 14.52 \frac{kg\ of\ air}{kg\ of\ fuel}}$$

∴ Air fuel ratio is 14.52:1

Now to check for T_f maximum discharge conditions,

$$T_f = 2\left[\frac{m+1}{m}\right] T_a = 2 \times \left[\frac{14.52+1}{14.52}\right] \times 303$$

$$\mathbf{T_f = 647.76\ K}$$

Since T_f actual given is 643 which is less than T_f for maximum discharge condtitions. So the chimney operates under maximum discharge conditions.

Example 8.8 *In a certain boiler installation a chimney of 42.5 m height produces a natural drought equivalent of 33.25 m of hot gas column. Boiler house temperature is 33°C. The mass of air supplied per kg of fuel is 19 kg, determine,*

(i) Temperature of gas leaving the chimney.

(ii) Temperature of flue gases and corresponding draught for maximum discharge.

Data: $H = 42.5$ *m* $\quad H' = 33.25$ *m of hot gas column*

$T_a = 33 + 273 = 306$ *K* $\quad m = 19$ *kg/kg of fuel*

$m + 1 = 20$ *kg of gases/kg of fuel* $\quad T_f = ?$

and T_f *for max discharge consitions = ?*

Solution

We know that, equivalent height of hot gas column,

$$H' = H\left\{\frac{m}{m+1} \times \frac{T_f}{T_a} - 1\right\} \text{ metre of hot gas column}$$

$$33.25 = 42.5\left\{\frac{19}{20} \times \frac{T_f}{306} - 1\right\} \text{ metre of hot gas column}$$

$$33.25 = 42.5\ \{3.1 \times 10^{-3}\ T_f - 1\}$$

$$T_f = 547.105\ \text{K}$$

We know that T_f for maximum discharge conditions,

$$T_f = 2\left[\frac{m+1}{m}\right] \times T_a$$

$$T_f = \left[\frac{20}{19}\right] \times 306$$

$$\mathbf{T_f = 644.211\ K}$$

Example 8.9 *Determine the height of chimney required for a boiler having an average coal consumption of 1500 kg/hr and producing 15 kg of dry flue gases per kg of coal fired. The pressure drop in mm of water at various stages in a boiler flue passes are (i) fuel bed 4.9 mm, (ii) Boiler flue 7.8 mm, (iii) Boiler tube passage 2.7 mm. (iv) Through bends, dampers etc. 0.6 mm.*

Actual draught may be taken as 25% greater than the theoretical draught. The flue gases have a mean te›perature of 277°C and ambient temperature is. 27°C.

Does this chimney fulfill the condition of maximum discharge?

Data: $H = ?$ $m_f = 1500$ *kg/hr*

$$m_{fg} = (m+1) = \frac{15\ kg\ of\ gases}{kg\ of\ coal}$$

Solution

Total pressure drop or Draught = 4.9 + 7.8 + 2.7 + 0.6 = 16 mm of H_2O

Since, Actual draught in 25% more than the theoretical draught,

$$\therefore \quad h_w = 16 \times 1.25 = 20 \text{ mm of } H_2O$$

$$T_f = 277°C + 273 = 500\text{ K} \quad \text{(Actual)}$$

$$T_a = 27 + 273 = 300\text{ K} \quad \text{(Mean)}$$

Now we know that the draught,

$$h_w = 353H\left\{\frac{1}{T_a} - \left[\frac{m+1}{m}\right] \times \frac{1}{T_f}\right\}$$

$$20 = 353 \times H\left\{\frac{1}{300} - \frac{15}{14} \times \frac{1}{550}\right\}$$

$$H = 40.91\text{ m}$$

And the condition for max. discharge is,

$$T_f = 2\left\{\frac{m+1}{m}\right\} \times T_a$$

$$= 2\left\{\frac{15}{14}\right\} \times 300$$

$$\mathbf{T_f = 642.8571\ K}$$

As T_f given in 550 K which is within T_f maximum. So the chimney satisfies the condition for maximum discharge.

Example 8.10 *Find the draught produced in mm of water by a chimney of 40 m height. Let the mass of flue gas be 20 kg/kg of fuel burnt in the combustion chamber. The temperature of flue gases and the ambient air are 270° C and 23° C. Assuming the diameter of the chimney is 150 cm and 30% of the theoretical draught is lost in friction, find the mass of the flue gases passing through the chimney/min.*

Data: h_w = ? $h = 40\ m$

$m + 1 = 20$ *kg of flue gases/kg of fuel*

$T_f = 543\ K$ $T_a = 296\ K$

Also mass of flue gases passing through the chimney/min = ?

Solution

We know that,

$$h_w = 353H\left\{\frac{1}{T_a} - \left[\frac{m+1}{m}\right] \times \frac{1}{T_f}\right\}$$

$$= 353 \times 40\left\{\frac{1}{296} - \frac{20}{19} \times \frac{1}{543}\right\}$$

$$h_w = 20.33 \text{ mm}$$

For the 2nd part from the continuity equation,

Mass of flue gases discharged

$$= \text{c/s area} \times P_f \times \text{velocity of flue gases} \quad \text{...(1)}$$

$$= A \times P_f \times C$$

$$\text{Units} = \text{m}^2 \times \frac{\text{kg}}{\text{m}^3} \times \frac{\text{m}}{\text{sec}} = \frac{\text{kg}}{\text{sec}} \quad \text{...(2)}$$

We also know that,

$$A = \frac{\pi}{4}D^2 = \frac{\pi}{4} \times 1.5^2 = 1.766 \text{ m}^2$$

$$P_f = \frac{m+1}{m} \times \frac{353}{T_f} = \frac{20}{19} \times \frac{353}{543}$$

$$P_f = 0.6843 \frac{\text{kg}}{\text{m}^3} \quad \text{...(3)}$$

and

$$c = \sqrt{2gH'}$$

To find H', we know that,

$$H' = H\left\{\frac{m}{m+1} \times \frac{T_f}{T_a} - 1\right\} \text{ metres of hot gas column.} \quad \text{...(4)}$$

$$= 40\left\{\frac{19}{20} \times \frac{543}{296} - 1\right\} = 40 \times 0.7427364$$

$$H' = 29.7095 \text{ m}$$

Since 30% of draught is lost frcition only 70% of the draught will be available

$\therefore$ Available draught $= 0.7 \times 29.7095$

H' available $= 20.7966$ m

∴ Substitute this in Eq. (4) we get,

$$C = \sqrt{2 \times g \times H'} = \sqrt{2 \times 9.81 \times 20.7966}$$

$$C = 20.199745 \text{ m/sec}$$

Now mass of the flue gases discharged from Eq. (1)

$$= 1.766 \times 0.6843 \times 20.199745$$

$$= 24.4109 \text{ kg/sec}$$

Mass of flue gases = 1464.65 kg/min

Example 8.11 *In a boiler trial following data were noted,*

Height of chimney : 25 m

Temperature of flue gases : 320°C

Ambient temperature : 22°C

Mass of air : 17 kg/kg of coal

Atmospheric pressure : 720 mm of Hg

Calculate h_w = ?

Solution

We know that draught produced in mm of H_2O column is given by,

$$h_w = 353 \times H \left\{ \frac{1}{T_a} - \frac{m+1}{m} \times \frac{1}{T_f} \right\} \frac{P_a}{P_N} \text{ mm of } H_2O$$

$$= 353 \times 25 \left\{ \frac{1}{295} - \frac{18}{17} \times \frac{1}{593} \right\} \frac{720}{760}$$

$$h_w = 13.40 \text{ mm of } H_2O$$

This problem can also be solved by using fundamental equations.

We know that, $\quad h_w = H[P_\alpha - P_f]$

Volume of 17 kg of air at NTP

$$P_0 V_0 = mRT_0$$

$$V_0 = \frac{17 \times 287 \times 273}{101325}$$

$$V_0 = 13.145 \text{ m}^3$$

Note. $P_0 = 760$ mm of Hg $= 101325$ N/m^2 and $T_0 = 273$

Volume of air at $\quad T_a = 22°\text{C} = 295$ K

and pressure of 720 mm of Hg is 95992.105 N/m^2

$$\therefore \quad \frac{P_0 V_0}{T_0} = \frac{P_a V_a}{T_a}$$

$$V_a = \frac{101325 \text{ N/m}^2 \times 13.145 \text{ m}^3 \times 295 \text{ K}}{273 \text{ K} \times 95992.105 \text{ N/m}^2}$$

$$V_a = 15\,\text{m}^3$$

$$\therefore \quad \rho_a = \frac{\text{Mass of air}}{\text{Volume of air at 295 K}}$$

$$\rho_a = \frac{17}{15} = 1.133\,\text{kg/m}^3$$

Note volume of flue gas at NTP =13.145 m^3

and to find volume of flue gases at 320ºC.

$$\frac{P_0 V_0}{T_0} = \frac{P_f V_f}{T_f}$$

$$V_f = \frac{101325 \times 13.145 \times 593}{273 \times 95992.105}$$

$$V_f = 30.137\,\text{m}^3$$

$$\rho_f = \frac{\text{Mass of flue gases}}{\text{Volume of flue gases}}$$

$$\rho_f = \frac{18}{30.137} = 0.5972\,\text{kg/m}^3$$

$$h_w = H[\rho_\alpha - \rho_\tau]\ 25\ [1.133 - 0.5972]$$

$$h_w = 13.40 \text{ mm of } H_2O$$

LIST OF FORMULAE

1. Heat transfer required to form the steam

$$= m_s (h_2 - h_1) \text{ kJ.}$$

m_s = Mass of steam generated in kg

h_2 = Specific enthalpy of steam formed in kJ/kg

h_1 = Specific enthalpy of feed water in kJ/kg

2. Energy received from fuel

$$= (m_f \times \text{C.V.}) \text{ kJ}$$

where m_f = mass of fuel in kg.

C.V. = Calorific value in kJ/kg

3. Boiler thermal efficiency,

$$\eta_{th} = \frac{m_s (h_2 - h_1)}{m_f \times \text{C.V.}}$$

4. Equivalent of Evaporation from and at 100ºC

$$= \frac{\dot{m}_s (h_2 - h_1)}{2256.9} \ \frac{\text{kg}}{\text{kg of coal}}$$

Note. Here $\dot{m}_s$ should be in kg/kg of fuel burnt and it is written as $\dot{m}_s$ or m_a.

(PROBLEMS ON BOILER PERFORMANCE)

Example 8.12 *A boiler working at a pressure of 14 bar evaporates 8 kg of water/kg of coal burnt from the feed water entering at 39°C. The steam at the steam stop valve is 0.95 dry. Determine the equivalent of evaporation from and at 2009°C.*

Data: $P = 14$ bar $\qquad \frac{m_s}{m_f} = 8$ kg of water/kg of coal $= m_s$

Feed water temperature = 39°C $\qquad x = 0.95$

Equivalent of evaporation?

Solution

We know that the equivalent of evaporation form and at 100°C

$$= \frac{\dot{m}_s (h_2 - h_1)}{2256.9} \quad \ldots(1)$$

where h_2 is specific enthalpy if steam formed at 14 bar pressure and 0.95 dry and the steam formed is wet,

$$\therefore \quad h_2 = h_f + x \cdot h_{fg}$$

From steam tables at 14 bar pressure

$$h_f = 830.07 \text{ kJ/kg}$$

$$h_{fg} = 1957.7 \text{ kJ/kg}$$

$$\therefore \quad h_2 = 830.07 + 0.95 \times 1957.7$$

$$h_2 = 2689.885 \text{ kJ/kg} \quad \ldots(2)$$

and h_1= specific enthalpy of feed water at 39°C from steam tables,

$$h_1 = h_f = 163.27 \text{ kJ/kg} \quad \ldots(3)$$

Also $\quad h_1 = h_f = m.C_p.\Delta t = 1 \times 4.187 \times 39°\text{C}$

$$h_1 = 163.29 \text{ kJ/kg}$$

∴ From Eq. (1),

Now, equivalent of evaporation $= \frac{\dot{m}_s \times (h_2 - h_1)}{2256.9}$

m_s should be in kg/kg of fuel burnt

$$= \frac{8 \times [2689.885 - 163.27]}{2256.9}$$

Equivalent of Evaporation = 8.9560548 kg of steam/kg of coal

Example 8.13 *A boiler plant supplies 5400 kg of steam /hr at 7502 and 0.98 dry from feed water at 41.5°C, when using 670 kg of coal/hr having a calorific value of 31000 kJ/kg.*

Determine,

(i) The efficiency of the boiler;

(ii) The equivalent of evaporation from and at 100°C.

Solution

We know that the thermal efficiency of the boiler is given by

$$\eta_{th} = \frac{m_s(h_2 - h_1)}{(m_f \times C.V.)} \times 100\%$$

Now Actual Evaporation

$$\text{or Steam raised/kg of coal} = m_a \text{ or } m_s = \frac{m_s}{m_f} = \frac{5400}{670}$$

$$m_a \text{ or } \dot{m}_s = 8.06 \text{ kg/kg of coal}$$

and h_2 = specific enthalpy of steam formed in kJ/kg at 7.5 bar and 0.98 dry,

$$\therefore \quad h_2 = h_f + x \cdot h_g$$

$$= 709.3 + 0.98 \times 2055.55$$

$$= 2723.74 \text{ kJ/kg}$$

Note h_f and h_{fg} are to be found at a pressure of 7.5 bar by interpolation, and h_1 = Specific enthalpy of feed water 41.5ºC = 173.7 kJ/kg

$$\therefore \quad \eta_{th} = \frac{8.06 \times (2723.74 - 173.7)}{31000} \times 100$$

$$\eta_{th} = \mathbf{66.30\ \%}$$

(ii) Equivalent of evaporation from and at 100ºC

$$= \frac{\dot{m}_s \times (h_2 \cdot h_1)}{2256.9}$$

$$= \frac{8.06 \times (2723.74 - 173.7)}{2256.9}$$

$$= \mathbf{9.11 \text{ kg/kg of coal.}}$$

Example 8.14 *Duration of a boiler trial the following data was obtained,*

During of trial = 8 hrs

Pressure of steam = 14 bar

Dryness fraction = 0.973

Feed water evaporated = 26700 kg

Temperature of water at inlet = 50°C

Coal used = 4260 kg

Calorific value of coal = 28900 kJ/kg

Air used / kg of coal = 17 kg

Temperature of flue gases = 344°C

Boiler room temperature = 21 °C

C_p of flue gases = 1.1 kJ / kg-K

Determine :

1. Boiler efficiency

2. Equivalent of evaporation

3. Heat lost to flue gases in kJ/kg of coal and in percentage

Solution

(i) Boiler efficiency

We know that, $$\eta_{th} = \frac{m_s[h_2 - h_1]}{m_f \times \text{C.V.}} \times 100\%$$

And steam raised/kg of coal = Actual evaporation

$$\dot{m}_s = \frac{m_s}{m_f} = \frac{26700}{4260}$$

$$= 6.2676 \frac{\text{kg}}{\text{kg of coal}}$$

Specific enthaply of steam formed at a pressure of 14 bar and 0.973 dry,

$$h_2 = h_f + x \times h_{fg}$$

$$= 830.3 + 0.973 \times 1957.7$$

$$h_2 = 2735.14 \text{ kJ/kg}$$

and h_1 = specific enthaply of feed water at 50°C

$$h_1 = 209.26 \text{ kJ/kg}$$

$$\therefore \quad \eta_{th} = \frac{6.2676 \times [2735.14 - 209.26]}{28900} \times 100$$

$$\boldsymbol{\eta_{th} = 54.77\%}$$

(ii) Equivalent of evaporation

$$= \frac{\dot{m}_s[h_2 - h_1]}{2256.9} \quad \left[\text{here } m_s \text{ should be in } \frac{\text{kg}}{\text{kg of coal}} \text{ i.e., } \dot{m}_s \text{ is to be used}\right]$$

$$= \frac{6.2676 \times [2735.14 - 209.26]}{2256.9}$$

$$= 7.01 \text{ kg/kg of coal}$$

(iii) Heat to flue gases = Mass of flue gases × C_p of flue gases × temperature rise

$$= m_g \times C_{pg} \times (T_f - T_a)$$

$$\left[\textbf{Note.} \text{ Units} = \frac{\text{kg}}{\text{kg of coal}} \times \frac{\text{kJ}}{\text{kg K}} = \frac{\text{kJ}}{\text{kg of coal}}\right]$$

$$= 18 \times 1.1 \times (344 - 21)$$

$$= 6395.4 \frac{\text{kJ}}{\text{kg of coal}}$$

Now to get heat to flue gases in % divide by C.V.

i.e.,
$$= \frac{m_g \times C_{pg}(T_f - T_a)}{\text{C.V.}} \times 100$$

$$= \frac{6395.4}{28900} \times 100 = 22.129\ \%$$

Example 8.15 *During a boiler trial following readings were noted,*

Chimney height = 90 m

Feed water = 1400 kg/hr

Temperature of feed water = 30°C

Boiler working pressure of dry and saturated steam = 20 bar

Coal used = 200 kg/hr

Calorific value of coal = 28000 kJ/kg

Flue gas temperature in chimney = 300°C

Flue gas formed = 18 kg/kg of coal burnt

Ambient temperature = 32° C

C_p of products of combustion = 1.028 kJ/kg K

Find

(1) η_{th}

(2) Equivalent of evaporation

(3) Heat carried away by flue gases

(4) Natural draught produced in mm of H_2O

Solution

(1) η_{th}

We know that,
$$\eta_{th} = \frac{m_s(h_2 - h_1)}{m_f \times \text{C.V.}} \times 100$$

From steam tables,

Specific enthalpy of dry and saturated steam at 20 bar

i.e.,
$$h_2 = h_g = 2797.2\ \text{kJ/kg}$$

And specific enthalpy of fluid at 30° C.

$$= h_1 = h_f \text{ at } 30°\text{C} = 125.7\ \text{kJ/kg}$$

And steam generator per kg of fuel

$$\frac{m_s}{m_f} = \frac{1400}{200} = 7\frac{\text{kg}}{\text{kg of fuel}} = m_a = \dot{m}_s$$

$$\therefore \quad \eta_{th} = \frac{7 \times (2797.2 - 125.7)}{28000} \times 100$$

$$\boldsymbol{\eta_{th} = 66.781\%}$$

(ii) Equivalent of evaporation

$$= \frac{\dot{m}_s (h_2 - h_1)}{2256.9}$$

$$\text{Equivalent of evaporation} = \frac{(2797.2 - 125.7)}{2256.9}$$

$$= 8.2859 \frac{\text{kg}}{\text{kg of coal}}$$

$$= 8.2859 \times 200 \text{ kg/hr}$$

Equivalent of evaporation = 1657.18 kg/hr

(iii) Heat to flue gases = $m_g \times C_{pg} \times (T_f - T_a)$

$$= 18 \frac{\text{kg}}{\text{kg of coal}} \times 1.028 \frac{\text{kJ}}{\text{kg K}} \times (300 - 32)\,\text{K}$$

$$= 4959.072 \frac{\text{kJ}}{\text{kg of coal}}$$

or Heat of the flue gases $= 4959.072 \dfrac{\text{kJ}}{\text{kg of coal}} \times 200 \dfrac{\text{kg of coal}}{\text{hr}}$

$$= 991814.4 \frac{\text{kJ}}{\text{hr}}$$

or $= 17.71\,\%$

(iv) Chimney draught

$$h_w = 353 \times H \left\{ \frac{1}{T_a} - \frac{m+1}{m} \times \frac{1}{T_f} \right\}$$

$$= 353 \times 90 \left\{ \frac{1}{305} - \frac{18}{17} \times \frac{1}{573} \right\}$$

$$h_w = 45.46 \text{ mm of } H_2O$$

Note. m = Mass of air supplied/kg of fuel

$m + 1$ = Mass of flue gases in kg/kg of fuel

Example 8.16 *In a boiler trial, following readings were noted,*

Mean temperature of feed water : 15°C

Mean boiler working pressure : 1.2 MPa

Mean dryness fraction : 0.95

Mass of coal burnt/hr : 250 kg

C. V. of coal burnt : 32400 kJ/kg

Mass of water supplied to the boiler in 7 hrs and 14 minutes = 16500 kg.

Mass of water in the boiler at the end of the test was less than at the commencement by 1000 kg. Calculate,

(i) Actual evaporation/kg of coal

(ii) Equivalent of evaporation from and at 100°C

(iii) η_{th}

Solution

(i) Water will be filled into the boiler, upto the prespecified level before firing.

Then the water is supplied continuously during the test period of 7 hrs and 14 minutes = 16500 kg and at the end of the trial water in the drum was less by 1000 kg.

∴ Total mass of steam raised in 7 hrs and 14 minutes

$$=16500+1000=17500\,\text{kg}$$

$$\text{Mass of steam raised/hr} = \frac{17500}{7\frac{14}{60}} = 2419.35\,\text{kg/hr (Note)}$$

∴ Actual evaporation or steam raised/kg of coal

$$m_a = \frac{m_s}{m_f} = \dot{m}_s = \frac{2419.35}{250} = 9.677\,\text{kg/kg of coal}$$

(ii) Equivalent of evaporation

$$= \frac{m_s(h_2 - h_1)}{2256.9}$$

h_2 at a pressure of 12 bar and 0.92 dry from steam tables

$$h_2 = h_f + \times h_{fg} = 798.43 + 0.95 \times 1984.3$$

$$h_2 = 2683.515\text{ kJ /kg}$$

And h_1 at 15°C from steam tables

$$h_1 = h_f$$

$$= 62.94\text{ kJ/kg}$$

∴ Equivalent of evaporation

$$= \frac{9.667[2683.515 - 62.94]}{2256.9}$$

$$= 11.2363\text{ kg/kg of coal}$$

(iii) Boiler thermal efficiency,

$$\eta_{th} = \frac{m_s(h_2 - h_1)}{m_f \times C.V.} \times 100$$

$$= \frac{9.677 \times [2683.515 - 62.94]}{32400} \times 100$$

$$\eta_{th} = \mathbf{78.2695\%}$$

Example 8.17 *A boiler is supplied with 15,900 kg of water at 15°C during a trial for a period of 8 hours 20 minutes. The mass of water at the end of trial was 680 kg less than that of the commencement of trial. The steam is* *produced at a pressure of 125 N/cm² and is 0.9 dry. The coal having calorific value of 30,000 kJ/kg was fired at the rate of 275.8 kg per hour. Calculate*

(i) *The actual evaporation*
(ii) *The equivalent evaporation*
(iii) *The thermal efficiency of boiler.*

Data: $$P = 125\frac{N}{cm^2} = 12.5\,bar$$

Note.
$$\left[\begin{array}{l} 100\,cm \times 100\,cm = 1000\,cm^2 = 1\,m^2 \\ P = 125\dfrac{N}{cm^2} = 125 \times 10000 \\ P = 1250000\dfrac{N}{m^2} - 12.5\,bar \end{array}\right]$$

$$m_s = 15900 \text{ kg of water in 8 hours 20 min.}$$

Solution

Mass of water at the end of test was less by 680 kg. So, this much water is also converted into steam . So total m_s = 15900 + 680 = 16500 kg of water in 8 hr and 20 min.

$$\text{C.V.} = 30000\frac{\text{kJ}}{\text{kg}}$$

$$m_f = 275.8\frac{\text{kg}}{h_r}$$

Calculate :

(i) Actual evaporation $= m_a = \dfrac{m_s}{m_f} = \dot{m}_s$

(ii) Equivalent of evaporation

(iii) η_{th}

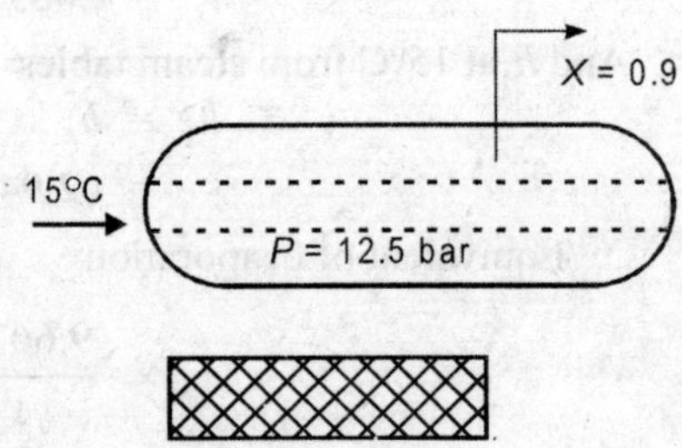

Fig. Ex. 8.17

Now, mass of steam produced/hr

$$= \frac{16500}{8\frac{20}{60}}$$

$$m_s = 1980\,\text{kg/hr}$$

(i) Now actual evaporation

$$m_a = \frac{m_s}{m_f} = \dot{m}_s$$

$$m_a = \frac{1980}{275.8}$$

$$m_a = 7.1791\frac{\text{kg of steam}}{\text{kg of fuel}}$$

(ii) Equivalent of evaporation

$$= \frac{\dot{m}_s(h_2 - h_1)}{2256.9}$$

and $$h_2 = h_f + \times h_{fg} \text{ at } 12.5 \text{ bar and } 0.9 \text{ dry}$$

$$= 806.7 + 0.9 \times 1977.5$$

$$h_2 = 2586.45 \text{ kJ/ kg}$$

and $$h_1 = h_f \text{ at } 15°\text{C} = 62.9 \text{kJ/kg}$$

Now from Eq. (1)
Equivalent of evaporation

$$= \frac{1980}{275.8} \times \frac{2586.45 - 62.9}{2256.9}$$

(iii) Thermal efficiency η_{th}

$$= \frac{m_s(h_2 - f_1)}{m_f \times \text{C.V.}} \times 100\%$$

$$= \frac{1980 \times (2586.45 - 62.9)}{275.8 \times 30000} \times 100$$

$$\boldsymbol{\eta_{th} = 60.389\%}$$

Example 8.18 *Following readings were obtained during a trial on two boilers.*

Boiler	*Pressure*	*Quality and steam*	*Evaporation rate*	*Feed water temp.*
I	*10 bar*	*0.9 dry*	*8.5 kg/kg of fuel*	*40°C*
I	*15 bar*	*300°C*	*8.0 kg/kg of fuel*	*60°C*

Fuel used has a clorific value of 30,000 kJ/kg. Compare these boilers in respect of equivalent evaporation and thermal efficiency.

Data. $$\frac{m_s}{m_f} = 8.5\frac{\text{kg of steam}}{\text{kg of fuel}} = m_a = \dot{m}_s$$

$$C.V. = 30{,}000\frac{kJ}{kg}$$

Equivalent of evaporation = ?

$$\eta_{th\,I} = ?$$

Solution

I –Boiler We know that equivalent of evaporation from and at 100°C

$$= \frac{\dot{m}_s(h_2 - h_1)}{2256.9} \quad \text{...(1)}$$

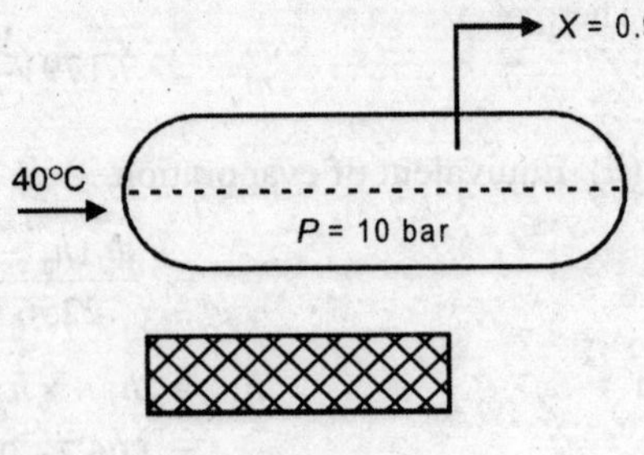

Fig. Ex. 8.18 (a)

$$h_2 = h_f + xh_{fg} \text{ at 10 bar and 0.98 dry}$$
$$= 762.6 + 0.98 \times 2013.6$$
$$h_2 = 2735.928 \text{ kJ/kg}$$

and $$h_1 = h_f \text{ at } 40°\text{C} = 167.5 \text{ kJ/kg}$$

Now from Eq. (1) $$= \frac{8.5 \frac{\text{kg of steam}}{\text{kg of fuel}} \times (2735.928 - 167.5)}{2256.9}$$

$$= \frac{21831.638}{2256.9}$$

Equivalent of evaporation $= \mathbf{9.673 \frac{kg\ of\ steam}{kg\ of\ fuel}}$

And Boiler thermal efficiency,

$$\eta_{th} = \frac{m_s (h_2 - h_1)}{m_f \times \text{C.V.}} \times 100$$

$$= \frac{m_s}{m_f} \times \frac{[h_2 - h_1]}{\text{C.V.}} \times 100\%$$

$$= \frac{8.5 \times (2735.928 - 167.5)}{30000} \times 100$$

$$\eta_{th} = \mathbf{72.772\%}$$

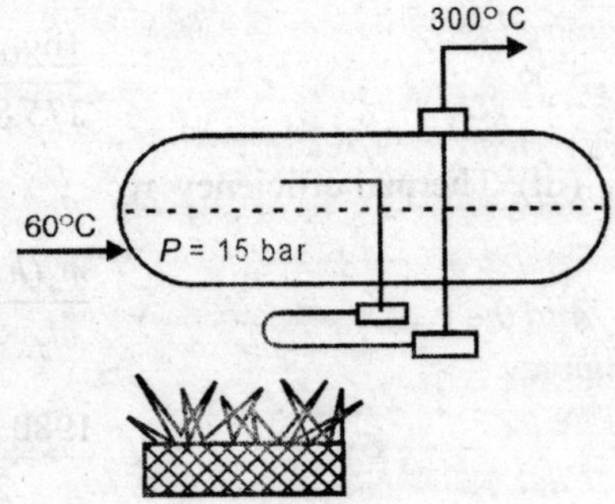

Fig. Ex. 8.18 (b)

II–Boiler $$\frac{m_s}{m_f} = \frac{8 \text{ kg of steam}}{\text{kg of fuel}} = m_a = \dot{m}_s$$

At 15 bar pressure T_{sat} is 198.3, and the given temperature of steam is 300° C, so the steam is superheated.

Now $$h_2 = h_{sub} \text{ at 15 bar and } 300°\text{C}$$
$$= 3038.9 \text{ kJ/kg}$$

and $$h_1 = h_f \text{ at } 60°\text{C} = 251.1 \text{ kJ/kg}$$

Now equivalent of evaporation from Eq. (1),

$$= \frac{\dot{m}_s (h_2 - h_1)}{2256.9}$$

$$= \frac{8 \frac{\text{kg of steam}}{\text{kg of fuel}} \times (3038.9 - 251.1)}{2256.9}$$

$$= \frac{22302.4}{2256.9}$$

Equivalent of evaporation $= \mathbf{9.8818 \frac{kg\ of\ steam}{kg\ of\ fuel}}$

and Thermal efficiency from Eq. (2),

$$\eta_{th} = \frac{\dot{m}_s (h_2 - h_1)}{m_f \times \text{C.V.}} \times 100\%$$

i.e.,

$$\eta_{th} = \frac{m_s}{m_f} \times \frac{(h_2 - h_1)}{\text{C.V.}} \times 100\%$$

$$= \frac{8 \times (3038.9 - 251.1)}{30000} \times 100$$

$$\eta_{th} = \mathbf{74.34\%}$$

So, second boiler is more efficient.

Example 8.19 *Design for principal dimensions i.e., determine the height and diameter of a chimney of a boiler which uses 18 kg of air/kg of coal burnt, when the coal is consumed at the rate of 1800 kg/hr. Actual draught required is 22 mm of H_2O taking into account all losses viz.,*

(i) Head loss due to fuel bed resistance

(ii) Head loss due to frictional resistance because of equipments and various bends/curves

Take, surrounding air temperature = 37°C

Temperature of flue gases = 294°C

And the velocity of gas flow to be 35% of the theoretical velocity because of roughness at the interior surface of chimney.

Data: $m = 18$ *kg/kg of coal*

Fuel consumed 1800 kg/hr

$h_w = 22$ *mm of* H_2O $\quad T_a = 37 + 273 = 310\ K$

$T_f = 294 + 273 = 567\ K$ $\quad C_{actual} = 0.35 \times C_{theoretical}$

$H = ?$ *and* $D = ?$

Solution

We know that

$$h_w = 353 \times H \left\{ \frac{1}{T_a} - \frac{m+1}{m} \times \frac{1}{T_f} \right\}$$

$$22 = 353 \times \left\{ \frac{1}{310} - \frac{19}{18} \times \frac{1}{567} \right\}$$

$\therefore$ $\quad$ **H = 45.68607 metres**

We also know that,

$$\rho_f = \frac{m+1}{m} \times \frac{353}{T_f}$$

$$\rho_f = \frac{19}{18} \times \frac{353}{567} = 0.6572 \text{ kg/m}^3$$

And note mass flow rate of flue gases = $(m + 1)$ × Mass of coal burnt/hr

[**ote.** Units $= \frac{\text{kg}}{\text{kg of coal}} \times \frac{\text{kg of coal}}{\text{hr}} = \frac{\text{kg}}{\text{hr}} = \frac{\text{kg}}{\text{hr}} \times \frac{1}{3600} = \text{kg/sec}$]

∴ Mass flow rate of the flue gases

$$= 19\times\frac{1800}{3600} = 9.5\,\text{kg/sec}$$

We also know that, $C = \sqrt{2gH'}$

To find H' $= H\left\{\frac{m}{m+1}\times\frac{T_f}{T_a}-1\right\}$

$$= 45:687\left\{\frac{18}{19}\times\frac{567}{310}-1\right\}$$

$$\textbf{H}' = \textbf{33.479 m of hot gas column}$$

Since $C_{actual} = 0.35$

$$C_a = 0.35\sqrt{2\times 9.81\times 33.4799}$$

$$C_a = 8.97008\,\text{m/sec}$$

From continuity equation, we know that,

Mass flow rate $= \rho_f\times A\times C_f$

$$9.5\frac{\text{kg}}{\text{sec}} = 0.6572\frac{\text{kg}}{\text{m}^3}\times A\times 8.97\frac{\text{m}}{\text{sec}}$$

$$A = 1.6115122\,\text{m}^2 = \frac{\pi}{4}D^2$$

$$\textbf{D} = \textbf{1.4324 m}$$

Example 8.20 *Following data was available in a boiler trial.*

Feed water = 2400 kg/hr

Ambient temperature = 33° C

Temperature of feed water = 42°C

Fuel used = 205 kg/hr

Constituents of fuel,

Carbon 84%, Hydrogen 9.27% and Oxygen = 6.73%

Calorific value of fuel = 39500 kJ/kg

Average chimney gas temperature = 307°C

Height of chimney = 32 m

Steam pressure = 11.054 bar (Gauge)

Barometric pressure = 710 mm of Hg

Air supplied = 50% in excess of stoichiometric (theoretical) air,

Steam condition = 0.96 dry.

Determine

(i) Boiler efficiency

(ii) Equivalent of evaporation from and at 100° and
(iii) Draught produced in min of H_2O.

Solution

Since in the problem Gauge pressure is given, we know that

$$P_{abs} = P_{atm} + P_g$$

Since $760 \text{ mm of Hg} = 1.01325 \text{ bar}$

$710 \text{ mm of Hg} = ?$

$$\therefore \quad \frac{710}{760} \times 1.01325 = P_{atm}$$

$$\therefore \quad P_{atm} = 0.9465888 \text{ bar}$$

$$\therefore \quad P_{abs} = 12.00 \text{ bar i.e., } (P_{atm} + P_g)$$

(i) We know that $\eta_{th} = \dfrac{m_s(h_2 - h_1)}{m_f \times \text{C.V.}} \times 100\%$

Now specific enthalpy of steam formed, h_2 at 12 bar and 0.96 dry

$$\therefore \quad h_2 = h_f + x h_{fg}$$

$$= 798.43 + 0.96 \times 1984.3$$

$$h_2 = 2703.3 \text{ kJ/kg}$$

And specific enthalpy of feed water at 42°C

$$h_1 = 175.81 \text{ kJ/kg}$$

Actual evaporation $ma = \dot{m}_s = \dfrac{m_s}{m_f} = \dfrac{2400}{205}$

$$= 11.7073 \frac{\text{kg of steam}}{\text{kg of fuel}}$$

$$\therefore \quad \eta_{th} = \frac{11.7073(2703.3 - 175.81)}{39500} \times 100$$

$$\boldsymbol{\eta_{th} = 74.912\%}$$

(ii) Equivalent of evaporation from and at 100°C

$$= \frac{\dot{m}_s(h_2 - h_1)}{2256.9} = \frac{11.7073(2703.3 - 175.81)}{2256.9}$$

$$= 13.1 \text{ kg/kg of coal}$$

(iii) Mass of theoretical air required to burn 1 kg of fuel. From fuels and combustion chapter,

$$= \frac{100}{23}\left[\frac{8}{3}\text{C} + 8\text{H} - \text{O}_2\right]$$

$$= \frac{100}{23}\left[\frac{8}{3}\times 0.84 + 8\times 0.0927 - 0.0673\right]$$

$$= 12.6709 \text{ kg/kg of coal}$$

∴ As given in the example

Actual air supplied $= 1.5 \times$ theoretical air

$$m = 12.6709 \times 1.5$$

$$m = 19.006 \text{ kg/kg of fuel}$$

$$\therefore \quad h_w = 353\times H\left\{\frac{1}{T_a} - \frac{m+1}{m}\times\frac{1}{T_f}\right\}$$

$$= 353\times H\left\{\frac{1}{306} - \frac{20.006}{19.006}\times\frac{1}{580}\right\}$$

$$\mathbf{h_w = 16.415 \text{ mm of } H_2O}$$

Example 8.21 *Equivalent of evaporation = 10.4 kg/kg of fuel. Boiler products 15 T/hr of steam at 2.00 MN/m². Feed water temperature is 40°C. Fuel consumption is 1650 kg/hr find boiler efficiency and condition of steam produced. Take calorific value = 29800 kJ/kg.*

Data: Equivalent of evaporation = 10.4 kg/kg of coal

m_s = *1500 kg/hr*

Pressure = 20 bar

Temperature of feed water at the inlet = 40°C

η_{th} = ? $\qquad x$ = ?

C.V. = 29800 kJ/kg

Solution

We know that,

$$\text{Equivalent of evaporation} = \frac{m_s(h_2 - h_1)}{2256.9}\frac{\text{kg}}{\text{kg of coal}}$$

$$10.4 = \frac{1500}{1650}\times\frac{(h_2 - h_1)}{2256.9}$$

$$(h_2 - h_1) = 2581.894\frac{\text{kJ}}{\text{kg}} \qquad \text{...(1)}$$

We also know that, $\eta_{th} = \dfrac{m_s(h_2 - h_1)}{m_f \times \text{C.V.}}\times 100$

$$\eta_{th} = 78.76\%$$

To determine the condition of steam.

We have already found that

$$(h_2 - h_1) = 2581.894$$

and h_1 is the specific enthalpy of feed water at 40°C = 167.45 kJ/kg from steam tables

$$\therefore \quad (h_2 - h_1) = 2581.894$$

$$h_2 = 2581.894 + 167.45$$
$$h_2 = 2749.33 \text{ kJ/kg}$$

At a pressure of 20 bar

$$h_2 = 908.6 \text{ kJ/kg}$$
$$h_{fg} = 1.888.7 \text{ kJ / kg}$$
$$\therefore \quad h_2 = h_f + xh_{fg}$$
$$2749.334 = 908.6 + \times 1887.7 \text{ kJ/kg}$$
$$\mathbf{x = 0.9746}$$

Example 8.22 *The observation recorded during a boiler trial are given below,*

Duration : 60 min

Steam generated : 5250 kg

Coal burnt : 695 kg

Calorific value of coal burnt : 30200 kJ/kg

Boiler pressure : 12 bar

Dryness fraction : 0.94

Temperature of steam leaving the superheater : 250 °C

Temperature of hot well : 45°C

Calculate without superheater

(i) Equivalent of evaporation (ii) η_{th} of boiler

Also calculate with superheater

(i) Equivalent of evaporation in kg/kg of coal (ii) η_{th} of boiler

(iii) Heat supplied by the superheater

Solution

(i) Without superheater

We know that equivalent of evaporation, $= \dfrac{\dot{m}_s (h_2 - h_1) \text{ kg of steam}}{2256.9 \quad \text{kg of coal}}$

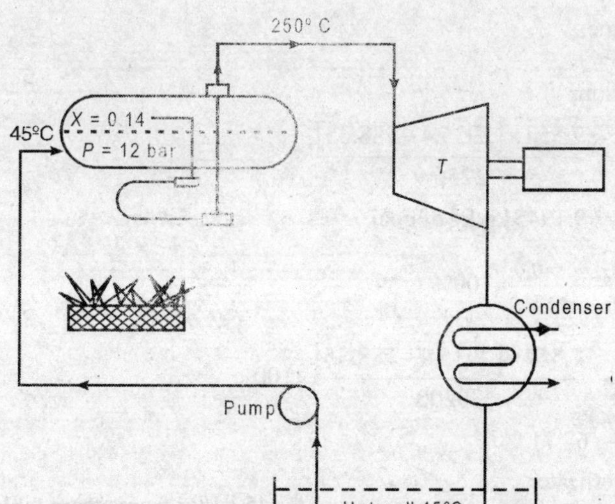

Fig. Ex. 8.22

Also steam raised/kg of coal $= \dfrac{m_s}{m_f} = m_a = \dot{m}_s$

$$= \frac{5250}{695} = 7.554 \text{ kg/kg of coal}$$

h_2, corresponding to a pressure of 12 bar and 0.94 dry

$$h_2 = h_f + x \cdot h_{fg}$$

$$= 798.43 + 0.94 \times 1984.3$$

$$h_2 = 2663.672 \text{ kJ/kg}$$

And h_1 corresponding to a temperature of hot well 45°C = 188.35 kJ/kg
(Since from the hot well water goes to the boiler drum)

∴ Equivalent of evaporation

$$= \frac{7.554 \times [2663.672 - 188.35]}{2256.9}$$

$$= 8.2851 \text{ kg/kg of coal}$$

(ii) Boiler Thermal Efficiency,

$$\eta_{th} = \frac{m_s(h_2 - h_1)}{m_f \times \text{C.V.}} \times 100$$

$$= \frac{7.554 \times [2663.67 - 188.35]}{30200} \times 100\%$$

$$\eta_{th} = 61.91\ \%$$

(iii) With superheater

Here h_2 should be corresponding to a pressure of 12 bar and temerature of 250°C from superheated steam tables,

$$h_2 = 2935.4 \text{ kJ/kg}$$

and Note h_1 will be the same.

∴ Now,

(i) Equivalent of evaporation,

$$= \frac{7.554 \times [2935.4 - 188.35]}{2256.9}$$

$$= 9.1945 \text{ kg/kg of coal}$$

(ii) η_{th} $= \dfrac{m_s(h_2 - h_1)}{m_f \times \text{C.V.}} \times 100\%$

$$= \frac{7.554 \times [2935.4 - 188.35]}{30200} \times 100$$

$$\eta_{th} = 68.71\ \%$$

(iii) Heat added in the superheater

= Enthalpy of superheated steam – Enthalpy of wet steam

$= h_{sub} - h_{wet}$

$$= (2935.4 - 2663.672)\ \text{kJ/kg}$$

$$= 271.728\frac{\text{kJ}}{\text{kg of steam}}$$

$\therefore$ Total heat added in the superheated/hr

$$= 271.728\frac{\text{kJ}}{\text{kg of steam}} \times 5250\frac{\text{kg of steam}}{\text{hr}}$$

$$= 1426572\ \text{kJ/hr}$$

Example 8.23 *The observation recorded during a boiler trial are as under:*

Duration : 60 min

Steam generated : 35,500 kg

Steam pressure : 12 bar

Steam temperature : 250°C

Temperature of water entering economiser = 17°C

Temperature of water leaving economiser = 77°C

Fuel burnt = 3460 kg

Calorific value of fuel used = 39500 kJ/kg

Calculate:

(1) Equivalent of evaporation from and at 100°C per in kg per kg of fuel.

(2) Thermal efficiency of the plant.

(3) The % of heat energy of the fuel utilised by the economiser.

Solution

We know that,

(i) Equivalent of evaporation from and at 100°C = $\dfrac{\dot{m}_s(h_2 - h_1)}{2256.9}\dfrac{\text{kg of steam}}{\text{kg of fuel}}$

Here h_2 is the specific enthalpy of steam formed at 12 bar and 250°C. The steam is superheated so h_2 from superheated steam tables = 2935.4 kJ/kg.

And h_1 at the feed water inlet temperature of economiser i.e.,

$17°C = 71.31\ \text{kJ/kg}$

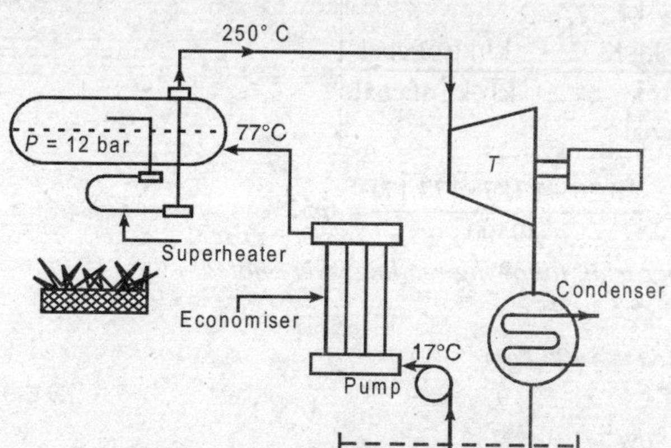

Fig. Ex. 8.23

And h_1 at the economiser outlet temperature i.e., inlet temperature of the water to the boiler

i.e., 77°C = 322.33 kJ/kg

$$\text{Now steam raised/kg of coal} = \frac{m_s}{m_f} = \frac{35500}{3460} = 10.26 = \dot{m}_s = m_a$$

$$\text{Equivalent of evaporation} = \frac{\dot{m}_s(h_2 - h_1)}{2256.9}$$

$$= \frac{10.26 \times (2935.4 - 322.33)}{2256.9}$$

$$= 11.879 \text{ kg/kg of coal}$$

Note. Here h_1 should be the enthaply of water entering the drum, from the definition of equivalent of evaporation.

(ii) Thermal efficiency of the plant

$$\eta_{th} = \frac{m_s(h_2 - h_1)}{m_f \times \text{C.V.}} \times 100$$

i.e.,

$$\eta_{th} = \frac{m_s(h_2 - h_1)}{m_f \times \text{C.V.}} \times 100\%$$

$$= \frac{10.26 \times (2935.4 - 71.31)}{39500}$$

$$= 74.395\%$$

Note. Since η_{th} of the plant is to be found, h_1 should correspond to the temperature of water entering the economiser

(iii) The % of used in the economiser

$$= \text{Mass of water passing through Economiser/kg of coal} \times \frac{C_p \text{ of water} \times \text{temperature rise}}{\text{C.V.}} \times 100\%$$

$$\left[\textbf{Note.} \text{ Unit} = \frac{\frac{\text{kg}}{\text{kg of coal}} \times \frac{\text{kJ}}{\text{kg K}} \times \text{K}}{\frac{\text{kJ}}{\text{kg of coal}}} = \frac{\text{kJ/kg of coal}}{\text{kJ/kg of coal}}\right]$$

$$= \frac{10.26 \times 4.187 \times (77.17)}{39500} = 6.52\%$$

Example 8.24. *During a boiler trial, following readings were noted*

Steam pressure = 15 bar

Amount of water evaporated = 5400 kg/hr

Condition of steam = 0.92

Feed water is supplied at 50°C

Determine the rate of coal consumption in kg/hr if η_{th} = 70%

Assume C.V. of coal = 40000 kJ/kg.

Also determine the equivalent of evaporation from and at 100°C in kJ/hr

If the superheater is used then the temperature of steam leaving the super heater is 250° C. Find η_{th} *= ? Assume* C_p *for superheated steam as 2.27 kJ/kg-K.*

Solution

We know that,

$$\eta_{th} = \frac{m_s\,(h_2 - h_1)}{m_f \times \text{C.V.}} \times 100 \qquad \ldots(1)$$

Now, h_2 = specific enthaply of steam formed at 15 bar and 0.92 dry.

$$= h_f + x \cdot h_{fg}$$

$$= 844.66 + 0.92 \times 1945.2$$

$$h_2 = 2634.244 \text{ kJ/kg}$$

$$\text{And } h_1 \text{ at } 50°\text{C} = 209.26 \text{ kJ/kg} = h_f$$

$$\therefore \text{ From Eq. (1)}, \quad \eta_{th} = \frac{\dot{m}_s\,(h_2 - h_1)}{m_f \times \text{C.V.}} \times 100$$

$$\therefore \quad m_f = \frac{5400 \text{ kg/hr } (2634.24 - 209.26)}{0.7 \times 40000}$$

$$m_f = 467.675 \text{ kg/hr}$$

(ii) Equivalent of evaporation,

$$= \frac{\dot{m}_s\,(h_2 - h_1)}{2256.9}$$

Note. Here m_s should be in kg/hr so as to get Equivalent of Evaporation in $\frac{\text{kg of steam}}{\text{hr}}$

$$= \frac{5400(2634.24 - 209.26)}{2256.9} = 5802.159 \text{ kg/hr}$$

(iii) With superheater, since C_p of superheated steam is given,

$$\therefore \quad h_2 = h_g + c_p\,(T_{\text{sup}} - T_{\text{sat}}) \text{ at } 15 \text{ bar}$$

$$h_2 = 2789.9 + 2.27\,(250 - 198.29)$$

$$h_2 = 2907.282 \text{ kJ/kg}$$

$$\eta_{th} = \frac{m_s\,(h_2 - h_1)}{m_f \times \text{C.V.}} \times 100 \quad \frac{5400(2907.28 - 209.26)}{467.675 \times 40000}$$

$$= 77.88\,\%$$

Example 8.25 *During a boiler trial on fire tube boiler, following readings were noted,*

Duration of trial = 24 hrs

Amount of water evaporated = 20 T

Feed water temperature = 50°C
Condition of steam = 0.92
Absolute pressure = 15 bar
Coal burnt = 6T
Grate area = 4 m²
Heating surface = 3200 m²
C.V. of coal = 20000 kJ/kg
Calculate:

(1) *Mass of coal burnt in kg per m² of grate area per hr* $\left[i.e., \frac{kg}{m^2\text{-}hr}\right]$

(2) *Equivalent of evaporation from and at 100° C in kg/kg of coal.*
(3) η_{th}
(4) *Equivalent of evaporation in kg/m² of heating surface per hour.*

Solution

Now, mass of coal burnt /hr $= \frac{6000\,\text{kg}}{24\,\text{hrs}} = 250\,\text{kg/hr}$

And steam reduced/hr $= \frac{20000}{24}$
$= 833.33$ kg/hr

Therefore,
(1) Mass of coal burnt in kg/m² of grate area/hr

$$= \frac{250\text{ kg/hr}}{4\text{m}^2} = 62.5\text{kg/m}^2\text{-hr.}$$

(2) Mass of steam produced/kg of coal

$$= \frac{20000}{6000} = 3.333\,\text{kg/kg of coal}$$

∴ Equivalent of evaporation from and at 100°C in kg/kg of coal

$$= \frac{\dot{m}_s(h_2 - h_1)}{2256.9} \qquad \text{...(1)}$$

Now from Eq. (1),
And h_2 = specific enthalpy of steam formed at 15 bar and 0.92 dry

$$= h_f + x \cdot h_{fg}$$
$$= 844.66 + 0.92 \times 1945.2 = 2634.244 \text{ kJ/kg}$$

$$h_1 \text{ at } 50°\text{C} = 209.26\,\text{kJ/kg}$$

Now from Eq. (1), Equivalent of Evaporation,

$$= \frac{3.333(2634.24 - 209.26)}{2256.9}$$

$$= 3.5815795 \text{ kg/kg of coal}$$

(3) $$\eta_{th} = \frac{m_s(h_2 - h_1)}{m_f \times \text{C.V.}} \times 100$$

$$= \frac{3.33(2634.24 - 209.26)}{20000} \times 100$$

$$= 40.41\ \%$$

(4) Mass of steam produced per m^2 of heating surface/hr

$$= \frac{833.33 \text{ kg/hr}}{3200\,\text{m}^2 \text{ of heating surface area}}$$

$$= 0.2604156 \frac{\text{kg}}{\text{hr - m}^2}$$

∴ Equivalent, of evaporation in $\frac{\text{kg}}{\text{hr - m}^2}$

$$= \frac{m_s(h_2 - h_1)}{2256.9}$$

∴ Equivalent, of evaporation in $\frac{\text{kg}}{\text{hr - m}^2}$

$$= \frac{0.2604156(2634.24 - 209.26)}{2256.9}$$

Equivalent of evaporation = 0.2798097 kg/hr-m^2

(PROBLEMS ON ENERGY BALANCE)

Example 8.26 *The following results were obtained in a boiler trial,*

Feed water/hr : 700 kg

Feed water inlet temperature : 27°C

Steam produced at a pressure : 8 bar

Dryness fraction : 0.97

Coal used : 100 kg/hr

C.V. of coal : 25000 kJ/kg

Ash and unburnt coal collected from beneath the grate bars = 7.25 kg/hr

C. V. of unburnt fuel = 2000 kJ/kg

Flue gases formed / kg of fuel = 17.3 kg

Flue gas temperature = 325° C

Temperature of air in the room = 16° C

C_p of flue gases = 1. 025 kJ/kg-K

Draw up energy balance on minute basis and find the boiler efficiency.

Solution

(1) Energy supplied/released due to the combustion of fuel

$= m_f \times$ C.V. of fuel
$= 100$ kg/hr $\times$ 25000 kJ/kg
$= 25{,}00{,}000$ kJ/hr
$= 41666.667$ kJ/min $= 100\%$ (Say)

(2) Energy utilised.

(i) Energy utilised to form the steam

$= m_s (h_2 - h_1)$

where,

h_2 = specific enthalpy of steam formed in kJ/kg corresponding to a pressure of 8 bar and since the steam is wet.

$h_2 = h_f + x \cdot h_{fg} = 720.9 + 0.97 \times 2046.5$

$\therefore \quad h_2 = 2706.005$

And h_1 specific enthaply of feed water at 27°C

$= 113.13$ kJ/kg

$\therefore$ Energy or Heat to form steam

$= m_s (h_2 - h_1)$
$= 700$ kg/hr $(2706.005 - 113.13)$ kJ/kg
$= 1815012.5$ kJ/kg
$= 30250.21$ kJ/min

Total energy released $= 41666.67$ kJ/min $= 100\%$
for 30250.21 kJ/min $= ?$

$$= \frac{30250.21}{41666.67} \times 100 = 72.60\% = \eta_{th}$$

(ii) Heat to flue gases

$= $ Mass of flue gases $\times C_p$ of flue gases $\times (T_f - T_a)$
$= 17.3 \times 1.025 \times (325 - 16)$
$= 5479.3425$ kJ/kg of coal

Since 100 kg of coal is burnt/hr

$$\therefore \quad = 5479.3425 \frac{\text{kJ}}{\text{kg of coal}} \times 100 \frac{\text{kg of coal}}{\text{hr}}$$

$= 547934.25$ kJ/hr
$= 9132.2375$ kJ/min

$$\therefore \quad \text{In } \% = \frac{9132.24}{41666.67} \times 100$$

(iii) Heat lost to unburnt fuel

$= m_{uf} \times$ C.V. of unburnt fuel
$= 7.25$ kg/hr $\times$ 2000 kJ/kg
$= 14500$ kJ/hr $= 241.667$ kJ/min
$= 0.58\%$

(iv) Heat unaccounted by difference

$=$ Total heat generated – Total heat utilised
$= 41666.67 - (30250.21 + 9132.24 + 241.67)$

$= 2024.55$ kJ/min

$= 4.1\%$

Energy Supplied	*kJ/min*	*%*	*Energy utilised*	*kJ/min*	*%*
	41666.67	100%	(i) To steam (ii) To flue gases (iii) To unburnt fuel (iv) To unaccounted	30250.21 9132.24 241.67 2042.55	72.60 21.92 0.58 4.10
	41666.67	100%		41666.67	100%

Example 8.27 *The following data were obtained in a boiler trial.*

Feed water supplied : 680 kg/hr

Temperature of feed water 20° C

Steam pressure : 15 bar

Steam temperature : 300° C

Coal used : 95 kg/hr

C.V. of coal used : 34000 kJ/kg

Ash and unburnt coal collected in the ash pit = 4 kg/hr

C.V. of unburnt coal = 2200 kJ/kg

Flue gas formed = 18 kg/kg of coal burnt

Flue gas temperature = 343° C

Ambient temperature = 28° C

Mean C_p of flue gases = 1.025 kJ/kg-K

Determine:

(1) η_{th}

(2) Equivalent of evaporation from and at 100°C

(3) % of heat unaccounted

(4) Drought produced in mm of H_2O, if the chimney height is 50 m.

Solution

(1) We know that,

$$\eta_{th} = \frac{m_s(h_2 - h_1)}{m_f \times \text{C.V.}} \times 100\%$$

At 15 bar and 300°C, from superheated steam tables,

$$h_2 = 3038.9 \text{ kJ/kg}$$

and

$$h_1 \text{ at } 20°\text{C} = 83.86 \text{ kJ/kg} = h_f$$

$$\eta_{th} = \frac{680 \times (3038.9 - 83.86)}{95 \times 34000} \times 100$$

$$\boldsymbol{\eta_{th} = 62.21\%}$$

(2) Equivalent of evaporation from and at 100°C

$$= \frac{\dot{m}_s (h_2 - h_1)}{2256.9}$$

$$= \frac{680}{95} \times \frac{(3038.9 - 83.86)}{2256.9}$$

Equivalent of Evaporation = $\mathbf{9.3721 \frac{kg}{kg\ of\ coal}}$

(3) (A) Heat supplied from coal

$= m_f \times \text{C.V.}$
$= 95 \text{ kg/hr} \times 34000 \text{ kJ/kg}$
$= 32{,}30{,}000 \text{ kJ/hr}$
$= 53833.333 \text{ kJ/min}$
$= 100\% \text{ (Say)}$

(B) (i) Heat to steam $= m_s (h_2 - h_1)$
$= 680 (3038.9 - 83.89) \text{ kJ/hr}$
$= 2009406.8 \text{ kJ/hr}$
$= 33490.113 \text{ kJ/min}$
$= 62.210\%$

(ii) Heat carried away by the flue gases,

$$= m_{\text{flue gas}} \times C_{p \text{ of flue gases}} \times (T_f - T_a)$$

$$= 18 \times 1.025 (343 - 28)$$

$$= 5811.77 \frac{\text{kJ}}{\text{kg of coal}}$$

$$= 5811.75 \text{ kJ/kg of coal} \times 95 \text{ kJ of coal/hr}$$

$= 552116.25 \text{ kJ/hr}$
$= 9201.9375 \text{ kJ/min}$
$= 17.09\%$

(iii) Heat lost to unburnt fuel

$$= m_{\text{uf}} \times \text{C.V.}_{\text{uf}}$$

$$= 4 \frac{\text{kg}}{\text{hr}} \times 2200 \frac{\text{kJ}}{\text{kg}}$$

$= 8800 \text{ kJ/hr}$
$= 146.667 \text{ kJ/min}$
$= 0.2724\%$

(iv) Heat unaccounted by difference

$$= 53833.333 - [33490.113 + 9201.938 + 146.667]$$

$= 10994.615 \text{ kJ/min}$
$= 20.423\%$

(4) Chimney height $= 50$ m

Mass of flue gases $= 18$ kg/kg of coal

$\therefore$ Mass of flue air $= 18 - 1 = 17$ kg/kg of coal

$$T_a = 28 + 273 = 301 \text{ K}$$

$$T_f = 343 + 273 = 616 \text{ K}$$

$$\therefore \quad h_w = 353 \times H \times \left\{\frac{1}{T_a} - \frac{m+1}{m} \times \frac{1}{T_f}\right\} \text{mm of } H_2O$$

$$h_w = 353 \times 50 \left\{\frac{1}{301} - \frac{18}{17} \times \frac{1}{616}\right\}$$

$$\mathbf{h_w = 28.3 \text{ mm of } H_2O}$$

Example 8.28 *Following observations were made during a boiler trial,*

Feed water = 1500 kg/hr

Inlet temp. of feed water = 20°C

Steam generated = 15 bar and 300°C

Coal consumed = 200 kg/hr

C.V. = 34000 kJ/kg

Ash in the ash pit contain unburnt products of 10 kg/hr of C.V. 2000 kJ/kg.

Flue gases formed = 18 kg/kg of coal burnt.

Flue gas temperature = 343° C

Atmospheric temp. = 28° C

Mean specific heat of fuel gases = 1.025 kJ/kg-K

Determine

(i) Boiler thermal efficiency

(ii) Equivalent of evaporation in $\dfrac{\text{kg of steam}}{\text{kg of coal}}$

(iii) Heat balance in kJ/min for the boiler and

(iv) Drought produced in mm of H_2O if chimney height H = 100 m.

Solution

$$m_s = 1500 \text{ kg/hr}$$

$$m_f = 200 \text{ kg/hr}$$

$$\text{C.V.} = 34000 \text{ kJ/kg}$$

$$m_{uf} = 10 \text{ kJ/kg}$$

$$\text{C.V.}_{uf} = 2000 \text{ kJ/kg}$$

$$m_{fg} = (m+1) = 18 \frac{\text{kg}}{\text{kg of coal}}$$

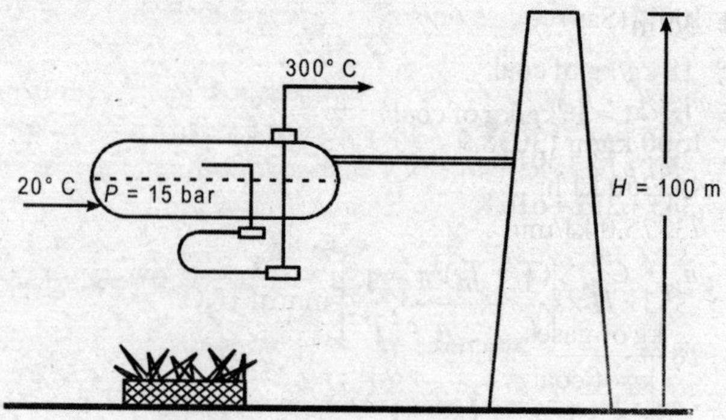

Fig. Ex. 8.28

$$T_f = 343 + 273 = 616 \text{ K}$$
$$T_a = 28 + 273 = 301 \text{ K}$$

Mean $C_{pg} = 1.025$ kJ/kg-K

At 15 bar pressure $T_{sat} = 198.3$°C and the steam temperature given is 300°C so the steam is superheated.
Now $h_2 = h_{sup}$ at 15 bar and 300°C from superheated steam tables

$$= 3038.9 \text{ kJ/kg}$$

and $h_1 = h_f$ at 20°C $= 83.9$ kJ/kg

(i) Now we know that Boiler thermal efficiency,

$$\eta_{th} = \frac{m_s(h_2 - h_1)}{m_f \times \text{C.V.}} \times 100\%$$

$$= \frac{1500\frac{\text{kg}}{\text{hr}}(3038.9 - 83.9)}{200\frac{\text{kg}}{\text{hr}} \times 34000}$$

$$= \frac{4432500}{200 \times 34000}$$

$$\eta_{th} = \mathbf{65.1838\%}$$

(ii) Equivalent of evaporation

$$= \frac{\dot{m}_s(h_2 - h_1)}{2256.9}$$

$$= \frac{1500}{200} \times \frac{(3038.9 - 83.9)}{2256.9}$$

Equivalent of evaporation = 9.8198 kg of st/kg of coal

(iii) Heat Balance sheet

(A) Heat energy released due to the combustion of fuel

$$= m_f \times \text{C.V.}$$
$$= 200 \text{ kg/hr} \times 34000 \text{ kJ/kg}$$
$$= 6800000 \text{ kJ/hr}$$
$$= 113333.33 \text{ kJ/min}$$

$= 100\%$ (Say)

(B) Heat Energy Utilised

(a) To form steam $= m_s (h_2 - h_1)$

$= 1500$ kg/hr $(3038.9 - 83.9)$ kJ/kg

$= 443250$ kJ/hr

$= 73875.0$ kJ/min

(b) Heat to flue gases $= m_{fg} \times C_{Pg} \times (T_f - T_a)$

$$= 18 \frac{\text{kg of gases}}{\text{kg of coal}} \times 1.025 \frac{\text{kJ}}{\text{kg-K}} (616 - 301)\,\text{K}$$

$= 5811.75$ kJ/kg of coal

$$= 5811.75 \frac{\text{kJ}}{\text{kg of coal}} \times 200 \frac{\text{kg of coal}}{\text{hr}}$$

$= 1162350$ kJ/hr

Heat to flue gases $= 19372.5$ kJ/min

(b) Heat lost due to unburnt fuel

$= m_{uf} \times C.V._{uf}$

$= 10$ kg/hr $\times$ 2000 kJ/kg

$= 2000$ kJ/hr

$= 333.333$ kJ/min

(d) Heat unaccounted by difference

$= \{113333.33\} - \{73875 + 5811.75 + 19375.5 + 333.33\}$

$= (113333.33 - 99392.58)$ kJ/min

$= 13940.75$ kJ/min

(iv) Since $H = 100$ m

We know that, $$h_w = 353 \times H \left\{ \frac{1}{T_a} - \frac{m+1}{m} \times \frac{1}{T_f} \right\}$$

$$= 353 \times 100 \{353 \times 100\} \left\{ \frac{1}{301} - \frac{18}{17} \times \frac{1}{616} \right\}$$

$$\mathbf{h_w = 56.61 \text{ mm of } H_2O}$$

Example 8.29 *A boiler generates 8 kg of steam/kg of coal at 30 bar and 350° C. Water is fed at 40°C. The mass of dry flue gas and water vapour/kg of coal burnt are 18.75 kg and 0.7 kg respectively. Boiler house temperature is 27°C and flue gas temperature leaving the boiler is 127°C.*

Draw-up heat balance sheet on kJ/kg of coal burnt basis and % basis.

Assume C_p of gases = 1.1 and C_p for the steam formed due to the combustion of H_2 = 2.1 kJ/kg K. Partial pressure of steam as 0.07 bar and C.V. of coal = 32000 kJ/kg.

Solution

(1) Heat energy liberated due to the combustion of 1 kg of fuel

$= m_f \times C.V.$

$= 1 \times 32000$

$= 32000$ kJ/kg of coal = 100% (say)

(II) Energy utilised

(i) Heat to steam $= m_s\,(h_2 - h_1)$

$= 8$ kg/kg of coal × (3117.5 – 167.45) kJ/kg

$= 23600.4$ kJ/kg of coal = 73.75%

(ii) Heat lost to the flue gases

$= m_g \times C_{pg} \times (T_f - T_a)$

$= 18.75 \times 1.1\,(127 - 27)$

$= 2062.5$ kJ/kg of coal

$= 6.45$ %

(iii) Heat lost due to the water vapour formed due to the combustion of H_2

$= m_1(h_2' - h_1')$

Here h_2' – is the specific enthalpy of steam formed due to the combustion of H_2 at flue gas temperature and at the given partial pressure.

Since C_p of steam formed is given, from steam tables corresponding to 0.07 bar pressure,

$$\therefore \quad h_2' = h_g + C_p(T_{sup} - t_{sat})$$

$$= 2572.6 + 2.1\,(127 - 39)$$

$$h_2' = 2557.4 \text{ kJ/kg}$$

and h_1' = specific enthalpy of water at room temperature

i.e., at 27°C = 113.13 kJ/hr

Heat lost to water vapour

$$= m_1(h_2' - h_1')$$

$$= 0.7\,\{2757.4 - 113.13\}$$

$$= 1850.99 \text{ kJ/kg}$$

$$= 5.79\%$$

(iv) Heat unaccounted by difference

$$= 32000 - \{23600.4 + 2062.5 + 1850.99\}$$

$$= 4486.11 \text{ kJ/kg}$$

$$= 14.02\%$$

THEORY QUESTIONS

8.1. What is a boiler? Explain how they are classified.

8.2. Explain the following boilers with the help of neat sketches.

(a) Landcashire boiler (b) Cornish boiler

(c) Locomotive boiler (d) Cochran boiler

(e) Bacock Wilcox boiler (f) Stirling boiler

8.3. Compare water tube and fire tube boilers.

8.4. What are the essentials of good boiler.

8.5. What is the difference between the Mountings and Accessories ? Mention the various Mountings and Accessories.

8.6. Explain with neat sketches the working of the following Mountings and Accessories

(a) Dead weight safety valve (b) High steam low water alarm

(c) Water level indicator (d) Fusible Plug

(e) Bourdon pressure gauge (f) Steam top valve
(g) Super heater (h) Economiser

8.7. What is boiler draught ?

8.8. Explain various types of draught used in a boiler.

8.9. Compare Artificial draught versus Natural (Chimney) draught.

8.10. Derive, $h_w = 353H\left[\frac{1}{T_a} - \frac{m+1}{m} \times \frac{1}{T_f}\right]$ mm of H_2O.

8.11. Derive the condition for maximum discharge of flue gases through the chimney.

8.12. What is a Steam Jet Draught ?

8.13. Explain,

(i) Equivalent evaporation from and at 100ºC

(ii) Boiler Efficiency

8.14. What are the draught losses ?

8.15. Write a note on IBR regulation.

8.16. State the essentials of a good boiler.

8.17. Draw a neat sketch of any one type of boiler you have studied. Label the parts and describe its working.

8.18. State the function and location of the following :

(a) Fusible plug (b) Feed check valve
(c) Blow-offcock (d) Pressure gauge

8.19. Staff the factors considered for selection of a boiler, for a specific application.

8.20. Describe with a neat indicator

(a) Water level indicator (b) Superheater
(c) Fusible plug (d) Economiser
(e) Spring loaded safety valve (f) Feed check valve

8.21. Describe briefly the advantages of incorporating an economiser, air-preheater and superheater in a steam boiler. Indicate the position of these in the passage of flue gases by a line diagram.

8.22. Define :

(i) Equivalent evaporation (ii) Boiler accessories

8.23. What is a boiler draught ? Compare natural draught versus artificial draught.

8.24. Define :

(i) Boiler efficiency,

(ii) Equivalent evaporation and explain the working parameters which influence them.

8.25. Define the term 'equivalent evaporation' and explain its significance.

8.26. Write a short note on 'classification of boiler'.

8.27. Write a short note on 'regulations for chimney height'.

8.28. State the working principle of

(a) Fusible plug, (b) Water level indicator, (c) Air preheater.

8.29. Prove that under maximum discharge condition of chimney, the draught produced in terms of the hot gas column is same as the chimney height.

8.30. State the advantage of high pressure boilers.

8.31. Explain Evaporative capacity of a boiler.

8.32. Explain Function and working of the following:

(a) Superheater

(b) Spring loaded safety valve

(c) Air preheater

8.33. What is IBR? What are the provisions made in IBR?

8.34. Compare water tube boilers with five tube boilers.

8.35. Explain:

(i) Grate
(ii) Shell
(iii) Mounting
(iv) Heating surface
(v) Manhole
(vi) Damper

8.36. Make a list of any five boiler mountings and write down their functions in brief.

8.37. Make 'a list of any five boiler accessories and write down their functions in brief.

8.38. With the help of neat sketch explain working of Benson boiler.

8.39. What is a boiler? How they are classified?

8.40. Explain the following brifers with the help of neat sketches.

(a) Lancashire Boiler
(b) Cornish Boiler
(c) Locomotive Boiler
(d) Cochram Boiler
(e) Babcock Wilcox Boiler
(f) Stirling Boiler

8.41. Compare water tube and fire tube brifers.

8.42. Whar are the essentials of a good brifer.

8.43. What is the difference between mountings and accessories mountings and accessories.

8.44. Explain with neat sketches working of following mountings and accessories.

(a) Dead water safety valve
(b) High-steam low water alarm
(c) Water level indicator
(d) Fusible plug
(e) Bourden pressure gauge
(f) Steam stop valve
(g) Superheater
(h) Economiser

8.45. What is boiler draught?

8.46. Explain various types of draught used in boiler?

8.47. Compare Artificial draught verses Natural draught.

8.48. Derive $h_w = 353 \times H \left[\frac{1}{T_a} - \frac{mel}{m} \times \frac{1}{T_f} \right]$ mm of H_2O.

8.49. Derive the condition for maximum discharge of the gases through the chimney?

8.50. What is a steam jet draught?

8.51. Explain

(i) Equivalent of evaporation from and @ 100°C

(ii) Boiler efficiency

8.52. What are the draught losses?

8.53. Write a note on IBR Regulations.

8.54. Explain the function of

(i) Scum valve (ii) Steam stop valve (iii) Steam trap

8.55. Define the Boiler accessories.

8.56. Explain

(i) Feed check valve (ii) Blow off cock

8.57. State the difference between

(i) Externally fired internally fired (ii) Forced circulation and Natural circulation

(iii) High pressure and Lower pressure (iv) Stationary Portable

(v) Single tube and multi tube.

Give at least one example of each type.

8.58. What do you understand by steam jet draught? Where is it generally employed. State advantages of it.

8.59. Draw neat pencil sketch of any one type of boiler you have studied. Label the parts and describe its working

8.60. Explain the function and working of

(i) Superheater (ii) Spring loaded safety valve

(iii) Air preheater

8.70. Bring out a comprehensive comparison between FT and WT Boilers.

8.71. Explain the following terms an applied, to a St. generator

(i) Grate (ii) Shell

(iii) Mounting (iv) Heating surface

(v) Marble (vi) Demper.

8.72. What is IBR? Explain some of its provisions.

8.73. Explain in detail classification of boilers.

8.74. Draw-neat sketches of following and explain their working

(i) Feed check valve (ii) Water level indicator

(iii) Economiser

8.75. Enumerate the factors to be considered in selections of a boiler.

8.76. Write the location and function of the following components used in a boiler

(i) Fusible plug (ii) Blow off cock

(iii) Superheater (iv) St. stop valve

8.77. What is equivalent of evaporation of a boiler? What is its importance?

8.78. What ave the factors involved in the selection of a boiler?

In what circumstances W.T boilers are used in preferance to F.T. boilers?

Give reasons.

Suggest a boiler for a small chemical industry which is in need of 500 kg/hrof steam at 12 atm pressure.

8.79. Write short note on :

(i) Classification of boilers (ii) IBR

PROBLEMS FOR PRACTICE

1. A boiler generates steam at 18 bar and 325° C feed water is supplied at 41.45° C. The thermal efficiency of the boiler is 80%. It uses furnace oil of C.V. 45500 kJ/kg. The steam is supplied to a turbine producing 100 kw power, having a specific steam consumption of 10 kg/hr-kW. Calculate the furnace oil consumption in kg/hr and equivalent of evaporation. (**Ans.** 395.71 kg/hr, 6382.55 kg/hr)

2. A boiler has a chimney of 35 m height. It produces a draught of 20 mm of water column. The mean temperature of flue gases is 370°C and the boiler house temperature is 34° C. Estimate the air:fuel ratio. Does this boiler satisfy the condition for maximum discharge? Also find the height of hot gas column under maximum discharge condition.

[**Ans.** (i) The chimney nearly fulfils the condition for maximum discharge,
(ii) $H' = H = 35$ m of hot gas columb]

3. A boiler generates 5000 kg/hr of steam at 16 bar and 300°C from feed water at 30°C. Coal used in 600 kg/hr of C.V. 30,000 kJ/kg. Find

(a) Equivalent of evaporation. (b) Boiler efficiency **[Ans.** (a) 10.75 kg/kg of coal (b) 80.9%]

4. The equivalent of evaporation of a boiler from and a 100°C is 10.4 kJ/kg of fuel. The C.V. of fuel is 29800 kJ/kg. Determine efficiency of boiler. If the boiler produces 15000 kg of steam per hour at 20 bar from a feed water at 40° C and the fuel used 1650 kg/hr. Determine the condition of steam produced.

[Ans. (a) 78.76 % (b) $x = 0.9734$]

5. The following readings were noted during an hour trial on a boiler.

Coal burnt : 750 kg

Feed water : 7000 kg

C.V. of coal : 33500 kJ/kg

Feed water temperature entering the economiser : 30ºC

Leaving the economiser : 90ºC

Boiler pressure : 10 bar

Steam dryness : 0.96

Temperature of steam leaving superheater : 250ºC

Determine the thermal efficiency of the plant and also calculate the enthalpy received by feed water in various components as a % of total enthalpy increase.

[**Ans.** (i) $\eta_{th} = 77.92\%$ (ii) Enthalpy received in economizer = 230.0 kJ/kg i.e., 8.23%
(iii) Enthalpy received in boiler = 2138.8 kJ/kg i.e., 82.91%
(iv) Enthalpy received in superheater = 247.4 kJ/kg i.e., 8.86%]

6. The following observations were obtained in a boiler trial:

Steam condition = 15 bar and 300°C

Coal used = 95 kg/hr having C.V. = 34,000 kJ/kg

Feed water supplied = 680 kg/hr at 20°C.

Ash and unburnt coal collected in ash pit = 4 kg/hr.

C.V. of unburnt coal = 2200 kJ/kg.

Flue gas formed = 18 kg/kg of coal burn

Flue gas formed = 18 kg/kg of coal burnt

Flue gas temperature = 343°C

Boiler room temperature = 28°C

Mean specific heat of flue gases = 1.025 kJ/kg-K

Determine:

(i) Boiler efficiency (**Ans.** 62.21%)

(ii) Heat carried away by the flue gases in kJ/min (**Ans.** 9201.9375 kJ/min)

(iii) Heat unaccounted in kJ/min and in percentage. (**Ans.** 10994.615 kJ/min, 20.4 %)

Chapter 9 High Pressure Boilers and Fluidised Bed Combustion Boilers

9.1 INTRODUCTION

9.1.1 Steam Generator

As we have studied, **boiler is a steam generator** or it is a closed vessel, in which water is heated and converted into steam.

Actually steam generator is a complex combination of boiler, economiser, air preheater, superheater, re-heater, etc. In addition it has various auxiliary equipments like stokers, pulverisers, burners, fans, emission control devices, water treatment plant, ash handling equipments etc.

9.1.2 High Pressure Boilers

These are used to increase the efficiency of the plant. These are developed because of rising cost of fuel and restrictions on air pollution. Modern power generation plants generally use high pressure boilers. These high pressure boilers have pressures above 140 bar and the temperature may be up to 540 – 610°C. High pressure boilers use both natural and forced circulations. Forced circulation is achieved with the help of centrifugal pumps. By using this forced circulation, because of high velocity of water scale formation in the tubes and boiler drums is reduced.

9.2 ADVANTAGES OF HIGH PRESSURE BOILERS

The advantages of high pressure boilers are as follows :

1. Because of high velocities, tendency of scale formation in the tubes is reduced.
2. Small light weight tubes of better heating surface arrangements can be used. This makes the unit compact and reduces erection time and cost of boiler.
3. Due to forced circulation, for heating of all the parts is uniform, which avoids overheating and development of thermal stresses.
4. There is more flexibility in the arrangement of furnace, tubes and other boiler components.
5. The steam can be raised quickly to meet the variable load requirements.
6. The efficiency of the plant is increased.
7. A very rapid start from cold is possible if an external power supply is available. Hence the boiler can be used for carrying peak loads or standby purposes with hydraulic power stations.

9.3 TYPES OF HIGH PRESSURE BOILER

Following are the types of high pressure boilers:

(i) La Mont

(ii) Benson
(iii) Schmidt–Hartmann boiler
(iv) Loeffler boiler
(v) Velox boiler.

9.3.1 La Mont Boiler

This is a modern high pressure, water tube boiler employs forced circulation. This was first introduced by La Mont in 1925. This is generally used in Europe and USA (Ref. Fig. 12.1)

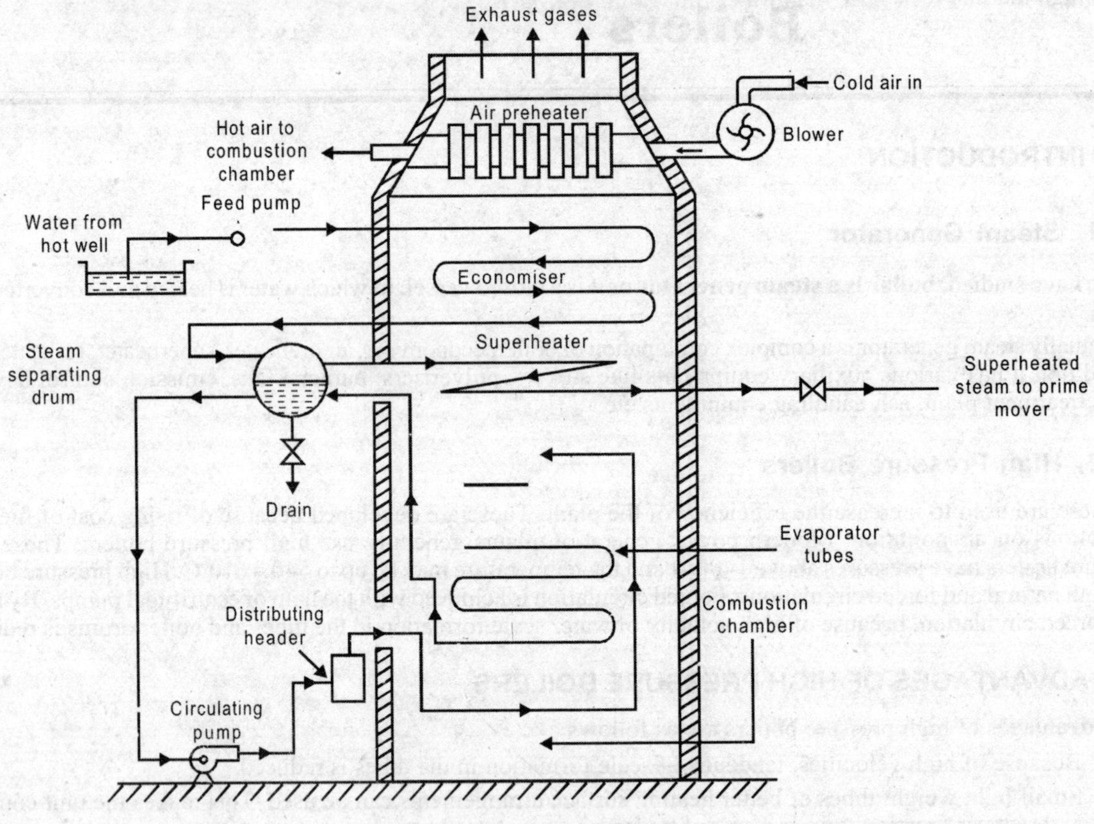

Fig. 9.1. *La Mont boiler.*

Working. This boiler consists of an economiser, boiler drum, superheater, evaporator, air preheater etc. as shown in Fig. 9.1.

The feed water from the hot well is pumped by the feed pump through the economiser and preheated water is supplied to the boiler drum.

The water is heated up by the hot gases flowing in the upward direction, which are later exhausted through the chimney high in the atmosphere to reduce air pollution.

This boiler drum also acts as steam separating drum. The circulating pump, which is operating at a pressure of about 2.5 bar above the boiler pressure, pumps the water to the distributor header.

This distributor header has nozzles, which distribute the feed water into the evaporator.

The water in the evaporator tubes gets evaporated. And the mixture of water and steam is forced into the boiler drum, where the steam is separated. This steam is passed through the superheater, so as to have superheated steam and the superheated steam goes to the steam turbine for power generation.

The air preheater is also provided in the flue gas passage for the economic combustion of fuel in the furnace. These boilers can generate steam up to critical pressure.

9.3.2 Benson Boiler

The main problem experienced with the La Mont boiler is the formation and attachment of bubbles on the inner surfaces of the tubes. These attached bubbles to the tube surfaces reduce the heat flow and steam generation, as it offers high thermal resistance than water film. (Ref. Fig. 9.2)

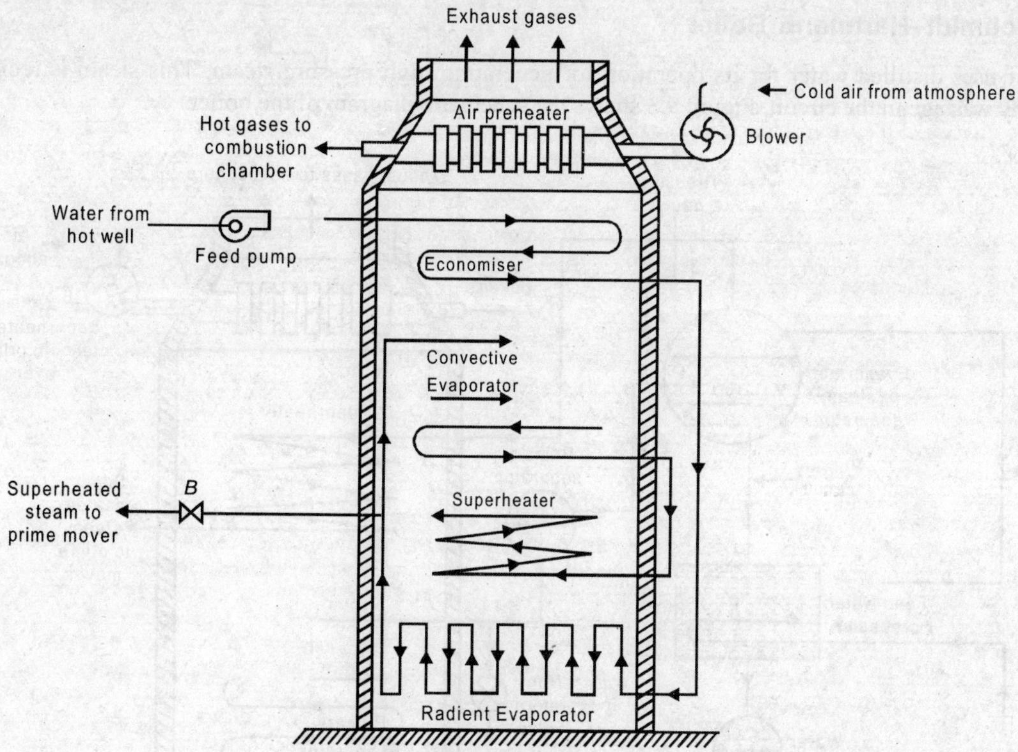

Fig. 9.2. *Benson boiler.*

In order to overcome this difficulty, Benson developed the boiler in 1927. In this boiler the above difficulty is overcome by raising the boiler pressure to critical pressure of steam. As at critical pressure steam and water co-exist at the same density and no bubble formation occurs. Further at critical pressure the enthalpy of evaporation (Latent heat) is zero. The water thus gets converted into steam directly without passing through evaporation process.

Figure 9.2 shows the Benson boiler. It is a high pressure, water tube, forced circulation boiler. It has unique characteristic of absence of drum and entire process of heating, steam formation and superheating takes place in a single continuous tube.

Working. Feed water pump feeds water to the economiser, from where it goes to the radient heating section. Here most of the heat is transmitted and water gets heated almost to the critical temperature. In the evaporative section i.e., in radient evaporator and connective evaporator, evaporation is completed and superheating begins. In the superheater the final desired temperature is obtained.

Advantages

1. Due to the absence of drum, initial cost is low.
2. Circulation pump is absent.
3. The boiler is light in weight because of absence of drum.
4. The boiler can be started within 15 minutes and can be put in service.
5. The boiler can be operated at supercritical pressure also.

Disadvantages

1. It requires specific means of washing the tubes to avoid solid deposition.
2. The tubes are likely to get overheated when the flow is not sufficient.
3. It requires skilled operators to co-ordinate the operations.

9.3.3 Schmidt–Hartmann Boiler

This boiler uses distilled water for its operation for generating high pressure steam. This steam is recirculated without any wastage in the circuit. Figure 9.3 shows the schematic diagram of the boiler.

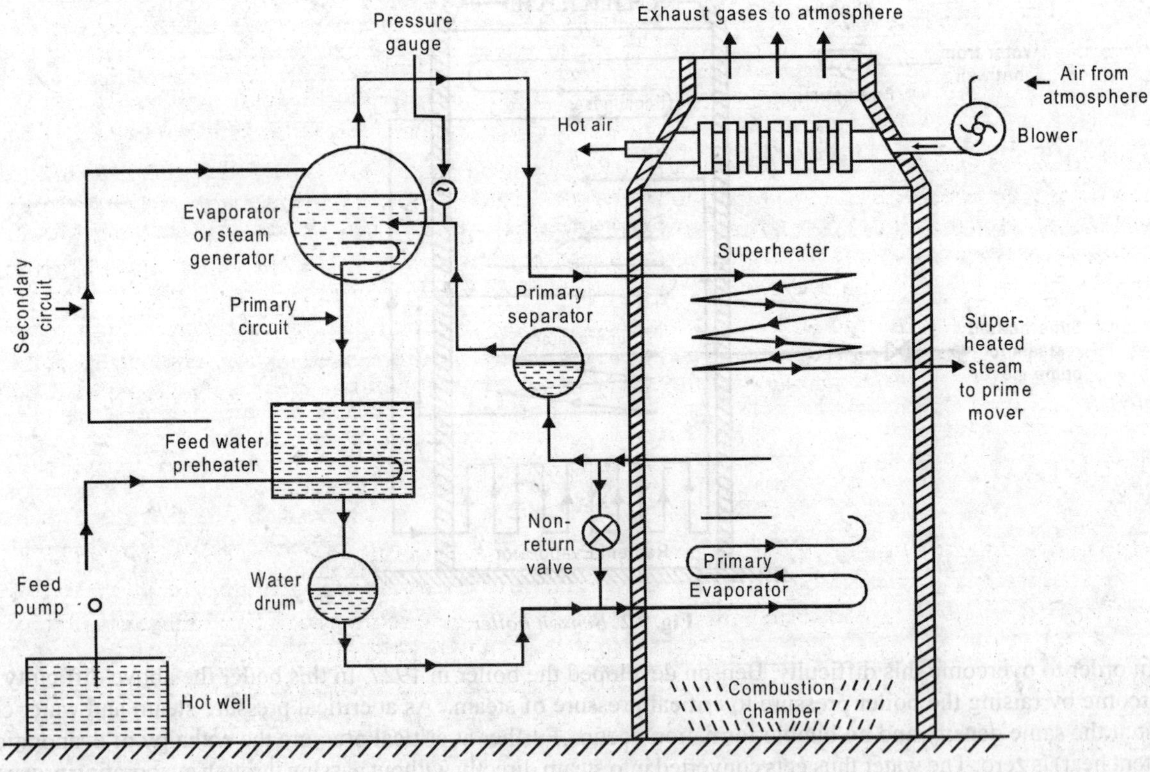

Fig. 9.3. *Schmidt-Hartmann boiler.*

In this case high pressure steam generated from pure water is used for generating low pressure steam from impure water or normal water and is used for power generation.

As shown, distilled water from the water drum flows into the primary evaporating tubes by natural circulation. The hot gases from the cumbustion chamber passes over the evaporator tubes and the steam upto a pressure of 100 bar is generated from distilled water. This steam then goes to the primary separator. And then this steam passes through the tubes provided in the steam generator, which are submerged in impure water in the steam generator. Here heat will be transferred from steam to impure water and the impure water is converted into steam at about 60 bar pressure.

This low pressure steam generated from the impure water is then passed through the superheater for superheating. Finally the superheated steam is supplied to the steam turbines for power generation.

Note that the condensate of high pressure steam from below the boiler drum passes into the feed water heater and then to the water drum.

It is also to be noted that the impure water from the hot well will be pumped by the feed pump and is made to pass through the tubes submerged. Where primary heating of this impure water takes place and preheated water is supplied to the steam generator.

As shown air preheater is also provided, which supplies preheated air for combustion.

Advantages

1. It can use impure water for steam generation.
2. Water softening plant is not required.
3. There is rare chance of overheating of components of primary circuit (distilled water circuit) as there is no danger of sạlt deposition. Hence long life of these parts.

9.3.4 Loeffler Boiler

It is a modern water boiler utilizing forced circulation. It differs from other boilers as it has two superheaters viz., Radient superheater and convective superheater. (Ref Fig. 9.4)

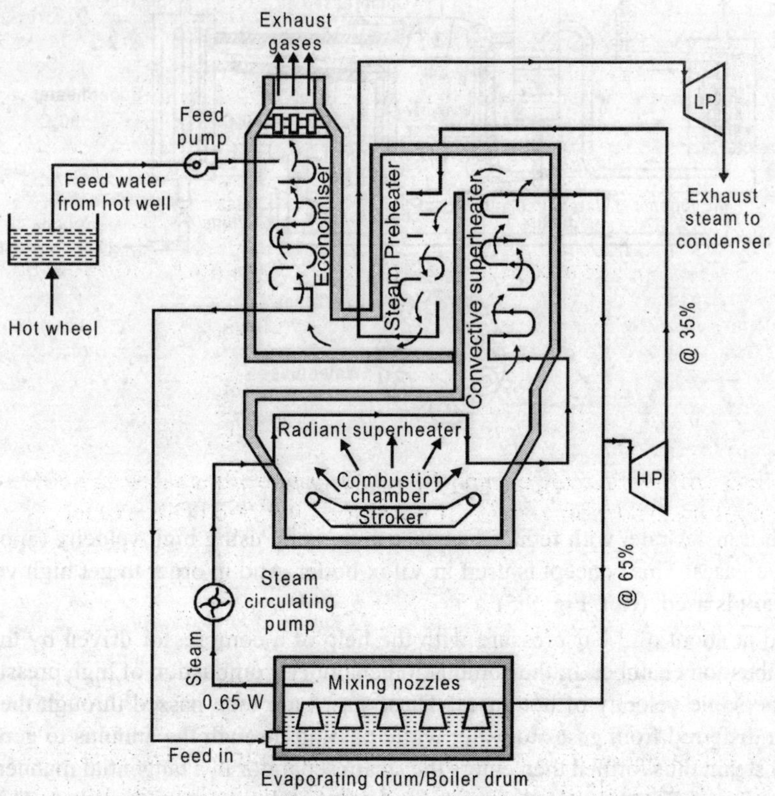

Fig. 9.4. *Loeffler boiler.*

Radient superheater is located above the combustion chamber, so that it receives intense radiations heat.

The feed pump supplies feed water to the economiser, where the water is heated by the hot gases and the hot water will be supplied to evaporating/boiler drum. Note that a part of superheated steam (@ 65%) generated is passed to the evaporating drum through the nozzles where it mixes with feed water and steam is generated.

The saturated steam from the boiler or evaporating drum is drawn by steam circulating pump and is passed through the radient superheater and then convective superheater. About 35% of steam coming out from the superheater is supplied to high pressure turbine. The steam coming out from the turbine is then passed through steam reheater before supplying to low pressure turbine as shown in Fig. 9.4.

As already mentioned about 35% of total superheated steam is used for power generation and remaining 65% is used in the evaporating drum for generating steam.

It has a steam-generating capacity of 90–95 T/hr and steam pressure upto 150–155 bar.

9.3.5 Velox Boiler

It is known fact that, the heat transfer rate from the flue gasses can be increased by increasing the velocity of gases.

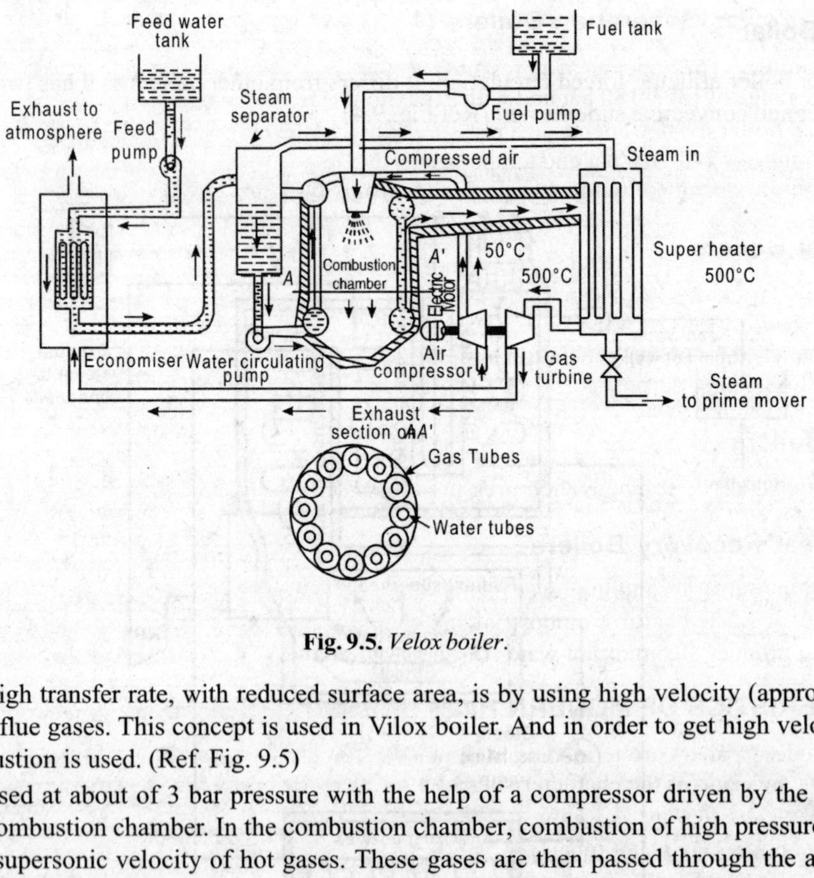

Fig. 9.5. *Velox boiler.*

Thus to have high transfer rate, with reduced surface area, is by using high velocity (approximately or above sonic velocity) of flue gases. This concept is used in Vilox boiler. And in order to get high velocity of flue gases, pressurised combustion is used. (Ref. Fig. 9.5)

Air is compressed at about of 3 bar pressure with the help of a compressor driven by the gas turbine before supplying to the combustion chamber. In the combustion chamber, combustion of high pressure air and fuel takes place to produce supersonic velocity of hot gases. These gases are then passed through the annulus of tubes as shown. The heat is transferred from gases to water while passing through the annulus to generate the steam. The mixture of water and steam thus formed then enters the steam separator in a tangential manner. It forms a vortex in the separator and due to centrifugal action, the moisture content of the steam is separated. This moisture separated is sent back to feed water line.

The dry steam so obtained in the steam separator is sent to the superheater, where it is further heated by means of hot flue gases coming out from combustion chamber.

The gases coming out of superheater are used to run the gas turbine, which drives the compressor.

The exhaust gases coming out from the gas turbine are passed through the economiser to preheat the water and finally they are exhausted through the chimney.

Advantages

(1) If can start quickly.

(2) The boiler design is compact and have flexibility towards varying load.

Disadvantages

Solid fuel can not be used. It operates only with liquid or gaseous fuel.

9.4 TYPES OF BOILERS

Boilers are broadly of the following types:

(1) Utility boilers or Power Generation boilers. (2) Industrial boilers. (3) Marine boilers.

(4) Waste heat recovery boilers.

9.4.1 Utility or Power Generation Boilers

(*a*) **Subcritical boilers:** They operate at a pressure below critical pressure of steam i.e., 221.2 bar. They usually operate in a pressure range of 130 – 180 bar and with a maximum steam temperature of 540 °C.

(*b*) **Supercritical boilers:** They operate at a pressure above critical pressure of steam. These usually operate in the pressure range of 240–300 bar and temperature 600°C.

The steam generation capacity of modern boilers range from 500 T/hr to 5000 T/hr.

9.4.2 Industrial Boilers

They are relatively low pressure, low capacity boilers which operate at a pressure range of a few bar to 100 bar and capacities ranging from few tons to 500 T/hr.

The steam is either used for power generation on small scale or for process heating in chemical, textile industries and in refineries or for heating purpose.

9.4.3 Marine Boilers

They are used on ships to drive turbines. Their pressure range is normally 60–100 bar and temperature 550 °C.

9.4.4 Waste Heat Recovery Boilers

These are used both in industries and in power plants. They use waste heat such as from a blast furnace, exhaust from the gas turbine, or waste heat from industrial and chemical processes. The waste heat is passed over a heat exchanger surface to produce steam or hot water for the required use.

9.5 CHARACTERISTICS OF MODERN HIGH PRESSURE BOILERS

Even though the modern boilers are termed as high pressure boilers, it does not mean them fully. Because high pressure operation is only one of the characteristics of modern boilers used for power generation.

Other aspects/characteristics are as follows,

(1) Modern power boilers are water tube.

(2) They use forced type of water circulation.

(3) They use feed water heaters.

(4) They use modern methods of firing viz. pulverised firing; cyclone furnace; fluidised bed combustion.

(5) They use F.D. and I.D. fans for creating the necessary draught.

(6) They use water treatment plant, demineralisation plant to purify boiler feed water to minimise scale formation and salt deposits in the boiler tubes.

(7) They use various controls for pressure temperature water level etc.

(8) They use complex assembly of evaporater superheaters, reheaters, economisers and air heaters etc.

9.6 FLUIDISED BED COMBUSTION SYSTEM (FBCS)

9.6.1 Introduction

(a) Oil and gas prices increased in recent decades.

(b) Coal is available in large quantities.

(c) Pulverized coal burning system is costly and the combustion takes place at 1400–1600ºC at which NO_x is produced.

(d) At high temperaure, ash forms a slag on superheater tubes and causes fouling, corrosion and reduces heat transfer rates.

To overcome these difficulties a new coal combustion system is developed and is called as Fluidized Bed Combustion System, (FBCS).

In this system any type of coal can be burnt efficiently in the temperature range of 900–1000ºC. At these temperatures production of pollutant like NO_x is less, and slag formation is almost nil.

It is to be noted that FBCS is introduced in 1950s and aims at removing sulphur from coal during combustion process itself and reducing to formation of NO_x.

9.6.2 Principle of Working of Fluidised Bed Combustion

Figure 9.6 shows the simple fluidised bed combustion which has a distributor or perforated support plate on which finely divided particles of fuel (6 mm–20 mm) are provided and the particles are in collapsed state. Air

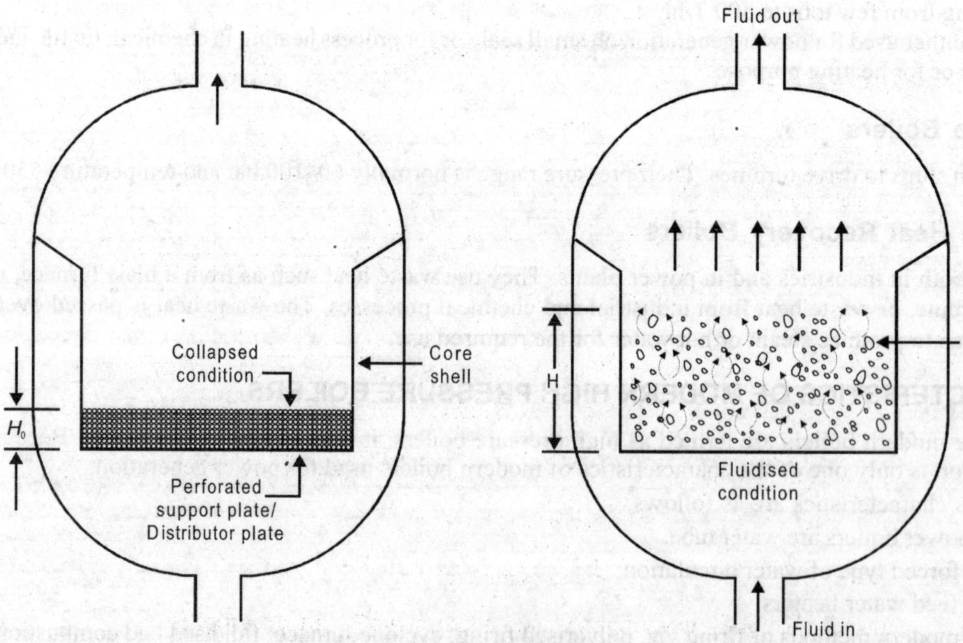

Fig. 9.6. *Schematic diagram of a fluidized bed.*

is passed in the upwards direction with low velocities. The pressure drop across the combustion chamber can be measured with the help of a manometer.

The velocities of air is gradually increased and the logarithmic plot of ΔP and velocities C is shown in Fig. 9.7.

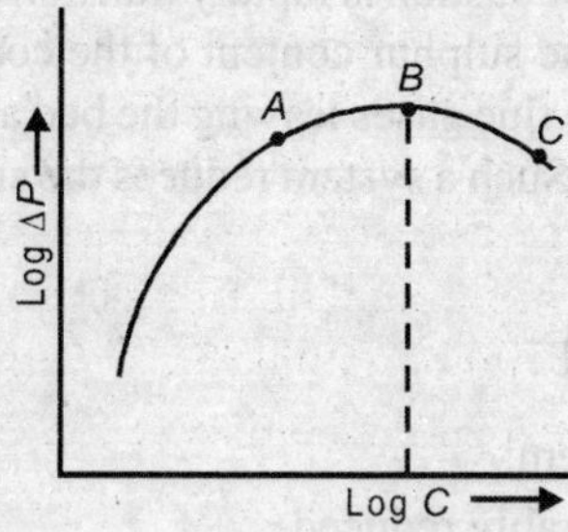

Fig. 9.7

Up to A — The bed is stable.

Between A and B — The bed is unstable particles adjust their position to present as little résistance to flow as possible.

At B — The particles are just in contact with each other.

Beyond B – Beyond this point – on further increase in velocity, the particles start moving freely in the bed with frequent collisions with each other. *Point B is called the point of fluidisations and the corresponding velocity is called the minimum fluidisation velocity.*

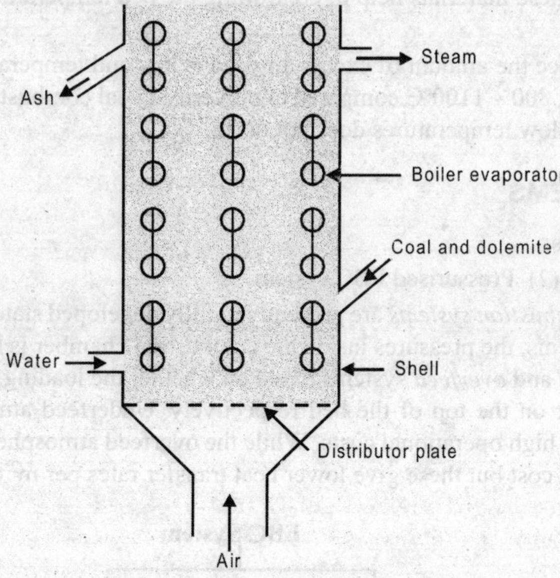

Fig. 9.8. *Arrangement of fluidised bed system.*

Beyond point B, on further increase in velocity, the resistance to flow decreceses as freely suspended in gas stream. In this state, the bed continues to expand and it becomes turbulent; also there is rapid mixing of particles. Under these conditions the mixuture of bed and gas behaves like a fluid and the burning of fuel in such a state is called **fluidized bed combustion.**

The simple schematic diagram of fluidised bed combustion system is shown in Fig. 9.8.

It consists of a shell with a distributor plate at its bottom with holes in it. The shell carries the boiler evaporator tubes which remain immersed in the fluidized bed as shown. Coal of 6 to 20 mm size with *dolemite* or lime stone are fed on the distributor plate. Air is supplied from the bottom at a velocity more than the minimum velocity of fluidisation. The heat energy released due to combustion is rapidly transferred at high rates to water in the evaporator tubes and the generated steam is taken out. The sulphur content of the coal is converted into SO_2 on burning and SO_2 is mostly absorbed by dolemite. Hence the flue gases leaving the bed are almost free from SO_2 gas. The molten ash formed is taken out from the top as shown. Such a system reduces the size of the boiler considerably due to high rates of heat transfer.

9.7 ADVANTAGES OF FBC SYSTEM

Following are the advantages of the FBC system:

1. Size of the plant and its cost is considerably reduced.
2. Coals with high ash contents can be used.
3. The low temperatures of combustion in FBC system allows ash to remain soft and less abrasive.
4. SO_2 emissions are much less due to addition of dolemite.
5. Combustion efficiencies obtained are as high as 99.5 %.
6. Its initial and operational costs are low.
7. The rates of heat transfer are much higher since the system behaves like a violently burning liquid.
8. Volatilisation of alkali compounds does not occur, therefore there deposits on boiler tubes is almost absent.
9. Coal used is 2 to 3 % of bed volume. Remained space is occupied by the inert material like sintered ash, fused alumina, sand etc. These materials help in controlling the bed temperatures in the range of 800ºC to 1000ºC.
10. NO_x emissions are low since the amount of excess air used is less and temperatures maintained are low in FBC system in the range of 800 – 1100ºC compared to pulverized coal combustion system. This is so as the fixation of free nitrozen at low temperatures does not occur.

9.8 TYPES OF FBC SYSTEMS

The FBC systems are classified as:

(1) Atmospheric FBC system. (2) Pressurised FBC system.

Atmospheric fluidized bed combustion systems are presently in fully developed state and these are being used in many power plants. In these systems, the pressures inside the combustion chamber is kept atmospheric. These are for further classified as *underfeed* and *overfeed* systems based on whether the loading of the fuel and supply of air is from the bottom of the bed or on the top of the bed respectively. Underfeed atmospheric FBC systems are compact in design, but, they have high operational costs. While the overfeed atmospheric FBC systems are simple in operation with less operational cost but these give lower heat transfer rates per m^2 of area of the bed.

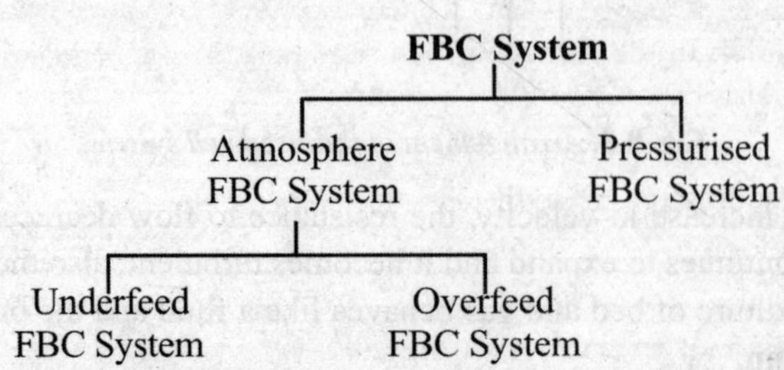

Pressurised FBC system is the latest development used for generation of power in which air is supplied under pressure of 10 bar approximately. With the help of an air compressor driven by a gas turbine as shown in Fig. 9.11. The supply of air under pressure results into better rates of heat transfer compared to atmospheric FBC system.

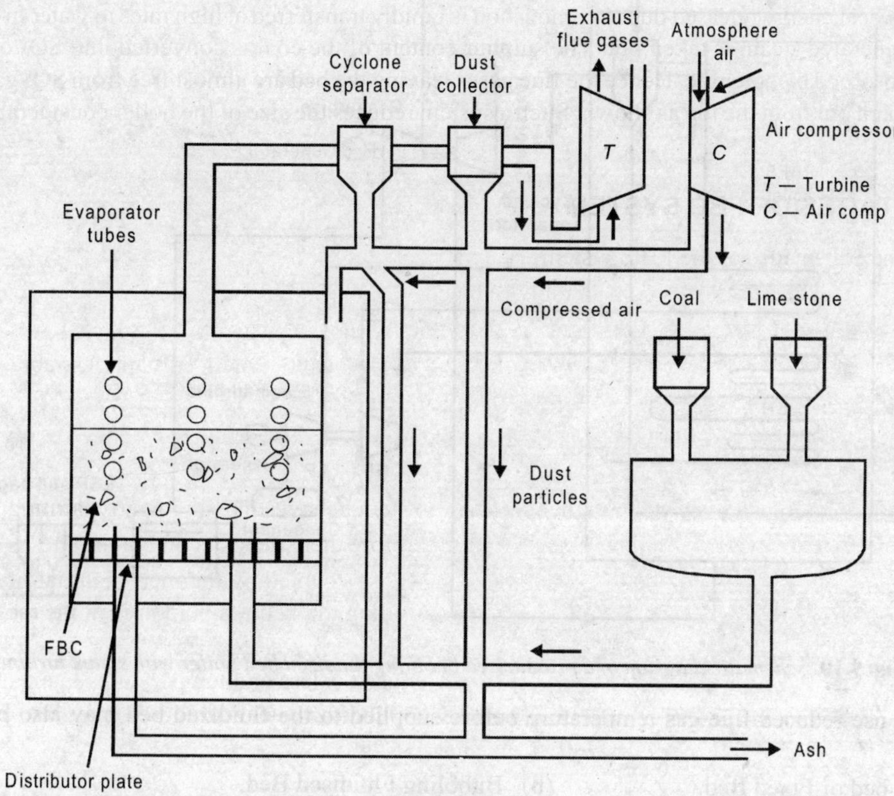

Fig. 9.9

The sketch is self explanatory. Note that the cleaned gas is supplied to the gas turbine producing power required by the air compressor which supplied pressurised air to the combustion chamber. FBC systems have various designs e.g.,

(a) FBC system with gas turbine (Ref. Fig. 9.9)

(b) FBC system with steam turbine as below.

9.9 PRESSURIZED BUBBLING FLUIDIZED BED BOILER (PBFB)

Figure 9.10 shows the PBFB boiler. In the furnace pressure is increased by the air compressor driven by the gas turbine. (Fig. 9.10 is self-explanatory)

Advantages of pressurized FBC system over atmospheric FBC system are as following

(a) It allows increased coal loading per m^2 in the combustion chamber since the burning rate of fuel is higher.

(b) It required reduced air velocity.

(c) No_x, emissions are reduced.

(d) It has improved overall thermal efficiency of the plant.

(e) The overall size of the plant is considerably reduced, hence the overall cost of the plant.
(f) Better desulphurisation is accomplished.

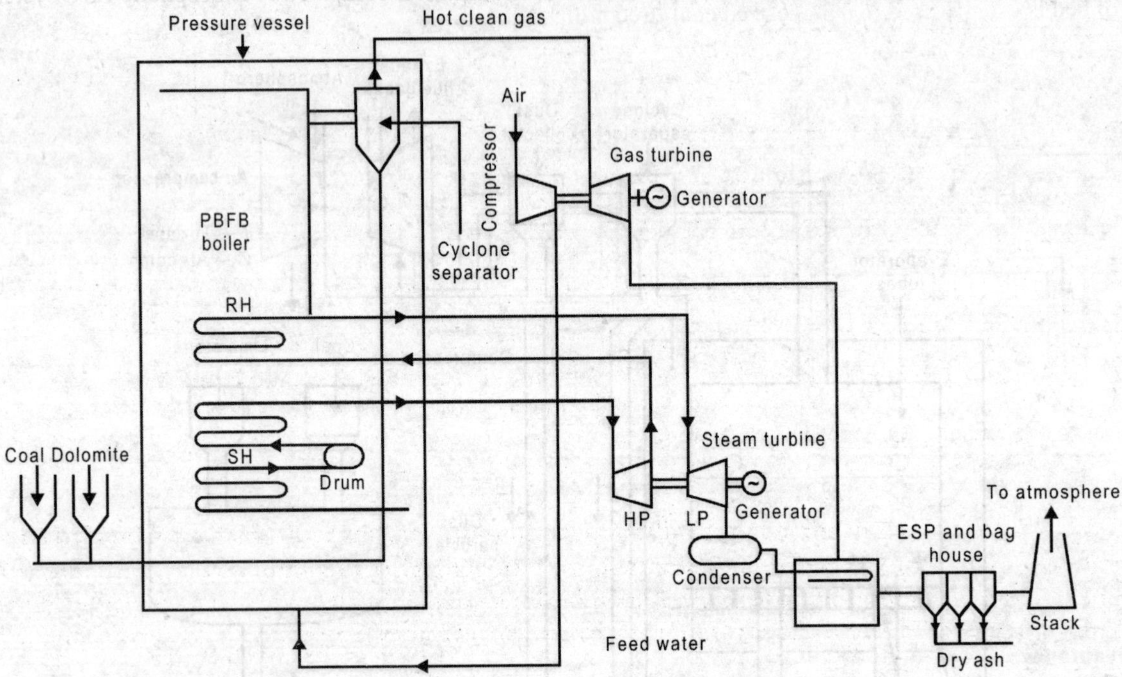

Fig. 9.10. *Schematic diagram of a pressurized bubbling fluidized bed boiler with steam turbine.*

It allows to use reduced flue gas temperature before supplied to the fluidized bed may also be classified or described as—

(a) Packed bed or Fixed Bed. (b) Bubbling Fluidised Bed.
(c) Fast or Circulating Fluidized Bed. (d) Transport Reactor Bed.

(a) Packed or Fixed Bed. It consists of a bed of stationary particles on a perforated grid or distributor plate, through which gas is flowing. In a moving packed bed which is used in stoker fired boilers, the solid particles move with respect to walls, but not with respect to each other. The velocity of air is 0.3 to 1.2 m/sec.

(b) Bubbling Fluidized Bed. When air velocity is more than the minimum fluidisation velocity, the bed behaves like a liquid fluid. The pressure drop across the bed equals the weight of the bed per unit area (m^2). The air flows through the bed in the form of bubbles, which pass through—what is called—the emulsion phase bed. The size of the bubbles increases as it rises up in the emulsion phase. They may carry some solid particles which they throw up into the space above the bed, called *Free Board*. Some particles fall back to the bed. Beyond a certain height above the bed called transport disengaging height (TDH), the particles do not full back and are carried away by the gas. This is called *Elutriation* of the solids. Air velocity in this zone is 1-1–4 m/s.

(c) Flast Fluidized Bed (Circulating Fluidized Bed). The air velocity in circulating bed combustion is between 4–10 m/s. In circulating bed the solid particles elutriated with are return back to the bed at sufficiently high rate. The bed column is divided into two regions. The core of column consists of ditute gas which carries the particles upwards. The walls or annulus of the bed consists of denser medium in which the solid particles form clusters and fell down to the bed. The cluster fall down a certain distance and disintegrate to be carried away upwards again. Thus a circulating motion of solids is established.

(d) Transport Reactor Bed. When the velociety of air exceeds 10 m/s the solid particles are entrained in, and elutriated with, flowing air. They do not fall back as the air velocity is more than the terminal velocity of the bed. This type of bed is usually used in a pulverized coal fired boilers.

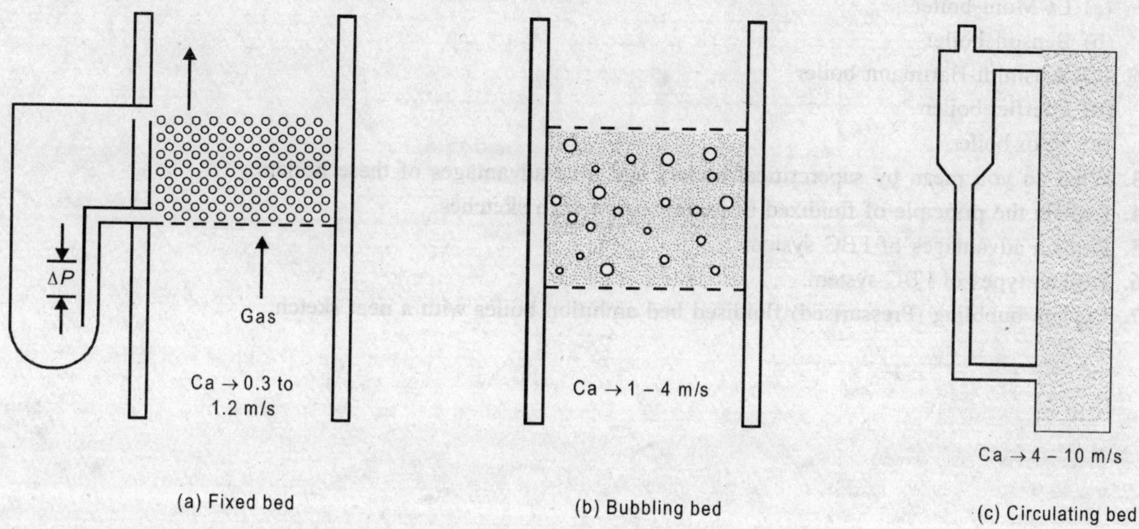

Fig. 9.11

Disadvantages

(a) Feeding of lime stone and coal into the bed in the required proportion is difficult especially under variable load conditions.

(b) Control of carbon carry over with the gases is also difficult.

(c) Erosion of furnace walls and immersed water tubes takes place.

(d) The cold water tube may quench the combustion process.

9.10 SPECIFICATIONS OF SOME BOILERS USED IN THERMAL POWER PLANTS IN INDIA

The specifications of some of the boilers used in India in thermal power plants are as follows.

Table 9.1

Name of the power plant	*Working pressure, bar*	*Maximum temperature, °C*	*Steaming capacity tons/hr*	*Electrical power output MW*
Bokaro (2 units)	60	500	150	50
Ramagundam	100	500	300	66
Chandrapura	135	540	500	140
Trombay old unit	175	670	520	150
Trombay (new unit)	178	540	1674	500

THEORY QUESTIONS

9.1. What do you mean by high pressure boiler? Give the advantages of high pressure boilers.

9.2. Explain following high pressure boilers with net schesmatic diagram.

(a) La-Mont boiler

(b) Benson boiler

(c) Schsmidt-Hartmann boiler

(d) Loeffler boiler

(e) Velox boiler.

9.3. What do you mean by supercritical boilers and give advantages of these boilers.

9.4. Explain the principle of fluidized bed combustion with sketches.

9.5. Explain advantages of FBC system.

9.6. Explain types of FBC system.

9.7. Explain bubbling (Pressurised) fluidised bed ambition boiler with a neat sketch.

Chapter 10 Simple Vapour Power Cycles

10.1 INTRODUCTION OR IDEAL STEAM PLANT CYCLES

Figure 10.1 shows the simple steam power plant. The heat energy released due to the combustion of fuel will be utilised in the boilers for converting water into steam (i.e., vapour) and this steam is then expanded into the steam engine/steam turbines to obtain useful work. The steam after producing work output, is generally condensed in the condensers. If the condensate is pure, then it is pumped as feed water to the boiler by means of a feed pump.

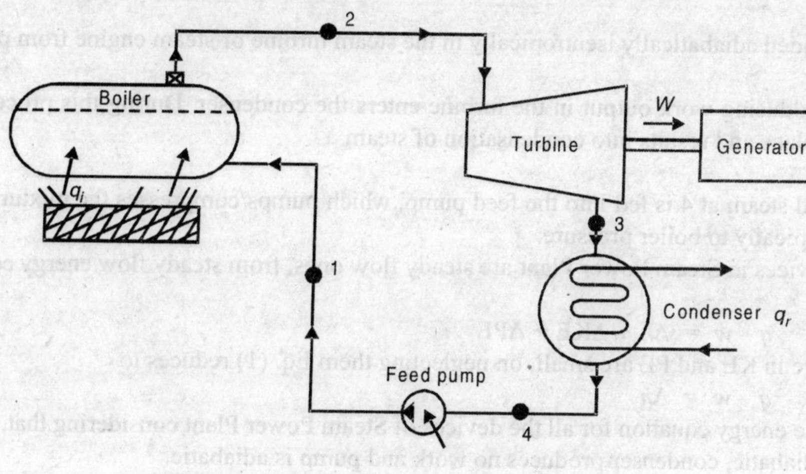

Fig. 10.1. *General arrangement of steam power plant.*

10.2 CARNOT CYCLE AND STEAM PLANT

As we know, the Carnot cycle has the greatest possible efficiency between any two given temperature limits T_1 and T_2 and is given by $\eta_{th} = \frac{T_1 - T_2}{T_1}$. It is important therefore to see whether Carnot cycle can be successfully applied to the steam plant.

Referring the *T–S* diagram as shown in Fig. 10.2.

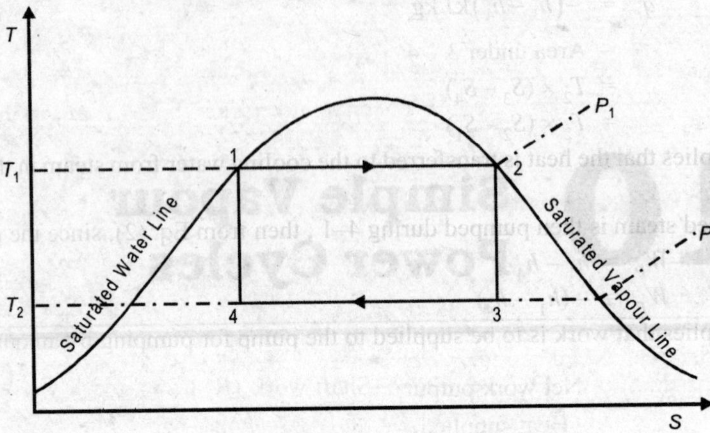

Fig. 10.2. *Carnot cycle on T–S diagram.*

Process 1–2:

This is an isothermal heat addition process and during this, water in the boiler gets converted into steam at constant temperature and pressure.

At point 2, steam is dry saturated.

Process 2–3:

The steam is expanded adiabatically isentropically in the steam turbine or steam engine from pressure P_1 to P_b.

Process 3–4:

The steam after producing work output in the turbine enters the condenser. During this process isothermal heat rejection q_r takes place and results into condensation of steam.

Process 4–1:

Partially condensed steam at 4 is fed into the feed pump, which pumps/compresses the mixture of wet steam and condensate isentropically to boiler pressure.

Since all the devices in Steam Power Plant are steady flow ones, from steady flow energy equation (SFEE) for unit mass,

$$q - w = \Delta h + \Delta \text{KE} + \Delta \text{PE} \quad \text{... (1)}$$

Since the change in KE and PE are small, on neglecting them Eq. (1) reduces to

$$q - w = \Delta h \quad \text{...(2)}$$

We can write the energy equation for all the devices of Steam Power Plant considering that, boiler produces no work, turbine is adiabatic, condenser produces no work and pump is adiabatic.

$\therefore$ Heat added in the boiler q_1 during 1–2

$$q_i = h_2 - h_1 \quad (\because W = 0 \text{ for boiler})$$

i.e.,

$$q_i = h_2 - h_1 \text{ kJ/kg}$$
$$= \text{Area under } 1\text{–}2$$
$$= T_1 \times (S_2 - S_1) \quad \text{...(3)}$$

Work produced by the turbine W_T during the expansion process 2–3 and since expansion adiabatic.

$$q = 0$$

$\therefore$

$$0 - W_T = h_3 - h_2$$
$$W_T = h_2 - h_3 \text{ kJ/kg} \quad \text{...(4)}$$

During the process 3–4 isothermal heat rejection takes place and results into condensation of steam (since $W = 0$)

$$q_r = h_4 - h_3$$

$$q_r = -(h_3 - h_4)\text{kJ/kg} \qquad \text{...(5)}$$
$$= \text{Area under } 3-4$$
$$= T_2 \times (S_3 - S_4)$$
$$= T_2 \times (S_2 - S_1)$$

Negative sign implies that the heat is transferred to the cooling water from steam in the condenser by an amount equal to $(h_3 - h_4)$.

Partially condensed steam is then pumped during 4–1 , then from Eq. (2), since the pump is adiabatic $q = 0$

$\therefore$
$$-W_P = h_1 - h_4$$

or
$$W_P = -(h_1 - h_4)$$

Negative sign implies that work is to be supplied to the pump for pumping by amount equal to $(h_1 - h_4)$

$$\eta_{th} = \frac{\text{Net work output}}{\text{Heat supplied}} = \frac{\text{Shaft work } (W_s)}{\text{Heat supplied } q_i}$$

$$\eta_{th} = \frac{\text{Difference between turbine and pump work}}{q_i}$$

$$= \frac{W_T - W_p}{q_i}$$

$$\eta_{th} = \frac{[h_2 - h_3] - [h_1 - h_4]}{[h_2 - h_1]} \qquad \text{...(7)}$$

$$= \frac{(h_2 - h_1) - (h_3 - h_4)}{h_2 - h_1}$$

$$= \frac{T_1(S_1 - S_2) - T_2(S_1 - S_1)}{T_1(S_2 - S_1)}$$

$$= \frac{T_1 - T_2}{T_1} \text{ as for ideal gas.}$$

Even though, the Carnot efficiency is maximum for a given value T_1 and T_2. But because of some inherent practical difficulties, as discussed below, Carnot cycle cannot be applied for the steam plant.

Carnot cycle when applied to S.P.P is practical upto a point that, Isothermal heat addition and the adiabatic expansion of steam in the turbine, which is quite reasonable.

The impractical part is in handling of steam in the condenser and pump. In the condenser, the steam is only partially condensed and the condensation must be stopped at 4. Also the feed pump must be capable of handling both wet steam and condensate.

A slight modification is made to this Carnot cycle which however produces a more practical cycle, called Rankine cycle. This cycle is accepted as, the practical but ideal cycle for the steam plant even though it has reduced thermal efficiency.

10.3 RANKINE CYCLE

To obtain Rankine cycle, modification made to Carnot cycle, is that, instead of stopping the condensation at some intermediate condition, the condensation is completed as shown in Fig. 10.3.

On *T–S* diagram Carnot cycle, would be 1'–2–3–4'. For Rankine cycle, however, condensation is continued until it is complete at 4. At this point there is all water. This can be successfully dealt with by the feed pump.

Process 4–1:

Pumping process, where water of condenser pressure P_b is raised to boiler pressure P_1. The pumping compression process is isentropic. At point (1) temperature is lower than T_{sat}. Subcooled state:

Process 1–2: (1 – 1' and 1' – 2)

During 1 – 1' feed water is heated to T_{sat} (may occur in economiser/boiler itself.) During 1' – 2 heat added q_t at constant temperature and pressure P_1 convert water into steam in the boiler.

Process 2–3:

Isentropic expansion of steam in the turbine from boiler pressure P_1 to back pressure P_b, which results into turbine work W_T.

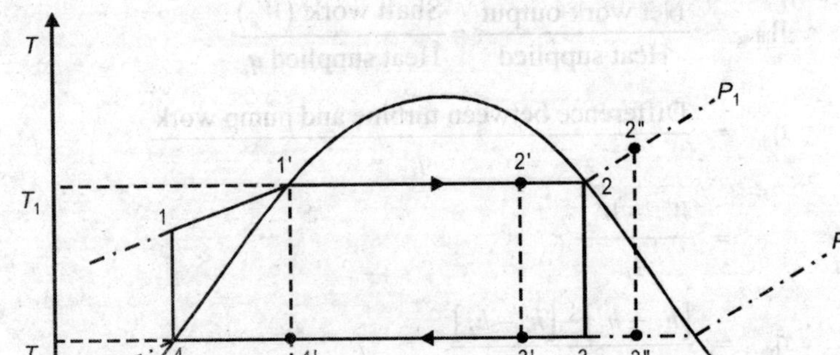

Fig. 10.3. *Rankine cycle.*

Process 3–4:

The steam is condensed at constant pressure P_b in the condenser. The steam rejects heat q_r to the cooling water.

It should be noted that the steam generated in the boiler or steam entering the turbine may be wet, dry-saturated and superheated. Correspondingly the expansion process in the turbine will be 2'–3', 2–3, 2"–3" respectively.

Analysis of Rankine Cycle (Assuming 1 kg of steam)

From Eq. (5) we get $q - w = \Delta h$

Process 1–2:

In the boiler heat q_i converts water into steam

$$q_i = h_2 - h_1 \text{ (since } q = 0)$$

Process 2–3:

In the turbine, when the steam expands, work will be produced.

$$W_T = h_2 - h_3 \text{ kJ/kg (since } q = 0)$$

Process 3–4:

In the condenser condensation will be completed and an amount of heat q_r is rejected to the cooling water.

$$q_r = h_4 - h_3$$

i.e., $q_r = -[h_3 - h_4]$ Negative sign implies heat rejection.

Process 4–1:

Feed pump pumps the water and increases the pressure of water from P_b to P_1.

$$0 - W_P = h_1 - h_4 \quad (\text{Since } q = 0)$$

or

$$W_P = -(h_1 - h_4)$$

Negative sign implies work to be supplied to pump.

Also since pump is a steady flow equipment, we know that work done in a steady flow system.

$$= \int vdP$$

$$\therefore \quad W_P = -\int_{P_b}^{P_1} vdP$$

The specific volume of water remains constant since the effect of pressure is negligible on liquids.

$$\therefore \quad W_P = -V\int_{P_b}^{P_1} dP$$

$$\therefore \text{ Magnitude of } W_P = V[P_1 - P_b]$$

We also know that

$$\text{Specific volume} = \frac{1}{\rho}$$

and for water

$$\rho_w = 1000\,\text{kg/m}^3$$

$\therefore$ Specific volume $= 0.001$ m³/kg. If the pressure are measured in bar then,

$$W_P = 0.001 \text{ m}^3/\text{kg } [P_1 - P_b]\ 10^5 \text{ N/m}^2$$

$$= 0.001\,(P_1 - P_b)\ 10^5 \text{ Nm/kg or J/kg}$$

$$= \frac{0.001(P_1 - P_b)10^5 \text{kJ}}{1000} \text{kJ/kg}$$

$$W_P = \left[\frac{P_1 - P_b}{10} \text{kJ/kg}\right] \quad \text{...(8)}$$

And the efficiency of Rankine cycle,

$$\eta_R = \frac{\text{Shaft work } W_s}{\text{Heat supplied } q_i} = \frac{W_T - W_P}{q_i}$$

$$\therefore \quad \eta_R = \frac{[h_2 - h_3] - [h_1 - h_4]}{[h_2 - h_1]}$$

Also we can write this as,

$$\therefore \quad \eta_R = \frac{[h_2 - h_3] - [h_1 - h_4]}{[h_2 - h_4] - [h_1 - h_4]}$$

Since the pump work $W_P = h_1 - h_4$ is very small quantity, it can be neglected,

$$\therefore \qquad \eta_{th} = \frac{[h_2 - h_3]}{[h_2 - h_4]}$$

Note: For solving the problems we can read the value of $h_4 = h_f$ from steam tables at back pressure P_b, h_2 and h_3 can be determined with the help of Mollier diagram or by calculations from steam tables. Since $W_P = h_1 - h_4$ the volume of h_1 can be determine by using Eq. (8) i.e., by finding W_P.

It should be noted that $\eta_R < \eta_{carnot}$ because the mean temperature of heat addition process in Rankine cycle is lower than the mean temperature of heat addition process in Carnot cycle. Whereas the mean temperature of heat rejection process in both the cycles is same.

In order to improve the efficiency of Rankine cycle, superheated steam has to be used, as represented by the process 2–2". It improves the cycle efficiency, as it increases the mean temperature of heat addition process in the boiler. Also when the superheated steam is used, it reduces the corrosion of turbines.

The Rankine efficiency has the same form as before, from Fig. 10.3 on similar lines

$$\eta_R = \frac{h_2'' - h_3''}{h_2'' - h_4''}$$

and when the steam is wet, then the cycle would be 1 – 1' – 2' – 3' – 4 –1

and $$\eta_R = \frac{h_2' - h_3'}{h_2' - h_4'}$$

10.4 EFFECT OF INLET PRESSURE (BOILER PRESSURE) AND BACK PRESSURE (EXHAUST PRESSURE) AND SUPERHEAT ON RANKINE EFFICIENCY

The effect of inlet pressure, back pressure and the superheat on the performance of Rankine cycle can be studied with the help of *T–S* diagram as shown in Fig. 10.4.

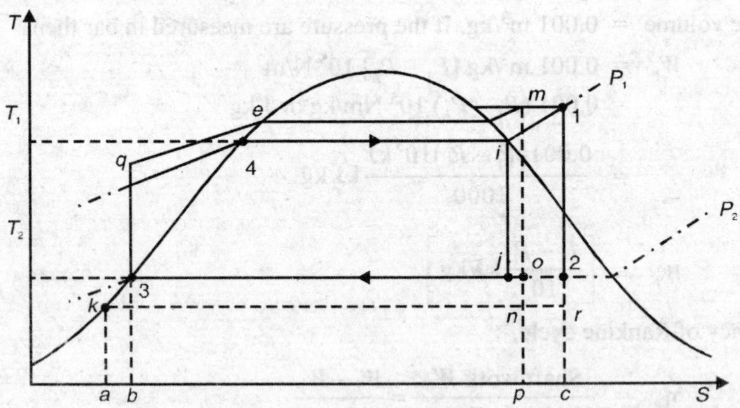

Fig. 10.4. *Rankine cylce varying working conditions.*

10.4.1 Boiler or Inlet Pressure

If the boiler pressure is raised, the evaporation line is raised as the saturation temperature is increased from 4–1 to *em*. The maximum temperature t_1 is kept constant i.e., $t_{m'} = t_1$. 1 – 2 – 3 – 4 –1 is the basic cycle taken for comparison.

With increased boiler pressure, keeping the condenser pressure and temperature constant, the new cycle will be m 03 qem. The amount of heat rejected is decreased given by C20 pc and difference in the heat added $-b341cb$ and *bqempb* is small and the thermal efficiency increases. It is experimentally found that at $p_1 = 170$ bar and $p_2 = 0.14$ bar, the efficiency is maximum and decrease with increase in p_1.

10.4.2 Condenser pressure and temperature is lowered

The new cycle wilt be 1*dk* 41. Work done is increased and heat rejected will be less so that the efficiency will be increased. Nature, however, precisely defines the limit of improvement that may be obtained by this means. Atmospheric temperature and therefore temperature of the cooling water puts the limits of the steam temperature and pressure.

10.4.3 Superheat

If the steam is dry and saturated at the inlet of turbine, and if the steam is superheated then the work is increased and the proportional rise of work is more than the heat supplied so that the efficiency of the Rankine cycle increases.

It is interesting to note that condensing plants commonly operate more efficiently in the winter than in summer because of the lower cooling water temperature in the winter.m

SOLVED EXAMPLES

Example 10.1 *A steam turbine operates on the Rankine cycle. It receives steam from the boiler at 30 bar, 250°C and it is exhausted into a condenser at 0.5 bar. Condensate is returned to the boiler by a feed pump. Calculate,*

(1) Heat supplied in the boiler drum.
(2) Dryness fraction of steam entering the condenser.
(3) Work done.
(4) Steam rate/kWh and
(5) Efficiency of the cycle.

Solution

Since it operates on Rankine cycle,

Refer to Rankine cycle 1–2–3–4, (Fig. Ex. 10.1)

h_4 from tables at 0.5 bar condenser pressure $h_4 = h_f = 340.56$ kJ/kg.

We know that,

$$W_P = \frac{[P_1 - P_b]}{10}\frac{\text{kJ}}{\text{kg}} = \left[\frac{30-0.5}{10}\right]\frac{\text{kJ}}{\text{kg}}$$

$$W_P = 2.95 \text{ kJ/kg}$$

Also, $$W_P = [h_1 - h_4]\,\text{kJ/kg}$$

or $$h_1 = W_P + h_4 = 2.95 + 340.56\,\text{kJ/kg}$$

$$h_1 = 343.51 \text{ kJ/kg}$$

From steam tables, h_2 at 30 bar and 250°C

$$h_2 = 2855 \text{ kJ/kg (2854.8)}$$

and $$s_2 = 6.28 \text{ kJ/kg-K}$$

For isentropic process (2–3)

Entropy before expansion s_2 = Entropy after expansion s_3

$$6.28 = s_{f3} + x_3 \,.\, s_{fg3} = 1.0912 + x_3 \,.\, (6.5035)$$

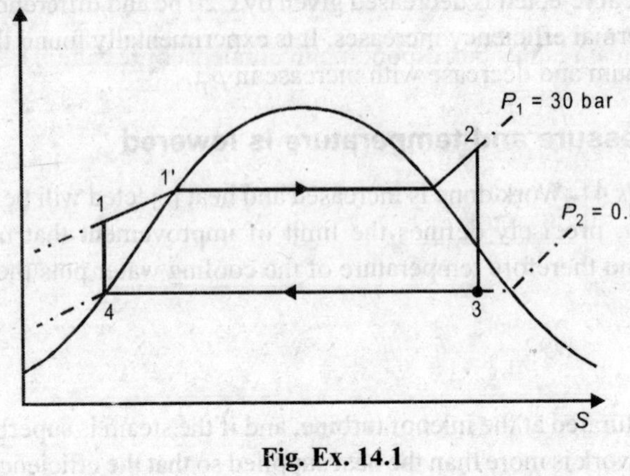

Fig. Ex. 14.1

∴ Dryness fraction of steam entering the condenser

$$x_3 = 0.797 \quad \text{...(A)}$$

∴ $h_3 = h_{f3} + x_3 \cdot h_{fg3}$ corresponding to condenser pressure

$$= 340.56 + 0.797 \times 2305.4$$

$$h_3 = 2177.9 \text{ kJ/kg}$$

Now,

(1) Heat supplied in the boiler drum

$$q_i = h_2 - h_1 = 2855 - 343.51$$

$$q_i = 2511.49 \text{ kJ/kg}$$

(2) Dryness fraction of steam entering condenser

$$x_3 = 0.797 \text{ as above}$$

(3) W.D. or shaft work

$$W_s = W_T - W_P = [h_2 - h_3] - [h_1 - h_4]$$

$$= (2855 - 2177.9) - 2.95 = 674.15 \text{ kJ/kg}$$

(4) Since 1 kWh = 3600 kJ

$$\text{Steam rate} = \frac{3600}{W_s(\text{kJ/kg})} = \frac{3600}{674.15}$$

$$= 5.34 \frac{\text{kg}}{\text{kWh}}$$

(5) $$\eta_R = \frac{W_s}{q_1} = \frac{674.15}{2511.49}$$

$$\eta_R = \mathbf{26.84\ \%}$$

Example 10.2 *Steam at 20 bar and 360°C expands in a steam turbine to 0.08 bar. It is then condensed in a condenser to saturated water. The pump feeds back the water to the boiler. Assume ideal Rankine cycle and determine,*

(i) The net work done/kg of steam *(ii)* η_R

Solution

Refer Fig. Ex. 10.2. From steam tables corresponding to condenser pressure 0.08 bar $h_4 = h_f = 173.86$ kJ/kg.

We know that, $$W_P = \left[\frac{P_1 - P_b}{10}\right]\frac{\text{kJ}}{\text{kg}} = h_1 - h_4$$

$$W_P = \left[\frac{20 - 0.08}{10}\right]\frac{\text{kJ}}{\text{kg}}$$

$$W_P = 1.992\frac{\text{kJ}}{\text{kg}} = h_1 - h_4$$

or $$h_1 = W_P + h_4 = 1.992 + 173.86$$

$$\mathbf{h_1 = 175.852\ kJ/kg}$$

or corresponding to 20 bar and 360° C, $h_2 = 3160.62$

At 350° C, $h = 3138.6$ kJ/kg and 400° $C = h$ 3248.7 kJ/kg

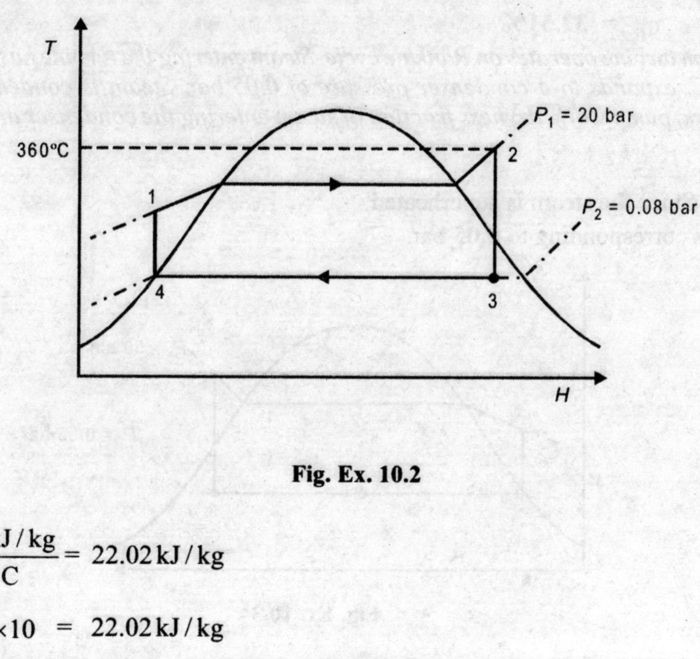

Fig. Ex. 10.2

Difference is $\dfrac{110.1\,\text{kJ/kg}}{50^\circ\text{C}} = 22.02\,\text{kJ/kg}$

$$\therefore \quad 2.202 \times 10 = 22.02\,\text{kJ/kg}$$

$$\therefore \quad \text{At } 360^\circ\text{C},\ h_2 = 3138.6 + 22.02 = 3160.62\,\text{kJ/kg}$$

This method of finding the values at intermediate points is known as Interpolation.

To find h_s:

For isentropic process 2–3

Entropy before expansion s_2 = Entropy after expansion s_3

To find $s_2 = 7.1296$ at 400° C

6.9596 at 350° C

$$0.17/50 = 0.0034 \text{ kJ/kg-K}$$

$$0.0034 \times 10 = 0.034 \text{ kJ/kg-K}$$

$$\therefore \quad 6.9596 + 0.034 = 6.9936 \text{ kJ/kg-K}$$

$$6.9936 = f_{f3} + x_3.s_{fg3} \text{ at condenser pressure}$$

$$= 0.5925 + x_3.7.6371$$

$$\therefore \quad x_3 = 0.8382$$

$$h_3 = h_{f3} + x_3\, h_{fg3} \text{ at condenser pressure}$$

$$= 173.86 + 0.8382 \times 2403.2$$

$$h_3 = 2188.15 \text{ kJ/kg}$$

Now, Work done $= W_s = W_T - W_P = (h_2 - h_3) - W_P$

$$= (3160.62 - 2188.15) - 1.992$$

$$= 970.478 \text{ kJ/kg}$$

and $\eta_R = \dfrac{W_S}{q_i} = \dfrac{970.478}{h_2 - h_1} = \dfrac{970.478}{3160.62 - 175.852}$

$$\eta_R = \mathbf{32.51\%}$$

Example 10.3 *A steam turbine operates on Rankine cycle. Steam entering the turbine has a pressure and temperature of 50 bar and 500°C, expands to a condenser pressure of 0.05 bar. Steam is condensed completely. Find heat supplied, turbine work, pump work, dryness fraction of steam entering the condenser and Rankine cycle efficiency.*

Solution

Refer Fig. Ex. 10.3. Since the steam is superheated,

From steam tables corresponding to 0.05 bar.

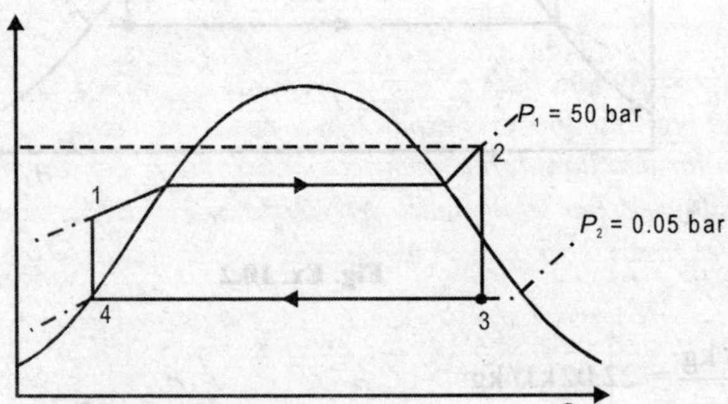

Fig. Ex. 10.3

$$h_4 = h_f = 137.8 \text{ kJ/kg}$$

Also pump work $W_P = \dfrac{P_1 - P_b}{10} = \dfrac{50 - 0.05}{10}$

$$W_P = 4.995 \text{ kJ/kg} = h_1 - h_4$$

$$\therefore \quad h_1 = 4.995 + 137.8$$

$$h_1 = 142.795 \text{ kJ / kg}$$

Now, h_2 corresponding to 50 bar and 500° C from superheated steam tables,

$$h_2 = 3433.7 \text{ kJ/kg}$$

To find h_3; from the isentropic process 2–3

Entropy before expansion = Entropy after expansion.

$$s_2 = s_3$$

$$s_2 = s_{f_3} + x_3 \cdot s_{fg_3} \qquad \text{...(A)}$$

Now s_2 at 50 bar and 500° C = 6.977 kJ/kg-K

∴ From (A), $\quad 6.977 = 0.4763 + x_3 \times (7.92)$

(At condenser pressure 0.05 bar)

$$\Rightarrow \quad x_3 = 0.8208 = 0.82$$

$$h_3 = h_{f_3} + x_3 h_{fg_3} = 137.77 + 0.82 \times 2423.8$$

$$\mathbf{h_3 = 2125.286 \text{ kJ/kg}}$$

Now,

(i) Heat supplied, $\quad q_i = h_2 - h_1 = 3433.7 - 142.795$

$$\mathbf{q_i = 3290.905 \text{ kJ/kg}}$$

(ii) Turbine work $W_T = h_2 - h_3 = 3433.7 - 2125.286$

$$\mathbf{W_T = 1308.414 \text{ kJ/kg}}$$

(iii) Pump work W_P = 4.995 kJ/kg, already found in the beginning.

(iv) Dryness fraction of steam entering the condenser.

x_3 = 0.8208 already found

(v) Rankine cycle efficiency

$$\eta_R = \frac{W_s}{q_i}$$

$$\eta_R = \frac{W_T - W_P}{q_i} = \frac{1308.414 - 4.995}{3290.905}$$

$$\mathbf{\eta_R = 39.606 \%}$$

Example 10.4 *Steam at 20 bar and 360°C is expanded in a steam turbine to 0.08 bar. It then enters a condenser where it is condensed to saturated liquid. It is then fed back to the boiler. Determine,*

(i) Net work (shaft work) per kg of steam. (ii) Dryness fraction of steam entering the condenser.

Solution

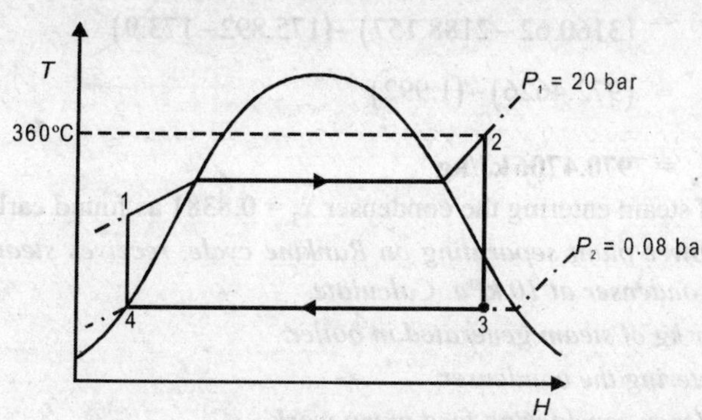

Fig. Ex. 10.4

From steam tables at 0.08 bar

$$h_4 = h_f = 173.9 \text{ kJ/kg}$$

And pump work $$W_P = \left[\frac{P_1 - P_b}{10}\right] = \left[\frac{20 - 0.08}{10}\right]$$

$$W_P = 1.992 \text{ kJ/kg} = h_1 - h_4$$

$$h_1 = 1.992 + h_4$$

$$h_1 = 1.992 + 173.9$$

$$h_1 = 175.892 \text{ kJ/kg}$$

h_2 corresponding to 20 bar and 360° C from superheated steam tables = 3160.62 kJ/kg (By interpolation)

To find h_3 from the isentropic process 2–3

Entropy before expansion = Entropy after expansion

$$s_2 = s_3$$

$$s_2 = s_{f_3} + x_3 \cdot s_{fg_3} \qquad \text{...(A)}$$

Now s_2 at 20 bar and 360° C = $6.994 \dfrac{\text{kJ}}{\text{kg K}}$ (By interpolation)

Now from (A) $$6.994 = 0.593 + x_3 \times 7.637 \text{ at } 0.08 \text{ bar}$$

$$6.994 - 0.593 = 7.637\, x_3$$

$$0.8381 = x_3$$

$$\therefore \quad h_3 = h_{f_3} + x_3 \,.\, h_{fg_3}$$

at $$0.08 \text{ bar} = 173.9 + 08381 \times 2403.2$$

$$\mathbf{h_3 = 2188.157\, kJ/kg}$$

(i) ∴ Net work or shaft work = $W_s = W_T - W_P$

$$W_s = [h_2 - h_3] - [h_1 - h_4]$$

$$= (3160.62 - 2188.157) - (175.892 - 173.9)$$

$$= (972.4626) - (1.992)$$

$$\mathbf{W_s = 970.4706\, kJ/kg}$$

(ii) Dryness fraction of steam entering the condenser x_3 = 0.8381 as found earlier.

Example 10.5 *A steam power plant separating on Rankine cycle, receives steam from boiler at 3.5 MPa and 350°C. It is exhausted to condenser at 10 kPa. Calculate,*

(i) Energy supplied per kg of steam generated in boiler.

(ii) Quality of steam entering the condenser.

(iii) Rankine cycle efficiency considering feed pump work.

(iv) Specific steam consumption.

Data: $P = 3.5$ *MPs* $= 35$ *bar,* $P_b = 10$ *kPa* $= 0.1$ *bar*

$$As\ 100\frac{kN}{m^2} = 1\,bar$$

(i) $q_i = ?$ (ii) $x_3 = ?$

(iii) $\eta_R = ?$ (iv) $SR = ?$

Solution

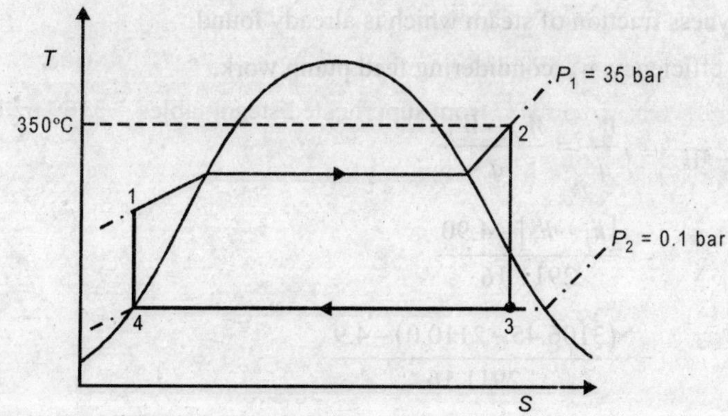

Fig. Ex. 10.5

From steam tables corresponding to 0.1 bar,

$$h_4 = h_f = 191.8 \text{ kJ/kg}$$

Also pump work $\quad W_p = \dfrac{P_1 - P_b}{10}$

$$= \frac{35 - 0.1}{10} = 3.490\,\text{kJ/kg} = h_1 - h_4$$

i.e.,

$$3.499 + h_4 = h_1$$
$$3.490 + 191.8 = h_1$$
$$195.29 = h_1$$

And h_2 corresponding to 35 bar and 350° C from superheated steam tables = 3106.45 kJ/kg by interpolation.

To find h_3 from the isentropic process 2–3

$$s_2 = s_3$$

$$s_2 = s_{f_3} + x_3 \cdot s_{fg_3} \qquad \text{...(A)}$$

Now s_2 at 35 bar and 350° C = $6.663\dfrac{\text{kJ}}{\text{kg-K}}$ by interpolation.

$\therefore$ From (A) $6.663 = 0.649 + x_3 \times 7.502$ at 0.1 bar

$$\frac{6.663 - 0.649}{7.502} = x_3$$

$$0.8016 = x_3$$

$$\therefore \quad h_3 = h_{f_3} + x_3 . h_{fg_3} \text{ at } 0.1 \text{ bar}$$

$$= 191.8 + 0.8016 + 2392.9$$

$$h_3 = 2110.075 \text{ kJ/kg}$$

Now,

(i) Heat supplied $q_i = h_2 - h_1 = 3106.45 - 195.29$

$$q_i = 2911.16 \text{ kJ/kg}$$

(ii) $x_3 = 0.8016$ dryness fraction of steam which is already found.

(iii) Rankine cycle efficiency η_R condidering feed pump work,

$$\eta_R = \frac{W_s}{q_i} = \frac{W_T - W_P}{q_i}$$

$$= \frac{[h_2 - h_3] - 4.90}{2911.16}$$

$$= \frac{(3106.45 - 2110.0) - 4.9}{2911.16}$$

$$\eta_R = 0.34226$$

$$\eta_R = 34.226\%$$

(iv) Steam Rate $= \dfrac{3600}{W_S} = \dfrac{3600}{[W_T - W_P]}$

$$= \frac{3600}{[h_2 - h_3] - 4.9} = \frac{3600}{996.375 - 4.9}$$

$$= \frac{3600}{991.475}$$

S.R. = 3.630 kg/kWh

Example 10.6 *Find heat supplied, turbine work, pump work, shaft work, efficiencies, steam rate of Carnot and Rankine cycle using steam between 30 bar and 0.04 bar.*

Steam is dry and saturated at 30 bar. Compare the results.

Data:

$q_i = ?$ $\quad W_T = ?$

$W_P = ?$ $\quad W_S = ?$

$\eta_c = ?$ $\quad \eta_R = ?$

$S.R. = ?$

$P_1 = 30 \text{ bar}$ $\quad P_b = 0.04 \text{ bar.}$

Solution

Refer Fig. Ex. 10.6

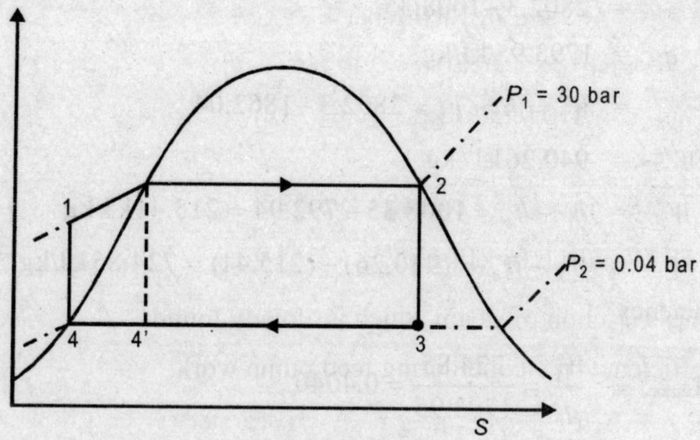

Fig. Ex. 10.6

First consider the Carnot cycle 1' – 2 – 3 – 4'

From steam tables at 30 bar,

$$h_{1'} = h_f = 1008.35 \text{ kJ/kg}$$

$$s_{1'} = s_f = 2.6455 \text{ kJ/kg-K}$$

and
$$h_2 = h_g \text{ at } 30 \text{ bar} = 2802.3 \text{ kJ/kg}$$

$$s_2 = s_g = 6.1837 \text{ kJ/kg-K}$$

For the isentropic process 2–3

$$s_2 = s_{f_3} + x_3 . s_{fg_3} \text{ at } 0.04 \text{ bar}$$

$$6.1837 = 0.423 + x_3 \times 8.053$$

$$x_3 = 0.7154$$

$$\therefore \quad h_3 = h_{f_3} + x_3 h_{fg_3}$$

$$= 121.41 + 0.7154 \times 2433.1$$

$$h_3 = 1862.04 \text{ kJ/kg}$$

Now for Isentropic Compression process 4'–1'

$$s_1' = s_4' = S'_{f_4} + x_4' s'_{fg_4}$$

$$2.6455 = 0.4225 + x'_4 \times 8.0530$$

$$x'_4 = 0.276 \quad \text{Nearer to liquid line}$$

$$\therefore \quad h_4' = h'_{f_4} + x'_4 h'_{fg_4}$$

$$= 121.41 + 0.276 \times 2433.1$$

$$h_4' = 792.94 \text{ kJ/kg}$$

Now,

(i) Heat supplied $\quad q_i = h_2 - h_{1'}$

$$= 2802.3 - 1008.35$$

$$q_i = 1793.95 \text{ kJ/kg}$$

(ii) Turbine work $= W_T = h_2 - h_3 = 2802.3 - 1862.04$

$$W_T = 940.26 \text{ kJ/kg}$$

(iii) Pump work $W_P = h_{1'} - h_{4'} = 1008.35 - 792.94 = 215.41 \text{ kJ/kg}$

(iv) Shaft work $W_s = W_T - W_P = (940.26) - (215.41) = 724.85 \text{ kJ/kg}$

(v) Carnot cycle efficiency

$$\eta_{carnot} = \frac{W_s}{q_i} = \frac{724.85}{1793.95} = 0.4040$$

$$\boldsymbol{\eta_{carnot} = 40.40\%}$$

(iii) Steam Rate $= \dfrac{3600}{W_s} = \dfrac{3600}{724.85}$

$$\mathbf{SR = 4.96654 \frac{kg}{kWh}}$$

II. Rankine Cycle 1 – 2 – 3 – 4

From Fig. Ex. 10.6 it is to be noted that

h_2 and h_3 will remain same.

$$\therefore \quad h_2 = 2802.3 \text{ kJ/kg}$$

$$h_3 = 1862.04 \text{ kJ/kg}$$

and $h_4 = h_f \text{ at } 0.04 \text{ bar} = 121.41 \dfrac{\text{kJ}}{\text{kg}}$

Now to find h_1,

We also know that, pump work

$$W_P = \frac{P_1 - P_b}{10} = \frac{30 - 0.04}{10} = 2.996$$

$$\cong 3.0 \text{ kJ/kg}$$

Also, $W_p = h_1 - h_4$

$$W_p + h_4 = h_1$$

$$3 + 121.41 = h_1$$

$$\Rightarrow \quad h_1 = 124.41 \text{ kJ/kg}$$

Now,

(a) Heat supplied $q_i = h_2 - h_1 = 2802.3 - 124.41$

$$q_i = 2677.89 \text{ kJ/kg}$$

(b) Turbine work $W_T = h_2 - h_3 = 940.26 \frac{\text{kJ}}{\text{kg}}$ · It will remain same.

(c) Pump work $W_P = h_1 - h_4 = 124.41 - 121.41$

$$\mathbf{W_P = 3 \text{ kJ/kg}}$$

(d) Shaft work $W_S = W_T = h_2 - h_3 = 940.26 - 3 = 937.26 \text{ kJ/kg}$

(e) Rankine Cycle Efficiency

$$\eta_R = \frac{W_S}{q_i} = \frac{937.26}{2677.89}$$

$$\eta_R = 0.34999$$

$$\eta_R = 34.999\%$$

(f) Steam Rate,

$$SR = \frac{3600}{W_s} = \frac{3600}{937.26}$$

$$\mathbf{SR = 3.8409 \text{ kh.kWh}}$$

Comparison:

(1) Pump work for Carnot Cycle is more, since it has to handle large amount of mixture of wet steam and condensate.

(2) Turbine work in both the cases is same as the state points are same.

(3) Carnot cycle efficiency is more than Rankine cycle efficiency.

Example 10.7 *Calculate the cycle efficiency and steam consumption in kg/kWh for Carnot cycle and Rankine cycle using steam between two pressure of 20 bar and 0.1 bar. The steam is dry and saturated at 20 bar.*

Find: $\eta_{carnot} = ?$ $\eta_R = ?$ $SR = ?$

Solution

Refer Fig. Ex. 10.7

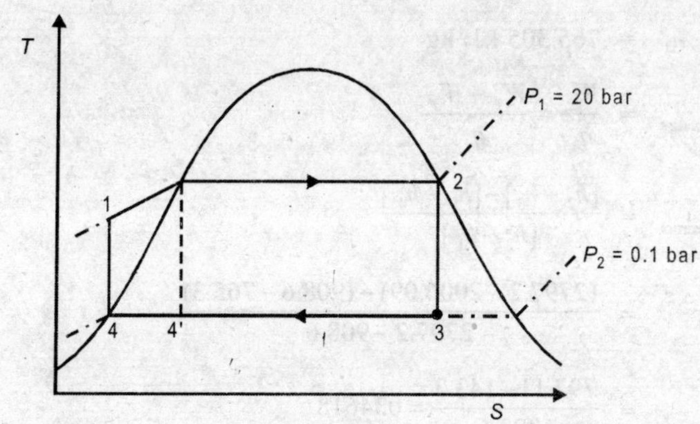

Fig. Ex. 10.7

(1) Consider carnot Cycle 1' – 2 – 3 – 4'

$$h_{1'} = h_f \text{ at 20 bar} = 908.6 \frac{\text{kJ}}{\text{kg}}$$

$$s_{1'} = s_f = 2.447 \frac{\text{kJ}}{\text{kg-K}}$$

and $$h_2 = h_g \text{ at 20 bar} = 2797.2 \frac{\text{kJ}}{\text{kg}}$$

$$s_2 = s_g \text{ at 20 bar} = 6.337 \frac{\text{kJ}}{\text{kg-K}}$$

For Isentropic Process 2–3

$$s_2 = s_3$$

$$s_2 = s_{f_3} + x_3 \cdot s_{fg_3}$$

$$6.337 = 0.649 + x_3 \times 7.502$$

$$x_3 = 0.75819 \cong 0.76$$

$$\therefore \quad h_3 = h_{f_3} + x_3 . s_{f_{g_3}} = 191.8 + 0.76 \times 2392.9$$

$$h_3 = 2006.09 \text{ kJ/kg}$$

Now for Isentropic Compression process 4' – 1'

$$s_{1'} = s_{4'} = s_{f4'} + x_{4'} \times s_{fg_{4'}}$$

$$2.447 = 0.649 + x_{4'} \, 7.502 \text{ at 0.1 bar}$$

$$0.2397 = x_{4'}$$

$$h_{4'} = h_{f_{4'}} + x_{4'} \times h_{fg_{4'}}$$

$$h_{4'} = 191.8 + 0.2397 \times 2392.9$$

$$h_{4'} = 765.305 \text{ kJ/kg}$$

Now $$\eta_{\text{carnot}} = \frac{W_s}{q_i} = \frac{W_T - W_P}{q_i}$$

$$\eta_{\text{carnot}} = \frac{[h_2 - h_3] - [h_{1'} - h_{4'}]}{[h_2 - h_{1'}]}$$

$$= \frac{(2797.2 - 2006.09) - (908.6 - 765.3)}{2797.2 - 908.6}$$

$$= \frac{797.11 - 143.3}{1888.6} = 034618$$

$$\eta_{\text{carnot}} = \mathbf{34.618\%}$$

And Steam Rate $= \dfrac{3600}{W_S} = \dfrac{3600}{W_T - W_P}$

$$= \frac{3600}{[h_2 - h_3] - [h_{1'} - h'_{4'}]}$$

$$SR = \frac{3600}{653.81} = 5.506 \text{ kg/kWh}$$

(II) Now Consider Rankine Cycle = 1 – 2 – 3 – 4

$$h_4 = h_f \text{ at } 0.1 \text{ bar} = 191.8 \frac{\text{kJ}}{\text{kg}}$$

From Fig. Ex. 10.7 h_2 and h_3 will remain same.

i.e., $h_2 = 2797.2$ kJ/kg

$h_3 = 2006.09$ kJ/kg

Now to find h_1, we also know that,

$$W_P = \frac{P_1 - P_b}{10} = \frac{20 - 0.1}{10}$$

$$W_P = 1.99 \text{ kJ/kg} = h_1 - h_4$$

$$\therefore \quad h_1 = 1.99 + 191.8$$

$$h_1 = 193.79 \text{ kJ/kg}$$

Now, $\eta_g = \dfrac{W_s}{q_i} = \dfrac{W_T - W_P}{q_i}$

$$\eta_R = \frac{[h_2 - h_3] - W_P}{[h_2 - h_1]}$$

$$= \frac{(2797.2 - 2006.09) - 1.99}{(2797.2 - 193.79)}$$

$$= \frac{791.11 - 1.99}{2603.41} = \frac{789.12 \text{ (i.e.,} W_s)}{2603.41}$$

$$= 0.30311$$

$$\eta_R = 30.311\%$$

and Steam Rate $= \dfrac{3600}{W_s} = \dfrac{3600}{[h_2 - h_3] - W_p} = \dfrac{3600}{789.12}$

SR = 4.562 kg/kWh

Example 10.8 *A steam plant operating on Rankine cycle generates superheated steam at 10 bar and 380°C. Condensation occurs at 0.06 bar. Calculate the thermal efficiency of the plant and compare this with the thermal efficiency of Carnot cycle working between the same temperature limits.*

Find: $\eta_R = ?$ $\eta_{car} = ?$

Solution

Refer Fig. Ex. 10.8.

(i) Rankine Cycle (1 – 2 – 3 – 4)

From steam tables, at 0.06 bar,

$h_4 = h_f = 151.5$ kJ/kg

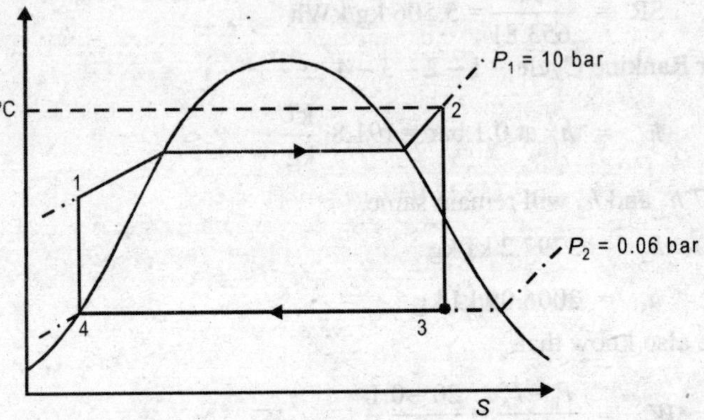

Fig. Ex. 10.8

Also Pump work

$$W_P = \frac{P_1 - P_b}{10} = \frac{10 - 0.06}{10}$$

$$W_P = 0.994 \text{ kJ/kg} = h_1 - h_4$$

$$0.994 + 151.5 = h_1$$

$$h_1 = 152.494 \text{ kJ/kg}$$

And h_2 at 10 bar and 380°C from superheated steam tables = 3222.04 kJ/kg by interpolation.

To find h_3; from the Isentropic process 2–3

$$s_2 = s_3$$

$$s_2 = s_{f_3} + x_3 s_{f_{g3}} \quad \text{...(A)}$$

Now s_2 at 10 bar and 380°C = 7.4.014 kJ/kg-K

Now from Eq. (A),

$$7.4014 = 0.521 + x_3 \times 7.81 \text{ at 0.06 bar}$$

$$\frac{7.4014 - 0.521}{7.81} = x_3$$

$$0.8809 = x_3$$

Now $\quad h_3 = h_{f_3} + x_3 . h_{f_{g3}}$ at 0.06 bar

$$= 151.5 + 0.8809 \times 2416$$

$$h_3 = 2279.93 \text{ kJ/kg}$$

Now, (i) $$\eta_R = \frac{W_S}{q_i}$$

$$\eta_R = \frac{W_T - W_P}{q_i}$$

$$\eta_R = \frac{[h_2 - h_3] - W_P}{[h_2 - h_1]}$$

$$\eta_R = \frac{(3222.04 - 2279.93) - (0.994)}{(3222.04 - 152.494)}$$

$$\eta_R = \frac{(942.108 - 0.994)}{3069.546}$$

$$\eta_R = \frac{941.1149}{3069.546}$$

$$\eta_R = 0.30659$$

$$\boldsymbol{\eta_R = 30.659\%}$$

(II) Carnot Cycle 1'–2–3–4'

From steam tables at 10 bar

$$h_{1'} = h_f = 762.6 \text{ kJ/kg}$$

$$s_1 = s_f = 2.138 \text{ kJ/kg-K}$$

And $h_2 = 3222.04$ and $h_3 = 2279.93$

These will remain same.

Now for Isentropic process 4'–1'

$$s_{1'} = s_{4'} = s_{f4'} + x_{4'} \times s_{fg_{4'}}$$

$$2.138 = 0.521 + x \times 7.81 \text{ at } 0.06 \text{ bar}$$

$$\frac{2.138 - 0.521}{7.81} = x_{4'}$$

$$0.2070 = x_{4'}$$

$$h_{4'} = h_{f_{4'}} + x_{4'} \times h_{fg_{4'}} \text{ at } 0.06 \text{ bar}$$

$$= 151.5 + 0.2070 \times 2416$$

$$h_{4'} = 651.714 \text{ kJ/kg}$$

Now, $$\eta_{carnot} = \frac{W_s}{q_i} = \frac{W_T - W_P}{q_i}$$

$$= \frac{[h_2 - h_3] - [h_{1'} - h_{4'}]}{h_2 - h_{1'}}$$

$$= \frac{(3222.04 - 2279.93) - (762.6 - 651.714)}{3222.04 - 762.6}$$

$$= \frac{(942.11) - (110.886)}{(2459.44)}$$

$$= 0.33797$$

$$\eta_{carnot} = \mathbf{33.797\%}$$

Hence Carnot Cycle efficiency is more than Rankine cycle for an engine working between same temperature limits.

THEORY QUESTIONS

10.1. Derive the expression for efficiency of Carnot cycle.

10.2. Draw Carrot cycle on P–V and T–S diagram.

10.3. Explain Rankine cycle with P–V and T–S diagram.

10.4. Derive an expression for efficiency of Rankine cycle.

10.5. Compare Carnot cycle with Rankine cycle.

10.6. Define the terms : (i) Work ratio (ii) SSC.

10.7. Write a short note on factors affecting performance of Rankine cycle.

10.8. Explain why Rankine cycle and not Carnot cycle is considered as the thermodynamic cycle for the steam power plant.

10.9. Discuss the effects of the following parameters on the performance of Rankine cycle:

(i) Steam pressure at turbine inlet.

(ii) Steam temperature at turbine inlet.

(iii) Steam pressure at turbine exhaust.

10.10. Draw a neat sketch of Rankine cycle of steam plant, label the parts and describe the working of each component, giving the formula for energy exchange for the same.

10.11. Enumerate the advantages of Rankine cycle over the Cannot cycle for steam power plant.

10.12. Explain the drawbacks of Cannot cycle for a steam power plant. How are they overcome in a Rankine cycle ?

10.13. In what respect does the Rankine cycle differ from the Carnot cycle. Draw Carnot and Rankine cycle on T–S plane.

10.14. Enumerate the advantages of Rankine cycle over the Carnot cycle for steam power plant.

10.15. Why is Carnot cycle not practicable for a steam power plant.

10.16. Draw a neat sketch of Rankine cycle steam plant, label the part and describe the working of each component, giving formula for energy exchange for the same.

10.17. Explain clearly how the following affects the performance of a Rankine cycle.

(i) Lowering the condensation pressure.

(ii) Superheating the steam to high temperature.

(iii) Increasing the boiler pressure.

10.18. Parameters affecting the performance of Rankine cycle.

10.19. Explain the factors affecting the efficiency of Rankine cycle.

10.20. Explain with *T–S* diagram, factors affecting performance of Rankine cycle.

10.21. Explain the effect of following on the efficiency of Rankine cycle.

(i) Increasing the average temperature at which heat is supplied.

(ii) Decreasing the temperature at which heat is rejected.

10.22. Show the Rankine cycle on

(i) Wet

(ii) Dry saturated

(iii) Superheated

PROBLEMS FOR PRACTICE

1. A steam power plant operating on Rankine cycle recieves steam from boiler at 35 bar and 350°C. It is exhausted to condenser at 0.1 bar, calculate

(i) Energy supplied/kg of steam generated in boiler

(ii) Quality of steam entering the condenser

(iii) Rankine cycle efficiency considering feed pump work.

[**Ans.** (i) 2911.16 kJ/kg, (ii) 0.8016, (iii) 34.226%]

2. Consider a steam power plant operating on the ideal Rankine cycle. The steam enters the turbine at 3 MPa and 350°C and is condensed in the condenser at a pressure of 10 kPa. Determine

(i) The thermal efficiency of this power plant.

(ii) The thermal efficiency if the steam is superheated to 600°C instead of 350°C.

(iii) The thermal efficiency if the boiler pressure is raised to 15 MPa while the turbine inlet temperature is maintained at 600° C.

Chapter 11 Steam Nozzles

11.1 INTRODUCTION

Turbomachines like steam turbines, water turbines and gas turbines produce power by utilising the kinetic energy of the jets produced by passing high pressure steam, water and gas through the devices called nozzles. Corresponding to the fluids used, the nozzles are called steam nozzles, water nozzles and gas nozzles.

These nozzles serves two purposes:

(1) To convert pressure energy and thermal energy into kinetic energy and

(2) To direct the fluid jet at the specific angle known as nozzle angle.

11.2 DEFINITION

A nozzle is a device, a duct of varying cross-section area in which a steadily flowing fluid can be made to accelerate by a pressure drop along the duct.

So when a fluid flows through a nozzle, its velocity increases continuously and pressure decreases continuously.

11.3 SOME APPLICATIONS OF A NOZZLE

There are following applications of a nozzle are:

1. Nozzles are used in steam turbines, gas turbines, water turbines and in jet engines, Jet propulsion.
2. Nozzles are used for flow measurement e.g., in venturimeter.
3. Nozzles are used to remove air from a condenser.
4. Injectors for pumping feed water to boilers.
5. Artificial fountains.

There are three types of nozzles:

1. Convergent nozzle,
2. Divergent nozzle, and
3. Convergent-Divergent nozzle.

If the cross-section of the nozzle decreases continuously from entrance to exit, it is called a convergent nozzle.

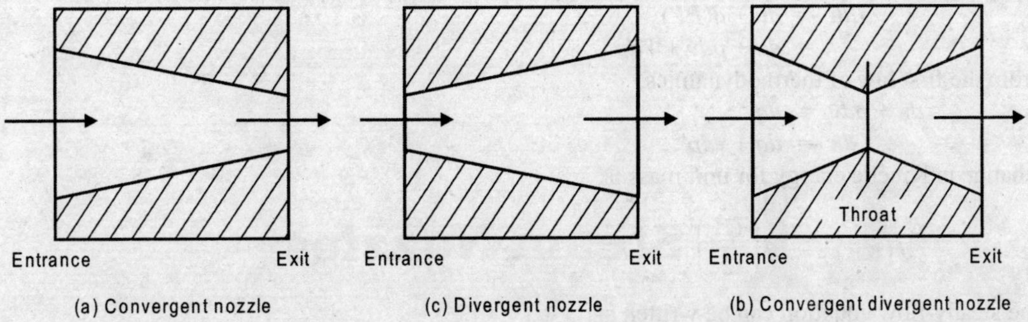

Fig. 11.1

If the cross-section of the nozzle increases continuously from entrance to exit, it is called a divergent nozzle.

If the cross-section of the nozzle first decreases and then increases, it is called a convergent-divergent nozzle. The section where cross-sectional area is minimum is called 'throat' of the nozzle.

Diffuser. When a steadily flowing fluid is decelerated in a duct causing rise in pressure along the stream, then the duct is called a diffuser.

So in a diffuser, velocity of the fluid decreases continuously and pressure increases continuously.

Similar to nozzle, there are three types of diffusers.

1. Convergent diffuser
2. Divergent diffuser
3. Convergent-divergent diffuser.

The given duct will work as a diffuser or a nozzle depending upon the fluid velocity at the inlet of a duct.

Some Applications

1. Diffusers are used in centrifugal compressor.
2. Diffusers are used in ram-jet engines to increase the pressure of incoming fresh-air.

11.4 GENERAL-FLOW ANALYSIS

Assumptions

1. The flow of the fluid is assumed to be one dimensional.
2. The change in area and curvature along the axis of the duct are gradual.
3. Thermodynamic and mechanical properties are uniform across planes normal to the axis of a duct.
4. The flow of the fluid is assumed to be steady flow.

The steady-flow equation is,

$$Q - W = \Delta H + \Delta PE + \Delta KE$$

or differential form of the equation is,

$$\partial q - \partial w = dh + d\,(PE) + d\,(KE)$$

We will assume heat the nozzle is horizontal,

$$\therefore \quad d\,(PE) = 0$$

The fluid is just flowing through a duct. So obviously work-done is zero

$$\therefore \quad \partial w = 0$$

The definition of enthalpy is

$$h = u + PV$$
$$\therefore \quad dh = du + d(PV)$$
$$= du + pdv + Vdp$$

But from the first law of thermodynamics,

$$du + pdv = dq$$
$$\therefore \quad dh = dq + vdp$$

The change in Kinetic energy for unit mass is,

$$d(\text{KE}) = d\left(\frac{C^2}{2}\right)$$

$\therefore$ The steady-flow equation can be written as,

$$\partial q = \partial q + vdp + d\left(\frac{C^2}{2}\right)$$

$$\therefore \quad d\left(\frac{C^2}{2}\right) = -vdp$$

$$\therefore \quad \frac{2C}{2} \cdot dC = -vdp$$

$$\therefore \quad CdC = -vdp \qquad \ldots(1)$$

The fluid flows through the duct under steady-flow conditions. Therefore the mass-flow rate is constant.

$$m = \rho A \cdot C = \text{Constant}$$

or

$$m = \frac{A \cdot C}{v} = \text{Constant}$$

where

A = cross-sectional area,
C = velocity of fluid
v = specific volume of the fluid.

$$\therefore \quad \frac{A \cdot C}{v} = \text{Constant}$$

$$\log A + \log C - \log v = \text{Constant}$$

Differentiating the above equation, we get

$$\frac{dA}{A} + \frac{dC}{C} - \frac{dv}{v} = 0$$

From Eq. (1)

$$dC = \frac{-vdp}{C}$$

$$\therefore \quad \frac{dA}{A} - \frac{vdp}{C^2} - \frac{dv}{v} = 0$$

$$\therefore \quad \frac{dA}{A} = \frac{vdp}{C^2} + \frac{dv}{v}$$

$$= vdp\left(\frac{1}{C^2} + \frac{dv}{v^2 dp}\right)$$

From the knowledge of fluid-dynamics, the velocity of sound wave is given by,

$$a^2 = -v^2\left(\frac{dp}{dv}\right)_s$$

where a is the velocity of sound (or sonic velocity)

$$\therefore \quad \frac{dA}{A} = vdp\left(\frac{1}{C^2} - \frac{1}{a^2}\right)$$

$$\therefore \quad \frac{dA}{A} = \frac{vdp}{C^2}\left(1 - M^2\right)$$

where M is mach number $= c/a$

$$\therefore \quad \frac{dA}{A} = \frac{-vdp}{C^2}\left(M^2 - 1\right)$$

But from Eq. (1) $CdC = vdp$

$$\therefore \quad \frac{dA}{A} = \frac{dC}{C}\left(M^2 - 1\right)$$

This equation gives information whether the given duct will act as a nozzle or a diffuser if the inlet fluid velocity is known. It also gives information which type of duct should be used for a particular application.

For a nozzle, velocity of the fluid should increase continuously from entrance to exit. So if the inlet fluid condition is known, we can select the nozzle as below.

1. When the velocity of fluid is less than sonic velocity (i.e., when the flow is sub-sonic), the mach no. (M) is less than 1.

$$\therefore \quad M < 1$$

$\therefore \quad \left(M^2 - 1\right)$ is negative.

For a nozzle change in velocity (dC) has to be positive.

$\therefore \quad \frac{dC}{C}\left(M^2 - 1\right)$ is negative.

$\therefore \quad \frac{dA}{A}$ is negative.

i.e., change in across sectional area along the duct should be negative.

This tells that, for *sub-sonic flow*, the duct must be convergent. By using this convergent nozzle, the flow of the fluid can be increased to sonic velocity.

But by using convergent nozzle we can not obtain super-sonic flow.

2. When the flow is super-sonic, $M > 1$

i.e., for super-sonic flow, the duct must be divergent.

3. The convergent-divergent nozzle is used for convert sub-sonic flow into super-sonic flow.

In the convergent part the velocity of fluid is increased from sub-sonic to sonic condition. At throat, the velocity is sonic. In the divergent part, the velocity is increased from sonic to super-sonic.

So only in convergent-divergent nozzle, the sub-sonic flow is converted into super-sonic flow.

For diffuser the velocity should decrease continuously so a diffuser is selected as below,

1. For sub-sonic flow

$$M < 1 \qquad \therefore \quad (M^2 - 1) < 0$$

$$\therefore \quad \frac{dC}{C}(M^2 - 1) > 0$$

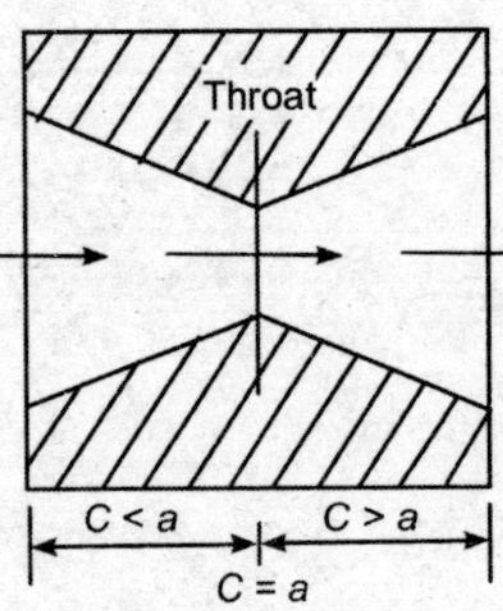

Fig. 11.2

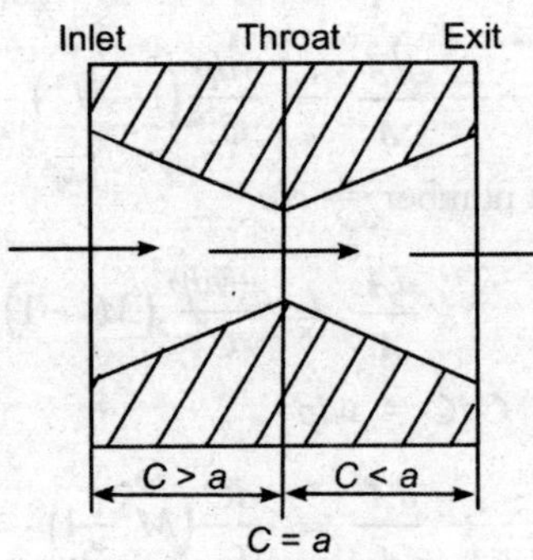

Fig. 11.3

$\therefore \quad \frac{dA}{A}$ is positive i.e., duct must be divergent.

2. For supersonic flow: $M > 1$

$$\therefore \quad \frac{dC}{C}(M^2 - 1) < 0$$

$\therefore \quad \frac{dA}{A}$ is positive i.e., duct must be divergent.

The convergent-diffuser will decrease the velocity of fluid to sonic velocity.

3. The convergent-divergent diffuser is used to convert super-sonic flow into sub-sonic flow.

11.5 VELOCITY IN A NOZZLE

For unit mass,

The steady flow equation is,

$$q - w = \Delta h + \Delta PE + \Delta KE$$

For a horizontal nozzle , $\Delta PE = 0$

There is no work-done in nozzle therefore $W = 0$. In the nozzle, the velocity of the fluid is so high that there is hardly any time available for fluid to exchange heat with the surroundings. Therefore for nozzle, it is assumed that heat transfer is zero i.e., flow in Isentropic

$$\therefore \quad q = 0$$

$$\therefore \quad \Delta h + \Delta \text{KE} = 0$$

$$\therefore \quad h_2 - h_1 + \frac{1}{2}\frac{\left(C_2^2 - C_1^2\right)}{1000} = 0$$

$$C_2 = \sqrt{2000\left(h_1 - h_2\right) + C_1^2}$$

where C_1 is velocity of fluid at inlet
h_1 is specific enthalpy of fluid at inlet
h_2 is specific enthalpy of fluid at exit condition.

For convergent-divergent nozzle, the velocity of fluid at throat is obtained by the equation,

$$C_t = \sqrt{2000(h_1 - h_t) + C_1^2}$$

where, h_t = specific enthalpy at the throat conditions.

For steam nozzles the values of enthalpy $(h_1, h_2, h_t \text{ etc.})$ are normally obtained by using Mollier chart.

If C_1, the initial or approach velocity is neglected, then

$$C_2 = \sqrt{2000\left(h_1 - h_t\right)} = 44.72\sqrt{h_1 - h_t}$$

$$= 44.72\sqrt{(h_1 - h_2) + \frac{C_1^2}{2000}} \text{ when } C_1 \text{ is considered.}$$

11.6 MASS-FLOW RATE IN A NOZZLE

The steady flow energy equation is

$$q - w = \Delta h + \Delta \text{PE} + \Delta \text{KE}$$

In the general analysis, we have obtained the equation,

$$d\left(\frac{C^2}{2}\right) = -vdp$$

Integrating the equation

$$\frac{C_2^2 - C_1^2}{2} = -\int_1^2 vdp$$

The simple relation, PV^n = const. holds good for any process. So we assume that the expansion in a nozzle is polytropic.

$$\therefore -\int_1^2 vdp \text{ will become } \frac{n}{n-1}\left(P_1 V_1 - P_2 V_2\right)$$

$$\therefore \quad \frac{C_2^2 - C_1^2}{2} = \frac{n}{n-1}\left(P_1 V_1 - P_2 V_2\right)$$

For nozzle, generally C_1 is too small.

$\therefore$ Neglecting C_1, we get

$$C_2 = \sqrt{\frac{2n}{n-1}(P_1 V_1 - P_2 V_2)}$$

$$C_2 = \sqrt{\frac{2n}{n-1} P_1 V_1 \left(1 - \frac{P_2 V_2}{P_1 V_1}\right)}$$

The mass flow rate is given by the continuity equation

$$m = \rho AC = \frac{AC}{v}$$

$$\therefore \quad \frac{m}{A_2} = \frac{C_2}{v_2} = \frac{\sqrt{\frac{2n}{n-1} P_1 V_1 \left(1 - \frac{P_2 V_2}{P_1 V_1}\right)}}{V_2}$$

For polytropic process, $P_1 v_1^n = P_2 v_2^n$

$$\therefore \quad \frac{m}{A_2} = \frac{\sqrt{\frac{2n}{n-1} P_1 v_1 \left(1 - \left(\frac{P_2}{P_1}\right)^{1-\frac{1}{n}}\right)}}{v_1 \cdot \left(\frac{P_1}{P_2}\right)^{1/n}}$$

$$\therefore \quad \frac{m}{A_2} = \sqrt{\frac{2n}{n-1} \cdot \frac{P_1}{v_1} \left[\left(\frac{P_2}{P_1}\right)^{\frac{2}{n}} - \left(\frac{P_2}{P_1}\right)^{\frac{n+1}{n}}\right]}$$

This is the equation for the mass-flow rate of fluid per unit area of nozzle.

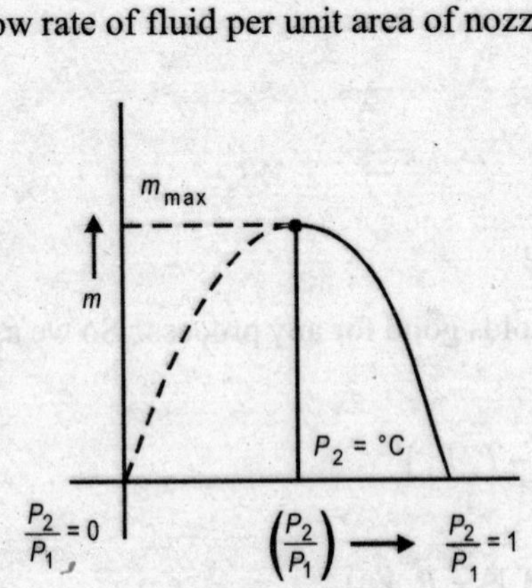

Fig. 11.4

The equation states that the mass flow-rate depends upon pressure-ratio $\left(\frac{P_2}{P_1}\right)$.

If $P_1 = P_2$ the mass-flow rate is zero.

Also if $P_2 = 0$, the mass flow rate is zero.

If the graph is plotted for mass flow rate v_s pressure ratio, it will be as shown in the figure. It is observed that at some value of (P_2/P_1) the velocity and mass-flow rate reaches to its maximum value.

11.7 CRITICAL PRESSURE RATIO

The nozzle is designed for maximum mass-flow rate per unit area. It is observed from the above discussion that the mass-flow rate depends on pressure-ratio. The equation for maximum mass-flow rate can be found out by differentiating the above equation w.r.t. $\left(\frac{P_2}{P_1}\right)$ and equating it to zero. The other parameters n, p, v are constant.

The equation is

$$\frac{m}{A_2} = \sqrt{\frac{2n}{n-1}\frac{P_1}{v_1}\left[\left(\frac{P_2}{P_1}\right)^{2/n} - \left(\frac{P_2}{P_1}\right)^{\frac{n+1}{n}}\right]}$$

$$\therefore \quad \frac{d\left(\frac{m}{A_2}\right)}{d\left(\frac{P_2}{P_1}\right)} = \frac{d}{d\left(\frac{P_2}{P_1}\right)}\left[\frac{2n}{n-1}\frac{P_1}{v_1}\left[\left(\frac{P_2}{P_1}\right)^{\frac{2}{n}} - \left(\frac{P_2}{P_1}\right)^{\frac{n+1}{n}}\right]\right]^{1/2}$$

$$= \frac{1}{\left[\frac{2n}{n-1}\frac{P_1}{v_1}\left[\left(\frac{P_2}{P_1}\right)^{\frac{2}{n}} - \left(\frac{P_2}{P_1}\right)^{\frac{n+1}{n}}\right]\right]^{1/2}} \times \left[\frac{2n}{n-1}\times\frac{P_1}{v_1}\left[\frac{2}{n}\left(\frac{P_2}{P_1}\right)^{\frac{2}{n}-1} - \frac{n+1}{n}\times\left(\frac{P_2}{P_1}\right)^{1/n}\right]\right]$$

$$= 0$$

$$\therefore \quad \frac{2}{n}\left(\frac{P_2}{P_1}\right)^{\frac{2-n}{n}} - \left(\frac{n+1}{n}\right)\times\left(\frac{P_2}{P_1}\right)^{1/n} = 0$$

$$\therefore \quad \frac{2}{n}\left(\frac{P_2}{P_1}\right)^{\frac{2-n}{n}} = \left(\frac{n+1}{n}\right)\left(\frac{P_2}{P_1}\right)^{1/n}$$

$$\therefore \quad \frac{2}{n}\times\frac{n}{n+1} = \left(\frac{P_2}{P_1}\right)^{\frac{1}{n}-\frac{2}{n}+1}$$

$$\therefore \quad \left(\frac{2}{n+1}\right) = \left(\frac{P_2}{P_1}\right)^{\frac{n-1}{n}}$$

$$\therefore \quad \left(\frac{P_2}{P_1}\right) = \left(\frac{2}{n+1}\right)^{\frac{n}{n-1}}$$

This is the equation for critical pressure ratio for maximum discharge. This pressure ratio will accelerate the flow to a velocity equal to sonic velocity.

For convergent-divergent nozzle, at throat the critical pressure is maintained to get maximum discharge. For steam nozzles value of n depends on condition of steam at inlet.

Initial condition	**value of n**
1. Superheated steam	1:3 $\quad \therefore \frac{P_2}{P_1} = 0.546$
2. Wet-steam	1.035 + 0.1 × x, where, x is the dryness fraction of steam.
3. Dry-saturated steam	1.135 $\quad \therefore \frac{P_2}{P_1} = 0.58$
4. For air	$\gamma = 1.4 \therefore \frac{P_2}{P_1} = 0.5283$

11.8 EFFECT OF FRICTION

The flow of steam through nozzle is assumed to be isentropic. The Mollier chart shows the isentropic flow $(1-t-2)$ of steam through a convergent-divergent nozzle. But in actual case, the friction losses occur. The frictional losses are due to,

(i) Friction between sides of nozzle (wall of nozzle) and fluid.
(ii) Internal fluid friction and
(iii) Due to eddies in the flow.

Most of the friction in convergent-divergent nozzle is assumed to occur between the throat and exit. The expansion upto throat is taken to be isentropic. This is due to low initial velocity. In Mollier chart, $1-t-2'$ is the actual expansion of steam through nozzle. Figure 11.5 shows the actual expansion of steam through nozzle.

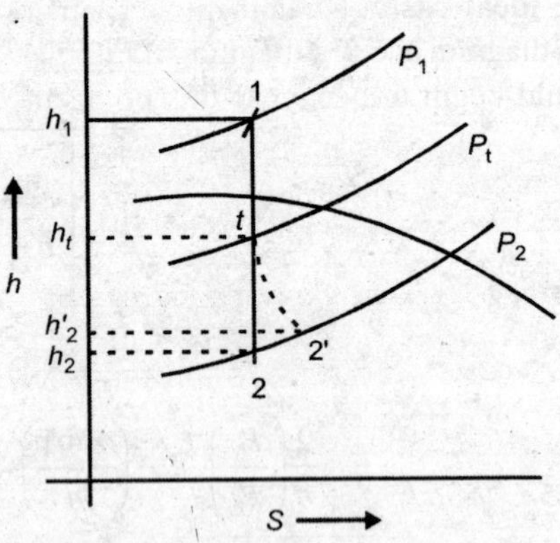

Fig. 11.5. *h-S diagram with friction.*

Effects

1. Reduction in actual enthalpy drop.
2. Reduction in exit velocity of fluid.
3. Increase in the dryness-fraction of steam.
4. Increase in the specific volume and
5. Decrease in the mass-flow rate.

11.9 NOZZLE EFFICIENCY

The nozzle efficiency is given as,

$$\text{Nozzle efficiency} = \frac{\text{Actual enthalpy drop}}{\text{Isentropic enthalpy drop}}$$

$$\eta_{nozzle} = \frac{h_1 - h_2'}{h_1 - h_2}$$

11.10 VELOCITY COEFFICIENT

Velocity coefficient is defined as the ratio of actual velocity of steam to an isentropic or theoretical velocity.

$$\text{Velocity coefficient } (C_V) = \frac{\text{Actual velocity}}{\text{Isentropic velocity}}$$

Velocity of steam is given as,

$$C_2 = \sqrt{2(h_1 - h_2)} \quad \text{when inlet velocity is neglected}$$

$$\therefore \quad C_V = \frac{\sqrt{2(h_1 - h_2')}}{\sqrt{2(h_1 - h_2)}} = \sqrt{\frac{h_1 - h_2'}{h_1 - h_2}}$$

$$\therefore \quad \text{Velocity coefficient} = \sqrt{\text{nozzle efficiency}}$$

11.11 SUPER SATURATED OR META-STABLE FLOW THROUGH NOZZLE

11.11.1 Normal or Equilibrium Expansion

The ideal case of isentropic expansion of a superheated vapour to a state in the wet region is shown in T–S diagram and h–s diagram of Fig. 11.6. At the point in the expansion where the pressure is P_s, a change of phase should begin to occur. At this point, the random kinetic energy of the molecules has fallen to a level which is

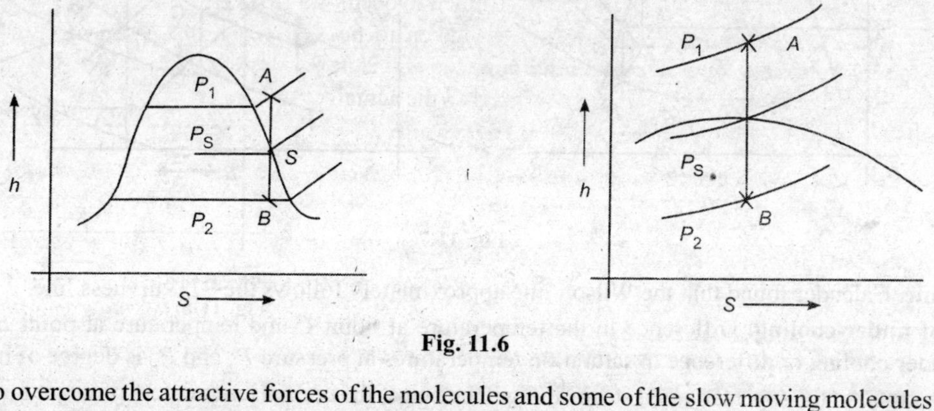

Fig. 11.6

insufficient to overcome the attractive forces of the molecules and some of the slow moving molecules start to form tiny droplets to condensate. Now although this process is rapid, it does not have time to occur in the nozzle where the flow velocity is very great. The achievement of equilibrium between the liquid and vapour phase is therefore delayed and vapour continues to expand in a superheated or dry state.

11.11.2 Super-saturated Expansion

Point *A* represents a steam in superheated region at pressure P_i. When at point *s*, due to the rapid expansion, instead of condensation commencing, the steam continues to behave as a super-heated vapour down to point *B*, at same intermediate pressure P_2. This non-equilibrium behaviour as a superheated vapour does not continue indefinitely and at point *B*, restoration of equilibrium quickly occurs and is after the throat in divergent portion of the nozzle. It is accompanied by a small increase in pressure.

In equilibrium flow, the energy released by condensing the molecules is provided for increasing the kinetic energy of the steam as it passes through the nozzle. If the steam does not condense, then the energy for this increase in kinetic energy come by reduction in the temperature and therefore the steam is called super-c ooled.

Thus we see that condensation does not start immediately after *S* is passed, no drops of liquid are formed until some state *B* is reached, where condensation suddenly occurs, a phenomenon sometimes called *Condensation Shock*. The steam in states between *S* and *B* is *supersaturated* or a *metastable state*.

Since the collaspe of the metastable state has been observed not in a converging nozzle, but always, in the diverging part of the De-Laval nozzle, one is probably safe assuming that the supersaturation, if it occurs at all, will persist to some point beyond throat. Of course the flow should be computed for the throat section since this is where it is limited.

In general, the velocity of supersaturated steam is less than the value computed for the equilibrium flow. The density of supersaturated steam is greater than the ewuilibrium density at the same pressure. However, the density increases by a large percentage that the velocity decreases, with the result that the computed mass flow of supersaturated steam is greater than the computed flow of equilibrium steam, given the initial state and throat area. When discharge from a nozzle is actually measured it is found to be from 2 to 5% greater than the calculated discharge.

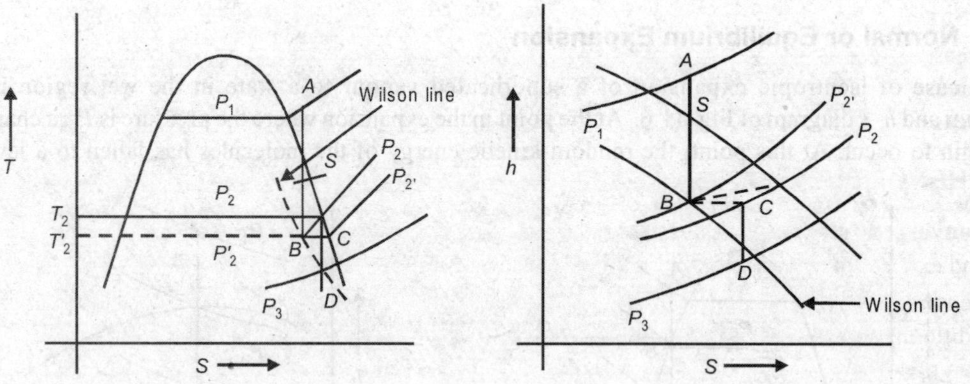

Fig. 11.7

Wilson line. Calender found that the Wilson line approximately follows the 97% dryness line.

Degree of under-cooling. Difference in the temperature at point *C* and temperature at point *B* is known as degree of under-cooling or difference in saturation temperatures at pressure P_2 end P_B is degree of under-cooling.

The temperature of the supersaturated steam at *B* will be below the normal temperature of steam for that pressure; this state is known as *undercooled state*, the amount of undercooling being the difference between the normal temperature and the actual temperature.

$\therefore$ Degree of under-cooling $= T_2 - T_2'$

$$= \begin{bmatrix}\text{Saturation temperature} \\ \text{corresponding to point } B\end{bmatrix} - \begin{bmatrix}\text{Actual temperature} \\ \text{at } B\end{bmatrix}$$

For this purpose we have

$$\frac{T_B}{T_A} = \left(\frac{P_B}{P_A}\right)^{\frac{0.3}{1.3}}$$

Degree of Supersaturation. Note that the actual pressure at the supersaturated state B is P_2, but the saturation pressure corresponding to the temperature t_B is P_2. The ratio of these pressures P_d / P_b is called the *Degree of Supersaturation* or *Supersaturation ratio.*

$$\text{Degree of super-saturation} = \frac{\text{Actual pressure}}{\text{Saturation pressure corresponding to actual temperature } (T_2')}$$

$$= \frac{P_2}{P_2'}$$

11.11.3 Wilson's Line

There is generally a limit to super-saturation. It is upto 96% dryness and beyond it, steam condensation occurs suddenly and irreversibly at constant enthalpy and remains in stable condition thereafter. This limit line is known as Wilson's line.

Effects of Super-saturated Flow

1. Increase in discharge by 2 to 5% due to increase in density due to super cooling.
2. Decrease in exit velocity.
3. Increase in final dryness-fraction and increase in enthalpy.

11.12 PHENOMENON IN NOZZLES OPERATING OFF THE DESIGN PRESSURE RATIO

(a) For Convergent Nozzle. Consider a convergent nozzle as shown in figure. The initial conditions are kept constant and exit pressure P_2 is reduced gradually from the initial pressure P_1 by a valve.

(i) When pressure P_2 is equal to P_t, there is no decrease in pressure and therefore mass-flow rate is zero. This condition is shown by curve (I) in Fig. 11.8.

(ii) When pressure P_2 is less than P_1, but more than critical pressure; distribution along the axis is shown by curve (II). The mass low rate increases as pressure P_2 is reduced as shown in figure.

(iii) When exit pressure P_2 is equal to critical pressure, the nozzle operates with maximum mass flow rate and the pressure distribution is shown by curve (III).
This state is called as 'chocked flow' or the nozzle is said to be 'chocked'.

(iv) When pressure P_2 is less than critical pressure, there is no change in mass-flow through nozzle and also pressure distribution along the nozzle is same. The pressure-drop from critical pressure to P_2 takes place after the nozzle. This expansion is irreversible and gives rise to pressure oscillations as shown by curve (IV).

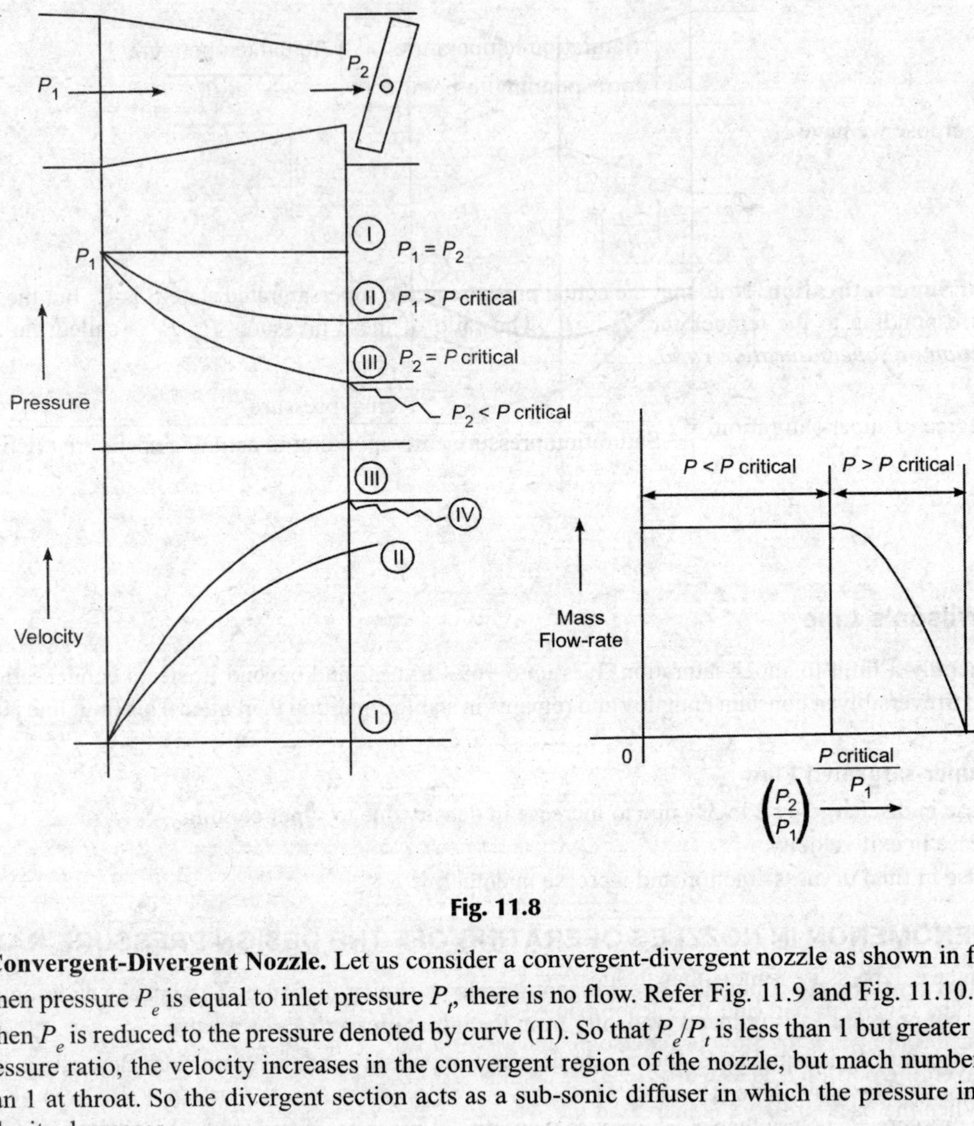

Fig. 11.8

(b) For Convergent-Divergent Nozzle. Let us consider a convergent-divergent nozzle as shown in figure.

1. When pressure P_e is equal to inlet pressure P_t, there is no flow. Refer Fig. 11.9 and Fig. 11.10.
2. When P_e is reduced to the pressure denoted by curve (II). So that P_e/P_t is less than 1 but greater than critical pressure ratio, the velocity increases in the convergent region of the nozzle, but mach number (m) is less than 1 at throat. So the divergent section acts as a sub-sonic diffuser in which the pressure increases and velocity decreases.
3. In case (IV), pressure is critical at throat and exit pressure P_e is design pressure. Therefore the flow is isentropic throughout the nozzle and velocity continuously increases along the nozzle. At throat, velocity is equal to sonic velocity. The divergent portion acts as a super-sonic nozzle with a continuous decrease in pressure and continuous increase in velocity.

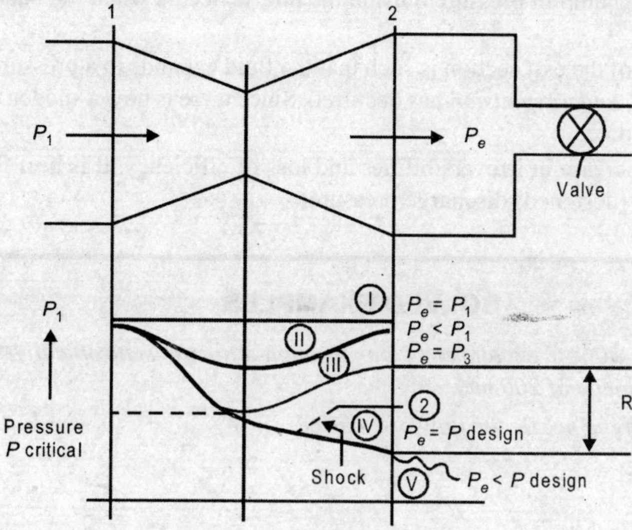

Fig. 11.9

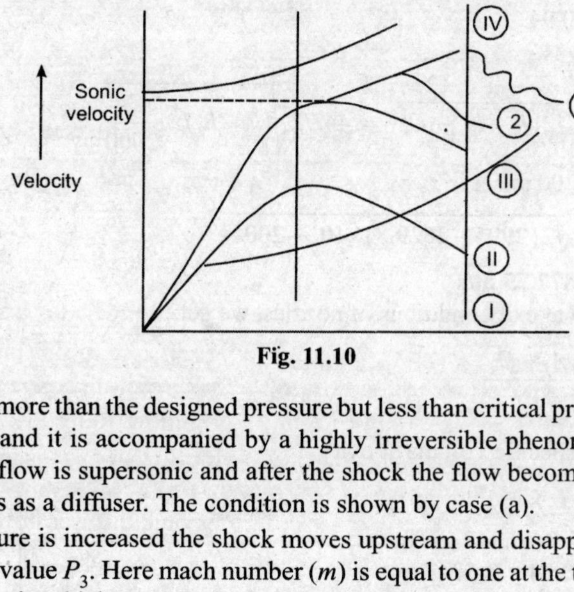

Fig. 11.10

4. If the exit pressure is more than the designed pressure but less than critical pressure, the flow is not isentropic in the divergent part and it is accompanied by a highly irreversible phenomena known as shocks. Shocks occur only when the flow is supersonic and after the shock the flow becomes sub-sonic and the rest of the diverging portion acts as a diffuser. The condition is shown by case (a).
5. When the back pressure is increased the shock moves upstream and disappears at the nozzle throat where pressure P_e has some value P_3. Here mach number (m) is equal to one at the throat but divergent portion acts as a sub-sonic diffuser in which pressure increases and velocity decreases.
6. If the pressure P_e is less than the design pressure, no further decrease in exit pressure occurs and drop of pressure from design pressure to P_e occurs outside the nozzle giving pressure fluctuations as shown by case (V).

11.12.1 Overexpansion and Underexpansion

If the area of the exit section of a nozzle is such that the fluid expands to a pressure at this section less than that of the discharge region. *Overexpansion* has occurred. As this lower pressure stream emerges into the higher pressure discharge region, there is a sudden increase in pressure, an act hat sets up compression pressure waves, much stronger than sound waves. This jump in pressure outside the nozzle occurs when the back-pressure is above the exit pressure.

If on the other hand the area of the exit section is such that the fluid expands to a pressure at this section greater than that in the discharge region, *underexpansion* has occurred. Since there is now a sudden decrease in pressure on the jet, expansion waves are initiated.

Both situations involve an increase in irreversibilities and loss of efficiency. It is best for the expansion in the nozzle to occur to just the right (designed) discharge pressure.

SOLVED EXAMPLES

Example 11.1 *Steam at 30 bar, 300° C expands to 5 bar through a nozzle isentropically. Calculate exit velocity. Assume steam is entering at velocity of 200 m/s.*

Also calculate area at exit of the nozzle for flow of 2 kg/sec.

Solution

From mollier chart,

$$h_1 = 2995 \text{ kJ/kg}$$
$$h_2 = 2629.8 \text{ kJ/kg}$$
$$x_2 = 0.94$$

The equation for exit velocity is,

$$C_2 = \sqrt{2(h_1 - h_2) + C_1^2} = 44.72\sqrt{(h_1 - h_2) + \frac{C_1^2}{2000}}$$

$$C_2 = 200 \text{ m/sec}$$

$$\therefore \quad C_2 = \sqrt{2(2995 - 2629.8)\times 10^3 + 200^2}$$
$$= \mathbf{877.75\ m/s}$$

Applying continuity equation at exit conditions of nozzles, we get,

$$m = \frac{A_2 - C_2}{V_2}$$

$$V_2 = \text{specific volume at exit}$$
$$= x_2 \times v_{g2}$$

From steam table, at 5 bar, $V_g = 0.347$ m³/kg

$$\therefore \quad v_2 = 0.94 \times 0.3747$$
$$= 0.3522$$

$$\text{Area at exit}\,(A_2) = \frac{m \times v_2}{c_2}$$

$$= \frac{2 \times 0.3522}{877.75}$$

$$= 8.025\times 10^{-4} \text{ m}^2$$
$$= \mathbf{8.025\ cm^2}$$

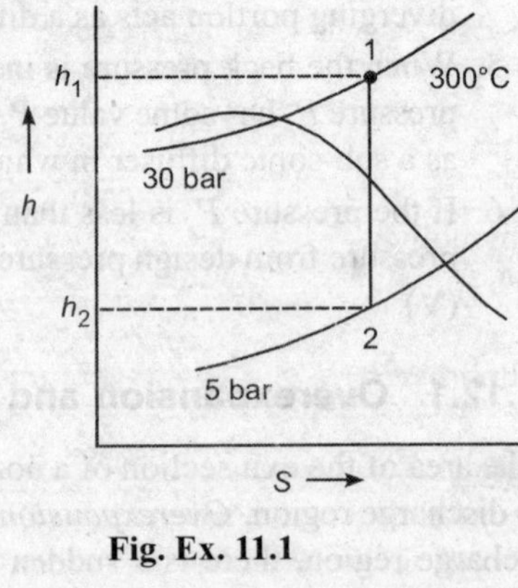

Fig. Ex. 11.1

Example 11.2 *Steam expands isentropically through a nozzle from 15 bar to 490 kPa with initial dry-saturated state.*

Calculate the exit velocity, neglecting initial velocity. If 10% heat is lost in friction, calculate the percentage reduction in velocity and improvement in the final quality of the steam.

Solution

$$P_1 = 15 \text{ bar}$$
$$P_2 = 4.9 \text{ bar}$$

From mollier chart,

$$h_1 = 2790 \text{ kJ/kg}$$
$$h_2 = 2590 \text{ kJ/kg}$$
$$C_2 = \sqrt{2(h_1 - h_2) + C_1^2}$$
$$C_1 = 0 \text{ (given)}$$
$$\therefore \quad C_2 = \sqrt{2(h_1 - h_2)} = \sqrt{2 \times (2790 - 2590) \times 10^3}$$
$$= 632.45 \text{ m/s}$$
$$= \text{Velocity without friction.}$$

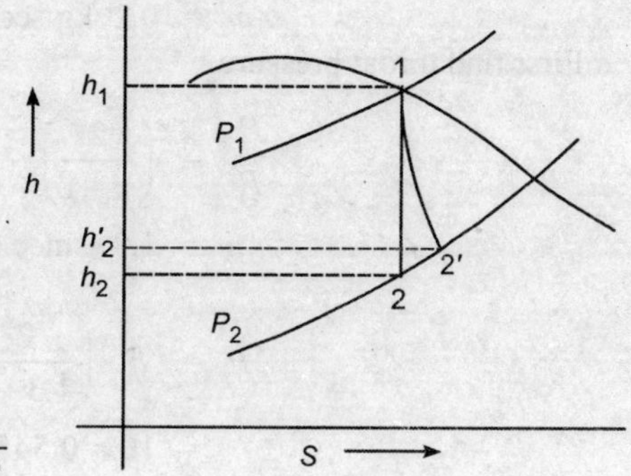

Fig. Ex. 11.2

When the friction is considered, the expansion will be similar to process $1 - 2'$ on *h–s* diagram of Fig. Ex. 11.2.

First find h_2'.

$$\text{Heat lost in friction} = 10\,\%$$
$$= 0.1\,(h_1 - h_2)$$
$$\text{So the enthalpy available} = 0.9\,(h_1 - h_2)$$
$$\therefore \quad (h_1 - h'_2) = 0.9\,(h_1 - h_2)$$
$$\therefore \quad h_1 - h'_2 = 0.9 \times (2790 - 2590)$$
$$= 180 \text{ kJ / kg}$$
$$C_2' = \sqrt{2(h_1 - h_2')}$$
$$= \sqrt{2(180) \times 10^3} = 600 \text{ m/s.}$$
$$\therefore \text{ Percentage reduction in velocity} = \left(\frac{C_2 - C_2'}{C_2}\right) \times 100$$
$$= \left(\frac{632.45 - 600}{632.45}\right) \times 100$$
$$= 5.13\ \%$$

From mollier-chart

$$x_2 = 0.92$$

and

$$x'_2 = 0.94$$

$\therefore$ Improvement in the final quality of steam, when friction considered

$$= x'_2 - x_2 = 0.02$$

Example 11.3 *Determine the throat area, exit velocity and exit area for convergent divergent steam nozzle to pass a mass of 0.2 kg/s, when the inlet conditions are 10 bar and 250° C and final pressure is 2 bar. Assume expansion is isentropic without friction and neglect initial velocity.*

Solution

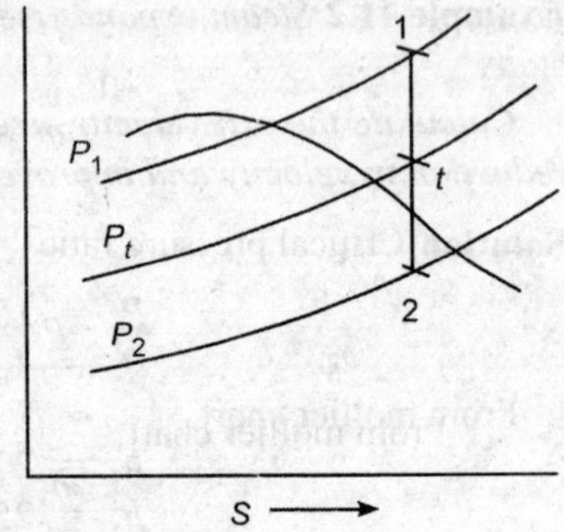

Fig. Ex. 11.3

$$P_1 = 10 \text{ bar}$$
$$P_2 = 2 \text{ bar}$$
$$P_t = \text{throat pressure}$$
$$m = 0.2 \text{ kg/sec}$$

First find throat pressure.

$$\frac{P_t}{P_1} = \left(\frac{2}{n+1}\right)^{n/n-1}$$

$$n = 1.3 \text{ (since steam is super-heated at inlet)}$$

$$\therefore \quad P_t = P_1 \times \left(\frac{2}{1.3+1}\right)^{\frac{1.3}{0.3}}$$

$$= 10 \times 0.5457 = 5.457 \text{ bar}$$

From mollier-chart,

$h_1 = 2944$ kJ / kg, $h_2 = 2628$ kJ/kg, $v_2 = 0.8537$ m³/kg

$h_t = 2813$.kJ / kg $V_t = 0.3735$ m³/kg

$$\text{Velocity at throat } (C_t) = \sqrt{2(h_1 - h_t)}$$

$$= \sqrt{2 \times (2944 - 2813) \times 10^3}$$

$$= 511.86 \text{ m/s}$$

$$\text{Area at throat } (A_t) = \frac{m \cdot v_t}{C_t}$$

$$= \frac{0.2 \times 0.3735}{511.86} = 1.46 \times 10^{-4} \text{ m}^2$$

$$= \mathbf{1.46 \text{ cm}^2}$$

$$\text{Exit velocity } (C_2) = \sqrt{2(h_1 - h_2)}$$

$$= \sqrt{2 \times (2944 - 2628) \times 10^3}$$

$$= 795 \text{ m/s}$$

$$\text{Exit area } (A_2) = \frac{m \cdot v_2}{C_2} = \frac{0.2 \times 0.8537}{795}$$

$$= 2.15 \times 10^{-4} \text{ m}^2 = \mathbf{2.15 \text{ cm}^2}$$

Example 11.4 *An adiabatic steam nozzle is to be designed for a discharge rate of 10 kg/sec at steam from 10 bar and 400° C to a back pressure of 1 bar. The nozzle efficiency is 92% and frictional loss is assumed to take place in the diverging portion of the nozzle only. Assume a critical pressure ratio of 0.545. Determine throat and exit areas.*

Solution

$$P_1 = 10 \text{ bar}$$
$$T_1 = 400°\text{C}$$
$$P_2 = 1 \text{ bar}$$

$$P_t = \text{throat pressure}$$
$$\eta_{nozzle} = 0.92$$
$$m = 10 \text{ kg/s}$$

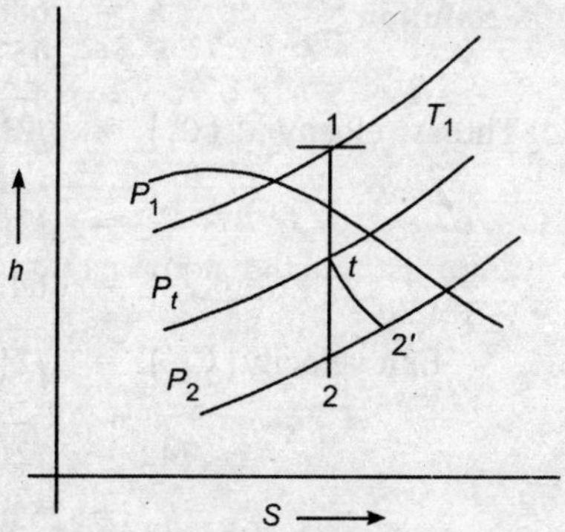

Fig. Ex. 11.4

$$\text{Critical pressure ratio} = \frac{P_t}{P_1}$$

$$\therefore \quad P_t = P_1 \times 0.545 = 10 \times 0.545$$
$$= 5.45 \text{ bar}$$

From mollier chart,

$$h_1 = 3263.9 \text{ kJ/kg}$$
$$h_t = 3110 \text{ kJ/kg}, \qquad V_t = 0.5 \text{ m}^3\text{/kg}$$
$$h_2 = 2700 \text{ kJ/kg} \qquad V_2 = 1.8 \text{ m}^3\text{/kg}$$

$$\text{Throat velocity}(C_t) = \sqrt{2(h_1 - h_t)}$$
$$= \sqrt{2(3263.9 - 3110) \times 10^3}$$
$$= \mathbf{554.8 \text{ m/s}}$$

$$\text{Throat area}(A_t) = \frac{(m \cdot v_t)}{C_t} = \frac{10 \times 0.5}{554.8} = 9.012 \times 10^3 \text{ m}^2$$

$$\text{Nozzle efficiency} = \frac{h_1 - h_2'}{h_1 - h_2}$$

$$h_1 - h_2' = 0.92\,(3263.9 - 2700)$$
$$= 518.79 \text{ kJ/kg}$$

$$\text{Exit velocity}(C_2) = \sqrt{2(h_1 - h_2')}$$
$$= \sqrt{2(518.79) \times 10^3} = \mathbf{1018.62 \text{ m/s}}$$

$$\text{Exit area}(A_2) = \frac{m \cdot v_2}{C_2} = \frac{10 \times 1.8}{1018.62}$$
$$= \mathbf{0.0176 \text{ m}^2}$$

Example 11.5 *A nozzle is supplied with steam at 7 bar and 180°C. The discharge takes place against back pressure of 1 bar. Approach velocity is 75 m/s. The pressure at throat is 4 bar. No frictional losses take place between inlet and throat. Determine velocity at throat.*

If exit velocity is 742 m/s, what is overall nozzle efficiency?

Solution

$$P_1 = 7 \text{ bar}$$
$$P_2 = 1 \text{ bar}$$
$$P_t = \text{throat pressure}$$
$$= 4 \text{ bar}$$

From mollier chart,

$$h_1 = 2800 \text{ kJ/kg}$$

$$h_t = 2680 \text{ kJ/kg}$$
$$h_2 = 2457 \text{ kJ/kg}$$

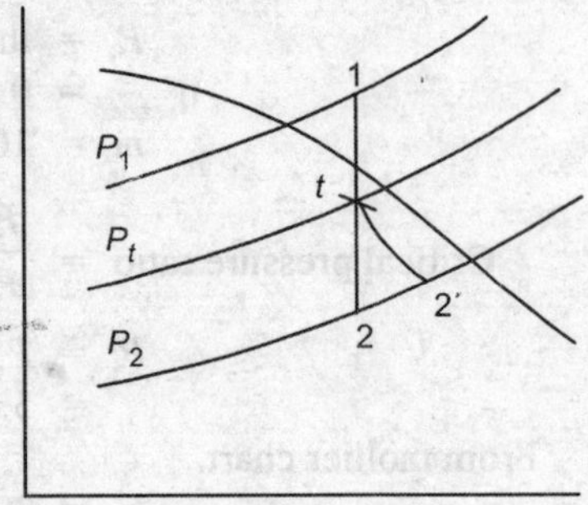

Fig. Ex. 11.5

$$\text{Throat velocity} \quad (C_t) = \sqrt{2(h_1 - h_t) + C_1^2}$$
$$= \sqrt{2(2800 - 2680) \times 10^3 + 75^2}$$
$$= 464 \text{ m/s}$$

$$\text{Exit velocity}(C_2') = \sqrt{2(h_1 - h_2') \times 10^3 + C_1^2}$$
$$\therefore \quad 742 = \sqrt{2(2800 - h_2') \times 10^3 + 75^2}$$
$$\therefore \quad h_2' = 2530 \text{ kJ/kg}$$

$$\text{Overall nozzle efficiency} = \left[\frac{h_1 - h_2' + \left(\frac{C_1^2}{2}\right)}{h_1 - h_2 + \left(\frac{C_1^2}{2}\right)}\right] \times 100$$

$$= \left[\frac{(2800 - 2530) + \frac{75^2}{2 \times 10^3}}{(2800 - 2457) + \frac{75^2}{2 \times 10^3}}\right] \times 100$$

$$= 78.89\ \%$$

Example 11.6 *Steam enters a convergent-divergent nozzle at 10 bar and 240ºC, with a velocity of 50 m/s. It is discharged at 0.5 bar with a velocity of 978 m/s. The expansion is isentropic upto throat and with friction in the diverging part.*

Determine:

1. *Final quality of steam.*
2. *Exit diameter required if the diameter at throat is 10 mm.*
3. *Number of nozzles required for a steam flow rate of 60 kg/min to a turbine.*
4. *Percentage of overall isentropic drop lost in the friction in the diverging part.*

Solution

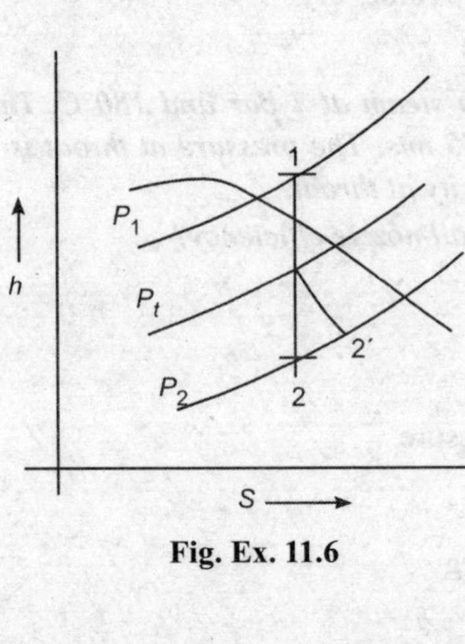

Fig. Ex. 11.6

$$P_1 = 10 \text{ bar} \qquad T_1 = 240°C$$
$$P_2 = 0.5 \text{ bar}$$
$$C_1 = 50 \text{ m/s}$$
$$C_2 = 978 \text{ m/s}$$

First found throat pressure (P_t)

$$\frac{P_t}{P_1} = \left(\frac{2}{n+1}\right)^{n/n-1}$$

$$n = 1.3 \text{ (since steam is superheated at inlet)}$$

$$\frac{P_t}{10} = \left(\frac{2}{1.3+1}\right)^{\frac{1.3}{0.3}}$$

$$\therefore \quad P_t = 5.45 \text{ bar}$$

From mollier chart,

$$h_1 = 2910 \text{ kJ/kg}$$
$$h_t = 2775 \text{ kJ/kg}, \qquad V_t = 0.4 \text{ m}^3\text{/kg}$$
$$h_2 = 2390 \text{ kJ/kg}, \qquad V_2 = 0.35 \text{ m}^3\text{/kg}$$

$$\text{Exit velocity } C_2' = \sqrt{2(h_1 - h_2') + C_1^2}$$

$$\therefore \quad 978 = \sqrt{2(2910 - h_2') + 50^2}$$

$$h_2' = 2433 \text{ kJ / kg}$$

Now by locating point 2′ mollier chart for $h_2' = 2433$ kJ/kg and $P_2' = 0.5$ bar, dryness fraction can be found.

$$\therefore \quad \mathbf{x_2' = 0.905}$$

$$\text{Throat velocity}(C_t) = \sqrt{2(h_1 - h_t) + C_1^2}$$

$$= \sqrt{2(2910 - 2775) \times 10^3 + 50^2}$$

$$= 522 \text{ m/sec}$$

Mass flow rate

$$m = \frac{A_t \cdot C_t}{V_t} = \frac{\frac{\pi}{4} \times (10^{-2})^2 \times 522}{0.4} = 0.1025 \text{ kg/s}$$

$$\text{Total steam flow rate} = 1 \text{ kg/s}$$

$$\therefore \text{ No. of nozzles required} = \frac{\text{Total flow rate}}{\text{Flow through a single nozzle}}$$

$$= \frac{1}{1.1025} = 9.756$$

$$\approx \mathbf{10 \text{ Nozzles}}$$

Now find exit diameter.

$$m = \frac{A_2' - C_2'}{V_2'}$$

$$\therefore \quad A_2' = \frac{m \cdot v_2'}{C'_2}$$

$$\therefore \quad \frac{\pi}{4}\left(d_2^2\right) = \frac{0.1025 \times 0.35}{978} \qquad \therefore \mathbf{d_2 = 21.6\ mm}$$

$$\text{Percentage enthalpy drop} = \frac{h_2' - h_2}{h_1 - h_2} \times 100$$

$$= \frac{2433 - 2390}{2910 - 2390} \times 100$$

$$= 8.27\ \%$$

Example 11.7 *A convergent-divergent nozzle is supplied with steam at a pressure of 1 MN/m² and 225°C. Supersaturated expansion occurs according to $pV^{1.3} = C$ in the nozzle down to an exit pressure of 0.32 MN/m². The exit diameter of the nozzle is 25 mm. If the flow through the nozzle is chocked, determine (i) exit velocity, (ii) mass flow, (iii) throat diameter.*

Solution

$$1\ \text{MN/m}^2 = \frac{1 \times 10^6}{10^5} = 10\ \text{bar}$$

$$0.32\ \text{MN/m}^2 = \frac{0.32 \times 10^6}{10^5} = 3.2\ \text{bar}$$

(i) Exit velocity

$$\text{From steam table, } h_1 = \frac{2826.8 + 2943}{2}$$

$$= \frac{5769.8}{2} = \mathbf{2884.9}$$

$$v_1 = \frac{0.2059 + 0.2327}{2} = \frac{0.4386}{2} = \mathbf{0.2193\ m^3/kg}$$

Work done/kg = Change in KE

= Area of pV diagram

$$= \frac{n}{n-1}\left[p_1 v_1 - p_2 v_2\right]$$

$$\therefore \quad \left(\frac{v_2}{v_1}\right)^{1.3} = \frac{p_1}{p_2} = \frac{10}{3.2}$$

$$\frac{v_2}{v_1} = \left(\frac{10}{3.2}\right)^{\frac{1}{1.3}} = (3.125)^{0.7692} = 2.4024$$

$$\therefore \quad v_2 = 0.2193 \times 2.4024 = \mathbf{0.5268\ m^3/kg}$$

$$\therefore \quad \text{Work done/kg} = \frac{1.3}{0.3}\times 100[10\times 0.2193 - 3.2\times 0.5258]$$

$$= 433.33\ [2.1930 - 1.6858]$$

$$= 433.33 \times 0.5072$$

$$= \mathbf{219.78\ kJ/kg}$$

$$\therefore \quad \text{Exit velocity} = 44.72\sqrt{219.78}$$

$$= \mathbf{662.93\ m/sec}$$

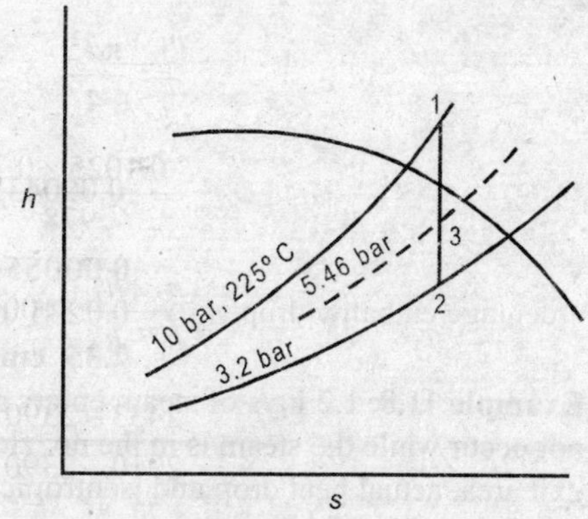

Fig. Ex. 11.7

(ii) Mass flow

$$\text{Mass flow at exit} = \frac{A_3V_3}{v_3} = \frac{A_2V_2}{v_2}$$

$$A_2 = \frac{\pi}{4}\times\frac{25\times 25}{1000\times 1000} = \frac{\pi}{4\times 1600}\ \text{m}^2$$

$$V_2 = 662.93\ \text{m/s},\ v_2 = 0.5268\ \text{m}^3/\text{kg}$$

$$\therefore \quad \dot{m} = \frac{\pi}{4\times 1600}\times\frac{662.93}{0.5268}$$

$$= \mathbf{0.6177\ kg/sec}$$

$$= \mathbf{2223.72\ kg/hour}$$

(iii) Throat diameter

$$\text{Throat pressure} = 0.546\ p_1 = 5.46\ \text{bar}$$

$$\therefore \quad v_3 = v_1\left(\frac{P_1}{P_3}\right)^{1/n} = 0.2193\left(\frac{10}{5.46}\right)^{0.7692}$$

$$v_3 = 0.2193 \times 1.5923$$

$$= \mathbf{0.3493\ m^3/kg}$$

$$\text{Change in K.E. upto throat} = \frac{n}{n-1}[p_1v_1 - p_3v_3]$$

$$= \frac{1.3}{0.3}\times\frac{10^5}{1000}[10\times 0.2193 - 5.46\times 0.3493]$$

$$= 433.33\ [2.193 - 1.907]$$

$$= \mathbf{123.93\ \ kJ/kg}$$

$$\therefore \quad V_3 = 44.72\sqrt{123.93}$$

$$= \mathbf{497.845\ \ m/s}$$

$$\therefore \quad \dot{m} = \frac{A_3V_3}{v_3} = \frac{A_3\times 497.845}{0.3493}$$

$$\therefore \quad A_3 = \frac{0.6177\times 0.3493}{497.845}$$

$$= 4.334 \times 10^{-4}$$

$$= 0.0004334\ \text{m}^2$$

$$= \frac{\pi d_3^2}{4}$$

$$\therefore \quad d_3^2 = 0.0004334 \times \frac{4}{\pi}$$

$$= 0.0005518$$

$$d_3 = 0.0235 \text{ m}$$

$$= \mathbf{2.35 \text{ cm}}$$

Example 11.8. 1.2 kg/s of steam enters a nozzle at 12 bar and 250°C and expands upto 2 bar. Condensation does not occur while the steam is in the nozzle. Find the throat area, degree of undercooling, degree of supersaturation, exit area, actual heat drop and isentropic heat drop.

Solution

p_1 = Inlet pressure = 12 bar

and temperature = 250°C

$\therefore \quad p_2$ = Throat pressure = 0.546 × 12

= 6.552 bar

p_3 = 2 bar

From diagram $(h - s)$,

h_1 = 2935 kJ/kg

h_2 = 2800 kJ/kg

The flow throughout is supersaturated and n = 1.3.

$$\therefore \quad \frac{T_1}{T_3} = \left(\frac{p_1}{p_3}\right)^{\frac{n-1}{n}} = \left(\frac{12}{2}\right)^{\frac{0.3}{1.3}} = 6^{0.2308}$$

$$= 1.531$$

$$\therefore \quad T_3 = \frac{250+273-523}{1.512} = 345.90 \text{ K} = \mathbf{72.90°C}$$

Saturation temprature at 2 bar = **120.23° C**

∴ Degree of undercooling = 120.23 – 72.9

= **47.33° C**

Pressure corresponding to 72.9° C = 0.3543 bar.

$$\therefore \text{Degree of supersaturation} = \frac{2}{0.3543} = \mathbf{5.645}$$

Upto throat the steam is superheated.

$$\therefore \quad h_1 - h_2 = 2935 - 2800$$

$$= \mathbf{135 \text{ kJ/kg}}$$

$$\therefore \quad V_2 = 44.72\sqrt{135}$$

$$= \mathbf{519.6 \text{ m/s}}$$

V_2= 0.35 m^3/kg

$$\therefore \quad m_2 = \frac{A_2 V_2}{V_2}$$

$$\therefore \quad 1.2 = \frac{A_2 \times 519.6}{0.35}$$

$$\therefore \quad A_2 = \frac{1.2 \times 0.35}{519.6} m^3 = 0.000808$$

$$= \mathbf{8.08\ cm^2}$$

$$h_1 - h_3 = \frac{n}{n-1} p_1 v_1 \left[1 - \left(\frac{p_3}{p_1}\right)^{\frac{n-1}{n}}\right]$$

$$= \frac{1.3}{0.3} \times \frac{12 \times 10^5 \times 0.1924}{1000}\left[1 - \left(\frac{2}{12}\right)^{\frac{0.3}{1.3}}\right]$$

$$= 1000.48\left[1 - \left(\frac{1}{6}\right)^{\frac{0.3}{1.3}}\right]$$

$$= 1000.48(1 - 0.6616)$$

$$= 1000.48 \times 0.3384$$

Actual heat drop $= \mathbf{338.56\ kJ/kg}$

Isentropic enthalpy drop $= 2935 - 2590$

$= \mathbf{345\ kJ/kg}$

$$\therefore \quad V_3 = 44.72\sqrt{338.56} = \mathbf{822.85\ m/s}$$

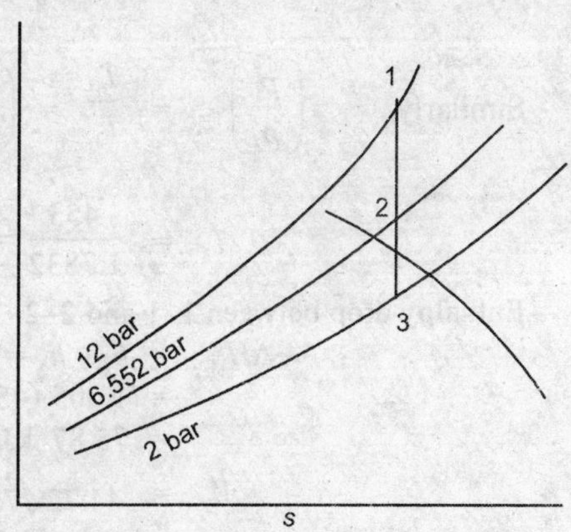

Fig. Ex. 11.8

$$\therefore \quad 1.2 = \frac{A_3 V_3}{v_3}$$

or

$$A_3 = \frac{1.2 \times v_3}{V_3}$$

$$= \frac{1.2 \times 0.779}{822.85}$$

$$= \mathbf{11.36\ cm^2}$$

Example 11.9 *(a) What do you understand by superstaurated flow and metastable state?*

(b) Derive the following property, ratios for isentropic flow in a duct in terms of specific heat ratio (γ) and Match number (M)

(i) Stagnation pressure/static pressure.

(ii) Stagnation density/static density.

(c) *Air at 7.8 bar, 180°C expands through a convergent-divergent nozzle into a space at 1.03 bar. The flow rate of air is 3.2 kg/s. Assuming isentropic flow throughout, calculate the throat and exit areas of the nozzle. Neglect inlet velocity.*

Solution

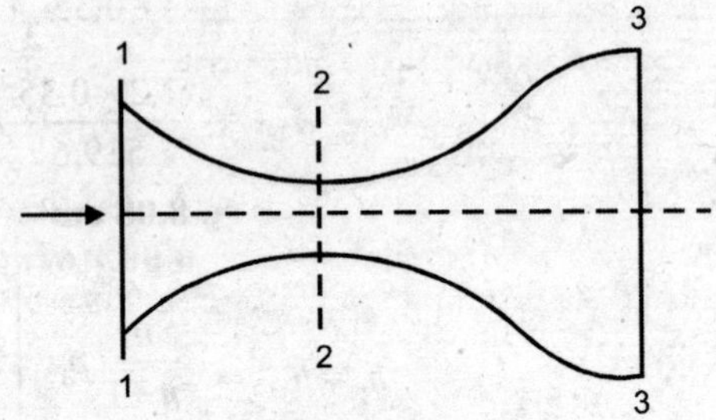

Fig. Ex. 11.9

$$p_1 = 7.8 \text{ bar}$$
$$p_3 = 1.03 \text{ bar}$$
$$T_1 = 180 + 273 = 453 \text{ K}$$

Critical pressure ratio,

$$\frac{p_2}{p_1} = 0.5283 \text{ for air}$$

$$\therefore \quad P_2 = 0.5283$$
$$= 4.12 \text{ bar}$$

$$\therefore \quad \frac{T_2}{T_1} = \left(\frac{p_2}{p_1}\right)^{\frac{\gamma-1}{\gamma}} = (0.5283)^{0.2857} = 0.8333$$

$$\therefore \quad T_2 = 0.8333 \times 453 = 377.51 \text{ K}$$

Similarly, $$\left(\frac{p_1}{p_3}\right)^{\frac{\gamma-1}{\gamma}} = \frac{T_1}{T_3} = \left(\frac{7.8}{1.03}\right)^{0.2857} = 1.7832$$

$$\therefore \quad T_3 = \frac{453}{1.7832} = \mathbf{254.04\ K}$$

Enthalpy drop between 1–1 and 2–2.

$$-\Delta H_{1-2} = h_1 - h_2 = C_{pa}\,(T_1 - T_2)$$
$$= 1.005\,(453 - 377.51) = 1.005 \times 75.49$$
$$= 75.87 \text{ kJ/kg}$$

$$\therefore \quad V_2 = 44.72\sqrt{75.87} = 44.72 \times 88.1$$
$$= 389.52 \text{ m/s}$$

Similarly, $$-\Delta H_{1-3} = h_1 - h_3 = C_{pa}\,(T_1 - T_3)$$
$$= 1.005\,(453 - 254.04)$$
$$= 1.005 \times 198.96$$
$$= \mathbf{199.955\ kJ/kg}$$

$$\therefore \quad V_3 = 44.72\sqrt{199.955}$$
$$= \mathbf{632.36\ m/s}$$

Specific volume of air at throat conditions

$$v_2 = \frac{RT_2}{p_2} = \frac{287 \times 377.51}{4.12 \times 10^5}$$

$$v_2 = 0.2631 \text{ m}^3/\text{kg}$$

Similarly, specific volume at exit conditions,

$$v_3 = \frac{RT_3}{p_3} = \frac{287 \times 254.04}{1.03 \times 10^5}$$

$$= \mathbf{0.7079\ m^3/kg}$$

$\therefore$ For throat, $\dot{m} = \dfrac{A_2 V_2}{v_2}$

$$\therefore \quad A_2 = \frac{3.2 \times 0.2631}{389.52}$$

$$= \mathbf{0.002161 m^2}$$

Let d_2 be the throat diameter in cm

$$\therefore \quad 21.61 = \frac{\pi}{4} \times d_2^2$$

$$d_2^2 = \frac{21.6}{\pi} \times 4 = 27.5$$

$\therefore d_2$ = throat diameter = **5.244 cm**

Similarly, exit area, $A_3 = \dfrac{\dot{m}\, v_3}{V_3} = \dfrac{3.2 \times 0.7079}{632.36}$

$$= 0.003582\ m^2$$

$$= \mathbf{35.82\ cm^2}$$

$$\therefore \quad d_3 = \left(\frac{35.82 \times 4}{\pi}\right)^{1/2} = (45.61)^{1/2}$$

$$= \mathbf{6.75\ cm}$$

= Exit diameter

Example 11.11 *(a) Prove that the maximum discharge in a steam nozzle per unit area at the throat is given by*

$$m_{max} = \left\{ 1000n. \frac{p_1}{v_1} \left(\frac{2}{n+1} \right)^{\frac{n+1}{n-1}} \right\}^{1/2}$$

(b) A convergent-divergent nozzle is supplied with dry saturated steam at 10 bar. The divergent portion of the nozzle is 5 cm along and the throat diameter is 6 cm. Calculate the semi cone angle of the divergent part of the nozzle so that the steam leaves the nozzle at 0.5 bar. Neglect friction.

Solution

(a) Discharge through any section is given by,

$$\dot{m} = \frac{\text{Area} \times \text{Velocity}}{\text{Specific volume}}$$

$$= \frac{AV}{v}$$

Consider the flow follows the law $pv^n = V$

$$\therefore \quad p_1 v_1^n = p_2 v_2^n$$

or
$$v_2 = \left(\frac{p_1}{p_2}\right)^{1/n} \cdot v_1 \quad \text{and} \quad v_1 = v_2 \times \left(\frac{p_2}{p_1}\right)^{1/n}$$

For the flow through throat, we have

$$\dot{m} = \frac{A_2 \cdot V_2}{v_2}$$

and
$$\frac{V_2^2 - V_1^2}{2 \times 1000} = h_1 - h_2 = \frac{n}{n-1} \cdot p_1 v_1 \left[\left(\frac{p_2}{p_1}\right)^{\frac{n-1}{n}} - 1\right]$$

$$\left(\frac{p_2}{p_1}\right) = \left(\frac{2}{n+1}\right)^{\frac{n}{n-1}} \quad \text{for maximum discharge}$$

$$\dot{m} = A_2 \left\{\frac{2000n}{n-1} \cdot \frac{p_1}{v_1}\left[\left(\frac{p_2}{p_1}\right)^{\frac{2}{n}} \cdot \left(\frac{p_2}{p_1}\right)^{\frac{n+1}{n}}\right]\right\}^{1/2}$$

$$\dot{m}_{\max} = \left\{\frac{2000 \cdot n}{n-1} \cdot \frac{p_1}{v_1}\left[\left(\frac{2}{n+1}\right)^{\frac{n}{n+1} \times \frac{2}{n}} - \left(\frac{2}{n+1}\right)^{\frac{n}{n-1} \times \frac{n+1}{n}}\right]\right\}^{1/2}$$

$$= \left\{\frac{2000\, n}{n-1}\, \frac{p_1}{v_1}\right\}\left[\left(\frac{2}{n+1}\right)^{\frac{2}{n-1}} - \left(\frac{2}{n+1}\right)^{\frac{n+1}{n-1}}\right]^{1/2}$$

$$= \left\{\frac{2000\, n}{n-1} \cdot \frac{p_1}{v_1} \cdot \left(\frac{2}{n+1}\right)^{\frac{n+1}{n-1}}\left[\left(\frac{2}{n+1}\right)^{\frac{2}{n-1} - \frac{n+1}{n-1}} - 1\right]\right\}^{1/2}$$

$$= \left\{\frac{2000\, n}{n-1} \cdot \frac{p_1}{v_1} \cdot \left(\frac{2}{n+1}\right)^{\frac{n+1}{n-1}}\left[\left(\frac{2}{n+1}\right)^{\frac{2-n-1}{n-1}} - 1\right]\right\}^{1/2}$$

$$= \left\{\frac{2000\, n}{n-1} \cdot \frac{p_1}{v_1} \cdot \left(\frac{2}{n+1}\right)^{\frac{n+1}{n-1}}\left[\left(\frac{2}{n+1}\right)^{\frac{1-n}{n-1}} - 1\right]\right\}^{1/2}$$

$$= \left\{ \frac{2000\,n}{n-1} \cdot \frac{p_1}{v_1} \cdot \left(\frac{2}{n+1} \right)^{\frac{n+1}{n-1}} \left[\left(\frac{2}{n+1} \right)^{\frac{1-n}{n-1}} - 1 \right] \right\}^{1/2}$$

$$= \left\{ \frac{2000\,n}{n-1} \cdot \frac{p_1}{v_1} \cdot \left(\frac{2}{n+1} \right)^{\frac{n+1}{n-1}} \left[\frac{n+1}{2} - 1 \right] \right\}^{1/2}$$

$$= \left\{ \frac{2000\,n}{n-1} \cdot \frac{p_1}{v_1} \cdot \left(\frac{2}{n+1} \right)^{\frac{n+1}{n-1}} \left[\frac{n+1-2}{2} \right] \right\}^{1/2}$$

$$= \left\{ \frac{2000\,n}{n-1} \cdot \frac{p_1}{v_1} \cdot \left(\frac{2}{n+1} \right)^{\frac{n+1}{n-1}} \times \frac{n-1}{2} \right\}^{1/2}$$

$$= \left\{ 1000\,n \cdot \frac{p_1}{v_1} \cdot \left(\frac{2}{n+1} \right)^{\frac{n+1}{n-1}} \right\}^{1/2}$$

Neglecting approach velocity V_1, we have

$$V_2 = \left\{ \frac{n}{n-1} \cdot p_1 v_1 \cdot \left[\left(\frac{p_2}{p_1} \right)^{\frac{n-1}{n}} + 1 \right] \right\}^{1/2}$$

$$= \sqrt{\frac{n}{n-1} p_1 v_1 \left[1 - \left(\frac{p_2}{p_1} \right)^{\frac{n-1}{n}} \right] \times 2000}$$

$$\therefore \qquad \dot{m} = \frac{A_2 \times \sqrt{2 \times 1000 \cdot \frac{n}{n-1} \cdot p_1 v_1 \left[1 - \left(\frac{p_2}{p_1} \right)^{\frac{n-1}{n}} \right]}}{\left(\frac{p_1}{p_2} \right)^{1/n} v_1}$$

$$= A_2 \times \sqrt{\frac{2000 \times n}{n-1} \cdot \frac{p_1}{v_1} \left[\left(\frac{p_2}{p_1} \right)^{2/n} - \left(\frac{p_2}{p_1} \right)^{\frac{n-1}{n'} \frac{2}{n}} \right]}$$

$$= A_2 \cdot \left\{ \frac{2000\,n}{n-1} \cdot \frac{p_1}{v_1} \cdot \left[\left(\frac{p_2}{p_1} \right)^{2/n} - \left(\frac{p_2}{p_1} \right)^{\frac{n+1}{n}} \right] \right\}^{1/2}$$

For maximum discharge,

$$\frac{d\dot{m}}{d\left(\frac{p_2}{p_1}\right)} = 0 = \frac{2}{n} \cdot \left(\frac{p_2}{p_1}\right)^{\frac{2-n}{n}} - \frac{n+1}{n}\left(\frac{p_2}{p_1}\right)^{\frac{n+1-n}{n}}$$

$$\therefore \quad \frac{2}{n} \cdot \left(\frac{p_2}{p_1}\right)^{\frac{2-n}{n}} = \frac{n+1}{n}\left(\frac{p_2}{p_1}\right)^{1/n}$$

$$\frac{2}{n+1} = \left(\frac{p_2}{p_1}\right)^{\frac{1}{n} - \frac{2-n}{n} = \frac{1-2+n}{n} = \frac{n-1}{n}}$$

$$\therefore \quad \frac{p_2}{p_1} = \left(\frac{2}{n+1}\right)^{\frac{n}{n-1}}$$

Condition for maximum discharge.

Substituting this condition in the above equation for $\dot{m}$, we get for unit area

$$\dot{m}_{max} = \left\{ \frac{2000 \cdot n}{n-1} \cdot \frac{p_1}{v_1} \left[\left(\frac{2}{n+1} \right)^{\frac{2}{n-1}} - \left(\frac{2}{n+1} \right)^{n+1} \right] \right\}$$

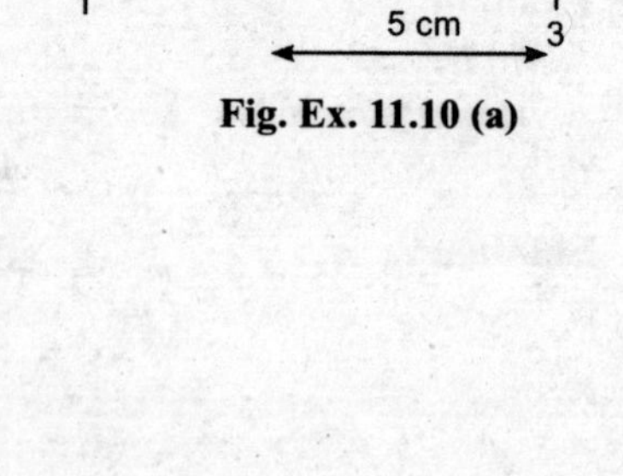

Fig. Ex. 11.10 (a)

(b) $p_1 = 10$ bar

$x_1 = 1$

$p_3 = 0.5$ bar

$p_2 = 0.58\ p_1 = 0.58 \times 10$

$= 5.8$ bar

From table, $h_1 = 2776.2$ kJ/kg

$s_1 = 6.5828$ kJ/kg-K, $s_2 = 1.9176 + x_2 \times 4.8514$

$$\therefore \quad x_2 = \frac{6.5828 - 1.9176}{4.8514} = \frac{4.6652}{4.8514} = \mathbf{0.9616}$$

$\therefore \quad h_2 = 664.69 + 0.9616 \times 2089.3$

$= 664.69 + 2009.1 = \mathbf{2673.79\ kJ/kg}$

and $v_2 = x_2\, v_{g2} = 0.9616 \times 0.32574 = \mathbf{0.3132\ m^3\ kg}$

$\therefore \quad V_2 = 44.72\sqrt{2776.2 - 2673.79} = 44.72\sqrt{102.41}$

$= \mathbf{452.56\ m/s}$

$s_1 = s_3 \quad \therefore\ 6.5828 = 1.0912 + x_3 \times 6.5035$

$$\therefore \quad x_3 = \frac{6.5825 - 1.0912}{6.5035} = \frac{5.4916}{6.5035}$$
$$= 0.8444$$
$$\therefore \quad h_3 = 340.56 + 0.8444 \times 2305.4$$
$$= 340.56 + 1946.7$$
$$= \mathbf{2287.26\ kJ/kg}$$
$$v_3 = x_3\, v_{g3} = 0.8444 \times 3.2402$$
$$= \mathbf{2.736\ m^3/kg}$$

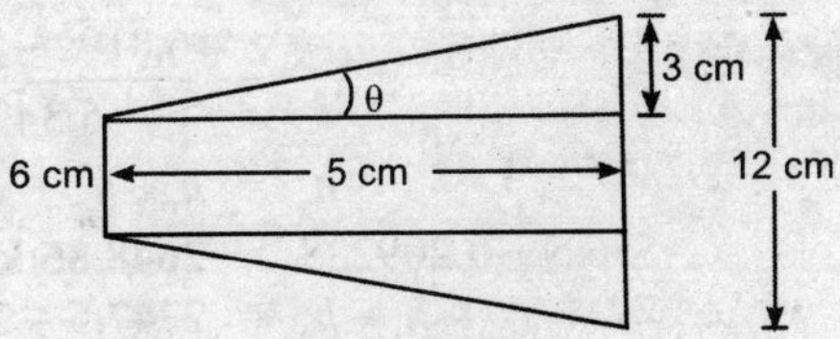

Fig. Ex. 11.10(b)

$$V_3 = 44.72\sqrt{(h_1 - h_3)} = 2776.2 - 2287.26$$
$$= 44.72\sqrt{488.94}$$
$$= \mathbf{988.85\ m/s}$$
$$\therefore \quad \dot{m}_S = \frac{A_2 V_2}{v_2} = \frac{A_3 V_3}{v_3}$$
$$\therefore \quad \frac{\pi d_2^2}{4} \times \frac{V_2}{v_2} = \frac{\pi d_3^2}{4} \times \frac{V_3}{v_3}$$
$$\frac{d_3^2}{d_2^2} = \frac{V_2}{V_3} \times \frac{v_3}{v_2} = \frac{452.56}{988.85} \times \frac{2.736}{0.3132}$$
$$= 4$$
$$\therefore \quad \frac{d_3}{d_2} = 2 \quad \therefore d_3 = 2 \times 6 = \mathbf{12\ cm}$$
$$\therefore \quad \tan\theta = \frac{3}{5} = 0.6$$
$$\therefore \quad \theta = \text{semi-cone angle} = \tan^{-1} 0.6$$
$$= \mathbf{30.964^\circ}$$
$$= \mathbf{30^\circ.58'}$$

Example 11.11 *(a) What do you understand by (i) critical pressure ratio and (ii) choked flow for a nozzle?*

(b) Dry saturated steam at 5 bar enters a converging-diverging nozzle at a velocity of 100 m/s. The exit pressure is 1.5 bar. The throat and exit area are 1200 m² and 1600 mm² respectively. Assuming isentropic flow upto the throat and taking critical pressure ratio as 0.58, estimate the mass flow rate of steam and nozzle efficiency.

Solution

$$V_1 = 100\ \text{m/s}$$
$$P_1 = 5\ \text{bar}$$
$$x_1 = 1$$
$$P_3 = 1.5\ \text{bar}$$
$$A_2 = 1200\ \text{mm}^2$$
$$A_3 = 1600\ \text{mm}^2$$

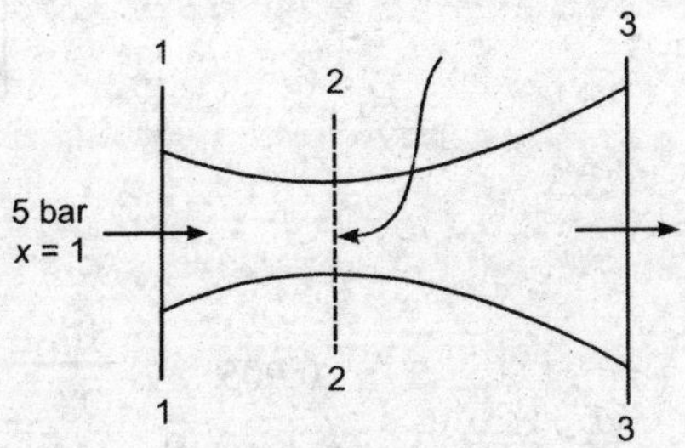

Fig. Ex. 11.11(a)

$$\frac{p_2}{p_1} = 0.58 \quad \therefore p_2 = 5 \times 0.58 = \mathbf{2.90\ bar}$$

$$h_1 = 2747.5 \text{ kJ/kg}, s_1 = 6.8192 = s_2 = 1.6595$$

$$\therefore \quad x_2 = \frac{6.8192 - 1.6595}{5.3428} = \frac{5.1597}{5.3428} = \mathbf{0.9657}$$

$$\therefore \quad h_2 = h_{f2} + h_{f2} + x_2 \times h_{fg2} = 556.5 + 0.9657 \times 2166.6$$

$$556.5 + 0.209235 = \mathbf{2648.85 \text{ kJ/kg}}$$

$$\therefore \quad h_1 - h_2 = 2747.5 - 2648.85 = 98.65 \text{ kJ/kg}$$

$$\therefore \quad V_2 = 44.72\sqrt{98.65}$$

$$= \mathbf{455.3 \text{ m/s}}$$

$$v_2 = x_2 v_{g2} = 0.9657 \times 0.62513$$

$$= 0.604 \text{ m}^3\text{/kg}$$

$$\therefore \quad \dot{m} = \text{mass flow rate} = \frac{A_2 V_2}{v_2}$$

$$= \frac{1200}{10^6} \times \frac{455.4}{0.604} = \mathbf{0.905 \text{ kg/s}}$$

Now, $p_3 = 1.5 \text{ bar} \therefore s_3 = 1.4336 + x_3 \times 5.7898$

$$\therefore \quad x_3 = \frac{6.8192 - 1.4336}{5.7898} = \frac{5.3856}{5.7898}$$

$$= 0.9302$$

$$\therefore \quad h_3 = 467.13 + 0.9302 \times 2226.2$$

$$= 467.13 + 2070.78$$

$$= 2537.91 \text{ kJ/kg}$$

Now, m is constant.

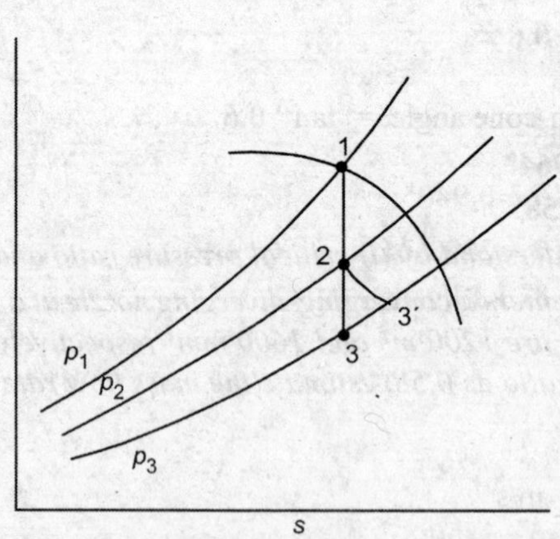

Fig. Ex. 11.11(b)

$$\therefore \quad 0.905 = \frac{A_3 V_{3'}}{v_{3'}}$$

$$= \frac{1600}{10^6} \times 44.72 \frac{\sqrt{2747.5 + 5 - 467.13}}{x_3' \times 1.59}$$

$$\therefore \frac{0.905 \times 100 \times 1.159\, x_3'}{0.16 \times 44.72} = \sqrt{2285.37 - 2226.2\, x_3'}$$

$$(14.66x_2')^2 = 2285.37 - 2226.2x_3'$$

$$214.9\, x_{3'}^2 = 2285.37 - 2226.2x_3'$$

$$x_{3'}^2 = 10.63 - 10.36x_3'$$

$$x_{3'}^2 = 10.36\, x_3' - 10.63 = 0$$

$$\therefore \quad x_3' = \frac{-10.36 \pm \sqrt{(10.36)^2 + 4 \times 10.63}}{2}$$

$$= \frac{-10.36 \pm \sqrt{107.32 + 42.52}}{2}$$

$$= \frac{-10.36 \pm \sqrt{149.84}}{2}$$

$$= \frac{-10.36 \pm 12.241}{2}$$

$$= \frac{12.241 - 10.360}{2}$$

$$= \frac{1.881}{2}$$

$$= \mathbf{0.9405}$$

$$\therefore \quad h_{3'} = 467.13 + 0.9405 \times 2226.2$$

$$= 467.13 + 2093.74$$

$$= \mathbf{2560.87\ kJ/kg}$$

$\therefore$ Nozzle efficiency,

$$\eta_{nozzle} = \frac{h_1 - h_{3'}}{h_1 - h_3}$$

$$= \frac{2747.5 - 2560.87}{2747.5 - 2537.91}$$

$$= \frac{186.63}{209.59}$$

$$= 0.89$$

or $$= \mathbf{89\%}$$

Example 11.12 *A steam turbine is developing 185 kW power with a specific steam consumption of 7 kg/kWh. The steam supplied to the turbine is dry and saturated and at a pressure of 13 bar and exausted at a pressure of 0.14 bar The diameter at the throat of each nozzle is 7 mm. Assume that 12% of the total heat drop is lost in friction. Expansion of steam is ideal upto throat and all the friction loses occur in diverging part of nozzles. Determine the total number of nozzles required and find the diameter of nozzles at their exit end. Assume the ratio of throat pressure to inlet pressure as 0.578.*

Solution

$$\text{Total steam required} = \text{SSC} \times \text{Power developed}$$

$$= 7 \times 185$$

$$= 1295 \text{ kg / hour}$$

$$= \mathbf{0.3597 \text{ kg/sec}}$$

$$h_1 = 2785.4 \text{ kJ / kg}$$

$$p_2 = 7.514 \text{ bar} = 13 \times 0.578 \text{ kJ / bar}$$

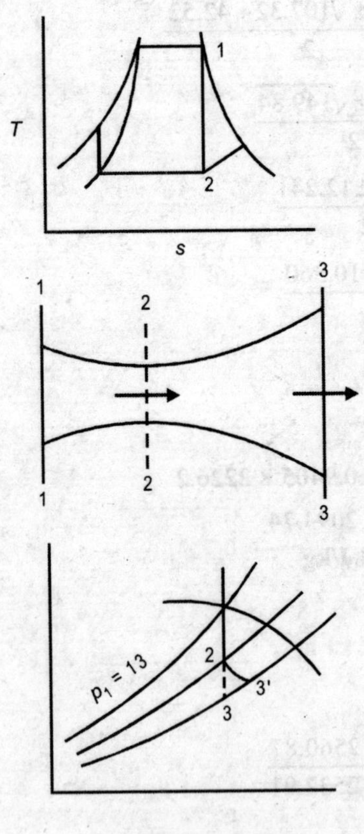

Fig. Ex. 11.12

$$s_1 = 6.4913 = s_2 = \frac{4.0390}{2} + x_2 \cdot \frac{9.3243}{2} = 2.0195 + 4.6621 \times x_2$$

$$x_2 = \frac{6.4913 - 2.0195}{4.6621} = \frac{4.4718}{4.6621} = \mathbf{0.9592}$$

$$\therefore \quad h_2 = \frac{706.9 + 711.67}{2} + x_2\left(\frac{2057.4 + 2053.7}{2}\right)$$

$$= 709.83 + 0.9592 \times 2055.55 = \mathbf{2681.5}$$

$$\therefore \quad v_2 = \left(\frac{0.25870 + 0.25224}{2}\right) \times 0.9592$$

$$= 0.9592 \times \left(\frac{0.51094}{2} = 0.25547\right) = \mathbf{0.245}$$

$$V_2 = 44.72\sqrt{h_1 - h_2} = 44.72\sqrt{2785.4 - 26875}$$

$$= 44.72\sqrt{103.9} \approx 455.8 \text{ m/s}$$

$$\text{Throat area } A_2 = \frac{\pi}{4} \times (0.007)^2\ m^2 = 0.0000385 \text{ m}^2$$

$$\therefore \text{ Mass of steam flowing/sec} = \frac{A_2 V_2}{v_2}$$

$$= \frac{0.0000385 \times 455.8}{0.245} = \mathbf{0.0716\ kg/s}$$

$\therefore$ No. of nozzles required

$$= \frac{\text{Total steam/sec}}{\text{Steam/nozzle}} = \frac{0.3597}{0.0716} = \mathbf{5.02\ say\ 5}$$

For divergent portion:

We have, $s_1 = s_3 = 6.4913$

$$= 0.7367 + x_3 \times 7.2967$$

$$\therefore \quad x_3 = \frac{6.4913 - 0.7367}{7.2967} = \mathbf{0.7886}$$

$$\therefore \quad h_3 = 220.02 + 0.7886 \times 2376.7$$

$$= \mathbf{2094.32\ kJ.kg}$$

$\therefore$ Total heat or enthalpy drop $= h_1 - h_3 = 2785.4 - 2094.32$

$$= \mathbf{691.08\ kJ/kg}$$

$$\therefore \quad \text{Heat lost in friction} = \frac{12}{100} \times 691.08 = \mathbf{73.33\ kJ/kg}$$

$$\therefore \quad h_{3'} = 2094.32 + 73.33 = \mathbf{2167.65\ kJ/kg}$$

$$= h_{f_3} + x_{3'} h_{fg_3} = 220.02 + x_{3'} \times 2376.7$$

$$\therefore \quad x_{3'} = \frac{2167.65 - 220.02}{2376.7} = \mathbf{0.824}$$

$$\therefore \quad v_{3'} = 0.824 \times 10.694 = \mathbf{8.812\ m^3/kg}$$

$$\therefore \quad V_{3'} = 44.72\sqrt{h_1 - h_{3'}} = 44.72\sqrt{2785.4 - 2167.65}$$

$$= \mathbf{1111.5\ m/s}$$

$$\therefore \quad m = \frac{A_3 V_{3'}}{v_{3'}}$$

$$\therefore \quad A_3 = \frac{0.0716 \times 8.812}{1111.5} = \frac{\pi d_3^2}{4}$$

$$\therefore \quad d_3 = \mathbf{2.73\ cm = 27.3\ mm}$$

Example 11.13 *(a) Explain why a subsonic nozzle is convergent while a supersonic nozzle is divergent.*

(b) A nozzle is supplied with dry steam at 12 bar and back pressure 1 bar. Whether the nozzle is convergent or divergent? Calculate the throat and exit diameters if it discharges 15 kg steam per minute. Solve by using steam tables only.

Solution

(a) Combining continuity and energy equations, we get

$$\frac{dA}{A} = \frac{dV}{V}\left(-1 + M^2\right) \quad \text{with usual notations.}$$

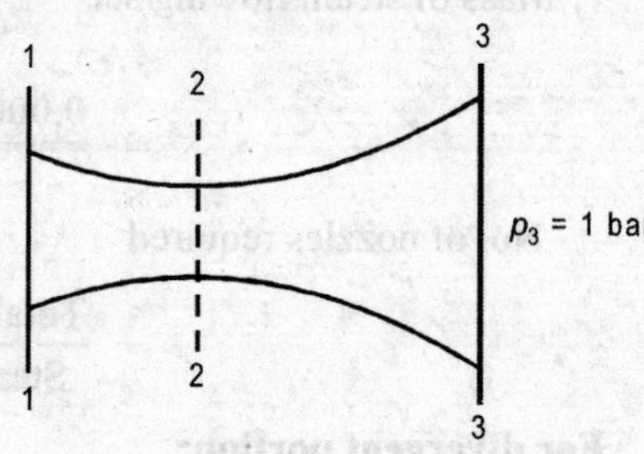

Fig. Ex. 11.13

In the subsonic region, $M < 1$

∴ $(-1 + M^2)$ is negative.

∴ dV is to be positive i.e., increase in velocity, dA has to be negative i.e., area is decreasing i.e., the nozzle is convergent.

∴ Subsonic nozzle is a convergent nozzle.

Similarly, a supersonic nozzle is a diverging nozzle.

(b) $p_1 = 12$ bar and $p_3 = 1$ bar.

∴ Critical pressure p_2 at the throat is given by

$$p_2 = 0.578, p_1 = 0.578 \times 12 = \mathbf{6.936\ bar}$$

which is greater than outlet pressure.

∴ The nozzle is **convergent-divergent.**

$$h_1 = 2782.7 \text{ kJ/kg}$$

$$s_1 = 6.5194 \text{ kJ/kg-°K}$$

For finding properties from steam tables, for p_2, p_2 is taken as 7 bar.

$$\therefore \quad s_2 = 1.9918 + x_2 \times 4.7134 \text{ and } s_1 = s_2$$

$$\therefore \quad x_2 = \frac{6.5194 - 1.9918}{4.7134} = \frac{4.5276}{4.7134}$$

$$\therefore \quad h_2 = 679.06 + 0.9606 \times 2064.9$$

$$= \mathbf{2681\ kJ/kg}$$

$$v_2 = \text{specific volume} = 0.9606 \times 0.27268$$

$$= \mathbf{0.26194\ m^3/kg}$$

$$h_1 - h_2 = \text{Enthalpy drop upto throat}$$

$$= 2782.7 - 2681 = 101.7 \text{ kJ/kg}$$

$$\therefore \quad v_2 = \text{Velocity of steam at throat}$$

$$= 44.72\sqrt{h_1 - h_2} = 44.72\sqrt{101.7}$$

$$= 450.99 \text{ m/s} \approx \mathbf{451\ m/s}$$

$$\therefore \quad m_2 = \frac{15}{60} = \frac{\text{Area} \times \text{Velocity}}{\text{Specific volume}} = \frac{A_2 \times 451}{0.26194}$$

$$= 1721.77\, A_2$$

$$\therefore \quad A_2 = \frac{151.26194}{60 \times 451} \times 14 = \mathbf{1.452\ cm^2} = \frac{\pi}{4} d^2$$

$$\therefore \quad d_2 = \text{Diameter at throat}$$

$$= \mathbf{1.36\ cm}$$

$$d_2 = 1.36 \text{ cm}$$

$$p_3 = 1 \text{ bar}$$

$$\therefore \quad s_3 = 1.3027 + x_3 \times 6.0571$$

$$\therefore s_1 = s_3 \quad \therefore 6.5194 = 1.3027 + 6.0571\, x_3$$

$$x_3 = \frac{6.5194 - 1.3027}{6.0571} = \frac{5.2167}{6.0571}$$

$$= \mathbf{0.8613}$$

$$\therefore \quad h_3 = 417.51 + 0.8613 \times 2257.9$$

$$= \mathbf{2362.13\ kJ/kg}$$

$$\therefore \quad v_3 = 0.8613 \times 1.6937 = 1.4588 \text{ m}^3$$

$$\therefore \quad V_3 = 44.72\sqrt{2782.7 - 2362.13}$$

$$= 44.72\sqrt{420.57} = \mathbf{917.11\ m/s}$$

$$\therefore \quad m_3 = \frac{15}{60} = \frac{A_3 V_3}{v_3} = \frac{A_3 \times 917.11}{1.4588}$$

$$\therefore \quad A_3 = \frac{1}{4} \times \frac{1.4588}{917.11} \text{ m}^2$$

$$= \frac{14588}{4 \times 917.11} \text{ cm}^2 = \mathbf{3.98\ cm^2}$$

$$\therefore \quad d_3 = \sqrt{3.98 \times \frac{4}{\pi}} = \mathbf{2.25\ cm}$$

$$d_3 = 2.25 \text{ cm}$$

Example 11.14 *(a) Explain the following terms as referred to a system nozzle*

(i) under expanding nozzle

(ii) chocked nozzle

(iii) over expanding nozzle.

(b) Steam enters a convergent-divergent nozzle at 10 bar and 240°C with the approach velocity of 50 m/s. It is discharged at 0.5 bar with a velocity of 978 m/s. The expansion is isentropic upto throat and with friction in the divergent part.

Determine:

(i) Final quality of steam.

(ii) Diameter at exit if the throat diameter is 10 mm.

(iii) Number of nozzles required for a steam flow rate of 60 kg/min.

(iv) % of overall isentropic enthalpy drop lost in the friction the diverging part.

Solution

(b) **(i) Final quality of steam:**

Initially steam is superheated.

$$\therefore \quad n = 1.3$$

$$\therefore \text{Critical pressure ratio} = \frac{p_2}{p_1} = 0.546$$

$$p_c = p_2 = 10 \times 0.546 = \mathbf{5.46\ bar \approx 5.5\ for\ calculation}\ 1\text{–}2 \text{ expansion upto throat.}$$

$$h_1 = 2925 \text{ kJ / kg}$$

For the divergent part :

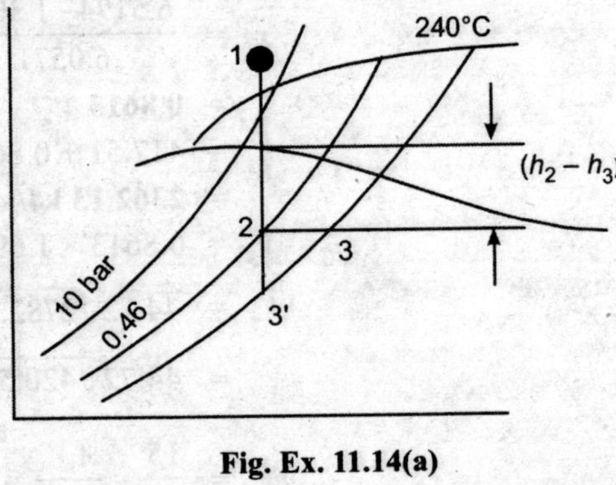

Fig. Ex. 11.14(a)

$$V_3 = 44.72\sqrt{h_1 - h_3}$$

$$\therefore \quad 978 = 4.72\sqrt{(h_1 - h_3)}$$

$$= 44.72\sqrt{(h_1 - h_3) + 0.25}$$

$$\therefore \quad \sqrt{h_1 - h_3} = \frac{978}{44.72} = 21.87$$

$$h_1 - h_3 + 1.25 = 478.27$$

$$\therefore \quad h_3 = h_1 - 478.27 = 2446.73$$

$$\therefore \quad x_3 = \text{Final quality of steam}$$

$$= \frac{2446.73 - 340.56}{2305.4} = \frac{2106.17}{2305.4} = \mathbf{0.914}$$

(ii) Exit diameter:

$$\therefore \quad h_1 + \frac{V_1^2}{2} = h_2 + \frac{V_2^2}{2}$$

$$\therefore \quad h_1 - h_2 = 2925 - 2800 = 125$$

$$= \frac{V_2^2 - V_1^2}{2 \times 1000}$$

$$\therefore \quad (V_2^2 - V_1^2) = 2000 \times 125$$

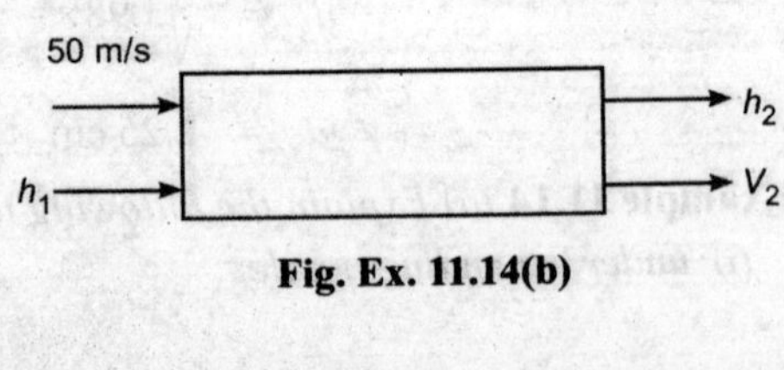

Fig. Ex. 11.14(b)

$$\therefore \quad V_2^2 = 2000 \times 125 + V_1^2$$

$$V_2 = \sqrt{2000 \times 125 + 2500} = 100\sqrt{0.2 \times 125 + 0.25}$$

$$= 100 \times \sqrt{25.25} = \mathbf{502.5\ m/s}$$

$$v_2 = \text{specific volume from chart} = 0.4046$$

$$v_3 = 0.0914 \times 3.2402 = 2.96$$

$$\therefore \quad \frac{A_2 V_2}{v_2} = \frac{A_3 V_3}{v_3} \quad \therefore A_3 = \frac{A_2 V_2}{v_2} \cdot \frac{v_3}{V_3}$$

$$\therefore \quad \frac{d_3^2}{d_2^2} = \frac{d_2^2 \times 502.5}{0.4046} \times \frac{2.96}{978} = 3.76$$

$$\therefore \quad \frac{d_3}{d_2} = 1.94 \ \therefore d_3 = 1.94 \times 10 = \mathbf{19.4\ mm}$$

(iii) Number of nozzles: Mass of steam flowing through one nozzle

$$= \frac{V_3 A_3}{v_3} = \frac{978 \times \frac{\pi}{4} \cdot \left(\frac{19.4}{1000}\right)^2}{2.96} = \mathbf{0.391\ kg/sec}$$

Actual flow is 60 kg/min i.e., **1 kg/sec.**

$$\therefore \text{Number of nozzles required} = \frac{1}{0.391}$$

$$2.56 \approx \mathbf{3\ nozzles}$$

(iv) % friction loss:

$$\text{Friction :} \quad h_3 - h_{3'} = 2446.73 - 2400 = \mathbf{46.73\ kJ/kg}$$

$$\therefore \quad \%\ \text{friction loss} = \frac{46.73}{525} \times 100$$

$$= \mathbf{8.9\ \%}$$

Example 11.15 *A steam nozzle is required to supply steam at rate 1 kg/sec from inlet condition of bar abs dry saturated and exit pressure of 1 bar. The isentropic efficiencies for the convergent section and divergent section are 0.95 and 0.9 respectively. Find:*

(1) the throat and exit diameters,

(2) length of divergent section if the cone angle of the divergent section is 14,

(3) Power in kW corresponding to the exit velocity of steam.

Solution

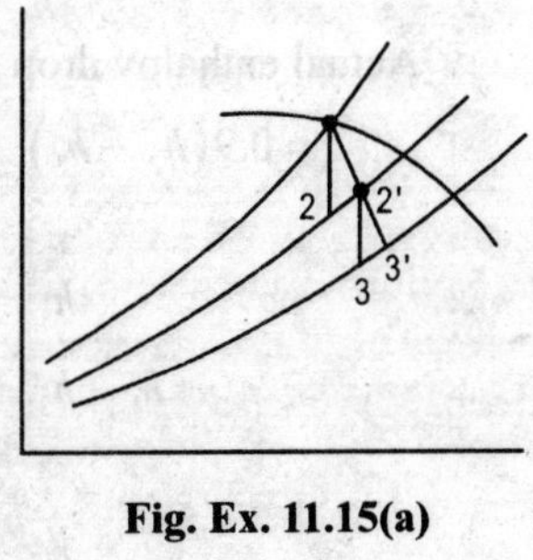

Fig. Ex. 11.15(a)

$$h_1 = 2776.2 \text{ kJ/kg}, \quad s_1 = 6.583 \text{ kJ/kg-K}$$

$$p_2 = 0.58\ p_1 = 10 \times 0.58 = 5.8 \text{ bar}$$

$$\therefore \quad s_2 = 1.918 + x_2 \times 4.851$$

$$\therefore \quad x_2 = \frac{6.583 - 1.918}{4.851} = \frac{4.665}{4.851} = \mathbf{0.9617}$$

$$h_2 = 664.7 + 0.9617 \times 2089.3$$
$$= \mathbf{2009.28 \ kJ/kg}$$

$\therefore$ Isentropic enthalpy drop $= h_1 - h_2 = 2776.2 - 2009.28 = 766.92$

$\therefore$ Actual enthalpy drop $= 7776.92 \times 0.95 = \mathbf{738.1 \ kJ/kg}$

$$h_{2'} = 2776.2 - 738.1 = 2038.1$$
$$= h_{f_2} + x_{2'} \, h_{fg_2}$$
$$= 664.7 + x_{2'} \times 2089.3$$

$$\therefore \quad x_{2'} = \frac{2038.1 - 664.7}{2089.3}$$
$$= 0.6573$$

$$\therefore \quad v_{2'} = 0.6573 \times 0.32572$$
$$\approx \mathbf{0.2157 \ m^3/kg}$$

$$V_2 = 44.72\sqrt{h_1 - h_{2'}} = 44.72\sqrt{738.1}$$
$$= \mathbf{1214.95 \ m/s}$$

$$\therefore \quad \dot{m} = \frac{A_2 V_2}{v_2} = \frac{A_2 \times 1214.95}{0.2157} = 1$$

$$\therefore \quad A_2 = \text{Throat area} = \frac{0.2157}{1214.95} \text{ m}^2$$

$$= \frac{2157}{1214.95} \text{ cm}^2 = \mathbf{1.78 \ cm^2}$$

$$\therefore \quad d_2 = \sqrt{\frac{1.78}{\frac{\pi}{4}}} = \mathbf{2.27 \ cm}$$

$$s_{2'} = 1.918 + 0.6622 \times 4.851$$
$$= \mathbf{5.130}$$
$$s_3 = 1.303 + x_3 \times 6.057$$

$$\therefore \quad x_3 = \frac{5.130 - 1.303}{6.057} = \mathbf{0.6318}$$

$$\therefore \quad h_3 = 417.5 + 0.6318 \times 2257.9 = 1844.04 \text{ kJ/kg}$$

$\therefore$ Actual enthalpy drop in divergent portion

$$= 0.9(h_{2'} - h_3) = 0.9(2048.13 - 1844.04)$$
$$= \mathbf{183.68 \ kJ/kg}$$

$$\therefore \quad h_{3'} = 1844.04 + 20.41 = \mathbf{1864.45 \ kJ/kg}$$

$$\therefore \quad h_1 - h_{3'} = 2776.2 - 1864.45$$
$$= \mathbf{911.75 \ kJ/kg}$$

$$\therefore \quad V_3 = 44.72\sqrt{911.75} = \mathbf{1350.33 \ m/s}$$

$$x_{3'} = \frac{1864.45 - 417.5}{2257.9}$$

$$= \frac{1446.96}{2257.9} = \mathbf{0.641}$$

$$\therefore \quad v_{3'} = x_{3'} v_{s_3} = 0.641 \times 1.6938 = 1.085 \text{ m}^3/\text{kg}$$

$$\therefore \quad A_3 = \text{Exit area} = \frac{1 \times 1.085}{1350.33} \text{ m}^2 = \frac{1.085 \times 10^4}{1350.33} \text{ cm}^2$$

$$= \mathbf{8.04 \text{ cm}^2}$$

$$\therefore \quad d_3 = \sqrt{\frac{8.04}{0.785}} = \mathbf{3.2 \text{ cm}}$$

$$\frac{d_3 - d_2}{2} = \frac{3.2 - 2.27}{2} = \frac{0.93}{2} = 0.465 \text{ cm}$$

$$\sin 7° = \frac{0.465}{L} = 0.1219$$

$\therefore \quad L$ = Length of the divergent portion

$$= \frac{0.465}{0.1219} = \mathbf{3.8 \text{ cm}}$$

(3) Power in kW corresponding to the exit velocity of steam

$$= \frac{\dot{m} V_3^2}{2} = \frac{1 \times 1350.33 \times 1350.33}{2 \times 1000}$$

$$= \mathbf{911.7 \text{ kW}}$$

THEORY QUESTIONS

11.1. What is a nozzle and what is its function?

11.2. Derive the expression for velocity and mass flow rate in a nozzle.

11.3. Derive the expression of critical pressure ratio for the maximum discharge through a nozzle.

11.4. What do you understand by supersaturated flow in case of nozzles.

11.5. Explain the various terms.
(a) Degree of supersaturation (b) Degree of undercooling.

11.6. An nozzle is supplied with steam at 7 bar and 180° C. The discharge take place against back pressure of 1 bar. The initial velocity is 75 m/s. The pressure at throat is 4 bar and there is no friction between inlet and throat. Determine the velocity at the throat. If exit velocity is 742 m/sec. What is overall nozzle efficiency.
(**Ans.** 78.62%)

11.7. Ten nozzles are required to discharge steam to an impulse turbine from a pressure of 8 bar and 350° C. The back pressure of the turbine is maintained at 0.14 bar. The nozzle efficiency is 90%. If the throat diameter for each nozzle is 0.8 cm find the total mass flow rate and exit diameter of each nozzle.

PROBLEMS FOR PRACTICE

1. The steam nozzle inlet pressure of steam is 7 bar and temperature 300°C and discharge at 1 bar. The diverging portion is 50 mm long and throat diameter is 5 mm. Assuming 10 % of the total enthalpy drop to be lost in friction in the diverging portion of nozzle, find the cone angle of the divergent portion. **(Ans.** 2° – 12')
2. The inlet condition of steam to a convergent-divergent nozzle is 15 bar pressure and 250°C temperature. The exit pressure is 0.4 bar. Assuming 85% nozzle efficiency and throat area of 30 cm², determine the exit area for the nozzle. **(Ans.** 221.5 sq.cm)
3. The nozzles in the stage of an impulse turbine receive steam at 12 bar, superheated by 62°C. The pressure in the wheel chamber is 4 bar. State whether the nozzle is convergent, divergent or convergent-divergent. Assuming negligible approach velocity and 10% frictional loss, suggest suitable number of nozzles to be used for a flow rate of 450 kg/min. Approximately, with the exit area of 2.4 cm². Also state the exact rate of discharge through the nozzle. **(Ans.** 24, 0.3125 kg/sec/nozzle)
4. Steam enters the nozzles of an impulse turbine at 10 bar, 250°C and leaves at 1bar. The turbine develops 225 kW with the specific steam consumption of 16. If the throat diameter is 8 mm, find the number of nozzles and the exit diameter, assuming that 10% of the total enthalpy drop is lost in friction in the divergent portion only, velocity of approach may be neglected. **(Ans.** 15, 12.3 mm)
5. A dry saturated steam is expanded in a nozzle from pressure 10 bar to 4 bar. If the expansion is supersaturated, find :

   (i) the degree of undercooling (ii) the degree of supersaturation.

   (Ans. (i) 10.4°C ; (ii) 1.354)

Chapter 12 Steam Turbines and Steam Turbine Systems

12.1 INTRODUCTION

Turbine is a device having rotating wheel to which is fastened a series of blades uniformly placed on its periphery. The expansion of the steam in nozzle or bucket converts its enthalpy into kinetic energy. The steam from nozzles or guide passage is directed continuously against the moving blades. The high velocity steam leaving the nozzle looses its velocity as it moves over moving blades causing rotation of moving blades.

The steam turbine has various applications. It is essentially a high speed device. It is used to run the electric generators, fans, blowers centrifugal pumps and with gear connections to run marine propeller for propulsion of ships. Turbines are also used in rolling mills.

12.1.1 Advantages

There are some advantages of turbines as given below:

1. Low initial cost smaller floor space required. Low expenses for maintenance. Light foundation is required.
2. The turbines have large over-load capacity.
3. The exhaust steam is free from oil.
4. The turbines have uniform velocity of rotation. So there is absence of the vibration.
5. High efficiency over a wide range.
6. Due to absence of reciprocating parts, perfect balancing is possible.
7. It can be built in much greater capacity than practical steam engine.
8. Due to uniform speeds, turbines are suitable for power generation.

12.2 COMPARISON BETWEEN STEAM TURBINE AND STEAM ENGINE

Steam turbines are compared with steam engines as described below.

1. In steam engine, reciprocating motion has to be converted into rotary motion by connecting rod and crank but in steam turbine direct rotational motion is available. Steam engines are inherently low speed engines becuase of the balancing problems. As against this, steam turbines being rotary ones there are no balancing problems.

2. It is difficult to couple steam engine to electric generator. Steam turbines are most suitable to be coupled to electric generators so turbines are used for power generation.

3. In steam-engine, steam gets mixed with oil, so condensate is not suitable as feed-water. For steam turbine, internal lubrication is not required so steam is free from oil and condensate can be used as boiler feed.

4. The wear and tear and maintenance cost of steam engine is more than turbine.

5. It is difficult to design large units for steam-engine. But for steam-turbines large units (250 MW to 500 MW) are feasible.

6. The thermal efficiency of steam turbine is more than that of steam engine.

12.3 CLASSIFICATION OF STEAM TURBINES

Steam turbines are classified as follows:

1. By principal of operation
   (a) Impulse turbine
   (b) Reaction turbine
   (c) Impulse Reaction Turbine

2. By the method of compounding
   (a) Pressure compounded turbine
   (b) Velocity compounded turbine
   (c) Pressure-velocity compounded turbine

3. By the pressure of inlet steam
   (a) High pressure turbine
   (b) Low pressure turbine

4. By the exhaust pressure
   (a) Condensing
   (b) Non-condensing
   (c) High back pressure
   (d) Bleeder

5. By shaft position
   (a) Vertical shaft
   (b) Horizontal shaft

6. By field service
   (a) Stationary, (b) variable speed, (c) locomotive, (d) ship, (e) industrial etc.

12.4 SIMPLE IMPULSE TURBINE

The impulse turbine depends completely upon the dynamic action of the steam. The direction of motion of the steam is changed as it passes across the blade. As a result of change of direction of steam across the blade, it will impart a force on the blade. Now if the blade is free, it will move in the direction of force.

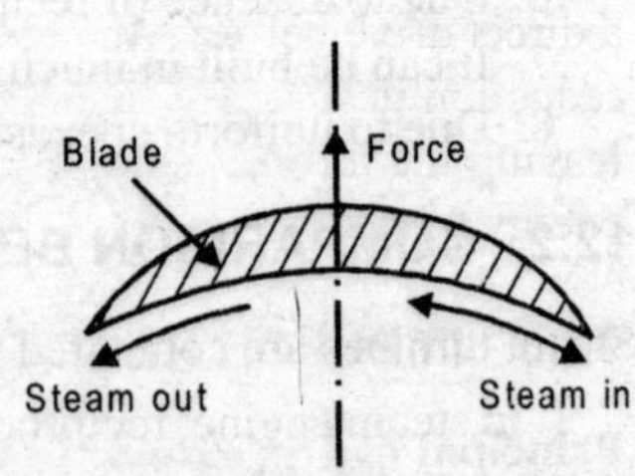

Fig. 12.1

In impulse turbine, steam first passes through the nozzle. In nozzle, the enthalpy is reduced and is converted into kinetic energy. The high velocity steam passes over moving blades and force is developed as shown in Fig. 12.1. The number of blades are fixed around the circumference of the shaft. So the steam will cause the turbine shaft to rotate. Thus mechanical work is obtained. The velocity of steam is reduced as it moves over the moving blades. But the pressure remain unchanged.

12.4.1 Working

The steam turbine shown in the Fig. 12.2 is simple impulse turbine or De-Laval turbine. There is only one set of nozzles. The complete expansion of steam from the steam chest pressure to the exhaust or condenser pressure takes place in the one set of nozzles. Thus the pressure in the moving blades chamber is approximately equal to condenser pressure.

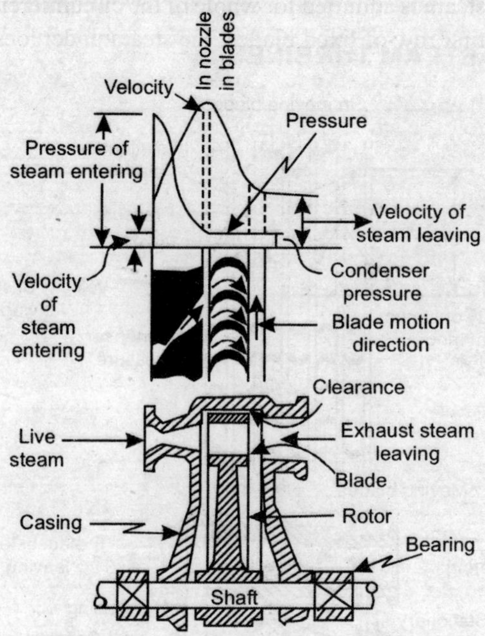

Fig. 12.2

The steam enters the moving blades chamber with a high velocity. The pressure in the set of moving blades remains constant. The velocity of steam is reduced as it passes over moving blades as some of the kinetic energy of the steam is used in producing work on the turbine shaft.

For example, impulse turbine, rotational speeds of the magnitudes of 20000 rpm may be obtained. This high speed of rotation will produce high centrifugal force which will restrict the size of the turbine. Thus the De-Laval type turbine is of relatively small size and hence has a small power-output. Also due to the high speed of rotation, a direct drive between the turbine disc and external equipment such as generator is not possible. For this reason, a reduction gear box is installed between the turbine-shaft and the external equipment. Also the velocity of steam leaving the turbine is quite appreciable resulting in a energy loss. It is as high as 10 to 12% of the initial kinetic energy of steam.

12.5 IMPULSE REACTION TURBINE

Principal of Operation. In reaction turbines, the pressure of the steam is reduced as it moves over the moving blades that gives rise to an increase in kinetic energy. This pressure drop also gives rise to a reaction in the direction opposite to that of increased velocity. Thus driving force is obtained. In impulse reaction turbine, impulse principle plus reaction principle is used.

12.5.1 Working

There are number of rows of moving blades fixed to the shaft and equal number of fixed blades attached to the casing. Refer Fig. 12.3. The fixed blades in this arrangement correspond to the nozzles referred to in the impulse turbine. Instead of a set of nozzles, steam is admitted for whole of the circumference and therefore there is all round admission. In passing through the first row of fixed blades, the steam undergoes a small drop in pressure and, its

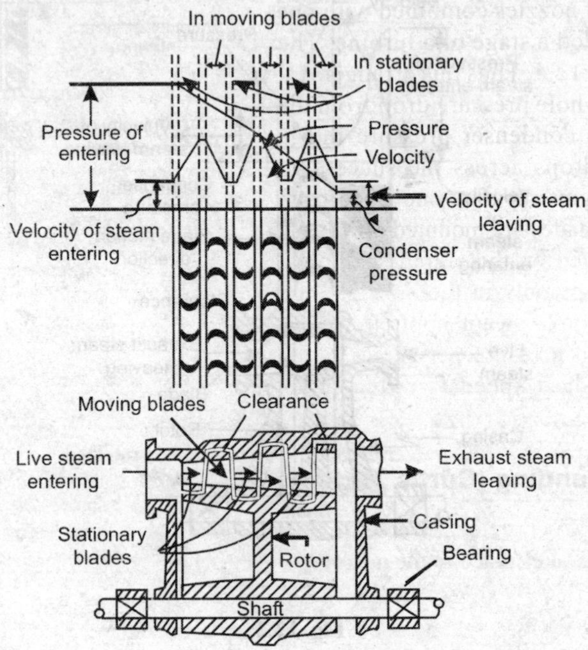

Fig. 12.3

velocity is increased. It then enters the first row of moving blades and just as in impulse turbine, it suffers change in direction and momentum. This gives an impulse to the blades. But here the passage of the blades is so designed that there is a small drop in pressure in the moving blades that gives rise to the reaction force. Thus the driving force is a vector sum of impulse and reaction forces. The steam velocities in this type of turbine are comparatively low.

12.6 COMPOUNDING OF STEAM TURBINES

Impulse turbines require nozzles and pressure drop of steam takes place in nozzles. The steam enters the turbine with a high velocity. There is no pressure drop in the turbine. The velocity reduces as some of the kinetic energy of steam is used up in producing power. The steam velocity entering the turbine is very high if the pressure drop from boiler pressure to the condenser pressure takes place in a single nozzle row. The speed of rotation becomes exceedingly high—upto 30000 r.p.m. — if single row of turbine blading is used. Hence, this speed needs reduction to practical speed.

Compounding of the steam turbines is the utilisation of the energy in stages. The methods used are:

(i) Velocity compounding.

(ii) Pressure compunding and

(iii) Pressure-velocity compounding

12.6.1 Pressure Compounding (Reateu Stage)

The high rotational speed can be avoided by combining the impulse stages in series and on the same output shaft. The exhaust steam from the rotor blades of the first stage enters the nozzle of the succeeding impulse stage. The nozzles in succeeding stage are built in and locked to fixed diaphragm. Each set of nozzles combined with one disc of moving blades is called a stage of a turbine. The arrangement is shown in Fig. 12.4. Thus this arrangement amounts to splitting up the whole pressure drop from the steam chest pressure to the condenser pressure into a series of smaller pressure drops across the successive impulse stages. The diaphragms separate one stage from the other and all the moving blades are mounted on wheels which are individually mounted on the output shaft. The expansion of steam takes place only in the nozzles. The blades of rotor change direction of steam, apply force and develop torque. With drop in pressure per stage, steam velocity is reduced. Thus shaft speed is reduced for maximum work done.

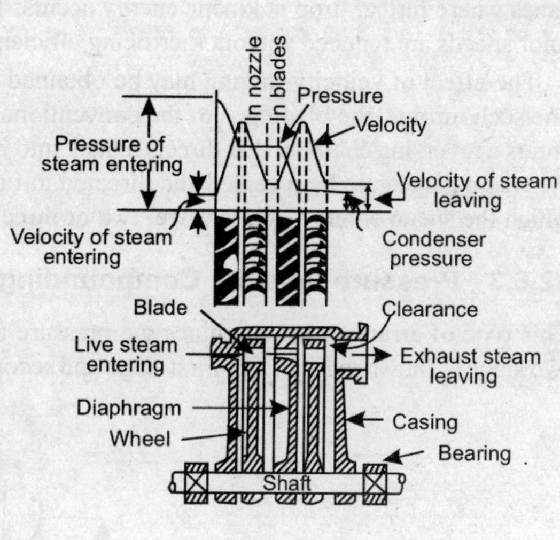

Fig 12.4

12.6.2 Velocity Compounding (Curtis Turbine)

This type of turbine consists of a set of nozzles and rows of moving blades fixed to the shaft and fixed blades fixed to casing. The entire expansion takes place in the nozzle and the high velocity steam parts with only a portion of the

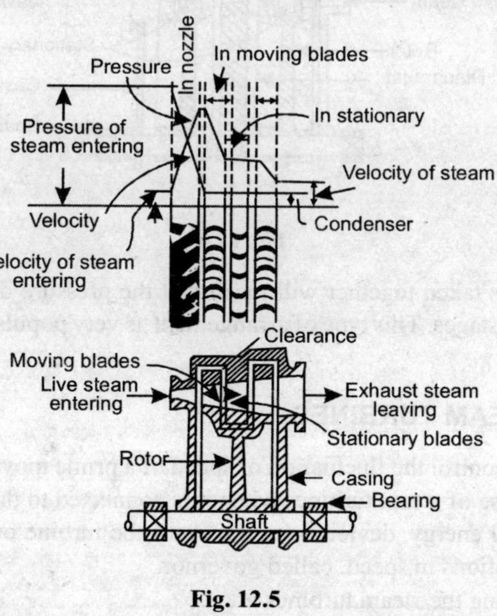

Fig. 12.5

kinetic energy in the first set of the moving blades and then passes onto fixed blades or guide vanes where only change in direction of jet takes place without appreciable loss in velocity. This jet then passes on to another set of moving vanes where further drop in kinetic energy occurs. The arrangement is shown in Fig. 12.5. By this method also high rotor speeds are reduced without sacrificing efficiency or work output.

The effect of velocity staging may be obtained with a single row of buckets by especially contrived means. In one such turbine, the blading is of the conventional type, but as the steam leaves the wheel the first time trough, it enters a reversing chamber that directs it back into the blades. Emerging for the second time from this row of blades, the steam may again be reversed and directed through the blades, thus providing three velocity stages. Turbines in which the steam enters a single wheel two or three times are called *re-entry turbines*.

12.6.3 Pressure Velocity Compounding

This type of arrangement combines the pressure compounding and velocity compounding. The arrangement for two stage is shown in Fig. 12.6. First stage and second stage taken separately are identical to a velocity compounding

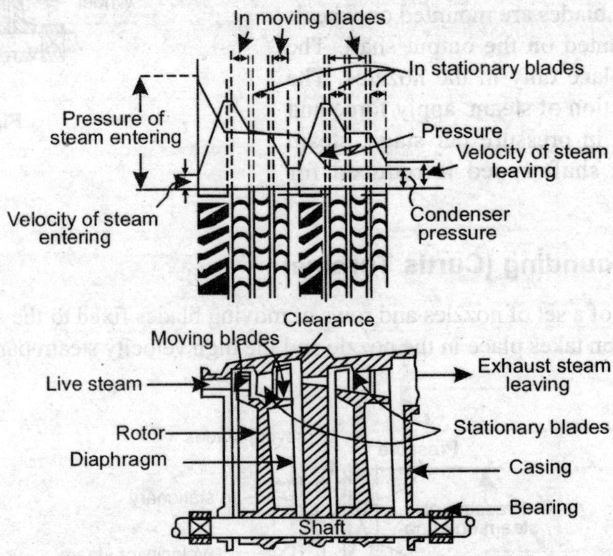

Fig. 12.6

explained earlier. But both stages taken together will mean that the pressure drop from the chest pressure to the condenser pressure occurs at two stages. This type of arrangement is very popular due to simple construction but its efficiency is very low.

12.7 GOVERNING OF STEAM TURBINES

The function of a governor is to control the fluctuation of speed of a prime mover within the prescribed limits with the variation of loads on it. In case of steam turbines when it is connected to drive an alternator for converting its mechanical energy into electrical energy, device is used to vary the turbine output according to the load on the alternator with very small fluctuations in speed, called governor.

The methods used for governing the steam turbines are:

(a) Throttle governing.

(b) Nozzle governing.

(c) By-pass governing.

(d) Combined throttle and nozzle governing.

(e) Combined throttle and by-pass governing.

12.7.1 Throttle Governing

In this type of governing the steam is throttled down to lower pressure according to the load on the turbine before it is supplied to the turbine. It reduces the enthalpy drop i.e., the available energy of the steam. Such a method is useful for small capacity power plants since the mechanism is simple with initial low cost.

The schematic arrangement employing the throttle governing is shown in Fig. 12.7. The throttling of steam is achieved with the help of balanced throttle valve which is controlled by a centrifugal governor. In case of small turbines the throttle valve can be actuated directly with the help of governor through linkages since the steam flow rates would be small and the valves needed are high. However, in case of medium and large power plants the effort of the governor may not be sufficient to actuate the throttle valve, due to this an oil operated relay is incorporated in the circuit as shown in Figure 12.7.This rely magnifies the small force produced by the governor for a small change of speed to produce a large force to move the throttle valve.

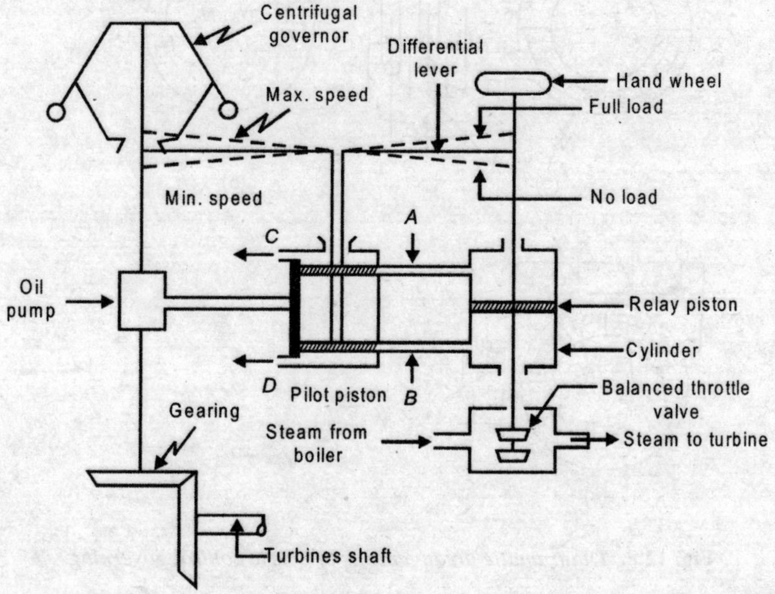

Fig. 12.7. *Throttle governing of steam turbines.*

The working of the mechanism is as follows. Consider the case when the load on turbine shaft is equal to the power developed by the turbine, the speed is constant and the system is in equilibrium. Let us assume that the load on the turbine is reduced suddenly. At this stage since the power developed is more than the load, the turbine and governor speeds will increase due to the excess energy developed by the turbine. The governor balls will fly out and it will raise the sleave of the governor, consequently, the different lever will cause the pilot piston to be raised. The oil which is supplied under a pressure of 3 – 4 bar will flow through the pipe *A* to the cylinder of relay piston and it would force the relay piston to move downwards, while the oil from relay piston cylinder is drained out through the pipe *B*. The downward movement of the relay piston operates the throttle valve which in turn closes the steam ports partially. It throttles the steam and the steam pressure at inlet to the turbine is reduced. When the power developed by the turbine equals to the load on the turbine, the oil ports $C - D$ are covered and the relay piston is locked.

It should be noted that this method of governing is through simple but it reduces the efficiency of power plants at part loads because a part of the available energy is lost in the irreversible throttling process. The Willan's straight line relationship between load and steam consumption is also followed as discussed in case of steam engines.

12.7.2 Nozzle-Control Governing

In this method, the first stage nozzles are split-up groups which are controlled by individual throttle valves. Various arrangements of valves and group of nozzles are employed. Three such arangements are shown digramatically in Fig. 12.8. An arrangement often adopted with large steam turbines using high-pressure steam is shown in Fig. 12.8(a). The number of nozzle groups may vary from three to five or more.

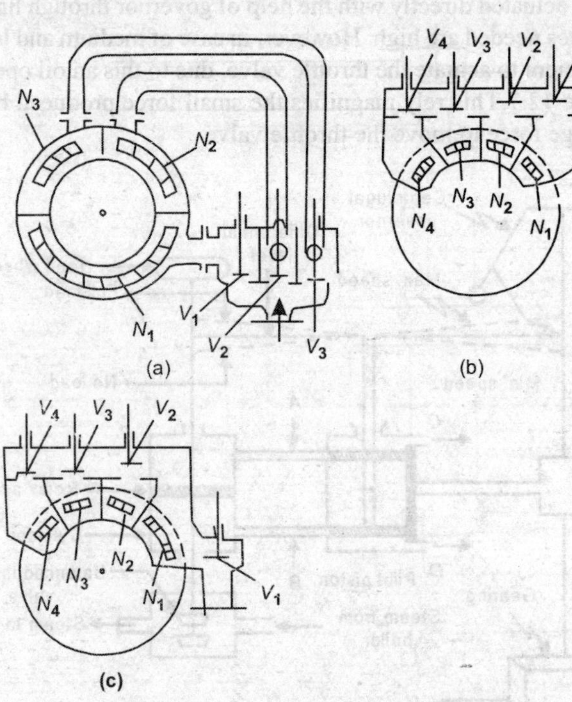

Fig. 12.8. *Diagramatic arrangements of nozzle control governing.*

Figure 12.8(b) shows an arrangement where the nozzle control valves are arranged in a casting forming part of the cylinder of bolted thereto, and containing passes loading to individual nozzle groups. The nozzles are confined to upper half of the cylinder and are of admission is limited to 180º or less.

The number of nozzle groups may vary from four to twelve.

Figure 12.8(c) shows an arrangement in which the group of nozzles N_i is under the control of valve V_1 through which all the steam entering the turbine passes. Further admission of stem is through the valves V_2, V_3 etc. in turn.

The nozzle-control is necessarily restricted to first stage and the absolute pressure of steam in front of the second stage nozzles will be directly proportional to the rate of steam flow through the turbine.

There is greater heat drop available when nozzle control is employed but the efficiency decreases when load is reduced. When there is a comparatively large heat drop in the first stage, nozzle control leads to reduction in steam consumption.

12.7.3 By-pass Governing

Modern steam power plants employing impulse turbines at very high pressure of steam at admission are usually designed to operate usually at an economical load which is bout 75–80% of the maximum load. In such cases it is desirable to have full admission of steam in first few stages of the turbine operating at very high pressure since the enthalpy drop is very small in the initial stages and the nozzle governing cannot be used effectively. Therefore the load regulation is achieved by throttle governing upto the stage of economical loads. However at the maximum loads the additional amount of steam required cannot be passed through the first-stages since the required additional number of nozzles are not available. This difficulty of steam regulation is overcome by employing by-pass governing as shown in Fig. 12.9.

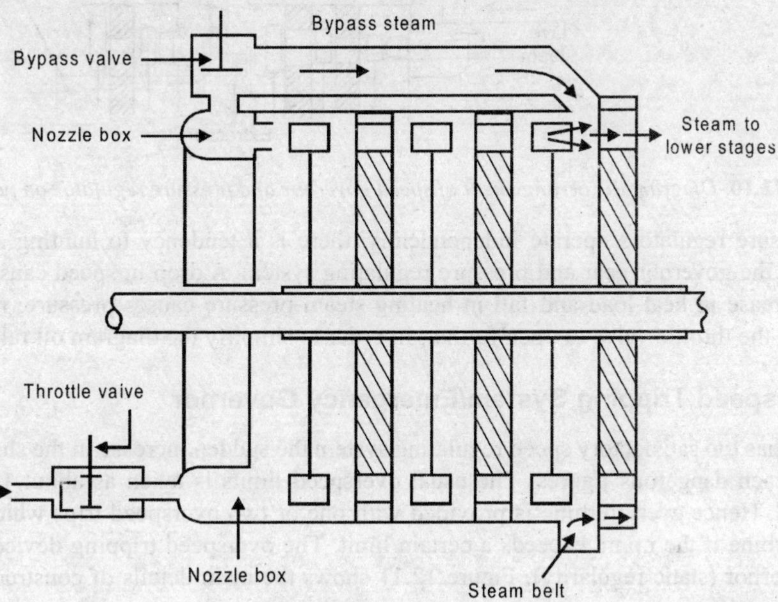

Fig. 12.9. *By-pass governing.*

Steam required upto economical loads is passed through the inlet valve and it is collected in the nozzle-chest. The governing is affected by throttle valve. However at loads more than economical load, the by-pass valve lifts and a part of the steam is by-passed into the steam belt and this steam mixes with the steam of high pressure turbine stages and it is supplied to lower stages of the turbine. This increases the power output of the turbine. The movement of the by-pass valve is controlled by the turbine governor.

12.7.4 Extraction Pressure Regulating System

Extraction pressure regulating system is used where steam at constant pressure is extracted for heating or process requirements from a condensing turbine which meets the power requirement. The control gear must be designed to maintain speed and pressure of extracted steam regardless of the variation in power and heating loads. These requirements are met by using a speed governor which controls the admission of h.p. steam, to the turbine and a pressure regulator controlling the admission of steam to the L.P. stages. A change of either power load or heating load will cause both governors to operate.

Figure 12.10 shows the diagramatic arrangement of the system. When the generator load increases the throttle valve opens causing a pressure rise in the extracted steam. The pressure regulator acts to open the regulating valve to maintain the steam pressure. If the heating load increases, the regulating valve closes and consequently the speed falls due to less flow of steam through L.P. stages. The speed governor acts to open the throttle valve. When the

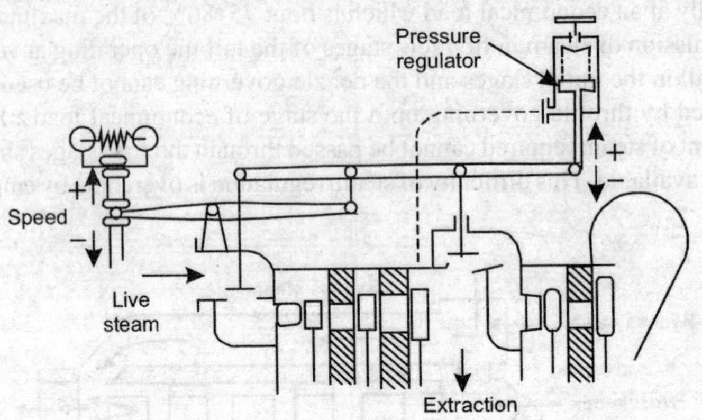

Fig. 12.10. *Diagramatic arrangement of speed governor and pressure regulator on pass-out turbine.*

speed and pressure regulators operate independently, there is a tendency to hunting and this is overcome by interconnecting the governor gear and pressure regulating system. A drop in speed causes both the valves to lift whereas an increase in heat load and fall in heating steam pressure causes pressure, regulating valve to close particularly and the throttle valve to open further. In order to simplify the diagram oil relays have been omitted.

12.7.5 Overspeed Tripping System/Emergency Governor

If a turbine not has too satisfactory speed regulating system the sudden increase in the shaft speed at times of load tripping may reach dangerous figures. The usual overspeed limits is taken as about 10 to 12% of the normal operating speed. Hence every turbine is provided with one or two overspeed trips which shut off the supply of steam to the turbine if the r.p.m. exceeds a certain limit. The overspeed tripping device consists of an unstable centrifugal governor (static regularity). Figure 12.11 shows the main details of construction of such a ring-type

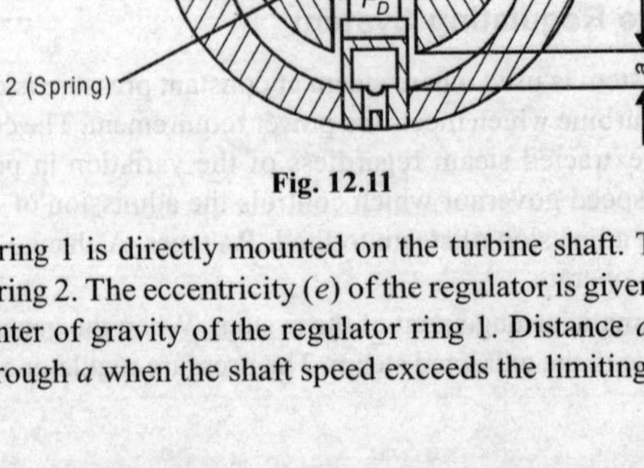

Fig. 12.11

static regulator. The eccentric ring 1 is directly mounted on the turbine shaft. The eccentric right is held in the position shown in Fig. 12.11 spring 2. The eccentricity (*e*) of the regulator is given by the distance between the axis of the turbine shaft and the center of gravity of the regulator ring 1. Distance *a* shows the regulator travel. The regulator ring 1 is displaced through *a* when the shaft speed exceeds the limiting speed. Overspeed regulators are made of various different constructions as well. One of the most widely used overspeed tripping devices as shown in Fig. 12.12. Regulator pin 1, when overspeeding, trips lever 7 disconnecting the interlock 8. The tensile force of the helical spring rotates lever 2 and segment 3 in the clockwise direction. Segment 3 now leaves its notch in sleeve 4 and allows the valve to close under the pressure of its own spring 5, thus shutting off steam supply to the turbine.

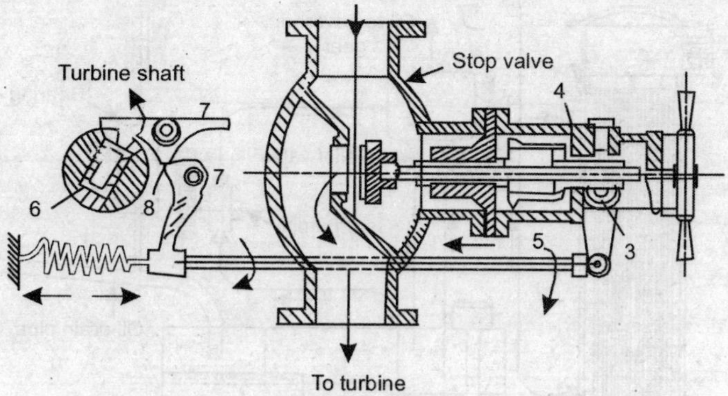

Fig. 12.12. *Overspeed trip with lever control.*

If the stop valve is to be re-opened the following operations have to be carried out. The hand wheel of the stop valve is rotated in the direction of closing. During this operation sleeve 4 is displaced upward. Levers 7, 2 and segments 3 are now brought back to their original position. The stop valve is opened by means of the handwheel.

12.8 STUDY OF STEAM TURBINE SYSTEMS

12.8.1 Steam Circuit

The steam piping to or from the turbine consists of:

1. Main steam supply to turbine. Valves in this steam piping are isolating valve, the stop valve, and the governor or throttle valve.
2. Drain piping which is provided to drain out the water condensed during the start-up of the turbine. This is normally provided at the lowest point,
   (a) before the isolating valve,
   (b) between the stop valve and the throttle valve, and
   (c) the turbine casing inlet. Valves are provided in the drain piping which will be opened during warming-up and closed after the turbine has been loaded to certain output.
3. Gland steam supply piping. Steam from the main steam piping is tapped off and supplied to the glands at the turbine shaft ends coming out of the cylinder to prevent the leakage of steam to the atmosphere and leakage of air into the turbine. The exhaust from the glands may be led to low pressure stages of turbine or condenser. The supply of steam to the glands is regulated by separate regulator.
   Steam may also be tapped off to steam ejector which maintains vacuum in the condenser.
4. Extraction steam Piping. Steam may be extracted from the turbine at intermediate stage for use as heating medium or for driving auxiliaries.

12.8.2 Oil Circuit

Oil circuit is provided to supply oil at a pressurre for lubriction of the bearings and also to operate the servo-motors through relays. A typical forced lubrication system as used on and turbines is shown in the Fig. 12.13. The pump which is normally of the positive action gear type usually delivers the whole of the oil required for lubrication and governing at the pressure required for the governor gear oil relays.

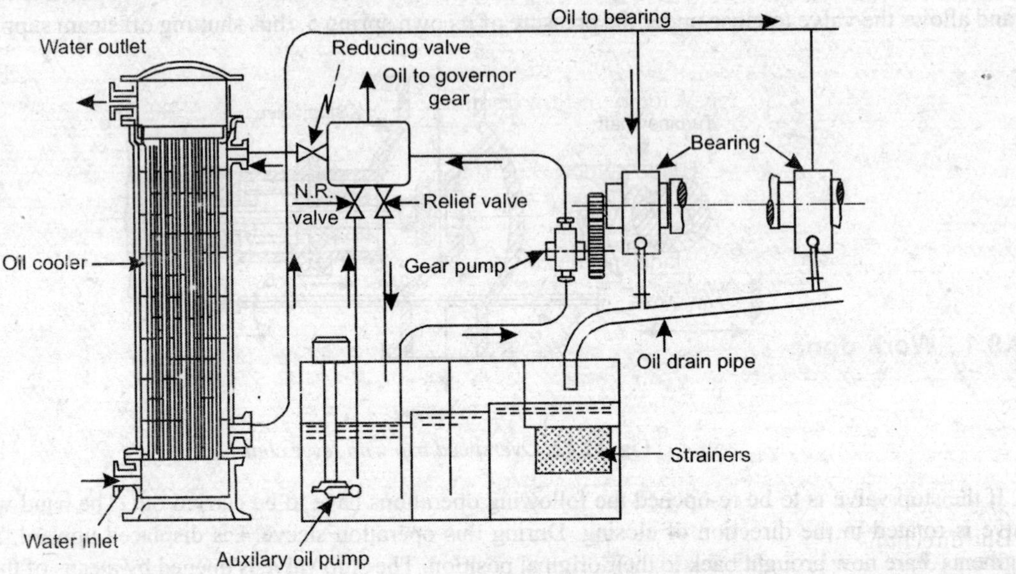

Fig. 12.13. *Diagramatic arrangement of forced lubrication system.*

As the oil for operating the relays is only required intermittently the surplus is allowed to passes through relief valve set to lift at the relay pressure. The oil used in the governor relays may or may not be cooled. Oil for lubrication is passed through a reducing valve which is so adjusted that the oil supply pressure at the bearings is about 0.5 to 1 bar. It is passed through all the bearings. Similarly a large bus pipe collects the oil from all the bearings and leads it to the oil drain tank. This pipe must be large enough to carry the oil with the pipe only partly filled so that air bubbles have a chance to escape. Oil strainers are placed on the oil inlet side of the tank.

The auxiliary oil pump comes into operation at starting. Stopping when the oil pressure falls below a certain valve to supply oil to the bearings when the shaft is being rotated by turning gear, and also as emergency stand-by. It is capable of supplying same quantity as the main pump. The type-most commonly used is electric/turbo driven centrifugal pump.

Water is used to cool the lubricating oil. Oil tanks should be large, enough to allow considerable time to elapse between the oil draining back into the tank and re-entering the pump.

12.9 VELOCITY DIAGRAM FOR SIMPLE IMPULSE TURBINE

Nomenclature used:

α = Angle of absolute velocity measured at inlet with the plane of rotation or Nozzles angle.

β = Angle of absolute velocity at outlet

θ = Blade inlet angle or angle of relative velocity at inlet

Φ = Blade outlet angle or angle of relative velocity at outlet

V_b = Linear blade speed or blade velocity

V_{ai} = Absolute velocity of steam at inlet

V_{ae} = Absolute velocity of steam at outlet

V_{ri} = Relative velocity of steam at inlet
V_{re} = Relative velocity of steam at outlet
V_{wi} = Tangential component of V_{ai} or velocity of whirl at inlet
V_{we} = Velocity of whirl at outlet
V_{fi} = Axial component of V_{ai} or velocity of flow at inlet
V_{fe} = Velocity of flow at outlet

$$K = \text{Blade velocity coefficient} = \frac{V_{re}}{V_{ri}}$$

$$\rho = \text{Blade speed ratio} = \frac{V_b}{V_{ai}}$$

12.9.1 Work-done

$$\text{Force} = \text{Rate of change of momentum in tangential direction}$$
$$= (\text{mass flow rate of steam}) \times (\text{Change in velocity of whirl})$$
$$= m \times (V_{wi} - V_{we})$$

But directions of V_{wi} and V_{we} are opposite. So change in velocity of whirl will be arithmetic addition of V_{wi} and V_{we}

$$\therefore \quad F = m \times (V_{wi} + V_{we})$$

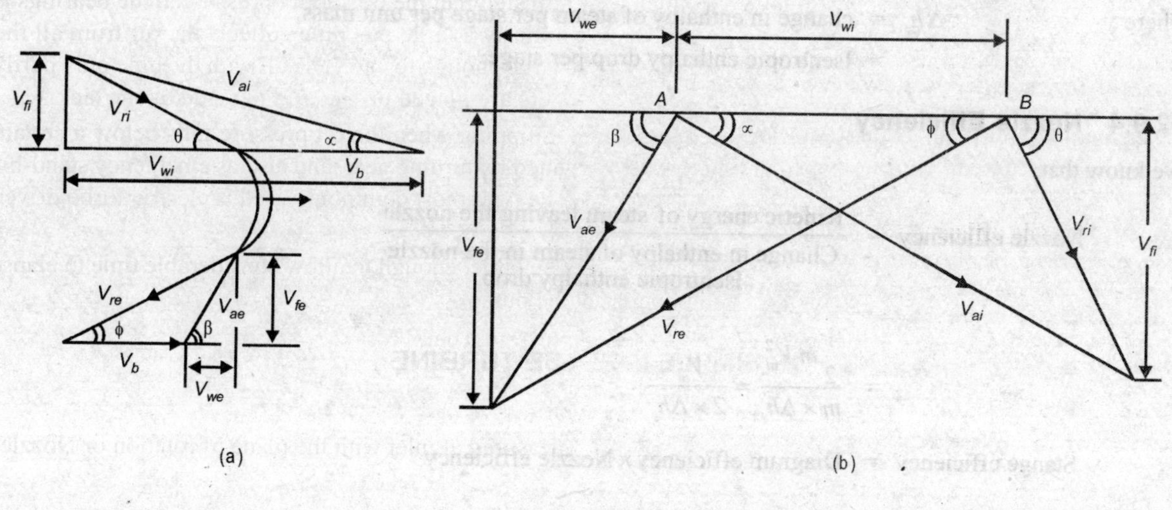

Fig. 12.14

$$\text{Work done} = \text{Force} \times \text{Velocity}$$

$$\therefore \text{Work done on blade} = m(V_{wi} + V_{we}) \times V_b$$

12.9.2 Diagram Efficiency (or Blade Efficiency)

$$\text{Diagram efficiency} = \frac{\text{Work done on the blade}}{\text{Kinetic energy supplied to the blade}}$$

$$\text{Kinetic energy supplied to the blade} = \frac{V_{ai}^2}{2} \times m$$

$$\therefore \ \text{Diagram efficiency} (\eta_{dia}) = \frac{m \times V_b (V_{wi} + V_{we})}{m \times \frac{V_{ai}^2}{2}}$$

$$= \frac{2 \times V_b \cdot (V_{wi} + V_{we})}{V_{ai}^2}$$

12.9.3 Stage Efficiency

$$\text{Stage efficiency} = \frac{\text{Work done per stage}}{\text{Energy supplied per stage}}$$

The stage consists of set of nozzles and set of moving blades and stage efficiency will also include the performance of nozzle.

$$\text{Stage efficiency} = \frac{m \times V_b \times \Delta V_w}{m \times \Delta h_s}$$

where Δh_s = change in enthalpy of steam per stage per unit mass.

= Isentropic enthalpy drop per stage.

12.9.4 Nozzle Efficiency

We know that,

$$\text{Nozzle efficiency} = \frac{\text{Kinetic energy of steam leaving the nozzle}}{\text{Change in enthalpy of steam in the nozzle}}$$

Isentropic enthalpy drop

$$= \frac{\frac{1}{2} m V_{ai}^2}{m \times \Delta h_s} = \frac{V_{ai}^2}{2 \times \Delta h_s}$$

$$\therefore \ \text{Stange efficiency} = \text{Diagram efficiency} \times \text{Nozzle efficiency}$$

12.9.5 End Thrust or (Axial Thrust)

= Change of momentum in axial direction

Axial thrust = mass flow rate × change of velocity in axial direction

$$= m \times (V_{fi} - V_{fe})$$

12.10 OPTIMUM OPERATING CONDITIONS

$$\text{Work done by blade} = m \times V_b \left(V_{wi} + V_{we}\right)$$

But

$$V_{wi} = V_{ai} \cos\alpha = V_b + V_{ri} \cos\theta$$

$$V_{we} = V_{ae} \cos\beta = V_{ro} \cos\Phi - V_b$$

$$\therefore \text{Work done by blade} = m \times V_b \times \left(V_{ai} \cos\alpha + V_{ae} \cos\beta\right)$$

$$= m \times V_b \times \left(V_{re} \cos\Phi + V_{ri} \cos\theta\right)$$

$$= m \times V_b \times V_{ri} \cos\theta \left(\frac{V_{re}}{V_{ri}} \times \frac{\cos\Phi}{\cos\theta} + 1\right)$$

$$\text{But } k = \frac{V_{re}}{V_{ri}}$$

and $$\text{let } c = \frac{\cos\Phi}{\cos\theta}$$

$$\therefore \quad \text{Work done} = m \times V_b \times V_{ri} \cos\theta \left(1 + kc\right)$$

$$= m \times V_b \left(V_{ai} \cos\alpha - V_b\right)\left(1 + kc\right)$$

$$\text{So the diagram efficiency} = \frac{m \times V_b \times \left(V_{ai} \cos\alpha - V_b\right)\left(1 + kc\right)}{m \times \dfrac{V_{ai}^2}{2}}$$

$$\rho = \text{blade speed ratio} = \frac{V_b}{V_{ai}}$$

$$\therefore \quad \eta_{dia} = \frac{2m}{m} \times \frac{V_b}{V_{ai}} \times \left(\cos\alpha - \frac{V_b}{V_{ai}}\right) \times \left(1 + kc\right)$$

$$= 2 \times \rho \left(\cos\alpha - \rho\right) \times \left(1 + kc\right)$$

Now if the impulse turbine has symmetric blades then $\theta = \Phi$ and $k = 1$ for frictionless flow

$$\therefore \quad \eta_{dia} = 4\left(\rho \cos\alpha - \rho^2\right)$$

$\therefore$ For fixed nozzle angle (α) the optimum value of diagram efficiency can be obtained by differentiating efficiency w.r.t. ρ and then equating it to zero.

$$\therefore \quad \frac{d}{d\rho}(\eta_{dia}) = 0$$

$$\therefore \quad \frac{d}{d\rho}\left[4\left(\rho\cos\alpha - \rho^2\right)\right] = 4\left(\cos\alpha - 2\rho\right) = 0$$

$$\therefore \quad \rho = \frac{\cos\alpha}{2},$$

which is the condition for maximum diagram efficiency

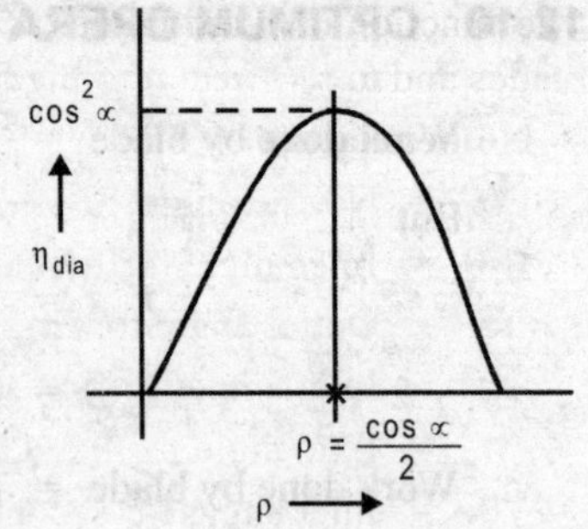

Fig. 12.15

$$(\eta_{dia})_{max} = 4\times\left(\frac{\cos\alpha\times\cos\alpha}{2} - \frac{\cos^2\alpha}{4}\right)$$

$$= \cos^2\alpha$$

$$\text{Maximum work} = 2\times m\times V_b^2$$

12.11 BLADE HEIGHT

For impulse turbine the blade height at a particular section of a turbine will be function of mass flow rate (m), specific volume (v), area (A), and velocity of steam

Let,

P = pitch = distance between two blades.
N = number of blades covered by nozzle.
H = blade height (in metres)

We know the equation,

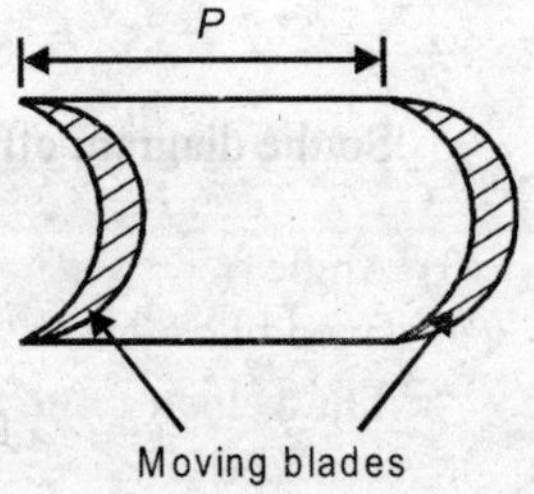

Fig. 12.16

$$\text{Mass flow rate } (m) = \frac{A\times C}{v}$$

Applying the same equation for a turbine, we get

$$m = \frac{N\times P\times H\times V_{fi}}{v}$$

$$\therefore \text{ Blade height } (H) = \frac{m\times v}{N\times P\times V_{fi}}, \text{ for impulse turbine}$$

For reaction turbine, there is full admission of steam through all nozzles

$$\therefore \quad N\times P = \text{Circumference at the blade mean diameter}$$

$$= \pi\times D$$

$$\therefore \text{ Blade height } (H) = \frac{m\times v}{\pi\times D\times V_{fi}}, \text{ for reaction turbine}$$

12.12 VELOCITY DIAGRAM FOR VELOCITY COMPOUNDED STEAM TURBINE CURTIS STAGE

The velocity compounded steam turbine has been already discussed. In this type of turbine, steam is first expanded in nozzles and is passed through set of moving blades fixed blades and then through second set of moving blades.

The function of fixed blades is to change the direction of motion of steam received from earlier set of moving blades and to re-direct it to the next set of moving blades. The complete expansion of steam takes place in nozzle.

Nomenclature:

α_1 = Nozzle angle

θ = Inlet angle for first set of moving blades

ϕ_1 = Outlet angle for first set of moving blades

β_1 = Inlet angle for first set of fixed blades

α_2 = Exit angle for fixed blades

θ_2 = Inlet angle for second set of moving blades

ϕ_2 = Outlet angle for second set of moving blades

β_2 = Inlet angle for second set of fixed blades

k = Friction coefficient

The velocity diagram for the first set of moving blades can be drawn easily if the required data is known. It will be similar to velocity diagram of simple impulse turbine.

Following points should be noted for drawing a velocity diagram for second set of moving blades.

(1) Angle α_2 is exit angle of fixed-blades. The steam leaving fixed blades will enter smoothly in second set of moving blades. So α_2 is similar to nozzle angle, if compared with velocity diagram of simple impulse turbine.

(2) If the friction is considered in blades, the relative velocity will decrease. If friction coefficient is k, we know that,

$$V_{re_1} = k \cdot V_{ri_1}$$

The steam leaving the first set of moving blades will enter into fixed blades. Due to friction relative velocity will decreased.

$$\therefore \quad V_{ri_2} = k \cdot V_{re_2}$$

and

$$V_{re_2} = k \cdot V_{ri_2}$$

(3) Angle β_1 is exit angle of steam leaving first moving blade. It is same as blade angle at inlet for the first set of fixed blades.

The velocity diagrams, for first set of moving blades is given by Fig. 12.17(a)

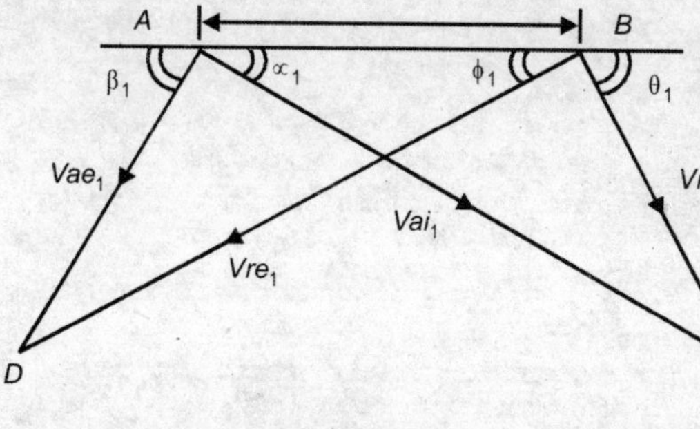

Fig. 12.17(a)

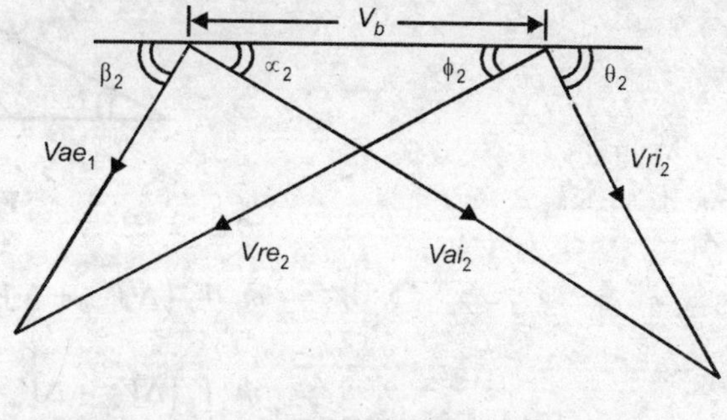

Fig. 12.17(b)

Blade velocity V_b is same in all sets of moving blades.

Total work done is work done in the first row of moving blades plus in the second set of moving blades. The velocity diagram for second stage is shown in Fig. 12.17(b).

$\therefore$ Total work (W) $= W_1 + W_2$

W_1 = Work done in the first set of moving blades.

$$= V_b\left(V_{wi_1} + V_{we_1}\right) \times m = m\,.\,V_b\,\Delta V_{w_2}$$

and $$W_2 = V_b\left(V_{wi_2} + V_{w_2}\right) \times m = m\,.\,V_b\,\Delta V_{w_2}$$

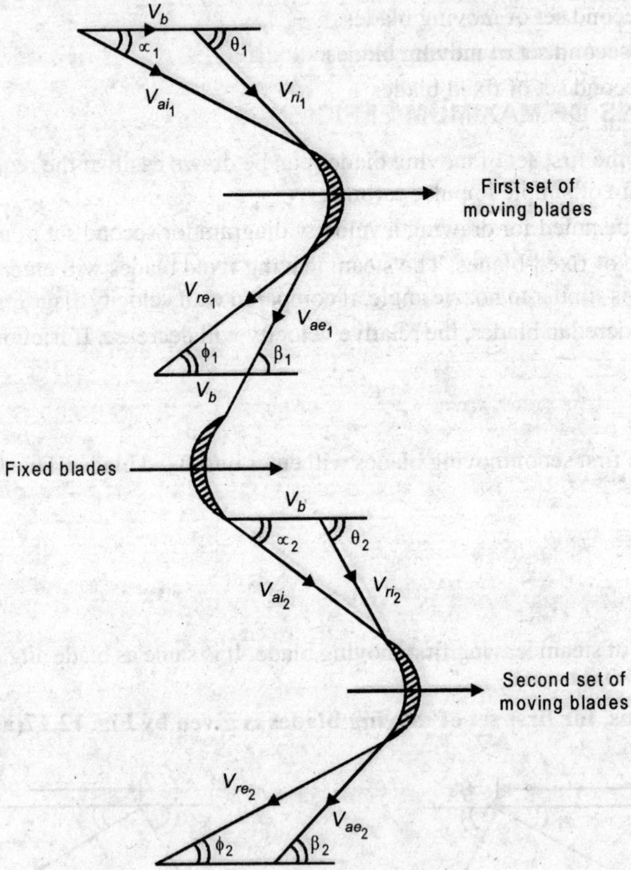

Fig. 12.18

$$\therefore \qquad W = m\,.\,V_b\left(\Delta V_{w_1} + \Delta V_{w_2}\right)$$

$$\therefore \text{ Blade Efficiency} = \frac{m\,.\,V_b\left(\Delta V_{w_1} + \Delta V_{w_2}\right)}{\frac{1}{2}m\,.\,V_{ai_1}^2}$$

$$= \frac{2 . V_b \left(\Delta V_{w_1} + \Delta V_{w_2}\right)}{V_{ai_1}^2}$$

$$\text{Stage Efficiency} = \frac{m . V_b \left(\Delta V_{w_1} + \Delta V_{w_2}\right)}{\Delta h \times m}$$

$$= \frac{V_b \left(\Delta V_{w_1} + \Delta V_{w_2}\right)}{\Delta h}$$

where, Δh = specific enthalpy drop in nozzle.

12.13 CALCULATIONS OF MAXIMUM EFFICIENCY

$$W = W_1 + W_2$$

For unit mass,

$$W_1 = V_b \left(V_{wi_1} + V_{we_1}\right)$$

$$= V_b \left(V_{ai_1} \cos \alpha_1 + V_{ae1} \cos \beta_1\right)$$

$$= V_b \left(V_{ri_1} \cos \theta_1 + V_{re_1} \cos \phi_1\right)$$

Similarly,

$$W_2 = V_b(V_{ai_2} \cos \alpha_2 + V_{ai_2} \cos \beta_2)$$

$$= V_b(V_{ri_2} \cos \theta_2 + V_{re_2} \cos \phi_2)$$

If the moving blades and fixed blades are symmetric, $\theta_1 = \phi_1$, $\theta_2 = \phi_2$ and $P_1 = \alpha_2$.

If the friction in moving blades and fixed blades is neglected, $V_{ri1} = V_{rel}$ and $V_{ri_2} = V_{re_2}$, $V_{ae_1} = V_{ai_2}$.

For maximum efficiency, energy loss to exit steam in the last stage has to be minimum. So angle has to be equal to 90º. Therefore, consider $\beta_2 = 90°$.

$$\therefore \quad W_1 = V_b \left(V_{ri_1} \cos \theta_1 + V_{re_1} \cos \phi_1\right)$$

$$= V_b . V_{ri_1} \cos \theta_1 \left(1 + \frac{V_{re_1}}{V_{ri_1}} \cdot \frac{\cos \phi_1}{\cos \theta_1}\right)$$

But $V_{rei} = V_{ri_1}$ and $\theta_1 = \phi_1$

$$\therefore \quad W_1 = V_b . \left(V_{ri_1} \cos \theta_1\right) \times 2$$

$$= 2 \cdot V_b \cdot \left(V_{ai_1} \cos \alpha_1 - V_b\right)$$

Similarly,

$$W_2 = 2 . V_b . \left(V_{ri_2} \cos \theta_2\right)$$

$$= 2\cdot V_b\cdot\left(V_{ai_2}\cos\alpha_2 - V_b\right)$$

But, $\alpha_2 = \beta_2$ and $V_{ai_2} = V_{ae_1}$

$$\therefore \quad W_2 = 2 - V_b\left(V_{ae_1}\cos\beta_1 - V_b\right)$$

But $V_{ae_1}\cos\beta_1 = \left(V_{re_1}\cos\phi_1 - 2V_b\right)$... see velocity diagram for first set of moving blades.

$$\therefore \quad W_2 = 2\,.\,V_b\left(V_{re_1}\cos\phi_1 - 2V_b\right)$$

But $V_{ri_1} = V_{re_1}$ and $\phi_1 = \theta_1$

$$\therefore \quad W_2 = 2\,.\,V_b\left(V_{ri_1}\cos\theta_1 - 2V_b\right)$$

$$V_{ri_1}\cos\theta_1 = V_{ai_1}\cos\alpha_1 - V_b$$

$$\therefore \quad W_2 = 2 - V_b\left(V_{ai_1}\cos\alpha_1 - 3V_b\right)$$

Total work done (W) $= W_1 + W_2$

$$= 2\,.\,V_b\left(V_{ai1}\cos\alpha_1 - V_b\right) + 2\,.\,V_b\left(V_{ai_1}\cos\alpha_1 - 3V_b\right)$$

$$= 2\,.\,V_b\left(2V_{ai_1}\cos\alpha_1 - 4V_b\right)$$

$$\text{Diagram efficiency} = \frac{\text{Total work done}}{\text{Kinetic energy of steam}}$$

$$= \frac{2\,.\,V_b\left(2V_{ai_1}\cos\alpha_1 - 4V_b\right)}{\frac{1}{2}\cdot V_{ai}^2}$$

$$= 4\cdot\frac{V_b}{V_{ai}}\cdot\left(2\cos\alpha_1 - \frac{4V_b}{V_{ai_1}}\right)$$

But, $\dfrac{V_b}{V_{ai}} = \rho$ = Blade speed ratio.

$$\therefore \text{Diagram efficiency} = 4\cdot\rho\cdot\left(2\cos\alpha_1 - 4\rho\right) = 8\,\rho\cos\alpha_1 - 16\,\rho^2.$$

For fixed nozzle angle (α_1), the maximum value of diagram efficiency can be obtained by differentiating efficiency w.r.t. ρ and equating to zero.

$$\therefore \quad \frac{d(\eta_{\text{dia}})}{d\rho} = \left(8\cos\alpha_1 - 32\,\rho\right) = 0$$

$$\therefore \qquad \rho = \frac{\cos\alpha_1}{4} \text{ – Condition for maximum efficiency.}$$

$$\text{Maximum diagram efficiency} = 4 \times \frac{\cos\alpha_1}{4} \times \left(2\cos\alpha_1 - \frac{4\cos\alpha_1}{4}\right)$$

$$(\eta_{dia})_{max} = \cos^2\alpha_1$$

The work done for maximum diagram efficiency

$$= 2 \times V_b \left(2\, V_{ai_1} \cos\alpha_1 - 4\, V_b\right)$$

$$\rho = \frac{V_b}{V_{ai_1}} = \frac{\cos\alpha_1}{4}$$

$$\therefore \qquad V_{ai_1} = \frac{4 \cdot V_b}{\cos\alpha_1}$$

$$\therefore \qquad W = 2 \times V_b \left(\frac{2 \times 4 \times V_b}{\cos\alpha_1} \cdot \cos\alpha_1 - 4\, V_b\right)$$

$$= 2 \cdot V_b \left(8\, V_b - 4\, V_b\right) = 8 \cdot V_{b2}$$

The work done in first stage $= 6\, V_b^2$.

The work done in the second stage $= 2\, V_b^2$.

The above analysis is for two rows of moving blades. Similarly analysis can be done for more number of rows. If there are *n* number of rows of moving blades, then optimum blade speeds ratio,

$$\rho = \frac{\cos\alpha_1}{2n}$$

and work done in the last row is $\frac{1}{2^n}$ of total work

$$\text{i.e.,} \qquad \rho_{optimum} = \frac{\cos\alpha_1}{2n}$$

$$\text{and work done by last row} = \frac{1}{2^n} \times (\text{Total work done})$$

It means that in two row turbine, the work obtained in second row is one-fourth of total work and if there are three rows of moving blades, work obtained in the row is one eighth of total work. Thus we see that the increase in number of rows the work obtained in the last decreases. So the number of rows are limited.

12.14 STEAM TURBINES PERFORMANCE

In steam turbines when steam flows over the blades, the friction takes place. It results into reduction in the useful enthalpy drop. So the useful or effective enthalpy-drop is less than isentropic enthalpy drop. The energy lost in friction reheats the steam and the quality of steam at exit from the stage gets improved.

Figure 19.19 shows $h_A - h_{\beta_1}$ is entropic enthalpy drop per stage for a pressure range form P_1 to P_2. The useful of enthalpy drop per stage is $\left(h_{A_1} - h_{c_1}\right)$ or $\left(h_{A_1} - h_{A_1}\right)C_1$.

So the stage efficiency is $\eta_{stage} = \dfrac{h_{A_1} - h_{C_1}}{h_{A_1} - h_{B_1}}$

Thus the quality of steam at exit from the first stage of a pressure compounded or reaction turbine is obtained as at point A_2 at pressure P_2 and enthalpy h_{c_1} or h_{A_2}. All such state points can be obtained for successive stages. The line joining such state points (i.e., A_1, A_2, A_3, A_4 and A_3) is called the condition-line.

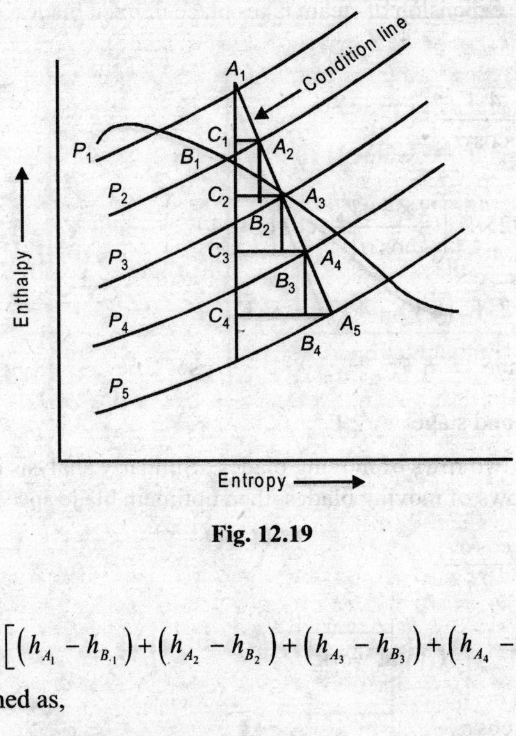

Fig. 12.19

Now,

$$\Sigma\left(h_A - h\beta\right) = \left[\left(h_{A_1} - h_{B_1}\right) + \left(h_{A_2} - h_{B_2}\right) + \left(h_{A_3} - h_{B_3}\right) + \left(h_{A_4} - h_{B_4}\right)\right]$$

The reheat-factor is defined as,

$$\text{Reheat factor (R.F.)} = \frac{\text{Cumulative enthalpy drop}}{\text{Isentropic enthalpy drop}}$$

$$= \frac{\Sigma\left(h_A - h_B\right)}{h_{A_1} - h_D}$$

The reheat factor is always greater than one. It depends on stage efficiency, initial temperature and pressure and exit pressure. The value normally lies from1.02 to 1.06

Now,

$$\eta_{normal} = \frac{\Sigma\left(h_A - h_C\right)}{\left(h_{A_1} - h_D\right)}$$

But

$$\left(h_{A_1} - h_D\right) = \frac{\Sigma\left(h_A - h_B\right)}{\text{R.F.}}$$

$$\therefore \qquad \eta_{\text{internal}} = \frac{\Sigma\left(h_A - h_C\right)}{\Sigma\left(h_A - h_B\right)} \times \text{R.F.}$$

$$\eta_{\text{internal}} = \eta_{\text{stage}} \times \text{R.F.}$$

12.15 VELOCITY DIAGRAM FOR REACTION TURBINE

In case of a reaction turbine, the expansion of steam takes place in fixed blades as well as in moving blades. Figure .... P_0. In fixed blades it expands to pressure P_1. So $(h_0 - h_1)$ is enthalpy drop in fixed blades. Then steam flows over the moving blades and expands from pressure P_1 to pressure P_2. The enthalpy drop in moving blades is $(h_1 - h_2)$. Due to expansion in moving blades, the relative velocity of steam continuously increases (i.e., $V_{r_e} > V_{r_i}$).

The velocity diagram for a reaction turbine is as shown in figure.

(i) Work-done $= m \times V_b \left(V_{wi} + V_{we}\right)$

(ii) Stage efficiency $= \dfrac{\text{Work done in a stage}}{\text{Enthalpy drop in a stage}}$

$$= \frac{m \times V_b \left(V_{wi} + V_{we}\right)}{m \times \Delta h}$$

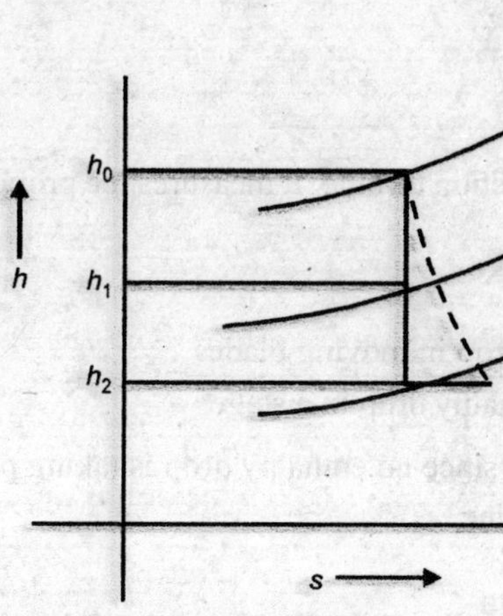

Fig. 12.20

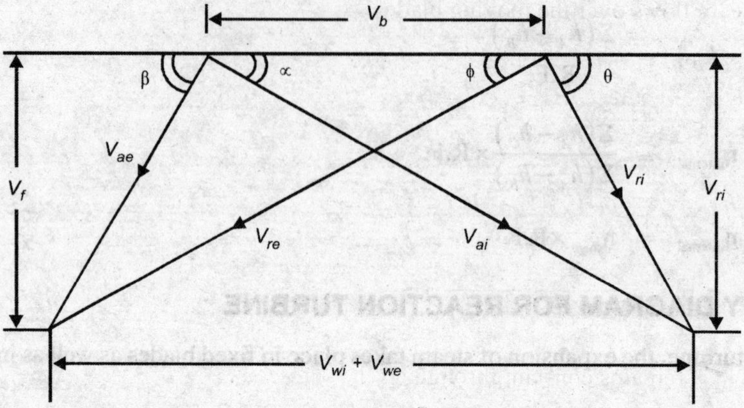

Fig. 12.21

$$= \frac{V_b\left(V_{wi}+V_{we}\right)}{\Delta h}$$

where Δh – Specific enthalpy drop in a stage.

(iii) Blade efficiency $= \dfrac{\text{Work done}}{\text{Energy supplied}}$

$$= \frac{m\,V_b\left(V_{wi}+V_{we}\right)}{\dfrac{V_{ai}^2}{2}+\left(\dfrac{V_{re}^2-V_{ri}^2}{2}\right)}$$

12.15.1 Degree of Reaction

The term 'degree of reaction' is used for reaction turbine. It measures the proportion of the work-done by the effect (due to pressure drop in moving blades)

The degree of reaction is defined as,

$$\text{D.R.} = \frac{\text{Enthalpy drop in moving blades}}{\text{Total enthalpy drop in a stage}}$$

When D.R. = 0, it is an impulse turbine, since no enthalpy drop is taking place in moving blades.

When D.R. = 1, it is 100% reaction turbine.

12.15.2 Parson's Reaction Turbine

For Parson's reaction turbine, the blades are symmetric and degree of reaction is 50%.

We know that,

$$\text{D.R.} = \frac{\text{Enthalpy drop in moving blades}}{\text{Enthalpy drop in a stage}}$$

Enthalpy drop in moving blades is equal to increase kinetic energy of the steam to corresponding to relative velocity while, the steam flows over the moving blades.

$$\therefore \quad (\Delta h)_{\text{moving}} = \left(\frac{V_{re}^2 - V_{ri}^2}{2}\right) \times m$$

The enthalpy in a stage

$$(\Delta h)_{\text{fixed}} = m \cdot V_b \cdot (V_{wi} + V_{we})$$

$$\therefore \quad \text{D.R.} = \frac{V_{re}^2 - V_{ri}^2}{2 \cdot V_b (V_{wi} + V_{we})}$$

Velocity of flow (V_f) remains constant throughout the stage

$$V_{ri} = V_f \text{ cosec } \theta \text{ and } V_{re} = V_f \text{ cosc } \Phi$$

$$(V_{wi} + V_{we}) = V_f (\cot \theta + \cot \Phi)$$

$$\therefore \quad \text{D.R.} = \frac{V_f^2 (\text{cosec}^2 \Phi - \text{cosec}^2 \theta)}{2\, v_f (\cot \Phi + \cot \theta)\, V_b}$$

$$= \frac{V_f}{2 \cdot V_b} (\cot \Phi - \cot \theta)$$

$$\left[\text{since} \left(1 + \cot^2 \theta\right) = \text{cosec}^2 \theta\right]$$

For 50 % reaction turbine, D.R. $= \dfrac{1}{2}$

$$\therefore \quad \frac{1}{2} = \frac{V_f}{2 \cdot V_b} (\cot \Phi - \cot \theta)$$

$$V_b = V_f (\cot \Phi - \cot \theta) \qquad \text{....(I)}$$

From velocity triangles of velocity diagram,

$$V_b = V_f (\cot \Phi - \cot \beta) \qquad \text{....(II)}$$

and

$$V_b = V_f (\cot \alpha - \cot \theta) \qquad \text{....(III)}$$

Comparing Eqs (I), (II) and (III), we get,

$$\alpha = \Phi \text{ and } \theta = \beta$$

Thus for 50 % reaction turbine i.e., for Parson's turbine, the fixed and moving blades have the same shape. This gives rise to symmetric velocity diagram.

12.15.3 Maximum Efficiency of Reaction Turbine

For 50 % reaction turbine, the velocity diagram is symmetric i.e., inlet velocity triangle is same as outlet velocity triangle.

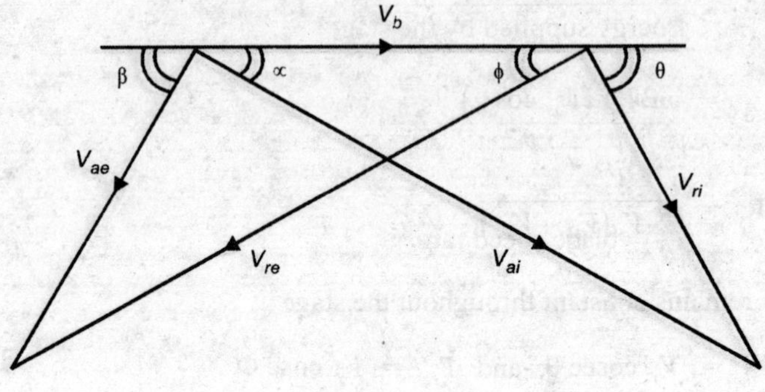

Fig. 12.22

Kinetic energy obtained in fixed blades $= \dfrac{V_{ai}^2}{2}$

Kinetic energy obtained moving blades $= \dfrac{V_{re}^2 - V_{ri}^2}{2}$

Total energy supplied $= \dfrac{V_{ai}^2}{2} + \dfrac{V_{re}^2 - V_{ri}^2}{2}$

From symmetric velocity triangles, we get $V_{re} = V_{ai}$

$\therefore$ Total energy supplied $= \dfrac{V_{ai}^2}{2} + \dfrac{V_{ai}^2}{2} - \dfrac{V_{ri}^2}{2} = V_{ai}^2 - \dfrac{V_{ri}^2}{2}$

From inlet velocity triangle,

$$V_{ri}^2 = V_{ai}^2 + V_b^2 - 2\,V_{ai} - V_b \cdot \cos\alpha$$

$\therefore$ Total energy supplied $= V_{ai}^2 - \left(\dfrac{V_{ai}^2 + V_b^2 - 2\,V_{ai}\,V_b\cos\alpha}{2}\right)$

$$= \left(\frac{V_{ai}^2 - V_b^2 + 2\,V_{ai} \cdot V_b \cos\alpha}{2}\right)$$

Work done per kg of steam $= V_b\left(V_{wi} + V_{we}\right)$

$$= V_b\left(V_{ai}\cos\alpha + V_{re}\cos\Phi - V_b\right)$$

But $\alpha = \Phi$

and $V_{ai} = V_{re}$

$\therefore$ Work done per kg of steam $= V_b\left(2V_{ai}\cos\alpha - V_b\right)$

Stage efficiency is given by,

$$\eta_{stage} = \frac{\text{Work done by the stage}}{\text{Energy supplied by the stage}}$$

$$= \frac{2V_b\left(2V_{ai}\cos\alpha - V_b\right)}{V_{ai}^2 - V_b^2 + 2V_{ai}\cdot V_b\cdot\cos\alpha}$$

Let us define, $\rho = \dfrac{V_b}{V_{ai}}$ = blade speed ratio

$$\therefore \quad \eta_{stage} = \frac{2\dfrac{V_b}{V_{ai}}\left[2\cos\alpha - \dfrac{V_b}{V_{ai}}\right]}{1-\left(\dfrac{V_b}{V_{ai}}\right)^2 + 2\left(\dfrac{V_b}{V_{ai}}\right)\cos\alpha} = \frac{2\rho\left(2\cos\alpha - \rho\right)}{1-\rho^2+2\rho\cos\alpha}$$

$$= \frac{2\left[\left(\rho\cos\alpha - \rho^2 + 1\right) - 1\right]}{\left(1-\rho^2+2\rho\cos\alpha\right)}$$

$$= 2 - \frac{2}{1-\rho^2+2\rho\cos\alpha}$$

The stage efficiency is maximum, when $\left(1-\rho^2+2\rho\cos\alpha\right)$ is maximum. So the required condition is,

$$\frac{d}{d\rho}\left(1-\rho^2+2\rho\cos\alpha\right) = 0$$

$$\therefore \quad -2\rho + 2\cos\alpha = 0$$

$$\therefore \quad \rho = \cos\alpha$$

This is the condition for maximum stage efficiency

$$\therefore \quad \left(\eta_{stage}\right)_{maximum} = \frac{2\cos\alpha\left(2\cos\alpha - \cos\alpha\right)}{1-\cos^2\alpha + 2\cos\alpha\cdot\cos\alpha}$$

$$= \frac{2\cos^2\alpha}{1+\cos^2\alpha}$$

SOLVED EXAMPLES

Example 12.1 *A De-Laval turbine has a nozzle angle of 20° and blade inlet angle 30° and blade outlet angle 25°. Enthalpy drop available is 350 kJ/kg. Nozzle efficiency is 82 % and blade friction coefficient is 0.8.*

Determine,

1. *Absolute velocity of steam at inlet (V_{ai})*
2. *Blade velocity (V_b)*
3. *Absolute velocity at outlet (V_{ae})*

Solution

Given

$$\alpha = 20°$$

$$\theta = 30°, \Phi = 25°$$

$$\Delta h = 350 \text{ kJ/kg}, \frac{V_{re}}{V_{ri}} = 0.8$$

$$\eta_{\text{nozzle}} = 0.82$$

$$\eta_{\text{nozzle}} = \frac{V_{ai}^2}{2 \times \Delta h}$$

$$V_{ai} = \sqrt{2 \times \eta_{\text{nozzle}} \times \Delta h} = \sqrt{2 \times 0.82 \times 350 \times 10^3} = 783 \text{ m/s}$$

Draw the velocity diagram to the scale

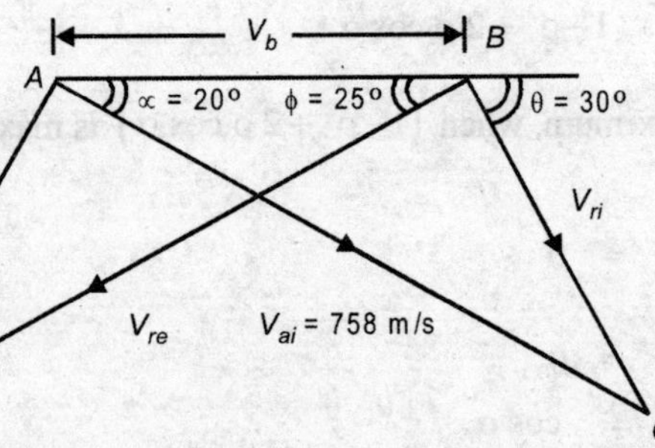

Fig. Ex. 12.1

From velocity diagram,

$$V_{ri} = 520 \text{ m/s}$$

$$V_{re} = 0.8 \times V_{ri} = 416 \text{ m/s}$$

Draw $BD = V_{re}$ to the scale so the velocity diagram is completed.

1. $V_{ai} = 578$ m/s
2. $V_b = AB = 265$ m/s
3. $V_{ae} = AD = 210$ m/s

Example 12.2 *A simple impulse turbine has a blade speed of 350 m/s and blade speed to steam velocity (inlet) ratio is 0.45. Nozzles are inclined at 20° to plane of rotation and steam leaves the stage at an angle of 70° to the plane of rotation.*

Determine,

1. *Blade inlet angle* (θ)
2. *Kinetic energy of steam at outlet.*

Solution

Given $V_b = 350$ m/s

$$\frac{V_b}{V_{ai}} = 0.45$$

$$\alpha = 20°, \beta = 70°$$

$$V_{ai} = \frac{V_b}{0.45} = 777 \text{ m/s}$$

Frictional losses are not given. Therefore assume it nil

$$\therefore \quad V_{re} = V_{ri}$$

Draw the velocity diagram.

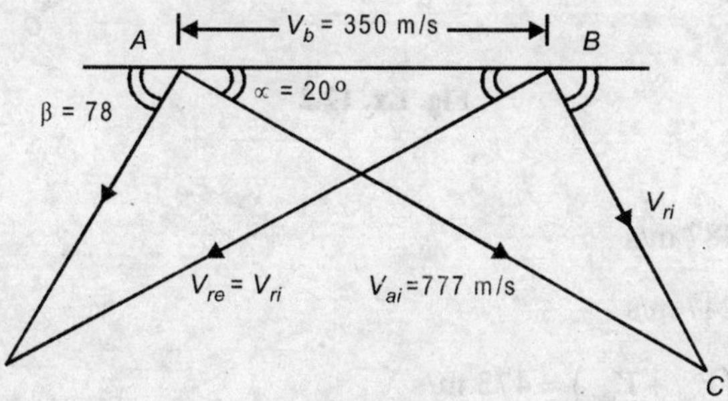

Fig. Ex. 12.2

From velocity diagram,

$$\theta = 35.5°$$

$$V_{ae} = 207 \text{ m/s}$$

$$\text{Kinetic energy of steam at outlet} = \frac{V_{ae}^2}{2} = \frac{(207)^2}{2 \times 1000} = 21.4 \text{ kJ/kg}$$

Example 12.3 *A De-Laval turbine has a blade speed of 200 m/s and relative velocity of steam at inlet is 300 m/s. The inlet and outlet blade angles are 30° each, while the steam leaves the wheel inclined at an angle of 85° to the plane of rotation. If the steam consumption is 4000 kJ/hr.*

Determine,

1. *Absolute velocity of steam at inlet,*

2. *Blade friction coefficient,*
3. *Power developed.*

Solution

Given,

V_b = 200 m/s

V_{ri} = 300 m/s

$\theta = \Phi = 30°$

β = 85°

m = 4000 kg/hr

Draw the velocity diagram

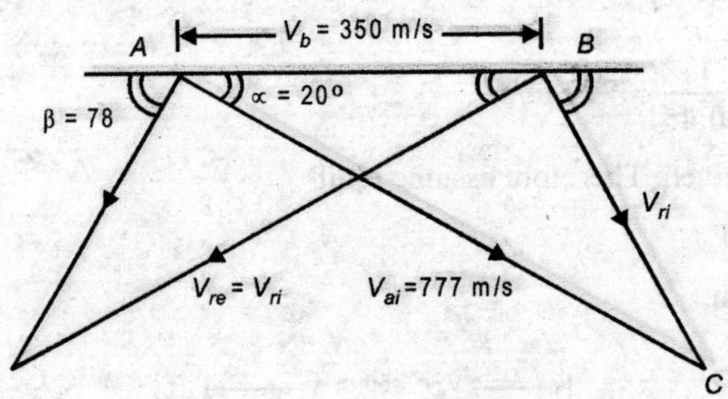

Fig. Ex. 12.2

From velocity diagram,

V_{ai} = 487 m/s

V_{re} = 247 m/s

$\Delta V_w = (V_{wi} + V_{we}) = 473$ m/s

Blade velocity coefficient $= \dfrac{V_{re}}{V_{ri}} = \dfrac{247}{300} = 0.823$

Power development $= m \times (\Delta V_w) \times V_b$

$$= \frac{4000}{3600} \times \frac{473 \times 200}{10^3} = 105 \text{ kW}$$

Example 12.4 *A De-Laval turbine is supplied with steam at pressure of bar and 250°C and back pressure is 0.12 bar.*

The nozzle efficiency is 0.9, blade velocity coefficient is 0.8. Mechanic efficiency of turbine is 0.9. The nozzle angle is 20°. The blades are symmetric and blade angle is 30°.

Draw the velocity diagram and calculate

1. *Speed of the rotor if mean diameter of wheel is 75 cm*
2. *Steam consumption per kW-hr.*

3. *Blade efficiency*
4. *Stage efficiency*
5. *Axial thrust per kg of steam*
6. *Percentage loss in exit*

Solution

Given,

$P_1 = 15$ bar and $T_1 = 250°C$, $P_2 = 0.12$ bar

$$\eta_{nozzle} = 0.9, \quad \frac{V_{re}}{V_{ri}} = 0.8, \quad \alpha = 20°$$

$$\eta_{mech} = 0.9 \qquad \theta = \Phi = 30°$$

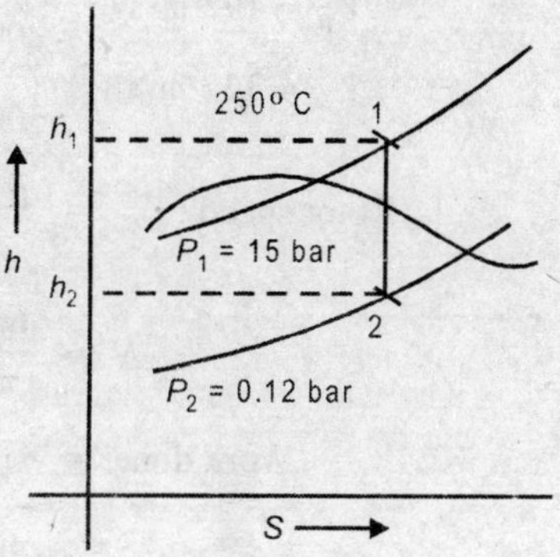

Fig. Ex. 12.4(a)

From Mollier-chart,

For $P_1 = 15$ bar and $T_1 = 250°C$. Mark point 1. Draw a vertical-downward line from point 1(since expansion in nozzle is isentropic). At exit pressure is 0.12 bar. So get point 2.

$$h = \Delta(h_1 - h_2) = 764.94 \text{ kJ/kg}$$

$$\eta_{nozzle} = \frac{V_{ai}^2}{2 \times \Delta h}$$

$$\therefore \quad V_{ai} = \sqrt{2 \times \eta_{nozzle} \times \Delta h}$$

$$V_{ai} = \sqrt{2 \times 0.9 \times 764.94 \times 1000}$$

$$= 1173.4 \text{ m/s}$$

Draw a velocity diagram,

$$k = \frac{V_{re}}{V_{ri}} = 0.8$$

$$\therefore \quad V_{re} = 0.8\, V_{ri}$$

$$\therefore \quad BD = 0.8 \times BC$$

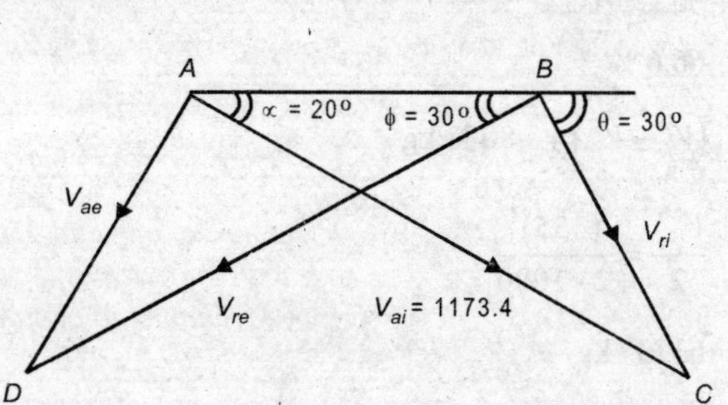

Fig. Ex. 12.4(b)

Use this while drawing the velocity diagram.

From velocity diagram,

$V_b = 407$ m/s , $(V_{wi} + V_{we}) = 1252$ m/s

$V_{fi} = 402$ m/s , $V_{ae} = 355$ m/s

$V_{fe} = 322$ m/s .

1. Rotor speed : $V_b = \dfrac{\pi D N}{60}$, where D = mean rotor diameter.

$$\therefore \quad N = \frac{60 \times V_b}{\pi \times D} = \frac{60 \times 407}{\pi \times 0.75} = 10364 \text{ rpm}$$

2. Work done $= m \times V_b \times \Delta V_w$

$$= \frac{407 \times 1252}{1000} = 509.564 \text{ kJ/kg}$$

Shaft output $= (\text{work done}) \times (\text{mechanical efficiency})$

$$= 509.564 \times 0.9$$

$$= 458.6 \text{ kJ/kg}$$

Steam consumption per kW-hr $= \dfrac{1}{458.6} \times 3600$

$$= 7.85 \text{ kg/kW-hr}$$

3. Blade efficiency $= \dfrac{\text{Work done by the blade}}{\text{Energy supplied to the blade}}$

$$= \frac{509.564}{\frac{(1173.4)^2}{2 \times 1000}} \times 100 = 74\ \%$$

4. Stage efficiency $= \eta_{\text{blade}} \times \eta_{\text{nozzle}} = (0.74 \times 0.9) \times 100$

$$= 66.6\ \%$$

5. Axial thrust per kg $= (V_{fe} - V_{fi}) = -80$ N/kg

6. Energy lost in exit $= \dfrac{V_{ae}^2}{2} = \dfrac{(355)^2}{2 \times 1000}$

$$= 63 \text{ kJ/kg}$$

$\therefore$ Percentage loss in exit $= \dfrac{\text{Energy lost in exit}}{\text{Energy available i.e., } \Delta h}$

$$= \frac{63}{764.94} \times 100 = 8.2\ \%$$

Example 12.5 *The impulse turbine has a covering nozzle and one ring of moving blades. The nozzles are inclined at 22° to the blades whose tip angles are both 35°. The velocity of steam leaving the nozzle is 660 m/s.*

Find,

1. *Blade speed (V_b) and diagram efficiency assuming no friction along the turbine blades.*
2. *If the relative velocity of steam to the blade is reduced by 15 % due to friction, find blade efficiency and end thrust.*

Assume power developed = 170 kW.

Solution

Given,

$$\alpha = 22°$$

$$\therefore \quad \theta = \Phi = 35°$$

$$V_{ai} = 660 \text{ m/s}$$

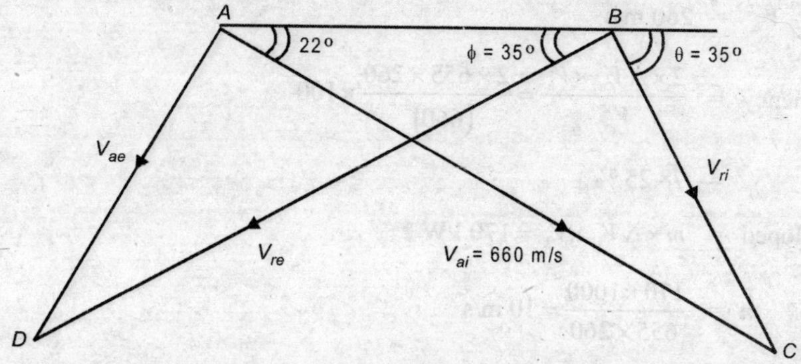

Fig. Ex. 12.5(a)

1. There is no friction

$$\therefore \quad V_{re} = V_{ri}$$

From velocity diagram, $V_b = 260$ m/s

$$\Delta V_w = (V_{wi} + V_{we}) = 715 \text{ m/s}$$

$$\text{Blade efficiency} = \frac{2 \times V_b \times \Delta V_w}{V_{ai}^2}$$

$$= \frac{2 \times 260 \times 715}{(660)^2} \times 100 = 85\ \%$$

2. When the blade friction is considered

$$k = \frac{V_{re}}{V_{ri}} = 0.85$$

Draw the velocity diagram,

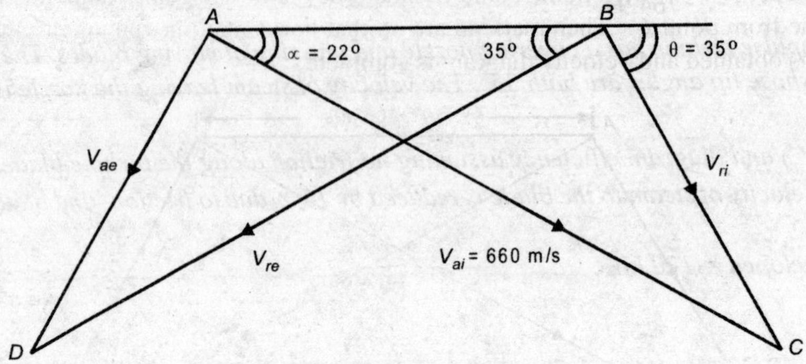

Fig. Ex. 12.5(b)

From the velocity diagram,

$$\Delta V_w = 655 \text{ m/s}$$

$$V_b = 260 \text{ m/s}$$

$$\text{Blade efficiency} = \frac{2 \times \Delta V_w \times V_b}{V_{ai}^2} = \frac{2 \times 655 \times 260}{(660)^2} \times 100$$

$$= 78.25\,\%$$

$$\text{Power developed} = m \times \Delta V_w \times V_b = 170 \text{ kW}$$

$$\therefore \quad m = \frac{170 \times 1000}{655 \times 260} = 10 \text{ m/s}$$

$$\text{End thrust} = m\left(V_{fi} - V_{fe}\right)$$

$$= 10 \times (250 - 205)$$

$$= 45 \text{ N}$$

Example 12.6 *In a simple impulse turbine, nozzles are inclined at an angle of 20° to the plane of rotation. Steam leaves the nozzle with velocity of 375 m/s. The blade speed is 165 m/s. Find the suitable inlet and outlet blade angles in order that there should be no axial thrust and allowing velocity reduction of 15 %, while steam flows over the blade. Determine also the power developed for now of 5 kg/s and kinetic energy of steam finally leaving the wheel.*

Solution

Given,

$$\alpha = 20°, \; V_b = 165 \text{ m/s}$$

$$V_{ai} = 375 \text{ m/s}, \; V_{fe} = V_{fi}$$

$$\frac{V_{re}}{V_{ri}} = 0.85, \; m = 5 \text{ kg/s}$$

$$BD = BC \times 0.85$$

For no-axial thrust $V_{fe} = V_{fi}$ i.e., point C and D should be on the same horizontal line parallel to AB so first drum a horizontal line from point C. Then mark an arc on that horizontal line, taking distance ($BC \times 0.85$) from point B. So point D is obtained and velocity diagram is complete.

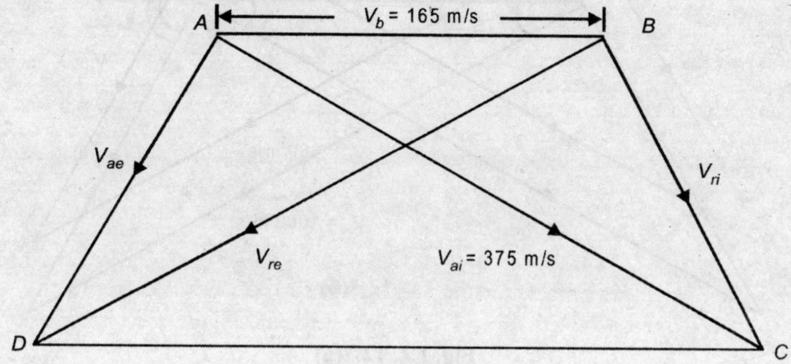

Fig. Ex. 12.6

From velocity diagram,

$$\theta = 34°, \ \Delta V_w = V_{wi} + V_{we} = 320 \text{ m/s}$$

$$\Phi = 41°, \ V_{ae} = 1325 \text{ m/s}$$

$$\text{Power developed} = m \times V_b \times \Delta V_w = \frac{5 \times 165 \times 320}{1000} = 264 \text{ kW}$$

$$\text{Kinetic energy of steam at exit} = \frac{m \times V_{ae}^2}{2}$$

$$= \frac{5 \times (1325)^2}{2 \times 1000} = 43.5 \text{ kW}$$

Example 12.7 *The following particulars refer to a velocity compounded impulse turbine with two rows of moving blades and a fixed row of guide blade between them. The steam velocity leaving the nozzle is 1000 m/s and nozzle angle is 20°. The blade speed is 200 m/s. The moving blades for the first row are symmetric. The exit angle for fixed blade is 23°. The exit angle of second row of moving blades is 30°. The blade friction coefficient 0.9 for each row. Determine power developed in kW.*

Solution

Given,

$$V_{ai_1} = 1000 \text{ m/s}, \quad V_b = 200 \text{ m/s} \quad _1 = \Phi_1$$

$$\alpha_1 = 20° \quad \alpha_2 = 23° \quad \Phi_2 = 30°$$

$$K = 0.9$$

The velocity diagram for first row of moving blades.

For drawing the velocity diagram for first row of moving blades, use $\theta_1 = \Phi_1$

and $$\frac{V_{re_1}}{V_{ri}} = 0.9$$

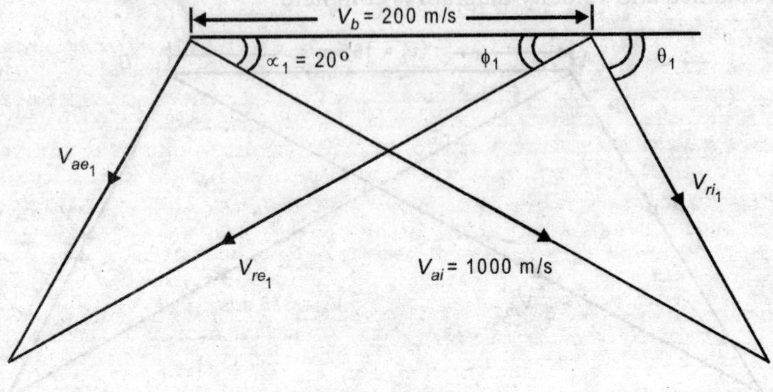

Fig. Ex. 12.7(a)

From the velocity diagram, we get,

$$V_{ri_1} = 813.33 \text{ m/s}$$

$$\therefore \quad V_{re_1} = 0.9 \times 813.33 = 732 \text{ m/s}$$

$$V_{ae_1} = 556.66 \text{ m/s}, \quad \Delta_{w_1} = V_{wi_1} + V_{we_2} = 1433.33 \text{ m/sec}$$

Also for fixed blades, K =0.9

$$\therefore \quad \frac{V_{ai_2}}{V_{ae_1}} = 0.9$$

$$\therefore \quad V_{ai_2} = 0.9 \times 556.66 = 501 \text{ m/sec}$$

Now draw the velocity diagram for second row of moving blades.

Use,

$$\frac{V_{re_2}}{V_{ri_2}} = K$$

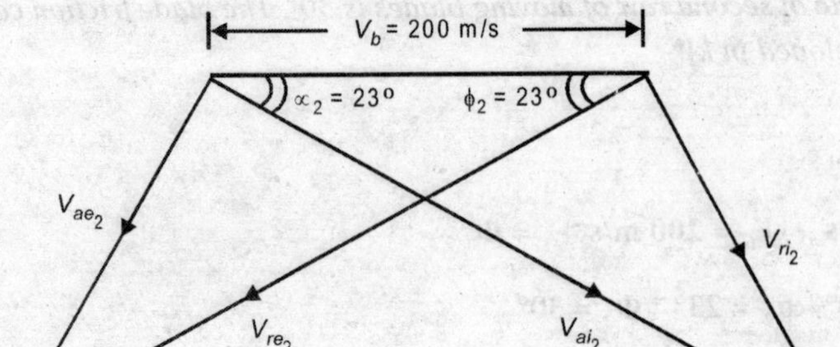

Fig. Ex. 12.7(b)

$$V_{ri_2} = 330 \text{ m/sec}$$

$$V_{re_2} = K \times V_{ri_2} = 0.9 \times 330 = 297 \text{ m/s}$$

$$\Delta V_{w_2} = \left(V_{wi_2} + V_{we_2}\right) = 520 \text{ m/sec}$$

Assume mass flow rate (m) = 1 kg/sec

$$\text{Power developed} = m \cdot V_b \left(\Delta V_{w1} + \Delta V_{w_2}\right)$$

$$= 1 \times 200 \times (1433.33 + 520)$$

$$= 390660 \text{ W}$$

$$= 390.66 \text{ kW}$$

Example 12.8 *The steam in a two row velocity compounded impulse turbine leaves the nozzle at 600 m/s. The blade speed is 120 m/s. The stage consists of two rows of moving and one fixed blades between the two. The nozzle angle is 16°, while the discharge angles are 18° for the first row of moving blades, 21° for fixed blades and 35° for the second row of moving blades. Assume 10% loss of velocity during the passage of steam through each ring of blades.*

Determine,

(a) Blade inlet angle for each row

(b) Driving force and axial thrust for stage in Newton

(c) Power developed in kW

(d) Diagram efficiency.

What would be maximum diagram efficiency for given steam inlet velocity and nozzle angle?

Solution

Given,

$$V_{ai_1} = 600 \text{ m/s}, \quad V_b = 120 \text{ m/s}, \quad _1 = \quad ^\circ$$

$$\Phi_1 = 18°, \ \alpha_2 = 21°, \ \Phi_2 = 35°$$

$$k = 0.9$$

Draw the velocity diagram for first row of moving blades

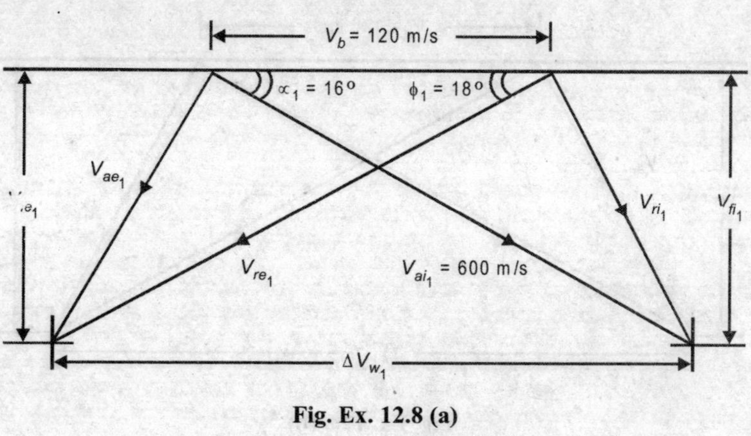

Fig. Ex. 12.8 (a)

Use,

$$\frac{V_{re1}}{V_{ri1}} = 0.9$$

From velocity diagram,

$$\Delta V_{w_1} = \left(V_{wi_1} + V_{we_1}\right) = 872 \text{ m/sec}$$

$$\theta_1 = 21°$$

$$\Phi_1 = 35°$$

$$V_{fi_1} = 40 \times \frac{872}{21.8} = 1600 \text{ m/s}$$

$$V_{ae_1} = 7.9 \times \frac{872}{21.8} = 316 \text{ m/s}$$

$$V_{fe_1} = 35 \times \frac{872}{21.8} = 1400 \text{ m/s}$$

$$\frac{V_{ai_2}}{V_{ae_1}} = 0.9$$

$$V_{ai_2} = 0.9 \times 316 = 284.4 \text{ m/s}$$

Draw the velocity diagram for second row of moving blades.

From the velocity diagram,

$$\Delta V_{w_2} = \left(V_{wi_2} + V_{we_2}\right) = 272 \text{ m/s}$$

(i) Power developed $= m \times V_b \left(\Delta V_{w_1} + \Delta_{w_2}\right)$

$= 1 \times 120 \times (872 + 272) = 137.28$ kW

(ii) Axial thrust :

For first row of moving blades

$$F_1 = m \times \left(V_{fi_1} - V_{fe_1}\right) = 1(20) = 20 \text{ N}$$

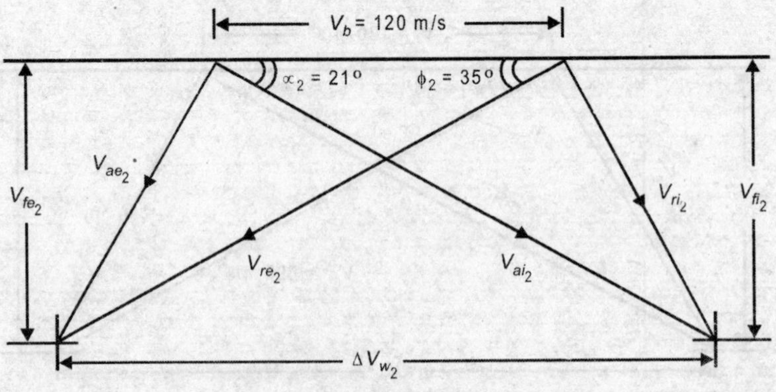

Fig. Ex. 12.8 (b)

For second row of moving blades,

$$F_2 = m \times \left(V_{fi_2} - V_{fe_2}\right) = 16 \text{ N}$$

Total axial thrust = 36 N

(iii) Diagram efficiency

$$\eta_{dia} = \frac{2 \times V_b \Delta V_w}{V_{ai_1}^2}$$

$$= \frac{2 \times 120 \times (872 + 272)}{(600)^2} = 76.27\ \%$$

(iv) Driving force : $\left[\left(V_{wi_1} + V_{we_1}\right) + \left(V_{wi_2} + V_{we_2}\right)\right] \times m$

$$= (872 + 272) \times 1 = 1144 \text{ N}$$

(v) Maximum diagram efficiency :

$$\eta_{dia} = (1 + KC) \times \frac{\cos^2}{2}$$

For maximum value, assume $K = 1$ and $C = 1$

$$\therefore \quad \eta_{dia} = \cos^2 =$$

Example 12.9 *Steam is supplied to a 3-stage turbine at 40 bar and 400° and the exhaust to the condenser takes place at 50 milli-bar, with a wetness fraction of 12%. The work developed in three stages is 1:1:2. The condition line may be assumed as straight line.*

Determine,

1. *Condition at entry of each stage*
2. *Stage efficiencies*
3. *Reheat factor*
4. *Internal efficiency of turbine*

Solution

$$(h_{A_1} - h_{A_4}) = (3215 - 2265) = 950 \text{ kJ/kg.}$$

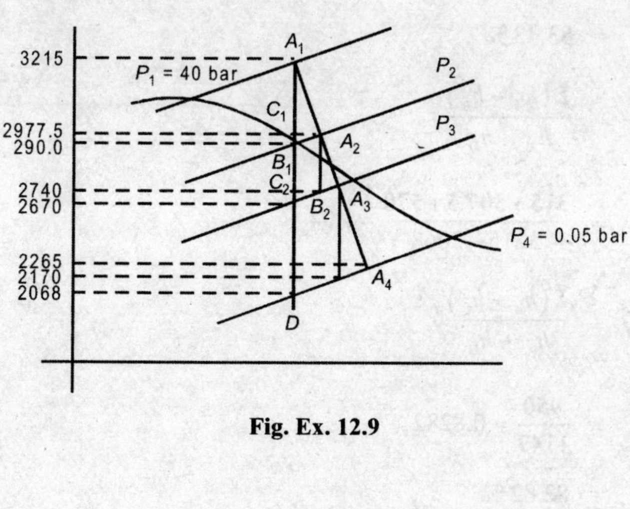

Fig. Ex. 12.9

$\therefore$ Enthalpy drop for first stage $= \dfrac{950}{4} = 237.5$ kJ/kg

Enthalpy drop for second stage = 237.5 kJ/kg

Enthalpy drop for third stage = 475 kJ/kg

1. Condition of steam at entry to each stage

From Mollier chart

At first stage the steam is superheated with pressure = 40 bar and temp = 400°C

Second stage, the steam is superheated

Pressure = 11.4 bar, Temp. = 265°C

for third stage, the steam is superheated and pressure = 2.5 bar

Temperature = 135°C

2. Stage efficiencies

$$\eta_{stage} = \frac{(h_A - h_C)}{(h_A - h_B)}$$

$$\eta_{1^{st}stage} = \frac{h_{A_1} - h_{C_1}}{h_{A_1} - h_{B_1}} = \frac{237.5}{315} = 0.754$$

$$= 75.4\ \%$$

$$\eta_{2^{nd}stage} = \frac{h_{A_2} - h_{C_2}}{h_{A_2} - h_{B_2}} = \frac{237.5}{307.5} = 0.7724$$

$$= 77.24\ \%$$

$$\eta_{3^{rd}stage} = \frac{h_{A_3} - h_{C_3}}{h_{A_3} - h_{B_3}} = \frac{475}{570} = 0.8333$$

$$= 83.33\ \%$$

3. Reheat factor $= \dfrac{\Sigma(h_A - h_B)}{h_{A_1} - h_D}$

$$= \frac{315 + 307.5 + 570}{3215 - 2068} = 1.04$$

$$\text{Internal efficiency} = \frac{\Sigma(h_A - h_C)}{h_{A_1} - h_D}$$

$$= \frac{950}{1147} = 0.8282$$

$$= 82.82\ \%$$

Example 12.10 *The three-stage steam turbine receives steam at 35 bar and 400°C. The steam is exhausted to condenser at 0.05 bar and dryness fraction = 0.9. The work developed by each stage is same.*

Determine,

(i) Condition of steam at entry to each stage

(ii) Stage efficiency

(iii) Reheat factor

(iv) Internal efficiency of turbine.

Solution

Given,

$$P_1 = 35 \text{ bar}$$

$$T_1 = 400°\text{C}$$

$$P_b = 0.05 \text{ bar}$$

$$X = 0.9$$

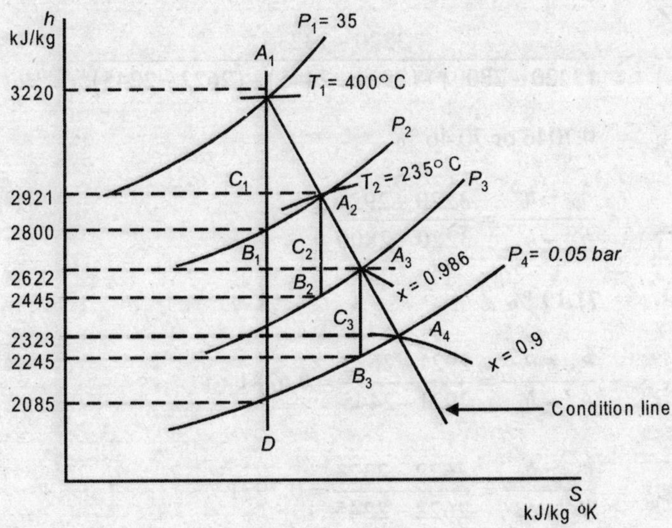

Fig. Ex. 12.10

It is given that work developed by each stage is same

$$\therefore \quad A_1C_1 = A_2C_2 = A_3C_3$$

$$\therefore \quad A_1C_1 + A_2C_2 + A_3C_3 = 8220 = 2323$$

$$\therefore \quad 3A_1C_1 = 897$$

$$\therefore \quad A_1C_1 = A_2C_2 = A_3C_3 = 299 \text{ kJ/kg}$$

$$\therefore \quad h_{A_2} = h_{C_1} = 3220 - 299 = 2921 \text{ kJ/kg}$$

$\therefore$ From chart $h_{B_1} = 2800$ kJ/kg

$$h_{C_2} = h_{A_3} = 2921 - 299 = 2622 \text{ kJ/kg}$$

$\therefore$ From chart $h_{B_2} = 2445$ kJ/kg

$$h_{A_4} = h_{C_3} = 2622 - 299 = 2323 \text{ kJ/kg}$$

From chart $h_{B_3} = 2245$ kJ/kg

(a) Condition at entry to each stage

1st stage superheated $P_1 = 35$ bar , $T_1 = 400°C$

2nd stage superheated $T_2 = 235°C$

3rd stage wet steam $x = 0.986$

(b) The stage efficiency

$$\text{Stage efficiency } \eta_s = \frac{\Sigma(h_A - h_C)}{\Sigma(h_A - h_B)}$$

$$= \frac{h_{A_1} - h_{C_3}}{\Sigma(h_A - h_B)} = \frac{3220 - 2323}{(3220 - 2800) + (2921 - 2445) + (2622 - 2245)}$$

$$\therefore \quad \eta_s = 0.7046 \text{ or } 70.46\ \%$$

For 1st stage $\quad \eta s_1 = \dfrac{h_{A_1} - h_{C_1}}{h_{A_1} - h_{B_1}} = \dfrac{3220 - 2921}{3220 - 2800}$

$\therefore \quad \eta s_1 =$ **71.19 %**

For 2nd stage $\quad \eta s_2 = \dfrac{h_{A_2} - h_{C_2}}{h_{A_2} - h_{B_2}} = \dfrac{2921 - 2622}{2921 - 2445} =$ **0.6281**

For 3rd stage $\quad \eta s_3 = \dfrac{h_{A_3} - h_{C_3}}{h_{A_3} - h_{B_3}} = \dfrac{2622 - 2323}{2622 - 2245} = 0.7931$

or 79.31 %

(c) Reheat factor

$$\text{R.F.} = \frac{\Sigma\, h_A - h_B}{h_{A_1} - h_D} = \frac{(3220 - 2800) + (2921 - 2445) + (2622 - 2245)}{3220 - 2085}$$

$\therefore$ **R.F. = 1.12**

(d) Internal turbine efficiency

$$\eta_i = \frac{\Sigma\, h_A - h_C}{h_{A_1} - h_D} = \frac{3220 - 2323}{3220 - 2085} = 0.7903$$

or **79.03 %**

Example 11.11 *At a stage in a reaction turbine, the mean blade ring diameter is 1 m and turbine runs at 50 rps. The blades are designed for 50% reaction with exit angle 30° and inlet angle 50°. The turbine is supplied with steam at the rate of 60,000 kg/hr and stage efficiency is 85%.*

Determine

(a) Power output

(b) Specific enthalpy drop in the stage in (kJ/kg)

(c) % increase in relative velocity.

Solution

$$V_b = \text{mean blade speed} = \pi D \cdot n = \pi \times 1 \times 50$$
$$= 157 \text{ m/s}$$

For 50 % reaction, $\alpha = \Phi$ and $\theta = \beta$

So draw the velocity diagram

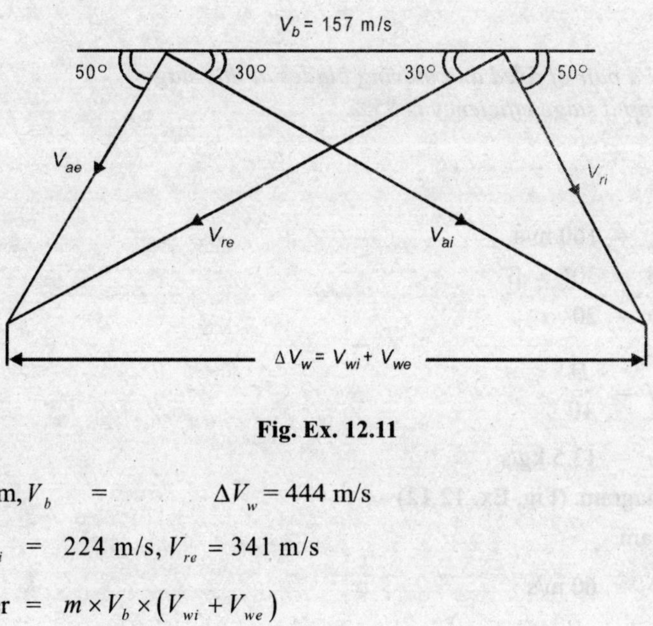

Fig. Ex. 12.11

For velocity diagram, $V_b = \Delta V_w = 444$ m/s

$$V_{ri} = 224 \text{ m/s}, V_{re} = 341 \text{ m/s}$$

$$\text{Power} = m \times V_b \times (V_{wi} + V_{we})$$

$$= \frac{60000}{3600} \times \frac{157 \times 444}{10^3}$$

$$= 11.6 \times 10^3 \text{ kW}$$

$$\eta_{stage} = \frac{V_b \times \Delta V_w}{\Delta h}$$

$$\therefore \quad \Delta h = \frac{V_b \times \Delta V_w}{\eta_{stage}}$$

$$= \frac{157 \times 444}{0.85} = 82 \times 10^3 \text{ J/kg} = 82 \text{ kJ/kg}$$

$$\% \text{ increase in relative velocity} = \frac{V_{re} - V_{ri}}{V_{ri}} \times 100$$

$$= \frac{341 - 224}{224} \times 100 = 52.2\,\%$$

Example 12.12 *At the particular stage of reaction turbine, the mean blade speed is 160 m/s and the steam is at the pressure of 350 kN/m² with the temperature of 175° C fixed and moving blades at the stage have inlet angle 30° and exit angle 20°. The blade height =* $\frac{D}{10}$.

The steam flow rate is 13.5 kg/s.

Determine,

1. *Blade height,*
2. *Power developed by a pair of fixed and moving blades at this stage*
3. *Specific enthalpy drop if stage efficiency is 85%.*

Solution

Given,

$$V_b = 160 \text{ m/s}$$

$$\theta = 30° = \beta$$

$$\Phi = 20° \ \alpha$$

$$H = \frac{D}{10}$$

$$m = 13.5 \text{ kg/s}$$

Draw the velocity diagram. (Fig. Ex. 12.12)

From velocity diagram,

$$V_{fi} = 60 \text{ m/s}$$

$$\Delta V_w = 270 \text{ m/s}$$

At 350 kN/m² and 175°C find the specific volume from Mollier-chart or steam-table.

$$V_s = 0.572 \text{ m}^3/\text{kg}$$

For reaction turbine, blade height is given as,

$$H = \frac{m \times V_s}{\pi D V f_i}$$

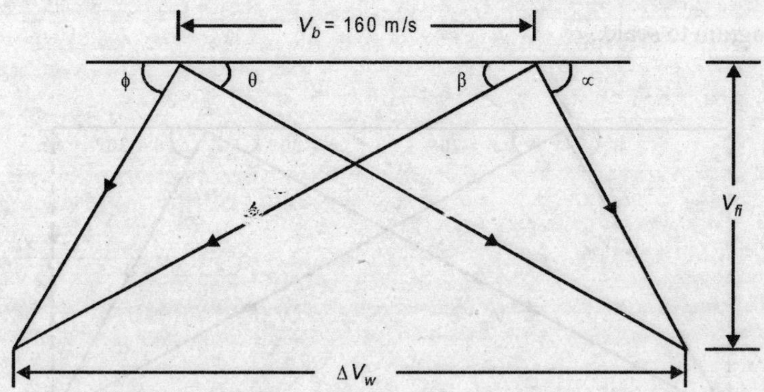

Fig. Ex. 12.12

$$H = \frac{D}{10} \text{ (Given)}$$

$$\therefore \quad D = 10H$$

$$\therefore \quad H = \frac{13.5 \times 0.572}{\pi \times 10 \times H \times 60}$$

$$\therefore \quad H = 0.064\ m = 64 \text{ mm}$$

$$\text{Power developed} = m \times V_b \times \Delta V_w$$

$$= \frac{13.5 \times 60 \times 270}{1000}$$

$$= 218 \text{ kW}$$

$$\eta_{stage} = \frac{V_b \times \Delta V_w}{\Delta h}$$

$$\therefore \quad \Delta h = \frac{V_b \times \Delta V_w}{\eta_{stage}}$$

$$\therefore \quad \Delta h = \frac{60 \times 270}{0.85 \times 10^3} = 19.1 \text{ kJ/kg}$$

Example 12.13 *Steam leaves the nozzle of a De-Laval turbine with a velocity of 1160 m/s. The nozzle is 20°. The blades are symmetrical with an angle of 30°. If the blade velocity coefficient is 0.8 and mechanical efficiency of 0.9. Calculate*

(i) Rotational speed if the rotor diameter is 75 cm,
(ii) Specific steam consumption in kg/kW-hr
(iii) Blade efficiency, and
(iv) Axial thrust.

Solution

Drawing the velocity diagram to scale.

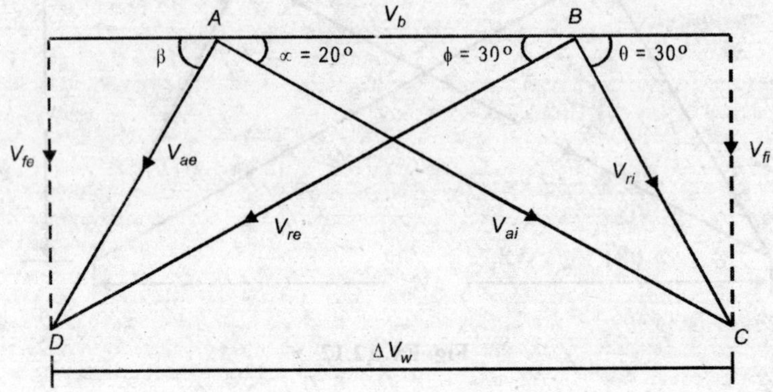

Fig. Ex. 12.13

From the velocity triangle,

$$V_{ai} = 1160 \text{ m/s}$$

$$V_b = 410 \text{ m/s}$$

$$V_{ri} = 800 \text{ m/s}$$

$$K = 0.8$$

$$\therefore \quad V_{re} = K\, V_{ri}$$

$$= 0.8 \times 800$$

$$= 640 \text{ m/s}$$

$$V_{fi} = 397.7 \text{ m/s}$$

$$V_{fe} = 320 \text{ m/s}$$

We have,

$$V_b = \frac{\pi\, DN}{60}$$

$$\therefore \quad N = \frac{60 \times V_b}{\pi \times D}$$

$$= \frac{60 \times 410}{\pi \times 7.5 \times 10^{-1}}$$

$$N = 10440 \text{ rpm}$$

$$\text{Power consumption} = \dot{m}\, V_b\, \Delta V_w$$

$$\Delta V_w = 1240 \text{ m/s}$$

$$\therefore \qquad \text{Power} = \frac{1 \times 410 \times 1240}{1000}$$

$$= 508.4 \text{ kW}$$

Specific steam consumption

$$= \frac{3600}{\text{Work done}}$$

$$= \frac{3600}{508.4}$$

$$= \mathbf{7.081 \text{ kg/kW.hr.}}$$

$$\text{Blade efficiency} = \frac{\dot{m}\, V_b\, \Delta V_w}{\frac{1}{2}\dot{m}\, V_{ai}^2}$$

$$= \frac{V_b\, \Delta V_w \times 2}{V_{ai}^2}$$

$$= \frac{410 \times 1240 \times 2}{(1160)^2} \times 100$$

$$\eta_{blade} = 75.56\ \%$$

$$\text{Axial thrust} = \dot{m}\left(V_{fi} - V_{fe}\right)$$

$$= 1 \times (397.7 - 320)$$

$$= 77.7 \text{ N}$$

Example 12.14 *In a stage of an impulse turbine, the mean diameter of blade is 1.5 m. It runs at 3000 rpm. The velocity of steam at inlet to blade is 500 m/s and nozzle angle is 20° moving blade angle at inlet and exit are same. If blade friction factor is 0.9, find power in kW. Axial thrust on the bearing is 80 N.*

Solution

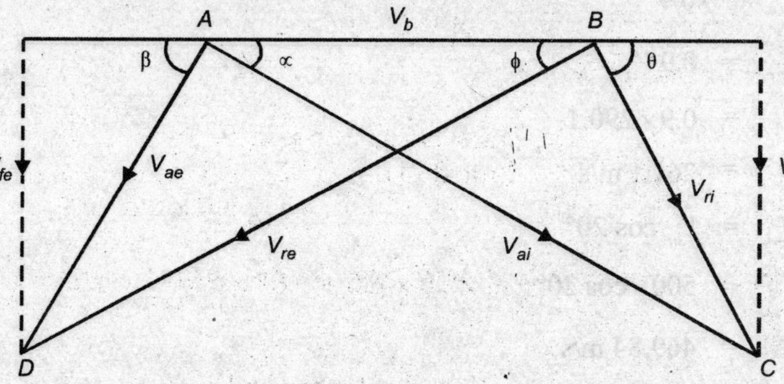

Fig. Ex. 12.14

Given,

$$N_b = 3000 \text{ rpm}$$

$$V_{ai} = 500 \text{ m/s}$$

$$\alpha = 20°$$

$$\phi = \theta$$

$$k = 0.9$$

$$\text{Axial thrust} = 180 \text{ N}$$

$$\text{Blade speed } V_b = \frac{\pi DN}{60}$$

$$= \frac{\pi \times 1.5 \times 3000}{60} = 235.62 \text{ m/s}$$

$$V_{ai} \cos\alpha = V_{ri} \cos\theta + C_b$$

$$V_{ai} \sin\alpha = C_{ri} \sin\theta$$

$$\therefore 500 \cos 20° - 235.6 = V_{ri} \cos\theta$$

$$\therefore 500 \sin 20° - 235.6 = V_{ri} \sin\theta$$

$$\therefore \tan\theta = \frac{500 \sin 20°}{500 \cos 20° - 235.6}$$

$$= 0.7299$$

$$\therefore \theta = 36.12° = \phi = \text{blade angles}$$

$$V_{ri} = \frac{500 \sin 20°}{\sin 36.12°} = 290.1 \text{ m/s}$$

We have

$$\frac{V_{ri}}{V_{re}} = \frac{1}{0.9} \qquad K = 0.9$$

$$\therefore V_{re} = 0.9 V_{ri}$$

$$= 0.9 \times 290.1$$

$$= 261.1 \text{ m/s}$$

$$V_{wi} = V_{ai} \cos 20°$$

$$= 500 \times \cos 20°$$

$$= 469.84 \text{ m/s}$$

$$V_{we} = V_{re} \cos\phi - V_b$$

$$= 261.1 \cos 36.12 - 235.9$$

$$= -23.8 \text{ m/s}$$

This shows that is greater than 90°, so the line *AD* will be inclined towards right side.

$$\therefore \quad V_{fi} = V_{ai} \sin \alpha$$

$$= 500 \sin 20$$

$$= 171 \text{ m/s}$$

$$V_{fe} = V_{re} \sin \phi$$

$$= 261 \sin 36.12$$

$$= 153.85 \text{ m/s}$$

Given axial thrust on the bearing is 180 N

$$\text{Axial thrust} = \dot{m}\left(V_{fi} - V_{fe}\right)$$

$$\therefore \quad 180 = \dot{m}(171 - 153.85)$$

$$\therefore \quad \dot{\mathbf{m}} = \mathbf{10.497\ kg/s}$$

$$\text{Power} = \dot{m}\, V_b \left(V_{wi} + V_{we}\right)$$

$$= 10.497 \times 235.62\left(469.84 + (-23.8)\right)$$

$$\therefore \quad \text{Power} = 1103320.4 \text{ J/s or N m/s}$$

$$= \mathbf{1103.32\ kW}$$

Example 12.15 *In a stage of an impulse turbine, steam leaves the nozzles at 1000 m/s. Nozzle angle is 20°. Mean blade speed is 400 m/s. Efficiency of nozzle is 0.95. Friction factor for blades is 0.8. Mass flow rate of steam is 0.5 kgs. Find: (i) Blade angle, (ii) Axial thrust, (iii) Output in kW, (iv) Diagram efficiency (v) Stage efficiency.*

Solution

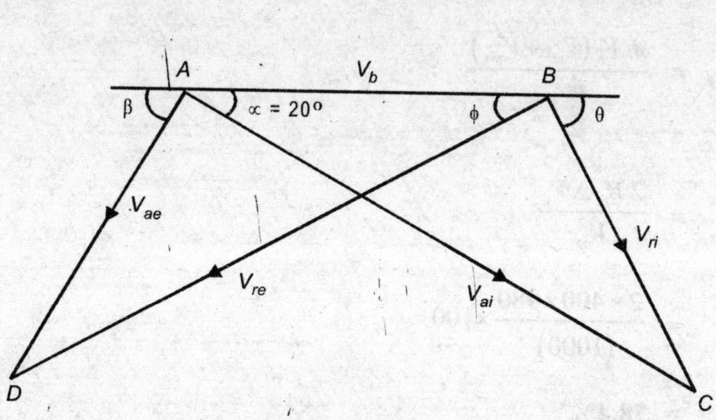

Fig. Ex. 12.15

Given,

$$\alpha = 20^\circ$$

$$V_{ai} = 1000 \text{ m/s}$$

$$V_b = 400 \text{ m/s}$$

$$k = 0.8$$

$$\dot{m} = 0.5 \text{ kg/s}$$

Assuming symmetrical blades and drawing velocity diagram to scale

Blades angles $\theta = \phi = 32°$

$$V_{ri} = 640 \text{ m/s}$$

$$V_{re} = 0.8\,V_{ri} = 512 \text{ m/s}$$

$$V_{fi} = 340 \text{ m/s}$$

$$V_{fe} = 270 \text{ m/s}$$

$$V_{wi} + V_{we} = 980 \text{ m/s}$$

$$\text{Axial thrust} = \dot{m}\left(V_{fi} - V_{fe}\right) = 0.5(340 - 270) = \mathbf{35\ N}$$

$$\text{Output in kW} = \frac{\dot{m}\, V_b\, \Delta V_w}{1000} \text{ kW} = \frac{0.5 \times 400 \times 980}{1000} = \mathbf{196\ kW}$$

$$\text{Diagram efficiency} = \frac{\dot{m}\, V_b\left(V_{wi} + V_{we}\right)}{\frac{\dot{m}}{2} V_{ai}^2} = \frac{2\, V_b\, \Delta V_w}{V_{ai}^2} = \frac{2 \times 400 \times 980}{(1000)^2} \times 100$$

$$\therefore \quad \eta_{dia} = \mathbf{78.4\%}$$

Stage efficiency = Diagram efficiency × Nozzle efficiency

$$= 0.784 \times 0.95$$

$$\eta_{stage} = \mathbf{74.48\%}$$

Example 12.16 *The data refers to single stage impulse turbine. Isentropic nozzle drop is 190 kJ/kg friction in nozzle is 10 % of isentropic heat drop. Nozzle angle is 20°. Ratio of blade speed to which component of steam speed is 0.4. Velocity component for blade is 0.9. The velocity of steam at entry to nozzle is 35 m/s. Find the blade angles so that steam may enter then without shock and leave them in axial direction.*

Solution

Given,

$$\alpha = 20°, \quad \beta = 90°$$

$$k = 0.9$$

$$\frac{V_b}{V_{wi}} = 0.4$$

Velocity at entry to nozzle = 35 m/s

Isentropic enthalpy drop in nozzle = 190 kJ/kg

Because of friction,

Actual enthalpy drop = 190×0.9

$$\Delta h = 171 \text{ kJ/kg}$$

Steam velocity at inlet

$$V_{ai} = \sqrt{2 \times \Delta h \times 1000 + (35)^2}$$

$$= 585.85 \text{ m/s}$$

Whirl velocity at inlet

$$V_{wi} = V_{ai} \cos \alpha$$

$$= 585.85 \times \cos 20$$

$$= 550.52 \text{ m/s}$$

$$V_b = 0.4\, V_{wi}$$

$$\therefore \quad V_b = 0.4 \times 550.52$$

$$= 220.21 \text{ m/s}$$

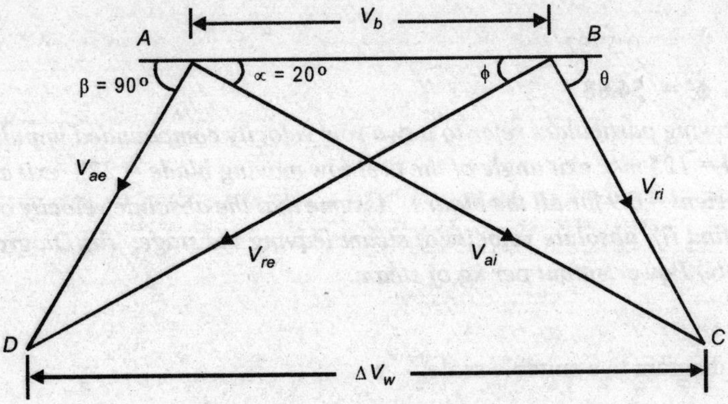

Fig. Ex. 12.16

From inlet triangle,

$$V_{ai} \sin 20 = V_{ri} \sin \theta$$

$$\therefore \quad V_{ri} \sin \theta = 585.52 \sin 20$$

$$= 200.37 \text{ m/s}$$

$$V_{ri} \cos \theta = V_{wi} - V_b$$

$$= 550.52 - 220.21$$

$$= 330.31 \text{ m/s}$$

$$\tan \theta = \frac{\sin \theta}{\cos \theta} = \frac{V_{ri} \sin \theta}{V_{ri} \cos \theta}$$

$$= \frac{200.37}{330.31}$$

$$= 0.6066$$

$$\therefore \quad \theta = 31.24°$$

$$V_{ri} = \frac{200.37}{\sin 31.24} = 386.33 \text{ m/s}$$

$$V_{re} = 0.9 V_{ri}$$

$$= 0.9 \times 386.33$$

$$= 347.7 \text{ m/s}$$

$$\cos \phi = \frac{V_b}{V_{re}} \quad \text{as } \beta = 90°$$

$$= \frac{220}{347.25}$$

$$= 0.5752$$

$$\therefore \quad \boldsymbol{\phi = 54.88°}$$

Example 12.17 *The following particulars refer to a two row velocity compounded impulse steam turbine : nozzle angle = 17°, blade speed = 125 m/s, exit angle of the first row moving blade = 22°, exit angle of the fixed blade = 30°, blade velocity coefficient = 0.9 for all the blades. Assume that the absolute velocity of steam leaving the stage is in the axial direction find (i) absolute velocity of steam leaving the stage, (ii) Diagram efficiency, (iii) Axial thrust per kg of steam, (iv) Power output per kg of steam.*

Solution

Drawing the velocity diagram to a suitable scale,

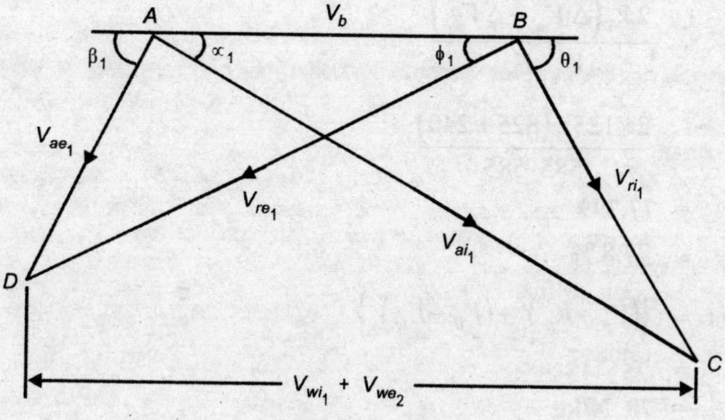

Fig. Ex. 12.17(a)

Given :

$$\alpha_1 = 17^\circ \qquad \alpha_2 = 26^\circ$$
$$V_b = 125 \text{ m/s} \qquad K = 0.9$$
$$\phi_1 = 22^\circ \qquad \phi_2 = 30^\circ$$
$$\beta_2 = 90^\circ$$

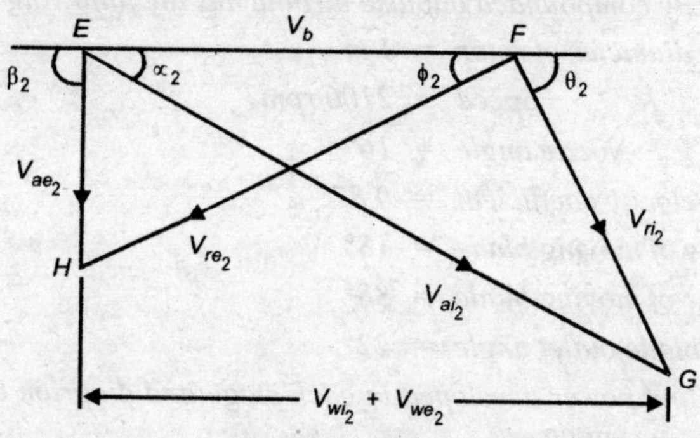

Fig. Ex. 12.17(b)

Second row

We get $V_b = 125$ m/s

From outlet triangle

$$V_{ae_2} = V_{fe_2} = 72 \text{ m/s}$$

i.e., exit velocity of steam at second stage.

Also we have

$$V_{ai_2} = V_{ae_1} \times 0.9 \qquad (\text{as } K = 0.9)$$

From the velocity triangles,

$$\text{Diagram efficiency} = \frac{V_b\left(\Delta V_{w_1} + \Delta V_{w_2}\right)}{\frac{1}{2}V_{ai_1}^2}$$

$$= \frac{2V_b\left(\Delta V_{w_1} + \Delta V_{w_2}\right)}{V_{ai_1} \cdot V_{ai_1}}$$

$$= \frac{2 \times 125 \times (825 + 240)}{585 \cdot 585}$$

$$= \mathbf{77.799}$$

$$\eta_{dia} \approx 77.8\,\%$$

$$\text{Axial thrust} = \left\{\left(V_{fi} - V_{fe}\right)_1 + \left(V_{fi} - V_{fe}\right)_2\right\}$$

$$= 25 + 45$$

$$= \mathbf{70\ N/kg}$$

Power output per kg/s of steam

$$= \frac{V_b\left\{\left(V_{wi} + V_{we}\right)_1 + \left(V_{wi}\right)_2\right\}}{1000}$$

$$= \frac{125\{825 + 240\}}{1000}$$

$$= \mathbf{133.125\ kW}$$

Example 12.18 *A two row velocity compounded impulse turbine has the following specifications*

Mean diameter of rotor = *1 m*

Speed = *2100 rpm*

Nozzle angle = *16°*

Velocity coefficient = *0.87*

Outlet angle of first row of moving blade = *18°*

Outlet angle of second row of moving blade = *38°*

Fixed blade outlet angle = *22°*

Draw velocity diagram and find power developed in each stage and diagram efficiency. Assume steam flow rate of 5 kg/s and steam jet velocity = 600 m/s.

Solution

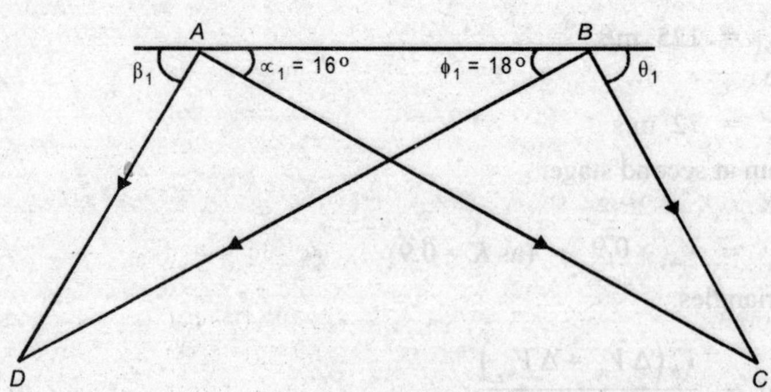

Fig. Ex. 12.18 (a)

Given,

$$V_{ai_1} = 600 \text{ m/s}$$

$$\alpha_1 = 16^\circ$$

$$\phi_1 = 18^\circ$$

$$\phi_2 = 38^\circ$$

$$\alpha_2 = 22^\circ$$

$$V_b = \frac{\pi DN}{60}$$

$$= \frac{\pi \times 1 \times 2100}{60}$$

$$= 109.95 \text{ m/s}$$

$$V_b \approx 110 \text{ m/s}$$

Velocity triangle for the first stage is drawn as shown.

We have

$$\Delta V_{w_1} = 875 \text{ m/s}$$

$$V_{re_1} = 0.87\, V_{ri_1}$$

Second row velocity triangle:

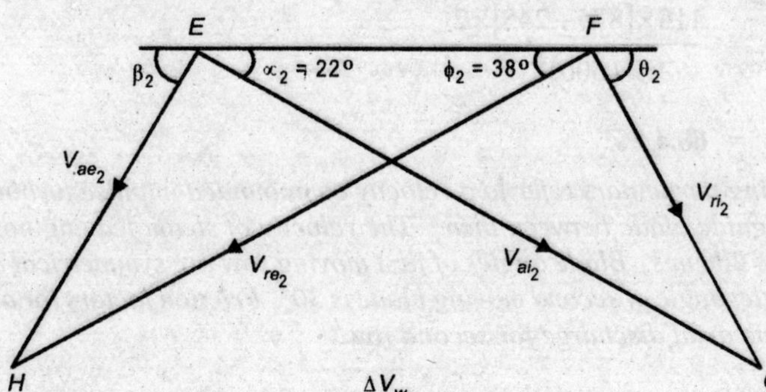

Fig. Ex. 12.18 (b)

$$\alpha_2 = 22^\circ \qquad \phi_2 = 38^\circ$$

$$V_{ai_2} = 0.87\, V_{ae_1}$$

From velocity diagram

$$\Delta V_{w_2} = 245 \text{ m/s}$$

$$V_{re_2} = 0.87\, V_{ai_2}$$

Power developed in first stage

$$P_1 = \dot{m} \cdot V_b \cdot \Delta V_{w_1}$$

$$= \frac{(5 \times 110 \times 875)}{1000}$$

$$= \mathbf{481.25\ kW}$$

Power developed in second stage

$$P_2 = \dot{m} \cdot V_b \cdot \Delta V_{w_2}$$

$$= \frac{(5 \times 110 \times 245)}{1000}$$

$$= \mathbf{134.7\ kW}$$

$\therefore$ Total power output

$$= 481.25 + 134.7$$

$$= \mathbf{615.95\ kW}$$

Diagram efficiency

$$= \frac{V_b \cdot \left(\Delta V_{w_1} + \Delta V_{w_2}\right)}{\frac{V_{ai_1}^2}{2}}$$

$$= \frac{110 \times [875 + 245] \times 2}{(600)^2}$$

$$\therefore \quad \eta_{dia} = \mathbf{68.4\ \%}$$

Example 12.19 *The following particulars refer to a velocity compounded impulse turbine with 2 rows of moving blades with a fixed row of guide blade between them. The velocity of steam leaving nozzle is 1000 m/s. Nozzle angle is 20°. Blade speed is 200 m/s. Blade angles of first moving row are symmetrical and blade outlet angle of fixed blade is 23°, blade outlet angle of second moving blade is 30°. Friction factors for all rows is 0.9. Determine power output in kW. Assume axial discharge for second row.*

Solution

Given, $V_b = 200$ m/s

$V_{ai} = 1000$ m/s

Nozzle angle $\alpha_1 = 20°$

Blade outlet angle $\phi_2 = 30°$

Fixed blade outlet angle $\alpha_2 = 23°$

Blade velocity coefficient

$k = 0.9$ for all blades

(i) Velocity triangles for first moving blades

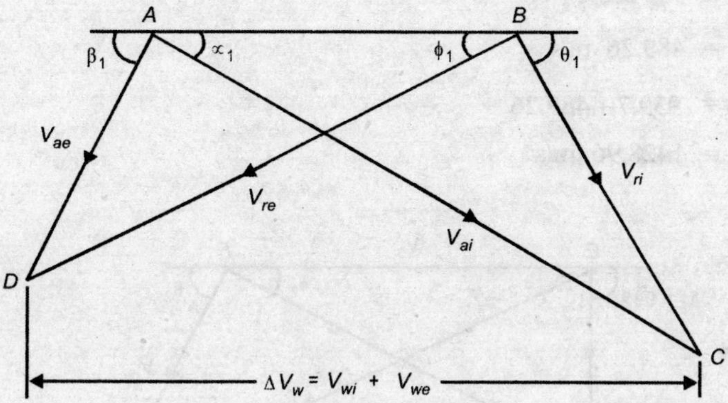

Fig. Ex. 12.19(a)

From velocity triangle of first row,

$$V_{ai_1} \cos \alpha_1 = V_{ri_1} \cos \theta_1 + C_b$$

and $$V_{ai_1} \sin \alpha_1 = V_{ri_1} \sin \theta_1$$

$$\therefore \quad \tan \theta_1 = \frac{V_{ai_1} \sin \alpha_1}{V_{ai_1} \cos \alpha_1 - C_b}$$

$$= \frac{1000 \sin 20}{1000 \cos 20 - 200}$$

$$= 0.462$$

$$\therefore \quad \theta_1 = 24.81° = \phi_1 \qquad (\because \text{ blades are symmetrical})$$

$$\therefore \quad V_{ri_1} = \frac{V_{ai_1} \sin 20}{\sin \theta_1}$$

$$= \frac{1000 \times \sin 20}{\sin 24.81}$$

$$= 815.1 \text{ m/s}$$

$$V_{re_1} = k \cdot V_{ri_1}$$

$$= 0.9 \times 815.1$$

$$= 733.5 \text{ m/s}$$

$$V_{wi_1} = V_{ai_1} \cos \alpha_1$$

$$= 1000 \cos 20$$

$$= 939.7 \text{ m/s}$$

$$V_{we_1} = V_{re_1} \cos \phi - V_b$$

$$= 489.26 \text{ m/s}$$

and $$\Delta V_{w_1} = 939.7 + 489.26$$

$$= 1428.96 \text{ m/s}$$

For second row:

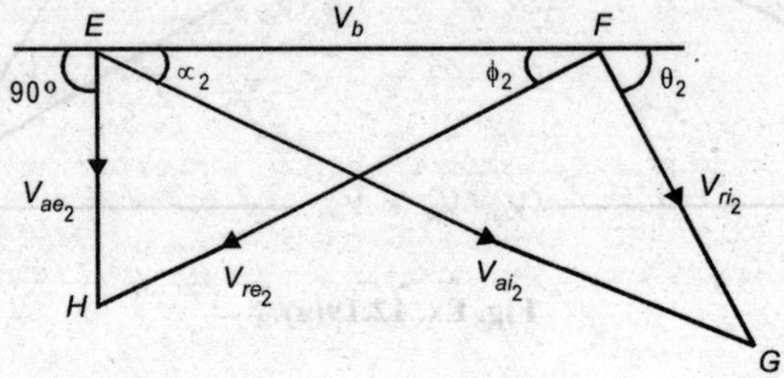

Fig. Ex. 12.19(b)

$$V_{ai_2} = V_{ae_1} \times 0.9$$

But $$V_{ae_1}^2 = V_{re_1}^2 + V_b^2 - 2\, V_{re_1}\, V_b \cos \phi_1$$

$$= 733.5^2 + 200^2 - 2 \times 733.5 \times 200 \times \cos 24.81°$$

$$V_{ae_1} = 558.3 \text{ m/s}$$

$\therefore$ $$V_{ai_2} = 0.9\, V_{ae_1}$$

$$= 502.47 \text{ m/s}$$

$$V_{wi_2} = V_{ai_2} \cos \alpha_2$$

$$= 502.47 \times \cos 23$$

$$= 462.53 \text{ m/s}$$

$$V_{we_2} = 0 \text{ as discharge is radial}$$

$\therefore$ $$\text{Power output} = \frac{V_b \times \left[\Delta V_{w_1} + \Delta V_{w_2}\right]}{1000}$$

$$= \frac{200 \times [1428.96 + 462.53]}{1000}$$

$$= \mathbf{378.3 \text{ kW}}$$

Example 12.20 *A reaction turbine has Identical blading. It delivers dry, saturated steam at 2.5 bar. Velocity of steam is 90 m/s. Mean blade height is 3 cm and exit angle of the moving blade is 20°. At the mean radius, the axial flow velocity equals 3/4 blade speed. For a steam flow rate of 9000 kg/hr, calculate (i) power output, (ii) rotor* *speed in rpm, (iii) diagram efficiency, (iv) enthalpy drop in a stage, (v) percentage increase in relative velocity in moving blade.*

Solution

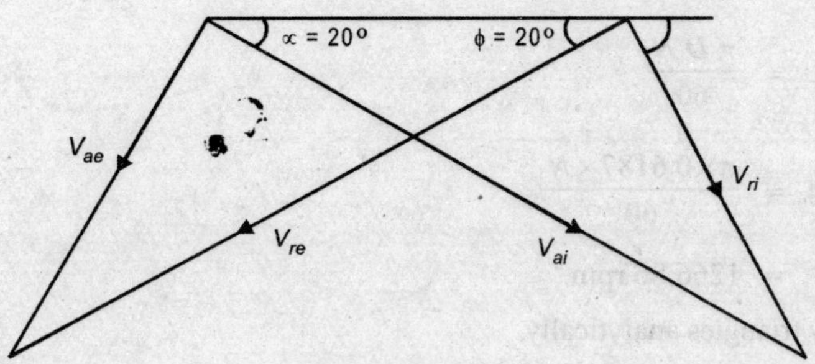

Fig. Ex. 12.20

Given,

$$V_{ai} = 90 \text{ m/s}$$

$$h = 3 \text{ cm} = 0.03 \text{ m}$$

$$\text{Stage pressure} = 2.5 \text{ bar}$$

$$\dot{m} = 9000 \text{ kg/hr} = \frac{9000}{3600} \text{ kg/s} = 2.5 \text{ kg/s}$$

$$V_{fi} = 90 \sin 20 = 30.78 \text{ m/s}$$

Steam is dry and saturated

$$\text{Specific volume } v = 0.718 \text{ m}^3\text{/kg at } p = 2.5 \text{ bar}$$

$$\dot{m} = \frac{\text{Area} \times V_f}{v} = \frac{\pi D h \times V_f}{v}$$

$$\therefore \quad 2.5 = \frac{\pi \times D \times 0.03 \times 30.78}{0.718}$$

$$\therefore \quad D = 0.6187 \text{ m}$$

$$V_b = \frac{4}{3} \cdot V_{fi}$$

$$= \frac{4}{3} \times 30.78$$

$$= 41.04 \text{ m/s}$$

But $$V_b = \frac{\pi D N}{60}$$

$$41.04 = \frac{\pi \times 0.6187 \times N}{60}$$

$$\therefore \quad N = 1266.86 \text{ rpm}$$

Solving the velocity triangles analytically,

$$\mathbf{V_{ai}\ cos} = \mathbf{V_{ri}\ cos} \quad \mathbf{b}$$

$$V_{ai} \sin \alpha = V_{ri} \sin \theta$$

$$\therefore \quad \tan \theta = \frac{V_{ai} \sin \alpha}{V_{ai} \cos \alpha - V_b}$$

$$= \frac{90 \times \sin 20}{90 \times \cos 20 - 41.04}$$

$$= 0.707$$

$$\therefore \quad \theta = 35.26° = \beta \quad \text{(as blades are identical)}$$

$$V_{ri} = \frac{V_{ai} \sin \alpha}{\sin \theta}$$

$$= 53.31 \text{ m/s}$$

$$V_{wi} = V_{ai} \cos \alpha = 90 \cos 20°$$

$$= 84.57 \text{ m/s}$$

$$V_{we} = V_{ae} \cos \beta = 90 \cos 35.26$$

$$= 73.48 \text{ m/s}$$

Blade speed ratio $$= \rho = \frac{V_b}{V_{ai}} = \frac{41.04}{90}$$

$$= 0.456$$

Diagram efficiency $$= \frac{2\rho(2\cos\alpha - \rho)}{1 - \rho^2 + 2\rho\cos\alpha}$$

$$= \frac{2 \times 0.456(2 \times 0.9396 - 0.456)}{1 - (0.456)^2 + 2 \times 0.456 \times 0.9396}$$

$$= 0.787$$

$$\boldsymbol{\eta_{dia} = 78.7\ \%}$$

Enthalpy drop per stage

$$= \frac{2\left(V_{ai}^2 - V_{ri}^2\right)}{2\times1000}$$

$$= \frac{2\times(90)^2 - (53.31)^2}{2\times1000}$$

$$= \textbf{5.258 kJ/kg.}$$

Percentage increase in relative velocity

$$= \left(\frac{V_{re} - V_{ri}}{V_{ri}}\right)\times100$$

$$= \left(\frac{90 - 53.31}{53.31}\right)\times100 \quad \left(\text{as } V_{re} = V_{ai}\right)$$

$$= 0.6888$$

$$= \textbf{68.88 \%}$$

$$\text{Power developed} = \frac{V_b\left(V_{wi} + V_{we}\right)}{1000}\times\frac{\dot{m}}{1}$$

$$= \frac{41.04(84.57 + 73.48)}{1000}\times2.5$$

$$= \textbf{16.216 kW}$$

Example 12.21 *In a stage of reaction turbine, the mean diameter of rotor is 1.4 m. Velocity ratio is 0.7. Turbine rotates at 3000 rpm. Blade outlet angle is 20°. Find diagram efficiency and also find value of maximum diagram efficiency.*

Solution

$$V_b = \frac{\pi DN}{60}$$

$$= \frac{\pi\times1.4\times3000}{60}$$

$$= 219.9 \text{ m/s}$$

$$V_{ai} = \frac{V_b}{0.7}$$

$$= 314.15 \text{ m/s}$$

$$\text{Blade efficiency} = \frac{2\rho(2\cos\alpha - \rho)}{1 - \rho^2 + 2\rho\cos\alpha}$$

$$= \alpha = 20°, \rho = 0.7$$

$$= \frac{2\times 0.7(2\times\cos 20-0.7)}{1-(0.7)^2+2\times 0.7\cos 20}$$

$$\therefore \quad \eta_{blade} = 90.4\ \%$$

Maximum diagram efficiency

$$= \frac{2\cos^2\alpha}{1+\cos^2\alpha}$$

$$= \frac{2\times\cos^2 20}{1+\cos^2 20}$$

$$= 93.78\ \%$$

Example 12.22 *At a stage of a reaction turbine, the mean ring diameter of 140 cm and speed ratio is 0.7. Speed is 3000 rpm. Determine required entrance angle to blading if exit angle is 20°. Also determine work done per kg of steam per second. Calculate percentage increase in diagram efficiency if the blades are desingned for and run at the best theoretical speed, the exit angle being kept 20°.*

Solution

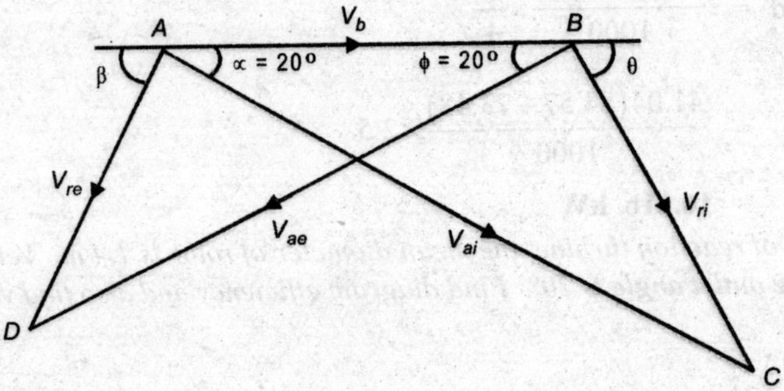

Fig. Ex. 12.22

Drawing velocity diagram, for 50% reaction, inlet and exit triangles are symmetrical

$$\alpha=\phi,\ \theta=\beta,\ V_{ai}=V_{re}$$

$$V_{fi} = V_{fe}$$

$$\text{Blade speed } V_b = \frac{\pi D N}{60}$$

$$= \frac{\pi\times 1.4\times 3000}{60}$$

$$= 220 \text{ m/s}$$

$$V_{ai} = \frac{V_b}{\text{Blade speed ratio}}$$

$$= \frac{220}{0.7}$$

$$= 314.16 \text{ m/s}$$

$$V_{ri}^2 = V_{ai}^2 + V_b^2 - 2V_b \cdot V_{ai} \cdot \cos\alpha$$

$$\therefore \quad V_{ri}^2 = 314.16^2 + 220^2 - 2\times 314.16 \times 220 \cos 20$$

$$\therefore \quad V_{ri} = 131.15 \text{ m/s}$$

Using sin rule,

$$\frac{V_{ri}}{\sin\alpha} = \frac{V_{ai}}{\sin(\pi-\theta)}$$

$$\therefore \quad \sin\theta = \frac{V_{ai}}{V_{ri}}\sin\alpha$$

$$= \frac{314.16}{131.15}\sin 20$$

$$\therefore \quad \theta = \mathbf{55^\circ}$$

$$\textbf{Work done per kg} = V_b\left(V_{wi} + V_{we}\right)$$

$$= V_b\left(V_{ai}\cos\alpha - V_b + V_{re}\cos\phi - V_b\right)$$

$$= 220\times 2\left(V_{ai}\cos\alpha - V_b\right)$$

$$= 440\times\left(314.16\cos 20 - 220\right)$$

$$= 33094.1 \text{ N.m}$$

$$= \mathbf{33.094 \ kJ}$$

$$\text{Diagram efficiency} = \frac{\text{Work done}}{\text{Energy supplied}}$$

$$= \frac{81}{\dfrac{V_{ai}^2 + V_{re}^2 - V_{ri}^2}{2\times 1000}}$$

$$= \frac{81\times 2\times 1000}{(314.16)^2 + (314.16)^2 - (131.15)^2}$$

$$= 0.899 \text{ i.e., } 89.9\,\%$$

For best theoretical condition

$$V_b = V_{ai}\cos\alpha$$

$$= 314.16\times\cos 20$$

$$= 295.2 \text{ m/s}$$

$$V_{ri} = \sqrt{(V_{ai})^2 + (V_b)^2 - 2 \cdot V_{ai} \cdot V_b \cos 20}$$

$$\therefore \quad V_{ri} = 107.45 \text{ m/s}$$

$$= V_{ae}$$

Maximum diagram efficiency

$$= \frac{2\cos^2\alpha}{1+\cos^2\alpha}$$

$$= \frac{2\cos^2 20°}{1+\cos^2 20°}$$

$$= 0.9378$$

$$= 93.78 \%$$

∴ Percentage increase in efficiency

$$= \frac{93.78 - 89.9}{89.9} \times 100$$

$$= \mathbf{4.324\ \%}$$

Example 12.23 *A 3 stage turbine is fed at 26 bar and 370°C. The exhaust takes place at 0.05 bar. Inter stage pressure are 5 bar and 1 bar. The stage efficiency for all the stages is 80%. Assuming a condition line to be straight.*

Determine:

(i) Rankine efficiency
(ii) Quality of steam leaving each stage
(iii) Reheat factor
(iv) Work dine per kg of steam in each stage
(v) Overall efficiency.

Solution

From the Mollier Chart

Pressure P_{bar}	*Temperature or Quality*	*Ise inthalpy drop kJ/kg*	*Actual enthalpy drop kJ/kg*
$P_1 = 26$	370°C	3170 – 2676	3170 – 2854.6
$P_2 = 5$	198°C at 2′		
$P_3 = 1$	$x_3 = 0.981$	2854.8 – 2568.8	2854.4 – 2626
$P_4 = 0.05$	$x_4 = 0.9$	2626 – 2200	2626 – 2285.2
		$\sum \Delta h_{ise} = 1106 \frac{\text{kJ}}{\text{kg}}$	$\sum \Delta h_{act} = 884.8 \frac{\text{kJ}}{\text{kg}}$

Also $\quad h_4'' = 2100$ kJ/kg

$$h_{f_4} = 137.7 \text{ kJ/kg}$$

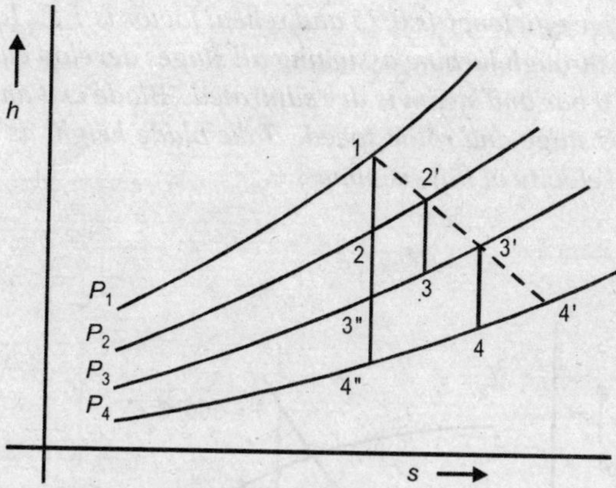

Fig. Ex. 12.23

We have

$$\text{Rankine efficiency} = \frac{h_1 - h_4^{"}}{h_1 - h_{f_4}}$$

$$= \frac{3170 - 2100}{3170 - 137.7} \times 100\%$$

$$= \mathbf{35.28\%}$$

$$\text{Reheat factor} = \frac{\Sigma\, \Delta\, h_{isen}}{h_1 - h_4^{"}}$$

$$= \frac{1106}{3170 - 2100}$$

$$= \mathbf{1.0336}$$

Overall efficiency = Stage efficierncy × Reheat factor

$$= \frac{\Sigma\, \Delta\, h_{act}}{h_1 - h_4^{"}} \times 100$$

$$= \mathbf{82.62\ \%}$$

Work done per stage :

W.D. in stage I $= h_1 - h_2^{'}$

$= 315.12$ kJ/kg

W.D. in stage II $= h_2^{'} - h_3^{'}$

$= 228.5$ kJ/kg

W.D. in stage III $= h_3^{'} - h_4^{'}$

$= 340$ kJ/kg

Example 12.24 *Steam at 15 bar and 300°C is supplied to a stage Parson's reaction turbine, the steam is exhausted from the turbine at 0.1 bar. The stage efficiency is 0.75 and reheat factor is 1.5. Total power developed by turbine is 11025 kW. Find steam flow rate through turbine assuming all stages develop equal power. At a certain stage of turbine, the pressure of steam is 1.0 bar and steam is dry saturated. Blade exit angle is 20° and blade speed ratio is 0.7. Find mean diameter of this stage and rotor speed. Take blade height as 1/10th of mean diameter of the rotor. Neglect thickness of blade. Velocity of flow is 60 m/s.*

Solution

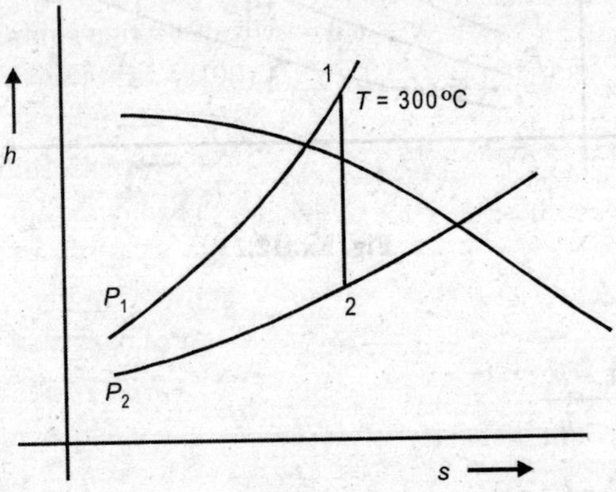

Fig. Ex. 12.24

$$P_1 = 15 \text{ bar}$$

$$T_1 = 300\ ^\circ\text{C}$$

$$P_2 = 0.1 \text{ bar}$$

$$\eta_{stage} = 75\ \%$$

From Mollier Chart,
We get isentropic enthalpy drop as

$$\Delta h = h_1 - h_2$$

$$= 3042 - 2185$$

$$= 857 \text{ kJ/kg}$$

We have

$$\text{Overall efficiency} = \text{Stage efficiency} \times \text{Reheat factor}$$

$$= 0.75 \times 1.05$$

$$= 0.7875 \text{ or } 78.75\ \%$$

$$\text{Actual enthalpy drop} = \text{Work done/kg}$$

$$= \Delta h_{ise} \times \text{Overall } \eta$$

$$= 857 \times 0.7875$$

$$= 647.88 \text{ kJ/kg}$$

Work done per kg of steam per stage

$$= \frac{647.88}{20}$$

$$= 33.74 \text{ kJ/kg}$$

$$\text{Power output/stage} = \text{Work done per stage} \times \dot{m}$$

$$\therefore \quad \frac{11025}{20} = 33.74 \times \dot{m}$$

$$\therefore \quad \dot{m} = 16.336 \text{ kg/s}$$

Work done/kg per stage

$$= (V_{wi} + V_{we}) \cdot V_b$$

We have $\frac{V_b}{V_{ai}} = 0.7$

$$\therefore \quad V_{ai} = 1.43\, V_b$$

$$33.74 \times 10^3 = V_b \{2 \times (1.43\, V_b \cos 20 - V_b)\}$$

$$= 2\, V_b \times 0.3424\, V_b$$

$$\therefore \quad V_b = 183.68 \text{ m/s}$$

Specific volume of steam at 1 bar of dry steam $= 1.694 \text{ m}^3 \text{ kg} = V_g$

$$\dot{m} = \frac{\pi \times D \times h \times V_{fi}}{V_g}$$

$$\therefore \quad 16.336 = \frac{\pi \times D \times \frac{D}{10} \times 60}{1.694}$$

$$\therefore \quad \mathbf{D = 1.211 \text{ m}}$$

$$V_b = \frac{\pi D N}{60}$$

$$183.68 = \frac{\pi \times 1.211 \times N}{60}$$

$$\therefore \quad N = \frac{183.68 \times 60}{\pi \times 1.211}$$

$$\therefore \quad \mathbf{N = 2896.8 \text{ rpm}}$$

THEORY QUESTIONS

12.1. Explain the principal of the operation of steam turbine?

12.2. Discuss the advantages of steam turbines?

12.3. Compare steam turbine and steam engine?

12.4. How the steam turbines are classified?

12.5. Differentiate between pressure and velocity compounding.

12.6. Write short notes on:

1. Governing
2. Types of steam turbines

12.7. State various methods of governing and describe any two methods of governing of steam turbines.

12.8. Derive the relation for condition for maximum efficiency of

(a) Impulse turbine

(b) Reaction turbine

(c) Two stage velocity compounded steam turbine.

PROBLEMS FOR PRACTICE

1. A simple impulse turbine has one ring of moving black running at 150 m/s the absolute velocity of steam at exit from stage is 85 m/s at an angle of 80° with tangent of wheel. The black velocity coefficient is 0.82. The steam rate is 2 kg/sec. Assume moving blade to be equiangular, find

(a) The blade angle

(b) Absolute velocity of steam at inlet

(c) Axial Thrust

(d)Power developed **[Ans.** (a) 26.5° (b) 365.59 m/s (c) 37 N (d) 108.9 kW]

2. At a stage in a reaction turbine the mean blade ring diameter is 1m and turbine runs at a speed 50 rpm. The blade are designed for 50% reaction with exit angle of 30° and inlet angle of 50°. The turbine is supplied with the rate of 6,00,000 kg/hr and $\eta_{isent} = 85\%$.

Determine

(a) Power developed

(b) Sp. enthalpy drop per stage per kg

(c) % increase in relative velocity **[Ans.** (a) 11.58 MW (b) 82.5 kJ/kg (c) 52.2%]

Chapter 13 Steam Condensers and Cooling Towers for Power Plants

13.1 INTRODUCTION: NECESSITY OF CONDENSERS

As we know steam will be generated in the boilers, after it is being superheated, it is expanded in the turbines for generating power. After generating the power, it is either exhausted into the atmosphere or it is exhausted into the condenser. In the condensers this steam will be condensed i.e., its phase will be changed from vapour to liquid by means of cooling water (Ref. Fig. 13.1).

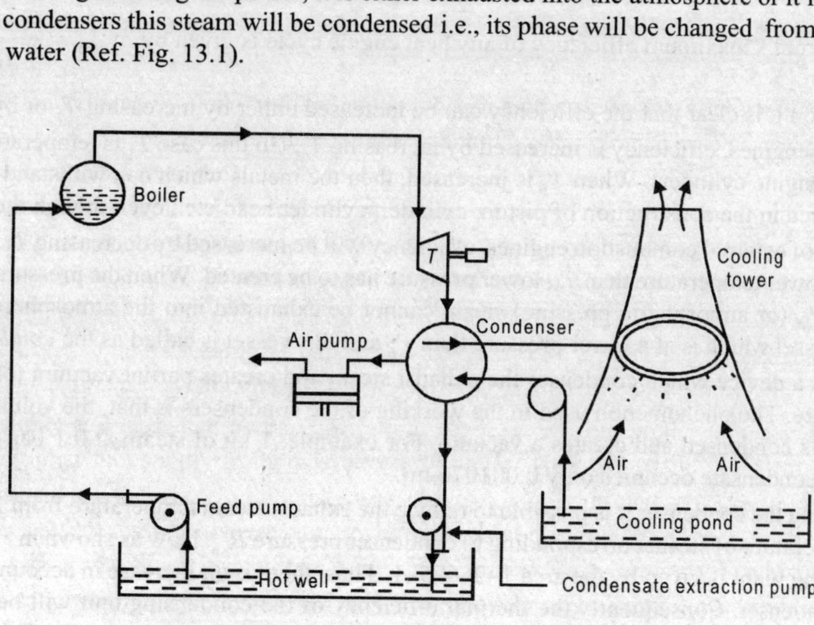

Fig. 13.1. *Elements of steam condensing plant.*

Figure 13.2 shows the theoretical indicator diagram for the steam turbine. Let the steam available in the turbine have the initial condition of pressure P_1 and temperature T_1. This steam is expanded in the turbines along 2–3 for generating work output and then it is exhausted into the atmosphere along 3–4 (slightly above atmospheric pressure). Let the final condition of steam be P_b and temperature T_2. Then the work output is given by the area 1–2–3–4–1 on the indicator diagram.

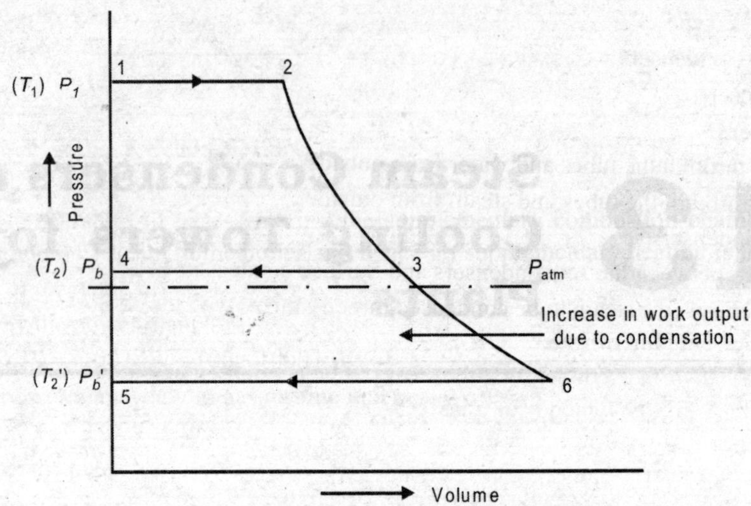

Fig. 13.2. *Theoretical/Ideal indicator diagram.*

According to Carnot's maximum efficiency of any heat engine cycle is given by $\eta_{carnot} = \frac{T_1 - T_2}{T_1}$.

From this equation it is clear that the efficiency can be increased either by increasing T_1 or by decreasing T_2.

In the case of IC engines, efficiency is increased by increasing T_1. (In this case T_1 is temperature of products of combustion in the engine cylinder). When T_1 is increased, then the metals which can withstand high temperature stresses are to be used in the construction of piston, cylinder, cylinder head etc., even though they are costly.

Whereas in case of external combustion engines, efficiency will be increased by decreasing T_2. By the properties of steam to obtain lower temperature than T_2, lower pressure has to be created. When the pressure of exhaust steam is lowered below P_b, (or atmospheric pressure) steam cannot be exhausted into the atmosphere. So, it has to be exhausted into a vessel which is at a lower pressure than P_b and this vessel is called as the *condenser*.

So, condenser is a device which condenses the exhaust steam and creates partial vacuum for the steam below atmospheric pressure. The phenomenon used in the working of the condensers is that, the volume of steam drops down heavily if it is condensed and creates a vacuum. For example, 1 kg of steam at 0.1 bar pressure occupies 14.93 m^3 while its condensate occupies only 0.001076 m^3.

Thus, by installing the condenser it is possible to reduce the exhaust steam temperature from T_2 to T_2' where T_2' is the saturation temperature of steam corresponding to condenser pressure P'_b. Now as shown in P–V diagram, work done in a condensing plant is given by the area 1–2–6–5–1. Thus, there is an increase in account of work by 3–4–5–6 by using a condenser. Consequently the thermal efficiency of the condensing unit will be higher than non-condensing unit. If the condensate is pure, then it can be used as a feed water to the boiler. This recovery of feed water is also an incidental advantage.

13.2 CLASSIFICATION

Steam condensers are broadly classified as,

I. Jet Condensers

(a) Low Level Jet Condensers

(i) Counter-Flow Type

(ii) Parallel-Flow Type

(b) High Level Jet Condensers

(c) Ejector Condensers

II. Surface Condensers

(a) Steam flowing through the tubes and water from outside

(b) Water flowing through the tubes and steam from outside

(c) Evaporative condensers

The main difference between the Jet condensers and Surface condensers is as given below.

In case of Jet condensers, the steam to be condensed directly mixes with the cooling water and the condensation will be effected with less amount of water.

Whereas in the case of surface condensers, the steam to be condensed and the cooling water are separated by means of a tube well and through which heat transfer takes place and the condensation will be effected. Here the amount of cooling water required will be more.

In case of Jet condensers, the temperature of condensate and cooling water are same, whereas, in case of surface condensers temperature of condensate will be more than the temperature of cooling water at the outlet.

In case of Jet condensers, the condensate collected will be impure. (Since the steam and cooling water mix with each other). So it has to be wasted. Whereas in case of surface condensers, it is pure and can be used as a feed water to the boiler.

The size, space required and cost of surface condenser is more, because it has to handle large quantity of water for the same capacity.

The most commonly used condensers are counterflow type Jet and Water tube type surface condensers.

(a) (i) Low level Counter-Flow Type Jet Condensers. In this type, the direction of flow of steam and cooling water are opposite.

In this condenser, the cold water is drawn up in the condenser from the cooling pond, due to the vacuum head created in the condenser shell. This condenser consists of a cylindrical shell arranged with 2–3 water trays with perforations to break up water into small jets. The exhaust steam and any mixed air enters the condenser at lower portion as shown in Fig. 13.3 and tries to ascend up through the falling sprays. Thus, when the steam comes in contact with the cold water particles gets condensed and the air gets cooled.

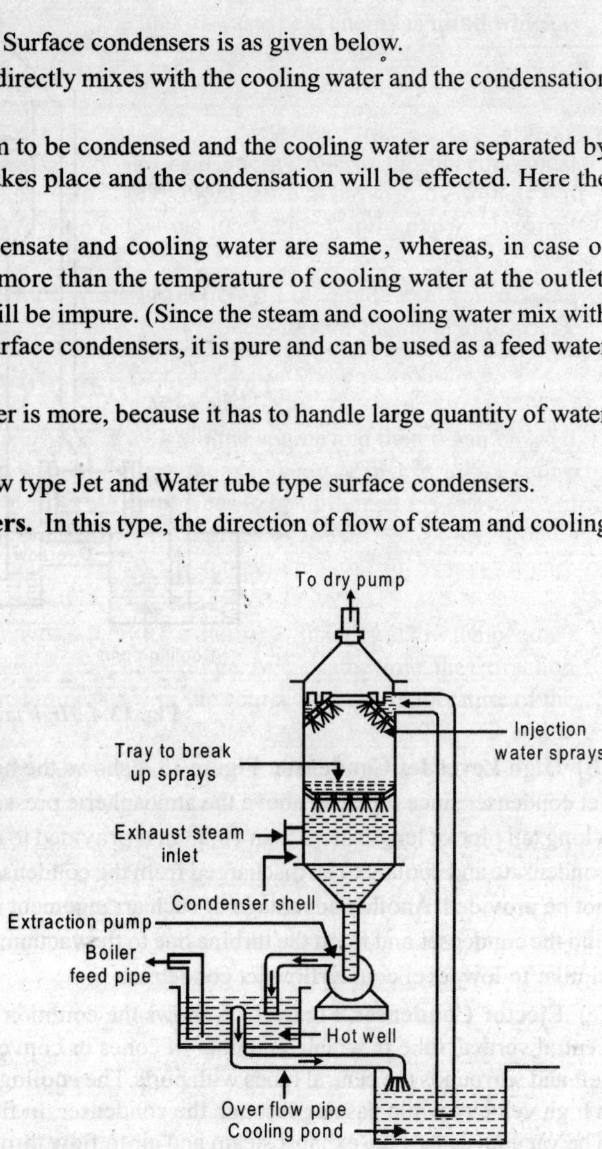

Fig. 13.3. *Low level counter flow type jet condenser.*

This cooled air will be removed through the air pump at the top. The mixture of condensate and cooling water descends down through the vertical pipe to the centrifugal extraction pump and is pumped to the hot well. If the condensate in the hot well is pure then it can be pumped to the boiler as a feed water otherwise it has to be wasted.

The capacity of air pump required is small, since it has to handle cold air and water vapour. These condensers can be directly installed below the turbines. Such condensers have a disadvantage of flooding the steam turbine, if the condensate extraction pump fails.

(ii) Low Level Parallel Flow Jet Condenser. In this case, the direction of flow of exhaust steam and cooling water are parallel. In this case, the steam enters from top of the condenser and the cooling water just below it from sides as shown in Fig. 13.4 and rest all the arrangements are similar to counter flow Jet condenser. A wet air pump is provided at the bottom which removes air, condensate and cooling water.

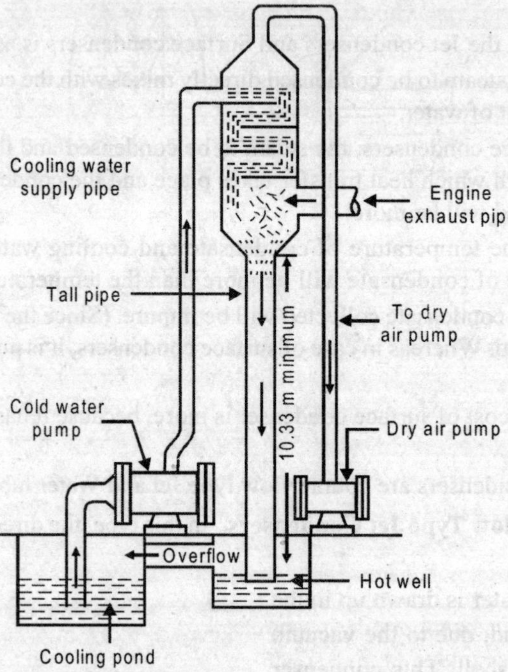

Fig. 13.4. *High level jet condenser.*

(b) High Level Jet Condenser. Figure 13.4 shows the high level Jet condenser. This is also called as barometric Jet condenser since it placed above the atmospheric pressure equivalent to 10.33 m of water pressure. In this case, a long tail pipe of length more than 10.33 m is provided to the bottom of the condenser as shown. This facilitates the condensate and coolant to be discharged from the condenser under gravity, hence condensate extraction pump need not be provided. Another advantage of such arrangement is that the water from the hot well will not be able to rise into the condenser and flood the turbine due to the vacuum head in the condenser. The other details and working are similar to low level counterflow jet condenser.

(c) Ejector Condenser. Figure 13.5 shows the common form of ejector condenser. This condenser consists of a central vertical tube in which a number of cones or converging nozzles are fitted. The exhaust steam enters at the left and surrounds the central tubes with ports. The cooling water enters at the top under a head of 5–6 m and attains a high velocity while passing through the condenser. In flowing port the steam ports, the water produces vacuum. The vacuum causes the exhaust steam and air to flow through the ports in the tube and mix with the cooling water. The exhaust steam gets condensed, as a result of which the vacuum is further increased. The condensate and air after passing through converging cones pass through the diverging cones and in doing so water gains the momentum which forces out condensate and air against the atmospheric pressure. So in this case, no air pump is required.

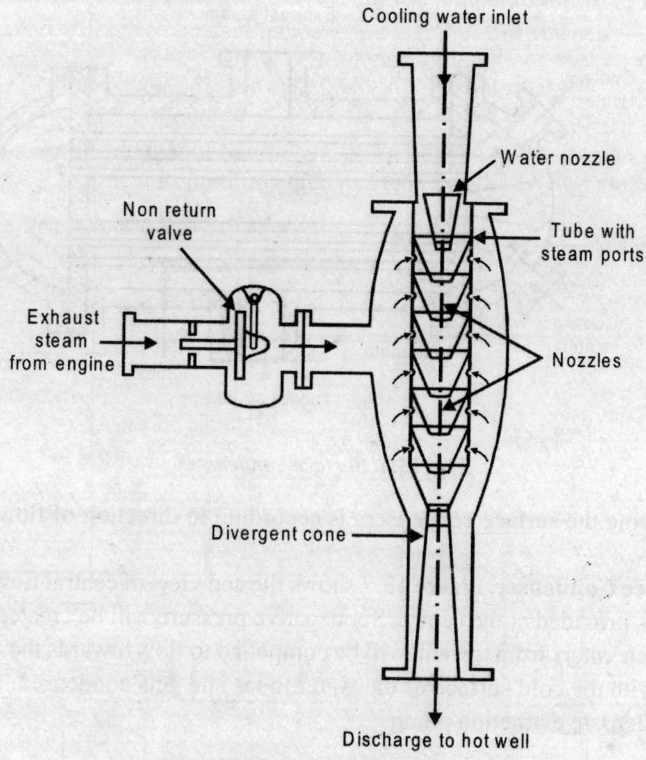

Fig. 13.5. *Ejector condenser.*

This type of condenser is usually fitted with a non-return value to prevent a sudden backward flow of water into the engine exhaust pipe, in case of sudden failure of water supply to the condenser.

III. Surface Condensers. (Cooling water flowing through tubes and steam from outside)

A surface condenser essentially consists of a Mild Steel rolled shell, closed at both the ends by dished ends (also called as water boxes) as shown. Ends of the shell and dishes are fitted with flanges. A tube sheet is provided in between the shell flanges and dish end flange on either side. A number of water tubes are fitted into these tube sheets. The ends of the tubes are either expanded or welded in order to have leak-proof joint.

The exhaust steam from the engine enters at the top of the condenser and gets condensed when it comes in contact with the cold surfaces of water tubes, through which cooling water is being circulated.

The cooling water enters at one end of the tubes (as shown in Fig. 13.6) situated at the lower half of the condenser and after flowing to the other end it returns in the opposite direction through the tubes situated in the upper half of the condenser. This is obtained by providing pass partition plates in the dished ends. In this type since the water traverses 2 times, it is called as 2-flow surface condenser. If the water traverses more than two times then it is called Multi-flow surface condenser.

As shown in Fig. 13.6 this condenser requires two pumps viz., wet air pump to remove air and condensate and secondly water circulating pump.

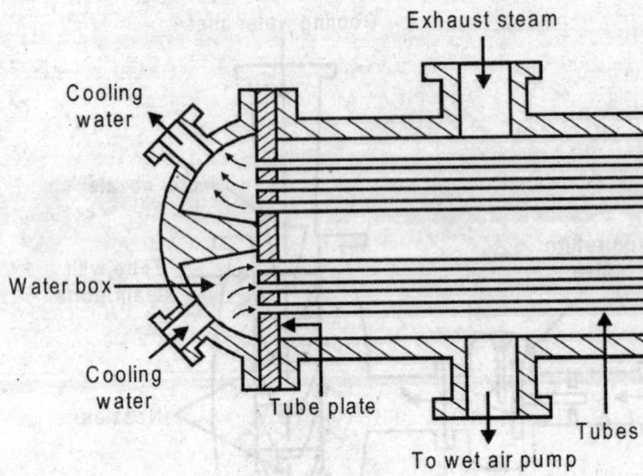

Fig. 13.6. *Surface condenser.*

Another way of classifying the surface condensers is according to direction of flow of exhaust steam viz., (i) central flow (ii) down flow.

(i) Central Flow Surface Condenser. Figure 13.7 shows the end view of central flow surface condenser. In this case, air extraction pump is provided at the centre. So, negative pressure will be created at the centre. In this case, the exhaust steam and any air enters from top and will be compelled to flow towards the centre as shown. During its flow, it comes in contact with the cold surface of the water tubes and gets condensed, and the condensate will be removed by means of condensate extraction pump.

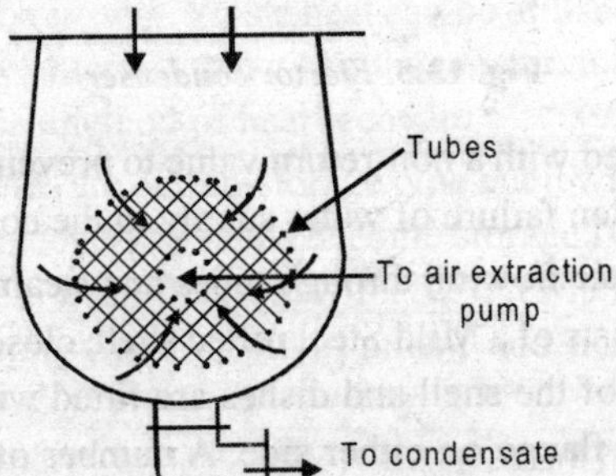

Fig. 13.7. *Central flow surface condenser.*

(ii) Down Flow Surface Condenser. Figure 13.8 shows the end view of the down flow surface condenser. In this case a baffles is provided throughout the length of the condenser. It consists of two pumps, one is for removing the condensate and other is the air pump. In this case exhaust steam mixed with any air enters from top and decends down as shown by arrows. The steam when comes in contact with the cold surfaces of tubes it gets condensed and the air gets cooled. Air is further cooled when it flows from below the baffle. So the air is cooled to the minimum temperature before it is extracted. Due to this cooling of air, its specific volume will be reduced, so it reduces the pump capacity. So, it reduces the energy consumption for running the air pump.

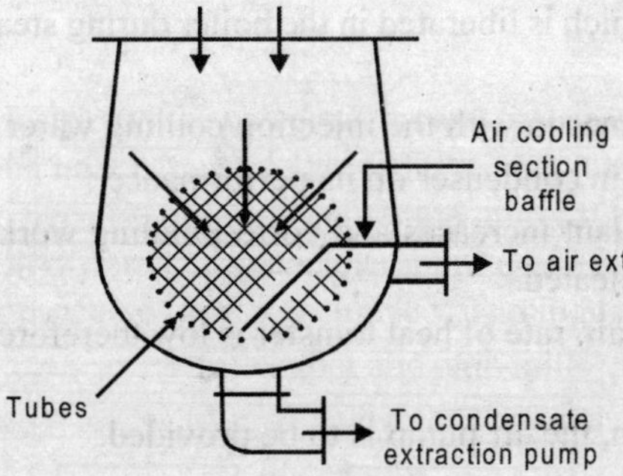

Fig. 13.8. *Down flow surface condenser.*

(iii) Evaporative Condenser. When the supply of cold water is extremely limited, then the evaporative condenser is used since it uses very small amount of circulating and make up water (since the coolant is used again and again). However these condensers are used with small power plants.

In this case the exhaust steam from the steam engine enters into a coil of gilled pipes or series of tubes, the outlet of which is connected to the wet air pump.

The water which is pumped up by means of a pump is sprayed from the top and decends down. The water falling downwards forms a thin film over the pipes as it falls from one level to another and remainder water will be collected in the water tank.

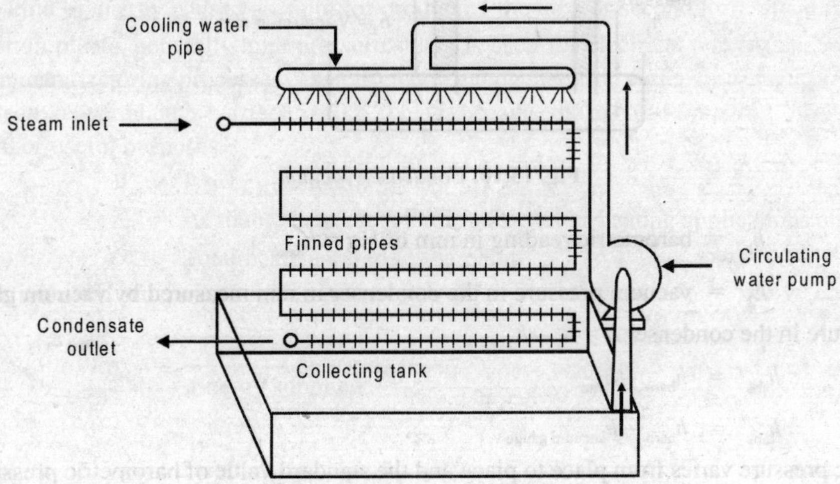

Fig. 13.9. *Evaporative condenser.*

A natural or forced air current causes rapid evaporation of this film of water. The rapid evaporation of film of water results into condensation of steam. The evaporated vapour will be removed by means of air.

13.3 SOURCES OF AIR IN THE CONDENSER

The sources of air in the condenser are as follows :

1. Leakage through packing glands and microscopic holes in shell joints since the pressure inside the condenser is less than the atmospheric pressure.

2. Leakage through relief valves and other accessories.
3. Feed water contains dissolved air, which is liberated in the boiler during steam formation, the exhaust steam from the steam engine carries air.
4. In case of jet condenser some air comes in with the injection/cooling water (in which it is dissolved).

Following are the effects of air, leakage in condenser on its performance :

1. Back pressure in the steam power plant increases and corresponding work output decreases and the very purpose of using the condenser is defeated.
2. Due to poor thermal conductivity of air, rate of heat transfer is low therefore more cooling water is required to be supplied.
3. For maintaining the required vacuum, the air pump is to be provided.

13.4 CONDENSER VACUUM

The vacuum in the condenser is usually expressed in mm of Hg. The absolute pressure in the condenser is equal to the difference of barometric pressure and the vacuum pressure as shown in Fig. 13.10.

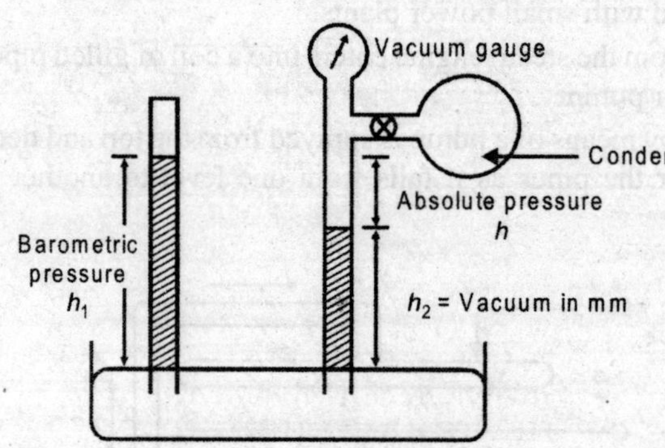

Fig. 13.10. *Condenser vacuum.*

Let h_1 = barometric reading in mm of Hg = h_{baro}

h_2 = vacuum pressure in the condenser in mm measured by vacuum gauge = h_{vac}

∴ Absolute pressure in the condenser $h = h_1 - h_2$

or
$$h_{abs} = h_{baro} - h_{vac}$$

$$h_{abs} = h_{baro} - h_{vacuum\ gauge}$$

The barometric pressure varies from place to place and the standard value of barometric pressure is 760 mm of Hg, at sea level. Then it is necessary to correct the vacuum gauge reading to a standard value.

The standard or corrected vacuum in mm of Hg.

$$h_s = 760 - (h_1 - h_2) = 760 - (h_{baro} - h_{vac})$$

Note 760 mm of Hg = 1.01325 bar

13.5 DALTON'S LAW OF PARTIAL PRESSURE

This law states that "Total pressure exerted by mixture of gases or mixture of gas (air) and vapour (steam) is equal to the sum of partial pressure of constituents, if they would occupy the same volume and are at the same temperature".

∴ Pressure of mixture in condenser = Partial pressure of steam + Partial pressure of air.

$$P_c = P_s + P_a$$

If the condenser temperature is known, P_s is found from steam tables and P_c can be read from the vacuum gauge.

∴ We can write, $P_a = P_c - P_s$

Knowing the partial pressure of air P_a its mass m_a can now be calculated with the help of ideal gas equation $P_a V_c = m_a R T_a$ where V_c = volume of condenser; T_a = temperature; R = characteristic gas constant of air.

Also note that : Each constituent is considered to occupy condenser volume V_c.

and V_c = specific volume × mass

∴ $V_c = m_s \times V_s = m_a \times V_a$

∴ Mass of air/m³ of condenser $= \dfrac{m_a}{V_c} = \dfrac{1}{V_a}$

∴ Mass of steam/m³ of condenser $= \dfrac{m_s}{V_c} = \dfrac{1}{V_s}$

By the conservation of mass

$$m_c = m_s + m_a$$

13.6 VACUUM EFFICIENCY

It may be defined as the ratio of actual vacuum as recorded by the vacuum gauge to the ideal vacuum.

∴ Vacuum Efficiency

$$\eta_{vacuum} = \frac{\text{Actual vacuum recorded by the gauge}}{\text{Ideal vacuum}}$$

The ideal vacuum, means the vacuum due to steam alone when air is absent. In that case total pressure in the condenser will approach the pressure of the steam corresponding to saturated temperature of steam.

∴ Ideal vacuum = barometric pressure – absolute pressure of steam

$$\eta_{vac} = \frac{\text{Actual vacuum recorded by the gauge}}{\text{Barometric Pressure} - \text{Absolute Pressure of Steam}}$$

Let P_c = Actual absolute pressure in the condenser.

$$P_c = P_s + P_a$$

and P_{baro} = Barometric pressure

Then Actual vacuum $= P_{baro} - P_c$

$$= P_{baro} - (P_s + P_a)$$

and Ideal vacuum $= P_{baro} - P_s$

$$\therefore \quad \eta_{vac} = \frac{P_{baro} - P_c}{P_{baro} - P_s}$$

$$\therefore \quad \boldsymbol{\eta_{vac} = \frac{P_{baro} - (P_s + P_a)}{P_{baro} - P_s}}$$

13.7 CONDENSER EFFICIENCY

There is no standard method of determining the condenser efficiency but a method adopted by the well-known makers of steam turbine M/s Parson & Co. has been widely used in engineering practice.

Let t_i = Inlet temperature of cooling water.

t_o = Exit temperature of cooling water.

t_s = Saturated temperature of steam corresponding to actual absolute pressure in the condenser.

Then condenser efficiency

$$\eta_{cond} = \frac{\text{Temperature rise of cooling water}}{\text{Temperature of steam} - \text{Inlet temperature of condenser}}$$

$$\therefore \quad \eta_{condenser} = \frac{\mathbf{t_o - t_i}}{\mathbf{t_s - t_i}}$$

13.8 COOLING WATER REQUIREMENTS

In a condenser, cooling water absorbs heat from steam to be condensed and the temperature of cooling water increases. In jet condensers steam to be condensed directly mixes with cooling water and hence temperature of condensate and cooling water are same.

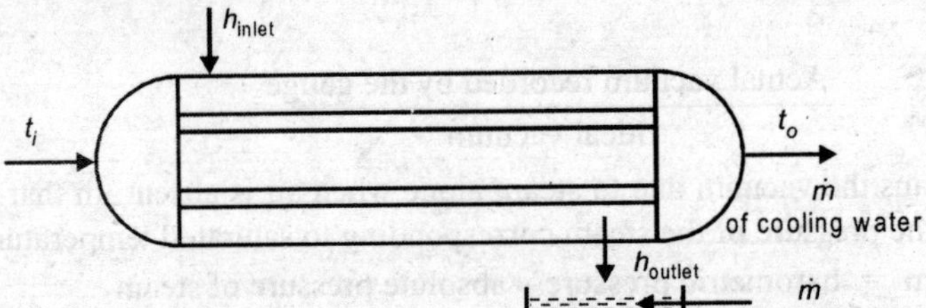

Fig. 13.11

Let t_i = Inlet temperature of cooling water.

t_o = Outlet temperature of cooling water.

$\dot{m}_{cw}$ = Mass flow rate of cooling water.

h_{inlet} = Enthalpy of steam entering condenser.

h_{outlet} = Enthalpy of condensate leaving condenser.

$\dot{m}_{condensate}$ = Mass flow rate of condensate.

Then heat lost by steam = Heat gained by cooling water

$$\dot{m}_{condensate}\left(h_{inlet} - h_{outlet}\right) = \dot{m}_{cooling\ water} \times C_{pw} \times (t_o - t_i)$$

$$\dot{m}_{condensate}\left\{\left(h_f + x\,h_{fg}\right) - h_{f_{outlet}}\right\} = \dot{m}_{cooling\ water} \times C_{pw} \times (t_o - t_i)$$

From this equation quantity of cooling water or condition of steam entering the condenser can be calculated if other quantities are given.

The quantity of water used varies from 20–60 kg/kg of steam. And temperature rise of circulating water may vary from 10 – 20 °C.

13.9 CAPACITY OF AIR EXTRACTION PUMPS

We know that, absolute pressure in the condenser

$$P_c = P_s + P_a \quad \text{(from Dalton's law)} \qquad \text{...(1)}$$

or

$$P_a = P_c - P_s \qquad \text{...(2)}$$

Also absolute pressure in the condenser P_c or $h_{abs} = h_{baro} - h_{vac}$ ...(3)

At the condenser temperature t°C, we can find P_s and volume of steam v_g from steam tables.

$v_g = v_a$ = volume of condenser if air and steam occupy condenser volume and are at the condenser temperature.

Then mass of air

$$m_a = \frac{p_a \cdot v_a}{RT_a}$$

$$m_a = \frac{p_a \times x \cdot v_g}{RT_a} \qquad \text{...(4)}$$

Note: Generally steam entering the condenser is wet.

Units

$$m_a = \frac{\dfrac{\text{kN}}{\text{m}^2} \times \dfrac{\text{m}^3}{\text{kg of steam}}}{\dfrac{\text{kJ}}{\text{kg of air K}} \times \text{K}}$$

$$m_a = \frac{\text{kg of air}}{\text{kg of steam}}$$

Note that along with condensate air is also removed by the wet air pump. If Z is the quantity of steam consumption of turbine in kg/hr then,

Mass of air to be handled by air pump

$$m_a = \frac{p_a \cdot x \cdot v_g}{RT_a} \times Z$$

$$= \frac{\text{kg of air}}{\text{kg of steam}} \times \frac{\text{kg of steam}}{\text{hr}}$$

Mass of air to be removed by air pump $m_a = \dfrac{\text{kg of air}}{\text{hr.}}$.

13.10 AIR PUMPS

Air pump is a device which is used to remove air from the condenser and to maintain necessary vacuum in the condenser. (The sources of air leakage and their effects on condenser have already been discussed in earlier articles).

Air pumps are of two types:

(i) Dry air pump —Which removes only cold air

(ii) Wet air pump — It is used to remove cold air and condensate.

These pumps may be reciprocating or rotary type. We shall discuss now a very common form of reciprocating air pump called Edward's air pump.

As shown in Fig. 13.12, it consists of head or delivery valves, one of them is shown at 1. These valves are provided in the cover 2, which on the top of the pump barrel 3. The piston 4 has the conical bottom 5 and the bottom of the pump casing also follows the conical shape as that of piston. Ports 6 are provided in the pump barrel near its bottom end. Connection with the condenser is made at 7.

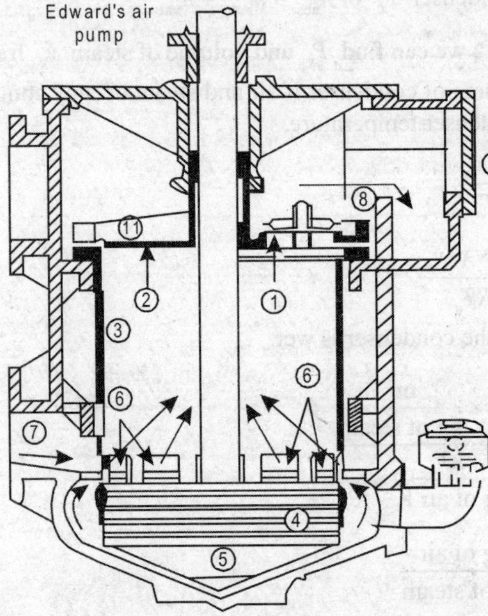

Fig. 13.12

During the downward stroke of the piston, a vacuum is produced in the barrel above the piston, the head valves being closed and as soon as the piston begins to uncover the ports 6, the air and water vapour from the condenser rush into the space above the piston. The condensate flows down in the pump casing. The conical end of downward moving piston displaces the condensate to the top of piston till the end of downward stroke.

Now during the upward stroke of the piston, some of the water flows down in the pump casing through the ports till the piston closes the ports. When the piston closes the ports the pressure of the charge above the piston will go on increasing and the charge of air, water vapour and condensate will be expelled through the head valves and then over the weir 8 on its way to the hot well. The weir ensures a head of water over the valves, so that if there is any leakage through these valves into the barrel, then it will be a leakage of water and not air.

The door 9 gives access to the head valves and 10 is the relief valve which opens to atmosphere, to release the pressure in case the pressure inside increase above the atmospheric pressure.

13.11 COOLING TOWERS

The function of the cooling tower is to cool the cooling water of the condenser, by the current of air flowing in the opposite direction.

A large amount of cooling water is required for condensation in large capacity power plants. If the water is freely available either from the river or lake, then the water can be directly pumped from the river to the condenser as shown in Fig. 13.13 (a).

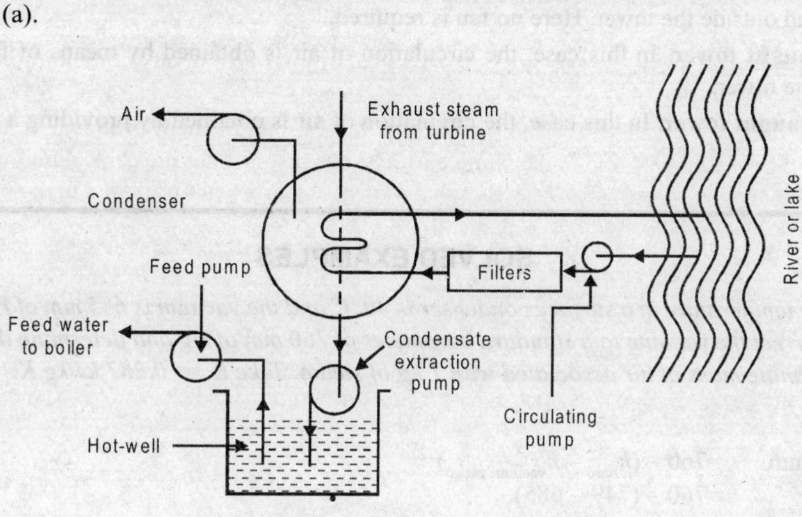

(a)

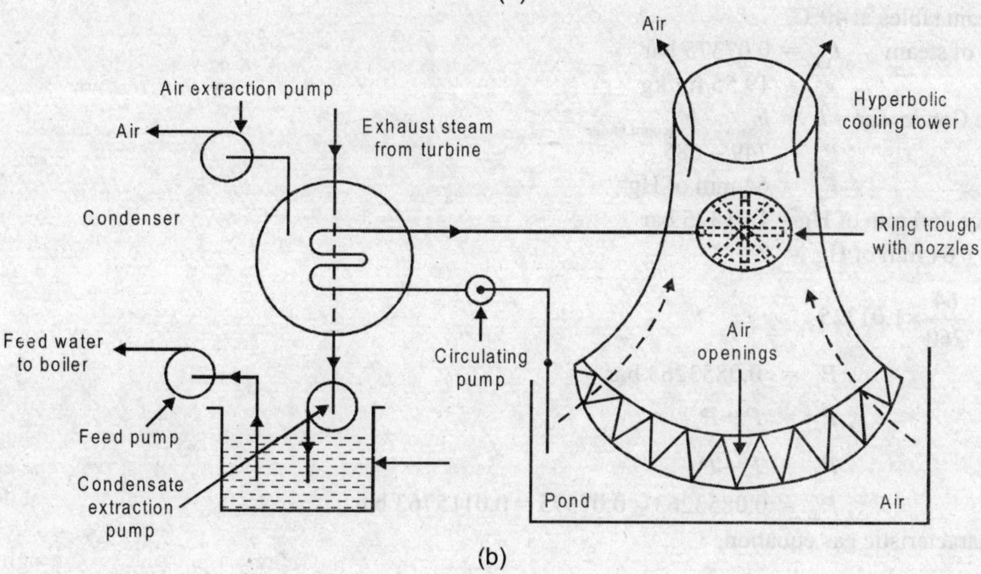

(b)

Fig. 13.13

If the water is not freely available, then a cooling tower has to be used for cooling the hot water of condenser. The cooled water can be used again as cooling water for the condenser.

Figure 13.13 (b) shows the Hyperbolic cooling tower. It is usually made of steel Reinforced cement concrete. The hot water from the condenser is supplied to the ring troughs which are placed at 8- 10 m above the ground level. The nozzles are provided on the bottom side of troughs to break up water into sprays. The air rises up from the pond in the opposite direction of water flow and absorbs heat from the falling water spray. The cooled water is collected into a pond built below the tower. This type of cooling tower is generally used since it is very efficient, however, it needs about 3–5 % of make up water for compensating the evaporation losses.

The other types of cooling towers are:

(i) **Natural draught towers.** In this type the circulation of air is obtained by virtue of pressure difference of the air inside and outside the tower. Here no fan is required.

(ii) **Forced draught tower.** In this case, the circulation of air is obtained by means of fans provided at the bottom of the tower.

(iii) **Induced draught tower.** In this case, the circulation of air is obtained by providing a fan at the top of the tower.

SOLVED EXAMPLES

Example 13.1 *The temperature in a surface condenser is 40 ºC and the vacuum is 685 mm of Hg. Barometer reads 749 mm of Hg. Correct the vacuum to a standard barometer of 760 mm of Hg and determine the pressure of steam and air. Also determine mass of air associated with 1 kg of steam. Take R = 0.287 kJ/kg K.*

Solution

Corrected vacuum $= 760 - (h_{baro} - h_{vacuum\ gauge})$

$= 760 - (749 - 685)$

$= 696$ mm of Hg

From steam tables at 40ºC,

Pressure of steam $P_S = 0.07375$ bar

and $v_g = 19.55\ m^3/kg$

Absolute Condenser $= h = h_{baro} - h_{vacuum\ gauge}$

or $P_c = 749 - 685$

$P_c = 64$ mm of Hg

Also since 760 mm of Hg= 1.01325 bar

64 mm of Hg = ?

$$\frac{64}{760} \times 1.01325$$

$$P_c = 0.0853263 \text{ bar}$$

Since, $P_c = P_s + P_a$

$$P_a = P_c - P_s$$

$$P_a = 0.0853263 - 0.07375 = 0.0115763 \text{ bar}$$

From characteristic gas equation,

$$m_a = \frac{P_a \times V_a}{R \times T_a}$$

From steam tables we have $v_g = 19.55\ m^3/kg$

By Dalton's law, air and steam occupy the same volume at their partial pressure and have same temperature, have air present/kg of steam will occupy a volume of 19.55 m³ at $P_a = 0.0115763$ bar and will at the temperature 40 ºC

i.e. $v_g = v_s = 19.55\ m^3/kg$

$$m_a = \frac{0.0115763 \times 100 \times 19.55}{0.287 \times 273 + 40} \qquad \text{(Since 1 bar = 100 kN/m}^2\text{)}$$

∴ Resultant unit will be,

$$= \frac{\frac{kN}{m^2} \times \frac{m^3}{kg}}{\frac{kJ}{kg\,K} \times K} = \frac{\frac{kJ}{kg\ of\ steam}}{\frac{kJ}{kg}} = \frac{kg}{kg\ of\ steam}$$

$$m_c = \frac{0.0115763 \times 100 \times 19.55}{0.287 \times 313}$$

$$\mathbf{m_a = 0.251936\ kg/kg\ of\ steam.}$$

Example 13.2 *In a surface condensating plant following data were obtained,*

Temperature of exhaust steam entering the condenser	=	*42.32 ºC*
Temperature of condensate leaving the condenser	=	*35 ºC*
Inlet temperature of cooling water	=	*16.5 ºC*
Outlet temperature of cooling water	=	*30 ºC*
Quantity of water/hour	=	*46250 kg*
Quantity of condensate/hour	=	*1190 kg*

Calculate the dryness fraction of steam entering the condenser.

Solution

Heat lost by steam = Heat gained by cooling water

$$\dot{m}_{condensate}\left[h_{inlet} - h_{outlet}\right] = \dot{m}_{cw} \times C_{p\ water} \times \Delta T$$

Now,

$$h_{inlet}\ \text{(at inlet it will be wet steam)} = h_f + x \times h_{fg}\ \text{at}\ 42.32°C$$

$$= 177.14 + x \times 2401.33\ \text{by interpolation}$$

and

$$h_{outlet}\ \text{(at outlet it will be water)} = h_f\ \text{at}\ 35°C$$

$$= 146.6\ \text{kJ/kg}$$

$$\therefore \quad = \frac{46250}{3600} \times 4.187 \times (30 - 16.5)$$

$$\therefore \quad [(177.14 + 2401.33\,x) - 146.6] = 2193.18$$

$$177.14 + 2401.3\,x = 2339.78$$

$$x = 0.9006$$

$$\mathbf{x \cong 0.9}$$

Example 13.3 ***The surface condenser has a vacuum of 685 mm of Hg and mean temperature 35ºC. The barometer reads 760 mm of Hg. Dryness fraction of steam entering the condenser is 0.9. The condensate collected is 15.5 kg/min. Calculate the vacuum efficiency and mass of air to be removed by the dry air pump/min.***

Solution

From steam tables at 35 ºC,

$$P_s = 0.5622\ \text{bar}$$

and

$$v_s = v_g = 25.24\ \text{m}^3/\text{kg}.$$

Absolute pressure in the condenser = $h_{baro} - h_{vac} = P_c$

$$= 760 - 685$$
$$= 75 \text{ mm of Hg}$$

$$760 \text{ mm of Hg} = 1.01325 \text{ bar}$$
$$75 \text{ mm of Hg} = ?$$

$$= \frac{75}{760} \times 1.01325$$

$$P_c = 0.0999917 \text{ bar}$$

We also know that,

$$\eta_{vac} = \frac{P_{baro} - P_c}{P_{baro} - P_s} = \frac{P_{baro} - (P_s + P_a)}{P_{baro} - P_s}$$

$$= \frac{760 - 75}{760 - P_s}$$

$$P_s = 0.05622 \text{ bar}$$
$$1.01325 \text{ bar} = 760 \text{ mm}$$
$$0.05622 \text{ bar} = ?$$

$$= \frac{0.05622}{1.01325} \times 760$$

$$P_s = 42.168468 \text{ mm of Hg}$$
$$= 42.17 \text{ mm of Hg}$$

$$\eta_{vac} = \frac{760 - 75}{760 - 42.17}$$

$$= \frac{685}{760 - 42.17}$$

$$= 0.9543 = 95.43\,\%$$

Also,

$$P_c = P_s + P_a$$
$$P_a = P_c - P_s$$
$$= 0.0999917 - 0.05622$$
$$P_a = 0.0437717 \text{ bar}$$

$\therefore$ Mass of air present/kg of steam = $\dfrac{P_a v_a}{R T_a}$

$$v_a = x \times v_g$$
$$= 0.9 \times 25.24$$

$$\therefore \quad m_a = \frac{0.0437717 \times 100 \, 25.24 \times 0.9}{0.287 \times (35 + 273)}$$

$$= 1.1248 \text{ kg of air/kg of steam.}$$

Given rate condensate $= 15.5$ kg/min

Air to be removed by air pump

$= 15.5 \text{ kg/min} \times 1.1243$ kg/kg of steam

$= 17.4351$ kg/min

Example 13.4 *A vacuum of 720 mm was obtained in a condenser when the barometer reads 765 mm. The temperature of condensate was 25°C. Determine the pressure of air and steam in the condenser and mass of air/kg of steam. Also determine the vacuum efficiency.*

Solution

Absolute pressure in the condenser $= h = h_{\text{baro}} - h_{\text{vac}} = P_c$

$$P_c = 765 - 720$$

$$= 45 \text{ mm of Hg}$$

Since 760 mm of Hg $= 1.01325$

45 mm of Hg $= ?$

$$\frac{45}{760} \times 1.01325 = P_c = 0.05995 \text{ bar}$$

Partial pressure of steam at saturation temperature of 25°C from steam tables,

$$P_s = 0.03166 \text{ bar}$$

Since 1.01325 bar $= 760$ mm of Hg

0.03166 bar $= ?$

$= 23.25$ mm of Hg

Since, $P_c = P_s + P_a$

$$P_a = P_c - P_s$$

$$P_a = 0.059995 - 0.03166$$

$$P_a = 0.028335 \text{ bar}$$

From steam tables, $v_g = v_s = v_a = 43.40 \text{ m}^3/\text{kg}$

$$\therefore \quad m_a = \frac{P_a v_a}{RT_a}$$

$$= \frac{0.028335 \times 100 \times 43.40}{0.287 \times (25 + 273)}$$

$$m_a = 1.4379 \text{ kg/kg of steam}$$

$$\eta_{\text{vac}} = \frac{P_{\text{baro}} - P_c}{P_{\text{baro}} - P_s}$$

$$= \frac{765 - 45}{765 - 23.75} = \frac{720}{741.25}$$

$$= 97.13\%$$

Example 13.5 *The following observations were made during a test on surface condenser*

Barometric reading $= 760$ *mm of Hg*

Condenser vacuum $= 705$ *mm of Hg*

Mean temperature of condensate = *35 °C*
Condensate collected = *2000 kg/hr*
Quantity of cooling water circulated = *60,000 kg/hr*
Rise in temperature of cooling water = *16 °C*
Hot well temperature = *28 °C*
Determine :

1. η_{vac}
2. $\eta_{condenser}$
3. *Quantity of steam entering the condenser*
4. *Mass of air present/m³ of condenser volume*

Assume inlet temperature of water as 20 °C

Solution

Absolute pressure in the condenser = $h_{baro} - h_{vac} = P_c$

$$P_c = 760 - 705$$
$$P_c = 55 \text{ mm of Hg}$$
$$P_c = 0.0733273 \text{ bar}$$

as $760 \text{ mm of Hg} = 1.01325 \text{ bar}$

From steam tables at 35°C

$$P_s = 0.05622 \text{ bar}$$
$$P_s = 42.168 \text{ mm of Hg}$$

as $1.01325 \text{ bar} = 760 \text{ mm of Hg}$

$0.05622 \text{ bar} = ?$

$$\therefore \quad P_a = P_c - P_s$$
$$= 0.0733273 - 0.05622 = 0.0171073 \text{ bar}$$

(i)
$$\eta_{vac} = \frac{P_{baro} - P_c}{P_{baro} - P_s}$$
$$= \frac{760-55}{760-42.168}$$
$$= \frac{705}{717.832}$$
$$= 98.21\%$$

(ii) Saturated temperature of steam corresponding to condenser pressure.

$P_c = 0.0733273$ bar $t_s = 39.88$ °C, by interpolation

$$\therefore \quad \text{Condenser efficiency} = \eta_{cond} = \frac{t_o - t_i}{t_s - t_i}$$
$$= \frac{16}{39.88-20}$$

$= 80.48\ \%$

(iii) Heat lost by steam = Heat gained by cooling water

$$\dot{m}_{\text{condensate}}\left[h_{\text{inlet}} - h_{\text{outlet}}\right] = \dot{m}_{\text{cooling water}} \times C_p \times \Delta T$$

$$\dot{m}_{\text{condensate}}\left[(h_f + x \times h_{fg}) - h_f\right] = \dot{m}_{\text{cooling water}} \times C_p \times \Delta T$$

$$\frac{2000}{3600}\left[(146.56 + x \times 2418.8) - 4.187 \times 28\right] = \frac{60000}{3600} \times 4.187 \times 16$$

Note: If hot well temperature is not given the corresponding to Mean temperature of condensate.

$$2000[(146.56 + x \times 2418.8) - 117.24] = 4019520$$

$$(146.56 + x \times 2418.8) - 117.24 = 2009.76$$

$$146.56 + 2418.8 \times x - 117.24 = 2009.76$$

$$2418.8x = 2009.76 - 146.56 + 117.24$$

$$= 1980.44$$

$$\therefore \quad \mathbf{x = 0.8187696}$$

(iv)

$$m_a = \frac{P_a\ v_a}{R T_a}$$

$$= \frac{0.0171073 \times 100 \times 1}{0.287 \times (35 + 273)}$$

$$= 0.019 \text{ kg/m}^3 \text{ of condensate volume}$$

Note Units

$$= \frac{\dfrac{\text{kN}}{\text{m}^2} \times \dfrac{\text{m}^3}{\text{m}^3}\left(\dfrac{\text{m}^3 \text{ of air}}{\text{m}^3 \text{ of cond. volume}}\right)}{\dfrac{\text{kJ}}{\text{kg K}} \times \text{K}}$$

$$= \frac{\text{kg}}{\text{m}^3 \text{ cond. volume}}$$

Example 13.6 *Following readings were noted during a trial on surface condenser*

Mean temperature of condenser = *30 ºC*

Condensate temperature = *23 ºC*

Barometer reading = *72 cm* = *720 mm of Hg*

Condenser vacuum = *66.375 cm* = *663.75 mm of Hg*

Condensate collected = *500 kg/hr*

Cooling water circulated = *18000 kg/hr*

Rise in temperature of cooling water = *14 ºC*

Calculate:

(i) Dryness fraction of steam entering the condenser

(ii) Capacity of air pump in m^3*/min.*

(iii) *Mass of air discharged in kg/min*

Solution

Absolute Pressure in the condenser = $h_{baro} - h_{vac} = P_c$

$$P_c = 720 - 663.75$$
$$= 56.25 \text{ mm of Hg.}$$
$$= 0.0749938 \text{ bar}$$

Since 760 mm of Hg $= 1.01325$ bar

(i) Heat lost by steam = Heat gained by cooling water

$$\dot{m}_{\text{condensate}}\left[h_{\text{inlet}} - h_{\text{outlet}}\right] = \dot{m}_{\text{cooling water}} \times C_p \times \Delta T$$

$$\frac{500}{3600} \times \left[(h_f + x \times h_{fg}) - h_f\right] = \dot{m}_{\text{cooling water}} \times C_p \times \Delta T$$

$$= \frac{500}{3600} \times \left[(125.66 + x \times 2430.7) - 4.187 \times 23\right]$$

$$\simeq \frac{1800}{3600} \times 4.187 \times 14°\text{C}$$

$\therefore$ $x = 0.856$

(ii) Rate of steam flow in condenser

$$\dot{V} = \frac{500}{60} \times 32.929 \times 0.856$$

$$= \frac{\text{kg}}{\text{min}} \times \frac{\text{m}^3}{\text{kg}} = \frac{\text{m}^3}{\text{min}}$$

Or note specific volume of steam/kg $v = x \times v_g$

$$= 0.856 \times 32.929$$
$$= 28.187 \text{ m}^3/\text{kg}$$

$$\therefore \quad \dot{v} = \frac{500}{60} \times 28.187 \text{ m}^3/\text{min}$$

$\therefore$ Capacity of air pump $= 234.89$ m^3/min.

(iii) Mass of air $m_a = \dfrac{P_a v_a}{R T_a}$

From steam tables at 30°C,

$$P_s = 0.04242 \text{ bar}$$

and P_c as we have found $= 0.0749938$ bar

Since $P_a = P_c - P_s$

$$= 0.0325738 \text{ bar}$$

$$\therefore \quad m_a = \frac{0.0325738 \times 100 \times 234.89}{0.287 \times 303}$$

$$= 8.7934 \text{ kg/min}$$

Note:

$$\frac{\frac{kN}{m^2} \times \frac{m^3}{min}}{\frac{kJ}{kg\,K} \times K} = \frac{kg}{min}$$

Example 13.7 *The absolute pressure in the condenser is 11.56 kPa. When the barometer reads 1 bar, the condenser temperature is 40 °C. Determine the partial pressure of air, η_{vac}, and the mass of air present in the condenser/kg of steam.*

Solution

Given Absolute pressure in the condenser = 11.56 kPa

$$P_c = 0.1156 \text{ bar}$$

∴ Vacuum in the condenser = Barometric pressure – Absolute pressure in the condenser

$$= 1 - 0.1156$$
$$= 0.8844 \text{ bar}$$

Partial pressure of steam at saturated temperature of 40°C,

From steam tables, $P_s = 0.07375$ bar

$$\therefore \quad P_a = P_c - P_s = 0.04185 \text{ bar}$$

$$\eta_{vac} = \frac{P_{baro} - P_c}{P_{baro} - P_s}$$

$$= \frac{1 - 0.1156}{1 - 0.07375}$$

$$= 95.59\ \%$$

Specific volume of steam at 40 °C = 19.546 m³/kg

= Specific volume of air = v_a

$$m_a = \frac{P_a\, v_a}{R\, T_a} = \frac{0.04185 \times 100 \times 19.546}{0.287 \times (40 + 273)}$$

$$\mathbf{m_a = 0.9106\ kg/kg\ of\ steam}$$

Example 13.8 *The following observations were made on a steam condensing plant,*

Barometer reading = *760 mm of Hg*

Recorded vacuum = *700 mm of Hg*

Mean temperature of condensation = *34 °C*

Hot well temperature = *27 °C*

Mass of condensate = *2120 kg/hr*

Mass of cooling water = *66000 kg/hr*

Rise in temperature of cooling water = *16 °C*

Find:

(i) *The state of steam entering the condenser*

(ii) *Mass of air present in kg/m³ of condenser*

(iii) *Vacuum efficiency*

Solution

(i) Heat lost by steam = Heat gained by cooling water.

$$\dot{m}_{\text{condensate}}\left[h_{\text{inlet}}-h_{\text{outlet}}\right] = \dot{m}_{\text{cooling water}}\times C_p\times\Delta T$$

$$\dot{m}_{\text{condensate}}\left[(h_f+x\times h_{fg})-h_{f\,(\text{hotwell})}\right] = \dot{m}_{\text{cooling water}}\times C_p\times\Delta T$$

$$\frac{2120}{3600}\left[(142.38+x\times 2421.2)-4.187\times 27\right] = \frac{66000}{3600}\times 4.187\times 16$$

$\therefore$ x = 0.8492 i.e., state of steam entering the condenser.

(ii) Absolute pressure in the condenser = $P_c = h_{\text{baro}} - h_{\text{vac}}$

$= 760 - 700$

$= 60$ mm of Hg

$P_c = 0.0799934$ bar

And P_s corresponding to condenser at temperature of 34 ºC = 0.05318 bar

$\therefore$ $P_a = P_c - P_s = 0.0268134$ barometer

$$\therefore \quad m_a = \frac{P_a v_a}{R T_a}$$

$$= \frac{0.0268134\times 100\times 1}{0.287\times(34+273)}$$

$$\mathbf{m_a = 0.0304\ kg/m^3\ of\ condenser\ volume}$$

(iii)
$$\eta_{\text{vac}} = \frac{P_{\text{baro}}-P_c}{P_{\text{baro}}-P_s}$$

$$= \frac{760-60}{760-P_s}$$

and $P_s = 0.05318$ bar = 39.89 mm of Hg

$$\therefore \quad = \frac{760-60}{760-39.89}$$

$$\boldsymbol{\eta_{\text{vac}} = 97.21\%}$$

Example 13.9 *The following observations were recorded during a test on a steam condenser*

Recorded vacuum = *71 cm of Hg* = *710 mm of Hg*
Barometer reading = *76.5 cm* = *765 mm of Hg*
Mean condenser temperature = *34 ºC*
Condensate collected = *1800 kg/hr*
Mass of cooling water = *57500 kg/hr*
Cooling water temperature rise = *17.5 ºC*

Calculate:

(a) Corrected vacuum to standard barometer of 76 cm of Hg

(*b*) *Vacuum efficiency*

(*c*) *Condenser efficiency*

(*d*) *State of steam entering the condenser*

Assume inlet temperature of cooling water = 8.5 ºC.

Solution

(a) Corrected vacuum $= 760 - [h_{baro} - h_{vac}]$

$= 760 - [765 - 710]$

$= 705$ mm of Hg

From steam tables at saturated temperature of 34 ºC, we get

$$P_s = 0.05318 \text{ bar} = 39.888 \text{ mm of Hg}$$

$$h_f = 142.4 \text{ and } h_{fg} = 2421.2 \text{ kJ/kg}$$

(b)
$$\eta_{vac} = \frac{P_{baro} - P_c}{P_{baro} - P_s}$$

$$P_c = h_{baro} - h_{vac}$$

$$= 765 - 710 = 55 \text{ mm of Hg}$$

$$P_c = 0.0733 \text{ bar}$$

$$\therefore \quad \eta_{vac} = \frac{765 - 55}{765 - 39.89}$$

$$\boldsymbol{\eta_{vac} = 97.92\%}$$

(c) Condenser Efficiency $= \eta_{cond} = \dfrac{t_o - t_i}{t_s - t_i}$

Now, saturated temperature corresponding to condenser pressure

$$P_c = 0.0733 \text{ bar}$$

From steam tables by interpolation $t_s = 39.88$ºC

$$\eta_{cond} = \frac{17.5}{39.88 - 8.5}$$

$$= 55.76\%$$

(d) Heat lost by steam = Heat gained by cooling water

$$\dot{m}_{condensate}\left[h_{inlet} - h_{outlet}\right] = \dot{m}_{cooling\ water} \times C_p \times \Delta T$$

$$\dot{m}_{condensate}\left[(h_f + x \times h_{fg}) - h_{f\,(hotwell)}\right] = \dot{m}_{cooling\ water} \times C_p \times \Delta T$$

$$= \frac{1800}{3600}\left[(142.4 + x \times 2421.2) - 4.187 \times 34\right]$$

$$= \frac{57500}{3600} \times 4.187 \times 17.5$$

$$\mathbf{x = 0.9667}$$

Note. Corresponding to condenser temperature since hot well temperature is not given.

Example 13.10 *Following observations recorded during a condenser test,*

Mean temperature of condenser = *35 ºC*
Hot well temperature = *30 ºC*
Condenser vacuum = *70 cm* = *700 mm of Hg*
Barometer reading = *76 cm* = *760 mm of Hg*
Condensate = *16.75 kg/min*
Cooling water = *39600 kg/min*
Rise in temperature of cooling water : 14 ºC
Inlet temperature of cooling water : 25 ºC

Find:

(a) Vacuum efficiency
(b) Condenser efficiency
(c) Dryness fraction of steam entering
(d) Mass of air/m³ of condenser volume

Solution

From steam tables at 35 ºC,

$$P_s = 0.05622 \text{ bar} = 42.17 \text{ mm of Hg}$$

$$v_g = 25.24 \text{ m}^3\text{/kg}, \quad h_f = 146.6, \quad h_{fg} = 2418.8$$

Absolute pressure in the condenser $P_c = h_{baro} - h_{vac}$

$$= 760 - 700 = 60 \text{ mm of Hg}$$

i.e., $P_c = 0.0799934$ bar

$\therefore$ $P_a = P_c - P_s$

$$= 0.0799934 - 0.05622$$

$$= 0.0237734 \text{ bar}$$

(a) Vacuum efficiency $= \eta_{vac} = \dfrac{P_{baro} - P_c}{P_{baro} - P_s}$

$$= \frac{760-60}{760-42.17}$$

$$= 97.52\%$$

(b) Condenser efficiency $= \eta_{cond} = \dfrac{t_o - t_i}{t_s - t_i}$

$$= \frac{14}{41.54-25}$$

$$= 84.64\ \%$$

Note. Saturated temperature corresponding to pressure $P_c = 0.0799934$ bar, $t_s = 41.54$, by interpolation

(c) $$m_a = \frac{P_a \times v_a}{R \times T_a}$$

$$= \frac{0.0237734 \times 100 \times 1}{0.287 \times (35+273)}$$

$$= 0.02715 \text{ kg/m}^3 \text{ of condenser volume}$$

(d) $$\dot{m}\left[h_{\text{inlet}} - h_{\text{outlet}}\right] = \dot{m} \times C_p \times \Delta T$$

$$\frac{16.75}{60}\left[(h_f + x \times h_{fg}) - h_{f\,(\text{houtlet})}\right] = \dot{m} \times C_p \times \Delta T$$

$$\frac{16.75}{60}\left[(146.6 + x \times 2418.8) - 4.187 \times 30\right] = \frac{39600}{3600} \times 4.187 \times 14$$

$$\therefore \quad x = 0.9469$$

THEORY QUESTIONS

13.1. Enumerate the advantage of using steam condenser in a steam power plant. Explain the significance of Vacuum efficiency and Condenser efficiency.

13.2. How the condenser are classified?

13.3. Give merits and demerits of a surface condenser over the Jet condenser.

13.4. How does air come into the condenser? What are its effects on the performance of a condenser? List the merits of air extraction.

13.5. Explain Dalton's law of Partial Pressures.

13.6. Write a note on cooling towers.

PROBLEMS FOR PRACTICE

1. The following observations were recorded during a test on a surface condenser.

Condenser vacuum	: 70 cm of Hg
Barometer	: 76.5 cm of Hg
Mean condenser temperature	: 35 ºC
Hot well temperature	: 28 ºC
Condensate collected	: 1800 kg/hr
Air leakage	: 1 kg/1000 kg of steam
Cooling water circulated	: 80,000 kg/hr
Inlet temperature of cooling water	: 120 ºC
Rise in temperature of cooling water	: 15 ºC

Calculate

(i) η_{vacuum}

(ii) $\eta_{\text{condenser}}$

(iii) Quality of steam entering the condenser.

(iv) Capacity of wet extraction pump required is m^3/hr **[Ans.** 96.84 %; 60 %; 0.8886; 53.985 m^3/hr]

2. The following data were obtained during a test on a steam condenser.

Condenser vacuum recorded	: 705 mm of Hg
Barometer reading	: 750 mm of Hg
Mean condenser temperature	: 35 ºC
Hot well temperature	: 30 ºC
Mass of cooling water circulated	: 54000 kg/hr
Inlet temperature of cooling water	: 9 ºC
Outlet temperature of cooling water	: 28.5ºC
Mass of condensate	: 1850 kg/hr

Find

(i) η_{vacuum} corrected to standard barometric pressure

(ii) $\eta_{condenser}$

(iii) Quality of steam

(iv) Mass of air present is per m^3 of condenser volume. **[Ans.** 99.6 %; 75 %; 0.9769; 0.004315 kg/m^3]

3. Following data refers to a surface condenser.

Vacuum in condenser	: 710 mm
Barometric reading	: 765 mm of Hg
Temperature in condenser	: 32 ºC
Hot well temperature	: 30 ºC
Cooling water circulated	: 48000 kg/hr
Cooling water inlet temperature	: 20 ºC
Cooling water outlet temperature	: 34 ºC
Condensate	: 1500 kg/hr

(a) Mass of air in kg/m^3 of condenser volume.

(b) Dryness fraction.

(c) η_{vacuum} **[Ans.** 0.029 kg/m^3; 0.7617; 97.35 %]

4. The data below refers to a surface condenser

Condensate temperature	: 30 ºC
Mean temperature of condenser	: 35 ºC
Barometer	: 76 cm of Hg
Condenser vacuum	: 70 cm of Hg
Condensate collected	: 930 kg/hr
Cooling water circulated	: 36800 kg/hr
Rise in temperature of cooling water	: 12.5 ºC

Calculate

(i) Dryness fraction.

(ii) Capacity of air pump in m^3/min.

(iii) Mass of air handled in kg/hr. **[Ans.** 0.845; 391.23 m^3/min; 259 kg/hr]

5. The following observations were made during the test on surface condenser.

Barometer reading	: 760 mm of Hg
Condenser vacuum	: 705 mm of Hg
Measure condenser temperature	: 35 ºC

Condensate collected	: 2000 kg/hr
Cooling water circulated	: 60000 kg/hr
Rise in temperature of cooling water	: 16 ºC
Hot well temperature	: 28 ºC
Inlet water temperature	: 15 ºC

Determine,

(i) η_{vacuum}

(ii) $\eta_{condenser}$

(iii) Mass of air present per m^3 of condenser volume. **[Ans.** 98.2 %; 80 %; 0.1933 kg/m^3]

Chapter 14

Improved Rankine Cycle

14.1 INTRODUCTION

In Chapter 10 we have observed that Carnot cycle is the most efficient thermodynamic cycle for all the heat engines utilising the heat energy and producing mechanical energy. We have also observed that there are limitations for the use of this cycle in practice.

The ideal practical cycles for IC Engines are Otto cycle (for petrol and gas engines) and Diesel cycle (for diesel or oil engines). Similarly the ideal practical cycle for steam—is the Rankine Cycle.

14.2 DIFFERENCE IN CARNOT AND RANKINE CYCLE

Figure 14.1 shows saturated cycle for both Carnot and Rankine cycle. Carnot cycle is 1'–2–3–4' and Rankine cycle 1–2–3–4.

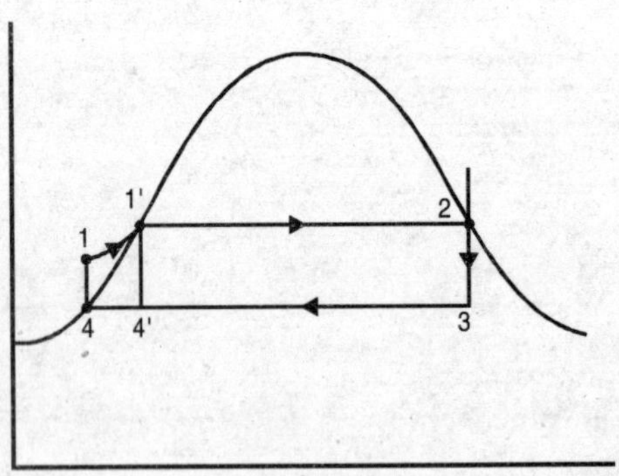

Fig. 14.1

Comparing these two cycles we observe that all the processes except water heating 1–1' are reversible processes. Rankine cycle efficiency is less than Carnot efficiency because of only one irreversible process 1–1'. If we could eliminate this irreversibility, we can get the Rankine cycle efficiency same as that of Carnot efficiency.

To eliminate this irreversible process, we have to heat water from 1–1' by some internal sources only. This type of the heating of water within the cycle is called Regenerative heating of water.

14.3 REGENERATIVE FEED HEATING CYCLE

Regenerative feed water heating can be achieved with the help of steam extracted or bled from the turbine, in the device called feed water heaters.

This irreversibility can be eliminated by the process of regeneration i.e., by internal and reversible heat exchange between the steam expanding in the turbine and water in the economizer. Figure 14.2 shows a flow diagram and a *T–S* diagram of the process for a saturated Rankine cycle.

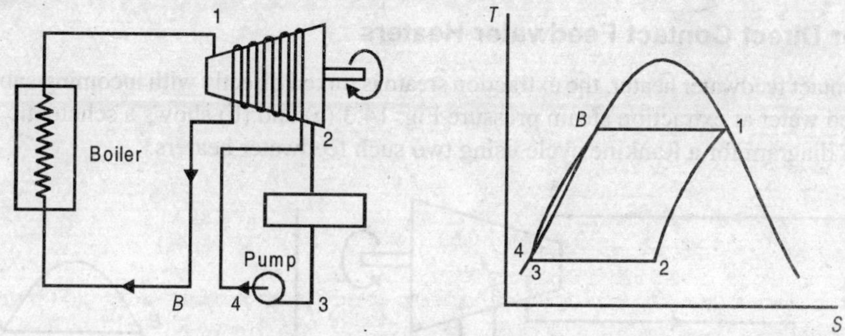

Fig. 14.2. *Ideal regeneration of Rankine cycle.*

The feed water after the condenser is carefully passed over the turbine, where it receives heat from the expanding steam reversibly at all times (i.e., with zero temperature difference). The water is progressively heated until it enters the steam generator at point *B*. The economizer and the irreversibilities introduced along with it, are thus eliminated. The cycle would receive and reject heat at constant temperature and would have same efficiency as that of Carnot cycle, working between same temperature limits.

This heating of feed water is called *ideal regenerative feed water heating*. It is however not employed as it has some serious limitations. They are as following:

Limitations of Ideal Regeneration

1. Water is flowing outside the turbine in tubes while the steam is expanding over turbine blades placed inside. In this situation, no adequate surface is available for effective heat transfer.
2. The mass flow rates of water and steam are very large. Hence effectiveness of heat exchange is very low.
3. The final dryness of steam leaving the turbine is very low and amount of moisture content is unacceptable for proper turbine operation and efficiency.

14.4 PRACTICAL–REGENERATIVE FEED WATER HEATING

In practical method of feed water heating, economizer is not eliminated totally but its size is reduced. The steam expands adiabatically through the turbine as usual. The feed water at 4 (Refer Fig. 14.2) is heated in steps and not continuously, by the steam bled from the turbine at selected stages. Heating of liquid takes place in heat exchangers, called *feed water heaters.*

Feedwater heating was introduced in early 1920's and since then has become an important and essential component of large thermal power plants, which use between five to eight such feed water heaters.

Feedwater heating is done in steps and not continuously. The feedwater therefore enters the boiler below point *B*. Also, feedwater heaters themselves are heat exchangers (they are practically very much like condensers) and have their own irreversibilities and losses. Thus we cannot attain an ideal situation of Fig. 14.2 and Rankine cycle cannot attain Carnot efficiency. A well designed Rankine cycle is however a closest practical cycle to Carnot and hence is widely accepted for most power plants.

14.5 TYPES OF FEEDWATER HEATERS

There are three types of feedwater heaters in use. They are,

1. Open or direct contact type
2. Closed type with drains cascaded backward
3. Closed type with drains pumped forward.

These types are now discussed and analysed in detail.

14.5.1 Open or Direct Contact Feedwater Heaters

In open or direct contact feedwater heater, the extraction steam is mixed directly with incoming subcooled feedwater; to produce saturated water at extraction steam pressure Fig. 14.3 (a) and (b) shows a schematic flow diagram and corresponding *T–S* diagram for a Rankine cycle using two such feedwater heaters.

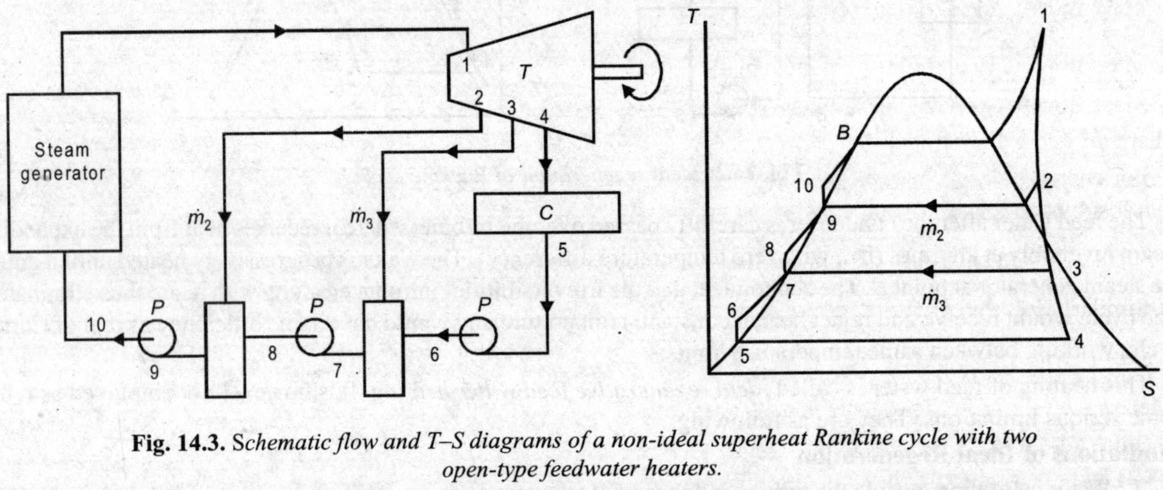

Fig. 14.3. *Schematic flow and T–S diagrams of a non-ideal superheat Rankine cycle with two open-type feedwater heaters.*

Condensate water leaves the condenser at 5 and is pumped to 6 to a pressure equal to that of the extraction steam at 3. The now-subcooled water at 6 and wet steam at 3 mix in the low pressure feed water heater to produce saturated water at 7.

Amount of bled steam $\dot{m}_3$ should essentially be such that subcooled water at 6, should be raised close to its saturation point. If it were much less, the water will be heated to below its saturation temperature and advantage of feedwater heating would be partially lost. If it were more there will be surplus steam at point 7. This would mean loss of turbine work and presence of steam-water mixture at 7 which is difficult to pump.

The pressure at 6–7 should be less than extraction steam pressure at 3 or else reverse flow of condensate water would enter the turbine at 3. The second pump pressurises saturated water at 7 to a subcooled condition at 8, which is at the pressure of extraction steam at 2. In the high pressure feed water heater superheated steam at 2, mixes with subcooled water at 8 to produce saturated water at 9. This is pressurised to 10 by a pump to enter the steam generator at its working pressure.

Steam at 2 or 3 loses its latent heat of vaporisation which is much larger than the gain of sensible heat by water at 8 or 6. Therefore, the quantity of extracted steam $\dot{m}_2$ and $\dot{m}_3$ is only a small fraction of the steam passing through the turbine.

In addition to a condensate pump, one pump for each open feedwater heater is required. This is a disadvantage. Open feed water heaters because of vigorous action of mixing remove non-condensible gases like Air, O_2, H_2, CO_2 from the feed water. They are therefore known as *deareating heaters* or *DA*.

14.5.2 Mass Flow Rates and Energy Balance

This is done on the basis of unit mass flow rate at the turbine inlet (point 1). The mass balance therefore is as follows:

Mass flow between points 1 and 2 = 1 kg/s

Mass flow between points 2 and 9 = $\dot{m}_2$

Mass flow between points 2 and 3 = $1-\dot{m}_2$

Mass flow between points 3 and 7 = $\dot{m}_3$

Mass flow between points 3 and 4 = $1-\dot{m}_2-\dot{m}_3$ ...(14.1)

Mass flow between points 4 and 7 = $1-\dot{m}_2-\dot{m}_3$

Mass flow between points 6 and 7 = $1-\dot{m}_2-\dot{m}_3$

Mass flow between points 7 and 9 = $1-\dot{m}_2$

Mass flow between points 9 and 1 = 1

where $\dot{m}_2$ and $\dot{m}_3$ are fraction of 1.

For energy balance, consider the low and high pressure feed water heaters. At low pressure feed water heater, heat lost by condensing steam equals heat gained by feed water heater, or,

$$\dot{m}_2(h_2-h_9) = (1-\dot{m}_2)(h_9-h_8) \quad ...(14.2)$$

Similarly for high pressure feed water heaters,

$$\dot{m}_3(h_3-h_7) = (1-\dot{m}_2-\dot{m}_3)(h_7-h_6) \quad ...(14.3)$$

where h_2, h_3, h_6, h_7, h_3, h_9 etc. are specific enthalpies at respective points.

Since admission, extraction and condenser pressures are known, all the enthalpy values are known. For two Eqs (14.2) and (14.3), therefore, there are only two unknowns $\dot{m}_2$ and $\dot{m}_3$. For any number of feedwater heaters there will be as many equations as there are unknowns, so solutions are always possible.

The Cycle Analysis

Based on the mass flow rates, cycle analysis can now proceed as follows:

1. Heat added $q_s = v(h_1-h_{10})$ kJ/kg

2. Turbine work $W_T = 1(h_1-h_2)+(1-\dot{m}_2)(h_2-h_3)+(1-\dot{m}_2-\dot{m}_3)(h_3-h_4)$ kJ/kg

3. Pump work $\left|\sum W_P\right| = (1-\dot{m}_2-\dot{m}_3)(h_6-h_5)+(1-\dot{m}_2)(h_8-h_7)+1(h_{10}-h_9)$ kJ/kg ...(14.4)

$= (1-\dot{m}_2-\dot{m}_3)v_5(p_6-p_5)+(1-\dot{m}_2)v_7(p_8-p_7)+1\times v_9(p_{10}-p_9)$ kJ/kg

4. Heat rejected $|q_r| = (1-m_2-m_3)(h_4-h_5)$ kJ/kg

5. Net cycle work $= \Delta W_{|net|} = |W_T - W_P|$ kJ/kg

6. Cycle thermal efficiency $= \eta_{th} = \dfrac{\Delta W_{net}}{q_s}$

7. Work ratio $= \dfrac{\Delta W_{net}}{W_T}$

8. Specific steam consumption, $= \dfrac{3600}{\Delta W_{net}}$ kg/k Wh

If turbine and pump efficiencies are considered then,

$$W_T\text{. actual} = \eta_{th} \times W_T$$

where η_t = turbine efficiency and η_p = pump efficiency

and $$W_P\text{. actual} = \frac{W_P}{\eta_P}$$

14.5.3 Closed Type Feed Water Heaters with Drains Cascaded Backward

This is the simplest and most commonly used type in power plants. It is a shell and tube type heat exchanger. The feedwater passes through the tubes, and the bled steam, on the shell side transfers its heat energy to it and condenses. Thus, they are like surface condensers, operating at higher pressures than those in the main condenser. Since feedwater does not mix with the steam and passes through the tubes of successive close feed water heaters, only one condensate pump is sufficient to pressurise the water from condenser to the boiler. However, to reduce the pressure rise in one pump, one condensate pump before the feed water heaters and one feed water pump after the feed water heaters are used. If deareater heater is used, one boiler feed pump is automatically required after the same.

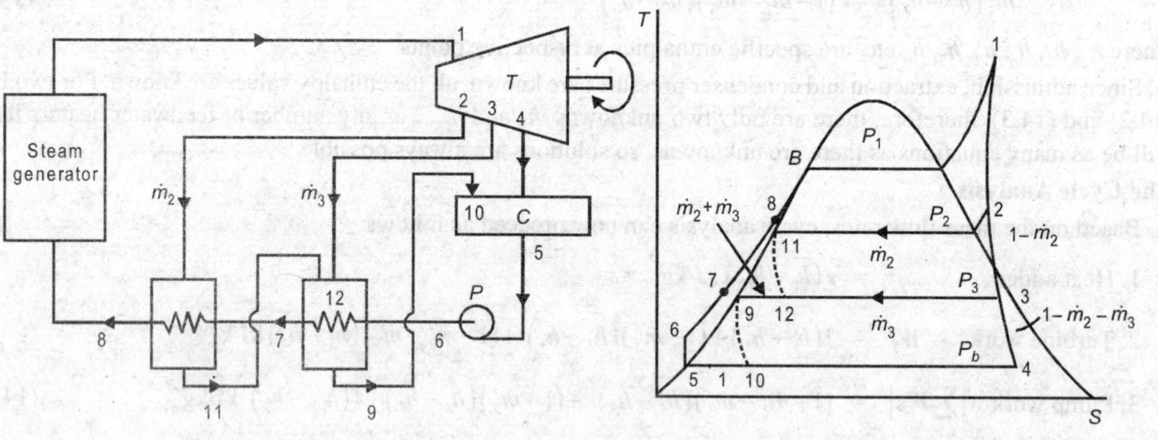

Fig. 14.4. *Schematic flow and T–S diagrams of a non-ideal superheat rankine cycle with two closed-type feed water heaters with drains cascaded backward.*

Figure 14.4 shows a simplified flow diagram and corresponding T–S diagram of a non-ideal superheat Rankine cycle. There are two feedwater heaters. One pump 5–6 pressurises the condensate, sufficient to pass through the two feedwater heaters and enters the steam generator at 8.

The bled steam condensed at each FWH is fed back to the next lower pressure FWH. The condensate from the last FWH is led back to the main condenser. One can imagine, then, a cascade from high pressure to lower-pressure heaters. Hence, the name.

14.5.4 Operation

Wet steam bled at 3 at the rate of $\dot{m}_3$ enters the low pressure FWH and heats the subcooled water at 6. The temperature of water at 7 is less than saturation temperature at 3. This difference is called the *terminal temperature difference* (TTD) and is defined for all closed FWHS.

TTD or TD = saturation temperature of bled steam – exit water temperature

The value of TTD for low pressure heaters is positive and is of the order of 3°C. For high pressure heaters it may be negative and lies between 0 and – 3°C. In case no data is available, temperature at 7 is taken equal to saturation temperature at 3.

The drain from low pressure FWH is led to the main condenser and enters it as two phase mixture at 10. The process 9-10 is throttling process from the pressure corresponding to 9 and the main condenser pressure. Due to throttling there is some loss of availability. Also enthalpy at 9 and 10 are equal, as the process is constant enthalpy one.

The high pressure FWH receives superheated steam bled from the turbine at 2 that flows on the shell side at the rate of $\dot{m}_2$ and heats the water entering the tubes at 7. The inlet steam at 2 is superheated, hence temperature at 8 may be higher than saturation temperature of that steam and TTD may be negative. The drain at 11 may be slightly subcooled; and is throttled to the low pressure FWH which it enters at 12 as a two phase mixture. There it joins with the steam bled at 3 and aids the heating of water in low pressure FWH. The drain from low pressure fwh is thus $\dot{m}_2 + \dot{m}_3$, which is throttled to the main condenser at 10. The higher pressure exit water at 8 is led to the steam generator.

14.5.5 Mass Flow Rate and Energy Balance

A mass balance based on unit flow rate at turbine inlet point 1, is given clockwise by

$$\begin{aligned}
\text{Mass flow between 1 and 2} &= 1\\
\text{2 and 3} &= 1-\dot{m}_2\\
\text{3 and 10} &= 1-\dot{m}_2-\dot{m}_3\\
\text{10 and 1} &= 1\\
\text{2 and 12} &= \dot{m}_2\\
\text{3 and 12} &= \dot{m}_3\\
\text{12 and 10} &= \dot{m}_2+\dot{m}_3
\end{aligned} \qquad \text{...(14.5)}$$

The energy balances on the high and low pressure heaters are given respectively by,

$$\dot{m}_2 (h_2 - h_{11}) = h_8 - h_7 \qquad \text{...(14.6)}$$

$$\dot{m}_3 (h_3 - h_9) + \dot{m}_2 (h_{12} - h_9) = h_7 - h_6 \qquad \text{...(14.7)}$$

For throttling processes, we have, $h_{12} = h_{12}$ and $h_{10} = h_9$.

All the relevant pressures and the enthalpies of the Eqs. (14.6) and (14.7) are known. Since there are only two unknowns $\dot{m}_2$ and $\dot{m}_3$, the equations can be solved.

14.5.6 Cycle Analysis

From mass flow rates, energy balance equations and enthalpy values, we can determine the amounts of steam extracted i.e., $\dot{m}_2$ and $\dot{m}_3$ and then the pertinent cycle parameters. We assume mass flow rate of 1 kg/s at turbine inlet (point 1).

1. Heat added $q_s = h_1 - h_8$
2. Heat rejected $q_r = (1 - \dot{m}_2 - \dot{m}_3)(h_4 - h_5) + (\dot{m}_2 + \dot{m}_3)(h_{10} - h_5)$
3. Turbine work $W_T = (h_1 - h_2) + (1 - \dot{m}_2)(h_2 - h_3) + (1 - \dot{m}_2 - \dot{m}_3)(h_3 - h_4)$
4. Pump work $|W_p| = h_6 - h_6 = v_5 (p_6 - p_5)$

5. Net cycle work $\Delta W_{net} = W_T - |W_P|$ ...(14.8)

6. Cycle thermal efficiency $= \eta_{th} = \dfrac{\Delta W_{net}}{q_s}$

7. Work ratio $= \dfrac{\Delta W_{net}}{W_T}$

8. Specific steam consumption, SSC $= \dfrac{3600}{\Delta W_{net}}$

14.5.7 Closed Type Feed Water Heaters with Drains Pumped Forward

This is a second closed type FWH. It avoids throttling and the irreversibilities therein but has to use a small additional pump. Like previous closed type FWH, it is also a shell and tube type heat exchanger. The feedwater passes through the tubes, the steam passes over them; transfers heat and condenses. They do not mix and only one pump is sufficient to carry water from the condenser to the boiler. If, however, *DA* is used one more pump would be necessary.

The drain from these FWHS is pumped forward into the main feedwater line. Figure 14.5 shows a simplified flow diagram and corresponding *T–S* diagram for a non-ideal superheat Rankine cycle with two such fwhs. Like open FWH these heaters also require an additional pump per heater. But the pump is small and carry only fractional flows corresponding to $\dot{m}_2$ and $\dot{m}_3$ while pumps in open FWH carry full steam and water flow.

14.5.8 Operation

The drain of low pressure FWH at 13 is pumped forward to the main feedwater line and enters it at 14. It mixes with exit water from l.p. heater at 7 and forms mixture at 8. Point 8 is closer to 7 than 1.4 on the *T–S* diagram because flow at 7 is much greater than the drain flow $\dot{m}_3$.

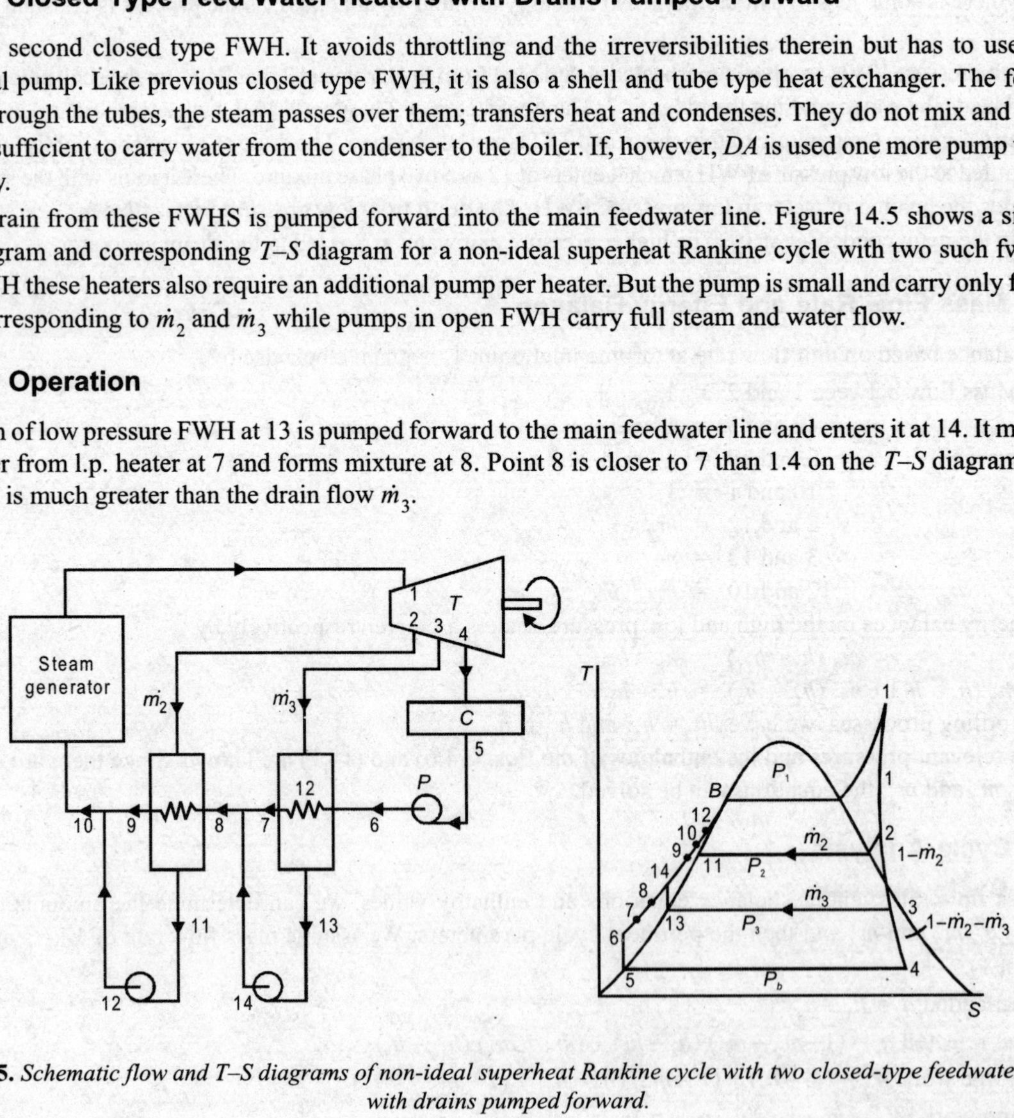

Fig. 14.5. *Schematic flow and T–S diagrams of non-ideal superheat Rankine cycle with two closed-type feedwater heaters with drains pumped forward.*

The water at 8 enters h.p. heater and is heated to 9. The drain leaves the heater at 11, is pumped to 12. It mixes with the feedwater at 9 to form full feedwater flow at 10, which crow goes to boiler.

14.6 MASS BALANCE AND ENERGY BALANCE

A mass balance based on a mass flow rate of 1 kg/s at turbine inlet (point 1) is given clockwise on T–S diagram by,

$$\begin{aligned}\text{Mass flow between 1 and 2} &= 1\\ \text{2 and 12} &= \dot{m}_2\\ \text{2 and 3} &= 1-\dot{m}_2\\ \text{3 and 14} &= \dot{m}_3\\ \text{3 and 7} &= 1-\dot{m}_2-\dot{m}_3\\ \text{at 14} &= \dot{m}_3\\ \text{8 and 9} &= 1-\dot{m}_2\\ \text{at 12} &= \dot{m}_2\\ \text{10 and 1} &= 1\end{aligned} \quad ...(14.9)$$

The energy balance equations are,

h.p. heater $\quad \dot{m}_2(h_2-h_{11}) = (1-\dot{m}_2)(h_9-h_8) \quad ...(14.10)$

and l.p. heater $\quad \dot{m}_3(h_3-h_{13}) = (1-\dot{m}_2-\dot{m}_3)(h_7-h_6) \quad ...(14.11)$

The values of h_9 and h_7 are obtained from the temperatures t_9 and t_7 which are equal to the saturation temperature of the steam at each heater minus its terminal temperature difference, or

$$t_9 = t_{11} - \text{TTD} \qquad \text{H. P. heater} \quad ...(14.12a)$$

and

$$t_7 = t_{13} - \text{TTD} \qquad \text{I. P. heater} \quad ...(14.12b)$$

h_{10} needed for q_A and h_8 to be used in Eq. (14.10) are obtained from h_{11} and h_{14} respectively, or

$$h_{12} = h_{11} + v_{11}\frac{p_{12}-p_{11}}{\eta_p} \quad ...(14.13a)$$

and

$$h_{14} = h_{13} + v_{13}\frac{p_{14}-p_{13}}{\eta_P} \quad ...(14.13b)$$

where η_P is pump efficiency.

Thus

$$h_{10} = m_2h_{12} + (1-m_2)h_9 \quad ...(14.14a)$$

and

$$(1-m_2)h_8 = m_3h_{14} + (1-m_2-m_3)h_7 \quad ...(14.14b)$$

14.6.1 Cycle Analysis

1. Heat added, $q_s = h_1 - h_{10}$
2. The turbine work, $W_T = (h_1-h_2)+(1-\dot{m}_2)(h_2-h_3)+(1-\dot{m}_2-\dot{m}_3)(h_3-h_4)$
3. Pump work, $W_P = (1-\dot{m}_2-\dot{m}_3)(h_6-h_5)+\dot{m}_3(h_{14}-h_{13})+\dot{m}_2(h_{12}-h_{11})$
4. Net work $\Delta W_{\text{net}} = |W_T - W_P|$ ...(14.15)

5. Thermal Efiiciency, $\eta_{th} = \dfrac{\Delta W_{net}}{q_s}$

6. Work Ratio W. R. = $\dfrac{\Delta W_{net}}{W_T}$

7. Specific steam consumption, SSC = $\dfrac{3600}{\Delta W_{net}}$

14.7 COMPARISON OF CLOSED FEED WATER HEATERS

Table 14.1 *Comparison of closed feed water heaters.*

Drains cascaded backward	*Drains pumped forward*
1. There is a loss of availability due to throttling.	1. No throttling, no loss of availability.
2. Number of pumps are less-one or two.	2. Number of pumps are more; one for each feedwater heater.
3. Simple and less expensive system.	3. Complex and expensive system.
4. Drain is slightly subcooled. This imparts more energy to the water and reduces loss of availability.	4. Is used as the lowest pressure feed water heater in an otherwise all cascaded system. This prevents throttling of the combined cascaded flows to the condenser.

14.7.1 The Placement of Feedwater Heaters

When number of FWHS are used, a natural question arises, as to where to place these FWHS in the cycle or what are the pressures at which steam is to be bled from the turbine, that will result in the maximum increase in the efficiency ?

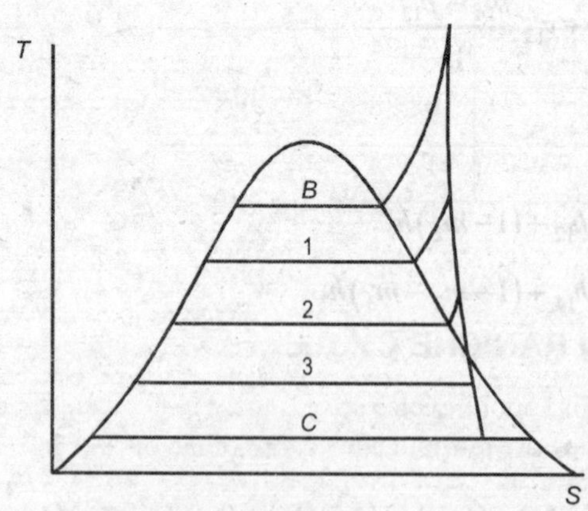

Fig. 14.6. *One feedwater heater in three possible positions.*

To answer this let us consider for the sake of simplicity that only one FWH is used. The aim of using any feedwater heater is to heat the feedwater as close to the boiler temperature as possible. We may now place the feedwater heater at position 1, 2 or 3. (Fig. 14.6) T_B and T_C are saturation temperatures at boiler and condenser pressures respectively. Now at position 1 the heat transfers are because of temperature differences $T_B - T_1$ and $T_1 - T_C$. At position 3 the heat transfers are because of $T_B - T_3$ and $T_3 - T_C$. In either case one of the temperature differences is very large. To minimise the temperature differences, we have to select position 2 which is in the middle i.e., $T_B - T_2 = T_2 - T_C$. Thus we find the temperature which is *halfway between* T_B and T_C and then obtain the saturation pressure at that temperature. Feed water placed at this pressure gives us optimum position of maximum efficiency. Note that the temperature at which steam is actually bled from the turbine may be superheated.

If two FWHS are used, the temperature range $T_B - T_C$ is divided into three equal parts and for three FWHS it is divided into 4 equal parts.

Thus, in general, for n FWHS the optimum temperature rise per heater would be given by

$$\Delta T_{opt} = \frac{T_B - T_C}{n+1} \quad \text{...(14.14)}$$

Consider for example a cycle where boiler pressure is 100 bar and condenser pressure is 0.08 bar and three FWHS are to be used. We find the pressures at which they are placed. From steam tables we get saturation temperatures at 100 bar and 0.08 bar. They are $T_B = 310.96°C$ and $T_C = 41.534°C$ respectively. From above Eq. (14.14) we have,

$$\Delta T_{opt} = \frac{310.96 - 41.534}{4} = 67.36$$

Thus the saturation temperatures of FWHS will be found by adding ΔT_{opt} successively from $T_C = 41.534$ onwards and then finding the corresponding saturation pressures from the steam tables. Table 14.1 summarises the results.

Table 21.2

Component	*Saturation temperature*	*Saturation pressures at sat. temperature*
The boiler	310.96°C	100 bar
The high pressure FWH	240.60°C	34 bar
The intermediate, pressure FWH	176.24°C	9.2 bar
The low pressure FWH	108.88°C	1.35 bar
The condenser	41.534°C	0.08 bar

Thus, the FWHS will be placed at 34 bar, 9.2 bar and 1.35 bar respectively.

14.8 IRREVERSIBILITIES IN RANKINE CYCLE

All ideal cycles whether gas cycles or vapour cycles are assumed to be made up of reversible processes only. In the actual cycles, however, irreversibilities are introduced because of factors like heat transfer through finite temperature difference. friction, mixing etc.

In Rankine cycle, we will discuss two kinds of irreversibilities:

1. **External Irreversibilities.** Introduced as a result of temperature differences between combustion gases in the furnace and feed water or steam in the boiler tubes and the temperature differences between the condensing steam and condenser cooling water.

2. **Internal Irreversibility.** Introduced because of fluid friction and throttling in the turbine and the pump and mixing.

14.8.1 External Irreversibilities

In Fig. 14.4 (a), line *ab* represents combustion gases in a counter-flow furnace and 4 – B – 1 represents water in a saturated Rankine cycle. Line *cd* represents condenser cooling water and 2 – 3 represents condensing steam.

The temperature differences between line *ab* and 4 – B – 1 – 1' and between 2 – 3 and line *cd* are not constant.

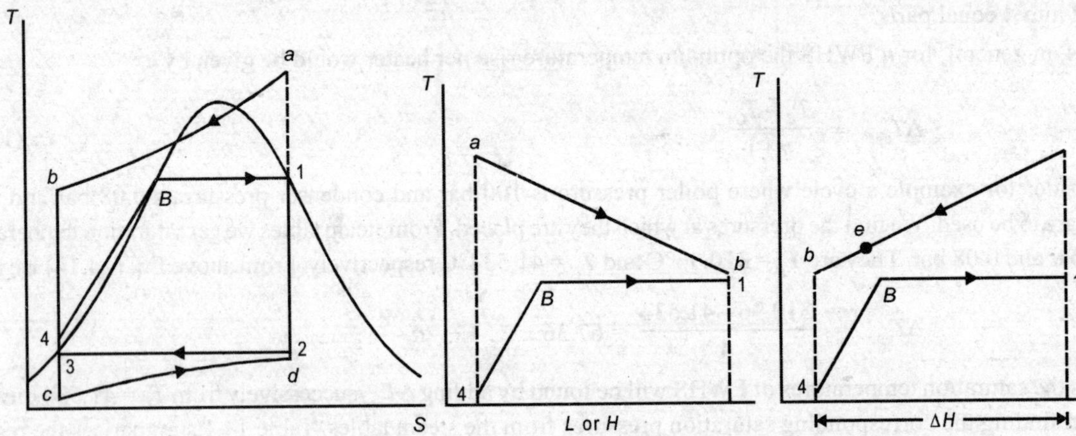

(a) External irreversibility with Rankine cycle

(b) Effect of flow direction on external irreversibility: (i) parallel flow, (ii) counter flow

Fig. 14.7

Figure 14.7 also shows temperature *vs* heat exchanger path length diagrams for (b) parallel flow and (c) counter flow steam generators and effect of flow directions in the steam generators. *L* – 1 and *e* – *B* represent minimum approach point called *pinch point*; and is finite. For very small pinch point temperature difference, the overall temperature differences are low and hence reversibilities are lower but the size and cost of steam generators are high. For very large pinch point temperature differences, size and cost of steam generators are low but overall temperature differences and irreversibilities are higher and plant efficiencies are lower. The most economical pinch point temperature difference is obtained by optimization of fixed cost (based on capital costs), and operating costs (based on fuel costs).

Figure 14.7 also shows that overall temperature differences between the combustion gases and water and steam are greater for parallel flow than counter flow heat exchange in steam generators. Heat transfer considerations also favour counter flow as it gives higher overall heat transfer coefficients. The counter flow is thus favoured over parallel flow.

14.8.2 Internal Irreversibilities

They are result of fluid friction, throttling and mixing in turbines, pumps, and pressure losses in pipes, bends, valves etc.

The actual expansion of steam in the turbine is adiabatic but not adiabatic reversible (isentropic) and during expansion entropy increases. Refer Fig. 14.8. The ideal expansion, which is isentropic is 1 – 2s but the actual expansion is 1 – 2.

The pump pressure P_4 has to be more than turbine inlet pressure P_1 because of pressure drop in heat exchangers, feed water heaters, pipes, bends and valves etc. Thus P_5 is the pressure of steam leaving the steam generator, while $P_{5'}$ is the pressure at turbine inlet. Heat loss in the turbine decreases entropy of steam from 5' to 1.

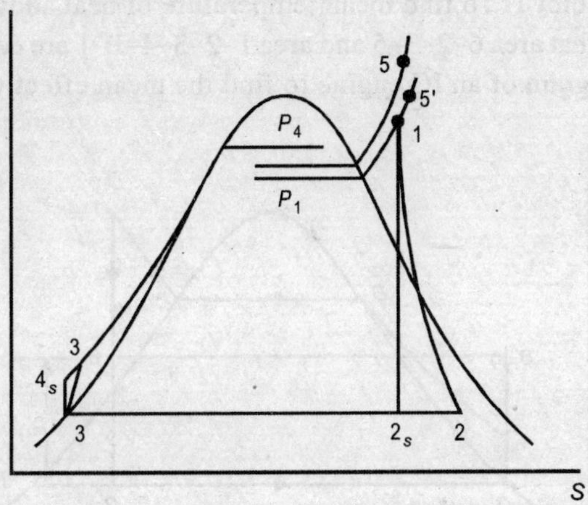

Fig. 14.8. *A T–S diagram of an internally irreversible superheat rankine cycle.*

As a result, dryness and therefore enthalpy at 2 is more than that at 2s. The irreversible losses in the turbine are represented by *turbine isentropic efficiency* and is given by the ratio of turbine actual work to the ideal isentropic work. Hence,

$$\eta_T = \frac{h_1 - h_2}{h_1 - h_{2s}} \qquad \ldots(14.15)$$

Well designed turbines give efficiencies of around 90%. η_T given above is not to be confused with *cycle thermal efficiency*. It is also overall isentropic efficiency. The individual turbine stages may have different efficiencies.

The compression of water in the pumps is also adiabatic but not isentropic. The entropy increases. The ideal compression is 3 – 4s while the actual one is 3 – 4. The losses are represented by *pump isentropic efficiency* which is the ratio of ideal isentropic work to the actual work taken by the pump. Hence

$$\eta_P = \frac{h_3 - h_{4s}}{h_3 - h_4} \qquad \ldots(14.16)$$

In both Eqs. (14.14) and (14.16), the smaller quantity is in the numerator. The actual pump work can be obtained by

$$|W_p| = \frac{h_{4s} - h_3}{\eta_P} = \frac{v_3(p_4 - p_3)}{\eta_P}$$

Thus we pay penalty for the turbine and pump irreversibilities. The turbine produces less work and the pump absorbs more work.

14.9 MEAN TEMPERATURE OF HEAT ADDITION

In Rankine cycle, heat is added at constant pressure but not at constant temperature. Refer Fig. 14.9. There is a large difference between temperature of water at the entry of the economiser at point 4 and that of steam at the exit of the superheater at point 1. To find mean temperature of heat addition T_m, construct a rectangle 6–2–3–2–5 on the base 2–3 such that area 6–2–3–5 and area 1–2–3–4–B-1 are equal. Recall a similar construction of a rectangle on indicator diagram of an IC engine to find the mean effective pressure P_m.

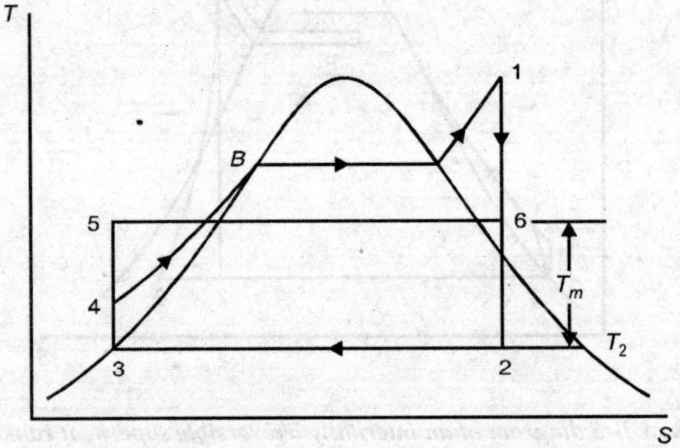

Fig. 14.9. *Mean temperature of heat addition.*

Now heat supplied is,

$$q_s = h_1 - h_4 = T_m(s_1 - s_4)$$

$$\therefore \quad T_m = \frac{q_s}{s_1 - s_4}\,°\mathrm{C} = \frac{h_1 - h_4}{s_1 - s_4}$$

Now heat rejected is

$$q_r = T_2(s_1 - s_4)$$

$$\therefore \quad \eta_{\text{Rankine}} = 1 - \frac{q_r}{q_s} = 1 - \frac{T_2}{T_m}$$

Since T_2 is fixed by condenser pressure for which very little improvement is possible over present practices, the efficiency largely depends on T_m. Larger the value of T_m greater is the cycle efficiency.

It will be realised now that superheating of steam will raise the value of T_m over the saturated cycle and hence will increase the cycle efficiency.

14.10 REHEAT CYCLE

Raising the turbine inlet pressure i.e., boiler outlet pressure or boiler working pressure as well as raising the temperature of superheat improves the thermal efficiency of the Rankine cycle. Higher working pressures and temperatures increase mean temperature T_m of heat addition; and takes T_m closer to gas line *ab* reducing the irreversibilities. Refer Fig. 14.7 (b).

Increasing the inlet pressure from P_1 to P_2 however results in the lowering of the final dryness fraction from 4 to 4' and all the stages after point *d* on the expansion line get progressively wet steam. The moisture present in the steam, in the form of suspended water particles hit the high speed turbine blades with a large relative velocities and erode them.

To improve the thermal efficiency of the cycle over and above that of superheat cycle as well as to improve the final dryness of steam, reheating of steam is used.

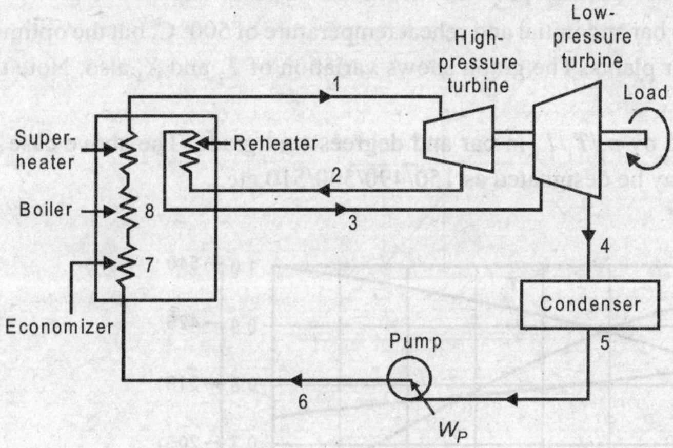

(a) Schematic of a Rankine cycle with superheat and reheat

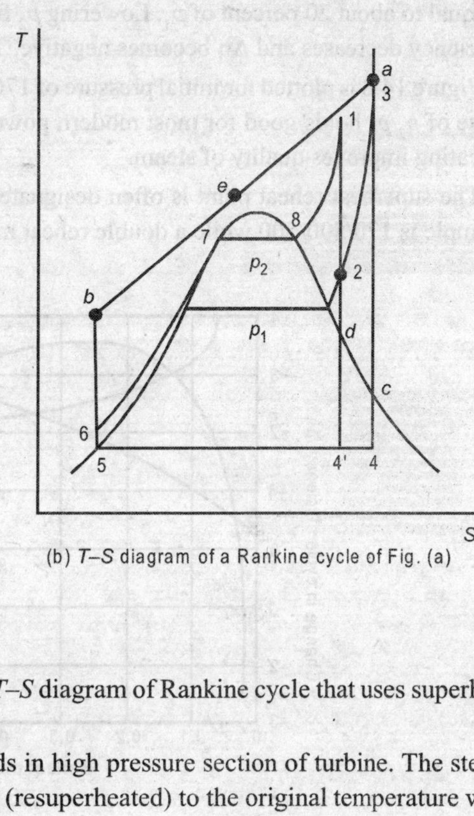

(b) *T–S* diagram of a Rankine cycle of Fig. (a)

Fig. 14.10

Figure 14.10. (a) and (b) show a schematic flow diagram and *T–S* diagram of Rankine cycle that uses superheat and resuperheat or reheat.

Reheating is resuperheating of steam at point 2 after it expands in high pressure section of turbine. The steam after expansion is sent to the steam generator where it is reheated (resuperheated) to the original temperature with the help of hot flue gases. The reheated steam now expands in the low pressure section of the turbine to the condenser pressure.

14.10.1 Advantages of Reheat Cycle

There are following advantages of reheat cycle is,

1. In reheat cycle, heat is added twice; from 6 to 1 and from 2 to 3. This increases the value of T_m—the temperature of heat addition and keeps the evaporator, superheat reheat portion from 7 to 3 closer to the flue gas line *ae*. This results in higher cycle efficiency.
2. It improves the final dryness of steam (from 4' to 4) and reduces turbine blade erosion.

3. It increases cycle work by increasing the cycle area. This increases the turbine power.

Modern fossil fuel power plants employ superheat and at least one stage of reheat. Some employ two. More than two reheats however, result in cycle complications and increased capital costs, that are not justified by improvements in efficiency.

14.10.2 Effect of Reheat Pressure Ratio on Cycle Efficiency

The pressure p_2 at which the steam is reheated affects the cycle efficiency. Figure 14.11 shows the relation between increase in the efficiency $\Delta\eta$ percent and the ratio of reheat pressure to initial pressure.

At $p_2/p_1 = 1$ there is no reheat and $\Delta\eta = 0$. As p_2 is lowered the efficiency improves and reaches the maximum at p_2 equal to about 20 percent of p_1. Lowering p_2 further decreases the efficiency and at p_2/p_1 less than say 0.025 the efficiency decreases and $\Delta\eta$ becomes negative. The optimum pressure ratio is between 0.2 and 0.25.

Figure 14.8 is plotted for initial pressure of 170 bar and initial and reheat temperature of 500° C, but the optimum value of p_2/p_1 holds good for most modern power plants. The graph shows variation of T_2 and X_4 also. Note that reheating improves quality of steam.

The superheat reheat plant is often designated by $p_1/T_1/T_3$ in bar and degrees centigrade. The above case for example is 170/500/500 while a double reheat may be designated as 150/490/500/510 etc.

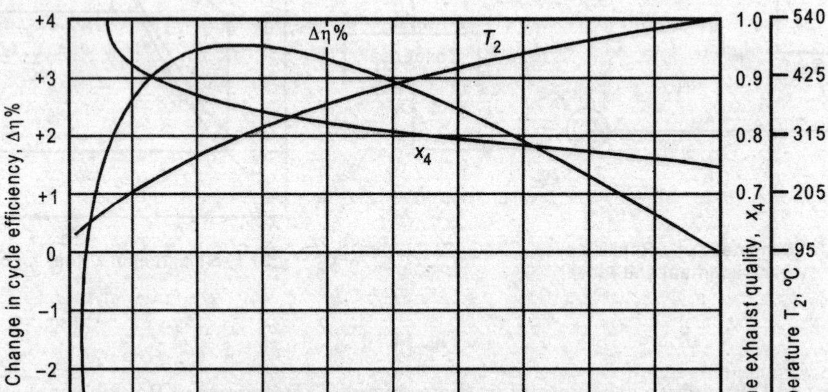

Fig. 14.11. *Effect of reheat-to-initial pressure ratio on efficiency, high-pressure turbine exit temperature, and low-pressure turbine exit quality.*

14.11 TYPICAL LAYOUT OF A STEAM POWER PLANT

Figure 14.12 shows a typical layout of a steam power plant. It shows steam and water circuits of the plant which includes reheaters and feedwater heaters. Note the positions of four closed feedwater heaters and one open feedwater heater (deaerator). Note also that there are two HP, two IP and one IP feedwater heater.

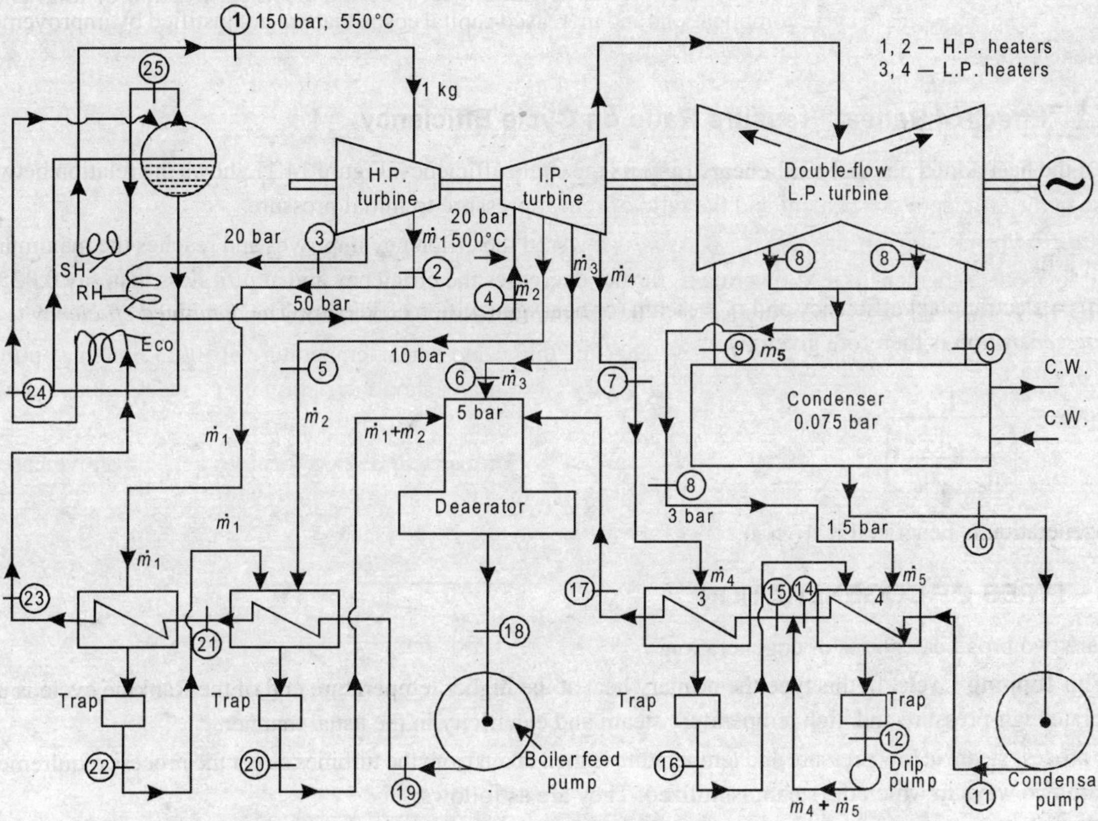

Fig. 14.12. *Schematic flow diagram of a modern steam power plant.*

14.12 COGENERATION

Cogeneration is the simultaneous generation of electricity and steam (or heat) in a single power plant. It has long been used by certain industries which need large quantities of steam or heating as well as electrical power. Examples are cement factories, paper mills, chemical industries, textile plants, fertilizer plants, extraction plants, edible oil industries, refineries etc. Cogeneration is used if the electricity thus produced is cheaper, than that purchased from the state electricity boards.

In a cogeneration plant if,

E = electrical energy generated

ΔH_s = heat energy generated or heat in the process steam

and Q_s = total heat added to the plant

then the cogeneration plant efficiency is given by

$$\eta_{co} = \frac{E + \Delta H_s}{Q_A} \quad \text{...(14.17)}$$

Now consider the generation of electricity and steam separately.

Total energy output of the plant = $E+\Delta H_s$ as stated earlier. Let e be the electrical fraction of total energy output,

$$\text{or} \quad e = \frac{E}{E+\Delta H_s}$$

Then the heat added per unit total energy output is,

$$\frac{e}{\eta_e}+\frac{1-e}{\eta_h}$$

where η_e = electric plant efficiency and η_h = steam (or heat) generation efficiency. The *combined efficiency* η_e *for separate generation* is therefore given by,

$$\eta_c = \frac{1}{\left(\dfrac{e}{\eta_e}+\dfrac{1-e}{\eta_h}\right)}$$

and cogeneration is beneficial if $\eta_{co} > \eta_c$.

14.13 TYPES OF COGENERATION

There are two broad categories of cogeneration.

1. The Topping Cycle. In this type the primary heat at the higher temperature end of the Rankine cycle is used to generate high-pressure and high temperature steam and electricity in the usual manner.

The process steam at low pressure and temperature is then taken from the turbines as per the process requirements. There are two ways in which the steam is utilized. They are as follows:

(a) By using steam at the turbine exhaust. The steam turbine in this case is called *back pressure turbine*. The condenser is removed, and turbine exhaust is led to the process plant. Since condenser is absent the exhaust pressure of the turbine is more (hence the name back pressure turbine) and its efficiency is lower. Figure 14.13 (a) shows a flow diagram of this type.

(b) By extracting steam from the turbine at an intermediate state; much as for feedwater heating. The turbine of this type is known as *extraction turbine* or *passout turbine*. Since condenser is used the efficiency of the cycle is higher; and steam extracted has much higher pressure and temperature. Figure 14.13 (b) shows a flow diagram of this type.

The main advantage of this cycle is its ability to use wide range of fuels including wood, coal, petroleum, coke, refuse etc. and hence is used where low quality refuse and byproduct fuels are readily available.

This type however is seldom used because of its high initial cost and low power-to-heat ratio; especially at high process-steam pressures.

2. Bottoming Cycle. Here first the primary heat at high temperature is directly used for process requirements. The low grade (low temperature and availability) waste heat from process plant is used to generate electricity. Figure 14.14. The efficiency of the cycle is obviously low, –10 percent for low and upto 20 percent at high temperature applications. Given that the waste heat is free, the cycle economics may be favourable.

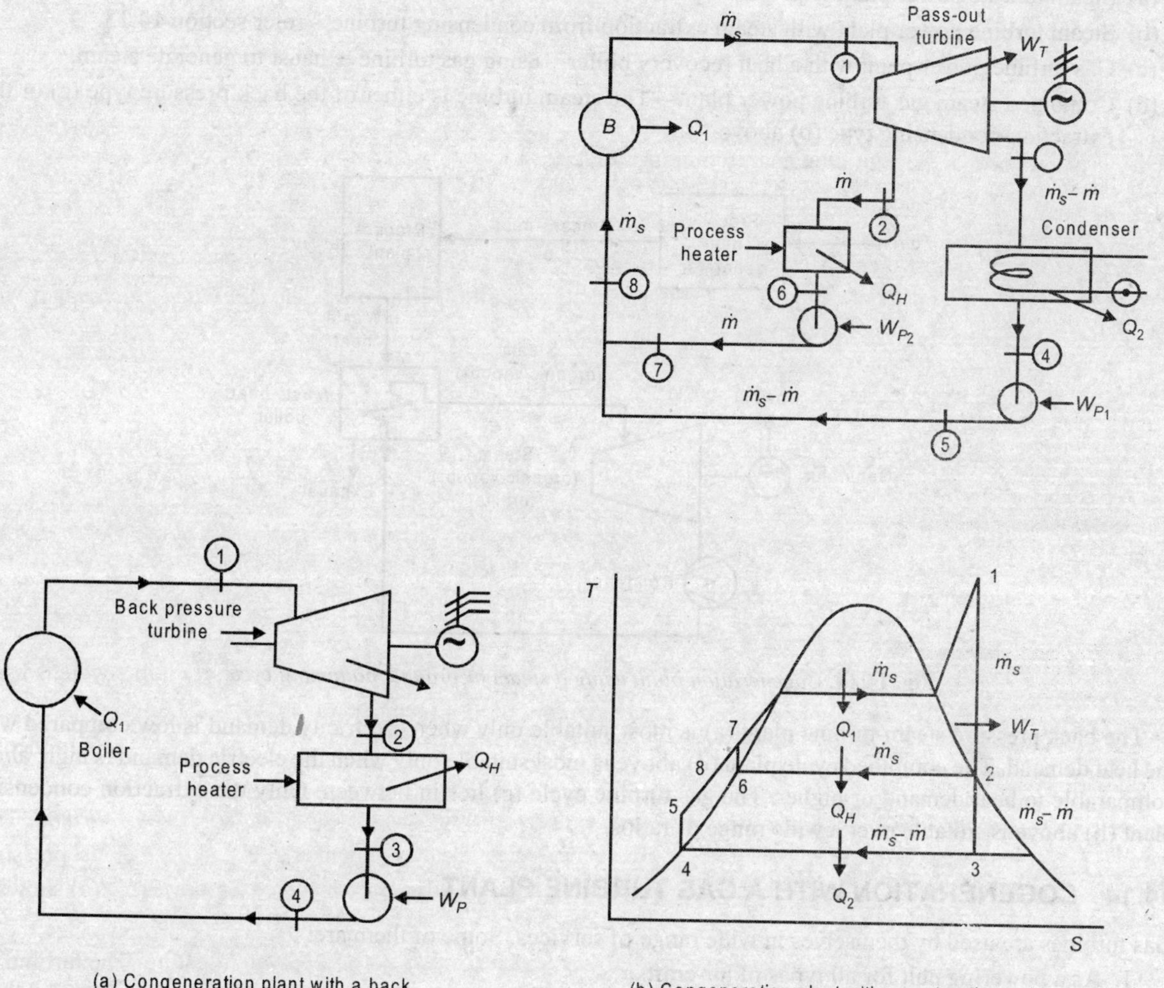

(a) Congeneration plant with a back pressure turbine

(b) Congeneration plant with an extraction turbine

Fig. 14.13

The main application of bottoming cycles are in the processes of the chemical and material processing industries such as cement, limestone and steel.

Since the temperature of waste heat is low, hydrocarbon fluids are often used as working medium.

14.13.1 Types of Topping Cycle

Since the bottoming cycle has very low combined efficiency; it is of little thermodynamic or economic interest and only the topping cycle therefore can provide true savings in primary energy. In addition, most process applications require a low grade (low temperature and availability) steam which can be conveniently produced in a topping cycle.

Some of the arrangements for cogeneration in topping cycle are as under.

(a) Steam turbine power plant with a back pressure turbine—refer section 14.13.

(b) Steam turbine power plant with steam extraction from condensing turbine—refer section 14.13.

(c) Gas turbine power plant with a heat recovery boiler—using gas turbine exhaust to generate steam.

(d) Combined steam gas turbine power plant—The steam turbine is either of the back pressure type (a) or the extraction-condensing type (b) above.

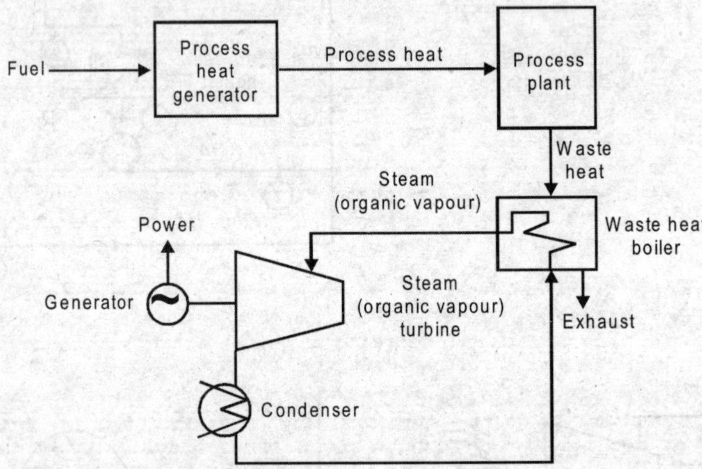

Fig. 14.14. *Cogeneration plant using a steam or organic bottoming cycle.*

The back pressure steam-turbine plant (a) is most suitable only when electricity demand is low compared with the heat demand. The combined cycle plant (d) above is most suitable only when the electric demand is high, about comparable to heat demand or higher. The gas turbine cycle (c) lies in between. Only the extraction condensing plant (b) above is suitable over a wide range of ratios.

14.14 COGENERATION WITH A GAS TURBINE PLANT

Gas turbines are used by themselves in wide range of services. Some of them are:

1. As a powering unit for all types of air-crafts.
2. For driving mechanical equipment such as pumps, compressors in industrial plants.
3. As a powering unit for railways and racing cars.
4. To drive electric generators in electric power plants — To produce power for peak loads.

There is also growing interest in using gas turbines for cogeneration and in combined cycle plants. The working temperatures of Gas Turbine are quite high (Max. temperature about 1200 – 1300°C) and hot exhaust gases leaving the turbine at about 500°C, still have a great deal of heat energy stored in them. This heat energy can be used for various purposes such as,

1. Drying and direct heating—In food industry.
2. Heating of fluids in process industries—to heat air, oil, water which is a working medium in a process plant.
3. Producing steam by using a waste recovery boiler—The steam produced may be used for process application or for generation of electricity by using steam turbines. This is known as "cogeneration" for the former and "combined cycle" for the latter.

Figure 14.15 shows a schematic flow diagram of a gas turbine plant used for cogeneration, with a heat recovery boiler.

A simple gas turbine cycle consisting of an air compressor (AC), combustion chamber (CC) and gas turbine (GT) is used. The turbine exhaust gases are passed through a heat recovery boiler (HRB) to generate process steam

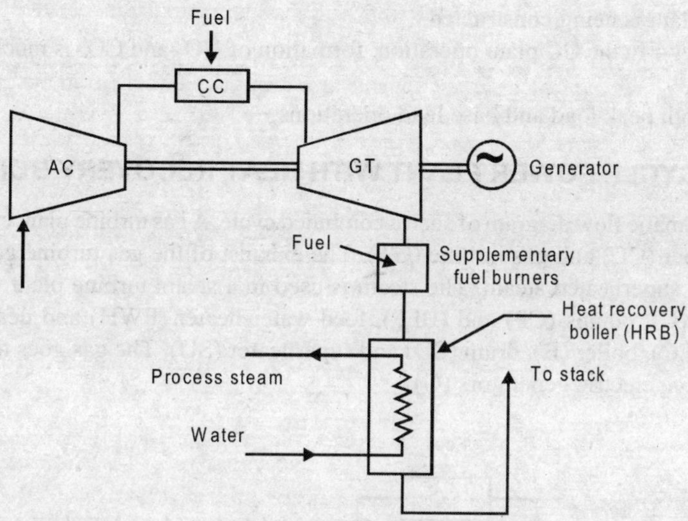

Fig. 14.15. *Cogeneration with gas turbine plant with heat recovery boiler (HRB).*

which is then taken out for process plant. Supplementary fuel burners are provided to burn additional fuel to increase the mass flow rate of process steam for a short period, if required.

14.15 COMBINED CYCLE POWER PLANTS

In *combined cycle* power plants both gas and steam turbines supply electrical power to the grid. The idea of combined cycle has grown out of the need to improve the simple Brayton-cycle efficiency by utilising the waste heat in the turbine exhaust gases. This can be done by regeneration also. The regeneration however has some serious drawbacks as given below.

1. Regeneration does not increase output. In fact regeneration reduces turbine pressure ratio and hence the net plant output by a few percent.
2. Because of large heat transfer surfaces and large gas and air piping; regeneration makes the plant costly.
3. With regeneration the optimum pressure ratio for maximum efficiency reduces sharply (from 20 – 30 to 7 – 10). This reduces the power output.

Thus raising the efficiency of a gas turbine plant by regeneration is costly. A means was saught whereby both efficiency and power are increased. A solution was found in using the large quantity of energy in the turbine exhaust, to generate steam for a steam turbine plant. This is a natural solution. Gas turbine is a high temperature machine (1100°C – 1625°C) while steam turbine is a low temperature machine (550 – 650°C). The joint operation with gas turbine at "hot end" and a steam turbine at "cold end" is called a *combined-cycle power plant.*

14.15.1 Advantages of Combined Cycle Power Plant

1. High overall plant efficiency—over 50% can be attained.
2. High total power–combined power of GT and ST is available.
3. Quick starting of plant—GT can be started quickly.
4. Low water requirement—GT needs no cooling. Only water required is for the condenser circulation.

Sometimes condensers also are air cooled. (e.g., Uran combined-cycle plant). Then water requirement is very low. Considering huge quantities of water required for condensers, this is a distinct advantage.

5. Phased installation—GT plant which takes less time to be installed, is installed first and stars producing power, while ST plant is being constructed.
6. Low pollution level—In the CC plant operation, formation of NO_x and CO_2 is much lower than that in coal fired plant.
7. Can be used for both peak load and base load operations.

14.16 COMBINED CYCLE POWER PLANT WITH HEAT RECOVERY BOILER

Figure 14.16 shows a schematic flow diagram of such a combined cycle. A gas turbine plant consists of air compressor (AC), combustion chamber (CC) and gas turbine (GT). The exhaust of the gas turbine goes to the heat recovery boiler (HRB) to produce superheated steam. The steam is used in a steam turbine plant which consists of steam turbine (ST), condenser (C), pumps (CP) and (BFP), feed water heater (FWH) and deareator (DA). The HRB consists of economizer (EC), boiler (B), drum (SD) and superheater (SU). The gas goes to stack after HRB. Both steam and gas turbine drive electric generators (G).

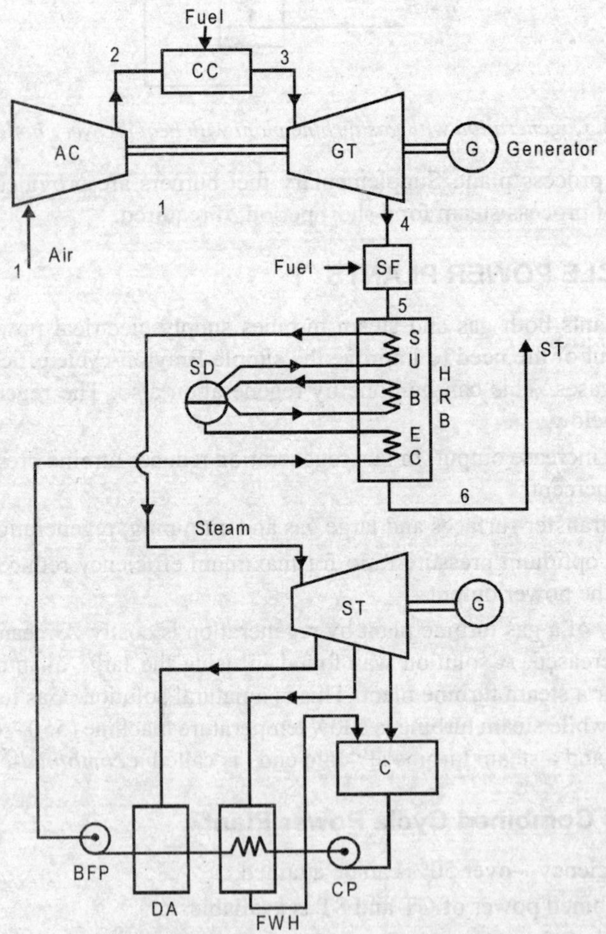

Fig. 14.16. *Schematic flow diagram of a combined cycle with heat-recovery boiler (HRB).*

The gas turbine is operated with a high air to fuel ratio (400 : 1) to ensure sufficient air in the gas turbine exhaust for further combustion. Plant output can be increased for short periods during load peaks by burning, additional full in supplementary fuel burners. HRB has no refractory lining nor water walls. Hence, its temperature is limited to 750°C.

In small plants steam turbine output is less than the gas turbine output by as much as 500 percent. In large plants, steam turbine output is greater than the gas turbine output by upto 8 : 1.

The supplementary firing equipment (SF) is interposed between the GT and HRB. The steam cycle is designed for high efficiency with reheat and full set of feed water heaters.

14.17 COMBINED CYCLE PLANT WITH MULTIPRESSURE STEAM

This plant allows the gas from MR to leave at a lower temperature and increases the efficiency of the plant.

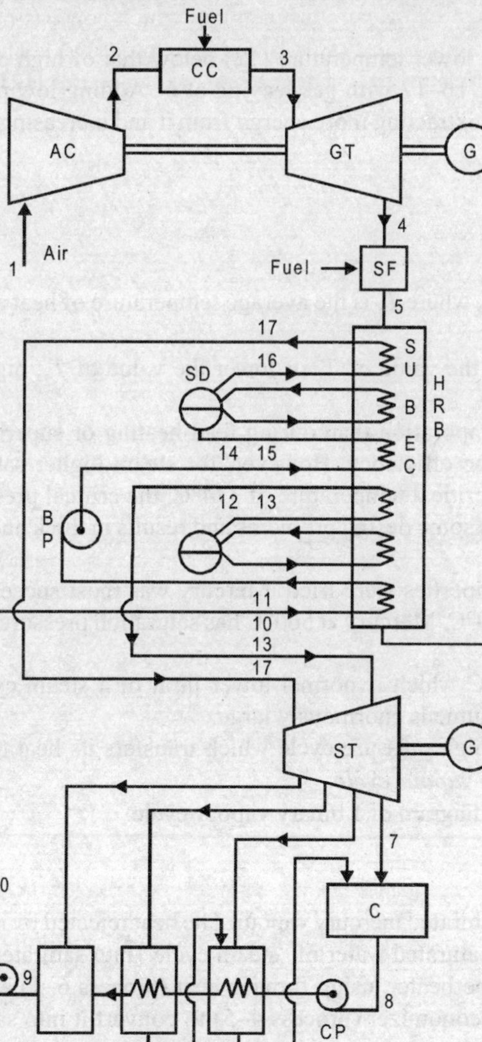

Fig. 14.17 (a) *A schematic diagram for a dual-pressure combined cycle.*

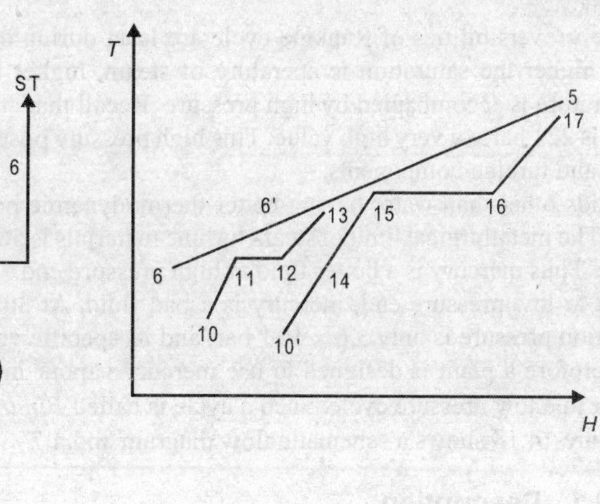

Fig. 14.17 (b) *Temperature-enthalpy (T-H) diagram of the heat recovery boiler of the dual-pressure combined cycle.*

Figure 14.17(a) shows a schematic flow diagram of the plant. The "dual-pressure cycle" has a HRB which has two steam circuits. The high pressure circuit supplies steam to the high pressure stages of the turbine and low pressure circuit supplies steam to the low pressure stages of the same turbine.

Figure 14.17(b) shows the temperature-enthalpy diagram of both gas and steam circuits in the HRB.

Exhaust gas leaving the gas turbine enters the supplementary firing (SF) at 4 and the HRB at 5, leaving it to the stack at 6. Condensate leaving the condenser at 8 enters the low pressure economizer at 10 via condensate pump, two feedwater heaters, one DA and boiler feed pump. Process 10 – 11 is feed heating of low pressure circuit; followed by evaporation to 12 and superheat to 13. Superheated steam at 13 enters the steam turbine at low pressure stage.

Water from 11 is pumped by booster pump (BP) to 14 and goes to the high pressure economizer. Evaporation occurs from 15 to 16 and superheat to 17. High pressure superheated steam at 17 enters the steam turbine first stage.

In *T-H* diagram we see that low pressure steam boils at a lower temperature (12) below that of high pressure steam (15). Single high pressure circuit is shown by 10'–15–16–17 with gas leaving at 6'. Adding low pressure circuit allows the gas to leave at a lower temperature (6), thus extracting more energy from it and increasing overall cycle efficiency.

14.18 BINARY VAPOUR CYCLE

The efficiency of a reversible cycle is given by $\eta_{rev.} = 1 - \frac{T_1}{T_2}$, where T_1 is the average temperature of heat addition and T_2 is the average temperature of heat rejection. Higher the value of T_1 or lower the value of T_2, higher the efficiency.

The irreversibilities of Rankine cycle are least during evaporation than during feed heating or superheating. Thus, higher the saturation temperature of steam, higher the efficiency. However, for steam higher saturation temperature is accompanied by high pressure. Recall that at critical temperature of 374°C, the critical pressure of steam is 221 bar—a very high value. This high pressure poses some design problems and results in thick and heavy boiler and turbine components.

Fluids other than water having better thermodynamic properties were tried. Mercury was most successful of them. The metallurgical limit of steam turbine materials is 560°C. Mercury at 560°C has saturation pressure of only 12 bar. Thus mercury is a better fluid at high pressure end.

But at low pressure end, mercury is a bad fluid. At 30°C which is normal lower limit of a steam cycle, its saturation pressure is only 3.6×10^{-6} bar; and its specific volume is enormously large.

Therefore a plant is designed to use mercury vapour in high pressure cycle which transfers its heat to water vapour in a low pressure cycle. Such a cycle is called *Binary vapour cycle*.

Figure 14.18 shows a schematic flow diagram and a *T–S* diagram of a binary vapour cycle.

14.18.1 Description

The "Mercury cycle" abcde is a simple Rankine cycle using saturated mercury vapour. The heat rejected by mercury during condensation (process *b–c*) is transferred to convert saturated water of "steam cycle" into saturated steam (process 5–6). The saturated steam is superheated in the superheater, using furnace heat (process 6–1) and then admitted to steam turbine. The condensate is heated in the economizer (process 4–5) to convert it into saturated liquid which is then admitted to steam boiler—mercury condenser, where it absorbs its latent heat.

Let $\dot{m}$ represent the flow rate of mercury in the mercury cycle per kg of steam circulating in the steam cycle. Then for 1 kg of steam,

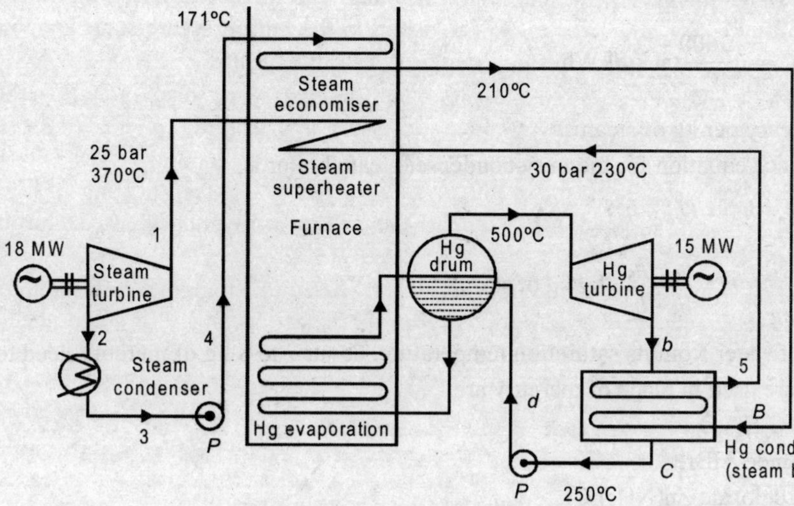

Fig. 14.18(a) *Binary vapour cycle.*

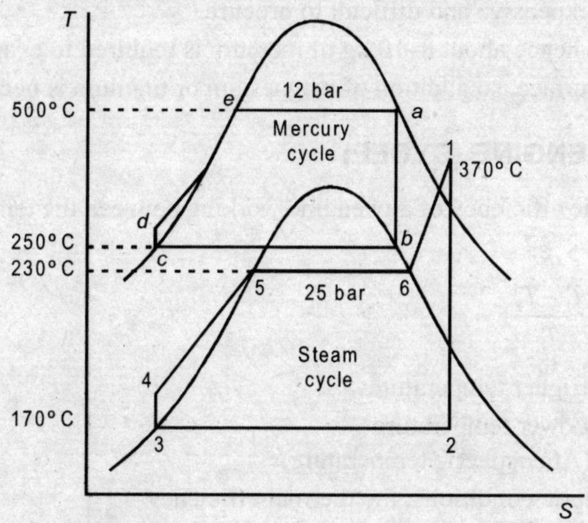

Fig. 14.18(b) *T–S diagram of binary vapour cycle.*

1. Heat supplied, $q_s = \dot{m}(h_a - h_d) + (h_1 - h_6) + (h_5 - h_4)$ kJ/kg
2. Heat rejected, $q_r = (h_2 - h_3)$ kJ/kg
3. Turbine work, $W_T = \dot{m}(h_a - h_b) + (h_1 - h_2)$ kJ/kg
4. Pump work, $W_P = \dot{m}(h_d - h_c) + (h_4 - h_3)$
5. Cycle efficiency,

$$\eta_{\text{cycle}} = \frac{q_s - q_r}{q_s} = \frac{W_T - W_P}{q_s} = \frac{W_{\text{net}}}{q_s} \quad \text{...(14.14)}$$

6. Specific steam consumption,

$$\text{S.S.C.} = \frac{3600}{W_{net}} = \text{kg/kWh}$$

7. Quantity of mercury per kg of steam.

The energy balance equation for mercury condenser steam boiler is,

$$\dot{m}(h_b - h_c) = (h_6 - h_5)$$

$$\dot{m} = \frac{h_6 - h_5}{h_b - h_c} \text{ kg/kg of steam}$$

To vapourise 1 kg of water from its saturation temperature, about 7 to 8 kg of mercury need to be condensed. Other fluids which are used in place of mercury are:

1. Diphenyl ether $(C_6H_5)_2O$
2. Aluminium bromide $AlBr_3$.
3. Zinc ammoniun chloride $Zn(NH_3)_2Cl$.
4. Liquid metals like sodium, potassium.

However, mercury vapour cycles are not very widely used because:

1. Mercury is toxic, highly expensive and difficult to procure.
2. It has low latent heat and hence about 8–10 kg of mercury is required to evaporate 1 kg of water.
3. It does not wet the tube surface, so addition of magnesium or titanium is necessary.

14.19 BINARY VAPOUR ENGINE (CYCLE)

We know that maximum possible efficiency of any engine working between the temperature limits of T_2 and T_1 is given by the equation

$$\eta = \frac{T_1 - T_2}{T_1}$$

where T_1 = Higher temperature

T_2 = Lower temperature (Atmospheric temperature)

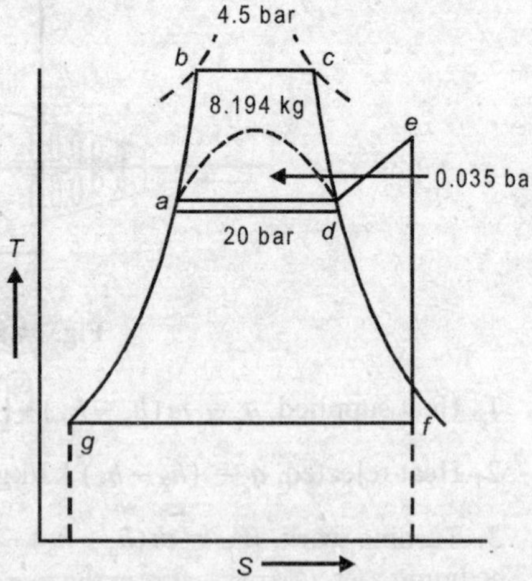

Fig. 21.19

Since T_2 is fixed by atmospheric conditions, the thermal efficiency of a plant can only be improved by increasing T_1. In a steam plant an increase of T_1 may entail the use of a corresponding higher pressure, which is one of the limiting factors in its design. By using mercury vapour in place of steam in the high temperature range of the cycle, an increase in T_1 is obtained without any increase in maximum pressure. The heat rejected by mercury in condensing can be utilized in raising superheated steam for the low temperature range of the cycle. A power plant using two vapours in this way is known as binary vapour plant.

On the T–S diagram; are shown the liquid and saturated steam lines for water and steam, meeting at the critical temperature which occurs at a steam pressure of 221 bar. The lines *ab* and *bc* represent the heating and evaporation of liquid mercury plotted to the same temperature scale as the steam, the scale of corresponding pressures is lower for mercury. The mercury vapour at *C* has a much higher temperature than steam at the same pressure. At *C* the mercury expanded adiabatically through a separate mercury vapour turbine to *d*; from *d* it is condensed to *a*, its latent heat being utilized for evaporating a corresponding amount of steam. The mercury has thus described the cycle *abcd*.

The steam cycle is represented by *gadef*, *ga* represents the heating of the feed water, *ad* its evaporation by the condensing mercury, and *de* the superheating of the steam by the flue gases. At *e* the steam is expanded adiabatically through a steam turbine to *f*; *fg* represents the condensing of the exhaust steam in the condenser. This completes the steam cycle.

To obtain the correct amount of heat from the condensing mercury for evaporating the steam, it is found that 8.196 (8.2 approx.) kg of mercury are required per kg of steam. It will be seen from the areas of *T–S* diagram that the use of mercury for the high temperature range of the cycle gives a higher efficiency than that obtainable from steam with the same addition of heat. Using pressure limits of 24 bar and. 0.07 bar, a steam plant gives an efficiency of 24–25% whilst that of the binary vapour plant is 50%.

A diagrametric view of a binary vapour plant is shown in Fig. 14.20. The hot gases from the furnace pass through the mercury boiler, through the liquid mercury heater, then through the steam superheater and finally through the feed water economiser, after which they are exhausted to the chimney.

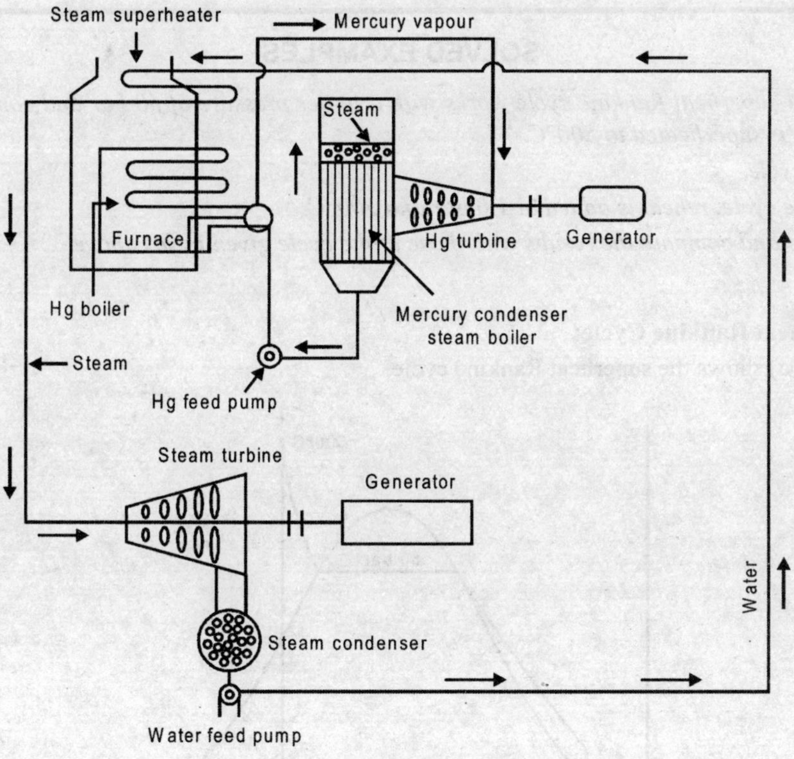

Fig. 14.20

The liquid mercury passes from the mercury liquid heater to the mercury vapour boiler where it is evaporated. It then flows to the mercury turbine through which it is expanded to its low pressure limit. From here it exhausts to the mercury condenser steam boiler where its latent heat is given upto the hot feed water; this operation condenses the mercury whilst at the same time the feed water is evaporated to steam. The mercury is then returned to the mercury liquid boiler, thus completing its cycle.

Referring to Fig. 14.20, the feed water from the economiser (not shown) is evaporated to steam in the mercury condenser steam boiler and then passes to the superheater where it is superheated by the hot flue gases. From the superheater the superheated steam passes to the steam turbine in which it is expanded down to the steam condenser pressure, it is then condensed in the steam condenser and passes back to the economiser, thus completing the cycle of the water and steam.

At the Hartford-USA–binary vapour plant, erected by the General Electric Co., the weight of mercury in use in the plant is 60 tonnes. A higher efficiency is obtained than is possible from a steam plant of the same output. Daily saving of 150 T of coal compared with the consumption of an equivalent steam plant. Five binary vapour plants have been erected by the General Electric Co.

In the early stages of the development of the mercury boiler, it was found that the liquid mercury did not wet the heating surface of the tubes. There was a tendency for a film of mercury vapour to form at the tube surface which prevented a good heat transfer to the liquid. The difficulty was overcome by a 0.002 percent solution of magensium and titanium being used in the boiler as a surface wetting agent.

The chief objection to the mercury vapour plant is the danger due to poisonous fumes if any leakage of mercury vapour occurs, and special pecautions are taken to prevent this.

SOLVED EXAMPLES

Example 14.1 *(a) A superheat Rankine cycle works with a boiler pressure of 50 bar and condenser pressure of 0.08 bar. The steam is superheated to 500°C.*

(a) Analyse the cycle.

(b) For the above cycle, reheat is added at 1.0 bar and 500°C.

Analyse the cycle and compare the results with those of the cycle given in (a) above.

Solution

Case (a) Superheat Rankine Cycle:

Figure Ex. 14.1 (a) shows the superheat Rankine cycle.

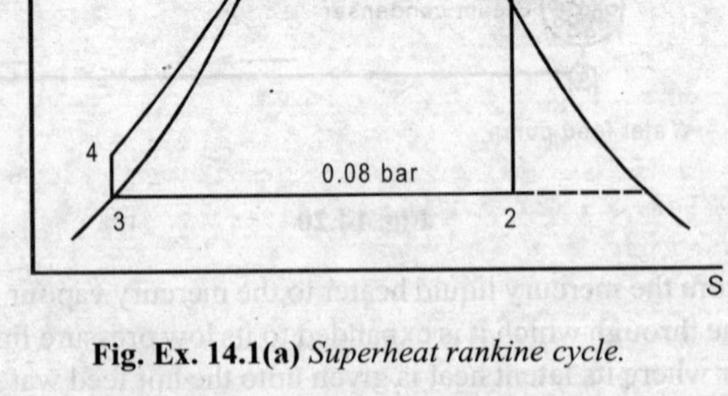

Fig. Ex. 14.1(a) *Superheat rankine cycle.*

Our first aim is to determine enthalpies at various salient points of the cycle. From the steam tables we have,

(i) at 50 bar 500°C

$$h_1 = 3433.7 \text{ kJ/kg}$$
$$s_1 = 6.977 \text{ kJ/kg-K}$$

(ii) at 0.08 bar

$$s_{f3} = 0.5925 \text{ kJ/kg-K}$$

$$s_{fg2} = 7.6371 \text{ kJ/kg-K}$$

$$h_{f3} = 173.86 \text{ kJ/kg}$$

$$h_{fg2} = 2403.2 \text{ kJ/kg}$$

1. To find X_2 equate entropies at point 1 and point 2.

$$6.977 = 0.5925 + x_2 (7.6371)$$
$$\therefore \quad x_2 = 0.386$$
$$\therefore \quad h_2 = 173.86 + 0.836 \times 2403.2 = 2182.93 \text{ kJ/kg}$$

2. Turbine work = $W_T = h_1 - h_2 = 3433.7 - 2182.93 = 1250.76$ kJ/kg
3. Pump work = $W_P = 0.001 [50 - 0.08] \times 100 = 5$ kJ/kg.
4. Net work = $W_{net} = W_T - W_P = 1250.76 - 5.0 = 1245.76$ kJ/kg
5. Heat supplied = $q_s = h_1 - h_4 = h_1 - (h_3 - W_P)$

$$= 3433.7 - 178.76 = 3254.84 \text{ kJ/kg}$$

6. Rankine efficiency = $h_r = \dfrac{W_{net}}{q_s} = \dfrac{1250.76}{3254.84} = 0.3843$ i.e., 38.43%

7. Work ratio = $\dfrac{W_{net}}{W_T} = \dfrac{1245.76}{1250.76} = 0.996$

8. Specific steam consumption = $\dfrac{3600}{1250.76} = 2.878$ kg/kWh

Case (b) Superheat Reheat Cycle: Fig. Ex. 14.1 (b) shows the cycle.

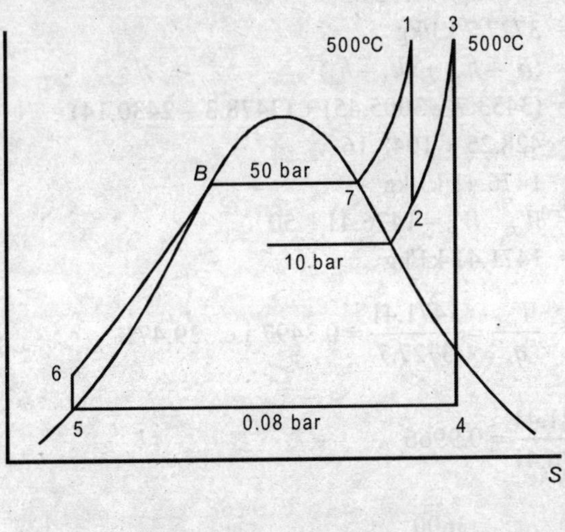

Fig. Ex. 14.1(b). *Superheat reheat rankine cycle.*

The steam expands from 50 bar to 10 bar before reheating. s_{g2} at 10 bar is 6.5828. Since $s_1 > s_{g2}$ the condition of steam at point 2 is superheated. To find T_{sup2} we equate entropies at point 1 and point 2.

or
$$s_1 = s_{g2} + C_p \ln \frac{T_{sup2}}{T_{sat2}}$$

$$\therefore \quad 6.997 = 6.5825 + 2.1 \ln \frac{T_{sup2}}{273 + 179.88}$$

$$\therefore \quad T_{sup2} = 551.62 \text{ K} = 278.62 \text{ °C}$$

∴ At 10 bar and 278.62°C, we have from steam tables (by interpolation)

$$h_2 = 3005.45 \text{ kJ/kg}$$

At 10 bar and 500 °C, we have from steam tables,

$$h_3 = 3478.3 \text{ kJ/kg and } s_3 = 7.7627$$

To find x_4 equate entropies at 3 and 4.

$$\therefore \quad s_3 = s_4$$

$$\therefore \quad 7.7627 = 0.5925 + x_4\, 7.6371$$

$$x_4 = 0.939$$

$$\therefore \quad h_4 = 173.86 + 0.939 \times 2403.2 = 2430.14 \text{ kJ/kg}$$

Let us analyse the cycle.

1. Heat supplied $= q_{s1} = h_1 - h_6 = 3433.7 - (h_5 + W_P)$ $\quad [W_P = 5 \text{ kJ/kg}]$

$$= 3433.7 - 178.86$$
$$= 3254.84 \text{ kJ/kg}$$

also heat supplied under reheat

$$= q_{s2} = h_3 - h_2$$
$$= (3478.3 - 3005.45)$$
$$= 472.85 \text{ kJ/kg}$$

∴ Total heat supplied $= q_s = q_{s1} + q_{s2}$

$$= 3254.84 + 472.85$$
$$= 3727.7 \text{ kJ/kg}$$

2. Turbine work $\quad W_T = (h_1 - h_2) + (h_3 - h_4)$

$$= (3433.7 - 3005.45) + (3478.3 - 2430.14)$$
$$= 428.25 + 1048.16$$
$$= 1476.41 \text{ kJ/kg}$$

3. Net work = $\quad W_{net} = W_T - W_P = 1476.41 - 50$

$$= 1471.41 \text{ kJ/kg}$$

4. Cycle efficiency $\quad h_r = \frac{W_{net}}{q_s} = \frac{1471.41}{3727.7} = 0.3497$ i.e., 39.47%

5. Work ratio $= \frac{W_{net}}{W_T} = \frac{1471.41}{1476.41} = 0.9966$

6. Specific steam consumption $= \frac{3600}{1471.41} = 2.447$ kg/kWh

Comments: Superheat reheat cycle shows following improvements,

1. Final dryness is improved from 0.836 to 0.939.
2. Rankine efficiency improves from 38.43% to 39.47% increased by 2.7%.
3. Specific steam consumption improves from 2.878 kg/kWh to 2.447 kg/kWh; reduction by 15%.

Example14.2 (*a*) *What is a combined cycle power plant? State the advantages.*

(*b*) *In a Rankine cycle, steam is supplied to a turbine at 50 bar. 400°C and is expanded to a pressure of 0.2 bar. Find the thermal efficiency of the plant if the reheat pressure is 7 bar and reheated steam temperature 400°C. Assume isentropic efficiency of the turbine as 80% throughout.*

Solution

The cycle is shown on *T–S* diagram.

Given:

$$P_1 = 50 \text{ bar}$$

$$T_1 = 673 \text{ K} = T_3$$

$$P_3 = 7 \text{ bar}$$

$$P_4 = 0.2 \text{ bar}$$

$$\eta_t = 80\ \%$$

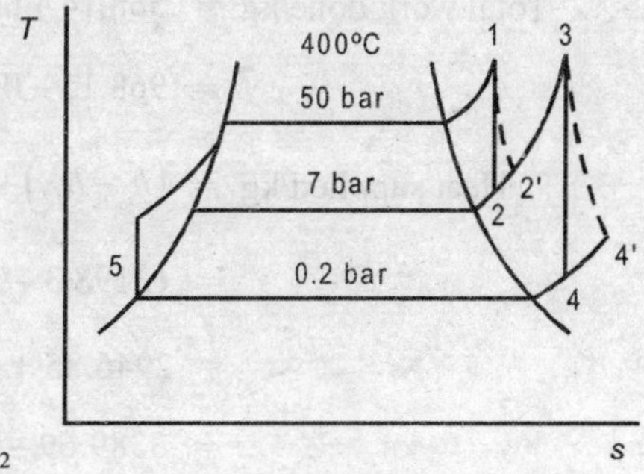

Fig. Ex. 14.2

From Steam tables,

$$h_1 = 3198.3 \text{ kJ/kg};\ h_3 = 3269 \text{ kJ/kg}$$

$$h_{f_5} = 251.45 \text{ kJ/kg}$$

$$s_1 = 6.6508;\quad s_3 = 7.6362$$

$$s_2 = 1.9918 + x_2 \times 4.7134 \qquad \because s_1 = s_2$$

$$x_2 = \frac{6.6508 - 1.9918}{4.7134} = \frac{4.6590}{4.7134}$$

$$= 0.9885$$

$$h_2 = 697.06 + 0.9885 \times 2064.9$$

$$= 697.06 + 2041.07$$

$$= 2738.13 \quad h_2 = 2738.13$$

From *T–S* diagram,

$$\eta_t = \frac{h_1 - h_{2'}}{h_1 - h_2} \qquad \therefore 0.8 = \frac{3198.3 - h_{2'}}{3198.3 - 2738.13}$$

$$\therefore \quad 0.8 \times 460.17 = \mathbf{368.14\ kJ/kg}$$

$$= \text{work of turbine I}$$

$$\therefore \quad h_{2'} = 3198.3 - 368.14 = 2830.16 \text{ kJ/kg}$$

Similarly, for 2nd stage turbine (after reheating)

$$s_3 = s_4 \qquad \therefore 7.6362 = 0.8321 + x_4 \times 7.0773$$

$$\therefore \quad x_4 = \frac{7.6362 - 0.8321}{7.0773} = \frac{6.8041}{7.0773}$$

$$= \mathbf{0.9614}$$

$$\therefore \quad h_4 = 251.45 + 0.9614 \times 2358.4$$

$$= 251.45 + 2267.36$$

$$= \mathbf{2518.81\ kJ/kg}$$

$$\therefore \quad 0.8 = \frac{h_3 - h_{4'}}{h_3 - h_4} = \frac{3269 - h_{4'}}{3269 - 2518.81}$$

$$\therefore \quad h_{4'} = 3269 - 0.8 \times 750.19$$

$$= 3269 - 600.15$$

$$= \mathbf{2668.85\ kJ/kg}$$

$$W_{t_2} = 600.15\ \text{kJ/kg}$$

$$\therefore \quad \text{Total work done/kg} = 368.14 + 600.15 - W_p$$

$$= 968.15 - W_p$$

$$\text{Heat supplied/kg} = \left(h_1 - h_{f_5}\right) + \left(h_3 - h_{2'}\right) - W_p$$

$$= (3198.3 - 251.45) + (3269 - 2830.16) - W_p$$

$$= 2946.85 + 439.84 - W_p$$

$$= 3389.69 - W_p$$

$$\therefore \quad W_p = \frac{p_1 - p_4}{10} = \frac{50 - 0.2}{10} = \frac{49.8}{10} = \mathbf{4.98\ kJ/kg}$$

$$W = (968.29 - 4.98) = \mathbf{963.31\ kJ/kg}$$

$$Q = 3386.69 - 4.98 = \mathbf{3381.71\ kJ/kg}$$

$$\therefore \quad \text{Thermal efficiency} = \frac{W}{Q} = \frac{963.31}{3381.71}$$

$$= \mathbf{0.285} \text{ or } \mathbf{28.5\%}$$

Example 14.3 *In a two stage steam turbine working on ideal Rankine cycle, steam enters the first stage at 90 bar and 550°C temperature. Second stage entry temperature after reheating is 550°C. Steam is never wet during expansion in both sides. Find the cycle efficiency and specific steam consumption. Also find the steam flow rate in the turbine in kg/sec if the turbine is directly connected to an alternator of 150 MW output. Assume 0.9 as alternate efficiency. Draw T-S diagram for the cycle. Neglect pump work.*

Solution

Assume dry and saturated steam after expansion.

$$h_1 = 3518.95$$

$$s_1 = 6.8087$$
$$P_2 = 5.5 \text{ bar}$$
$$h_2 = 2750$$
$$h_3 = 3600$$
$$h_4 = 2590$$
$$p_1 = 90 \text{ bar}, \ p_2 = 5.5 \text{ bar} = p_3; \ p_4 = 0.1 \text{ bar.}$$

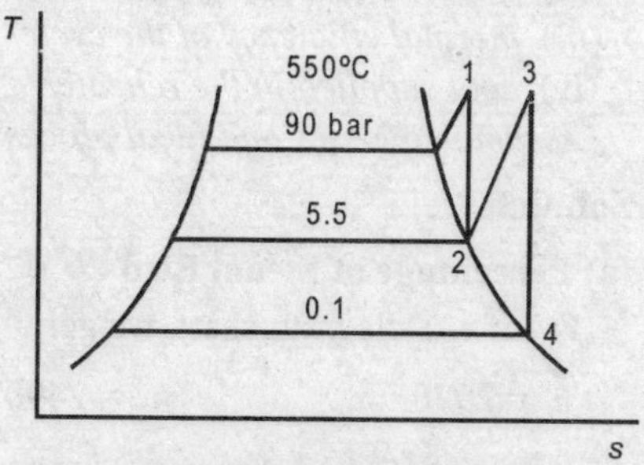

Fig. Ex. 21.3

$$\text{Work done/kg} = (h_1 - h_2) + (h_3 - h_4)$$
$$= (3518.95 - 2750) + (3600 - 2590)$$
$$= 768.95 + 1010$$
$$= \mathbf{1778.95 \text{ kJ/kg}}$$

$$\text{Heat supplied/kg} = h_1 - h_{f4}) + (h_3 - h_2)$$
$$= (3518.95 - 191.83) + (3600 - 2750)$$
$$= 3397.12 + 850$$
$$= \mathbf{4247.12 \text{ kJ/kg}}$$

$$\therefore \quad \text{Cycle efficiency} = \frac{W}{Q_A} = \frac{1778.95}{4247.12}$$
$$= 0.4189$$
$$= \mathbf{41.89\%}$$

$$\text{Specific steam consumption} = \frac{3600}{\text{work/kg}}$$
$$= \frac{3600}{1778.95}$$
$$= \mathbf{2.027 \text{ kg/kWh}}$$

$$\text{Alternator output} = 150 \text{ MW}$$

$$\therefore \quad \text{Alternator input} = \frac{150}{0.9} = 166.67 \text{ MW} = \mathbf{166670 \text{ kW(kJ/s)}}$$

$$\text{Steam flow rate} = \frac{166670}{1778.95} = \mathbf{93.69 \text{ kg/s}}$$

Example 14.4 (*a*) *How can a combined cycle plant be used for co-generation ? What is its thermodynamic advantage?*

(*b*) *Why is the efficiency of a condensing steam power plant is less in a warm region than in a cold region?*

(*c*) *A steam power plant of 120 MW capacity is equipped with regenerative as well as reheat arrangement. The steam is supplied to the turbine at 80 bar and 50°C superheat. Steam is extracted at 7 bar for feed heating and the remaining steam is reheated to its initial temperature before it is expanded to 0.36 bar in the L.P. stage. Assuming indirect contact type of feed water heater, determine*

(*i*) *percentage of steam bled*

(ii) generating capacity of the boiler
(iii) thermal efficiency of the cycle
(iv) heat supplied in the reheater.
Assume no losses and ideal processes of expansion.

Solution

(a) Percentage of Steam Bled

From *h–S* diagram/chart, we get

$h_1 = 2970$ $\quad h_2 = 2490$

$h_3 = 2880$ $\quad h_4 = 2380$

$h_{f_2} = 697.06$ kJ/kg $\quad h_{f_4} = 307.12$ kJ/kg

Let m_2 be the weight of steam bled for feed heating

$$\therefore \quad m_2\left(h_2 - h_{f_4}\right) = 1\left(h_2 - h_{f_4}\right)$$

$$\therefore \quad m_2(2490 - 307.12) = (697.06 - 307.12)$$

$$= 389.94$$

$$m_2 = \frac{389.94}{2182.88}$$

$$= 0.1786 \text{ kg/kg supplied}$$

$$\therefore \quad \% \text{ of steam bled} = \mathbf{17.86\ \%}$$

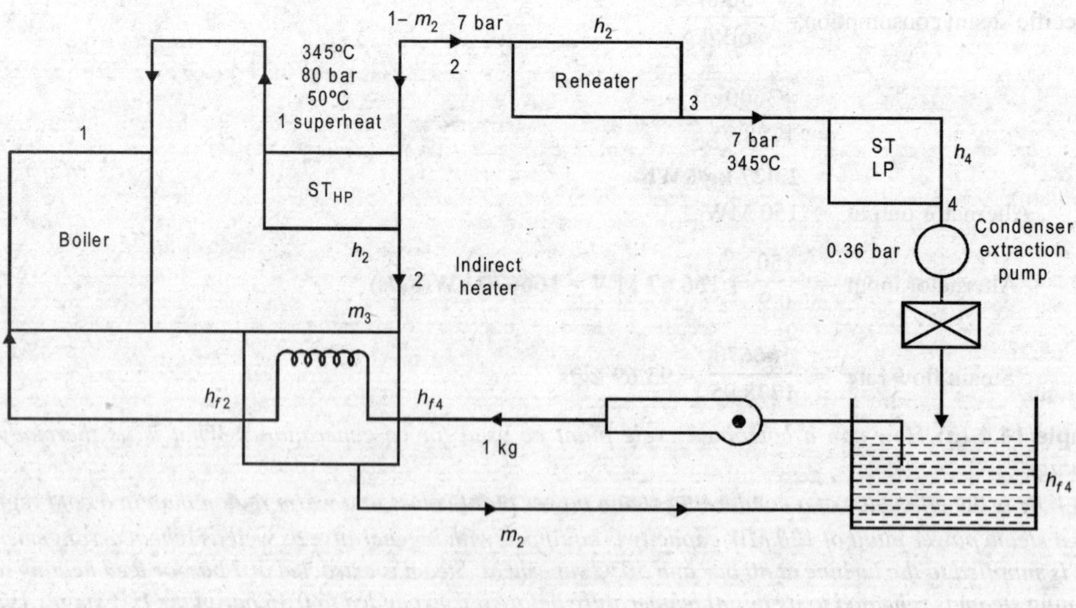

Fig. Ex.14.4 (a)

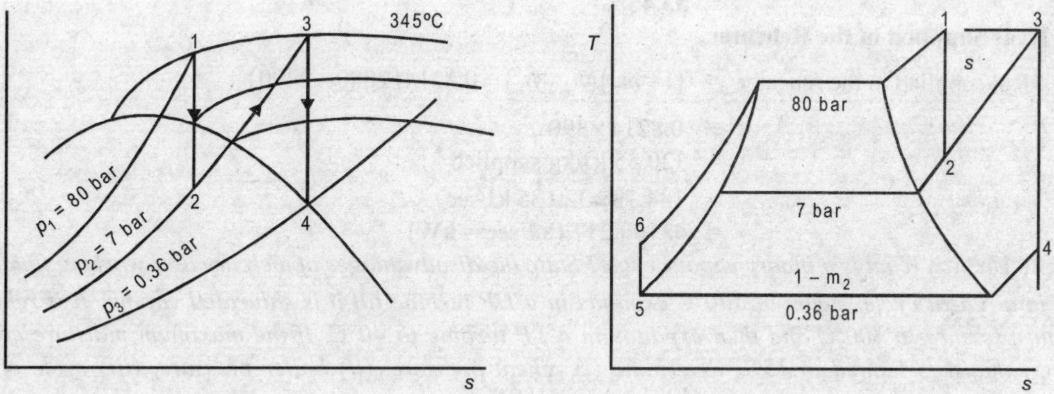

Fig. Ex. 14.4 (b)

(b) Generating Capacity of Boiler

Weight of steam expanded in HP turbine = 1 kg

$\therefore$ Work done in HP turbine $= 1(h_1 - h_2) = 2970 - 2490 =$ **480 kJ/kg**

Weight of steam expanded in LP turbine = 1 – 0.1786 = **0.8214 kg**

$\therefore$ Work done in LP turbine $= 0.8214(h_3 - h_4)$

$= 0.8214(2880 - 2380) = 0.8214 \times 500$

= 410.70 kJ/kg supplied

$\therefore$ Total work done/kg steam supplied

$= 480 + 410.7 =$ **890.7 kJ/kg**

Capacity of the plant = 120 MW = **120,000 kW**

$\therefore$ Steam consumption/sec

$$= \frac{120000}{890.7} = 134.73 \text{ kg/sec}$$

= **485012 kg/hour**

= Generating capacity of boiler

(c) Thermal Efficiency of Cycle

Heat supplied in the cycle/kg $= (h_1 - h_{f_2}) + (1 - m_2)(h_3 - h_2)$

$= (2970 - 697.06) + 0.8214(2880 - 2490)$

$= 2272.94 + 0.8214 \times 390$ kJ/kg steam supplied

$= 2272.94 + 320.35 = 2593.29$

Thermal efficiency of the cycle

$$\frac{\text{Work done/kg}}{\text{Heat supplied/kg}} = \frac{890.7}{2272.04 + 390}$$

$$= \mathbf{0.3345}$$
$$= \mathbf{33.45\%}$$

(d) Heat Supplied in the Reheater

$$\text{Heat supplied in the reheater} = (1-m_2)(h_3-h_2) = 0.8214(2880-2490)$$
$$= 0.8214 \times 390$$
$$= 320.35 \text{ kJ/kg supplied}$$
$$= 134.73 \times 320.35 \text{ kJ/sec}$$
$$= \mathbf{43160.217 \text{ (kJ/sec = kW)}}$$

Example 14.5 (*a*) *What is a binary vapour cycle? State the disadvantages of mercury as a working fluid.*

(*b*) *In a reheat cycle, steam at 500°C expands in a HP turbine till it is saturated vapour. It is reheated at constant pressure to 400°C and then expands in a LP turbine to 40°C. If the maximum moisture content at turbine exhaust is limited to 15%, determine* (*i*) *reheat pressure,* (*ii*) *boiler pressure,* (*iii*) *cycle efficiency,* (*iv*) *steam rate.*

Assume all processes are ideal. What would have been the quality at turbine exhaust, the net work output and cycle efficiency without reheating of steam?

Solution

Locate point 4. Draw vertical to point 3 on 400°C temperature line. Point 3 will give the pressure of reheating. Follow this pressure line upto saturation curve. We will get point 2. From point 2, draw a vertical line to meet a temperature line of 500°C and we will get point 1 which will give the boiler pressure.

From diagram:

$h_4 = 2220$ kJ/kg, $p_3 = 20$ bar, $h_3 = 3240$

$h_2 = 2800$ kJ/kg; 0 $h_1 = 3300$ kJ/kg

$p_1 = 150$ bar $h_{f_4} = 167.45$ kJ/kg

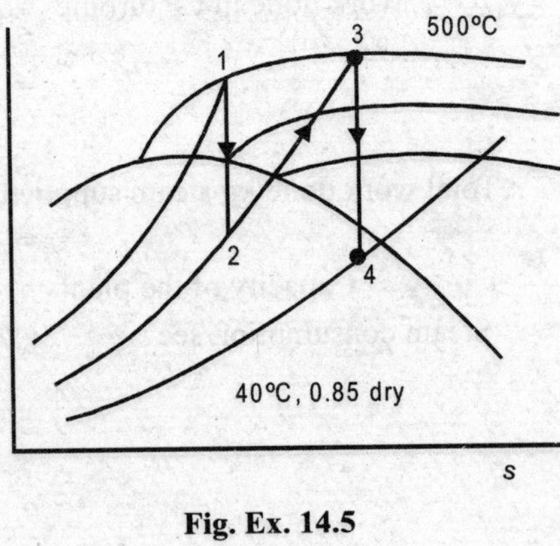

Fig. Ex. 14.5

(i) Reheat pressure

$$\text{Reheat pressure} = p_2 = \mathbf{20 \text{ bar}}$$

(ii) Boiler pressure

$$\text{Boiler pressure} = p_1 = \mathbf{150 \text{ bar}}$$

(iii) Cycle efficiency

$$\text{Cyclic work/kg} = (h_1 - h_2) + (h_3 - h_4)$$
$$= (3300 - 2800) + (3240 - 2220)$$
$$= 500 + 1020$$
$$= \mathbf{1520 \text{ kJ/kg}}$$

$$\text{Heat supplied/kg} = (h_1 - h_{f_4}) + (h_3 - h_2)$$
$$= (3300 - 167.45) + (3240 - 2800)$$
$$= 3132.55 + 440$$
$$= \mathbf{3572.55 \text{ kJ/kg}}$$

$$\text{Cycle efficiency} = \frac{\text{Work}}{Q_A} = \frac{1520}{3572.55} = 0.4255$$

$$= \mathbf{42.55\ \%}$$

(iv) Steam rate

$$\text{Steam rate} = \frac{3600}{\text{Work done/kg}} = \frac{3600}{1520}$$

$$= \mathbf{2.368\ kg/kWh}$$

Example 14.6 *The net power output of an ideal reheat regenerative steam cycle is 80 MW. Steam enters the h.p. turbine at 80 bar, 500°C and expands till it becomes saturated vapour. Some of the steam then goes to an open feed water heated and the balance is reheated to 400°C, after which it expands in an LP turbine to 0.07 bar.*

Compute: (i) the reheat pressure, (ii) the steam flow rate to h.p. turbine, (iii) the cycle efficiency, (iv) the rate of flow of cooling water in the condenser if the temperature rise of water is 8°C, (v) if the velocity of steam flowing from the turbine to the condenser is limited to 130 m/s, find the diameter of the connecting pipe.

Solution

The schematic diagram for the arrangements of the equipments (plant) is shown in Fig. Ex. 14.6.

From *h–s* diagram, we get the following ethalpies.

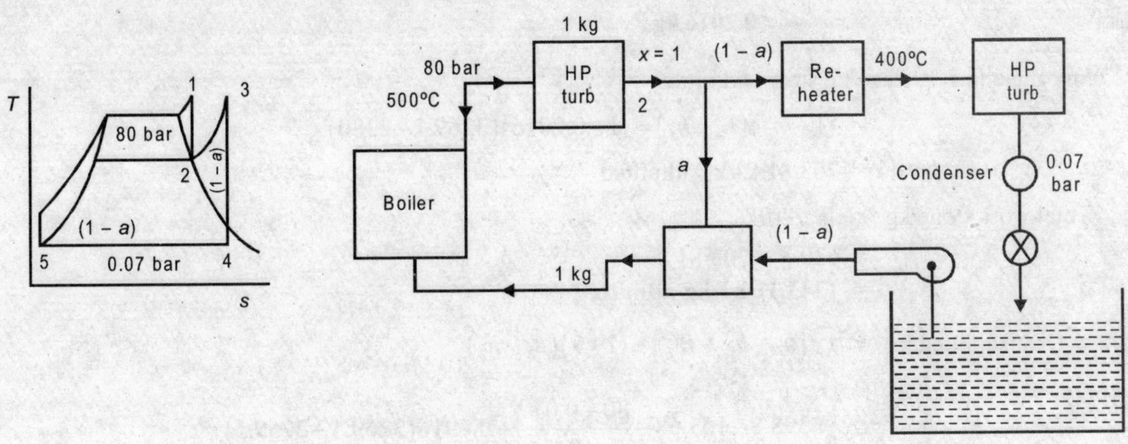

Fig. Ex. 14.6

$h_1 = 3398.8$ kJ/kg $\quad \therefore \quad p_2 = 6.6$ bar

$h_2 = 2759.5$ kJ/kg $\quad s_1 = 6.7262 = s_{g_2}$

$h_3 = 3269.1$ kJ/kg $\quad h_{f_2} = 686.78$ kJ/kg

$h_4 = 2380$ kJ/kg $\quad h_{f_4} = 163.38$ kJ/kg

$x_4 = 0.914$

Specific volume at LP turbine outlet = 0.914 × 20.531 = 18.77 m^3

(i) Reheat Pressure = 6.6 bar

(ii) Work done/kg in HP turbine

$$= W_{HP} = h_1 - h_2 = 3398.8 - 2759.5$$

$$= \mathbf{639.3\ kJ/kg}$$

∴ Steam flow rate to H.P. turbine

$$= \frac{3600}{639.3} = \mathbf{5.631\ kg/kWh}$$

(iii) Cycle efficiency : Consider the open heater

$$m_a \times (h_2 - h_{f_2}) = (1 - m_a)(h_{f_2} - h_{f_4})$$

$$m_a (2759.5 - 686.78) = (1 - m_a)(686.78 - 163.38)$$

$$m_a \times 2072.72 = (1 - m_a) 523.4$$

$$1 - m_a = m_a \times \frac{2072.72}{523.4}$$

$$= m_a$$

$$1 = 4.96\, m_a$$

or

$$m_a = \frac{1}{4.96} = \mathbf{0.2016\ kg}$$

∴ Work done in L.P. stage/kg steam admitted

$$= (1 - m_a)(h_3 - h_4) = (1 - 0.2016)(3269.1 - 2380)$$

$$0.7916 \times 889.1 = 703.81 \text{ kJ/kg admitted}$$

∴ Total work/done/kg $= W_{HP} + W_{LP}$

$$= 639.3 + 703.81$$

$$= \mathbf{1343.11\ kJ/kg\ admitted}$$

$$\text{Heat supplied/kg} = 1 \times (h_r - h_{f_5} - W_p) + (1 - 9)(h_3 - h_2)$$

$$= \left(3398.8 - 163.68 - \frac{80 - 0.07}{10}\right) + 0.7916(3269.1 - 2759.5)$$

$$= 3235.12 - 7.993 + 0.7916 \times 509.6$$

$$= 3227.127 + 403.4$$

$$= 3630.527 \text{ kJ/kg}$$

$$\therefore \quad \text{Cycle efficiency} = \eta_{cyc} = \frac{\text{Work done/kg}}{\text{Heat supplied/kg}}$$

$$= \frac{1343.11}{3630.527}$$

$$= \mathbf{0.37\ or\ 37\%}$$

(iv) Flow rate of cooling water at condenser

0.7916 kg if steam leaves the LP turbine and enters the condenser per kg of steam admitted to the HP turbine or the system.

$$\text{But, steam flow/sec to the system} = \frac{80\times100}{1343.11}$$

$$= \mathbf{59.56\ kg/sec}$$

$$\therefore \text{Steam to condenser/sec} = (1-9)\times59.56$$

$$= 0.7916\times59.56$$

$$= \mathbf{47.15\ kg/s}$$

∴ Heat gained by cooling water/sec. = Heat given by steam/sec

$$\therefore \quad M_W C_{PW}\Delta T = 47.15\times\left(h_4 - h_{f_4}\right)$$

$$\therefore \quad M_W = \text{Mass of cooling water/sec}$$

$$= \frac{47.15\times(2380-163.38)}{4.187\times8}$$

$$= \mathbf{3120\ kg/s}$$

(v) Total volume of steam coming to condenser/sec

$$= \text{Mass of steam/s} \times \text{Specific volume of steam}$$

$$= 47.15 \times 18.77\ \text{m}^3/\text{s}$$

$$= 885\ \text{m}^3/\text{s}$$

$$= \text{Velocity} \times \text{Area of pipe}$$

$$= 130\times\frac{\pi}{4}d^2$$

$$d^2 = 8.666$$

$$\mathbf{d} = \mathbf{2.944\ m}$$

Example 14.7 (*a*) *Write a short note on combined gas steam power cycle.*

(*b*) *A steam power plant is operating on the ideal reheat Rankine cycle. Steam enters the high pressure turbine at 15 MPa and 600°C and is condensed in the condenser at F pressure of 1.0 kPa. If the moisture content of the steam at the exit of the low pressure turbine is not to exceed 10.4%, determine:*

(*i*) *the pressure at which the steam should be reheated before supplying the same to the low pressure turbine, and*

(*ii*) *the thermal efficiency of the cycle.*

Assume the steam is reheated at constant pressure to a temperature of 600°C. Assume both steam turbines are mounted on same shaft.

Solution

Neglecting pump work,

$$15\ \text{MPa} = 150\ \text{bar}$$

$$10\ \text{kPa} = \frac{10\times1000}{10^5}$$

$$= \frac{1}{10} = \mathbf{0.1\ bar}$$

10.4% wetness means 89.6% dryness

$$x_4 = 0.896$$

$$= \mathbf{0.9\ Approx.}$$

$$h_1 = 3580,\ h_2 = 3160,\ h_3 = 3675,\ h_4 = 2340 \text{ kJ/kg}$$

(i) Reheat Pressure = 40 bar

(ii) Thermal Efficiency

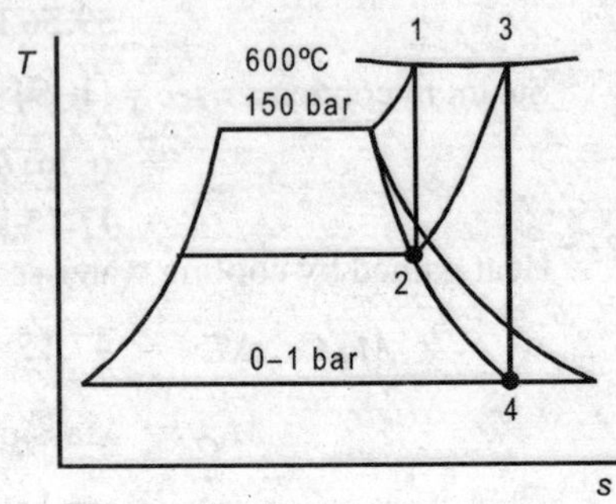

Fig. Ex. 14.7

$$\text{Work done/kg} = (h_1 - h_2) + (h_3 - h_4)$$

$$= (3580 - 3160) + (3675 - 2340)$$

$$= 420 + 1335$$

$$= \mathbf{1755\ kJ/kg}$$

$$\text{Heat added/kg} = (h_1 - h_{f_4}) + (h_3 - h_2)$$

$$= (3580 - 191.83) + (3675 - 3160)$$

$$= 3388.17 + 515$$

$$= \mathbf{3903.17\ kJ/kg}$$

$$\therefore \quad \text{Thermal efficiency} = \frac{\text{Work done/kg}}{\text{Heat added/kg}}$$

$$= \frac{1755}{3903.17}$$

$$= 0.4496$$

$$= \mathbf{or\ 0.45\ or\ 45\%}$$

Example 14.8 (*a*) *Write a short note on cogeneration.*

(*b*) *In a single feed water, regenerative cycle, the steam enters the turbine at a pressure of 30 bar and 400°C. The exhaust pressure of steam is 0.1 bar. The exhaust pressure of steam is 0.1 bar. The feed water heater is direct-contact type which operates at a pressure of 5 bar. Find the thermal efficiency of the cycle and steam rate. Show the flow diagram, this regenerative cycle on h–s and T–S diagram.* (*Neglect pump work*).

Solution

The flow diagram and *T–S, h–s,* diagrams are shown. Given data:

$p_1 = 30$ bar, $p_2 = 5$ bar $\quad p_3 = 0.1$ bar, $T_1 = 400$°C

For steam table, $\quad h_{f_2} = 640.12$ kJ, $h_{f_4} = 191.83$ kJ

From chart, $\quad h_1 = 3225$ kJ, $p_2 = 2780$ kJ, $h_3 = 2200$ kJ

Consider the heater:

Heat given by system = Heat gained by water

$$\dot{m}_2 (h_2 - h_{f_2}) = (1 - \dot{m}_2)(h_{f_2} - h_{f_3})$$

$$\dot{m}_2 (2780 - 640.12) = (1 - \dot{m}_2)(640.12 - 181.83)$$

$$\dot{m}_2 (2139.88) = (1 - \dot{m}_2)(458.29)$$

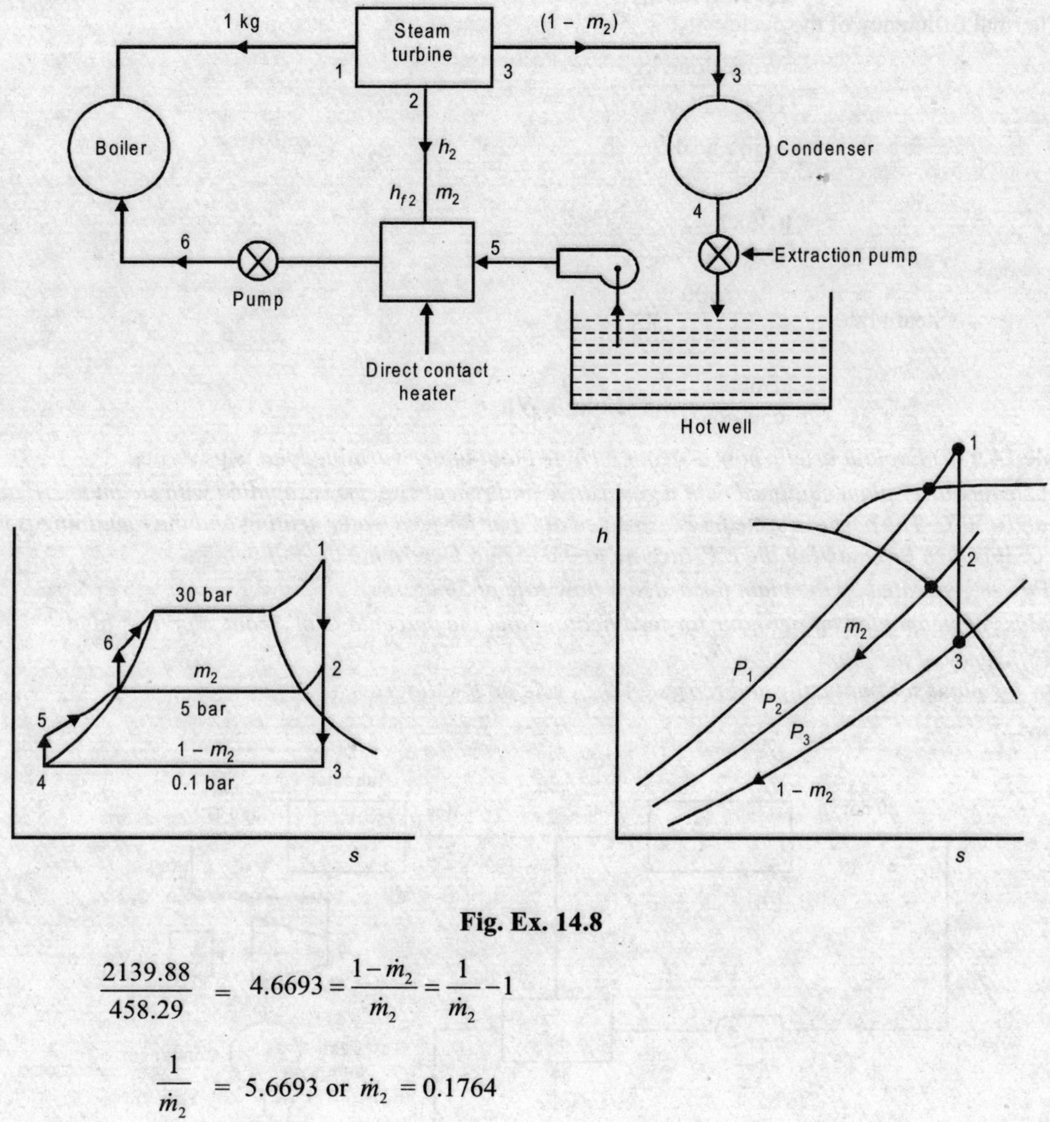

Fig. Ex. 14.8

$$\frac{2139.88}{458.29} = 4.6693 = \frac{1-\dot{m}_2}{\dot{m}_2} = \frac{1}{\dot{m}_2} - 1$$

$$\frac{1}{\dot{m}_2} = 5.6693 \text{ or } \dot{m}_2 = 0.1764$$

$$\dot{m}_2 = \mathbf{0.1764}$$

Neglecting pump work,

$$\text{Work done/kg} = 1(h_1 - h_2) + (1 - \dot{m}_2)(h_2 - h_3)$$

$$= (3232 - 2780) + (1 - 0.1764) \times (2780 - 2200)$$

$$= 452 + 0.8236 \times 580$$

$$= 452 + 477.69$$

$$= \mathbf{929.69 \text{ kJ/kg}}$$

$$\text{Heat supplied/kg} = h_1 - h_{f_2}$$

$$= 3232 - 640.12$$
$$= \mathbf{2591.88\ kJ/kg}$$

∴ Thermal Efficiency of the cycle

$$= \frac{\text{Work done/kg}}{\text{Heat added/kg}}$$

$$= \frac{929.69}{2591.88}$$
$$= \mathbf{0.3587}$$
$$= \mathbf{35.87\%}$$

$$\text{Steam rate} = \frac{3600}{\text{Work done/kg}}$$

$$= \frac{3600}{929.69} = \mathbf{3.8722\ kg/kWh}$$

Example 14.9 (*a*) *Explain briefly how a steam turbine plant under variable load is governed.*

(b) A steam power plant equipped with a generative and reheat systems is supplied with steam to HP turbine at 80 bar and 470°C. The part of the steam is extracted at 7 bar for feed water heating and the remaining is reheated to 350°C. It is then expanded in the L.P. turbine to 0.035 bar. Determine the following:

(i) Power generated by the plant for a steam flow rate of 50 kg/s.

(ii) Mass of steam bled off per hour for feed heating and the percentage of steam supplied, bled.

(iii) Efficiency of the plant.

Draw the plant schematically and represent the cycle on h-s diagram.

Solution

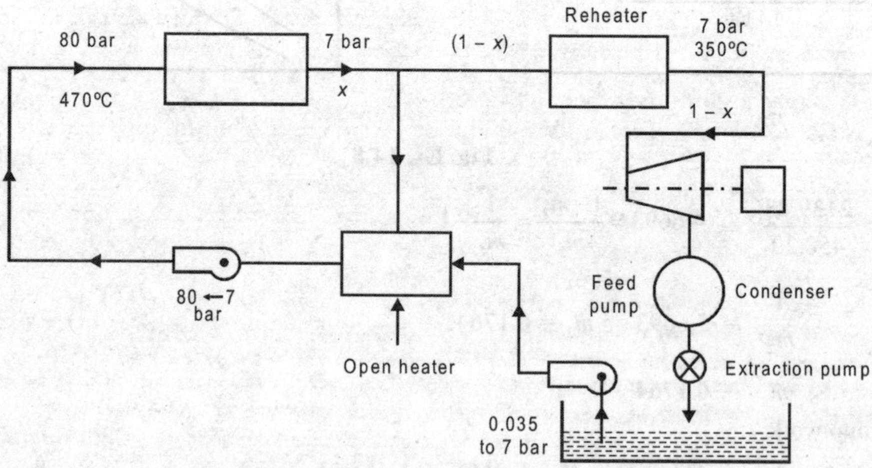

Fig. Ex. 14.9 (a)

Plant in equation is shown on a schematic diagram.

$p_1 = 80$ bar

$T_1 = 470°$ C

$p_2 = 7$ bar

$$T_3 = 350° \text{ C}$$

$$p_4 = 0.035 \text{ bar}$$

Steam flow rate = 50 kg/s

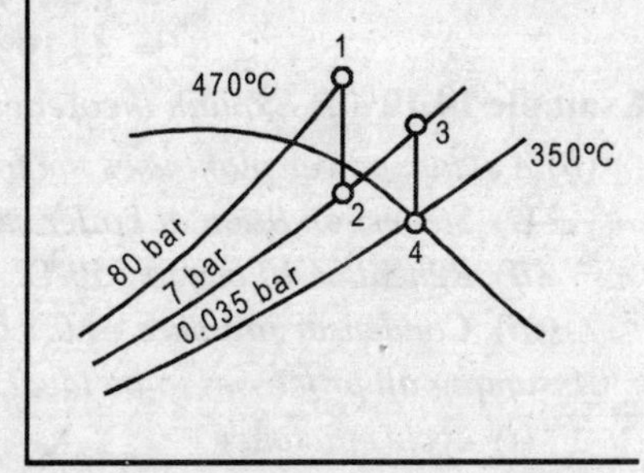

Fig. Ex.14.9 (b)

(i) Power generated : (ii) Steam bled

$h_1 = 3325$ kJ/kg $\qquad h_2 = 2725$ kJ/kg

$h_3 = 3170$ kJ/kg $\qquad h_4 = 2250$ kJ/kg

$h_{f_4} = 111.85$ kJ/kg

$h_{f_2} = 697.06$ kJ/kg

Assume 1 kg steam supplied to H.P. turbine and x kg is bled for feed water heating. Again assume a open (direct) feed heater (steam and water are mixed together).

For the feed heater

Heat lost by steam = Heat gained by feed water

$$x\left(h_2 - h_{f_2}\right) = (1-x)\left(h_{f_2} - h_{f_4}\right)$$

$$\therefore \quad (2725 - 697.06)x = (1-x)(697.06 - 111.85)$$

$$2027.94x = (1-x)585.21$$

$$= 585.21 - 585.21\,x$$

$$\therefore x\,(2027.94 + 585.21) = 585.21$$

$$x = \frac{585.21}{2613.15}$$

$$= \mathbf{0.224 \text{ kg/kg steam} = 22.4\%}$$

$$\text{Power generated} = (h_1 - h_2) + (1-x)(h_3 - h_4)$$

$$= (3325 - 2725) + (1 - 0.224)(3170 - 2250)$$

$$= 600 + 0.776 \times 920$$

$$= 600 + 713.92$$

$$= \mathbf{1313.92 \text{ kJ/kg}}$$

$$= \mathbf{65696 \text{ kJ/s} = kW}$$

$$= \mathbf{66 \text{ MW}}$$

(iii) Efficiency of the plant

$$\text{Heat added/kg} = \left(h_1 - h_{f_2}\right) + (1-x)(h_3 - h_2)$$

$$= (3325 - 697.06) + (1 - 0.224)(3170 - 2725)$$

$$= 2627.94 + 0.776 \times 445$$

$$= 2627.94 + 345.32$$

$$= \mathbf{2973.26 \text{ kJ/kg}}$$

$$\therefore \text{Efficiency of the plant} = \frac{W}{Q_A} = \frac{1313.92}{2973.26}$$

$= 0.4419$
$= \mathbf{44.19\%}$

Example 14.10 (*a*) *Explain theoretical and actual regenerative feed heating as applied to Rankine cycle plant.*

(*b*) *A steam power plant uses the following cycle:*

(*i*) *Steam condition at boiler stop valve — 150 bar, 550°C.*

(*ii*) *Reheat at 40 bar to 550°C.*

(*iii*) *Condenser pressure — 0.1 bar.*

Assuming all processes to be ideal, find

(*i*) *cycle efficiency*

(*ii*) *specific steam rate (kg/kWh of the plant).*

Solution

From Mollier chart,

$$h_1 = 3445.2$$

$$h_2 = 3050$$

$$h_3 = 3558.9$$

$$h_4 = 2280$$

$$h_{f_4} = 191.83$$

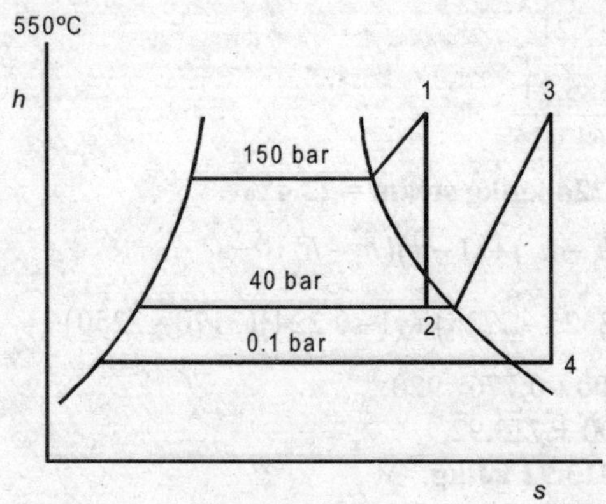

Fig. Ex.14.10

$$\text{Expansion work/kg} = (h_1 - h_2) + (h_3 - h_4) - W_p$$

$$= (3445.2 - 3050) + (3558.9 - 2280) - \frac{149.9}{10}$$

$$= 395.2 + 1278.9 - 14.99$$

$$= \mathbf{1659.11\ kJ/kg}$$

$$\text{Heat supplied} = (h_1 - h_{f_4}) + (h_3 - h_2) - W_p$$

$$= (3445.2 - 191.83) + (3558.9 - 3050) - 14.99$$

$$= 3253.37 + 508.9 - 14.99$$
$$= \mathbf{3747.28\ kJ/kg}$$

$$\therefore \quad \text{Cycle efficiency} = \frac{\text{Work}}{\text{Heat supplied}} = \frac{1659.11}{3747.28}$$
$$= 0.4428$$
$$= \mathbf{44.28\%}$$

$$\text{Specific steam rate} = \frac{3600}{1659.11} = \mathbf{2.17\ kg/kWh}$$

Example14.11 (*a*) *Explain the working of a modern reheat regenerative cycle used for steam power plants with the help of T–S and h–s diagrams and plant schematic.*

(*b*) *A steam power plant working on reheat regenerative cycle has the following data:*

(*i*) *Steam inlet to HP turbine 80 bar 470°C.*

(*ii*) *Steam extraction for feed heating at 7 bar.*

(*iii*) *Remainder steam reheated to 350°C in reheater.*

(*iv*) *Exhaust pressure from LP turbine = 0.035 bar.*

Determine: (*i*) *Fraction of steam bled off for reheating.* (*ii*) *Heat supplied/kg of steam in boiler and in reheater.* (*iii*) *Power output of the turbine.* (*iv*) *Thermal efficiency.*

Solution

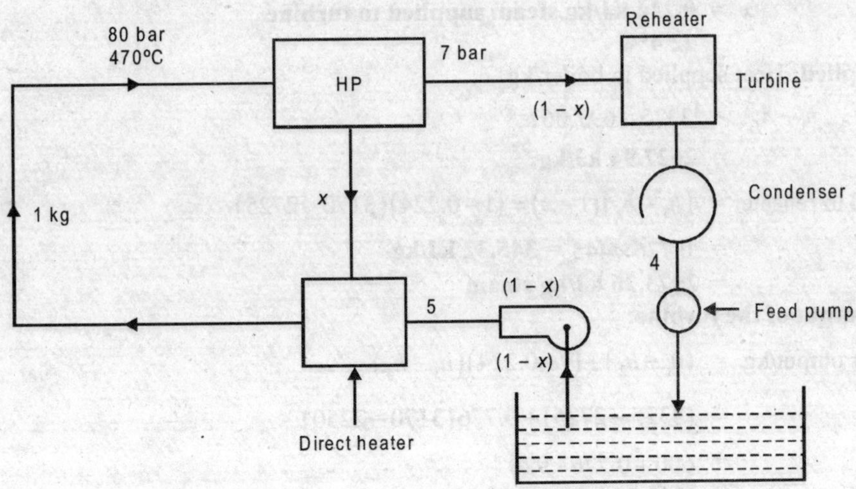

Fig. Ex. 14.11(a)

Assume direct heater.

Referring to Mollier chart,

$h_1 = 3325$ kJ/kg, $h_2 = 2725$, $h_{f_2} = 697.06$, $h_{f_3} = 3170$, $h_4 = 2250$, $h_{f_4} = 111.85$

(i) Fraction of steam bled off or for feed heating: Let x be the steam bled per kg of steam supplied to turbine.

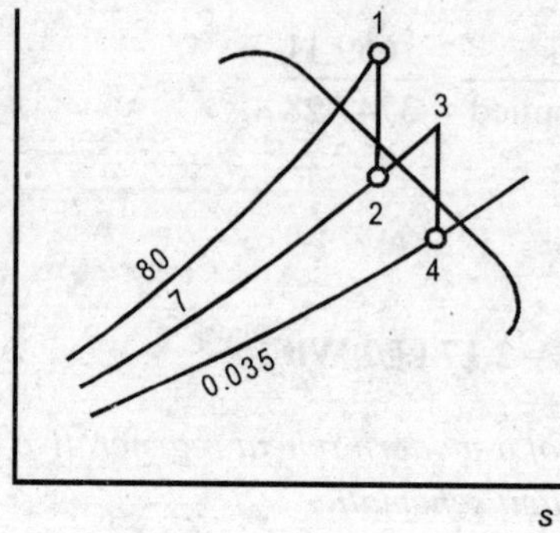

Fig. Ex. 14.11 (b)

$\therefore$ Heat given by steam = Heat gained by water

$$x\left[h_1 - h_{f_2}\right] = (1-x)\left(h_{f_2} - h_{f_4}\right)$$

$$\therefore \quad x(2725-697.06) = (1-x)(697.06-111.85)$$

$$2027.94x = 585.21 - 585.21x$$

$$x = \frac{585.21}{2613.15} = \mathbf{0.224}$$

$$x = \mathbf{0.224\ kJ/kg\ steam\ supplied\ to\ turbine}$$

i.e. **22.4%**

(ii) Heat supplied: Heat supplied in boiler/kg

$$h_1 - h_{f_2} = 3325 - 697.06$$

$$= \mathbf{2627.94\ kJ/kg}$$

$$\text{Heat supplied in reheater} = (h_3 - h_2)(1-x) = (1-0.224)(3170-2725)$$

$$= 0.776 \times 445 = \mathbf{345.32\ kJ/kg}$$

$$= \mathbf{2973.26\ kJ/kg\ steam}$$

(iii) Power putput of the turbine:

$$\text{Power output/kg} = (h_1 - h_2) + (1-0.224)(h_3 - h_4)$$

$$= (3325-2725) + 0.776(3170-2250)$$

$$= 600 + 0.776 \times 920$$

$$= \mathbf{1313.92\ kJ/kg}$$

(iv) Thermal efficiency: Thermal efficiency of the cycle

$$= \frac{W}{Q_A} = \frac{1313.92}{2973.26} = \mathbf{0.4419 = 44.19\%}$$

Example 14.12 *A binary vapour plant uses mercury 1 between a temperature range of 204°C and 537°C, the mercury being just dry at the higher limit. The pressure range for the steam is between 73.66 cm vacuum and 17 bar and it is superheated to 371°C at its higher limit. The feed water is raised to 204°C in economiser and is* *evaporated to dry steam in Hg-H_2O boiler; it is superheated by the flue gases. Assuming adiabatic expansion in Hg and steam turbine, find, (1) the required weight of Hg per kg of steam used, (2) work done by mercury, (3) work done by steam and (4) the ideal efficiency of the plant. The following are the required values of total heats and entropies of Hg and steam.*

Mercury:

Temperature, °C	*Enthalpy, kJ/kg*			*Entropy, kJ/kg.K*	
	h_f	h_{fg}	h_g	s_f	s_g
204°C	*29*	*300*	*329*	*0.044*	*0.3883*
537°C	*75.15*	*288*	*363.15*	*0.08413*	*0.2820*

Steam:

Pressure bar	*Enthalpy*			*Entropy*	
	h_f	h_{fg}	h_g	s_f	s_g
0.03 bar 73.66 cm Hg	*101.0*	*2444*	*2545*	*0.354*	*8.576*
16.9 bar	*8.72*	*1923*	*2792*	*2.372*	*6.4*

Solution

Steam at 371°C and 17 bar, h = 3200 kJ/kg, s = 7.1259.

Refer the *T–S* diagram drawn.

$$s_c = 0.2820 = s_d = 0.044 + x_d \times (0.3883 - 0.044)$$

$$= 0.044 + x_d \times 0.3443$$

$$\therefore \quad x_d = \frac{0.2820 - 0.044}{0.3443}$$

$$= \frac{0.2380}{0.3443}$$

$$= \mathbf{0.6913}$$

(i) Weight of mercury

Let, W_m = Weight of mercury required per kg of steam

Then, Loss of heat of Hg = Gain of latent heat of steam from *a* to *d* betwen *d* and *a*

$$W_m \times x.h_{f_g} = L_s \text{ at 17 bar}$$

$$W_m \times 0.6913 \times 300 = 1923$$

$$\therefore \quad W_m = \frac{1923}{0.6913 \times 300}$$

$$= \mathbf{9.27\ kg}$$

(ii) Work done by mercury

$$\text{Work} = W_m[H_c - H_d]$$
$$= 9.27[363.15 - 237.29]$$
$$= 9.27 \times 125.86$$
$$= \mathbf{1166.72\ kJ/kg\text{-}steam}$$

where, $H_c = 363.15$

$$H_d = 29.9 + 0.6913 \times 300$$
$$= 29.9 + 207.39$$
$$= 237.29$$

(iii) Work done by steam

$$H_e = 3200 \text{ kJ/kg}, \quad s_e = 7.1259$$
$$s_f = 0.354 + x_f\,(8.576 - 0.354)$$
$$= 0.354 + 8.222\ x_f$$
$$\therefore \quad x_f = \frac{7.1259 - 0.354}{8.222} = \frac{6.7719}{8.222} = \mathbf{0.8236}$$
$$\therefore \quad H_f = 101 + 0.8236 \times 2444$$
$$= 101 + 2012.88 = 2113.88$$
$$\therefore \quad \text{Work done/kg} = 3200 - 2113.88$$
$$= \mathbf{1086.12}$$

(iv) Total work done/kg of steam = 1166.72 + 1086.12 = **2252.84**

$$[\text{Heat supplied by fuel}] = \begin{bmatrix}\text{Heat supplied}\\ \text{to mercury}\end{bmatrix} + \begin{bmatrix}\text{Heat to}\\ \text{superheated steam}\end{bmatrix} + \begin{bmatrix}\text{Heat of}\\ \text{feed water}\end{bmatrix}$$
$$= W_m[H_c - H_a] + [H_e - H_d] + [H_a - H_g]$$
$$= 9.27(363.15 - 29) + (3200 - 2795) + (872 - 101)$$
$$= 3097.57 + 405 + 771$$
$$= \mathbf{4273.57\ kJ/kg\ steam}$$

$$\therefore \text{Ideal efficiency of the plant} = \frac{\text{Total work done}}{\text{Total heat supplied}}$$
$$= \frac{2252.84}{4273.57} = \mathbf{0.5272 \text{ or } 52.72\%}$$

Example 14.13 *A cyclic steam power plant is to be for a steam temperature at the turbine inlet of 360°C and an exhaust pressure of 0.08 bar isentropic expansion of steam in the turbine, the moisture content at the turbine exhaust is not to exceed 15%. Determine the greatest allowable steam pressure at the turbine inlet and calculate the Rankine cycle efficiency for these steam conditions. Also estimate the mean temperature of heat addition.*

Solution

$$T_1 = 360^\circ \text{ C}$$
$$P_2 = 0.08 \text{ bar} \quad x_2 = 0.85$$

From steam tables – $t_{sat} = 42.689$

$$\therefore \text{Mean temperature of heat addition} = \frac{360+41.5}{2} = \frac{401.5}{2} = 200.75°C$$

Corresponding to this temperature as saturation temperature, pressure is given as

$$p = \frac{15.6 \text{ bar}}{16} = \mathbf{0.975\ bar}$$

$$\text{Rankine efficiency} = \frac{h_1 - h_2}{h_1 - h_2} \text{ neglecting } W_p$$

$$h_1 = 2795.0 - 2.1\,(360 - 200.75)$$
$$= 2795.0 - 334.425$$
$$= 3129.425 \text{ kJ/kg}$$
$$h_2 = 174 + 0.85 \times 2402$$
$$= 174 + 2041.7$$
$$= 2215.7 \text{ kJ/kg}$$
$$h_{f_2} = 174 \text{ kJ/kg}$$
$$\therefore \quad \eta_f = \frac{3129.425}{3129.425 - 174}$$
$$= \frac{913.725}{2955.425}$$
$$= \mathbf{0.3092 \text{ or } 30.92\%}$$

Example 14.14 *Find the ideal cycle efficiency and steam consumption of a reheat cycle operating between the pressures of 30 bar and 0.04 bar with initial superheated temperature of 400°C, expanded to the point where the steam is dry and then it is reheated at that pressure upto initial temperature of 400°C.*

The feed pump work may be neglected.

Solution

The cycle represented on *T–s* diagram as shown.
Reffering to *h-s* diagram,

$$h_1 = 3230 \text{ kJ/kg}$$
$$h_2 = 2730 \text{ kJ/kg}$$
$$h_3 = 3270 \text{ kJ/kg}$$
$$h_4 = 2395 \text{ kJ/kg}$$

$$\text{Ideal work done/kg steam} = (h_1 - h_2) + (h_3 - h_4)$$
$$= (3230 - 2730) + (3270 - 2395)$$
$$= 500 + 875$$
$$= \mathbf{1375 \text{ kJ/kg}}$$

$$\text{Heat supplied/kg of steam} = (h_1 - h_{f_5}) + (h_3 - h_2)$$
$$= (3230 - 121.4) + (2370 - 2730)$$
$$= 3108.6 + 540 = \mathbf{3648.6 \text{ kJ/kg}}$$

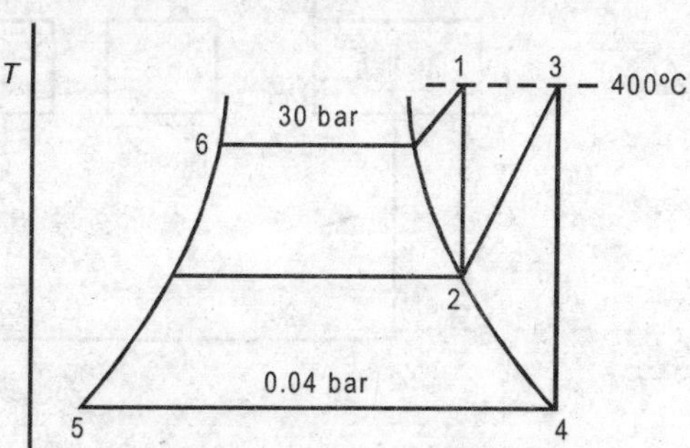

Fig. Ex. 14.14

$$\therefore \text{Ideal cycle efficiency} = \frac{\text{Work done}}{\text{Heat supplied}}$$

$$= \frac{1375}{3648.6} = 0.3769 = \mathbf{37.69\%}$$

$$\text{Specific steam consumption} = \frac{3600}{1375} = \mathbf{2.618\ kg/k\ Wh}$$

Example 14.15 *In a thermal power plant of 60 MW capacity, boiler supplies steam at 100 bar and 400°C. The steam expands to 20 bar in the HP turbine. Steam is then reheated to 380°C in a reheater. Thereafter, steam expands to a condenser pressure of 0.05 bar. The pressure loss in the reheater is 1 bar. Assuming isentropic efficiency in both stages of expansion is 80%, find the steam generating capacity required in ton/hour. Take transmission efficiency from turbine to generator = 98% and generator efficiency of 95%. Neglect feed pump work. Also find the overall efficiency of the plant.*

Draw a neat line diagram of the plant and represent the cycle on T-s diagram.

Solution

From Mollier diagram, we have

$$h_1 = 3095 \text{ kJ/kg}$$
$$h_2 = 2730 \text{ kJ/kg}$$
$$h_3 = 3200 \text{ kJ/kg}$$
$$h_4 = 2140 \text{ kJ/kg}$$
$$h_{f_5} = 137.77 \text{ kJ/kg}$$

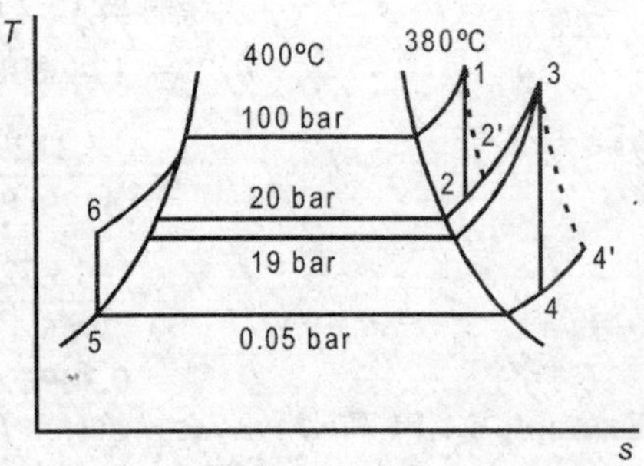

Fig. Ex.14.15 (b)

We have, $\eta_s = 0.8 = \dfrac{h_1 - h_{2'}}{h_1 - h_2} = \dfrac{3095 - h_{2'}}{3095 - 2730} = \dfrac{3095 - h_{2'}}{365}$

$\therefore \quad 3095 - h_{2'} = 0.8 \times 365 = 292.0$

$\therefore \quad h_{2'} = 3095 - 292 = \mathbf{2803\ kJ/kg}$

Again $\quad 0.8 = \dfrac{h_3 - h_{4'}}{h_3 - h_4} = \dfrac{3200 - h_{4'}}{3200 - 2140} = \dfrac{3200 - h_4}{1060}$

$\therefore \quad 3200 - h_{4'} = 0.8 \times 1060 = \mathbf{848\ kJ/kg}$

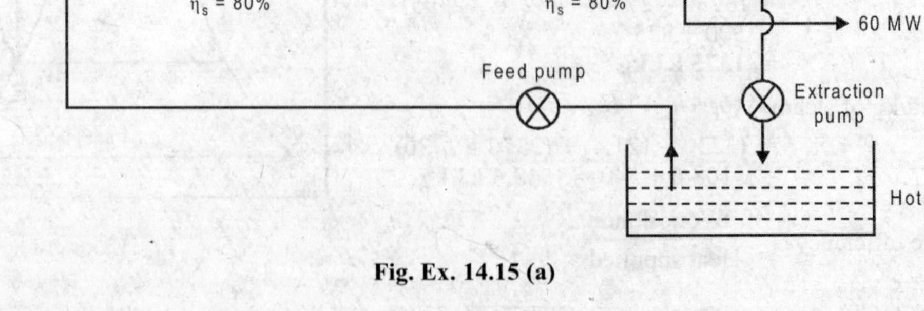

Fig. Ex. 14.15 (a)

$\therefore$ Work done in HP turbine $= h_1 - h_2 = 292\,\text{kJ/kg}$

$\therefore$ Work done in LP turbine $= h_3 - h_4 = 848\,\text{kJ/kg}$

$\therefore$ Net work/kg $= 292 + 848 = \mathbf{1140\ kJ/kg}$

Work transmitted to generator $= 1140 \times 0.98 = \mathbf{1117.2\,kJ/kg}$

$\therefore$ Work delivered by generator $= 1117.2 \times 0.95 = 1061.34\,\text{kJ/kg}$

Capacity of the plant is 60 MW= 60 × 1000 kW

$$\therefore \quad \text{Mass flow rate} = \frac{60 \times 1000}{1061.34}$$

$$= 56.53 \text{ kg/sec}$$

$$= \frac{56.53 \times 3600}{1000} \text{ tons/hour}$$

$$= \mathbf{203.5\ tons/hour}$$

= Capacity of steam generation

$$\text{Heat supplied/kg} = (h_{1-h_{f5}}) + (h_3 - h_{2'}) = (3095 - 137.77) + (3200 - 2803)$$

$$= 2957.23 + 397 = \mathbf{3354.23\ kJ/kg}$$

$\therefore$ Overall efficiency of the plant

$$= \frac{\text{Work delivered/kg}}{\text{Heat supplied/kg}} = \frac{1061.34}{3354.23}$$

$$= \mathbf{0.3164 = 31.64\ \%}$$

Example 14.16 *The steam at 392 N/cm² and 300°C is supplied to steam turbine. The steam is bled for regenerative heating at 137.2 N/cm² and 33.3 N/cm². The condenser pressure is 0.686 N/cm². Calculate*

(i) *At each stage the optimum weight of bled beam per kg of steam supplied to the turbine and*

(ii) *the cycle efficiency.*

Assume turbine efficiency of 0.8 for each portion of the expansion and that the feed water leaving each heater is raised to the steam temperature entering the heater.

The heaters used are indirect heaters and the condensate coming out from the first heater is passed through the second heater is passed to the hot well. Neglect pump work.

Solution

Indirect heater

$\therefore$ Steam does not mix with feed water

$$p_1 = 40 \text{ bar } (39.2)$$

$$p_2 = 14 \text{ bar } (13.72)$$

$$p_3 = 3.50 \text{ bar } (3.33)$$

$$p_4 = 0.07 \text{ bar } (0.0686)$$

$$h_1 = 2965\,\text{kJ/kg}$$

$$h_a = 1740\,\text{kJ/kg}$$

$$h_{2'} = 2740 + (2965 - 2740) \times 0.2$$

$$= 2740 + 45 = \mathbf{2785\ kJ/kg}$$

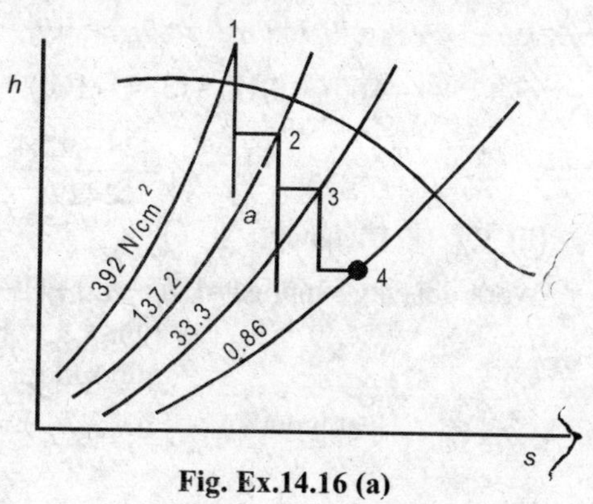

Fig. Ex.14.16 (a)

$$h_b = 2540 \text{ kJ/kg}$$

$$h_3 = (2785 - 2540) \times 0.2 + 2540$$
$$= 2540 + 0.2 \times 245 = 2540 + 49 = \mathbf{2589}$$

$$h_{f_3} = 584 \text{ kJ/kg}$$

$$h_c = 2330 \text{ kJ/kg}$$

$$h_4 = 2330 + 0.2(2589 - 2330)$$
$$= 2330 + 51.8$$
$$= \mathbf{2381 \text{ kJ/kg}}$$

$$h_{f_4} = 160 \text{ kJ/kg}$$

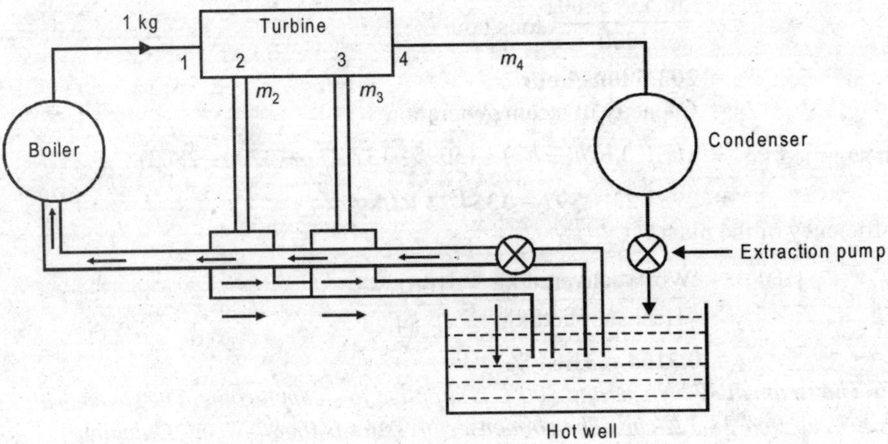

Fig. Ex.14.16 (b)

(i) Heater at 2

$$\dot{m}_2(h_2 - h_{f_3}) = 1(h_{f_2} - h_{f_3}) = 830 - 584 = 246$$

$$\therefore \quad \dot{m}_2 = \frac{246}{2785 - 584} = \frac{246}{2201} = \mathbf{0.1118\,kg}$$

Heater at 3

$$\dot{m}_3(h_3 - h_{f_4}) + \dot{m}_2(h_{f_3} - h_{f_4}) = 1(h_{f_3} - h_{f_4})$$

$$\dot{m}_3\,(2589 - 160) + 0.1118\,(584 - 160) = 584 - 160 = \mathbf{424}$$

$$\therefore \quad \dot{m}_3 = \frac{424 - 47.4}{2429} = \mathbf{0.155\ kg}$$

(ii) Cycle efficiency

$$\text{Work done/kg supplied} = 1(h_1 - h_2) + (1 - \dot{m}_2)(h_2 - h_3) + (1 - \dot{m}_2 - \dot{m}_3)(h_3 - h_4)$$
$$= (2965 - 2785) + (1 - 0.1118)\,(2785 - 2589) + (1 - 0.1118 - 0.155)\,(2589 - 2381.8)$$
$$= 506 \text{ kJ/kg}$$

$$\text{Heat input/kg} = 1(h_1 - h_{f_2}) = 2965 - 830 = 2135 \text{ kJ/kg}$$

$$\therefore \quad \text{Cycle efficiency} = \frac{\text{Work output}}{\text{Heat input}} = \frac{506}{2135} = 0.237$$

$$= \mathbf{23.7\%}$$

Example 14.17 *The steam turbine of a power station using regenerative cycle receives dry and saturated steam at 32 bar. The steam is bled at 2.8 bar and the condenser vacuum is 723.1 mm of Hg. The heating of feed water is done in the direct contact heater. Find (i) the quantity of steam bled per kg of steam supplied to the turbine, (ii) cycle efficiency and (iii) specific steam consumption.*

Solution

$$\text{Condenser pressure} = 760 - 723.1 = 36.9 \text{ mm Hg}$$
$$= 0.04916 \text{ bar} = \mathbf{0.05 \text{ bar}}$$

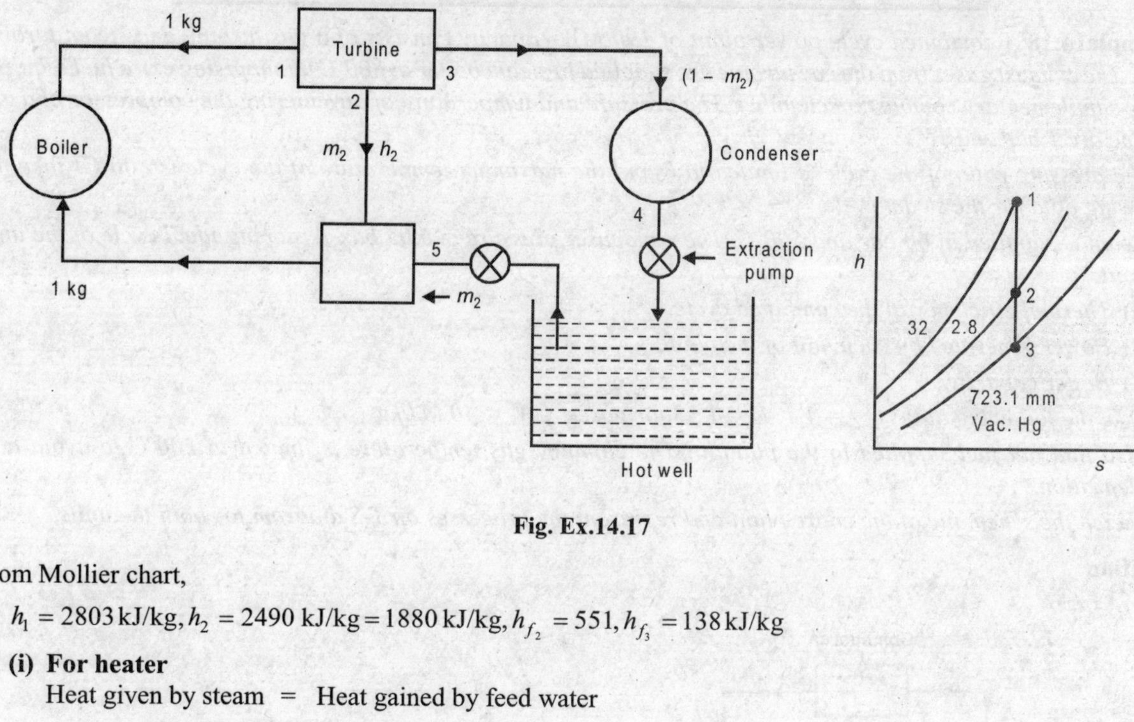

Fig. Ex.14.17

From Mollier chart,

$$h_1 = 2803 \text{ kJ/kg}, h_2 = 2490 \text{ kJ/kg} = 1880 \text{ kJ/kg}, h_{f_2} = 551, h_{f_3} = 138 \text{ kJ/kg}$$

(i) For heater

$$\text{Heat given by steam} = \text{Heat gained by feed water}$$

$$\dot{m}_2(h_2 - h_{f_2}) = (1 - \dot{m}_2)(h_{f_2} - h_{f_3})$$

$$\therefore \quad \dot{m}_2 \,(2490 - 551) = (1 - \dot{m}_2)(551 - 138)$$

$$1939\, \dot{m}_2 = 413\,(1 - \dot{m}_2)$$

$$\therefore \quad \dot{m}_2 = \frac{413}{1939 + 413} = \frac{413}{2352}$$

$$= \mathbf{0.1756 \text{ kg}}$$

(ii)

$$\text{Work done/kg} = (h_1 - h_2) + (1 - m_2)(h_{f_2} - h_{f_3})$$
$$= (2803 - 2490) + (1 - 0.1756)\,(2490 - 1880)$$
$$= 313 + 0.8244 \times 610 = 313 + 502.88$$
$$= \mathbf{815.88 \text{ kJ/kg}}$$

$$\text{Heat supplied/kg} = h_1 - h_2 = 2803 - 551$$

$$= \mathbf{2232\ kJ}$$

$$\therefore \text{Cycle thermal efficiency} = \frac{\text{Work output}}{\text{Heat input}} = \frac{815.88}{2252}$$

$$= \mathbf{0.3623\ or\ 36.23\%}$$

$$\text{(iii) Specific steam consumption} = \frac{3600}{815.88} = \mathbf{4.412\ kg/k\ Wh}$$

EXAMPLES ON COMBINED GAS-VAPOUR POWER CYCLE

Example14.18 *A combined cycle power plant of 190 MW capacity consists of a gas turbine and steam turbine units. The exhaust gases from the gas turbine are supplied to steam boiler at 800°C by supplying extra fuel required in the supplementary combustion chamber. The pressure and temperature of air entering the compressor of a gas turbine are 1 bar and 15°C.*

The pressure ratio of the cycle is limited to 8 and the maximum temperature of the cycle is 800°C. Take the following data for the steam cycle.

Steam is supplied at 60 bar and 500°C. The condenser pressure = 0.05 bar. Assuming ideal cycle of the unit, find out

(1) Thermal efficiency of the combined cycle.

(2) Power generated by each unit of the plant.

(3) Boiler capacity.

Take the following data: $C_{pg} = 1.1$, $K = 1.33$ *for gases, C.V. = 40 MJ/kg*

Also find out fuel supplied to the plant/hr. The chimney gas temperature is limited to 200°C to avoid acid condensation.

Sketch the schematic of the entire plant and represent the processes on T-S diagram for both the units.

Solution

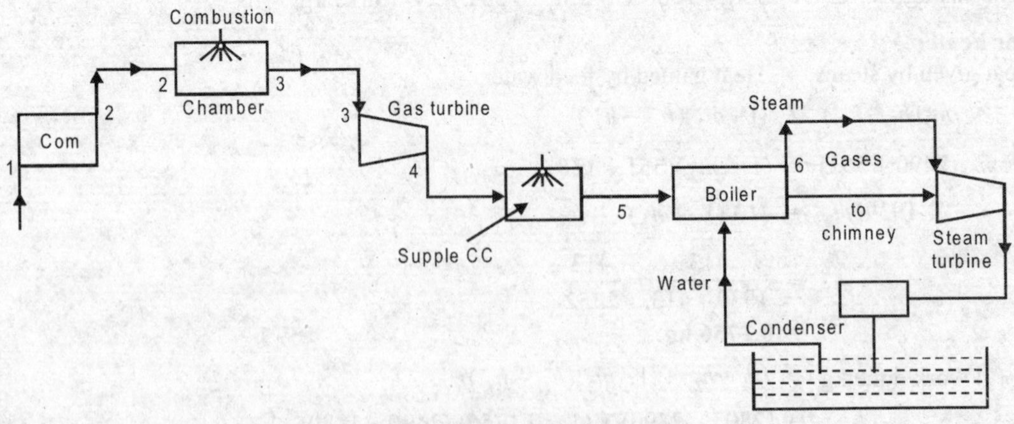

Fig. Ex.14.18 (a)

Gas turbine unit:

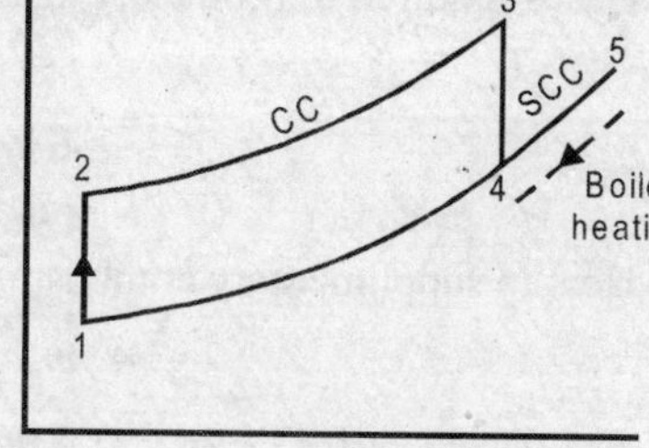

Fig Ex. 14.18 (b)

$$T_1 = 288 \text{ K}$$

$$p_1 = 1 \text{ bar}$$

$$p_2 = 8 \times 1 = 8 \text{ bar}$$

$$T_3 = 1073^\circ \text{ K}$$

Assume 1 kg air, $C_{p_a} = 1.005$ kJ/kg

γ for air $= 1.4$

$$\frac{T_2}{T_1} = \left(\frac{p_2}{p_1}\right)^{\frac{y-1}{y}} = 8^{\frac{1.4-1}{1.4}} = 8^{0.286} = 1.8125$$

$\therefore$ $T_2 = 288 \times 1.8125 = \mathbf{522\ ^\circ K}$

$\therefore$ $$T_4 = \frac{T_3}{8^{\frac{133-1}{1.33}}} = \frac{1073}{(8)^{0.24812}}$$

$$= \frac{1073}{1.67523}$$

$$= \mathbf{640.51^\circ\ K}$$

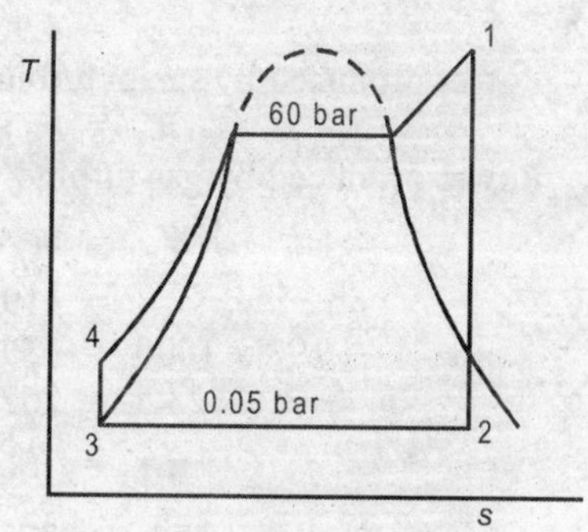

Fig Ex.14.18 (c)

$\therefore$ Compressor work/kg $= C_{p_a}(T_2 - T_1) = 1.005\ (522 - 288)$

$= \mathbf{235.17\ kJ/kg}$

Turbine work/kg $= C_{p_g} = (T_3 - T_4) = 1.1\ (1073 - 640.51)$

$= 1.1\ (432.49)$

$= \mathbf{475.74\ kJ/kg}$

$\therefore$ Net work of gas/turbine/kg $= W_t - W_c = 475.74 - 235.17$

$= \mathbf{240.57\ kJ/kg}$

Heat supplied air $= C_{p_g} T_3 - C_{p_a} T_2 = 1.1 \times 1073 - 1.005 \times 522$

$= 1180.3 - 524.61$

$=$ **655.69 kJ/kg** i.e., **0.01164 kgf/kg air**

Steam turbine $h_1 = 3422.2$ kJ/kg from table and 3410 from chart,

$h_{f_3} = 137.77$ kJ/kg

Heat required/kg steam $= 3410.2 - 137.77$

$= \mathbf{3272.43\ kJ/kg}$

$h_2 = 2080$ kJ/kg from chart

$\therefore$ Work done/kg steam $= h_1 - h_2$

$= 3410 - 2080$

$= \mathbf{1330\ kJ/kg}$

Gases required to produce 1 kg steam

$$= \frac{3274.43}{1.1\times(800-200)} = \frac{3274.43}{660}$$

$= \mathbf{4.961\ kJ/kg\ steam}$

Heat in supplimentary combustion chamber/kg steam

$$= m_g c_{pg}(T_5 - T_4) = 4.961 \times 1.1(1073 - 640.51)$$

$= 4.961 \times 1.1 \times 432.49$

$= \mathbf{2360.14}$

i.e. $= \mathbf{0.05918\ kgf/kg\ steam}$

Total power generated $= 190 \times 10^6$

$= 190 \times 1000$ kW ...(I)

Let m_s is mass of steam/sec

$\dot{m}_a = \dot{m}_g$ = mass of air required/sec

$= \mathbf{4.961 \times m_s}$

Power produced by steam turbine

$W_s = \dot{m}_s \times 1330$ kW

Power produced by gas turbine

$W_g = \dot{m}_a \times 240.57 = 4.961 \times \dot{m}_s \times 240.57$

$= \mathbf{1193.46\ \dot{m}_s\ kW}$ ...(II)

$\therefore \quad 190 \times 1000 = 1330\ \dot{m}_s + 1193.46\ \dot{m}_s$

$2523.5 = 2527.58\ \dot{m}_s$

$\therefore \quad \dot{\mathbf{m}}_s = \mathbf{75.29\ kg/sec}$

$\dot{\mathbf{m}}_a = \mathbf{373.52\ kg/sec}$

Heat supplied/sec $= 2360.14 \times 75.29 + 373.52 \times 655.69$

$= 178235.5 + 244913.32$

$= 423148.82$ kJ/s

$$\therefore \quad \text{Fuel supplied/hour} = \frac{\text{Heat supplied/sec} \times 3600}{\text{C. V. of fuel}} = \frac{423148.82 \times 3600}{40000}$$

$= 38083.4$ kg/hour = **38.083 T/hour**

$\therefore$ Thermal efficiency of the plant

$$= \frac{190 \times 100}{423148.82} = 0.449 = \mathbf{44.9\%}$$

$$\text{Power produced by GT} = \frac{1197.08 \times 75.29}{1000}$$

$= 90.13 = \mathbf{90\ MW}$

Steam turbine $= 75.29 \times 1330$

$= 99.87 = \mathbf{100\ MW}$

Heat supplied in GT plant= $\dot{m}_a \times 655.69 = 373.52 \times 655.69 = 244913.32$ kJ/sec

Example 14.19 *In a combined cycle power plant, the air is supplied at the rate of 2000 × 10³ kg/hr at 20°C. The pressure ratio used in gas turbine plant is 7. The suction pressure of the compressor and the exhaust pressure of turbine is 1 bar. The maximum temperature in the gas turbine is limited to 1000°C. The isentropic efficiencies of* *the gas turbine and compressor are 85% and 80% respectively. The following data is given for the steam power plant.*

t_g of exhaust from G.T. is increased to 1200°C by burning fuel before passing into the boiler. The steam is generated at 50 bar and 500°C. The condenser pressure is maintained at 0.1 bar. Temperature of gases entering the chimney is 200°C.

Find the following:

(i) Total power generating capacity of the combined plant.

(ii) Overall efficiency of the plant.

(iii) Fuel consumption of two combined plant/hr.

(iv) Air to fuel ratio used for gas turbine plant. Use the following data:

C.V. of fuel = 45 MJ/kg, C_{pa} = 1 kJ/kg °K, γ_a = 1.4, C_{pg} = 1.1 γg = 1.33

Do not neglect the mass of the fuel used.

Draw neat line diagram for the components.

Represent the process on T-S diagram.

Solution

Gas turbine

$$\frac{T_2}{T_1} = \left(\frac{p_2}{p_1}\right)^{\frac{\gamma-1}{\gamma}} = 7^{0.286} = 1.745$$

$$\therefore \quad T_2 = 293 \times 1.745 = \mathbf{511.17^\circ\ K}$$

$$\frac{T_3}{T_4} = (7)^{\frac{1.33-1}{1.33}} = 7^{0.2481} = 1.621$$

$$\therefore \quad T_4 = \frac{1273}{1.621} = 785.32^\circ K$$

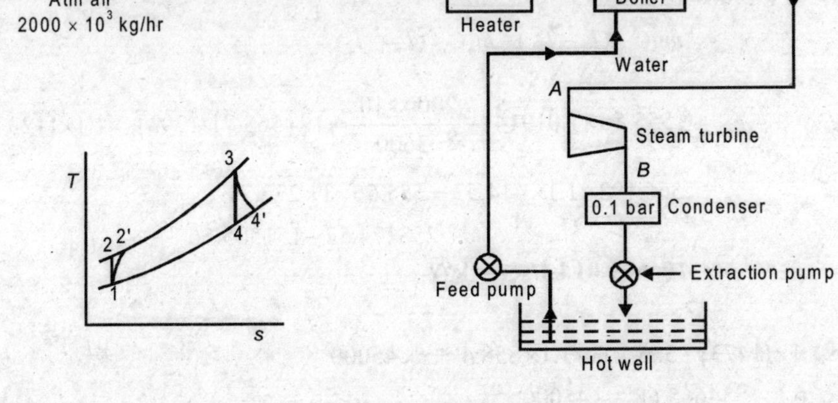

Fig. Ex.14.19 (a)

$$\eta_c = 0.8 = \frac{T_2}{T_{2'}} - \frac{T_1}{T_1}$$

$$0.8 = \frac{511.17 - 293}{T_{2_1} - 293} = \frac{218.17}{T_{2_1} - 293}$$

$$\therefore \quad T_{2'} = 293 + \frac{218.17}{0.8} = 293 + 272.71$$

$$= \mathbf{565.71^\circ\ K}$$

Similarly $\quad \eta_t = 0.85 = \frac{T_3 - T_{4'}}{T_3 - T_4} = \frac{1273 - T_{4'}}{1273 - 785.32}$

$$\therefore \quad T_{4'} = 1273 - 0.85 \times 487.68 = 1273 - 414.53$$

$$= \mathbf{858.47\ ^\circ K}$$

Consider combustion chamber. Let x be the mass of fuel per kg of air.

$$C_{pg}(1-x)T_3 = C_{pa}T_{2^1} + x \times \text{C.V.}$$

$$(1+x) \times 1.1 \times 1273 = 1 \times 1 \times 565.71 + 45000x$$

$$1400 + 1400\,x = 565.71 + 45000\,x$$

$$1400 - 565.71 = x\,(45000 - 1400)$$

$$834.29 = x \times 43600$$

$$x = \frac{834.29}{43600} = 0.01914$$

$$\text{Air to fuel ratio} = \frac{1}{0.01914} = \mathbf{52.25}$$

$\therefore$ Fuel supplied to gas turbine plant

$$= 2000 \times 10^3 \times 0.01914$$

$$= \mathbf{38280\ kg/hour = 10.63\ kg/s}$$

Power produced by gas turbine plant

$$p_1 = \dot{m}gC_{pg}(T_3 - T_{4^1}) - \dot{m}_a C_{pa}(T_{2^1} - T_1)$$

$$= (555.56 \times 1.01914) - \frac{2000 \times 10^3}{3600} \times 1 \times (565.71 - 293) \ \times 1.1 \times (12373 - 858.47)$$

$$= 566.193 \times 1.1 \times 414.53 = 555.56 \times 1 \times 272.71$$

$$258174.57 - 151505.56$$

$$= \mathbf{106669.01\ kJ/sec = kW} \qquad \ldots(\text{I})$$

For Heater

$$(566.193 + x) \times 1.1 \times (1473) - 566.193 \times 1.1 \times 858.4 = x \times 45000$$

$$917402.52 + 1620\,x - 534665.68 = 45000\,x$$

$\therefore$ $382736.84 = (45000 - 1620) \times 43380$

$$x = \frac{382736.84}{43380}$$

$= 8.823$ kg/sec

$= 31762.8$ kg/hour

$\therefore$ Total fuel supplied to the plant

$= 38280 + 31762.8$

$=$ **70042.8 kg/hr**

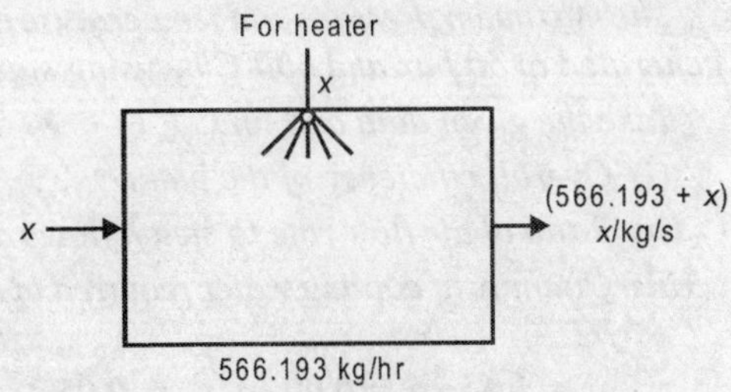

Fig. Ex. 14.19 (b)

Steam turbine:

$$p_1 = 50\,\text{bar}, t_1 = 500°\,\text{C}, p_2 = 0.1\,\text{bar}$$

$\therefore$ From *H-s* chart, $h_1 = 3420$ kJ/kg, $h_2 = 2190$

$h_{f_2} = 191.83$ kJ/kg

$\therefore$ Heat required for 1 kg steam

$h_1 - h_{f_2} = 3420.0 - 191.83$

$=$ **3228.17 kJ/kg**

Available heat for steam generation

$= \dot{m}_g C_{pg} (t_5 - t_6)$

$= (555.56 + 10.63 + 4.87) \times 1.1 \times (1200 - 200)$

$= 571.056 \times 1100$ kJ/sec

$= 628161.6$ kJ/sec

$\therefore$ Amount of steam generated/sec

$$= \frac{628161.6}{3228.17} = \textbf{194.59 kg/sec}$$

$\therefore$ Work done in steam turbine/plant

$= \dot{m}_s \times (h_1 - h_2) = 194.59 \times (3420 - 2190)$

$= 194.59 \times 1230$

$=$ **239342.65 kJ/sec = kW**

Total power produced in the combined plant

$= 106669.01 + 239342.65$

$=$ **346011.66 kW**

$$\text{Overall efficiency of the plant} = \frac{\text{Total work produced}}{\text{Fuel used} \times \text{Calculated value}}$$

$$= \frac{36011.66}{\frac{70042.8}{3600} \times 45000} = \frac{7.689}{19.46}$$

$=$ **0.3951 or 39.51%**

Example14.20 *A combined cycle power plant is to be designed to develop 100 MW. Out of this 60 % power is to be developed in a gas turbine plant and the remaining in a steam power plant. The pressure and temperature of the air entering the compressor are 1 bar and 27°C.*

The maximum pressure and temperature in the gas turbine cycle are limited to 8 bar and 1000°C. Steam is to be generated at 50 bar and 600°C by using supplementary fuel. The condenser pressure is maintained at 0.05 bar.

Take the given data and find:

(i) Overall efficiency of the plant.

(ii) Ratio of air flow rate to steam flow rate.

(iii) Quantity of cooling water required in ton/hour if the rise in temperature of the cooling water is limited to 10 °C.

$\eta_c = 0.85,\ \eta_t = 0.90,\ \eta_{comb} = 0.95.$

$C_{pa} = 1$ kJ/kg-°K, $\gamma_a = 1.4$, $C_{pg} = 1.1$ kJ/kg-°K, $\gamma_g = 1.33$

C.V. of fuel used = 40 MJ/kg

Gas temperature going to chimney = 200°C.

Neglect W_p. Consider mass of fuel in gas turbine cycle.

Solution

(i) First consider gas turbine plant.

$$p_1 = 1\,\text{bar},\ p_2 = 8\,\text{bar},\ T_3 = 1273\,^\circ\text{K}$$

$$\frac{T_2}{T_1} = 8^{0.286} = 1.813$$

$$\therefore \quad T_2 = 1.813 \times 300 = 543.76^\circ\text{K}$$

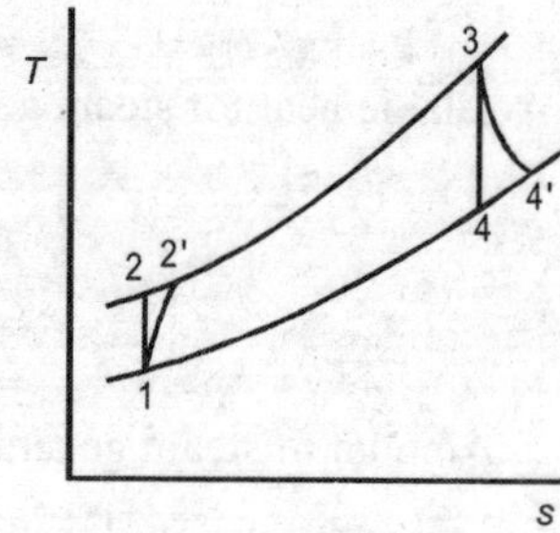

Fig. Ex.14.20 (a)

$$\therefore \quad 0.85 = \eta_c = \frac{T_2 - T_1}{T_{2'} - T_1}$$

$$= \frac{543.76 - 300}{T_{2'} - 300}$$

$$\therefore \quad T_{2'} = 300 + \frac{243.76}{0.85}$$

$$= 300 + 286.8 = \mathbf{586.8\ ^\circ K}$$

For combustion chamber –

$$1 \times C_{pa} \times t_{2'} + x_f \times \text{C.V.} = (1 + x_f) C_{pg} \times t_3$$

$$1 \times 1 \times (586.8 - 273) + (x_f \times 40000 \times \eta_{\text{comb}}) = (1 + x_f) \times 1.1 \times 1000$$

$$313.8 + x_{\frac{1}{2}} \times 0.95 \times 40000 = (1 + x_{\frac{1}{2}})1100 = 1100 + 1100 x_{\frac{1}{2}}$$

$$38000\, x_{\frac{1}{2}} - 1100\, x_{\frac{1}{2}} = 1100 - 313.8 = 786.2$$

$$\therefore \quad x_{\frac{1}{2}} = \frac{786.2}{36900} = \mathbf{0.0213\ kg/kg\ air}$$

Compressor work/kg air $= C_{pa}(T_{2'} - T_1) = 1\ (586.6 - 300)$

$= \mathbf{286.8\ kJ/kg}$

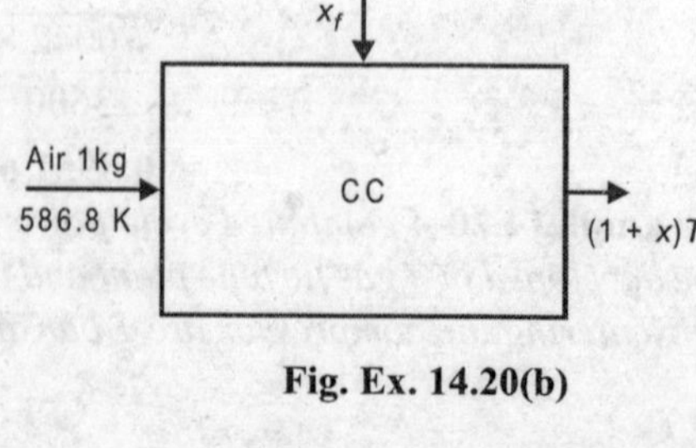

Fig. Ex. 14.20(b)

Now consider gas turbine

$$\left(\frac{T_3}{T_4}\right) = 8^{0.2481} = 1.675$$

$$\therefore \quad T_4 = \frac{1273}{1.675} = \mathbf{760\ °K}$$

$$\therefore \quad \eta_t = 0.9 = \frac{T_3 - T_{4'}}{T_3 - T_4} = \frac{1273 - T_{4'}}{1273 - 760} = \frac{1273 - T_{4'}}{513}$$

$$\therefore \quad T_{4'} = 1273 - 461.7 = \mathbf{811.3\ °K}$$

$$\therefore \quad \text{Turbine Work} = W_t = (1 + x_f) \times C_{pg}(T_3 - T_{4'})$$

$$= 1.1\ (1273 - 811.3)$$

$$= 1.1 \times 461.7 \times 1.0213 = \frac{507.87}{1.0213}\ \text{kJ/kg}$$

$$= \mathbf{518.7\ kJ/kg}$$

∴ Net gas turbine work/kg air admitted

$$= 518.7 - 286.8 = \mathbf{231.9\ kJ/kg}$$

To produce 60 MW, quantity of air required

$$\dot{m}_a = \frac{60 \times 1000}{231.9} = 258.73\ \text{kg/sec}$$

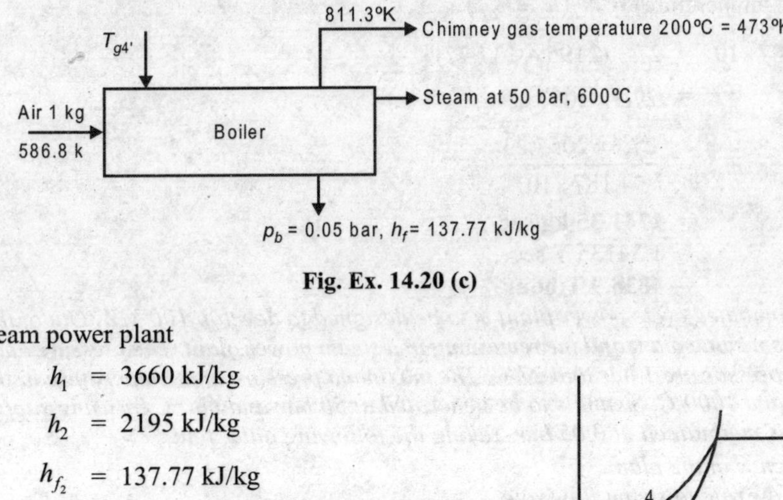

Fig. Ex. 14.20 (c)

Now, consider steam power plant

$$h_1 = 3660\ \text{kJ/kg}$$

$$h_2 = 2195\ \text{kJ/kg}$$

$$h_{f_2} = 137.77\ \text{kJ/kg}$$

$$\text{Work done/kg steam} = h_1 - h_2 = 3660 - 2195$$

$$= \mathbf{1465\ kJ/kg}$$

$$\therefore \quad \text{Mass of steam/sec} = \frac{40 \times 1000}{1465}$$

$$= \mathbf{27.3\ kg/sec}$$

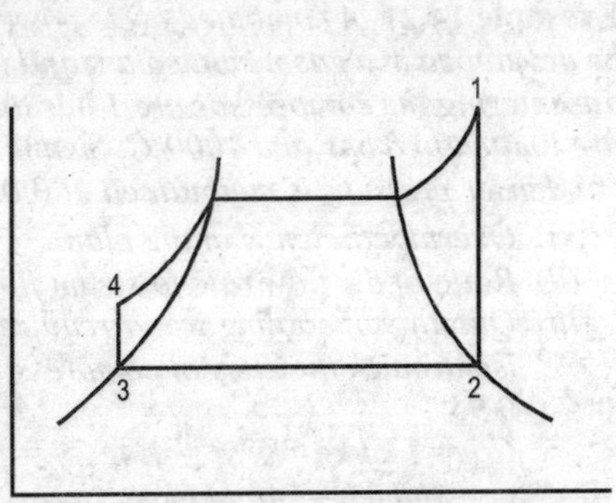

Fig. Ex. 14.20 (d)

Let t_g be the temperature of the gases coming out of the boiler combustion chamber.

∴ Heat given by gases/sec = Heat given by steam/sec

$$m_g C_{pg}(T_{g_4'} - T_{chim}) = \dot{m}_s = (h_1 - h_{f_3})$$

$$(258.73 \times 1.0213) \times 1.1 \times (811.3 - 473) = 27.3(3660 - 137.77)$$

$$98331.98 = 96156.9$$

Temperature of gases coming from gas turbine is sufficient to produce steam in the boiler and therefore, no extra fuel is required to be burned in boiler combustion chamber.

$$\therefore \quad \text{Heat supplied/sec} = \dot{m}f \times \text{C.V.}$$

$$= 258.73 \times 0.0213 \times 40000$$

$$= 220438 \text{ kJ/sec}$$

$$\text{Net work/sec} = 100{,}000 \text{ kW}$$

$$\therefore \quad \text{Overall efficiency} = \frac{100{,}000}{220438}$$

$$= 0.4536$$

$$= \mathbf{45.36\%}$$

(ii) Ratio of air flow to steam flow rate

$$= \frac{\dot{m}_a}{\dot{m}_s} = \frac{258.73}{27.3}$$

$$= \mathbf{9.48}$$

(iii) Quantity of cooling water: We have

$$\dot{m}_w \times C_{pw} \text{ (Rise in temperature)} = \dot{m}_s (h_2 - h_{2f})$$

$$\dot{m}_w \times 4.187 \times 10 = 27.3\ (2195 - 137.77)$$

$$= 27.3 \times 2057.23$$

$$\therefore \quad \dot{m}_w = \frac{27.3 \times 2057.23}{4.187 \times 10}$$

$$= \mathbf{1341.35 \text{ kg/sec}}$$

$$= 1.34135 \text{ T/sec}$$

$$= \mathbf{4828.9 \text{ T/hour}}$$

Example 14.21 *A combined cycle power plant is to be designed to develop 100 MW. Out of this, 60% power is to be developed in a gas turbine plant and the remaining in a steam power plant. The pressure and temperature of the air entering the compressor are 1 bar and 27°C. The maximum pressure and temperature in the gas turbine cycle are limited to 8 bar and 1000°C. Steam is to be generated at 50 bar and 600°C by using supplementary fuel. The condenser pressure is maintained at 0.05 bar. Taking the following data, find*

(i) Overall efficiency of the plant.

(ii) Ratio of air flow rate to steam flow rate.

(iii) Quantity of cooling water required in tons/hr. If the rise in temperature of cooling water is limited to 10°C, isentropic efficiency of compressor = 85%, isentropic efficiency of turbine = 90%, combustion efficiency = 0.95.

C_{pa} = 1 kJ/kg-°K, γ_a = 1.4, C_{pg} = 1.1 kJ/kg-°K, γ_g = 1.33.

C.V. of fuel = °40 MJ/kg.

Gas temperature going to chimney = 200°C.

Neglect feed pump work.

Consider mass fuel in gas turbine.

Solution

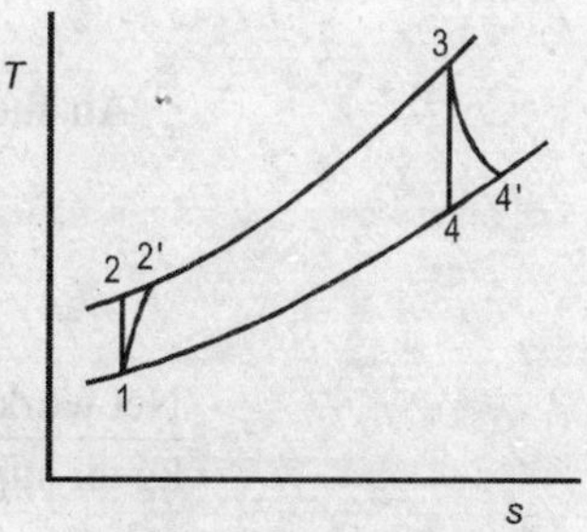

Fig. Ex.14.21 (a)

$$p_1 = 1\,\text{bar} = p_4$$

$$T_1 = 300^\circ\,\text{K}$$

$$C_{pa} = 1\,\text{kJ/kg}^\circ\text{K}$$

$$\gamma_a = 1.4$$

$$p_2 = p_3 = 8\,\text{bar}$$

$$\eta_o = 0.85$$

$$C_{pg} = 1.1\,\text{kJ/kg-}^\circ\text{K}, \gamma_g = 1.33, \eta_t = 90\%$$

$$\text{C.V. of fuel} = 40\,\text{mJ/kg} = 40000\,\text{kJ/kg}$$

$$\eta_{\text{comp.}} = 95\%,\ T_3 = 1273^\circ\text{K}$$

$$\frac{T_2}{T_1} = \left(\frac{p_2}{p_1}\right)^{\frac{\gamma_a - 1}{\gamma_a}} = (8)^{0.287} = 1.8163$$

$$T_2 = 300 \times 1.8163 = 544.9\,^\circ\text{K}$$

$$\eta_c = \frac{T_2 - T_1}{T_{2^1} - T_1} = 0.85 = \frac{544.9 - 300}{T_{2^1} - 300}$$

$$\therefore \quad T_{2^1} = 300 + \frac{244.9}{0.85} = 300 + 288.12 = \mathbf{588.12\,^\circ K}$$

Similarly,

$$\frac{T_3}{T_4} = \left(\frac{p_3}{p_4}\right)^{\frac{\gamma - 1}{\gamma}} = (8)^{\frac{0.33}{1.33}} = 8^{0.24812} = \mathbf{58.12\,^\circ K}$$

$$\therefore \quad T_4 = \frac{1273}{1.6752} = \mathbf{759.91^\circ K}$$

$$\therefore \quad \eta_t = 0.9 = \frac{T_3 - T_{4'}}{T_3 - T_4} = \frac{1273 - T_{4'}}{1273 - 759.91} = \frac{1273 - T_{4'}}{513.09}$$

$$\therefore \quad T_{4^1} = 1273 - 0.9 \times 513.09$$

$$= 1273 - 461.78$$

$$= \mathbf{811.22^\circ K}$$

Now, we will consider combustion chamber of gas turbine. Let x be fuel supplied per kg of air.

$$\therefore \quad C_{pg} \times \dot{m}_g \times t_3 - \dot{m}_a C_{pa} T_{2^1} = x \times \text{C.V.} \times \eta_{\text{comb}}$$

$$(1 + x) \times 1.1 \times 1273 - 1 \times 1 \times 588.12 = x \times 40000 \times 0.95$$

$$\therefore \quad 1400.3 + 1400.3\,x - 588.12 = 40000x \times 0.95$$

$$812.18 = x(40000 \times 0.95 - 1400.3) = 36599.7$$

$$\therefore \quad x = \frac{812.18}{36599.7} = 0.022191 \frac{\text{kg of fuel}}{\text{kg air}}$$

$$\therefore \quad \text{Air fuel ratio} = \frac{1}{0.022191}$$

$$\frac{A}{F} = 45.06 \text{ kg air/kg fuel}$$

$$\therefore \quad \frac{\text{Net work done}}{\text{kg air supplied}} = W_t - W_c = m_g C_{pg}(T_3 - T_4) - m_a C_{pa}(T_{2'} - T_1)$$

$$= 1.021041 \times 1.1(1273 - 811.22) - 1 \times 1 \times (588.12 - 300)$$

$$= 1.123145 \times 461.78 - 288.12$$

$$= 518.65 - 288.12 = \mathbf{230.53\ kJ/kg}$$

Power produced by gas turbine

$$= 0.6 \times 100 \times 1000 \text{ kW} = 60{,}000 \text{ kW}$$

∴ Mass of air required/sec

$$\frac{60000}{230.53} = \mathbf{260.27\ kg/sec}$$

In gas-turbine plant, fuel used per second

$$= 260.27 \times 0.022191$$

$$= \mathbf{5.7756\ kg/s}$$

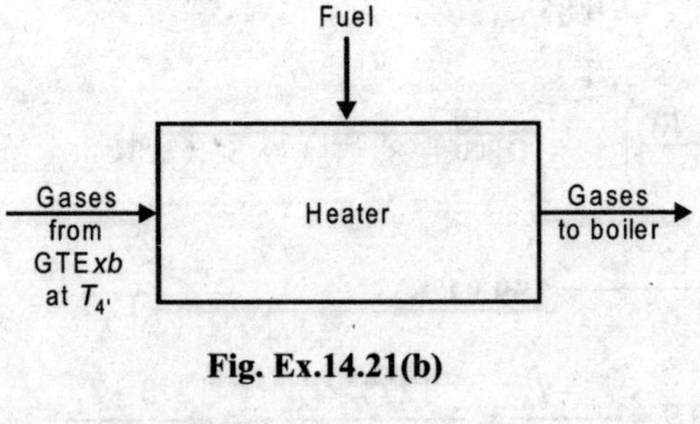

Fig. Ex.14.21(b)

From chart, $h_1 = 3680$ kJ/kg

$h_2 = 2200$ kJ/kg

$$\therefore \quad \text{Work done/kg steam} = h_1 - h_2$$

$$= 3680 - 2200$$

$$= \mathbf{1480\ kJ/kg}$$

Power produced by steam turbine

$$= 0.4 \times 100 \times 1000 = 40000 \text{ kW}$$

∴ Amount of steam required per sec

$$= \frac{40000}{1480} = \mathbf{27.03\ kg\ steam/sec}$$

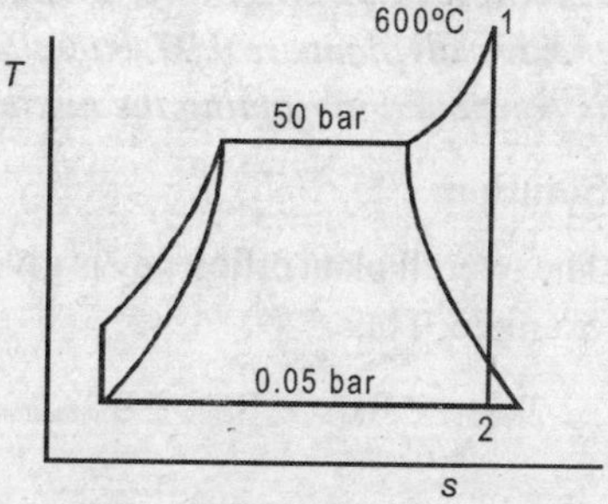

Fig. Ex. 14.21 (c)

$\therefore$ Amount of heat required/sec $= 27.03\left[h_1 - h_{f_2})\right]$

$= 27.03\ [3680 - 137.77]$

$= 27.03 \times 3542.23 = \mathbf{95746.5\ kJ/sec}$

$$\therefore\ 266.0456 \times 1.1 \times \left[\begin{pmatrix}\text{Temperature at}\\ \text{boiler net}\end{pmatrix} - \begin{pmatrix}\text{Temperature of gases}\\ \text{at chimney inlet}\end{pmatrix}\right] = 95746.5$$

$$\therefore \begin{bmatrix}\text{Temperature of}\\ \text{gases at boiler inlet}\end{bmatrix} - 200 = \frac{95746.5}{265.75 \times 1.1}$$

$= 327.53°\text{C}$

$\therefore$ Temperature of gasesat boiler inlet or heater exhaust $= 327.53 + 200$

$= 537.53\ °\text{C}$

$= 810.53\ °\ \text{K}$

This is more or less equal to the temperature of gases leaving the turbine and hence no heat is added in the heater.

$\therefore$ Ratio of air flow rate to steam flow rate

$$= \frac{260.27}{27.03} = 9.63$$

$$\frac{m_a}{m_s} = 9.63$$

$$\text{Overall efficiency of the plant} = \frac{100 \times 1000}{5.48 \times 40000} = \frac{\text{Work}}{\text{Heat supplied}}$$

$$= \frac{10}{5.7756} = 0.4328$$

$$\eta_o = \mathbf{43.48\%}$$

Heat to be rejected to cooling water in condenser/sec

$= 27.03 \times DF \times h_{fg}$

$= 27.03 \times 0.852 \times 2423.8$

$= \mathbf{55819\ kJ/sec}$

$= 55819 \times 3600\,\text{kJ/hr}$

$= \dot{m}_w \times C_{pw} \times \text{Rise in temperature}$

$= \dot{m}_w \times 4.187 \times 10$

$$\dot{m}_w = \frac{55819 \times 3600}{4.187 \times 10}\text{kg/hr}$$

$$= \frac{55819 \times 3600}{4.187 \times 10 \times 1000}\ \text{tons/hr}$$

$= 4799.34$ tons/hr

$= 4800$ tons/hr

$\dot{m}_w = \mathbf{4800\ tons/hr}$

Example14.22 *In a power plant, the efficiencies of the electric generator, turbine (mechanically) boiler cycle and the overall plant are 0.97, 0.95, 0.92, 0.42 and 0.33 respectively. What percentage of the total electricity generated is consumed in running the auxiliaries?*

Solution

The overall plant efficiency is given by by the multiplication of all the other efficiencies mentioned in the text of the example. Thus,

$$\eta_{plant} = \eta_{boiler} \times \eta_{cycle} \times \eta_{turbine(mech)} \times \eta_{generator} \times \eta_{auxiliaries}$$

$$\therefore \quad \eta_{auxiliaries} = \frac{0.33}{0.92 \times 0.42 \times 0.95 \times 0.97}$$

$$= 0.9268$$

∴ The losses in the auxiliaries = 1 – 0.9268 = 0.0732 i.e., **7.32%**

or 7.32% of total electrical power generated is consumed in running the accessories.

Example 14.23. A cyclic steam power plant is to be designed for a steam temperature at turbine inlet of 360°C and an exhaust pressure of 0.08 bar. After isentropic expansion of steam in the turbine, the moisture content at the turbine exhaust is not to exceed 15%. Determine the greatest allowable steam pressure at the turbine inlet, and calculate the Rankine cycle efficiency for these steam conditions. Estimate also the mean temperature of heat addition.

Solution

Refer Fig. Ex. 14.23 (a)

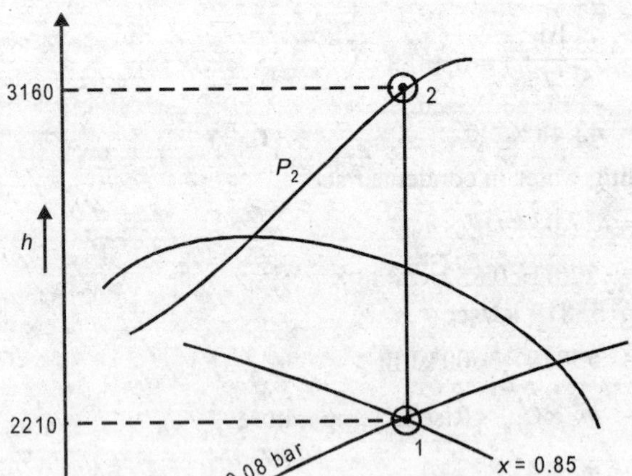

Fig. Ex. 14.23 (a)

The example is best solved with the help of the Mollier diagram. Locate point 1 on 0.08 bar and $x = 0.85$. Go straight upon constant entropy line till you get temperature equal to 360°C and locate point 2. Read off pressure on pressure line passing through point 2. It is 16.5 bar which is required pressure.

$\therefore$ Inlet pressure = **16.5 bar**

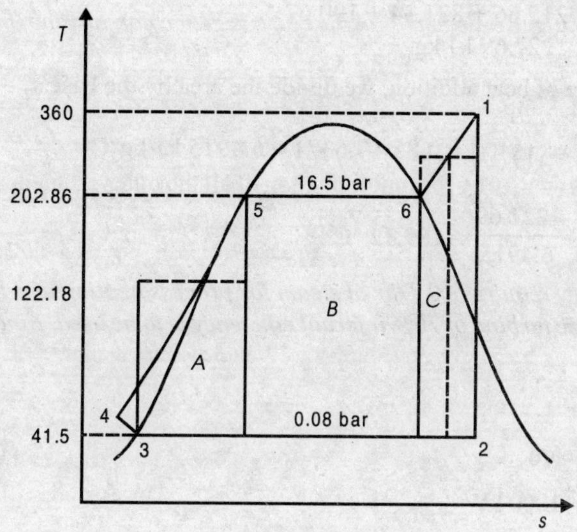

Fig. Ex.14.23 (b)

Now to find cycle efficiency read off h_2 at point 2 = 3160 kJ/kg and h_1 at point 1 = 2210 kJ/kg then work done or turbine work,

$$W_T = 3160 - 2210 = 950 \text{ kJ/kg}$$

To find heat supplied, find h_{f1} = 173.86 at 0.08 bar from the steam tables then, neglecting the pump work,
Heat suuplied during the cycle, $q_s = 3160 - 173.86 = 2986.14$ kJ/kg

The cycle efficiency, $\eta_R = \frac{W_T}{q_s} = \frac{950}{2986.14} = \mathbf{0.3181}$ or **31.81%**

To find mean temperature of heat addiction:

We first find average temperatures of $T_5 - T_3, T_6 - T_5$ and $T_1 - T_6$.

Average temperature of $T_5 - T_3$

$$= \frac{202.86 + 41.5}{2}$$

$$= 122.18 \text{ °C}$$

Average of $T_5 - T_6$ is 202.86°C

Average of $T_6 - T_1$

$$= \frac{202.86 + 360}{2}$$

$$= 281.43 \text{ °C}$$

We now find total area of *A*, *B* and *C*. Fig. Ex. 21.23(b)

$$\text{Total Area} = \text{Area } A + \text{Area } B + \text{Area } C$$

$$= 122.18(s_5 - s_3) + 202.86(s_6 - s_5) + 281.43(s_1 - s_6)$$

$$= 122.18(2.3576 - 0.5925) + 202.86(4.0489) + 281.43(7.084 - 6.4065)$$

$$= 215.66 + 821.34 + 190.67$$
$$= 1227.69 \text{ kJ/kg}$$

$\therefore$ To find mean temperature of heat addition, we divide the area by the base $s_2 - s_3$

$$\therefore \quad s_2 - s_3 = x_2\left(s_{f_{g2}}\right) = 0.85 \times 7.6371 = 6.4915 \text{ kJ/kg°C}$$

$$T_m = \frac{1227.69}{6.4915} = \mathbf{189.12\ °C}$$

Example 14.24 *A textile factory requires 10 T/hr of steam for process heating at 3 bar saturated and 1000 kW of power, for which a back pressure turbine of 70% internal efficiency is to be used. Find the steam condition required at inlet of the turbine.*

Solution

If $\dot{m}$ =mass flow rate of steam,

$$\dot{m}\left(h_1 - h_{2^1}\right) = 1000 \text{ kW}$$

$$\therefore \quad h_1 - h_{2^1} = \frac{1000 \times 3600}{10,000} = 360 \text{ kJ/kg}$$

With reference to Fig. Ex. 14.24

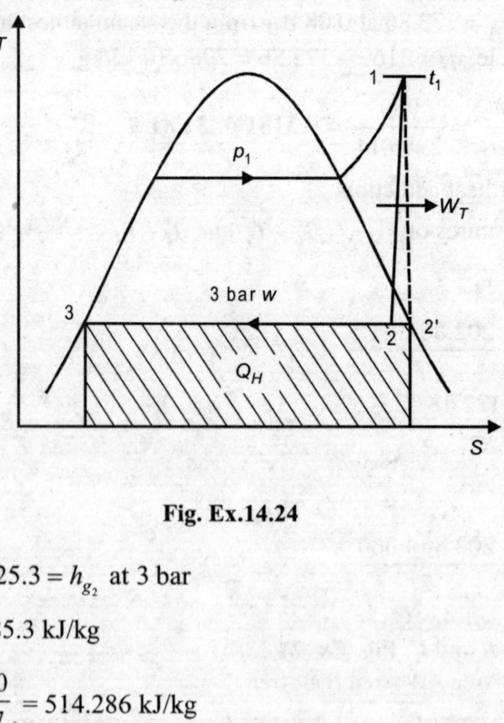

Fig. Ex.14.24

$$\therefore \quad h_{2^1} = 2725.3 = h_{g_2} \text{ at 3 bar}$$

$$\therefore \quad h_1 = 3085.3 \text{ kJ/kg}$$

Now, $$h_1 - h_2 = \frac{360}{0.7} = 514.286 \text{ kJ/kg}$$

$$h_2 = 3085.3 - 514.286$$
$$= 2571.014 \text{ kJ/kg}$$

$$\therefore \quad h_2 = h_f + x_2 h_{f_g} = 561.47 + x_2 \times 2163.8 = 2571.014$$

$$x_2 = 0.9287$$

$$s_2 = 1.6718 + 0.9287\ (6.9919 - 1.6718) = 6.6125 \text{ kJ/kg °K}$$

$$= s_1$$

Corresponding to $h_1 = 3085.3$ kJ/kg

and $s_1 = 6.6125$ kJ/kg °K, from the mollier diagram,

$p_1 = \mathbf{37.3}$ bar, $t_1 = \mathbf{344°C}$, which is the state of steam at turbine inlet.

Example 14.25 *A steam power plant operates between a boiler pressure of 42 bar and condenser pressure of 0.035 bar. The steam is superheated to 500°C. Neglecting feed pump work, compare the performance of the plant if reheat is included. The steam condition at inlet to the turbine is same in reheat cycle and also condenser pressure. Assume that steam is just dry saturated on leaving the first turbine and is reheated to its initial temperature.*

Solution

Here performance means cycle efficiency and specific steam consumption with and without reheat.

Case (a) without reheat:

At 42 bar and 500°C, we have

$$h_1 = 3442.7 \text{ kJ/kg and } s_1 = 7.0662 \text{ kJ/kg°K}$$

At 0.035 bar we have

$$s_{f_2} = 0.3907 \text{ and } s_{f_{g2}} = 8.1325$$

Also $h_{f_2} = 111.85$ and $h_{f_{g2}} = 2438.5$

To find x_2, we equate entropies

$$\therefore \quad s_1 = s_2 = s_{f_2} + x_2 s_{f_{g2}}$$

$$\therefore \quad 7.0662 = 0.3907 + x_2 8.1325$$

$$x_2 = 0.821$$

$$\therefore \quad h_2 = 111.85 + 0.821 \times 2438.5 = 2113.5 \text{kg}$$

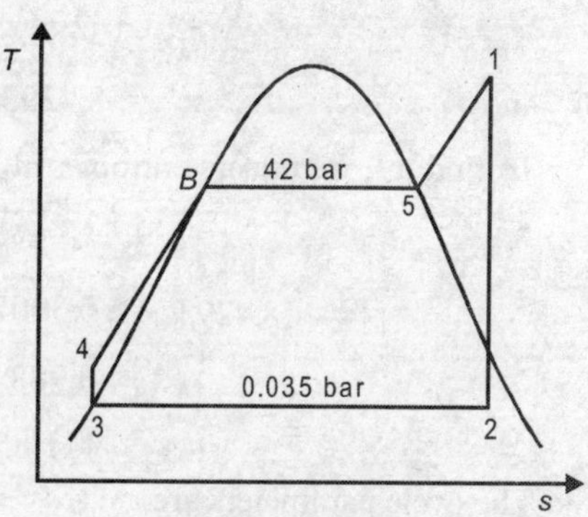

Fig. Ex. 14.25 (a)

We now determine cycle parameters. For one kg of steam

1. Heat supplied, $q_s = h_1 - h_3$ (Pump work neglected)

$$= 3442.7 - 111.85 = 3330.85 \text{ kJ/kg}$$

2. Turbine work, $W_T = h_1 - h_2 = 3442.7 - 2113.5 = 1329.2$ kJ/kg

3. Cycle efficency, $\eta_R = \dfrac{W_T}{q_s} = \dfrac{1329.2}{3330.85} = \mathbf{0.3991 = 39.91\%}$

4. Specific steam consumption, S. S. C.

$$= \frac{3600}{W_T} = \frac{3600}{1329.2} = \mathbf{2.708 \text{ kg/kWh}}$$

Case (b) with reheat:

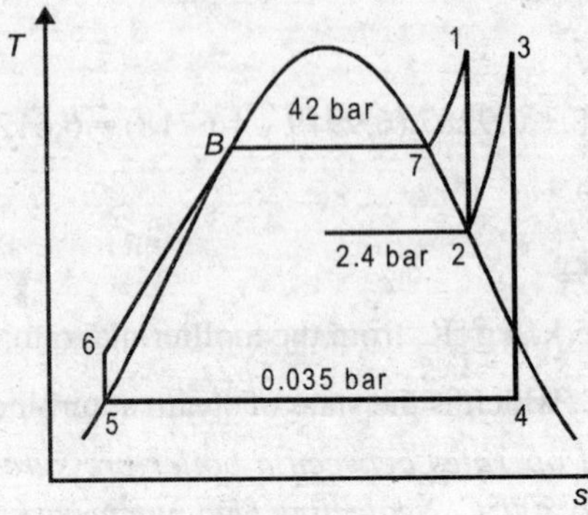

Fig. Ex. 14.25 (b)

Steam condition at 2 is dry saturated.

$$\therefore \quad s_1 = s_2 = s_{g_2} = 7.0662$$

$\therefore$ From steam tables, $P_2 = 2.4$ bar

Also from steam tables,

$$h_2 = h_{g_2} = 2714.5 \text{ kJ/kg}$$

At 2.4 bar and 500ºC

$$h_3 = 3486.6 \text{ kJ/kg}$$

and $$s_3 = 8.4293 \text{ kJ/kg-K}$$

To find x_4, we equate entropies at 3 and 4.

$$\therefore \quad s_3 = s_4 = s_{f_4} + x_4 s_{f_{g4}}$$

$$\therefore \quad 8.4293 = 0.3907 + x_4 8.1325$$

$$x_4 = 0.988$$

$$\therefore \quad h_4 = 111.85 + 0.988 \times 2438.5 = 2522.2$$

The cycle parameters are,

1. Heat supplied during the cycle,

$$q_s = (h_1 - h_5) + (h_3 - h_2)$$
$$= (3442.7 - 111.85) + (3486.6 - 2714.5)$$
$$= 3330.85 + 772.1 = 4102.95 \text{ kJ/kg}$$

2. Turbine work, $W_T = (h_1 - h_2) + (h_3 - h_4)$

$$= (3442.7 - 2714.5) + (3486.6 - 2522.2)$$
$$= 728.2 + 964.4 = 1692.6 \text{ kJ/kg}$$

3. Cycle efficiency $= \dfrac{W_T}{q_s} = \dfrac{1692.95}{4102.95}$

$$= 0.4125 = \mathbf{41.25\%}$$

4. Specific steam consumption. S. S. C.

$$= \frac{3600}{1692.6} = \mathbf{2.127\ kg/kWh}$$

Comments:

With reheat following improvements are observed.

1. Cycle efficiency increases from 39.91 % to 41.25 %. An increase of 3.36 %. The increase in efficiency is small as reheat pressure is very low.
2. S.S.C. is reduced considerably from 2.708 to 2.127 kg/kWh reduction of 21.45 %.

Example 14.26 *An ideal steam power plant operates between 70 bar, 550°C and 0.075 bar. It has seven feedwater heaters. Find the optimum pressure and temperature at which the heaters operate.*

Solution

With reference to Fig. Ex. 14.26

$$T_B = \text{Saturation temperature at 70 bar}$$
$$= 285.9\ °C$$

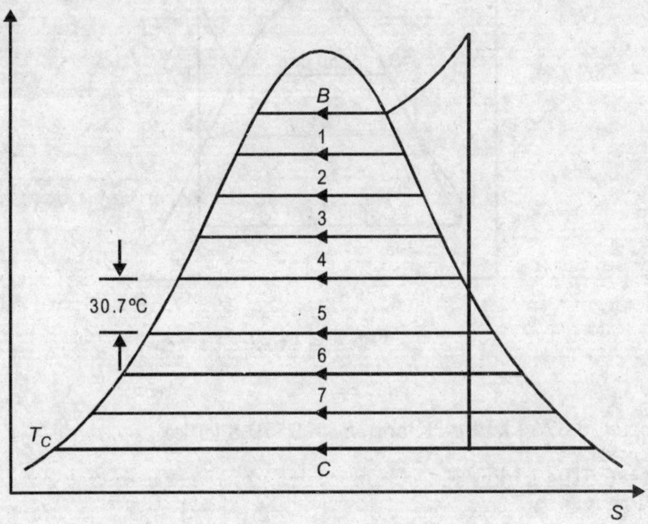

Fig. Ex.14.26

$$T_C = \text{Saturation temperature at 0.075 bar}$$
$$= 40.3\ °C$$

The temperature rise per heater for maximum cycle efficiency

$$= \frac{285.9 - 40.3}{7+1}$$
$$= 30.7°C$$

Heater 1 : $t_1 = 285.9 - 30.7 = 255.2°\text{C}$

: $p_1 = 43.3 \text{ bar (from steam tables)}$

Heater 2 : $t_2 = 255.2 - 30.7 = 224.5°\text{C}; p_2 = 25.32$ bar

Heater 3 : $t_3 = 193.8°\text{C}; p_3 = 13.67$ bar

Heater 4 : $t_4 = 163.1°\text{C}; p_4 = $ **6.71 bar**

Heater 5 : $t_5 = 132.4°\text{C}; p_5 = 2.91$ bar

Heater 6 : $t_6 = 101.7°\text{C}; p_6 = 1.1$ bar

Heater 7 : $t_7 = 71.0°\text{C}; p_7 = 0.327$ bar

Example 14.27 *A steam power plant operating on the ideal regenerative Rankine cycle with one open feed water heater. Steam enters the turbine at 150 bar and 600°C and is condensed at a pressure of 10 kPa. Steam is bled at 12 bar and enters the open feed water heater. Determine the fraction of steam extracted from the turbine and the thermal efficiency of the cycle and specific steam consumption. Consider pump work.*

Solution

First we determine enthalpies at various points. See Fig. Ex. 14.27

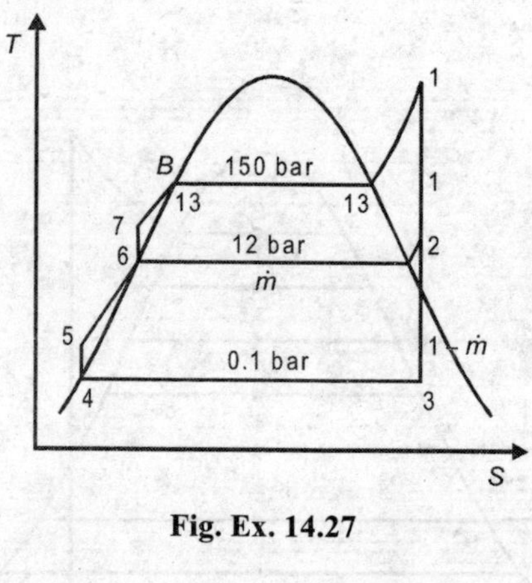

Fig. Ex. 14.27

State 1: At 150 bar 600°C

$$s_1 = 6.6764 \text{ kJ/kg-°K and } h_1 = 3579.8 \text{ kJ/kg}$$

State 2: At 12 bar, $s_{g_2} = 6.5194$ kJ/kg-°K

Since $s_1 > s_{g_2}$, steam at 2 is superheated.

To find T_{sup} 2, equate s_1 and s_2.

$$6.6764 = 6.5194 + 2.1 \ln \frac{T_{sup} 2}{273 + 188}$$

$$T_{sup} 2 = 496.78°\text{K} = 223.78°\text{C} \sim 224°\text{C}$$

At 12 bar 224°C, we find h_2 by interpolation.

$$h_2 = 2814.4 + \frac{2935.4 - 2814.4}{50} \times 24$$

$$= 2872.48 \text{ kJ/kg}$$

State 3: To find s_3 equate $s_1 = s_3$ (at 0.1 bar).

$$\therefore \quad 6.6764 = 0.6493 + x_3 7.5018$$

$$\therefore \quad x_3 = 0.803$$

$$\therefore \quad h_3 = 191.83 + 0.803 \times 2392.9 = 2114.3 \text{ kJ/kg}$$

State 4: $h_4 = 191.83$ kJ/kg

State 5: W_P at 4 = $0.001(12 - 0.1) \times 100 \sim 1.2$ kJ/kg

$$\therefore \quad h_5 = h_4 + W_P = 193.83 + 1.2 = 193 \text{ kJ/kg}$$

State 6: $h_6 = 798.43$

State 7: W_P at 6 = $0.001(150 - 12) \times 100 = 13.8$ kJ/kg

$$\therefore \quad h_7 = h_6 + W_P = 798.43 + 13.8 = 812.23 \text{ kJ/kg}$$

Let $\dot{m}$ be the quantity of extracted. To find $\dot{m}$, we write energy balance for the open FWH.

$$\therefore \quad \dot{m}(h_2 - h_6) = (1 - \dot{m})(h_6 - h_5)$$

and $\quad \dot{m}(2872.48 - 798.43) = (1 - m)(798.43 - 193)$

$$\therefore \quad \frac{1 - \dot{m}}{\dot{m}} = \frac{2074.05}{605.43} = 3.426$$

$$\therefore \quad \dot{\mathbf{m}} = \mathbf{0.226 \text{ kJ/kg of steam supplied}}$$

We now determine the cycle parameters.

1. Heat supplied in the cycle, $q_s = h_1 - h_7$

$$= 3579.8 - 812.23 = 2767.57 \text{ kJ/kg}$$

2. Turbine work, $W_T = (h_1 - h_2) + (1 - \dot{m})(h_2 - h_3)$

$$= (3579.8 - 2872.48) + (0.774)(2872.48 - 2114.3)$$

$$= 707.32 + 586.83 = 1294.15 \text{ kJ/kg}$$

$$\text{Net work} = \Delta W_{\text{net}} = W_T - (W_{P_1} + W_{P_2}) = 1294.15 - (1.2 + 13.8)$$

$$= 1279.15 \text{ kJ/kg}$$

3. Cycle efficiency $= \dfrac{\Delta W_{\text{net}}}{q_s} = \dfrac{1279.15}{2767.57} = 0.4621 \sim$ **46.21%**

4. Specific steam consumption, S. S. C.

$$= \frac{3600}{1279.15} = \mathbf{2.814 \text{ kg/kWh}}$$

Example 14.28 *In a cogeneration plant, the power load is 5.6 MW and the heating load is 1.163 MW. Steam is generated at 40 bar and 500°C and is expanded isentropically through a turbine to a condenser at 0.06 bar. The heating load is supplied by extracting steam from the turbine at 2 bar, which condensed in the process heater to saturated liquid at 2 bar and then pumped back to the boiler. Compute (a) the steam generation capacity of the boiler in t/h, (b) the heat input to the boiler in kW, (c) the fuel burning rate of the boiler in t/h if a coal of calorific* *value 25 MJ/kg is burned and the boiler efficiency is 88%, (d) the heat rejected to the condenser, (e) the rate of flow of cooling water in the condenser if the temperature rise of water is 6°C. Neglect pump work.*

Solution

With reference to Fig. Ex. 14.28

$$h_1 = 3445.3 \text{ kJ/kg}, \; s_1 = 7.0901 = s_2 - s_3$$

$$7.0901 = 1.5301 + x_2 \times 5.5970 \text{ or } x_2 = \frac{5.56}{5.597} \approx 1.0$$

$$\therefore \quad h_2 = h_g = 2706.7 \text{ kJ/kg}$$

$$h_2 - h_6 = h_{f_g} = 2201.9 \text{ kJ/kg}$$

If $\dot{m}$ is the rate of steam extraction for process heating,

$$\dot{m}(h_2 - h_6) = 1.163 \times 10^3$$

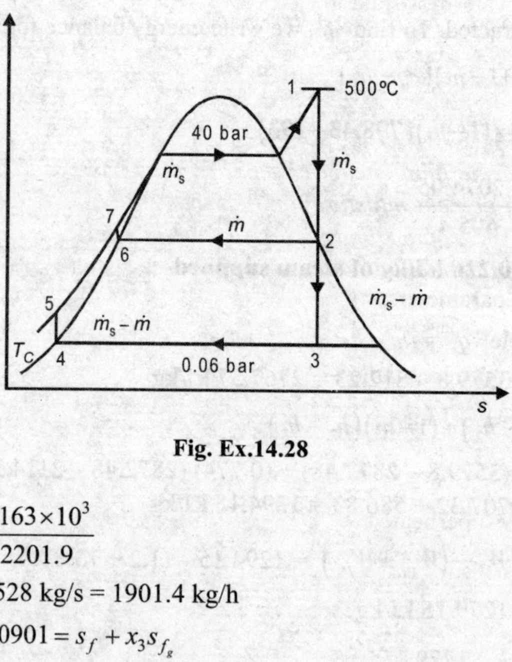

Fig. Ex.14.28

$$\dot{m} = \frac{1.163 \times 10^3}{2201.9}$$

$$= 0.528 \text{ kg/s} = 1901.4 \text{ kg/h}$$

$$s_1 = 7.0901 = s_f + x_3 s_{f_g}$$

$$= 0.52 + x_3 \times 7.815$$

$$\therefore \quad x_3 = 0.84$$

$$\therefore \quad h_3 = 149.79 + 0.84 \times 2416 = 2180.59 \text{ kJ/kg}$$

Total work output, $W_T = \dot{m}_s(h_1 - h_2) + (\dot{m}_s - \dot{m})(h_2 - h_3)$

$$\therefore \quad 5.6 \times 10^3 = \dot{m}_s \times 738.6 + \dot{m}_s \times 526.11 - 277.8$$

$$\therefore \quad \dot{m}_s = \frac{5877.8}{1264.7} = 4.648 \text{ kg/s} = 16731 \text{ kg/h} = \mathbf{16.73 \text{ t/h}}$$

$$h_7 = 5.407+1.061\times10^{-3}(40-2)\times100 = 508.73 \text{ kJ/kg}$$

$$h_5 = 149.79+1.006\times100\times40\times10^{-3} = 153.8 \text{ kJ/kg}$$

$$Q_1 = (\dot{m}_s-\dot{m})(h_1-h_5)+\dot{m}(h_1-h_7)$$

$$= (4.648-0.528)(3445.3-153.8)+0.528(3445.3-508.73)$$

$$= 4.120\times3291.5+0.528\times2936.57 = 15111.5 \text{ kJ/s}$$

$$= \mathbf{15.111\ MW}$$

$$\eta_{\text{boiler}} = \frac{Q_1}{\dot{m}f\times\text{C.F.}} = \frac{15.111}{\dot{m}f\times25} = 0.88$$

$$\dot{m}f = 0.687 \text{ kg/s} = 2473.2 \text{ kg/h} = \mathbf{2.473\ t/h}$$

$$Q_2 = (\dot{m}_s-\dot{m})(h_3-h_4) = 4.12\times2030.8 = 8367 \text{ kW}$$

$$= \mathbf{8.367\ mW}$$

If m_w = water flow rate in condensor,

then, $Q_2 = \dot{m}_w C_p (t_2-t_1)$

$$\dot{m}_w = \frac{8367}{4.187\times6} = 333.05 \text{ kg/s} = \mathbf{0.333\ m^3/s}$$

Example 14.29 *In a Rankine cycle plant steam is supplied to turbine at 32 bar, 400°C and is expanded to a pressure of 0.16 bar. Calculate percent moisture and thermal efficiency if the steam is reheated at 6 bar to a temperature of 350°C. Assume isentropic efficiency of turbine 80% throughout. If the turbine is put up with one extraction point (for optimum results) and an open feed heater, recalculate thermal efficiency.*

Solution

Case (I) Reheat cycle

At 32 bar 400°C, we have from steam tables

$S_1 = 6.8912$ and $h_1 = 3229.2$ kJ/kg

at 6 bar $s_{g_2} = 6.7575$ kJ/kg-K

as $s_1 > s_{g_2}$, the steam at 2 is superheated

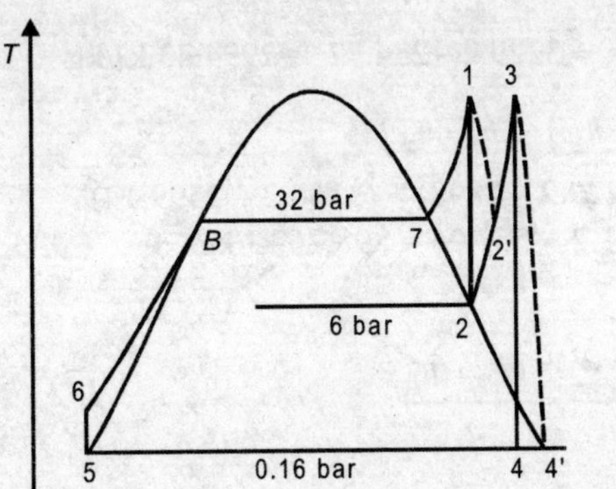

Fig. Ex.14.29 (a)

To find T_{sup} at 2, we have

$$6.8192 = 6.7545 + 2.1 \ln \frac{T_{sup^2}}{273 + 158.84}$$

$$\therefore \quad T_{sup^2} = 460.23 - 273 = 187.23°C$$

$$\therefore \quad h_{sup^2} = 2755.5 + 2.1\,(187.23 - 158.84)$$

$$\therefore \quad h_2 = 2815.1 \text{ kJ/kg}$$

At 6 bar 350°C $h_3 = 3166.2$ kJ/kg and $s_3 = 7.5479$ kJ/kg-°K

To find x_4 equate $s_3 = s_4$

$$7.5479 = 0.7721 + x_4 7.2148$$

$$\therefore \quad x_4 = 0.939 \qquad \therefore \textbf{Moisture} = \mathbf{1 - 0.939 = 0.061 \text{ or } 61\%}$$

$$\therefore \quad h_4 = 231.5 + 0.939 \times 2370 = 2457.38 \text{ kJ/kg}$$

Considering turbine efficiencies, we find h_{2^1} and h_{4^1} as follows

$$0.8 = \frac{h_1 - h_{2^1}}{h_1 - h_2} \quad \therefore \quad h_{2^1} = h_1 - 0.8(h_1 - h_2)$$

$$= 3229.2 - 0.8\,(3229.2 - 2815.1)$$

$$\therefore \quad h_{2^1} = 2898 \text{ kJ/kg}$$

also $$0.8 = \frac{h_3 - h_{4^1}}{h_3 - h_4} \quad \therefore \quad h_{4^1} = h_3 - 0.8(h_3 - h_4)$$

$$= 3166.2 - 0.8\,(3166.2 - 2457.38)$$

$$\therefore \quad h_{4^1} = 2599.14 \text{ kJ/kg}$$

Let us determine cycle parameters.

1. Pump work $= |W_p| = (32 - 0.16)100 \times 0.001 = 3.184$ kJ/kg
2. Heat supplied $= q_s = h_1 - h_6 = 3229.2 - 234.77 = 2994.43$ kJ/kg
3. Turbine work $= W_T = (h_1 - h_{2^1}) + (h_3 - h_{4^1})$

$$= (3229.2 - 2898) + (3166.2 - 2599.14)$$

$$= 331.2 + 567.06 = 898.26 \text{ kJ/kg}$$

4. Thermal efficiency, η_{th}

$$= \frac{W_{Tnet}}{q_s} = \frac{W_T - W_P}{q_s}$$

$$= \frac{898.26 - 3.18}{2994.43} = 0.2989 \text{ or } \mathbf{29.89\%}$$

Case (II) Regenerative cycle:

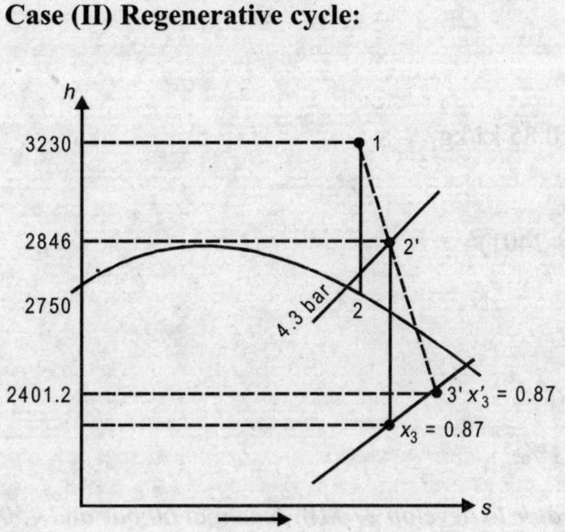

Fig. Ex.14.29 (b)

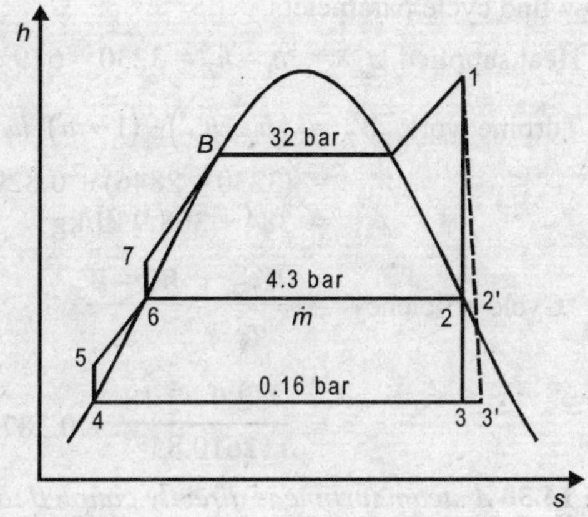

Fig. Ex.14.29 (c)

For optimum results the saturation temperature at the point of extraction is mean of saturation temperatures at inlet and condenser pressures.

$$\therefore \quad T_{opt} = \frac{T_{sat} \text{ at } 32 \text{ bar} + T_{sat} \text{ at } 0.16 \text{ bar}}{2}$$

$$= \frac{237.45 + 55.341}{2} = 146.4°\text{C}$$

Saturation pressure at 146.4 is ~ 4.3 bar (from steam tables)
Thus extraction pressure is 4.3 bar.
The example is best solved by the mollier chart Fig. Ex. 14.29 (b).

To find $h_{2'}$ with turbine efficiency = 0.8

$$h_{2'} = 3230 - 0.8\,(3230 - 2750)$$

$$= 2846 \text{ kJ/kg}$$

$$h_3 = 2290 \text{ kJ/kg}$$

$$h_{3'} = 2846 - 0.8\,(2846 - 2290)$$

$$= 2401.2 \text{ kJ/kg}$$

From Mollier chart,

$$\mathbf{x'_3} = \mathbf{0.918}$$

To find $\dot{m}$ we write energy balance equation,

$$\dot{m}(h_{2'} - h_6) = (h_6 - h_5)$$

$$h_5 = h_3 + W_P = 231.59 + 3.18$$

$$= 234.77 \text{ kJ/kg}$$

$$\therefore \quad \dot{m}(2846 - 615.97) = (615.97 - 234.77)$$

$$\therefore \quad \dot{m} = 0.171 \text{ kg/kg of steam}$$
$$1-\dot{m} = 0.829$$

We now find cycle parameters,

1. Heat supplied $q_s = h_1 - h_7 = 3230 - 619.15 = 2610.85$ kJ/kg

2. Turbine work, $W_T = (h_1 - h_{2'}) - (1-m)(h_2 - h_{3'})$

$$= (3230 - 2846) + 0.829\,(2846 - 2401)$$
$$= 384 + 368.9 \text{ kJ/kg}$$

3. Cycle efficiency $= \dfrac{W_{Tnet}}{q_s} = \dfrac{W_T - W_P}{q_s}$

$$= \frac{752.9 - 3.18}{2610.85} = 0.2871 = \mathbf{28.71\%}$$

Example 14.30 *A steam turbine is directly coupled to a generator to develop 27 MW. Steam at 60 bar and 450°C is supplied to the turbine and condensed in a surface condenser. The condenser vacuum is 73 cm of Hg when the barometer reads 76 cm of Hg.*

To enhance the efficiency of the system, a regenerator feed heating is used by extracting the steam at 3 bar pressure, and directly mixing with the condensate. Assuming turbine efficiency for each portion of expansion as 87%, find:

(i) mass of steam bled, per kg of steam supplied to the turbine.

(ii) steam flow rate supplied to the turbine in ton per hour.

(iii) assuming 10 % generator power is used to run the pumps and accessories, find overall efficiency of the plant.

Take the following:

Boiler efficiency = 0.9, Alternator efficiency = 0.95, Mechanical efficiency = 0.98

Draw a neat line diagram of all components rents of plant and represent the cycle on h-s diagram.

Solution

A schematic flow diagram and the cycle on *h-s* diagram are drawn. refer Fig. Ex. 14.30 (a) and (b) . The example is solved by using Mollier chart. Refer Fig. Ex. 14.30. (c).

1. The condenser vacuum is 73 cm of Hg and barometer is 76 cm of Hg.

∴ Absolute pressure in the condenser

$$= \frac{76-73}{75} = 0.04 \text{ bar.}$$

2. The inlet conditions are 60 bar and 450°C. Mark point 1 and read $h_1 = 3300$ kJ/kg. Refer Fig. Ex. 14.30 (c).

3. Draw vertical line to 3 bar. The pressure at which steam is bled for open FWH. Read $h_{2'} = 2610$ kJ/kg. The turbine efficiency for this expansion is 0.87. We find h_2 by $h_{2'} = 3300 - 0.87\,(3300 - 2610) = 2699.7 - 2700$ kJ/kg.

4. Further expansion is upto 0.04 bar. Draw line 2' – 3 from 3 bar to 0.04 bar and read $h_3 = 2085$ kJ/kg. Find h_3' by $h_3' = 2700 - 0.87\,(2700 - 2085) = 2164.95 - 2615$ kJ/kg. Other enthalpies can be found out from the steam tables.

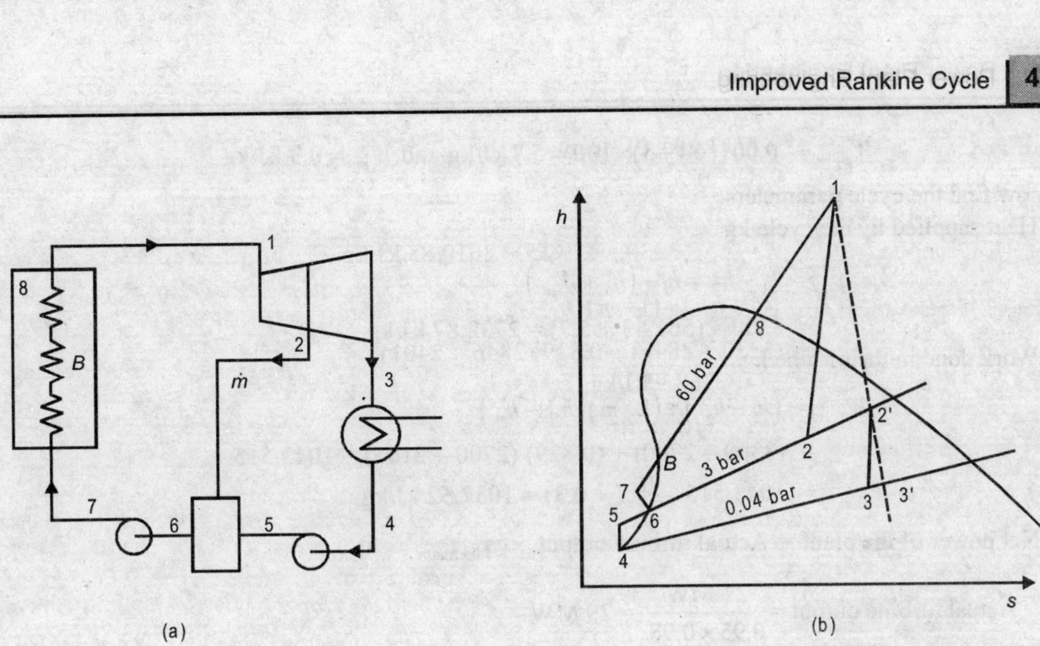

Fig. Ex.14.30

5. To find $\dot{m}$ we write energy balance for open feed water heater.

$$\dot{m}\left(h_{2'}-h_6\right) = (1-\dot{m})(h_6-h_5)$$

Now $h_5 = h_4+\left(W_{p_1}\right)$ since W_{p_1} is too small, it is neglected.

and $h_5 = h_4 = 121.43 \text{ kJ/kg} \left[W_{P_1} = 0.001(3.0-0.04)100 = 0.3 \text{ kJ/kg}\right]$

$$\therefore \quad \dot{m}(2700-561.43) = (1-\dot{m})(561.43-121.43)$$

Solving $\dot{m}$ = **0.171 kg/kg of steam**

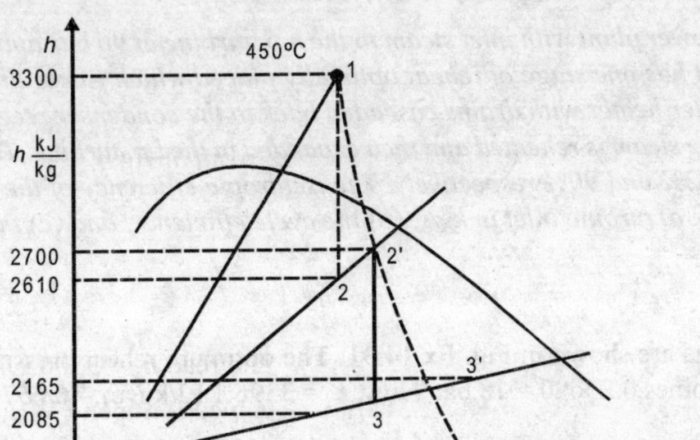

Fig. Ex. 14.30

6. $W_{p_2} = 0.001(60-3)\times 100 = 5.7$ kJ/kg and $W_{p_1} = 0.3$ kJ/kg

We now find the cycle parameters.

7. Heat supplied by the cycle/kg

$$q_s = h_1 - h_7 = h_1 - \left(h_6 + W_{p_2}\right)$$

$$= 3300 - (561.43 + 5.7) = 2732.87 \text{ kJ/kg}$$

8. Work done in the turbine/kg

$$W_T = \left(h_1 - h_{2^1}\right) + (1-m)\left(h_{2^1} - h_{3^1}\right)$$

$$= (3300 - 2700) + (0.829)(2700 - 2165) = 1043.515$$

$$W_{T\text{net}} = 1043.515 - (5.7 + 0.3) = 1037.52 \text{ kJ/kg}$$

9. Net power of the plant = Actual turbine output $\times\ \eta_{\text{alternator}} \times \eta_{\text{mech}}$

$$\therefore \quad \text{Actual turbine output} = \frac{27 \text{ MW}}{0.95 \times 0.98} = 29 \text{ MW}$$

But actual turbine output = Steam flow through the turbine kg/s × Work done/kg

$$\therefore \quad \text{Steam flow rate, } \dot{m}_s = \frac{29 \times 100}{1043.515} = \mathbf{27.79\ kg/s}$$

$$= \mathbf{100\ T/h}$$

10. Net power available deducting 10% pump power

$$= 27000 \times 0.9 = 24300 \text{ kW}$$

Total heat supplied in the boiler with 90% efficiency

$$= \frac{27.79 \times 2732.87}{0.9} = 84347.84 \text{ kW}$$

11. Overall efficiency $= \dfrac{\text{Net power}}{\text{Heat supplied}} = \dfrac{24300}{84347.84} = 0.2881 = \mathbf{28.81\%}$

Example 14.31 *A steam power plant with inlet steam to the h.p. turbine at 90 bar and 500°C, and condensation at 40°C produces 500 MW. It has one stage of reheat optimally placed which raises the steam temperature back to 500°C. One closed feedwater heater with drains cascaded back to the condenser receives bled steam at the reheat pressure, and the remaining steam is reheated and then expanded in the l.p. turbine. The h.p. and l.p. turbines have isentropic efficiencies of 92% and 90% respectively. The isentropic efficiency of the pump is 75%. Calculate* (*a*) *the mass flow rate of steam at turbine inlet in kg/s,* (*b*) *the cycle efficiency, and* (*c*) *the cycle work ratio. Use TTD = – 1.6°C.*

Solution

The flow and *T–S* diagrams are shown in Fig. Ex.14.31. The optimum reheat pressure is taken to be 20% of teh boiler pressure, which becomes 0.2 × 90 = 18 bar. Now, h_1 = 3396.1 kJ/kg, ε_1 = 6.6576 kJ/kg-K = s_2, h_2 = 2915 kJ/hg, h_3 = 3469.8 kJ/kg.

$$s_3 = 7.4825 = s_4 = s_{f_4} + x_4 s_{f_4}$$

$$\therefore \quad x_4 = \frac{7.4825 - 0.5725}{7.6845} = 0.8992$$

$$\therefore \quad h_4 = 167.57 + 0.8992 \times 2406.7 = 2331.7 \text{ kJ/kg}$$

$$h_5 = 167.57 \text{ kJ/kg}, \; h_7 = 883.42 \text{ kJ/kg}$$

$$W_{p_5} = \int vdp = 0.001008 \times 90 \times 10 = 9.072 \text{ kJ/kg}$$

$$h_6 = 176.64 \text{ kJ/kg}$$

$$h_1 - h_2 = 0.92\,(3386 - 2915) = 433.3 \text{ kJ/kg, or } h_2 = 3386.1 - 433.3$$

$$= 2952.8 \text{ kJ/kg}$$

$$h_3 - h_{4^1} = 0.9\,(3469.8 - 2331.7) = 1024.29 \text{ kJ/kg}$$

$$h_{4^1} = 3169.8 - 1024 = 2445.5 \text{ kJ/kg}$$

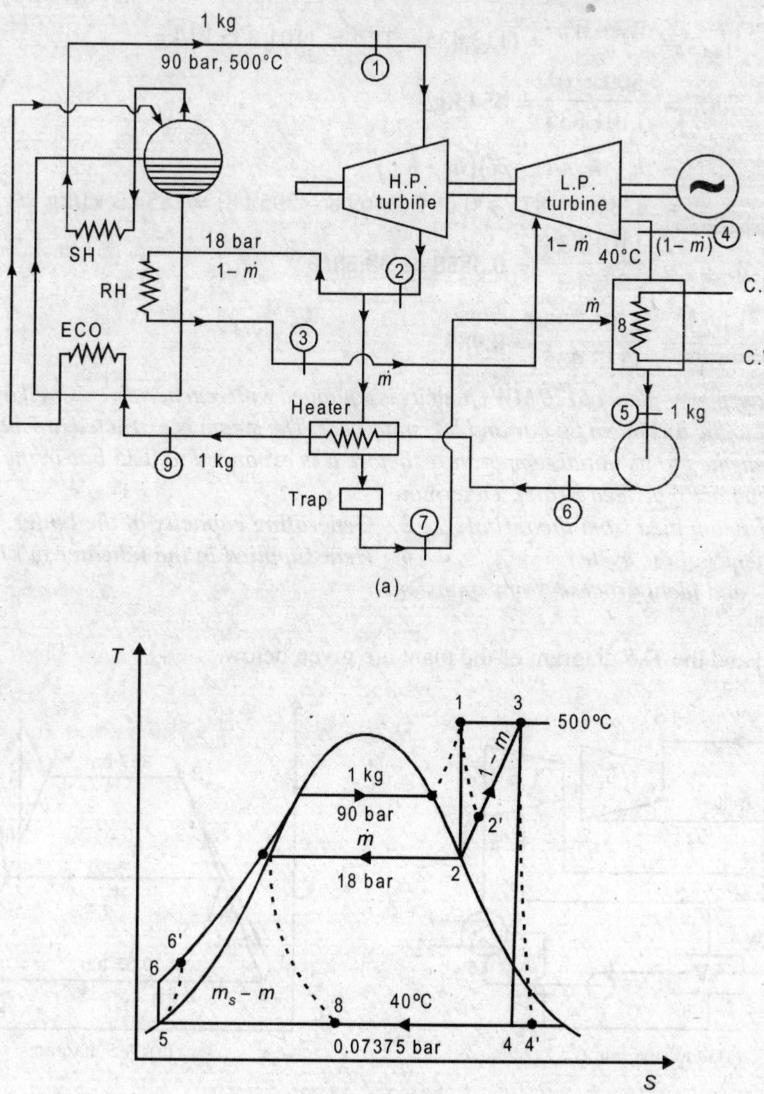

Fig. Ex. 14.31

$$W_P = h_6' - h_5 = \frac{9.072}{0.75} = 12.1 \text{ kJ/kg, or } h_6' = 167.15 + 12.1$$

$$= 179.67 \text{ kJ/kg}$$

$$t_{sat} \text{ at 18 bar} = 207.15°\text{C} \qquad t_9 = 207.15 + 1.6 = 207.75°\text{C}$$

$$h_9 = 875 \text{ kJ/kg}$$

$$1(h_9 - h_{6'}) = \dot{m}(h_2 - h_7) \quad \dot{m} = \frac{875-179.67}{2952.8-883.4} = 0.336 \text{ kg}$$

$$W_T = (h_1 - h_2') + (1-\dot{m})(h_3 - h_4') = 433.3 + 0.664 \times 1024.3$$

$$= 1113.435 \text{ kJ/kg}$$

$$W_{net} = W_T - W_P = 1113.435 - 12.1 = 1101.335 \text{ kJ/kg}$$

$$\dot{m}_s = \frac{500 \times 10^3}{1101.335} = \mathbf{454 \text{ kg/s}}$$

$$q_s = h_1 - h_9 + (1-\dot{m})(h_3 - h_{2'})$$

$$= 3386.1 - 875 + 0.664\,(3469.8 - 2952.8) = 2854.3 \text{ kJ/kg}$$

$$\eta_{cycle} = \frac{1101.335}{2854.3} = \mathbf{0.3858 \text{ or } 38.58\%}$$

$$\text{Work ratio} = \frac{W_{net}}{W_T} = \frac{1101.335}{1113.435} = \mathbf{0.989}$$

Example 14.32 *A steam power plant of 120 MW capacity is equipped with regenerative as well as reheat arrangements. The steam is supplied to the turbine at 86 bar and 50° superheat. The steam is extracted at 7 bar for feed heating and remaining steam is reheated to its initial temperature, before it is expanded to 0.35 bar in the LP stage.*

Assume indirect contact type feed heater. Determine :

1. Percentage of steam bled from the turbine *2. Generating capacity of the boiler*

3. Thermal efficiency of the cycle *4. Heat supplied in the reheater in kJ/min.*

Assume no losses and ideal processes of expansion.

Solution

The flow diagram and the *T–S* diagram of the plant are given below.

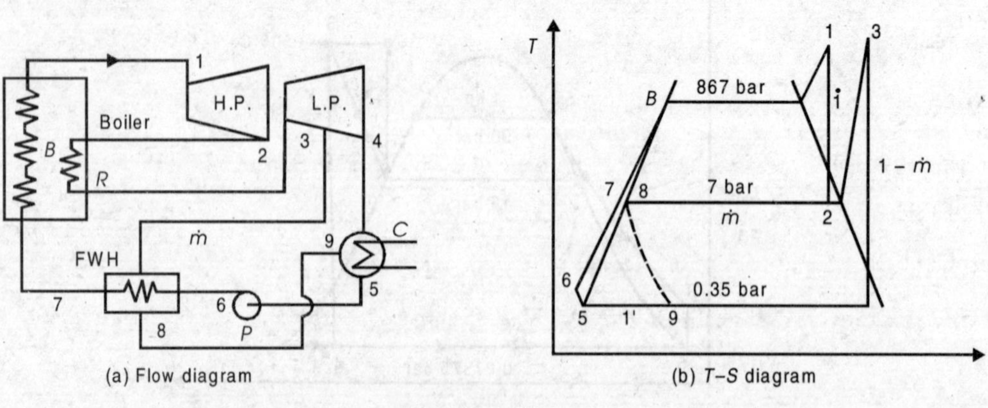

Fig. Ex. 14.32

At point 1, pressure is 86 bar and temperature = 350ºC

1. From steam table, we have $h_1 = 2971.6$ kJ/kg and $s_1 = 6.0781$ kJ/kgºK
2. Since s_{g_2} at 7 bar = 6.7052 > s_1 condition at 2 is wet.

   We have at 7 bar, $s_{f_2} = 1.9918$ and $sf_{g_2} = 4.7134$

$$\therefore \quad x_2 = \frac{6.0781-1.9918}{4.7134} = 0.867$$

$$\therefore \quad h_2 \text{ at 7 bar} = 697.06 + 0.867 \times 2064.9 = 2487.33 \text{ kJ/kg}$$

3. Steam at 7 bar is reheated to 350ºC at 7 bar 350ºC from steam tables,

$$h_3 = 3164.3 \text{ kJ/kg and } s_3 = 7.4745 \text{ kJ/kgºK}$$

4. To find dryness at point 4 equate s_3 and s_4

$$\therefore \quad 7.4745 = 0.98765 + x_4 \times 6.72915$$

$$\therefore \quad x_4 = 0.964$$

$$\therefore \quad \text{At 0.35 bar } h_4 = 304.3 + 0.964 \times 2327.2 = 2547.72 \text{ kJ/kg}$$

5. To find quantity of steam $\dot{m}$ extracted at 7 bar, we write the energy balance for feedwater heater. Thus,

$$\dot{m}[h_2 - h_8] = h_8 - h_6$$

Now, $$W_P = 0.001 \times (86 - 0.35) \times 100 = 8.25 \text{ kJ/kg}$$

$$\therefore \dot{m}[2487.33 - 697.06] = 697.06 - [304.03 + 8.25]$$

$$\therefore \quad \dot{m} = 0.215 \text{ i.e., } \mathbf{21.5\%}$$

6. Turbine work is given by,

$$W_T = (h_1 - h_2) + (1 - \dot{m})(h_3 - h_4)$$

$$\therefore \quad W_T = (2971.6 - 2487.33) + 0.785\,(3164.3 - 2547.72)$$

$$= 484.27 + 484.01 = 968.28 \sim 968.3 \text{ kJ/kg}$$

7. $$W_{net} = W_T - W_P = 968.28 - 8.25 = 960.03 \sim 960 \text{ kJ/kg}$$
8. Net turbine × work steam rate/s = net power

$$960 \text{ kJ/kg} \times \dot{m}_s \text{ kg/s} = (120 \times 10^3) \text{ kJ/s}$$

$$\therefore \quad \dot{m}_s = \frac{120 \times 10^3}{960} = \mathbf{125 \text{ kg/s i.e., } 450 \text{ T/H}}$$

$$= \textbf{Generating capacity of the boiler}$$

9. Now heat supplied in the boiler and reheater/kg,

$$q_s = (h_1 - h_8) + (1 - \dot{m})(h_3 - h_2)$$

$$= (2971.6 - 697) + 0.785(3164.3 - 2487.3)$$

$$= 2274.6 + 531.45 = 2806 \text{ kJ/kg}$$

10. $$\text{Cycle efficiency} = \frac{W_{net}}{\text{heat supplied}} = \frac{960}{2806} = \mathbf{0.3421 \text{ or } 34.21\%}$$

11. Total heat supplied in reheater = $531.45 \times 125 = 66.43$ MW

$\therefore$ Heat supplied in reheater/min. = $66.43 \times 60 =$ **3985.875 MJ/min**

Example 14.33 *In a thermal power plant of 60 MW capacity, boiler supplies steam at 100 bar and 400°C. The steam expands to 20 bar in the H.P. turbine. Steam is then reheated to 380°C in a reheater. Thereafter, steam expands to a condenser pressure of 0.05 bar. The pressure loss in the reheater is 1 bar. Assuming isentropic efficiency in both stages of expansion is 80%, find the steam generating capacity required in ton/h. Take transmission efficiency from turbine to generator = 98%, and generator efficiency of 95%. Neglect feed pump work. Also find the overall efficiency of the plant.*

Draw a neat line diagram of the plant and represent the cycle on T–S diagram.

Solution

The schematic flow diagram, the *T–S* diagram and *h–s* diagram are given below.

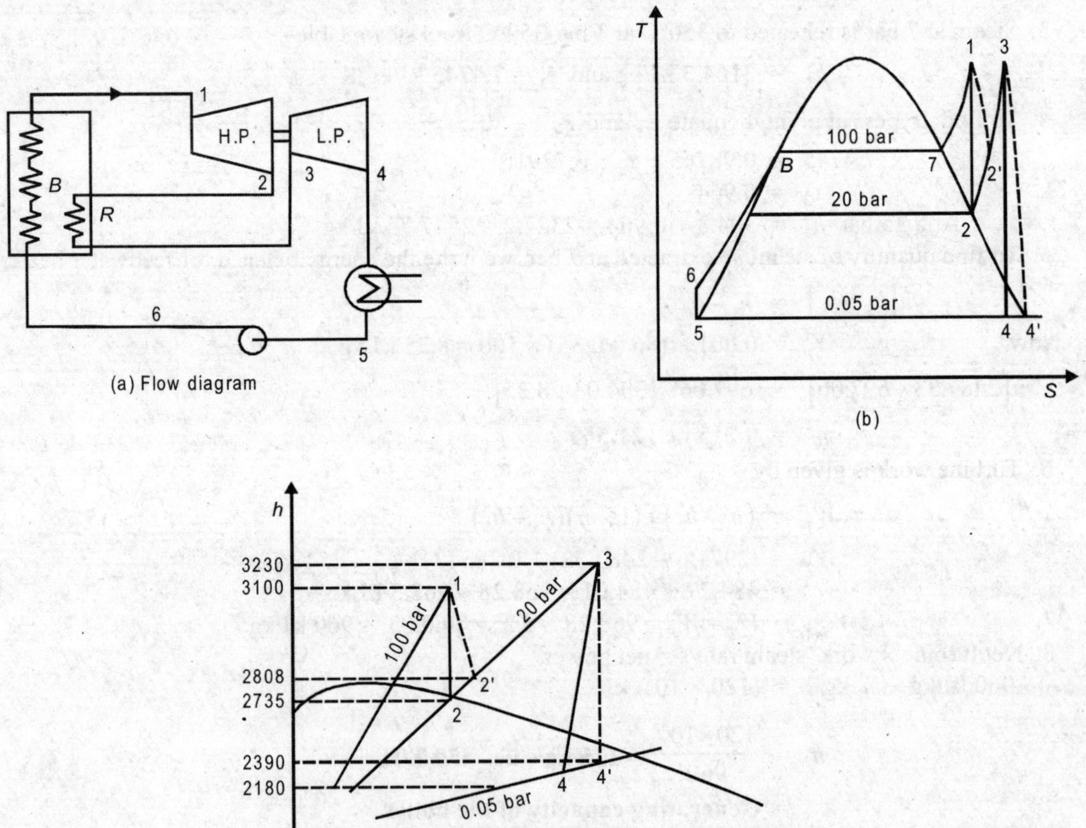

(a) Flow diagram

(b)

(c)

Fig. Ex. 14.33

The steam expands in hp turbine from 100 bar to 20 bar. From *h-s* diagram, we have h_1@100 bar = 3100 kJ/kg and h_2@20 bar = 2735 kJ/kg.

The turbine efficiency is 80%. We find $h_{2'}$ by

$$h_{2'} = h_1 - 0.8(h_1 - h_2) = 3100 - 0.8(3100 - 2735) = 2808\,\text{kJ/kg}$$

The steam is then reheated at 20 bar to 350°C

$$h_3 \text{ at 20 bar 350°C} = 3230\text{ kJ/kg}$$

The steam then expands to 0.05 bar.

$$h_4 = 2180 \text{ kJ/kg. We find } h_{4'} \text{ by}$$

$$h_{4'} = h_3 - 0.8(h_3 - h_4) = 3230 - 0.8(3230 - 2180) = 2390\,\text{kJ/kg}$$

We now find cycle parameters:

1. Work done in turbirre,

$$W_T = (h_1 - h_2) + (h_3 - h_4')$$

$$= (3100 - 2808) + (3230 - 2390) = 1132\,\text{kg}$$

Pump work is neglected $\therefore W_p = 0$ and $W_{\text{net}} = W_T$

2. Work transferred to the generator,

$$= 1132 \times 0.98 = 1109.36\,\text{kJ/kg}$$

3. Work delivered by the generator,

$$1109.36 \times 0.95 = 1053.9 \text{ kJ/kg}$$

4. Capacity of the plant = 60 MW

$\therefore$ Mass flow rate of steam

$$= \frac{60 \times 1000}{1053.9} = 56.93\,\text{kg/s} = \mathbf{204.95\ T/h}$$

= **Capacity of the steam generation**

5. Heat supplied in the boiler/kg

$$= (h_1 - h_5) + (h_3 - h_{2'})$$

$$= (3100 - 138) + (3230 - 2808)$$

$$= 2962 + 422 = 3384 \text{ kJ/kg}$$

6. Overall efficiency of the plant

$$= \frac{\text{Work delivered/kg}}{\text{Heat supplied/kg}} = \frac{1053.9}{3384} = 0.3114 = \mathbf{31.14\%}$$

Example14.34 *Steam at 392 N/cm² and 300°C is supplied to steam turbine. The steam is bled for regenerative heating at 137.2 N/cm² and 33.3 N/cm². The condenser pressure is 0.686 N/cm². Calculate*

1. *At each stage the optimum mass of bled steam per kg of steam supplied, to the turbine, and*
2. *The cycle efficiency.*

Assume turbine efficiency of 0.8 for each portion of the expansion and that the feed water leaving each heater, is raised to the steam temperature, entering the heater.

The heaters used are indirect heaters and the condensate coming out from the first heater is passed through the second heater is passed to the hot well. Neglect pump work.

Solution

The schematic flow diagram, the *T–S* diagram and *h–s* diagram are given on next page.

1. The turbine η is 0.8

$$\therefore \quad h_2' = 2740 + (2907 - 2740) \times 0.2 = 2773\,\text{kJ/kg}$$

$$\text{also} \quad h_3' = 2540 + (2773 - 2540) \times 0.2 = 2587\,\text{kJ/kg}$$

also $\quad h_4' = 2120 + (2587 - 2120) \times 0.2 = 2218\,\text{kJ/kg}$

also $\quad h_8 = h_9 = h_{10} = 826.28\,\text{kJ/kg}$ from steam tables by interpolation

and $\quad h_{11} = h_7 = 577.69\,\text{kJ/kg}$ from steam tables by interpolation

and $\quad h_6 = h_5 = W_p = 160.51\,\text{kJ/kg}$

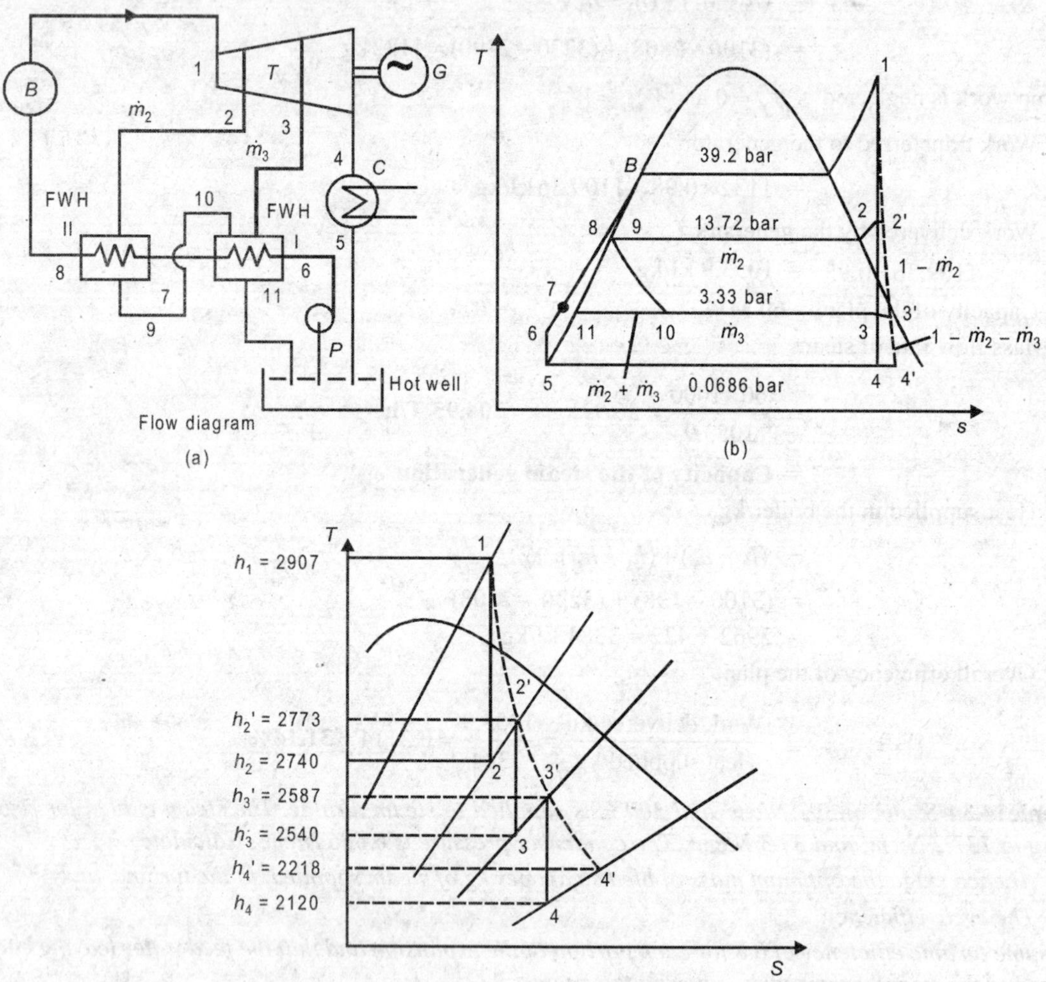

Fig. Ex. 14.34

2. The energy balance equations for feedwater heaters are,

(i) $\quad \dot{m}_2 (h_2' - h_9) = h_8 - h_7$

(ii) $\quad \dot{m}_3 (h_3' - h_{11}) + \dot{m}_2 (h_{10} - h_{11}) = h_7 - h_6$

$$\therefore \quad \dot{m}_2 = \frac{826.28 - 577.69}{2773.4 - 826.28} = \mathbf{0.128\,kg/kg\ of\ steam}$$

and $\dot{m}_3(2587-577.79)+0128(826.28-577.69)$

$$= 577.69 - 160.51 - 31.82$$

$$\therefore \quad \dot{m}_3 = \frac{385.36}{2009.2} = \mathbf{0.19\ kJ/kg\ of\ steam}$$

3. Turbine Work, W_T $= (h_1 - h_{2'}) + (1 - m_2)(h_{2'} - h_{3'})(1 - \dot{m}_2 - \dot{m}_3)(h_{3'} - h_{4'})$

$$= (2907 - 2773) + (1 - 0.128)(2773 - 2587) + (1 - 0.128 - 0.19)(2587 - 2218)$$

$$= 134 + 162.2 + 251.8 = 548 \text{ kJ/kg}$$

4. Heat supplied/kg, q_s $= h_1 - h_8 = 2907 - 826.28 = 2080.72$ kJ/kg

5. Cycle efficiency $\eta_{cycle} = \dfrac{W_t}{q_s} = \dfrac{548}{2080.72} = 0.2634 = \mathbf{26.34\%}$

Example 14.35 *A combined cycle power plant of 190 MW capacity consists of a gas turbine and steam turbine units. The exhaust gases from the gas turbine are supplied to a boiler at 800°C by supplying extra fuel in the supplementary combustion chamber. The pressure and temperature of air entering the compressor of a gas turbine are 1 bar and 15°C. The pressure ratio of the cycle is limited to 8, and maximum temperature of the cycle is 800°C. Take the following data for the steam cycle.*

Steam supplied at 60 bar, 500°C. The condenser pressure is 0.05 bar. Assuming the ideal cycle for the unit find out,

1. *Thermal efficiency of the combined cycle.*
2. *Power generated by each unit of the plant.*
3. *Boiler capacity.*
4. *Fuel supplied to the plant/hr.*

The chimney gas temperature is limited to 200°C to avoid acid condensation. Take C_p = 1.1 kJ/kg-°K and γ = 1.33 for gas C.V. of fuel = 40000 kJ/kg.

Sketch schematic plant and T–S diagrams for both the units.

Solution

The schematic flow diagram and *T–S* diagrams for both the units are shown on next page.

(i) Gas turbine unit:

1. $T_a = 288\,K, P_1 = 1 = 1$ bar and pressure ratio = 8. To find T_b, we have

$$\frac{T_b}{T_a} = \left(\frac{P_b}{P_a}\right)^{\frac{r-l}{r}}$$

$$\therefore \quad T_b = 288(8)^{0.4/1.4} = 288 \times 1.8125 = 522°K$$

also $T_c = 800 + 273 = 1073°K$

$$\therefore \quad T_d = \frac{T_c}{(r_p)^{\gamma-1/\gamma}} = \frac{1073}{(8)^{1.33-1/1.33}} = \frac{1073}{1.67523} = 640.51°K$$

2. Compressor work per kg,

$$W_c = C_{pa}(T_b - T_a) = 1.005(522 - 288) = 235.17 \text{ kJ/kg}$$

and turbine work per kg

$$W_{GT} = C_{pg}(T_c - T_d) = 1.1(1073 - 640.51) = 475.74 \text{ kJ/kg}$$

∴ Net work produced by the gas turbine.

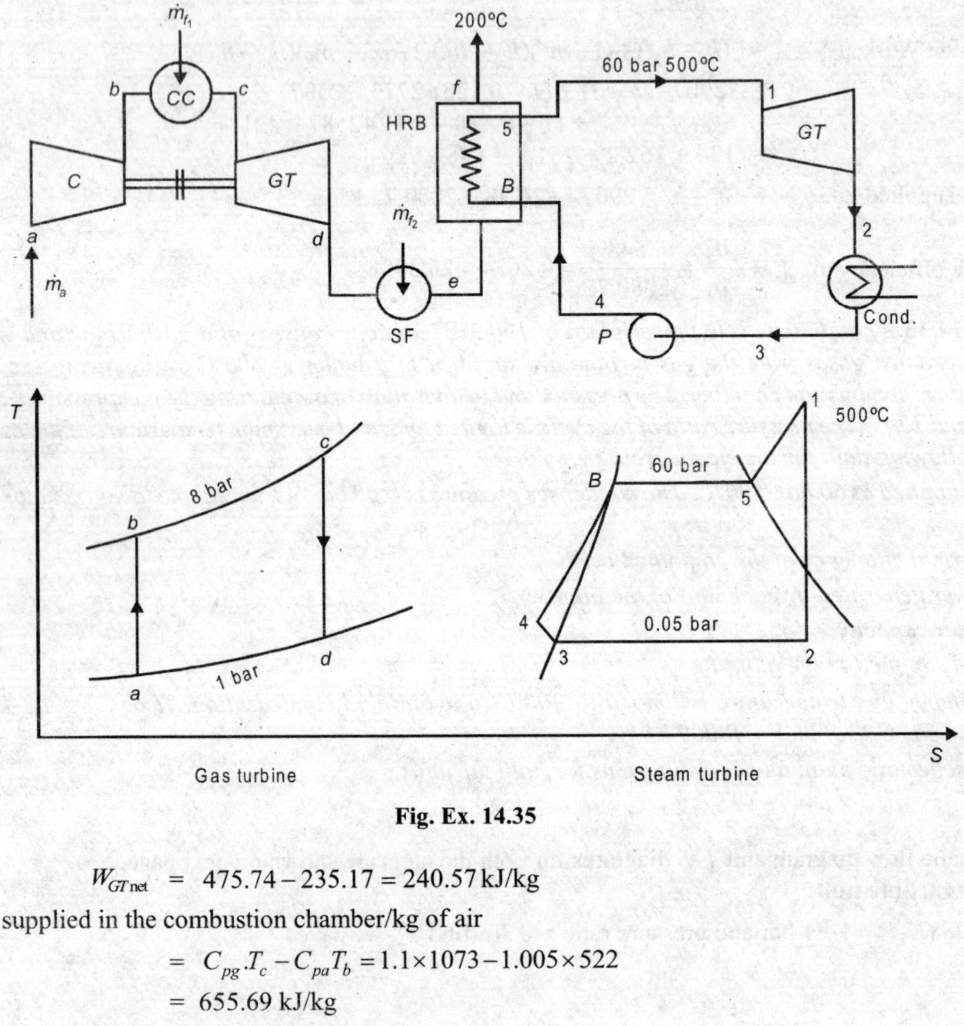

Fig. Ex. 14.35

$$W_{GT\,net} = 475.74 - 235.17 = 240.57 \text{ kJ/kg}$$

3. Heat supplied in the combustion chamber/kg of air

$$= C_{pg}.T_c - C_{pa}T_b = 1.1 \times 1073 - 1.005 \times 522$$

$$= 655.69 \text{ kJ/kg}$$

∴ Fuel supplied/kg of air $\dot{m}_{f_1} = \dfrac{655.69}{40000}$

$$= 0.0164 \text{ kg/kg of air}$$

Steam Turbine Unit:

4. At 60 bar and 500°C,

$$h_1 = 3422.2 \text{ kJ/kg and}$$

$$s_1 = 6.8818 \text{ kJ/kg}$$

∴ $$x_2 = \frac{6.8818 - 0.4763}{7.9197} = 0.809$$

$\therefore \qquad h_2 = 137.77 + 0.809 \times 2423.8$

$= 2098.62$ kJ/kg

$\therefore$ Work done by steam turbine/kg

$= 3422.2\ 0 - 2098.62 = 1323.58$ kJ/kg

5. Also heat supplied per kg of steam

$= 3422.2 - 137.77$

$\therefore \qquad q_s = 3284.43$ kJ/kg

To find gases required to produced 1 kg of steam, we have

$$\dot{m}_g \times C_{pg} \times (T_e - T_f) = \dot{m}_s \times q_s$$

$$\therefore \qquad \dot{m}_g = \frac{1 \times 3284.43}{1.1(800-200)} = \frac{3284.43}{660} = 4.976 \text{ kg/kg of steam}$$

$\therefore$ Heat supplied in the supplementary combustion chamber per kg of steam,

$$= \dot{m}_g \times C_{pg}(T_c - T_d) = 4.976 \times 1.1(1073 - 640.51) = 2367.28 \text{ kJ}$$

$\therefore$ Mass of fuel supplied per kg of steam in supplementary firing

$$\dot{m}_{f_2} = \frac{2367.28}{40000} = 0.05918 \text{ kg/kg of steam}$$

6. Total power generated =190 MW = 190 × 10³ kW

Let $\qquad \dot{m}_s$ = mass flow rate of steam, kg/s

and $\qquad \dot{m}_a = \dot{m}_g$ = mass flow rate of gases, kg/s (neglect mas of fuel)

$= 4.976 \times \dot{m}_s$ kg/s

$\therefore$ Power produced by the steam turbine

$= \dot{m}_2 \times 1323.58$ kW ...(I)

and Power produced by the gas turbine

$= \dot{m}_a \times 240.57 = 4.976 \times \dot{m}_s \times 240.57$ kW ...(II)

$= 1197.08\ \dot{m}_s$ kW

To find $\dot{m}_s$, we have

$190 \times 1000 = 1323.58\ \dot{m}_s + 1197.08\ \dot{m}_s$

$\therefore \qquad \dot{m}_s = 75.38$ kg/s

$\therefore \qquad \dot{m}_a = 4.976 \times 75.38 = 375$ kg/s

7. Heat supplied/sec for the total plant

$= \dot{m}_s \times$ heat supplied/kg of steam + $\dot{m}_a \times$ heat supplied/kg of air

$= 75.38 \times 23.67.28 + 375 \times 655.69$

$= 178445.57 + 245883.75$ kJ/s

8. Thermal efficiency of the combined plant

$$= \frac{190 \times 1000}{424329.32} = 0.4478 \text{ i.e., } \mathbf{44.78\%}$$

9. Power produced by gas turbine

$= 1197.08 \times 75.38 = 90.236$ MW $-$ **90 MW**

and Power produced by steam turbine

$= 75.38 \times 1323.58 = 99.764$ MW $-$ **100 MW**

10. Fuel supplied for total plant/hr

$$= \frac{\text{total heat supplied}}{W} = \frac{424329.32 \times 3600}{40000}$$

$$= 38189.64 \text{ kg/hr} = \mathbf{3819\ T/hr}$$

Example14.36 *In a combined cycle power plant the air is supplied at the rate of 2000 × 10³ kg/hr at 20°C. The pressure ratio used in the gas turbine plant is 7. The suction pressure of the compressor and the exhaust pressure of the turbine is 1 bar. The maximum temperature in the gas turbine is limited to 1000°C. The isentropic efficiencies of the gas turbine and the compressor are 85% and 80% respectively. The following data is given for the steam power plant.*

The temperature of exhaust gas from the gas turbine is increased to 1200°C by burning fuel before passing into the boiler. The steam is generated at 50 bar and 500°C. The condenser pressure is maintained at 0.1 bar. Temperature of the gases entering the chimney is 200°C. Find the following :

1. *Total power generating capacity of the combined plant*
2. *Overall efficiency of the plant*
3. *Fuel consumption of the combined plant/hour*
4. *Air/fuel ratio for the gas turbine plant. Use the following data,*

Calorific value of the fuel = 45 MJ/kg

C_{pa} = 1 kJ/kg-°K γ for air = 1.4

C_{pg} = 1.1 kJ/kg-°K γ for gas = 1.33

Do not neglect the mass of the fuel used.

Draw neat line diagram for the components.

Represent the process on T–S diagram.

Solution

The schematic flow diagram and *T–S* diagrams of the combined plant are given below.

1. The GT plant: We find the temperature $T_{2'}$ and $T_{4'}$. We have $T_1 = 20 + 273 = 293$°K

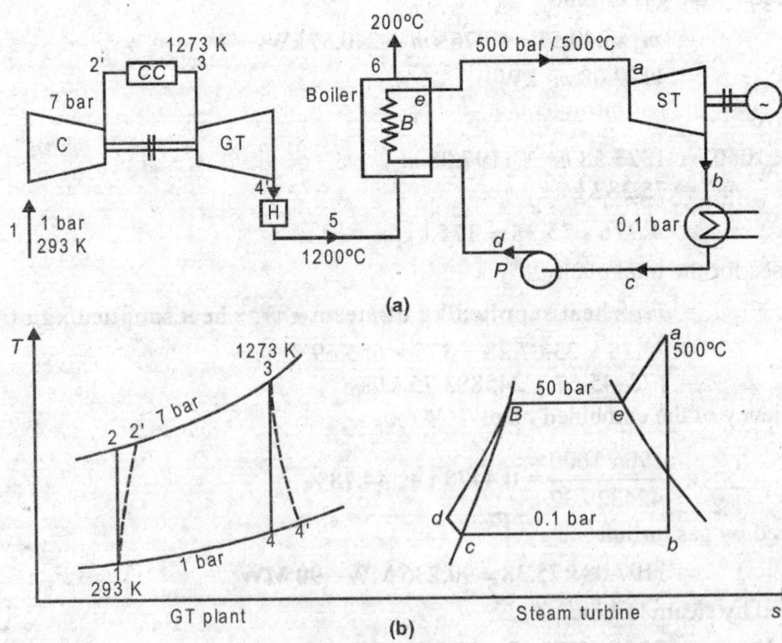

Fig. Ex. 14.36

$$\therefore \quad T_2 = T_1\left(\frac{P_2}{p_1}\right)^{\frac{y-1}{y}} = 293(7)^{0.4/1.4} = 510.886°K$$

$$\eta_c = 0.8 = \frac{T_2 - T_1}{T_{2'} - T_1} = \frac{510.886 - 293}{T_{2'} - 293} \qquad \therefore T_{2'} = 565.36°K$$

For GT, $\quad T_3 = 1000 + 273 = 1273°K$

We have, $\quad \frac{T_4}{T_3} = \left(\frac{P_4}{P_3}\right)^{\frac{\gamma-1}{\gamma}} \qquad T_4 = 1273\left(\frac{1}{7}\right)^{\frac{0.33}{1.3}} = 785.49°K$

As GT has efficiency of 0.85

$$\therefore \quad T_{4'}T_3 - 0.85(T_3 - T_4) = 1273 - 0.85(1273 - 785.49)$$

$$\therefore \quad T_{4'} = 858.62°K$$

2. To find A/F ratio, we write energy balance for the combustion chamber. Let *t* be the mass of fuel per kg of air

$$\therefore \quad C_{ga} \times T_{2'} \times C.V. \times f = C_{pg}(1+f)T_3$$

Substituting the values,

$$\therefore \quad 1 \times 1 \times (565.36) + 45000 f = 1.1(1+f)1273$$

$$43599.7 f = 834.94$$

$$\therefore \quad f = 0.01915 \text{ kg/kg of air}$$

$$\therefore \quad \text{Air/Fuel ratio} = \frac{1}{f} = \mathbf{52.22\ kg/kg\ of\ fuel}$$

3. Fuel supplied to the gas turbine plant,

$$= 0.01915 \times 2000 \times 10^3 = 38300 \text{ kg/hr i.e., } 10.64 \text{ kg/s}$$

4. Power produced by the gas turbine plant,

$$P_{GT} = \dot{m}_g C_{pg}(T_3 - T_{4'}) - \dot{m}_a C_{pa}(T_{2'} - T_1)$$

Now $\quad \dot{m}_g = \dot{m}_a + \dot{m}_f - (52.22 + 1)\dot{m}_f = 53.22 \times 10.64 = 566.26 \text{ kg/s}$

and $\quad \dot{m}_a = 52.22 \times 10.64 = 555.62 \text{ kg/s}$

$$\therefore \quad P_{GT} = 566.26 \times 1.1(1273 - 858.62) - 555.62 \times 1 \times (565.36 - 293)$$

$$= 258111.50 - 151328.66 = 106782.84 \text{ kW (I)}$$

5. To find amount of fuel supplied to the heater, we write the heat balance for the heater. Let *x* be the quantity of fuel in kg/s.

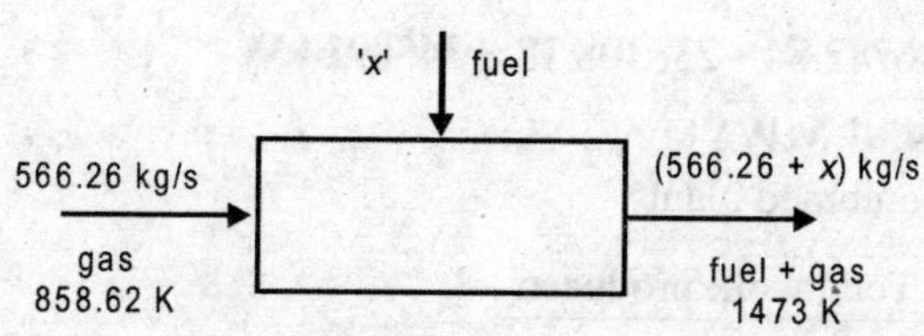

Fig. Ex. 21.36 (c)

$$\therefore \quad 566.26\times1\times858.62+45000x = (566.26+x)1.1\times1473$$

$$\therefore \quad (45000-1620.3)x = 566.26(1620.3-858.62)$$

$$\therefore \quad x = \frac{431308.92}{43379.7}$$

$$= 9.9426 \text{ kg/s}$$

$$= 35793.52 \text{ kg/hr (II)}$$

6. Total fuel supplied to the plant,

$$= 38300 + 35793.52 = \mathbf{74093.52\ kg/hr}$$

$$= \mathbf{74.1\ T/hr}$$

7. For steam turbine

At 50 bar 500°C $\quad h_1 = 3433.7\,\text{kJ/kg and } s_1 = 6.977\,\text{kJ/kg°K}$

At 0.1 bar $\quad s_{f_2} = 0.6493\,\text{kJ/kg°K and } s_{f_{g^2}} = 7.5018\,\text{kJ/kg°K}$

also $\quad h_{f_2} = 191.83 \text{ and } h_{f_{g2}} = 2392.9\,\text{kJ/kg}$

$$\therefore \quad x_b = \frac{6.977-0.6493}{7.5018} = 0.846$$

$$\therefore \quad h_b = 191.83 + 0.846 \times 2392.9 = 2216.6 \text{ kJ/kg}$$

Pump work, $\quad W_p = 0.0001(50 - 0.1) \times 100 = 5 \text{ kJ/kg}$

$$\therefore \quad h_d = h_e + w_p = 191.83 + 5 = 196.83\,\text{kJ}$$

Heat required per kg of steam $= h_a - h_d = 3433.7 - 196.83 = 3236.87\,\text{kJ/kg}$

8. Heat supplied in the boiler $= m_g \times C_{pg}\ (1200 - 200)$

$$= (566.26 + 9.94)\ 1.1 \times 1000$$

$$= 633820 \text{ kJ/s}$$

9. Amount of steam generated/sec.

$$= \frac{633820}{3236.87} = 195.8\,\text{kg/s or } 704.9\,\text{T/h}$$

10. Power produced by the steam turbine plant

$$= 195.8\,(h_a - h_b)$$

$$= 195.8\,(3433.7 \times 2216.6)$$

$$= 238308.18 \text{ kW}$$

11. Total power produced by combined plant

$$= 06782.84 + 238308.18 = \mathbf{345091\ kW}$$

$$= \mathbf{345.1\ MW}$$

12. Overall efficiency of the combined plant

$$= \frac{\text{Total work produced}}{\text{Fuel used} \times \text{Cal. value}}$$

$$= \frac{345091}{74093.52/3600 \times 45000} = 0.3726 \text{ or } \mathbf{37.26\%}$$

Example 14.37 *A combined cycle power plant is to be designed to develop 100 MW. Out of this 60% power is to be developed in a gas turbine plant, and the remaining in a steam turbine plant. The pressure and temperature of the air entering the compressor are 1 bar and 27ºC.*

The maximum pressure and temperature in the gas turbine cycle are limited to 3 bar and 1000°C. Steam is to be generated at 50 bar and 600°C, by using supplementary fuel. The condenser pressure is maintained at 0.05 bar.

Take the given data and find:

1. *Overall efficiency of the plant*
2. *Ratio of air flow rate to steam flow rate*
3. *Quantity of cooling water required in tons/hr, if the rise in temperature of cooling, water is limited to 10°C.*

Data:

Isentropic epic efficiency of the compressor = 85%

Isentropic efficiency of the turbine = 90%

Combustion efficiency = 95%

Solution

$C_{pa} = 1 \text{kJ/kg°K}, \quad \gamma_a = 1.4$

$C_{pg} = 1.1 \text{kJ/kg°K}, \quad \gamma_g = 1.33$

C. V. of fuel = 40 mJ/kg

Gas temperature going to chimney = 200°C.

Neglect the pump work. Consider the mass of fuel in the gas turbine.

The schematic diagram and the *T–S* diagrams are given on next page.

(I) Gas turbine plant:

1. We find $T_{b'}$ and $T_{d'}$ by usual method.

$P_a = 1 \text{bar}, t_a = 300°\text{K} \; P_d = 1 \text{bar}, P_b = P_c = 8 \text{bar}.$

$\eta_c = 0.85, C_{pa} = 1 \text{kJ/kg °K}, \gamma_a = 1.4, T_c = 1273 °\text{K}$

$\eta_t = 0.9, C_{pg} = 1.1 \text{kJ/kg °K}, \gamma_g = 1.33$

$$\frac{T_b}{T_a} = \left(\frac{p_b}{p_a}\right)^{\frac{\gamma-1}{\gamma}} \qquad \therefore \; T_b = 300 \times (8)^{0.4/1.4} = 543.43 \text{K}$$

$$T_{b'} = 300 + \frac{543.43 - 300}{0.85} = 586.4°\text{K}$$

Similarly,
$$T_d = \frac{T_c}{(8)^{0.33/1.33}} = \frac{1273}{1.675} = 759.89°\text{K}$$

Now, $T_{d'} = 1273 - 0.9(1273 - 759.89) = 811.20°\text{K}$

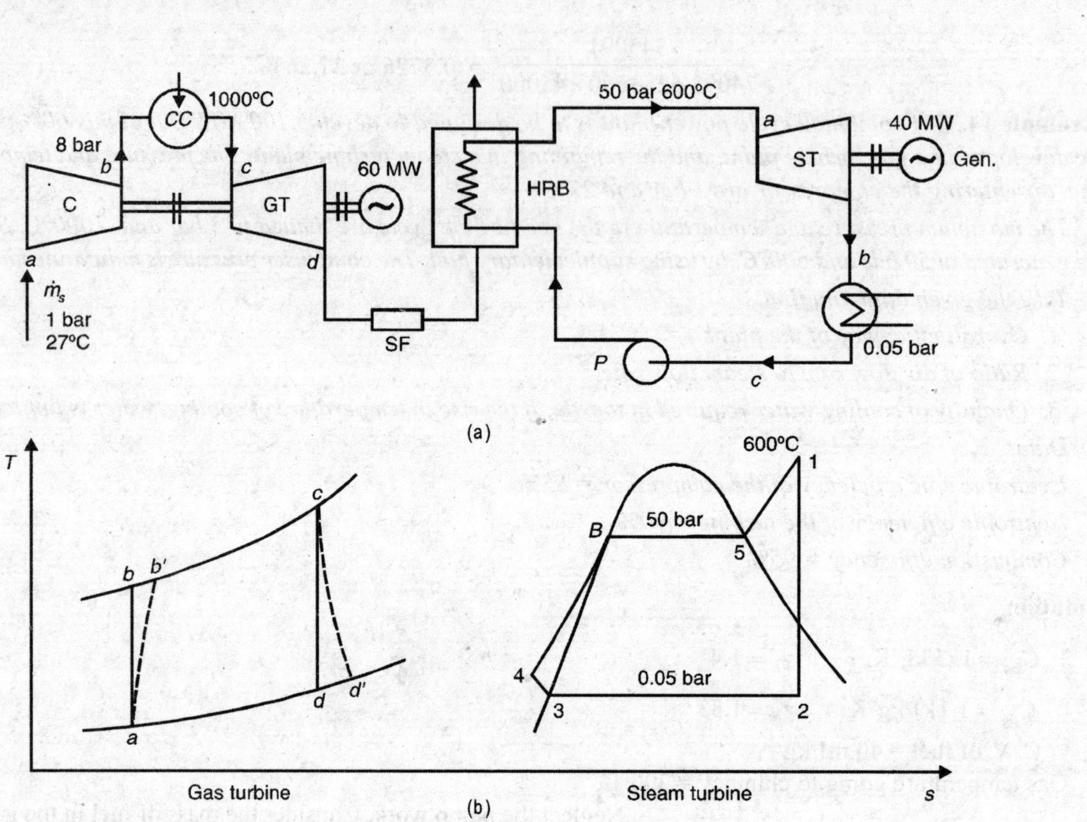

Fig. Ex. 14.37

2. Now consider combustion chamber of Gas turbine.
Let x be fuel supplied per kg of air.
Writing energy balance for combustion chamber, we get

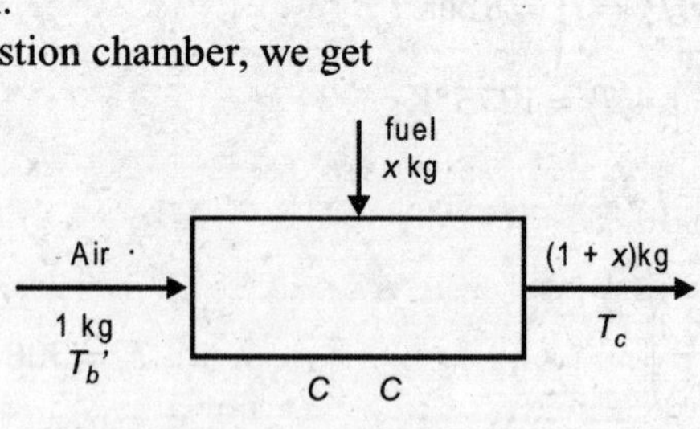

Fig. Ex. 14.37 (c)

$$C_{pg} \times \dot{m}_g \times (T_3) - \dot{m}_a C_{pa} T_{2'} = \text{C.V.}x.\eta_{\text{comb}}$$

$$1.1\ (1+x)\ 1273 - 1 \times 1 \times 586.4 = 40000x \times 0.95$$

$$x = \frac{813.9}{36600} = 0.02224 \text{ kg of fuel/kg of air}$$

$$\text{Air fuel ratio} = \frac{1}{x} = \frac{1}{0.2224} = 44.969 - 45:1\text{kg of air/kg of fuel}$$

3. Net work done by gas turbine,

$$W_{GT\,net} = W_T - W_C = \dot{m}_g C_{pg}\left(T_c - T_{d'}\right) - \dot{m}_a C_{pa}\left(T_{b'} - T_a\right)$$

$$= 1.02224 \times 1.1 \times (1273 - 811.20) - 1 \times 1 \times (586.4 - 300)$$

$$= 519.277 - 286.4 = 232.877 \text{ kJ/kg}$$

Power produced by GT $= 0.6 \times 100 \times 1000 = 60{,}000\,\text{kW} = 60\,\text{MW}$

$$\therefore \text{Mass of air required/sec} = \frac{60000}{232.877} = 257.877 \text{ kg/sec}$$

$\therefore$ Fuel used/sec. in GT plant $= 0.02224 \times 257.877$

$= 5.73$ kg/sec.

(II) Steam Turbine Plant

At 50 bar 600ºC, $h_1 = 3664.5$ and $s_1 = 7.2578$

$$\therefore \quad x^2 = \frac{7.2578 - 0.4763}{7.8626} = 0.863$$

$$\therefore \quad h_2 = 144.91 + 2419.8 \times 0.863 = 2232 \text{ kJ/kg}$$

WD by turbine/kg $= h_1 - h_2 = 3664.5 - 2232 = 1432.5\,\text{kJ/kg}$

1. Power generated by steam turbine = 0.4 × 100 = 40 MW = 40000 kW

Quantity of steam required/sec

$$= \frac{40000}{1432.5} = 27.923 \text{ kg of steam/sec}$$

2. Amount of heat required to generate steam/sec

$$27.923(h_1 - h_4) = 27.923\,[3664.5 - (137.7 + 5)]$$

$$= 98337.267 \text{ kW}$$

3. We find inlet temperature of gases to the boiler.

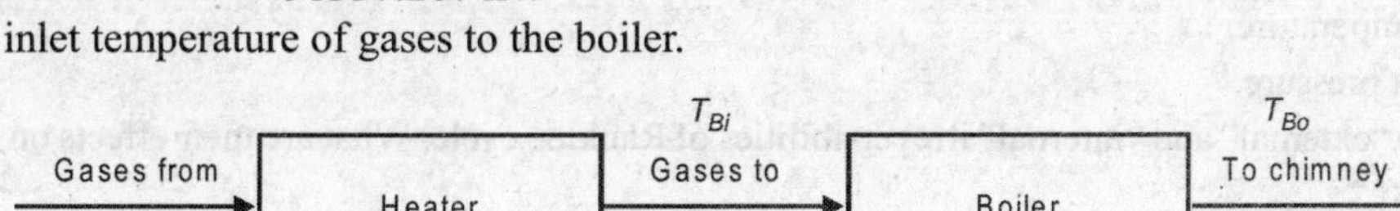

Fig. Ex. 14.37 (d)

$$\therefore \quad \dot{m}_g \times C_{pg}\left(T_{Bi} - T_{B0}\right) = \text{Total heat required by steam}$$

$$\left(\dot{m}_a + \dot{m}_f\right) C_{pg}\left(T_{Bi} - T_{B0}\right) = \text{Total heat required by steam}$$

$$(257.877 + 5.73)1.1\left(T_{Bi} - 200\right) = 98337.267$$

$$\therefore \quad T_{Bi} = 539.132$$

$$= 812.132 \text{ ºK}$$

This is very nearly equal to the inlet temperature of gases to the heater and hence no heat added in the heater.

4. Ratio of air flow to steam flow rate,

$$\frac{\dot{m}_a}{\dot{m}_s} = \frac{257.877}{27.923} = \mathbf{9.235}$$

5. Overall efficiency of the plant

$$= \frac{\text{Power developed}}{\text{Total heat supplied}} = \frac{\text{Power}}{m_f \times \text{C.V.}}$$

$$= \frac{100 \times 1000}{5.73 \times 40000} = \mathbf{0.436 \text{ i.e., } 43.60\%}$$

6. Heat rejected by steam in the condenser to its cooling water

$$= \dot{m}_s \times \text{dryness fraction} \times h_{f_g}$$

$$= 27.923 \times 0.863 \times 2423.8 = 58407.639 \text{ kW}$$

This is equal to heat gained by water

$$\therefore \quad \dot{m}_w C_{pw} \Delta T = 58407.639$$

$$\dot{m}_w = \frac{58407.639}{4.1867 \times 10} = 1395.076 \text{ kg/s}$$

$$= \mathbf{5022.273 \text{ ton/hr.}}$$

THEORY QUESTIONS

14.1 Explain. Why Rankine cycle rather than Carnot cycle is used as a standard cycle for steam power plants.

14.2 What is work ratio? What is its importance in power cycles?

14.3 Explain the effect of following parameters on the performance of Rankine cycle:

(a) admission pressure,

(b) inlet temperature,

(c) exhaust pressure.

14.4 Explain the "external" and "internal" irreversibilities of Rankine cycle. What are their effects on the performance of the cycle?

14.5 What is 'reheating' employed in Rankine cycle? How does reheating improve the cycle efficiency?

14.6 What are the advantages of superheat-reheat cycle?

14.7 How Rankine cycle efficiency is affected by reheat pressure ratio?

14.8 Explain ideal regenerative feed heating. Why it is not used in practice?

14.9 What is practical method of regenerative feed heating?

14.10 What are the advantages and limitations of regenerative feed heating?

14.11 What are different types of feed water heaters?

14.12 What is DA? Where and why it is used in power plants?

14.13 Compare following closed feed water heaters.

(i) With drains cascaded backwards

(ii) With drains pumped forward.

14.14 How feed water heaters are placed ideally in a power plant?

14.15 Sketch a typical layout of modern power steam power plant with superheating, repeating and regenerative feed heating.

14.16 What is cogeneration? What are the categorise of cogeneration?

14.17 What are the basic types of topping cycles in cogeneration?

14.18 Explain with a sketch the gas turbine plant with cogeneration.

14.19 Explain the concept of a combined cycle operation. What are the advantages of combined cycle power plant?

14.20. Explain with a sketch the combined cycle power plant with a heat recovery boiler.

14.21 What is a multipressure combined cycle plant? What are its advantages?

14.22 What is binary vapour cycle ? Explain with a sketch the working of binary vapour power plant. What are its limitations?

PROBLEMS FOR PRACTICE

1. Find the ideal cycle efficiency and steam consumption of reheat cycle operating between pressure of 30 and 0.04 bar with a superheat temperature of 450°C. Assume that the first expansion is carried out to the point where the steam is dry saturated and that the steam is reheated to the original temperature. The feed pump work may be neglected. **[Ans.** η = 0.387, S.S.C. = 2.39 kg/kWh]

2. Steam is supplied to a two stage turbine at 40 bar and 350°C. It expands in the first turbine until it is just dry saturated, then it is reheated to 350°C and expanded through the second stage turbine. The condenser pressure is 0.035 bar. Calculate the work output and the heat supplied per kg of steam for the plant. Calculate also the specific steam consumption and the cycle efficiency.

[Ans. (i) 1290 kJ, (ii) 3362 kJ, (iii) 2.79 kg/kWh, (iv) 38.4 %]

3. In a reheat cycle, the initial steam pressure and the maximum temperature are 150 bar and 550°C respectively. If the condenser pressure is 0.1 bar and the moisture at the condenser inlet is 15%, and assuming ideal processes, determine (a) the reheat pressure, (b) the cycle efficiency and (c) the steam rate.

[Ans. (a) 13.5 bar, (b) 43.6 %, (c) 2.05 kg/kWh]

4. In a reheat steam cycle, the maximum steam temperature is limited to 500°C. The condenser pressure is 0.1 bar and the quality at turbine exhaust is 0.8778. Had there been no reheat the exhaust quality would have been 0.7592. Assuming ideal processes, determine (a) the reheat pressure, (b) the boiler pressure, (c) the cycle efficiency, and (d) the steam rate. **[Ans.** (a) 30 bar, (b) 150 bar, (c) 50.51 %, (d) 1.9412 kg/kWh]

5. Find the cycle efficiency and steam consumption of regenerative cycle with one open feed heater. The steam leaves the boiler at 30 bar and superheated to 450°C. The bleed pressure is so that the temperature difference is divided into approximately equal steps. **[Ans.** η = 0.373, S.S.C. = 4.29 kg/kWh]

6. Calculate the efficiency of regenerative cycle using a single feed water heater, neglecting the feed pump work. The steam leaves the boiler at 30 bar and superheated to 450ºC and the condenser pressure is 0.035 bar. The heater is of closed type and bled steam is flashed into the condenser.

[Ans. Bleeding pressure = 2.7 bar, Steam extracted = 0.165 kg, Efficiency = 0.401]

7. A steam turbine is supplied with steam at 40 bar and 450°C and exhausted at 0.035 bar. Neglecting the feed pump work, find the ideal Rankine cycle efficiency of the plant and compare it with the ideal efficiency, which could be obtained if regenerative feed heating were employed. Assume the use of open feed water heater at one bleeding point at 3 bar. Also calculate the amount of steam extracted per kg of steam supplied.

[Ans. Ideal efficiency = 0.391, Ideal efficiency with feed water heater = 0.415,
Steam extracted = 0.173 kg of steam supplied]

8. In a single-heater regenerative cycle the steam enters the turbine at 30 bar, 400°C and the exhaust pressure is 0.10 bar. The feedwater heater is a direct contact type which operates at 5 bar. Find (a) the efficiency and the steam rate of the cycle, and (b) the increase in mean temperature of heat addition, efficiency and steam rate, as compared to the Rankine cycle (without regeneration). Neglect pump work.

[**Ans.** (a) 35.36 %, 3.93 kg/kWh, (b) 27.4°C, 1.18%, 0.47 kg/kWh]

9. A regenerative steam cycle employing three closed heaters the steam is supplied to the turbine at 42 bar and 500°C and is exhausted to the condenser at 0.035 bar. The bled steam for heating is taken at pressure of 15.4 and 0.5 bar. Assuming the ideal processes and neglecting pump work, calculate:

(i) the fraction of steam bled at each stage

(ii) power output per kg of steam flow rate

(iii) the cycle efficiency. [**Ans.** (1) 0.0969, 0.0902, (ii) 1133.6 kW/kg, (iii) 43.6%]

10. A simple steam power cycle uses solar energy for the heat input. Water in the cycle enters the pump as a saturated liquid at 40°C, and is pumped to 2 bar. It then evaporates in the boiler at this pressure, and enters the turbine as saturated vapour. At the turbine exhaust the conditions are 40°C and 10% moisture. The flow rate is 150 kg/h. Determine (a) the turbine isentropic efficiency, (b) the net work output, (c) the cycle efficiency, and (d) the area of the solar collector needed if the collectors pick up 0.58 kW/m².

[**Ans.** (a) 76.7 %, (b) 15.5 kW, (c) 12.78 %, (d) 182 m²]

11. The net power output of an ideal reheat regenerative steam cycle is 80 MW. Steam enters the h.p. turbine at 80 bar, 50W and expands till it becomes saturated vapour. Some of the steam then goes to an open feedwater heater and the balance is reheated to 400ºC, after which it expands in an L.P. turbine to 0.07 bar. Compute: (a) the reheat pressure, (b) the steam flow rate to the h.p. turbine, (c) the cycle efficiency, and (d) the rate of flow of cooling water in the condenser if the temperature rise of water is 8ºC. (e) If the velocity of steam flowing from the turbine to the condenser is limited to 130 in/s, find the diameter of the connecting pipe

[**Ans.** (a) 6.5 bar, (b) 58.4 kg/s, (c) 43.7%, (d) 3146.5 kg/s, (e) 2.97 m]

12. A binary vapour cycle operates on mercury and steam. Saturated mercury vapour at 4.5 bar is supplied to the mercury turbine, from which it exhausts at 0.04 bar. The mercury condenser generates saturated steam at 15 bar which is expanded in a steam turbine to 0.04 bar. (a) Find the overall efficiency of the cycle. (b) If 50,000 kg/h of steam flows through the steam turbine, what is the flow through the mercury turbine? (c) Assuming that all processes are reversible, what is the useful work done in the binary vapour cycle for the specified steam flow? (d) If the steam leaving the mercury condenser is superheated to a temperature of 300ºC in a superheater located in the mercury boiler, and if the internal efficiencies of the mercury and steam turbines are 0.85 and 0.87 respectively, calculate the overall efficiency of the cycle. The properties of saturated mercury are given below: [**Ans.** (a) 53%, (b) 59.35 × 10⁴ kg/h, (c) 28.5 MW, (d) 46.2%]

$p(bar)$	t(°C)	h_f	h_g	s_f	s_g	v_f	v_g
		(kJ/kg)		(kJ/kg-°K)		(m^3/kg)	
4.5	450	62.93	355.98	0.1352	0.5397	79.9×10^{-6}	0068
0.04	216.9	29.98	329.85	0.0808	0.6925	76.5×10^{-6}	5.178

13. In a combined cycle, the gas turbine portion has a pressure ratio of 16. Air enters the compressor at 300 °K at a rate of 14 kg/s and heated to 1500 °K in the combustion chamber. The combustion gases leaving the gas turbine are used to heatsteam to 460°C at 100 bar in a heat exchanger. The combustion gases leave the heat exchanger at 420°C. The steam leaving the turbine is condensed at 15 kPa. Assuming all the compression and expansion processes to be isentropic, determine (1) the mass flow rate of the steam, (2) the net power output and (3) the thermal efficiency of the combined cycle. Assume constant specific heat for air at room temperature.

[**Ans.** (1) 1.275 kg/s, (2) 7881 kW, (3) 66.3%]

14. A pass-out two-stage turbine receives steam at 50 bar, 350°C. At 1.5 bar, the h.p. steam exhausts and 12,000 kg of steam per hour are taken at this stage for process purposes. The remainder is reheated at 1.5 bar to 250°C and then expanded through the i.p. turbine to a condenser pressure of 0.05 bar. The power output from the turbine unit is 3750 kW. Take isentropic efficiency for h.p. and i.p. stages as 0.84 and 0.81 respectively.
Calculate the boiler capacity required. [**Ans.** 18.5 t/h]

15. Steam is generated at 70 bar, 500°C and expands in a turbine to 30 bar with an isentropic efficiency of 77%. At this condition, it is mixed with steam twice its mass at 30 bar, 400°C. The mixture then expands with an isentropic efficiency of 80% to 0.06 bar. At a point in the expansion where the pressure is 5 bar, steam is bled for feedwater heating in a direct contact heater, which raises the feedwater to the saturation temperature of the bled steam. Calculate the mass of steam bled per kg of high pressure steam and the cycle efficiency. Assume that the L.P. expansion condition line is straight. [**Ans.** 0.53 kg, 31.9 %]

PART 2

Chapter 15 Nuclear Power Plant

15.1 FUNDAMENTALS OF NUCLEAR ENERGY

15.1.1 Introduction

Today the demand of power increasing at a tremendous rate day by day and this growing energy needs of mankind has forced us to go in search of alternative sources of energy. This is because of the diminishing quantity of the conventional source—crude petroleum. We have experienced by this time that the development of a country depends on the per capita energy consumption. Industry, transport and agriculture sector need energy.

So far man has used various forms of solar energy. Coal and petroleum have been derived from solar energy accumulated by plants in millions of years. Hydropower is also derived from solar energy, as solar energy causes evaporation of water from sea and oceans which comes back as rain.

The reserves of solar energy (coal and petroleum) through tremendous, are limited to a few hundred years. The quantity of fuel required is enormous leading to transportation and ash disposal problems.

Nuclear energy is one such source which requires a very less quantity of fuel. Electricity was generated for the first time by nuclear reactor on December 20, 1951 at EBR-I experimental station near Arco, Idaho in the United States. On June 27, 1954, world's first nuclear power plant to generate electricity for a power grid started operation at Obninsk, Russia (USSR). The world's first commercial scale power station, Calder Hall in England opened on 17th October 1956.

Nuclear energy enlarges the world's power resources enormously. The fuel uranium alone has a huge potential and its reserves contain more energy than coal and petroleum reserves of the world put together. A unique feature of nuclear energy is that, it has an exceptionally high degree of concentration which exceeds by millions of times the concentration of energy in the conventional fuels. The energy obtained from 1 kg of uranium is nearly equal to 200 tonnes of high grade coal i.e., 2,00,000 times.

As of 2006, there are 442 licensed nuclear power reactors in operation in the world, operating in 31 countries.

15.2 NEED OF NUCLEAR POWER

From the discussion above, we come to the conclusion that the required power for the development of our country, has to be generated with the help of nuclear power plants in addition to the other existing power plants like thermal, hydraulic and gas turbine power plants.

At present, in our country, to meet the ever growing demand of power, generation of nuclear power seems to be the major solution, and this nuclear power generation should be on the top priority because,

(a) concentration of energy in nuclear fuel gives the advantage regarding the consumption of the fuel

(b) transport cost will be very less

(c) there is no danger of the quality of nuclear fuel being vanished.

15.3 ADVANTAGES AND DISADVANTAGES OF NUCLEAR POWER

Following are the advantages of nuclear power :

(a) During normal operation of the power plant no greenhouse gases are emitted e.g., carbon dioxide.

(b) There is no pollution of air or atmosphere i.e., zero production of dangerous and polluting gases such as carbon monoxide, carbon dioxide, aerosols, oxides of nitrogen, particulates and photochemical smog.

(c) Very little solid waste is generated.

(d) Less fuel consumption as very little fuel is needed (1 kg of uranium = 200 tonnes of coal).

(e) Power plants are not affected by adverse weather conditions.

(f) Nuclear power generation is reliable.

Following are the disadvantages of the nuclear power plants :

1. Capital cost is very high (initial cost).

2. Maintenance cost is very high.

3. The cost of decommissioning of plants is very high.

4. There is a risk of major accidents e.g., worst possible situation is what happened at the Cherobyl (Russia). Nuclear Power Plant which did not have a containment building.

5. It can help to produce Nuclear Bomb. All designs of sectors can produce plutonium, aiding nuclear proliferation.

6. Nuclear waste disposal is a big problem as high level reactor waste produced can remain dangerous for thousands of years.

15.4 FUNDAMENTALS OF NUCLEAR ENERGY

Before knowing the production of useful nuclear energy, it will be better to know the different terms that are used in **Nuclear power production**. Here, the different terms that are generally used can be listed as belows :

(a) Atomic structure

(b) Atomic number and mass number

(c) Energy equivalence of mass

(d) Nuclear fission

(e) Nuclear fusion

(f) Fast neutrons

(g) Thermal neutrons

(h) Isotopes

(i) Chain reaction (self-sustaining)

(j) Fissile materials

(k) Fertile materials

(l) Moderator

(m) Reflector

(n) Radiations

(o) Biological shield

(p) Coolant

(q) Control rods.

Now all these terms will be discussed in brief.

15.4.1 Atomic Structure

An element is defined as a substance which cannot be decomposed into other substances. The smallest particle of an element which takes part in chemical reactions is known as an **atom**. The atom is a planetary system. The centre of the system consists of a nucleus made up of protons and neutrons (both being referred to as neutrons) and revolving round the nucleus are the electrons. The radius of the proton and neutron is about 10^{-14} m and the radius of the electron's orbit is 10^{-8} m.

Protons are positively charged and neutrons are neutral, thus making the complete nucleus as positively charged. The electrons are negatively charged. As the positive charge of the proton particle is equal to the negative charge on the electron particle, the number of electrons is equal to the number of protons and the atom is a neutral element. Any addition of electron to the neutral atom will make it negatively charged. Also removal of any electron will make the atom positively charged. Atoms which are positively or negatively charged, are known as **ions** and this process is known as **ionisation.** The atomic structure is shown in Fig. 15.1.

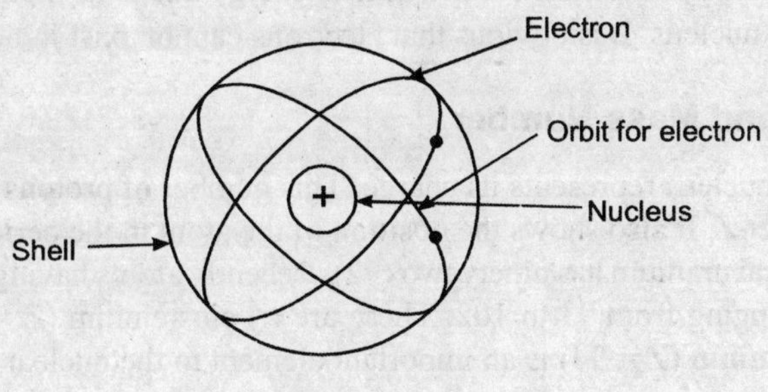

Fig. 15.1 *Atomic stucture.*

Protons and neutrons are having mass of about 1837 times and 1839 times the mass of an electron.

Nuclear power engineering is connected with the variation of nucleons in the nucleus.

The modern atomic theory tells that the atom has a diameter of about 10^{-7} mm. In a neutron atom the electrons are bound to the nucleus by the electrostatic forces, which follows the Coulomb's Law of forces i.e., like charges repel and unlike charges attract each other. The function of electrostatic force is similar to the gravitational force.

The study of the atomic spectrum has revealed that every electron in a atom is in one group of specific states of motion which is corresponding to its total energy. In a atom the electrons are spinning around the nucleus in orbits.

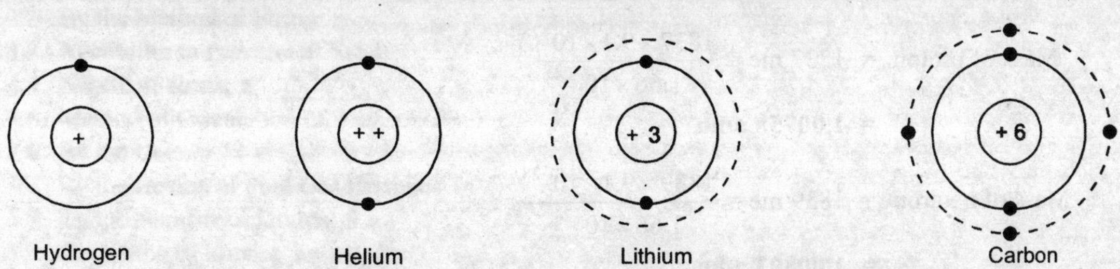

Fig. 15.2 *Atomic stucture of H_2, He, Li and carbon.*

These orbits are called shells, which represents the energy levels for the electrons. All the electrons having very nearly the same total energy are said to be in the samel shell. The shells have been named as K, L, M, N etc. Each shell consists of the specific maximum number of electrons. The K shell (inner shell) contains two electrons, L shell has 8 (eight) electrons, M shell is limited to eighteen (18) and the N shell possesses thirty two electrons. It is obvious from the study that amplitude difference in energy between two shells is much more than the difference in between energy levels in one shell. In a shell the number of electrons may be less than the specified number of electrons but not a large number. The inner shell is filled up first and then the other successive shells are completed.

The chemical properties of the atom varies with the composition of number of electrons in various shells and the state of energies within the shells determine the electrical characteristics of the atom. For example, hydrogen consists of one electron in first shell, helium has two electrons in first shell, lithium has two electrons in first shell and one in second shell, carbon contains two electrons in the first shell and four electrons in the second shell. See Fig. 15.2. The electrons posted in the outermost shell are termed **valence electrons**. If the outermost shell is completely filled, the **atom** is **stable** and will not take any electron to fill up the gap.

However, the incomplete outer shell will try to snatch the required number of electrons from the adjacent atom in a matter. The binding force between the electrons and nucleus is the electrostatic force of attraction. To emit one electron, energy required is more than the electrostatic force of attraction. When the energy is supplied, the electron jumps from one discrete energy level to another permissible level. The process starts from the outermost shell.

The electron possesses the energy in two forms, i.e., kinetic energy due to its motion and potential energy due to its position with respect to the nucleus. It is obvious that electrons cannot exist in between the permissible orbits.

15.4.2 Atomic Number and Mass Number

The number of protons in the nucleus represents its charge. This number of protons is known as **Atomic Number** and it is designated by the letter Z. It also shows the position of the atom in the periodic table. Hydrogen has only one atomic number while natural uranium has ninety two (92). Recently atoms having **higher atomic number** have been developed artificially, ranging from 93 to 102. These are (i) einsteinium (Z = 99), ferinium (Z = 100) and mendelevium (Z = 101). **Platinium (Z = 94)** is an important element to the nuclear power field.

The mass number (A) is the sum of total number of protons and neutrons in the nucleus. The number of neutrons is represented by the letter N.

$$\therefore \quad A = Z + N \text{ or } N = A - Z.$$

The mass of the atom is expressed in terms of the mass of the electron. The unit of mass has been considered as one-sixteenth of the mass of neutral oxygen atom which contains eight (8) protons and eight neutrons. The atomic mass unit is equal to one sixteenth $\left(\frac{1}{16}\right)$ the mass of oxygen neutral atom.

$$\text{One amu} = 1.66 \times 10^{-24} \text{ grams.}$$

$$\text{Mass of proton} = 1837 \text{ me} = \frac{1837 \times 9.1 \times 10^{-28}}{1.66 \times 10^{-24}}$$

$$= \mathbf{1.00758 \text{ amu}}$$

$$\text{Mass of neutron} = 1839 \text{ me} = \frac{1839 \times 9.1 \times 10^{-28}}{1.66 \times 10^{-24}}$$

$$= \mathbf{1.00893 \text{ amu}}$$

The scientific approach has concluded that the density of matter in a nuclei is enormous. It has been investigated that the radius of nucleus is equal to $1.57 \times 10^{-13} \times 3\sqrt{A}$ cm. Here A is the number of nucleons in nucleus.

15.5 ENERGY EQUIVALENT OF MASS

Einstein has given the theoretical phenomenon which says that there is acute relationship between the matter and energy. The phenomenon says that the matter is convertible into energy and vice versa.

The first proposed theory of relativity of Einstein says that the mass of the moving body varies with velocity and is expressed by the formula

$$m = \frac{m_0}{\sqrt{1-\frac{V^2}{C^2}}}$$

where,

m = mass of the moving body at any velocity

m_0 = mass of the body at cost

C = velocity of light

It can be seen that by substitution of the values in the above equation, for low velocities, the difference between m and m_0 is not much whereas at high velocities, slight change in velocity will cause significant change in mass.

The relationship between the mass and energy is expressed by the below noted equation, given by Einstein

$$E = mC^2$$

where

m = mass in gms

E = energy in ergs

C = velocity of light in vacuum in cm/sec

Complete conversion of mass to the energy has not been observed so far, and only small fraction of the mass is reduced after the energy is released. Even 0.5 kg of matter is converted completely to the energy, a large quantity of energy will be calculated as is clear from the calculations –

$$E = mC^2$$

$$= 500 \times (3 \times 10^{10})^2 \text{ ergs.}$$

$$= 45 \times 10^{22} \text{ ergs.}$$

$$= 45 \times 10^{15} \text{ Joules}$$

$$= \mathbf{45 \times 10^{12}} \textbf{ kilo-joules}$$

15.6 ELECTRON VOLT (eV)

In nuclear engineering, the common unit of energy is **Electron Volt** (eV). This electron volt is the amount of energy required to raise the potential of an electron by one volt.

One electron volt = 1.6021×10^{-12} erg

$= 1.6021 \times 10^{-21}$ joules.

15.7 BINDING ENERGY

The formation of the nucleus of atom takes place when the nucleons come closer to each other and this distance between the two nucleons is of the order of nearly 10^{-12} mm. At the moment of combination there is a release of energy and is known as nuclear binding energy. Further if it is required to separate out or to disintegrate two nucleons the equivalent amount of binding energy is to be supplied from the external source to overcome the force of attraction. The binding energy can be expressed as the energy required to overcome the binding forces of nucleus.

It has been observed that when two nuclear particles are combined to form a nucleus, there is a difference in the mass of the resultant nucleus and the sum of the masses of two parent nuclear particles. This decrement of mass is known as **mass defect**. The amount of mass defect is directly proportional to the amount of energy released. This statement can be verified by studying the following illustrations.

The three basic particles are having the below noted masses :

Mass of proton = 1.007595 amu

Mass of neutron = 1.008933 amu

Mass of electron = 0.00055 amu.

Example 1. Combination of one neutron and one proton forms a nucleus of heavy hydrogen with a mass of 2.0142 amu.

$$\begin{aligned}\text{Mass defect} &= -2.0142 + (1.007595 + 1.008983)\\ &= -2.0142 + 2.016578\\ &= \mathbf{0.002378\ amu}\end{aligned}$$

The binding energy corresponding to the mass defect can also be calculated by putting the values in the equation $E = mC^2$

$$\begin{aligned}\therefore \qquad E &= mC^2\\ &= 1.66 \times 10^{-24} \times (3 \times 10^{10})^2\\ &= 1.66 \times 9 \times 10^{-4}\\ &= \mathbf{1.495 \times 10^{-3}\ ergs.}\\ &= 1.495 \times 10^{-10}\ \text{joules}\\ &= 1.495 \times 10^{-10} \times 6.242 \times 10^{8}\ \text{eV}\\ &= 931.25 \times 10^{6}\ \text{eV}\\ &= \mathbf{931.25\ MeV}\end{aligned}$$

($\because$ 1 amu = 1.66×10^{-24} and 1 joule = 6.242×10^{18} eV)

$\therefore$ Binding energy of heavy hydrogen

$$\begin{aligned}&= 0.002378 \times 931.25\\ &= \mathbf{2.218\ MeV}\end{aligned}$$

Example 2. Helium nucleus is the resultant of the combination of two protons and two neutrons. Mass of He^4 nucleus is 4.00387 amu.

$$\begin{aligned}\therefore \qquad \Delta m &= \text{mass defect}\\ &= (2 \times 1.007595 + 2 \times 1.008983) - 4.00387\\ &= (2.015190 + 2.017966) - 4.00387\\ &= 4.033156 - 4.00387\\ &= \mathbf{0.03028\ amu}\end{aligned}$$

$\therefore$ Binding energy of helium nucleus

$$\begin{aligned}&= 931.25 \times 0.03028\\ &= \mathbf{28.2\ MeV}\end{aligned}$$

It has been observed that the nuclear binding energy per nucleon increases with the increase of the number of nucleons in the nucleus. For example, the binding energy per nucleon for H^2 is 1.109 MeV and for He^4 it is 28.2/4 = 7.05 MeV. A curve representing the variation of nuclear binding energy per nucleon with the mass number is shown in Fig. 15.3. Here the average value of binding energy per nucleon is considered. The curve indicates that the average binding energy per nucleon increases as the mass number increases initially with the peak value of about

8.8 MeV at nearly 60 mass number. The elements falling in this region are nickel and iron. As the mass number increases further, the binding energy curve falls gradually to 7.6 MeV for U^{238}. For U^{235}, the binding energy per nucleon is 7.7 MeV.

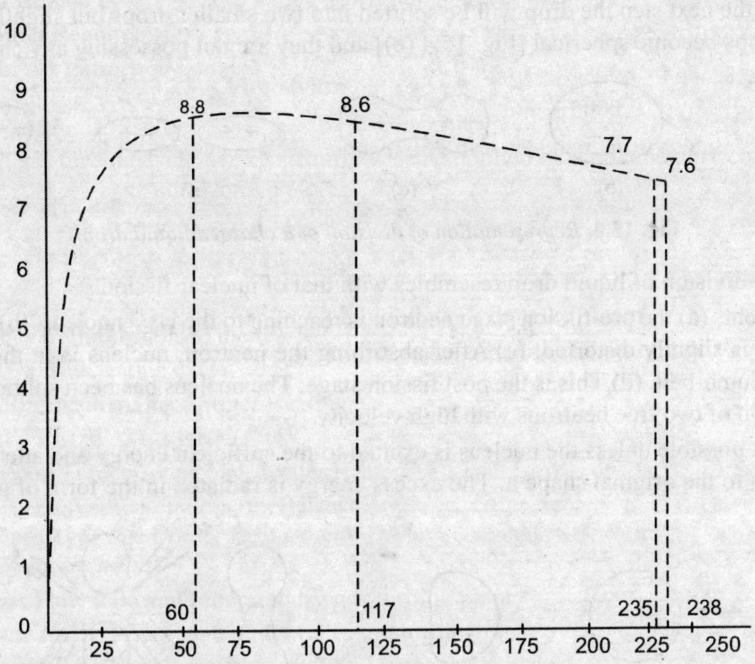

Fig. 15.3 *Variation of average binding energy per nucleon with mass number.*

U^{235} nucleus is splitted into two approximately equal nuclei. The formation of two nuclei will release the energy of above 0.9 MeV per nucleon. There is a release of energy as the mass number decreases within the range of 60 to 250 mass number. This release of energy is corresponding to the increase of mass defect. In fission process, U^{235} nuclei is splitted to two other nuclei and the liberation of energy takes place. Thus the nuclear transformation of other nuclei is also possible such as U^{235}, U^{233} and Pu^{239}. These are the important fuels used in the production of nuclear power.

It has been observed that an atom with even number of protons and of mass number, is more stable because of the pairing of protons and neutrons. This type of atom is also possessing higher binding energy per nucleon. This atom is also represented as even type of atom.

U^{235} is fissionable with a slow neutrons (neutrons having less energy) but U^{238} is fissionable only when the neutrons are having energy more than 1 MeV.

15.8 NUCLEAR FISSION

In 1938 it was discovered that the unstable heavy nucleus when bombarded with the neutrons splits into fragments of more or less the equal mass. This process is known as **nuclear fission**. The heavy nuclei breaks into two isotopes which are located in the middle of the periodic table. It is already seen that in this region the nuclear binding energy per nucleon is more than that of heavy nuclei. Thus it is evident that there is considerable release of energy during the process. Besides the above advantage the process is followed by the emission of several neutrons which will prove to be the essential condition to make the fission reactions self sustaining.

The fission phenomenon can be explained very clearly by the division of a charged liquid drop. A neutral drop is having a spherical shape [Fig. 15.4 (a)] but as soon as it is charged it is subjected to the excited state and the shape is slightly distorted because of the vibrations [Fig. 15.4 (b)] and becomes spherical. If this force is sufficient enough the drop will take the shape of dumb bell [Fig. 15.4 (c)] otherwise because of the surface tension it will again become spherical. In the next step the drop will be splitted into two smaller drops but slightly distorted [Fig. 15.4 (d)]. Finally these drops become spherical [Fig. 15.4 (e)] and they are not possessing any charge.

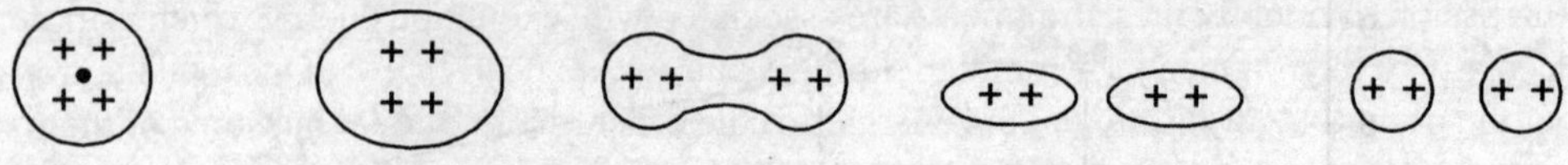

Fig. 15.4 *Representation of division of a charged liquid drop.*

The process of the division of liquid drop resembles with that of nuclear fission.

Figure 15.5 represents (a) the pre-fission stage neutron is reaching to the U^{235} nucleus. (b) The neutron collides with the nucleus and is slightly distorted. (c) After absorbing the neutron, nucleus is in the excited state and it acquires the form of dumb-bell. (d) This is the post fission stage. The nucleus has been splitted into two fragments and there is an ejection of two free neutrons with high velocity.

This process is not possible unless the nucleus is excited to the sufficient energy and attended the state (c) and the nucleus will come to the original shape a. The excess energy is radiated in the form of gamma quantum. The

Fig. 15.5 *Representation of fission process in steps.*

energy must be able to overcome the electrostatic repulsive forces i.e., nuclear binding energy (forces) working between the two spheres of the dumb-bell and it breaks into fragments.

The excitation energy required to split the nucleus is called the **critical energy**.

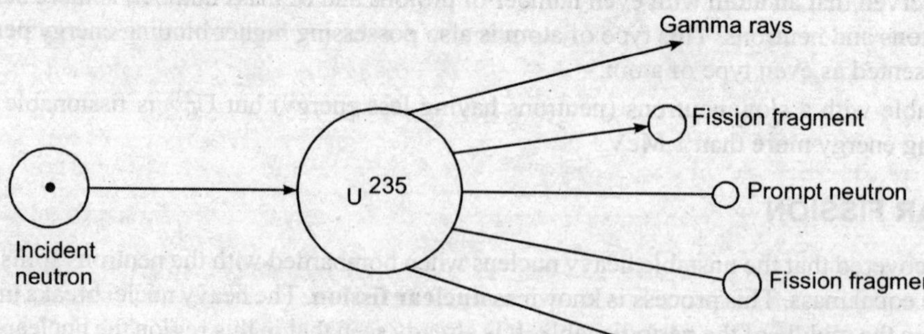

Fig. 15.6 *Fission of U^{235}*

Figure 15.6 shows the fission of a uranium-235 (U^{235}). The energy released due to fission is the basis for nuclear power generation. 2–3 (2.5 ± 0.5) neutrons are released which make it possible to produce sustained energy due to fissioning. The neutrons emitted in fission are of two types — prompt neutrons and delayed neutrons.

15.9 PROMPT NEUTRONS

The prompt neutrons are those neutrons which are ejected within a period of 10^{-14} to 10^{-12} sec from the fission fragments after the fission. Evidently these fragments are associated with excitation energy which is more than the neutrons binding energy of the fragment. The emitted prompt neutrons are nearly 99% of the total neutrons ejected in fission. Most of the prompt neutrons are having energies of the order of 1 to 2 MeV but some of them may also have energies beyond 10 MeV.

15.10 DELAYED NEUTRONS

After the fission has occurred the delayed neutrons are emitted for several minutes and the intensity decreases exponentially with time. In U^{235} about 0.775% neutrons are the delayed neutrons, which is about three times than that of Pu^{239}. These delayed neutrons are important because they permit the chain reaction to be controlled easily.

The critical energy is more than the neutron binding energy and it is possible to fission the nucleus by a fast neutron which can supply extra kinetic energy. Th^{232} (Thorium) and U^{238} (natural uranium) are even type of nuclei and can be fissioned by neutrons possessing kinetic energy of 1.1 MeV. In case the critical energy is less than the neutron binding energy the nucleus is fissionable with slow neutrons. Even odd type of nuclei can be fissioned by slow neutrons such as U^{235}, U^{233}, and Pu^{239}.

15.11 THE SELF-SUSTAINING CHAIN FISSION REACTION

A chain reaction is one in which the number of neutrons keeps on increasing (multiplying) in geometric progression during fission till all the fissionable material is disintegrated. The chain reaction will sustain (by itself) only if, for every neutron absorbed, at least one fission neutron is available to cause fission of another nucleus. The ratio of

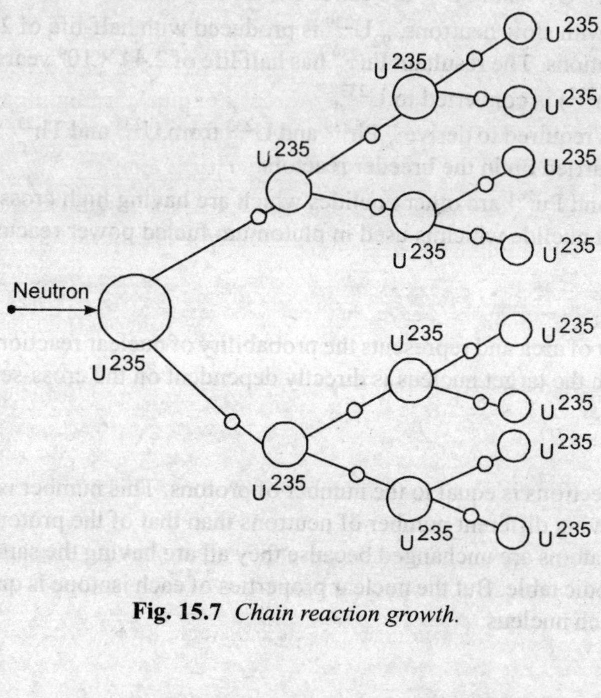

Fig. 15.7 *Chain reaction growth.*

number of neutrons in any fission to the number of neutrons generated in the preceding generation is known as **multification factor K**. If K is less than 1, the number of fissions decreases rapidly and the process dies down (stops). If K is greater than 1, then the rate of reaction is very fast and may result in sudden explosion as it occurs in an atom bomb. Hence it is necessary to keep the value of K = 1. But there will be loss of neutrons due to leakage, capture in control rods, etc. which requires to keep the value of K around 1.04. Maintaining the value of K at the required value is the most difficult part of reactor control. A fission chain reactor is shown in Fig. 15.7.

15.12 REQUIREMENTS OF FISSION PROCESS

The requirements of a fission process can be summarised as under.

1. The neutrons emitted in fission must have adequate energy to cause fission of another nuclei.
2. The produced number of neutrons must be able not only to sustain the fission process but also to increase the rate of fission. Certain loss of neutrons during the process is also to be accounted.
3. The process must be followed by the liberation of energy.
4. It must be possible to control the rate of energy liberation i.e., the rate of fission by some means.

The time required between the absorption of neutron and the emission of fission fragments is of the order of 10^{-14} sec. The fission fragments gain their most of the kinetic energy in nearly 10^{-20} sec. After the fission the fragments take nearly 10^{-9} sec before they collide with other adjacent nuclei.

Any reactor requires minimum quantity of fuel and this is called **critical mass**. The size associated with this mass is called **critical size**. When the nuclear fuel is assembled to the point of critical mass i.e., reaches the point of self-sustaining chain reaction, the reactor is said to go **critical**.

15.13 FISSILE AND FERTILE MATERIALS

A elements which undergo a nuclear fission process with a neutron are known as **fissile materials**. The examples of these materials are U^{235}, U^{233} and Pu^{239}. Of these only U^{235} occurs in nature are 0.7% of natural uranium.

Uranium-233 (U^{233}) and Plutonium (Pu^{239}) are not naturally available, but can be obtained from U^{238} (natural uranium) and Th^{232} (Thorium). U^{233} and Pu^{239} are called the fertile materials.

When U^{238} is bombarded with slow neutrons, ${}_{92}U^{239}$ is produced with half-life of 23.5 days and is unstable and undergoes two beta disintegrations. The resultant Pu^{239} has half life of 2.44×10^4 years and is a good alpha emitter.

Similarly thorium-232 (Th^{232}) is converted to U^{233}.

U^{235} is the source neutrons required to derive ${}_{94}Pu^{239}$ and U^{233} from U^{238} and Th^{232} respectively. This process of conversion is performed or carried on in the breeder reactors.

Th^{227}, Pa^{232}, U^{231}, Np^{238} and Pu^{241} are other nuclides which are having high cross-section for neutron thermal fission. Pu^{241} is the important nuclide which is used in plutonium fueled power reactors.

15.13.1 Cross-section

The cross-section has the unit of area and represents the probability of nuclear reaction. The probability of reaction of the impinging particle with the target nucleus is directly dependent on the cross-section of the target nucleus.

15.14 ISOTOPES

In any atom the number of electrons is equal to the number of protons. This number is independent of the neutrons in the nucleus. The atoms having different number of neutrons than that of the protons is known as **Isotopes**. The chemical properties of these atoms are unchanged because they all are having the same atomic number, occupying the same position in the periodic table. But the nuclear properties of each isotope is quite different depending upon the number of neutrons in each nucleus.

Isotope is not identified chemically but the difference of mass is the main basis for its identification. In addition, some isotopes are radioactive which is another sign of the identification of isotopes.

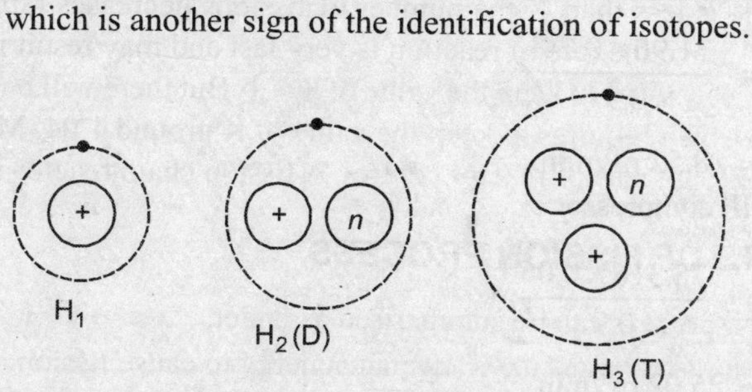

Fig. 15.8 *Isotopes of Hydrogen.*

The isotopes of some of the elements have been dealt here. Hydrogen has no neutrons and is known as Hydrogen (H_1). Hydrogen with one neutron is called heavy hydrogen or deuterium (D or H_2) and hydrogen with two neutrons is called heavy hydrogen or tritium (T or H_3). Thus hydrogen has three isotopes and shown in Fig. 15.8.

In case of oxygen, the variation of neutrons in the nucleus is ranging from six to eleven i.e., O_2 isotopes vary from O_{14} to O_{19} as such O_2 nucleus consists eight protons. Figure 15.9 shows the isotopes of oxygen.

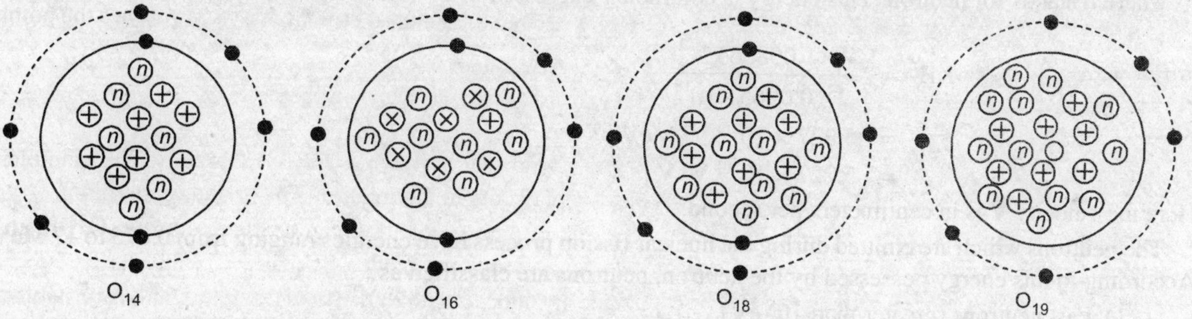

Fig. 15.9 *Isotopes of oxygen.*

Total compressor power $= P_1 + P_2 = 14.382 + 35.75$

$= 50.1364$ kW

Cooling load on condenser,

$$Q_c = M_3 (h_4 - h_5)$$

$$= \frac{75.316}{60}(638 - 458)$$

$$= 225.948 \text{ kW}$$

$$\text{Overall COP} = \frac{(RE_1 + RE_2) \times 3.517}{P_1 + P_2}$$

$$= \frac{(20 + 30) \times 3.517}{50.1364}$$

$$= 3.5074$$

Displacement volume of CP compressor

$$= \frac{m_1 V_1}{n_v} = \frac{31.96 \times 0.24}{0.9}$$

$$= 8.52 \text{ m}^3/\text{min.}$$

Displacement volume of HP compressor :

$$= \frac{m_3 V_3}{n_v} = \frac{75.316 \times 0.06}{0.9}$$

$$= 5.02 \text{ m}^3/\text{min.}$$

15.15 NEUTRON ENERGIES : FAST AND THERMAL NEUTRONS

For the nuclear fission process to take place, it is essential that neutrons are present with certain energy. This energy is the kinetic energy and with usual notations. This kinetic energy is given by

$$KE_n = \frac{1}{2} m_n V^2$$

where n stands for neutron. This energy is commonly expressed in eV (electron volt) or MeV.

$$\therefore \quad KE_n = \frac{1 \times 1.008665 \text{ V}^2}{2 \times 0.0965 \times 10^{18}}$$

$$= 5.227 \times 10^{-19} \text{ V}^2 \text{ MeV}$$

$$= 5.227 \times 10^{-13} \text{ V}^2 \text{ V}$$

Here the velocity V is in centimeters per second.

The neutrons which are emitted during the nuclear fission process have energies ranging from 0.075 to 17 MeV. According to this energy possessed by the neutron, neutrons are classified as :

(1) Fast neutrons (greater than 10^5 eV)

(2) Intermediate neutrons ($1 - 10^5$ eV)

(3) Slow neutrons (less than 1 eV).

Corresponding to these energies, the velocities of the neutrons are :

(1) Fast neutrons — velocity $> 4.4 \times 10^6$ m/s

(2) Intermediate neutrons — $(1.38 \text{ to } 4.4) \times 10^6$ m/s

(3) Slow neutrons — $< 1.38 \times 10^6$ m/s.

15.15.1 Fast Neutrons

Fast neutrons are those neutrons which have lost relatively very little energy since being produced in the fission process. As noted above, the lower limit of their energy is taken as 0.1 MeV and more. Fast neutron is required for the operation of **Fast Reactor** (Nuclear fission in the reaction).

15.15.2 Thermal Neutrons

When the neutrons are emitted during the fission process, they travel through the matter surround them and collide with the nuclei and thus decelerated, mainly by the higher nuclei, and in doing so, some of the energy is given and hence the neutrons are decelerated with each collision. This process is called **scattering**. Generally an effective scattering medium is **moderator**. This is one having small nuclei with high scattering and low neutron-absorption probabilities. In scattering, they attend the lowest energies and the neutrons are in thermal equilibrium with the surrounding. Such neutrons which are in thermal equilibrium with the surrounding, are called **thermal neutrons** and is a special category of **slow** neutrons. Thermal neutrons possess a mean energy of about 0.025 eV, at normal temperature of 15°C.

Thermal neutron speeds and energies are tabulated as given in Table 15.1

The most probable speed V_m is given by

$$V_m = \frac{(g_c 2KT)^{1/2}}{m}$$

where

K = Boltzmann's constant

m = mass of particle.

T = Abs. temperature

$V_m = 128.39\ \sqrt{T}$.

Table 15.1

Temp °C	*Speed* *m/s*	*Corresponding energy, eV (Probable)*
20	2200	0.0252
260	2964	0.0459
537.8	3656	0.0699
1000.0	4580	0.1097

15.16 SCATTERING

It is seen above that the deceleration of the neutrons because of the collisions of the fast neutrons with the lighter nuclei is called **scattering**.

There are two types of scattering: (i) Inelastic scattering and (ii) elastic scattering.

15.16.1 Inelastic Scattering

Incident particle and the ejected particle remains unchanged. When the neutron interacts with the nuclei, its kinetic energy is reduced and this reduced energy is utilized in exciting the nuclei. Thus the reduction in neutron kinetic energy appears partly as excitation energy of the struck nucleus and partly as kinetic energy. Therefore, inelastic scattering occurs with high-energy neutrons. The excitation energy is released in the form of gamma emission. This transformation of energy is known as **inelastic scattering**. Thus, to excite the target nucleus, the neutron should have minimum energy. Then if

K_{n1} = Kinetic energy of nucleus before collision

K_{n2} = Kinetic energy of nucleus after collision

E_{n1} = Excitation energy of neutron before collision

E_{n2} = Excitation energy of neutron after collision

E_c = Excitation energy of struck nucleus

∴ Balancing the energies of colliding particles before and after collision, we get

$$(E_{n1} + K_{n1}) = (E_{n2} + K_{n2} + E_c)$$

∴ A neutron has to possess an initial high kinetic energy greater than E_c. This is the condition for a neutron to engage in **inelastic scattering**.

15.16.2 Elastic Scattering

In contrast to inelastic scattering in elastic scattering, striking neutron does not possess the minimum energy to excite the nucleus and in the equation given above. $E_c = 0$. Therefore, the kinetic energy loss of the neutron is equal to the kinetic energy gain of the nucleus. Slowing down the neutron will depend upon the mass of nucleus and the direction of incidence. Loss will be maximum when head-on collision takes place.

In elastic scattering, the striking of the neutron will the nucleus will cause only displacement of the nucleus in the crystal lattice by a significant distance and can change the structural properties of the material. Elastic scattering is also called **elastion collision**.

In elastic collision, no nuclear reaction takes place but only redistribution of kinetic energy takes place.

15.17 RADIOACTIVITY

Radioactivity is the property exhibited by certain types of matter of emitting **radiation** spontaneously. This phenomenon was first reported in 1896 by Henri Becquerel for a uranium salt, and it was soon found that all uranium compounds are radioactive due to the uranium's radioactivity. In 1898 Marie Curie and her husband discovered two other naturally occurring, strongly radioactive elements—Radium and Polonium. The radiation is emitted by **unstable atomic** nuclei as they attempt to become more stable. In 1934 it was discovered that radioactivity could be induced in ordinary matter by artificial transmutation.

Many isotopes occurring in nature are stable. Examples of these stable isotopes are :

(a) thallium (Z = 81), (b) lead (Z = 82), and (c) bismuth (Z = 83). Isotopes, which are unstable, of heavier elements like polonium (Z = 84) and above (Z > 84) emit radiation till they become nearly stable and spontaneous disintegration process takes place. This process of spontaneous disintegration is called Radioactivity or **Radioactive decay**.

In addition to the heavier elements being radioactive, there are a few low-mass isotopes which are naturally radioactive. Examples of these isotopes are K^{40}, Rb^{87} and Ln^{115}.

Radioactive isotopes—natural or man-made are generally called **radioisotopes**. **Radioactivity is always accompanied by a decrease in mass or liberation of energy**. It is observed that the rate of radio radiations from unit mass is fixed and cannot be changed by any method. Therefore, the quantity of radiation per unit time for any isotope can be found out.

The radioactive change is accompanied by the emission of the following particles and radiations :

(a) α-particles
(b) β-particles and
(c) γ-radiations.

In addition to these radiations, following emission takes place :

(d) positrons
(e) K capture
(f) neutrons.

(a) α-particles (Alpha particles). The α-particles contain helium nuclei which contains 2-protons and 2-neutrons. [$_2H^4$]. A particular isotope will require a definite amount of energy to emit α-particles. 4.3 MeV is required to emit α-particles from uranium $_{92}U^{238}$.

$$_{92}U^{238} \rightarrow {_{90}Th^{234}} + {_2He^4}$$

The decay of plutonium $_{94}Pu^{239}$ also emits α-particles, reaction of which is written as

$$_{94}Pu^{239} \rightarrow {}_{92}U^{235} + {}_{2}He^{4}.$$

Thus we see that α-particles or emission produces new isotopes such that mass number is reduced by 4 and atomic number is reduced by 2. This is verified by the two equations noted above.

(b) Beta particles (β-particles). It is observed that the nucleus of the isotope will undergo a spontaneous change when the number of neutrons and protons in the nucleus are such that the ratio of neutron to proton lies outside the stability range for the mass number. Here, the neutron will be spontaneously changed or converted into a proton and at the same time a negative electron. A negative β-particle will be driven out as given by the following equation :

	Neutron	→	Proton	+	Negative β-particle.
Charge	0		+1		– 1
Mass	1		1		0

This is given in the equation form as

$$n \rightarrow p + -e^0$$

$$\therefore \quad n \rightarrow p + \beta^-$$

In this way neutron is replaced by a proton so that atomic number is increased by 1 and the mass number is unchanged. Thus in the newly formed element, the number of neutron is decreased by one and the number of protons is increased by one resulting the decrease in the ratio of neutron to proton. Generally, after one or more stages of disintegration, a stable element is formed.

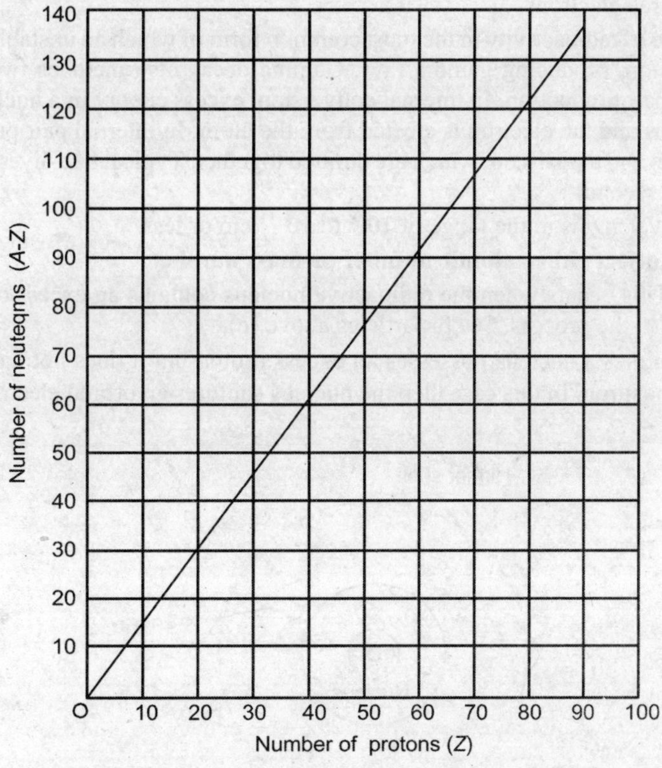

Fig. 15.10 *Stability range for neutrons and protons.*

A representative example of emission of β-particles (–vely charged) from a particular isotope is given by :

$$_{56}Ba^{140} \rightarrow {_{57}La^{140}} + {_{-1}e^{0}} \rightarrow {_{58}Ce^{140}} + {_{-1}e^{0}} \rightarrow {_{58}Ce^{140}} + \beta^{-}$$

Here $_{58}Ce^{140}$ is stable.

Again if the number of neutrons is too small and the number of protons is too large, then also a nucleus will be unstable. In this case, proton is converted into neutron and a positive electron are emitted. This positive electron is called **positron**.

	Proton	→	Neutron	+	Positive β change.
Charge	+ve		0	+	+ve
Mass	1		1	+	0

Therefore, the new nucleus will have atomic number less by one than the original (parent) nucleus, and the new nucleus—called daughter nucleus—will be unstable and will be radioactive. Thus after one or more stages of positive β decay, a stable nucleus will be formed, as the neutron to proton ratio will be with the stability range.

Again the emission of β (+vely charged electrons) particle by the reaction given as

$$_{7}N^{13} \rightarrow {_{6}C^{13}} + {_{1}e^{0}}$$

or

$$_{7}N^{13} \rightarrow {_{6}C^{13}} + \beta^{+}$$

(c) γ-radiation. γ-rays or radiations are similar to an *x*-ray but of higher energy. These radiations are highly penetrating and are of very short wavelength electromagnetic radiation. These are emitted spontaneously by some radioactive substances. Gamma radiation also originates in the decay of certain **subatomic** particles and in particle-antiparticle Annihilation. Gamma rays can initiate nuclear fission and can be absorbed by ejection of an electron and can be scattered by free electrons.

Gamma decay is a type of radioactivity in the most common form of which an unstable atomic nucleus dissipates energy by gamma emission, producing gamma rays. Gamma decay also includes two other processes, internal conversion and internal pair production. In internal conversion, excess energy in a nucleus is transferred to one of its own orbitting electrons and the electron is ejected from the atom. In internal pair production, excess energy is converted into an electron and a **positron**, which are emitted together. Typical half-lives for gamma emission range from about 10^{-9} to 10^{-14} second.

Gamma rays have wavelengths in the range of 10^{-8} to 10^{-11} cm or less.

γ-radiation does not affect either atomic number or mass number.

(d) Positron decay. This occurs when the radioactive nucleus contains an excess of protons. In this proton is converted into a neutron in the process. See β-particles above (c).

(e) K-capture. Sometimes a nucleus possesses an excess proton but it does not have the sufficient energy of 1.024 MeV to emit that positron. In this case then the nucleus captures an orbital electron from the rear most shell (or K-shell) (Fig. 15.11)

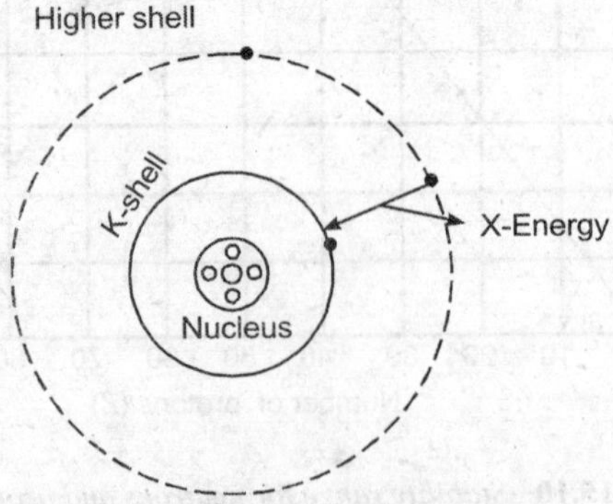

Fig. 15.11 *K-capture.*

Because of this K-capture, proton in the nucleus changes into neutron and the vacancy in K-shell is filled by another electron falling from a higher orbit. This K-capture is accompanied by X-ray emission from the atom.

(f) Neutron emission. When the nucleus of the radioactive material (isotope) is excited to the extent that the binding energy per nucleon is of the order of 7.8 MeV, then the neutron is emitted and the mass number is reduced by unity.

15.18 RADIOACTIVE DECAY AND HALF-LIFE

It is noted above that a spontaneous disintegration process, in case of radioactive materials, takes place and this process is called **radioactive decay**. The rate of decay is a function of the radioactive nuclei present at a time irrespective of the temperature, pressure or physical and chemical states of matter. Also the rate of decay is called **activity** A and the dimensions of this activity is disintegration per second.

Let N be the number of radioactive nuclei of a certain species at any time θ, the rate of decay is given by

$$\frac{dN}{d\theta} = -\lambda N$$

where λ is a constant of proportionality and this constant is called Decay constant or disintegration constant or the radioactive decay constant. Negative sign indicates that luring disintegration the number of nuclei is decreasing.

∴ Integration of the above equation between the proper limits we get

$$\frac{dN}{d\theta} = -\alpha N$$

$$\therefore \quad \frac{dN}{N} = -\alpha.\, d\theta$$

$$\therefore \quad \int_{N_0}^{N} \frac{dN}{N} = -\lambda \int_0^{\theta} d\theta$$

$$\therefore \quad \log_e N - \log_e N_0 = -\lambda\theta$$

$$\therefore \quad \log_e \frac{N}{N_0} = -\lambda\theta.$$

$$\text{or} \quad \frac{N}{N_0} = e^{-\lambda\theta}$$

$$\text{or} \quad N = N_0 \times e^{-\lambda\theta}$$

$$\therefore \quad \frac{dN}{d\theta} = -\lambda N = -\lambda . N_0 e^{-\lambda\theta}$$

$$\boxed{\frac{dN}{d\theta} = -\lambda \cdot N_0 e^{-\lambda\theta}}$$

This equation indicates that decay follows the exponential law.

Activity A can be expressed by an equation given below :

$$A = K\left(-\frac{dN}{d\theta}\right)$$

$$= K \lambda N$$
$$= K \lambda N_0 e^{-\lambda \theta}$$

where K is called the detection coefficient.

The unit of activity is **Curie**.

1 curie = 3.7×10^{10} disintegrations/sec.

15.18.1 Half-Life

Half-life is the time interval required for one-half of the atomic nuclei of a radioactive sample to decay or sample to change spontaneously into other nuclear species by emitting particles and energy. In other words, it is the time required for the number of disintegrations per second of a radioactive material to decrease by one half. Also, half-lives are characteristic properties of the various unstable atomic nuclei and the particular way in which they decay. It is found that Alpha decay and Beta decay are generally slower processes than Gamma decay.

Substituting $N = \dfrac{N_0}{2}$ and $\theta = \theta_{1/2}$ in the above equations we have

$$\frac{N_0}{2} = N_0 e^{-\lambda \theta 1/2}$$

or
$$\frac{1}{2} = e^{-\lambda \theta 1/2}$$

or
$$\log_e 0.5 = \ln 0.5 = 0.693 = \lambda \theta_{1/2}$$

∴
$$\theta_{1/2} = \frac{0.0693}{\lambda}.$$

$$\boxed{\theta_{1/2} = \frac{0.693}{\lambda}}$$

It is found that half life is from fraction of second to millions of years. If the **half life** process is repeated in nature, activity comes to stand still and no further spontaneous changes are possible.

Thorium Th-232 has half-life of nearly 1.4×10^{10} years and that of uranium U^{238} is approximately 4.5×10^9 years.

Half-life is important factor in selecting an isotope for various purposes. For example, to avoid frequent replacements of nuclear fuels used in power generation, isotopes having longer half-life are required and for medicine and agricultural uses, isotopes having shorter half-life are desired.

Regarding human body, half-life considered is frequently called as biological half-life to distinguish it from radioactive half-life. This biological half-life is defined as the time required by the human body to eliminate or reduce half of the nuclei to which the body has been subjected.

These two half-lives are related with each other as

$$\frac{1}{t_{eff}} = \frac{1}{t_{bio}} + \frac{1}{t_{1/2}}$$

where

t_{eff} = Effective half-life time

t_{bio} = Biological half-life time

$t_{1/2}$ = Radioactive half-life time

15.19 ARTIFICIAL RADIOACTIVITY

In 1934 Curie and Joliot discovered the artificial radioactivity. They found out the technique by which the light elements were artificially transformed to produce radioactivity. Such light, radioactive elements are not found in nature. Artificial radioactivity is **induced radioactivity** in the element. The artificial radioactivity follow the same laws of decay of natural radioactivity.

Curie and Joliot took the light elements like boron, and aluminium etc. and they studied the effect of the bombarment of α particles on nucleus. It was observed that there was an emission of position after the removal of α-particles source. The rate of emission was found decreasing following an exponential decay law $N = N_0 e^{-\lambda t}$.

15.20 MODERATOR

The moderator is a substance which is used to reduce the neutron energy (fast neutron). These fast neutrons collide with the nuclei of the moderator material and slow down by losing their energy. The moderator can be in the form of solid, liquid or gas.

The following properties are required for a good moderator material.

1. It should not react with neutrons because the neutrons reacted in nuclear reactions, no neutron will be available for the fission reaction, and the reactor becomes inefficient or inactive.
2. It should not be costly and should be readily available.
3. It should be chemically stable, inert and should not erode or corrode.
4. The moderating material should have high moderating ratio to slow down the neutrons in the reactor core. [Ratio of moderating power to the microscopic neutron capture coefficient.]

The moderators that are used in nuclear reactor are

(i) Graphite
(ii) Water
(iii) Heavy water (deuterium)
(iv) Helium
(v) Baryllium
(vi) Baryllium oxide.

(i) Graphite. Among all these materials, graphite is heavier and therefore neutrons have to make more collisions to achieve thermal energy. Graphite does not react with other materials and therefore the problem of corrosion is of little importance. Graphite is, also, not toxic unlike beryllium and beryllium oxide.

This has the moderating ratio of 1>1.

Graphite can be produced artificially by partial recrystallization at high temperature of suitable carbon acqueous materials. The graphite to be suitable for use in reactor is obtained from petroleum coke mixed with coal tar pitch and graphitized at a temperature of 2500 to 2800°C. To get graphite in its pure form it is bathed in a purifying gas containing chlorine and fluorine at nearly 2450–2500°C and in doing so, the neutron absorbing elements are volatilize.

The major disadvantage with reactor using graphite as moderator is that the graphite is subjected to radiation damage.

(ii) Water. Amongst all the moderating materials, water has the least moderating ratio of >1.

Lighter water (ordinary water) is the cheapest moderating material but it must be in the pure form (no solution salt should be present). In the pure form, decomposition of water is reduced to a large extent.

Water corrodes the metal and at high temperature and pressure the metal is corroded seriously. Similarly, radiation damage is also severe with water as moderator.

Zirconium alloy is affected less by corrosion.

(iii) Beryllium and Beryllium Oxide. *Beryllium*, as a moderator, has the following advantages :

(a) Excellent chemical and corrosion resistant properties

(b) No radiation damage

(c) Lighter than carbon.

As against these advantages, beryllium has the following disadvantages :

1. High cost
2. Poor ductility
3. Highly toxic.

Mainly because of the high cost, it is not used as a moderator in the power reactor because the cost is nearly equal to the total cost of a small reactor.

Moderating ratio for beryllium is nearly 15%

Similarly, baryllium oxide is as good as beryllium itself. Its advantages are,

(a) It is chemically inert.

(b) It has high thermal conductivity.

(c) It is heat resistant.

(d) High melting point 2520°C.

Major disadvantages are —

1. It has poor mechanical properties. It is brittle.
2. Easily affected by radiations.

(iv) Heavy Water (Deuterium). Heavy water has nearly the same thermodynamic, physical and chemical properties as ordinary water. For example, boiling point or temperature is 101–102°C as compared to 100°C for ordinary water, melting point is nearly 3.8°C as compared to 0°C for ordinary water and density of heavy water is 1100 kg/m^3 as against 1000 kg/m^3 for ordinary water. These properties are at atmospheric conditions.

When heavy water is to be used as moderator in a power reactor, it should work at high temperature and consequently high pressure. It has different nuclear characteristics. It requires a neutron to travel double as against ordinary water so that the same energy will be cost. It is found that heavy water absorbs practically no neutrons and ordinary water will absorb neutron. Moderating ratio is about 160 times greater for heavy water than ordinary water. [Moderating ratio = 11900 for D_2O and >1 for H_2O.]

Because of the particular property regarding absorption of neutron (nearly zero absorption) heavy water can also be used as coolant and moderator with natural uranium U^{238} fuels.

(v) Helium. It is a chemical element and its atomic number is 2. It is a noble gas. It is coloureless, odourless, tasteless completely unreactive and non-toxic. It was first found by spectroscopy of the sun's atmosphere in 1868. It is the second most abundant and second lightest element in the universe (after hydrogen). It is the *product of radioactive decay*. Liquid helium exists only below – 268.9° C, about 4° above absolute zero. Helium is used as moderator in nuclear reactor.

15.21 REFLECTOR

There are various methods used to control the reactor. One of these methods is the use of the materials called reflectors. Reflectors are generally made of the same material as that of moderator. Reflectors are generally placed round the core to reflect some of the neutrons that leak out from the surface of the core. To increase or decrease the power of the reactor, these reflectors are swung away or toward or moved axially with respect to the core. Of course, this method can be used only for small-power reactors or in special individual cases.

15.22 COOLANTS

In the nuclear reactor, a medium is required to take out the heat of fission and for this some medium is required. This medium is called the **coolant**. Thus coolant is a heat transfer media to take away heat from the reactor core. These coolants may be in the liquid form or in the gaseous form.

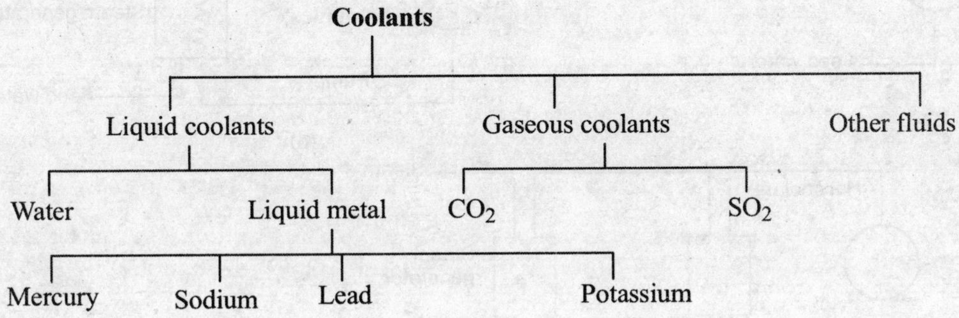

The coolant transfers heat from the core in two ways :

(i) coolant transfers heat to another medium which then transfers heat to water for generation of steam and

(ii) direct heat exchange when water is used as coolant.

Here the water takes heat from the core, get converted into steam which is supplied to steam turbine.

The desirable properties required for coolants are :

(a) It should have a good heat transfer coefficient.

(b) It should not corrode the metal.

(c) It should not absorb more neutrons i.e., it should have low cross-section. Because of this, more amount of coolant can be used to carry or take away more heat from the reactor core.

(d) It should not be toxic.

(e) It should be free from radiation damage.

(f) It should have low viscosity.

(g) It should have high density and specific heating so as to carry more heat. It will require less power for pumping.

(h) Liquid metal coolants should have low melting point.

(i) It should be possible to operate coolant at high temperature.

(j) It should have a low moderation ratio.

15.22.1 Water as Coolant

Because of its high density and good heat transfer coefficient, water is the better coolant than gaseous coolants. For gas, because of its very low density, large pumps (circulators/blowers) and heat exchanges are required. In addition to this the boiling temperature of water is low and can not be used in heterogeneous reactor core. The maximum temperature for water is 374°C (critical temperature) but the pressure is high and the system has to work under pressure. When this water is used as coolant, then the vapours are not tolerated because of the heat transfer problems.

As against these disadvantages of water as coolant, there are some advantages for the use of water as coolant in nuclear reactors. Light water is a good moderator for neutrons. Reactivity in the core of a nuclear reactor varies with the density of water. Density of water is a function of temperature. At lower temperatures the density is higher and reactivity is decreases.

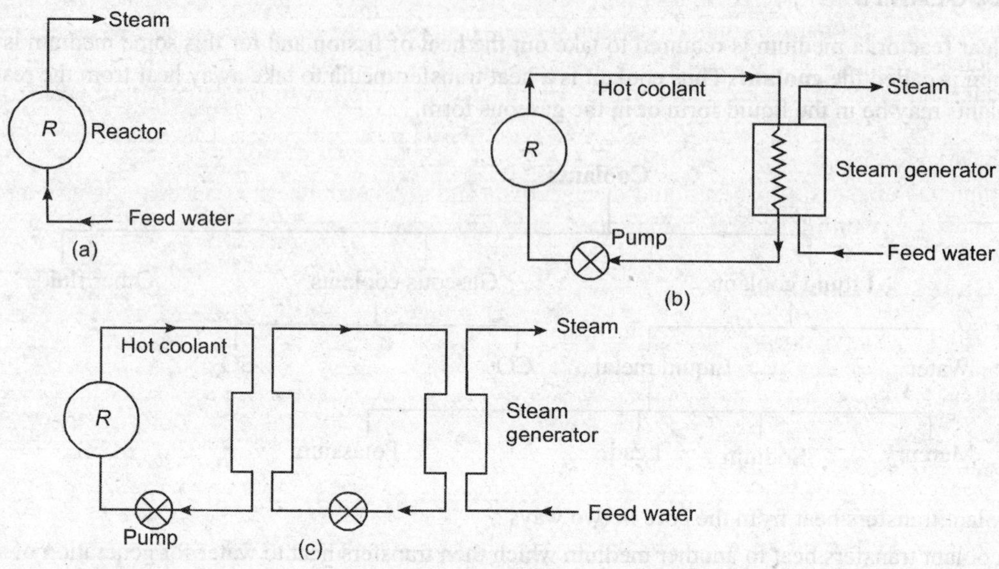

Fig. 15.12 *Coolant cycles.*

There are three different coolant cycles as shown in Fig. 15.12.

(a) Direct cycle

(b) Single circuit system

(c) Double circuit system

In direct cycle, the water is used as a coolant and it takes heat from the core and steam is take out for use in steam turbines, Fig. 15.12 (a).

In single circuit system, one heat exchanger is used while in double circuit system, two heat exchangers are used as shown in Figs. 15.11(a), (b) and (c).

15.22.2 Liquid Metals as Coolant

It is known that the efficiency of the plant cycle depends on the temperature of heat addition in the cycle, and therefore, the plant should be operated at the highest temperature. For this purpose, the metals in the liquid form can be used at high temperatures and at the same time at normal pressure.

The main advantages of liquid metals are high heat transfer coefficients and less radiation damage. The metals that can be used are :

(i) Sodium

(ii) Aluminium

(iii) Bismuth

(iv) Lead and

(v) Magnesium.

These metals have small cross-sections as compared to water and therefore can be used as coolant in thermal reactors. Mercury also is liquid at room temperature but because of its bigger cross-section, mercury is not suitable as coolant. The melting points of the liquid metal coolants should be low and many times the alloys can be prepared to suit the requirement. Along with melting points, cross-section of the resultant alloy should also be within limits so that it can be used as coolant in the nuclear reactors.

The alloy of sodium and potassium (Na-K) is found to be the most satisfactory liquid metal coolant because small cross-section, high thermal conductivity and specific heat. With the liquid metal coolants in the reactors, the thermal efficiency is high and the cost per kW is less. Graphite is found to be the best moderator with this liquid-metal coolant. This is because of the retention of its mechanical strength and purity at high temperatures. Sodium is cheap and is available in sufficient quantity. It does not react with stainless steel upto 80°C. Sodium should not come in contact with air or water reaction with them.

Pure sodium is found to be the best liquid-metal coolant and gives better performance. As against pure sodium, potassium undergoes explosive reaction and super oxide of potassium is the product.

15.22.3 Gases as Coolants

Air being available in large quantity, it was the usual practice to use air for many purposes. Similarly air was tried as the coolant in nuclear reactors. Also air was used at low pressures. But air contains nitrogen as a major constituent and that it has high cross-section and was not suitable as coolant to be used in nuclear reactors. Similarly helium gas was tried but as the heat transfer capacity is poor, it also could not be used as coolant for the required power demands from economical considerations.

With gas coolants, large quantities of gas are required and for that purpose large capacity-blowers are necessary power required is also large. For a good gas coolant, the heat transfer capacity should be large so that less quantity and subsequently low power will be required. Gas coolants can be used at only high temperatures came across the nuclear reactors. Reactor can be operated at different pressures when the gas coolant temperature is kept constant and in this case the thermal efficiency is high because of high temperatures. As the power consumption for gas coolant circulation is high, the overall efficiency is low.

When gas is used as coolant, the neutron absorption is almost nil or it is negligible and therefore reactivity is independent of the quantity of coolant used. Gas coolants are used when sufficient quantity of water is not available.

Carbon dioxide gas and helium gas do not react with the materials used for the structures of the reactor core.

Other fluids that can be used as coolants in nuclear reactors are as follows :

(i) Organic fluids such as hydrocarbons. They are stable when subjected to heat and radiation.

(ii) Polyphenyls

(iii) Diphenyl

(iv) Terphenyl.

These are corrosion resistant.

The main disadvantage of the organic coolant is the radioactive decomposition. In addition to this, organic fluids have less heat transfer capacity and therefore to carry the given quantity of heat, large amount of liquids is required and consequently large power is required for circulation for this fluids.

Organic liquids used as coolants give somewhat toxic effect.

15.23 REACTOR CONTROL

It is understood that the reactivity and the reactor control are closely related with each other. Also, it is then clear that the reactor will be unstable when the reactivity becomes excessive and in this case the multiplication factor plays a very important role. The value of multiplication factor by which it is greater than unity ($K_{eff} - 1$) is expressed as excess multiplication factor (K_{exc}). The ratio of K_{exc} to K_{eff} is termed as reactor reactivity ε. [K_{eff} = effective multiplication factor]

$$\therefore \quad \varepsilon = \frac{K_{exc}}{K_{eff}} = \frac{K_{eff} - 1}{K_{eff}}$$

The value of K_{exc} decides the transient behaviour of the reactor. If $\varepsilon > 0$, the reactor is said to be supercritical and the power level increases, with $\varepsilon < 0$, the reactor is subcritical and power level reduces.

Generally, the reactivity is measured inhouse unit. It is defined as the amount of reactivity which makes the stable reactor period equal to one hour and one in hour is equal to 2.595×10^{-5} of reactivity. Reactivity, sometimes, is expressed, in dollar. One dollar reactivity is that which makes the reactor prompt critical and one cent is one hundredth of the dollar.

Again, neutron flux is the reactor is a function of multiplication factor and the effective multiplication factor can be defined as the ratio of the number of neutrons produced in successive generations considering the reactor to be critical.

The prompt and delayed neutrons are of major importance specially when the reactor is shut down.

The delayed neutrons are very effective in the control of the reactor. The delayed neutrons increase the pile period for better control of flux and consequently power. Generally, the delayed neutrons are 0.755 to 0.8% of one neutron generation. The delay neutrons reduce the rate of power decrement. Similarly with prompt neutrons the positive change of reactivity is similar to the negative change of reactivity effects.

When the heavy water or beryllium is used as a moderator in the reactor, the shut down of the reactor is delayed further because the fission of uranium produces radioactive fragments and emit high energy gamma rays interacting with heavy hydrogen or beryllium nuclei and producing neutrons. The effect of this is to increase the multiplication factor which causes the increase in delay and consequently there is a delay in damping out the chain reaction.

Reactor control is also affected by the temperature and density of the materials considerably.

15.23.1 Methods of Reactor Control

After studying the facts underlying the control of the reactors, it is the proper place to study the different methods that are used to control the reactors. There are many methods that are used to control the reactor working. The different functions required are as following :

(i) Starting of the reactor
(ii) Increase the power generation
(iii) Decrease the power generation
(iv) Stop or turn-off the reactor.

To have these functions, following means are used to (a) control rods and (b) reflectors.

(a) Control Rods. Generally, **control rods** are used for controlling the thermal and breeder reactors. For thermal reactors, the material used for the rods should have absorption cross-section for thermal neutrons. These materials are : (i) Boron and (ii) Cadmium and they are used because they are very cheap as compared to other materials available. The boron forms the alloy either with steel or aluminium and commonly used in the shape of rod. Similarly cadmium is plated on steel or aluminium rod or it is used as a sandwich material with steel or aluminium.

If the control rods are used in the core of the reactor, then it is possible that the burning of the fuel may be non-uniform and a good number of valuable neutrons will be absorbed by the control rods, a great disadvantage. To avoid this the reactors are operated with the control rods out of the reactor. Also it is very important that the core should be properly loaded.

Generally, each reactor is equipped or provided with three types of control rods :

(i) Shim rods
(ii) Fine control rods
(iii) Safety rods.

(i) Shim Rods. The **shim rods** absorb the excess reactivity during normal operation of the reactor. When the reactors are to be started, then the rods are taken out of the reactor. Once the reactor is started, then the rods are positioned such that the constant power level is maintained. After certain time of interval, the positions of the rods are adjusted to take care of the poisoning and other effects of fission products in the core.

(ii) Fine Control Rods. Small changes in pressure, temperature and density of the fuel material may take place during operation and these change causes reactivity. Very small percentage of reactivity is absorbed by the **fine control rods**. These rods can be operated automatically or manually. In the modern nuclear reactor practice, automatic operation is preferred.

(iii) Safety Rods. The safety rods give significant poisoning effect and therefore it is very difficult or rather impossible to start-up the reactor when the safety rods are inside the core. In case of emergency i.e., in the event of potential hazards or dangerous pressure surges, the safety rods enter the reactor core instantaneously reducing the multiplication factor well below unity. Similarly, in the event of earthquake or excessive power, the safety rods come into action in nearly 3–5 seconds and prevent the damage to the plant. The sudden insertion of the rod is sometimes called **scram**.

Generally, when only one safety rod or control rod is used, it is to be placed at the centre of the core as any other position of the rod will not be so effective. In large power reactors, a bank of control rod is provided for uniform control of flux. In that case the effectiveness of the complete bank is slightly less than the sum of the effectiveness of individual control rod. This is because each rod is placed at a certain distance apart and it distorts the thermal neutral flux.

In small reactors, generally two rods are placed instead of one rod. In the case of two rods, the distance between them is one-third the dimension of the core so that maximum effectiveness is obtained. Also, the symmetrical arrangement of rods restricts the distortion of the flux and thus the thermal stresses are reduced. Special care is to be taken to arrange the shim rods.

All the rods, namely, shim rods, fine control rods and safety rods are inserted in or taken out of the reactor by the same distance.

15.24 BIOLOGICAL SHIELDING

Nuclear reactors are provided with biological shielding. Shielding the radioactive zones in the reactor from possible radiation hazard is essential and very important to protect the operating men from the harmful effects. During fission of the nuclear fuel, alpha (α) particles, beta (β) particles, deadly gamma rays and neutrons are produced. Out of these, neutrons and gamma rays are of main importance or significant. The protection must be provided against these products of fission. The thick layers of plastics, thin metal plates for alpha and beta rays are provided. Similarly thick layers of lead or steel or concrete are provided all round the reactor for preventing the gamma rays which are deadly rays and causes the effects similar to x-rays. Sometimes concrete of the thickness of about 6 to 10 feet or 2 to 3 metres is used for the concrete walls.

The whole reactor system is made completely leak-proof.

15.25 REACTOR POISONING

Out of the total products of fission, some have high cross-section and that causes the poisoning of the reactor. Among these products of fission, Xe^{135} and Sm^{149} are of significance. Thermal energy utilization is very much affected because of Xe^{135} (zenon) poisoning.

The poisoning of the reactor is a matter of concentration of fuel in the reactor which gives flux of the order of 10^{14}. The poisoning of the reactor is expressed as the ratio of number of thermal neutrons absorbed due to poisoning to the number of neutrons absorbed by the total fissionable material. It is observed that poisoning of the reactor by 5% will reduce the reactor activity between 3 to 5%. When the power is being produced by the reactor, the effect of poisoning is subsided as Xe^{135} decays to Xe^{136} which is stable and some of its percentage undergo thermal neutron capture. When the reactor is shut down, the flux reduces very rapidly the loss of xenon by neutron capture reduces to zero. On the other hand, the concentration of xenon increases to a maximum value within twelve hours after shut down. Afterwards, if the reactor is to be started, it is to be supplied by excess amount of reactivity to overcome the xenon effect. The flux under these conditions is very large, of the order of 10^{14}, it becomes impracticable to provide sufficient reactivity and hence the reactor is started within 1–2 days.

Another poisoning product is rare **earth** specifically Sm^{149} (Samarium) which gives the same poisoning effect as xenon.

15.26 SITE REQUIREMENTS FOR NUCLEAR POWER PLANTS

The following points or factors have to be considered while selecting the site or location for a nuclear power plant:

(a) Load Centre. To minimise the transmission losses in lines, the nuclear power generating station should be as near the load centre as possible.

(b) Cooling Water Availability. Large quantity of water is required for :

(i) Boiler feed
(ii) Condenser circulating (cooling) water and
(iii) Make-up water.

It is estimated that for a 500 MW nuclear power plant, about 2800 billion litres of water is required. For this purpose, the nuclear power plants are located very near to sea or large rivers that are flowing for throughout the year. In India nuclear power plants like Tarapur, Kaiga and Kalpakam power plants are near the sea while Rana Pratap Sagar Plant is near the river.

(c) Accessibility to Site. The site should be easily accessible either by rail or road for transporting the materials required during construction and during maintenance.

(d) Safeguards against Earthquakes or Seismology. The site should not be prone to earthquake. Generally the site is classified into its respective seismic zones like 1, 2, 3, 4, 5. The zone 5 is most dangerous seismic and unsuitable for nuclear power plant and for any power plants. For this purpose, once the site is selected, then the area of around 300 km radius is studied for its history of tremors and earthquakes. Accordingly, the foundation of the buildings, machinery and equipments supports can be designed for severest earthquakes. This will ensure that the plant will retain integrity of structure, piping and equipments should an earthquake occur.

(e) Geology Soil Quality. The bearing capacity of the soil has to be tested as large and heavy structures have to be supported. Sudden earth movement will damage the plant severely. Therefore, proper care should be taken as far as the soil quality is concerned.

(f) Nuclear Waste Disposal. The final stage of the nuclear fuel cycle is the management of the still highly radioactive 'spent' fuel, which constitutes the most problematic component of the nuclear waste stream. After fifty years of nuclear power the question of how to deal with this material remains fraught with safety concerns and technical problems, and one of the most important lines of criticism of the industry is based on the long-term risks and costs associated with dealing with the waste.

Management of the spent fuel can include various combinations of storage, reprocessing and disposal.

Another, more important method of disposal of high-level nuclear waste call for the material to be buried deep underground in certain geological formations. The Canadian government, for example, was seriously considering this method of disposal, known as the **Deep Geological Disposal** concept. In this plan, a vault is to be dug 500 to 1000 metres below ground and the high-level nuclear waste is stored.

Storing high level nuclear waste above ground for century or so is considered appropriate by many scientists. This allows for the material to be more easily observed and any problems detected and managed, while the decay over this time period significantly reduces the level of radioactivity and the associated harmful effects to the container material. It is also considered likely that over the next century newer materials will be developed which will not break down as quickly when exposed to a high neutron flux thus increasing the longevity of the container once it is permanently buried.

(g) Population Distribution. While considering the proximity to the load centre, we have to consider the **population density** around the nuclear power plant. The safety of the people working in the plant, outside the plant and at a certain distance from the nuclear power plant, in the event of small and major accidents take place, must be given a major priority while selecting the site for nuclear power plant. While considering this aspect of the people safety, the possible extension of the capacity of the power plant can also be given proper thought.

(h) Other Considerations. There are many other minor points that are to be given considerations. These points are :

(i) External natural events like floods (heavy rains) in rivers
(ii) Accendental up-stream dam-failures (floods in the rivers)
(iii) Tropical cyclones
(iv) Wind direction (for radioactive gases) and velocity and many other small incidents.

(i) Hydrology. 'Local hydrology' is another important point to be considered while selecting a site for the nuclear power plant. In this regard, the characteristics of the ground water and the level of water-table at the site must be taken into consideration so that contamination of local water sources by the discharge of radioactive liquid wastes does not occur.

For some of these points, the local history of the last 10–20 years will be useful while selecting the site for nuclear power plant.

15.27 METHODS TO START THE NUCLEAR FISSION AND FUSION REACTIONS

If we require to start the nuclear fission process, we have to excite the nuclei by given sufficient energy required to overcome the static force of repulsion. There are—in general — three methods to initiate or start the nuclear fission process.

15.27.1 Use of Accelerated Particles

The particles used here are α-particles (Hydrogen H_2 or helium (He)). These particles are highly accelerated by means of an accelerator and are used to strike or bombard the target nuclei and the fission takes place. The particles used have less repulsive forces and hence they are very much required for the purpose. In practice this method is rarely used because the accelerated particles, after striking, lose their energy in joining the atom and therefore are unable to fulfil the requirements.

15.27.2 Heating to High Temperature

In this method, the substance is heated to a very high temperature at which the thermal energy possessed by the substance is high enough to overcome the electrostatic repulsive forces. The temperatures that are required for this process is million degrees centigrade. This high temperature is actually required for fusion process which is not practical as the materials are not available which will sustain this high temperature. This type of the reaction because of high temperature is called thermo-nuclear reaction. In the central core of the sun, this thermo-nuclear reaction takes place and hydrogen is the gas (lighter substance) present there.

15.27.3 Use of Neutron

In this method, the particles are bombarded by the neutrons. Neutrons are neutral i.e., have no charge—positive or negative. Therefore, with low energy also, they can make their way through the shells of the electrons and then through the nucleus. This is the practical method and is used in all modern nuclear reactors (fission). This type of the reaction is known as neutron-induced uranium fission reaction.

THEORY QUESTIONS

15.1. What is the need of nuclear power? Explain.

15.2. What are the advantages and disadvantages of nuclear power?

15.3. Write short note on Atomic Structure.

15.4. Define atomic number and mass number.

15.5. What is energy equivalent of mass? What is electron volt eV?

15.6. Write in short about Binding energy.

15.7. What do you understand about the 'nuclear fission'? What are the products of fission? What are the requirements of fission process?

15.8. Write briefly about the self-sustaining chain fission reaction.

15.9. Explain what are prompt neutrons and delayed neutrons. What are their effects?

15.10. What do you understand by critical mass, critical size and critical reactor?

15.11. Write shortly on (a) Fissile material and (b) Fertile material.

15.12. What do you understand by 'Isotope'? Give examples of different isotopes.

15.13. Write short notes on :

(a) Fast and Thermal neutrons

(b) Scattering

(c) Radioactivity and radioisotopes

(d) Positrons and Neutrons

(e) K-capture

(f) Radioactive decay and Half-life

15.14. What is artificial radioactivity?

15.15. What is a 'Moderator'? What is the functions of moderator? What are the different materials that are used as moderator? What are the desirable properties of a good moderator?

15.16. What are the advantages and disadvantages of the following materials used as moderators?

(a) Graphite

(b) Water

(c) Heavy water

15.17. Where the coolant is used in nuclear reactor? What are the desirable properties of a good coolant?

15.18. What different coolants are used in nuclear reactors?

15.19. To what extent, the liquid metals can be used as coolants? Name such coolant materials.

15.20. What are the disadvantages of gases used as coolants? Which gases are generally used as coolants in nuclear reactors? Are there any other fluids that can be used as coolants in nuclear reactors? What are they? Give any advantages and disadvantages of such fluids.

15.21. What is the purpose of reactor control? What are the different methods employed to control the reactors?

15.22. Write in short about the reflectors used in reactors.

15.23. What is poisoning of the reactors? Are there any remedies for reactor poisoning?

Chapter 16 Nuclear Reactors and Nuclear Power Plants

16.1 NUCLEAR REACTORS

Nuclear reactor is the device in which nuclear reactions are taking place. These nuclear reactions produce energy by nuclear fission process which is a controlled process. From the discussions in the previous chapter regarding fundamentals of nuclear energy, it is now clear that **the nuclear reactor will consist of the following parts or components.**

16.1.1 Nuclear Fuel

Natural uranium U^{238} consists of $U^{238} + U^{235} + U^{234}$. In this natural uranium, $U^{238} = 99.3\%$, $U^{235} = 0.7\%$, U^{234} Traces.

Artificially produced fuel are U^{233} and P^{239}.

U^{233} is produced from Th^{232} and Pu^{239} is produced from U^{238}. Out of these fuels U^{235} is most unstable and is capable of sustaining chain reaction and is called as the primary fuel. U^{233} and Pu^{239} called secondary fuel. These secondary fuels are produced from Th^{232} and U^{238} and undergoes fission process by thermal neutrons.

All nuclear fuels should not be expensive to fabricate. Also they should be able to operate at high temperatures and should be resistant to radiation damage.

The fuel should be protected from corrosion and erosion of the coolant and for this it is always enclosed in metal cladding generally stainless steel or aluminium.

For continuous operation of the reactor, fuel supply should be regular and adequate and the arrangements for charging or discharging and storing of the fuel should be made.

For safety and economical operation of the reactor and nuclear power plant arrangements should be made to reprocess the spent (burnt) fuel elements to recover the unconsumed fuel. The burnt fuel elements are intensively radioactive and emits some neutrons and very harmful gamma rays and hence should be handled carefully.

Generally care is taken to prevent the contamination of fission products with coolant by providing cladding which separate the fuel from the coolant steam. Fuel cladding should possess the following properties :

(i) It should have high corrosion resistance

(ii) It should withstand high temperature within the reactor

(iii) It should have high conductivity

(iv) It should not absorb neutrons

(v) It should have strength.

In addition to the natural uranium as nuclear fuel, other important nuclear fuel element is **uranium oxide (UO_2).**

It has the advantages, as given below, over natural uranium.

(i) **Corrosion.** Better corrosion resistant than natural uranium.

(ii) **Stability.** It is more stable than natural uranium.

(iii) **Melting point.** Its melting point is very high than uranium.

(iv) **Dimensional stability.** It has greater dimensional stability during operation.

(v) It is not attacked by H_2 and N_2.

(vi) It is a brittle ceramic and is produced as a powder and then sintered to form pellets.

Disadvantages. Uranium oxide has the following disadvantages :

(i) Low conductivity

(ii) It is very much brittle and breaks due to thermal stresses

(iii) Its enrichment is a must.

Because of the brittleness, UO_2 can be produced in the powder form and the sintered to form fuel pellets.

In addition to these fuels, uranium carbide – UC – is another fuel that is used in nuclear reactor in the form of pellets.

Density and melting temperatures of natural uranium, uranium oxide and uranium carbide are :

$$\text{U-238} \rightarrow 19000 \text{ kg/m}^3 \text{ and } 1130°\text{C}$$
$$UO_2 \rightarrow 11000 \text{ kg/m}^3 \text{ and } 2750°\text{C}$$
$$\text{UC} \rightarrow 13600 \text{ kg/m}^3 \text{ and } 2350°\text{C}.$$

Another fuel to be used in nuclear reactor is uranium carbide. Uranium carbide is a black ceramic and is used in the form of pellets. Advantages of uranium carbide as nuclear fuel are :

(i) High density

(ii) High melting point

(iii) High thermal conductivity

(iv) Free from phase change

Uranium carbide is not economically suitable as compared to other fuels.

16.1.2 Moderator and Reflector

A moderator is used to slow down the fast neutrons. The fast neutrons collide with the nuclei of moderator material and slow down by losing their energy. The moderator can be in the form of solid, liquid or gas. Thus moderator is used (i) to slow down the neutrons from the high energy level (velocity) which they have when released from the fission process, (ii) to slow down the neutrons without absorbing them.

The desirable properties of a good moderator are as follows :

1. Should have high power to slow down the neutrons
2. It should be corrosive resistant
3. It should be abundance in nature in the pure form
4. It should be chemically stable
5. It should have radiation stability
6. It should have high thermal conductivity
7. It should have low parasite capture
8. It should have high melting point for solids and low melting point for liquids
9. It should have high machinability in case of solid
10. It should be inert and should not erode or corrode
11. It should not be costly.

Moderators generally used are :

(a) Water (H_2O)
(b) Heavy water (D_2O)
(c) Helium gas (He)
(d) Beryllium (Be)
(e) Graphite (C)
(f) Beryllium oxide

Amongst these materials, Beryllium is costly.

Helium is used in the pressurised state.

Reflector is generally placed round the core to reflect some of the neutrons that leak out from the surface of the core. Reflectors are generally made of the same material as that of moderator.

16.1.3 Coolant

Heat is generated in the core because of the fission process and it is necessary to limit the temperature to the safe functioning of the reactor. For this purpose, heat generated has to be removed from the core. For this purpose, coolants are used. The coolants used are either liquid or gas.

The coolants used in reactors should have the following properties :

1. It should not absorb more neutrons. The advantage is that more amount of coolant can be used at a time to take away more heat.
2. It should have a good heat transfer coefficient.
3. It should be free from radiation damage.
4. It should not corrode the metals.
5. Liquid metal coolants should have low melting points.
6. It should have high density and high specific heat so as to carry more heat (It reduces the pumping power).
7. It should have low viscosity.
8. It should be non-toxic.
9. It should have low parasite capture.
10. It should be chemical stable and it should have radiation stability.
11. It should have high boiling point.
12. It should have a low moderation ratio.
13. It should be costly as it is one of the basic requirements of the nuclear reactor.
14. It should have high thermal conductivity.
15. It must be non-oxidising.

The coolants used in power reactors are :

(a) Water (H_2O)
(b) Liquid metal — Sodium (Na), Lead (Pb), Mercury (Hg), Potassium (K)
(c) Gas coolants — Carbon dioxide (CO_2), Sulphuric dioxide (SO_2)
(d) Organic liquids.

(a) Water as Coolant. Water is naturally available in ample quantity and is definitely a better coolant than gas because (1) it has good heat transfer coefficient, (2) it can be used for high power operation, (3) it has low boiling point, and (4) light water is a good neutron moderator.

Water cannot be used in heterogeneous reactor. Because of the boiling water, whole system of heat transfer from the core will be impaired because of the conversion of water into vapours. However, if the pressure is increased, water can be used upto the temperature of 374°C which is the critical temperature.

The reactivity in the core will vary with the density of water i.e., weight content of water. If the amount of water is increased in the core the reactivity will decrease.

(b) Liquid Metal as Coolant. To increase the efficiency of the power plant it is necessary to operate it at high temperature. Metals in the liquid form can be used at high temperatures at normal pressures. Metals are less subjected to radiation damage and also have high heat transfer coefficients. Because the liquid metals like sodium, aluminium, lead and magnesium have small cross-sections as compared to water, they are used as coolants in thermal reactors.

Liquid metals used as coolants should have low melting point.

Sodium–Potassium alloy brings down the melting point even at room temperature but the cross-section of the alloy is higher than pure sodium (0.52 barns) and that of potassium is (1.98) barns.

Lithium has high cross-section and corrodes the metals. Because of high specific heat, small quantity is required to transfer the given heat.

Thus most suitable or satisfactory metal coolant is sodium–potassium alloy because of low boiling point, small cross-section, high thermal conductivity and specific heat. With moderate pressure the temperature of 540°C can be achieved.

Pure sodium gives better performance as coolant.

Potassium has disadvantage as it undergoes explosive reaction and forms potassium superoxide.

Sodium is cheap metal and is available in sufficient quantity. Sodium does not react with stainless steel upto 580°C.

Sodium should come in contact with air as it reacts with it.

(c) Gas Coolants. Powerful blowers are required if gas is to be used as coolant and power consumption will be high. Carbon dioxide gas is used as coolant in natural uranium reactor with graphite moderator. The advantages for gas as coolant are :

(i) It can be used at any high temperature.
(ii) Reactor can operate at different pressures.
(iii) Thermal efficiency is high.
(iv) CO_2 and He do not react with metal.

As the gas pumping power is high, overall efficiency of the system is low.

(d) Organic Liquids.

(i) Less subjected to corrosion.
(ii) The hydrocarbons are sufficiently stable to heat and radiation.
(iii) Organic liquids can also be used as moderator.
(iv) Do not corrode steel or aluminium and hence costly structural materials need not be used (stainless steel is not required at all).

Possible coolants are **diphenyl** and **terphenyl**.

(i) Radioactive decomposition.
(ii) Heat transfer capacity is less.
(iii) Large volume is needed to take away the given amount of heat.
(iv) Larger pumping set is required.
(v) Larger heat exchanger is required.
(vi) Organic liquid gives somewhat toxic effect.

At high temperatures fused salts are also used as coolants. Both alkali and alkaline earth halides — fluorides — possess low cross-section. Fused salts are used in homogeneous reactors because they form solutions with uranium. The disadvantage is that the salts are subjected to radiation damage and corrosion.

16.1.4 Control Rods

Control rods or mechanism is the most important mechanism in the working of reactor. The reaction rate, starting and shutting down of a reactor is controlled by the control rods. The thermal and breeder reactors use control rods. Boron steel and cadmium strips are used for absorbing excess neutrons. They are good absorbers of slow neutrons and have the advantage of not becoming radioactive due to neutron capture. By pushing these rods deep into the central core, any amount of excess neutron can be absorbed. To start the reactor, it is necessary to carefully withrod the control rods and adjust them till required output is attained. The movement of control rods can be made automatic with the help of servomechanism.

The use of control rods in the reactor core is disadvantageous because of two reasons. They give rise to non-uniform burning of fuel and good percentage of valuable neutrons are absorbed by the rods. For small reactivity this loss is not significant. It can be reduced by properly loading the core and the reactor is operated with the control rods out of the reactor.

Each reactor is equipped with three types of control rods :

(i) Shim rods
(ii) Fine control rods
(iii) Safety rods.

The shim rods absorb the excess reactivity during normal operation of the reactor. At the time of starting, the rods are taken out of the reactor. Afterwards they are positioned in the reactor to maintain the constant power level. These rods are adjusted after certain interval of time to compensate for the poisoning and other effects of fission products in the core.

The fine control rods are designed to absorb small percentage of reactivity caused by the small changes in pressure, temperature and density of fuel material. The rods can be operated manually or automatically.

The safety rods give significant poisoning effect and it is impossible to start-up the reactor with the safety rods inside. In case of an emergency the rods enter the reactor core instantaneously reducing the multiplication factor well below unity. The safety rods also operate in nearly 3–5 seconds when the reactor is subjected to earthquake or excessive power.

In large power reactors a bank of control rods is provided for uniform control of flux. In small reactor also in place of one rod, two rods are placed at a distance of 1/3 the dimension of the core to obtain maximum effectiveness.

The control rod is properly calibrated before being used in the reactor.

Thus control mechanism serves the following purposes :

(a) For starting the reactor.
(b) To maintain the power-level at the safety levels.
(c) For shutting down the reactor under normal or emergency conditions.

The control system is also necessary to prevent the chain fission reaction from becoming violent and consequently damaging the reactor.

16.1.5 Reactor Core

Reactor core is the heart of the reactor of nuclear power plant where fission chain reaction is taking place and thereby fission energy is liberated in the form of heat energy to be utilized for generating steam which is used in steam turbine producing power. The core of the reactor consists of fuel elements, control rods, coolant and moderator. Generally, reactor core have a shape approximating to a right circular cylinder with diameters ranging from 0.5 to 15 m. Many times the pressure vessel which houses the reactor core is also considered a part of the core. The schematic diagram of the simple nuclear reactor is shown in Fig. 16.1.

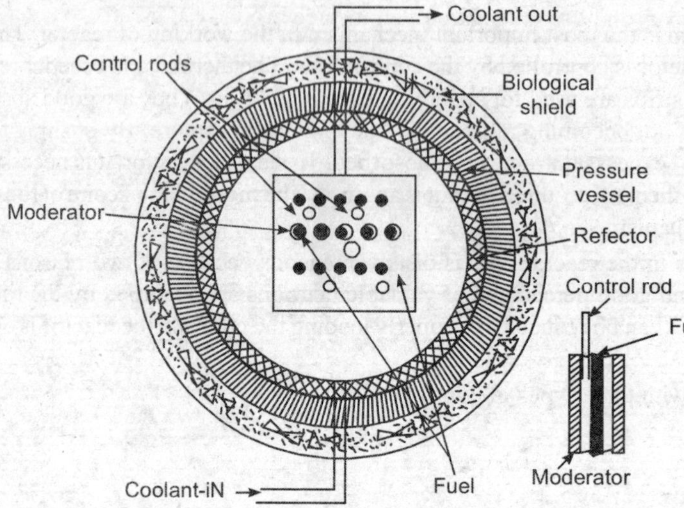

Fig. 16.1 *Nuclear Reactor.*

Canning Materials. The fuel element in the nuclear reactor is put in cans (canned) so that the fuel does not contaminate the coolant. Hence canning eliminates radiation hazards. Selection of the canning material depends on fuel used in the reactor. Aluminium, magnesium, beryllium and stainless steel are the canning materials used.

16.2 WORKING OF NUCLEAR REACTOR

Several commercial reactor designs are currently in use in United States and other developed countries. The most widely used design consists of a heavy steel pressure vessel surrounding a reactor core. The reactor core contains the uranium fuel. The fuel is formed into cylindrical ceramic pellets about one-half inch (10 to 15 mm) in diameter, which are sealed in long metal tubes called fuel tubes. The pins are arranged in groups to make a fuel assembly. A group of fuel assemblies forms the core of the reactor.

Heat is produced in a nuclear reactor when neutrons strike Uranium atoms causing them to fission in a continuous chain reaction. Control elements, which are made of materials that absorb neutrons, are placed among the fuel assemblies. When the control elements or control rods as they are often called, are pulled out of the core, more neutrons are available and the chain reaction speeds up, producing more heat. When they are inserted into the core, more neutrons are absorbed and the chain reaction slows down or stops, reducing the heat.

Most commercial nuclear reactors use ordinary water to remove the heat generated the fission process. These are called light water reactors. The water also serves to slow down, or 'moderate' the neutrons. In this type of reactor, the chain reaction will not occur without water to serve as a moderator.

16.3 POOL-TYPE REACTOR

Pool-type reactors are a type of reactor that has a core immersed in an open pool of water. The reactor core consisting of the fuel elements and the control rods is situated in an open water pool. The water acts as neutron moderator, cooling agent and radiation shield. The layer of water above the reactor core shields the radiation so completely that operators may work above the reactor in total safety. This design has two major advantages : (1) the reactor is easily accessible and (2) the whole primary cooling system i.e., pool water, is under normal pressure. This avoids high temperatures and great pressures of nuclear power plants. Popularly, called a "swimming pool reactor", they are used as a source of neutrons and for training, and in rare instances for process heat but not for electrical power generation. Life boats and life savers are often located around the facility to rescue personnel that may fall

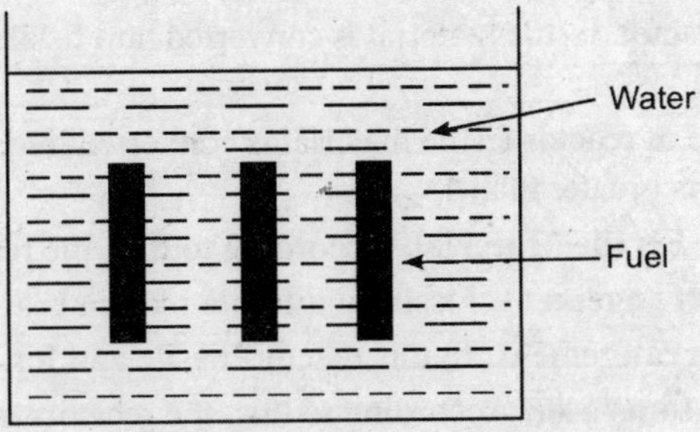

Fig. 16.2 *Pool type reactor.*

into the pool, adding further to the appearance of a pool-like environment. Most research reactors are of the pool type.

16.4 CLASSIFICATION OF NUCLEAR REACTORS

Nuclear reactors are classified based on :

(i) Purpose
(ii) Type of fission
(iii) Fuel material
(iv) Moderators
(v) Coolants.

The main heads in which the classification is done are explained as below.

1. Purpose. According to "Purpose", the reactors are classified as :

(a) Research reactor
(b) Power reactor
(c) Production reactor

(a) Research Reactor. For research purposes and for testing new reactor designs.

(b) Power Reactor. For generation of electric energy.

(c) Production Reactor. For converting fertile materials into fissile materials.

2. Type of fission. According to the type of fission, the reactors are classified as :

(a) **Slow.** In this neutron kinetic energy is less than 0.1 eV.
(b) **Intermediate.** Neutron kinetic energy is between 0.1 eV and 0.1 MeV.
(c) **Fast.** Neutron kinetic energy is 1 MeV and more.

3. Fuel used. According to fuel used, the reactors are classified as :

(a) Natural uranium
(b) Enriched uranium
(c) Plutonium.

4. State of fuel. According to the state of the fuels, the reactors are classified as :

(a) Solid fuel and
(b) Liquid fuel.

5. Fuel cycle. According to fuel cycle, the reactors are classified as :

(a) Burner (Thermal). The reactor is designed for producing heat only without any recovery of converted fertile material.

(b) Converter. In this type of reactor, fertile material is converted into fissile material different from the initially fed into reactor. γ is less than 1.

(c) Breeder reactor. In this type of reactor fertile material is converted into fissile material which is the same as that initially fed into the reactor. γ is greater than 1.

6. Arrangement of Fissile and Fertile Material. According to this, the reactors are classified as :

(a) Reactor with one region arrangement. Fissile and fertile materials are mixed.

(b) Reactor with two region arrangement. In this reactor, fissile and fertile materials are separate.

7. Arrangement of Fuel and Moderator. According to this, the reactors are classified as :

(a) Homogeneous. In this type of reactor, fuel and moderator are fixed together homogeneously.

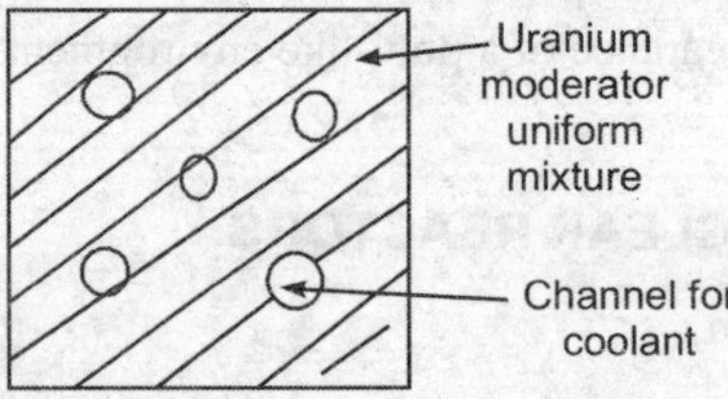

Fig. 16.3 *Homogeneous reactor core.*

(b) **Heterogeneous.** In this type of reactor the fuel is in discrete lumps in moderator.

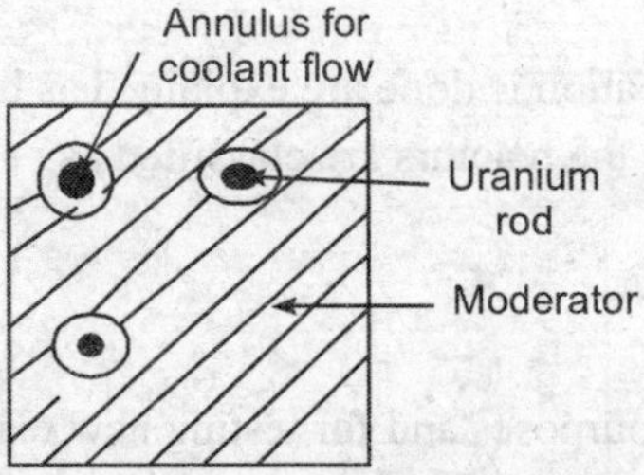

Fig. 16.4 *Cross-section of hetrogeneous reactor core.*

8. Moderator Material. According to the moderator material used, the reactors are classified as following :

(a) Heavy water

(b) Ordinary water

(c) Graphite

(d) Beryllium

(e) Organic

9. Cooling System. According to this, the reactors are classified as below

(a) Direct Cooling. In direct cycle, the reactor coolant is used as the working fluid in the power cycle.

(b) Indirect Cooling. In this cycle, coolant is passed through the reactor core and through the heat exchange to transfer heat to separate working fluid for generation of steam.

(c) Indirect Cooling with Intermediate Loop. In this case, the reactor coolant transfers heat through intermediate loop to working fluid for generation of steam.

These three systems are shown schematically in Figs. 16.5, 16.6 and 16.7.

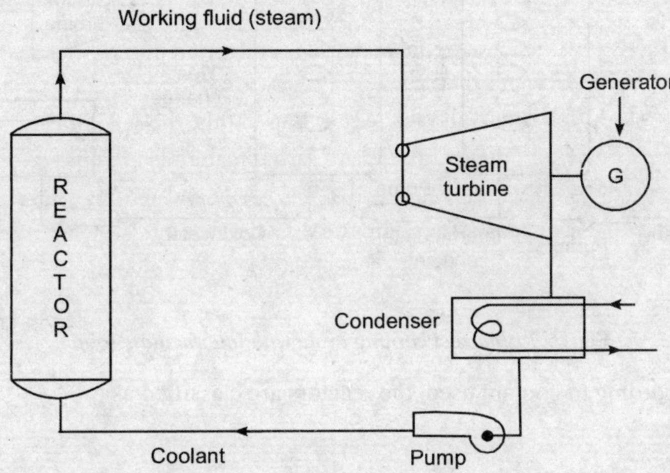

Fig. 16.5 *Direct Cooling Cycle.*

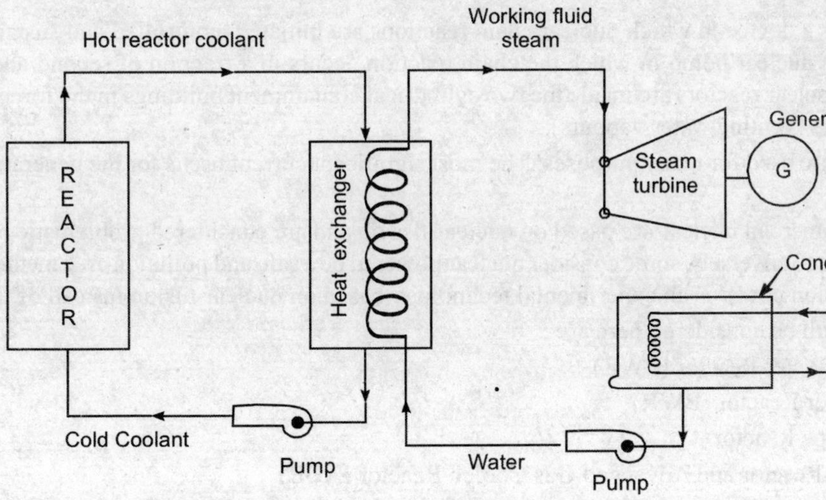

Fig. 16.6 *Indirect cooling cycle.*

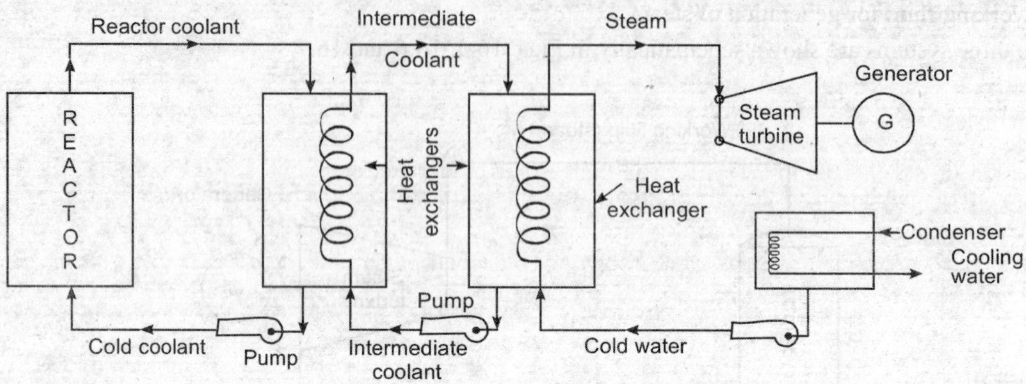

Fig 16.7 *Indirect cooling cycle with intermediate loop.*

10. Coolant used. According to coolant used, the reactors are classified as :

(a) Water cooled reactor

(b) Gas cooled reactor

(c) Liquid metal cooled reactor

(d) Heavy water cooled reactor.

16.5 TYPES OF REACTORS

A **nuclear reactor** is a device in which nuclear chain reactions are initiated, controlled and sustained at a steady rate (as opposed to a nuclear bomb in which the chain reaction occurs in a fraction of second and is completely uncontrolled). The nuclear reactors are inside the two cylindrical containment buildings in the foreground – behind are the cooling towers (venting water vapours).

Nuclear reactors are used for many purposes. The most significant current use is for the generation of electrical power.

Currently, all commercial nuclear are based on nuclear fission and are considered problematic by some of their safety and health risks. Conversely, some consider nuclear power to be a safe and pollution-free method of generating electrical power. Fusion power is an experimental technology based on nuclear fusion instead of fission.

Reactors which will be considered here are :

(a) Pressurised Water Reactor (PWR)

(b) Boiling Water Reactor (BWR)

(c) CANDU Type Reactor

(d) Gas Cooled Reactor and Advanced Gas Cooled Reactor (AGR)

(e) Liquid Metal Cooled Reactor (Sodium Cooled Reactor)

(f) Organic Moderated and Cooled Reactor

(g) Fast Breeder Reactor

(h) Advanced Reactors

(i) 3rd Generation Reactors

(j) 4th Generation Reactors.

16.5.1 Pressurised Water Reactor (PWR)

In 1954 the first reactor of this type was built in USSR. Two water moderated, light water cooled, enriched fuel reactors are under operation in USSR (Voronezh Atomic Power Station). Fuel is enriched upto 1.5%. Produce 760 MW (Thermal power) and 125 MW electricity. Overall efficiency is 25.78%.

USA has contructed its first land-based pressurised water reactor for power generation in 1957 at shipping port in Western Pennsylvania on Ohio river. It is light water moderated and cooled heterogeneous thermal reactor producing 60–68 MW (Electrical power output) at an thermal efficiency of 29.30%.

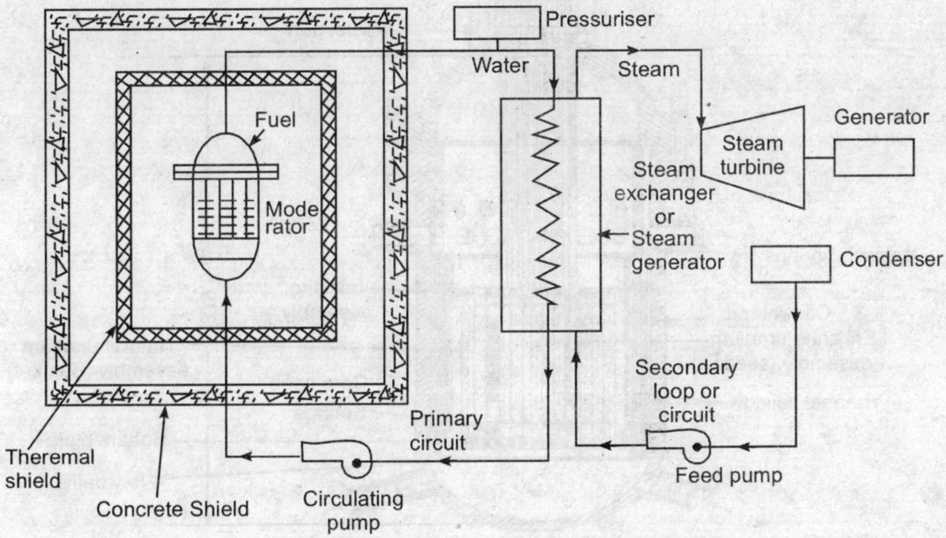

Fig. 16.8 *Schematic of PWR power plant.*

The schematic diagram for PWR is shown in Fig. 16.8. In a pressurised water reactor system, heat generated in the nuclear core (fission energy) is removed by water-reactor coolant-circulating at high pressure through the primary circuit. The water in the primary circuit cools and moderates the reactor. The heat is transferred from the primary to the secondary circuit in the heat exchanger which acts as a boiler. In this boiler, the water is converted into steam. In this reactor, the pressure of the coolant water has to be greater than the saturation pressure of steam generated in the heat exchanger (boiler) – sat at 300°C and 85.927 bar, so that the bulk boiling is prevented. The limit of temperature of coolant (water) is its critical value of 374°C but the actual temperature is much less and is generally upto 300°C.

Following are the parts of the reactor.

(a) Pressure Vessel or Reactor Vessel. The pressure vessel is cylindrical in shape provided with hemispherical domes on the two sides. The hemispherical dome is secured to the vessel by means of studs. These studs are typical in construction as they incorporate a heater so that they can be heated and expanded before installation, which results in their being highly stressed when cooled. This is necessary for the same operation of the vessel at elevated working temperatures.

The height of the pressure vessel is of the order of 10 m (nearly of the double-decker bus height). It is constructed of carbon steel plates cladded with 0.65–0.7 cm thick stainless steel. General dimensions of the cylinder is –ID = 2.75 m and 21.5 cm thick. Again, the top hemispherical portion is a forged piece of 250–260 mm thick. Provision is made for nearly 24 control rods and 9 fuel port tubes assembly.

For this typical PWR, there are nearly 42 studs each of 15 cm to fix the top head with the vessel. The net weight of the reactor vessel is 254 tonnes and flanges 34 tonnes.

(b) Reactor Thermal Shield. When the reactor is in operation a large neutron and gamma flux leaks from the core. In order to minimise the absorption of gamma rays and neutrons in the pressure vessel wall a thermal shield is interposed between the reactor core and the pressure vessel wall. To minimise the thermal stresses the vessel is packed with nearly 10 cm thick glass wool insulation having density of 0.145 gm/cc.

(c) Fuel Elements. The fuel in PWR is slightly enriched uranium in the form of thin rods or plates. The cladding is either of stainless steel or zircaloy. The fuel elements are incorporated in speed assemblies and blanket assemblies.

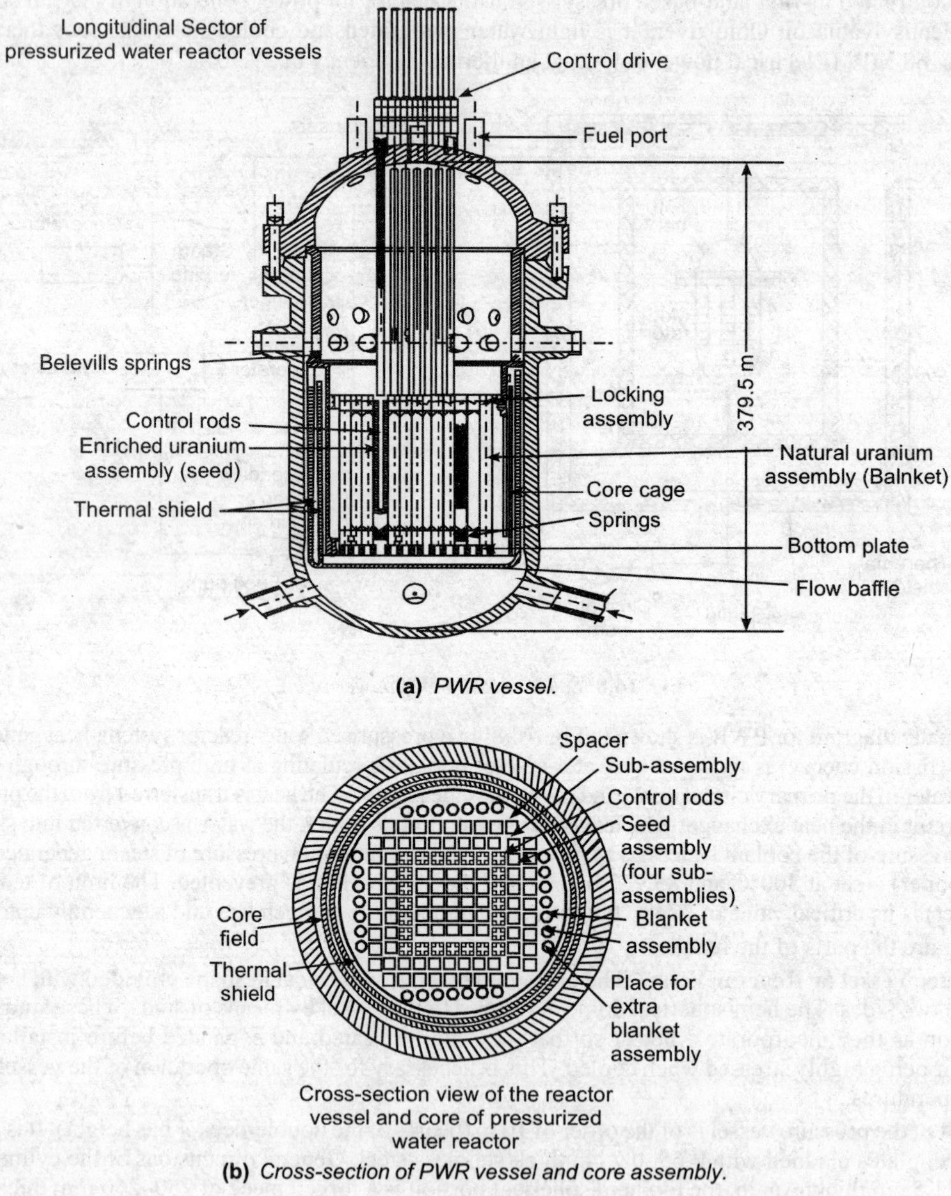

(a) *PWR vessel.*

(b) *Cross-section of PWR vessel and core assembly.*

Fig. 16.9

Speed assemblies consist of a number of plates welded together to form a square cross-sectional arrangement with passages left between them for water to pass through. A typical PWR contains about 200 fuel assemblies, each assembly being an array of rods.

The blanket assemblies are filled with natural uranium oxide rods with the inside zircaloy-2 tubing with OD = 1 cm and wall thickness = 0.7 mm. One bundle of rod = 100 rods and one fuel assembly = seven bundles. The total weight of enriched U^{235} required = 52 kg and 12.3 tonnes is filled in the blanket assembly.

(d) Control Rods. The control rods in this type of reactor are made of **hafnium** in a cruciform shape. Hafnium possesses good mechanical properties and is also highly corrosion resistant. The control rods are driven by canned rotor type electric motor through the seed elements where power density is highest.

(e) Reactor Containment. Usually there are a number of reactor coolant loops. Some loops are meant for regular service with the choice of the number of loops in operation depending on the load on the plant. Some standby loops are also provided for emergency. Each loop is provided with separate heat exchanger. In addition to the heat exchanger each loop is having centrifugal canned motor pump and gate type isolation valve.

(f) Reactor Pressurizer. The function of the pressuriser is to maintain high pressure of the order of 150–155 bar in the primary loop of the coolant system. This is greater than the pressure corresponding to the saturation temperature of steam generated in the secondary loop or circuit. This is to prevent the bulk boiling. A pressurizer usually remains half-filled with water and partly with steam. This acts as a **surge tank** or **accumulator** for the system.

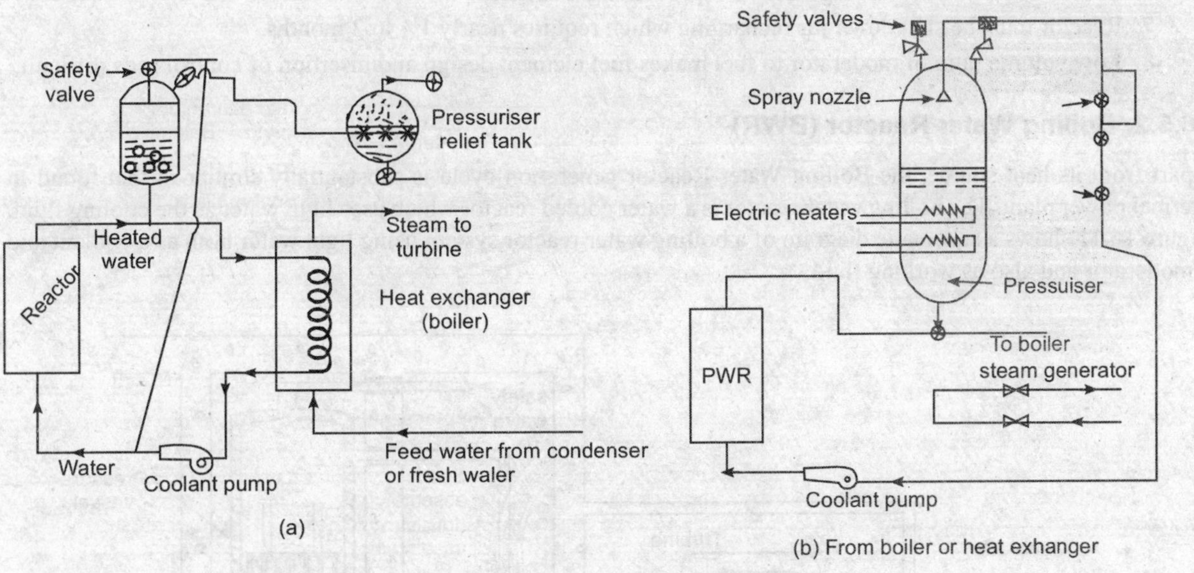

Fig. 16.10 *Vapour type pressurizer system.*

A schematic of the pressuriser is shown in Fig. 16.10(a). If the reactor transient occurs and the pressure in the reactor system goes up, coolant is forced through the line to the pressuriser, and condenses there. If the pressure in the primary system drops, some of the steam in the pressuriser flashes, forcing the coolant from the pressuriser into the primary coolant loop. From this it is observed that the pressurizer works as a system pressure regulator and its upper and lower limits of working are set by the hot spot temperature and safety and relief valves settings for a particular pressure. Another version of pressuring system is shown in Fig. 16.10(b).

Advantages and disadvantages of PWR are as below

Advantages

1. The reactor makes use of single fluid as coolant, moderator and reflector. Water is cheap and is available in plenty.

2. The reactor is compact and the size is minimum. The power density is higher and is 65 kW/litre.
3. Being compact, it is suitable for naval propulsion units (submarine).
4. Fission products remain contained in the reactor and are not circulated.
5. It provides complete freedom to inspect and maintain the turbine, feed heaters and condenser during operation.
6. A small number of control rods is required.
7. Fuel cost is less because more energy per unit weight of fuel can be extracted.
8. Because the fuel used is enriched uranium, the reactor can be compact in size.

Disadvantages.

1. High primary loop pressures require strong pressure vessel.
2. Because of the strong vessel requirement, the capital cost is high.
3. Low pressure and temperature in secondary loop result in poor thermodynamic efficiency.
4. Use of water under pressure at high temperature creates the problem of corrosion which calls for use of stainless steel.
5. Fuel element fabrication is expensive.
6. Fuel suffers radiation damage and its reprocessing is difficult.
7. Reactor must be shut down for recharging which requires nearly 1½ to 2 months.
8. Low volume ratio of moderator to fuel makes fuel element design and insertion of control rods difficult.

16.5.2 Boiling Water Reactor (BWR)

Apart from its heat source, the Boiling Water Reactor generation cycle is substantially similar to that found in thermal power plant. The boiling water reactor is a water cooled reactor which uses light water as the cooling fluid. Figure 16.11 shows a schematic diagram of a boiling water reactor system using light water both as a coolant and a moderator and also as working fluid.

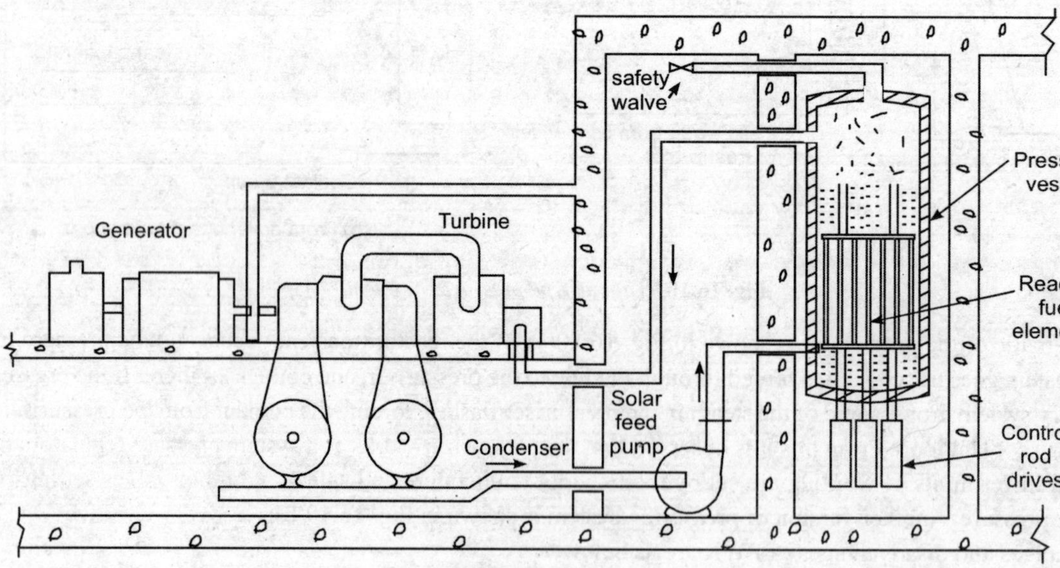

Fig. 16.11 *BWR plant.*

The fuel used is enriched uranium oxide (oxide of uranium with additional U^{235} content to that contained in natural uranium) canned in zirconium alloy. Another simple schematic diagram of Boiling Water Reactor is shown in Fig. 16.12.

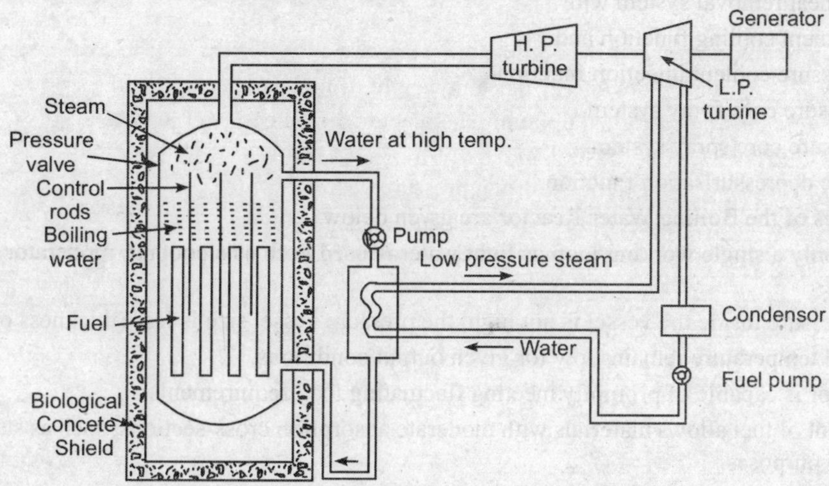

Fig. 16.12 *Boiling water reactor.*

Normally a reactor operating at 80 bar can produce steam at 60–65 bar pressure so that a conventional thermal plant cycle could be used on secondary side.

The steam from such reactors is of course radioactive. This radioactivity of the steam system is short-lived and exists only during power generation. Extensive generating experience has fully demonstrated that shut down maintenance on BWR turbine, condensate and feed water components can be performed essentially as at a thermal plant.

The reactor core, the source of nuclear heat, consists of fuel assemblies and control rods contained within the reactor vessel and cooled by the circulating water system. The power level is maintained or adjusted by positioning control rods up and down within the core. The BWR core power level is further adjustable by changing the circulation flow rate without changing control rod position, a feature that contributes to the superior load following capacity of the BWR. Generally the active or fueled core of a BWR consists of about 800 fuel assemblies, each containing 8 × 8 array of fuel rods. The zircaloy channel around the fuel rods prevent cross flow in the core. BWR fuel rods are slightly larger than PWR fuel rods. The average fuel enrichment varies from 1.9 to 2.6%.

The BWR operates at constant pressure and maintains constant steam pressure as in thermal plants. The integration of the turbine pressure regulator and control system with reactor water circulation flow control system permits automated changes in steam flow to accommodate varying load demands on the turbine. Power changes upto 25% can be accomplished automatically by circulation flow control alone, at rates of 15% per minute increasing and 60% per minute decreasing. This provides a load-following capability that can track rapid changes in power demand.

Following auxiliary systems are used for normal plant operation.

1. Reactor water clean-up system.
2. Shut down cooling function of residual heat removal system.
3. Fuel and containment pools cooling and filtering system.
4. Closed cooling water system for reactor service.
5. Radioactive waste treatment.

The following auxiliary systems are used as back-up (standby) or emergency systems :

1. Standby liquid control system.
2. Reactor core isolation cooling system.
3. Residual heat removal system with
   - containment cooling function and
   - low pressure coolant injection function.
4. High pressure core spray system.
5. Low pressure core spray system.
6. Automatic depressurisation function.

The **advantages** of the Boiling Water Reactor are given below.

1. There is only a single working loop as light water is used both as a coolant, moderator and as a working fluid.
2. As the pressure inside the vessel is not high, the pressure vessel size is less (thickness of plates is less).
3. The metal temperature remains low for given output conditions.
4. The reactor is capable of promptly meeting fluctuating load requirements.
5. Enrichment of fuel allows materials with moderate absorption cross-sections, such as stainless steel, to be used for structural purposes.

However, there are some disadvantages as explained below :

1. Activation of water (used as coolant, moderator and working fluid) and steam, involves the risk of radioactive contaminator of the steam turbine used. Half-life is about 15 minutes.
2. More biological shield or protection is required.
3. Boiling limits the power density, only 3 to 5% by mass, can be converted to steam per pass through the boiler.
4. Part of steam is wasted at low loads.
5. Enrichment of fuel for the reactor is an extremely costly process.
6. Unless properly designed, the BWR will have negative power demand coefficient. So that when more power is demanded from the reactor, it may produce less.
7. The possibility of burn-out of fuel is more in this reactor.

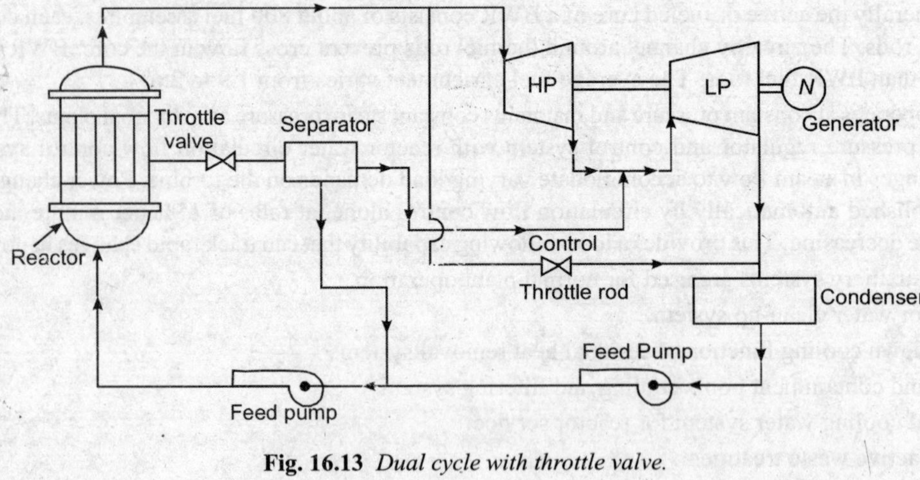

Fig. 16.13 *Dual cycle with throttle valve.*

Because of the possibility of negative power demand coefficient, the single circuit cycle does not adjust heat output (or steam output) corresponding to the power requirement. The required steam demand can be supplied by using the dual cycle as shown in Figs. 16.13 and 16.14.

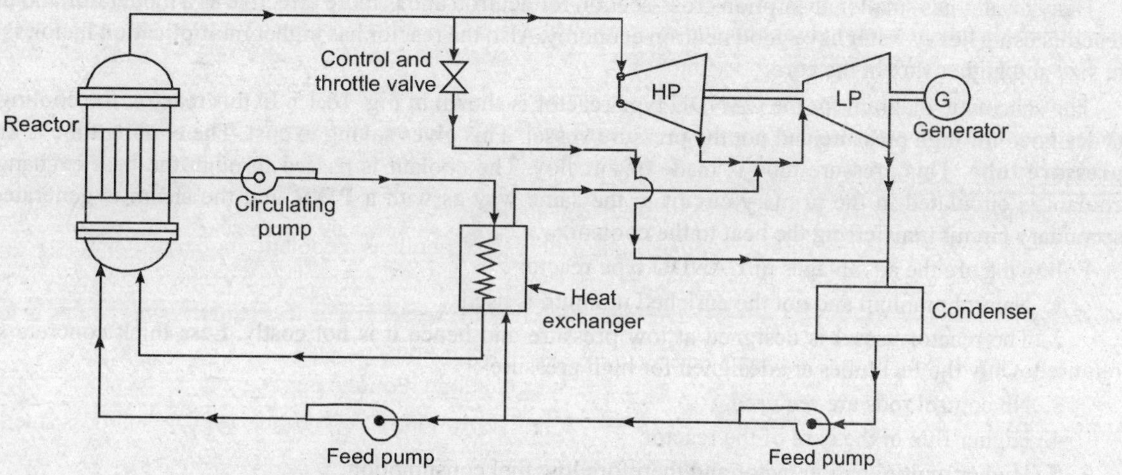

Fig. 16.14 *Dual cycle with secondary heat exchanger.*

As shown in the Figs. 16.13 and 16.14, the high pressure steam from the reactor is directly supplied to the HP turbine and the LP steam formed by throttling the high temperature water or generating steam with the use of high temperature water, is fed to the LP turbine. The dual supply arrangement is used to govern the speed of the turbine by low pressure steam. This enables to adjust the increased power demand without shutting down the reactor which is common in single circuit.

16.5.3 CANDU Type Reactor

A reactor designed in Canada and using natural uranium as fuel, heavy water as moderated, and water as coolant and reflector is very well known as **CANDU** reactor. This type of the reactor is used at Douglas Point Ontario,

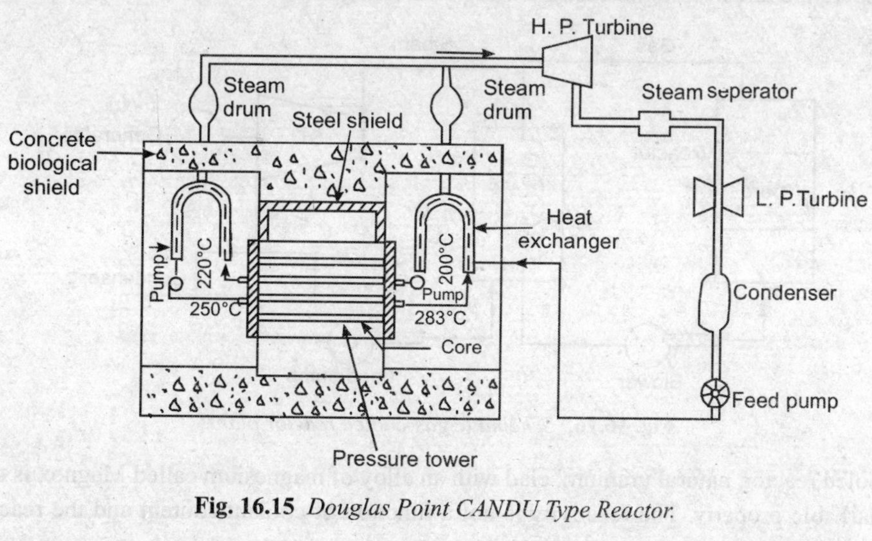

Fig. 16.15 *Douglas Point CANDU Type Reactor.*

Canada and is known as The Douglas Point Candu Reactor. In this reactor, natural uranium oxide as fuel and heavy water are used. India is using similar types of the reactors at Rajasthan Atomic Power Project at Rana Pratap Sagar and Madras Atomic Power Project at Kalpakam.

Heavy water has smaller absorption cross-section for neutron and is more effective as a moderator and therefore reactors using heavy water have good neutron economy. Also the reactor has higher multiplication factor, is smaller in size and higher flux in the core.

The schematic diagram for the CANDU type reactor is shown in Fig. 16.15. In this reactor, the cooling system is designed for high pressure and not the pressure vessel. This gives saving in cost. The cooling tube is known as **pressure tube**. This pressure tube is made of zircaloy. The coolant is passed through the heat exchanger. The coolant is circulated in the primary circuit in the same way as with a **PWR** and the steam is generated in the secondary circuit transferring the heat to the coolant.

Following are the advantages of CANDU type reactor :

1. Natural uranium and not the enriched uranium is used.
2. The reactor vessel is designed at low pressure and hence it is not costly. Less thick concrete shield is required. Only the fuel tubes are designed for high pressure.
3. No control rods are required.
4. Higher flux in the core of the reactor.
5. Higher multiplication factor and therefore low fuel consumption.

Against these advantages, following are the disadvantages of this CANDU type reactor.

1. The cost of heavy water is extremely high (nearly Rs. 550/kg).
2. The power density is extremely low and therefore reactor size for the same power is very large.
3. High standard of inspection and maintenance is required.

16.5.4 Gas Cooled Reactor and Advanced Gas Cooled Reactor (AGR)

Under the British nuclear power programme the gas-cooled reactor was developed from Fermi's Chicago Pile. Actually the first gas cooled reactor using CO_2 gas at a pressure of 16 bar, as coolant and graphite as moderator was developed in Britain during 1956–69. Because of the large variation in temperature and pressure of the gas, it was preferred in place of water. Depending on the maximum permissible temperature of the fuel slugs, the gas can attain any working temperature and keeping this temperature constant, gas can be subjected to any pressure.

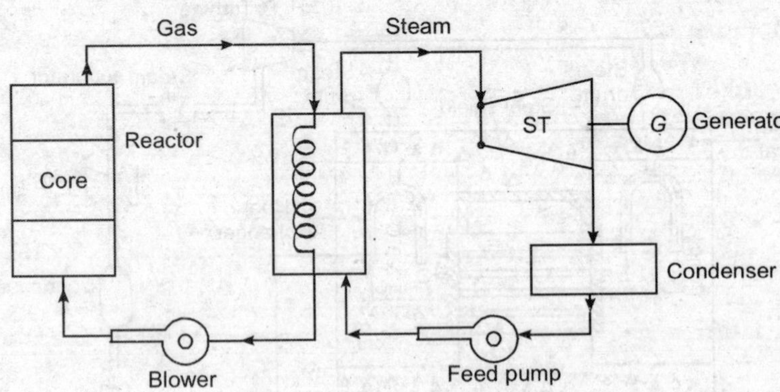

Fig. 16.16. *Schematic gas-cooled reactor plant.*

In this gas cooled reactor, natural uranium, clad with an alloy of magnesium called Magnox is used. The gas as coolant has remarkable property. The reactivity is not a function of coolant content and the reactor operation is inherently safe. In the reactor, the weight of the gas is very small and hence the neutron absorption by gas is negligible, and therefore, any leakage of the gas will not affect the reactivity. Instead, the leakage of gas will affect the heat transfer system.

At present several types of gas-cooled reactors have been designed and built. UK Atomic Energy Commission has developed Calder Hall and Hinkley point reactors. Also England developed Advanced Gas Cooled Reactor (AGR). In this category, Germany and the USA developed Helium-cooled, graphite-moderated system (HTGR).

A simple schematic diagram of gas-cooled reactor is shown in Fig. 16.16. In this reactor (AGR), UO_2 is used as fuel clad in stainless steel tubes with CO_2 gas as coolant and graphite as moderator.

In HTGR, Helium is used as coolant, graphite is a moderator and U^{233} as fissile material and thorium as fertile material. In HTGR, U^{235} is initially fuelled until sufficient U^{233} is available for make up fuel. Graphite has a very high melting point and therefore the fuel elements can operate at very high temperatures generating steam at the condition equivalent to those in modern coal fired power plant. Helium is chemically inert and therefore it is a suitable coolant. Also helium has good heat transfer characteristics and low neutron absorption. It is a monoatomic gas and produces more power for the given temperatures, in the Brayton cycle.

A schematic diagram of a direct cycle HTGR gas turbine plant is shown in Fig. 16.17. It includes multi stage compressor with intercooling, a regenerator with a reactor in between compressor and gas turbine. This reactor with U^{233}/Th^{232} fuel may be looked upon as a thermal breeder reactor.

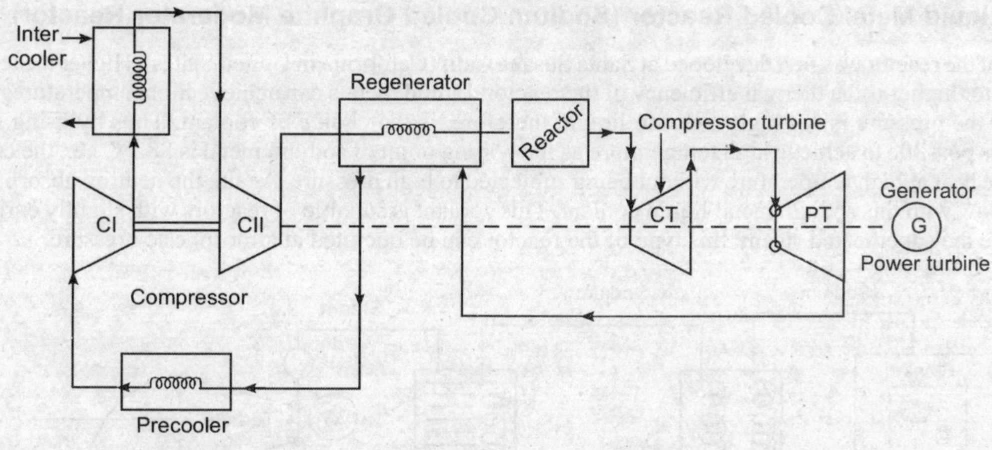

Fig. 16.17. *Schematic of HTGR direct cycle gas turbine plant using helium.*

Reason for Selecting CO_2 as Coolant.

1. The neutron economy is the main consideration in selecting the coolant. CO_2 gas in cheap and its neutron capture cross-section is also satisfactory.

2. It is chemically stable under all operating conditions of the reactor and it does not react with graphite moderator.

3. CO_2 does not attract nitrogen. Nitrogen is having high neutron capture cross-section.

4. It can attain any temperature but the maximum temperature is decided by the permissible working temperature of the fuel.

Only disadvantages is its poor heat transfer coefficient.

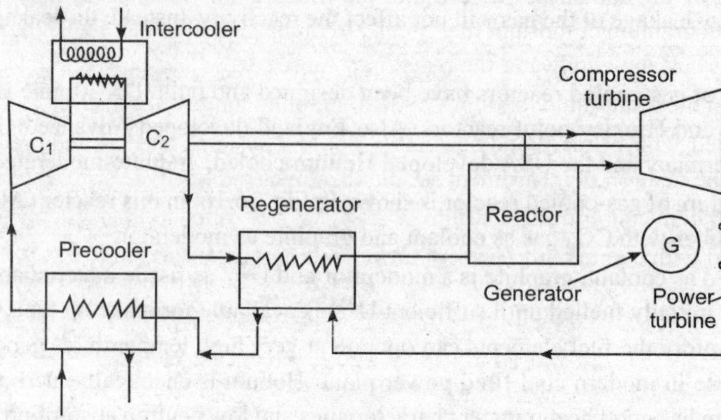

Fig. 16.18 *Direct cycle gas-turbine plant with reactor.*

Another schematic diagram of a simple, high temperature reactor—gas turbine power plant is shown in Fig. 16.18.

16.5.5 Liquid Metal Cooled Reactor (Sodium Cooled Graphite Moderator Reactor)

This type of the reactor was first developed at Santa Susana Calif (California in United States). Higher the temperature of the reactor higher is the thermal efficiency of the reactor. Liquid metals can achieve high temperatures and at the same time the pressure is not high. Sodium liquid, therefore, is the choice of coolant. Thus by using sodium as coolant it is possible to achieve high temperature as the boiling point of sodium metal is 882°C i.e., the coolant can transfer the heat at high temperature without being subjected to high pressure. Again, the neutron absorption cross-section is low with this sodium metal liquid coolant. This coolant is suitable of reactors with slightly enriched fuel. To produce the superheated steam, this type of the reactor can be operated at atmospheric pressure.

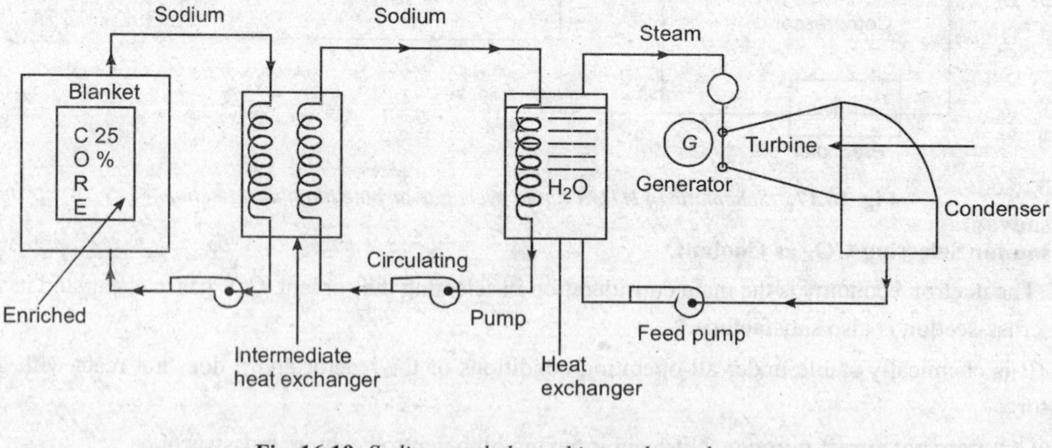

Fig. 16.19 *Sodium cooled graphite moderated reactor.*

Higher efficiency of the reactor reduces the cost per kW. Graphite is used as moderator having high mechanical strength at high temperature. The mixing of water and sodium is hazardous and they are thermally insulted. Sodium and graphite are to be kept separate in this type of the reactor. Both can be separated either by canning each block of graphite by zirconium or the coolant is run through the tubes. As a coolant sodium in the pure form is not subjected to corrosion with uranium or thorium fuel, steel, zirconium and graphite. In the impure form, the reaction of sodium with O_2 is vigorous and with water the reaction is more severe and cause rapid corrosion.

The schematic diagram of metal liquid (sodium) cooled, graphite moderator reactor is shown in Fig. 16.19.

This reactor is provided with primary and secondary cooling systems and are required to be shielded with concrete because due to the neutron bombardment sodium becomes radioactive.

It has been studied that conversion from Thorium Th^{232} to U^{233} is suited to this system.

The reactor is provided with eight control rods of nickel and boron.

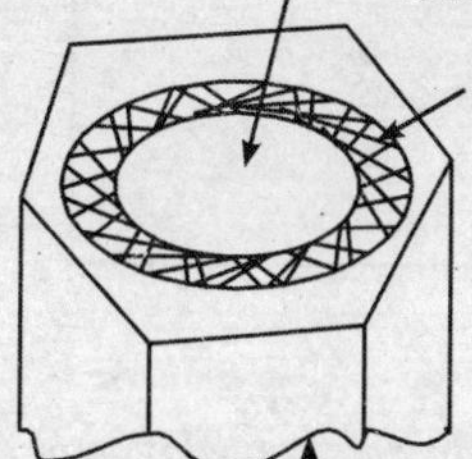

Fig. 16.20 *Fuel, moderator and coolant arrangement.*

Figure 16.20 shows fuel rods assembled in the hexagonal graphite moderator blocks.

Some advantages and disadvantages of this reactor are as following.

Advantages

1. High temperature available at low pressure at low cost and high thermal efficiency.
2. Sodium need not be pressurised because of its high melting point at low pressure.
3. Excellent heat removal.
4. High conversion ratio.
5. Superheating of steam is possible.
6. Reactor size is small.
7. Graphite moderator is cheap and it has high mechanical strength at high temperature.
8. Sodium cooled reactor is safe from the point of view of reactivity.

Disadvantages

1. Sodium reacts violently with water and actively with air.
2. Heat exchangers must be leak-proof.
3. Thermal stress are a problem.
4. Intermediate circuit is necessary to separate active sodium from water.
5. Separation of sodium and graphite is necessary as porous graphite may absorb sodium. The penetration of sodium between the layers of graphite can cause mechanical failure.

16.5.6 Organic Cooled and Moderated Reactor

As the name of the reactor suggests the organic fluids are used as coolant and also as moderator. In this reactor, instead of sodium, some hydrocarbon is used as coolant as well as moderator. The coolant and moderator is a commercial mixture of ortho-meta terphenyl and paratherphenyl operated at a pressure of about 8.5 bar. The same organic material also acts as a reflector. The flux control is achieved with the help of boron control rods numbering 13-14.

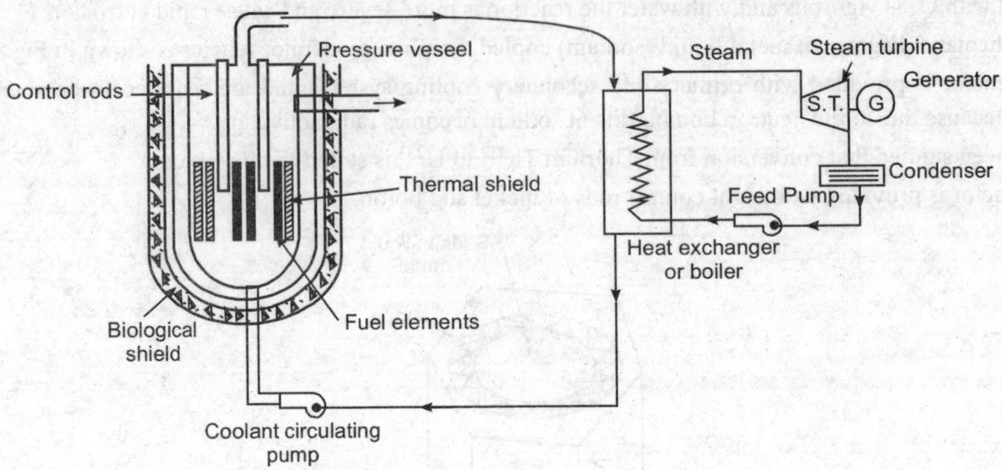

Fig. 16.21 *Organic cooled reactor (Piqua)*

Figure 16.21 shows the schematic diagram for Piqua organic cooled reactor and the steam power plant.

The organic fluid working as coolant and also as moderator has some **advantages**. They are :

1. The hydrocarbon contains carbon and hydrogen which have super moderating properties. No separate moderator is required and therefore the design will be such that compact core is possible. Hence the construction will be economical.

2. Hydrocarbons have low vapour pressure at high temperature. Therefore higher thermal efficiency is the result. The reactor is designed for low pressure.

3. With organic fluid uranium can be used in the form of oxide, alloy or carbide.

4. With organic coolant, the steam generated is subjected to very small radioactivity. The water circuit is designed for high temperature and pressure as there is no radiation damage.

5. Corrosion is absent with organic coolant and the pressure vessel and piping can be made of low carbon steel. This further reduces the cost of construction.

Some **disadvantages** are also present with organic fluid used as coolant and moderator. They are :

1. The organic fluid has very low heat transfer coefficient than water.
2. The organic fluids catch fire and therefore care must be taken to avoid accidents.

Near Piqua city, Ohio in USA, there is a nuclear power plant known as Piqua Nuclear Power Plant of 11.5 MW capacity with 25% overall efficiency. This power plant uses an organic cooled. Moderator nuclear reactor and is called Piqua organic cooled reactor.

16.5.7 Fast Breeder Reactor

Natural uranium U^{238} is found in nature and it contains fissionable material or fuel U^{235} of the order of 0.7%. Many nuclear scientist were thinking that in the same way as natural uranium U^{238}, the fissil fuel U^{235} may be in the highly enriched condition in nature. But in vain. It is known that when U^{235} is fissioned by a slow neutrons, then 2.5 neutrons are one of the product of the fission process. It is known that out of 2.5 neutrons released, one neutron is used to maintain the further fission process. Out of 1.5 neutrons remaining, 0.9 neutrons are captured by U^{238} and 0.6 neutrons are absorbed by the coolants. The absorption of neutron by U^{238} converts U^{238} to Pu^{239} which is again a fissile material. This Pu^{239} is a man-made fuel and is called **secondary fissionable fuel**. Similarly, thorium Th^{232} can be converted into U^{233} which is also a fissionable fuel and is again man-made or secondary fissionable fuel. Both these secondary fuels can be fissioned by a slow neutrons.

The efficiency of natural uranium is low because of poor concentration of U^{235}. When U^{235} is used to produce Pu^{239} or U^{233} from U^{238} or Th^{232}, the factor called as **Conversion Factor** comes into picture. The conversion factor is defined as the ratio of the number of secondary fuel atoms produced to the number of primary fuel atoms consumed. Normal reactor using uranium-graphite gives conversion factor as 0.9 i.e., 90% recovered and 10% consumed.

A reactor is known as **Breeder Reactor** when conversion ratio is unity or greater than 1. The process of conversion is represented as :

$$_{90}Th^{232} + _0n^1 \rightarrow _{90}Th^{233} \xrightarrow{23.26\,\text{min}} _{91}Pa^{233} \xrightarrow{27.38\text{ days}} _{92}U^{233}$$

For getting Pu^{239}, the process is as follows :

$$_{92}U^{238} + _0n^1 \rightarrow _{92}U^{239} + \gamma$$

$$_{92}U^{239} \xrightarrow{24\,\text{min}} _{93}Np^{239} + _{-1}e^0 \xrightarrow{23\text{ days}} _{94}Pu^{239} + _{-1}e^0 .$$

Always, the breeder reactor can only be started by U^{235} or by Pu^{239} or U^{233} produced in some other reactor because both the latter isotopes are not available in nature.

When the fission is due to **fast neutrons** and when the moderator is not required, the reactor is called **Fast Breeder Reactor**. In this case the energy produced is enormous and therefore, only liquid metal (sodium) coolant is suitable to transfer the heat.

Doubling Time. The rate of accumulation of fissionable fuel is expressed in **Doubling Time**. The rate of accumulation is a function of breed gain. Doubling time is the time needed for the initial investment of fissionable material to be doubled.

Fast Breeder Reactor. The fast breeder reactor derive its name from its ability to breed, that is to create/ produce more fissionable material than it consumes; and from the fact that its neutron travel is faster than they do in a thermal reactor.

In fast breeder reactor Pu^{239} or enriched uranium is used as a fuel and U^{238} as a fertile material without a moderator, in the casing or the core. The central portion of the reactor is generally a stainless steel vessel in which the core of fissile material is surrounded by a blanket of fertile material. It is necessary to use a coolant which does not slow the neutrons or capture them as they travel through the coolant. One such coolant is sodium and it is widely used in fast reactors.

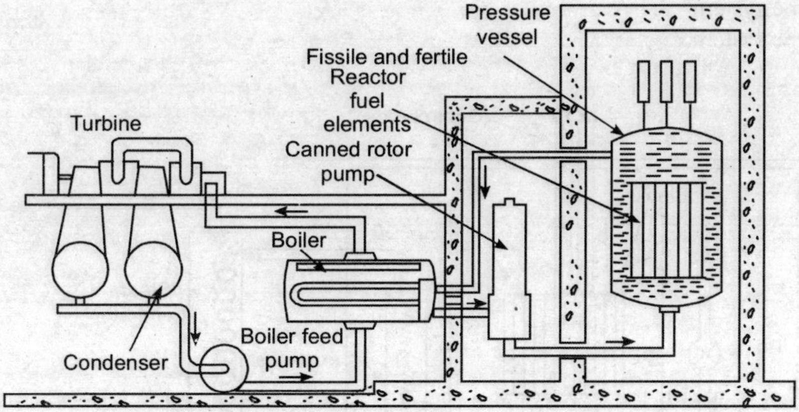

Fig. 16.22 *Fast breeder reactor.*

The general arrangement of the plant is shown in Fig. 16.22. Sodium, used a reactor coolant, exchanges heat in primary heat exchanger. The secondary cooling circuit also used sodium, which must not become radioactive, and a neutron shield is placed in between the vessel and the primary heat exchanger.

Characteristics of sodium due to which it is used as coolant (Helium and CO_2 are also available) are :

1. Its cross-section for absorbing and thermalizing neutron is very low.
2. Its ability to transfer heat is excellent.
3. It has high boiling point (893°C).
4. It has very low vapour pressure at most temperatures.
5. It does not react on any of the structural materials used in primary circuits.
6. It has good emergency cooling characteristics.

Major or serious disadvantage of sodium is its chemical reactivity with water and air, highly induced radioactivity and very expensive engineering.

Because of this, nowadays the gas-cooled reactors concept is being promoted.

The helium as an alternative coolant for fast breeder reactor has the following advantages :

1. The metallurgical problems are not severe.
2. Chemically inert, does not become radioactive and does not change place.
3. Heat transfer characteristics are equally good as sodium.
4. Doubling time is 10–12 years as compared to 20 years for sodium cooled reactors.

Only disadvantage of helium is that high pressure (70–80 bar) is required.

It is predicted that in future helium cooled reactors may be used.

Some **advantages** of the fast breeder reactor are :

1. High breeding gain is possible.
2. High power density.
3. Absorption of neutrons is low.
4. Efficient utilization of uranium fuel than that could be achieved in thermal reactors.
5. It does not require moderator.

Disadvantages are as follows :

1. Specific power is not high as in thermal reactor.
2. Enriched fuel is required (15%).
3. Neutron flux is high at the centre of the core and therefore control is difficult.
4. Safe for melt down of material.

There are two systems of faster breeder reactor. They are

1. the pool type system
2. the loop type system.

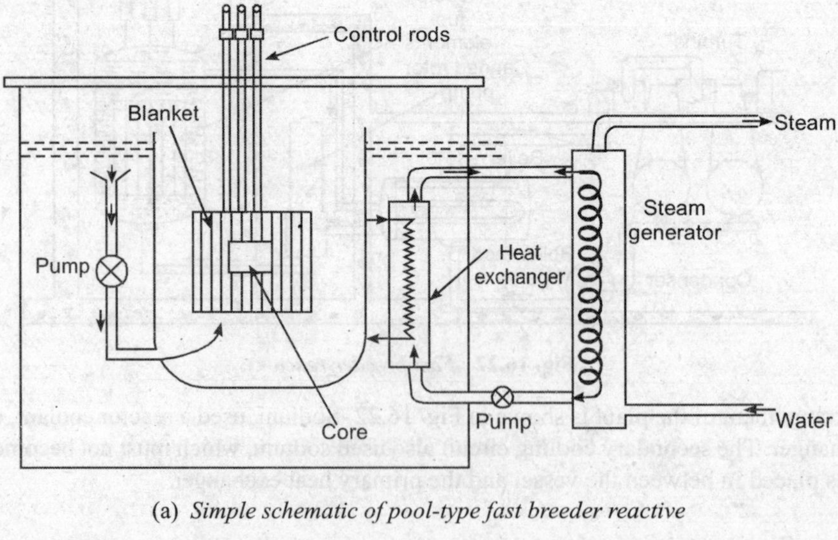

(a) *Simple schematic of pool-type fast breeder reactive*

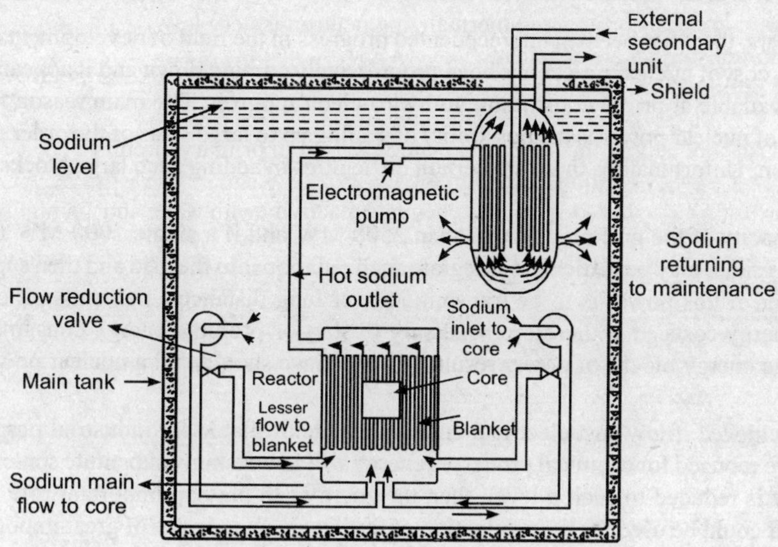

(b)

Fig. 16.23 *Pool type fast breeder reactor.*

Schematic diagram of pool type fast breeder reactor is shown in Fig. 16.23(a) and (b). In this system the reactor core, primary pump and the primary heat exchangers are all placed in a large pool of liquid sodium contained in a reactor vessel. The continuity of cooling can be assured by this system.

In the loop system which is shown in Fig. 16.24 the heat exchanger in Fig. 16.23(a) and (b) is located outside the reactor vessel.

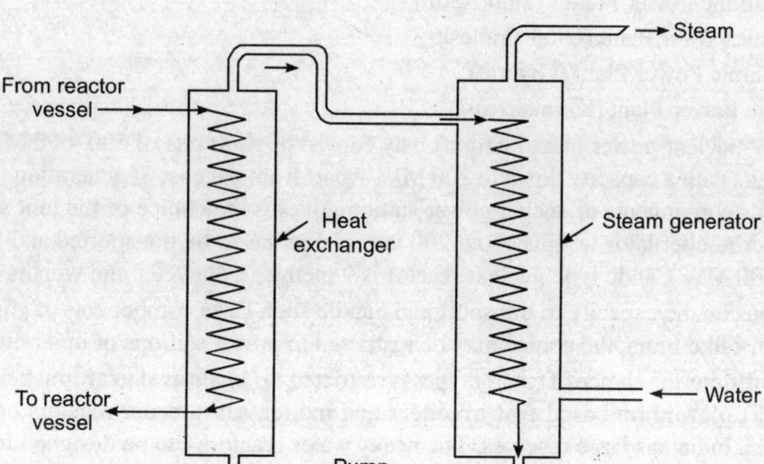

Fig. 16.24 *Loop-type fast breeder reactor.*

16.6 NUCLEAR POWER PLANTS IN INDIA

During the last 45 years, there has been an unprecedented progress in the field of development of nuclear power for commercial use. The cost of nuclear power has been progressively coming down and it appears that nuclear power may soon become available at prices comparable to hydro-electric power. The main reason for this considerable reduction in the cost of nuclear power is the possibility of scaling up to large sizes of the order of 1000 MW or even more for single station. Unfortunately, there are certain difficulties in adding such large blocks of power to grids in India.

Generally the capacity of the grid may be less than 2500 MW and if a single 1000 MW unit in one station is added to any of such grid, it may be difficult to integrate such a station into the grid and then supply the power to the consumers. Moreover, if this power is to be transmitted over long distances, transmission costs will nullify the advantage of low energy costs. It is therefore necessary to start or promote energy consuming industries in the vicinity of these large energy blocks of power resulting in what we should call a nuclear powered agro-industrial complex.

Much could be achieved if low cost electrical energy is made available for industrial purposes. It can be seen that if power costs are reduced for chemical processes energy will increasingly substitute some of the raw materials. If the cost of energy is reduced to such a level, then the energy can play an important role in the production of gasoline from coal. It could be used for the production of fertilizers, an industry of great importance to the country.

Apart from fertilizer industry, there are other power consuming industries of greater importance to India. Aluminium is one of them. The production of aluminium requires 18000 to 20000 kWh of power per tonne of ingot and about 20% of the cost of production is the cost of power.

The government of India, keenly aware of the potential of agro-industrial complexes based on nuclear power, constituted working group to study the probale location of such complexes. Accordingly nuclear power plants are set-up at :

1. Tarapur Atomic Power Station (Maharashtra)
2. Rana Pratap Sagar Atomic Power Station (Rajasthan)
3. Kalpakam Atomic Power Plant (Tamil Nadu)
4. Narora Atomic Power Plant (Uttar Pradesh)
5. Kakrapar Atomic Power Plant (Gujarat)
6. Kalga Atomic Power Plant (Karnataka)

A 1000–1200 MW nuclear power plant (station) may consist of two units of 500–600 MW reactors or a large number of smaller units with a capacity down to 200 MW. Apart from the cost of generation the ability to fabricate and handle the various components of such a power station governs the choice of the unit size. For instance, the rotor of a 600 MW turbo-alternator weighs about 200 tonnes and has to be transported and handled in one piece. The Calandria of a 600 MW Candu type nuclear reactor is 9 metres in diameter and weighs about 100 tonnes.

Until the limitations on the capacity to transport and handle such large components of considerable weight are removed in the country like India, the choice may be restricted to power stations of units smaller than 600 MW.

Under **Indian conditions** the choice of reactor types is restricted to (1) **natural uranium heavy water moderated thermal reactors,** (2) **plutonium** based **fast breeders** and molten salt breeder reactors breeding uranium U^{233} from thorium of which India has large reserves. The heavy water reactors can be designed for on load charge and discharge of fuel thereby reducing down-time whereas fast breeders will require reactor shut down for a fuel change.

The molten salt breeder reactor concept has the lowest capital cost and also generates power at the lowest cost per kWh. The concept is further attractive to Indian conditions on account of the potential for utilization of large resources of thorium available in the country.

Financial aspects play leading role in deciding the size of the plant. In the case of heavy water based thermal reactors and fast reactors, the size is of considerable importance as the cost of generation decreases rapidly with the increase in the size.

Further, in the case of fast reactors, due to necessity for a shut down for fuel change the grid has to be firm and of suitable size to be able to meet the requirements of the industry during shut down period. In the early concepts of molten salt breeders the possibility of a moderator set up consisting of a large number of reactors of smaller size (say 250 MW) operating in parallel was visualised. Such a scheme will permit the development of the complex in stages commensurate with financial resources.

16.6.1 Tarapur Atomic Power Station (TAPS)

The 420 MW Tarapur Atomic Power Station was set up at a total cost of Rs. 68.4 crores and is one of the largest operating nuclear power stations in Asia (Price reference in 1964).

The work on TAPS was started in October 1964 after the selection of site for the plant in 1960. All aspects namely (a) adequate cooling water, (b) foundation, (c) health and safety conditions, (d) power transmission possibilities and (e) transport facilities were considered. Atomic Energy commission investigated all these factors and selected the site at Tarapur situated 65 (105 km) miles north of Mumbai. The plant was synchronised with the grid in Oct. 1969. The gestation period was one of the lowest for similar plants elsewhere.

TAPS consists of two Mark I Boiling Water Reactors supplied by General Electric Co. of USA. Each has a basic capacity of 210 MWe and feeds steam at nearly 0 bar pressure and 280°C to specially designed turbo-generator. The total generating capacity of 380 MWe because of the marginal changes common to all similar reactors, arising from safety factors. Two more plants of 500 MWe capacity (TAPP 3 and 4 PHWR) are under commissioning stage construction started in 1998.

In the BWR reactors, **enriched uranium** is used in preference to natural uranium because of the financial consideration. Quotation given by USA was the lowest and other main consideration was the availability of loan from USA.

Sea water provides **condenser cooling**.

The reprocessing plant is designed to reprocess the spent fuel for extracting plutonium and collecting residual enriched uranium from it. Both could be used for refuelling.

Other details of the TAPS are :

1. Weight of each pressure vessel = 200 T
2. Diameter of each pressure vessel = 3.6 m
3. Height of each pressure vessel = 16.2 m (54 ft. long)
4. Cooling water required = 240×10^7 kg/day through 910 m long channel into sea.
5. Fresh water for steam generation from reservoir. A special dam is built for the purpose for this purpose.

Because of the nuclear blast or test carried out by India in 1974 (Pokharan I), USA stopped the supply of enriched uranium to TAPS. Then France started supplying this requirement.

India then developed indigenously the production of mixed oxide (MOX) and started using this after France stopped supplying as per agreement. For the production of MOX, a reprocessing plant is set up at BARC (Bhabha Atomic Research Centre) Mumbai.

With the bad experience with USA supply of enriched uranium, India decided to develop indigenously the nuclear power plants where natural uranium (available in India) as fuel can be used and heavy water can be used as moderator. This type of the reactor is then known as PHWR and many nuclear power stations (plants) are being set-up.

16.6.2 RANA PRATAP SAGAR ATOMIC POWER STATION (RAPS OR RAPP)

The second atomic power plant in India is RAPP.

The station or plant is situated on the right bank of Rana Pratap Sagar lake which is formed between two dams Gandhi Sagar dam upstream and Rana Pratap Sagar dam downstream of the Chambal river. The station is located at a distance of 64 km from the city Kota.

This **station** is designed to work as **base-load** plant. Power is supplied to Rajasthan grid. Heavy water plant is set up close to the reactor site.

Two reactors of **CANDU** type are used in the station because :

(a) Natural uranium fuel used which is available in the country.

(b) CANDU type reactor gives high power rating.

In RAPP, the reactor is characterised by the following things :

(i) It uses natural uranium as fuel

(ii) It is a horizontal pressure tube reactor

(iii) Low pressure heavy water used as moderator

(iv) High pressure heavy water as coolant and

(v) Bidirectional fueling on power.

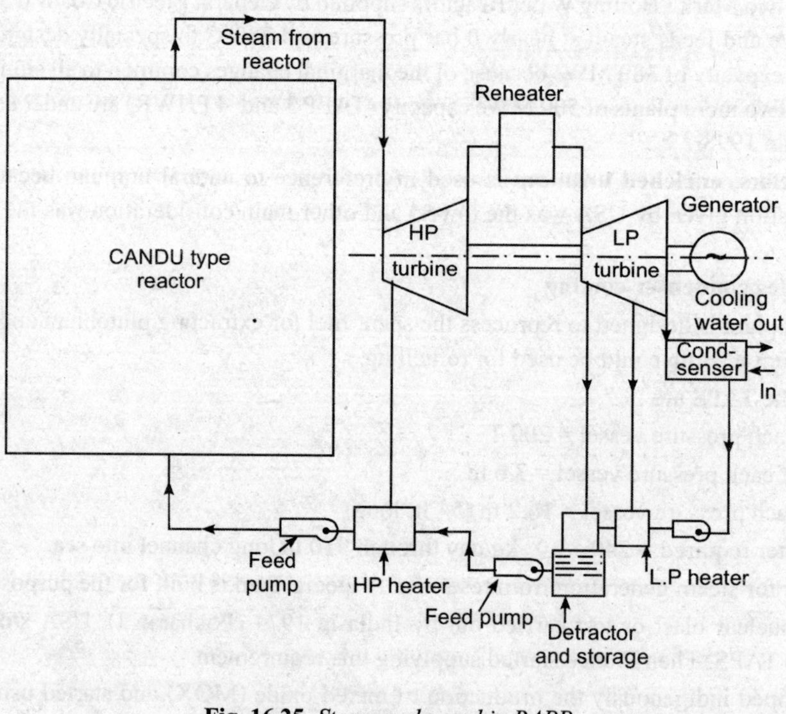

Fig. 16.25 *Steam cycle used in RAPP.*

Figure 16.24 shows the schematic diagram steam cycle used in Rana Pratap Sagar Atomic Power Plant.

16.6.3 Kalpakam Atomic Power Station

The plant at Kalpakam near Chennai in Tamil Nadu state is totally designed by Indian Engineers using indigenous materials. The capacity of this Kalpakam Atomic Power Plant is 470 MW. In this power plant, the reactors that are used are of CANDU type similar to those in RAPP. The actual thermal energy generation capacity of both the reactors is about 700 MWe.

16.6.4 Narora Atomic Power Plant

The Narora atomic power plant is situated in UP (Uttar Pradesh). Two CANDU type reactors similar to those at Rajasthan and Kalpakam atomic power stations are used in this power plant. The capacity of the plant is 2 × 235 = 470 MW.

16.6.5 Kakrapar Atomic Power Plant

This plant is situated at Kakrapar Surat district in Gujarat state. Similar to RAPS and MAPS, this power plant has two CANDU type reactors of 2 × 220 capacity. Four more units of 235 MWe capacity are proposed.

16.6.6 Kaiga Atomic Power Plant

Kaiga is situated at 56 km east of Karvar and 13 km upstream of Kadra dam on the left bank of Kali river in Karnataka state. For the site selected, it is estimated that six units of 220 MW each can be accommodated. The reactors used in this power plants are similar to those at RAPP, MAPP, NAPP and KAPS and started delivering electrical power in Dec. 1999 (Unit I) and Sept. 2000 (Unit II).

16.6.7 KUNDANKULAM ATOMIC POWER PLANT

In the Ninth Five Year Plan, it was proposed to have a nuclear power plant of Kundankulam in Tamil Nadu for generating power from 2 × 1000 MWe.

The information given above can be tabulated in a table given below :

Nuclear Power Plants location	*Installed capacity MW*	*Year of work commencement*	*Commissioned*
1. Tarapur (BWR)	2 × 210 = 420	1964	1969
Maharashtra TAPS	2 × 500 = 1000	1997	1998
2. Rajasthan (1, 2) RAPS	2 × 220 = 440	1973	1981
3. Madras MAPS	2 × 235 = 470	—	1983, 85
4. Narora NAPS (UP)	2 × 235 = 470	—	1991, 92
5. Kakrapar KAPS (Gujarat)	2 × 235 = 470	—	1993, 95
6. Kaiga KA 1, 2	2 × 235 = 470	—	1996, 2000
(Karnataka) 3, 4, 5, 6	4 × 235 = 940	—	
7. Tarapur (Maharashtra)	2 × 500 = 1000	—	1997, 98
8. RAPS 3, 4 5, 6, 7, 8	2 × 235 = 470	—	1995, 96
	4 × 500 = 2000	—	1998, 2000
9. Kudankulam (TN) PWR	2 × 1000 = 2000	—	1999, 2000
New projects proposed	6 × 500 = 3000	—	1998, 2000

16.7 COMPARISON OF NUCLEAR POWER PLANT WITH OTHER TYPES OF POWER PLANT

Comparison of Nuclear Power Plant with other types of power plants is tabulated in the table 16.1 :

TABLE 16.1 *Comparison of various power plants*

Sr. No.	*Point of comparison*	*Steam power plant*	*Nuclear power plant*	*Diesel power plant*	*Gas turbine power plant*	*Hydro power plant*
1.	Site	Located near load centre	Location depends on availability of water and away from populated area	Installed anywhere	Located near gas supply points	Away from load centre. Depends on head of water available Higher than steam power plant
2.	Capital cost (Initial cost)	Low in comparison with nuclear plants	Very high	Low	Low	Higher than steam power plant
3.	Cost of fuel	High	Very low	Less than steam power plant	Low	Zero.
4.	Operating cost	Very high compared to nuclear plant	Low	Very high	High	Practically zero
5.	Maintenance cost	High w.r.t. hydro and diesel plants	Higher	Low	Higher than all except diesel plants.	Very low
6.	Source of fuel	Limited	Unlimited	Limited	Limited	Unlimited
7.	Transmission and distribution cost	Comparatively low	Comparatively low	Very low	Less than hydro plants and more than other plants	High due to location in remote place
8.	Reliability	Less reliable	Reliable.	Less reliable	Less reliable.	Highly reliable
9.	Working (load)	Base load station	Base load station	Peak or standby power plant	Peak or standby power plant	Base load or peak load plant

The major nuclear fuel resources are :

1. Singhbum region in Bihar
2. Dakshina Kannada in Karnataka
3. Uttara Kannada in Karnataka
4. MP
5. Rajasthan
6. Central part of India.

16.8 NUCLEAR POWER IN FUTURE

Prior to the commission of nuclear power plant at Tarapur in 1969, power generation in India was mainly Hydro-Electric Power and Thermal Power. India has good hydro-electric power potential and so far all the economically feasibly hydropower schemes have been developed. For thermal power generation, good quality coal is the basic necessity. In India, the quality of coal is not very good and secondly, the low quality coal reserves are not uniformly distributed over the entire India but are concented in one or two parts of the country. Another fact is that these coal reserves are being depleted at a fast rate.

India has good and adequate deposits of uranium and thorium which are fissionable (U^{238}) ($Th^{232} \rightarrow U^{233}$) and can eventually be used for generation of power. Therefore, to meet the ever growing demand of electric power, nuclear power development is necessary for India. Thus, the future of nuclear power is quite bright. However, the following three factors must be considered seriously.

16.8.1 Safety of the Nuclear Power Plants

Nuclear power plants are to be designed such that the working will be safe for the public and the staff working in the plant premises.

16.8.2 Availability of Nuclear Fuel

The problem of availability of large amount of fissionable nuclear fuel, can be overcome to a great extent by using fast-breeder reactors and producing the new fissionable fuel while consuming the nuclear fuel.

16.8.3 Cost of Power Generation

At present the cost of power generation in a nuclear power plant is comparatively more but with the development of nuclear technology and design, the cost of power generation in nuclear power plant—it is expected—will come down to the value comparable with those of other plants.

16.9 ADVANCED REACTORS

16.9.1 Advanced Nuclear Power Reactors

Under this heading, at present there are two generations of Nuclear Power Reactors. They are

1. 3rd generation of nuclear power reactor
2. 4th generation of nuclear power reactors.

These reactors are under development stage in various countries like Japan, America, Russia, Britain etc.

It is known that

1. Generation-I reactors were developed in 1950–60s and outside the UK none are still running today.
2. Generation-II reactors are typified by the present US fleet and most in operation elsewhere.

The designers of the third and fourth generation nuclear reactors (newer advanced reactors) claim that these reactors are simple in designs which will reduce the capital cost. Also these reactors will be more fuel efficient and will be safer to operate. The third generation reactors are the advanced reactors and are in operation in Japan.

Designs of the 4th generation are still on the drawing board and it is predicted that before 2020 at the earliest, the reactors will not be operational. The second generation nuclear power units have been found to be safe and reliable, but they are being superseded by better designs. Fourth generation reactors are at concept stage.

Third generation reactors have :

1. a standardised design to reduce capital cost and reduce constructional time

2. a simpler and more rugged design, making them easier to operate and less vulnerable to operational upsets
3. longer operating life typically 60 years
4. reduced possibility of core-melt accidents
5. resistance to serious damage that will allow radiological release from an aircraft impact
6. higher burn-up to reduce fuel use and the amount of waste
7. burnable absorbers (poisons) to extend fuel life.

Traditional reactor safety systems are 'active' in the sense that they involve electrical or mechanical operation on command. Some engineered systems operate passively e.g., pressure relief values. They operate or function without operator control and despite any loss of auxiliary power. Both require parallel redundant systems. Inherent or full passive safety depends only on physical phenomena such as convection, gravity or resistance to high temperatures, not on functioning of engineered components.

The greatest departure from second generation designs is that many incorporate passive or inherent safety features, which require no active controls or operational intervention to avoid accidents in the event of malfunction, and may rely on gravity, natural convection or resistance to high temperatures.

Another departure is that some will be designed for load-following. While most French reactors today are operated in that mode to some extent, the EPR design has better capabilities. It will be able to maintain its output at 25% and then ramp up to full output at the rate of 2.5% of rated power per minute upto 60% output and 5% of the rated output per minute upto full rated power. This meant that potentially the unit can change its output from 25% to 100% in less than 30 minutes, though this may be at some expense of wear and tear.

16.10 LIGHT WATER REACTORS

Department of Energy (DoE) of USA and commercial nuclear industry in the 1990s developed four **advanced reactor** types. These reactors are in the 1300 Megawatts range. One is an advanced boiling water reactor (ABWR) derived from General Electric design. The other type, **SYSTEM 80+**, is an advanced pressurised water reactor, ready for commercialisation but is now being promoted for sale.

In USA :

(a) Advanced reactor, 600 MWe — AP 600 (AP–Advanced Passive)

(b) Large Advanced BWR ABWR

(c) AP–1000. The first generation 3+ type. Has a 60 year operational life. 1300 men-year and $ 440 million design and testing programme.

4 units are being built in China.

GE Hitachi Nuclear Energy's **ESBWR** (Economic and Simplified BWR) is a Generation III+ technology that utilizes passive safety features and natural circulation principles and is essentially an evolution from its predecessor, the SBWR at 670 MWe. The ESBWR will produce approximately 1520 MWe net.

IRIS (International Reactor Innovative and Secure) is a modular 335 MWe – advanced 3rd Generation project. Fuel is initially similar to present LWRs (Light Water Reactors) with 5% enrichment and burnable poison, with fuel assemblies similar to AP-1000.

In Japan 2 ABWRs (1350 MWe) are working since 1996 and several of the 1350 MWe units are under construction and planned in Japan and Taiwan.

To complement this ABWR, Hitachi-GE has completed system design for 600, 900 and 1700 MWe versions of 1350 MWe design.

Mitsubishi's large APWR (1538 MWe) advanced PWR. The **US-APWR** will be 1700 MWe, due to longer (4.3 m) fuel assemblies, higher thermal efficiency (39%) and has 24 month refuelling cycle.

In South Korea, the APR-1400 – Advanced PWR design has evolved from the US system 80+ with enhanced safety and seismic robustness (Earlier known as the Korean Next Generation Reactor) 1450 MWe. Expected to be operating in 2013.

In Europe Areva NP (ANP) has developed large 1600–1750 MWe. European Pressurised Water Reactor (**EPR**).

Areva NP together with German utilities and safety authörities developed the **SWR-1000** – a 1250 MWe BWR known as Kerena.

Toshiba has been developing its evolutionary **advanced BWR** (1500 MWe) design.

In Russia several advanced reactor designs have been developed – advanced PWR with passive safety features. Gidropress late model **VVER-1000** with enhanced safety are being built in India and China.

A third generation standardised **VVER-1200** reactor of 1150–1200 MWe is an evolutionary development of VVER-1000. The **VVER-1500** model being developed by Gidropress.

OKBM's VBER-300 PWR is 295–325 MWe unit developed from naval power plants and was originally envisaged in pairs as a floating nuclear power plant.

16.11 HEAVY WATER REACTORS

Canada has had two designs under development which are based on its reliable **CANDU-6** reactors, the most recent of which are operating in China.

The **CANDU-9** (925–1300 MWe) was developed from this also as a single-unit plant. It has flexible fuel requirements ranging from natural uranium through slightly-enriched uranium, recovered uranium from reprocessing spent PWR fuel, mixed oxide (U and Pu) fuel, direct use of spent PWR fuel to thorium. It may be able to burn military plutonium or actinides separated from reprocessed PWR/BWR waste. In spite of this, the design has been shelved.

Some of the innovations of this, along with experience in building recent Korean and Chinese units, was then put back into the **Enhanced CANDU-6** – built as twin units with power increase to 750 MWe and flexible options. This is under consideration for new build in Ontario.

The Advanced Candu Reactor (ACR), a 3rd generation reactor, is a more innovative concept. While retaining the low pressure heavy water moderator, it incorporates some features of the pressurised water reactor. Adopting light water cooling and a more compact core reduces capital cost, and because the reactor is run at higher temperature and coolant pressure, it has higher thermal efficiency.

The **ACR-700** design was 700 MWe but is physically much smaller, simpler and more efficient as well as 40% cheaper than the CANDU-6. But the **ACR-1000** of 1080–1200 MWe is now the focus of attention by AECL. It has more fuel channels. The ACR will run on low-enriched uranium (about 1.5–2.0% U-235) with high burn-up, extending the fuel life by about three times and reducing high level waste volumes accordingly. It will also efficiently burn MOX fuel, thorium and actinides.

The first ACR-1000 unit is expected to be operating in 2016 in Ontario.

The **CANDU X** or **SCWR** is a variant of the ACR, but with supercritical light water coolant (25 MPa and 625°C) to provide 40% thermal efficiency. The size range envisaged is 350 to 1150 MWe, depending on the number of fuel channels used. Commercialisation envisaged after 2020.

India is developing the Advanced Heavy Water Reactor (**AHWR**) as the third stage in its plan to utilise thorium to fuel its overall nuclear power programme. The AHWR is a 300 MWe reactor moderated by heavy water at low pressure. The Calandria has about 450 vertical pressure tubes and the coolant is boiling light water circulated by convection. A large heat sink – "Gravity-driven water pool" with 7000 cu.m of water is near the top of the reactor building. Each fuel assembly has 30 Th-U_{233} oxide pins and 24 Pu-Th oxide pins around a central rod with burnable absorber. Burn-up of 24 GW$^{d}/_{t}$ is envisaged. It is designed to be self-sustaining in relation to U-233 breed from Th-232 and have a low Pu inventory and consumption, with slightly negative void coefficient of reactivity. It is designed for 100 year plant life and is expected to utilise 65% of the energy of the fuel, with two thirds of the energy coming from thorium.

Once it is fully operational, each AHWR fuel assembly will have the fuel pins arranged in three concentric rings arranged.

Inner. 12 pins Th-U_{233} with 3.0% U-233.

Intermediate. 18 pins Th-U_{233} with 3.75% U-233.

Outer. 24 pins Th-Pu_{239} with 3.25% Pu.

The fissile plutonium content will decrease from an initial 75% to 25% at equilibrium discharge burn-up level.

In 2009 an export version of this design was announced.

The AHWR-LEU. This will use low enriched uranium plus thorium as a fuel, dispensing with the plutonium input. About 39% of the power will come from thorium (via in situ conversion to U-233), and burn-up will be 64 GW $^{d}/_{t}$. Uranium enrichment level will be 19.75%, giving 4.21% average fissile content of the U-Th fuel. While designed for closed fuel cycle, this is not required. Plutonium production will be less than in light water reactors, and the fissile proportion will be less and the Pu-238 portion three times as high, giving inherent proliferation resistance. The **AEC** says that "the reactor is manageable with modest industrial infrastructure within the reach of developing countries".

In AHWR-LEW, the fuel assemblies will be configured :

Inner ring. 12 pins Th-U with 3.555% U-235

Intermediate ring. 18 pins Th-U with 4.345% U-235

Outer ring. 24 pins Th-U with 4.444% U-235.

16.12 HIGH TEMPERATURE GAS COOLED REACTORS

These reactors use helium as a coolant which at upto 950°C drives a gas turbine for electricity and a compressor to return to the gas to the reactor core. Fuel is in the form of TRISO particles less than a millimetre in diameter. Each has a kernel of uranium oxycarbide, with the uranium enriched upto 17% U-235. This is surround by layers of carbon and silicon carbide, giving a containment for fission products which is stable to 1600°C or more.

These particles may be arranged in blocks as hexagonal prisms of graphite, or in billiard ball-sized pebbles of graphite encased in silicon carbide.

South Africa's **PBMR (Pebble Bed Modular Reactor)** production units will be of 165 MWe. They will have a direct cycle gas turbine generator and thermal efficiency of about 42%. The pressure vessel is lined with graphite and there is a central column of graphite as reflector. Control rods are in the side reflectors and cold shut down units in the central column.

Performance includes great flexibility in loads (40–100%) with rapid change in power settings. Each unit will finally discharge about 19 tonnes/year of spent pebbles to ventilated on-site storage bins. A demonstration plant is due to be built in 2009, with fuel loading expected in 2013.

A larger US design, the **Gas Turbine - Modular Helium Reactor** (GT-MHR) will be built as modules of 285 MWe each directly driving a gas turbine at 48% thermal efficiency. The cylindrical core consists of 102 hexagonal fuel element columns of graphite blocks with channels for helium and control rods. Graphite reflector blocks are both inside and around the core. Half the core is replaced every 18 months. Burn-up is about 100,000 MW $^{d}/_{t}$. It is being developed by General Atomics in partnership with Russia's Minatom, supported by Fuji (Japan).

16.12.1 Fast Neutron Reactors

Several countries have research and development programs for improved Fast Breeder Reactors (FBR), which are a type of Fast Neutron Reactor. These use the uranium-238 in reactor fuel as well as the fissile U-235 isotope used in most reactors.

Research Continues in India. At the Indira Gandhi centre for Atomic Research at 40 MWt fast breeder test reactor has been operating since 1985. In addition, the tiny Kamini there is employed to explore the used of thorium as nuclear fuel, by breeding U-233 in 2004 construction of a 500 MWe prototype fast breeder reactor started at Kalpakam. The unit is expected to be operating in 2011, fuelled with uranium-plutonium carbide and with a thorium blanket to breed U-233. This will take India's ambitious thorium program to stage 2, and set the scene for eventual full utilization of the country's abundant thorium to fuel reactors.

(a) **Japan Standard Fast Reactor (JSFR)** with breeding ratio less than 1:1. It could be of any size from 500 to 1500 MWe.

(b) The Russian BN-600 fast breeder reactor at Beloyarsk.

(c) The BN-800–a new larger 880 MWe FBR from OKBM.

(d) **Russia** has experimented with **lead-cooled reactor designs**. Lead-Bismuth cooling for 40 years in 7 Alfa class submarines.

(e) The European Lead-cooled System (ELSY) of 600 MWe in Europe.

(f) Korea's KALIMER (Korea Advanced Liquid Metal Reactor) is 600 MWe pool type sodium-cooled fast reactor. Future development of KALIMER as a Generation IV type is envisaged.

US Nuclear Regulatory Commission deaft policy–May 2008 expects from the designers of reactors the following points :

1. Reactor should be highly reliable, less complex safe shutdown systems.
2. Reactor should have simplified safety systems that allow more straightforward engineering analysis, operate with fewer operator actions and increase operator comprehension of reactor conditions.
3. Overall security system that requires fewer human actions.
4. Features that maintain spent fuel pool integrity following an aircraft impact.
5. Features that prevent a simultaneous breach of containment and loss of core cooling from an aircraft impact, or than inherently delay any radiological release.

Reference : Advanced Nuclear Power Reactors (September 2009).

Table 16.2 giving information regarding Advanced Nuclear Power Reactors in different countries of the world.

Country and developer	*Reactor*	*Size MWe*	*Design Progress*	*Main Features (Improved safety in all)*
US-Japan (GE-Hitachi, Toshiba)	ABWR	1300	Commercial operation in Japan since 1996-7. In US: NRC certified 1997. FOAKE.	• Evolutionary design • More efficient less waste • Simplified constructions (48 months) and operation
	AP-600	600	AP-6000: NRC certified 1999, FOAKE. AP-1000 NRC	• Simplified constructions and operation

Contd.

Contd. Table 16.2

Country and developer	*Reactor*	*Size MWe*	*Design Progress*	*Main Features (Improved safety in all)*
USA (Westinghouse)	AP-1000	1100 (PWR)	Certification 2005, first units being built in China, many more planned	• 3 years to build • 60-year plant life
France-Germany (Areva NP)	EPR US-EPR (PWR)	1600	Future French standard. French design approval. Being built in Finland and France, planned for China, US version developed.	• Evolutionary design • High fuel efficiency. • Flexible operation
USA (GE-Hitachi)	ESBWR	1550	Developed from ABWR. Under certification in USA, likely construction there.	• Evolutionary design • Short construction time.
Japan (utilities, Mitsubishi)	APWR US-APWR EU-APWR	1530 1700 1700	Basic design in progress, planned for Tsuruga US design certification application 2008	• Hybrid safety features. • Simplified Construction and operation
South Korea (KHNP, derived from Westinghouse)	APR-1400 (PWR)	1450	Design certification 2003, First units expected to be operating c 2013	• Evolutionary design. • Increased reliability. • Simplified construction and operation.
Germany (Areva NP)	SWR-1000 (BWR)	1200	Under development, pre-certification in USA	• Innovative design • High fuel efficiency.
	VVER-		Replacement under	• Evolutionary design

Contd.

Contd. Table 16.2

Country and developer	*Reactor*	*Size MWe*	*Design Progress*	*Main Features (Improved safety in all)*
Russia (Gidropress)	1200 (PWR)	1200	construction for Leningrad and Novovoroneh plants	• High fuel efficiency. • 50-year plant life • Evolutionary design.
Canada (AECL)	CANDU-6 CANDU-9	750 925+	Enhanced model Licensing approval 1997	• Flexible fuel requirements. • C-9: Single stand-alone unit. • Evolutionary design.
Canada (AECL)	ACR	700 1080	Undergoing certification in Canada	• Light water cooling. • Low-enriched fuel
South Africa (Eskom, Westinghouse)	PBMR	170 (module)	Prototype due to start building (Chinese 200 MWE counterpart under const.)	• Modular plant, low cost. • High fuel efficiency. • Direct cycle gas turbine.
USA-Russia et al (General Atomics-OKBM)	GT-MHR	285 (module)	Under development in Russia by multinational joint venture	• Modular plant, low cost. • High fuel efficiency. • Direct cycle gas turbine.

16.13 GENERATION IV REACTORS (GEN IV)

Generation IV reactors are a set of theoretical nuclear reactor designs currently being researched. Most of these designs are generally not expected to be available for commercial construction before 2030, with the exception of a version of Very High Temperature Reactor (VHTR) called the **Next Generation Nuclear Plant (NGNP)**. The **NGNP** is to be **completed by 2021**. Research into these reactor types was officially started by the Generation IV International Forum (GIF) base on eight technology goals, primary goals being :

1. Improve nuclear safety
2. Improve proliferation resistance
3. Minimise waste
4. Natural resource utilization
5. Decrease the cost to build and
6. Decrease the cost to run such plants.

The reactors are intended for use in nuclear power plants to produce nuclear power from nuclear fuel.

16.13.1 Advantages of Generation IV Reactors

Relative to current nuclear power plant technology, the claimed benefits for 4th generation reactors include :

(a) Nuclear waste that lasts decades instead of millennia.

(b) 100–300 times more energy yield from the same amount of nuclear fuel.

(c) The ability to consume existing nuclear waste in the production of electricity.

The systems that meet the goals of Gen IV initiative are :

(i) Thermal reactors and

(ii) Fast reactors.

(i) Thermal Reactors. There are three thermal reactor types :

1. Very High Temperature Reactor (VHTR)
2. Supercritical Water Cooled Reactor (SCWR)
3. Molten Salt Reactor (MSR).

1. Very High Temperature Reactor (VHTR). This utilizes a graphite-moderated core with a once through **uranium fuel cycle**. This reactor designs expect an outlet temperature of 1000°C. The reactor core can be either a **prismatic block** or a **pebble-bed reactor design**. The high temperatures enable applications such as process heat or hydrogen production via the thermochemical iodine-sulphur process. It would also be passively safe.

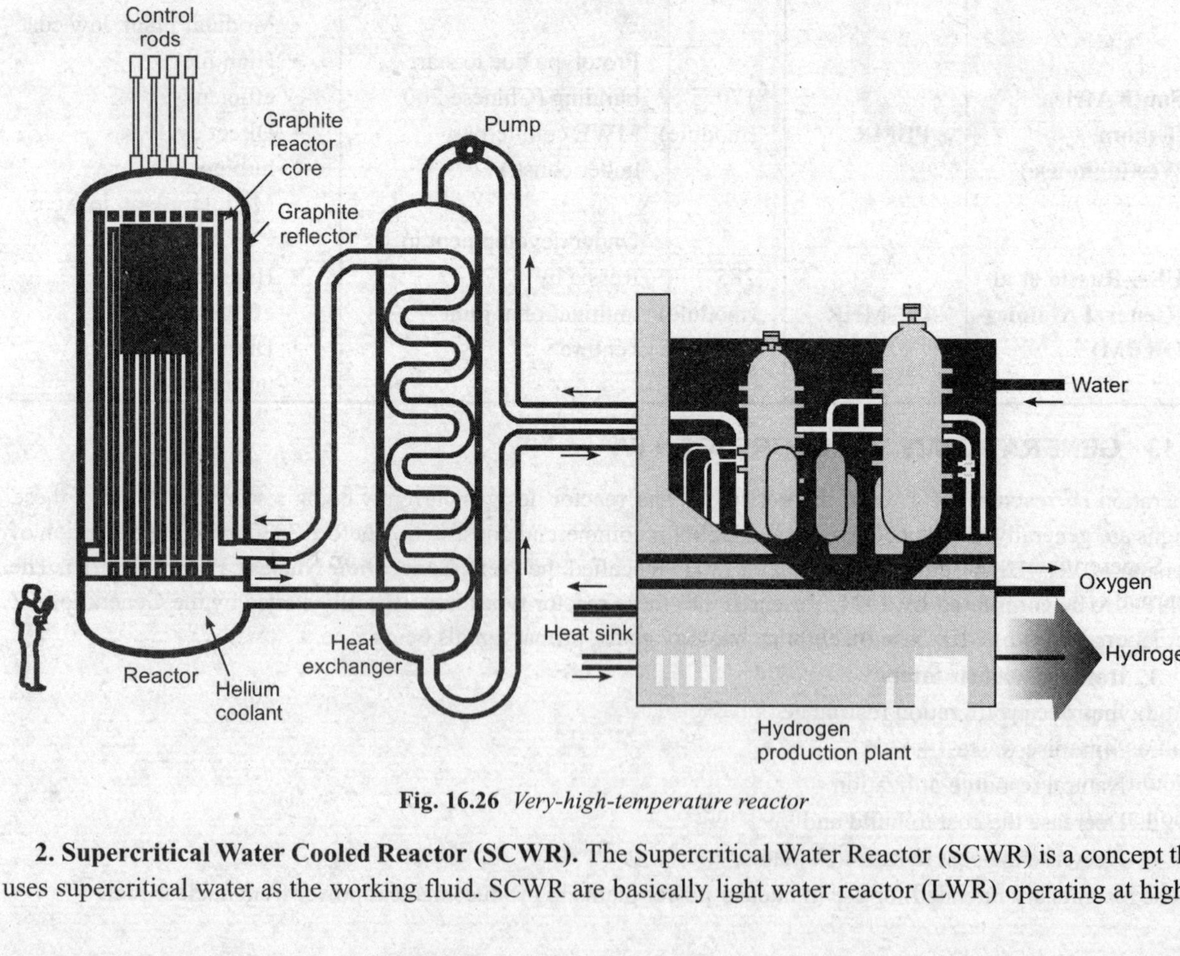

Fig. 16.26 *Very-high-temperature reactor*

2. Supercritical Water Cooled Reactor (SCWR). The Supercritical Water Reactor (SCWR) is a concept that uses supercritical water as the working fluid. SCWR are basically light water reactor (LWR) operating at higher pressure and temperatures with a direct once-through cycle. In normal course, it would operate on a direct cycle, much like a Boiling Water Reactor (BWR), but since it uses supercritical water as the working fluid, would have only one phase present, like the Pressurised Water Reactor (PWR). It could operate at much higher temperatures than both current PWRs and BWRs.

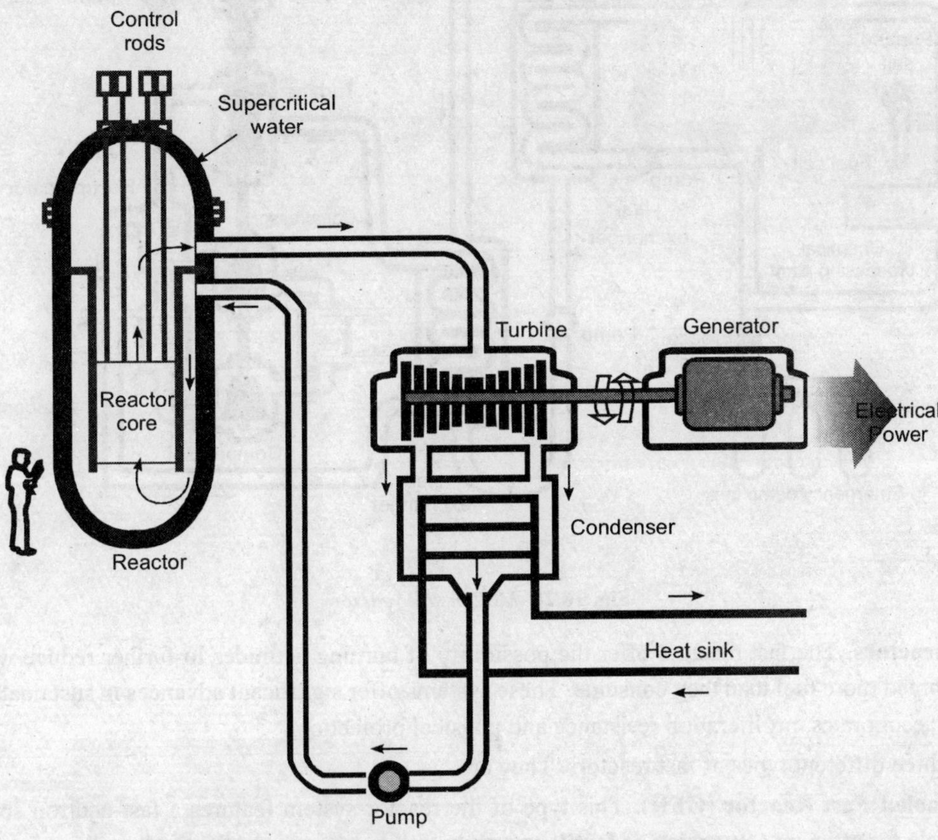

Fig. 16.27 *Supercritical-water-cooled reactor*

Supercritical water cooled reactors (SCWRs) are promising advanced nuclear systems because of their high thermal efficiency (i.e., about 45% vs about 33%) for current (LWRs) and considerable plant simplification.

32 organisations in 13 countries are investigating the concept of SCWR.

3. Molten Salt Reactor (MSR). This is a type of nuclear reactor where the **coolant** is a molten salt. Few prototypes are built. The early concepts and many current ones had a **nuclear fuel** dissolved in the **molten fluoride** salt as **uranium tetrafluoride** (UF_4), the fluid would reach **criticality** by flowing into a **graphite core** which would serve as the **moderator**. Many current concepts rely on fuel that is dispersed in a graphite matrix with the molten salt providing low pressure, high temperature cooling.

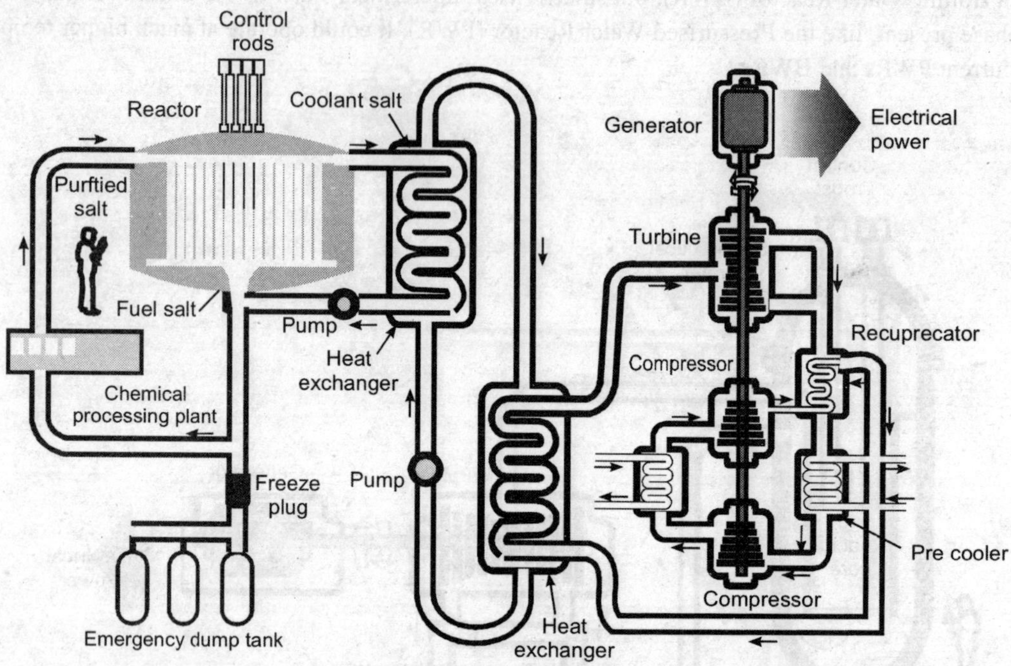

Fig. 16.28 *Molten salt reactor*

(ii) Fast Reactors. The fast reactors offer the possibility of burning actinides to further reduce waste and of being able to breed more fuel than they consume. These systems offer significant advances in sustainability, safety and reliability, economics, proliferation resistance and physical protection.

There are three different types of fast reactors. They are :

(a) Gas Cooled Fast Reactor (GFR). This type of the reactor system features a fast-neutron spectrum and closed fuel cycle for efficient conversion of **fertile uranium** and management of actinides. The reactor is **helium cooled,** with an outside temperature of 850°C and using a direct **Brayton cycle gas turbine** for high thermal efficiency. Several fuel forms are being considered for their potential to operate at very high temperatures to ensure an excellent retention of fission products : (1) composit ceramic fuel (2) advanced fuel particles or (3) ceramic clad elements of actinide compounds. Core configurations are being considered based on pin or plate-based fuel assemblies or prismatic blocks (Show Fig. 16.29).

(b) Sodium Cooled Fast Reactor (SFR). Under this (SFR) project, there are two closely related projects are :

1. Liquid metal fast breeder reactor and
2. Integral Fast Reactor (IFR).

The goals are to increase the efficiency of uranium usage by breeding plutonium and eliminating the need for transuranic isotopes ever to leave the site. The reactor design uses an unmoderated core running on fast neutrons, designed to allow any transuranic isotope to be consumed (as in some cases used as a fuel). In addition to the benefits of removing the long **half-life** transuranics from the waste cycle, the SFR fuel expands when the reactor overheats, and the chain reaction automatically slows down. In this manner, it is passively safe.

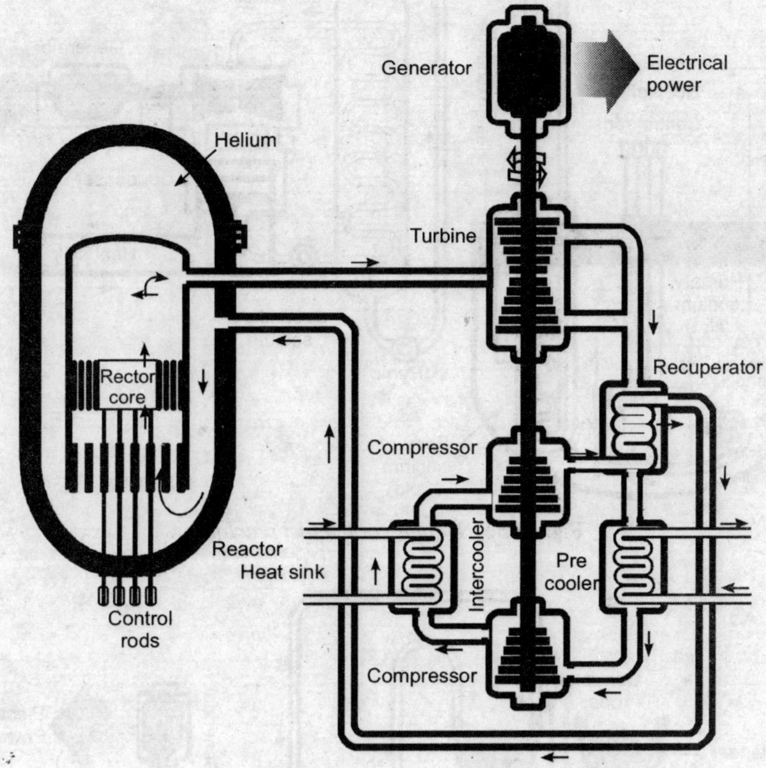

Fig. 16.29 *Gas-cooled fast reactor*

The SFR reactor is cooled by liquid sodium and fueled by a metallic alloy of uranium and plutonium. The fuel is contained in steel cladding with liquid sodium filling in the space between the clad elements which make up the fuel assembly. One of the design challenges of **SFR** are the risks of handling sodium, which reacts explosively if it comes into contact with water. However, the use of liquid metal instead of water as coolant allows the system to work at atmospheric pressure, reducing the risk of leakage.

(c) Lead Cooled Fast Reactor (LFR). The **lead cooled fast reactor** features a fast-neutron spectrum lead or lead/bismuth eutectic (LBE) liquid-metal cooled reactor with a closed fuel cycle. Options include a range of plant ratings, including battery of 50 to 150 MW of electricity that features a very long refueling intervel, a modular system rated at 300 to 400 MW, and a large monolithic plant option at 1200 MW (The term "battery" refers to the long life, factory fabricated core, not to any provision for electrochemical energy conversion). The fuel is a metal or nitride-based containing **fertile uranium** and **transuranics**. The **LFR** is cooled by natural convention with a reactor outlet coolant temperature of 550°C, possibly ranging upto 800°C with advanced materials. The high temperature enables the production of hydrogen by thermochemical processes.

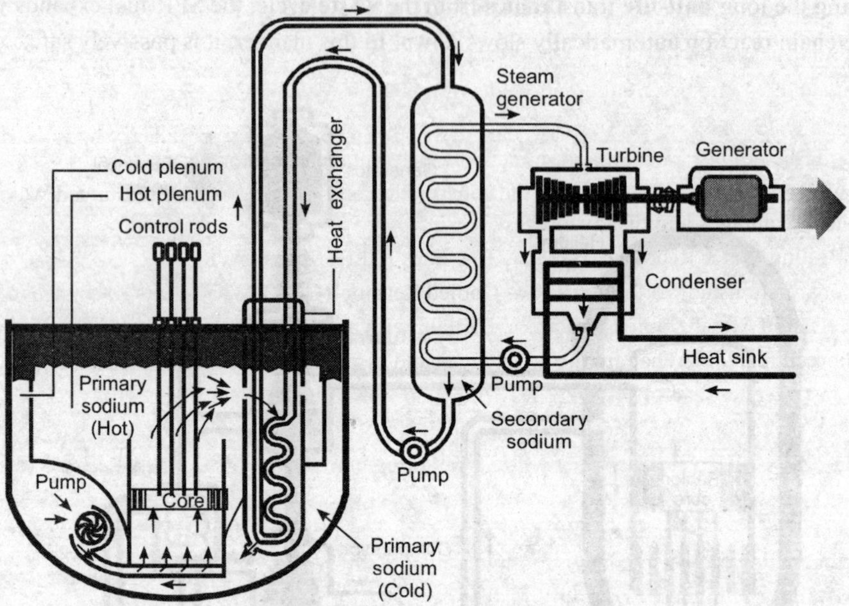

Fig. 16.30 *Sodium-cooled fast reactor*

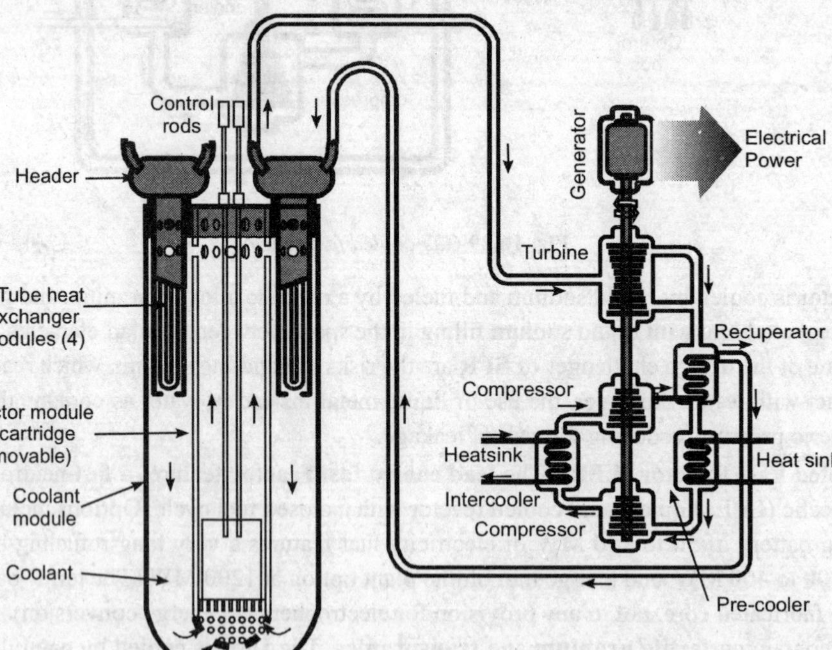

Fig. 16.31 *Lead cooled reactor*

THEORY QUESTIONS

16.1. What is a nuclear reactor?

16.2. Enumerate and explain essential components of a nuclear reactor.

16.3. How are nuclear reactors classified?

16.4. Explain with the help of a neat diagram the working of nuclear reactor.

16.5. Describe with the help of a neat diagram the construction and working of the Pressurised Water Reactor (PWR). What are its advantages and disadvantages?

16.6. What is "Boiling Water Reactor" (BWR)? How does it differ from PWR?

16.7. Give the construction and working of a 'Gas Cooled Reactor' (GCR). What are its advantages and disadvantages?

16.8. What is a "Liquid Metal Cooled Reactor"? Explain with a neat sketch a typical liquid metal cooled reactor.

16.9. What is a breed reactor? What are the advantages and disadvantages of breeder reactor?

16.10. What is a fast breeder reactor? In what way it differs from breeder reactor?

16.11. Explain with neat sketch the working of CANDU reactor.

16.12. Write a short note on Pressurised Heavy Water Reactor (PHWR).

16.13. Explain the help of neat sketch the working of organic moderated and cooled reactors. What are the organic materials used?

16.14. Compare the nuclear power plants with other conventional type of power plants.

16.15. What is the future of nuclear power in India?

Reference

(i) "4th Generation Nuclear Power" http/www.ossfoundation,usprojects/energy/nuclear

Chapter 17 Nuclear Waste and Its Disposal

17.1 INTRODUCTION

Whatever may be the state or type of the waste occurring in industry or for that matter domestic (household) waste, the main problem for everybody is to get rid off this waste material. In nuclear power plants, the fission products are the outcome of the nuclear reactor operation and requires proper disposal. These fission products are radioactive and create radioactive hazard and require strong control to ensure that radioactivity is not released into atmosphere to avoid atmospheric pollution.

The fission products are the mixture of various beta and gamma emitting radioactive isotopes with various half-lives. After disintegration which depends upon their half-life, the radioactive isotopes are transformed into stable isotopes. Some of the high energy radioactive isotopes are used for industrial application. Radioactive isotopes cannot be neutralised by any chemical method. Drinking water and atmospheric air will be contaminated causing biological damage if the nuclear waste is discharged into atmosphere. If a particular limit of radioactivity is above 100 tons of radium daily then no harm is there for the human and other living ones. If these products are exposed to atmosphere, all the living organisms within the area of about 160 square kilometres will be killed and therefore these wastes material has to be disposed of safely.

17.2 EFFECTS OF NUCLEAR RADIATION

It was observed in the previous chapter 15 that radioactive materials emit three different kinds of rays i.e., alpha, beta and gamma. These rays can also be released when atoms are bombarded by charged particles or neutrons. In addition neutrons and protons are released by nuclear bombardment of atoms. It is experienced that the radioactive products of 400 MW power station.

Nuclear radiations may be grouped into two classes, **ionising** and **non-ionising**. Ionising radiations readily react with shielding material and are easily stopped, the main nuclear shielding problem is to stop the non-ionising radiations. Whenever radiations of any kind are stopped by matter, the energy of the radiations is turned into heat and the temperature of the material rises.

17.2.1 Ionising Radiations

Alpha radiations are heavy, energetic and strongly ionising. These can be quickly stopped even by an aluminium foil. They cannot penetrate the human skin and constitute a negligible external hazard but, because they are strongly ionising, constitute a serious internal hazard in the body when alpha emitting substances are ingested. For one ion pair the energy needed is nearly 32.5 eV.

Beta rays are less ionising than alpha rays but are more penetrating. Energy associated is of the order of 1 MeV and will produce 31000 ion pairs.

Protons are intensely ionising and are consequently stopped.

17.2.2 Non-ionising Radiations

Gamma radiations consist of electromagnetic waves, which can penetrate greater thickness of material. Neutron emission, which also occurs in fission process, has tremendous penetrating power.

Radioactivity is almost always accompanied by a chemical change, new elements appearing, alpha, beta and gamma rays being shot off and the radioactivity gradually becoming smaller. The rate of decay of activity of a substance is measured by the half life or time for the activity of fall to half value. Half lives vary enormously being 4.5×10^9 years for the 238 isotope of uranium and only a few seconds for some unstable elements.

Reactor Shielding. While the reactor is operating, neutron and gamma radiations from the fission process escape from the core. These radiations can cause serious injury to human beings and the reactor steel pressure vessel is therefore surrounded by a biological shield, the function of which is to reduce the radiations to levels than ensure safe working conditions around the plant. The reactor shield must satisfy the following conditions :

1. No leakage of active dust or gas
2. Adequate attention of neutron and gamma beams
3. No serious restraint on the pressure vessel
4. Adequate shield cooling
5. No deterioration of the shield under irradiation or thermal stress.

Health Hazards. Human beings are continuously exposed to radiations from cosmic rays and various radioactive materials in the earth and air. Small amounts of radiations can be tolerated but exposure to radiations above certain levels is dangerous to health and life.

The uniquely dangerous aspect of radiation is that no human sense recognises it and hence great damage can be caused without any knowledge of its happening. Human beings can be exposed to danger from radiations in two ways :

(a) *Externally* by direct bombardment by penetrating gamma rays or neutrons. This hazard is checked by providing shielding in nuclear power plants.

(b) *Internally* through breathing, eating, drinking, cuts. This implies loose radioactive contaminants. In nuclear power plants these hazards are minimised providing enclosures on radioactive materials and also by rigorous control of the gaseous, liquid and solid effluents.

The ultimate effect of all these hazards on human beings is to damage the living cells of the body by ionisation. The result of such damage may be immediate, like burns or even death, or delayed effects like leukaemia, anaemia or cancer and may be genetic giving hereditary effects.

In order to guarantee protection from these unseen dangers of radiations from reactors and their products and to ensure that nobody is injured, the nuclear power plants have to fulfil the following obligations :

A. Personnel Monitoring. Under this photographic films which are sensitive to radiations are used to measure and record the radiation doses received by plant staff.

B. Periodical Check-up. The routine check-up on the medical condition of the staff are carried out.

C. Environmental Control. The dangers of radiations are constantly analysed and estimates of probable exposure of staff to radiations are carried out. Precautionary measures including the use of shielding, distance and time (limiting the time spent by individuals in radiation zones) are taken. In addition to this survey camps are organised to periodically check the radiation levels both in the surrounding areas and also around the plant.

There are certain maximum limits of radiations to which an individual can be exposed safely. Another consideration is the intensity of radiation. The biological damage is greater for a given radiation dose spread over a short period as compared to that spread over longer period. Also type of radiation matters a lot. For example, one unit of alpha radiation does twenty times as much biological damage as one unit of gamma radiations.

Units of Radiation. Biological damage is taking place due to —

(i) ionization

(ii) absorption of neutrons.

The rate of absorption is a function of flux or field intensity.

The unit of radiation dose is roentgen (γ) and is defined as the amount x-rays or gamma rays which will ionize the dry air of 0.001293 gram i.e., 1 cu cm of dry air to one electrostatic unit of charge of either sign of 0°C and 760 mm Hg. It is a large unit and a small unit is milliroentgen (1 mr = 10^{-3} γ) is in common use.

To produce 1 ion pair, energy needed is nearly 32.5 eV. Thus to ionize 0.001293 gm of dry air (i.e., 1 cu cm air) the energy required is about 6.85×10^4 MeV/cm^3 which is equivalent to 83.9 ergs/g of air.

It is to be noted that the relative absorptivities for tissues for different radiations differ and naturally are not same as that of air. It is, therefore, clear that the limit of the radiation dose expressed in Roentgen units will be different for different nuclear radiations, for different energies of radiation and different types of tissues. To avoid this, a new unit called Rem (Roentgen equivalent man) is used. Unit of Rem is the dose of absorbed radiation that will have the same effect that exposure to one Roentage of γ-radiation will have. This is also same as absorption of 100 ergs/gm of tissue. Table given below shows the various radiations and the rem values.

Type of radiation	*X or γ rays*	*β-particles*	*Slow neutrons*	*Fast Neutrons*	*Proton*	*α-particles*
Rem (value)	1	1	3–5	10	10	10–20

The maximum permissible exposure is that dose of the whole body exposure to external radiation extending many years which will not cause any permanent injurious effects. It is estimated or found that the total dose received during a person's life time should not exceed 200 rem and weekly dose should not exceed 0.3 rem which is accepted by health physicians. The exposure over number of years may also shorten the life or may come out in the form of **leukaemia**.

17.2.3 Permissible Radiation Dose Limits

Mean Lathal Dose is the radiation dose which will kill fifty percent of the population exposed to such radiation and is equivalent to nearly 450 γ total body dose.

Rad is the unit adopted by The International Commission on Radiological Unit and is defined as amount of radiation which after absorption by the tissue of body emits 100 ergs per gram.

Relative Biological Effectiveness (RBE) is defined as :

$$\text{RBE} = \frac{\text{No. of Rads due to gamma rays to cause give biological damage}}{\text{No. of Rads of radiation to cause the same damage}}$$

The RBE of all gammas ⊔ 1 at all energies.

The human body can withstand the radiations upto a certain limit. Generally, the human body is subjected to radiation damage which is due to radioactive impurities in atmosphere, earth, cosmic rays, luminous watch dials etc. The blood generating organs in the human body is situated at the depth which is greater than 5 cm and to protect the human body from the radiations, care has to be taken such that the penetration due to radiations will not reach that depth.

The NCRP (National Committee on Radiation Protection) US has recommended the maximum permissible weekly dose of 3γ of purely gamma radiations.

The exposure of the human body to radiation reduces the life span. Table given below gives the maximum permissible external radiation dose.

Age limit	*Dose rem/week*	*Feet, hands, head*
Below 18 years	0.01	0.05
18 and above 18 years	0.03	0.50

The presence of radioactivity in water, air or eatables is also harmful.

17.2.4 Effect of Radiation on Materials

Radiation has different effects on different materials. When the materials are exposed to radiations, displacement of electrons, displacement of nuclei or addition of foreign atoms may take place.

Some effects are experienced and they are :

(a) On glass the effect is of colouration.

(b) Plastic materials are subjected to cracks.

17.3 TYPES OF RADIOACTIVE WASTES

There are three main types of the wastes produced in a nuclear power plant. They are :

(a) Solid radioactive waste

(b) Liquid radioactive waste and

(c) Gaseous radioactive wastes.

17.3.1 Solid Radioactive Waste

Solid wastes include the scrap material or discarded materials contaminated with radioactive matter. The solid waste may be classified as (a) combustible waste and (b) non-combustible waste. Combustible solid wastes include wood, paper, clothes etc. The non-combustible wastes include in cinerator ash, floor sweepings, evaporator bottoms and contaminated process plant.

17.3.2 Liquid Radioactive Waste

Liquid materials are of three types or categories :

(i) Metal waste

(ii) Process waste

(iii) Chemical waste material.

Metal waste contains good amount of thorium, uranium or plutonium with high energy beta and gamma particles.

In liquid wastes, the condenser water, floor drainage, roof drainages etc. are contaminated with the radioactive materials.

Chemical waste materials generally consist of tank drainage, hot sink drainage and discarded or unused process solutions.

17.3.3 Gaseous Waste

There are two basic types of gases produced as wastes.

1. Hot-off gases from slug dissolvers, process tank etc.
2. Ventilation gases coming from cells, exhausts etc.

The gases is the mixture of inert gases i.e., xenon, argon and krypton, volatile fission products like iodine, ruthenium etc. and some solid and liquid particles. Radioactive iodine is not harmful to health of living beings. The inert gases have short half-lives and they are not absorbed by the system. Generally the inert gases can be absorbed by silica gel or charcoal.

Any solid particles present in the gaseous waste are most harmful and are to be taken care of.

Gaseous wastes can easily result in atmospheric pollution.

17.4 DISPOSAL OF RADIOACTIVE WASTE MATERIALS

It has been studied that fissile and fertile materials give radiations. Also when fission takes place in the reactor, fission products are the outcome of nuclear reactor operation. These materials are radioactive and are the mixture of various beta and gamma emitting radioactive isotopes with various half-lives. Depending on their half-life, these materials are transformed into stable isotopes after disintegration. Radioactive isotopes having high energy have industrial applications.

One thing is to be noted that radioactive isotopes can not be neutralized by any chemical method or process. Biological damage will be caused if the **waste** is discharged into atmosphere without taking into proper precautions and hence these waste materials are to be disposed off properly. The waste disposal system should be designed so that air and water contamination should be well below the prescribed limits.

Following are the basic methods of storage or disposal of radioactive waste materials.

(i) Storage Tanks. Generally the tanks are underground for storing the waste materials and are built at site or in the remote locations. These tanks should be corrosion resistant and the leakages are to be avoided. Generally solid and liquid wastes are stored in concrete or stainless steel tanks. The tank is situated in a large prohibited area.

(ii) Dilution. Disposal of liquids after dilution to safe limits is done in the rivers or sea. In some cases the waste material is stored for a short time and is diluted either in liquid, solid or gaseous materials. Gases are also let off in air after dilution. Before dispersal in the dilutent the radioactivity of the gas or liquid being discharged is reduced to acceptable levels by, for example, gas filteration and treatment of liquid by ion exchange.

Generally low energy wastes can be diluted in river water or sewage.

(iii) Sealed Containers. Radioactive liquids and solids are put in sealed containers which prevent the radioactive contamination. These sealed containers are then disposed off at sea where they are quickly and completely covered with mud in the bottom. Radioactivity can only get into the sea by slow diffusion through the mud.

This has been in use by British.

(iv) Underground Burial. Another alternative is the burial of waste direct in the ground. However, burial grounds must be isolated from the public and water must not be able to seep through as it may cause radioactive contamination of drinking water supplies. The solid waste is buried at the depth of nearly 10 ft or 3 m or slightly more.

(v) Atmospheric Dilution. This method is suitable for gaseous radioactive wastes.

(vi) Absorption by Soil. Large quantities of fission products are disposed by this method. In this case the radioactive particles are absorbed on the soil particles. This is the cheapest method of disposing the nuclear wastes.

(vii) National Storage or Burial Ground. The disposal of the nuclear wastes and its control can be simplified if the national storage system is developed. In this case, the deep craters created by the underground atomic blasts can also be utilized for disposing the nuclear wastes. The blast in a silica containing material forms a big hollow gap with a thick over-bearing lining. The wastes can be stored in it but there is a likelihood of contamination of ground water in the event of earthquake.

The nuclear wastes can also be classified according to the radioactivity levels, as

1. High level waste
2. Medium level waste
3. Low level waste.

The high level waste has radioactivity above 1000 curie. The medium level waste has radioactivity between 100 and 1000 curie while the low level waste has radioactivity below 100 curie.

There is a water pond in which the spent fuel is placed so that the heat is first removed and the pond water is continuously treated so that the activity due to defective cladding if any is removed. After the removal of heat the spent fuel is taken to reprocessing plant where the cladding containing fuel is removed. This fuel is then dissolved in nitric acid. From this solution, U^{235} and Pu^{239} are removed and nearly 99% non-volatile fission products is left behind. This liquid solution is known as "Highly Active Liquid Water (HALW). U^{235} and Pu^{239} are then purified and used in reactors as fuel. The liquid waste from the pond is then transferred to **storage** where it is required to be kept for nearly 30 years and more. So that the radioactivity is considerably reduced to low level. After this the waste is transferred to the permanent storage.

17.4.1 Low Level Solid Waste Disposal

The nuclear low level solid waste is cast in cement and is stored in steel drums. These drums are buried either below soil (5–10 m) or kept at the bottom or bed of the ocean. When the dilution of the activity is reduced due to the dispersion at the bottom of the sea, the risks are reduced to the minimum.

17.4.2 Medium Level Solid Waste Disposal

The nuclear medium level solid state wastes are mainly coming in contact with neutron activation product isotopes. These wastes are kept into cement cylinders because cement is non-combustible and provides a good shielding against external exposure. Another advantage of cement is the resistance to leaching by ground water.

17.4.3 High Level Liquid Waste Disposal

As seen in the previous sections, the high level liquid waste is obtained after U^{235} and Pu^{239} are removed from the solution of the spent fuel and nitric acid. This high level liquid is then stored on the site in special steel tanks in concrete vaults. They are water cooled and then taken to the permanent storage area.

There are two new methods to treat high level wastes.

(i) Harvest Process (Highly Active Residues Vitrification Engineering Study). In 1970, this process is developed. In this process, liquid waste together with glass forming materials are fed continuously into a heated stainless steel vessel in which the mixture fuses to a homogenous glass. When the vessel is full, it is removed from the furnace, cooled, sealed and discharged to the storage place.

The schematic for such a process is shown in Fig. 17.1.

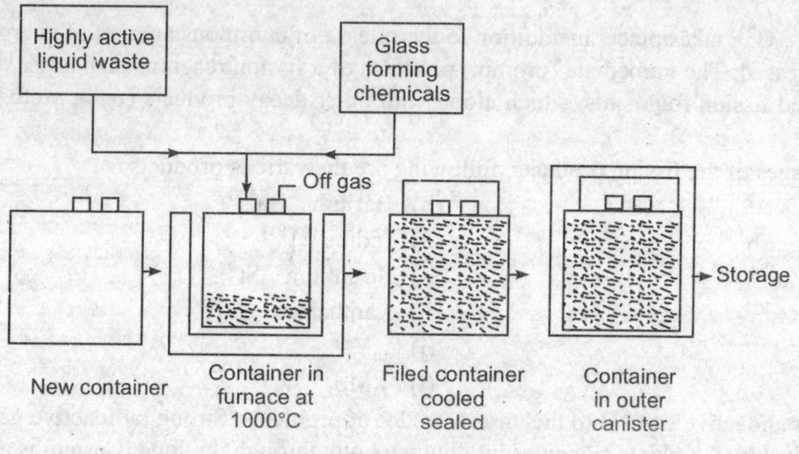

Fig. 17.1 *Harvest process.*

(ii) AVM (Atelier de Vitrification Marcoule) Process. This method is developed in France. The schematic diagram for this process method is shown in Fig. 17.2. In this process, the liquid waste is first dried to powder form in the rotary kiln. This powder is fed along with the glass powder into a glass making furnace.

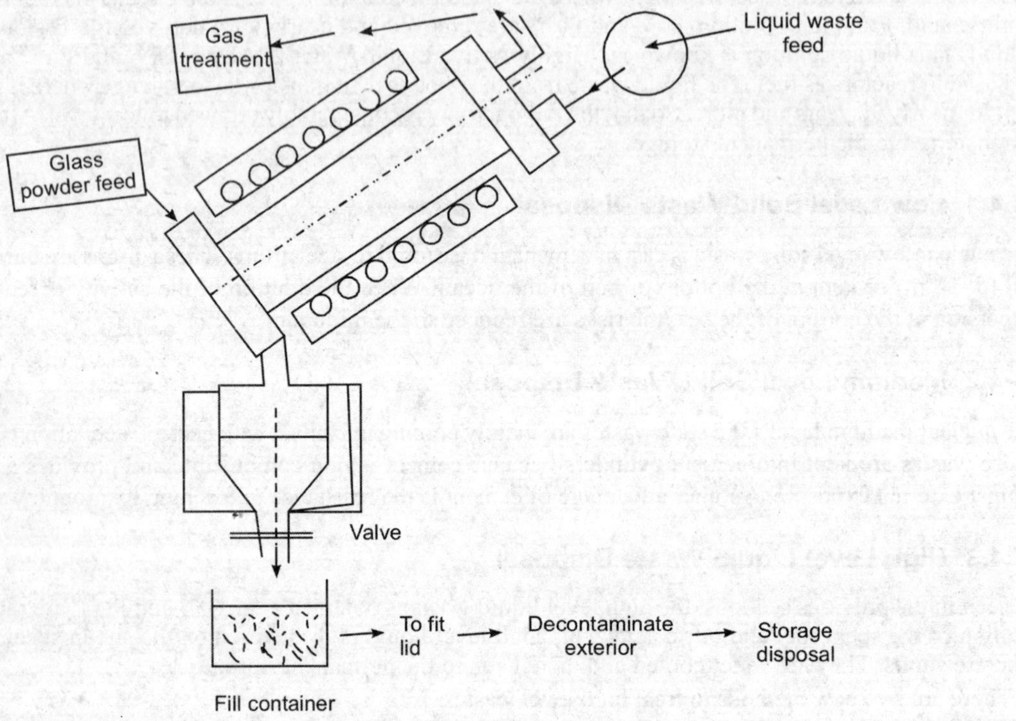

Fig. 17.2 *AVM Process.*

The advantage of this system is that there is higher production rate.

17.5 GASEOUS WASTE DISPOSAL

When the fission of ${}_{92}U^{235}$ takes place, in addition to the release of enormous amount of energy, fission products and neutrons are released. The immediate (prompt) products of a fission reaction such as Ze^{140} (xenon) and Sr^{94} (strontium) are called fission fragments, which along with other decay products (α, β, γ etc.) are called fission products.

At the various stages of the fission products, following are the various products :

(a) Krypton ${}_{36}Kr^{92}$
(b) Barium ${}_{56}Ba^{140}$
(c) Antimony ${}_{51}Sb^{133}$
(d) Iodine ${}_{53}I^{133}$
(e) Xenon ${}_{54}Xe^{133}$
(f) Strontium ${}_{38}Sr^{94}$
(g) Cesium ${}_{55}Cs^{133}$
(h) Lanthanum ${}_{57}La^{140}$
(i) ${}_{58}Ce^{140}$
(j) ${}_{39}Y^{94}$
(k) ${}_{40}Zr^{94}$
(l) ${}_{42}Mo^{95}$

The disposal of radioactive gases into the air creates lot of problems. Strong radioactive gases like iodine and strontium are absorbed by the plants and enter into human body through the food. Cesium is absorbed by muscle and strontium in bones and paralyses the health.

Hence generally radioactive gases are collected and stored in the tank buried in the ground and disposed off to the atmosphere when activity level sufficiently comes down.

THEORY QUESTIONS

17.1. What are the different types of nuclear waste? Which are more dangerous and why?

17.2. What is Nuclear Radiation? What are the different types of radiations? What are the effects on human body? Explain.

17.3. What is the unit of Nuclear Radiation? Mention the permissible radiation dose limits. Explain.

17.4. Explain the different methods of disposal of nuclear wastes.

Chapter 18 Nuclear Materials and Equipments

18.1 INTRODUCTION

The basic factors to be considered while studying the nuclear reactor are the components of the reactor. These components are :

(a) Reactor vessel

(b) Nuclear fuel to be used

(c) Coolant

(d) Moderators to be used

(e) Control system.

In addition to these components, it is necessary to consider the cladding and structural components to be used. The requirements of the materials that are used will be

1. mechanical and metallurgical adequate strength and corrosion resistance during operation.
2. machining and fabrication requirement of ductibility.
3. nuclear properties requirement.

18.2 REACTOR VESSEL

In thermal power plant, boiler furnace where heat is generated by burning the fuel thereby producing steam to be used in steam turbines is the heart of the plant. Similarly combustion chambers of the gas turbine power plant and diesel power plant are the main components where the burning of the fuel takes place thereby producing heat to be used for further operation of the gas turbine and diesel power plants. As compared to these components, the nuclear power plant has a component called **Nuclear Reactor** which may be regarded as the substitute for the fire box or furnace and combustion chambers in diesel power plant and gas turbine power plant.

Thus nuclear reactor is a device in which energy by nuclear fission process can be generated at the control rate. The nuclear reactor vessel encloses the following elements as :

(i) Reactor core

(ii) Reflector and

(iii) Shield.

The reactor vessel has to provide inlet and outlet passages for the coolant. Again it has to withstand the pressure as high as 200 bar and above. For inserting the control rods from the top of the vessel, holes are to be provided at the top of the vessel.

Stainless steel is an alloy consisting of steel, chromium, nickel, niobium, titanium, tungsten etc. This alloy possesses good structural properties and is also corrosion resistant. Because of its excellent properties it is used for the containers of reactors using metallic and aqueous solutions of uranium, cooling tubes, fabrication of pipes of heat exchangers etc.

18.3 NUCLEAR FUEL

Fuel required for nuclear reactor is nuclear fuel which is one of the isotopes of uranium U. This nuclear fuel can be one or all of the following U^{233}, U^{235}, U^{239}. Natural uranium found in nature (earth crust) contains three isotopes U^{234}, U^{235}, U^{238}. The percentages of these materials are

$U^{238} = 99.3\%$, $U^{235} = 0.7\%$, U^{234} = Traces.

Following are the resources of nuclear fuel :

(i) Granite Rocks. It is found that these rocks contain 4 ppm uranium and uranium concentration in the upper crest is as high as 5%.

(ii) Volcanic Rocks. Sediments originated from volanic rocks.

(iii) Sand Stone. This is another source of uranium. This may contain maximum 3% uranium.

The main locations of these resources are :

(i) USA (30–35%)

(ii) South Africa (18–20%)

(iii) Australia (20%)

(iv) Canada (20–22%)

(iv) **Sea Water.** It is estimated that the uranium content of the sea water is only 3–3.5 micrograms/litre.

The method has been developed by Japanese scientists to extract U^{235} from sea water by using Ca (OH_2) [200 gms/ton of sea water].

Uranium is used either in the form of pure metal or as a solution of uranium salts in ordinary or heavy water or as a metallic uranium solution in some low melting point metal.

At room temperature natural uranium is chemically stable but oxidation is severe even at low temperature of nearly 100°C. Uranium in the powder form ignites even at room temperature. With water, uranium gives strong corrosion effect.

Uranium melts at 1100–1130°C. High melting point alloys can be developed with beryllium, aluminium, copper, iron, tungsten, magnesium, zinc, vanadium and thorium. Some of the engineering properties required by the uranium fuel are as follows :

1. High tensile strength at high temperature to prevent buckling of fuel elements and to bear thermal stresses.
2. To reduce the cost of machining, higher ductility and better machinability.
3. High conductivity.
4. It should be corrosion resistant so that coolant will not corrode.
5. It should have high radiation stability. The fuel elements will not buckle due to distortion because of nuclear radiation.

The **uranium** exists in **alpha phase** with the density of matter 19.1 gm/cm^3 **upto 660°C**. Also the mechanical strength is nearly 27 kg/mm^2 which can be increased to 138 kg/mm^2 by cold rolling process.

For the temperatures from 660°C to 770°C, the uranium is stable in Beta-phase. Above 770°C the uranium exists in gamma phase with cubic crystal lattice structure.

In addition to the natural uranium as a fuel for nuclear reactor, oxides and carbides of uranium are also used as nuclear fuels.

Uranium Oxide. Another nuclear fuel used in nuclear reactor is uranium oxide (UO_2). Uranium oxide is a brittle ceramic. It is produced in a powder form and then sintered to form fuel pellets.

There are advantages and disadvantages of uranium oxide as fuel for nuclear reactor.

Advantages. The advantages of UO_2 are as follows :

(i) UO_2 is more stable than natural uranium because it is oxidised and therefore presents less problems.

(ii) It is not attacked by N_2, He and H_2 and most of the coolants.

(iii) It can be used for higher temperatures because UO_2 does not change phases.

(iv) The fuel rating is high as 8.5 MW/te. For uranium it is 3 MW/te.

(v) Cladding failure is not a problem because UO_2 does not corrode as natural uranium.

(vi) As irradiation effects are absent, there will be greater dimensional stability. Therefore, buckling, swelling, wrinking are absent.

As against these advantages, the **disadvantages** are :

1. Because it is brittle than uranium, it tries to crack under thermal stresses. Canning provides the required strength.

2. If the can fails, the fuel UO_2 may disintegrate into powder.

3. The powder thus formed is oxidised to U_4O_9. This is far inferior in physical properties compared with UO_2 and further deterioration of the fuel assembly takes place.

4. Low thermal conductivity and it is undesirable property. Heat removal is poor and therefore thermal stresses are much greater. And hence, fuel element has small diameter.

5. When UO_2 is used, the enrichment is necessary because the presence of two oxygen atoms for every uranium atom reduces the proportion of fissile uranium U^{235}.

Uranium Carbide. Another fuel used in nuclear reactor is uranium carbide UC. Uranium carbide is a black ceramic and is used in the form of pellets. Its density is high, melting point also is high, and possesses good (high) thermal conductivity. There is no trouble of phase change. Uranium carbide is more stable under irradiation. It is, at present, not economical to use. Its melting point is 2350°C and thermal conductivity is 82.4 $\frac{\text{kW}}{\text{m}-°\text{C}}$

In natural uranium, the fissile part proportion (U^{235}) is only 0.7% and therefore in some types of the reactors it is very difficult to maintain a chain reaction. Therefore, the percentage of U^{235} in the fuel has been increased so that the chain reaction can be maintained. The process of increasing the percentage of U^{235} is known as **enrichment**. The enrichment may be from 1.5 to 90% depending on the different types of the reactors used.

Without discussing any details of the processes involved in enriching the uranium, the different methods involved are :

1. Gaseous diffusion method
2. Thermal diffusion method
3. Centrifugal method
4. Electromagnetic method
5. Separation nozzle method.

18.3.1 Coolants

For coolants Refer See 16.1.3

18.3.2 Moderators and Reflectors

For Moderators Refer See 16.1.2

18.3.3 Control System

Refer See 16.1.4

18.4 CLADDING AND STRUCTURAL MATERIALS

The fuel elements that are generally used in reactors are usually long cylindrical rods or rectangular plates of uranium or thorium enclosed by cladding. The cladding material is necessary to prevent the fuel elements from corrosion by the coolant. Basically cladding serves three functions :

(a) to provide structural support and strength for the fuel and prevent distortion

(b) to prevent the release of radioactive fission products into the coolant steam

(c) in certain types of the reaction (mainly gas-cooled reactor) to provide extended surfaces in the form of fins to promote more heat transfer to the coolant.

The desirable properties of a good cladding material are as follows :

(a) It should have high residence to corrosion during operation under high pressure and temperature.

(b) It should have low neutron capture cross-section.

(c) It should have high thermal conductivity to facilitate the quicker heat transfer and low thermal stress.

(d) It should have good ductility and high machinability to reduce the fabrication or production cost of the element.

(e) It should have high melting point.

(f) It should be non-toxic.

(g) It should maintain high strength against high pressure and temperature.

The requirement for canning and structural material will be same as those for cladding materials. Sometimes cost of the material will also have to be taken into account while deciding the materials for all these purposes.

The materials which are generally used for these purposes are :

(i) Stainless steel

(ii) Magnesium

(iii) Beryllium and

(iv) Zirconium.

(i) Stainless Steel. Stainless steel is an alloy which is obtained by alloying the steel with chromium, nickel, niobium, titanium, tungsten etc. The alloy possesses good structural properties and is also corrosion resistant. Because of its excellent properties it is used for the containers of reactors using metallic and aqueous solutions of uranium, coding tubes, fabrication of pipes for liquid metal heat exchange systems and also for other piping system. Stainless steel can be decontaminated easily.

Stainless steel offers high resistance to oxidation and it can be used at a temperature as high as 800°C. All these properties are in favour of **stainless steel as can material**.

(ii) Magnesium. Magnesium has low-neutron absorption cross-section and high resistance to oxidation. The magnesium can be easily sealed so that welding with arc welding process can be done at any time. Magnesium has high thermal expansion and oxidised at about 350°C in the presence of air and once the magnesium melts, magnesium catches fire above the melting point.

Magnesium can be used as cladding material.

(iii) Beryllium. Beryllium is having excellent chemical and corrosion resistant properties. It is not effected by radiation damage. It has low neutron absorption section. Also its melting point is high and has high thermal conductivity. Beryllium has high strength. All these factors are in favour of material for cladding.

Following points are, also, to be considered in case of beryllium material to be used for cladding, can and structural.

(i) Toxic (ii) High ductility
(iii) Low ductility (iv) Costly
(v) Accelerated corrosion at 600°C and above (with moisture).

(iv) Zirconium. *Zirconium* is used in the reactor as construction material. It has good mechanical strength, small neutron absorption cross-section and high melting point (1860°C). Zirconium is less effected by corrosion. In nature zirconium is accompanied with hafnium which has high cross-section of 105 barns. Zirconium has good thermal conductivity and high resistance to corrosion at high temperature and pressure. This is a costly material and the scrap cannot be reused. Zircaloy-2 is an alloy of zirconium having nearly 1.5 percent tin, and smaller percentage of iron, nickel and chromium. Zirconium is preferred as can material.

18.5 ALUMINIUM

Aluminium is widely used in the reactor for jacketing the uranium slug, coolant pipes upto the temperature of 300–400°C, pipes for the heat exchangers etc. Aluminium is comparatively cheaper material and alloys are being developed to replace the other metallic structure by aluminium alloy. The aluminium metal is not subjected to radiation damage but it is affected by corrosion. The presence of inorganic acids and salt solutions corrode the metal. Above 100°C the effect of corrosion becomes significant. The alloy of aluminium with silicon has proved to be better corrosion resistant. The alloy produced should have low cross-section.

The aluminium in pure form is not subject to contamination.

18.6 HEALTH MONITORING INSTRUMENTS (DOSIMETERS)

Because of the possibilities of radiations present in the viscinity of the nuclear power plant, health of the workers or every person working in the plant or power station is of utmost important. To measure the intensity of radiation in the vicinity of buildings inside and outside both and the intensity reaching to workers in the power plants, health monitoring instruments have been developed. These instruments are dosimeters and are of following types :

(i) survey dosimeters
(ii) personal dosimeter
(iii) dosimeters for monitoring the radiation does
(iv) dosimeters for monitoring radiation dose in water and air.

Out of these survey dosimeters and personal dosimeter are quite common in nuclear station.

18.6.1 Survey Dosimeter

Figure 18.1 shows the schematic of survey dosimeter. This dosimeter consists of one ionisation chamber, one amplifier and an ammeter calibrated in terms of dose rate units.

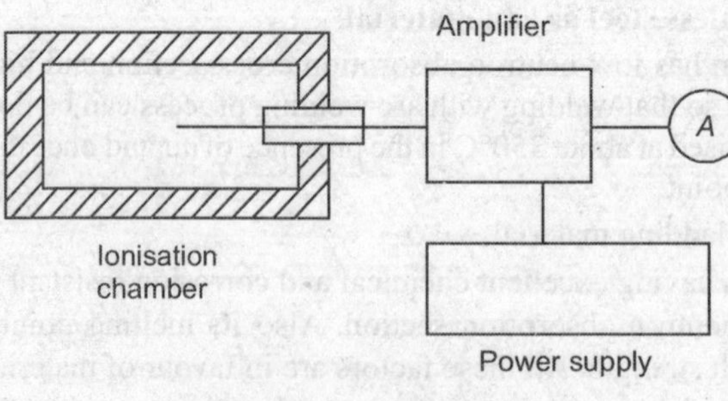

Fig. 18.1 *Survey dosimeter.*

The unit of dose is **Roentgen**. It is a international unit of x-ray or gamma radiation. Unit of Roentgen is defined as the quantity of radiation which results in the absorption of 83.9 ergs of energy by one gram of air. Other units of radiation are **Rem, Rep** and **Rad**. These are similar to each other but each one is used for specific purpose.

Rem is **Roentgen equivalent for man.** One **Rem** is equivalent to 93 ergs energy per gram absorbed due to gamma radiation and this related to the biological effectiveness.

Rep is **Roentgen equivalent for physical effects.** This is also equivalent to 93 ergs of energy absorbed by one gram of material cause by the gamma radiation.

The dosimeter measures intensity of gamma radiations in Roentgen per hour. This determines for how long the man can work in the specified compartment without being affected by the health hazards.

Working. The ionisation chamber produces current which is proportional to the intensity of incident gamma radiations. The amplifier amplifies the current and the ammeter reads the dose rate. The safe working time in any room can be calculated as given below :

Let the dose rate in the room is 0.006 γ/hour. The tolerance dose for the working day is 0.05 γ. Then the working time of a man should not be more than $\frac{0.05}{0.006} = 8.33$ hours in that room.

18.6.2 Personal Dosimeter

Another instrument used in nuclear power station is a **Personal Dosimeter**. This is used to monitor the radiation doses coming to the individual person in the power station. This dosimeter gives direct measurement. The principle of working of this instrument is that the photoconductivity increases the electrical conductivity of some nonmetal when subjected to gamma radiation. In this case the effect of gamma radiation on photographic film emulsions is the basis of radiography and this phenomenon is used in several industrial devices and also in film badges. In the dosimeter, the photographic film is enclosed in light-tight wrapping. When the worker or any person in the power station is working he or other bears the badge in the chest pocket and thus the dosimeter is exposed to gamma radiation. Due to the exposure of the film to gamma radiation, the film is blackened. The film dosimeter is calibrated in terms of degree of blackness and irradiation dose. After some interval, the exposed film is developed and the effect of accumulated radiation dose. Generally, the film is developed at the end of each working day or once in a week depending on the requirement. Instant measurement of the dose is not possible.

18.6.3 Pocket Chamber or Fountain Pen Dosimeter

Working Principle. There is an electroscope in which metal leaves (two) electrodes has been suspended in the centre. When the electrode is charged by gamma radiations, the thin leaves are subjected to motion because of the development of the coulomb forces. Because of the similar charges the leaves will repel each other and then the

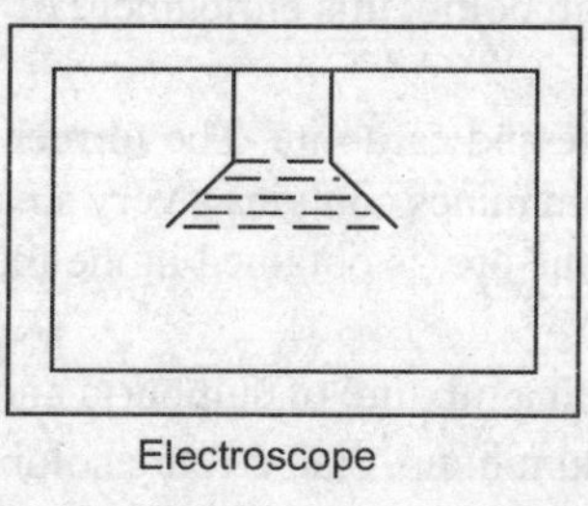

Electroscope

Fig. 18.2 *Fountain Pen Dosimeter.*

degree of separation is directly proportional to the magnitude of charge. If the electroscope is well insulated, it will be possible to maintain the charge for few days.

The pocket chamber or fountain pen dosimeter works on the principle given above. The dosimeter is shown in Fig. 18.2 and it consists of two loops. One of these two loops is fixed one and the other is moving. Moving loop is made of quartz fibre. These two loops are connected to form a conducting system. We can observe the position of loops through the eye-piece of the microscope. A high voltage battery or some similar source of voltage will charge the meter and positive charge on the loops will be produced. Because of this, the moveable loop will deflect to its maximum distance and coincides zero position on the transparent scale which is visible through an eye-piece. The gas in the chamber is ionized when the meter is subjected to gamma radiation. Thus both the loops are positively charged, and the negatively charged ions will be attracted towards the loops. Therefore, this negative charge will decrease the charge on the loops. Thus the moveable loop will move towards the fixed loop. This movement will be proportional to the negative charge with the ions. The movement is measured either by reading directly on the scale or by electronic voltmeter. In these both cases the calibration is in terms of **Roentgens**. Generally, the range of normal fountain pen dosimeter is upto 0.3 γ and is normally charged after six days or once in a week (Normal radiation dose is 0.05 γ per day).

The meter is made shockproof and is extensively used in installations subjected to gamma radiations.

18.7 OTHER INSTRUMENTS

In addition to the above dosimeters, there are a few detectors that are used in nuclear power installations to protect the workers working there. Without giving the details of these instruments, following are the equipments used :

(a) Electromagnetic Pump. These pumps are used for pumping radioactive coolants and play important role in transferring the heat. There are two types of pumps :

(i) Direct Current or Alternating Current Electromagnetic liquid metal pump

(ii) Induction type electromagnetic liquid metal pump.

(b) Radiation Detectors. In these types of instruments the radiation to be detected interacts with the specific material and this effect is measured. The possible ways of interactions are :

(i) Charge accumulation (i.e., electrostatic charge)

(ii) Ionization

(iii) Transformation of radiation energy into visible light, heat and electricity

(iv) Photoconductivity

(v) Nuclear transformation.

18.8 PRODUCTION OF ENRICHED URANIUM

It is observed that the natural uranium consist of, mainly, fissional fuel U^{235} only 0.7% which is insufficient for many reactor-breeder reactor. These reactors require the nuclear fuel upto the tune of 90% enrichment of U^{235}. If we consider the use of nuclear fuel for atom bomb, this enrichment of U^{235} is about 90–95 or more percent. Thus natural uranium has to be enriched.

The ores available from mines is uranite and carnotite. The former is a pitch blend and the latter is a uranium vanadium complex. Other ore obtained from mines contains a very small percentage of uranium. These ore are the basic sources of uranium. Uranium, from the ore, is obtained in the metallic form. The procedure followed to get uranium from the ore is as follows:

1. Ore from the mines is treated with a mixture of sulphuric and nitric acid.
2. In this mixture, uranium, iron, aluminium, zinc etc. are soluble and forms another mixture.
3. From the ore, barium and lead are insoluble and forms the insoluble sulphate precipitate.
4. The acidic solution is mixed with a good quantity of soda so that the solution becomes alkaline in nature.
5. Iron, aluminium and zinc are precipitated in the alkaline solution.

6. The solution in (5) is filtered and the solution containing uranium is obtained.
7. The solution in (6) is treated with sufficient quantity of nitric acid to make the solution acidic.
8. Uranium in solution in (7) is transformed to uranyl nitrate.
9. Add ether to the solution and uranium passes into either solution.

This particular process is called as **selective extraction**.

10. From this process, high quality (purity) uranium is obtained and this is used to get metallic uranium.

Neutron absorbing material like cadmium, indium and boron are to be separated while producing uranium fuel. Again when uranyl nitrate is subjected to firing, uranous-uranic oxide (U_3O_8) is obtained in the pure form.

To produce the enriched uranium from natural uranium, following five commercial methods have been developed.

1. Gas diffusion process
2. Thermal diffusion process
3. Separation nozzle method
4. Electromagnetic method
5. Centrifugal method.

The very basic principle on which these all methods are based is the **Principle of mass difference**.

The mass of natural uranium U^{238} is slightly more than that of U^{235}.

1. Gas Diffusion Process. To produce enriched uranium on an industrial scale, gas diffusion process is used. This process is the most economical process, and was developed in US in 1932 by Hertz. It was proved by scientist Graham that the rate of gas flow through porous barrier in inversely proportional to the square root of molecular weight. In case of a gas having two isotopes, it was found that the lighter isotope will diffuse at a fast rate. This process is not reversible.

Uranium hexafluoride (UF_6) is the volatile compound which is in solid state at room temperature but at 64.5°C but transformed into vapour at about 1100 mm Hg pressure. Unsaturated UF_6 vapours are used for diffusion. Figure 18.3 shows a separating stage in gas diffusion plant. It consists of many diffusion tubes. UF_6 is compressed and is

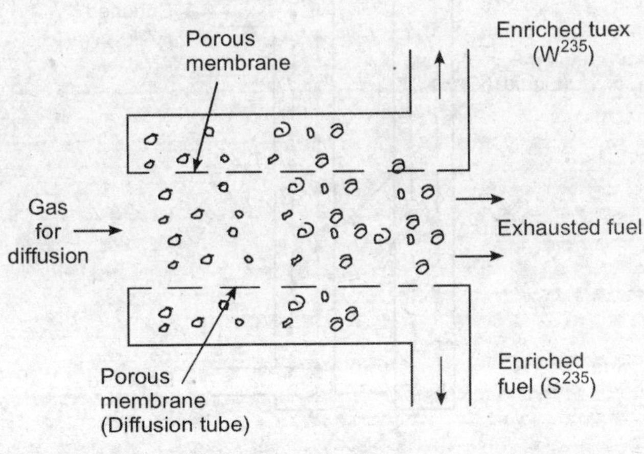

Fig. 18.3 *Schematic of Gas Diffusion Plant*

fed the plant. Compressed gas is cooled by coolers so that the compression heat is removed. The diffused gas is $U^{235}F_6$. To achieve good percentage of U^{235}, the cascading is used. It is found that for the enrichment of the fuel to

3 to 5%, there are nearly 1500 to 1800 diffusion stages. The plant must be air sealed otherwise humidity will decompose UF_6 which will plug the membrane pores.

2. Thermal Diffusion Process. Thermal diffusion process is very simple. The schematic showing the phenomenon is shown in Fig. 18.4. As shown, the container contains two gases and the two walls are at the temperatures T_1 and T_2 such that $T_2 > T_1$. The difference $(T_2 - T_1)$ is maintained constant. In such circumstances, it was observed that one

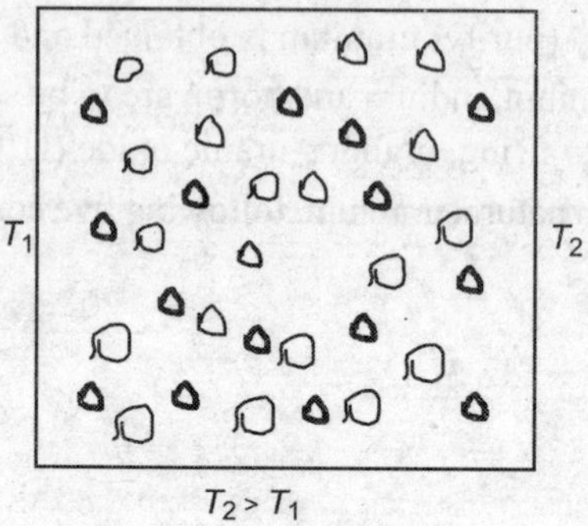

Fig. 18.4 *Separation of molecules by Thermal diffusion.*

type of the molecule concentrates in the vicinity of hot wall but in the vicinity of cold wall it is poor. This phenomenon is called **Thermal Diffusion**. This is also applicable to liquids. It is observed that light molecules concentrate near the hot wall. This phenomenon is also applicable to a isotopic mixtures.

In the present case, UF_6 is in the liquid form. The schematic plant is shown in Fig. 18.5. This plant consists of two concentric pipes as shown, and they act as columns.

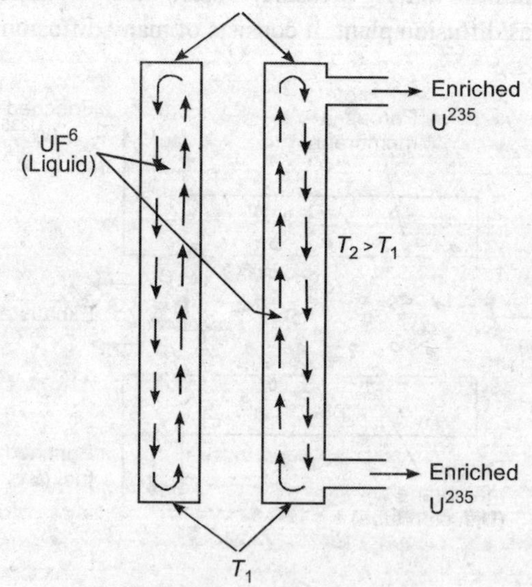

Fig. 18.5 *Thermal diffusion plant.*

Liquid UF_6 is filled in the gap and the pipes are kept at different temperatures. These two temperatures are kept constant. The difference in temperature makes the fluid flow in the direction shown in Fig. 18.5. Particularly, the inner pipe is kept at high temperature and the fluid will rise upwards along the hot pipe wall. The liquid collected at the high temperature wall is of lower density. Here U^{235} is collected in the vicinity of hot wall. The fuel is concentrated in the upper part of the column because of lower density. Number of columns are connected in succession (cascading type) so that high order of enrichment is achieved.

3. Separation Nozzle Method. Another simple and efficient method for enriched uranium is **separation nozzle method**. In this case the mixture of gases is passed through a curved passage made of curved strips forming a nozzle passage and the two gases having a mixture of two different masses, are separated as shown in Fig. 18.6.

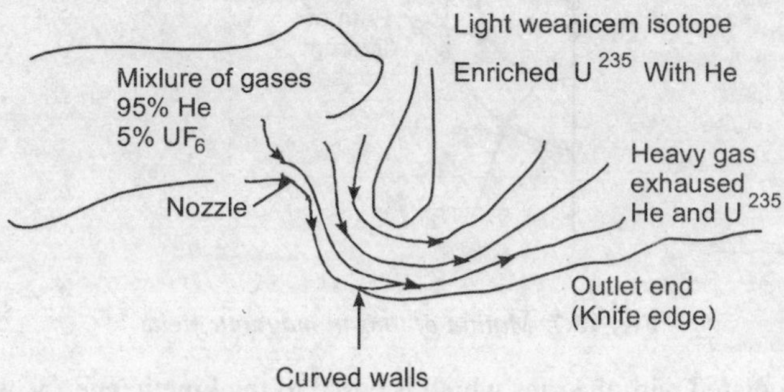

Fig. 18.6 *Separation nozzle schematic.*

It can be shown that the radius of curvature is directly proportional to the mass of molecules. Therefore, the heavy gas will be enriched at the inner surface of deflection grove and the light gas will be collected at the outer surface and the knife edge divides the flow into light and heavy fractions of the gases.

In this plant set up, the required mixture [UF_6–5% and He–95%] is fed to the nozzle. The heavy fraction contains small percentage of Helium and UF_6 concentration is more. It is generally observed that this method is three times more effective than that of any diffusion method. To increase the effectiveness of the process, cascade arrangement is used in practice.

4. Electromagnetic Method. It is known that magnetic field affects the direction of motion or flow of ions. If the ions are moving in one particular direction and are subjected to magnetic field, which is at perpendicular to the direction of flow then the forces created will reflect or deviate the path of motion of the ions. In this case the ions follow the semi-circular path which is decided by the mass of the ion.

Let F = Magnetic force at right angles to the flow direction

H = Magnetic field strength

V = Velocity of ion

e = Charge of ion

Then $F = e_v H$

$$= \frac{mv^2}{r} \text{ centripetal force}$$

$$\therefore \quad evH = mv^2$$

or

$$r = \frac{mv^r}{eH}$$

From Fig. 18.7 it is clear that, the radius of curvature is different for different ionic masses. Similarly it is clear that the ionic mass is directly proportional to the radius of curvature of the ion path.

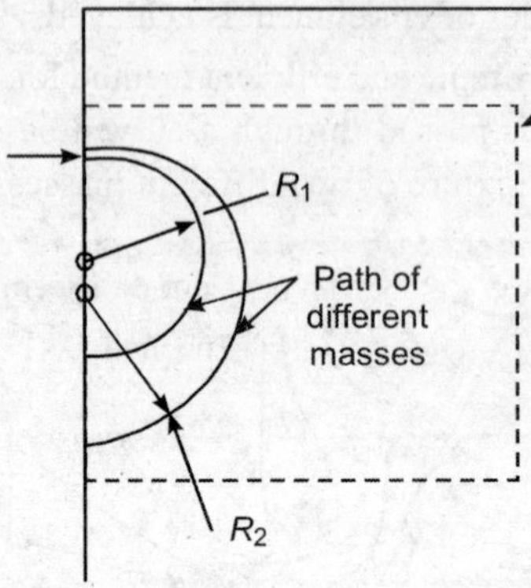

Fig. 18.7 *Motion of ions in magnetic field.*

Also the total energy associated with the ions which is equal to the kinetic energy, we have

$$E = \frac{mv^2}{2}$$

$$\therefore \quad v = \sqrt{\frac{2E}{m}}$$

$$\therefore \quad eHv = \frac{mv^2}{r}$$

$$\therefore \quad eH = \frac{mv}{r}$$

$$r = \frac{mv}{eH}$$

$$= \frac{m \times \frac{\sqrt{2E}}{m}}{eH}$$

$$\boxed{r = \sqrt{\frac{2Em}{eH}}}.$$

From this equation it is clear that the radii of circumferences of different masses having same kinetic energy are directly proportional to the square root of their ion masses when passing through the magnetic field. This principle is used in separating $U^{235}F_6$ gas from the mixture of $U^{235}F_6$ and $U^{238}F_6$. The schematic diagram of the arrangement is shown in Fig. 18.8.

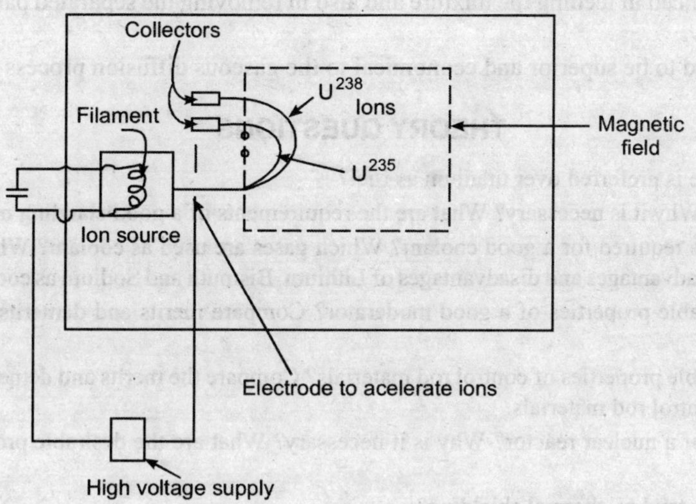

Fig. 18.8 *Electromagnetic plant for enrichment of uranium of U^{235}.*

A gaseous mixture is introduced into the ion source as shown in Fig. 18.8. Here neutral atoms are ionized. Then the ion beam is passed through an electric field where the ions are highly accelerated. The accelerated ions of the mixture are then passed through magnetic field where $U^{235}F_6$ and $U^{238}F_6$ move along the different radii and are collected in different collectors as shown in Fig. 18.8. The collected ions are neutralised and deposited on the wall as the coating of uranium compounds. These collectors are removed to collect the uranium compounds when the enrichment process is over.

Very high enrichment even 100% is possible by using this method.

5. Centrifugal Method. During the Second World War, this centrifugal method used for enrichment of the uranium fuel, was developed in USA and Germany. The principle of working for this method is that if a mixture of

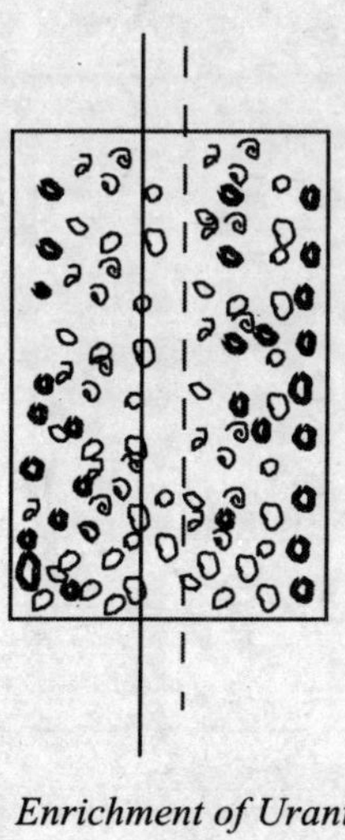

Fig. 18.9 *Enrichment of Uranium U^{235}.*

two gases of different molecular weights are enclosed in a centrifuge and rotated at high speed, the heavier gas is concentrated near the outer periphery because of the centrifugal forces acting on the heavier particles. This is shown in Fig. 18.9.

Difficulty was experienced in feeding the mixture and also in removing the separated parts one enriched $U^{235}F_6$ and the other $U^{238}F_6$.

This method was found to be superior and economical to the gaseous diffusion process described earlier.

THEORY QUESTIONS

18.1. Why uranium oxide is preferred over uranium as fuel?

18.2. What is cladding? Why it is necessary? What are the requirements of a good cladding material?

18.3. What properties are required for a good coolant? Which gases are used as coolant? Which is more preferable? Why? What are the advantages and disadvantages of Lithium, Bismuth and Sodium as coolant for nuclear reactor?

18.4. What are the desirable properties of a good moderator? Compare merits and demerits of H_2O, D_2O and C as moderators.

18.5. What are the desirable properties of control rod materials? Compare the merits and demerits of Boron, Cadmium and Hafnium as control rod materials.

18.6. What is shielding of a nuclear reactor? Why is it necessary? What are the desirable properties of the shielding material used?

18.7. What do you understand by thermal shielding?

18.8. Why concrete is considered as best shielding material? Why drying out of concrete is undesirable?

18.9. What are the merits and demerits of stainless steel and zirconium used as fuel element cladding?

18.10. Name the various methods commercially used to produce the enriched uranium.

18.11. Describe the most efficient and economical method suited to produce highly enriched U^{235}.

18.12. Describe the problem of health hazard from nuclear radiation. Briefly enumerate different instruments used in nuclear power plant to measure radioactive dose.

PART 3

Chapter 19 Elements of Hydrology

19.1 INTRODUCTION

The economic development of the country depends on the per capita energy consumption. Developing countries like India need to produce more power as there is a large gap between demand and supply. Energy is available in various forms viz. mechanical, electrical, thermal energy etc. Out of these, electrical energy is the most preferred, because of certain inherent advantages.

The main resources of electrical energy generation in the country are :

(i) Hydro-power plants
(ii) Thermal power plants based on coal
(iii) Thermal power plants based on nuclear fuel
(iv) Diesel power plants
(v) Gas turbine power plants.

Electrical energy generation has certain inherent advantages :

(a) It can be easily transported from one place to another.
(b) Transmission and distribution losses are minimum.
(c) It can be easily converted into other forms.
(d) Controlling and regulation is easy to suit the requirement.

At present, USA accounts for about 6 to 8% of total world population but consumes about 30% of electrical energy consumed in the world.

Hydroelectric power plants account for nearly 26% of electrical power produced in India. The first hydroelectric power station started in Karnataka at Shivanasamudram in 1902, which was generating 4.5 MW. These plants require either high head or high discharge.

The energy of water can be utilized for the generation of electrical energy. When water flows through a height, the kinetic energy of water is used to rotate a turbine which is coupled to an alternator. Higher the potential energy or head (it is the difference in level of water between two points). Hence, it is necessary to have a continuous flow of water to generate electrical energy. For this, water has to be collected and stored in reservoirs or lakes at high altitudes. Water is artificially stored in dams which are constructed over rivers in which large quantity of water is stored in a wide area known as catchment area. Rainfall is the main source of water and depends on several factors such as temperature, wind, cloudiness, humidity etc. Not all the water stored in dams can be utilized for power generation. Some of the water is evaporated, some seeps into the soil and some of it is used for the vegetation. The remaining water flows on the ground surface of the catchment area to form a stream and is known as run-off. This water is utilized for hydro-projects i.e., power generation and irrigation.

The factors affecting run-off are :

(a) Rainfall pattern
(b) Geology of the area
(c) Size and shape of the catchment area
(d) Topography and nature of soil in the catchment area
(e) Vegetation and
(f) Weather condition in catchment area.

19.2 HYDROLOGY AND THE HYDROLOGIC CYCLE

Hydrology may be defined as the science that deals with the processes governing the depletion and replenishment of the water resources of the land areas of the earth. In its broadest sense, it is the study of water in all its phases and includes hydraulics, the physics and chemistry of water, meteorology and various other allied natural sciences.

Hydrology may also be defined as the science which deals with the rainfall and run-off study.

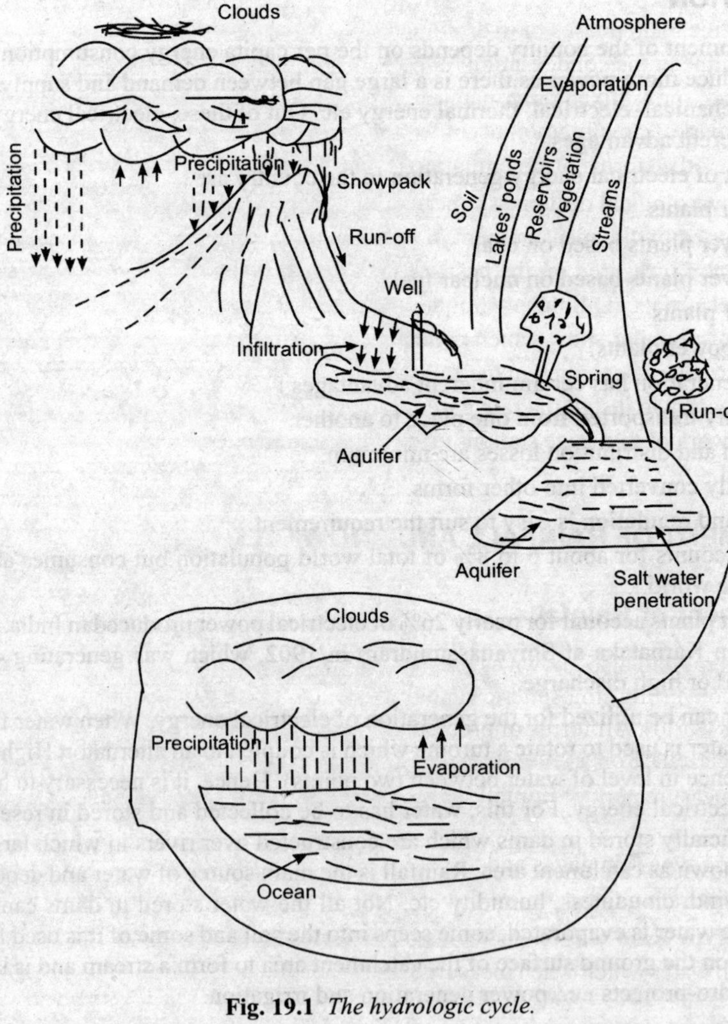

Fig. 19.1 *The hydrologic cycle.*

The hydrologic cycle is the descriptive term applied to the general circulation of water from the oceans to the atmosphere, to the ground and back to the oceans again. Figure 19.1 shows the various phases of hydrologic cycle. The cycle may be considered to begin with the water of the oceans. Water from the ocean surface is evaporated into atmosphere. The vapour is condensed by various processes and falls to the earth as precipitation. Some of this precipitation fall directly on the oceans, and some falls on the land surfaces. A portion of that falling on the land, is retained temporarily in the soil, in surface depressions and on vegetation and other objects until it is returned to the atmosphere by evaporation and transpiration. Another portion runs off from the ground surface into the streams and is returned to the ocean. Still another portion percolates into the ground and join the ground water which also slowly finds its way to the streams as ground water flow. However, some portion of the ground water which percolates the great depths appears after long intervals as springs, artesian wells and geysers. Further the entire quantity of water that reaches the stream does not flow directly to the oceans, because, throughout its travel a portion of it returns to the atmosphere by evaporation and transpiration and some portion seeps into the ground. The hydrologic cycle provides the ground work upon which the science of hydrology is used.

19.3 PRECIPITATION

Precipitation is one of the main phases of hydrologic cycle. It includes all moisture that reaches the earth's surface in liquid or solid form due to the condensation of the atmospheric vapour. The principal forms of the precipitation are rain, snow, sleet (i.e., snow mixed rain), hail (i.e., frozen rain drops), dew and frost. Out of the various forms of precipitation, a major part of precipitation occurs in the form of rain and a minor part occurs in the form of snow. The contribution of the other forms of precipitation is very little and hence the same is generally ignored.

For almost all the hydrological designs it is necessary to have the precipitation records for a long period. Such records may be developed by the measurement of precipitation occurring at any place from time to time. All the forms of precipitation are measured on the basis of the vertical depth of water or water equivalent which would accumulate on the level surface if all the precipitation remained where it fell. Many types of gauges are developed for the measurement of rain and snow which constitute the major part of the precipitation. Some of these gases which are commonly used are described in the following section. Since the amount of rainfall varies from place to place, it would be necessary to install the rain gases at the various key points in the area. Especially in the drainage basin of a stream a network of rain gage stations evenly distributed over the entire area of the drainage basin should invariably be provided.

19.4 MEASUREMENT OF RAINFALL AND SNOWFALL

19.4.1 Measurement of Rainfall

As it is seen above that rainfall also known as precipitation is one of the main phases of the hydrologic cycle and is the natural process of converting atmospheric vapour into water. The rainfall varies widely from location to location (may be the country or world). All forms of precipitation are measured in practice on the basis of vertical depth of water which would accumulate on a level surface if all precipitation remained where it fell.

Generally, for the measurement of rainfall, gages known as **Rain gages** are used. There are two main types of rain gages :

1. Non-recording type rain gages and
2. Recording type rain gages.

1. Non-recording Type Rain Gages. As the name indicates these rain gages do not record the rainfall directly but only collect the rain water which when measured gives the total amount of rainfall at the rain gage station during the measuring interval. The India Meteorological Department has adopted the Standard Symon's rain gage stations in India for measuring the rainfall. It is a non-recording type rain gage which is used at most of the rain gage stations in India.

As shown in Fig. 19.2 the Symon's rain gage consists of a cylindrical metal case of internal diameter 127 mm with base enlarged to 203.2 mm diameter. At the top of the case a funnel is fixed which is provided with a brass rim measuring exactly 127 mm inside diameter. The funnel shank is inserted in a glass bottle placed inside the case. The

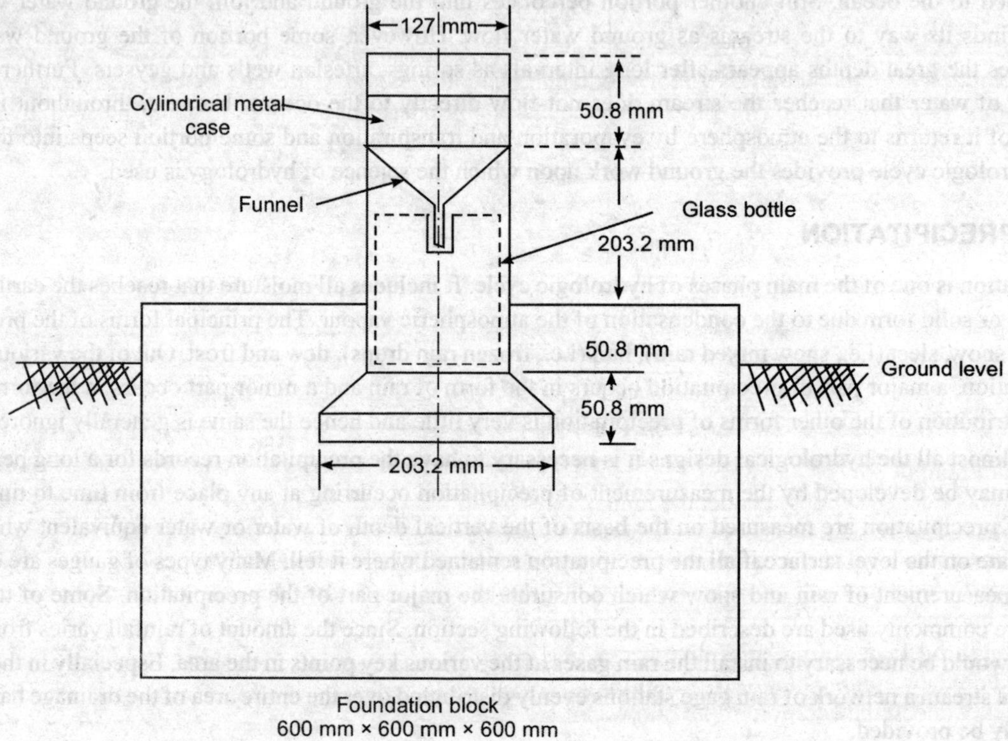

Fig. 19.2 *Symon's rain gage.*

case of the rain gage is fixed in a masonary or concrete foundation such that the funnel rim is exactly 305 mm above the ground level. The rain water enters the bottle through the funnel and gets collected in the bottle. The rain water collected in the bottle is measured with the help of standard measuring glass supplied with each rain gage which indicates the millimeters of rain that has fallen at the rain gage station. At each gage station the observations for the rainfall are taken daily two times at 8.30 a.m. and 5.30 p.m. (IST). However, if the rainfall is likely to exceed the capacity of the bottle then a few intermediate observations are also taken. The sum of the observations taken will represent the total rainfall of the past 24 hours of the day on which the observation at 8.30 am is taken.

2. Recording Type Rain Gages. These rain gages automatically record the intensity of rainfall and the time of its occurrence in the form of a pentrace on a clock driven chart, from the total amount of rainfall for the desired duration may also be determined. These gages may be operated over extended periods without attention and the recorded observations may be collected from the gages each time after a certain fixed duration. The most widely used recording rain gages are as given below :

(a) Tipping bucket rain gage (b) Weighing type rain gage (c) Float type rain gage.

(a) Tipping Bucket Rain Gages. It consists of a 300 mm diameter funnel which collects rain water and conducts it to a pair of small buckets pivoted just below the funnel as shown in Fig. 19.3. The buckets are so designed that

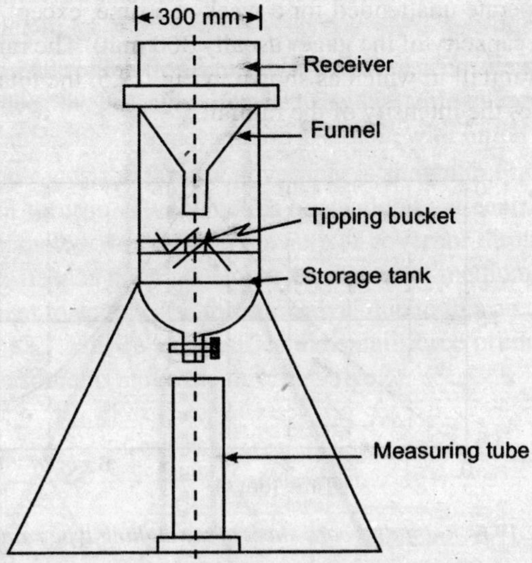

Fig. 19.3 *Tipping bucket rain gage.*

0.25 mm of rainfall collects in one bucket, it tips and empties its water into the storage tank below and at the same time the other bucket is brought under the funnel. The tipping of the bucket actuates an electric circuit which causes a pen to make a mark on a record sheet mounted on a clock driven revolving drum. Since each mark on the record sheet corresponds to 0.25 mm of rainfall, by counting the same the intensity of rainfall may be determined. The total rainfall as determined from the record at the end of day may also checked by measuring the rain water collected into the storage tank in the same manner as in the case of a non-recording gage.

(b) Weighing Type Rain Gages. In this rain gage the rain water passes through a funnel into a bucket which is supported on the weighing platform of a spring or lever balance as shown in Fig. 19.4. The increase in the weight of the bucket due to the addition of the rain water causes the platform to move. The movement of the platform is

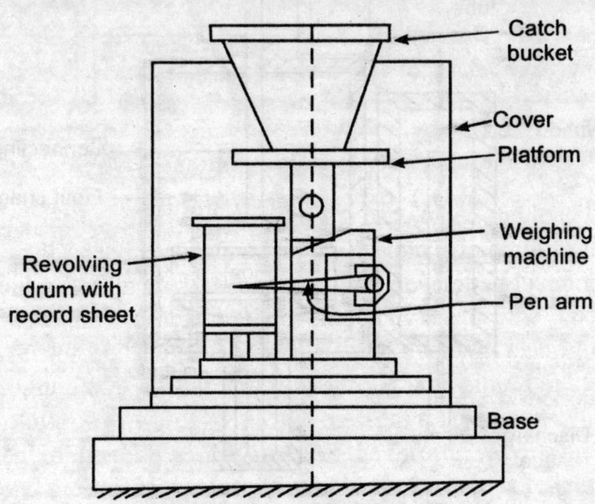

Fig. 19.4 *Weighing type rain gage.*

transmitted through a system of links and levers to a pen which makes trace of the accumulated amount of rainfall on a suitably graduated chart wrapped around a clock driven revolving drum. The mechanism is arranged to reverse the travel of the pen after a certain rainfall (say 150 mm) has accumulated and reverse again after another equal amount, so that the gage may operate unattended for a week at a time, except in regions of very intense rainfall, where the totals may exceed the capacity of the gage (usually 300 mm). The rainfall record produced by this gage is in the form of mass curve of rainfall in which as shown in Fig. 19.5, the total rainfall is plotted with respect to time. The slope of the curve gives the intensity of the rainfall.

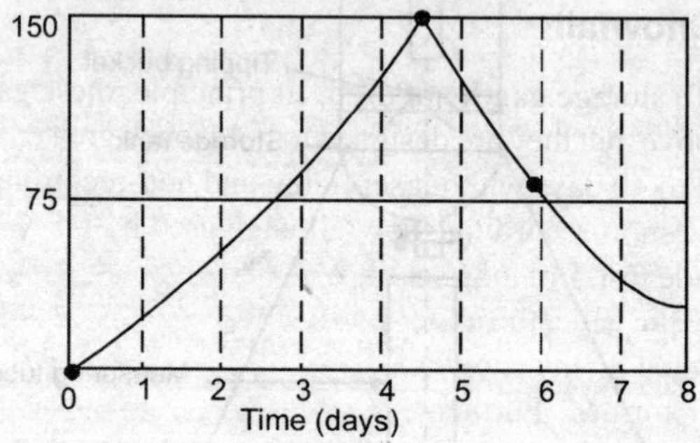

Fig. 19.5 *Rainfall record sheet of a weighing type rain gage.*

(c) Float Type Rain Gage. As shown in Fig. 19.6 in this rain gage the rain water enters a chamber which contains a float. As the level of the rain water collected in the float chamber rises, the flows moves up, which actuates a pen

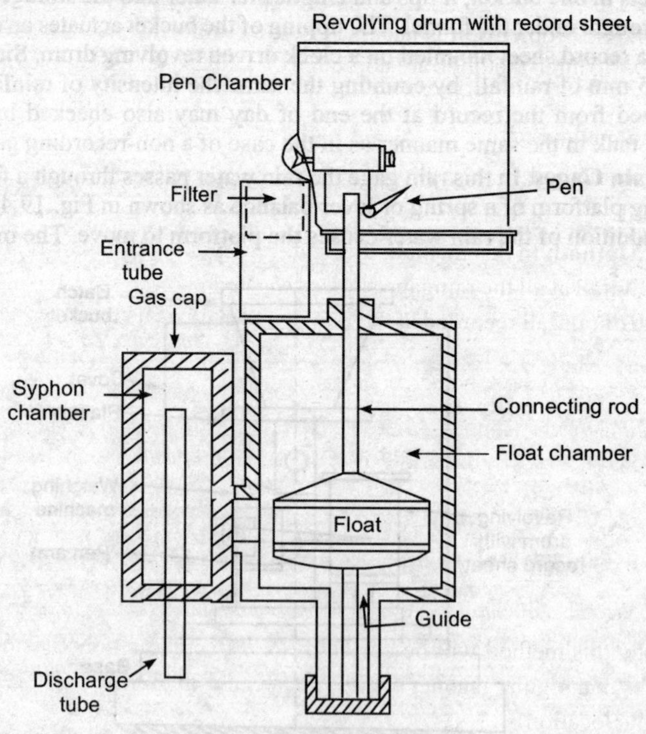

Fig. 19.6 *Float type rain gage.*

connected to it through a connecting rod. The pen makes a trace of accumulated amounts of rainfall on a suitably gradated chart wrapped round a clock driven revolving drum. Thus in the case of this gage also the rainfall record is in the foem of mass curve of rainfall, which is same as in the case of weighing type rain gage. When the float chamber gets completely filled, it is automatically emptied by means of siphon as shown in Fig. 19.6.

In addition to the above noted gages, in the modern times the use of radar as an aid in the measurement of rainfall is being made. However, the main use of radar is in the determination of the real extent, orientation and movement of rain storms.

19.4.2 Measurement of Snowfall

For the measurement of snowfall, storage gages are used. In principle, these gages are same as the non-recording and recording gages described above, but they are designed to hold much greater amounts of precipitation. Recording gages designed to operate for 30 to 40 days without servicing and non-recording gages with sufficient capacity for an entire season are now in use. A non-recording snowfall gage used by US Weather Bureau is made for 300 mm thin-walled pipe. The gage is made in 1.5 m long sections so that any height in multiples of 1.5 m may be obtained. To permit free fall of snow inside the gage, the upper portion of the gage is in the shape of a frustrum of a cone. The equivalent depth of water contained in any snowfall may be found by melting the snow collected in the gage and measuring the quantity of water formed. Further to provide gages with capacity for large amounts of snowfall without making the container excessively large, arrangements are made to melt the snow in the gage itself as it gets collected there. This is generally-usually accomplished by placing in the gage certain chemicals such as calcium chloride, ethelene glycol etc.

19.5 MEAN DEPTH OF RAINFALL OVER AN AREA

For the most of the hydrologic problems, it is necessary to determine the mean or average depth of rainfall over a particular area. The following three methods are commonly used for computing the mean depth of rainfall over an area :

(i) Arithmetic mean method
(ii) Theissen polygon method
(iii) Isohyetal method.

(i) Arithmetic Mean Method. In this method, the average depth of rainfall over an area is obtained by dividing the sum of the depths of rainfall at all the raingage stations located in the area by the number of stations. Thus, if P_1, $P_2, P_3, ..., P_n$ are the depth of rainfall recorded at the n raingage stations distributed over an area then average depth of rainfall P is given by

$$P = \frac{P_1 + P_2 + P_3 + ... + P_n}{n}$$

$$= \frac{\sum_{i=1}^{i=n} P_i}{n} \qquad (19.1)$$

The results obtained by this method will be correct if the raingage stations are uniformly distributed over the area and the rainfall varies in a regular manner. This is so because in this method every rain gage station has equal weightage regardless of its location.

(ii) Theissen Polygon Method. In this method the adjacent rain gage stations are joined by straight lines thus dividing the entire area into a series of triangles as shown in Fig. 19.7. On each of these lines perpendicular bisectors are erected, thereby forming a series of polygons, each containing only one rain gage station. The entire

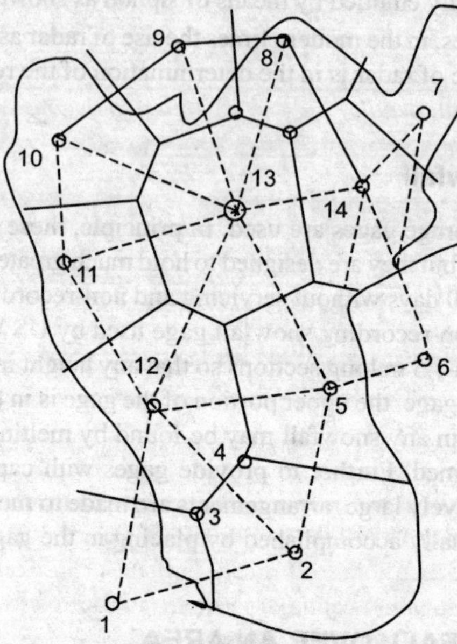

Fig. 19.7 *Theissen polygon method.*

area within any polygon is nearer to the rain gage station contained therein than to any other station, and hence it is assumed that the rainfall recorded at that station is the representative rainfall for than area. If $P_1, P_2, P_3, ..., P_n$ represent the depths of rainfall recorded at the rain gage stations enclosed by polygons, the areas of which are respectively $A_1, A_2, A_3, ..., A_n$, then the average depth of rainfall P for this entire area A is given by

$$P = \frac{P_1A_1 + P_2A_2 + A_3P_3 + ... + P_nA_n}{A}$$

where

$$A = A_1 + A_2 + A_3 + ... + A_n$$

and

$$P = \frac{\sum_{i=1}^{i=n} P_i A_i}{A}.$$

$$\mathbf{P} = \frac{\sum_{i=1}^{i=n} \mathbf{P_i A_i}}{\mathbf{A}} \quad (19.2)$$

In the Theissen Polygon method each rain gage station is given weightage according to its position with respect to the boundary of the area under consideration, and this method is better than the arithmetic mean method.

The area of each polygon enclosed by the boundary of the area is obtained by using planimeter.

The measured gauge readings and the polygon areas enclosing the gage locations are generally tabulated as given in the Table 19.1.

Table 19.1

Station No. (a)	*Rainfall depths* *mm* (b)	*Area of polygon* *sq km* (c)	*Total volume of rainfall in polygon* (d) = b × c
1	P_1	A_1	P_1A_1
2	P_2	A_2	P_2A_2
3	P_3	A_3	P_3A_3
— — —	— — —	— — —	— — —
n	P_n	A_n	P_nA_n

(iii) Isohyetal Method. Isohyetals are the contours of equal rainfall. In the isohyetal method, isohyetals are drawn on the map of the area under consideration. For drawing the isohyets the depths of the rainfall recorded at each of the rain gage stations are noted on the map of the area, at the respective stations. Then assuming a linear variation between the two stations, the probable positions of rainfall values at some interval are interpolated between the stations. The points of equal values of rainfall are connected by smooth curves which forms the isohyetal pattern for the area as shown in Fig. 19.8. The area between the two adjacent isohyets is measured with the help planimeter and for this area the rainfall is presumed to be equal to the mean of the two isohyet values. The average depth of rainfall P for the area A may be found by the following equation.

$$P = \frac{\Sigma(\text{Area between two adjacent isohyets} \times \text{mean of the two isohyet values})}{A}$$

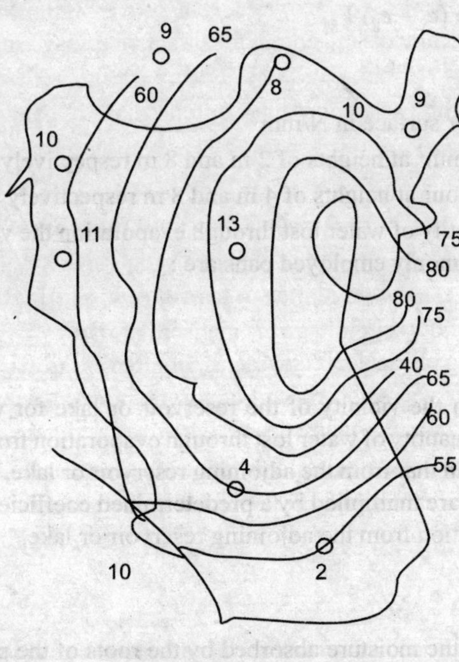

Fig. 19.8 *Isohyetal method.*

This isohyetal method is more accurate method for computing the average depth of rainfall. This is so because the isohyetal method takes into account the actual spatial relationship of the rain gage stations.

It is observed from Fig. 19.8 that :

(a) two isohyets do not cross each other

(b) each isohyet closes on itself

(c) isohyete of higher value indicates the higher rainfall.

This method is generally used for the basin area above 500 sq km.

19.6 EVAPORATION, TRANSPIRATION AND EVAPO-TRANSPIRATION

19.6.1 Evaporation

It is the process by which water is converted into gaseous state and is returned to the atmosphere as vapour. The water is continuously evaporating from the water surfaces of natural and artificial reservoirs, lakes etc., snow and soil surfaces and also from the surfaces of vegetation. As such a part of the precipitation reaching the earth's surface is lost through evaporation. Evaporation depends upon several factors such as temperature, wind velocity, humidity, salinity of water and surface area. The rate of evaporation increases with the increase of temperature as well as with the increase of wind velocity, but it decreases with the increase of humidity as well as with the increase of dissolved salts in water. Further the evaporation increases with the increase in the open surface area of water and soil.

In order to determine the quantity of water lost through evaporation from the water surfaces of natural and artificial reservoirs, lakes etc. Several methods have been developed, but those commonly adopted are (i) the use of empirical formulae, and (ii) the direct measurement with the help of evaporation pans.

A large number of empirical formulae have been developed for determining the quantity of water lost through evaporation but one which is widely used is the Lake Hefner formula which is :

$$E = 6.9317\,(e_s - e_2)\,V_4$$

and

$$E = 5.5046\,(e_s - e_8)\,V_8$$

where

E = evaporation in mm/day

e_s = vapour pressure at water surface in N/mm²

e_2 and e_8 = vapour pressures in N/mm² at heights of 2 m and 8 m respectively above the water surface and

V_4 and V_8 = wind velocities in km/hour at heights of 4 m and 8 m respectively above the water surface.

For the direct measurement of quantity of water lost through evaporation the various types of evaporation pans have been designed, but the most commonly employed pans are :

(a) USWB Class A Land Pan

(b) Colorado Sunken Pan and

(c) Floating Pan.

The evaporation pan is installed in the vicinity of the reservoir or lake for which the loss of water through evaporation is to be determined. The quantity of water lost through evaporation from the pan is measured. Since the evaporation from the pan is greater than that from the adjoining reservoir or lake, the observed values of the water lost through evaporation from the pan are multiplied by a predetermined coefficient of the pan to obtain the actual quantity of water lost through evaporation from the adjoining reservoir or lake.

19.6.2 Transpiration

Transpiration is the process by which the moisture absorbed by the roots of the plants and circulated through the plant body, returns back to the atmosphere as water vapour breathed out by the stromata of the leaves of the plants.

Since transpiration is essentially the evaporation of water from leaf cells, the rate of transpiration is also influenced by all the factors which affect the evaporation. Thus high temperature, low humidity and strong wind all tend to increase transpiration. But in addition to these factors transpiration is also affected by a few other factors such as the available amount of sunshine, available soil moisture, type and extent of plant growth etc.

It is not possible to measure the quantity of water lost through transpiration from a large area under natural conditions, and hence the measurement of the transpiration losses is restricted to studies of small samples under laboratory conditions. One such practical method of measuring the transpiration loss is with the help of instruments called potometers and phytometers. A potometer is a small vessel containing water and sealed so that the only escape of moisture is by transpiration from a small plant with its cut end inserted in the water. A phytometer is a large vessel filled with soil in which one or more plants are rooted. The soil surface is sealed to prevent evaporation so that the only escape of moisture is by transpiration. Thus in each of the above devices the loss in the weight of plant and vessel is a measure of transpiration loss for the specimen tested.

The land surfaces which are partially covered with vegetation, loose moisture simultaneously through evaporation from the land surface and through transpiration from vegetation. As such the common practice is to estimate the combined loss of moisture from the land surface through evaporation as well as through transpiration. This combined loss of moisture from the land surface is known as evapo-transpiration or consumptive use of water. The term **evapo-transpiration** is defined as the total volume of water per unit area used by the vegetal growth of a given area in transpiration and building of plant issue and that evaporated from adjacent soil or intercept precipitation on the area in a specified time. In the above definition the term vegetal growth includes the natural vegetation as well as the crops grown on the area. Evidently the rate of evapo-transpiration is influenced by all the factors which affect both evaporation and transpiration. Further in order to determine evapo-transpiration several methods have been developed.

19.7 INFILTRATION

Infiltration is the process by which water enters the surface of the soil and moves downward to join the ground water. This water first replenishes the soil moisture deficiency and the excess water moves downwards to become a part of the ground water. The rate at which the water actually infiltrates through a soil at a given time is known as the **infiltration rate**. Further the maximum rate at which a soil in any given condition is capable of absorbing water is called its **infiltration capacity**. Initially when the soil is dry and the rainfall occurs some rain water is consumed in wetting the soil particles and hence the infiltrate is high. But as the rain continues, the capacity of the soil to absorb water progressively decreases. Thus if infiltration rate is plotted with respect to time then a falling curve is obtained which is known as **infiltration capacity curve**. Horton found that infiltration capacity curves approximate the form :

$$f_p = f_c + (f_0 - f_c)e^{-kt}$$

where f_p = Infiltration capacity in mm/hr at any time t

f_0 = Initial infiltration capacity in mm/hr

f_c = Final infiltration capacity in mm/hr

t = Time in hours from the beginning of rainfall

k = An empirical constant.

Infiltration capacity defends on several factors such as rainfall characteristics, surface condition of the soil, soil characteristics, moisture content of soil, climate condition, man's activities etc.

For determining infiltration capacity of a soil, the following two methods are adopted :

(i) by infiltrometer and

(ii) by analysis of rainfall and run-off data.

The first method is commonly used and the same is briefly discussed here. There are two types of infltrometers:

(a) Ring and tube type infiltrometers and (b) Rain simulators.

Ring type infiltrometer consists of two shallow concentric rings of sheel metal, usually ranging from 225 mm to 900 mm in diameter. The rings are placed with their lower edges a few centimetres below the ground surface and with the upper portion projecting above the ground surface. Water is then applied in the inner ring as well as in the annular space between the two rings and is always kept at the same level in both. The function of the outer ring is to prevent the water within the inner ring from spreading over a large area after penetrating below the bottom of the ring. The infiltration capacity of the soil is determined directly as the rate at which water must be added to maintain a constant level.

Tube type infiltrometer consists of a single tube about 225 mm in diameter and 450 mm to 600 mm long, which is placed in the ground upto a depth at least equal to that to which the water penetrates during the experiment. Thus

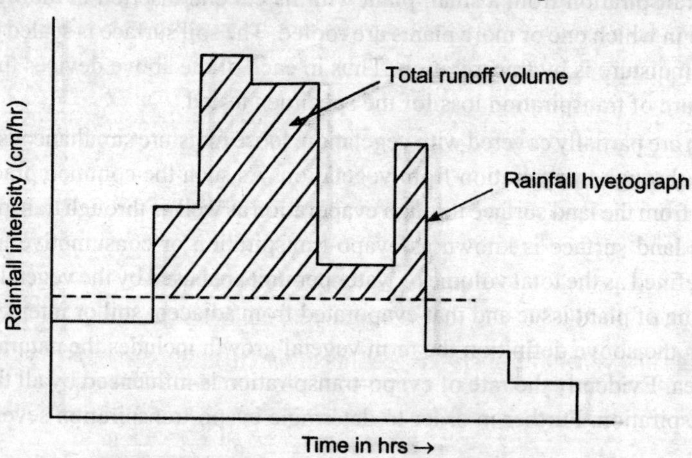

Fig. 19.9 *Diagrammatic representation of ϕ-index.*

in this case no spreading of water can occur. Again the rate at which water must be added to maintain a constant depth within the tube is then observed which directly gives the infiltration capacity of the soil.

In case of rain simulators, on a small area of land ranging from 0.1 m² to 50 m², water is applied by sprinkling at a uniform rate that is in excess of infiltration capacity. The resulting surface runoff is measured and by subtracting it from the rate of application of water, the infiltration capacity of the soil undermined.

The infiltration concept can be applied to the rational computation of surface runoff. For this purpose certain infiltration indices have been devised as indicated below.

ϕ-index. ϕ-index may be defined as the average rainfall intensity above which the volume of rainfall equals the volume of runoff (Fig. 19.9).

Since the difference between the rainfall and the runoff volumes is equal to the sum of the volume of water intercepted, depression storage and that infiltrates into the soil, the ϕ-index represents the combined effect of interception, depression storage and infiltration, all of which are combinedly designed as basin recharge. Further the ϕ-index is based on the assumption that, for a specified storm with given initial conditions, the rate of basin recharge remains constant throughout the storm period.

W-index. W-index is defined as the average rate of infiltration during the time rainfall intensity exceeds the infiltration capacity. The value of W-index may thus be obtained from the following equation :

$$W = \frac{F_i}{t_f} = \frac{1}{t_f}(P - Q_s - S_e)$$

where F_i = Total infiltration

t_f = Time during which rainfall intensity exceeds infiltration capacity

P = Total precipitation

Q_s = Total surface runoff

S_e = Retention by interception and depression storage

Thus W-index is equivalent to ϕ-index minus the retention by interception and depression storage.

W_{min}-index. With very wet conditions, when the infiltration capacity is essentially equal to the final infiltration capacity f_c and the retention rate is a minimum, the values of W and ϕ indices are almost identical. Under these conditions, W-index becomes the W_{min}-index by definition.

19.8 RUNOFF

Runoff is that part of precipitation as well as any other flow contribution, which is transmitted through natural surface channels or streams or rivers. In the general sense runoff includes :

(a) surface runoff or overland flow received in the stream immediately after a heavy rain.

(b) interflow which is a portion of soil moisture that flows laterally through the upper soil layers and joins the stream before joining the ground water.

(c) delayed runoff or ground water flow that enters the stream after passing through deeper portions of the earth and

(d) other delayed runoff that has been temporarily detained as snow cover or stored in natural lakes and swamps.

Thus runoff is the total quantity of water received by a stream from its drainage basin or catchment. The runoff is generally classified as direct runoff and base flow or base runoff. The direct runoff comprises the overland flow and the interflow which are generally grouped together, while the ground water flow that enters a stream is termed as base flow or (base runoff).

The runoff is generally considered in terms of the total flow carried by the stream during a month, season or year and accordingly it is termed as monthly, seasonal or annual runoff. The runoff is expressed either in millimeters or centimeters of water depth over the entired drainage basin or, in hectare-metre or cubic-metre per unit area of the drainage basin. Further the terms stream flow, discharge of a stream and rate of runoff are generally used to mean one and the same thing.

19.8.1 Factors Affecting Runoff

The runoff from the drainage basin or catchment area of a natural stream depends on several factors as noted below:

(a) Characteristics of Precipitation. The pertinent characteristics of precipitation which may considerably affect the runoff are the type of precipitation, its intensity extent and duration. The precipitation may occur either in the form of rain or snow. If the precipitation occurs in the form of rain, it will immediately produce the runoff, but if the precipitation occurs in the form of snow, it will produce the runoff much later, only when the snow melt due to rise in temperature. The runoff increases with the increase in the intensity of rainfall. This is so because an intense rainfall in a short duration results in less evaporation and infiltration losses. On the other hand, a mild rainfall extending over a longer duration would produce less runoff. Further a greater areal extent of the rainfall over the drainage basin and also the rainfall for a longer duration would produce greater runoff.

(b) Characteristics of Drainage Basin. The runoff is considerably affected by the characteristics of the drainage basin such as size, shape, surface, orientation, altitude, topography and geology of the drainage basin.

In general larger the drainage basis greater will be the runoff, but it will not be always true. This is so because for a larger drainage basin it is very seldom that the precipitation of same intensity occurs over the entire basin. Further a larger drainage basin may have larger storage effect resulting in the reduction of the runoff. On the other hand even for a smaller drainage basin with intense rainfall, uniformly spread over the entire basin, greater runoff rate may be produced.

The drainage basin may be either fan-shaped or fern-shaped. A fan-shaped drainage basin produces a greater runoff rate than a fern-shaped drainage basin. This is so because in a fan-shaped drainage basin all the tributaries of a stream area of comparable length and hence the runoff from each tributary reaches almost simultaneously. But with a fern-shaped drainage basin reverse is the case.

The surface of the drainage basin also has substantial effect on the runoff. For example a bare surface gives more runoff as compared with a surface having vegetal covering. Similarly a paved surface also produces a greater runoff.

For the drainage basin located on the windward side of the mountains heavy precipitation occurs and hence greater runoff is produced. On the other hand if the drainage basin is located on the leeward side it receives less precipitation and hence less runoff is produced. Similarly the drainage basins located at higher altitude receive more precipitation and hence yield greater runoff.

The topography of the drainage basin also has significant effect on the runoff. A steeply sloping drainage basin produces greater runoff because a steep slope helps quicker drainage and there being virtually no depression storage, less opportunity for evaporation and infiltration losses is provided. On the contrary a drainage basin with mild slope or with undulating land will yield less runoff.

The rain off is also significantly affected by the geology of the drainage basin. A drainage basin with previous soil will yield less runoff due to larger infiltration losses, while the one with impervious soil or having a hard straum will produce large runoff due to less infiltration losses. Further if a drainage basin comprises unfissured rock it will produce greater runoff, but the runoff will be considerably reduced if the rock is fissured.

(c) Meteorological Characteristics. The runoff is significantly affected by the meteorological characteristics such as temperature, humidity, wind velocity, pressure variation etc. Higher temperature and high wind velocity cause more losses due to evaporation and transpiration and hence reduce the runoff. On the other hand greater humidity decreases evaporation and transpiration losses, resulting in greater runoff. The variation of the pressure in the atmosphere helps the movement of storms causing the precipitation. If the direction of the storm is the same as the direction of flow of water in the drainage basin, then water will remain in the basin for lesser period and hence more runoff will be produced in a short duration, as compared to the case when the storm is moving in the opposite direction.

(d) Storage Characteristics. The storage characteristics of a drainage basin also have significant effect on the runoff. If a drainage basin has a larger natural depressions, pools, lakes etc. which store a part of precipitation and the same will not appear as runoff, then the runoff at the outflow point of the basin will be reduced. Similarly the creation of the artificial reservoirs or tanks will also increase the storage capacity of the drainage basis, resulting in the reduction of the runoff. However, if from these surface storage, the water is released subsequently, it will appear in the form of delayed runoff. Further on drainage basins having previous deposits, large ground water storage may be created, which may also appear in the form of delayed runoff.

In short the factors affecting the runoff are :

1. **Rainfall Pattern.**
   (i) Heavy rainfall;
   (ii) Lean rainfall;
   (iii) Duration of rainfall.
2. **Character of Catchment Area.** *Topography*. The steep and rocky surface gives more runoff.
3. **Size of Catchment Area.** Large area gives more runoff.
4. Vegetation
5. Geology of the area
6. Weather conditions

19.8.2 Measurement of Runoff

The runoff may be determined either by direct measurement of the stream flow or it may be estimated in an indirect manner. The stream flow may be measured either by providing a weir across stream or by velocity area method. Out of these two methods the velocity area method is commonly adopted for measurement of the stream flow. Further the elevation of the water surface in the stream measured above an arbitrary datum at a section of the stream is known as the **stage**. Corresponding to the measured discharge of the stream the stage is also recorded with the help of a gage provided at the section of the stream where the discharge is measured. By plotting the stage against the corresponding discharge a stage-discharge curve may be prepared for a site, which may be used for determining the discharge corresponding to a known stage. During high floods it is often impossible to make the discharge measurement. In such cases after the flood subsides, the stage corresponding to the flood may be obtained by noting the elevation of high water marks on the banks of the stream and corresponding to this value of the stage the maximum or peak value of flood discharge may be determined from the extrapolated stage-discharge curve. Alternatively the maximum flood discharge may also be determined by the slope area method. In this method by noting the high water marks along the banks of the stream the cross-sectional area of the flow section as well as the water surface slope may be computed and using the Manning's formula, with suitably assumed value of roughness coefficient, the corresponding maximum value of the flood discharge may be determined.

The runoff may be indirectly estimated by the following methods :

(a) by empirical formulae relating rainfall and runfall

(b) by infiltration method

(c) by rational method

(d) by unit hydrograph method.

(a) Empirical Formulae Relating Rainfall and Runoff. Several empirical formulae, curves and tables relating the rainfall and runoff have been developed which may be used for the estimation of runoff. Some of the empirical formulae are as indicated below :

(i) Inglis and DeSouza's formula

For ghat area,

$$R = 0.85\,P - 305$$

For plains,

$$R = \frac{(P-178)\,P}{2540}$$

where R = Average annual runoff in mm and P is the average annual rainfall in mm over the entire drainage basin.

(ii) Lacey's formula for Indo-Gangetic Plain

$$R = \frac{P}{1+\dfrac{3048}{P}} \times \left(\frac{F}{S}\right)$$

where R = average runoff in mm

P = average annual rainfall in mm over the entire drainage basin

F = Monsoon duration factor varying between 0.5 to 1.5

S = catchment factor (drainage basin)

= 0.25 for flat lands to 3.45 for hilly areas.

(iii) *A.N. Khosla's formula*

$$R = P - \frac{T}{0.208}$$

where R = average annual runoff in mm
P = average annual rainfall in mm
T = mean annual temperature °C.
} For the entire drainage basin

(iv) *Formulae for some of the drainage basins in India*

Ganga basin $R = 4.9025\ P^{0.64}$

Yamuna basin $R = 0.1112\ P^{1.1}$

Rihand basin $R = P - 1.6150\ P^{0.86}$

Chambal basin $R = 120\ P - 49450$.

In addition to the empirical formulae, several investigators have presented the rainfall and runoff relationships in the form of curves and tables such as Binnie's percentage tables, Strange's curves and tables, Barlow's tables etc. which may also be used to determine the average annual runoff.

Several empirical formulae have also been developed for estimating the maximum or peak value of the flood discharge. In these formulae the maximum flood discharge Q of a river is expressed in terms of one or more of the flood producing factors, the most common being the catchment area A. Some of the important formulae used in India are listed in Table 19.2.

TABLE 19.2 *Empirical formulae for maximum flood discharge of rivers*

S.No.	*Author*	*Formula*	*Remarks*
1.	Dicken	$Q = CA^{3/4}$ C = 11.4 for Northern India = 13.9 to 19.5 for Central India = 22.5 to 25 for Western Ghats. Max. value = 35	Generally for moderate size river basins in Northern and Central India
2.	Ryves	$Q = CA^{2/3}$ C = 6.75 with 24 km of coast = 8.45 between 24 and 161 km. = 10 for limited area near hills	Generally applicable for river basins in South India
3.	Inglis	$Q = \frac{123A}{\sqrt{A+10.4}}$	Derived using data of rivers of Maharashtra.
4.	Ali Nawaz Jung	$Q = C\,[0.386A]^{(0.925-K)}$ C = 48 to 60. Max = 85 $K = \left(\frac{1}{14}\right) \log 0.386\,A$.	For former Hyderabad state. With lower values of C for rivers in South India and with higher values of C for river basins of Northern India

(b) Infiltration Method. By this method the runoff may be estimated either by the use of the infiltration capacity curve or by the use of infiltration indices. In the first method the infiltration capacity curve for a given soil and moisture conditions is subtracted from the curve of rainfall pattern to derive the excess rainfall which represents the runoff. In the second method by subtracting the predetermined values of the infiltration indices from the rainfall intensity the excess rainfall representing the runoff may be obtained.

The infiltration capacity method is rational and would give good results if there is high degree of areal uniformity in rainfall pattern and in infiltration capacity, and if reasonably, accurate infiltration curves are available. On the other hand the infiltration index methods are largely empirical and the derived values of the indices are applicable when the rainfall characteristics and initial soil moisture conditions are identical to those for which these are derived. However, the infiltration index methods are best suited for estimating runoff which would be produced by a rainfall occurring on a drainage basin with its soils being saturated either by a preceding rainfall, or shortly after the starting of the rainfall of higher intensity and longer duration, so that the rate of infiltration might be assumed to have reached a final constant rate prior to or shortly after the starting of the rainfall.

(c) Rational Method. In the rational method the basic equation commonly known as rational formula is used which correlates runoff and rainfall and is as follows :

$$Q = CIA$$

where Q flow in hectare-metre per hour (or cubic metre per hour);

I = Intensity of rainfall in metres per hour

A = Area of drainage basin in hectares (or square metres)

C = Runoff coefficient.

The values of runoff coefficient C depends on the characteristics of the drainage basin such as soil type, vegetation, geological features etc. For different types of drainage basin values of C are as given below :

Types of drainage basin	*Values of C*
1. Rocky and impermeable	0.8–1.0
2. Slightly permeable	0.6–0.8
3. Cultivated or covered with vegetation	0.4–0.6
4. Cultivated absorbent soil	0.3–0.4
5. Sandy soil	0.2–0.3
6. Heavy forest	0.1–0.2

In this method the entire area of the drainage basin lying on the upstream of the point where the runoff is estimated, is divided into a number of sub-areas such that the time taken for the flow to reach the reference point (i) from the extreme dividing line of the nearest sub-area is one hour; (ii) from the extreme dividing line of the next sub-area is two hours and so on. Then knowing the intensity and duration of the rainfall and the values of the runoff coefficient for each sub-area at different times, and using the rational formula the runoff contributed by each sub-area at different times can be obtained. By adding the runoff contributed by each of the sub-areas during every hour, the direct runoff obtained from the entire drainage basin during every hour at the reference point is determined.

The rational method can also be used to estimate the maximum flood discharge for which the following formula is used.

$$Q = CI_cA$$

where

Q = Maximum flood discharge in hectare-metre per hour (or m^3/hr)

A = Area of the drainage basin in hectares (or m^2)

I_c = Critical intensity (design intensity) of rainfall in metres per hour corresponding to time concentration t_c

C = Runoff coefficient.

The time of concentration is defined as the time which would be required for the surface runoff from the remotest part of the drainage basin to reach the point where the runoff is estimated. This time vary, generally depending on the slope and character of the drainage basin.

The critical intensity of rainfall I_c can be obtained from the equation :

$$I_c = \frac{P}{t_R}\left(\frac{t_R + 1}{t_c + 1}\right)$$

where P = Maximum precipitation in metres that occurs during a strom period t_R hours.

However, if the time concentration t_c is not known then

$$I \simeq \frac{P}{t_R}$$

The value of the runoff coefficient C depends on the characteristics of the drainage basin and the same for different types of terrain are as given below :

Types of drainage basin	*Values of C*
1. Flat residential, 30% impervious area	0.4
2. Moderately steep residential, 50% impervious area	0.6
3. Built-up area, 70% impervious	0.8
4. Flat cultivated land, sandy soils	0.2
5. Rolling cultivated land, clay-loamy soil	0.5
6. Hilly area, forests, clay and loamy soil	0.5

(d) Unit Hydrograph Method. The unit hydrograph method was first proposed by L.K. Sherman in 1932. It is one of the widely used methods for the estimation of the flood discharge of a stream. By this method the entire flood hydrograph corresponding to any anticipated rainfall may be developed. A brief outline of the method is given below.

A unit hydrograph may be defined as a hydrograph of direct surface runoff resulting from one centimeter of effective rainfall falling uniformly over the drainage basin or catchment in space as well as in time for a specified or unit duration. The effective rainfall also called rainfall excess in that part of the total rainfall that enters the stream as direct runoff. The specified or unit duration is the period within which the effective rainfall is considered to be uniformly distributed. A unit hydrograph is designated according to the unit duration. For example if a unit hydrograph results from a rainfall of 3 hour unit duration it is known as a 3 hour unit hydrograph, meaning thereby a hydrograph produced by surface runoff from a storm lasting for 3 hours and yielding a rainfall excess of 1 centimetre.

Since the area under the hydrograph represents the total runoff, it is evident from the definition of unit hydrograph that the area under unit hydrograph represents a direct runoff a 1 centimetre.

The theory of unit hydrograph is based on the following assumptions :

(i) The effective rainfall is uniformly distributed within its duration or specified period of time.

(ii) The effective rainfall is uniformly distributed over the entire area of the drainage basin.

(iii) For a given drainage basin, the base period of the hydrographs of direct runoff corresponding to effective rainfalls of different intensities but same unit duration is constant.

(iv) The ordinates of all the direct runoff hydrographs of a common base period are directly proportional to the total amount of direct runoff represented by each hydrograph.

(v) For a given drainage basin, the hydrograph of direct runoff corresponding to a given period of rainfall reflects all the physical characteristics of the basin.

For the application of the unit hydrograph method a unit hydrograph of suitable unit duration is derived from an observed hydrograph of the drainage basin under consideration. Then using the unit hydrograph so derived the flood hydrograph corresponding to any rainfall of the same unit direction may be obtained for the drainage basin on the basis of the above noted assumptions. In order to derive the unit hydrograph from an observed hydrograph the following steps may be adopted.

(i) Select a hydrograph resulting from an isolated, intense, short duration rainfall of nearly uniform distribution over the entire drainage basin as well as within its duration.

(ii) Separate the base flow from the total runoff represented by the hydrograph, in order to obtain the direct runoff. For base flow separation several methods are adopted but all the methods are empirical. One of the methods to separate the base flow is by drawing a line AC tangential to both the limbs of the hydrograph at its lower portion as shown in Fig. 19.10(a). The shaded portion below the line AC represents the base flow. In another method as shown in Fig. 19.10(b), the recession curve existing prior to the occurrence of the rainfall is extended to a point D on a vertical line passing through the peak B of the hydrograph. From the point D a straight line DE is drawn to the hydrograph where E is a point on the hydrograph N days after the peak of the hydrograph. The value of N days is approximately given by

$$N = 0.826\, A^{0.2}$$

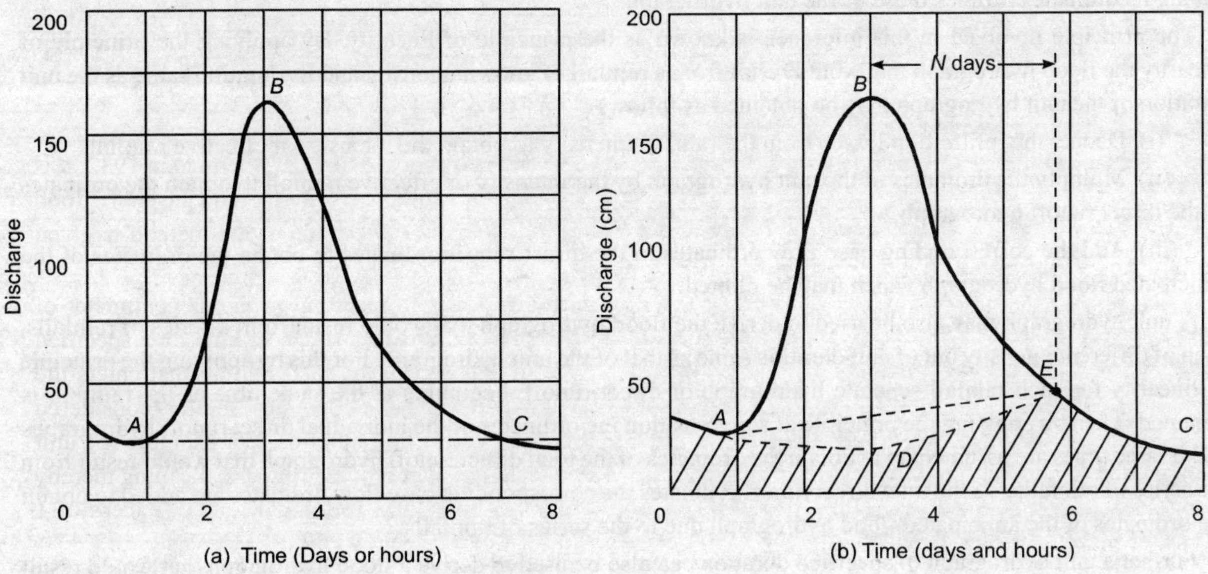

Fig. 19.10

where A is the area of the drainage basin in sq. km. The shaded portion below $ADEC$ represent the base flow. Many a time points A and E are joined by a straight line as shown in Fig. 19.10(b) and the area below this line is taken as the base flow.

(iii) Draw the series of ordinates at some uniform interval for the entire hydrograph.

(iv) From these ordinates of the hydrograph by deducting the corresponding ordinates of base flow obtain the ordinates of direct runoff.

(v) Find the volume of direct runoff which is equal to ($\Sigma 0 \times t \times 3600$) m^3, where $\Sigma 0$ is the sum of the direct runoff ordinates in cusec and t is time interval in hours between the successive ordinates.

(vi) Divide the volume of the direct runoff by the area of the drainage basin to obtain the equivalent depth of direct runoff which is given by

$$\text{Direct runoff depth} = \frac{[\Sigma 0 \times t \times 3600] \times 100}{A \times 10^6} \text{ cm.}$$

$$= \frac{0.36 \times \Sigma 0 \times t}{A} \text{ cm.}$$

where A is the area of the drainage basin in square kilometres.

(vii) By dividing each of the ordinates of direct runoff by the depth of direct runoff obtain the ordinates of the unit hydrograph. By plotting these ordinates at their respective times and drawing a smooth curve passing through the unit hydrograph so derived is same as the duration of these, unit hydrogram is obtained. The unit duration of the unit hydrograph so derived is same as the duration of effective rainfall corresponding to which the resulting hydrograph has been used for deriving the unit hydrograph. The use of unit hydrograph for deriving a flood hydrograph is indicated below.

From the above noted assumptions of the unit hydrograph theory it may be inferred that an effective rainfall of intensity R cm occurring uniformly over the entire drainage basin in a unit duration of T hours unit duration but having its ordinates R times those of the unit hydrograph.

The principle involved in this inference is known as the principle of linearity. By applying the principle of linearity the flood hydrograph that would result from a rainfall of known intensity and its duration same as the unit duration of the unit hydrograph may be obtained as follows :

(i) Deduct the infiltration losses from the rainfall intensity to obtain the intensity of effective rainfall.

(ii) Multiply the ordinates of the unit hydrograph by the intensity of effective rainfall to obtain the ordinates of the direct runoff hydrograph.

(iii) Add the corresponding base flow ordinates to the direct runoff ordinates to obtain the ordinates of the anticipated flood hydrograph which may be plotted.

A unit hydrograph may also be used to derive the flood hydrograph that would result from a series of rainfalls, each of different intensity but of unit duration same as that of the unit hydrograph. For this by applying the principle of linearity for each rainfall separate hydrograph of direct runoff, beginning at the same time as the rainfall, is obtained. Then by applying the principle of superposition the ordinates of the individual direct runoff hydrographs at the same times are assumed up to obtain the ordinates of the total direct runoff hydrograph that would result from the series of rainfalls. To the total direct runoff ordinates the corresponding base flow ordinates are added to obtain the ordinates of the anticipated flood hydrograph due to the series of rainfalls.

Further a unit hydrograph of specified duration can also be used to derive a flood hydrograph that would result from a rainfall of uniform intensity but duration longer than specified duration of the unit hydrograph. This may be accomplished by dividing the total rainfall into a series of rainfalls, each of duration equal to the specified duration of the unit hydrograph. Then adopting the same method as indicated above the anticipated flood hydrograph for this case may also be obtained.

19.9 HYDROGRAPH

A hydrograph is a graphical plot or representation of discharge of a natural stream or river versus time. Discharge is usually expressed in cu. m (m^3/sec) per second, or hectare-metre per day (or per month) and time expressed in hours, days or months. Discharge is plotted on Y-axis and the corresponding time is plotted on X-axis. Figure 19.11 shows a hydrograph of a typical stream resulting from rainfall of intensity as indicated by a hydrograph plotted at the top left corner. At the beginning there is only a base flow i.e., the ground water contribution to the stream as indicated at *A* in Fig. 19.11. After the commencement of rain which lasts for some time when the initial losses like interception and infiltration are met, the surface flow begins and there is progressive increase in the stream flow.

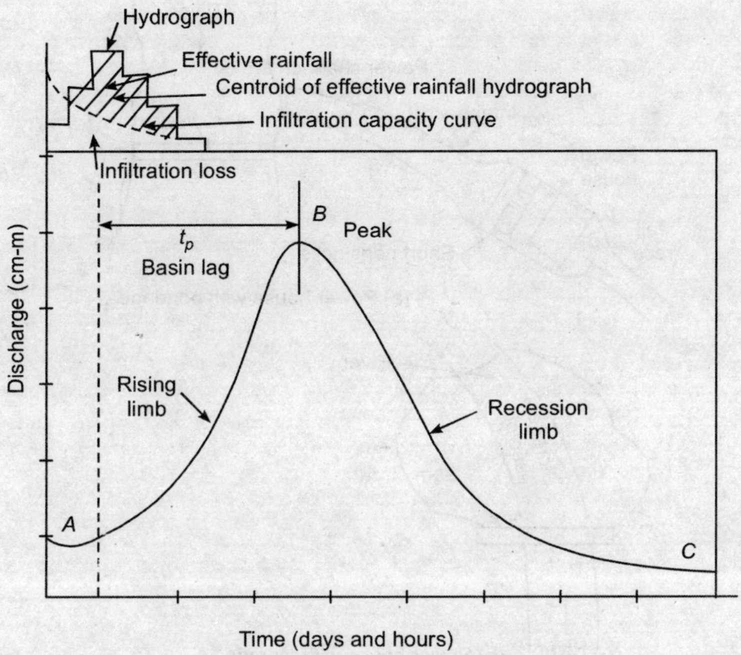

Fig. 19.11 *Typical hydrograph.*

The hydrograph therefore gradually rises and reaches a peak value at *B* at time t_p, called time lag or basin lag or time to peak, measured from the centroid of the hydrograph of effective rainfall (defined as that portion of that rainfall which enters the stream as direct runoff). Therefore the flow in the stream recedes and hence the hydrograph declines. Thus a hydrograph consists of the rising limb *AB* and recession limb *BC*. Discharge at *B* corresponds to the maximum or peak flood discharge and the area under the curve *ABC* would give the volume of the total runoff during the flood.

A single peaked hydrograph as shown in Fig. 19.11 is developed when there is a single storm causing precipitation of more or less intensity. However, in actual practice the hydrograph may not be a smooth single peaked curve as shown in Fig. 19.11, but it may be a multiple peaked curve known as complex hydrograph, due to several factors such as the occurrence of precipitation of different intensities with some interval, the characteristics the drainage basin etc.

19.10 STORAGE AND PONDAGE

As stated earlier the flow rate of a stream varies considerably with time. For instance, during rainy season when the stream is in floods it carries a huge quantity of water as compared to other times of the year when the quantity of water carried by it is considerably less. However, the demands of the power ordinarily do not correspond to such variations of the natural flow of the stream. As such some arrangement in the form of storage and pondage of water is required to be made to regulate the flow of water so that it may be made available to requisite quantity at time when the generation of power is required.

Storage may be defined as impounding of a considerable amount of excess runoff during seasons of surplus flow for use during dry seasons. It is achieved by constructing a dam across a stream at suitable site, thus forming a storage reservoir on the upstream side of the dam.

Pondage may be defined as a regulating body of water in the form of a relatively small pond or reservoir provided just at the hydropower plant. The pondage is necessary to regulate the flow to suit the variation in daily or

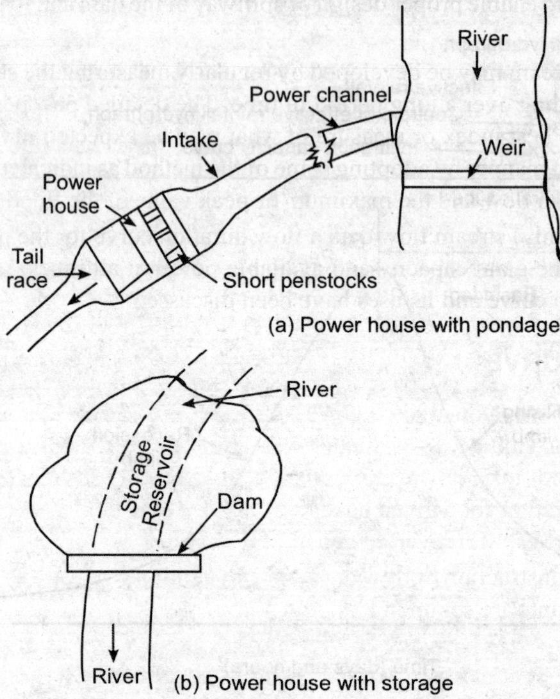

Fig. 19.12 *Hydropower houses with pondage and storage.*

weakly power (or load) demand. It is required to be provided either when the hydropower station is located at some distance away from the storage reservoir, or when the storage reservoir is not provided. When the hydropower station is located at some distance away from the storage reservoir and water is supplied to the hydropower station through a power channel or a natural stream, then the regulation of outflow of water from the storage reservoir, to take care of the variation in daily or weekly power demand, may not be possible and a regulating body or pondage is needed directly at the hydropower station. Similarly when the storage reservoir is not provided and the hydropower station is located directly on the natural stream, then the pondage is necessary to regulate the natural flow of the stream to take care of the variation in daily or weekly power demand.

Thus in general the term **storage** is used to indicate the building up of the low natural flow of the stream to a uniform discharge and the term **pondage** is used to indicate the regulation of the resulting uniform flow, or the natural flow, if there is no storage, to suit the variation in daily or weekly load demand. If the hydropower station is close to the storage reservoir then the storage reservoir itself will also provide the necessary pondage. Figure 19.12 shows the location of power houses with pondage and storage.

19.11 ESSENTIAL STREAM FLOW DATA FOR WATER POWER STUDIES

The essential stream flow data required for the assessment of the hydropower that would be available from any stream are as follows :

(i) Daily (weekly or monthly) flow over a period of years (the longer is the data better it is) to determine the plant capacity and estimated output, both of which are a function of the average flow of the stream as well as its distribution during the year.

(ii) Minimum or low water flow to determine the amount of primary or firm or dependable power and also the size of thermal or stream auxiliary plant if used.

(iii) Maximum or flood flow to enable proper design of spillway of the dam and to provide for the safety of the plant.

The above noted data for any stream may be developed by regularly measuring the stream flow and maintaining the records of its daily flow extending over a long period of time. The detailed past performance of a stream thus determined by measurement is the best index or measure of what may be expected in the future. The stream flow may also be estimated in an indirect manner by adopting some of the method as indicated earlier. However, by these methods only average annual stream flow and the maximum or peak value of the flood discharge can be obtained.

From the daily (weekly or monthly) stream flow data a flow duration curve for the power site may be prepared which may be used to determine the plant capacity and available power at all times. In the following section the method of drawing a flow duration curve and its uses have been discussed.

19.12 FLOW DURATION CURVE

The **flow duration curve** as ordinarily constructed consists of a plot of values of stream flow (daily, weekly or monthly) in increasing order of magnitude, as ordinates and the percent of time under consideration the flow equalled or exceeded, as abscissas. It may, however, be noted that instead of discharge, any function of discharge such as power in kW per metre of fall or for a given head, may also be used as ordinates. The curves thus obtained are termed as **power duration curves**. Moreover, in general these curves are also termed as **duration curves**.

Computations involved in the construction of flow duration curve of Fig 19.13 are shown in Table 19.3. Rates of flow, differing by convenient amounts are set up in Col. 1 in ascending order. In Col. 2 is indicated the number of

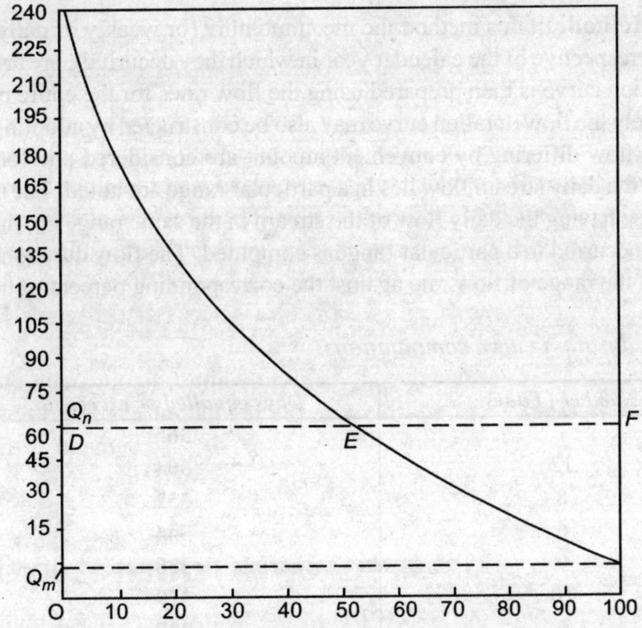

Fig. 19.13 *Daily flow duration curve.*

times the daily flow of the stream equalled a value between the corresponding flow in Col. 1 and the flow next below. For example, during the year, a daily flow between 28 and 30 cusec occurred 13 times. A summation of occurrences is tabulated in Col. 3. It shows that a flow of 28 cusec was equalled or exceeded on 303 days during the year. Col. 3 is then reduced to percent of time and recorded in Col. 4. Columns 1 and 4 are used to plot the flow duration curve. For a flow duration curve of 1 year period, Col. 2 will total 365 days and for a longer period than a year it will be more than 365 days.

A flow duration curve may also be constructed in the same manner with the help of mean weekly or monthly flow rates instead of daily flow rates. But the flow duration curve computed from mean monthly flows are only approximate, because in each month there is a considerable variation in flow not indicated in the duration curve. The error involved in the use of mean monthly flows for the construction of the duration curves usually ranges from 5 to 15% depending upon the characteristics of the stream and the extent of utilization of the flow. The difference between the daily and monthly flow duration curves would be negligible for a stream with a steady flow, but much greater for very flashy streams.

The flow duration curve may either be constructed only for that year which is having the lowest flow rate out of the entire stream flow record, or it may be constructed by considering the entire stream flow record. For the later case two methods are in general use for constructing the flow duration table for the plotting of the curve, which are as indicated below :

(i) Calender Year Basis Method. In this method the mean monthly (or weekly or daily) flow rates of each year of the record are separately arranged in increasing order of magnitude and tabulated as first, second, third etc. month (or week or day). There will thus be as many vertical columns in the table as the number of years of record considered. Then the mean value of the flow rates noted in each of the horizontal columns of the table is computed. Thus the mean of the first horizontal column of the table will give the flow rate for the driest month (or week or day) of the average year, the mean of the second horizontal column that for the next driest month (or week or day) etc. In this manner monthly (or weekly or daily) flow rates for the average year are obtained and the same are used to plot the flow duration curve as indicated earlier.

(ii) Total Period Basis Method. In this method the mean monthly (or weekly or daily) flow rates for the entire period under consideration (irrespective of the calendar year in which they occurred), are arranged in increasing order of magnitude. The flow duration curve is then prepared using the flow rates for the entire period in the same manner as indicated above. Alternatively the flow duration curve may also be constructed by adopting the following procedure. Continuous ranges of stream flow differing by convenient amount are considered and for each year separately the number of days during which the daily stream flow lies in a particular range are noted. The number of days, for all the years are considered, and those having the daily flow of the stream in the same range are then added, and the percent of time the daily stream flow occurred in a particular range is computed. The flow duration curve is then constructed by plotting the lower value of the range of flow rate against the corresponding percent of time.

TABLE 19.3 *Daily flow duration curve computations.*

Flow rate m^3/s	*Number of days*	*Days equalled or exceeded*	*Percent of time*
10	1	365	100
12	12	364	99.7
14	8	352	96.4
16	6	344	94.2
18	9	338	92.6
20	10	329	90.1
22	6	319	87.4
24	7	313	85.7
26	3	306	83.8
28	9	303	83.0

Contd.

Contd. Table 19.3

Flow rate m³/s	*Number of days*	*Days equalled or exceeded*	*Percent of time*
30	13	294	80.5
35	6	281	77.0
40	14	275	75.3
45	21	261	71.5
50	14	240	63.7
55	10	226	61.9
60	10	216	59.2
65	5	206	56.4
70	5	201	55.0
75	18	196	53.7
80	11	178	48.7
85	17	167	45.7
90	13	150	41.1
95	13	137	37.5
100	10	124	34.0
105	2	114	31.2
110	8	112	30.7
115	13	104	28.5
120	13	91	25.0
130	14	78	21.4
140	4	64	17.5
150	5	60	16.5
160	0	55	15.1
170	6	55	15.1
180	10	49	13.4
190	10	39	10.7
200	8	29	7.9
210	7	21	5.7
220	5	14	3.8
230	9/365	09	2.4

Out of these two methods the total period basis method gives a relatively true representation of the flow duration curve. This is so because in this method the actual flow rates of stream appear at appropriate places in the curve, but in the calendar year basis method the stream flow rates might be averaged up with the values for the other years.

The area under the flow duration curve (Fig. 19.13) would evidently represent the average yield from the stream. As stated ealier for a hydropower project the flow duration curve can be easily converted into power duration curve by changing the ordinate, in which case the area under the curve would represent the average yield of power from the project. It can be noted that without storage, Q_m is the maximum flow rate that would be available for all the times (100% of time), and the area *OABC* would represent the firm yield of water or power. However, if a flow rate of Q_n is required to be provided which is indicated by the flow demand line *DEF*, then it is evident that it would be possible to meet this uniform demand of flow rate (or power) for all the times only if storage equal to area *BEF* is provided. However, an alternative to storage, for power projects, is to instal a thermal power unit of maximum capacity equal to *BF* (on power duration curve ordinate in kW), to work as supplementary to hydropower unit. The curve also shows that natural flow sufficient to meet the flow demand Q_n is available for 53.5% of time or 195 days in this year of the lowest flow of the record.

19.13 MASS CURVE

In order to facilitate the storage computations mass curves are commonly used. **A mass curve is a part of accumulated flow (in hectare-metre or any convenient unit) against time**. It is usually made from the records of mean monthly flows of a stream. The mean monthly flows are summed up consecutively and each sum is plotted above the corresponding time upto which the summation has been done. Figure 19.14 shows the mass curve for a typical stream. Obviously a mass curve must continuously rise as it shows accumulated flows. The slope of the curve at any point indicates the rate at that particular time. If the curve is horizontal, the flow is zero, and if there is high rate of flow the curve rises steeply. Thus relatively dry periods are indicated as concave depressions on the mass curves.

As shown in Fig. 19.14, the slope of the straight line *AB* joining the end points of the mass curve represents the average discharge over the period for which the mass curve has been plotted. Further if two straight lines *A'B'* and *A"B"* are drawn parallel to *AB* and tangent the mass curve at the lowest tangent point *C* and the highest tangent point *D* respectively, then the vertical intercept between these two straight lines represents the storage volume required to permit continuous release of water at this average discharge rate for the entire period. If the reservoir having this capacity is assumed to contain a volume of water equal to *AA'* at the beginning of the period, then the reservoir would be full at *D* and it would be empty at *C*. However, if the reservoir was empty in the very beginning, then it would be empty again at the point *E* and also during the period from *F* to *B*. On the other hand, if the

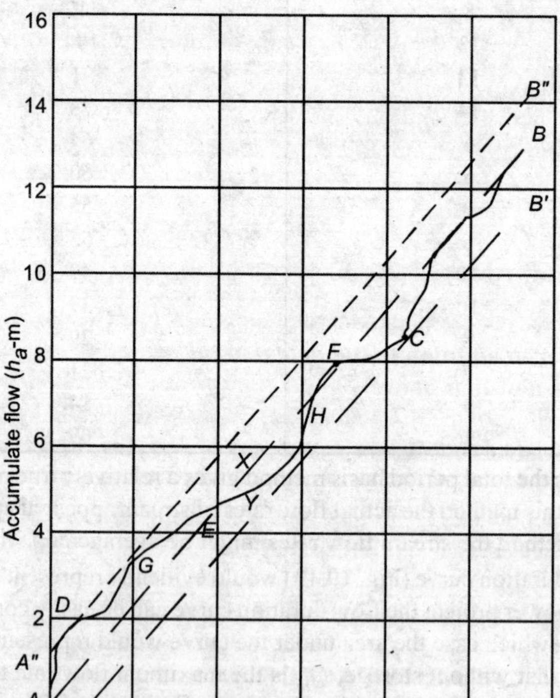

Fig. 19.14 *Mass curve.*

reservoir was full in the very beginning, if would be full at points *F* and *B*, and between points *F* and *B* and between points *A* and *E* there will be spill of water from the reservoir.

In order to obtain the storage required for a given rate of depand, a demand line corresponding to the given rate of demand is obtained. The lines such as *GH* shown in Fig. 19.14, are drawn parallel to the demand line and tangential to the high points of the mass curve. The maximum vertical intercept, such as *X–Y*, between such lines and the mass curve gives the storage required to satisfy the given rate of demand. This is so because, assuming the reservoir to be full at *G*, there is a total flow in the stream represented by the vertical distance between *G* and *Y* and there is a total demand represented by the vertical distance between *G* and *X*, leaving a gap of volume represented by *XY* which must be met from the storage.

The tangential line drawn parallel to the demand line when extended forward must intersect the mass curve, such as at *H*, so that the reservoir which was full at *G* will be filled again at *H*. However, if the line does not intersect the mass curve, the reservoir will not be filled again. Further if the reservoir is very large the time interval between the points *G* and *H* may be several years.

The rate of demand may not be always constant, in which case the demand line becomes a curve with its slope varying from point in accordance with the variable rate of demand at different times. Such a demand line is actually a mass curve of demand. In this case also the storage requirements can be obtained in the same way superimposing the mass curve of demand on the mass curve of steam flow from the beginning of the dry period till the two meet again. The maximum vertical intercept between the two curves gives the requirement of storage. It is however essential that the demand line for variable demand coincide chronologically with the mass curve of stream flow i.e., June demand must coincide with the June inflow and so on.

In both the above noted cases the requirement of the storage so obtained would be the net storage which must be available for utilization and it must be increased by the amount of water lost by evaporation and seepage.

Mass curve may also be used to determine the maximum possible uniform rate of flow which may be obtained from the given storage capacity. For this sloping tangential lines are drawn at the high points of the mass curve in such a manner that the maximum intercept between these lines and mass curve is equal to the given storage capacity. The slope of these lines would give the uniform rate of flow which may be obtained from the given storage capacity. Further the slope of the flattest line would represent the firm rate of flow which may be obtained for the entire period from the given storage capacity.

19.14 PRIMARY (FIRM) AND SECONDARY POWER

The **minimum power** which can be generated throughout the year from a hydroelectric plant is called **the firm or primary power**. The **power in addition to the firm power** which would be generated for only a part of the year is known as **secondary** or **non-firm power**. For a run-of-river plant without any storage the firm power will correspond to the minimum flow of the river which would be available throughout the year. However, by providing the storage the firm power can be considerably increased.

A flow duration curve can be used to determine the firm and secondary powers that would be generated from a stream. Thus shown in Fig. 19.13 the flow rate Q_m would correspond to the firm power that would be available from the river without any storage being provided. But if the provision for the storage equal to area *BEF* is made then the firm power for the same river may correspond to a higher flow rate Q_n represented by line *DEF*. However, in the absence of any storage, area *BCDE* would correspond only to the secondary power that would be available from the river.

19.15 LOAD FACTOR, CAPACITY FACTOR, UTILISATION FACTOR

Load factor is defined as the ratio of the average load during the certain period to the maximum or peak load during that period. Load factor may be **daily load factor** or **weekly load factor** or **monthly load factor**.

If the load factor of a power plant is low, a large proportion of the generating capacity remains idle for most of the time and the cost of generation is high. Conversely, if the load factor of a power plant is high the generating capacity is better utilized and the cost of generation is relatively less.

Capacity factor or **plant factor** may be defined as the ratio of the energy that the plant actually produces during any period to the energy that it might have produced if operated at full capacity throughout this period. Capacity factor and load factor will be identical when the maximum or peak load just equals the plant capacity. For a hydroelectric plant, capacity factor commonly varies from about 0.25 to 0.7 or more, depending on the load factor, plant capacity, available pondage and storage etc.

Utilization factor or **plant use factor** is defined as the ratio of the peak load developed during a certain period to the installed capacity of the plant. For a hydroelectric plant, utilization factor commonly varies from 0.4 to 0.9, depending on the plant capacity, load factor, available pondage and storage etc.

SOLVED EXAMPLES

Example 19.1 *Two turbogenerators each of capacity 25000 kW have been installed at a hydel power station. During a certain period the load on the hydel plant varies from 15000 kW to 40000 kW. Calculate (i) the total installed capacity (ii) the load factor (iii) plant factor and (iv) utilization factor.*

Solution

(i) *Total installed capacity* :

Two turbogenerators each of capacity of 25000 kW.

$$\therefore \quad \text{Installed capacity} = 2 \times 2500$$

$$= \mathbf{50{,}000\ kW}$$

(ii) *Load factor* :

$$\text{Load factor} = \frac{\text{Average load}}{\text{Peak load}}$$

where

$$\text{Average load} = \frac{15000+40{,}000}{2}$$

$$= \frac{55000}{2}$$

$$= 27500 \text{ kW}$$

$$\text{Peak load} = 40000 \text{ kW}$$

$$\therefore \quad \text{Load factor} = \frac{27500}{40000}$$

$$= 0.6875$$

$$= \mathbf{68.75\%}$$

(iii) *Plant factor* :

$$\text{Plant factor} = \frac{\text{Energy actually produced}}{\text{Maximum energy which can be produced}}$$

$$= \frac{\text{Average load} \times \text{Time period}}{\text{Capacity} \times \text{Time period}}$$

$$= \frac{\frac{15000+4000}{2} \times T}{50000 \times T}$$

$$= \frac{27500}{50000}.$$

$$= 0.55$$

$$= \mathbf{55\%}$$

(iv) Utilization factor (Capacity factor) :

$$\text{Utilization factor} = \frac{\text{Maximum power developed}}{\text{Total power that can be developed}}$$

$$= \frac{40000 \times T}{50000 \times T}$$

$$= 0.8$$

$$= \mathbf{80\%}$$

Example 19.2 *A run-of-river hydroelectric power station is proposed across a river at a site where the net head of 25 m is available on the turbine. The river carries a sustained minimum flow of 30 m³/sec in dry weather and behind the power station sufficient pondage is provided to supply daily peak load of demand with a load factor of 71%. Assuming the plant efficiency of 56.4%, determine the maximum generating capacity of the generator to be installed at the power house. If the daily load pattern indicates 21½ hours average load and 2½ hours of peak load, determine the volume of pondage to be provided to supply the daily demand.*

Solution

The average power developed corresponding to the dry weather is given by

$$P = 9.81Q(H - h_f)\eta$$

$$= 9.81 \times 30 \times 25 \times 0.564$$

$$= \mathbf{4150\ kW}$$

$$\text{Load factor} = \frac{\text{Average load}}{\text{Peak load}} = 0.71.$$

$$\therefore \quad \text{Peak load} = \frac{4150}{0.71}$$

$$= \mathbf{5845\ kW}$$

∴ The maximum capacity of the generators to be installed

$$= \mathbf{5845\ kW}$$

Excess power to be developed during 2½ hours

$$= 5845 - 4150$$

$$= \mathbf{1695\ kW}$$

The discharge required for developing this excess power during the peak hours is

$$Q = \frac{1695}{9.81 \times 25 \times 0.564}$$

$$= \mathbf{12.25\ m^3/s}$$

$$\therefore \quad \text{Pondage} = 12.25 \times 2.5 \times 3600$$

$$= 11 \times 10^4\ m^3$$

$$= \mathbf{11\ hectare\ m}$$

Example 19.3 *The following data gives the average monthly flow in m³/sec in a dry year and it is intended to use it for designing a hydropower plant :*

July – 17.0; Aug – 18.4; Sept – 17.6; Oct – 13.6
Nov – 13.3; Dec – 11.3; Jan – 9.9; Feb – 9.6; Mar – 8.5
Apr – 11.8; May – 12.7; June – 14.2.

Assuming the net head on the turbine as 91.5 m and efficiency 80% draw a power duration curve. It is intended to develop electric power at a firm rate of 9000 kW either by providing a storage or by providing standby diesel unit and no storage. In the two cases, determine the minimum capacities of the reservoir or the diesel unit.

Solution

First we calculate the power developed in each month and then draw the power or load curve and from that power duration curve as shown in following tables. $P = 9.81Q(H - h_f)\eta$ kW.

Table *Power curve*

July – 12208 kW; Aug – 13213 kW; Sept – 12638 kW; Oct – 9766 kW
Nov – 9551 kW; Dec – 9120 kW; Jan – 7109 kW; Feb – 6894 kW
Mar – 6104 kW; Apr – 8473 kW; May – 12638 kW; June – 10177 kW.

Table *Power/flow duration curve*

Power	6104	6894	7109	8114	8473	9120	9551	9766	10197	12208	12638	13213
% time	100	91.7	83.3	75	66.7	58.3	50.0	41.7	33.3	25	16.7	8.3

As seen from Power-Duration curve, in order to obtain a firm power at the rate of 9000 kW, the power corresponding to area *ABC* will be required to be developed from the storage. Further it is seen that for 60% the time the power developed from the natural flow in the stream in an excess of the required power. But for the 40% of the time the natural flow will be required to be supplemented from the storage to develop the required power.

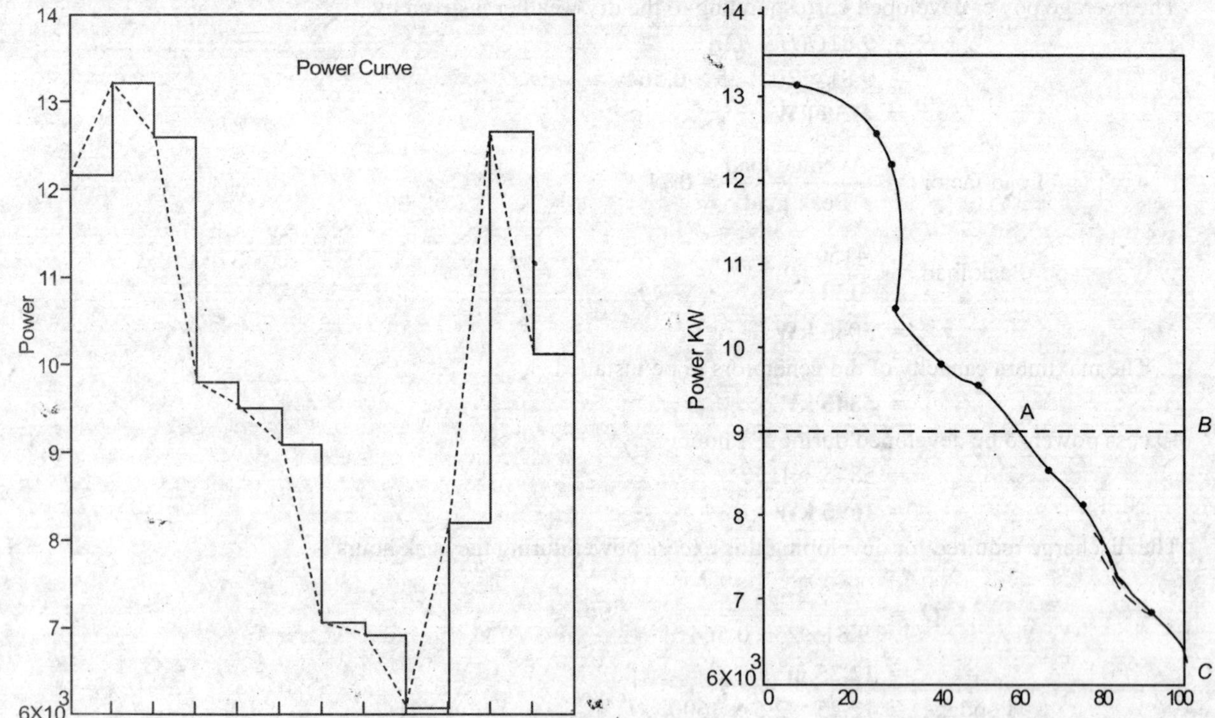

(a) *Power (load) curve* (b) *Load duration curve*

Fig. Ex. 19.3

The area of the portion ABC is obtained by counting the squares as 565 mm^2 which is equivalent to

$$= \frac{565 \times 1000 \times 10 \times 12}{10 \times 10 \times 100} \text{ kW month}$$

$$= \mathbf{6780 \text{ kW month}}$$

$$\therefore \text{ The storage required } = \frac{6780 \times 30 \times 24 \times 3600}{9.81 \times 91.5 \times 0.8}$$

$$= 2.45 \times 10^7 \text{ m}^3$$

$$= 2450 \times 10^4$$

$$= \mathbf{2450 \text{ hectare metre}}$$

With no pondage, the firm power available

$$= 6104 \text{ kW}$$

∴ Capacity of the diesel unit required

$$= 9000 - 6104$$

$$= \mathbf{2896 \text{ kW}}$$

$$\simeq \mathbf{2900 \text{ kW}}$$

Example 19.4 *A run-of-river plant is installed on a river having a minimum flow of 15 m³/s. If the plant is used as a peak load plant operating only for 6 hours a day, determine the firm capacity of the plant (a) without pondage (b) with pondage but allowing 10% of the water to be lost in evaporation and other losses. Head at the plant is 10 m and the plant efficiency may be assumed to be 85%.*

Solution

(a) *Without pondage* :

$$= 9.81\, Q(H - h_f)\eta$$

The firm capacity = 9.81 × 15 × 10 × 0.85

$$= \mathbf{1251 \text{ kW}}$$

(b) *With pondage* :

The total volume of water stored during the 18 hours (24 – 6) when the plant is not operating

$$= 15 \times 18 \times 3600 \text{ m}^3$$

$$= 9.72 \times 10^5 \text{ m}^3.$$

The loss of water due to evaporation etc.

$$= \frac{10}{100} \times 9.72 \times 10^5$$

$$= 0.972 \times 10^5 \text{ m}^3$$

∴ Net amount of water available from the pondage for the generation of power during 6 hours when the plant is operating

$$= (9.72 - 0.972) \times 10^5$$

$$= 8.748 \times 10^5 \text{ m}^3$$

The corresponding rate of flow available from the pondage

$$= \frac{8.748 \times 10^5}{6 \times 3600}$$

$$= \mathbf{40.5 \text{ m}^3/\text{sec}}$$

∴ Total flow available for the power generation

$$= 40.5 + 15$$

$$= \mathbf{55.5 \text{ m}^3/\text{s}}$$

∴ Firm capacity of the plant

$$= 9.81\, Q(H - h_f)\eta$$

$$= 9.81 \times 55.5 \times 10 \times 0.85$$

$$= \mathbf{4638 \text{ kW}}$$

Example 19.5 *A run-of-river hydel power plant with a installed capacity of 15000 kW operates at 20% load factor when it serves as a peak load station. What should be the minimum discharge in the stream so that it may serve as a base load station? The plant efficiency may be taken as 80% when working under a head of 15 m. Also calculate the maximum load factor of the plant when the discharge in the stream is 30 m³/s.*

Solution

When the plant operates as a peak load station with a load factor of 20%, the average power developed is given by

$$\text{Load factor} = 0.2 = \frac{\text{Average load}}{\text{Max. load}}$$

$$\therefore \quad \text{Average load} = 0.2 \times 15000$$

$$= \mathbf{3000\ kW}$$

If Q is the minimum flow necessary for the plant to run as a base load plant, the power developed is given by

$$P = 9.81\, Q(H - h_f)\eta$$

$$= 9.81\, Q \times 15 \times 0.8$$

$$= \mathbf{117.72\ Q\ kW}$$

$$\therefore \quad 3000 = 117.72\, Q$$

$$Q = \frac{3000}{117.72}$$

$$= \mathbf{25.48\ m^3/s}$$

If the stream flow is 30 m³/s, then the power developed is

$$P = 9.81 \times 30 \times 15 \times 0.8$$

$$= \mathbf{3532\ kW}$$

$$\therefore \text{ Maximum load factor} = \frac{3532}{15000}$$

$$= 0.2355$$

$$= \mathbf{23.55\%}$$

Example 19.6 *What is the volume of rainfall in day sec-metre if 6.2 cm occurs over an area of 1000 sq. km.? Also find out in terms of km²-m.*

Solution

Total rainfall = Area × Height

$$= 1000 \times 10^6 \times \frac{6.2}{100}$$

$$= \mathbf{6.2 \times 10^7\ cu.\ m.\ (m^3)}$$

$$1 \text{ Day} = 24 \times 3600 \text{ sec.}$$

$$\therefore \quad 6.2 \times 10^7 = \frac{6.2 \times 10^7}{3600 \times 24} \text{ day sec-m}$$

$$= \frac{6.2 \times 10^7}{86400}$$

$$= \frac{6.2 \times 10^3}{8.64}$$

$$= 0.7176 \times 10^3$$

$$= \mathbf{717.6\ day\ sec\text{-}m}$$

$$\text{Similarly rainfall in km}^2\text{-m} = \frac{6.2 \times 10^7}{10^6}$$

as $1 \text{ km} = 1000 \text{ m}$

$1 \text{ km}^2 = 10^6$

$= \mathbf{620\ km^2\text{-}m}$

Example 19.7 *A lake behind dam has a capacity of 3000 km²-m approximately. For how many days would this water supply be sufficient to a nearby city of 20 × 10⁶ population if daily requirement per person is 300 litres.*

Solution

$$\text{Available water} = 3000 \times 10^6 \times 1 \text{ m}^3$$

Daily requirement of the city

$$= \text{No. of persons} \times \text{daily requirement}$$

$$= 20 \times 10^6 \times \frac{300}{1000} \text{ cu. m (m}^3)$$

$$= \mathbf{6 \times 10^6\ m^3/day}$$

∴ Number of days water can be supplied

$$= \frac{3000 \times 10^6}{6 \times 10^6}$$

$$= \mathbf{500\ days}$$

Example 19.8 *The average daily stream flows for seven days of a week, resulting from a heavy storm on a basin of 3000 sq. km is given in the table shown below. Find the total flow volume in day-sec-meter, km²-m, cm and millions of cu. m.*

Days	*1st*	*2nd*	*3rd*	*4th*	*5th*	*6th*	*7th*
Mean daily flow cu. m/s	*100*	*300*	*200*	*110*	*40*	*30*	*20*

Solution

Total flow volume in 7 days is given by

$$\text{Total volume} = 24 \times 3600 \,[100 + 300 + 200 + 110 + 40 + 30 + 20]$$

$$= 24 \times 3600 \times 800 \text{ cu. m}$$

$$= 6912 \times 10^4 \text{ cu. m}$$

$$= 69.12 \times 10^6$$

$$= 69.12 \text{ million m}^3$$

Again flow in day-sec-metre

$$= \frac{69.12 \times 10^6}{24 \times 3600}$$

$$= \mathbf{855\ day\text{-}sec\text{-}metre}$$

$$\text{Flow} = 69.12 \times 10^6 \text{ cu. m} = \text{Area in m}^2 \times \text{height in m}$$

$$= 10^6 \times \text{m}$$

$$\therefore \quad \text{km}^2\text{-m} = \frac{69.12 \times 10^6}{10^6}$$

$$= \mathbf{69.12\ km^2\text{-}m}$$

Again

$$\text{Flow} = 69.12 \times 10^6 \text{ m}^3 = \text{Area} \times \text{Height}$$

$$= 3000 \times \text{cm} \times 10^4$$

$$\therefore \quad \text{cm} = \frac{69.12 \times 10^6}{3000 \times 10^4} = \frac{69.12 \times 10^3}{3 \times 10^4}$$

$$= \frac{69.12}{3 \times 10}$$

$$= \frac{23.04}{10} = \mathbf{2.304\ cm}$$

Example 19.9 *Monthly discharge for 12 months at a particular location of the river is given below :*

April – 400 m³, May – 150 m³, June – 1400 m³, July – 2600 m³

August – 3000 m³, Sept. – 2500 m³, Oct – 1800 m³, Nov – 1500 m³

Dec – 1400 m³, Jan – 1000 m³, Feb – 600 m³, Mar – 600 m³.

Draw

(i) The hydrograph and find average monthly flow

(ii) The power available at mean flow of water if the available head is 75 m and overall efficiency of generation is 80%.

Solution

(i) The average monthly flow $= \dfrac{\text{Total flow}}{\text{No. of months}}$

Total flow (m³) × 10^6 = 400 + 150 + 1400 + 2600 + 3000 + 2500 + 1800 + 1500 + 1400 + 1000 + 600 + 600 = 16950 m³ × 10^6

$$\therefore \text{ Average flow/month } = \frac{16950}{12} \times 10^6$$

$$= \mathbf{1412.5 \times 10^6\ m^3}$$

$$= \mathbf{1412.5\ millions\ of\ m^3/month}$$

Hydrograph is shown in Fig. Ex. 19.9(a)

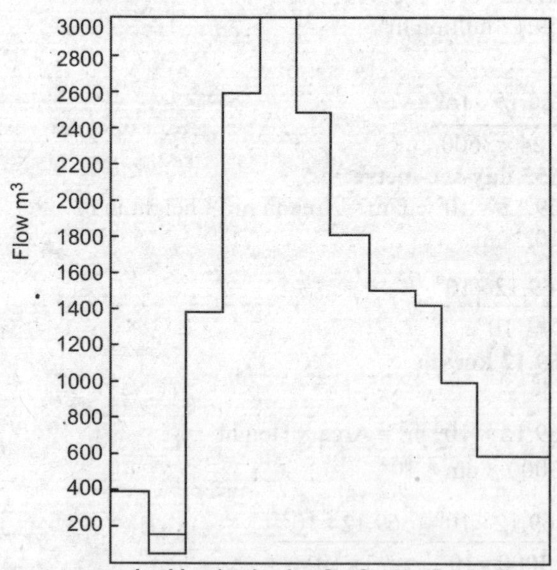

Fig. Ex. 19.9 (a) *Hydrograph*

For constructing a flow-duration curve, it is necessary to find the duration of time during which a certain flow is available. The following table will give data for flow-duration curve.

Flow	Duration	Percentage
150	12 Months	100
400	11 Months	91.8
600	10 Months	83.4
1000	8	66.6
1400	7	58.4
1500	5	41.67
1800	4	33.3
2500	3	25.0
2600	2	16.65
3000	1	8.32
—		

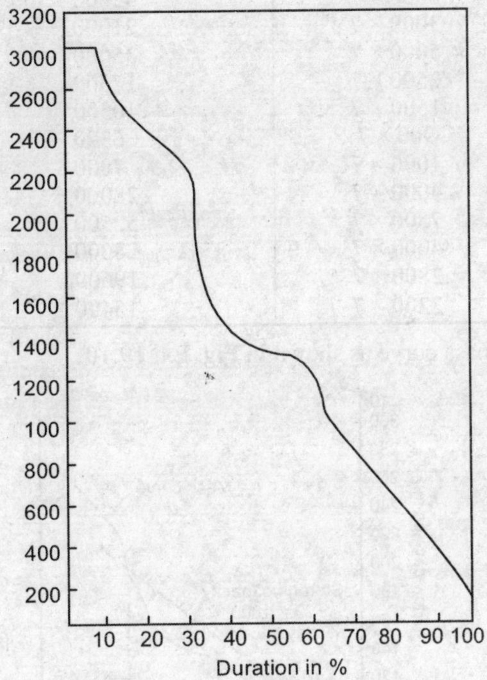

Fig. Ex. 19.9 (b) *Flow duration curve.*

The average flow = 1412.5 × 10^6 cu. m/month

$$= \frac{1412.5 \times 10^6}{30 \times 24 \times 3600} \text{ m}^3/\text{s}$$

$$= 544.95 \text{ m}^3/\text{s}$$

∴ Average power (kW) available at the site is

$$P = Q.P.H \times \eta$$

$$= 544.95 \times 1000 \times 75 \times 0.8 \text{ kW}$$

$$= 32697000 \text{ kW}$$

$$= 32697 \text{ MW}$$

Example 19.10 *At a particular site the weekly flow rate for 12 weeks is given as follows :*

^{3}/s, 2 – 3000 m^3/s, 3 – 5000 m^3/s, 4 – 2500 m^3/s

5 – 1500 m^3/s, 6 – 900 m^3/s, 7 – 1000 m^3/s, 8 – 4000 m^3/s

9 – 7500 m^3/s, 10 – 4000 m^3/s, 11 – 2800 m^3/s, 12 – 2200 m^3/s.

Find out the size of the reservoir and the possible rate of available flow after the reservoir had in-built with the mass curve.

Solution

For drawing the mass curve, find out the cumulative volume of water that can be stored week after week. The data for curve is given below :

Note : If the mean flow is available in the week at the given rate, then the total flow in the week is given by Total flow = 7 (days) × m^3/sec = 7 × day-sec-meter

Week	*Weekly flow m³/s*	*Weekly flow-day-sec-m*	*Cumulative volume*
1	6000 × 7	42000	42000
2	3000 × 7	21000	63000
3	5000 × 7	35000	98000
4	2500 × 7	17500	105500
5	1500 × 7	10500	116000
6	900 × 7	6300	122300
7	1000 × 7	7000	129300
8	4000 × 7	28000	157300
9	7500 × 7	52500	209800
10	4000 × 7	28000	237800
11	2800 × 7	19600	257400
12	2200 × 7	15400	272800

By using this data, draw the mass curve as shown in Fig. Ex. 19.10

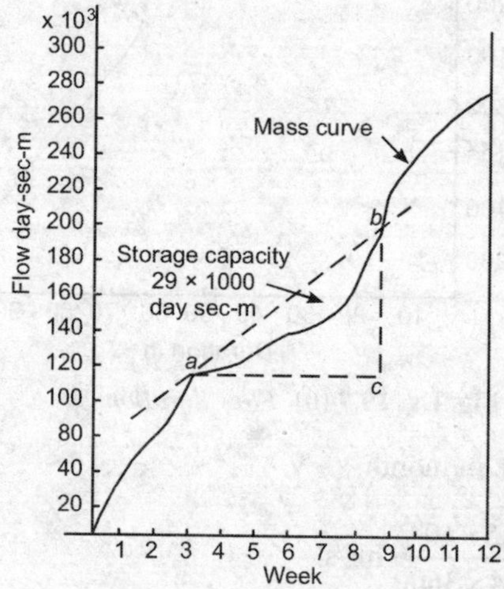

Fig. Ex. 19.10 *Mass curve.*

The tangent is drawn at the highest point on the mass curve from *a* and measure the highest distance between the tangent drawn and the mass curve which given the capacity of the reservoir. In this case, the capacity of the reservoir is **29,000 day-sec-m.**

The slope of the line *ab* gives the flow rate available for the given capacity of the reservoir.

From graph,

$$\text{Flow rate available} = \frac{bc}{ac}$$

$$= \frac{880 \times 1000}{5.8 \times 7(\text{days})}$$

$$= 21674.9 \text{ m}^3/\text{sec}$$

Example 19.11 *The run-off data from the particular location of the river for ten months is given in the table given below :*

Month	*Discharge m³/month × 10⁶*	*Month*	*Discharge m³/month × 10⁶*
1	*100*	*6*	*90*
2	*50*	*7*	*20*
3	*10*	*8*	*140*
4	*10*	*9*	*30*
5	*130*	*10*	*60*

Find the minimum flow available throughout the year if the storage capacity at the location is 50×10^6 cu. m.

Solution

Cumulative volume in ($m^3 \times 10^6$) is tabulated as given below :

Month	*Discharge*	*Cumulative Discharge*	*m³/m × 10⁶*
1	100	100	Draw the mass curve (cumulative volume). Draw the tangents from the points *a*, *b* and *c* in such away that the distance *x*, *y* and *z* each will be equal to 50×100^6 m³.
2	50	150	
3	10	160	
4	10	170	
5	130	300	
6	90	390	
7	20	410	
8	140	550	
9	30	580	
10	50	640	

Measure the slopes at the point *a*, *b* and *c* which will directly give flow rates.

Flow rate at point $a = 36.3$ m³ × 10⁶/month

Flow rate at point $b = 83.2 \times 10^6$ m³/month

Flow rate at point $c = 68.8 \times 10^6$ m³/month

The lowest rate 36.3×10^6 m³ among the three is the minimum available flow rate throughout the period considered.

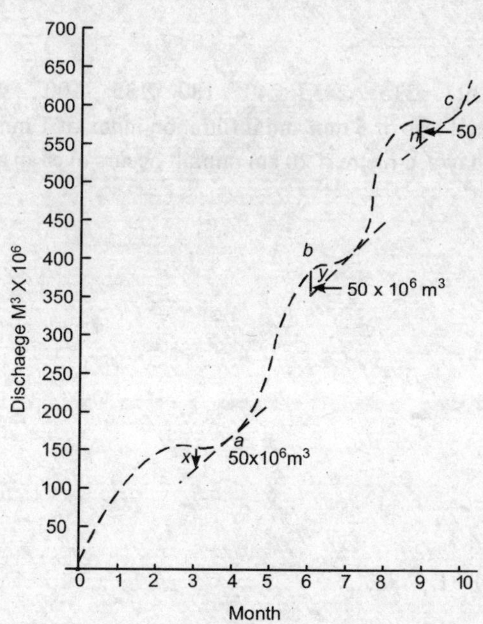

Fig. Ex. 19.11

THEORY QUESTIONS

19.1. Describe briefly with a neat sketch the hydraulic cycle.

19.2. What do you understand by term precipitation?

19.3. Describe with neat sketches the different types of rain-gages used for measuring the rainfall.

19.4. Discuss : (i) Theissen polygon method and (ii) Isohyetal method for determining the average rainfall.

19.5. Define evaporation, transpiration and evapo-transpiration.

19.6. Differentiate between infiltration rate and infiltration capacity.

19.7. How can infiltration capacity be determined in the field?

19.8. What are the infiltration indices?

19.9. What is runoff? On what factors does runoff depend?

19.10. What are the different methods of estimating runoff from a catchment?

19.11. What is a hydrograph? Indicate the various components of hydrograph?

19.12. How can runoff be estimated by infiltration method?

19.13. What is a unit hydrograph? How can it be constructed from a flood hydrograph of a drainage basin? Indicate the various assumptions involved in the unit hydrograph theory.

19.14. Indicate the use of unit hydrograph for developing a flood hydrograph that would result from a series of rainfalls of different intensities but duration same as the unit duration of the unit hydrograph.

19.15. What is base flow? Describe briefly how base flow may be separated from the total runoff.

19.16. What is the importance of rainfall and runoff data in the design of hydroelectric power plant?

19.17. Explain the construction of flow-duration curve and discuss its importance in comparing the power potentiality of different storages used for power generation.

19.18. Compute and plot the ordinates of a flood hydrograph resulting from a 4 hour rainfall of intensity 10.5 cm. The ordinates of 4 hour unit hydrograph are given below.

Time in hours	03	06	09	12	15	18	21	24	27	30	33	36	39	42	45	48
Ordinates of UH cu. m/s	0	120	370	505	397	315	225	240	180	135	100	70	45	25	15	0

Assume base flow is m³/s. An initial loss of 5 mm and infiltration index of 1 mm/hr.

19. What is the volume of rainfall in day-sec-metre if 20 cm rainfall occurs over an area of 2000 km^2? Also find out the rainfall in km^2-m.

Chapter 20 Hydroelectric Power Plants

20.1 INTRODUCTION

Water power hydropower is one of the major sources of energy. The energy of water can be utilized for the generation of electrical energy. When water flows through a height, the kinetic energy of water is used to rotate a turbine which is coupled to an alternator. Higher the potential energy as head—it is the difference in level of water between two points—greater will be the kinetic energy available (which is a function of mass of water and velocity or head available). Hence, it is necessary to have continuous flow of water to generate electrical energy. For this, water has to be collected and stored in reservoirs or lakes at high altitudes. Water is artificially stored in dams. Dams are constructed over rivers in which large quantity of water is stored in a wide area (known as catchment area). Rainfall is the main source of water and depends on several factors such as temperature, wind, cloudiness, humidity etc. Not all the water stored in dams can be utilized for power generation. Some of the water is evaporated, some seeps into soil and some of it is used by the vegetation. The remaining water flows on the ground surface of the catchment area to form the stream and is known as run-off. This water is utilized for hydro-projects i.e., power generation and irrigation. The factors affecting run-off are :

1. Rainfall pattern
2. Geology of the area
3. Size and shape of catchment area
4. Topography and nature of soil in catchment area
5. Vegetation and
6. Weather condition in catchment area.

Apart from power generation, water is utilized for irrigation and drinking purposes. **Hydropower is a renewable source of energy which is clean, free from pollution and has a good environment effect**.

20.2 SITE SELECTION FOR HYDRO ELECTRIC POWER PLANTS

Apart from the availability of water, a number of factors have to be considered while selecting a site for a hydro-electric plant. The most economical option has to be considered for selecting the site. There is no easy answer to the question raised "which is the suitable site for power plant?" as several points are involved as dam, conduits, intakes, surge tanks, power house and many others.

The following points are to be considered for the selection of site for a hydro electric power plant :

I. Hydrological investigation
II. Topographical investigation

III. Geological investigation
IV. Load centre
V. Access to site
VI. Water pollution effects.

20.2.1 Hydrological Investigation

The hydrological investigations include :

1. Availability of Water. Capacity of the hydro electric power plant will be decided once the data or information regarding the maximum, minimum and average quantity of water is available throughout the year. For this purpose, hydrograph will be useful. Here run-off data for over a period of years should be available. Also the data regarding the total rainfall and the catchment area will also required to study the feasibility of the project.

2. Water Storage. There will always a wide variation in rainfall during the year and therefore it is necessary to store water and then utilize for generation of power. From the data of rainfall and the catchment area, information regarding the total quantity of water that can be made available for generation of power throughout the year. Thus it is essential to store water to get the uniform output. For this purpose a storage reservoir is constructed at the site.

Generally the reservoir constructed will be such that after providing the required water for the generation of power throughout the year, there will be some water in case the rains are not there in time or this water will be useful during dry periods.

3. Water Head. The topography of the area will decide the maximum height for the storage dam. As the height of the dam increases, the cost of construction will increase as the base width will have to be more. As the head of water is increased, the quantity of water required will be reduced for the generation of same power. The compromise will have to be exercised as far as the capital cost and power generation are concerned ($P \propto HQ$).

20.2.2 Topographical Investigation

While deciding the location of dam, it is necessary to know the topographical features of the catchment area. These features are mapped sufficiently for alternative locations or schemes for comparison purposes. Generally for the area where dam is to be constructed, the mapping is done at 1 to 1.5 m intervals and for catchment area this interval may be 3 to 6 m.

20.2.3 Geological Investigations

Geological investigations are required for (i) reservoir area i.e., catchment area and the water storage area and (ii) dam site and other structures.

The basic factors required to be studied for the selection of site are :

1. Ample area with tight, hard base.
2. A strong foundation able to support the structures (dam and other structures).
3. To disposal of the surplus water, sufficient and safe spillway.
4. The impounded or stored water should not submerge valuable minerals and agricultural land.
5. The area should not be prone to earthquakes.
6. The rocks in the foundation of the dam should be reasonably impervious.

20.2.4 Load Centre

It is always desirable that the power station should be set up near the load centres so that cost of the transmission lines and their maintenance are low. But, generally the hydro electric plants are located in hilly or mountaineous region much away from the load centres. Hence long transmission lines are required which will increase cost and transmission losses are increased.

20.2.5 Access to Site

The site, where the hydro electric power plant is to be construscted, should be easily accessible. The site should have transportation facilities through rail or road.

20.2.6 Water Pollution Effects

After the construction of the hydro electric plant, many time it is found or felt foul smelling in the plant and vicinity, corrosion of metallic parts including electrical equipments. All these effects are due to very poor quality of water. This is due to :

(a) Soil under the reservoir may contain mineral deposits of harmful nature.

(b) If sulphates are present, the pungent gas H_2S is formed because of the action of micro organisms.

(c) The presence of alkaline and acidic deposits largely increase the corrosive effects of water.

(d) Submerged vegetation also causes badily smelling gases. This is because of the decay of twigs, leaves, logs etc. under water are allowed to decay.

(e) The presence of algae is another cause of water pollution.

Apart from the smelling of foul gas, corrosion and errosion of the metallic plants, the presence of H_2S and SO_2 can cause health hazards also. Headache, dizziness, dryness and nose, throat and chest pains are the effects of H_2S. The high concentration (> 300 ppm) of H_2S even causes paralysis.

In addition to all these considerations, one more important fact is to be considered and this to preserve important historic, cultural and natural aspects of the site.

20.3 CLASSIFICATION OF HYDRO ELECTRIC POWER PLANTS

Hydro electric power plants may be classified according to the following points :

1. The head of water available. Hydroplants may be classified on the basis of head under which they operate as

(a) Low head plants

(b) Medium head plants and

(c) High head plants.

20.3.1 Low Head Plants

When the operating head of water is less than 30 m, the plant is known as **Low Head Plant**. These low head plants, a dam is built across river to create the necessary head of water. Run-of-river plants are usually of this type. The excess water is allowed to flow over the dam itself. Figure 20.1 shows the schematic arrangement of low water head plant.

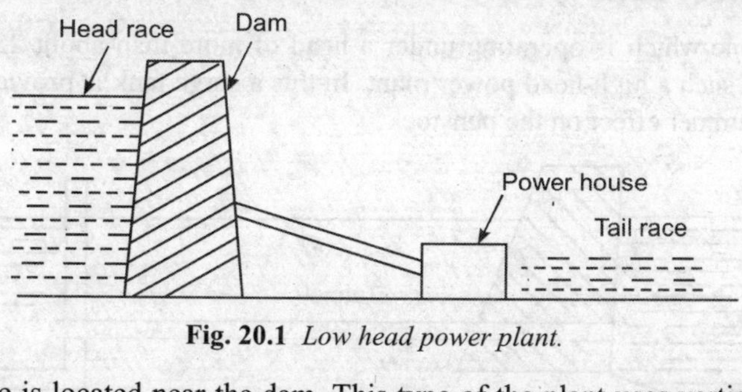

Fig. 20.1 *Low head power plant.*

In this, the power house is located near the dam. This type of the plant uses vertical shaft Francis turbine or Kaplan turbine. Figure 20.2 shows the another arrangement for the low head plants.

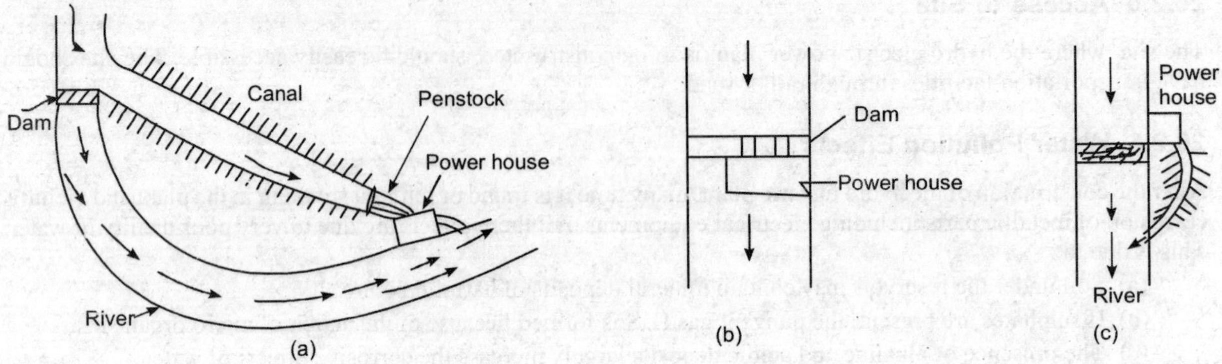

Fig. 20.2 *Various arrangements of power houses, and dam in low head power plant with dam.*

20.3.2 Medium Head Power Plants

A medium head power plant is the one which is operating under a head between 30 and 250 m. In this case a forebay is provided at the beginning of penstock. This forebay acts as a reservoir. In these plants the water is taken from the reservoir to the forebay through canals. From the forebay the water is taken to the power house through the penstock. In this type of the plant, the prime movers used are Francis turbines or Kaplan turbines.

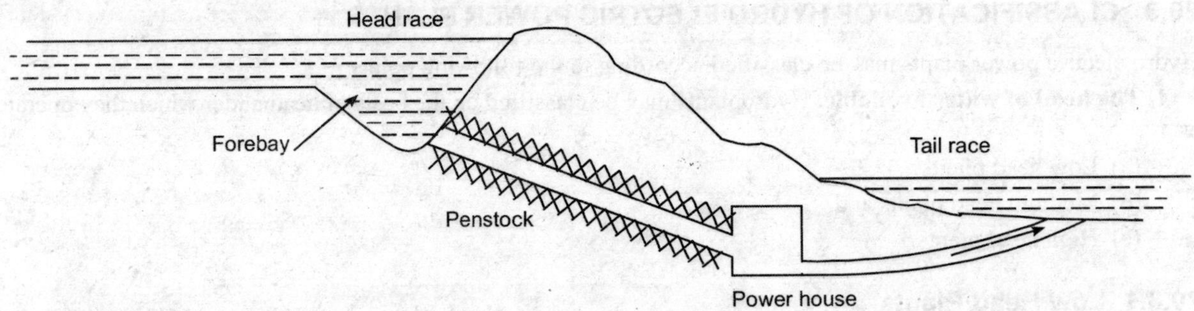

Fig. 20.3 *Medium head power plant.*

20.3.3 High Head Power Plants

A high head plant is the one which is operating under a head of more than about 250 m. Fig. 20.4 shows the schematic arrangements of such a high-head power plant. In this a surge tank is provided as shown surge tank is used to reduce the water hammer effect on the penstock.

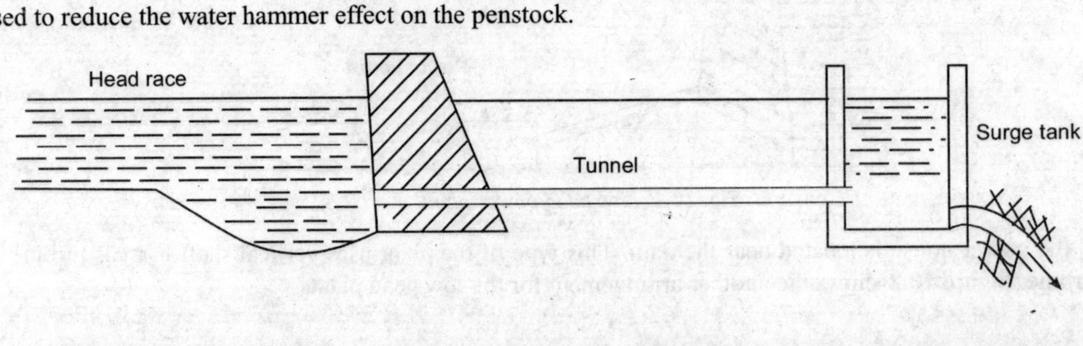

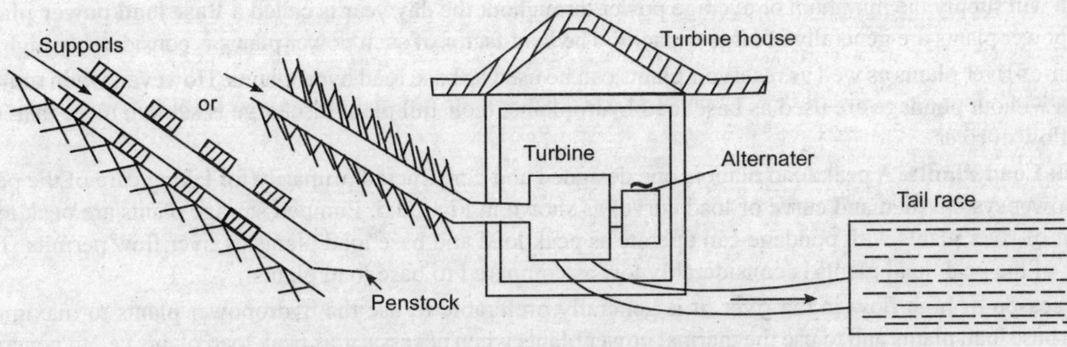

Fig. 20.4 *High head power plant.*

Penstock provided is generally a steel pipe. The water from the main reservoir is first taken upto the surge tank through a tunnel and then it is carried through the penstock to the power house. In high head plants, Francis turbine is used upto 300 m head and beyond that Pelton wheel is used as the prime mover.

Low heads as well as the lower ranges of medium heads may be made available by utilizing a steep stope or a fall in river or a canal. On the other hand the higher ranges of medium heads as well as high heads may be obtained by constructing dams of sufficient heights and locating the power house either at the toe of the dam close to it, or in a deep depression if available at a certain distance away from the dam.

The above noted head ranges for the different types of hydropower plants are, however, arbitrary and it is customary to associate these head ranges with the types of the turbines used. Thus hydropower plants for which **impulse turbines** are used are called **high head plants**. Similarly **mixed flow turbines** are used for **medium head plants** and **axial flow turbines** are used for **low head plants**. With the advances in the turbine design it has become possible to use axial and mixed flow turbines for higher heads. Consequently, the ranges of head indicated above also move up.

2. Hydro-electric power plants may also be classified on a functional basis as :

(a) Base load power plants and

(b) Peak load power plants.

(a) Base Load Power Plants. It is very well experienced that an individual consumer of electric power does not use all the appliances in his possession all the time. Whatever is true for a single consumer, it is true for all the consumers supplied with power from a power station. If the load curve as shown in Fig. 20.5 is drawn then the power

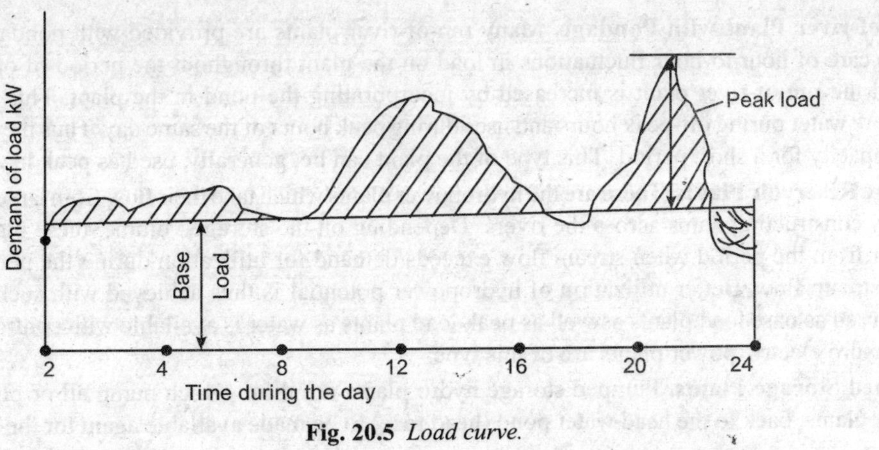

Fig. 20.5 *Load curve.*

plant which will supply the minimum or average power throughout the day/year is called a **Base load power plant.** Base load power plants are generally of large capacity. The load factor of such power plants is considerably high.

Both run-of-river plants as well as reservoir plants can be used as base load hydroplants. However, when run-of-river plants without pondage are used as base load hydroplants, their full plant discharge is seldom more than the minimum flow of river.

(b) Peak Load Plants. A peak load plant is one designed and constructed primarily for taking care of the peak load of a power system (demand curve or load curve) as shown in Fig. 20.5. Pumped storage plants are peak load plants. Run-of-river plants with pondage can operate as peak load and base load plants as river flow permits. The load factor of the peak load plants is considerably low as compared to base load plants.

During season of high flow in the river, it is generally preferable to use the hydropower plants to maximum capacity as base load plants and to use the thermal power plants when necessary, as peak load plants i.e., to generate the extra power needed beyond the capacity of the hydropower plants at peak load periods. Conversely during the periods of low flow, it would be better to use the thermal power plants to take the baseload and use hydropower plants with storage for taking the peak load. However, out of the hydropower and thermal power plants, the former have better peaking characteristics. Hydropower plants can be started at a notice seldom exceeding 3-4 minutes and there is absolutely no wastage when they are idle. But if thermal power plants are used for peak loads, the boiler fire must be kept lighted and thus there is a continuous loss of fuel, and even then it usually requires at lease 30 minutes before a steam unit can generate to full capacity.

3. Hydropower plants can be classified according to the quantity of water available as :

(a) Run-of-river plant without pondage
(b) Run-of-river plant with pondage
(c) Storage reservoir plants
(d) Pumpstorage plants
(e) Tidal plants
(f) Mini- and micro- hydel plants.

(a) Run-of-river Plant without Pondage. Run-of-river plants without pondage are those plants which utilize the flow as it comes, without any storage being provided. As such these plants would be feasible only on such rivers which have a minimum day weather flow of such magnitude which makes the development worthwhile. A weir or barrage may be constructed across the river close to the power plant to maintain a given water level. These are generally low head plants and often at times of flood, tail water rises to such an extent that the plants are inoperative. The capacity of the plant becomes less in dry weather and hence the utility of this plant is less as compared with other plants.

(b) Run-of-river Plant with Pondage. Many run-of-river plants are provided with pondage, which enables them to take care of hour-to-hour fluctuations in load on the plant throughout the period of operation. Thus the usefulness of the run-of-river plant is increased by incorporating the pond in the plant. This type of the plants permits to store water during off-peak hours and uses during peak hours of the same day. Thus the pondage increases the stream capacity for a short period. This type of the plant can be, generally, used as peak load plants.

(c) Storage Reservoir Plants. These are the hydropower plants which take their flow from large storage reservoirs developed by constructing dams across the rivers. Depending on the storage volume, these plants can hold over surplus water from the period when stream flow exceeds demand for utilization during the period when demand exceeds the stream flow. Better utilization of hydropower potential is thus achieved with such plants. Thus this plant can be used as base load plants as well as peak load plants as water is available with control as required. The majority of hydro electric power plants are of this type.

(d) Pumped Storage Plants. Pumped storage hydro plants are those which pump all or portion of the water used by these plants, back to the head water pond (head race) to be made available again for the power generation.

Essentially they consist of a tail water pond and a head water pond. During times of peak load, water is drawn from the head water pond through the penstocks to operate hydro electric generating units. This water is collected in the tail water pond and during the off-peak hours pumps are operated to pump the water back from the tail water pond to the head water pond. The power for operating the pumps is provided by some off-peak thermal or hydropower plants.

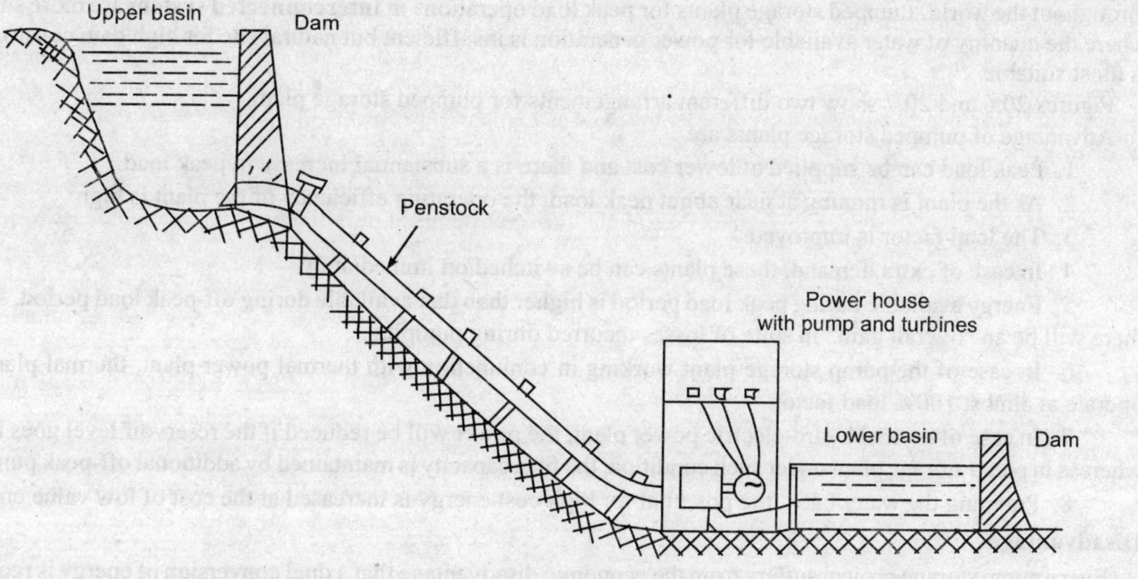

Fig. 20.6 *Simple arrangement of pumped storage hydroelectric power plant*

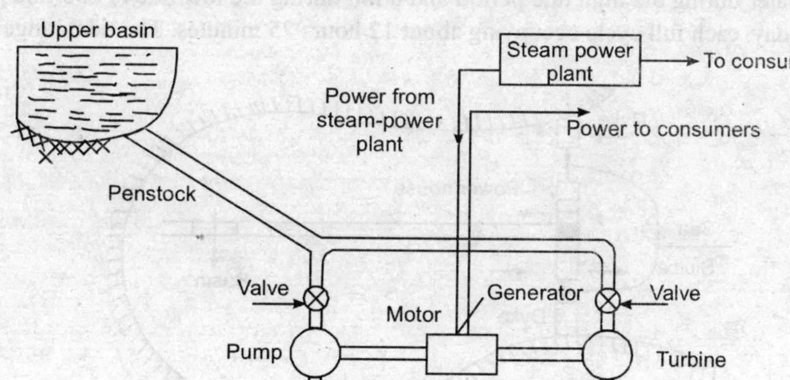

Fig. 20.7 *Pumped storage plant (peak load) and steam power plant (base load)*

For heads upto about 120 m special Francis turbines have been developed for the pumped storage plants, the runners of which are so shaped that they can be used both as turbine as well as pump. With such turbines which are known as **reversible turbines (Deriaz turbine)**, it is feasible to use the same unit for both pumping water and generating power. Alternatively pumped storage plants may have separate pumping and power generating units, especially for high heads for which multistage centrifugal pumps are used for pumping water and high head Francis turbines are installed for power generation.

When steam or thermal power plant is used in combination with hydropower plant, then the pumped storage plant decreases the operating cost of steam plant because it serves to increase the load factor of the steam plant and provides the added capacity to meet peak loads. Nowadays, the pumped storage plants gained high popularity throughout the world. Pumped storage plants for peak load operations in **interconnected system** are more suitable where the quantity of water available for power generation is insufficient but natural site for high dam construction is most suitable.

Figures 20.6 and 20.7 show two different arrangements for pumped storage plant.

Advantage of pumped storage plants are :

1. Peak load can be supplied at lower cost and there is a substantial increase in peak load.
2. As the plant is running at near about peak load, the operating efficiency of the plant is high.
3. The load factor is improved.
4. In case of extra demand, these plants can be switched on immediately.
5. Energy available during peak load period is higher than that available during off-peak load period, so that there will be an "overall gain" in spite of losses incurred during pumping.
6. In case of the pump storage plant working in conjunction with thermal power plant, thermal plant can operate at almost 100% load factor.
7. In case of normal hydro-electric power plant, the power will be reduced if the reservoir level goes to low whereas in pump storage plant under such condition, the firm capacity is maintained by additional off-peak pumping.
8. Pumping the water back, the potential for high cost energy is increased at the cost of low value energy.

Disadvantage

Every pump storage project suffers from the economic disadvantage that a duel conversion of energy is required.

(e) **Tidal Hydro Plants.** Tidal plants do not involve any storage of water. These plants work on the principle that there is a rise in sea water during the high tide period and a fall during the low tide or ebb tide period. Sea water rises and falls twice a day, each full cycle occupying about 12 hours 25 minutes. The tidal range or the difference

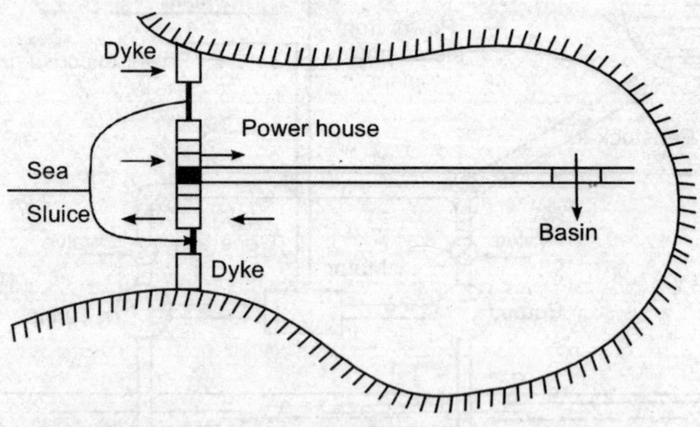

Fig. 20.8 *Simple tidal power plant (two-way generation)*

between the high tide and low tide levels is utilized to generate power. This is accomplished by constructing a basin separated from the sea by a wall and installing turbines in openings through this wall. During high tide water passes from the sea to the basin thus running turbines and generating power. During low tide water from the basin flows back to the sea which can also be utilized to generate power provided another set of turbines operating in the opposite direction is installed or special turbines which can generate power for either direction of flow are installed. Thus depending on the types of installation either one or both directions of flow can be utilized to generate power.

Tidal plants are the recent innovations in the field of hydropower generation. However, these may be used only where the tidal range is high. Rance power plant in France is an example of this type of plant where the tidal range is about 11 m and it has 9 units of 38000 kW each.

Figure 20.8 shows a schematic diagram of simple tidal power plant.

20.4 SCHEMATIC ARRANGEMENTS OF HYDRO ELECTRIC PLANTS AND ELEMENTS OF HYDRO ELECTRIC POWER PLANTS (STORAGE TYPE)

The reservoirs or the storage will have large quantity of water collected during the rainy season and the water is supplied to hydro electric plant during wet as well as dry period of the year. The capacity of the reservoirs will be very large so that the hydro-electric power plant will work to its full capacity throughout the year. This storage capacity will be much more as compared to that for other types of the hydro electric power plant.

Figure 20.9 shows the general arrangement for the reservoir type hydro electric plant showing different major parts or elements.

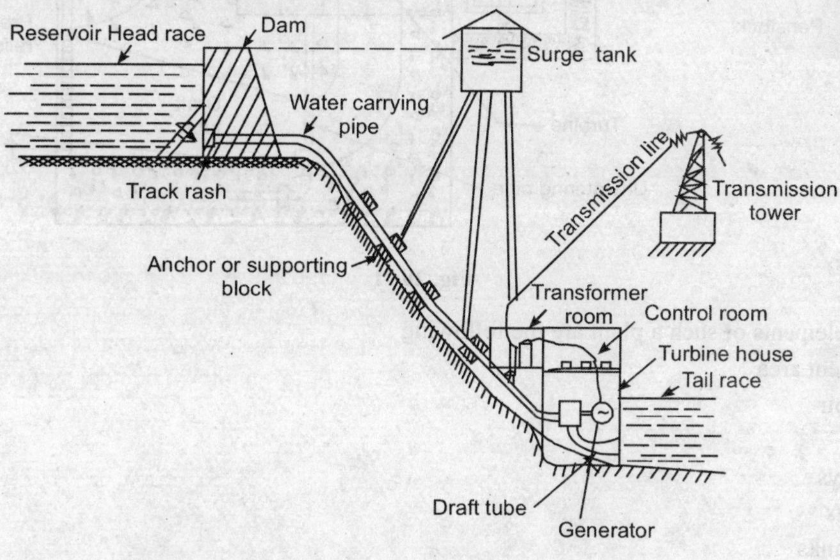

Fig. 20.9 *Reservoir type hydro-electric power plant.*

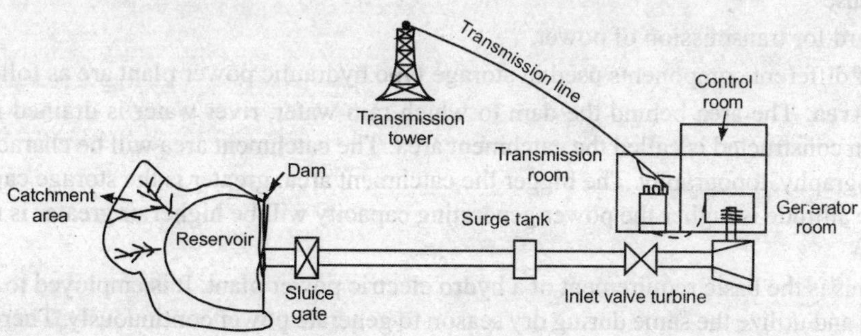

Fig. 20.10 *Plan for another arrangement of the hydro electric power plant.*

Figure 20.11 shows the typical layout of hydroelectric power plant.

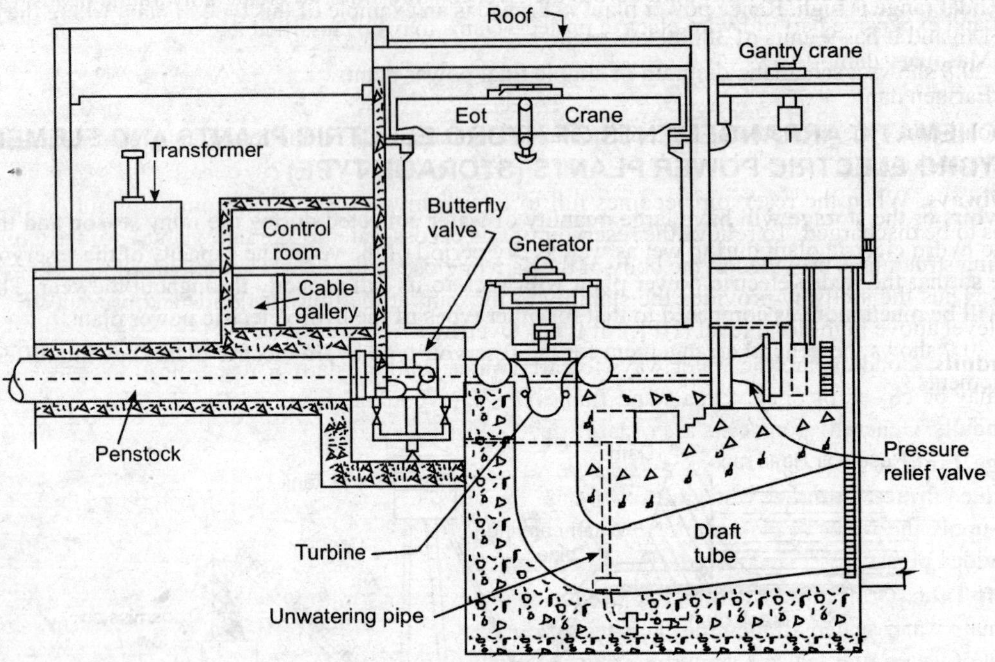

Fig. 20.11

The essential elements of such a plant are the following :

1. Catchment area
2. Reservoir
3. Dam
4. Spillways
5. Conduits
6. Surge tanks
7. Draft tubes
8. Power house
9. Switch yard for transmission of power.

The functions of different components used in storage type hydraulic power plant are as follows :

1. Catchment Area. The area behind the dam to which rain water, river water is drained into stream across which dam has been constructed is called the catchment area. The catchment area will be characterised by its size, shape, altitude, geography, topography. The bigger the catchment area, greater is the storage capacity. If the slope is steeper and if the altitude is higher the power generating capacity will be higher or greater is the total run-off of water.

2. Reservoir. This is the basic requirement of a hydro electric power plant. It is employed to store water during wet or rainy season and utilize the same during dry season to generate power continuously. There are basically two types of reservoirs :

(a) Natural e.g., lake in high mountains

(b) Artificial e.g., dam across a river.

3. Dam. A dam is a barrier (wall) built across the river to store water to create a hydraulic head. It is the most important and expensive part of hydro electric project. Dams may be classified as :

(a) Masonary dam

(b) Earthen dam

Selection of the type of the dam, for a particular location depends on the topography of the site, sub-soil and geological conditions.

4. Spillways. When the reservoir becomes full to its capacity, excess water coming in during rainy season, floods, has to be discharged. To relieve the reservoir of this excess water to keep the stability of the dam intact, the safeguarding structure is provided in the body of the dam or close to the dam. This safeguarding structure is called a spillway. Thus the spillway provides the stability to the dam at the time of floods and prevent the raising the reservoir level above high flood level (HFL). It is an 'Overflow'.

5. Conduits. Conduits are the water ways to carry water from the dam or reservoir to the power house. The conduits may be closed or open. Canals and flumes are open conduits and tunnels, pipelines and penstocks are closed conduits. Generally penstocks are of large diameter and each turbine has its own separate penstock.

6. Surge Tank. A surge tank is a small reservoir of water which is used to prevent the bursting or damaging of penstock due to **water hammer** when the load on the turbine decreases and thereby raising the water level in surge tank and supply the water to the turbine when the load on the turbine increases. Thus, the surge tank is used to prevent sudden pressure rise and also decrease in pressure.

7. Draft Tube. Draft tube or tail race is used to discharge water into the river (downstream). The draft tube should remain water seated all the time. The design and size of the draft tube should be such that water has a free exit and jet of water after leaving the turbine has no obstruction.

8. Power House. A power house is a stable building which includes the equipments having the layout such that adequate space is provided around the equipment for convenient dismantling and repairs. These equipments are as follows :

(a) Hydraulic turbines

(b) Electric generators

(c) Governors

(d) Gate valves

(e) Relief valves

(f) Water circulating pumps

(g) Air duct

(h) Switch board and instruments

(i) Storage batteries

(j) Cranes.

20.5 ADVANTAGES AND DISADVANTAGES OF HYDRO ELECTRIC POWER PLANTS

Following are the advantages and disadvantages of the hydro electric power plants :

Advantages. Hydro electric power plants have certain distinct advantages over other power plants. They are :

(a) No fuel changes, as no fuel, in its real sense, is required.

(b) The life of the hydro electric plant is around 50 years as compared to steam power plant which is 25–30 years.

(c) The running cost of the plant is low.

(d) Maintenance cost of the hydro electric plant is low.

(e) There are no standby losses. They can be started in short time and synchronised in few minutes.

(f) Efficiency of these plants does not decrease with age.

(g) The number of personnel (staff) is much less as compared to other power plants.

(h) There are no fuel and ash handling problems.

(i) Though a large area is required for storage, the cost of land is not a major problem because these are located in remote areas where cost of land is low.

(j) The hydro projects are multipurpose projects because in addition to power generation, they are useful for flood control and irrigation in addition of drinking water.

Disadvantages. There are few disadvantages of hydro electric power plants as follows :

(a) The time required for erection of hydro electric plants is long. This is because much time is required for evaluation of people (staying in the catchment area) and their rehabilitation.

(b) The total cost of the plant is high. Hence, cost of power generation per MW is quite high.

(c) Power generation is affected by the availability of water. If monsoons are not good, then the power generation is reduced.

(d) As these plants are located in hilly areas (generally), the cost of laying transmission lines (high tension lines) is high.

(e) The hydro plants disturb the ecology of the area as it destroy vegetation and deforestation takes place. Particularly because of the evaluation or uprooting the people from the area and disturbance in their life, they are against the erection of such plants and is a major deterrent factor.

Table 21.1 gives the comparison of all the power plants (Nuclear power plant, Thermal power plant, Hydro-electric power plant, Diesel power plant and Gas turbine power plant).

Table 20.1 *Comparison of various power plants*

Sr. No.	*Point of comparison*	*Steam power plant*	*Nuclear power plant*	*Diesel power plant*	*Gas turbine power plant*	*Hydro power plant*
1.	Site	Located near load centre	Location depends on availability of water and away from populated area	Installed anywhere.	Located near gas supply points.	Away from land centre. Depends on head of water available. Higher than steam power plant.
2.	Capital cost (Initial cost)	Low in comparison with nuclear plants	Very high	Low	Low	Higher than steam power plant
3.	Cost of fuel	High	Very low	Less than steam power plant	Low	Zero
4.	Operating cost	Very high compared to nuclear plant	Low.	Very high	High	Practically zero.
5.	Maintenance cost	High w.r.t. hydro and diesel plants.	Higher	Low	Higher than all except diesel plants	Very low

Contd.

Contd. Table 20.1

Sr. No.	*Point of comparison*	*Steam power plant*	*Nuclear power plant*	*Diesel power plant*	*Gas turbine power plant*	*Hydro power plant*
6.	Source of fuel	Limited	Unlimited	Limited	Limited	Unlimited (depends on rain)
7.	Transmission and distribution cost	Comparatively low	Comparatively low	Very low	Less than hydro plants and more than other plants	High due to location in remote place
8.	Reliability	Less reliable	Reliable	Less reliable.	Less reliable	Highly reliable.
9.	Working (load)	Base load station	Base load station	Peak or standby power plant	Peak or standby power plant.	Base load or peak load plant

20.6 MINI AND MICRO HYDRO ELECTRIC POWER PLANTS

There are many instances when electrification of the villages or the isolated pockets where transmission lines or systems cannot reach. In such cases, exploitation of mountain streams, rivulets and canal drops is done and the generation of power on the very small scale is possible. **The plants which can generate power with small head (upto 5 metres) is called a Microhydro plants and the generated power is upto 100 kW. If the water head available is from 5 to 20 m, a hydrel power plants can be installed generating power upto 1 MW. Such plants are called Mini hydel power plant.**

There are some plants generating power upto 15 MW. They are called as small power plants.

The consumers of the power generated by these plants are nearby the plant and hence the cost of transmission lines etc. is negligible. Such development of micro and mini hydropower plants is very much economical and hence it is a need of rural regions of the country. Small hydro power plants (SHP) which generate electricity in small scale are now being developed. These power plants are an attractive renewable source of energy in remote and hilly regions isolated from the main grids.

The advantages of these plants are :

1. Source of renewable energy.
2. Water heads from 2 m and above can be used to generate power in either micro or mini or small hydro plants.
3. No need of dams in its present meaning.
4. No pollution.
5. Limited investments.
6. Transmission and other losses negligible.

The Indian Renewable Energy Development Agency (IREDA) requires such plants to be developed and encourages organisations in private as well as public sectors for the same. By 2002 AD, the total capacity of mini/micro/small hydro power generation was estimated to 85 MW from nearly 120 schemes. It is estimated that the total potential of SHP is nearly 5000 MW majority of which is untapped.

In India, such sites are available in Himachal Pradesh, Arunachal Pradesh, UP, West Bengal, Karnataka. The potential of such sites is about 15000 MW. So far (till 2007) about 10% of this (1500 MW) has been installed.

THEORY QUESTIONS

20.1. What are the different factors involved while selecting the sites for hydro-electric power plant?

20.2. What are the points of classification of the hydro-electric power plants?

20.3. What do you understand by base load and peak load power plants? Explain this with the help of load curve.

20.4. What type of power plants are used as base load plant and why?

20.5. Which type of the power plants are used as peak load power plants? Why?

20.6. What do you understand by the term "run-of-river" power plant?

20.7. What do you understand by the term "pondage"? Why is it beneficial in the performance of the hydro-electric power plant?

20.8. What do you understand by storage reservoir type hydro power plants? What are its advantages? In what way it is different from other hydro electric power plants?

20.9. Write briefly about 'Pump Storage Plant'.

20.10. Draw a schematic sketch or diagram of storage type hydro electric power plant. Describe the function of each component used in this plant.

20.11. Compare hydro electric power plant with thermal power plant.

20.12. "Combined operation of hydro electric power plant with thermal power plant is economical than individual operation of the plant." Justify.

Chapter 21 Components of Hydro-electric Power Plants

21.1 INTRODUCTION

Hydraulic or water turbines are the machines which use the energy of water (Hydro-power) and convert it into mechanical energy. As such these may be considered as hydraulic motors or prime movers. The mechanical energy developed by a turbine is used in running an electric generator which is directly compled to the shaft of the turbine. The electric generator thus develops electric power, which is known as **hydro-electric power**.

21.2 ELEMENTS OF HYDRO-ELECTRIC POWER PLANTS

One of the essential requirements of the hydro-electric power generation is the availability of a continuous source of water with a large amount of hydraulic energy. Such a source of water can be available if a natural lake or a reservoir have been found at a high elevation or an artificial reservoir may be formed by constructing a dam across a river. Figure 21.1 shows a general layout of a hydro-electric power plant in which an artificial storage reservoir formed by constructing a dam has been shown. The water surface in the storage reservoir is known as **head-race level** or simply **head race**. The water from the reservoir is carried through **penstock** or **canals** to the power house. **Penstocks** are the pipes of large diameters, usually made of steel, wood or reinforced concrete, which carry water under pressure from the storage reservoir to the turbine. In some installations smaller reservoirs known as forebays are also provided. A **forebay** is essentially a storage reservoir at the head of penstocks. The purpose of a forebay is to temporarily store water when it is not required by the turbine and supply the same when required. When the power house is located just at the base of the dam no forebay is required to be provided since the reservoir itself serves the same purpose. However, if the power house is situated away from the storage reservoir and water is carried to the power house through a canal, then a forebay may be provided. In that case water from the reservoir is first led into the forebay which in turn distributes it to penstocks through which it is supplied to the turbines. Furthermore, where the power house is located across a canal, a forebay may be provided by enlarging the canal just ahead of the power house.

The water passing through the turbine is discharged to the tail race. The **tail race** is the channel which carries water (known as a tail water) away from the power house after it has passed through the turbine. It may be a natural stream at some point downstream from the power house. The water surface in the tail race channel is known as tail race level or simply tail race.

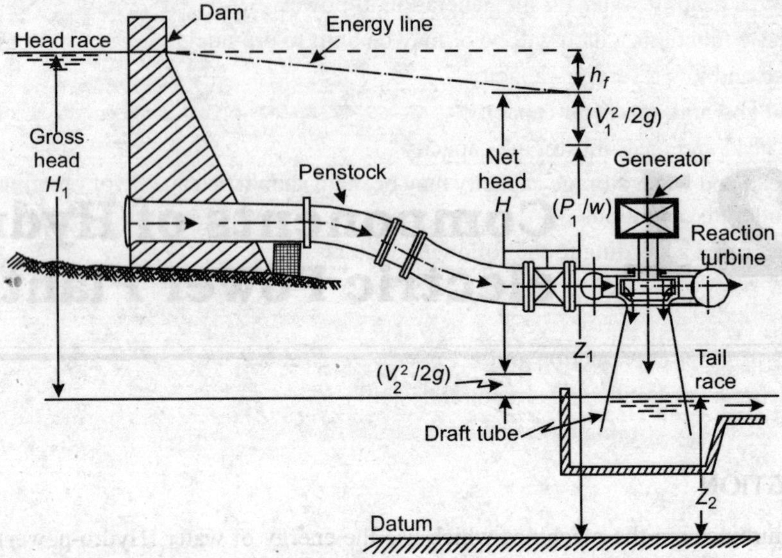

(a)

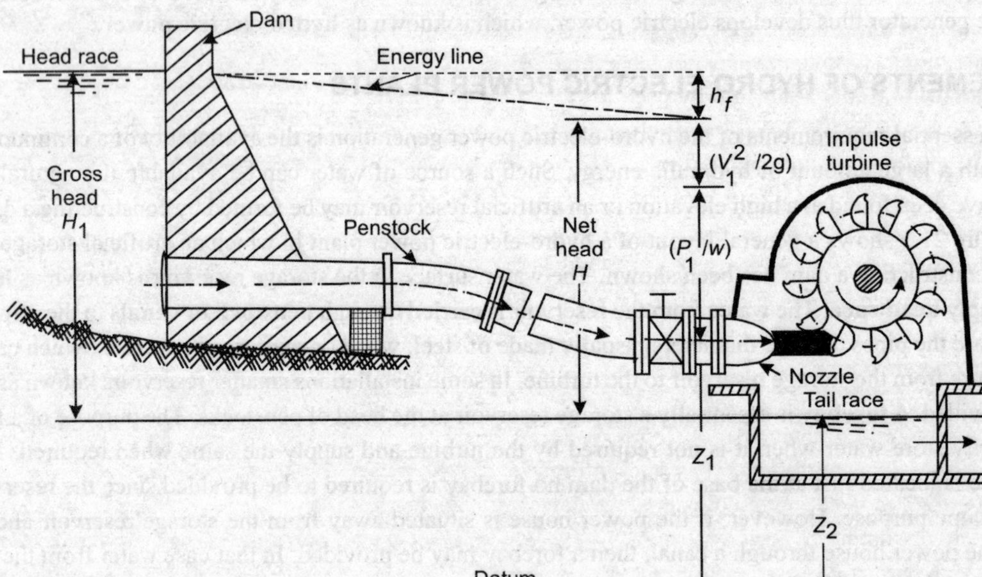

(b)

Fig. 21.1 *General layout of a hydro-power plant.*

21.2.1 Dam

The structure constructed across a river is called a dam. The materials used for construction may be masonry or any other suitable one. The functions of the dam are :

(a) To develop a reservoir of the desired storage capacity and

(b) To provide a head of water for the generation of power.

According to these functions a dam will be or may be built to provide

(i) High head and low reservoir capacity

(ii) Low head and high reservoir capacity

(iii) Medium head and medium storage capacity.

Whatever the head and water storage capacity may be, dam should be capable of resisting pressure of water and should be stable under all conditions.

Dams can be classified according to the following factors :

(a) Function

(b) Shape

(c) Materials used for construction and

(d) Design criteria (hydraulic and structural design).

(a) Dams are classified according to their function as follow

(i) Storage dams

(ii) Diversion dams and

(iii) Detention dams.

(i) **Storage dams** are mainly constructed for storing water and use of water when and where required. Storage may be required for generating power, irrigation or water supply.

(ii) **Diversion dams**, as its name suggests, are constructed to raise the water level and to divert the river flow in another direction. These dams may not have large storage capacities.

(iii) **Detention dams** are primarily used to store flood waters.

(b) Dams are classified according to their shape as :

(i) Trapezoidal section and

(ii) Arch type.

(c) Dams are classified according to the materials used for the construction of dams as :

(i) earthen dam – Earth is used

(ii) Rock piece used as material

(iii) Stone masonary dam

(iv) Concrete dam

(v) RCC dam

(vi) Timber and rubber.

The concrete dams plain as well as steel inforced, earthen and rockfit dams are most popular.

(d) According to the structural designs, the dams may be classified as :

(i) Gravity dam

(ii) Arch dam

(iii) Buttress dam

(iv) Earthen dam

(v) Rockfill dam.

(i) Gravity Dam. The ratained water thrust is resisted by gravity action in case of a gravity dam. Gravity dams are constructed either in stone masonary or in concrete. With better quality control, concrete dams can be built faster. Figure 21.2 shows the cross-section of a solid gravity dam. This gravity dam is subjected to the following static forces :

(a) Water pressure P

(b) Weight of dam W

where R is the resultant static pressure

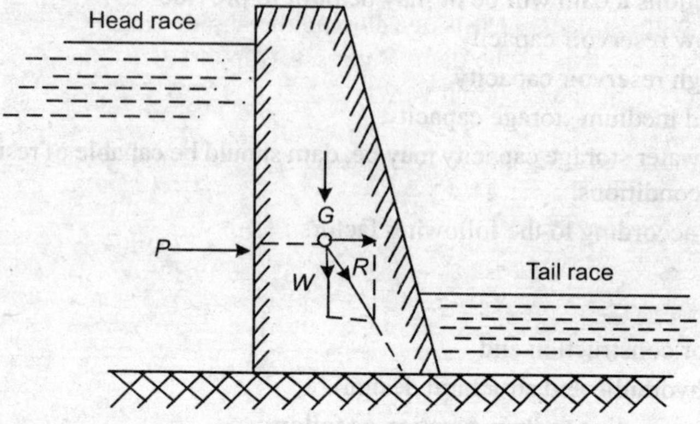

Fig. 21.2 *Solid gravity type masonary dam.*

While designing the cross-section and stability of the dam, in addition to the force P and W, following forces are required to be considered.

(a) Wind pressure

(b) Earthquake pressure

(c) Reaction of the foundation

(d) Uplift pressure.

(ii) Arch Dam. In arch type dam, the dam resists the pressures of water partly due to its weight and partly due to its arch action. Here all forces are acting perpendicular to the surface so that they tend to compress the material in

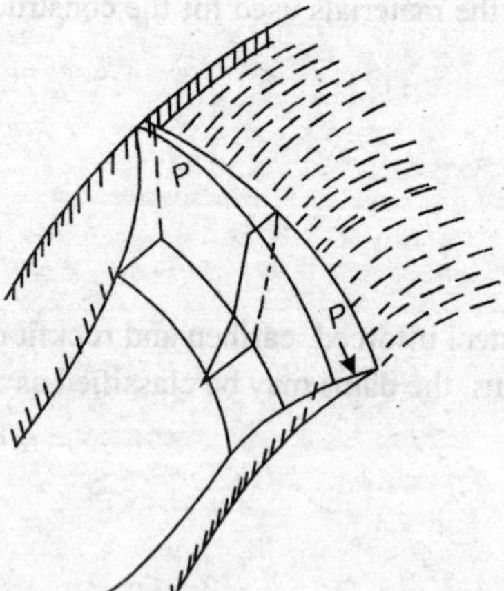

Fig. 21.3 *Arch type of masonary dam.*

the arch. This type of the dam is located in relatively narrow valley with steep slopes suitable for arch abutment. The main advantage using an arch dam is that the amount of masonary or concrete required is much less than a gravity dam of the same height. And as such the cost of such dam is much less. Over fall type spillway can not be used in arch type dam. The spillway has to be separately provided which increases the cost of outlet works and hence the overall cost of the project.

(iii) Buttress Dam. The buttress dam consists of a sloping membrane which transmits the water load to a series of buttresses at right angles to the axis of the dam. Buttresses support a flat stab or reinforced concrete. Figure 21.4(a) and (b) shows buttress dam and sectional view.

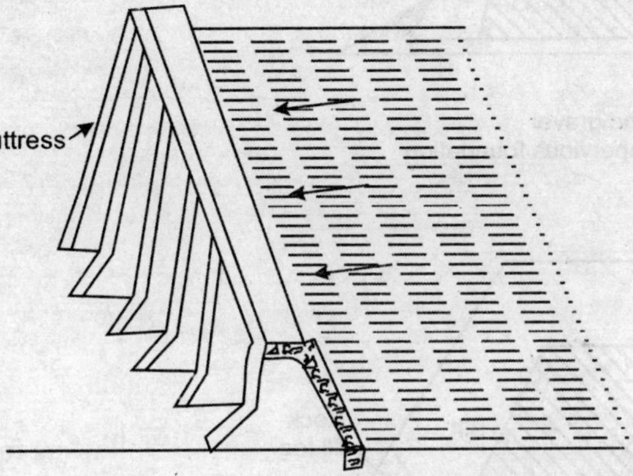

(a) Buttress dam

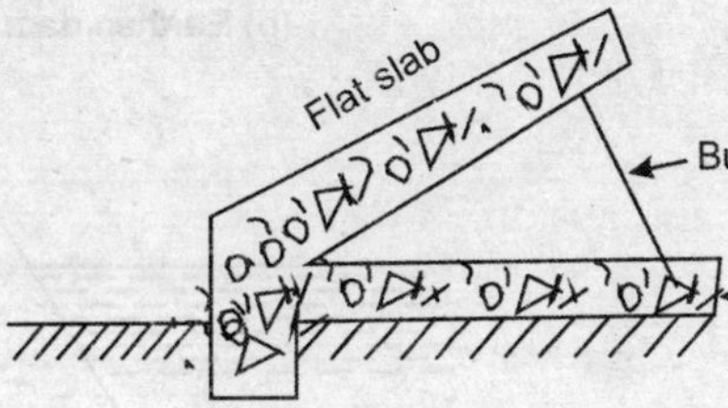

(b) Section of buttress dam

Fig. 21.4

Flat slab and buttress dams are particularly adopted to wide valleys where a long dam is required and foundation is week. Masonary dams have the advantages of maximum height, longest life, most economical in water conservation and lowest maintenance cost as compared to earthen dam.

(iv) Earthen Dam. With the development in soil mechanics and the earth moving machinery, earthen dams are nowadays used for the smaller projects (upto 70 m height dams). These dams can be built safely and economically on all types foundations of earth and rocks. These dams are used when river banks are not steep and where the site cannot take the weight of gravity dam. Figure 21.5 shows the schematic diagram of earthen dam.

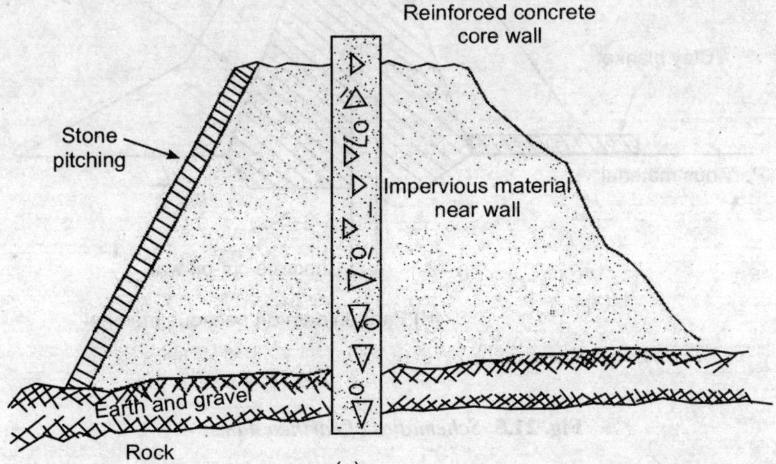

(a)

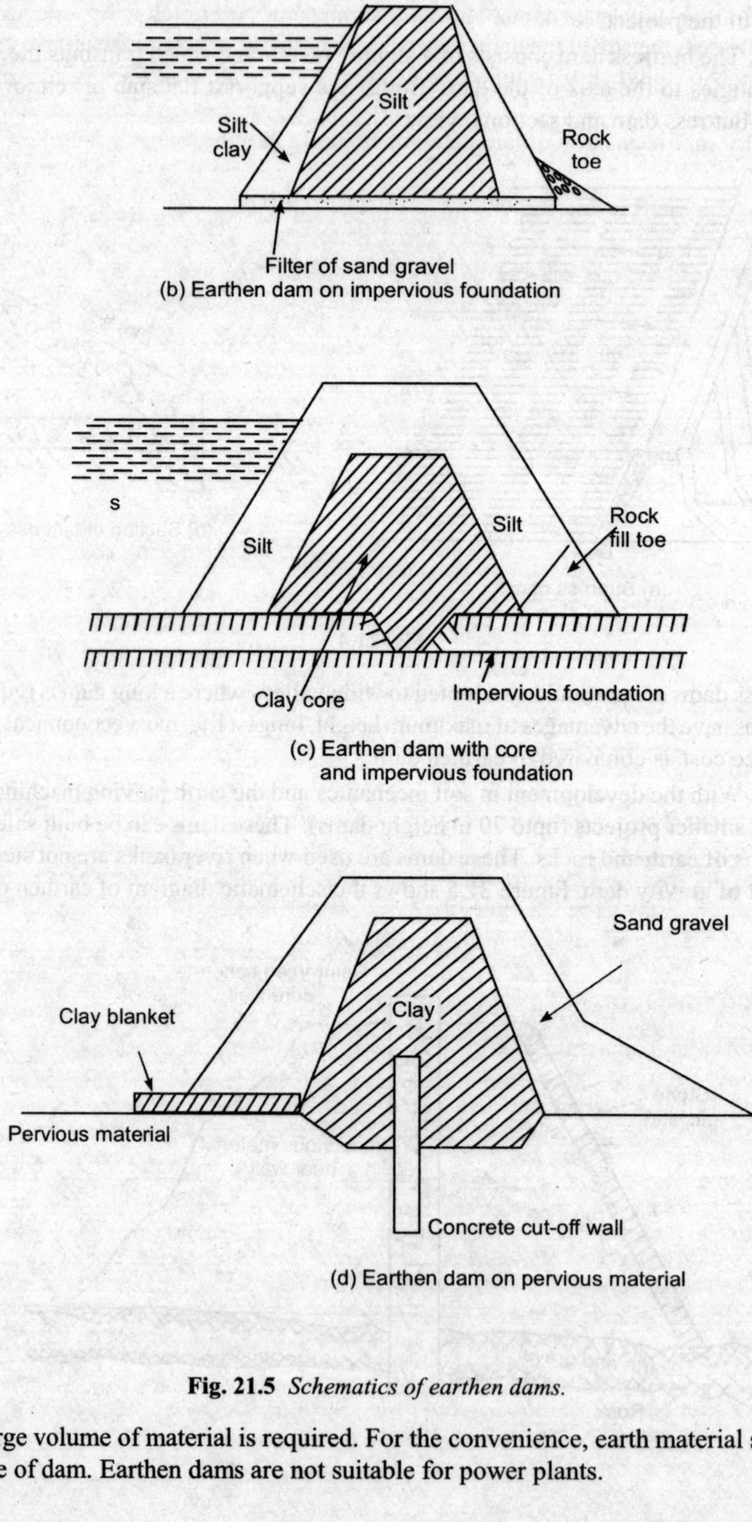

Fig. 21.5 *Schematics of earthen dams.*

For these dams large volume of material is required. For the convenience, earth material should be available in the vicinity of the site of dam. Earthen dams are not suitable for power plants.

(v) Rock fill Dams. This type of the dam has characteristics midway between gravity dams and earthen dams. These dams are not as common as earthen dams. This type of the dam is preferred where adequate quantity of good quality rocks are available near the site of the dam. Figure 21.6 shows the schematic diagram of the simple rock-fill dam. The main components of the rock fill dam are as follows :

(a) Loose rock fill

(b) An upstream dry rubble cushion of laid up stone bonding into the dumped rock

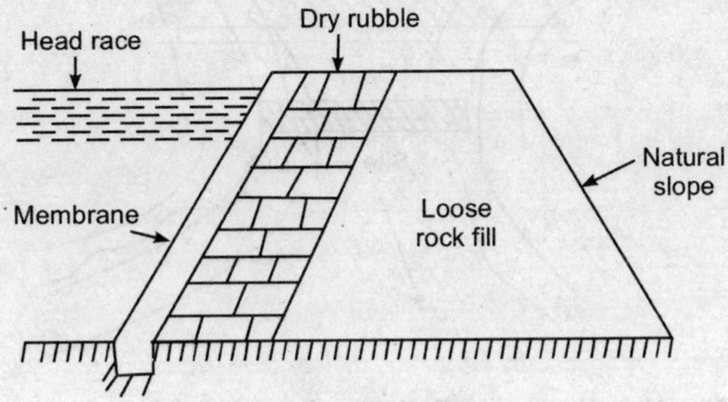

Fig. 21.6 *Rock fill dam.*

(c) An upstream impervious membrane resting on the dry rubble cushion.

Advantages of the rock fill dam are :

(i) High resistance to earthquake (ii) Foundation rock need not be as strong as for gravity dams.

21.2.2 Water Reservoir

Storage of water is called water reservoir. Reservoirs are the basic requirements of a hydro-electric power plant. The water in the reservoir is stored during the wet season and use the same during the dry periods of the year. The water stored in the reservoir is supplied to the water turbines to produce electric power. Reservoirs may be classified as natural such as lake or artificial reservoir which is built by erecting or constructing a dam across the river. The management of water supply from the reservoir is of great importance for the power plants. In addition to power generation, water from reservoirs can be used for irrigation purposes, for controlling the floods, water supply for drinking and other uses for the public and lastly for navigation.

21.2.3 Site Selection for Dam

In addition to the geology of foundations, availability of materials near the site, economy and safety, cost of construction of the dam and the storage capacity is to be considered in selecting the site of the dam. Smaller the length of dam, less will be the cost. For this the site has to be one where river valley has a neck formation as shown in Fig. 21.7.

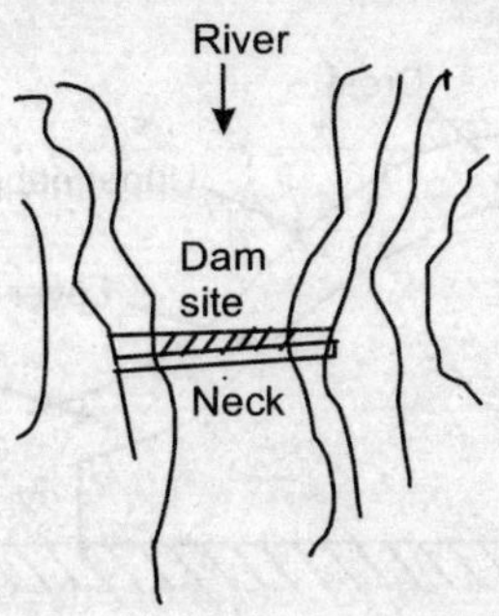

Fig. 21.7 *Neck for dam across river.*

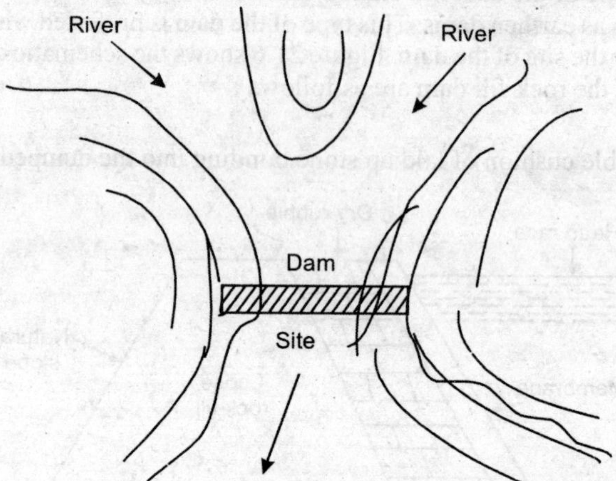

Fig. 21.8 *Confluence for dam at the merging of two rivers.*

Many times the dam is located after the confluence of rivers as shown in Fig. 21.8. In this case, the capacity of storage will be large.

21.2.4 Spillways and Control Gates

During rainy period, the water level in the dam reservoir start rising. The dam height is designed for specific water level in the reservoir. If the water level increases beyond this safe level, the stability of the dam structure is endangered. Some provisions are made in the design of a dam to allow the discharge of water downstream. For this a structure is provided in the body of the dam or close to it. This safeguarding structure is called **spillway**. This provides structural stability to the dam under conditions of flood without raising reservoir level above High Flood Level (HFL).

Following are the types of spillways which are commonly used for different dams :

(a) Overflow spillways (b) Chute or trough spillways
(c) Side channel spillways (d) Shaft spillways
(e) Saddle spillways (f) Siphon spillways.

(a) Overflow Spillways. This type of the spillway is widely used on gravity, arch and buttress dam. This type of the spillway is also called solid gravity spillway. A section of dam designed to permit water to pass over its crest during flood period is known as **Overflow Spillway**.

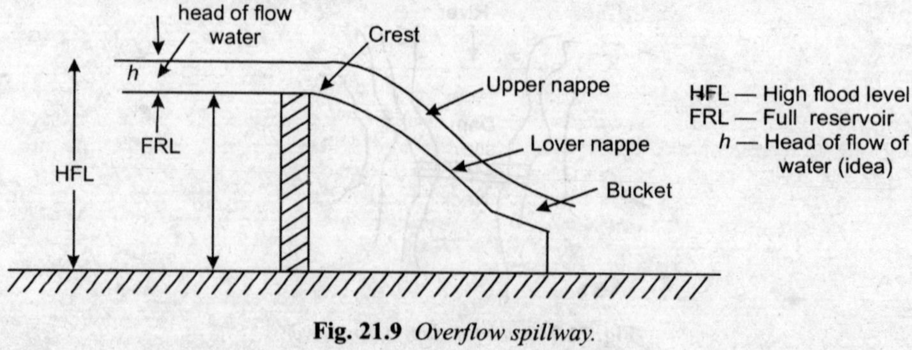

Fig. 21.9 *Overflow spillway.*

Figure 21.9 shows the overflow spillway. In this case, water spills and flows over the crest in the form of a rolling sheet of water. It is always important that the head of flow water over the spillway crest should be minimum, and it should flow with minimum turbulence. The ideal overflow type spillway is called **Ogee Spillway**.

(b) Chute or Trough Spillways. This is the spillway similar to the overflow spillway but isolated from the dam. It is suitable when the valley is too narrow to accommodate the solid gravity spillway in the body of the dam. Its crest is normal to the centre line and it has a discharge channel to the river downstream in an *excavated trench*. This type of spillway is constructed for the Obra and Ramganga dams in India.

(c) Side Channel Spillway. This type of the spillway is provided for non-rigid dams where the flood water is not desired to flow over the dam.

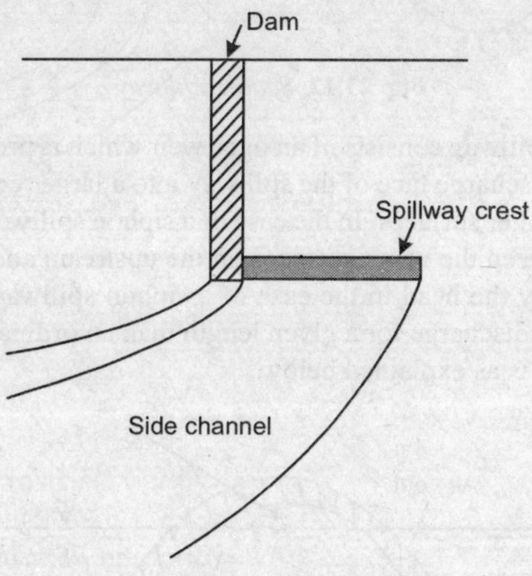

Fig. 21.10 *Side channel spillway.*

Side channel spillway is generally adopted for earthen and rock fill dams because overflow of flood water over the dam is not allowed for the safety of the dam.

When the valley is too narrow and where sufficient length is not available, side channel spillway is used.

(d) Shaft Spillway. Shaft spillway is used where no adequate space for types of the spillways is available. This type of spillway is shown in Fig. 21.11. This spillway is like a hollow shaft to which water is admitted through the intakes at the sides as shown in Fig. 21.11. This shaft is turned at right angles and then taken out below the dam horizontally. If the topography prevents the use of other spillways, a shaft spillway through the foundation is the only solution.

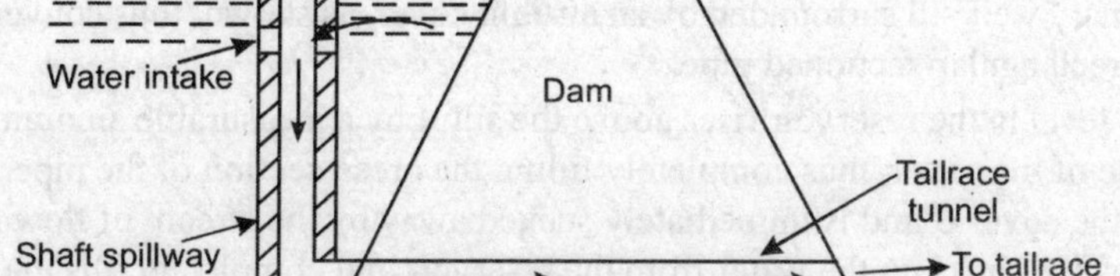

Fig. 21.11 *Shaft spillway.*

(e) Saddle Spillways. Many times there are natural depressions or saddles on the periphery of the reservoir basin away of the dam, and when the conditions are not favourable for any of the spillways, then the saddle spillway is used. Here the bottom of the depression is at the full reservoir level as shown in Fig. 21.12.

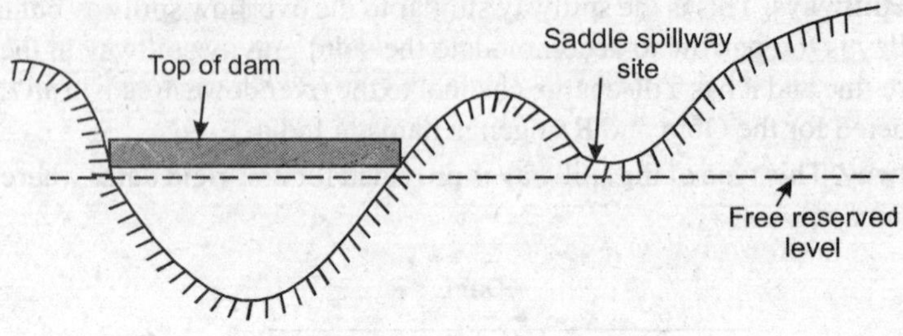

Fig. 21.12 *Saddle spillway.*

(f) Siphon Spillway. A siphon spillway consists of an ogee weir which is provided with air tight cover as shown in Fig. 21.13, thus converting the discharge face of the spillway into a large rectangular sectioned pipe connecting the upstream and the downstream water surfaces. In the case of a siphon spillway the head *H* under which the water flow is equal to the difference between the water surfaces on the upstream and the downstream sides. As such as compared with an ordinary spillway the head in the case of a siphon spillway is more, on account of which the siphon spillway has a much greater discharge for a given length than an ordinary open spillway. The working of a siphon spillway is automatic which is as explained below.

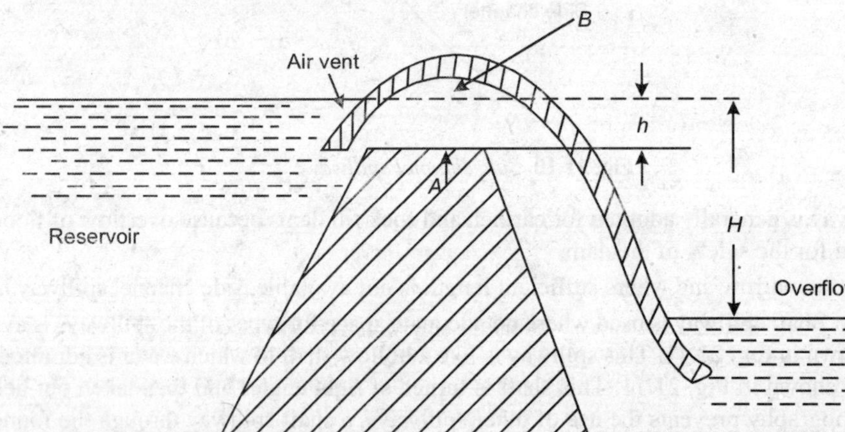

Fig. 21.13 *Automatic siphon spillway.*

It consists of an ordinary weir sill surrounded by an airtight cover, as shown, thus converting the discharge face of the weir into a large rectangular-sectioned pipe.

As soon as the water level in the reservoir rises above the sill *A* by a measurable amount, the water flows over a sill and strikes the inside of the cover, thus completely filling the cross-section of the pipe. The air is now trapped in the upper portion of the cover *B* and is immediately sucked away by the stream of flowing water. This causes a negative pressure at *B* which sucks up the water from the reservoir and completely fills the pipe, the siphon action is thus started. The water will now rush down the pipe to waste, with a velocity caused by the total head *H*, thus causing a large discharge. Had this operation been performed by an ordinary open weir, the velocity of discharge would have been due to the small head h only, and would have been consequently very small. When the level of water in the reservoir reduces and becomes almost equal to the crest level of the spillway, then the siphonic action should stop, otherwise the water in the reservoir may be lowered to a level much below the crest of the spillway. For this purpose air vents are provided in the cover at a level slightly above the crest of the spillway so that when the water in the reservoir drops below this level the air vents are exposed and air enters the upper portion thereby stopping the siphonic action. By such an arrangement the working of a siphon spillway may be made automatic and hence often it is called **automatic siphon spillway**.

21.3 CONTROL GATES

If the capacity of the reservoir is to be increased beyond spillway cress, it can be achieved by providing movable gates. During run-off period, the capacity can be increased with the help of these gates. But during flood period, the gates are required to be opened.

The different types of gates commonly used are :

1. Vertical lift gate
2. Radial gate
3. Rolling gate
4. Drum gate
5. Bear trap gate
6. Tilting gate.

Floods may come in rainy season due to the heavy rains in the catchment area or due heavy rains in the upstream regions other than catchment area. The latter type of the flood may or may not any warning and in that case the operations of the gates should be **automatic**. For this automatic control, methods given below are used :

(a) Float operation control
(b) Balance weight control
(c) Hydraulic control.

21.3.1 Waterways – Tunnels, Canals and Penstocks

A waterway which includes tunnels, canals or penstock is used to carry water from the dam to the power house. **Tunnel** is made by cutting the mountains where topography prevents the use of a canal or a pipe line. Tunnel is made to save the distance.

Penstocks are made up of steel, reinforced concrete and wood. For smaller developments cast iron pipes may be used. The intake of penstock at the dam or forebay of canal should be at a level low enough to provide an adequate water seal under all conditions particularly at low water. It is desirable that the penstock should be sloping towards the power house and its grade may be varied as desired to suit the topography. Sharp bends in the penstock should be avoided because they cause loss of head and require special anchorages. Generally penstocks are not covered because exposed pipes are cheaper and easily accessible for repairs and maintenance. Covered penstocks should be used where topography is such that there is danger from slides of snow, rock, earth etc. The velocity of water in penstock is about 2 m/s in low head power plants and about 4 m/s for medium head power plants and 7 m/s for high head power plants.

There are two types of penstocks :

(i) Exposed penstock
(ii) Burried penstock.

Exposed type penstocks are less expensive to install, easy maintenance and inspection. Its life is more. Burried penstocks are preferred where landslide, snow slide or falling rocks may occur now and then. In case of burried penstocks, corrosion is more and life is less.

Canal is an open waterway. Waterway excavated in natural ground. Flume is an open channel erected on a surface above the ground.

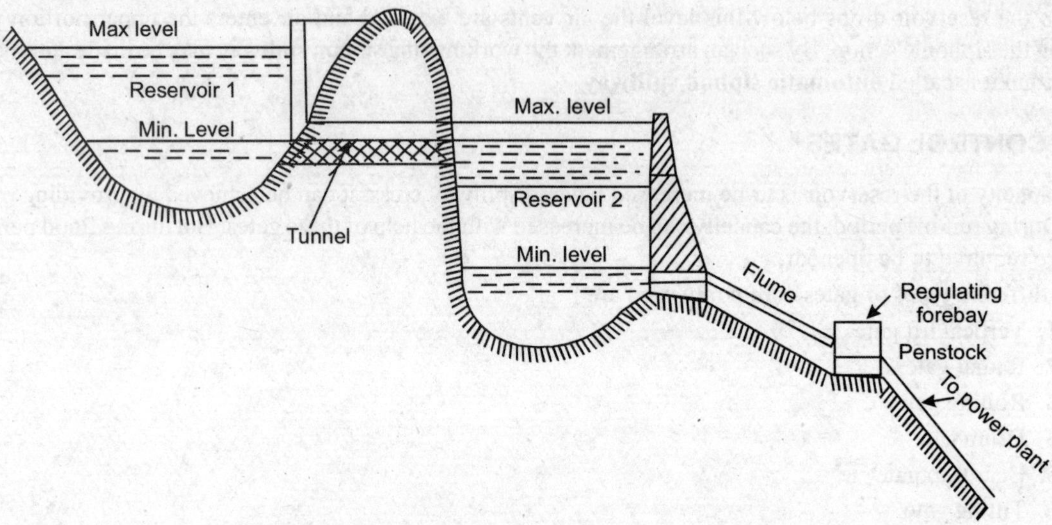

Fig. 21.14 *Waterway including tunnel, flume, penstock.*

21.3.2 Water Hammer and Surge Tank

Major difficulty in water turbine governing is the regulation of the increase of pressure due to inertia effect of the column of water in the supply pipe. On the governor partly closing the gate there will be a slowing down of the water in the supply pipe; this will cause an increase of pressure at the guide vanes which may tend to speed up the turbine. In order to prevent this, a pressure regulator in the form of a spring relief valve is fitted at the turbine end of the supply pipe.

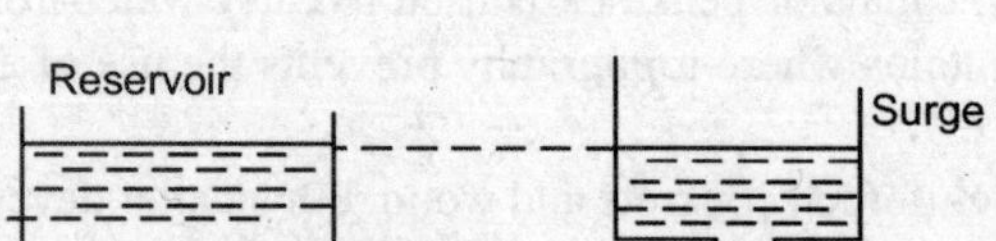

Fig. 21.15 *Surge tank.*

Another method of overcoming the inertia effect of the water column in the supply pipe is to fit a vertical pipe and tank, known as a "surge tank" or stand pipe, on the supply pipe as near to the turbine as possible. This tank in open to the atmosphere at the top. When the turbine gates are closing, the slowing down of the water column in the supply pipe will cause a rise of pressure, and water will flow into the surge tank, thus reducing the shock. When the turbine gates are opening, water will flow from the surge tank into the turbine whilst the water column in the supply pipe is accelerating.

For turbines with very high heads the surge tank is closed at the top, the air trapped-in being compressed and expanded by the closing and opening of the turbine gates. This is the same in principle as the air vessel on the reciprocating pump.

The sudden change of pressure and its fluctuations in the pipeline during reduction of load on turbine is known as **water hammer.**

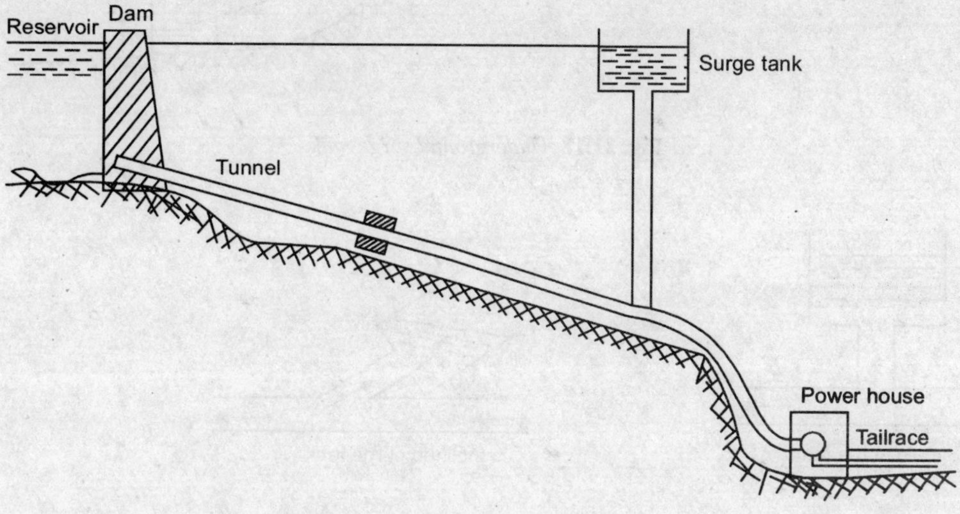

Fig. 21.16 *Surge tank for nearly ground level.*

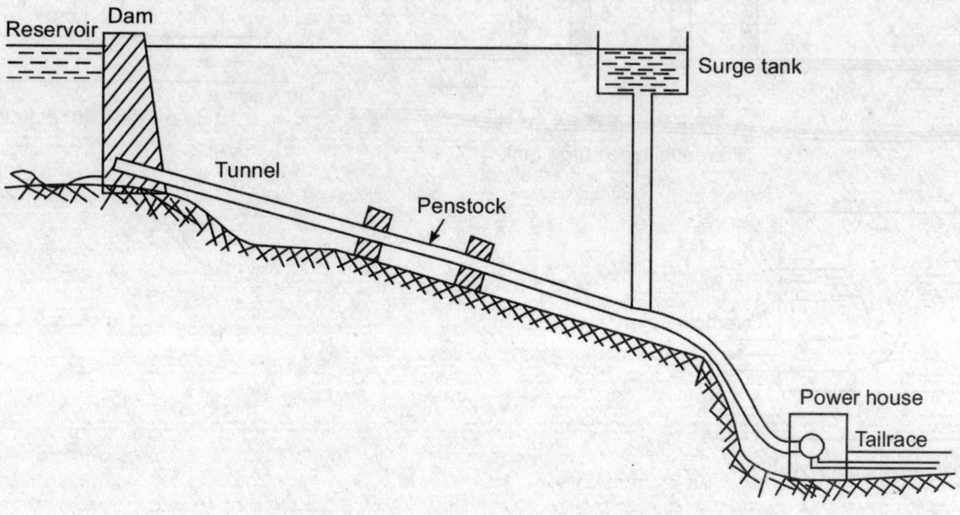

Fig. 21.17 *Surge tank for steep topography.*

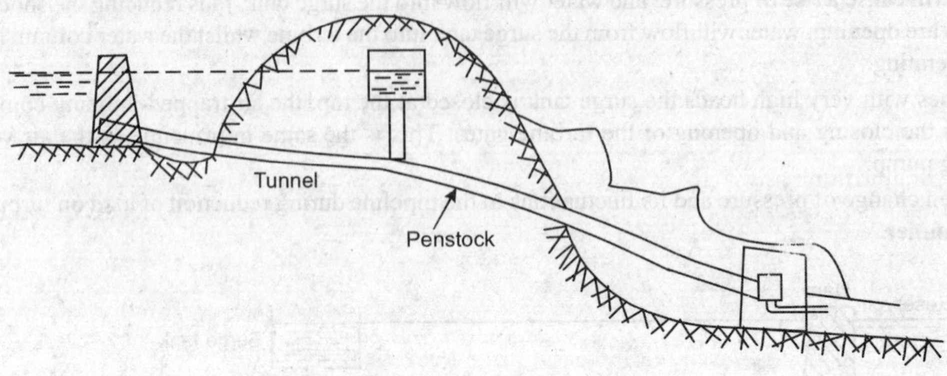

Fig. 21.18 *Underground surge tank.*

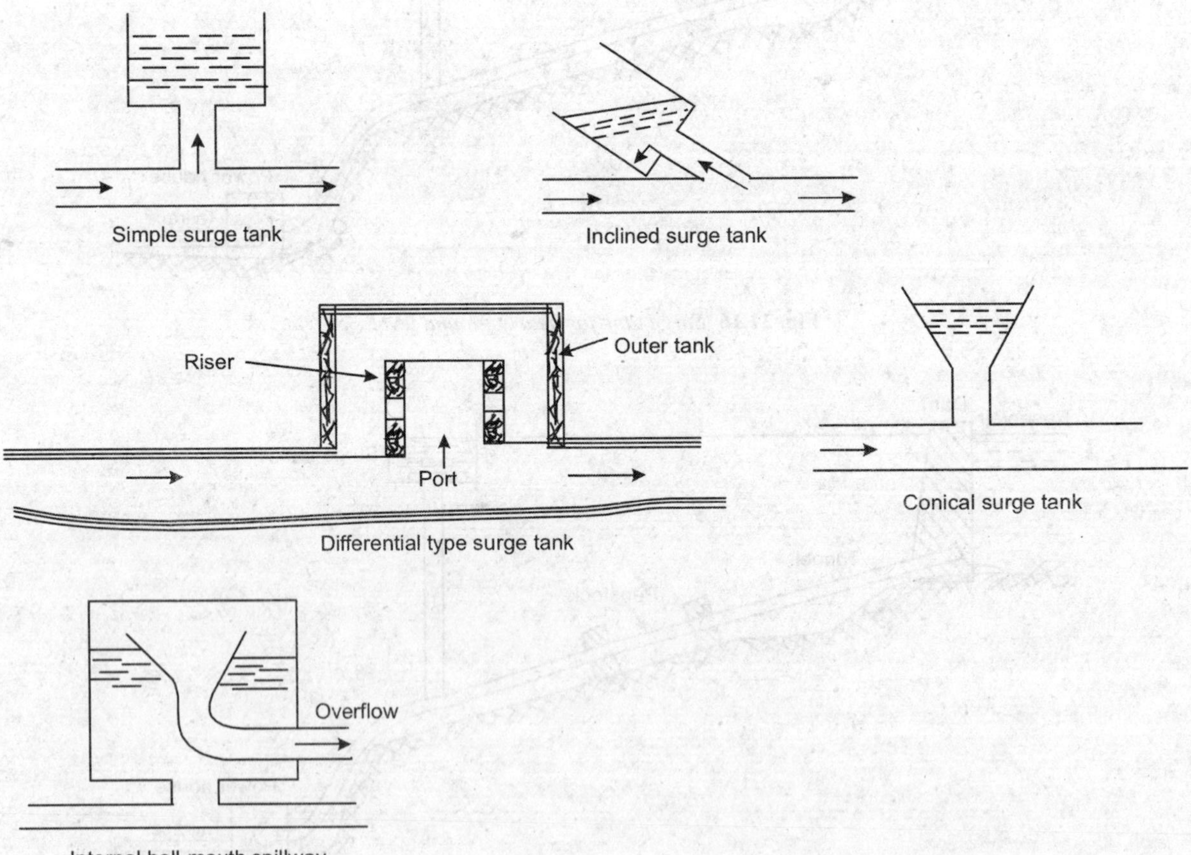

Fig. 21.19 *Different types of surge tank.*

The schematic diagrams for different arrangements are shown in Figs. 21.15–21.19.

21.3.3 Power House

A power house is basically a stable structure and is a main building where the conversion of water energy is converted into electric energy. The various equipments are arranged in the building in such a way that there will be sufficient space for convenient maintenance — dismantling and repair — of the equipment. The equipments provided in the power house are as follows :

(a) Hydraulic turbines
(b) Electric generators
(c) Governors
(d) Relief valves
(e) Gate valves
(f) Water circulating pump and flow measuring equipments
(g) Air duct
(h) Transformers
(i) Switch board equipment and instruments
(j) Cranes
(k) Storage batteries
(l) Oil circuit breakers
(m) Outgoing connections
(n) In some cases shops and offices.

(a) Hydraulic Turbines as Prime Movers. Power was formerly obtained from water by means of water wheels which were revolved either by the weight of the water or by the impulse of the stream. In middle ages water wheels were used for flows mills — particularly for wheat grinding or ore crushing etc. At present there is one four hundred years' old water wheel (Pan Chakki) in Aurangabad (Maharashtra) — working by the action of steam of water. Such wheels are now obsolete and have been replaced by the water turbine.

21.4 CLASSIFICATION OF WATER TURBINES

Hydraulic or water turbines may be classified according to several considerations explained as below :

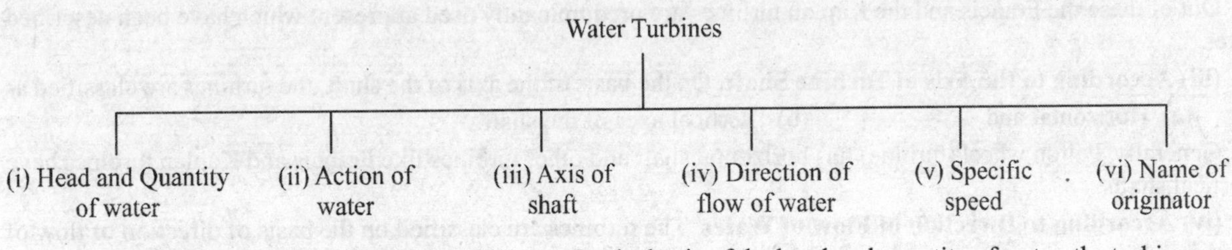

(i) According to Head and Quantity of Water. On the basis of the head and quantity of water, the turbines may be classified as :

1. High head and low quantity turbines,
2. Medium head and medium quantity turbines,
3. Low head and high quantity turbines.

1. High head turbines are those which are capable of working under very high heads ranging from several hundred metres to few thousand metres. These turbines thus require relatively less quantity of water. Generally impulse turbines are high head turbines. In particular Pelton wheel has so far been used under a **highest head** of about (1770 m). Turbines using water head from 71 to 500 and more are called **High head turbines.**

2. Medium head turbines are those which are capable of working under medium heads ranging from 60 m to 250 m. These turbines require relatively large quantity of water. Modern Francis turbines may be classified as **Medium head turbines.**

3. Low head turbines are those which are capable of working heads less than 60 m. These turbines thus require a large quantity of water. Kaplan and other propeller turbines may be classified as **Low head turbines.**

(ii) According to the Action of the Water. According to the action of water flowing the turbine runners the turbines may be classified as :

1. Impulse turbines and
2. Reaction turbines.

1. Impulse Turbine. In an impulse turbine, all the available energy of water is converted into kinetic energy or velocity head by passing it through a convergent nozzle provided at the end of the penstock. The water coming out of the nozzle is formed into a free jet which impinges on a series of buckets or blades of the runner thus causing it to revolve. The runner revolves freely in air. The water is in contact with only a part of the runner at a time, and throughout its action on the runner and in its subsequent flow to the tail race, the water is at atmospheric pressure. A casing is however provided on the runner to prevent splashing and to guide the water discharged from the bucket to the tail race. Some of the impulse turbines are Pelton wheel, Turg impulse wheel, Girard turbine, Banki turbine, Jon Val turbine etc. Out of these turbines only Pelton wheel is predominantly used at present, which has been described later.

2. Reaction Turbine. In a reaction turbine, at the entrance to the runner, only a part of the available energy of water is converted into kinetic energy and a substantial part remains in the form of pressure energy. As water flow through the runner the change from pressure to kinetic energy takes place gradually. As such the pressure at the inlet to the turbine is much higher than the pressure at the outlet and it varies throughout the passage of water through the turbine. For this gradual change of pressure to be possible the runner in this case must be completely enclosed in an air-tight casing and the passage is entirely full of water throughout the operation of the turbine. The difference of pressure (or pressure drop) between the inlet and the outlet of the runner is called **reaction pressure**, and hence these turbines are known as reaction turbines. Some of the reaction turbines are :

(a) Fourneyon
(b) Thomson
(c) Francis
(d) Propeller
(e) Kaplan etc.

Out of these the Francis and the Kaplan turbines are predominantly used at present which have been described later.

(iii) According to the Axis of Turbine Shaft. On the basis of the axis of the shaft, the turbines are classified as

(a) Horizontal and
(b) Vertical axes of the shaft.

Generally, Pelton wheel (turbine) has horizontal shaft and other turbines like Francis and Kaplan turbines have vertical shafts.

(iv) According to Direction of Flow of Water. The turbines are classified on the basis of direction of flow of water as :

(a) Tangential flow turbine
(b) Radial flow turbine
(c) Axial flow turbine and
(d) Mixed flow turbine.

In a **tangential flow turbine** water flows along the tangent to the path of rotation of the runner. Pelton wheel is a tangential flow turbine.

In a radial flow turbine water flows along the radial direction and remains wholly and mainly in the plane normal to the axis of rotation, as it through the runner. A radial flow turbine may be either inward radial flow type or outward radial flow type. In an **inward radial flow** turbine the water enters at the outer circumference and flows radially inwards towards the centre of the runner. Old Francis turbine, Thomson turbine, Girard radial flow turbine etc. are some of the examples of inward radial flow turbine. In an **outward radial flow** turbine water enters at the centre and flows radially outwards toward the outer periphery of the runner. Fourneyron turbine is an example of outward radial flow turbine.

In an **axial flow** turbine the flow of water through the runner is wholly and mainly along the direction parallel to the axis of rotation of the runner. Jonval turbine, Girard axial flow turbine, Propeller turbine, Kaplan turbine etc. are some of the examples of axial flow turbines.

In the **mixed flow** turbine, water enters the runner at the outer periphery in the radial direction and leaves it at the centre in the direction parallel to the axis of rotation of the runner. Modern Francis turbine is an example of the mixed flow type turbine.

(v) According to Specific Speed. The turbines are also classified according to their specific speed. The specific speed of a turbine is the speed of a geometrically similar turbine that would develop one kilowatt power when working under a head of one metre. However, in metric units the specific speed of a geometrically similar turbine that would develop one metric horse-power when working under a head of one metre. On the basis of the specific speed the various turbines may be considered in the following three groups in which the values given in the brackets represent the range of specific speed in metric units.

(i) Specific speed varying from 8.5 to 30 (10–35) – Pelton wheel with single jet and upto 43 (50) for Pelton wheel with double jet.

(ii) Specific speed varying from 50 to 340 (60 to 400) – Francis turbine.

(iii) Specific speed varying from 255 to 860 (300 to 1000) – Kaplan and other propeller turbines.

(vi) According to the Originator

(a) Pelton wheel or turbine — Named after Lester Allen Pelton of USA.

(b) Francis turbine — named after James B. Francis.

(c) Kaplan turbine — named after De Victor Kaplan.

(d) Deriaz turbine — named after the Swiss Engineer Deriaz.

21.5 PELTON WHEEL

This is the only impulse type of hydraulic turbine now in common use. It is named after Lester A Pelton (1829–1908), the American engineer who contributed much to its development in about 1880. It is well suited for operating under high heads.

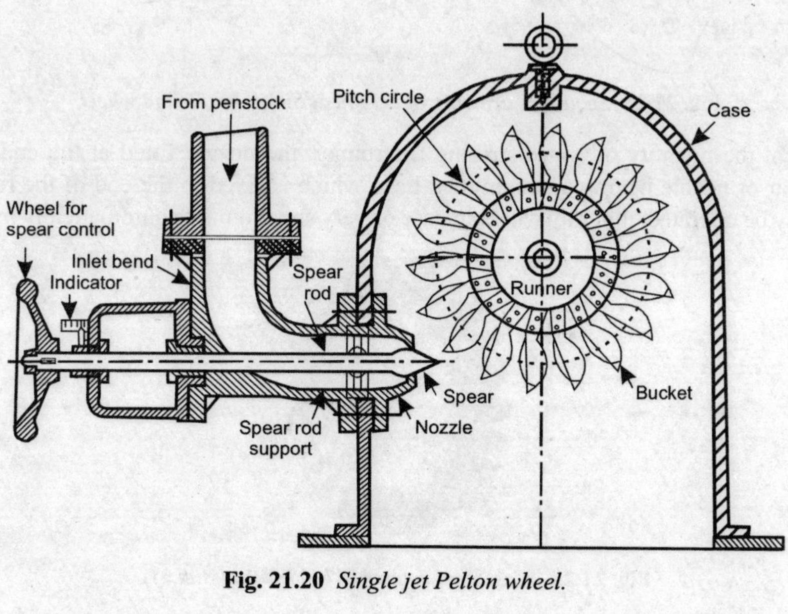

Fig. 21.20 *Single jet Pelton wheel.*

Figure 21.20 shows the elements of a typical Pelton wheel installation. The runner consists of a circular disc with a number of buckets evenly spaced round its periphery. The buckets have a shape of double semi-ellipsoidal cups. Each bucket is divided into two symmetrical parts by a sharp-edged ridge known as splitter. One or more nozzles are mounted so that directs a jet along a tangent to the circle through the centres of the buckets called the pitch circle. The jet of water impinges on the splitter, which divide the jet into two equal portions, each of which after flowing round the smooth inner surface of the bucket leaves it at its outer edge. The buckets are so shaped that the angle at the outlet tip varies from 10° to 20° (usually kept at 15°) so that the jet of water gets deflected through 160° to 170°. The advantage of having a double cup-shaped buckets is that the axial thrust neutralise each other, being equal and opposite, and hence the bearings supporting the wheel shaft are not subjected to any axial or end thrust. The back of the bucket is so shaped that as it swings downward into the jet no water is washed by splashing. Further at the lower tip of the bucket a notch is cut which prevents the jet striking the preceding bucket being intercepted by the next bucket very soon, and it also avoids the deflection of water towards the centre of the wheel as the bucket first meets the jet. For low heads, the buckets are made of cast iron, but for higher heads, they are made of cast steel, bronze or stainless steel.

Figure 21.21 shows a schematic drawing of a typical single-jet Pelton wheel.

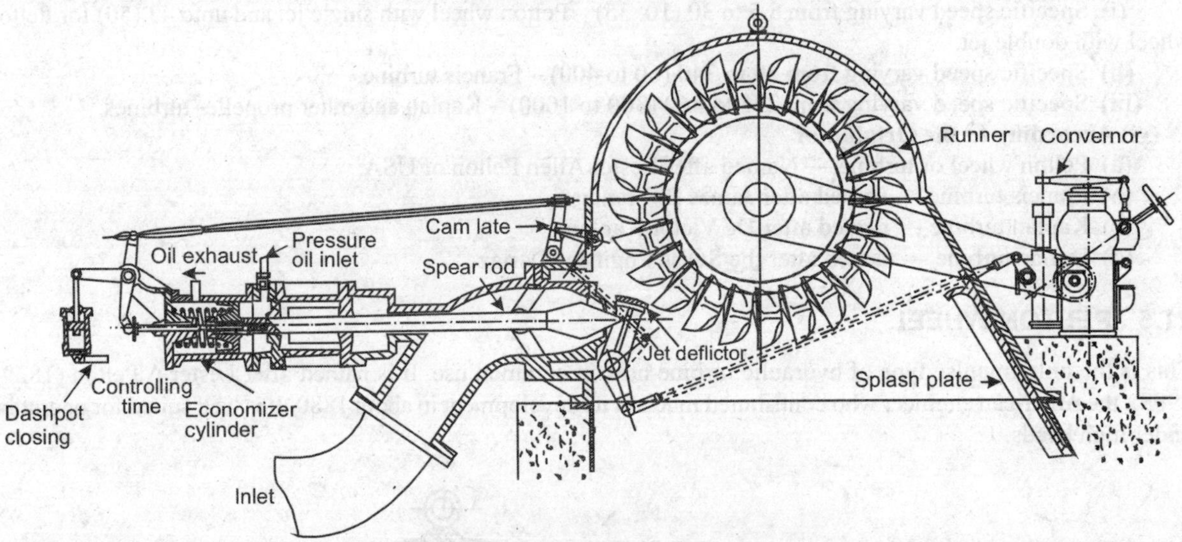

Fig. 21.21 *Sectional drawing of a Typical Single-jet Pelton wheel.*

In order to control the quantity of water striking the runner, the nozzle fitted at the end of the penstock is provided with a spear or needle having a streamlined head which is fixed to the end of the rod as shown in Fig. 21.20. The spear may be operated either by wheel in case of very small units or automatically by a governor in case

Fig. 21.22 *Cast steel runner (6700 kW and 800 M).*

of almost all the bigger units. When the shaft of the Pelton wheel is horizontal then not more than two jets are used. But if the wheel is mounted on a vertical shaft a larger number of jets (upto six) is possible.

Figure 21.22 shows a cast steel runner for 6700 kW head Pelton wheel operating on 800 m head. A casing made of cast iron or fabricated steel plates is usually provided for a Turgo Impulsion Wheel, as shown in Fig. 21.23. It has no hydraulic function to perform. It is provided only to prevent splashing of water, to lead water to the tail race and also to act as a safeguard against accidents.

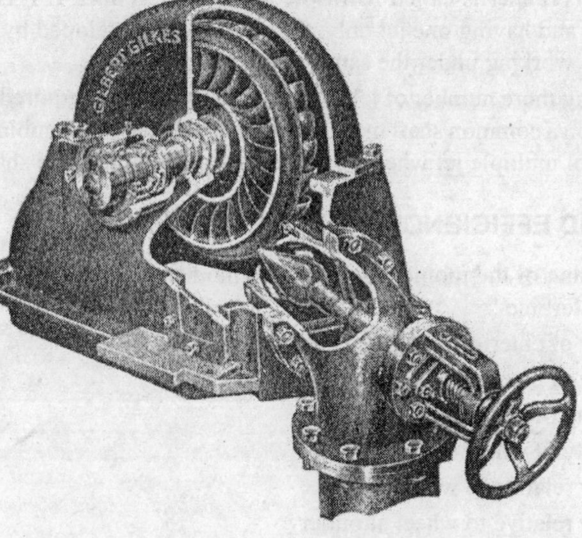

Fig. 21.23 *Sectional view of a typical Turgo Impulsion wheel.*

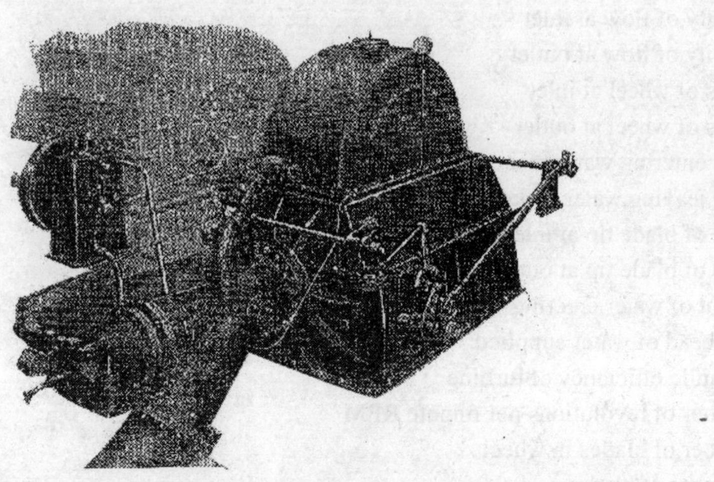

Fig. 21.24 *A 3200 kW Pelton wheel working on a head of 350 m.*

Larger Pelton wheel are usually equipped with small brake nozzle which when opened directs a jet of water on the back of the buckets, thereby bringing the wheel quickly to rest after it is shut down. Figure 21.24 shows a large Pelton wheel generating 3200 kW power and working under a head of 350 m.

21.6 MULTIPLE JET PELTON WHEEL

The power developed by a Pelton wheel provided with a single jet is usually quite low. This is so because on account of the restrictions of the jet velocity, wheel speed and the jet ratio, a single jet cannot be made big enough to develop any desired power. The amount of power developed by a single runner of a Pelton wheel turbine may however be increased by providing more than one jet spaced evenly around the same runner. The nozzles must never be spaced so closely that water issued from one jet after striking runner interferes with another jet. As such the maximum number of jets so far used with a single runner of some large units is six. A Pelton wheel having more than one jet spaced around its runner is called **Multiple Jet Pelton Wheel**. If P is the power, by a Pelton wheel when working under head H and having one jet only, then the power developed by the same Pelton wheel will be (nP), if n jets are used for its working under the same head.

Sometimes even if by using more number of jets for a single runner, the required power is not developed then a number of runners mounted on a common shaft may be used. In some cases a combination of the above two systems may be used, i.e., a number of multiple jet wheels may be mounted on the same shaft.

21.7 WORK DONE AND EFFICIENCIES OF PELTON WHEEL

In the design of turbine blading of the momentum type, the momentum equation is used. The following notation will be used for types of the turbine :

V_1 = absolute velocity of entering water
V_2 = absolute velocity of entering water
V_{b1} = tangential velocity of wheel at inlet
V_{b2} = tangential velocity of wheel at outlet
V_{r1} = velocity of water relative to wheel at inlet
V_{r2} = velocity of water relative to wheel at outlet
V_{w1} = velocity of whirl at inlet
V_{w2} = velocity of whirl at outlet
V_{f1} = velocity of flow at inlet
V_{f2} = velocity of flow at outlet
r_1 = radius of wheel at inlet
r_2 = radius of wheel at outlet
α = angle entering water makes with wheel's tangent
β = angle leaving water makes with wheel's tangent
θ = angle of blade tip at inlet
ϕ = angle of blade tip at outlet
m_w = weight of water entering wheel in kg/second
H = total head of water supplied
η_H = hydraulic efficiency of turbine
N = Number of revolutions per minute RPM
n = Number of blades in wheel
t = thickness of blades
b_1 = breadth of blades at inlet
b_2 = breadth of blades at outlet.

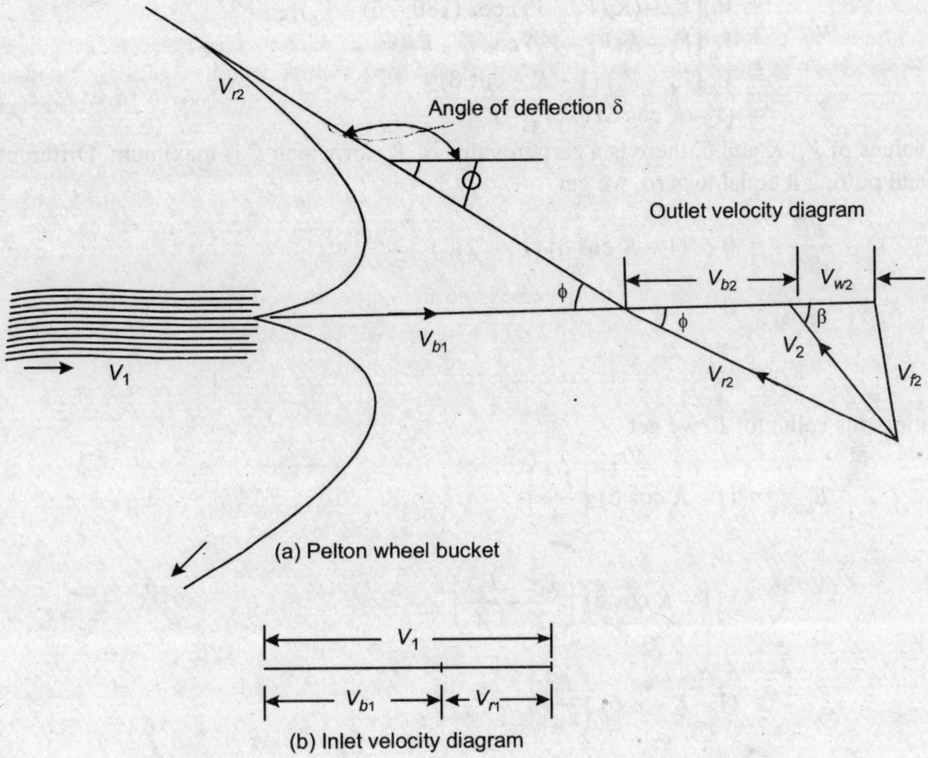

Fig. 21.25 *Pelton wheel bucket and velocity diagrams.*

Figure 21.25 shows a Pelton wheel bucket and the velocity diagrams for the same. In this case,

$V_{w1} = V_1 =$ Velocity of whirl at inlet.

$V_{r1} =$ Velocity relative to blade or bucket at inlet.

$= V_1 - V_{b1}$

$= V_{r2}$ if friction is absent

$V_{r2} = K.V_{r1}$ if friction or loss of velocity due to friction and K is friction coefficient.

From Fig. 21.25,

$V_{w2} =$ Velocity of whirl at outlet

$= -V_{r2} \cos \phi + V_{b2}$ with proper sign.

$= -K V_{r1} \cos \phi + V_{b2} = V_b + K V_{r1} \cos \delta$

For Pelton wheel, $V_{b1} = V_{b2} = V_b$.

∴ Total energy transferred to the wheel is given by Euler equation.

∴ $E =$ Work done $= (V_{w1} V_{b1} - V_{w2} V_{b2})$

$= V_b(V_{w1} - V_{w2})$

Let $\delta =$ angle of deflection of the jet

∴ $\phi = 180 - \delta$.

∴ $E = V_b[V_1 - (V_{r2} \cos \phi - V_b)]$

$$= V_b\,[V_1 - (KV_1 \cos\phi - V_b)]$$
$$= V_b\,[V_1 - (K\,(V_1 - V_b)\cos(180 - \delta) - V_b)]$$
$$= V_b\,[V_1 - K(V_1 - V_b)\cos\delta - V_b]$$
$$= V_b\,(V_1 - V_b)\,(1 - K - \cos\delta)$$
$$= (1 - K\cos\delta)\,(V_1 V_b - V_b^2)$$

For given values of V_1, K and δ, there is a certain value of V_b for which E is maximum. Differentiating E with respect to V_b and putting it equal to zero, we get

$$\frac{dE}{dV_b} = 0 = (1 - K\cos\delta)\,(V_1 - 2V_b)$$

$$\therefore \quad V_1 - 2V_b = 0$$

or $$V_b = \frac{V_1}{2}$$

$\therefore$ Substituting this value for E, we get

$$E_{max} = (1 - K\cos\delta)\left(\frac{V_1^2}{4}\right)$$

$$=(1 - K\cos\delta)\left(\frac{V_1^2}{2} - \frac{V_1^2}{4}\right)$$

$$= (1 - K\cos\delta)\left(\frac{V_1^2}{4}\right).$$

$$\boxed{E_{max} = (1 - K\cos\delta)\left(\frac{V_1^2}{4}\right)}$$

Kinetic energy of the entering jet $= \dfrac{V_1^2}{2}$

$\therefore$ The blade or diagram or hydraulic efficiency of the wheel is given by

$$(\eta_b)_{max} = \frac{E}{V_{1/2}^2} = \frac{(1 - K\cos\delta)\left(\frac{V_1^2}{4}\right)}{\frac{V_1^2}{2}}$$

$$= \frac{1 - K\cos\delta}{2}$$

or $$\eta_b = (1 - K\cos\delta)\,(V_1 V_b - V_b^2) \times \frac{2}{V_1^2}$$

$$= 2\,(1 - K\cos\delta)\left(\frac{V_b}{V_1} - \frac{V_b^2}{V_1^2}\right)$$

Let $\frac{V_b}{V_1}$ = Velocity ratio = x

$\therefore$ $\eta_b = 2(1 - K\cos\delta)(x - x^2)$

$\therefore$ For maximum efficiency,

$$\frac{d\eta_b}{dx} = 0 = 2(1 - K\cos\delta)(1 - 2x)$$

or $1 - 2x = 0$

$$x = 1/2$$

$$(\eta_b)_{max} = 2(1 - K\cos\delta)\left(\frac{1}{2} - \frac{1}{4}\right)$$

$$= \frac{1}{2}(1 - K\cos\delta).$$

$$\boxed{(\eta_b)_{max} = \frac{1}{2}(1 - K\cos\delta)}$$

$\therefore$ **For maximum efficiency, blade velocity is half the jet velocity.**

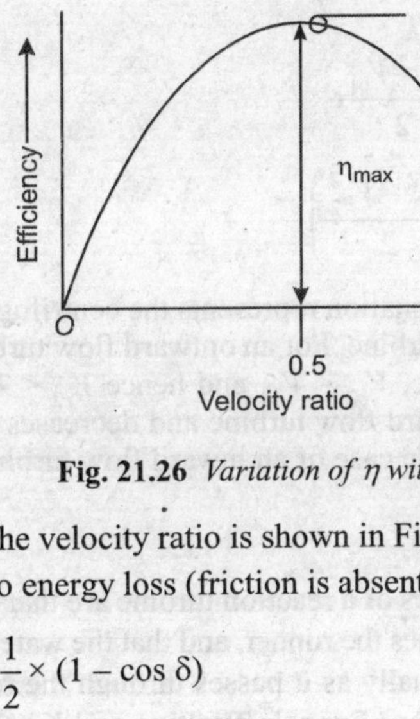

Fig. 21.26 *Variation of η with velocity ratio.*

Variation of efficiency with the velocity ratio is shown in Fig. 21.26.

If $K = 1$, i.e., when there is no energy loss (friction is absent) then the efficiency is given by

$$\eta_{max} = \frac{1}{2} \times (1 - \cos\delta)$$

Again if $\delta = 180$ (semi-spherical bucket cups)

$$\eta_{max} = \frac{1}{2}(1 + 1) = \frac{2}{2}$$

then $= \mathbf{1}$ i.e. **100%**

Generally, value of K lies between 0.8 and 0.85 and $\delta = 165°$, so that the water at outlet does not strike the following bucket.

$$\therefore \quad \eta_{max} = \frac{1-0.8\cos 165°}{2}$$

$$= \mathbf{0.886.}$$

In practice, the velocity ratio $x = 0.46$ instead of 0.5 and $V_1 = C_v\sqrt{2gH}$ where C_v = velocity coefficient.

21.8 RADIAL FLOW IMPULSE TURBINE

For a radial flow impulse turbine the inlet velocity triangle is not a straight line and hence

$$\text{Work done} = m\,(V_{w1}\,V_{b1} \pm V_{w2}V_{b2})$$

Further if there is no loss of energy in the runner vanes then the work done may also be expressed by

$$m\,(V_{w1}\,V_{b1} \pm V_{w2}V_{b2}) = \frac{m}{2}\left(V_1^2 - V_2^2\right).$$

or

$$V_{w1}V_{b1} \pm V_{w2}V_{b2} = \frac{V_1^2 - V_2^2}{2}$$

but

$$V_{w1}V_{b1} \pm V_{w2}V_{b2} = \frac{V_1^2 - V_2^2}{2} + \frac{V_{b1}^2 - V_{b2}^2}{2} + \frac{V_{r2}^2 - V_{r1}^2}{2}$$

$$\therefore \quad O = \frac{V_{b1}^2 - V_{b2}^2}{2} + \frac{V_{r2}^2 - V_{r1}^2}{2}$$

or

$$\frac{V_{r2}^2}{2} = \frac{V_{r1}^2}{2} + \frac{V_{b2}^2 - V_{b1}^2}{2}$$

$$= \frac{V_{r1}^2}{2} - \left(\frac{V_{b1}^2 - V_{b2}^2}{2}\right)$$

The second term on the right hand side of this equation represents the centrifugal head impressed on water as it flows through the runner of a radial flow impulse turbine. For an outward flow turbine $V_{b2} > V_{b1}$ and then from this equation, $V_{r2} > V_{r1}$ and for an inward flow turbine, $V_{b2} < V_{b1}$ and hence $V_{r2} < V_{r1}$. That is the centrifugal head increases the relative velocity of water in an outward flow turbine and decreases it in an inward flow turbine. As such a better control of speed can be enforced in the case of an inward flow turbine.

21.9 REACTION TURBINES

As stated earlier the principal distinguishing features of a reaction turbine are that the only part of the total head of water is converted into velocity head before it reaches the runner, and that the water completely fills all the passage in the runner. The pressure of water changes gradually as it passes through the runner. The two reaction type of turbines which are predominantly used these days are **Francis Turbine** and Kaplan turbine, which are described below.

21.10 FRANCIS TURBINE

Figure 21.27 illustrates a Francis turbine which is a mixed flow type reaction turbine. It is named in honour of James B. Francis (1815–92), an American Engineer, who was the first to develop an inward radial flow type of reaction turbine in 1849. Later on it was modified and the modern Francis turbine is a mixed flow type, in which water enters the runner radially at its outer periphery and leaves axially at its centre.

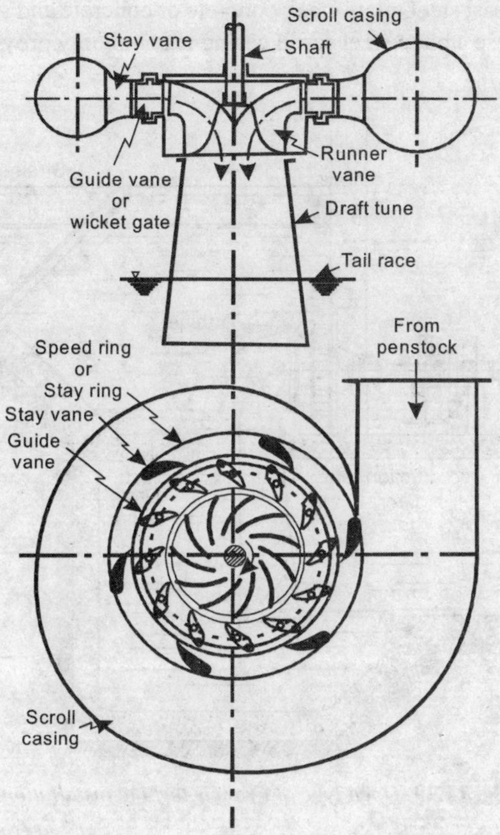

Fig. 21.27 *Sectional arrangement of Francis turbine.*

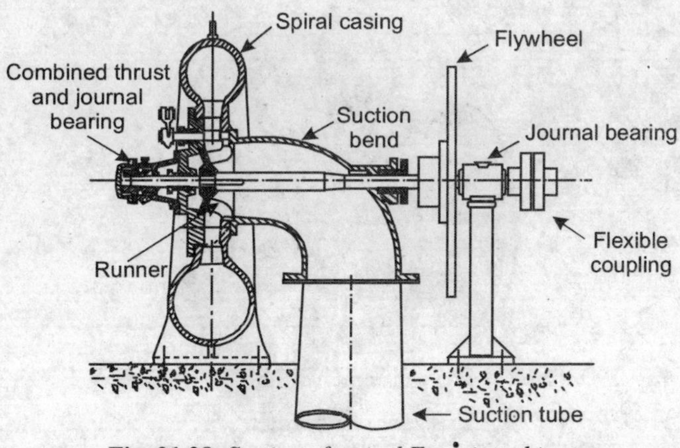

Fig. 21.28 *Section of typical Francis turbine.*

The water from the penstock enters a **scroll easing** (also called spiral casing) which completely surround the runner. The purpose of the casing is to provide an even distribution of water around the circumference of the turbine runner, maintaining an approximately constant velocity for the water so distributed. In order to keep the velocity of water constant throughout its path around the runner, the cross-sectional area of the casing is generally decreased. The casing is made of cast steel, plate steel, concrete or concrete and steel depending upon the pressure to which it is subjected. Out of these a plate steel scroll casing is commonly provided for turbines operating under 30 m or higher heads.

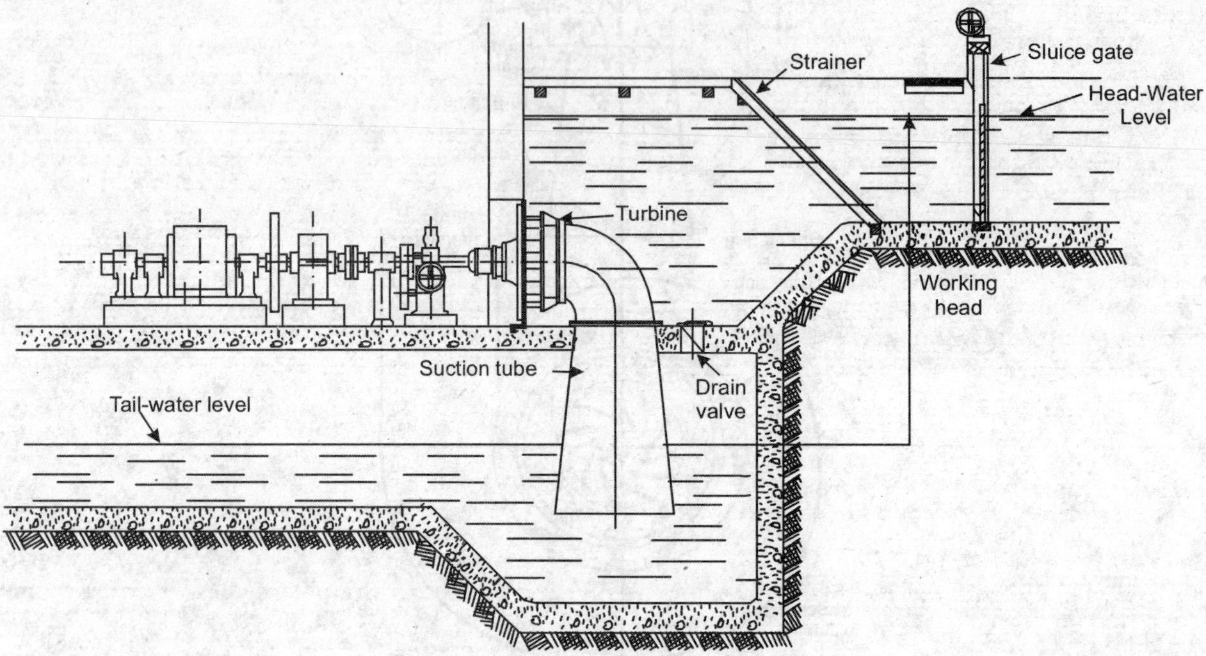

Fig. 21.29 *Open type of Francis turbine installation.*

From the scroll casing the water passes through a speed or stray ring. The speed ring consists of an upper and a lower ring held by series of fixed vanes called **stay vanes**. The number of stay vanes is usually taken as half the

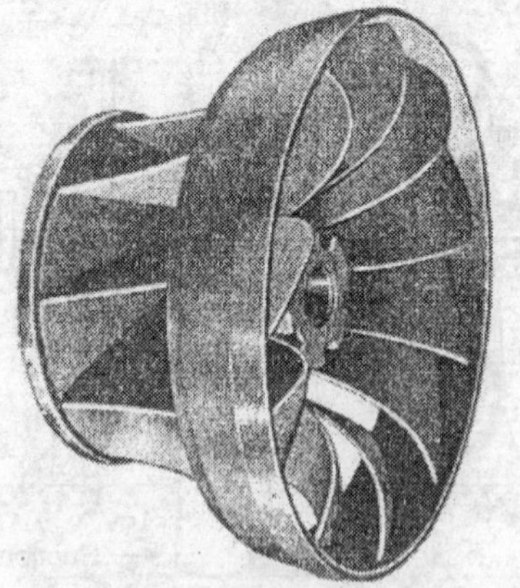

Fig. 21.30 *Francis turbine runner.*

number of guide vanes. The speed ring has two functions to perform. It directs the water from the scroll casing to the guide vanes or wicket gates. Further it resists the load imposed upon it by the internal pressure of water and the weight of the turbine and the electrical generator and transmits the same to the foundation. The speed ring may be either of cast iron or cast steel or fabricated steel.

From the speed ring the water passes through a series of **guide vanes** or **wicket gates** provided all around the periphery of the turbine runner. The function of guide vanes is to regulate the quantity of water supplied to the runner and to direct water on to the runner at an angle appropriate to the design. The guide vanes are airfoil shaped

Fig. 21.31 *Guide-vane operating gear of a spiral-cased Francis turbine.*

and they may be made of cast steel, stainless steel or plate steel. Each guide vane is provided with two stems, the upper stem passes through the head cover and the lower stem seats in a bottom ring. By a system of levers and links, all the guide vanes may be turned about their stems, so as to alter the width of the passage between the adjacent guide vanes, thereby allowing a variable quantity of water to strike the runner. The guide vanes are operated either by means of a wheel or automatically by a governor.

The main purpose of the various components so far described is to lead the water to the runner with a minima loss of energy. The runner of the Francis turbine consists of a series of a curved vanes (about 16 to 24) evenly arranged around the circumference in the annular space between two plates. The vanes are so shaped that the water enters the runner radially at the outer periphery and leaves it axially at the inner periphery. The change in the direction of flow of water, from radial to axial, as it passes through the runner, produces a circumferential force on the runner which makes the runner to rotate and thus contributes to the useful output of the runner. The runners are usually made up of cast iron, cast steel, mild steel or stainless steel. Often instead of making the complete runner of stainless steel, only those portions of the runner blades, which may be subjected to cavitation erosion, are made of stainless steel. This reduces the cost of the runner and at the same time ensures the operation of the runner with a minimum amount of maintenance. The runner is keyed to a shaft which is usually of forged steel. The torque produced by the runner is transmitted to the generator through the shaft which is usually connected to the generator shaft by a bolted flanged connection.

Figures 21.28 to 21.32 are the schematic diagrams showing the parts and plants for Francis turbine.

The water after passing through the runner flows to the tail race through a draft tube (See section 21.15).

For Francis turbine

$$\text{Work done/kg} = (V_{w_1} V_{b_1} - V_{w_2} V_b)$$

For axial outlet, $V_{w_2} = 0.$

$$\therefore \quad \text{Work done/kg} = V_{w_1} V_{b_1}.$$

If H = Net head

$$\therefore \quad \frac{\text{Input energy}}{\text{kg}} = H$$

∴ Hydraulic efficiency of the Francis turbine is given by

$$\eta_h = \frac{V_{w_1} V_{b_1}}{H}$$

or if $V_{w2} \neq 0$, then

$$\eta_n = \frac{V_{w_1} V_{b_2} - V_{w_1} V_{b_2}}{H}$$

Generally, hydraulic efficiency of Francis turbine ranges from 0.85 to 0.95.

Again if P = Power available at the runner shaft, then

$$\eta_m = \text{Mechanical efficiency}$$

$$= \frac{P}{m\left(V_{w_1} V_{b_1}\right)}$$

where m = Mass or weight of water flowing through runner.

and

$$\eta_o = \text{Overall efficiency of the Francis turbine}$$

$$= \eta_h \times \eta_m$$

$$= \frac{P}{m\,H}$$

Generally this overall efficiency ranges from 0.8 to 0.9.

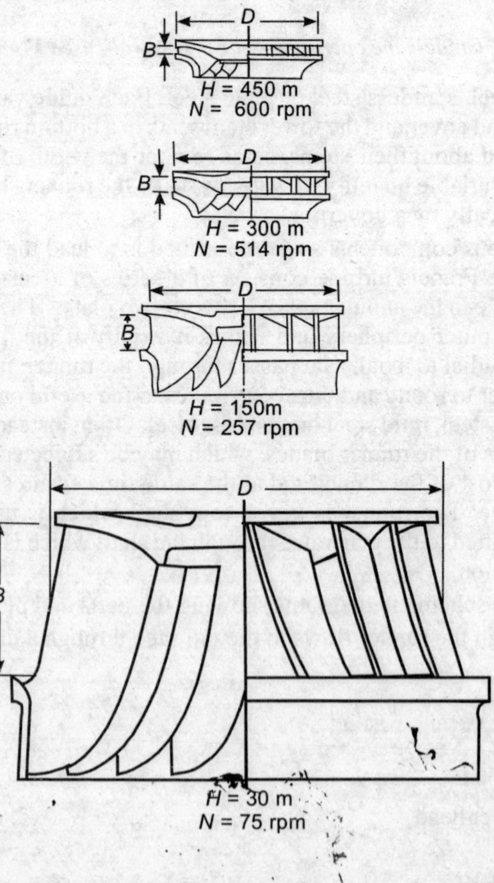

Fig. 21.32 *Relative sizes of Francis turbine to produce 73500 kW.*

Degree of Reaction

Degree of reaction R, is defined as the ratio of pressure drop in the runner to the hydraulic work done on the runner. Thus if P_1 and P_2 are the pressures at the inlet and outlet of the runner, then

$$R = \frac{\frac{P_1}{w} - \frac{P_2}{w}}{V_{w_1}V_{b_1} - V_{w_2}V_{b_2}}$$

$\therefore$ If $V_{w_2} = 0$, then

$$R = \left(\frac{\frac{P_1}{w} - \frac{P_2}{w}}{V_{w_1}V_{b_1}} \right)$$

In Francis turbine

(i) **n = B/D** $= \frac{\text{Breadth of runner}}{\text{Diameter of runner}}$

ranges from **0.1 to 0.45**

(ii) Speed ratio, $x = \frac{V_{b_1}}{V_1}$

$$= \frac{V_{b_1}}{\sqrt{2gH}}$$

$$= \mathbf{0.15 \text{ to } 0.3}$$

(iii) The ratio of the velocity of flow V_{f_1} at the inlet tip of the vane to the spouting velocity $\sqrt{2gH}$ is known as **flow ratio** ψ is given by

$$\psi = \frac{V_{f_1}}{\sqrt{2gH}}$$

and ranges from 0.15 to 0.3

21.11 KAPLAN TURBINE

A Kaplan turbine is a type of propeller turbine which was developed by the Austrian engineer V. Kaplan (1876–1934). It is an axial flow turbine, which is suitable for relatively low heads, and hence requires a large quantity of water to develop large amount of power. It is also a reaction type of turbine and hence it operates in an entirely closed circuit from the head race to the tail race.

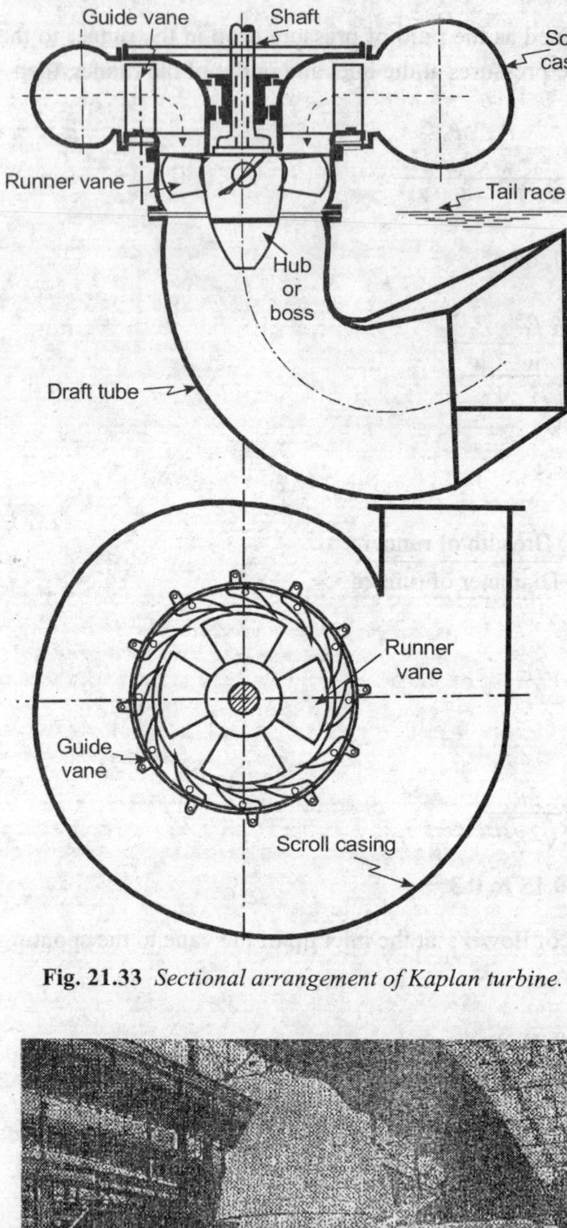

Fig. 21.33 *Sectional arrangement of Kaplan turbine.*

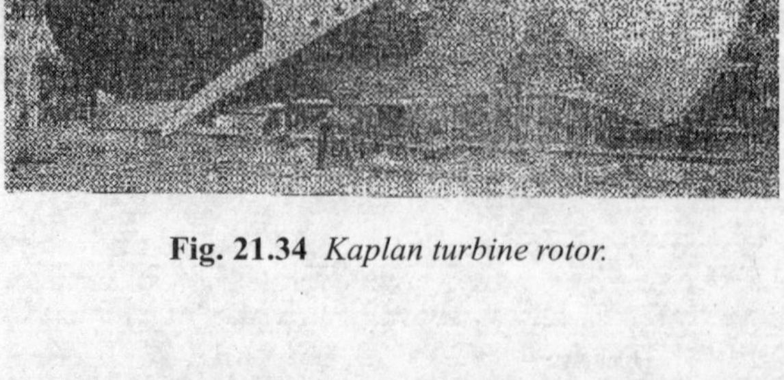

Fig. 21.34 *Kaplan turbine rotor.*

Figures 21.33 to 21.35 shows the various arrangements of Kaplan turbine. From these figures it will be seen that the main components of a Kaplan turbine such as scroll casing, stay ring, arrangement of guide vanes, and the draft tube are similar to those of Francis turbine. Between the guide vanes and the runner the water in Kaplan turbine

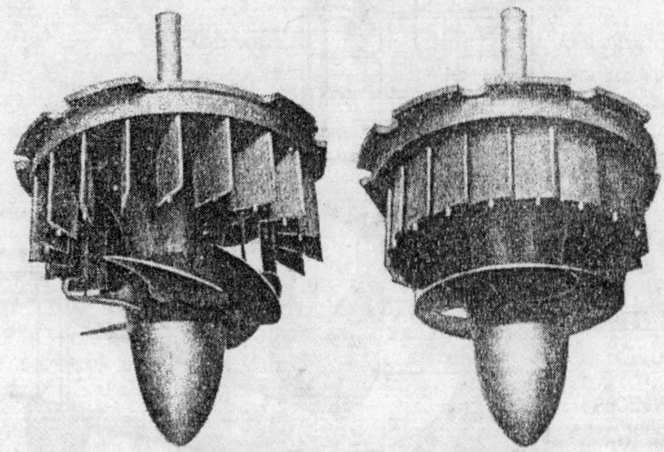

Fig. 21.35 *Kaplan turbine runner.*

turns through a right-angle into the axial direction and then passes through the runner. The runner of a Kaplan turbine has four or six or eight blades and it closely resembles a ship's propeller. The blades or vanes attached to a hub or boss are so shaped that water flows axially through the runner. Ordinarily the runner blades of a propeller turbine are fixed, the Kaplan turbine runner blades can be turned about their own axis, so that their angle of inclination may be adjusted while the turbine is in motion. This adjustment of the runner blades is usually carried out automatically by means of a servomotor operating inside the hollow coupling of turbine and generator shaft. When both guide vane angle and runner blade angle may thus may be varied, a high efficiency can be maintained over a wide range of operating conditions. In other words even at part load, when a lower discharge is flowing through the runner, a high efficiency can be attained in the case of a Kaplan turbine.

If, d = hub or boss diameter

D = runner outside diameter

then $\frac{d}{D}$ varies from 0.35 to 0.60

Again the discharge Q flowing through the runner is given by

$$Q = \frac{\pi}{4}(D^2 - d^2)\, V_f$$

$$= \frac{\pi}{4}(D^2 - d^2)\, \psi\, \sqrt{2gh}$$

where $\psi = 0.7$ = flow ratio for a Kaplan turbine.

The runner blades of Kaplan turbine runner are wraped or twisted, the blade angle being greater at the outer tip than at the hub.

The expressions for the work done and the efficiencies of Kaplan turbine are same as those for Francis turbine.

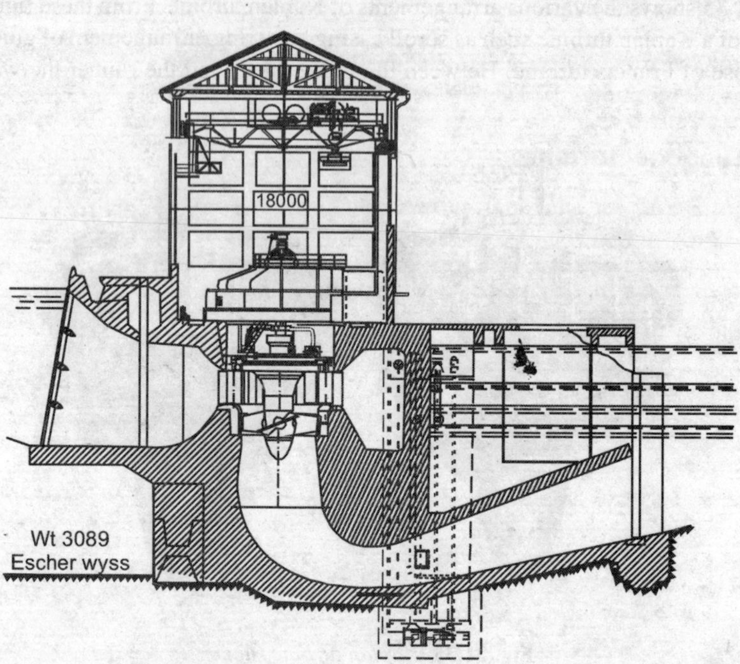

Fig. 21.36 *Cross-section through the power plant.*

Fig. 21.37 *Longitudinal section of a turbine of the power station.*

Figures 21.36 and 21.37 shows the two power plants arrangements.

21.12 NEW TYPES OF TURBINES

In addition to the conventional water turbines described above, there are three new types of water turbines :

(i) Deriaz or Diagonal turbine (ii) Tubular turbine (iii) Bulb turbine.

21.12.1 Deriaz or Diagonal turbine

Fig. 21.38 is a diagram of Deriaz (or Diagonal) turbine which is a reaction turbine. It is named in honour of its inventor P Deriaz. This type of the turbine is intermediate between the mixed flow and the axial flow turbines,

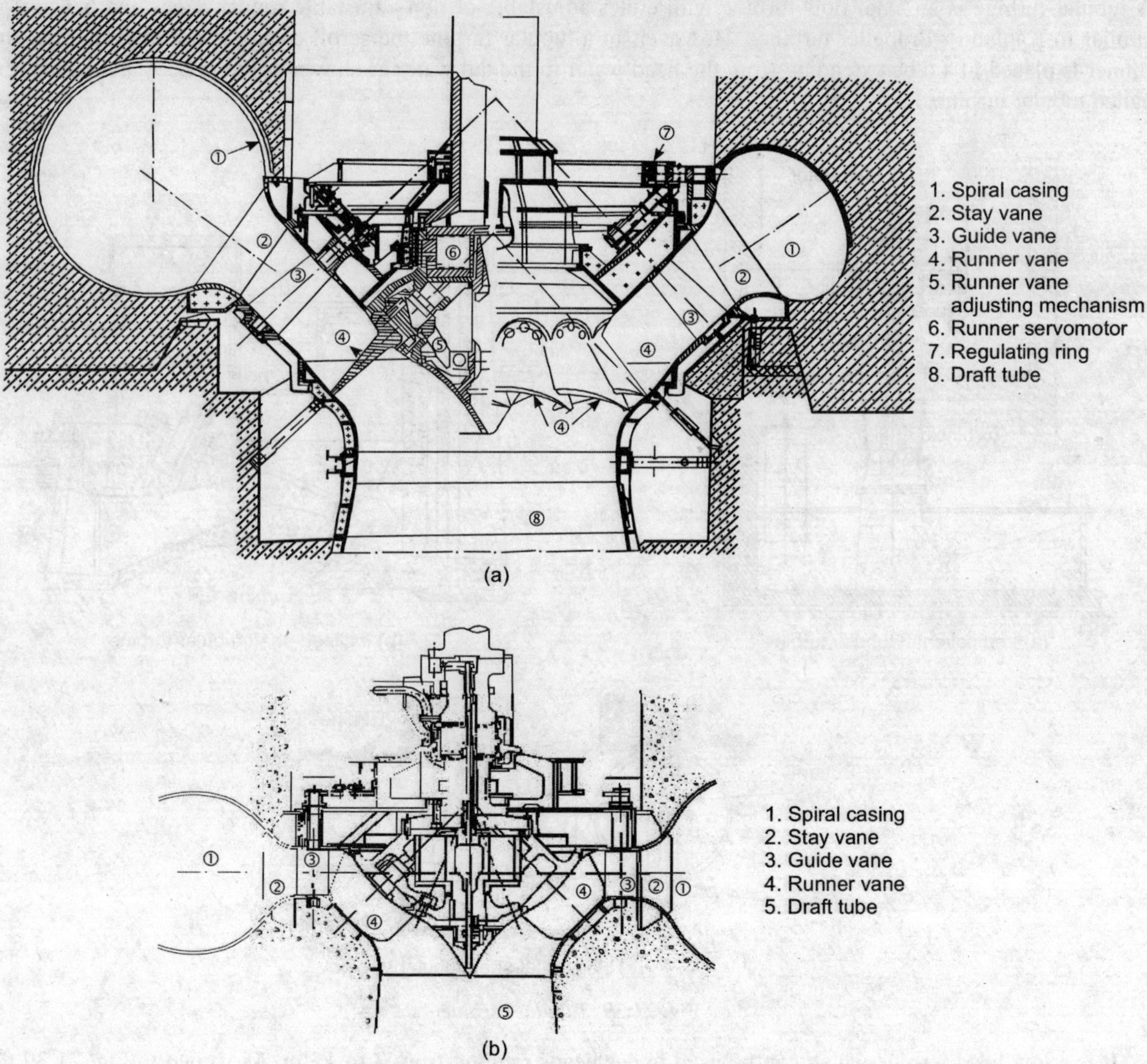

Fig. 21.38 *Deriaz or Diagonal turbine.*

because the flow of water as it passes through the runner is at an angle of 45° to the axis and it is also known as Diagonal turbine. It can be used under heads ranging from 30 to 150 m. Alike Kaplan turbines the runner blades of these turbines are also adjustable. The other components such as scroll or spiral casing, stay vanes, guide vanes and draft tube are same as those in the case of other reaction turbines. Figure 21.38(a) shows the passages are straight unlike in conventional Kaplan turbines. However, some Deriaz turbines are also provided with stay vanes and guide vanes similar to those provided for conventional Kaplan turbines as shown in Fig. 21.38(b). The Deriaz runner is so shaped that it can be used both as a turbine as well as a pump and hence it may be classified as a reversible type machine. As such Deriaz turbines are quite suitable for pumped storage hydro-power plants.

21.12.2 Tubular turbine

A tubular turbine is an axial flow turbine with either adjustable or non-adjustable runner vanes and hence it is similar to Kaplan or Propeller turbines. However, in a tubular turbine the scroll casing is not provided but the runner is placed in a tube extending from the head water to the tail water as shown in Fig. 21.39, and hence it is called tubular turbine.

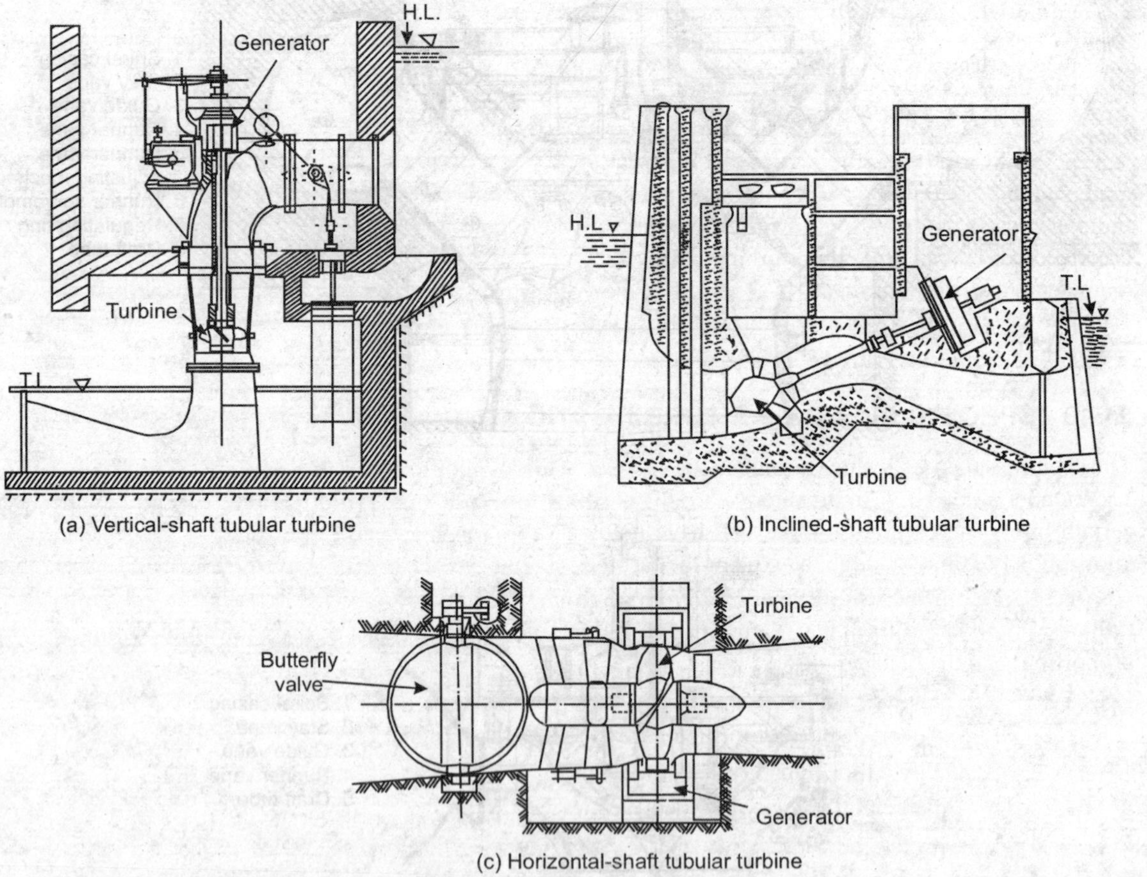

Fig. 21.39 *Tubular turbine.*

It is a low head turbine which can be used under heads ranging from 3 to 15 m. As shown in Fig. 21.39 the tubular turbines may have either vertical or inclined or horizontal disposition of shaft. The tubular turbines are quite suitable for tidal power plants.

21.12.3 Bulb turbine

It is same as tubular turbine with horizontal disposition of shaft in which the generator is housed in a bulb shaped water tight casing as shown in Fig. 21.40 and hence it is called **bulb turbine**. The bulb together with the turbine runner form an integral unit. The outer surface of the bulb is streamlined so as to form a passage for the flow of water from the head pond to the turbine with minimum loss of energy. The bulb could be either upstream or downstream of the runner but in most of the cases it is on the upstream of the runner. The bulb turbines are also

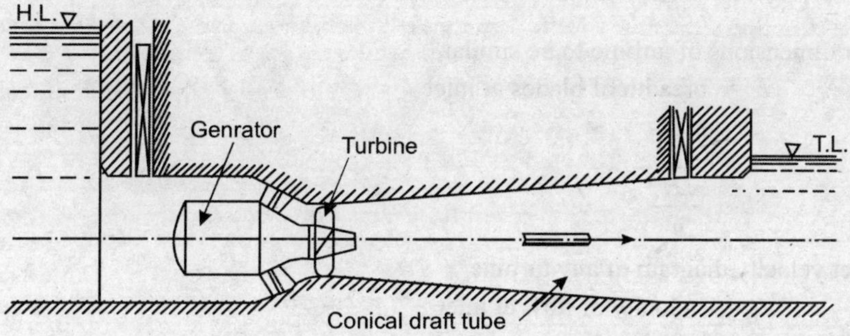

Fig. 21.40 *Bulb turbine.*

quite suitable for tidal power plants. A number of bulb turbines are operating successfully all over the world. In India at present there are only two power stations where bulb turbines have been installed. At Gandak Western Canal Power House three bulb turbine units each of 5.5 MW are operating under a head of 6.1 m. The runner diameter is 4.1 m and it run at 107.2 rpm. The total discharge rating is about 103.8 m^3/sec per unit. The Kosi East Canal Power House has four bulb turbine units each of 5.6 MW operating under a head of 6.1 m and has discharge rating about 104.5 m^3/s per unit. The runner diameter in this case is 4.5 m.

21.13 SPECIFIC SPEED OF A TURBINE AND COMPARISONS OF TURBINES

The specific speed of a water turbine is the speed at which a geometrically similar turbine would run if producing 1 kW under a head of 1 m of water (n_S). This is sometimes called the **type characteristic** of the turbine. Within certain limits each type of turbine will have its own value for the specific speed; hence if the specific speed is known it is possible to judge the type of the turbine.

An equation for the specific speed of a turbine can be obtained by applying the principle of similarity to water turbines. It will be assumed that all turbines are geometrically similar; that is, that all their linear dimensions are in proportion, and the blade angles are constant.

Let, D = Diameter of turbine in metre

n_S = Specific speed of turbine in revolutions per minute (RPM)

P = Power developed (kW)

v = tangential velocity of wheel at inlet

$$= \frac{\omega D}{2}$$

$$\therefore \quad D \propto \frac{v}{\omega} \text{ and } \omega \propto n$$

where n = number of revolutions per minute.

Also, from inlet velocity diagram of any turbine,

$$v \propto V \text{ and } V \propto \sqrt{H}$$

where H = Total head of water supplied.

$$\therefore \quad v \propto \sqrt{H}$$

$$\therefore \quad D \propto \frac{\sqrt{H}}{n} \qquad (1)$$

Assuming linear dimensions of turbine to be similar,

$$b = \text{breadth of blades at inlet}$$

$$\propto D$$

$$\therefore \quad b \propto \frac{\sqrt{H}}{n} \qquad (2)$$

Again, from inlet velocity diagram of any turbine

$$V_f = \text{Velocity of flow at inlet}$$

$$\propto V$$

$$\propto \sqrt{H} \qquad (3)$$

Again, quantity per second passing through turbine

$$= \text{area of flow} \times \text{velocity of flow}$$

$$= \pi D h \times V_f$$

$$\therefore \text{ Quantity per second } Q \propto \frac{\sqrt{H}}{n} \times \frac{\sqrt{H}}{n} \times \sqrt{H}$$

$$\propto \frac{H^{3/2}}{n^2}$$

$\therefore$ Weight of water per second = Density × Quantity Q

$$W \propto Q$$

$$W \propto \frac{H^{3/2}}{n^2} \qquad (4)$$

$$\therefore \text{ Power of turbine} = \frac{\text{Weight} \times \text{Head}}{1000} = \frac{WH}{1000}$$

$$\therefore \quad P \propto \frac{H^{3/2} \times H}{n^2}$$

$$\propto \frac{H^{5/2}}{n^2}$$

$$\text{or} \quad n \propto \sqrt{\frac{H^{5/2}}{P}}$$

$$\alpha \frac{H^{5/4}}{\sqrt{P}}$$

$$\therefore \quad n = K \times \frac{H^{5/4}}{\sqrt{P}} \quad (5)$$

where

$$K = \text{Constant depending on the type of turbine} = \frac{n\sqrt{P}}{H^{5/4}}$$

It will be noticed that the constant K is a function of the constants of Eqs. (1), (2), (3), (4). Hence K is a function of the linear proportions of the turbine.

When the turbine is developing 1 kW power under a head of 1 m, it will be noticed that K is equal to n which, under these conditions, is known as the specific speed n_s.

$$K = n = n_s$$

$$\therefore \quad n = n_s \times \frac{H^{5/4}}{\sqrt{P}}$$

or

$$\boxed{n_s = \frac{n\sqrt{P}}{H^{5/4}}} \quad (6)$$

It is found that for impulse turbines n_s lies between 6 and 60 and for Francis reaction turbines n_s lies between 50 and 400 and for Kaplan and propeller turbine it lies between 280 and 1100.

There is a large variation in the efficiency of a turbine when the gate and speed are varied; for small gate openings and low speeds the efficiency is very low. To obtain the conditions for the maximum efficiency for a turbine a diagram is plotted showing the efficiencies for all conditions of running; from this diagram the condition for maximum efficiency may be obtained.

Before plotting the diagram certain characteristics for the turbine are calculated from the results of the tests; these characteristics are known as **unit power, unit speed** and **unit quantity**.

Unit Power (Pu) : The unit power of any particular turbine may be defined as the power developed under a head of 1 m, or under unit head if any other system of dimensions be used.

Let

$$\begin{aligned} P &= \text{Power developed} \\ &= \text{Density} \times \text{Volume} \times \text{Head} \\ &= \rho \times aV \times H \\ &\alpha \sqrt{H} \times H \\ &\alpha H^{3/2} \\ &= K_1 . H^{3/2} \end{aligned}$$

where K_1 is a coefficient which will vary with the efficiency of the turbine, that is, with the gate opening and speed.

When,

$$H = 1 \text{ m}$$

$$P = K_1 = \text{unit power.}$$

$\therefore$ Unit power of a turbine

$$P_u = K = \frac{P}{H^{3/2}}$$

$$P_u = \frac{P}{H^{3/2}}$$

Unit Speed (N_u)

The unit speed for a particular turbine is the speed when running under a head of 1 m.

We have,

$$V_D = \frac{\pi DN}{60}, \text{ but } V_b = \sqrt{2gH}$$

$$\alpha \sqrt{H}$$

$$\therefore \quad N = \frac{60}{\pi D}.V_b$$

$$\alpha \sqrt{H}$$

$$\therefore \quad N = K_2 \sqrt{H}$$

where K_2 is a coefficient which will vary with the conditions of running.

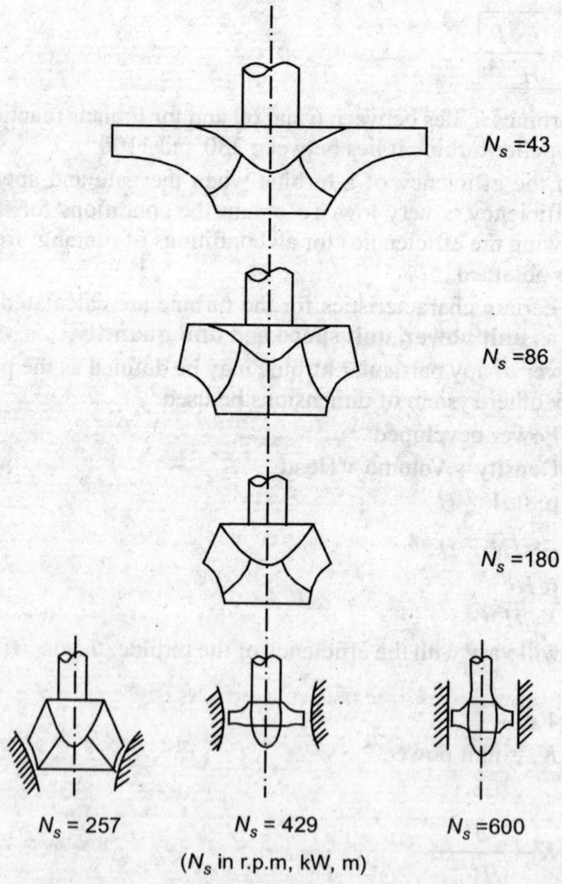

Fig. 21.41 *Relative sizes of turbine runners developing same power under same head for different specific speeds*

When $H = 1$ m,

$N = K_2 =$ unit speed N_u

Hence $N = N_u \times \sqrt{H}$

$$\therefore \quad N_u = \frac{N}{\sqrt{H}}$$

$$\boxed{N_u = \frac{N}{\sqrt{H}}}$$

Unit Quantity (Q_u)

Unit quantity of water passing through the turbine when the head is 1 m.

$$Q = aV$$

$$\alpha\ a.\sqrt{H}$$

$$\alpha\ \sqrt{H}$$

$$\therefore \quad Q = K_3\sqrt{H}$$

where K_3 is a coefficient depending on the conditions of running.

∴ When $H = 1, Q = K_3 =$ Unit Quantity Q_u.

$$\therefore \quad Q = Q_u.\sqrt{H}$$

or

$$Q_u = \frac{Q}{\sqrt{H}}$$

$$\boxed{Q_u = \frac{Q}{\sqrt{H}}}$$

Example 1 *A turbine develops 16000 kW power working under a head of 400 m when running at 400 rpm. Calculate the specific speed of the turbine. Specify the type of the turbine to be used.*

Solution

Head $H = 400$ m, $N =$ RPM $= 400$, $P =$ Power $= 16000$ kW

$$\therefore \text{ Specific speed} \quad N_s = \frac{N\sqrt{P}}{H^{5/4}}$$

$$= \frac{400\sqrt{16000}}{400^{5/4}} = \frac{400\sqrt{16000}}{400 \times 400^{1/4}}$$

$$= \frac{\sqrt{16000}}{4.472} = \frac{126.49}{4.472}$$

$$= \mathbf{28.28}$$

For this specific speed, Pelton wheel is used.

Example 2 *Calculate the specific speed of a turbine and suggest the type of turbine required for a river having a discharge of 240 litres/sec. with an available head of 60 metres. Assume efficiency of the turbine as 80% and speed 450 rpm.*

Solution

Given:

$$\text{Discharge} = Q = 240 \text{ litres/sec.}$$
$$\text{Head } H = 60 \text{ metres}$$
$$\text{Efficiency } \eta = 80\%$$
$$\text{Speed } N = 450 \text{ rpm}$$

∴ Power produced is given by

$$P = \rho Q \times H \times \eta$$
$$= \frac{1 \times 240 \times 60 \times 0.8}{1000} \times \frac{9.81}{1}$$
$$= \mathbf{113\ kW}$$

(Density of water is taken as 1 kg/lit. = 9.81 *N*/litre.)

$$\therefore \quad N_s = \text{Specific speed}$$
$$= \frac{N\sqrt{P}}{H^{5/4}}$$
$$= \frac{450\sqrt{113}}{60^{5/4}} = \frac{24.453}{2.783}$$
$$= \mathbf{8.79}$$

For this specific speed, Pelton turbine is suitable.

Example 3 *It was proposed to build a hydro-electric power plant across a river having a discharge of 30,000 litre/ second at a head of 10 m. Assuming turbine efficiency of 75% and speed ratio as 0.84, determine the following :*

(i) Is it possible to use two turbines with a speed not less than 120 rpm and specific speed not more than 350 rpm?

(ii) Specify the type of the runner than can be used. Also calculate the diameter of the runner.

Solution

Power that can be obtained from discharge and head available from river is given by

$$P = \frac{WH\ \eta \times 9.81}{1000}$$
$$= \frac{1 \times 9.81 \times 30{,}000 \times 10}{1000} \times 0.75$$
$$= 9.81 \times 30 \times 7.5$$
$$= \mathbf{2207.25\ kW}$$

Here two possibilities are possible.

(a) Assume number of units and then decide the type of the turbine.

(b) Assume the specific speed and then decide the number of units to be used.

(i) In the present case specific speed limit for the turbine is given.

$$\therefore \quad N_s = \frac{N\sqrt{P}}{H^{5/4}}$$

$$350 = \frac{120\sqrt{P}}{10^{5/4}}.$$

$$\therefore \quad \sqrt{P} = \frac{350 \times 10^{5/4}}{120} = \frac{35}{12} \times (10 \times 1.778)$$

$$= 51.86$$

$$\mathbf{P = 2689\ kW}$$

$\therefore$ 1 turbine can be used.

Now, if specific speed is assumed to be 200 then

$$\sqrt{P} = \frac{200 \times 10^{5/4}}{120}$$

$$= 29.63$$

$$\therefore \quad P = 878 \text{ kW}$$

$\therefore$ No. of units used

$$= \frac{2207.25}{878}$$

$$= \mathbf{2.51}$$

Say **3**

(ii) In both these cases, Francis turbines are to be used.

Blade velocity = Speed ratio × $\sqrt{2gH}$

$$= 0.83 \times \sqrt{2 \times 9.81 \times 10}$$

$$= \mathbf{11.626\ m/s}$$

$$= \frac{\pi DN}{60}$$

$\therefore \quad D$ = Diameter of the runner

$$= \frac{11.626 \times 60}{\pi \times 120}$$

$$= \mathbf{1.85\ metres}$$

Example 4 *Deduce an expression for the specific speed of a reaction turbine. Under a head of 12 m the maximum feasible specific speed is 440. If under this head, the installation of 15000 kW is required and if the speed is to be 150 rpm, how many units should be used?*

Solution

The specific speed is given by

$$N_s = \frac{N\sqrt{P}}{H^{5/4}}$$

$$\therefore \quad 440 = \frac{150 \times \sqrt{P}}{12^{5/4}}$$

$$\therefore \quad \sqrt{P} = \frac{440 \times 12^{5/4}}{150} = 65.51$$

$$P = \mathbf{4292\ kW/unit}$$

$\therefore$ Total power = 15000 kW

$\therefore$ Number of units (turbines) required

$$= \frac{15000}{4292}$$

$$= \mathbf{3.495}$$

Say **4**

21.14 SELECTION OF TURBINES

The selection of a suitable type of turbines is usually governed by the following factors :

(i) Head and Specific Speed. It has been found that there is a range of head and specific speed for which each type of turbine is most suitable which is given below :

S.No.	*Head in metres*	*Type of turbine*	*Specific speed*
1	300 and more	Pelton wheel	8.5 to 47 (SI unit)
		Single or Multiple Jet	10 to 55 (Metric units)
			30 to 85 (SI units)
2	150–300 m	Pelton or Francis	35–100 (Metric units)
3	60–150 m	Francis or	85 to 188 (SI units)
		Deriaz (or Diagonal)	100 to 220 (Metric units)
4	Less than 60 m	Kaplan or propeller	188 to 860 (SI units)
		or Deriaz or Tubular	220 to 1000 (Metric units)

However, as a general rule, it may be stated that as far as possible a turbine with highest permissible specific speed should be chosen, which will not only be the cheapest in itself but is relatively small size and high rotational speed will reduce the size of the generator as well as the power house. But the specific speed can not be increased indefinitely, because higher specific speed turbine is generally more liable to cavitation. The cavitation, however, be avoided by installing the turbine at a lower level with respect to the tail race.

(ii) Part Load Operation. The turbines may be required to work with considerable load variations. As the load deviates from the normal working load, efficiency would also vary. In Fig 21.41 (Load vs Efficiency) a plot between η_0 and percentage of full load has been shown. At part load the performance of Kaplan and Pelton turbines is better in comparison to that of Francis and Propeller turbines. The variability of load will influence the choice of type of turbine if the head lies between 150 m to 300 m or lies below 30 m. For high range of heads Pelton wheel is preferable for part load operation in comparison to Francis turbine, though the former involves high initial cost. For heads below 30 m, Kaplan turbine is preferable for part load operation in comparison to propeller turbine.

In addition to above mentioned factors there are certain other factors to be considered for the selection of the suitable type of the turbine. The **overall cost** which includes the initial cost and the running cost should be considered. The **cavitation** characteristics of the turbine should also be considered since it affects the installation of a reaction turbine.

21.15 DRAFT TUBES

In reaction turbines—Francis and Kaplan—the water after passing through the runner flows to the tail race through a **Draft Tube**. A draft tube is a pipe or passage of gradually increasing cross-sectional area which connects the runner exit to the tail race. It may be made of cast or plate steel or concrete. It must be straight and under all conditions of operation its lower end must be submerged below the level of water in the tail race. The draft tube has two purposes as follows :

(i) It permits a negative or suction head to be established at the runner exit, thus making is possible to instal the turbine above the tail race level without loss of head.

(ii) It converts a large proportion of velocity energy rejected from the runner into useful pressure energy i.e., it acts as a recuperator of pressure energy.

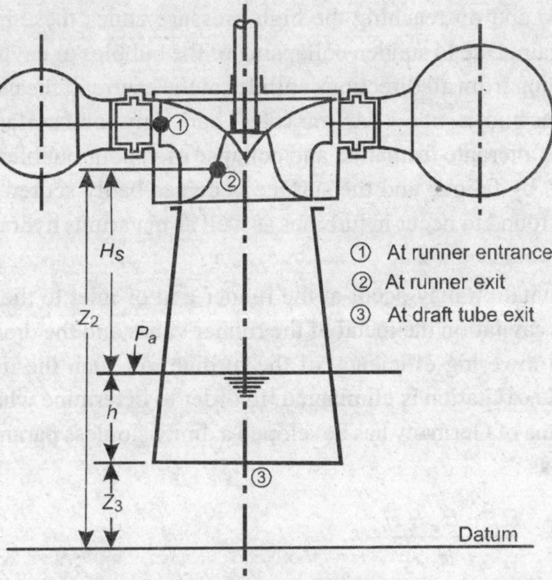

(a) Draft tube

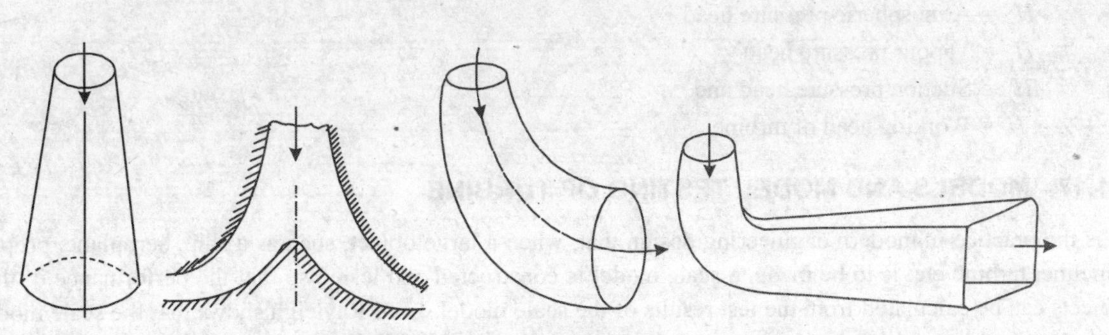

(b) Different types of draft tubes

(a) Straight divergent tube. (b) Moody spreading tube (or hydraucone tube) (c) Simple elbowtube (d) Elbow tube having circular cross section at inlet and rectangular at outlet

Fig. 21.42

Figure 21.42 shows the different types of draft tubes which are employed in the field to suit particular conditions of installation. Of these the types (a) and (b) are the most efficient, but the types (c) and (d) have an advantage that they require lesser excavation for their installation. It has been observed that for straight divergent type draft tube the central cone angle should not be more than 8°. This is so because if this angle is more than 8° the water flowing through the draft tube will not remain in contact with its inner surface, with the result that eddies are formed and the efficiency of the draft tube is reduced.

21.16 CAVITATION IN TURBINES

When the pressure in any part of the flow passage reaches the vapour pressure of the flowing liquid, it starts vaporizing and small bubbles of vapour form in large numbers. These bubbles (or vapour filled pockets or cavities) are carried along by the flow, and on reaching the high pressure zones these bubbles suddenly collapse as the vapour condenses to liquid again. Due to sudden collapsing of the bubbles or cavities the surrounding liquid rushes in to fill them. The liquid moving from all directions collides at the centre of the cavity, thus giving rise to very high local pressure, which may be as high as 686.7 MN/m^2 (6867 bar). Any solid surface in the vicinity is also subjected to these intense pressures. The alternate formation and collapse of vapour bubbles may cause severe damage to the surface which ultimately fails by fatigue and the surface becomes badly scored and pitted. This phenomenon is known as **cavitation** which is found to occur in turbines as well as in various hydraulic structures such as penstocks, gates, valves, spillways etc.

In reaction turbines the cavitation may occur at the runner exit or inlet to the draft tube where the pressure is considerably reduced. Due to cavitation the metal of the runner vanes and the draft tube is gradually eaten away in these zones, which results in lowering efficiency of the turbine. As such the turbine components should be so designed that as far as possible cavitation is eliminated. In order to determine whether cavitation will occur in any portion of the turbine, D. Thoma of Germany has developed a dimensionless parameter called **Thoma's Cavitation Factor σ which is expressed as :**

$$\sigma = \frac{H_a - H_v - H_s}{H}$$

where

H_a = Atmospheric pressure head
H_v = Vapour pressure head
H_s = Suction pressure head and
H = Working head of turbine.

21.17 MODELS AND MODEL TESTING OF TURBINE

It is the practice in modern engineering design that, when a large object, such as a ship, aeroplane, propeller or pipeline, turbine etc. is to be made, a scale model is constructed and tested so that the performance of the large objects can be calculated from the test results of the scale model. Lord Rayleigh shows that the scale model tests give comparable results only when the non-dimensional factor of the model test is equal to that of the large object, when working under its design conditions. By equating the non-dimensional factor of the large object to that of the model, the test speed of the model is obtained. This is known as the **corresponding speed**, and the comparison of the two conditions between the large object and the test results of a scale moded, at its corresponding speed, is known as the **principle of dynamical similarity**.

In order to have an idea about the performance of the actual turbine in advance, a small scale model of the turbine which is geometrically similar to the actual turbine is first prepared. The various linear dimensions of the model turbine bear the same proportion to their corresponding dimensions of the actual turbine. The model turbine is then tested under a known head, speed and flow rate and its output as well as the efficiency are determined. From these test results it will be possible to predict the performance of the actual turbine. Moreover on the basis of the test results if the design is to be modified it may be so done in the model turbine without incurring much expenditure. Thus model testing of turbines assists in obtaining the perfect design for the actual turbines, as well as in the development of the new types of turbines with higher specific speed and better efficiency.

Generally the model of a turbine and its prototype are in definite geometric ratio depending on their respective heads and rotative speeds. The ratio of blade velocity V_b and the water velocity V which is called as the **speed ratio** has a definite values for different turbines. For Pelton wheel this ratio varies from 0.42 to 0.47 while for Francis and Kaplan propeller turbine it is from 0.55 to 1 or more and from 1.5 to 3 or more respectively.

$\therefore\ V_b \propto V$ But $V_b \propto DN$ and $V \propto H$

Therefore, if *m* is the subscript for model turbine and *P* is the subscript for prototype, then we can write :

$$\frac{D_m N_m}{D_P N_P} = \sqrt{\frac{H_m}{H_P}}$$

and $$\frac{D_m}{D_P} = \sqrt{\frac{H_m}{H_P}} \times \frac{N_P}{N_m}.$$

$\frac{D_m}{D_P}$ is called the **scale ratio**.

21.18 PERFORMANCE CHARACTERISTIC CURVES OF TURBINES

The turbines are generally designed to work at particular values of H, Q, P, N and η_0 which are known as the design conditions. But often the turbines are required to work at conditions different from those for which they have been designed. Therefore, it is essential to determine the exact behaviour of the turbines under the varying conditions by carrying out tests either on the actual turbines or on their small scale turbines. The result of these tests are usually graphically represented and the resulting curves are known as **Characteristic Curves**. For the sake of convenience the characteristic curves are plotted in terms of unit quantities. These characteristic curves are of the following three types :

(i) Constant head characteristic curves (or Main Characteristic Curves)

(ii) Constant speed characteristic curves (or Operating Characteristic Curves)

(iii) Constant efficiency curves.

(i) Constant Head Characteristic Curves. In order to obtain these curves the tests are performed on the turbine by maintaining a constant head and a constant gate opening and the speed is varied by changing the load on the turbine. A series of values of N are thus obtained and corresponding to each value of N, discharge Q and the output power P are measured. The series of such tests are performed by varying the gate opening, the head being maintained constant at the previous value. From the data of the tests the values of Q_u, P_u, N_u and η_o are computed for each gate opening. Then N_u as abscissa the values of Q_u, P_u and η_o for each gate opening are plotted. The curves thus obtained for Pelton wheel and the reaction turbines for four different gate openings are shown in Fig. 21.43.

For Pelton wheels since Q_u depends only on gate opening and is independent of N_u, Q_u vs N_u plots are horizontal straight lines. However, for low specific speed Francis turbines Q_u vs N_u are drooping curves, thereby indicating

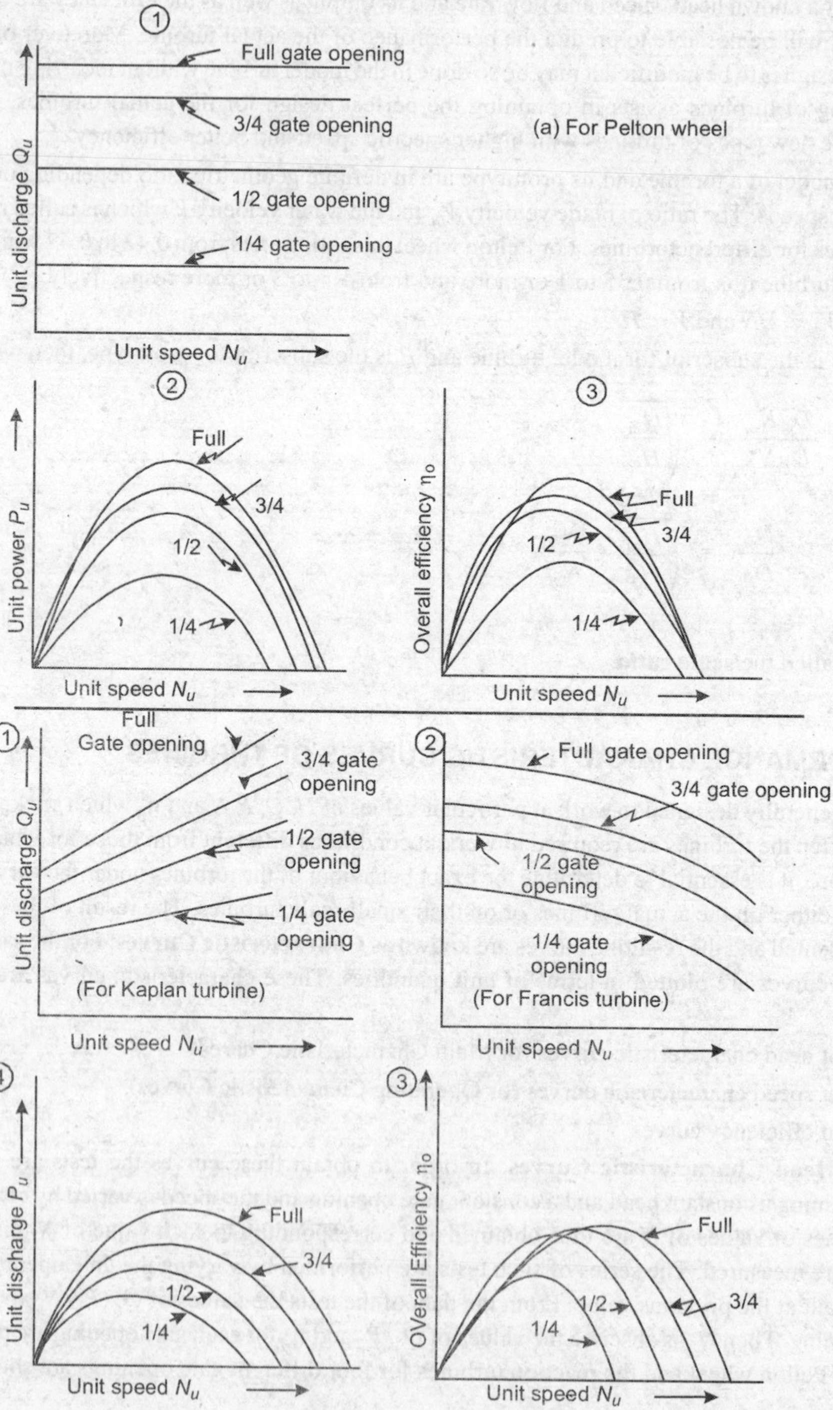

Fig. 21.43 *Constant head characteristics for Pelton wheel and reaction turbines.*

that as the speed increases the discharge through the turbine decreases. This is so because in these turbines a centrifugal head is developed which retards the flow. Since the centrifugal head increases with the speed, the flow through the turbine is reduced as the speed increases. On the other hand, for high specific speed Francis turbines as well as Kaplan turbines, since the flow is axial there is no such centrifugal head developed which may cause the retardation of the flow.

(ii) Constant Speed Characteristic Curves. In order to obtain these curves the tests are performed on the turbines by operating them at constant speed. The constant speed is attained by regulating the gate opening thereby varying the discharge flowing through the turbine as the load varies. The head may or may not remain constant. The power developed corresponding to each setting of the gate opening is measured and the corresponding values of η_o are computed. Further knowing the total load capacity of the turbine the percentage of full load may be computed from the measured power, and a plot of η_o vs % of full load is prepared. The curves thus obtained for the four different types of turbines working at constant speed under constant head at varying gate openings are shown in Fig. 21.44 from which it may be seen that as the % full load increases η_o also increases. Further it may be observed that the Kaplan turbine and the Pelton wheel maintain a **high efficiency over a longer range of the part load** as compared with either the Francis or the fixed blade propeller turbine. This is an **important factor in the selection of turbines**.

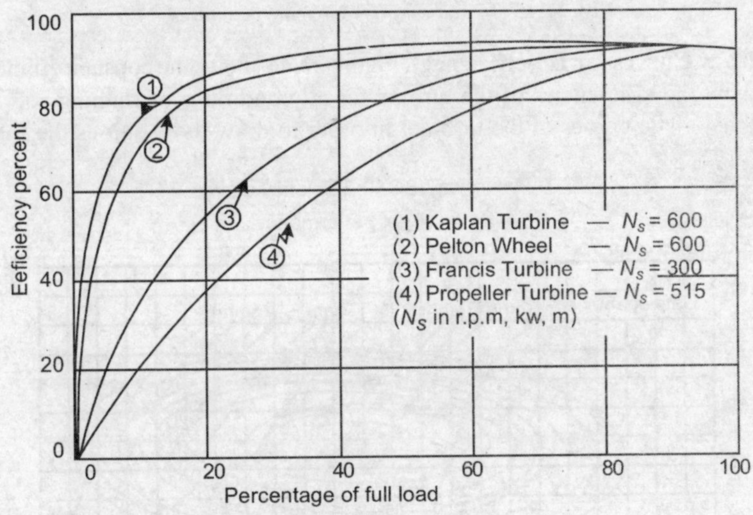

Fig. 21.44 *Percent of full load vs η_o for different type turbines.*

In addition to the above graph, output power vs discharge and η_o vs discharge graphs are also plotted which are shown in Fig. 21.45. Both these graphs are plotted from a certain minimum discharge Q_o which is required to initiate the motion of the turbine runner from its state of rest. Further since output power is directly proportional to discharge if head is constant, output vs discharge graph is a straight line. On the other hand η_o vs discharge graph is curvilinear and as shown in Fig. 21.45, the overall efficiency η_o increases with discharge and remains more or less constant beyond a particular value of discharge.

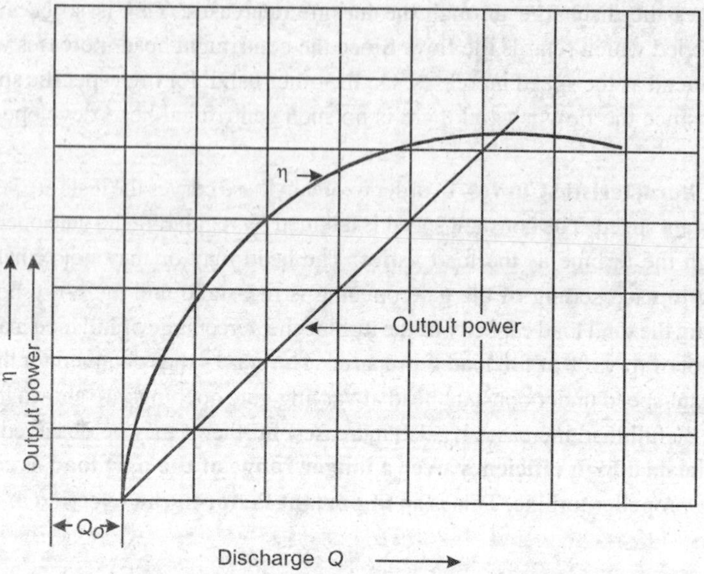

Fig. 21.45 *η_o and output vs discharge curves.*

(iii) Constant Efficiency Curves (or Iso-efficiency). Figure 21.46 shows the constant efficiency or iso-efficiency curves. These curves show the efficiencies of the turbine for all conditions of running and hence these are also known as universal characteristic curves of the turbine. In order to draw these curves the following procedure is

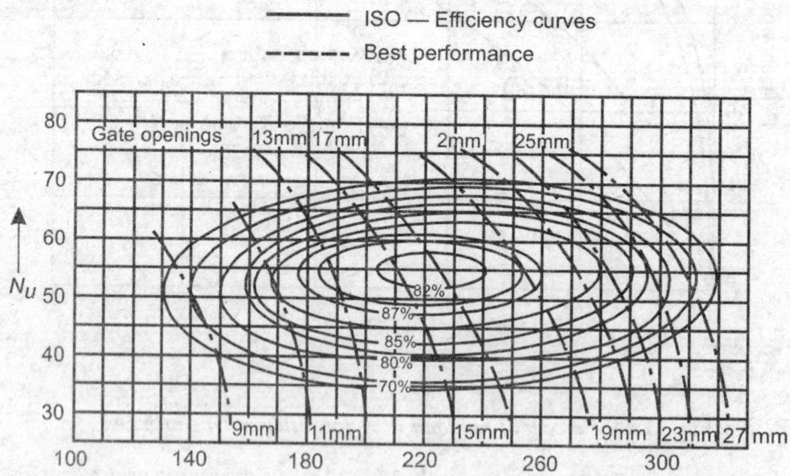

Fig. 21.46 *Iso-efficiency curves of a Francis turbine.*

adopted. By operating the turbine at about 8 to 10 gate openings, the corresponding number of η_o vs N_u and N_u vs Q_u (or P_u) curves are plotted. On the η_o vs N_u curves set of horizontal lines (each line representing the same efficiency) are drawn which will cut the curves corresponding to each gate opening at different points. These points are projected on the corresponding N_u vs Q_u (or P_u) curves for each gate opening and the points of the same efficiency are joined by smooth curves which are the iso-efficiency curves. It is, however, clear from the graph that the innermost iso-efficiency curve represents the highest efficiency of the turbine and the outer curves represent the lower efficiences. Further it will be seen from Fig. 21.46 that for a given unit discharge (or unit power) if a vertical line is drawn it will cut some iso-efficiency curve at two points and it will also touch some other inner iso-efficiency curve of higher efficiency, which may not have been drawn, just at one point only. Thus for a given unit discharge (or unit power) the vertical line touches the curve of maximum efficiency at only one point. It, therefore, means that if the peak points of the various iso-efficiency curves are joined together by a smooth curve then we obtain the best performance curve for the turbine, because the peak points of each iso-efficiency curve are the points of maximum efficiency for some value of the unit discharge. With the help of these characteristic curves it is possible to predict the best performance of the turbine.

21.19 GOVERNING OF TURBINES

All the modern hydraulic turbines are directly coupled to the electric generators. The generators are always required to run at constant speed irrespective of the variations in the load. This constant speed N rpm of the generator is given by the expression

$$f = \frac{pN}{60}$$

where

f = frequency for power generated in cycles per second and

p = number of pair of poles for the generator.

Usually $f = 50$ and hence

$$N = \frac{60 \times 50}{P}$$

$$= \frac{3000}{P}.$$

The speed of the generator can be maintained constant only if the speed of the turbine runner is constant equal to the one given by the above noted equation. It is then known as the **Synchronous Speed** of the turbine runner for which it is designed.

The load on the governor will, however, go on varying and now if the input for the turbine remains same then the speed of the turbine runner will tend to either increase or decrease depending on the load on the generator is reduced or increased. This in turn will cause the speed of the generator also to vary accordingly, which is, however, not desirable because it may result in varying the frequency of power generation. As such the speed of the turbine runner is required to be maintained constant so that the generator always runs at constant speed under all conditions of working. It is usually done by regulating the quantity of water flowing through the runner in accordance with the variations in the load. Such an operation of regulation of speed of the turbine runner is known as **governing of turbine** and is usually done automatically by means of a **governor**. One of the common types of governor which is predominantly used with modern turbines is **oil pressure governor**. As shown in Fig. 21.47 its component parts are as follows :

(i) Servomotor also known as relay cylinder

(ii) Relay valve also known as control valve or distribution valve

(iii) Actuator or pendulum which is belt or gear driven from the turbine main shaft

(iv) Oil sump

(v) Oil pump which is driven by belt connected to turbine main shaft

(vi) A system of oil supply pipes connecting the oil sump with the relay valve and the relay valve with the servometer.

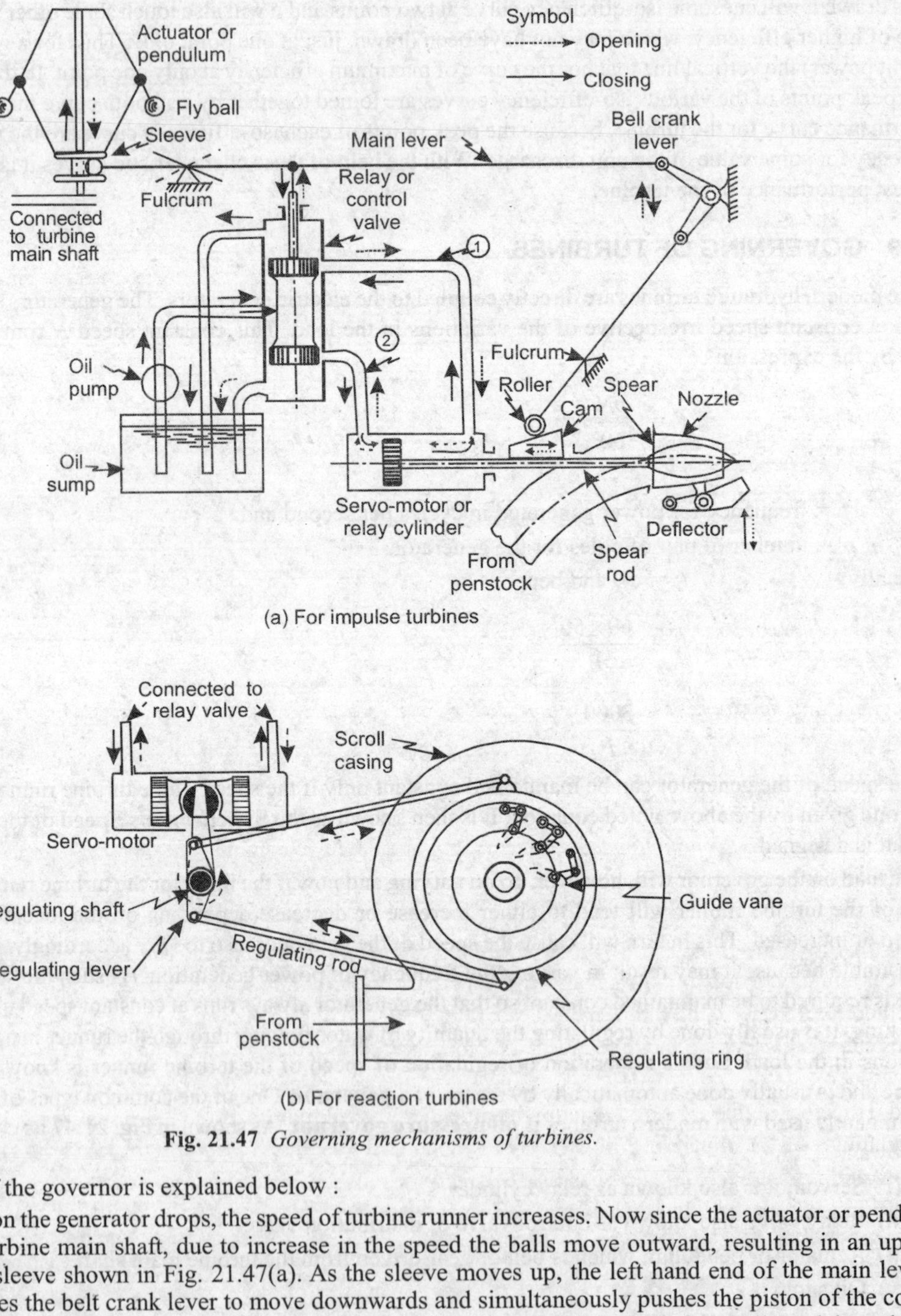

Fig. 21.47 *Governing mechanisms of turbines.*

The working of the governor is explained below :

When the load on the generator drops, the speed of turbine runner increases. Now since the actuator or pendulum is driven by the turbine main shaft, due to increase in the speed the balls move outward, resulting in an upward movement of the sleeve shown in Fig. 21.47(a). As the sleeve moves up, the left hand end of the main lever is raised, which causes the belt crank lever to move downwards and simultaneously pushes the piston of the control valve down in its cylinder. In the case of Pelton wheel (turbine) the downward motion of the bell crank lever beings the **deflector** in front of the jet thereby diverting a portion of the jet away from the buckets. On the other hand in the case of reaction turbine such as Francis or Kaplan turbine a relief valve shown in Fig. 21.47(b) is provided. A relief valve consists of a spear, which under normal conditions of operation of turbine, is held by oil or water under pressure in such a position that it keeps the opening of the bypass from the spiral casing to the tail race closed. The

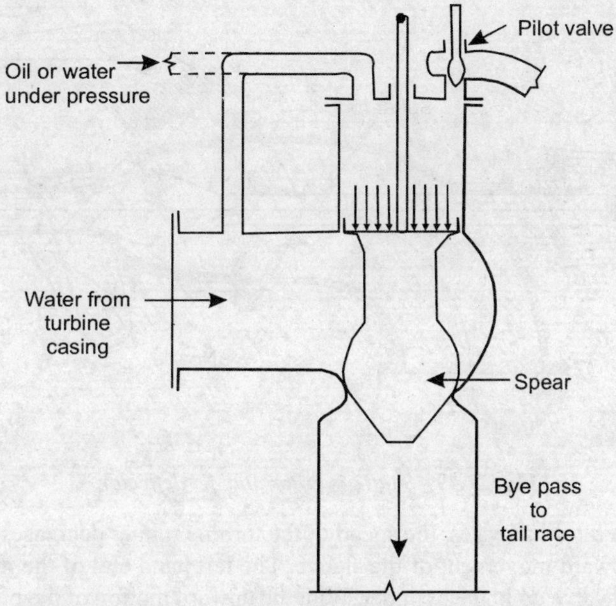

Fig. 21.48 *Relief valve.*

downward motion of the bell-crank lever (or other suitable arrangement) opens the pilot valve of the pressure chamber so that the pressure on the spear is reduced, thereby permitting the spear to be lifted up and allowing a portion of water to flow directly from the spiral casing to the tail raçe through the bypass without striking the turbine runner. Thus both these devices viz. deflactor and relief valve have similar function to perform. These eliminate the necessity of rapid closure of the nozzle opening or the guide vanes, at the same time the quantity of water striking the runner is reduced. The rapid closure of the nozzle opening or the guide vanes is not desirable because a sudden reduction of the rate of flow in the penstock may result in serious **water hammer** problem.

With the downward motion of the piston of the control valve the passage for pipeline 2 opens and oil under pressure is admitted from the control valve cylinder to the servomotor on the left side of the piston. The servomotor piston, therefore, move to the right. In the case of Pelton wheel, the servomotor piston being connected to the spear rod, thus causes the spear to move forward. The forward motion of the spear reduces the area of the nozzle outlet and thus decreases the rate of flow (as required for the decreased load) whereby the normal turbine speed is restored. However, in the case of a reaction turbine the forward motion of the servomotor piston is transmitted to the regulating ring as shown in Fig. 21.50 (b) which causes all the guide vanes to move simultaneously in one direction, and thus the area of flow passage between the adjacent guide vanes is reduced and the rate of flow of water striking the runner is also reduced. When once the normal turbine speed is restored the main lever attains its normal position and through a suitable can arrangement the deflector is brought back to its original position or the relief valve stops bypassing of water to the tail race.

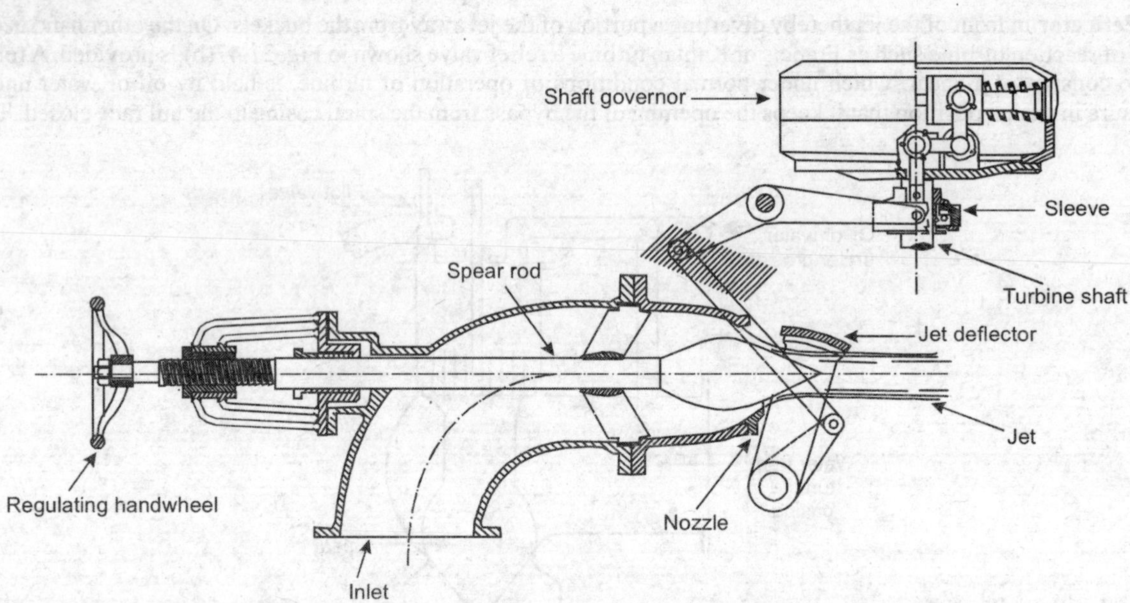

Fig. 21.49 *Shaft governor and jet deflector.*

When the load on the generator increases, the speed of the turbine runner decreases. Due to this the balls move inward resulting in the downward movement of the sleeve. The left hand end of the main lever is lowered which pulls the piston of the control valve up in the cylinder. With the upward motion of the piston of the control valve the passage of the pipe line 1 opens and oil under pressure rushes from the control valve cylinder to the servomotor on the right side of the piston. The servomotor piston then moves to the left. This increases area of the nozzle outlet or the passage between the adjacent guide vanes, thereby allowing a large quantity of water to strike the runner (as required for the increased load) and the normal speed for the turbine runner is thus restored.

In the case of Kaplan turbine since in addition to guide vanes the runner vanes are also adjustable, the governor is required to operate both sets of vanes simultaneously. The runner vanes are also operated by a separate servometer and a control valve, and the servomotors and the control valves for both, the runner and the guide vanes, are interconnected to ensure that for a given guide vane opening there shall be a definite runner vane inclination.

However, the large hydro-electric units are always provided with **electrohydraulic governors**. In general, an electrohydraulic governer consists of a hydro-mechanical actuator and an electrical equipment cabinet, both mounted in the machine room. The electrical equipment cabinet contains the electrical part of the governor only, while the actuator cabinet contains the hydro-mechanical devices. A special techogenerator connected to the mainshaft transmits current at a frequency of 50 cycles to the electric measuring circuits of the governor. When the speed of the unit changes, the frequency of the techogenerator supplies also changes, whereupon a descriminator responding to frequency variations emits an electrical signal.

After being amplified, this signal is transmitted to the electro-mechanical final control element, where it is transformed into a mechanical displacement transmitted through the lever system of the amplifier valve to the main governor valve. Being moved from its mid (normal) position in direct proportion to the strength of the electric signal, the main governor feeds oils under pressure to the servomotor of the turbine distributor. The spear in the nozzle or the guide vanes thus turn in the direction corresponding to the sign of electric signal.

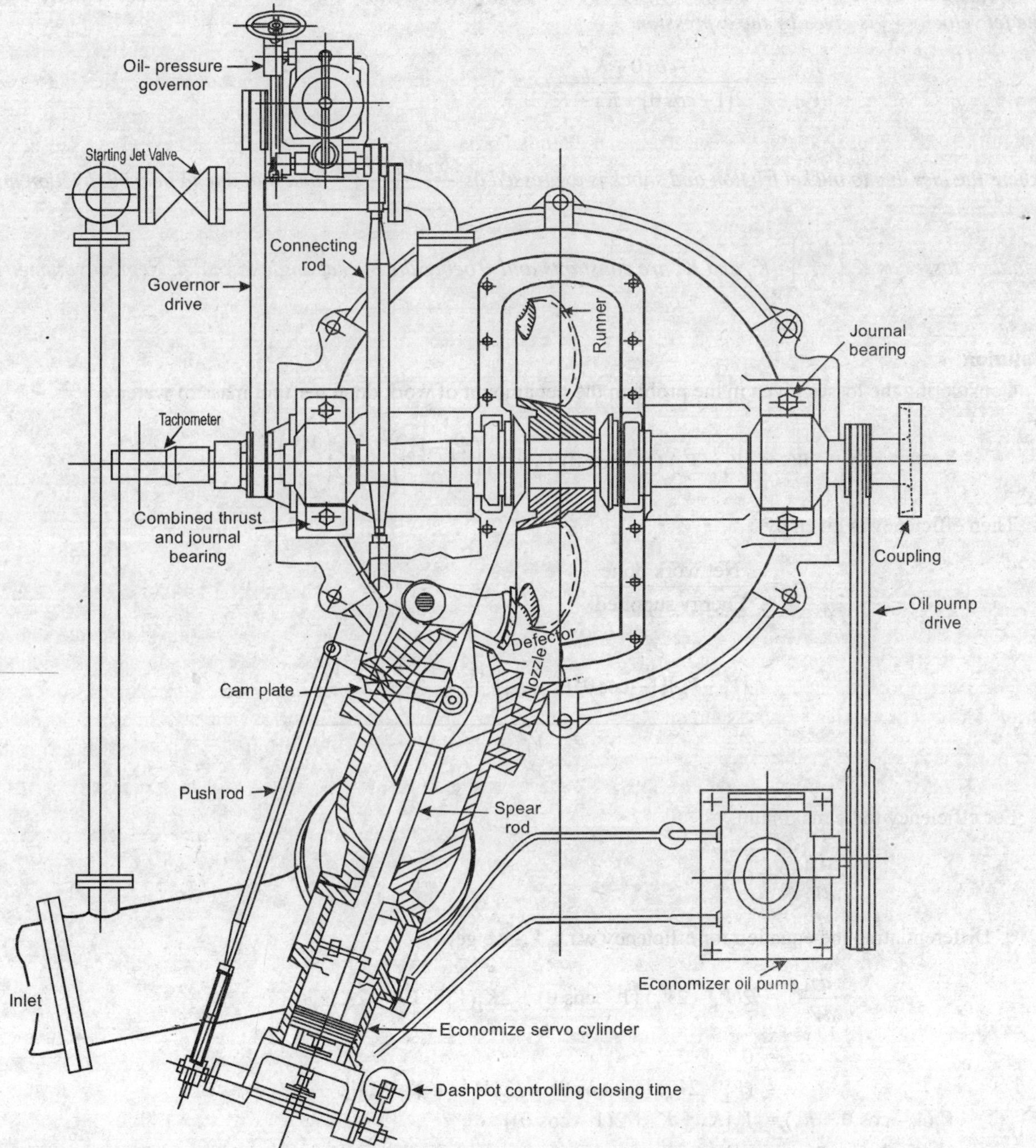

Fig. 21.50 *Oil pressure governor & automatic spear control.*

SOLVED PROBLEMS

Example 21.1 *Prove that the maximum efficiency of Pelton wheel occurs when the ratio of bucket velocity V_b to the jet velocity V_1 is given by the expression*

$$\frac{V_b}{V_1} = \frac{1-\cos\theta + K_1}{2(1-\cos\theta)+K_1+K_2}$$

where the loss due to bucket friction and shock is expressed as $\frac{K_1(V_1-V_b)^2}{2}$ *and that due to bearing friction and windage losses as* $K_2\left(\frac{V_b^2}{2}\right)$*; K_1 and K_2 are constants and θ being the bucket angle at outlet. Neglect volumetric losses.*

Solution

Considering the losses given in the problem the net amount of work done per unit mass of water

$$W = (V_4 - V_b)(1-\cos\theta)\,V_b - K_1\left(\frac{V_1-V_b}{2}\right)^2 - K_2\left(\frac{V_b^2}{2}\right)$$

Then efficiency is given by

$$\eta = \frac{\text{Net work done}}{\text{Energy supplied}}$$

$$= \frac{(V_1-V_b)(1-\cos\theta)V_b - K_1\left(\frac{V_1^2-V_b^2}{2}\right) - K_2\frac{V_b^2}{2}}{\frac{V_1^2}{2}}$$

For efficiency to be maximum,

$$\frac{d\eta}{dV_b} = 0$$

∴ Differentiating the equation for efficiency w.r.t. V_b, we get

$$\frac{d\eta}{dV_b} = 2(V_1 - 2V_b)(1-\cos\theta) + 2K_1(V_1 - V_b) - 2K_2V_b$$

$$= 0$$

$$= (V_1 - 2V_b)(1-\cos\theta) + K_1(V_1 - V_b) - K_2V_b.$$

$$\therefore \quad V_1(1-\cos\theta + K_1) = V_b[K_2 + K_1 + 2(1-\cos\theta)]$$

or

$$\frac{V_b}{V_1} = \frac{1-\cos\theta + K_1}{K_1+K_2+2(1-\cos\theta)}$$

Example 21.2 *In a Francis turbine, prove that hydraulic efficiency η_h of the turbine can be expressed as*

$$\eta_h = \frac{2}{2+\dfrac{K_1+K_2+K_3+K_4}{(\cot\alpha-\cos\theta)\left[\cot\alpha\left(1+n^2\right)-n\left(\cot\phi+n\cot\theta\right)\right]}}$$

where K_1, K_2, K_3 and K_4 represent the fraction the losses in the guide vanes, runner vanes, draft tube and at exit respectively expressed in terms of velocity of flow head, α, θ and ϕ are the guide vane angle, the runner vane angle at inlet and the runner vane angle at outlet respectively and n is the ratio of the inter to outer diameter of the runner. Assume the velocity of flow to remain constant in the runner.

Solution

If H is the net head acting on the turbine then the work done by the runner per unit mass of water may be expressed by

H – losses = Work done = $(V_{w1}V_{b1} - V_{w2}V_{b2})$

The losses may be combined and expressed by

$$(K_1+K_2+K_3+K_4)\,\frac{V_{f1}^2}{2} = (K_1+K_2+K_3+K_4)\,\frac{V_f^2}{2}$$

$$\because \quad V_{f1} = V_{f2} = V_f$$

$$\therefore \quad (V_{w_1}V_{b_1} - V_{w_2}V_{b_2}) = H-(K_1+K_2+K_3+K_4)\,\frac{V_f^2}{2} \tag{1}$$

From velocity diagrams (Fig. 21.51)

$$V_{w_1} = V_{f1}\cot\alpha$$

and
$$\tan\theta = \frac{V_{f_1}}{V_{w_1}-V_{b_1}}$$

$$= \frac{V_{f_1}}{V_f\cot\alpha - V_{b_1}} \text{ or } \frac{V_{f_1}}{\tan\theta} = V_{f1}\cot\alpha - V_{b_1} = V_{f1}\cot\theta$$

or
$$V_{b_1} = V_{f1}(\cot\alpha - \cot\theta).$$

Similarly
$$\phi = 180 - \theta.$$

and
$$\tan\phi = \frac{V_{f_1}}{V_{w_2}+V_{b_2}}$$

Also
$$V_{b_1} = \frac{\pi D_1 N}{60} \text{ and } V_{b_2} = \frac{\pi D_2 N}{60}$$

$$\therefore \quad \frac{V_{b_2}}{V_{b_1}} = \frac{D_2}{D_1} \quad \text{or } V_{b_2} = V_{b_1}\times\frac{D_2}{D_1}$$

$$= nV_{b_1}$$

$$= nV_{f1}(\cot\alpha - \cot\theta)$$

Introducing this value of V_{b_2}, we get

$$\tan\phi = \frac{1}{\cot\theta} = \frac{V_{f1}}{V_{w_2} + nV_{f_1}(\cot\alpha - \cot\theta)}$$

$$\therefore \quad V_{w2} = V_{f1}[\cot\theta - n(\cot\alpha - \cot\theta)]$$

The hydraulic efficiency is given by

$$\eta_h = \frac{V_{w_1}V_{b_1} - V_{w_2}V_{b_2}}{gH}$$

$\therefore$ From Eq. (1), we have

$$gH = (V_{w1}V_{b1} - V_{w2}V_{b2}) + (K_1 + K_2 + K_3 + K_4)\frac{V_{f_1}^2}{2}$$

$$\therefore \quad \eta_h = \frac{V_{w_1}V_{b_1} - V_{w_2}V_{b_2}}{\left(V_{w_1}V_{b_1} - V_{w_2}V_{b_2}\right) + \left(K_1 + K_2 + K_3 + K_4\right)\frac{V_{f_1}^2}{2}}$$

$$= \frac{2}{2 + \frac{\left(K_1 + K_2 + K_3 + K_4\right)V_{f_1}^2}{V_{w_1}V_{b_1} - V_{w_2}V_{b_2}}}$$

and substituting the values of V_{w1}, V_{b1}, V_{w2} and V_{b2} we get

$$\eta_h = \frac{2}{2 + \frac{K_1 + K_2 + K_3 + K_4}{(\cot\alpha - \cot\theta)\left\{\cot\alpha\left(1 + n^2\right) - n\left(\cot\theta + n\cot\theta\right)\right\}}}$$

Example 21.3 *A Pelton wheel has to be designed for the following data : Power to be developed = 6000 kW. Net head available = 300 m; speed = 550 rpm; ratio of jet diameter to wheel diameter = 1/10 and overall efficiency = 85%. Find the number of jets, diameter of the jet, diameter of the wheel, and the quantity of water required.*

Solution

Assuming velocity coefficient to be 0.98 and $\frac{u}{V_1} = 0.46$,

$$V_1 = 0.98\sqrt{2gH} = 0.98\sqrt{2 \times 9.81 \times 300}$$

$$= \mathbf{75.19\ m/s}$$

$$V_{b1} = 0.46\sqrt{2gH} = 0.46\sqrt{2 \times 9.81 \times 300}$$

$$= \mathbf{35.29\ m/s}$$

$$\eta_o = \frac{P}{wQH} = \frac{6000 \times 10^3}{(1000 \times 9.81)Q \times 300}$$

$$\therefore \quad Q = \frac{6000 \times 10^3}{0.85 \times 300 \times 1000 \times 9.81}$$

$$= \mathbf{2.399\ m^3/s}$$

and
$$V_{b_1} = V_b = \frac{\pi DN}{60} = \frac{\pi \times D \times 550}{60}$$

$$= 35.29 \text{ m/s}$$

$$\therefore \quad D = \frac{35.29 \times 60}{550 \times \pi}$$

$$= \mathbf{1.225\ m}$$

$$\therefore \quad d = \text{diameter to jet}$$

$$= \mathbf{0.1225\ m = 122.5\ mm}$$

$$\therefore \quad \text{Area of jet} = a = \frac{\pi}{4} \times (0.1225)^2$$

$$= \mathbf{0.0118\ m^2}$$

Total jet area required

$$= \frac{Q}{V}$$

$$= \frac{2.399}{75.19}$$

$$= \mathbf{0.0319\ m^2}$$

$$\therefore \quad \text{No. of jets required} = \frac{0.0319}{0.0118}$$

$$= 2.7$$

$$\simeq \mathbf{3}$$

Example 21.4 *A Pelton wheel has a mean bucket speed of 12 m/s and is supplied with water at the rate of 750 litres per second under a head of 35 m. If the bucket deflects, the jet through an angle of 160°, find the power developed by the turbine and its hydraulic efficiency. Take the coefficient of velocity as 0.98. Neglect friction in the bucket. Also determine the overall efficiency of the turbine if its mechanical efficiency is 80%.*

Solution

$$\text{Jet velocity} = V_1 = C_v = \sqrt{2gH}$$

$$= 0.98\sqrt{2 \times 9.81 \times 35}$$

$$= \mathbf{25.68\ m/s}$$

$$\text{Blade velocity} = V_b = 12 \text{ m/s}$$

Power developed by the turbine is given by

$$P = m[(V_1 - V_b)(1 + K\cos\phi)]V_b$$

where $m = 750$ kg/s

$$\therefore \quad P = (750)\,[(25.68 - 12)(1 + \cos 20°)] \times 12$$

$= 750\ [13.68 \times 1.9397]12$

$= \mathbf{238816\ W}$

$= \mathbf{238.816\ kW}$

Hydraulic efficiency of the turbine is given by

$$\eta_h = \frac{2V_b(V_1 - V_b)(1 + K\cos\theta)}{V_1^2}$$

$$= \frac{2 \times 12 \times 13.68 \times 1.9397}{(25.68)^2}$$

$= \mathbf{0.966}$ or $\mathbf{96.6\%}$

Mechanical efficiency $= 0.8 = \eta_m$

$\therefore$ Overall eff. $= \eta_o = \eta_m \times \eta_h$

$= 0.8 \times 0.966$

$= 0.773$

or $= \mathbf{77.3\%}$

Example 21.5 *The following are the design particulars of a large Pelton wheel (turbine) :*

Head at the distributor $= 630$ *m*

Discharge $= 12.5\ m^3/s$

Power developed $= 65$ *MW*

Speed of rotation $= 500$ *rpm*

Runner diameter $= 1.96$ *m*

Number of jets $= 4$

Jet diameter $= 0.192$ *m*

Angle through which the jet is deflected by the bucket $= 165°$

Mechanical efficiency of the turbine $= 96\%$

(a) Determine the hydraulic power losses in the distributor – nozzle assembly and the buckets.

(b) If the loss in the buckets is given to be proportional to V_{r1}^2 *where* V_{r1} *is the relative velocity at inlet, determine the best speed of rotation for this head and discharge.*

Solution

(a) *Hydraulic Power Loss :*

Mechanical efficiency of the Pelton wheel is given as

$$\eta_m = \frac{P}{m\left[\dfrac{(V_1 - V_{b_1})V_{b_1}(1 + K\cos\theta)}{g}\right]}$$

V_b^2

where $\eta_m = 96\%$

$P = 65\ \text{MW} = 65 \times 10^6$ W

Substitution of these values in the equation, we get

$$0.96 = \frac{65 \times 10^6}{m\left[\left(V_1 - V_{b_1}\right)V_{b_1}\left(1 + K\cos\theta\right)\right]} / g$$

$$\therefore \qquad m = \frac{\left[\left(V_1 - V_{b_1}\right)V_{b_1}\left(1 + K\cos\theta\right)\right]}{9.81} = \frac{65 \times 10^6}{0.96}$$

$$= 67.71 \times 10^6 \text{ watts}$$

Now power supplied to the turbine runner at the distributor is

$$P_1 = \omega QH$$
$$= (9.81 \times 1000) \times 12.5 \times 630$$
$$= 77.25 \times 10^6 \text{ watts}$$

∴ Power losses in the distributor – nozzle assembly and the buckets is given by

$$(P_1 - P) = 77.25 \times 10^6 - 67.71 \times 10^6$$
$$= \mathbf{9.54 \times 10^6 \text{ watts}}$$
$$= \mathbf{9.54 \times 10^3 \text{ kW}}$$

(b) The head loss in the buckets is given to be proportional to V_{r1}^2 i.e.,

Head loss $\qquad h_L = C\,V_{r_1}^2 = C(V_1 - V_{b_1})^2$

where C is the constant of proportionality.

Total discharge = 12.5 m³/s

and number of jets = 4

∴ Discharge through each jet = $\frac{12.5}{4}$ m³/s

Diameter of jet = 0.192 m.

$$\therefore \qquad V_1 = \frac{12.5/4}{\frac{\pi}{4} \times (0.192)^2}$$

$$= \mathbf{107.93 \text{ m/s}}$$

and $\qquad V_{b_1} = \frac{\pi DN}{60} = \frac{\pi \times 1.96 \times 500}{60}$

$$= \mathbf{51.31 \text{ m/s}}$$

$$\therefore \qquad h_L = C(107.93 - 51.31)^2$$
$$= 3206C$$

The best speed of rotation is the one which would provide maximum hydraulic efficiency for which we have

$$V_b = 0.5V_1$$

$$\therefore \qquad h_L = C(2V_b - V_b)^2 = C\,V_b^2$$

$$\therefore \qquad 3206C = CV_b^2$$

$$\therefore \qquad V_b = \sqrt{3206}$$
$$= \mathbf{56.62 \text{ m/s}}$$

$$= \frac{\pi DN}{60}$$

$\therefore$ Best speed of rotation

$$N = \frac{56.62 \times 60}{\pi \times 1.96}$$

$$= \mathbf{552\ rpm}$$

Example 21.6 *A Pelton wheel produces 746 kW under a gross head of 200 m. Its nozzle has a diameter of 10 cm and the losses in the pipeline due to friction amount to 90 Q^2 where Q is the discharge in m^3/s. Assume the gross head and efficiency of the wheel to be constant and CV for the nozzle is 0.98. Find the discharge and overall efficiency.*

Solution

Net head = $(200 - 90\,Q^2)$

$\therefore$ Actual velocity $V_1 = C_v \times \sqrt{2 \times 9.81 \times \left(200 - 90Q^2\right)}$

Assuming, jet diameter = Nozzle diameter, we get

$$Q = \text{water flow}$$

$$= \frac{\pi}{4} \times (0.1)^2 \times 0.98 \sqrt{2 \times 9.81\left(200 - 90Q^2\right)}$$

Solving these equations we get

$$Q = \mathbf{0.465\ m^3/s}$$

$\therefore$ Net head $= H = [200 - 90 \times (0.465)^2]$

$$= \mathbf{180.54\ m}$$

$$\text{Overall efficiency} = \frac{P}{wQH}$$

$$= \frac{746 \times 1000}{9810 \times 0.465 \times 180.54}$$

$$= \mathbf{0.906} \text{ or } \mathbf{90.6\%}$$

Example 21.7 *An inward flow turbine (reaction type with radial discharge) with an overall efficiency of 80% is required to develop 150 kW. The head is 8 m; peripheral velocity of the wheel is 0.96 $\sqrt{2gH}$; the radial flow velocity is 0.36 $\sqrt{2gH}$. The wheel is to make 150 rpm and the hydraulic losses in the turbine are 22% of the available energy. Determine :*

(i) the angle of the guide blade at inlet

(ii) the wheel vane angle at inlet

(iii) diameter of the wheel

(iv) the width of the wheel at inlet.

Solution

Given = η_o = 80%, H = 8 m, P = 150 kW, N = 150 rpm.

Peripheral velocity $V_{b_1} = 0.96\sqrt{2 \times 9.81 \times 8} = \mathbf{12.03\ m/s}$

Velocity of flow at inlet $V_{f_1} = 0.36\sqrt{2 \times 9.81 \times 8} = 4.51$ m/s

Hydraulic losses = 22%

$$\therefore \quad \eta_h = (100 - 22) = 78\%$$

For radial discharge

$$\eta_h = \frac{V_{w1}V_{b1}}{gH}$$

$$\therefore \quad 0.78 = \frac{V_{w1} \times 12.03}{9.81 \times 8}$$

$$\therefore \quad V_{w_1} = 5.09 \text{ m/s}$$

From inlet velocity triangle, we have

$$\tan\alpha = \frac{V_{f_1}}{V_{w_1}} = \frac{4.51}{5.09}$$

$$= 0.8861$$

$$\therefore \quad \alpha = \text{the angle of the guide blade at inlet}$$

$$= \mathbf{41°–33'}$$

$$\text{Also } \tan(180 - \theta) = \frac{V_{f_1}}{V_{b_1} - V_{w_1}}.$$

$$= \frac{4.51}{12.03 - 5.09} = \mathbf{0.6499}$$

$$\therefore \quad \theta = \text{Wheel vane angle at inlet}$$

$$= \mathbf{147°}$$

Again

$$V_{b_1} = \frac{\pi D_1 N}{60} = \frac{\pi \times D_1 \times 150}{60}$$

$$\therefore \quad D_1 = \text{Diameter of the wheel}$$

$$= \frac{12.03 \times 60}{\pi \times 150}$$

$$= \mathbf{1.532 \text{ m}}$$

$$\text{Overall efficiency} = \eta_o = \frac{P}{\omega QH}$$

$$0.8 = \frac{150 \times 10^3}{9810 \times Q \times 8}$$

$$\therefore \quad Q = \text{Water flow/s}$$

$$= \frac{150 \times 10^3}{9810 \times 8 \times 0.8}$$

$= \mathbf{2.389\ m^3/s}$

Neglecting the vane thickness, we have

$$Q = \pi D_1 B_1 V_{f1}$$

$$2.389 = \pi \times 1.532 \times B_1 \times 4.51$$

$$\therefore \quad B_1 = \text{Wheel width at inlet}$$

$$= \frac{2.389}{\pi \times 1.532 \times 4.51}$$

$$= 0.11 \text{ m}$$

$$= \mathbf{110\ mm}$$

Example 21.8 *An inward flow reaction turbine discharges radially and the velocity of flow is constant and equal to the velocity of discharge from the turbine. Show that the hydraulic efficiency can be expressed by*

$$\eta_h = \frac{1}{1 + \dfrac{\frac{1}{2}\tan^2\alpha}{1 - \dfrac{\tan\alpha}{\tan\theta}}}$$

where α and θ are respectively the guide vane angle and wheel vane angle at inlet.

Solution

From the inlet velocity triangle,

$$V_{f_1} = V_{w_1} \tan\alpha$$

$$V_{b_1} = V_{w_1} - \frac{V_{f1}}{\tan\theta}$$

$$= V_{w_1} - \frac{V_{w1}\tan\alpha}{\tan\theta}$$

$$= V_{w_1}\left[1 - \frac{\tan\alpha}{\tan\theta}\right] \quad \text{(i)}$$

For radial discharge at outlet,

$$V_{w_2} = 0 \text{ and } V_2 = V_{f_2}$$

$$\therefore \quad \frac{V_{w_1}V_{b_1}}{g} = H - \frac{V_2^2}{2g}$$

$$\therefore \quad H = \frac{V_{w_1}V_{b_1}}{g} + \frac{V_2^2}{2g}$$

$$= \frac{V_{w_1}V_{b_1}}{g} + \frac{V_{f_2}}{2g}$$

$$= \frac{V_{w_1}V_{b_1}}{g} + \frac{V_{f_1}}{2g}$$

$$= \frac{V_{w_1} V_{b_1}}{g} + \frac{V_{w_1}^{\ 2} \tan^2 \alpha}{2g}$$

Substituting the value of V_{b1}, we get

$$H = \frac{V_{w_1}^2 \times V_{w_1}\left(1 - \dfrac{\tan\alpha}{\tan\theta}\right)}{g} + \frac{V_{w_1}^{\ 2} + \tan^2 \alpha}{2g}$$

$$= \frac{V_{w_1}}{g}\left[1 - \frac{\tan\alpha}{\tan\theta} + \frac{\tan^2\alpha}{2}\right]$$

$$\therefore \qquad \eta_h = \frac{V_{w_1} V_{b_1}}{gH} = \frac{V_{w_1}^{\ 2}\left[1 - \dfrac{\tan\alpha}{\tan\theta}\right]}{gH}$$

or

$$\eta_h = \frac{\left(1 - \dfrac{\tan\alpha}{\tan\theta}\right)}{\left[1 - \dfrac{\tan\alpha}{\tan\theta} + \dfrac{\tan^2\alpha}{2}\right]}$$

$$= \frac{1}{1 + \dfrac{\frac{1}{2}\tan^2\alpha}{1 - \dfrac{\tan\alpha}{\tan\theta}}}$$

Example 21.9 *Design a Francis turbine runner with the following data : Net head H = 68 m, Speed N = 750 rpm, Output power P = 330 kW, η_h = 94%, η_o = 85%, flow ratio = 0.15 n = Breadth, ratio = 0.1, inner diameter of runner is $\frac{1}{2}$ outer diameter. Also assume 6% of the circumferential area of the runner to be occupied by the thickness of the vanes. Velocity of flow remains constant throughout and flow is radial at exit.*

Solution

$$\eta_o = \frac{P}{\omega Q H}$$

$$0.85 = \frac{330 \times 10^3}{9810 \times Q \times 68}$$

$$\therefore \qquad Q = \frac{330 \times 10^3}{0.85 \times 9810 \times 68}$$

$$= \mathbf{0.582\ m^3/s}$$

Also $V_{f_1} = 0.15\sqrt{2gH} = 0.15\sqrt{2 \times 9.81 \times 68}$

$= \mathbf{5.48\ m/s}$

Again $Q = (K\pi n\, D^2{}_1)V_{f1}$

$0.582 = 0.94\pi \times 0.1 \times D^2{}_1 \times 5.48$

$\therefore \quad D_1 = 0.6\ \text{m} = \mathbf{600\ mm}$

$B/D = n = 0.1$

$\therefore \quad B = \mathbf{60\ mm}$

$$V_{b_1} = \frac{\pi D_1 N}{60} = \frac{\pi \times 0.6 \times 750}{60}$$

$= \mathbf{23.56\ m/s}$

$$\eta_n = \frac{V_{w_1} V_{b_1}}{gH}$$

$$0.94 = \frac{V_{w_1} \times 23.56}{9.81 \times 68}$$

$\therefore \quad V_{w1} = \mathbf{26.62\ m/s}$

From inlet velocity diagram, we have

$$\tan\alpha = \frac{V_{f_1}}{V_{w_1}} = \frac{5.48}{26.62}$$

$= 0.2059$

$\therefore \quad \alpha = \mathbf{11°38'}$

$$\tan\theta = \frac{V_{f_1}}{V_{w_1} - V_{b_1}}$$

$$= \frac{5.48}{26.62 - 23.56}$$

$= \mathbf{1.7908}$

$\therefore \quad \theta = \mathbf{60°49'}$

$$D_2 = \frac{D_1}{2} = 0.3\ \text{m} = 330\ \text{mm}$$

Also $$\frac{V_{b_1}}{D_1} = \frac{V_{b_2}}{D_2}$$

$$\therefore \quad V_{b_2} = \frac{D_2}{D_1} \times V_{b_1}$$

$= \mathbf{11.78\ m/s}$

Since the flow is radial at exit and $V_{f_1} = V_{f_2}$,

$$\tan\theta = \frac{V_{f_2}}{V_{b_2}} = \frac{5.48}{11.78}$$

$$= \mathbf{0.4652}$$

$$\therefore \quad \phi = \mathbf{24°57'}$$

Example 21.10 *A Kaplan turbine produces 60000 kW under a net head of 25 m with an overall efficiency of 90%. Taking the value of speed ratio Ku as 1.6, flow ratio ψ as 0.5 and the hub diameter as 0.35 times the outer diameter, find the diameter and speed of the turbine.*

Solution

$$\eta_o = \text{Overall efficiency} = 0.9.$$

$$= \frac{P}{\omega QH} = \frac{60000\times 10^3}{9810\times Q\times 25}$$

$$\therefore \quad Q = \frac{60000\times 10^3}{9810\times 25\times 0.9}$$

$$= \mathbf{271.83\ m^3/s}$$

$$= \frac{\pi}{4}(D^2 - d^2)\,\psi\,\sqrt{2gH}$$

$$= \frac{\pi}{4}D^2\left[1-\frac{d^2}{D^2}\right]\times\psi\times\sqrt{2gH}$$

$$= \frac{\pi}{4}D^2\,[1-(0.35)^2]\times 0.5\times\sqrt{2\times 9.81\times 25}$$

$$= \frac{\pi}{4}D^2\,[1-0.1225]\times 0.5\times\sqrt{50\times 9.81}$$

$$= \frac{30.54}{4}D^2$$

$$D = \sqrt{\frac{271.83\times 4}{30.54}} = \mathbf{5.97\ m}$$

$$= \text{Diameter of the turbine.}$$

Speed ratio $\quad Ku = 1.6 = \dfrac{V_{b1}}{\sqrt{2gH}}$

$$\therefore \quad V_{b_1} = 1.6\sqrt{2\times 9.81\times 25}$$

$$= \mathbf{35.44\ m/s}$$

$$= \frac{\pi DN}{60}$$

$$\therefore \quad N = \frac{35.44 \times 60}{\pi \times 5.97}$$

$$= \text{Speed of the turbine}$$

$$= \mathbf{113.38\ rpm}$$

Example 21.11 *Determine the efficiency of a Kaplan turbine developing 3000 kW under a net head of 5 m. It is provided with a draft tube, with its inlet diameter 3 m, is set 1.6 m above the tail race level. A vacuum gage connected to the draft tube indicates a reading of 5 m of water. Assume draft tube efficiency as 78%.*

Solution

Neglecting the head loss in the draft tube, we have

$$\frac{P_2}{W} = -5 = -\left[H_s + \frac{V_2^2 - V_3^2}{2g}\right]$$

$$= -\left[1.6 + \frac{V_2^2 - V_3^2}{2g}\right]$$

$$3.4 = \frac{V_2^2 - V_3^2}{2g}$$

Again, the draft tube efficiency is given by

$$\eta_d = \frac{\left(V_2^2 - V_3^2\right)/2g}{V_2^2/2g}$$

$$\therefore \quad 0.78 = \frac{3.4}{V_2^2/2g}$$

$$\therefore \quad V_2^2 = \frac{3.4 \times 2 \times 9.81}{0.78}$$

$$= 85.523$$

$$\therefore \quad V_2 = \mathbf{9.25\ m/s}$$

$$\therefore \quad Q = \left(\frac{\pi}{4} \times d_i^2\right) \times V_2$$

$$= \frac{\pi}{4} \times 9 \times 9.25$$

$$= \mathbf{65.38\ m^3/s}$$

$$\therefore \quad \text{Overall efficiency} = \eta_o$$

$$= \frac{P}{wQH}$$

$$= \frac{3000 \times 1000}{9810 \times 65.38 \times 5}$$

$$= 0.9355$$

$$= \mathbf{93.55\%}$$

Example 21.12 *A propeller turbine runner has outer diameter of 4.5 m and the diameter of the hub 2 m. It is required to develop 20600 kW when running at 150 rpm, under a head of 21 m. Assuming hydraulic efficiency of 94%, and overall efficiency of 0.88%, determine the runner vane angles at inlet and exit at the mean diameter of the vanes. Also determine the runner vane angles at inlet and exit of two sections near the hub and the outer periphery.*

Solution

Overall efficiency $\eta_o = \dfrac{P}{\omega QH} = \dfrac{20600 \times 10^3}{9810 \times Q \times 21}$

$\therefore \quad Q = \dfrac{20600 \times 10^3}{9810 \times 0.88 \times 21}$

$= \mathbf{113.63\ m^3/s}$

Again $\quad Q$ = Area of flow × velocity of flow

$= \dfrac{\pi}{4}[D^2 - d^2]\, V_{f1}$

$\therefore \quad 113.63 = \dfrac{\pi}{4}[(4.5)^2 - (2)^2]\, V_{f1}$

$V_{f1} = \dfrac{113.63 \times 4}{\pi \times 16.25}$

$= \mathbf{8.9\ m/s}$

I. Computation for Mean Diameter

Mean diameter $= D_m = \dfrac{4.5 + 2}{2}$

$= \mathbf{3.25\ m}$

$\therefore \quad V_{bm}$ = Blade mean velocity

$= \dfrac{\pi D_m N}{60}$

$= \dfrac{\pi \times 3.25 \times 150}{60}$

$= \mathbf{25.53\ m/s}$

∴ Hydraulic efficiency

$\eta_h = \dfrac{V_{w_1} V_{b_1}}{gH}$

$\therefore \quad 0.94 = \dfrac{V_{w_1} \times 25.53}{9.81 \times 21}$

$\therefore \quad V_{w_1} = \dfrac{0.94 \times 9.81 \times 21}{25.53}$

$$= \mathbf{7.59\ m/s}$$

From inlet velocity diagram, we have

$$\tan(180 - \theta) = \frac{V_{f_1}}{V_{w_1} - V_{b_1}}$$

$$= \frac{8.9}{25.53 - 7.59}$$

$$= 0.4961$$

$$\therefore \quad \theta = \mathbf{153°37'}$$

Similarly from outlet or exit velocity diagram,

$$\tan\phi = \frac{V_{f_2}}{V_{b_2}} = \frac{V_{f_1}}{V_{b_1}}$$

$$= \frac{8.9}{25.53}$$

$$= \mathbf{0.3486}$$

$$\therefore \quad \phi = 19°13'$$

II. Computation for Section Near the Hub

$$V_{b_1} = \frac{\pi D_1 N}{60} = \frac{\pi \times 2 \times 150}{60}$$

$$= \mathbf{15.71\ m/s}$$

and
$$\eta_h = 0.94 = \frac{V_{w_1} V_{b_1}}{gH}$$

$$\therefore \quad V_{w_1} = \frac{0.94 \times 9.81 \times 21}{15.71}$$

$$= \mathbf{12.33\ m/s}$$

From inlet velocity diagram, we have

$$\tan(180 - \theta) = \frac{V_{f_1}}{V_{w_1} - V_{b_1}}$$

$$= \frac{8.9}{15.71 - 12.33} = 2.6331$$

$$\therefore \quad \theta = \mathbf{110°48'}$$

From exit velociy diagram, we have

$$\tan\phi = \frac{V_{f_2}}{V_{b_2}} = \frac{8.9}{15.71}$$

$$= \mathbf{0.5665}$$

$$\therefore \quad \phi = \mathbf{29°32'}$$

III. Computations for Section at the Outer Periphery

$$V_{b_1} = \text{Blade speed}$$

$$= \frac{\pi D_0 N}{60} = \frac{\pi \times 4.5 \times 150}{60}$$

$$= \mathbf{35.34\ m/s}$$

$$= V_{b_2}$$

$$\eta_h = 0.94 = \frac{V_{w_1} V_{b_1}}{gH}$$

$$= \frac{V_{w_1} \times 35.34}{9.81 \times 21}$$

$$\therefore \quad V_{w_1} = \frac{0.94 \times 9.81 \times 21}{35.34}$$

$$= \mathbf{5.48\ m/s}$$

∴ From inlet velocity diagram, we have

$$\tan(180 - \theta) = \frac{V_{f_1}}{V_{a_1} - V_{b_1}}$$

$$= \frac{8.9}{35.34 - 5.48}$$

$$= \mathbf{0.2981}$$

$$\therefore \quad \theta = \mathbf{163°24'}$$

Similarly, from exit velocity diagram, we have

$$\tan\theta = \frac{V_{f_2}}{V_{b_2}}$$

$$= \frac{8.9}{35.34} = \mathbf{0.2518}$$

$$\therefore \quad \phi = \mathbf{14°8'}$$

Example 21.13 *The velocity of whirl at inlet to the runner of an inward flow reaction turbine is (3.15 $\sqrt{H}$) m/s and the velocity of flow at inlet is (1.05 $\sqrt{H}$) m/s. The velocity of whirl at exit is $\left(0.22\sqrt{H}\right)$ m/s in the same direction as an inlet and the velocity of flow at exit is (0.83 $\sqrt{H}$) m/s, where H is the head in metres. The inner diameter of the runner is 0.6 times the outer diameter. Assuming hydraulic efficiency of 80% compute the angles of the runner vanes at inlet and exit.*

Solution

$$\text{Hydraulic efficiency } \eta_h = \frac{V_{w_1} V_{b_1} - V_{w_2} V_{b_2}}{gH}$$

$$\therefore \quad 0.8 = \frac{3.15\sqrt{H} \times V_{b_1} - 0.22\sqrt{H} \times 0.6V_{b_1}}{9.81 \times H}$$

$$= \frac{\sqrt{H}}{9.81H}[3.15\, V_{b_1} - 0.22 \times 0.6\, V_{b_1}]$$

$$= \frac{V_{b_1}}{9.81\sqrt{H}} = [3.15 - 0.132]$$

$$\therefore \quad V_{b_1} = \frac{0.8 \times 9.81\sqrt{H}}{3.018}$$

$$= 2.6\sqrt{H}$$

$$\therefore \quad V_{b_2} = 0.6 \times 2.6\sqrt{H} = 1.56\sqrt{H}.$$

From inlet velocity diagram, we have

$$\tan\theta = \frac{V_{f_1}}{V_{w_1} - V_{b_1}} = \frac{1.05\sqrt{H}}{3.15\sqrt{H} - 2.6\sqrt{H}}$$

$$= \frac{1.05}{3.15 - 2.6} = \frac{1.05}{0.55}$$

$$= 1.9091$$

$$\therefore \quad \theta = \text{Blade angle at inlet}$$

$$= \mathbf{62°21'}$$

Similarly, from outlet velocity diagram, we have

$$\tan\phi = \frac{V_{f_2}}{V_{b_2} - V_{w_2}}$$

$$= \frac{0.83\sqrt{H}}{1.56\sqrt{H} - 0.22\sqrt{H}}$$

$$= \frac{0.83}{1.56 - 0.22} = \frac{0.83}{1.34}$$

$$= \mathbf{0.6194}$$

$$\therefore \quad \phi = \text{Vane outlet angle}$$

$$= \mathbf{31°46'}$$

PERFORMANCE OF TURBINES

Example 21.14 *A Francis turbine working under a head of 5 m at a speed of 210 rpm develops 75 kW when the rate of flow is 1.8 m³/s. The runner diameter is 1 m. If the head on this turbine is increased to 16 m, determine the new speed, discharge and power.*

Solution

We have $N_1 = \frac{N\sqrt{H_1}}{\sqrt{H}}$ = New speed

$$= \frac{210 \times \sqrt{16}}{\sqrt{5}}$$

$$= \mathbf{375.7\ rpm}$$

Similarly Q_1 = New discharge

$$= \frac{Q\sqrt{H_1}}{\sqrt{H}}$$

$$= \frac{1.8\sqrt{16}}{\sqrt{5}}$$

$$= \mathbf{3.22\ m^3/s}$$

and P_1 = New power

$$= \frac{PH_1^{3/2}}{H^{3/2}}$$

$$= 75 \times 16^{3/2} / 5^{3/2} = \mathbf{429.33\ kW}$$

Example 21.15 *A hydraulic turbine has an output of 6600 kW when it works under a head of 25 m and runs at 100 rpm. What is the type of the turbine? What would be its speed and what power will it develop when working under a head of 16 m?*

Solution

Specific speed is given by

$$N_s = \frac{N\sqrt{P}}{H^{5/4}} = \frac{100\sqrt{6600}}{(25)^{5/4}}$$

$$= \mathbf{145.33\ rpm}$$

In this case, $85 < N_s < 188$

∴ Turbine will be a Francis turbine.

$$\therefore \quad N_1 = \frac{N\sqrt{H_1}}{\sqrt{H}} = \frac{100\sqrt{16}}{\sqrt{25}}$$

$$= \mathbf{80\ rpm}$$

New power developed is given by

$$P_1 = \frac{PH_1^{3/2}}{H^{3/2}}$$

$$= \frac{6600 \times 16^{3/2}}{25^{3/2}}$$

$$= \mathbf{3379.2\ kW}$$

Example 21.16 *A turbine develops 7460 kW under a head of 24.7 m at 135 rpm. What is the specific speed? What would be its normal speed and output power under a head of 19.5 m?*

Solution

Specific speed is given by

$$N_s = \frac{N\sqrt{P}}{H^{5/4}}$$

$$= \frac{135\sqrt{7460}}{(24.7)^{5/4}}$$

$$= \textbf{211.75 rpm}$$

If N_1 is the speed under a head of 19.5 m, then

$$N_1 = \frac{N\sqrt{H_1}}{\sqrt{H}}$$

$$= \frac{135\sqrt{19.5}}{\sqrt{24.7}}$$

$$= \textbf{119.95 rpm}$$

Similarly if P_1 is the output under a head of 19.5 m, then

$$P_1 = \left(\frac{H_1}{H}\right)^{3/2} \times P$$

$$= 7460\left(\frac{19.5}{24.7}\right)^{3/2}$$

$$= \textbf{5232.94 kW}$$

Example 21.17 *A Pelton wheel develops 5520 kW under a head of 240 m at an overall efficiency of 80% when revolving at a speed of 200 rpm. Find the unit discharge, unit power and unit speed. Assume peripheral coefficient = 0.46. If the head on the same turbine falls during the summer season to 150 m, find the discharge power and speed for this head.*

Solution

$$\text{Overall efficiency} = \eta_o = \frac{P}{\omega QH}$$

$$0.8 = \frac{5520}{9810 \times 240Q}$$

$$\therefore \quad Q = \text{Flow/sec}$$

$$= \frac{5520}{9810 \times 240 \times 0.8}$$

$$= \textbf{2.931 m}^3\textbf{/s}$$

$$\therefore \text{ Unit discharge} \quad Q_u = \frac{Q}{\sqrt{H}} = \frac{2.931}{\sqrt{240}}$$

$$= \textbf{0.189}$$

Similarly, unit power $= P_u = \frac{P}{H^{3/2}} = \frac{5520}{240^{3/2}}$

$= \mathbf{1.485}$

Unit speed $N_u = \frac{N}{\sqrt{H}} = \frac{200}{\sqrt{240}}$

$= \mathbf{12.91}$

As it is assumed that peripheral coefficient is constant and equal to 0.46, the efficiency may be assumed to be constant under different heads and hence the unit quantities will be same under different heads. Thus

$$Q_u = 0.189 = \frac{Q_1}{\sqrt{150}}$$

$$\therefore \quad \mathbf{Q_1 = 2.315\ m^3/s}$$

$$P_u = 1.489 = \frac{P_1}{(150)^{3/2}},\ P_1 = 2735.47 \text{ kW.}$$

$$N_u = 12.91 = \frac{N_1}{\sqrt{150}},\ \therefore N_1 = 158.1 \text{ rpm}$$

Example 21.18 *A model of Francis turbine one-fifth of full size, develops 3 kW at 306 rpm under a head of 1.77 m. Find the speed and power of full size turbine operating under a head of 5.7 m if*

(a) the efficiency of the model and the full size turbine are same.

(b) the efficiency of the model turbine is 76% and the scale effect is considered.

Solution

(a) For the same efficiency of the model and the prototype, we have

$$\frac{D_m N_m}{\sqrt{H_m}} = \frac{D_P N_P}{\sqrt{H_P}}$$

$$\therefore \quad \frac{306}{5 \times \sqrt{1.77}} = \frac{N_P}{\sqrt{5.7}} \text{ as } \frac{D_m}{D_P} = \frac{1}{5}$$

$$\therefore \quad \mathbf{N_P = 109.8\ rpm}$$

Similarly, $\frac{P_m}{D_m^2 H_m^{3/2}} = \frac{P_P}{D_P^2 H_P^{3/2}}$

$$\frac{3}{(1.77)^{3/2}} = \frac{P_P}{5^2 \times (5.7)^{3/2}}$$

$\therefore \quad P_P =$ Power of the prototype turbine

$$= \left(\frac{5.7}{1.77}\right)^{3/2} \times \frac{3}{25}$$

$= \mathbf{433.43\ kW}$

(b) Moody's equation gives

$$\frac{1-\eta_{oP}}{1-\eta_{om}} = \left(\frac{D_m}{D_P}\right)^{0.2}$$

$$\therefore \quad \frac{1-\eta_{oP}}{1-0.76} = \left(\frac{1}{5}\right)^{0.2}$$

$$\therefore \quad 1-\left(\frac{1}{5}\right)^{0.2} \times (1.076) = \eta_{oP}$$

$$= \mathbf{0.826}$$

Similarly,

$$\frac{D_m N_m}{\sqrt{H_m \cdot \eta_{om}}} = \frac{D_P N_P}{\sqrt{\eta_{oP} . H_P}}$$

or

$$\frac{306}{5\times\sqrt{1.77\times 0.76}} = \frac{N_P}{\sqrt{0.826\times 5.7}}$$

$$\therefore \quad \mathbf{N_P = 114.5\ rpm}$$

$$\therefore \quad \frac{P_m}{\eta_{om}^{3/2} . H_m^{3/2} . D_m^2} = \frac{P_P}{\eta_{oP}^{3/2} H_P^{3/2} D_P^2}$$

$$\therefore \quad \frac{3}{(0.76)^{3/2}\times(1.77)^{3/2}} = \frac{P_P}{(0.826)^{3/2}\times(5.7)^{3/2}\times 5^2}$$

$$\therefore \quad P_P = \text{Prototype turbine power.}$$

$$= \mathbf{491.09\ kW}$$

Example 21.19 *Estimate the maximum height of straight conical draft tube of 13240 kW Francis turbine running at 150 rpm under a head of 27 m. The turbine is installed at a station where the effective atmospheric pressure is 10.6 m of water. The draft tube must sink at least 0.77 m below the tail race.*

Solution

Specific speed of the turbine is given by

$$N_s = \frac{N\sqrt{P}}{H^{5/4}} = \frac{150\sqrt{13240}}{(27)^{5/4}}$$

$$= \mathbf{280.43}$$

Again

$$\sigma_c = 0.625\left[\frac{N_s}{380.78}\right]^2 = 0.625\left[\frac{280.43}{380.78}\right]^2$$

$$= \mathbf{0.339}$$

$$= \frac{H_a - H_v - H_s}{H}. \text{ But } H_a - H_v = 10.6 \text{ and } H = 27 \text{ m}$$

$$\therefore \quad 0.339 = \frac{10.6 - H_s}{27}$$

or $$H_s = 1.45 \text{ m}$$

Hence the maximum length of the draft tube

$$= (1.45 + 0.77)$$

$$= \mathbf{2.22 \text{ m}}$$

Example 21.20 *Calculate the value of Thoma's cavitation parameter for a Kaplan turbine having non-dimensional specific speed of 19.36 and suction specific speed of 17.6.*

Solution

The non-dimensional specific speed n_s is given by

$$n_s = \frac{N_s}{17.362}$$

$$N_s = (17.362 \times n_s) = 17.362 \times 19.36$$

Similarly the non-dimensional suction specific speed is given by

$$s = \frac{S}{17.362}$$

$$\therefore \quad S = 17.362 \times s$$

$$= 17.362 \times 17.6$$

$$\therefore \quad \sigma = \left(\frac{N_s}{S}\right)^{4/5}$$

$$= \left(\frac{17.362 \times 19.36}{17.362 \times 17.6}\right)^{4/5}$$

$$= \mathbf{1.08}$$

Example 21.21 *A model of a turbine 0.3 m diameter when running at 900 rpm under a head of 12 m gave the following results :*

Percentage gate opening	*40*	*60*	*80*	*100*
Output power kW	*8.38*	*13.24*	*17.50*	*19.71*

The actual turbine is required to develop 7360 kW at 85% gate opening under a head of 35 m. Assuming same efficiency at corresponding gate opening for the model and actual turbines, find

(i) diameter and speed of the actual turbine

(ii) output of the actual turbine at 50% gate opening.

Solution

(i) From the plot of percentage gate opening v/s output power for the given data, for 85% gate opening the output power of the model is obtained as

$$P_m = 18.20 \text{ kW}$$

$$\therefore \quad \left(\frac{P}{H^{3/2}D^2}\right)_m = \left(\frac{P}{H^{3/2}D^2}\right)_P$$

or $\left(\dfrac{18.20}{12^{3/2} \times 0.3^2}\right)_m = \left(\dfrac{7360}{35^{3/2} \times D^2}\right)_P$

$\therefore \quad D_P$ = Diameter of actual turbine

$= \mathbf{2.7\ m}$

Also $\left(\dfrac{H}{N^2 D^2}\right)_m = \left(\dfrac{H}{N^2 D^2}\right)_P$

$$\left(\frac{12}{(900)^2 \times (0.3)^2}\right)_m = \left(\frac{35}{N^2 \times (2.7)^2}\right)_P$$

$\therefore \quad N_P$ = Speed of the actual turbine

$= \mathbf{171\ rpm}$

(ii) From the same plot for 50% gate opening, the output power of the model is

$\mathbf{P_m = 11.03\ kW}$

$$\therefore \quad \left(\frac{P}{H^{3/2} D^2}\right)_m = \left(\frac{P}{H^{3/2} D^2}\right)_P$$

$$\left(\frac{11.03}{(12)^{3/2} \times (0.3)^2}\right)_m = \left(\frac{P}{(35)^{3/2} \times (2.7)^2}\right)_P$$

$\therefore \quad P_P$ = Power of the actual turbine at 50% gate opening

$= \mathbf{4450.3\ kW}$

Example 21.22 *A vertical shaft Kaplan turbine under a head of 9.8 m has a runner diameter of 9.3 m. At the optimum regime the turbine runs at 51.7 rpm and develops 45000 kW power discharging 535 m³/s water. Determine the values of unit speed, unit discharge, unit power and specific speed of the turbine.*

If the critical value of cavitation coefficient for the runner is 0.95, determine the location of the runner with respect to the tail race water level. The atmospheric and saturated vapour pressures may be taken as 0.91 kg_f/cm^2 and 0.25 kg(f)/cm² respectively.

Solution

Given : $H = 9.8$ m, $D = 9.3$ m, $N = 51.7$ rpm, $P = 45000$ kW and $Q = 535\ m^3/s$.

Unit speed $\quad N_u = \dfrac{ND}{\sqrt{H}} = \dfrac{51.7 \times 9.3}{\sqrt{9.8}}$

$= \mathbf{153.59}$

Unit discharge $\quad Q_u = \dfrac{Q}{D^2\sqrt{H}} = \dfrac{535}{(9.3)^2 \times \sqrt{9.8}}$

$= \mathbf{1.98}$

Unit power $\quad P_u = \dfrac{P}{D^2 H^{3/2}} = \dfrac{45000}{(9.3)^2 \times (9.8)^{3/2}}$

$$= \mathbf{16.96}$$

Specific speed $$N_s = \frac{N\sqrt{P}}{(H)^{5/4}} = \frac{51.7\sqrt{45000}}{(9.8)^{5/4}}$$

$$= \mathbf{632.51}$$

Again we have $$\sigma = \frac{H_a - H_v - H_s}{H}$$

$\therefore$ $$\sigma = \sigma_c = 0.95$$

$$H_a = \frac{0.98 \times 10^4}{1000} = 9.8 \text{ m}$$

and $$H_v = \frac{0.25 \times 10^4}{1000} = 2.5 \text{ m}$$

$\therefore$ $$0.95 = \frac{9.8 - 2.5 - H_s}{9.8}$$

$\therefore$ $$\mathbf{H_s = -2.01 \text{ m}}$$

= Suction head is negative and hence the turbine runner must be located 2.01 m below the tail race water level.

Example 21.23 *A water turbine develops 130 kW at 230 rpm under a head of 16 m. Determine the scale ratio and the speed of a similar machine which will generate 660 kW when working under a head of 25 m.*

Solution

Since $$\frac{P_1}{D_1^2 H_1^{3/2}} = \frac{P_2}{D_2^2 H_2^{3/2}}$$

We have $$\frac{130}{D_1^2 \times 16^{3/2}} = \frac{660}{D_2^2 \times 25^{3/2}}$$

$\therefore$ $$\left(\frac{D_2}{D_1}\right)^2 = \frac{660}{130} \times \frac{16^{3/2}}{25^{3/2}} = \frac{660}{130} \times \frac{64}{125}$$

$\therefore$ $$\frac{D_2}{D_1} = \text{scale ratio}$$

$$= \left(\frac{660 \times 64}{125}\right)^{1/2}$$

$$= \mathbf{1.612}$$

Also, $$\frac{N_1 D_1}{\sqrt{H_1}} = \frac{N_2 D_2}{\sqrt{H_2}}$$

$$\frac{230 \times D_1}{\sqrt{16}} = \frac{N_2 \times D_2}{\sqrt{25}}$$

$$\therefore \quad N_2 = \frac{230 \times D_1}{4} = \frac{5}{D_2}$$

$$= \frac{230 \times 1.25}{1.612}$$

$$= \mathbf{178.35\ rpm}$$

Example 21.24 *A turbine is to operate under a head of 25 m at 200 rpm. If the discharge is 9 m³/s and turbine efficiency is 90%, calculate power generated by the turbine, specific speed of the turbine and performance of the turbine under a head of 20 m. Also state the type of the turbine.*

Solution

We know, $$\eta_o = \frac{P}{\omega QH} = P = \eta_o = wQH$$

$$= 0.9 \times 9810 \times 9 \times 25/1000$$

$$= \mathbf{1986.5\ kW}$$

= Power generated by turbine

$$N_s = \frac{N\sqrt{P}}{H^{5/4}} = \frac{200 \times \sqrt{1986.5}}{(25)^{5/4}}$$

$$= \mathbf{159.46}$$

∴ For this value of N_s, the turbine is Francis turbine.

The performance of the turbine under a head of 20 m is given by the following expressions :

(i) $$N_1 = \frac{N\sqrt{H_1}}{\sqrt{H}} = \frac{200\sqrt{20}}{\sqrt{25}}$$

$$= \mathbf{178.89\ rpm}$$

(ii) $$Q_1 = \frac{Q\sqrt{H_1}}{\sqrt{H}} = \frac{9\sqrt{20}}{\sqrt{25}}$$

$$= \mathbf{8.05\ m^3/s}$$

(iii) $$P_1 = \frac{PH_1^{3/2}}{H^{3/2}} = \frac{1986.5 \times (20)^{3/2}}{(25)^{3/2}}$$

$$= \mathbf{1421.4\ kW}$$

Example 21.25 *The following data were obtained from the main characteristics of a Kaplan turbine of runner diameter 1 m; $P_u = 30.695$; $Q_u = 108.6$; $N_u = 63.6$. Estimate for runner diameter, the discharge and the speed of a similar runner working under a head of 30 m and developing 2000 kW. Determine the specific speed of the runner.*

Solution

We known that the two runners will be similar if

$$\left(\frac{Q_1}{N_1 D_1^3}\right) = \left(\frac{Q_2}{N_2 D_2^2}\right); \left(\frac{N_1 D_1}{\sqrt{H_1}}\right) = \left(\frac{N_2 D_2}{\sqrt{H_2}}\right)\left(\frac{N_2 D_2}{\sqrt{H_2}}\right)$$

and
$$\left(\frac{N_1\sqrt{P_1}}{H_1^{5/4}}\right) = \left(\frac{N_2\sqrt{P_2}}{H_2^{5/4}}\right)$$

From these conditions the following equation can be written

$$\frac{P_1}{D_1^2 H_1^{3/2}} = \frac{P_2}{D_2^2 + H_2^{3/2}}$$

$$\frac{30.695}{1^2 \times ?} = \frac{2000}{D_2^2 \times (30)^{3/2}}$$

$$\therefore \qquad \mathbf{D_2 = 0.6297\ m}$$

Again
$$\frac{N_1 D_1}{\sqrt{H_1}} = \frac{N_2 D_2}{\sqrt{H_2}}$$

$$\therefore \qquad 63.6 \times 1 = \frac{N_2 \times 0.6297}{30^{1/2}}$$

$$\therefore \qquad \mathbf{N_2 = 553.2\ rpm}$$

Similarly following conditions will be obtained

$$\frac{Q_1}{D_1^2\sqrt{H_1}} = \frac{Q_2}{D_2^2\sqrt{H_2}}$$

or
$$\frac{108.6}{1^2} = \frac{Q_2}{(0.6297)^2 \times 30^{1/2}} \text{ or } \mathbf{Q_2 = 235.86\ m^3/s}$$

Specific speed is given by

$$N_s = \frac{N\sqrt{P}}{H^{5/4}} = N_u \cdot \sqrt{P_u}$$

For the initial or first runner

$$N_s = 63.6\sqrt{30.695}$$
$$= \mathbf{352.36}$$

For the second runner

$$N_s = \frac{553.2\sqrt{2000}}{30^{5/4}}$$
$$= \mathbf{352.36}$$

THEORY QUESTIONS

21.1. Enumerate the essential elements of a hydro-electric power plant.

21.2. What is a catchment area? Why a reservoir is required?

21.3. State the function of a dam.

21.4. What factors are considered in selecting the site for a dam?

21.5. How the dams are classified? Explain briefly the various types of dams that are used for various purposes.

21.6. What is the function of a spillway? Discuss the various types of spillways commonly used.

21.7. Explain the following terms :

(a) Head race

(b) Tail race

(c) Canal

(d) Flume

(e) Tunnel

(f) Penstock.

21.8. What is water hammer? In what way a surge tank helps in reducing water hammer?

21.9. What is a surge tank and a forebay and what are their functions? Describe with neat sketches different types of surge tanks.

21.10. What are uses of a draft tube? Describe with neat sketches different types of draft tubes.

21.11. Enumerate the various equipments provided in a power house.

21.12. What do you understand by Mini and Micro-hydel plants? In what way they are important these days?

21.13. What is a water wheel?

21.14. How the hydraulic turbines are classified?

21.15. Classify the hydro-turbines according to head, power, size and specific speed.

21.16. Explain with a neat sketch the principle of operation of a Pelton wheel. Why spear and deflector plate are used in Pelton wheel?

21.17. Derive the relation for maximum power and maximum efficiency of Pelton wheel.

21.18. How the number of jets in a Pelton wheel are determined?

21.19. Define degree of reaction of a hydraulic turbine? Derive the relation for degree of reaction R for the hydraulic turbine. What do you understand by $R = 0$, $R = 0.5$ and $R = 1$?

21.20. Explain with a neat schematic diagramatic sketch, the working of a Francis turbine. What are its advantages?

21.21. Derive, with the help of velocity diagrams of inward flow Francis turbine, the relation for blade or diagram efficiency in terms of blade angles.

21.22. What is a Kaplan turbine? How it is different from Propeller turbine? What are the main features of Kaplan turbine?

21.23. Write in short the new developments of turbines as

(i) Deriaz turbine

(ii) Tubular turbine

(iii) Bulb turbine.

21.24. Explain (a) unit speed (b) unit discharge (c) unit power of a hydraulic turbine. Derive the expressions for each of them.

21.25. Explain the terms specific speed, unit speed and unit power as applied to hydraulic turbines. Deduce expressions to indicate their values.

21.26. How are specific speed, capacity and head related in a turbine? Explain the significance of specific speed in selecting the type of hydro-electric turbine in a hydro-electric power station. Derive the relation for specific speed in terms of speed, power and head.

21.27. What is meant by cavitation? What is Thoma's cavitation factor, and what is its significance for water turbines?

21.28. Write short notes on :

(a) Pitting of turbine blades and its prevention

(b) Servo motors.

21.29. What do you understand by governing of hydraulic turbines? Explain with neat sketches the working of an oil pressure governor.

21.30. How is the governing of a reaction turbine carried out? What is the function of a Relief Valve?

21.31. What do you understand by a term "synchronous speed"? How is it determined?

21.32. What is 'Principle of Similarity' as applied to water turbines? Where is it used? What is a scale ratio? What is its importance?

21.33. What is the performance of the hydraulic turbine? Explain the characteristic curves related to the performance of the hydraulic turbines.

21.34. Write short notes on :

(a) Scroll casing

(b) Draft tube

(c) Wicket gates

(d) Speed ratio

(e) Flow ratio

(f) Turbine efficiencies

(g) Relief valve

(h) Oil pressure governor

(i) Electro-hydraulic governor.

21.35. Explain the purpose of providing (i) scroll casing (ii) stay vanes (iii) guide vanes for a reaction turbines.

21.36. How you will select a turbine in a hydro-electric power plant at a particular site selected?

21.37. Design a Pelton wheel which is required to develop 1500 kW, when working under a head of 160 m at a speed of 420 rpm. The overall efficiency may be taken as 85% and assume other data required.

[**Ans.** D = 1.17 m, No. of jets = 3, d = 9.32 cm, No. of buckets = 21]

21.38. An inward flow reaction turbine works under a head of 22.5 m. The external and internal diameters of the runner are 1.35 m and 1 m respectively. The angle of guide vanes is 15° and the moving vanes are radial at inlet. Radial velocity of flow through runner is constant and there is no velocity of whirl at outlet. Determine the speed of the runner in rpm and the angle of vane at outlet. If the turbine develops 275 kW, find the specific speed. Neglect friction losses. [**Ans.** 206.5 rpm, 19°53′, 69.88 rpm]

21.39. An inward flow reaction turbine is required to develop 300 kW at 200 rpm. The active head at the turbine is 18 m. Determine the outside and inside diameters, the inlet and exit angles for the vanes and the exit angle for the guide vanes. Assume the inlet diameter equal to twice the outlet diameter, the hydraulic efficiency as 80%, the constant radial velocity of flow of 3.6 m/s through the runner, mechanical efficiency as 95% and the width ratio as 0.1. Water leaves the runner radially. [**Ans.** 1.406 m, 0.703 m, 144°56′, 26°4′, 20°34′]

21.40. Find the leading dimensions of the runner of an inward flow Francis turbine producing 625 kW at 1000 rpm under a head of 100 m. Assume the guide vane angle of 16°, the axial length of the blade at inlet 0.1 times the outer diameter, inner radius 0.6 times the outer radius, radial flow velocity constant final discharge radial, hydraulic efficiency 87%, mechanical efficiency 97%, allowance for the blade thickness 5%.

[**Ans.** 541.3 mm, 324.8 mm, 54.13 mm, 90.22 mm, 78°26′, 26°55′]

21.41. A reaction turbine runner has inlet diameter 3.5 m, exit diameter 2.1 m, width 0.5 m both at inlet and exit, and vane angles with the direction of motion 60° and 150° at inlet and exit respectively. If the guide blade angle is 15° and the rate of flow of water is 15 m^3/s, calculate the required speed of the runner for smooth flow at inlet. For this condition, also compute the torque exerted on the runner, the power developed, the energy extracted per unit weight of water, and the pressure drop through the runner.

[**Ans.** 47 rpm, 309.903 kN-m, 1525.3 kW, 10.366 N-m/N, 60.15 kN/m^2]

21.42. A Kaplan turbine runner develops 9300 kW under a net head of 7.4 m. Mechanical efficiency of the wheel is

86%, speed ratio based on outer diameter is 2.2 and the flow ratio 0.66. Diameter of the boss is 0.35 times the external diameter of the wheel. Determine (i) diameter of the runner, (ii) its synchronous speed, (iii) specific speed of the runner. Assume mechanical efficiency equal to overall efficiency.

[Ans. (i) 5.21 m (ii) 97.12 rpm (iii) 767.4 rpm**]**

21.43. A Kaplan turbine develops 2100 kW under a net head of 7.2 m with an overall efficiency of 87%. It is to be fitted with an elbow-type draft tube (with efficiency 88%) having its inlet 1.8 m diameter. Determine how much above or below the tail race level should the draft tube inlet be set so that vacuum pressure there does not exceed 460 mm of mercury. **[Ans.** 1.83 m below the tail race**]**

21.44. A Kaplan turbine develops 2575 kW under a net head of 7.5 m. It is provided with an elbow type draft tube with a circular inlet of 2.5 m diameter. The inlet is set at a height of 1.5 m above the tail race level. A vacuum gage connected to draft tube inlet records a reading of 36.3 kN/m^2. If the efficiency of the draft tube is 78%, calculate the efficiency of the Kaplan turbine. If the ratio of area of circular inlet and rectangular exit of the draft tube is 1:5, determine the power lost due to friction in the tube. If the turbine output were reduced to 1250 kW, head and speed remaining unchanged, determine the vacuum gage reading. **[Ans.** 96%, 181.9 kW, 19.78 kN/m^2**]**

21.45. A reaction turbine 0.5 m diameter when running at 600 rpm, develops 265 kW, the flow through the turbine being 0.75 m^3/s. Under the above operating condition the pressure head at the entrance to the turbine is 28 m, the elevation of the turbine above the total water level is 1.3 m and the velocity of flow at entrance to the turbine runner is 3.7 m/s. Assuming the runner vane angle at inlet as 90°, calculate the effective head on the turbine and the efficiency. **[Ans.** 42.58 m, 84.6%**]**

PERFORMANCE OF TURBINES

21.46. A turbine develops 7355 kW under a head of 24.7 m at 210 rpm. What is the specific speed? Indicate the type of the turbine suitable for the purpose. If this turbine is tested in the laboratory where the flow of water available is only 7.5 m, what power it will develop and at what speed?

[Ans. 327 rpm, Kaplan or Deriaz, 1231 kW, 115.7 rpm**]**

21.47. A Francis turbine runner having a diameter of 2.92 m operates at 163.5 rpm under 54 m head and develops 19900 kW at an efficiency of 87%. Find the other characteristics if this turbine is operated under 60 m head.

[Ans. 172.3 rpm, 45.51 m^3/s, 23307 kW**]**

21.48. An impulse turbine at best speed produces 92 kW under a head of 63 m. By what percent should the speed be increased for a head of 87 m? **[Ans.** 17.5%**]**

21.49. In an inward flow reaction turbine the speed ratio is Ku, the flow ratio is y and the breadth ratio is n. If the overall efficiency of the turbine is 85% and the area taken up by the vanes at inlet is 5% of the peripheral area at inlet, then show that $N_s = 888.23\, Ku\left(\sqrt{yn}\right)$.

A runner of the above type having $N_s = 185$ is required to develop 6620 kW under a head of 90 m. Take $y = 0.18$, $n = 0.2$, calculate speed and the diameter of the runner. **[Ans.** 630.3 rpm, 1.4 m**]**

21.50. A model turbine has a runner of diameter 0.61 m. It develops 50 kW under a head of 30 m at a speed of 4000 rpm. Compute N_s and N_u for this model. It is required to build a similar turbine to develop 155 kW under a head of 36 m. Calculate the required diameter. **[Ans.** 402.85 rpm, 730.3 rpm, 0.937 m**]**

21.51. A $\frac{1}{4}$ scale turbine model is tested under head of 12 m. The prototype turbine is required to work under a head of 30 m at a speed of 425 rpm. At what speed the model be run and if it develops 100 kW and uses 1.14 m^3/s of water at this speed, what power will be obtained from prototype turbine, assuming that its efficiency is 3% better than that of the model? Also indicate the type of runner used in this turbine.

[Ans. 1054 rpm, 6710 kW, Kaplan ($N_s = 496$)**]**

21.52. Estimate the maximum height of a conical draft tube for a Kaplan turbine developing 14710 kW under a net head of 20 m and running at 145 rpm. The turbine is installed at a station where the effective atmospheric pressure is

10.4 m of water. The draft tube must sink at least 1 m below the tail race level.

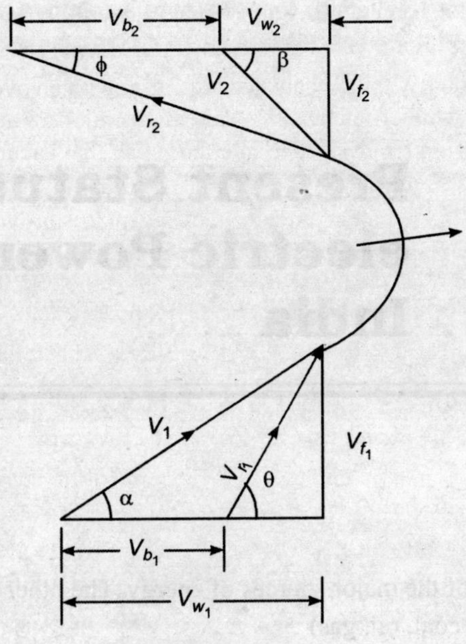

Fig. 21.51

21.53. A model having scale ratio $\frac{1}{10}$ is constructed to determine the best design of Kaplan turbine to develop 7355 kW under the net head of 10 m at a speed of 100 rpm. If the head available at the laboratory is 6 m and the model efficiency is 88% find (i) running speed of the model, (ii) the flow required in the laboratory, (iii) the output power of the model, (iv) the specific speed in each case. Assume same efficiency for the model and prototype turbine. **[Ans.** (i) 774.6 rpm (ii) 0.66 m^3/s, (iii) 34.19 kW, (iv) 482.3 rpm]

Chapter 22 Present Status of Hydro-electric Power Plants in India

22.1 INTRODUCTION

Water power or hydropower is one of the major sources of energy. The other sources of energy being :

(i) The thermal power (fuels like coal, oil, gas)

(ii) Nuclear power (Atomic power)

Hydro or water power developed next to thermal power.

22.2 HYDRO ELECTRIC POWER DEVELOPMENT IN INDIA

India has a vast hydro power potential which is estimated to be about 90 million kW at 60% load factor. So far very little percentage of this vast potential has been harnessed. As such in order to develop this vast potential of power, efforts are being made and several projects have been undertaken to generate more and more hydro electric power.

The major portion of power in India is generated by hydro-electric power plants and thermal power plants in the ratio 2 : 3. A start has, however, been made to develop large scale atomic power in the country with Tarapur Atomic Power Station having installed capacity of 0.44 million kW, Rana Pratap Sagar Atomic Power Station having installed capacity of 0.22 million kW and few others. By 1982–83, the total installed capacity in India was only about 35.34 million kW. This is made of 13.04 million kW hydropower, 21.44 million kW thermal power and 0.86 million kW nuclear power.

Table 22.1 gives the salient features of some of the major hydropower projects of India upto 2006.

Table 22.1 *Major hydro power projects of India*

S.No.	*Name of Power Station*	*River*	*State*	*Installed Capacity*	*Year of*	
					Commencement	*Completion*
1	Bhakra	Govind Sagar	Himachal Pradesh	1050	1948	1968
2	Ganguwal	Sutlej	Punjab	77	1948	1968
3	Kotla	Sutlej	Punjab	77	1948	1968
4	Gandhi Sagar	Chambal	M.P.	115	1954	1960

Contd.

Contd. Table 22.1

S.No.	Name of Power Station	River	State	Installed Capacity	Year of Commencement	Year of Completion
5	Rana Pratap Sagar	Chambal	Rajasthan	172	1961	1970
6	Jawahar Sagar	– do –	– do –	99	1962	1973
7	Rihand	Rihand	U.P.	300	1952	1966
8	Obra	– do –	– do –	99	1963	1971
9	Ukai	Tapti	Gujarat	300	1960	1974
10	Kayna Stage I and II	Koyna	Maharashtra	540	1954	1968
11	Kayna – Stage III	– do –	– do –	320	1966	1975
12	Machkund	Machkund	A.P.	114.75	1946	1961
13	Upper Sileru	Sileru	– do –	120	1958	1968
14	Lower Sileru Stage I	– do –	– do –	460	1964	—
15	Lower Sileru Stage II	– do –	– do –	230	—	—
16	Sharavathy	Sharavathy	Karnataka	891	1956	1974
17	Sabarigiri	Pampa and Kakki	Kerala	300	1961	1968
18	Balimela Stage I	Sileru	Orissa	360	1961	1968
19	Balimela Stage II	– do –	– do –	480	—	1977
20	Ramganga	Ramganga	U.P.	198	1961	1978
21	Srisailam Stage I	Krishna	A.P.	440	1963	1984
22	Idukki	Periyar	Kerala	390	1963	1976
23	Kalinadi	Kali	Karnataka	910	1968	—
24	Nagarjuna Sagar	Krishna	A.P.	440	1956	1974
25	Nagarjuna Sagar (Pumped Storage)	– do –	– do –	110	1970	—
26	Yamuna Stage I	Tons	U.P.	84.75	1960	1970
27	Yamuna Stage II	– do –	– do –	240	1966	1975
28	Kundah	Kundah	Tamil Nadu	435	1956	1967
29	Mettur (Tunnel Power House)	Cauvery	– do –	200	1960	1966
30	Hirakud	Mahanadi	Orissa	270	1948	1964
31	Beas Unit I (Beas Sutlej Link)	Beas	HP	990	1961	1977
32	Beas Unit II	– do –	– do –	360	1960	1974
33	Salal	Chenab	J & K	540	1969	1976
34	Baira Suil	Baira	H.P.	201	1969	1976
35	Loktak	Manipur	Manipur	105	1969	—
36	Umiam Hydel Power Project	Umiam	Assam	39	—	—
37	Subarnarekha Hydel-cum-water supply water	Subarnarekha	Bihar	120	—	—
38	Likar Project	Tapi	Gujarat	160	—	—
39	Chonani Hydel Project	Tawi	J & K	37	—	—

Contd.

Contd. Table 22.1

S.No.	Name of Power Station	River	State	Installed Capacity	Year of Commencement	Year of Completion
40	Iddiki Project	Iddiki	Kerala	800	—	—
41	Tungabhadra Project	Tungabhadra	Karnataka	126	—	—
42	Doab Canal Hydro Project	– do –	Punjab	45	—	—
43	Yamuna Hydro-electric Project	Yamuna	U.P.	444	—	—
44	Jaldala Hydro-electric Project	Jaldala	W.B.	36	—	—
45	Dehar	Beas, Satluj	Rajasthan	990	—	—
46	Ranjit Sagar	Ranjit	Orissa	600	—	—
47	Ranganadi	Ranganadi, Dikrong	Arunachal Pradesh	405	—	—
48	Kadampari	Cauvery	Tamil Nadu	400	—	—
49	Koteshwar	Bhagirathi	U.P.	400	—	—
50	Pong	Beas	H.P.	360	—	—
51	Upper Kolab	Kolab	Orissa	320	—	—
52	Bansagar	Sone	M.P.	315	—	—
53	Rengali	Brahmani	Orissa	250	—	—
54	Chibro	Yamuna	Uttaranchal	240	—	—
55	Kanada	Mahi	Gujarat	240	—	—
56	Upper Sileru	Godavari	A.P.	240	—	—
57	Varahi	Varahi	Karnataka	230	—	—
58	Mukerian	Beas	Punjab	207	—	—
59	Kopili	Umrong	Assam	200	—	—
60	Gerusuppa	Sharavati	Karnataka	180	—	—
61	Pench	Narmada	M.P.	160	—	—
62	Kadra	Bethi, Kalinadi	Karnataka	150	—	—
63	Chillia	Ganga	Uttaranchal	144	—	—
64	Mahi Bajaj Sagar	Mahi	Rajasthan	140	—	—
65	Anandpur Sahib	—	Punjab	134	—	—
66	Bhira	—	Maharashtra	132	—	—
67	Upper Sindh	Sindh	J & K	127	—	—
68	Kuttiadi	Kuttiadi	Kerala	125	—	—
69	Hasdeobango	—	Chattisgarh	120	—	—
70	Jog	Sharavathi	Karnataka	120	—	—
71	Khodri	Tons	Uttaranchal	120	—	—
72	Kodasalli	Kalinadi	Karnataka	120	—	—
73	Lower Mettur	Cauvery	Tamil Nadu	120	—	—
74	Sanjaya Bhabha	Bhaba Khad	H.P.	120	—	—
75	Tatanpur	Sarda	Uttaranchal	120	—	—
76	Umiam	Umiam	Meghalaya	114	—	—
77	Shanan (Uhl I)	Uhl	Punjab	110	—	—
78	Lower Jhelum	Jhelum	J & K	105	—	—
79	Kodayar	Kodayar	T.N.	100	—	—

At present stage, the primary role of hydro-electric power stations in most areas of the country is to provide maximum possible capacity to the grid in coordination with the base load thermal or nuclear power stations. This is because the hydropower projects are not providing sufficient water throughout the year so that the power station will work as base load station.

Following are the major hurdles in developing hydropower projects in India :

1. Varying climatic features
2. Vast geographical regions
3. Topographic features
4. Long duration for the completion of the project
5. Investigations regarding hydrological, topographical, geological and lastly economical features
6. Seismological survey or history
7. Lack of suitable surface storage particularly in Himalaya though India has large river flows from Himalaya Indus, Ganges and Brahmaputra. Also Narmada, Tapi and Mahanadi from central India and Godavari, Krishna and Cauvery from Deccan plateau
8. For Brahmaputra basin, lack of enough protection against floods.

22.3 TYPICAL HYDRO-ELECTRIC DEVELOPMENTS OF INDIA

22.3.1 Bhakra-Nangal Hydro-electric Project

Figure 22.1 shows a schematic section of Bhakra-Nangal hydro-electric power project which is one of the major hydro-electric projects in India. It consists of Bhakra dam constructed across the river Sutlej and Nangal barrage across the same river about 13 km downstream of Bhakra dam. There are two power houses, one on the left bank and other on the right bank, located near the dam. The left bank power house has 5 units of 90 MW each and the right bank power house has 5 units of 120 MW each. Francis turbines have been provided for these power plants which are classified as medium head plants and also they come under the category of storage or reservoir plants.

In addition to the above noted power houses, there are two more power houses in this project, which are located on a power channel taking off from the Nangal barrage. One of the power houses is located at Ganguwal which has in all 3 units of which 2 units are 24 MW each and one unit of 29 MW. The other power house is located at Kotla on the downstream of Ganguwal and has in all 3 units of capacities exactly same as at Ganguwal. The power plants at Ganguwal and Kotla may be classified as low head plants.

22.3.2 Chambal Valley Development Scheme

The Chambal Valley Development scheme is another major hydro-electric scheme of India. The full development consists of three dams with a power house at each dam site and barrage. The three dams constructed in series on the Chambal river are Gandhi Sagar Dam in Madhya Pradesh and Rana Pratap Sagar Dam and Jawahar Sagar Dam in Rajasthan. At Gandhi Sagar Dam there are 5 units of 23 MW each. At Rana Pratap Sagar Dam there are 4 units of 43 MW each. At Jawahar Sagar Dam there are 3 units of 33 MW each. Francis turbines have been provided for each of these power plants which may be classified as medium head plants and also they come under the category of storage plants. A unique feature of Rana Pratap Sagar project is that since the power house is located in a pit below the river bed, the water after issuing from the draft tubes of the turbines is returned to the river at downstream through a standard horse-shoe shaped tunnel 1466 m long and having 12 m internal diameter.

A barrage has been constructed across the Chambal river near Kota city in Rajasthan to divert the water released from the storage and power dams into irrigation canals.

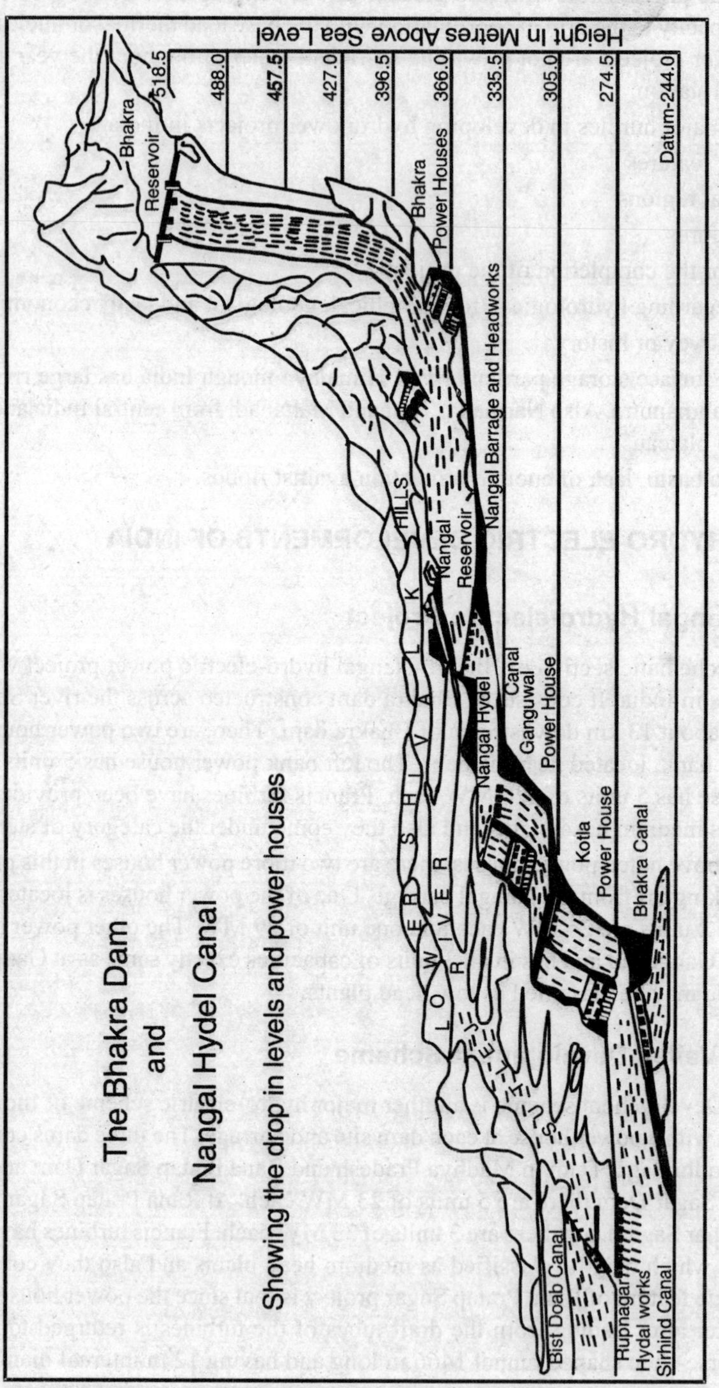

Fig. 27.4 *Schematic Section of Bhakra-Nangal Hydroelectric Project*

22.3.3 Koyna Hydro-electric Power Project KHEP (Maharashtra)

The Koyna Hydro-electric project was commissioned in 1962 to generate power from the potential Koyna river, due to its geographical significance. The Sahyadri ranges with its unique feature of suddenly tapering down to sea-level from peak within about 50 km on the western side aid to this cause.

There are total four stages for this project.

Intake structure includes lake tap, stop log gates, track racks and service gates. Koyna lake is commonly used to cater needs of irrigation through slices provided in the body of the dam and power through existing intakes in Koyna stage I and II withdrawn for the utilisation through Koyna IV needs water conductors arrangements suitable for 260 m^3/s discharges. Storage has been impounded from 1962 onwards and reaches very rarely to the low level during working seasons. This necessitated separately lake piercing of Koyna IV. Such gestation being first of its kind in Asiatic subcontinent was meticulously studied. Advice was also resorted to, from eminent continental consultants. It was unanimously opined to have a smaller sizes shafts to pierce the lake instead of one shaft of larger diameter.

Turbine is provided with spiral case of rolled, welded steel plates designed for transient stresses during starting and stopping of units. Spiral case is provided with a manhole, a drain valve, piping, taps for measuring discharge and water pressure.

Draft tube and cone consists of a rolled steel welded structure with niches for water level detection, complete with drain valve piping, manhole with a door.

22.4 GUIDE VANES AND REGULATING APPARATUS

Guide vanes are made of stainless steel casting of high rigidity, cavitations and corrosion resistant and are 24 in numbers. Each guide vane is provided with self lubricating bearings. Guide vanes are connected through couplings, links and levers to the gate regulating ring. Double acting hydraulic servomotor (accumulators) is provided to operate and hold guide vanes through the regulating ring.

22.4.1 Runner

Runner consists of splitter vanes of chromium-Nickel stainless steel casting, cavitationless and is a non-corrosive alloy of chromium-Nickel steel provided with water injection device to prevent seizure of runner.

22.4.2 Turbine Main Shaft

Turbine shaft is forged, provided with integrally forged coupling flanges top and bottom connecting to the generator shaft and to runner hub.

22.4.3 Spherical Valve and Butterfly Valve

Main spherical valve with a hollow spherical plug, with service and maintenance seals connecting to inlet pipe of spiral case and steel lined shaft, along with motor pump, oil accumulator. A butterfly valve with lattice type disc, with service and maintenance seals, hydraulic drive, motor pump for servomotor generation, closing weight will be provided. The butterfly valves are located in the emergency valve tunnel. Butterfly valve is capable of shutting down the total discharge of turbine at all heads safety and reliably under balanced conditions.

22.4.4 Generator

Hydro-generator with rated continuous apparent output 335 MVA power factor 0.9, 50 Hz, 30 m vertical shaft AC with closed air circuit ventilation is provided. Along with its thrust and guide bearings brakes and proper air cooling systems are also provided. Each turbine is provided with a separate generator. They are manufactured by GEC,

Alstom Electro mechanique Belfort – France. 12 single phase, 100 MVA transformers are also provided.

Starting and change over time

Standstill to rated speed = 50 sec.

No load to full load = 40 sec.

Complete unit slopping = 548 sec.

Emergency slopping = 280 sec.

22.5 CONCLUSION

With rapid growth and clusters of industrial new born belts, the peak demand in the western region is being partially met with, expensive to run, thermal plants. But with completion of Koyna IV the load is being shared, and Koyna IV at present acts to serve the peak demand. But after total completion of all the long term expansion plan Koyna (IV) will be parallel to Koyna (I – II) and Koyna III will acts as the peak load station.

TABLE 22.2 *Salient features of K.H.E.P.*

	Stage I and II	*Stage III*	*Stage IV*
(a) Dam : Catchments's area	891.78 km^2	25.40 km	891.78 km
Max. height above function	103.02 m	66.30 m	103.02 m
Length	807.72 m	497.0 m	807.72 m
(b) Head Race tunnel length	3748 m	4551 m	4230 m
Intake tunnel diameter	6.4 m circular	7.47 m 'D' shape	7 × 9.5 m Horse
Discharge capacity	164 m^3/s	170 m^3/s	260 m^3/s
(c) Pressure shaft nos.	4	4	4
Pressure shaft length	616 m	192 m	590 m
(d) Power house approach tunnel	864 m	780 m	988 m
Trail race tunnel	2215 m	4543 m	1886 m
(e) Switchyard	45.72 × 29.5 m	340 × 137 m	130 × 40 m
(f) Generator capacity	I – 65 MW × 4 II – 75 MW × 4 Damfoot – 20 MW × 2	80 MW × 4	250 MW × 4
Load factor	60%	24%	18% (1 + 2 and 4)

Figures 22.2 to 22.5 show the schematics about KHP.

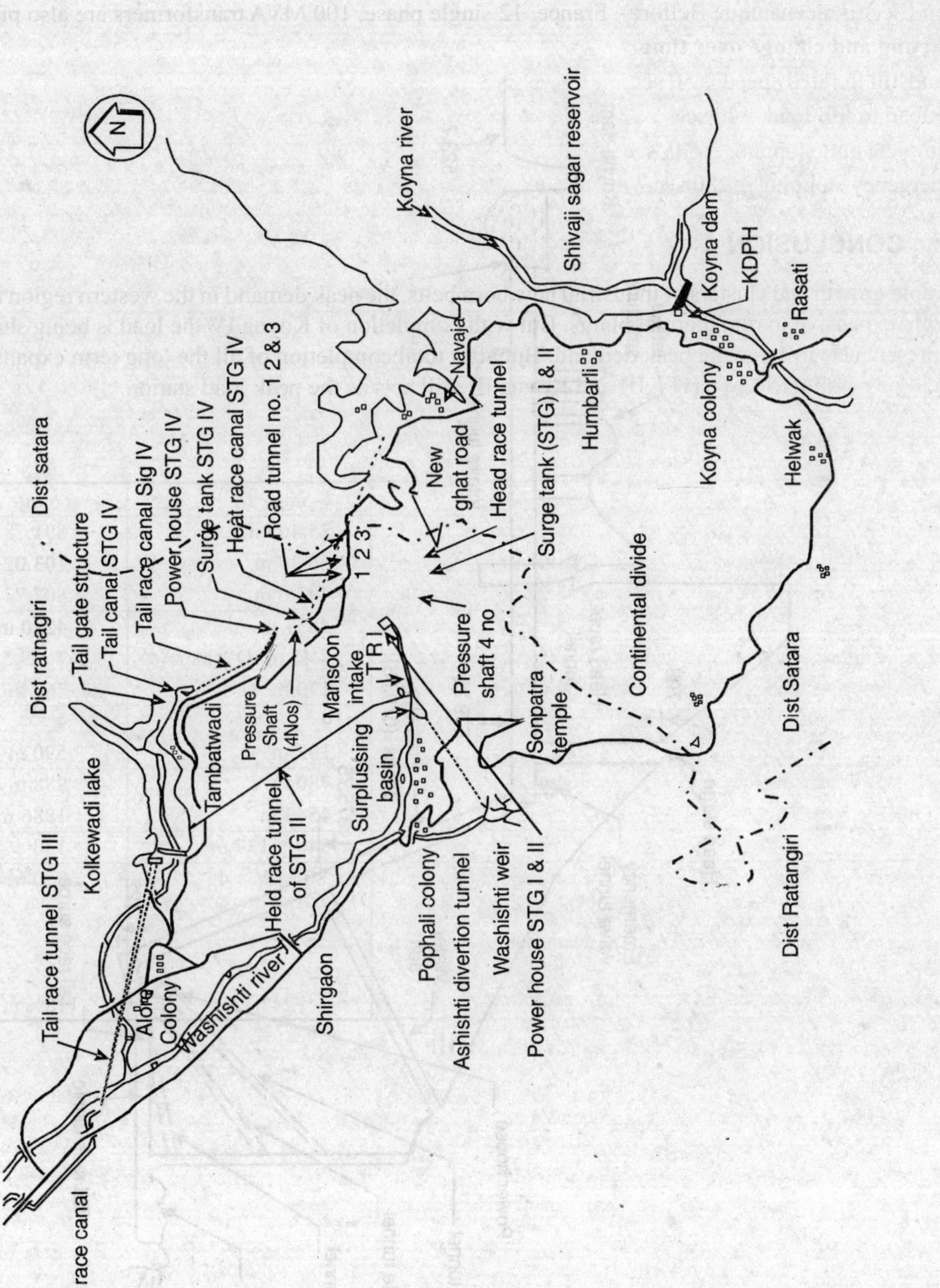

Fig. 22.2 *Sechematic layout of Koyna Hydroelectric project state I, II, III and IV.*

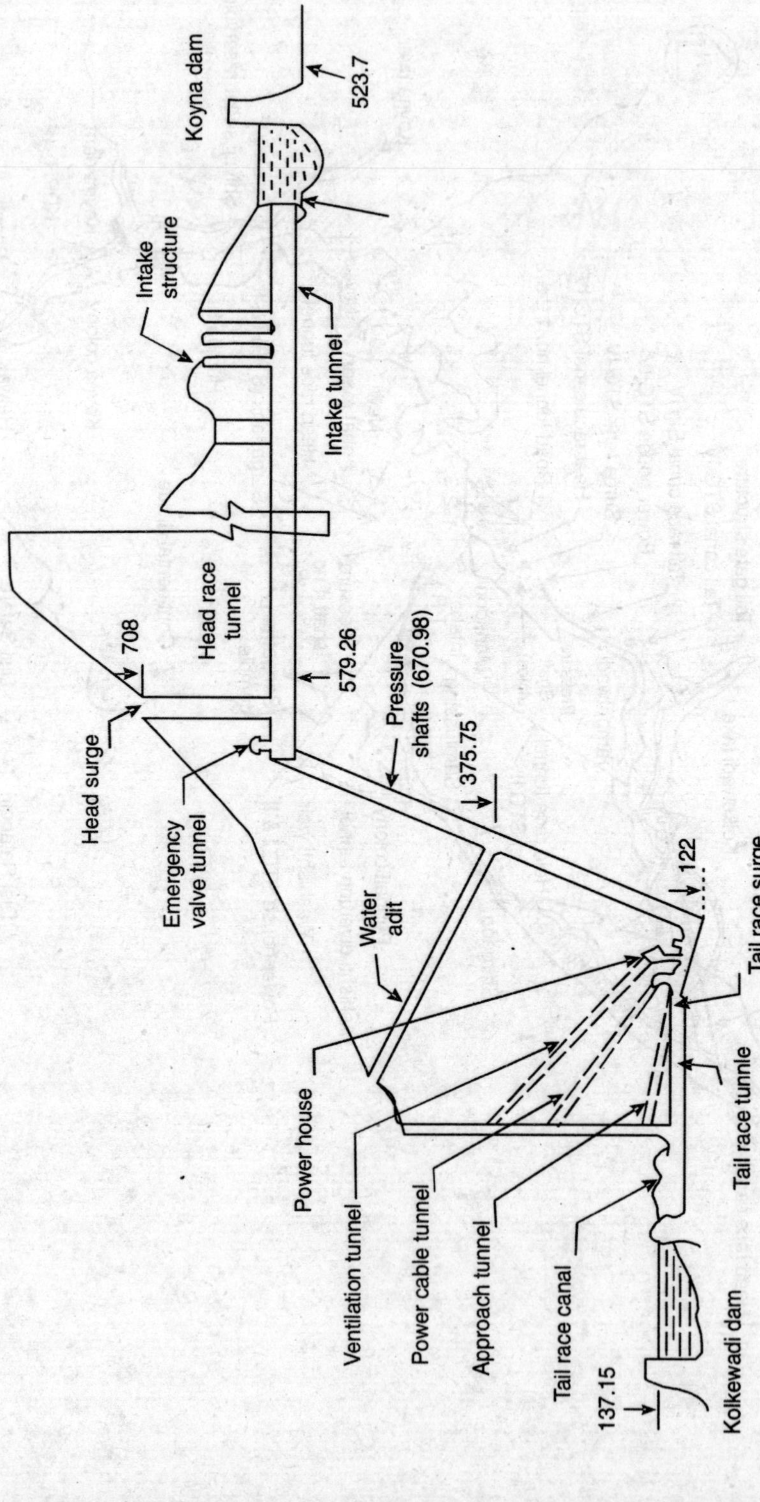

Fig. 22.3 *Cross section of K.H.P. stage–IV*

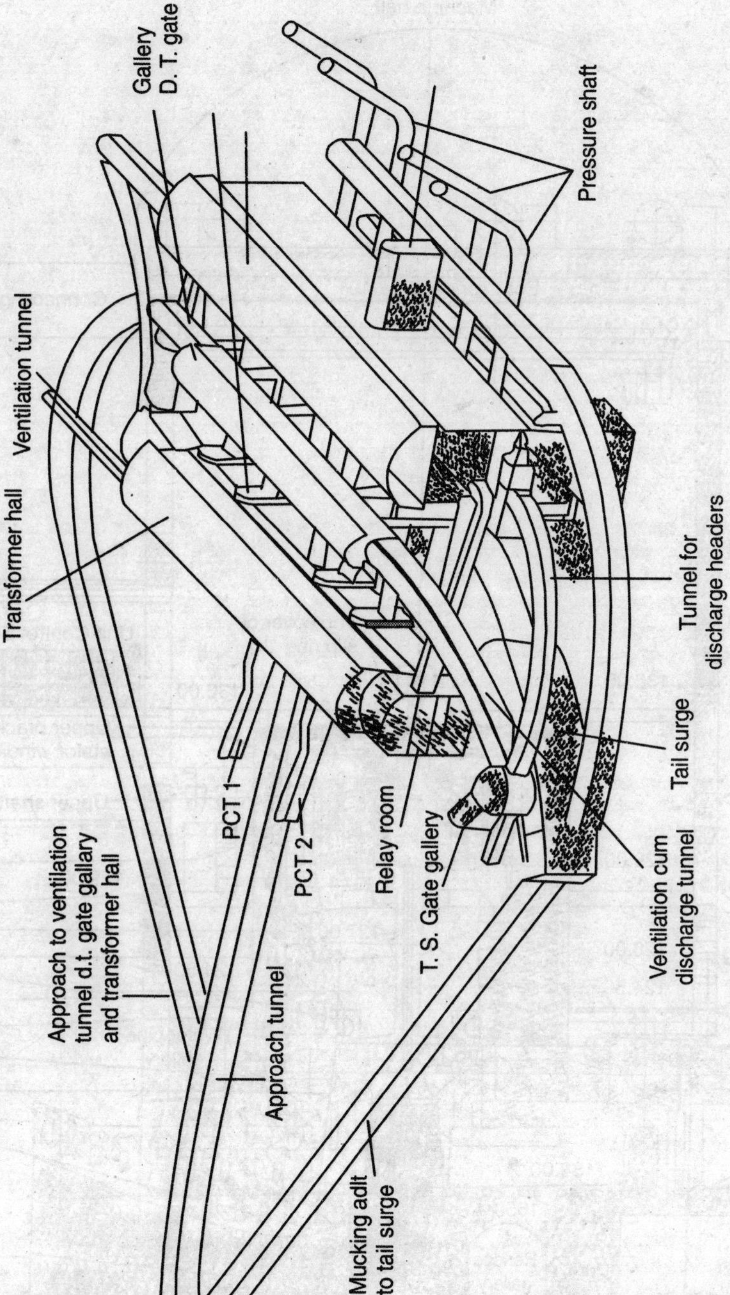

Fig. 22.4 *Under ground power house complex of K.H.E.P stage–IV*

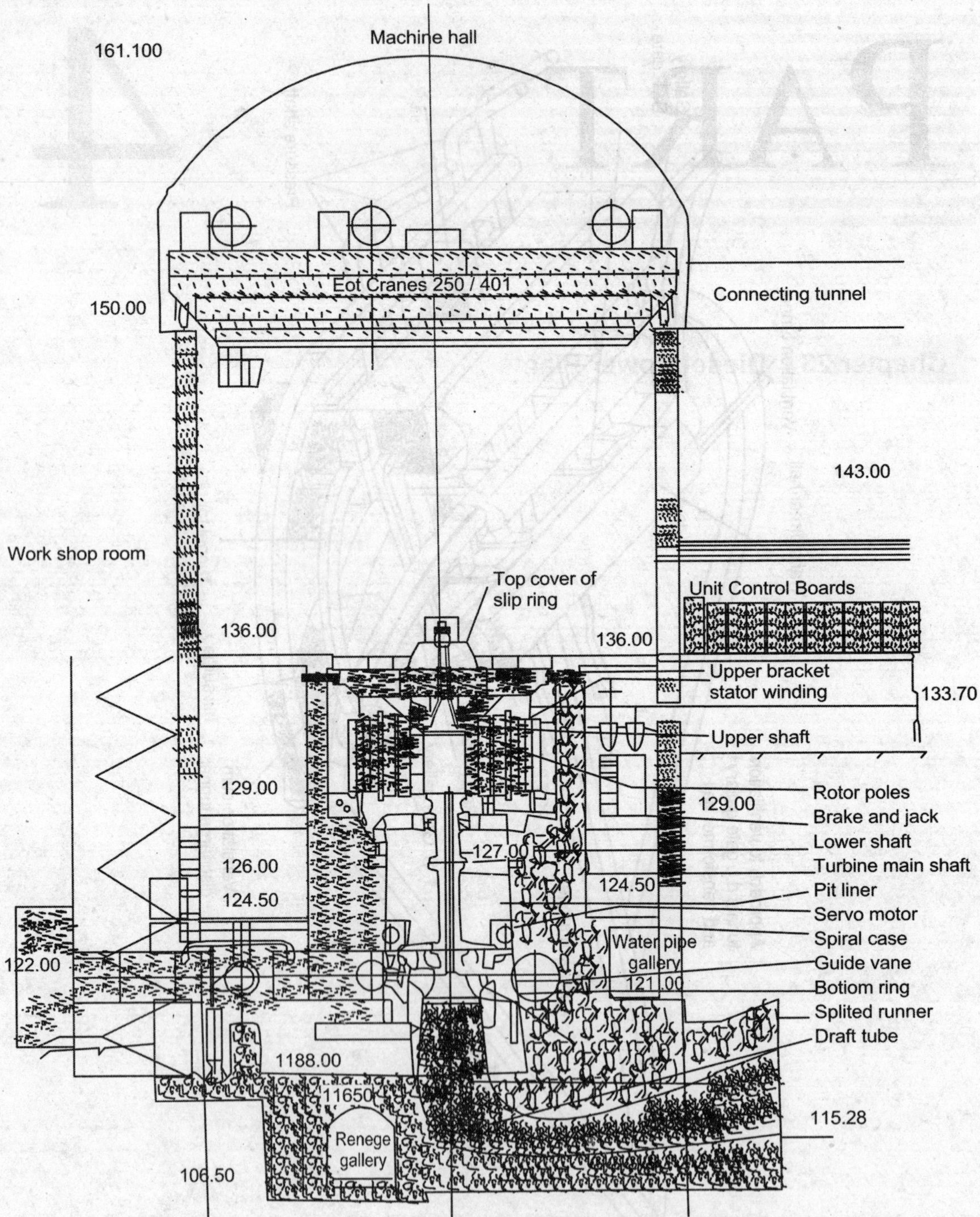
161.100
Machine hall
Eot Cranes 250 / 401
150.00
Connecting tunnel
143.00
Work shop room
Top cover of slip ring
Unit Control Boards
136.00
136.00
Upper bracket
stator winding
133.70
Upper shaft
129.00
Rotor poles
129.00
Brake and jack
Lower shaft
127.00
126.00
Turbine main shaft
124.50
124.50
Pit liner
Servo motor
Spiral case
Water pipe gallery
122.00
Guide vane
121.00
Botiom ring
Splited runner
Draft tube
1188.00
11650
115.28
Renege gallery
106.50

Fig. 22.5

PART 4

Chapter 23 : Diesel Power Plants

Chapter 23 Diesel Power Plants

23.1 INTRODUCTION

It is known that power can be developed or generated by heat engines by utilising heat energy in the products of combustion acting as the working substance in the engine. It is also known that the engines in which the products of combustion are obtained are called **Internal combustion engines**. Again these internal combustion engines are further classified as spark ignition (SI) engines and compression ignition (CI) engines. Spark ignition engines are low compression engines and compression ignition engines are high compression engines.

These engines are largely responsible for the self-propelled vehicles and have their major application in that field, and therefore, the influence of these Internal combustion engines on life in a technically advanced country like ours is profound. Small IC engines power mobile equipment or vehicles and stationary units where electrical service is not convenient.

Although, increasingly, stationary power needs are being met by electrical motors, an important exception is where the mechanical power is employed to drive generators originating the supply of electrical power. It is this use of the internal combustion engine that we are considering in this chapter.

There are a variety of thermodynamic cycles of working of different types of internal combustion engines and it is natural then that one type may vary from another in characteristics such as fuels used, weight, cost, flexibility and so many others. But all these various types of I.C. engines will have many features in common such as :

1. Use of piston and cylinder mechanism
2. Working medium such as gas
3. Open cycle
4. Two or four stroke operating cycle
5. For more power, multicylinders.

Generally, IC engines are directly connected to generators of relatively low speeds. The speeds — 200 to 1000 rpm — necessitate a large diameter, salient pole generator of rather short axial length. High reliability is obtained by the heavy, rugged slowspeed construction and though costly, the use of engines can be justified in central station.

23.2 APPLICATION OF DIESEL POWER PLANTS

The diesel-electrical power plants find wide applications in various fields. These fields of applications are :

(a) Peak Load Plant. The main or chief requirement for a power plant to be used as peak load plant is the quick starting within a very short time and diesel-electric power plant can be started with a very short notice. Therefore, diesel-electric power plant can be used as a peak load power plant in combination with thermal power plant and with hydro-electric power plants.

(b) Stand-by Power Plants. The diesel-electric power plants can be used as standby units when it is required to supply part load power for short duration. Generally this can be associated with hydro-electric power plants. When the available water is not in sufficient quantity — may be due to less rainfall etc. The capacity of this standby plant is comparatively very less than the capacity of the hydro-electric power plant. Standby diesel power plants can also be used in case of emergency.

(c) Mobile Plants. Generally the capacity of these plants is very small. The plant is generally mounted on trolleys or skids and is required to supply electric power at construction sites like buildings, bridges, roads etc.

(d) Starting Auxiliaries in the Base Load Power Plants. The starting of thermal power plants requires that all the auxiliaries be run first and diesel-electric power unit is used for this purpose.

(e) Base Load Power Station (Central Power Plant). For a remote small town or the group of villages, if the total requirement of power is limited to 10–15 MW, diesel-electric power plant can be used to supply the power. This limit may generally be decided by the cost of the plant, availability of fuel and water and the space for the station. Such a power station will be constructed when no grid is available or will not be available in future.

(f) Nursery Station. The main grid cannot be extended — many times — to every corner of the country unless there is enough load. Also, many times the main grid cannot be extended because of the constructional difficulties. In such cases, diesel electric plant of small capacity can be installed to supply the load. When such a locality or region is well developed, then the diesel-electric power plant can be transferred to another locality which needs power. This plant will supply the power during the developmental stage and after development, the plant can be or will be shifted to another region.

Such power plants are called **nursery power plants**.

23.3 MAIN COMPONENTS OF DIESEL ELECTRIC POWER PLANTS

For a diesel-electric power plant, apart from the engine which is the main component, following equipments or systems must be provided. These systems are :

(a) Air intake system
(b) Fuel system
(c) Engine cooling system
(d) Engine exhaust system
(e) Engine lubrication system
(f) Engine starting system
(g) Electrical system
(h) Governing system.

These systems are discussed in the subsequent sections.

23.3.1 Air Intake System

This system includes air filter, inkate ducts or suction manifold and super charger.

(i) Air Filter. In large diesel engines, generally air-fuel ratio is about 15 : 1. In natural atmospheric state the air will vary in temperature and in dust content. The air system begins with an intake located away from the engine. The intake is provided with a filter to catch dirt which would otherwise cause excessive wear in the engine. At some locations, the outside air may sometimes reach such low temperatures that it may promote misfiring at low roads; consequently, the air intake system may need to have a heating element using exhaust gases.

The filters used may be conveniently classified as

1. Dry air filters
2. Oil-impingment filters
3. Oil-bath filters.

The dry type of filter is made of felt, glass-wool, wire mesh or gauge etc. The effect of such filters is to catch dirt by causing it to cling to the surface of the filter material. This type also needs to be cleaned at intervals, the period between cleanings being governed by the amount of air which has been used by the engines and the dust concentrations in it.

Certain dry filters are either cleanable filters or throw away filters.

Throw away filters are made of glass wool, plastic fibre, steel wool, animal hair or vegetable fibre. If resin power is added with dry wool electrostatic charge develops which attract and retain small aerosols. Addition of resin powder increases filtering efficiency but has limited dust holding capacity. These types of the filters can not be used when the air is dusty.

Dry filters are shown in Figs. 23.1 (a – c)

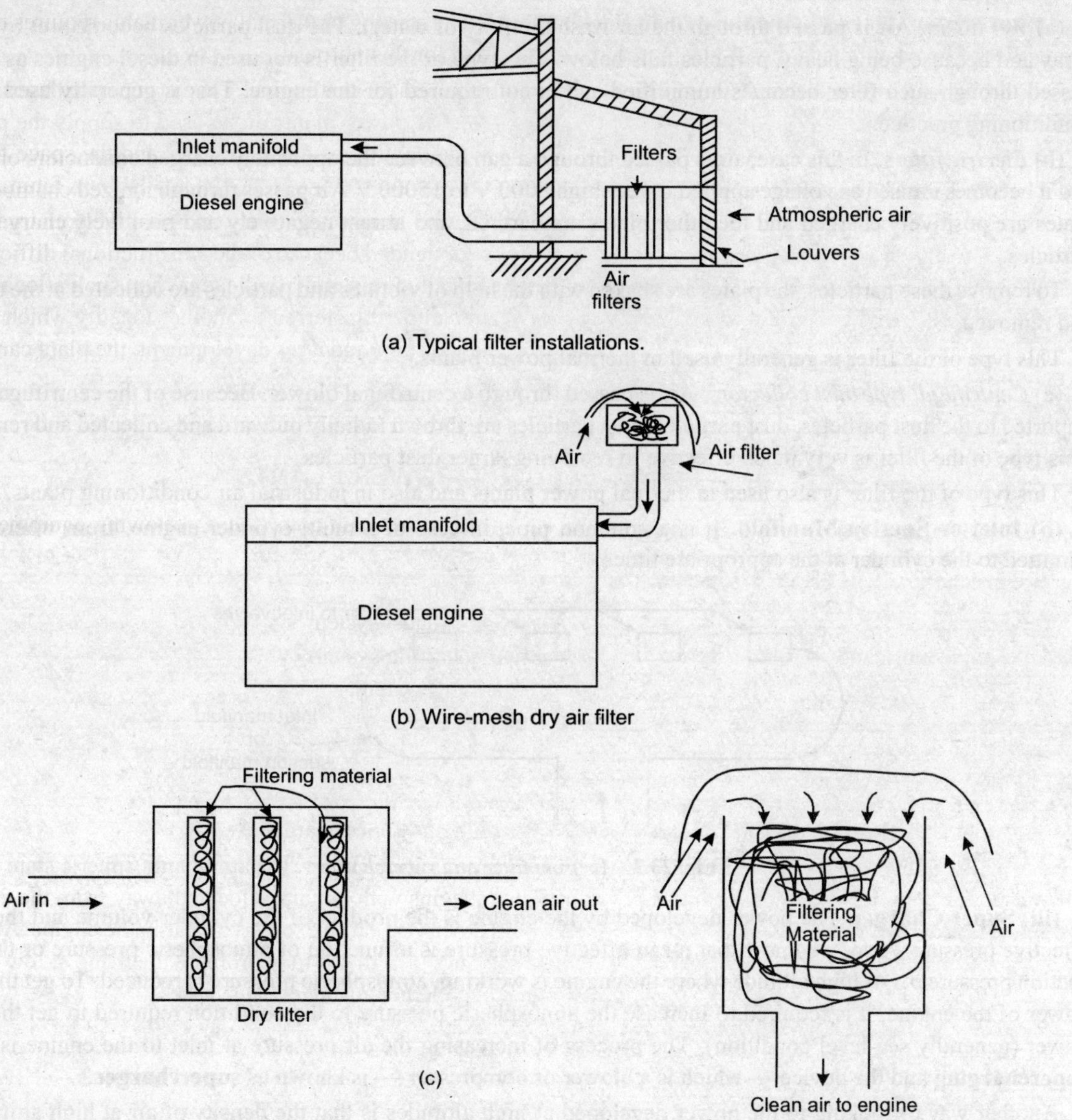

Fig. 23.1 *Dry air filters.*

The impingement type filter consists of a frame filled with crimped wire or metal shavings. These are coated with a special oil so that the air in passing through the frame and being broken up into numbers of small filaments comes into intimate contact with the oil, whose property is to seize and hold any dust particles being carried by the air. When in service, the cleansing capacity of this type of filter becomes progressively exhausted, and it is necessary periodically to refresh and clean in by removing, washing and re-oiling.

In the oil-bath type of cleaner, the air is swept over or through a pool of oil so that the particles of dust become coated. The air is then caused to flow through a filter, which retains the oil coated dust particles.

Light weight steel pipe is the material for intake ducts.

Other types of air filters are :

(a) *Wet filters.* Air is passed through the air washer (spray of water). The dust particles become wet by water spray and because being heavy, particles falls below. This type of the filter is not used in diesel engines as the air passed through such filter becomes humidified and is not required for the engine. This is generally used in air-conditioning practice.

(b) *Electric filters.* In this case, air is passed through a gap between the oppositely charged conductors or plates and it becomes ionised as voltage applied is very high 8000 V to 15000 V. Air passes through ionized chamber. The plates are positively charged and the other plates are earthed, and attract negatively and positively charged dust particles.

To remove these particles, the plates are shaken with the help of vibrates and particles are collected at the bottom and removed.

This type of the filter is generally used in thermal power plants.

(c) *Centrifugal type dust collectors.* Air is passed through a centrifugal blower. Because of the centrifugal force imported to the dust particles, dust particles, dust particles are thrown radially outward and collected and removed. This type of the filter is very much effective in removing larger dust particles.

This type of the filter is also used in thermal power plants and also in industrial air conditioning plants.

(ii) Inlet or Suction Manifold. It is a common pipe, in case of a multi-cylinder engine, from where air is admitted to the cylinder at the appropriate time.

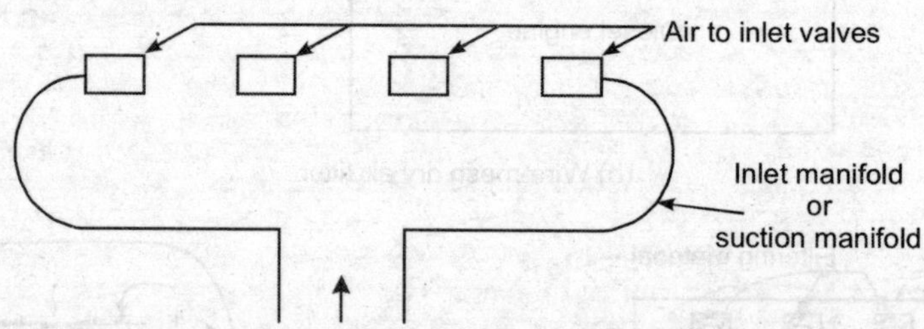

Fig. 23.2 *Air from filter and supercharger.*

(iii) Super Charger. The power developed by the engine is the product of the cylinder volume and the mean effective pressure. Also, we know that mean effective pressure is a function of atmospheric pressure or the inlet suction pressure p_1. At high altitude where the engine is working, atmospheric pressure is reduced. To get the rated power of the engine, it is required to increase the atmospheric pressure to the condition required to get the rated power (generally sea-level condition). The process of increasing the air pressure at inlet to the engine is called **supercharging** and the device — which is a blower or compressor — is known as **supercharger**.

Another way of looking to the power developed at high altitudes is that the density of air at high altitudes is reduced because of decreased pressure. Less density means for the same stroke volume of the cylinder, less quantity (kg) of air is admitted. Less quantity of air means less quantity of oxygen required for combustion of the fuel. Thus less amount of fuel will be burnt and consequently less energy will be generated. Thus the power developed will be reduced. To increase the power developed, we have to increase the density of surrounding air at high altitudes by increasing the pressure with the help of a compressor or blower so that the density of air is same that is required to develop the rated power.

Also, the rated power of the engine can be increased by increasing the inlet pressure with the help of blower or compressor.

Thus **supercharging** is employed either to obtain more power from the same engine cylinder capacity or to compensate for the power lost at high altitudes due to the reduction in air density.

Basically the superchargers are air compressors, driven either by engine shaft through gearing or by exhaust gas turbines. Engine driven superchargers absorb a considerable proportion of the engine power, while the exhaust turbines run on the energy of the exhaust gases which otherwise go waste. The exhaust gases from the engine carry away about one-third of the total heat liberated in the cylinders and as such any useful work obtained from them is an asset to the engine performance.

The air compressors used as superchargers are chiefly of two types, (i) reciprocating compressors and (ii) rotory compressors. Rotory compressors running at very high speeds, include (i) Roofs blower (ii) Vane type (iii) Centrifugal compressors.

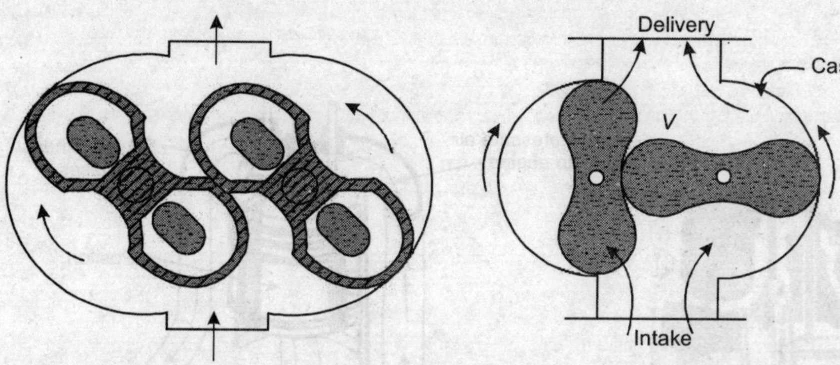

(a) (b)

Roots blower

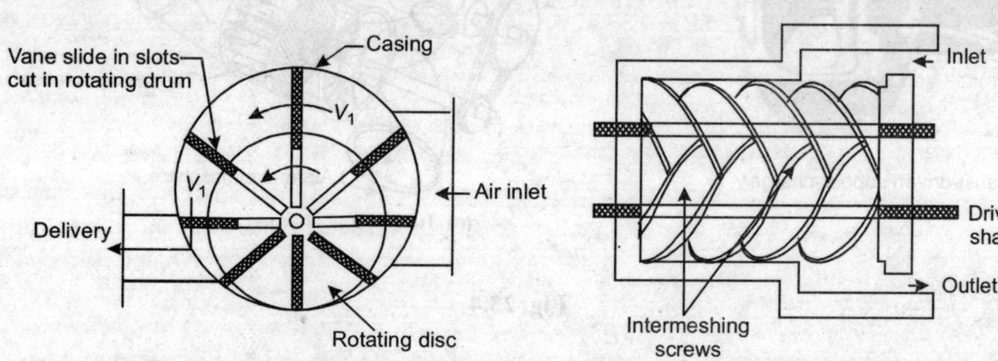

(a) Vane type compressor (b) Screw type comprssor

Fig. 23.3 *Different compressors used as super chargers*

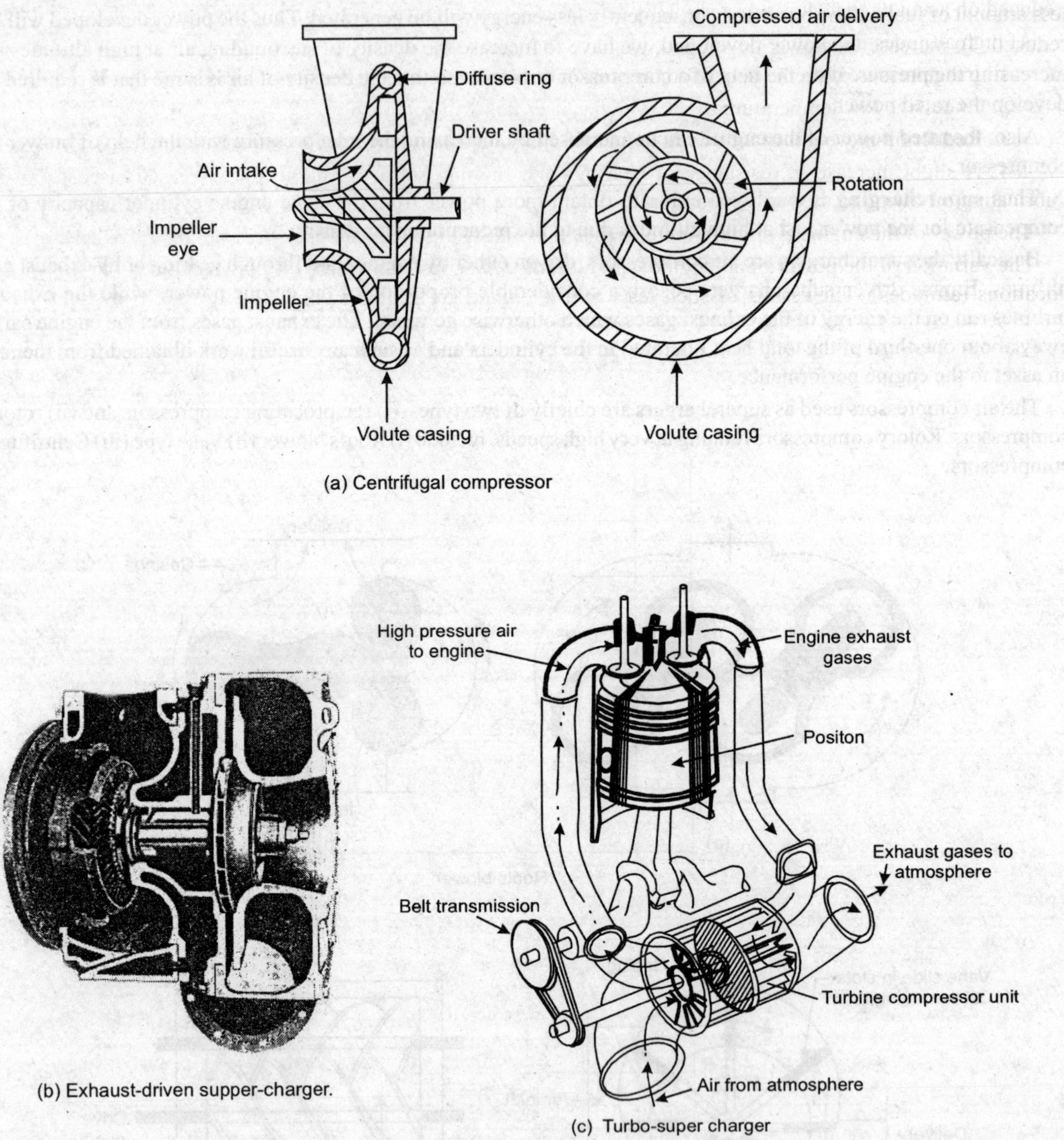

Fig. 23.4

The power of diesel engines is found to be considerably reduced when working in the rarified air at high altitudes. This reduction in power is mainly due to the lack of oxygen, owing to the low pressure of the atmosphere. Although the volume of air drawn in the carburettor or cylinder in case of diesel engine remains constant, its weight is greatly reduced on account of its low pressure. This deficiency of air is corrected by supercharging.

At high altitudes the engine power is reduced by the following causes :

1. Irregularities of fuel evaporation, causing failure to evaporate all the fuel injected.
2. Excessive low temperature of air, which causes delay in combustion.
3. Reduction in weight of air owing to its low pressure; this is corrected by supercharging.
4. A slight increase in resistance of bearings and rubbing surfaces, due to the increased viscosity of the lubricating oil owing to the low temperature.

The power of the engine is slightly increased by the effect of the reduced back pressure on the exhaust gases.

The variation of the power of an engine with altitude can be tested by means of a dynamometer at the various locations (altitudes). The results of such tests on the engine are shown in Fig. 23.5. In this figure, the power

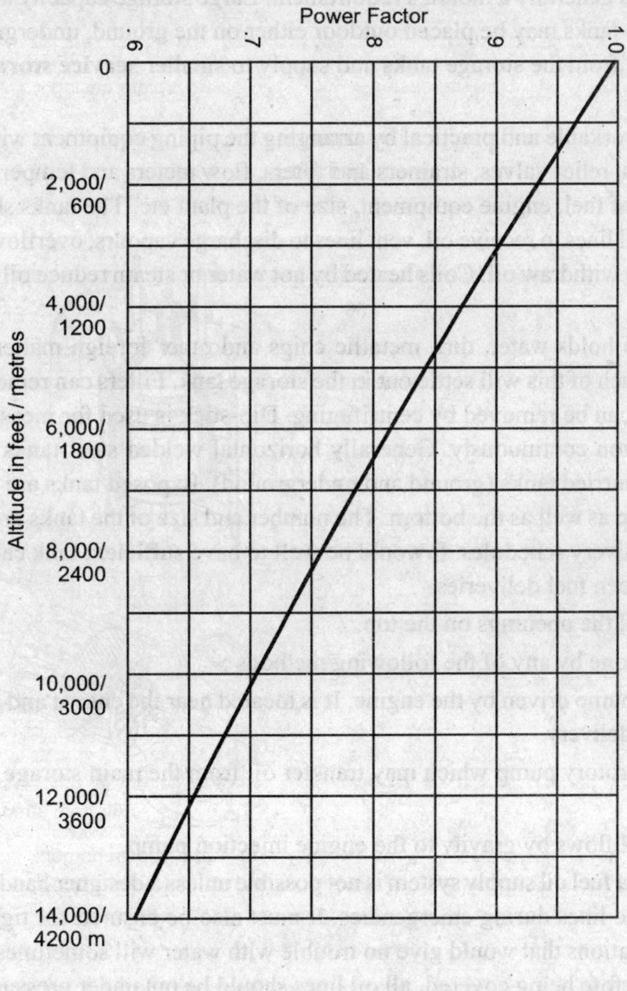

Fig. 23.5 *Effect for altitude*

factor is plotted on a base representing altitude. It will be noticed from the curve that the engine power fell to 60% of its full value at an altitude of 4200 m.

23.3.2 Fuel System

Fuel system includes :

(i) Fuel storage tank
(ii) Pumps
(iii) Strainers
(iv) Oil filters
(v) Oil heaters
(vi) Piping and pressure gauges
(vii) Fuel injection system.

The fuel oil may be delivered at the plant site by trucks, railroad tank cars or barge and tankers. From tank car or truck the delivery is through the unloading facility to **main storage tanks**. The capacity of these tanks depends on the location of the site and is generally a month's requirement. Large storage capacity allows purchasing fuel when prices are low. Bulk storage tanks may be placed outdoor either on the ground, underground or as overhead tanks. Fuel oil pumps draw the oil from the storage tanks and supply to smaller **service storage tanks** known as **engine day tanks**.

The main flow is made workable and practical by arranging the piping equipment with the necessity heaters, by-passes, shut-offs, drain lines, relief valves, strainers and filters, flow meters and temperature indicators. The actual flow plans depend on type of fuel, engine equipment, size of the plant etc. The tanks should contain manholes for internal access and repair, fill lines to receive oil, vent lines to discharge vapours, overflow return lines for controlling oil flow and a suction line to withdraw oil. Coils heated by hot water or steam reduce oil viscosity to lower pumping needs.

Delivered oil sometimes holds water, dirt, metallic chips and other foreign matter that must be removed by filtering or centrifuging. Much of this will settle out in the storage tank. Filters can remove the contamination of oil and heavier contamination can be removed by centrifuging. Dip-stick is used for measuring the tank contents and the meter gives the indication continuously. Generally horizontal welded steel tanks are used. These tanks can either be exposed tanks or burried tanks (ground and underground). Exposed tanks are set on concrete saddles and can have openings on the top as well as the bottom. The number and size of the tanks are established from the plant size, capacity factor and delivery schedules. It would be well to have sufficient tank capacity to carry the plant for twice the normal time between fuel deliveries.

The buried tanks have all the openings on the top.

The oil transfer can be done by any of the following methods :

1. By simple suction pump driven by the engine. It is located near the engine and a few metres below it. Such pumps are content volume delivery.

2. By a motor driven rotory pump which may transfer oil from the main storage tank to an elevated service tank or day tank.

From the day tank the oil flows by gravity to the engine injection pump.

Satisfactory operation of a fuel oil supply system is not possible unless a designer handle provisions for cleanliness and for changing over of the lines during emergencies. It must also be ensured for tight pipe joints in all suction lines. Imperfect pipe installations that would give no trouble with water will sometimes suck air at the joints when carrying oil at a vacuum. Before being covered, all oil lines should be put under pressure and the joints tested with soap solution. Small air leaks into the line can be the source of exasperating operating difficulties and are hard to remedy once the plant is in operation. The piping between filter and the engine should be thoroughly oil flushed before being first placed in service. Great importance should be given for cleanliness in handling bulk fuel oil. Dirt particles will ruin the fine lap of injection pumps or plug the injection nozzle orifices. So high grade filters are of great importance to the diesel oil supply system. Engines are equipped with one or more external filters in the piping system. They should be in duplicate.

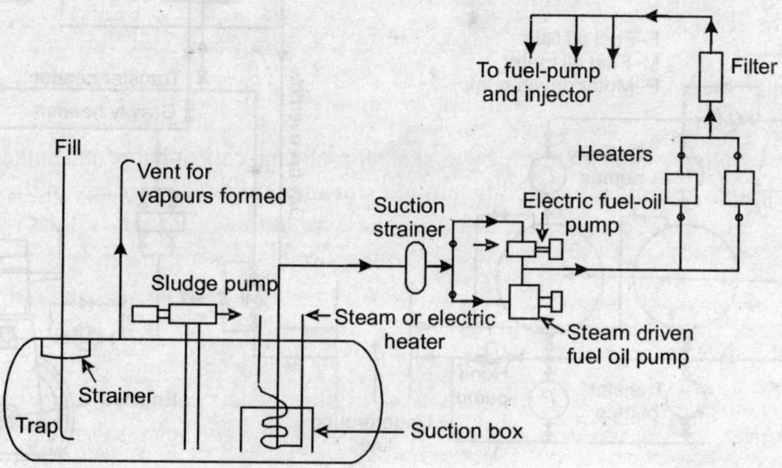

Fig. 23.6 *Fuel-oil storage tank.*

Figures 23.6, 23.7 and 23.7(a) show fuel-oil storage tank, fuel service tank or day tank and multiple unit fuel handling system.

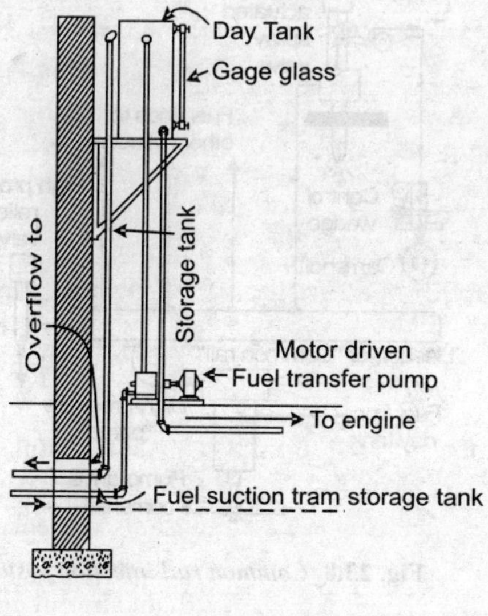

Fig. 23.7 *Fuel service tank.*

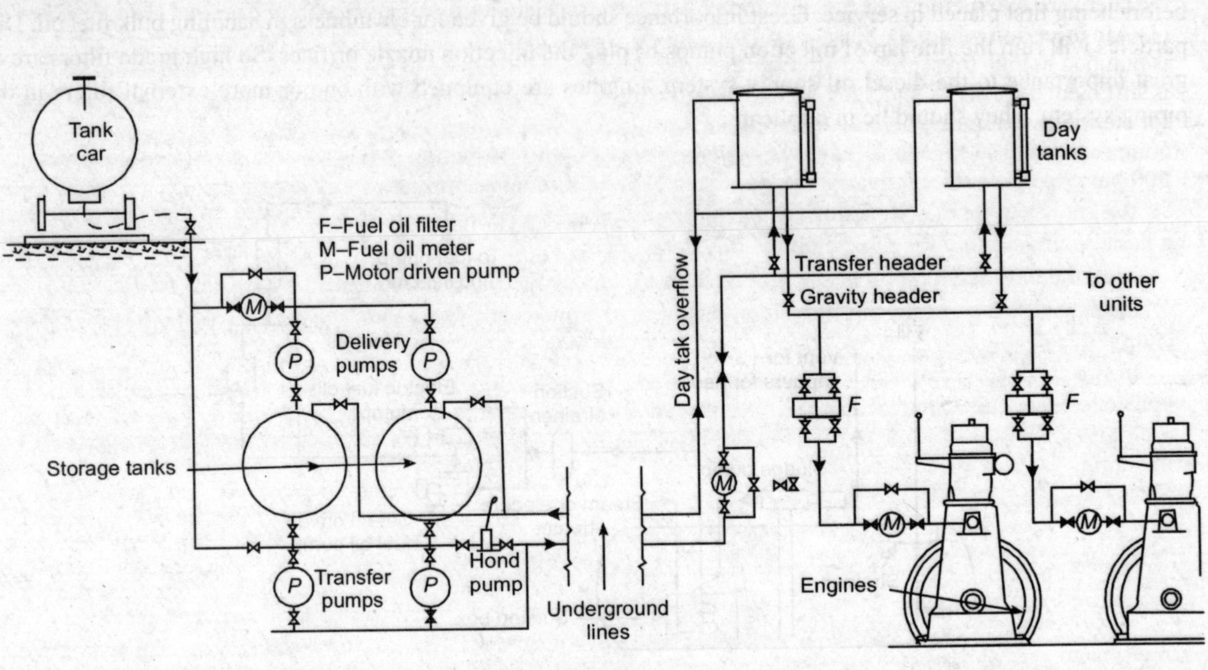

Fig. 23.8 *Fuel-oil storage tank, fuel service tank or day tank and multiple unit fuel handling system*

Fuel injection system of the diesel engine is the mechanical heart of the engine. The performance of the engine depends mainly on its fuel injection system. A very small quantity of fuel must be measured out, injected, atomized

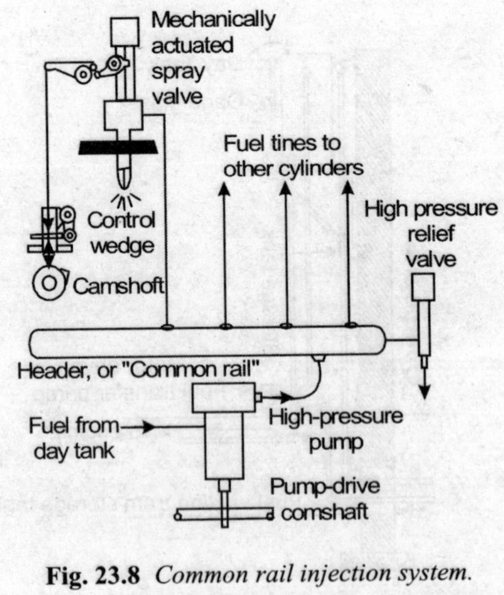

Fig. 23.8 *Common rail injection system.*

and mixed with combustion air. The mixing problem becomes more difficult the larger the cylinder and the faster the rotative speed. Special combustion arrangements such as precombustion chambers, air cells etc. are necessary to secure good mixing. For lower speed engines simple combustion chambers are provided.

There are two methods of fuel injection by means of which atomization of fuel oil has been secured. They are : (1) **Air injection** where air blast or compressed air is used to atomize the fuel and (2) solid injection where fuel oil is pumped to high pressure and atomize the oil. Early diesel engines used air for fuel injection at about 70 bar. This is sufficient not only to inject the oil, but also to atomize it for a rapid and thorough combustion. This type was used for stationary engines where the compressor was used to supply the high pressure air. The cost of providing an air compressor and storage tank led to the development of **'solid' injection**, using a liquid pressure of between 100 to 200 bar which is sufficiently high to atomize the oil by forcing oil through spray nozzles.

We know that the fuel injection system must pressurise, meter and time the fuel. These functions are achieved in a variety of ways, but the most employed systems are :

(a) the individual pump or the unit injector system and (b) the common rail system.

From this, we can find the fuel storage capacity required and these systems are usually employed on power station diesel engines.

In Fig. 23.8 a system uses a single pump supplying high pressure fuel to header. A relief valve holds pressure constant. The control wedge adjust lift of mechanically operated valve to set amount and time of injection.

Figure 23.9 shows another common rail injection system in which controlled pressure system has pump which maintains set header pressure. Pressure and relief and timing valves regulate injection time and amount of fuel. Spring loaded spray valve acts merely as a check.

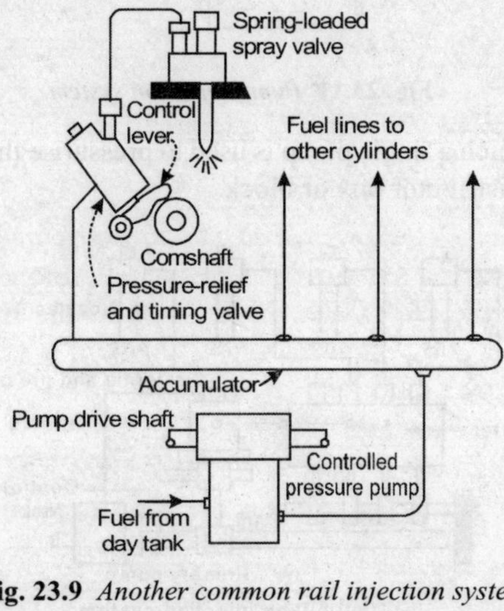

Fig. 23.9 *Another common rail injection system.*

Figure 23.10 shows a unit injector system. Mechanically actuated pump plunger raises fuel to a high pressure, meters the quantity and controls injection timing. Nozzle contains a spring loaded delivery valve actuated by change in the fuel oil pressure.

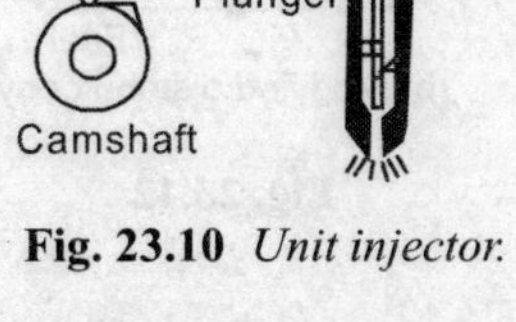

Fig. 23.10 *Unit injector.*

Figure 23.11 shows an individual pump or pump cylinder connects directly to each fuel nozzle. Pump meters charge and control injection timing. Nozzles contain a delivery valve actuated by fuel-oil pressure.

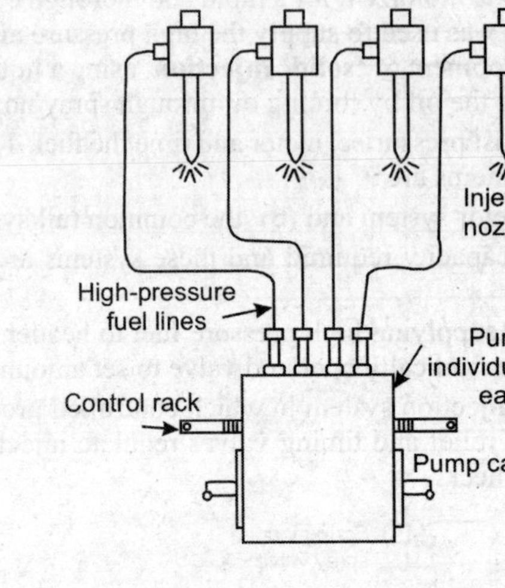

Fig. 23.11 *Pump injection system.*

Figure 23.12 shows system in which a single pump is used to pressurise the fuel which is correctly distributed to the individual injectors through a distributor box or block.

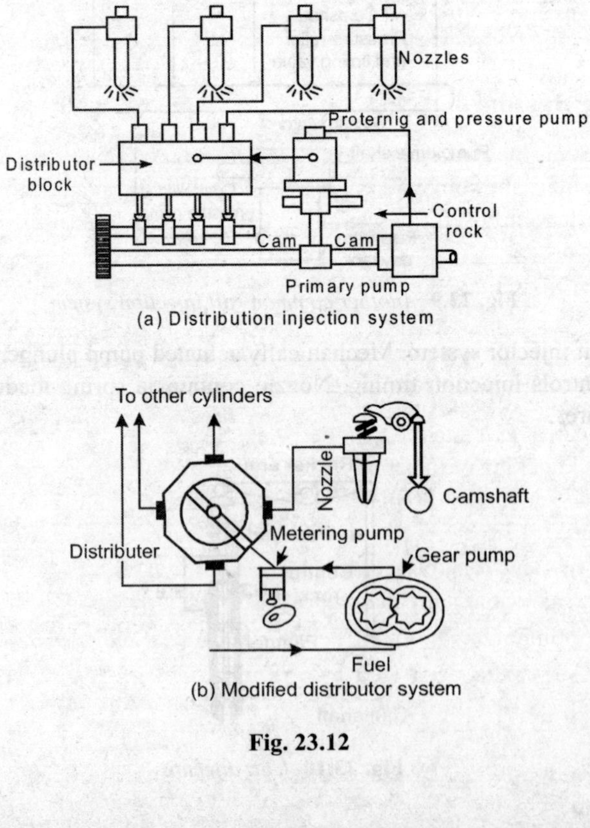

(a) Distribution injection system

(b) Modified distributor system

Fig. 23.12

Figure 23.13 shows the modified distributor system. In this case fuel at low pressure is pumped to a distributor block which supplies fuel at low pressure to the various individual injectors and this unit injector pressurises the fuel further and inject the fuel at proper time.

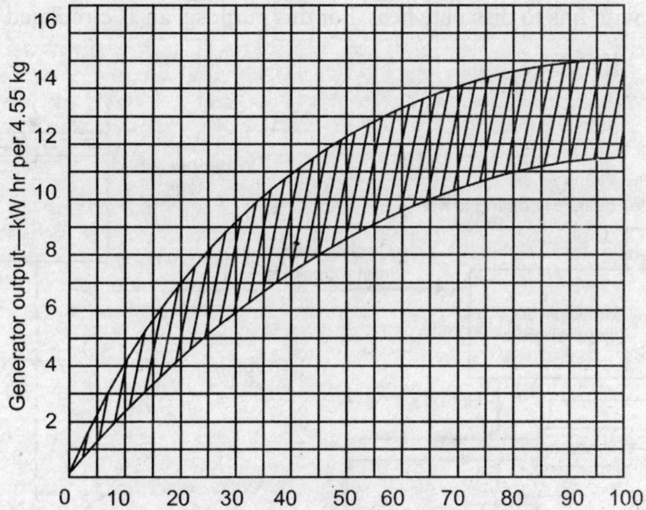

Fig. 23.13 *Relation between plant capacity factor and generator output.*

Figure 23.13 shows the relation between plant capacity factor and the generator output.

23.3.3 Engine Cooling System

During the process of converting thermal energy to mechanical energy, high temperatures are produced in the cylinders of the engine as a result of the combustion process. All the heat produced by the combustion of fuel in the engine cylinders is not converted into useful power at the crankshaft. A typical distribution of the fuel energy is given below :

Useful work at the crankshaft	20–25%
Loss to the cylinder walls	30–35%
Loss in exhaust gases	30–35%
Loss in friction	10%

It is seen that the quantity of heat given to the cylinder walls is considerable and if this heat is not removed from the cylinders it would result in knocking or the combustion will not be proper. In addition, the lubricant would also burn away, thereby causing the seizing of the piston. Excessive heating will also damage the cylinder material, gaskets etc.

Keeping the above factors in view, it is observed that suitable means must be provided to dissipate the excess heat from the cylinder walls, so as to maintain the temperatures below certain limits.

However, cooling beyond optimum limits is not desirable, because it decreases the overall efficiency due to the following reason :

(i) thermal efficiency is decreased due to more loss of heat to the cylinder walls

(ii) the vaporization of the fuel is less, this results in fall of combustion efficiency

(iii) low temperature increases the viscosity of lubricant and hence more piston friction is encountered, thus decreasing mechanical efficiency.

The method used for cooling diesel power plant engine is water cooling. If this power plant is situated at the place where water is not available, then air cooling is an alternative to water. In that case, cylinders, cylinder head etc. will have to be provided with fins to dissipate heat. For this purpose, air is circulated with the fans or blowers.

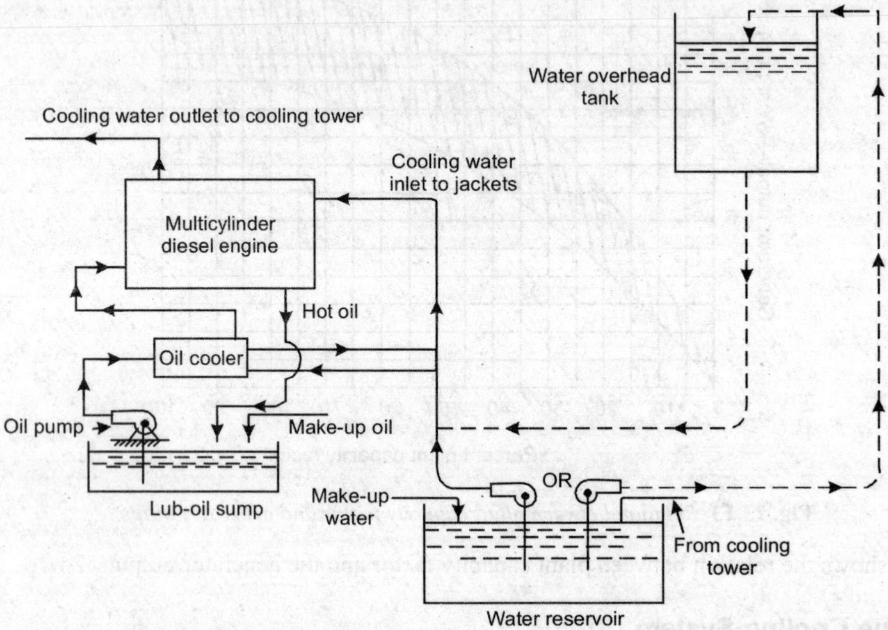

Fig. 23.14 *Simple engine water cooling system.*

Again if the water supply is inadequate, then the water circulated through the engine jackets is not exposed to evaporation loss, but is indirectly cooled by flowing through pipe coils over which either raw water drips or air is circulated.

We know that the atmosphere is a mixture of air and water vapour and this water vapour in air is termed humidity. Air rarely contains vapour it is capable of holding, and when it does its vapour-holding capacity can be further increased by warming it. In any process wherein more water vaporized into the air the proper latent heat of evaporation must be supplied from some source. Under certain conditions this source can be the internal energy of the liquid water from which the vapour is being produced. What liquid remains after the vapour has passed off is found to be considerably cooled. This is called **Evaporative cooling**.

The practical application of this type of evaporative cooling to the diesel plant is used in cooling apparatus which is designed to expose a large surface of warm water to an air flow, thereby humidifying the air and cooling the remaining water. This apparatus is called the **cooling tower**. There are three basic types of cooling towers. They are :

(a) atmospheric cooling towers

(b) mechanical draft cooling towers and

(c) evaporative coolers.

The cooling action is the same in all of them. Air is humidified and some of the warm water goes off into the atmosphere and fresh water must be added to the system to take its place. Generally this make-up water is 2–5% of the water flow.

(*a*) *Atmospheric towers* are long, narrow structures of considerable height placed with the long axis normal to the prevailing wind. They are built to utilize horizontal wind components. Decks of wooden lattice inside the structure serve to break up a stream of water released at the top of the tower so that it rains down through the air currents and is evaporatively cooled. This type is not often used with diesel power plant.

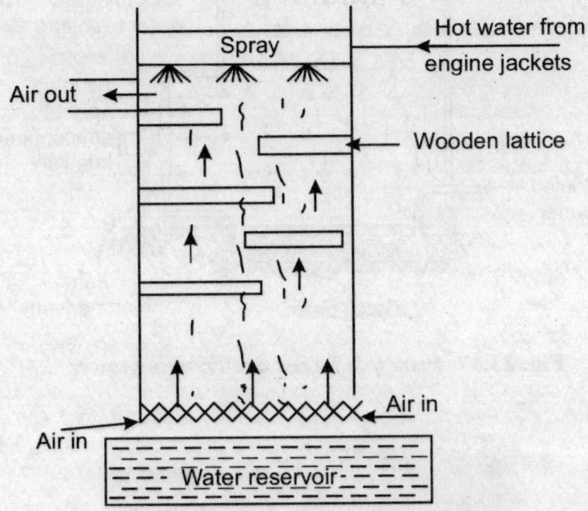

Fig. 23.15 *Atmospheric cooling tower*

(*b*) *Natural draught cooling tower*. Figure 23.16 shows the cooling tower which works on natural draft. Air circulation is obtained through the levity of a confined column of warm air. The performance of these towers is very good with high water temperature and if built as a high structure. These towers are very common to small plants.

Fig. 23.16

(*c*) *Mechanical draft cooling tower.* If we use a fan for the dependable air movement or air flow, the cooling tower is called a **mechanical draft cooling tower** and it is necessary to be as high as with natural draft. According, to the **fan location**, the tower is classified as **forced** or **induced draft cooling towers**.

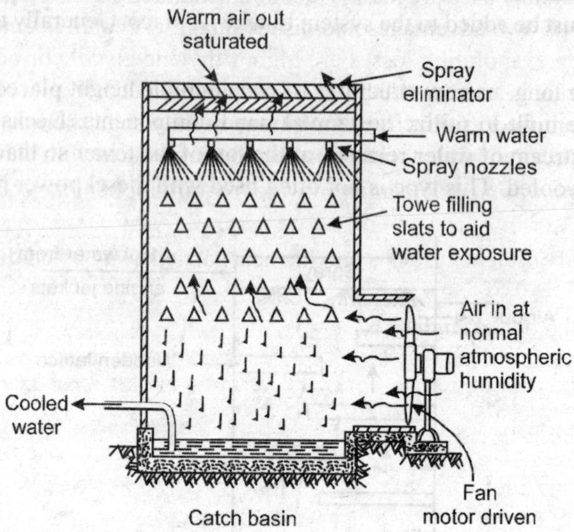

Fig. 23.17 *Principle forced draft cooling tower.*

Fig. 23.18 *Induced draft cooling tower.*

Forced and induced draft cooling towers are shown in Fig. 23.17 and Fig. 23.18.

Induced fan is placed at the top of the tower and the motor is take away and mounted as shown. This motor should not come in contact with the warm and humid air.

Mechanical draft cooling towers are efficient than natural cooling towers.

(*d*) *Evaporative cooling towers.* When the cooling tower is built entirely of steel, including a heat exchange for cooling water of the engine circuit, it is called an **evaporative cooler**. This type of the cooler is having rare use in diesel plant.

Most coolers are either natural or forced draft outdoor coolers.

In cold climates, there is always a danger that the water in the cooling system may get frozen. As the volume of water, when converted into ice, increases by about 10% and as a result the damage of the entire system including in extreme cases the bursting of the pipes etc. may take place. To avoid this, some additives are used, which when mixed with water in suitable proportions, lower the freezing point of water. Such additives are called antifreezers. Even for hot climates the use of antifreezers has the incidental advantage that boiling point of the cooling water is also raised.

The antifreezers most commonly used are methyl alcohol, denatured alcohol (ethyl alcohol), glycerine, ethylene glycol etc. Each of these has its own advantages and disadvantages. Alcohol is quite effective, but it is very much volatile and due to this reason evaporation losses are high. Ethylene glycol corrodes copper, aluminium and tin-lead solder alloys. The glycerine is less volatile but it is comparatively costly and also it attacks rubber hose pipes. Calcium chloride is another antifreezer. Antifreeze solutions usually spoil the finish of the paint work. Therefore, adequate precaution should be taken while topping up the cooling system with antifreeze solutions.

(*e*) *Engine cooling water requirement.* Quantity of water required for engine cooling can be calculated as given below :.

Assumption, brake thermal efficiency is 30% and the heat carried away by cooling water is 30–32% of the heat supplied by fuel.

Let *BP* is the capacity of the diesel engine.

∴ Brake thermal efficiency

$$= 0.3 \text{ say}$$

$$= \frac{BP \times 3600}{\frac{\text{Fuel}}{\text{hr}} \times \text{Cal. value}}$$

$$= \frac{\text{Work/hour}}{\text{Heat supplied/hr}}$$

∴ Heat supplied/hour

$$= \frac{\text{Work done/hour}}{\text{Brake thermal eff.}}$$

$$= \frac{BP \times 3600}{0.3}$$

Again heat carried away by engine cooling water/hr.

$$= \frac{\text{Mass of water}}{\text{hr}} \times \text{Sp. heat} \times \text{Rise in temp.}$$

$$= m_w \times C_{pw} \times (t_{w_2} - t_{w_1})$$

where

m_w = wt. of water/hour

C_{pw} = Sp. heat of water

= 4.187 kJ/kg/K.

t_{w_1} = water inlet temperature °C

t_{w_2} = water outlet temperature °C.

∴ Heat given to cooling water/hr.

$$= 0.32 \times \text{heat given by fuel.}$$

$$= 0.32 \times BP \times 3600.$$

$$\therefore \quad \frac{0.32 \times BP \times 3600}{0.3} = m_w C_{pw}(t_{w_2} - t_{w_1})$$

∴ m_w = wt. of water/hour

$$= \frac{BP \times 3600 \times 0.32}{0.3 \times 4.187\left(t_{w_2} - t_{w_1}\right)}$$

$$\therefore \quad = \frac{3600 \times 0.32}{0.3 \times 4.187\left(t_{w_2} - t_{w_1}\right)} \frac{\text{kg}}{\text{hr. kW}}$$

$$= \frac{917.12}{\left(t_{w2} - t_{w1}\right)} \text{ kg/kr/BP.}$$

If we assume the rise of water temperature as 40°C, then

$$m_w = \frac{917.12}{40}$$

$$= \mathbf{22.93\ kg/hour/BP.}$$

Another simple method of cooling of engine cooling water is to use a spray pond. Figure 23.19 shows such a spray pond. It is a very good example of evaporative cooling. In this case water coming out of the engine jacket is sprayed with the help of spray nozzles, into air. Evaporating cooling takes place and this water is collected in the pond and hence the name **spray pond**. From spray pond water is recirculated through the **diesel engine and the cycle is completed**.

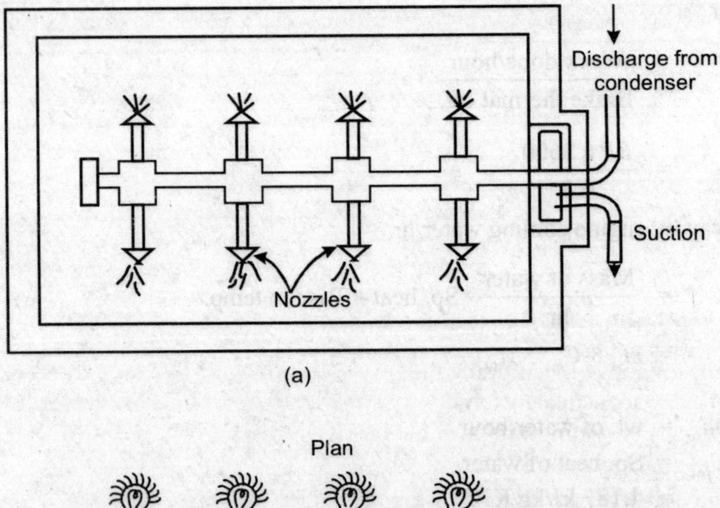

(a)

Plan

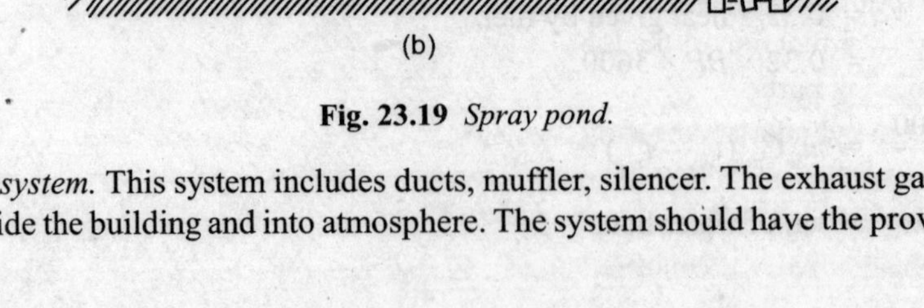

(b)

Fig. 23.19 *Spray pond.*

(*f*) *Engine exhaust system.* This system includes ducts, muffler, silencer. The exhaust gas system is used to lead the exhaust gases outside the building and into atmosphere. The system should have the provision for the following:

1. Silencing of the exhaust noise to the required degree.
2. Discharge of the exhaust sufficiently high above the ground level.
3. Expansion and contraction due to changes in temperature between the extremes of full load and cold.
4. Arrangement of the exhaust system to minimise the back pressure created by the exhaust system itself.
5. Isolation of engine vibration from building and muffler system by the use of a flexible section of the exhaust pipe.
6. High temperature of the exhaust gas which may require water cooled exhaust lines or special high temperature materials.

Muffling of the exhaust noise is the main problem which should be dealt in the exhaust system. This is done in a silencer usually located outside the building. Generally individual silencers are provided for each engine. They may be of cast iron or sheel steel. A pipe or stack slightly smaller than the exhaust line should extend vertically from

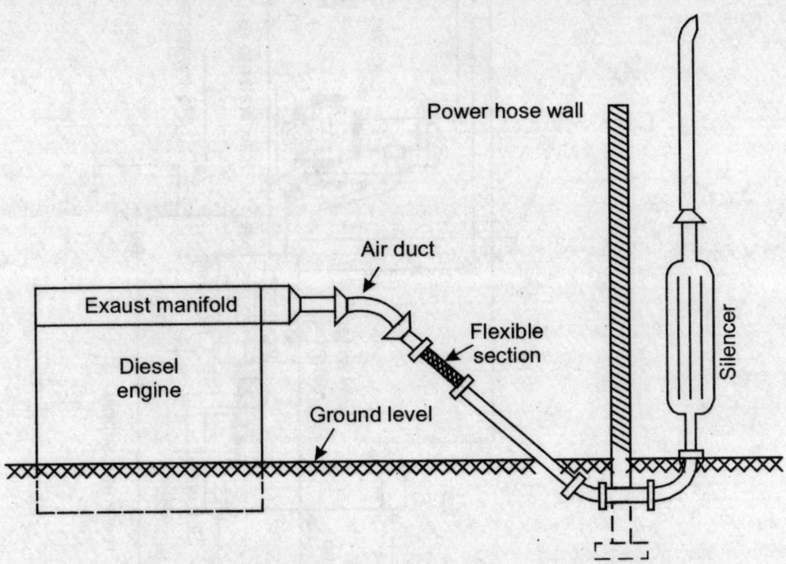

Fig. 23.20 *Exhaust gas system of a diesel power plant with silencer.*

the silencer and carry the gases above the building roof. Several principles are used in silencers. Some operate by absorption of pressure waves by acoustical material, some by expansion through carefully designed inner chambers in the silencer body and some by providing baffles or in other ways creating internal friction. All designs of silencers must necessarily create some back pressure.

It is customary to use superchargers in all modern engines to get the increased power output. The exhaust gases may be used to drive such superchargers. An exhaust gas turbine will drive the centrifugal compressor which will act as the supercharger.

Sometimes, the exhaust gas may be used in heat exchangers, where steam could be generated. This is the case of exhaust gas heat recovery boilers.

(g) *Lubrication system.* The function of the lubricating oil is :

1. to lubricate the moving parts
2. to remove heat from cylinder and bearings
3. to carry away solid matter from rubbing moving parts and help piston rings to seal gases in the cylinder.

Lubrication is vital to diesel engines because of high pressures and small clearances. The life of the engine and its efficiency depend on the quality of the lubricating oil.

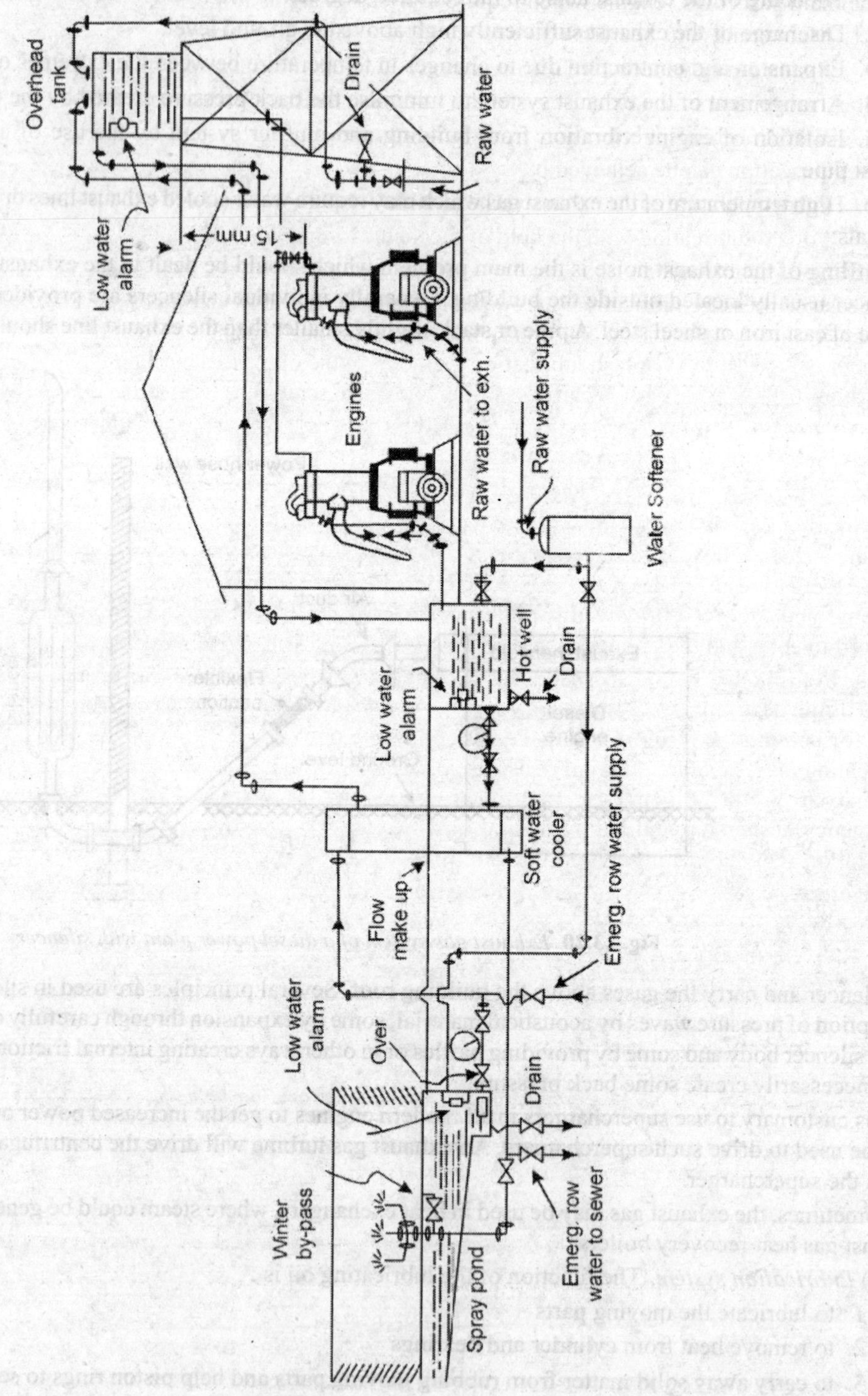

Fig. 23.21 *Closed cooling water system using water coolers and spray pond.*

The main parts of IC engine which require lubrication are as follows :

1. Main crankshaft bearings
2. Big end bearings
3. Piston pin or gudeon pin bearings
4. Piston rings and cylinder walls
5. Timing gears
6. Camshaft and camshaft bearings
7. Valve mechanism etc.

The lubrication can be achieved by :

1. Gravity lubrication from an overhead tank
2. Forced lubrication with the help of mechanical force feed lubricator
3. Splash lubrication.

In case of pressure lubrication, an oil pump supplies the lubricant under pressure to many parts of the engine through a duct system and to the crankshaft and wrist pin bearings by drilled passages in the shaft and rods. Cylinder is lubricated by splash lubrication method. As the oil passes through the lubrication cycle it accumulates impurities in the form of carbon particles, water and metal scrap. The lubricating oil may be purified by the following methods :

(i) Settling (ii) Centrifuging
(iii) Filtering (iv) Chemical reclaiming.

Typical of mechanical filters are cloth bags, wood felt pads, paper discs and cartridges of porous material. The oil can be cleaned by high speed centrifuges, where hot oil should be admitted into the centrifuge. This method will yield excellent results thus giving rise to excellent purification of oil. Chemical reclaiming uses a combination of heat and activated clay followed by filtering. In case of settling method, large tanks are used where hot oil is allowed to enter and the impurities settled down. Clean oil is then tapped from the top of the tank.

Due to friction of moving parts, the lubricating oil gets heated and should be cooled before recirculation. About 2-3% of the heat in the fuel is dissipated as heat which is taken by the lubricating oil. Hence tube coolers can be used for cooling the lubricating oil. The cooling water used for cooling the engine can be used to cool this hot lubricating oil. The lubricating oil consumption is about 1% of fuel consumption. The qualities, film strength and viscosity are produced by the selection and processing of the mineral oil base. Resistance to oxidation and foaming, detergency, dispersion and anti-corrosion qualities are provided by adding detergents and inhibitors to straight mineral oil. The lubricating oils are graded by SAE numbers.

Figure 23.22 shows the splash lubrication system. This is used in light duty engines. The system is self explainatory.

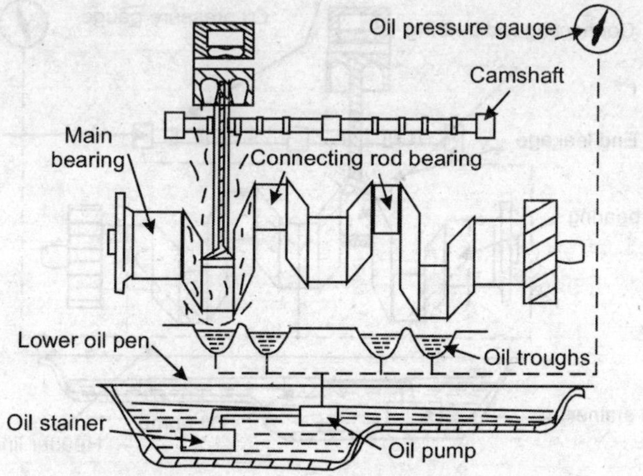

Fig. 23.22 *Splash lubricating system.*

Figure 23.23 shows the splash and pressure lubrication system. Here the lubricating oil is supplied under pressure to main and camshaft bearings. Oil is also supplied under pressure to pipes which direct a stream of oil against the

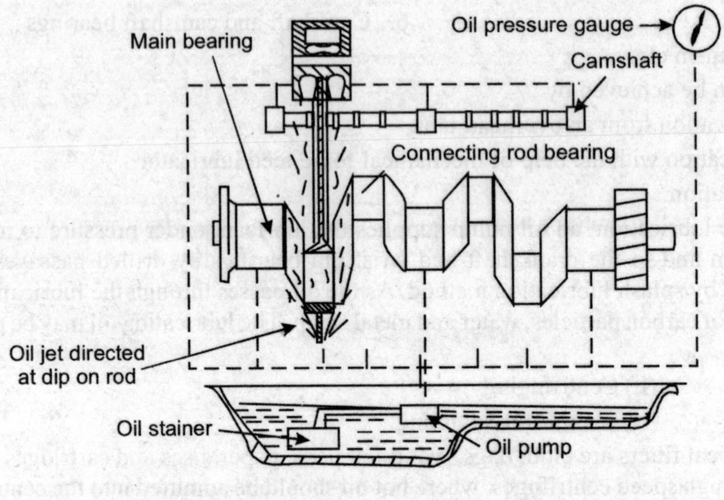

Fig. 23.23 *Splash and pressure lubrication system.*

dippers on the big end of connecting rod bearings cup and thus the crank pin bearings are lubricated by splash or spray of oil thrown up by the dipper.

The pressure feed system is illustrated in Fig. 23.24 in which oil is drawn in from the sump and forced to all the main bearings of the crankshaft through distributing channels. A pressure relief valve will also be fitted near the delivery point of the pump which opens when the pressure in the system attains a predetermined value. An oil hole is drilled in the crankshaft from the centre of each crankpin to the centre of the adjacent main journal, through

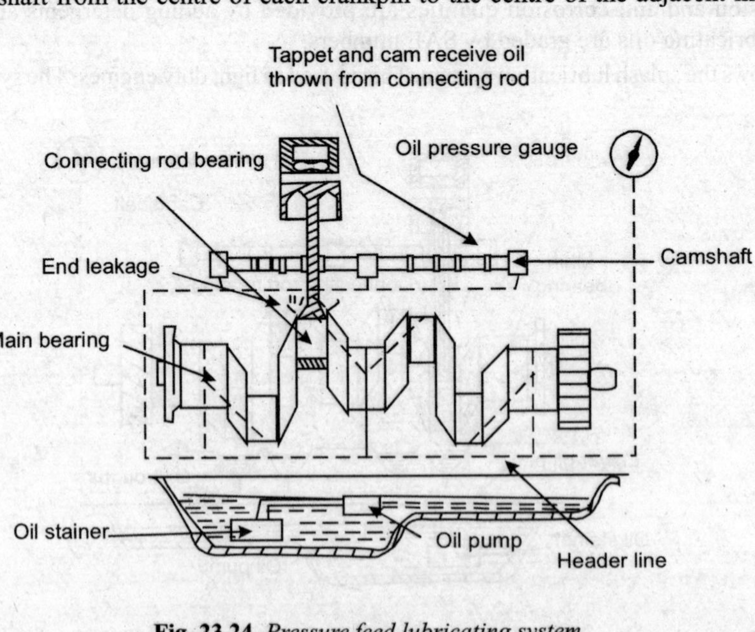

Fig. 23.24 *Pressure feed lubricating system.*

which oil can pass from the main bearings to the crankpin bearing. From the crankpin it reaches piston pin bearing through a whole drilled in the connecting rod.

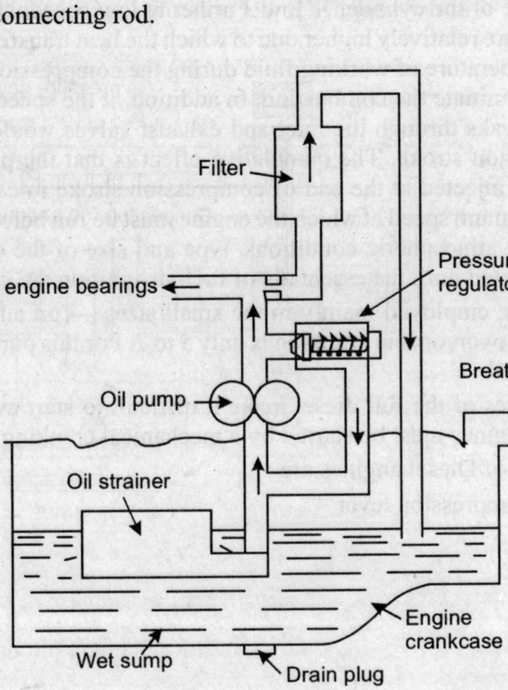

Fig. 23.25 *Wet sump.*

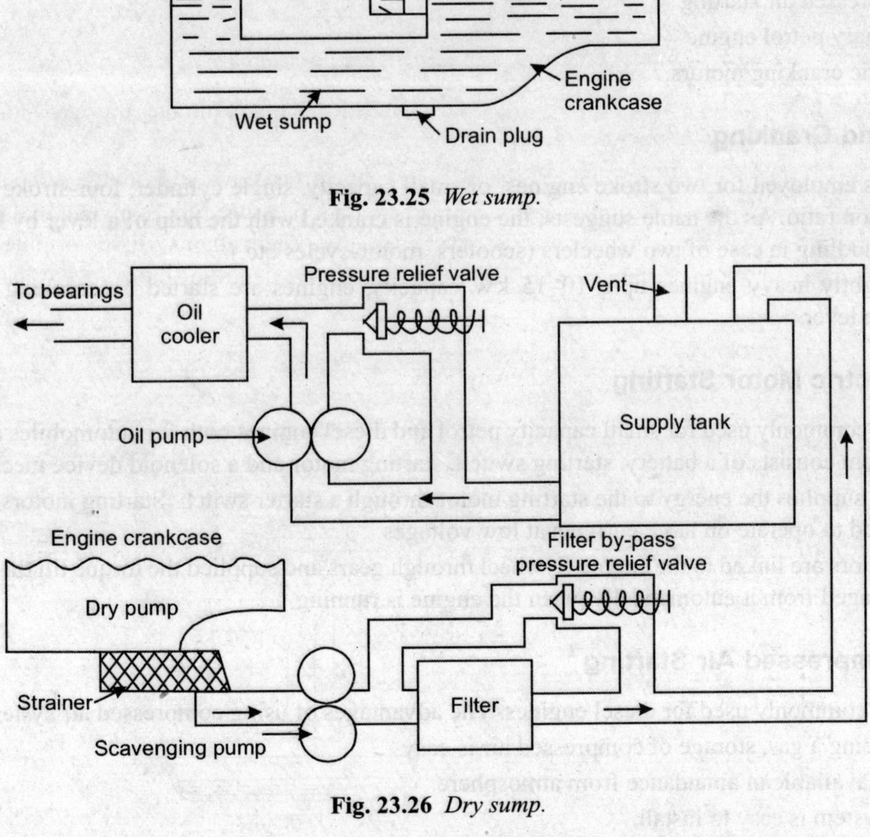

Fig. 23.26 *Dry sump.*

Other systems shown in Fig. 23.25 and Fig. 23.26 are wet sump lubrication system and dry sump lubrication system.

(*h*) *Engine starting systems.* The condition of the engine at the time of starting is not conducive for initiation of combustion since the temperature of the cylinder is low. Further at low speeds the time required to complete the suction and compression strokes are relatively higher due to which the heat transfer from the working substance are high. This further lowers the temperature of working fluid during the compression stroke. Therefore, temperature and pressure are not sufficient to initiate the combustion. In addition, if the speeds are low, the unavoidable leaks past the piston rings and likely leaks through the inlet and exhaust valves would further lower the pressure and temperature during the compression stroke. The cumulative effect is that the pressure and temperature are not sufficient to ignite the fuel to be injected at the end of compression stroke in case of compression ignition (CI) engines. Therefore, there is a minimum speed at which the engine must be run before the engine can start firing. The required speed depends upon the atmospheric conditions, type and size of the engine. Thus the cycles of these engines must be externally motivated until the essentials of fuel-air-ignition are correlated and power ensues.

Spark ignition (SI) engines are employed mainly in the small sizes — on automatic and portable equipment where the compression ratio to be overcome in cranking is only 5 to 7. For this purpose hand and electric motor (6–12 volts d.c.) cranking are practical.

The high compression pressures of the full diesel make it difficult to start even the smallest diesels by hand cranking and in general, diesel engines must be started by a mechanical cranking system.

Thus, the methods for starting of Diesel engines are :

1. Hand cranking with decompression lever
2. Compressed air starting
3. Auxiliary petrol engine
4. Electric cranking motors.

23.3.4 Hand Cranking

This method is employed for two stroke engines, or small capacity, single cylinder, four-stroke diesel engines of low compression ratio. As the name suggests, the engine is cranked with the help of a lever by hand in case of 4 stroke or by paddling in case of two wheelers (scooters, motor-cycles etc.).

For the slightly heavy engines upto 10–15 kW capacity, engines are started by cranking with the help of decompression lever.

23.3.5 Electric Motor Starting

This system is commonly used for small capacity petrol and diesel engines both for automobiles and land-engines. The basic system consists of a battery, starting switch, starting motor and a solenoid device mechanism.

The battery supplies the energy to the starting motor through a starter switch. Starting motors are series wound motors designed to operate on large currents at low voltages.

Starting motors are linked to the engine flywheel through gears and supplied the torque till the engine starts and then be disengaged from it automatically when the engine is running.

23.3.6 Compressed Air Starting

This system is commonly used for diesel engines. The advantages of using compressed air system are :

(i) Air being a gas, storage of compressed air is easy.

(ii) Air is available in abundance from atmosphere.

(iii) The system is easy to install.

(iv) System can quickly and surely start the engine.

Compressor, air starting valve, compressed air reservoir are the requirements of this system.

In addition to compressed air for starting, diesel engines with air injection of fuel require an air injection system.

A mechanical injection engine, having no air compressor as an integral part, requires the separately driven compressor for starting air.

Two or more compressed air storage tanks, or 'bottles' are provided. A small compressor is installed either as a normal or emergency supply of compressed air to recharge the storage tanks. An air injection engine has an integral compressor, and air may be bled from the injection system for the purpose of recharging the starting tanks. Starting air pressures need to be between 15 to 50 bar.

During the starting period, compressed air is admitted to one or more of the engine cylinders. This powers the engine much the same as steam works in a steam engine and this air working may taken as an air engine where air at pressure, instead of steam, is admitted and in that case it may be called air motor. Then air is stopped and oil injection is started and the engine starts running.

23.3.7 Auxiliary Petrol Engine

Instead of using a battery and a motor as in case of electric starting, a small capacity petrol engine is used to crank the diesel engine with the help of bendix or over-running clutch drive. The petrol engine is coupled to diesel engines through reduction gear. The petrol engine is usually started by cranking. When the diesel engine picks up adequate speed and it starts firing, the drive disengages the petrol engine and subsequently shut down.

(*i*) *Electric power system.* This system consists of the following elements :

1. Alternator or generator
2. Switch gear
3. Exciters
4. Transformer
5. Auxiliary power and
6. Light.

(*ii*) *Governing system.* The purpose of governor is to keep the engine running at a desired speed regardless of the changes in the load carried by the engine. If the load decreases, the speed of the engine will begin to increase, because the fuel supply is more for the decreased load. As the speed of the engine increases, the centrifugal force on the rotating weights of the governor will also increase and will move the control sleeve, together with the fuel regulating mechanism in the direction of less fuel supply, thereby the speed is brought to the rated value. If, on the other hand, the load on the engine increases the engine will begin to slow down because the fuel supply is not sufficient for the increased load. As the speed of the engine decreases, the centrifugal force on the rotating weights of the governor will also decreases and will move the control sleeve, together with the fuel regulating mechanism, in the direction of more fuel supply.

There are three methods of governing of CI engines as following :

(a) Quality governing

(b) Quantity governing

Quality governing is employed for compression ignition (CI) engines (oil engines) while quantity governing is employed for petrol engines.

Governing is accomplished by control of the quantity of fuel oil spray. Fuel oil pump may have a constant stroke with variable by-pass or suction or will have a variable stroke. Governors are sensitive enough to mermit exact load adjustment. For electrical generation even the multi-cylinder engines must be equipped with a heavy flywheel to prevent cyclic variation of speed.

23.4 PRINCIPLES OF WORKING OF IC ENGINES

Internal combustion engines are built around a variety of thermodynamic cycles. One type may vary from another in characteristics such as flexibility, fuel used, ease of starting, weight, cost etc., but all have many features in common such as :

(a) Use of the piston and cylinder to create a variable volume chamber in which to work out the cycle.

(b) Gaseous working medium.

(c) Open thermal cycles, which means a flow of air as well as fuel into the engine as a system and a discharge of gaseous products of combustion from it.

(d) Two or four stroke mechanical cycle.

(e) Instead of uniform power output as in case of gas turbines, to get uniform output from cyclic working, multiple cylinders and/or heavy flywheels for steady delivery.

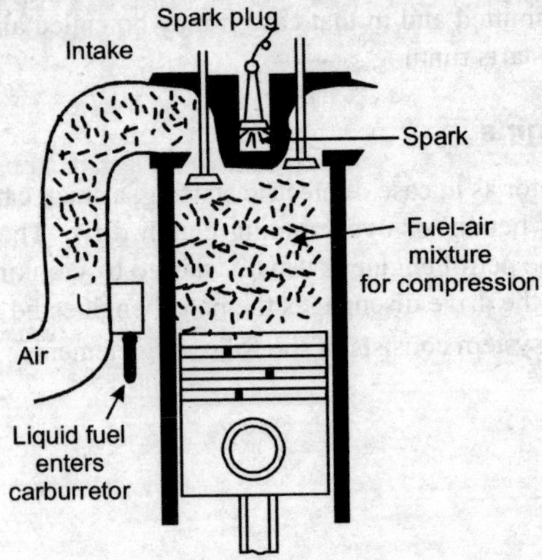

Fig. 23.27

There are different nine principles and features of commercial variations used nowadays and are given below :

(i) Fuel is introduced by carburretor.

(ii) Fuel and air mix outside the cylinder and before compression.

(iii) Compression ratio is low.

(iv) Ignition method used is spark ignition.

23.4.1 Cycle of Events

1. Carburretor vaporizes fuel and mixes it with air in proper proportion.
2. Suction stroke begins with inlet valve open, fills the cylinder with mixture.
3. Then the compression stroke begins. The pressure of the mixture increases to 7 to 15 bar.
4. At the end of compression or at near the end of compression, spark ignites the mixture. This is ideally at constant volume or instantaneous explosion of the mixture. Pressure increases.
5. The ignited mixture expands pushing piston outwards.
6. At the end of expansion, exhaust valve opens and the pressure in the cylinder is released instantly.
7. Then the burnt gases are cleared off the cylinder by the piston moving inside the cylinder.

All these events are taking place in a **Gasoline or Petrol engine**.

In this type of engine :

1. Fuel is introduced by mixing valve (Gas fuel).
2. Fuel and air mix outside the cylinder and before compression.

3. Compression ratio is low.
4. Method of ignition is spark ignition.

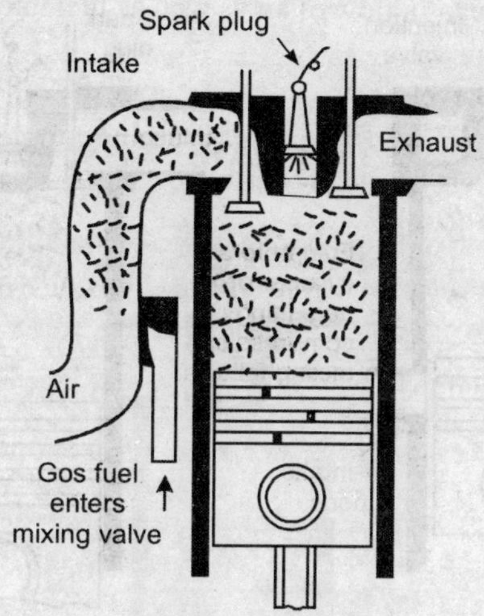

Fig. 23.28

Cycle of Events

1. Mixing valve blends air and gas fuel in proper proportion.
2. Suction takes place during suction stroke with intake valve open, fills the cylinder with mixture.
3. Then compression stroke takes place and the pressure of the mixture is increased 5 to 20 bar.
4. At the end of compression stroke, the mixture is ignited by spark.
5. This ignition is instantaneous and takes place at constant volume.
6. The fired, high pressure mixture expands ideally and the piston moves outwards.
7. At the end of expansion stroke, the exhaust valve opens, and the pressure in the cylinder is released instantly.
8. The piston moves inwards and the spent gases are driven out. This is exhaust.

All these events are taking place in a **Gas Engine**.

In the third type of the engine :

(i) The fuel is introduced into the engine cylinder by L.P. injection. Fuel is gas.
(ii) Fuel and air mix inside the cylinder and early in compression.
(iii) Compression ratio is low.
(iv) Method of ignition is spark ignition.

Cycle of Events

1. Air enters the cylinder through inlet ports in the cylinder wall.
2. Piston moves inwards. Inlet and exhaust ports are closed because of the motion of the piston. After the valves are closed, compression of air starts.
3. At early in the compression stroke, gas fuel is injected at low pressure.
4. The resulting mixture of air and fuel gas is compressed to 5 to 20 bar pressure.
5. At the end of compression, speak ignites the mixture increasing the pressure of the mixture because the ignition takes place instantly i.e., at constant volume.

6. Next the stroke is an working stroke or power stroke and the expansion of the mixture gases takes place.

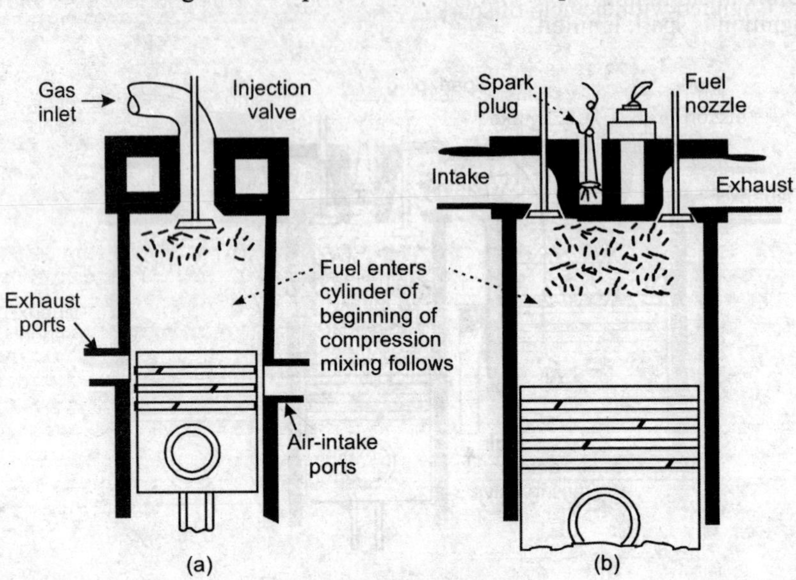

Fig. 23.29

7. At the end of the expansion, the piston uncovers the exhaust port and the pressure in the cylinder is released instantly and the spent charge is exhausted by the incoming fresh air which process is called scavenging.

All these events are taking place in a two-stroke gas engine where the gas is admitted under low pressure.

In this type of engine :

(i) Fuel is introduced by high pressure injection system.

(ii) Fuel and air mix together in the cylinder and during early in compression.

(iii) In this case compression ratio is low.

(iv) Ignition method used is spark ignition.

Cycle of Events

1. During suction stroke, inlet or suction valve is open and the cylinder is filled with air.

2. During the next stroke, inlet valve closes because of the inward motion of the piston. At the early stage of compression fuel is injected at high pressure.

3. During compression, a mixture of air and fuel is formed in the cylinder.

4. Compression of the mixture raises its pressure 7 to 15 bar.

5. At the end of compression, the mixture is ignited by means of a spark.

6. Ignition of the mixture increases the pressure further and the mixture of products of combustion expand driving the piston outwards.

7. At the end of expansion, the exhaust valve opens and the pressure in the cylinder is released and the piston start moving inwards and the gases are driven out or exhausted through an exhaust valve.

This is a case of **Injection Gasoline Engine.**

In this type of engine :

(i) Fuel is introduced by high pressure injection system.

(ii) Fuel and air mix together inside the cylinder and at the end of compression.

(iii) Compression ratio is low.

(iv) The mixture is ignited with the help of spark.

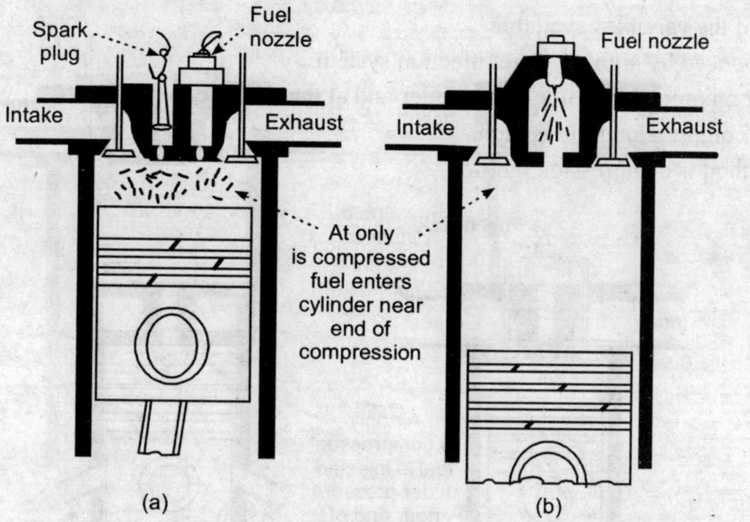

Fig. 23.30

Cycle of Events

1. Piston moves outwards during the suction stroke and inlet valve is opened admitting air into the cylinder.
2. Then inlet valve is closed. The inward motion of the piston, compresses the air and the pressure increases to 8 to 10 bar.
3. Just before the end of compression stroke, fuel injection is complete.
4. Just before the end of compression, spark is struck the fuel is ignited by the spark. Fuel gets vaporized during compression.
5. Pressure is increased and the piston moves forward, (outward) thus causing the expansion of the gases.
6. At the end of expansion, exhaust valve opens and the cylinder pressure is released instantly.
7. Piston moves inwards driving the spent gases out of the cylinder. It is an exhaust stroke.

This is a case of **Hesselman Engine.**

In this variety of engine :

(i) Fuel is introduced by high pressure injection system.

(ii) Fuel and air mix together inside the cylinder and at the end of compression.

(iii) Compression ratio is low.

(iv) The mixture is ignited by **hot surface**.

Cycle of Events

1. During the suction stroke, piston moves outwards, inlet valve is opened and air fills the cylinder.
2. Ignition surface is heated by the exhaust gases and the fuel is injected in the combustion chamber and gets vaporized. This mixes with the exhaust gases in the chamber but does not burn.
3. During compression stroke the pressure rises to 4 to 10 bar and piston forces air into the vaporizer.
4. Heated surface of the vaporizer ignites the mixture near the end of compression stroke.
5. Combustion of the mixture is explosive and the piston is pushed outwards. Expansion of the gases takes place.

6. At the end of expansion, exhaust valve is opened and the pressure in the cylinder is released instantly.
7. Piston moves inwards forcing the gases out of the cylinder.

This method of ignition is used with dual-combustion cycle. The engine is called **vaporizing oil engine**.

VII. This is one of the variables available

(i) Fuel is introduced by high pressure injection system.
(ii) Fuel and air mix together inside the cylinder and at the end of compression.
(iii) In this case compression ratio is high.
(iv) Ignition method is compression ignition.

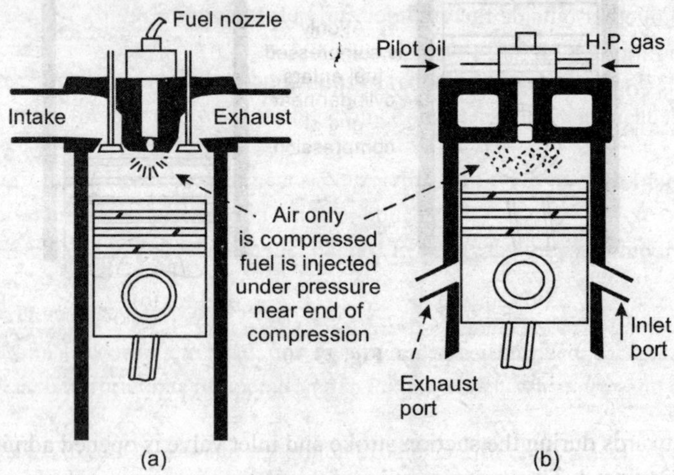

Fig. 23.31

Cycle of Events

1. Suction stroke with inlet or suction valve open. Air fill the cylinder when the piston moves outward.
2. Piston moves inwards. Inlet and outlet or exhaust valves are closed. Air inside the cylinder is compressed. The pressure increases to about 35–40 bar.
3. At the end of compression, fuel injection starts.
4. During compression, air temperature increases and at the end of compression it is sufficient to ignite the fuel injected.
5. Burning mixture expands, pushing piston outwards on the expansion stroke (power stroke, working stroke).
6. At the end of expansion stroke, exhaust valve opens and the pressure in the cylinder is released instantly.
7. Piston moves inwards driving the gases out of the cylinder. This is the exhaust stroke.

This is the case of the familiar **Diesel Engine**.

This type of the variation in engine has the following events :

(i) Fuel is introduced by high pressure injection system.
(ii) Fuel and air mix together inside the cylinder and at the end of compression.
(iii) In this case the compression ratio is high.
(iv) Ignition method is compression plus pilot oil.

Cycle of Events

1. As the piston moves out during the suction, air enters the cylinder through inlet port.

2. During upwards or inwards motion piston closes or covers inlet and exhaust ports. Compression of air takes place to pressure of 30–35 bar.

3. Near the end of compression, fuel gas and pilot oil are injected at high pressure.

4. Because of the high temperature of compression, ignition of fuel takes place. Pilot oil stabilizes combustion.

5. Burning mixture expands and the piston moves outwards.

6. Near the end of expansion stroke, piston uncovers the exhaust port and inlet port.

This is a case of **Gas-Diesel Engine** (Two stroke).

This type of variation includes :

(i) Gas fuel is introduced by induction.

(ii) Fuel and air mix together outside the cylinder and before compression.

(iii) The compression ratio is high.

(iv) Ignition method is compression ignition plus pilot oil.

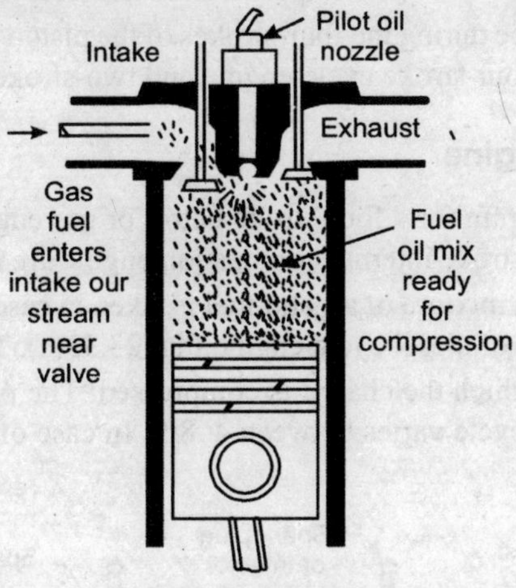

Fig. 23.32

Events in the cycle are :

1. During suction stroke inlet valve opens and the cylinder is filled with the air and gas mixture.

2. Compression stroke. Pressure of the mixture increases to about 35–40 bar.

3. Near the end of compression, pilot oil is injected to initiate combustion.

4. Heat of compression ignites the pilot oil which causes the mixture to burn.

5. Mixture of gases expands pushing the piston outwards. At the end of expansion exhaust valve opens and the pressure in the cylinder is released to atmospheric pressure.

6. Further inward motion of piston drives out the spent gases. This is exhausted stroke.

This is the case of **Dual-Fuel Engine**.

23.5 ENGINES USED IN DIESEL POWER PLANT

It is known that energy is supplied to the internal combustion engine by the combustion of fuel within the cylinder. Because of the widespread use of internal combustion engines in automobiles, on the farm, in industrial plants, on ships, in **Power Plants** etc. the ideal cycles for them are particularly significant. The fuels used are natural or manufactured gas, gasoline, kerosene oil, fuel oil, alcohol and others. The common fuels are natural gas, gasoline and **fuel oil** (Diesel oil).

From operational point of view, there are two types of engines used in practice :

1. Four-stroke engines and
2. Two-stroke engines.

The main thermodynamic process that are required for a cycle to be completed are :

(a) Admission of oxidant (air) for combustion of fuel
(b) Admission of fuel
(c) Compression of the charge
(d) Combustion of the charge or fuel
(e) Expansion of the burnt charge
(f) Exhaust of the spent charge.

In short for a thermodynamic cycle

(i) Heat addition Q_A (ii) Heat rejection Q_R
(iii) Compression (iv) Expansion are necessary.

All these operations are taking place during the four-strokes of the piston or during the two-strokes of the piston. In such cases the engines are called four-stroke cycle engine and two-stroke cycle engines.

23.5.1 Four Stroke Cycle Engine

Figure 23.33 shows a schematic diagram for a four-stroke petrol or gas engine. The sequence of events shown in Fig. 23.33 are the same for any four-stroke internal combustion engine are aa following

1. *A suction stroke*, during which a mixture of air and fuel is taken in case of an Otto-cycle engine, or only air is taken in during suction stroke in case of diesel cycle engine Fig. 23.33(a). This is the **charge** taken.

2. *A compression stroke*, during which the charge is compressed. The pressure and temperature of the charge increase. Compression ratio in Otto-cycle varies between 4–8/9. In case of diesel cycle, the compression ratio is between 12 to 21.

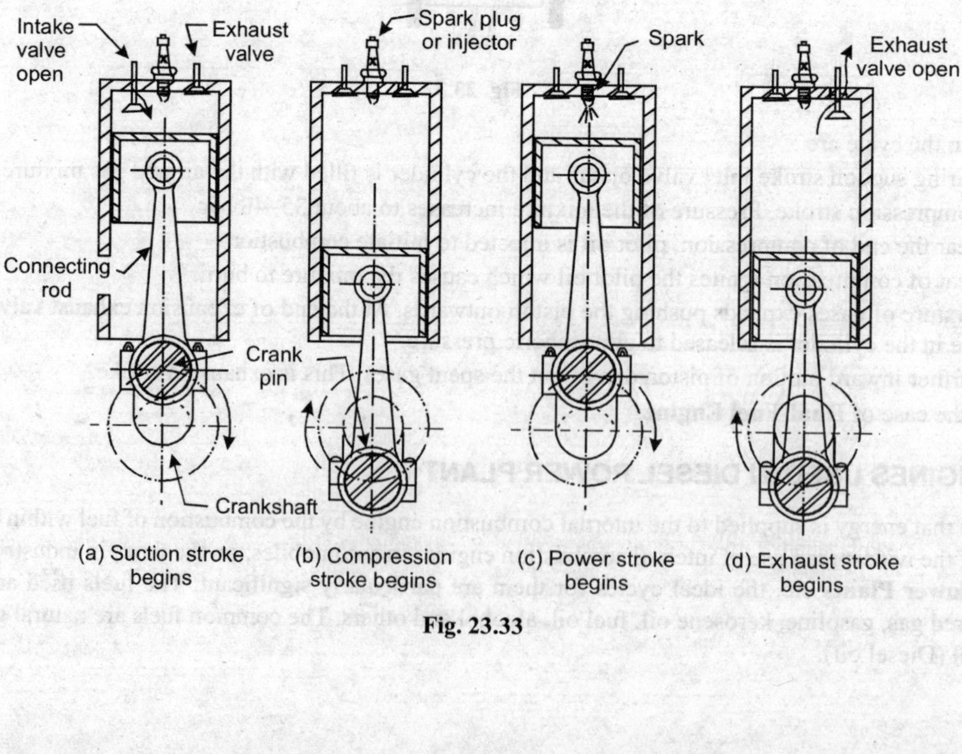

Fig. 23.33

3. At the end of compression stroke, the spark is struck and the fuel, present in the cylinder, is ignited instantly so that the pressure and temperature increase at constant volume.

In case when only air is taken in during suction stroke, because of the high compression ratio, the pressure and temperature increases high enough to cause the fuel — injected at the end of compression stroke — to burn without spark ignition. This burning or combustion of the fuel is taking place ideally at constant pressure.

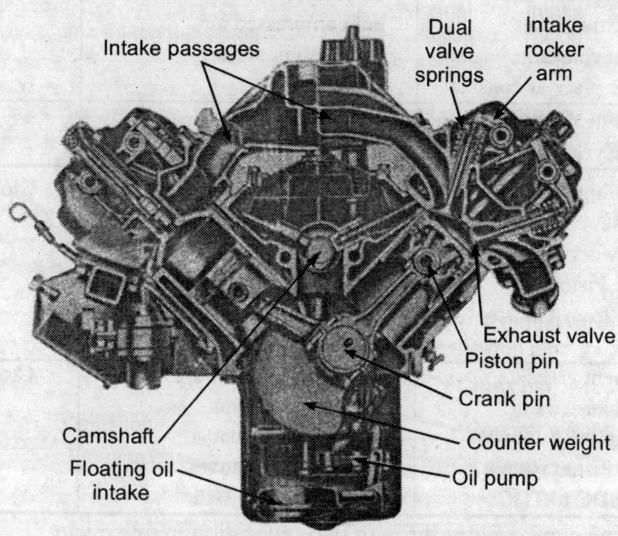

Fig. 23.34 *V-type, 4 stroke cycle automotive engine.*

The burning of the fuel releases energy for use by the system.

4. *An expansion stroke* or *power stroke* during which expansion work or positive work is done Fig. 23.33(c).

5. *An exhaust stroke* during which most of the products of combustion are push out of the cylinder and the cycle is complete and is repeated.

Figure 23.34 shows a V-type, 4-stroke cycle automotive engine.

Stroke	*Petrol/Otto cycle SI engine*	*Diesel engine CI engine*	*Inlet valve*	*Exhaust valve*
First suction stroke	The piston moves down from TDC air fuel mixture is sucked or induced from carburettor	The piston moves from TDC down to BDC and sucks only air	Open	Closed
Second stroke compre-ssion stroke	Piston moves from BDC to TDC and the charge is comp-ressed. Pressure and temperature rise	Piston moves from BDC to TDC and air is compressed. Pressure and tempera-ture increased	Closed	Closed

Contd.

Contd.

Stroke	*Petrol/Otto cycle SI engine*	*Diesel engine CI engine*	*Inlet valve*	*Exhaust valve*
Igni-tion	Near the end of compression the spark is struck and combustion takes place at constant volume.place at constant	Near the end of comp-ression the diesel oil is injected into hot compressed air combustion takes pressure	Closed	Closed
	Constant volume CycleCycle	**Constant pressure**		
Third stroke power or working stroke	Due to high pressure the gases expand doing work on the piston. Piston moves from TDC to BDC	Due to high pressure the gases expand doing work on the piston. Piston moves from TDC to BDC	Closed	Closed
Fourth stroke exhaust stroke	The spent gases are exhausted out through the exhaust valve. Piston moves from BDC to TDC	The spent gases are exhausted out through the exhaust valve. Piston moves from BDC to TDC.	Closed	Open

In four-stroke cycle engine crank rotates through two revolution in one cycle.

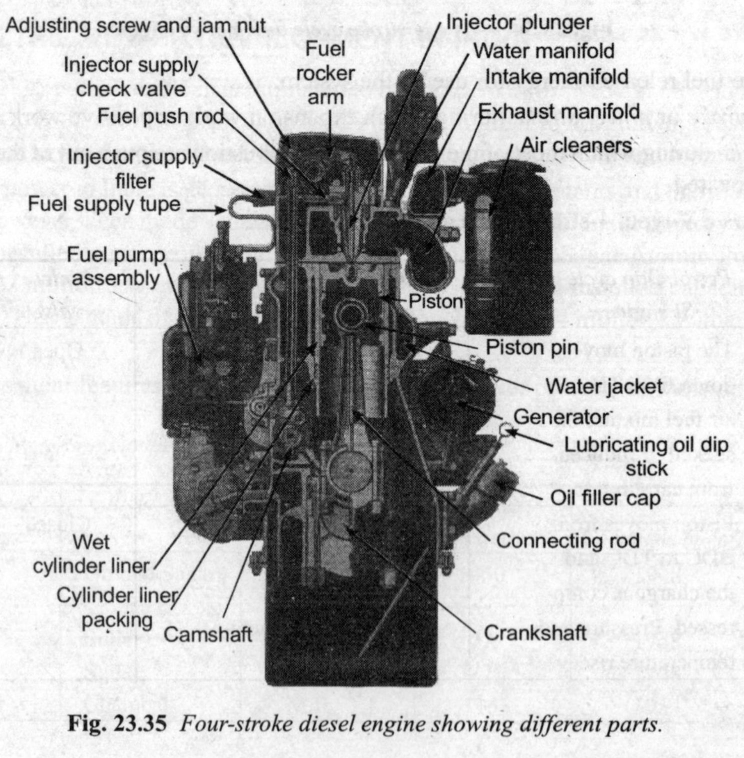

Fig. 23.35 *Four-stroke diesel engine showing different parts.*

Figure 23.35 shows a four-stroke diesel engine where the different parts are shown. Figure 23.38 is another diagram of a diesel engine showing lubrication system of the engine.

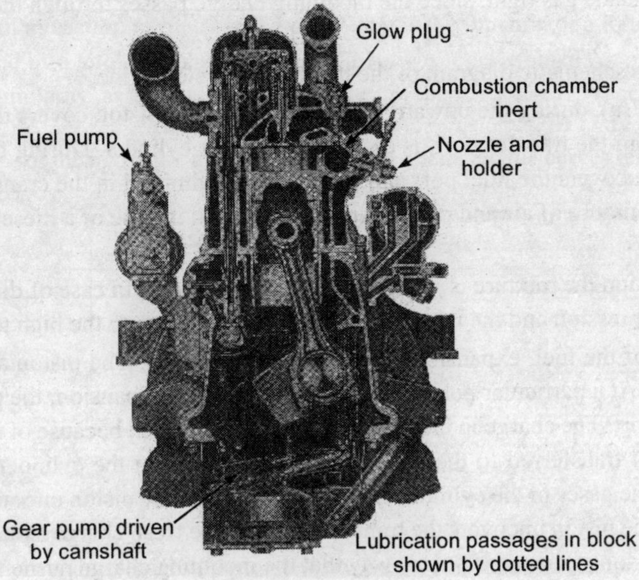

Fig. 23.36 *Four-stroke diesel engine with lubrication.*

23.5.2 Two Stroke Cycle Engine

In two-stroke cycle engine, all the basic processes for a thermodynamic cycle namely, suction or induction, compression, combustion, expansion and exhaust take place in two-strokes or in one revolution of the crankshaft.

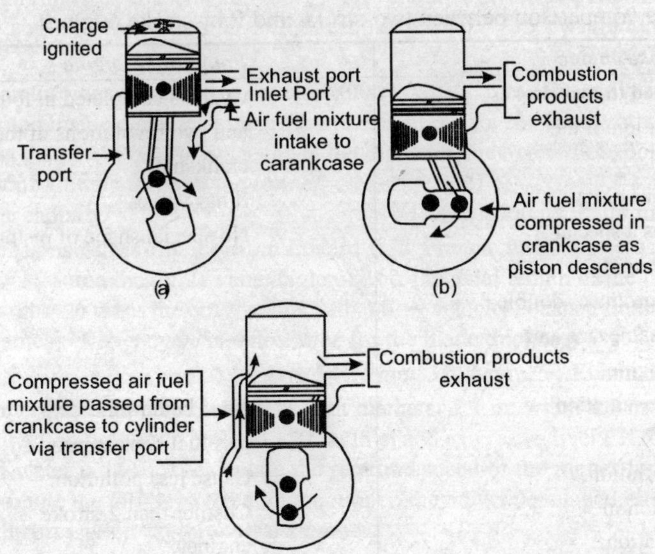

Fig. 23.37

Two-stroke engines are provided with ports instead of valves. These ports are inlet ports, exhaust port and transfer port. These ports are opened and closed by the piston itself (Ports are covered or uncovered by piston).

The crankcase will be made gas tight, since the incoming charge passes through the crankcase units way to the engine cylinder.

Figure 23.37 shows the schematic diagram of the two stroke cycle engine.

As shown in Fig. 23.37(a), during the upward stroke of the piston, piston covers the transfer port and exhaust port. The fresh charge from the transfer port, is compressed in the cylinder. During the compression stroke, at a certain point, the piston uncovers the inlet port and the charge is admitted in the crankcase. In case of Otto-cycle engine, the charge is the mixture of air and petrol vapour or gas and in case of a diesel cycle engine, the charge is only air.

At the end of compression the mixture is ignited by means of a spark. In case of diesel cycle engine, fuel oil is injected at the end of compression and the ignition of fuel takes place due to the high temperature of compression.

After the combustion of the fuel, expansion of the gases takes place. The piston starts moving towards BDC doing work on the piston. At a particular position of the piston during expansion, the piston uncovers the exhaust port and covers the inlet port. The charge in the crankcase gets compressed because of the downward motion of the piston, and is ready to get transferred to the upper side of the piston in the cylinder. When the exhaust port is uncovered by the piston, the gases in the cylinder are exhausted. After the piston uncovers the exhaust port, further downward movement of the piston uncovers the transfer port and the fresh charge from the crankcase comes in the cylinder. The top of the piston is shaped in such a way that the incoming charge pushes out the burnt gases through the exhaust port. The displacing the combustion gases or products out of the engine cylinder with the help of fresh incoming charge is called **"scavenging"**. This is shown in Figs. 23.37(b) and Fig. 23.37(c).

When the piston is at BDC, the conditions of the ports are as (i) transfer port is uncovered, exhaust port is open or uncovered and the piston is on the verge of compressing the fresh charge. The cycle thus, is complete and is repeated.

23.5.3 Comparison of Two-stroke and Four-stroke Engines

Following table shows the comparison between two-stroke and four-stroke engines.

Two stroke engine	Four stroke engine
1. Cycle completed in two strokes and one revolution of the crankshaft.	Cycle is completed in four strokes and two revolutions of the crankshaft.
2. No valves. It has ports.	It has valves.
3. Piston head has a special shape for scavenging.	No special shape of piston head.
4. Theoretically, produces double power for the same size and speed of the engine.	—
5. Higher fuel consumption less mileage.	Have better mileage less fuel consumption.
6. Cause more pollution.	Cause less pollution.
7. Simple construction use mist lubrication.	Costlier than 2-stroke engines.

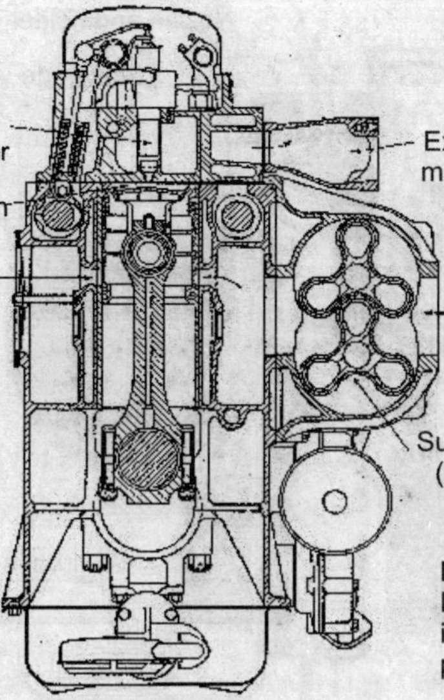

Fig. 23.38 *Section of 2-stroke engine.*

Figure 23.38 shows the section of the 2-stroke diesel engine.

Figure 23.39 shows a 2-stroke diesel power plant with a 2-stroke engine section for the plant shown.

As employed in stationary power plants, engines are directely-connected to generators of relatively low speeds. The speeds employed — 200 to 1000 rpm — necessitate a large diameter, salient generator of rather short axial length. The heavy, rugged, slow speed construction is productive of high reliability and although costly, can be justified in central station practice.

Engines using gaseous fuels are less frequently seen that the liquid fuel types, as liquids are more readily transported. But in certain cases like some manufacturing company like aluminium producer where electricity is a principal manufacturing expenses, factory can be located in a gas field territory and built a power supply plant in which engine-generator units driven by 2-stroke cycle, outside mixing gas engines can be installed.

The Diesel engine is an excellent prime mover for electrical generation in capacities of from 75 to 4000 kW. As such, diesel engines are widely used by private industry large hotels, utility companies etc. It may also be used as peak load plants along with thermal power plants.

The advantages of the Diesel engines are :

(a) Low fuel cost
(b) No long warming up period
(c) No standby losses
(d) Uniformly high efficiency of all sizes
(e) Simple plant layout
(f) Needs no large water supply.

Diesel engines become attractive prime mover wherever first cost is written off slowly enough so that operating costs are influential. But the Diesel is not the answer to all power supply problems. In case of Diesel power plants, an increase of capacity increases the plant floor area and cost nearly in the same proportion. This fixes an economic limit to the size of Diesel central stations. Another objection to the Diesel is the exhaust noise. Proper exhaust silencers can minimise this fault. Exhaust smoke is also objectionable. For that purpose, the exhaust is let off to the atmosphere at an elevated height.

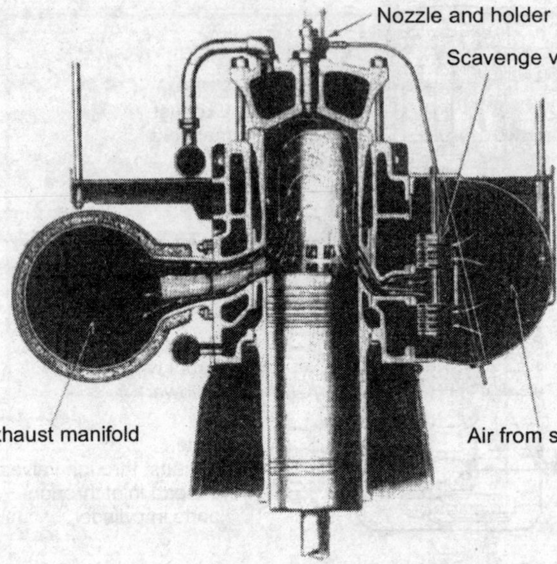

(a)

(b)

Fig. 23.39 *Two-stroke diesel power plant.*

Thus,

(i) where fuel prices or reliability of supply of oil over coal
(ii) where water supply is limited
(iii) where loads are relatively small and
(iv) where electric line services is unavailable or is available at too high rates.

There one will frequently find **Diesel Engine Power** as a solution.

23.6 THERMODYNAMIC CYCLE FOR DIESEL POWER PLANT AND ITS ANALYSIS

The ideal, thermodynamic cycle used for diesel engine is Diesel cycle. This cycle of operations was introduced by Dr. R. Diesel in 1897. In this cycle, all processes are ideal processes. This ideal cycle is called **Air Standard Cycle** in which we are assuming the following events :

1. Air is the working substance throughout.
2. Effect of fuel is neglected.
3. Compression and expansion processes are assumed to be isentropic processes.
4. Heat addition is assumed to take place at constant pressure.
5. Heat rejection is assumed to take place at constant volume.
6. Suction and exhaust are taking place at atmospheric pressure.

Taking this into consideration, the *P–V* and *T–S* diagrams for the Diesel cycle are shown in Fig. 23.40(a) and (b).

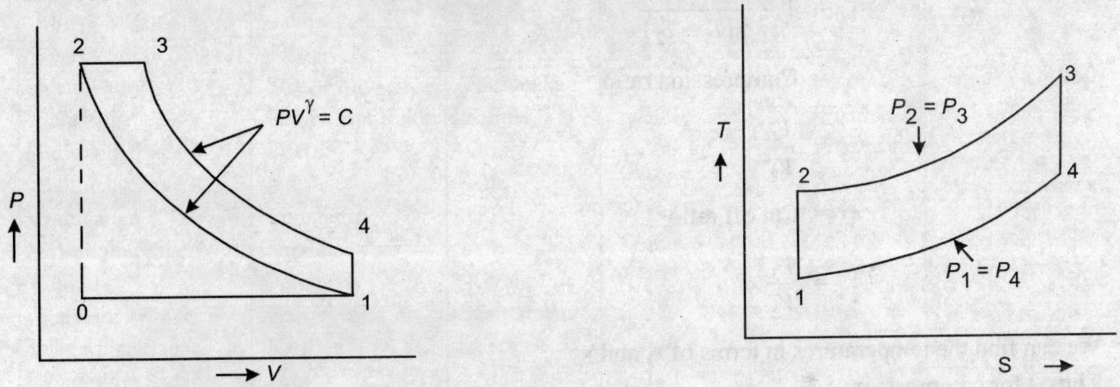

(a) *P–V* diagram (b) *T–S* diagram

Fig. 23.40 *Diesel cycle*

In Fig. 23.40,

0–1	Suction of the charge (air)
1–2	Isentropic compression
2–3	Constant pressure heat addition Q_A
3–4	Isentropic expansion
4–1	Pressure release at constant volume
	Heat is rejected at constant volume Q_R
1–0	Exhaust of the spent gases.

For a cycle, we know that

$$\text{Net work done/kg} = Q_A - Q_R$$

$$\text{Heat added/kg} = Q_A$$

∴ Thermal efficiency of the cycle

$$\eta_{th} = \frac{\text{Net work/kg}}{\text{Heat added/kg}}$$

$$= \frac{Q_A - Q_R}{Q_A}$$

$$= 1 - \frac{Q_R}{Q_A}$$

$$= 1 - \frac{C_V(T_4 - T_1)}{C_P(T_3 - T_2)}$$

$$= 1 - \frac{T_4 - T_1}{\frac{C_P}{C_V}(T_3 - T_2)}$$

$$= 1 - \frac{T_4 - T_1}{\gamma(T_3 - T_2)} \quad \text{(I)}$$

Let r_k = Compression ratio

$$= \frac{V_1}{V_2}$$

r_c = Cut off ratio

$$= \frac{V_3}{V_2}$$

We can find the temperatures in terms of r_k and r_c

Thus – for compression

$$\frac{T_2}{T_1} = r_k^{\gamma-1} \text{ and } T_2 = T_1 \; r_k^{\gamma-1} \quad \text{(i)}$$

For heat addition process,

$$\frac{T_3}{T_2} = \frac{V_3}{V_2} = r_c \therefore T_3 = T_2 . r_{c\gamma-1}$$

$$= T_1 . r_k^{\gamma-1} . r_c \quad \text{(ii)}$$

For isentropic expansion process,

$$\frac{T_3}{T_4} = \left(\frac{V_4}{V_3}\right)^{\gamma-1} = \left(\frac{V_1}{V_3}\right)^{\gamma-1} \text{ as } V_4 = V_1$$

$$\therefore \quad \frac{T_3}{T_4} = \left(\frac{V_1}{V_2} \times \frac{V_2}{V_3}\right)^{\gamma-1} = \left(r_k \times \frac{1}{r_c}\right)^{\gamma-1}$$

$$\therefore \quad T_4 = \frac{T_3}{\left(\frac{r_k}{r_c}\right)^{\gamma-1}} = T_3 \times \left(\frac{r_c}{r_k}\right)^{\gamma-1}$$

$$= T_1 \times r_k^{\gamma-1}.r_c \times \frac{r_c^{\gamma-1}}{r_k^{\gamma-1}}$$

$$= T_1 \times r_c^{\gamma} \qquad \text{(iii)}$$

Substituting these values of T_2, T_3 and T_4 in Eq. (I) we get

$$\eta_{th} = 1 - \frac{T_4 - T_1}{\gamma(T_3 - T_2)}$$

$$= 1 - \frac{T_1 r_c^{\gamma} - T_1}{\gamma\left[T_1.r_k^{\gamma-1}.r_c - T_1 - r_k^{\gamma-1}\right]}$$

$$= 1 - \frac{T_1\left(r_c^{\gamma} - 1\right)}{\gamma.T_1.\,r_k^{\gamma-1}\left(r_c - 1\right)}$$

$$\boxed{\eta_{th} = 1 - \frac{r_c^{\gamma} - 1}{\gamma r_k^{\gamma-1}\left(r_c - 1\right)}} \qquad \text{(II)}$$

Mean Effective Pressure (MEP) is defined as that constant pressure, when acting on the piston throughout the stroke, produces the work equal to the work of the cycle. It is given by p_m

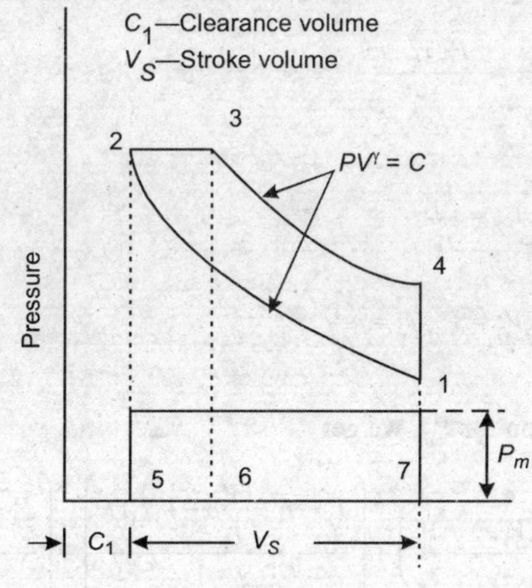

Fig. 23.41

Mean Effective Pressure can be shown on P–V diagram in Fig. 23.41.

$$\text{Work of the cycle} = \text{Area of } P\text{–}V \text{ diagram}$$

$$= \text{Area 1–2–3–4–1}$$

$$= p_m \times V_s$$

But
$$V_s = V_1 - V_2$$
$$= V_1\left(1 - \frac{V_2}{V_1}\right)$$
$$= V_1\left(1 - \frac{1}{r_k}\right)$$
$$= V_1\left(\frac{r_k - 1}{r_k}\right) \quad (1)$$

Area 1–2–3–4–1 = Area (2–3–6–5–2) + Area (3–4–7–6–3) – Area (1–2–5–6–7)
$$= P_2(V_3 - V_2) + \frac{P_3V_3 - P_4V_4}{\gamma - 1} - \frac{P_2V_2 - P_1V_1}{\gamma - 1}$$

Here $P_2 = P_3$ and $V_1 = V_4$

Substitute all volumes in terms of V_1 and all pressures in terms of P_1.
$$\frac{P_2}{P_1} = r_k^\gamma \qquad \therefore P_2 = P_1 . r_k^\gamma = P_3$$

Again $\dfrac{P_3}{P_4}$ = (Expansion ratio)$^\gamma$ = $\left(\dfrac{V_4}{V_3}\right)^\gamma$
$$= \left(\frac{V_1}{V_2} \times \frac{V_2}{V_3}\right)^\gamma = \left(\frac{r_k}{r_c}\right)^\gamma$$

$\therefore$
$$p_4 = \frac{P_3 \times r_c^\gamma}{r_k} = \frac{P_1 . r_k^\gamma . r_c^\gamma}{r_k^\gamma}$$
$$= P_1 . r_c^\gamma.$$

Similarly,
$$r_k = \frac{V_1}{V_2} \therefore V_2 = \frac{V_1}{r_k}$$
$$r_c = \frac{V_3}{V_2} \therefore V_3 = V_2 . r_c = \frac{V_1}{r_k} . r_c$$
$$V_4 = V_1$$

Substituting these values in equation for P_m, we get
$$p_m \times V_1\left(\frac{r_k - 1}{r_k}\right) = P_1 r_k^\gamma\left(\frac{V_1 . r_c}{r_k} - \frac{V_1}{r_k}\right) + \left(\frac{P_1 - r_k^\gamma . \frac{V_1 r_c}{r_k} - V_1 P_1 r_c^\gamma}{\gamma - 1}\right) - \left(\frac{P_1 - r_k^\gamma . \frac{V_1}{r_k} - P_1 V_1}{\gamma - 1}\right)$$

$$p_m = \left[\frac{r_k}{r_k - 1}\right]\left[P_1 . r_k^{\gamma-1}(r_c - 1) + \frac{\left(P_1 r_k^{\gamma-1} . r_c - P_1 r_c^\gamma\right)}{\gamma - 1} - \frac{P_1}{\gamma - 1}\left(r_k^{\gamma-1} - 1\right)\right]$$

$$= \frac{r_k}{r_k - 1} P_1 \left[r_k^{\gamma-1}(r_c - 1) + \frac{r_c r_k^{\gamma-1} - r_c^{\gamma} - r_k^{\gamma-1} + 1}{\gamma - 1} \right]$$

$$= \frac{P_1 . r_k}{(\gamma-1)(r_k-1)} \left[(\gamma-1)(r_c-1) r_k^{\gamma-1} + r_k^{\gamma-1}(r_c-1) - (r_c^{\gamma}-1) \right]$$

$$\boxed{p_m = \frac{p_1}{(\gamma-1)} \cdot \frac{r_k}{(r_k-1)} \left[\gamma . r_k^{\gamma-1}(r_c-1) - (r_c^{\gamma}-1) \right]}$$

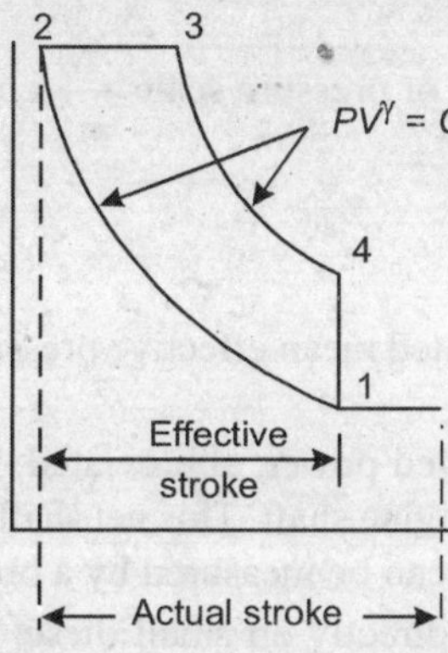

Fig. 23.42 *P–V diagram for 2-stroke engine.*

Figure 23.42 shows the *P–V* diagram for a 2 stroke diesel engine.

23.7 ENGINE PERFORMANCE

The rapid cyclic action in an actual engine and the irreversibilities of the process in the cycle modify the ideal cycle of Fig. 23.39. The diagram corners are no longer sharp, fuel must be injected ahead of dead centre and friction (fluid) and many other losses are the different actual events in the cycle.

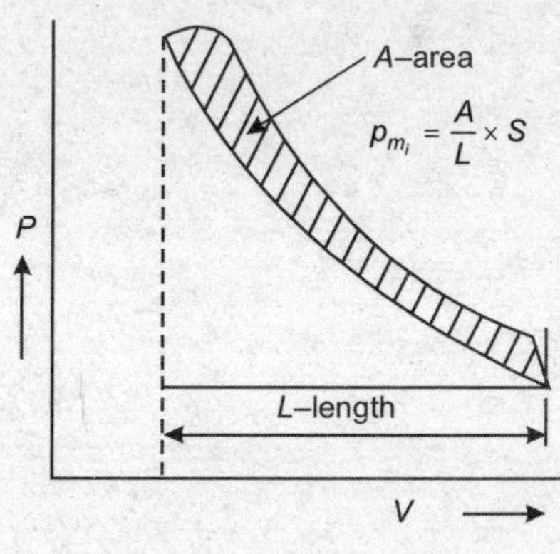

Fig. 23.43

Figure 23.43 shows the actual *P–V* diagram of diesel cycle engine.

With the help of indicator — may be mechanical or electronic as CRO — we can get an **indicator card** from the running engine.

Knowing the pressure scale of the spring used in case of the mechanical indicator or the pressure used in electronic indicator as CRO, we can find the actual mean effective pressure from the relation given below.

$$p_{mi} = \frac{\text{Area of diagram}}{\text{Length of diagram}} \times \text{Spring strength}$$

Units will be

$$A = \text{Area of indicator diagram cm}^2$$

$$L = \text{Length of diagram cm}$$

$$S = \text{Spring strength or pressure scale } \frac{N}{m^2} \text{ cm or bar/cm}$$

$$\therefore \quad p_{mi} = \frac{A}{L} \times S$$

This mean effective pressure is called indicated mean effective pressure p_{mi}. This is because it is obtained from indicator diagram.

The developed cylinder power is the indicated power, abbreviated, ip or IP. After considering engine friction etc., we can get the net power available at the engine shaft. This net shaft power generally is known as brake power abbreviated bp or BP. It is so called because it can be measured by a brake dynamometer.

Indicated power IP is difficult to measure directly on small diesel engines with great accuracy, because the volume of indicator lead through the cylinder head appreciably lowers the normal compression ratio. Also, the cylinder heads of small-bore high speed engines have special chambers, valves, and other equipment, leaving little or no space available for an indicator lead. Since none of these difficulties exists with the large stationary Diesel engine, indicator openings are provided as standard equipment.

Indicator power is given by

$$IP = p_{mi} \times \frac{LAn}{1000} \text{ kW}$$

where

$$p_{mi} = \text{Indicated mep } \frac{N}{m^2}$$

$$L = \text{Stroke of piston } m$$

$$A = \text{Cylinder diameter } m^2$$

$$n = \text{Number of working cycles/sec.}$$

$$= \frac{N}{2} \times \frac{1}{60} \text{ for a 4-stroke engine}$$

$$= \frac{N}{60} \text{ for a 2-stroke engine}$$

$$N = \text{Speed of the engine RPM.}$$

If $\quad X = \text{Power output of the power station}$

and $\quad \eta_g = \text{Efficiency of the generator}$

Then BP = Shaft power of the engine

= Power supplied to generator

$$= \frac{\text{Power Output}}{\text{Generator Efficiency}}$$

$$= \frac{X}{\eta_g}$$

For the engine, this brake power can also given as

$$\text{BP} = p_{mb} \times \frac{LAn}{1000}$$

where p_{mb} = brake mean effective pressure (bmep).

$$\therefore \quad p_{mb} = \frac{\text{BP} \times 1000}{LAn}.$$

At the engine manufacturer's end, *BP* can be found or determined by the relation :

$$\text{BP} = \frac{2\pi N \times Wr}{1000}$$

where

N = Engine speed rpm

r = Dynamometer arm. m

W = Net dynamometer load

As is true of all prime movers, there are a number of efficiency expressions applied to diesel engines.

$$\text{Mechanical efficiency} = \frac{\text{Brake Power}}{\text{Indicated Power}} = \frac{\text{BP}}{\text{IP}}$$

Indicated thermal efficiency

$$\eta_{thi} = \frac{\text{IP} \times 1000}{m_f \times \text{CV}}$$

where

m_f = mass of fuel used/sec

CV = Calorific value of fuel kJ/kg.

Again, η_{thb} = Brake thermal efficiency

$$= \frac{BP \times 1000}{m_f \times CV}$$

One more term which is very much important for the engine performance is the brake specific fuel consumption.

BSFC. It is defined as the mass of fuel consumption per hour per BP. Units of BSFC is kg/kWH.

The energy supplied to IC engines in the form of Q_h of the fuel input is generally broken down into the following items for heat balance purposes.

A **heat balance** is an accounting of all energy units involved in a system, a cycle, or an individual piece of equipment. It is the book keeping of energy credits and debits with kJ as the unit of currency.

Items for heat balance sheet are as given below :

1. **Useful work.** The actual net shaft output.

2. **Cooling.** Heat absorbed by water jackets. Sometimes this is not separable from some of the friction and exhaust loss due to merging of stream of water that cool cylinders, exhaust manifolds and the lub oil heat exchangers. This is $m_w\ C_{pw}\ \Delta t$ type of loss.

3. **Exhaust gas loss.** Heat carried away as sensible heat and latent heat in the products of combustion sensible heat is a $m_g\ C_{pg}\ \Delta t$ loss; latent heat is of the form gH_2 × Latent heat.

4. **Mechanical friction.** Crankshaft and connective rod bearings, piston etc. are taken to include energy supplied to engine-mounted auxiliaries such as pump, governor etc.

5. **Radiation and unaccounted for.** Heat radiated from engine and incomplete combustion loss due to presence of fuel in the exhaust gases.

Generally item 4 and 5 are not separable and also cannot be measured. There are usually combined and then the balance is made by subtracting items 1, 2 and 3 from the heat of the fuel and considering the difference as **unaccounted**.

Typical full-load heat balances are given below :

	Item	*SI engine (Otto cycle)*	*CI engine (Diesel cycle)*
1.	Useful work	25	34
2.	Cooling	30	30
3.	Exhaust	37	26
4.	Unaccounted	8	10
	Input-fuel heat	100	100

Heat balance sheet (On minute basis)

Input %	*Output %*	
Heat supplied/min.	1. Heat equivalent of BP or shaft work $\dfrac{\text{BP} \times 60}{Q_A} \times 100$	(a)
$= \dfrac{m_f}{\text{min}} \times$ Cal. value 100%	2. Heat given to cooling water	
$= Q_A$	$= \dfrac{\dfrac{m_w}{\text{min}} C_{pw}\ (t_{w_2} - t_{w_1})}{Q_A} \times 100$	(b)
	3. Heat with exhaust gases $= \dfrac{\dfrac{m_g}{\text{min}} \times C_{pg} \times (t_g - t_{room})}{Q_A} \times 100$ (Latent heat neglected)	(c)
	4. Unaccounted for $100 - (a + b + c)$	(d)

23.8 COMPRESSION IGNITION ENGINE PERFORMANCE CURVES

Compression ignition engines are not controlled by throttling but by adjusting the amount of fuel supplied to the engine, and hence are quality governed. When adjusting the fuel supplied to a C.I. engine the limiting condition is given by the smoke limit, which is the appearance of black smoke in the exhaust. Engines should not be operated

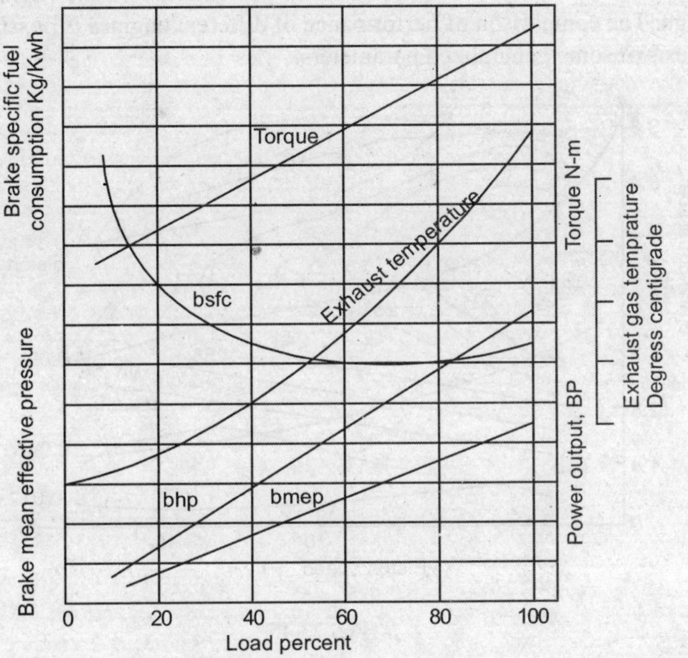

Fig. 23.44

with mixture rich enough to produce smoke, although such a mixture may give more power output. The efficiency under these conditions is low and the engine soon becomes dirty. The smoke limit occurs at air: fuel ratio of about 16 : 1. The engine is tested at different speeds to the smoke limit which can be observed visually or measured by smoke meter. The values of torque, BP, fuel consumption etc. are then plotted against engine speed in rpm.

Figure 23.44 represents the performance curves of a six-cylinder, four-stroke cycle automotive type, CI engine at constant speed. Engines used in the power plants are to run at constant speed because for constant voltage generation by the generator, the generator has to run at constant speed only. And hence the performance curves are plotted at constant speed and variable load on the engine.

By inspection of the BSFC curve it may be observed that the most economical operating point at the particular speed, is in the vicinity of 70% load. By obtaining such families of curve (i.e., one curve for one speed), it is possible to determine the most economical load to be used at any speed throughout the engine speed range. The BSFC curve rises sharply with a decrease in load since the percent heat loss at low loads, and at constant speed is greater than at higher loads which affect the shape of the BSFC curve. At high loads, the BSFC curve rises again. The curve is reasonably flat over a wide range of value of power output which shows that CI engine is better than SI engine. The reduction in thermal efficiency at part load is less for CI engine as compared to SI engine.

Exhaust gas temperature increases with load due to increase in the amount of fuel used and heat loss remain almost constant on percentage basis.

23.9 PERFORMANCE MAPS

The speed characteristics and load characteristics or performance curves give the relationship between the power output, torque, fuel consumption per hour and BSFC covering useful range of speed and power output. These relationships are usually sufficient for the user of a given engine to assess the performance of the engine. However, these performance curves are not suitable for comparing the performance of one engine, with that of an engine of different size and design. The comparison of performance of different engines is possible if the performance data are presented on the basis of some generalized parameters.

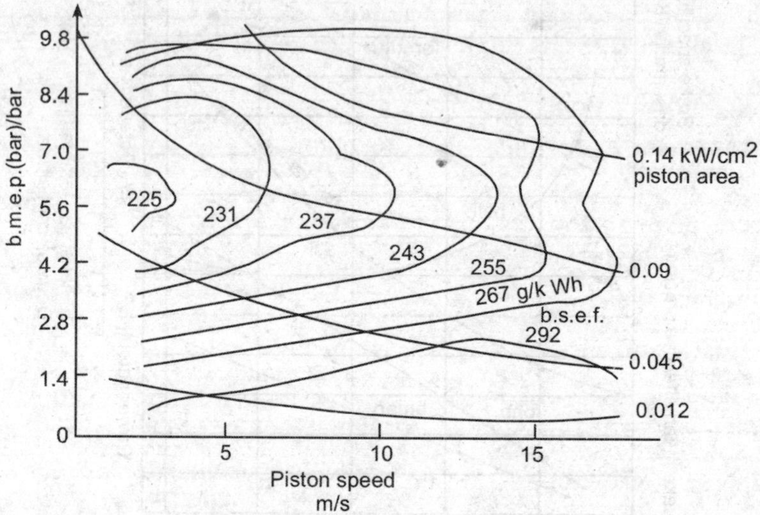

Fig. 23.45

A method has been presented where the performance characteristics are plotted on the basis of bmep (brake mean effective pressure), bsfc (brake specific fuel consumption) and specific power output (power per unit piston area). In this method the data are reasonably independent of cylinder size and are therefore directly comparable between engines of different cylinder size. The effect of change in design of the engine on the performance can also be compared in this method. Figure 23.45 shows the form of performance map for a diesel engine. In such maps bmep is plotted against piston speed and includes the curves of constant specific fuel consumption and power per unit area of piston. In diesel engines the bsfc increases at high load owing to the increased fuel waste (smoke) associated with high fuel air ratios. At lower bmep, the bsfc increases due to reduced mechanical efficiency.

23.10 SELECTION OF SITE FOR A DIESEL POWER PLANT

After knowing about the diesel engines to be used for power generation, their different systems provided, and the testing, it is now time to think about the location of such a plant. Following are the main factors to be considered while selecting a site for a diesel engine power plant.

23.10.1 Load Centre

Diesel engine power plant does not require much of space and water as is required for thermal power plant. Therefore, Diesel engine power plant should be preferably located close to the load centre in order to minimise the cost of transmission lines and losses occurring in the plant.

23.10.2 Availability of Land and Water

Although the space requirement of a diesel engine power plant is less, still reasonable space is definitely required. Particularly when the diesel plants are located at the load centres, the cost of the land may be excessively high.

Similarly, when the cooling tower is used for recirculation of jacket cooling water, provision must be made for the make-up water and for this the source of water must be available.

23.10.3 Foundations

Diesel engines used in power plants are reciprocating engines and there are always unbalanced forces which give rise to severe vibrations. Therefore, good foundations must be provided for these engines. Subsoil conditions at the site selected for the plant should be such that a foundation at the reasonable depth should serve the purpose.

23.10.4 Transportation of Fuel

The cost of transportation of the fuel from the nearest source of fuel should not be high.

23.10.5 Local Conditions

At some locations, the conditions may be such that the operation of the diesel engine power plant may not be safe, for example, in some places where wind flow is always is large and air is always accompanied by dust, and small sand-like particles, the problem of air filteration becomes severe. Also, filteration of fuel oil, lubricating oil may prove to be a costly affair.

Also, the power developed by a diesel engine is affected by the location of the plant (height or altitude from sea level). The power developed by the engine decreases with the increase of height from sea level.

23.10.6 Nuisance for Neighbours

The site selected for a diesel engine plant should not create problem due to noise, smoke, fire hazards etc. for the neighbouring area.

23.11 DIESEL-ELECTRIC POWER PLANT LAYOUT

Before considering the actual layout plans for the diesel-generator sets, we may think of the building which will house these sets. The central electric power station exhibits a great variation of ideas from the purely functional to the impressively refined architecture seen in so many of the modern plants. The buildings should be good-looking

Fig. 23.46 *Typical building of the diesel power plant.*

and it should always be kept in mind that economics cannot always govern. Even from public point of view, the buildings should be shabby but clean and tidy from inside as well as outside. If the power plant building is not to be purely functional, the exterior should receive a treatment appropriate to the field.

Diesel-electric power plants may receive very simple treatment. The auxiliaries to the Diesel engine are few and small in size. No coal piles hinder the landscape treatment. It is possible to locate the oil tanks underground, although the reason for doing so is more often for convenience in unloading from tank-car or protection against the fire hazard than for a desire to hide the tank. If the water for the jackets must be recirculated, a cooling tower or spray pond will, of necessity, be located near the plant. The capacity of the diesel engine plant rarely exceeds 10,000 kW, the design of the building being more or less directly determined by the dimensions of the engine units it houses. The rectangular block shape illustrated in Fig. 23.46 is the usual result.

As the diesel engine power plants require very few auxiliaries beyond those built onto the engine by the manufacturer, these plants are almost the self-contained prime movers. For the stationary type and high capacity factor service in power plants, generally all the systems of the power plants are connected or provided on the location of the power plants. Figure 23.47 shows the usual auxiliary equipments that must be selected and worked into complete plant. These auxiliaries are grouped into systems as :

1. Fuel system
2. Intake and exhaust systems
3. Cooling system
4. Lubrication system and
5. Starting system.

All these systems are considered in the previous sections.

Generally, it is a common practice to set the engine-generator units in a power plant on parallel-centre lines. Again, as the average plant has two or more engines, the parallel arrangements leads to a building of somewhat square shape. There should be ample clearance in the layout for maintenance of the engine, generator and exciter. This clearance must be sufficient for dismantling of these equipments. Sometimes the manufacturers' prints (drawings) indicate the extreme dimensions for removal of rotors, piston etc.

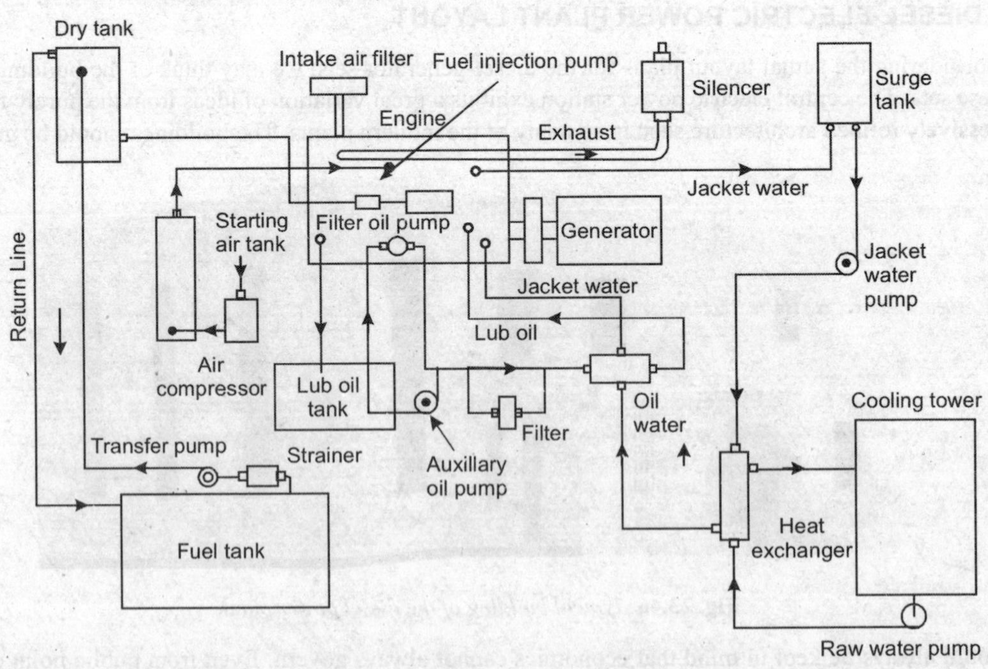

Fig. 23.47 *Diesel engine plant with auxiliary systems.*

It is known that reciprocating masses of the diesel engine cannot be completely balanced and the heavy reciprocating parts of the diesel engine will cause objectionable vibrations and must be taken care of by providing heavy, sufficient foundation. It is always better to have heavy foundation so that no vibration complaints will be present, as nothing can be done about the foundation once poured and the engine is in place.

If the small diesel power plants are required in the hotels, lodges and many other such locations, the vibrations transmitted from the foundations are undesired. In these cases, the foundations are isolated from the rest of the building by insulating by providing corkboard or vibration dampers.

Diesel engines in power station are always directly connected to their generators. Flywheels are provided with jacks for rotating the engine to its starting position and for use when repairing or adjusting.

Voltages preferred for ac generators from about 100 to 600 kW are either 480, 600, or 2400 V. Excitation is, generally, at 125 V dc with exciters being belt-driven to increase speed, thereby reducing physical size.

The relation between number of electrical poles n, the frequency f and the rotating speed N is given by the equation :

$$N = \frac{120 f}{n} \text{ rpm}$$

Figure 23.48 shows a typical layout of a multi-engine Diesel power plant.

Generally, diesel plants have basements excavated which will provide room for piping, conduits, storages, workshops etc.

As far as types of Diesel power plants on the basis of capacity, are concerned, the plant may be a low capacity plant (below 25 MW), or a medium capacity plants (25 to 50 MW) or high capacity plants (above 50 MW).

Depending on the capacity of the plant and the maximum capacity of the single engine (generally 5 MW and more), the layout can be decided. Basically, the engine sets will be layed in parallel as shown in Fig. 23.48. There may be two or more rows for the high capacity power plants. Again according to space available, each row may contain 4 to 8 engine-generator sets and the positions or locations of the storage, workshops will be conveniently decided. For the high capacity plants and where the number of rows are more than the cooling spray pond and the pump house will be located in between the rows. If cooling tower is to be provided instead of spray pond, it will be away from the engine room so that the atmosphere in the vicinity of engine air-intake is not humid.

The switch board should be at the centre on one side of the generating sets.

In short, the plant layout shown in Fig. 23.48 may be extended in both direction and the other auxiliaries required may be arranged accordingly.

If cooling spray ponds are to be used, there are two systems to be employed. One is a single circuit cooling circuit [Fig. 23.48(a)] and a double circuit cooling system. In case of a double circuit cooling circuit one heat exchanger is included as shown in Fig. 23.48(b).

The disadvantage of a single circuit cooling system is that the cylinder jackets may be subjected to corrosion because of the dissolved gases in the cooling water. Such thing is absent in double circuit cooling system.

Present status of diesel power generation in India

Power plant	*City*	*State*	*Fuel*	*MW*
Chennai Vasai	Chennai	Tamil Nadu	Oil	200
White field Ind. Park	Bangalore	Karnataka	Oil	158
Velahanka	Bangalore	Karnataka	Oil	132
Kozhikode	Kozhikode	Kerala	Oil	128
Brahmapuram	Kochi	Kerala	Oil	110
Samayanallur	Samayanallur	Tamil Nadu	Oil	106
Samalpatti	Samalpatti	Tamil Nadu	Oil	105

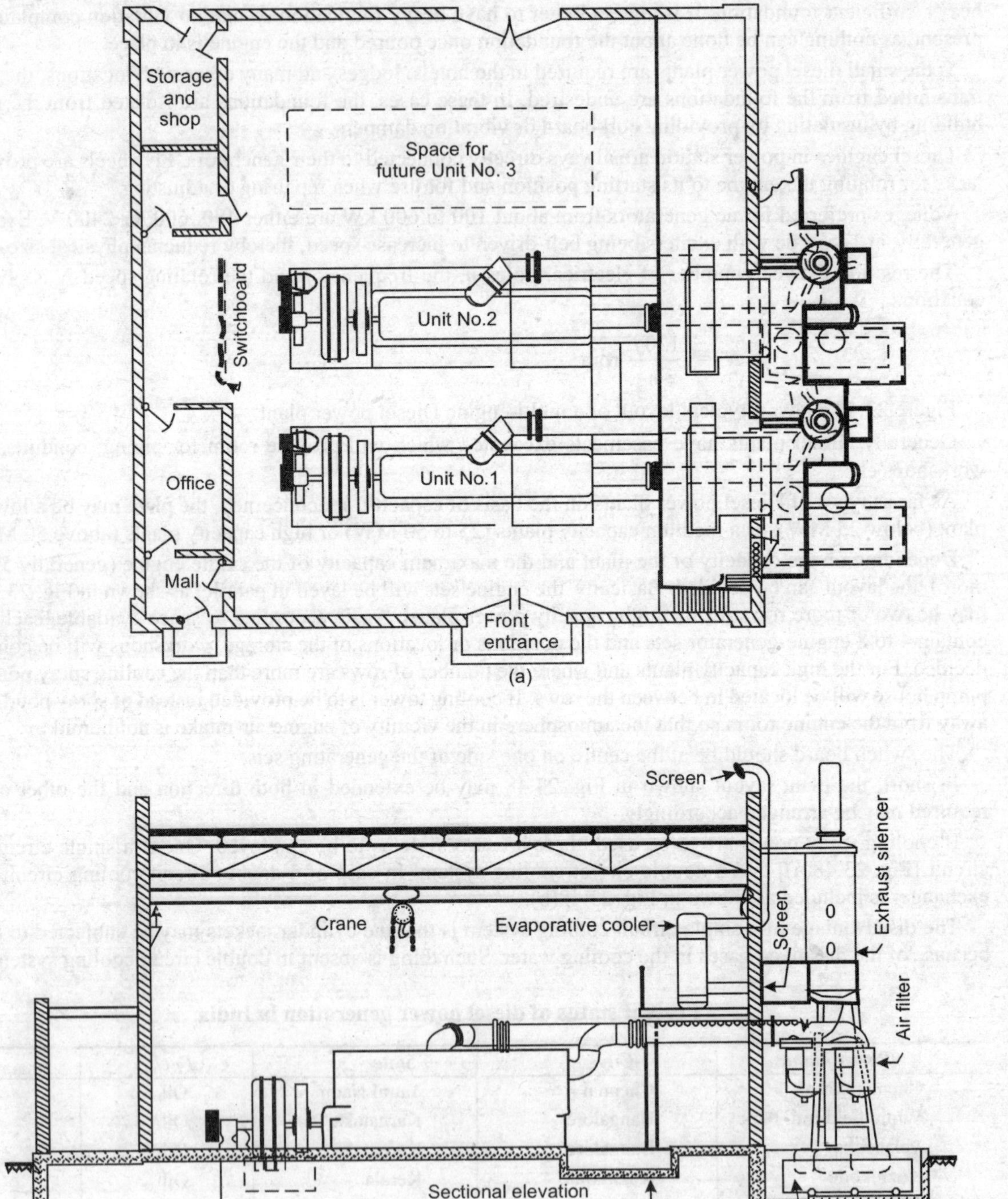

Fig. 23.48 *Typical diesel power plant layout.*

SOLVED EXAMPLES

Example 23.1 *The following observations were made during a test on a two-stroke cycle oil engine :*

Bore 20 cm; Stroke 25 cm; Speed 6 bps

Effective brake drum diameter	*1.2 m*
Net brake load	*440 N*
Indicated m.e.p.	*280 kPa*
Fuel oil consumption	*3.6 kg/hour*
Calorific value of oil	*42500 kJ/kg*
Mass of jacket cooling water	*468 kg/hour*
Rise in temperature of jacket cooling water	*28°C*
Air used	*34 kg/kg of fuel oil*
Temperature of air in test house	*30°C*
Temperature of exhaust gases	*400°C*
Mean specific heat of exhaust gases	*1 kJ/kg/K*

Calculate :

(i) BP

(ii) IP

(iii) Mechanical efficiency

(iv) BMEP

(v) BSFC in kg/kWH

Draw up a heat balance sheet in kJ/min and as percentage of the heat supplied to the engine.

Solution

(i) *Brake Power BP*

$$\text{BP} = \frac{2\pi NT}{1000} = \frac{2\pi \times W \times R \times N}{1000}$$

$$= \frac{2\pi \times 440 \times \frac{1.2}{2} \times 6}{1000}$$

$$= \mathbf{9.948\ kW}$$

(ii) *Indicated Power IP*

$$\text{IP} = p_{mi} \times \frac{LAn}{1000}$$

$$= \frac{280 \times 10^3 \times \frac{25}{100}}{1000} \times \frac{p \times 20 \times 20}{4 \times 10^4} \times \frac{6}{1}$$

$$= \mathbf{13.2\ kW}$$

(iii) *Mechanical Efficiency*

$$\eta_m = \frac{\text{BP}}{\text{IP}} = \frac{9.948}{13.2}$$

$$= 0.7541 = \mathbf{75.4\%}$$

(iv) *Brake Mean Effective Pressure*

$$\text{BMEP} = (\text{Mech. Eff.}) \times \text{IMEP}$$
$$= 0.7541 \times 280$$
$$= \mathbf{211.15\ kPa}$$

(v) *Brake Specific Fuel Consumption (BSFC)*

$$\text{BSFC} = \frac{\text{Fuel consumption/hour}}{\text{Brake Power}}$$
$$= \frac{3.6}{9.948} = \mathbf{0.3616\ kg/kW}$$

Heat Balance Sheet

1. *Heat supplied/min.*

$$= \frac{\text{Fuel used}}{\text{min}} \times \text{Cal. value.}$$
$$= \frac{3.6}{60} \times 42500$$
$$= \mathbf{2550\ kJ/min.\ (100\%)}$$

Heat Out of the System

2. *Heat equivalent of BP.*

$$= \text{BP} \times 60 \text{ kJ/min}$$
$$= 9.948 \times 60$$
$$= \mathbf{596.88\ kJ/min.}$$
$$= \mathbf{23.41\%}$$

3. *Heat given to jacket cooling water/min.*

$$= m_w C_{pw}(t_{w2} - t_{w1})$$
$$= \frac{\text{Mass of water}}{\text{min}} \times \text{Sp. heat of water} \times \text{Rise in temperature}$$
$$= \frac{468}{60} \times 4.187 \times 28$$
$$= \mathbf{914.5\ kJ/min.}$$
$$= \mathbf{35.86\%}$$

4. *Heat carried away by exhaust gases/min.*

$$= \text{Mass of gases/min.} \times \text{Sp. heat of gases} \times (t_{ge} - t_R)$$
$$= m_g C_{pg} \times (t_g - t_R)$$

$$\text{Mass of gases/min.} = \frac{\text{Mass of fuel}}{\text{min}} + \frac{\text{Mass of air}}{\text{min}}$$
$$= \left(\frac{3.6}{60} + 34 \times \frac{3.6}{60}\right)$$

$$= \frac{3.6}{60}(1+34) = \frac{3.6}{60} \times 35$$

$$= \mathbf{2.1\ kg/min.}$$

∴ Heat carried away by exhaust gases/min.

$$= 2.1 \times 1 \times (400 - 30)$$

$$= 2.1 \times 370$$

$$= \mathbf{777\ kJ/min.}$$

$$= \mathbf{30.47\%}$$

5. *Heat unaccounted for (by difference)*

$$= 2550 - (596\text{-}88 + 914.5 + 777)$$

$$= 2550 - 2288.38$$

$$= \mathbf{261.62\ kJ/min.}$$

$$= \mathbf{10.26\%}$$

Heat balance can be given in the tabular form as below :

Heat supplied/min.		*Heat expanded/min.*	
Heat supplied/min.		1. Heat equivalent of BP	
$= \frac{\text{mass of fuel}}{\text{min}} \times \text{C.V.}$		= 596.88 kJ/min.	23.41%
$= \frac{3.6}{60} \times 42500$		2. Heat to cooling water/min.	
= 2550 kJ/min.	100%	= 914.5 kJ/min.	35.86%
		3. Heat carried away by exhaust gases min	
		= 777 kJ/min.	30.47%
		4. Heat unaccounted/min.	
		= 216.62 kJ/min.	10.26%

Example 23.2 *In a trial of a single cylinder oil engine working on dual cycle, the following observations were made :*

Compression ratio	*15*
Oil consumption	*10.2 kg/hr*
Cal. value of fuel	*43890 kJ/kg*
Air consumption	*3.8 kg/min.*
Speed	*1900 rpm*
Torque on brake drum	*186 Nm*
Quantity of cooling water used	*15.5 kg/min.*
Temperature rise of cooling water	*36°C*
Exhaust gas temperature	*410°C*
Room temperature	*20°C*
Sp. heat for exhaust gases	*1.17 kJ/kg/K*

Calculate:

(i) BP

(ii) BSFC

(iii) *Brake thermal efficiency*

Draw up a heat balance on minute basis.

Solution

(i) *Brake Power (BP)*

$$BP = \frac{2\pi NT}{1000}$$

$$= \frac{2\pi \times 1900}{60} \times \frac{186}{1000}$$

$$= \mathbf{37\ kW}$$

(ii) *Brake Specific Fuel Consumption (BSFC)*

$$BSFC = \frac{\text{Fuel used/hour}}{BP} = \frac{m_f}{BP}$$

$$= \frac{10.2}{37}$$

$$= \mathbf{0.2756\ kg/kWh}$$

(iii) *Brake Thermal Efficiency* (η_{BTH})

$$\eta_{BTH} = \frac{BP}{\frac{\text{mass of fuel}}{\text{sec}} \times \text{Cal. value}}$$

$$= \frac{37}{\frac{10.2}{3600} \times 43890}$$

$$= 0.2975$$

$$= \mathbf{29.75\%}$$

Heat Balance Calculations

1. *Heat supplied by fuel/min.*

$$= \frac{10.2}{60} \times 43890$$

$= \mathbf{7461\ kJ/min.}$ **(100%)**

2. *Heat equivalent of BP (useful work)*

$$= BP \times 60$$

$$= 37 \times 60$$

$= \mathbf{2220\ kJ/min.}$ **(29.75%)**

3. Heat carried away by cooling water/min.

$$= \frac{\text{Mass of water}}{\text{min.}} \times \text{Sp. heat} \times \text{Rise in temperature}$$

$$= m_w C_{pw} \times \Delta T$$

$= 15.5 \times 4.187 \times 36$

$=$ **2336.46 kJ/min.** **(31.31%)**

4. *Heat carried away by exhaust gases/min.*

$$= \frac{\text{Mass of gases}}{\text{min.}} \times \text{Sp. heat} \times (t_g - t_R)$$

$$= \left(\frac{\text{mass of air}}{\text{min.}} + \frac{\text{mass of fuel}}{\text{min.}}\right) \times C_{pg} \times (t_g - t_R)$$

$$= \left(3.8 + \frac{10.2}{60}\right) \times 1.17 \times (410 - 20)$$

$$= (3.8 + 0.17) \times 1.17 \times 390$$

$=$ **1911.5 kJ/min.** **(24.27%)**

5. *Heat unaccounted for* (*by difference*)

$= 7461 - (2220 + 2336.46 + 1911.5)$

$= 7461 - 6467.96$

$=$ **994.04 kJ/min.** **(14.67%)**

Heat balance sheet on minute basis

Heat supplied/min	*Heat expenditure*	*kJ/min*	%
$m_f \times$ CV	1. Heat equivalent of BP	2220	29.75
= 7461 kJ/min	2. Heat to cooling water	2336.46	31.31
= 100%	3. Heat with exhaust gases	1911.5	24.27
	4. Heat unaccounted for	994.04	14.67

Example 23.3 *In a constant speed CI engine operating on four-stroke cycle and fitted with band-broke, the following observations were taken :*

Brake wheel diameter	*60 cm*
Band thickness	*5 mm*
Speed	*450 rpm*
Load on band	*210 N*
Spring balance reading	*30 N*
Area of indicator diagram	*4.15 cm²*
Length of indicator diagram	*6.25 cm*
Spring No.	*11 i.e. 11 bar/cm*
Bore	*10 cm*
Stroke	*15 cm*
Brake specific fuel consumption	*0.3 kg/kWh*
Calorific value of fuel	*41800 kJ/kg*

Determine :

(*i*) *BP*

(*ii*) *IP*

(*iii*) *Mechanical efficiency*

(*iv*) *Indicated and brake thermal efficiency*

Solution

(i) *BP* (*Brake Power*)

$$\text{BP} = \frac{2\pi NT}{1000}$$

where N = revolutions per second

$$= \frac{450}{60}$$

$$T = \text{Torque}$$

$$= (W - S) \times R$$

$$= (210 - 30) \times \frac{60 + 0.5}{2 \times 100}$$

$$= \frac{180 \times 60.5}{200}$$

$$\therefore \quad \text{BP} = \frac{2\pi}{1000} \times \frac{450}{60} \times \frac{180 \times 60.5}{200}$$

$$= \mathbf{2.566\ kW}$$

(ii) *IP* (*Indicated Power*)

$$\text{IP} = p_{mi} \frac{LAn}{1000}$$

where $L = \frac{15}{100}$ m, $A = \frac{\pi}{4} \times \left(\frac{10}{100}\right)^2$

n = No. of working cycles/sec.

$$= \frac{450}{2 \times 60}$$

p_{mi} = Actual indicated mean effective pressure

$$= \frac{\text{Area of diagram}}{\text{Length of diagram}} \times \text{Spring number}$$

$$= \frac{4.15}{6.25} \times 11$$

$$= \mathbf{7.3\ bar.}$$

$$\therefore \quad \text{IP} = 7.3 \times \frac{15}{100} \times \frac{\pi}{4} \times 0.01 \times \frac{450}{2 \times 60 \times 1000}$$

$$= \mathbf{3.225\ kW}$$

(iii) *Mechanical efficiency*

$$\eta_m = \frac{\text{BP}}{\text{IP}} = \frac{2.566}{3.225}$$

$$= 0.795 \text{ or } \mathbf{79.5\%}$$

(iv) *Brake thermal efficiency*

$$\eta_{bth} = \frac{\text{BP} \times 3600}{\frac{m_f}{\text{hour}} \times \text{Cal. value}}$$

$$= \frac{3600}{\frac{m_f}{\text{hour}} \times \frac{1}{\text{BP}} \times \text{Cal. value}}$$

$$= \frac{3600}{\text{Specific fuel consumption} \times \text{Cal. value}}$$

$$= \frac{3600}{0.3 \times 41800}$$

$= 0.2849$ or **28.49%**

Similarly, indicated thermal efficiency is given by

$$\eta_{ith} = \frac{\eta_{bth}}{\text{Mech. eff.}} \times \frac{0.2849}{0.795}$$

$= 0.3611$ or **36.11%**

Example 23.4 *The following particulars were obtained in a trial on a single cylinder 4-stroke oil engine :*

Cylinder diameter	*250 mm*
Stroke length	*400 mm*
IMEP	*6.5 bar*
Engine speed	*250 rpm*
Net load on the brake	*1080 N*
Effective diameter of the brake	*1.5 m*
Fuel used	*10 kg/hr*
Calorific value of fuel	*44300 kJ/kg*

Calculate :

(i) IP

(ii) BP

(iii) Mechanical efficiency

(iv) Indicated thermal efficiency

Solution

(i) *Indicated Power (IP)*

$$\text{IP} = p_{mi} \times \frac{L \times A \times n}{1000}$$

where p_{mi} = Indicated mean effective pressure

$= 6.5$ bar

$= 6.5 \times 10^5$ N/m^2

$L = 400$ mm $= 0.4$ m

$$A = \frac{\pi}{4} \times \left(\frac{250}{1000}\right)^2 = \frac{\pi}{4} \times \frac{1}{16} \text{ m}^2$$

$$n = \frac{250}{2 \times 60} = \frac{25}{12}$$

$$\therefore \quad \text{IP} = \frac{6.5 \times 10^5}{1000} \times 0.4 \times \frac{\pi}{4} \times \frac{1}{16} \times \frac{25}{12}$$

$$= \mathbf{26.6 \text{ kW}}$$

(ii) *Brake Power* (*BP*)

$$\text{BP} = \frac{2\pi NT}{1000} = \frac{2\pi N \times \text{Net load } (W - S) \times R}{1000}$$

$$= \frac{\pi DN \times \text{Net load}}{1000}$$

$$= \pi \times 1.5 \times \frac{250}{60} \times 1080$$

$$= \mathbf{21.2 \text{ kW}}$$

(iii) *Mechanical efficiency*

$$\eta_m = \frac{\text{BP}}{\text{IP}} = \frac{21.2}{26.6}$$

$$= 0.797 \text{ or } \mathbf{79.7\%}$$

(iv) *Indicated thermal efficiency*

$$\eta_{ith} = \frac{\text{IP}}{\frac{\text{fuel}}{\text{sec}} \times \text{Cal. value}}$$

$$= \frac{26.6}{\frac{10}{3600} \times 44300}$$

$$= \mathbf{0.216} \text{ or } \mathbf{21.6\%}$$

Example 23.5 *Following data relates to 4 cylinder, 4-stroke CI Engine :*

Air : Fuel ratio	*16 (by weight)*
Calorific value of fuel	*45200 kJ/kg*
Mechanical efficiency	*82%*
Air standard efficiency	*60%*
Relative efficiency based on IP	*70%*
Volumetric efficiency	*78%*
Stroke to Bore ratio L/D	*1.25*
Suction conditions	*1 bar and 25°C*

RPM *2400*

Power at brake *72 kW*

Fuel cut-off ratio *2.2*

Calculate

(i) Compression ratio

(ii) Indicated thermal efficiency

(iii) Brake specific fuel consumption

(iv) Brake and stroke

Solution

(i) *Compression ratio*

Air standard efficiency of a CI engine working on Diesel cycle is given by :

$$\text{ASE} = 1 - \frac{1}{\gamma . r_k^{\gamma-1}}\left[\frac{r_c^{\gamma}-1}{r_c-1}\right]$$

where γ = specific heat ratio

$= 1.4$ for air

r_c = cut-off ratio

$= 2.2$

$$\therefore \quad \text{ASE} = 0.6 = 1 - \frac{1}{1.4 \times r_k^{0.4}}\left[\frac{2.2^{1.4}-1}{2.2-1}\right]$$

$$= 1 - \frac{1}{r_k^{0.4}}\left[\frac{3.016-1}{1.4 \times 1.2}\right]$$

$$= 1 - \frac{1}{r_k^{0.4}}\left[\frac{2.6}{1.68}\right]$$

$$\therefore \quad 0.4 = \frac{1}{r_k^{0.4}} \times \frac{2.016}{1.68}$$

$$\therefore \quad r_k^{0.4} = \frac{1}{0.4} \times \frac{2.016}{1.68} = \frac{2.016}{0.672} = 3.87$$

$$r_k = (3.87)^{1/0.4} = (3.87)^{2.5}$$

$$= (3)^{2.5}$$

$$= \mathbf{15.588} \simeq 16$$

(ii) *Indicated thermal efficiency*

BP = Power at brake

$= 72$ kW

Mech. eff. $= 82\%$

$$= \frac{\text{BP}}{\text{IP}}$$

$$\therefore \quad IP = \text{Indicated Power}$$

$$= \frac{72}{0.82}$$

$$= \mathbf{87.8\ kW}$$

Heat supplied/sec

$$= \frac{\text{mass of fuel}}{\text{sec}} \times \text{Cal. value}$$

Also, Relative efficiency on IP basis

$$= 0.7$$

$$= \frac{\text{Indicated thermal efficiency}}{\text{Air standard efficiency}}$$

$\therefore$ Indicated thermal efficiency = 0.7 × 0.6

$$= \mathbf{0.42}$$

$$\therefore \quad = \frac{IP}{\frac{\text{mass of fuel}}{\text{sec}} \times \text{Cal. value}}$$

$$\text{Fuel used/sec} = \frac{IP}{0.42 \times 45200}$$

$$= \frac{87.8}{0.42 \times 45200}$$

$$= 0.004625 \text{ kg/sec}$$

$$= \mathbf{16.65\ kg/h}$$

(iii) *Brake specific fuel consumption*

$$BSFC = \frac{\text{Fuel used/hour}}{BP}$$

$$= \frac{16.65}{72}$$

$$= \mathbf{0.23126\ kg/kWh}$$

(iv) *Bore and stroke*

Air to fuel ratio = 16

$\therefore$ Weight of air/hour = 16 × 16.65

= 266.4 kg/h

= 4.44 kg/min.

$$\text{Volume of this air/min} = V = \frac{mRT}{P}$$

$$= \frac{4.44 \times 287 \times 298}{1 \times 10^5}$$

$$= \mathbf{3.7973\ cu.\ m/min.}$$

This is the actual volume of air supplied.

But volumetric efficiency is given as

$$\eta_{voi} = \frac{\text{Actual volume}}{\text{Stroke volume}}$$

$$\therefore \quad 0.78 = \frac{3.7973}{\text{Stroke volume/min.}}$$

$$\therefore \quad \text{Stroke volume/min.} = \frac{3.7973}{0.78}$$

$$= \mathbf{4.8684\ m^3/min.}$$

$$\text{Stroke volume/min.} = \frac{\pi}{4} d^2 \times L \times N$$

$$= \frac{\pi}{4} \times \frac{d^3 \times d \times 1.25 \times 2400}{1}$$

$$= \frac{\pi}{4} \times 3000 d^3$$

$$\therefore \quad d^3 = \frac{4.8684 \times 4}{\pi \times 3000}$$

$$= \mathbf{0.002065}$$

$$= 2.065 \times 10^{-3}$$

$$\therefore \quad d = 10^{-1} \times 1.274 = 0.1274 \text{ m}$$

$$= \mathbf{12.74\ cm}$$

$$L = \text{Stroke} = \mathbf{16\ cm}$$

Example 23.6 *A laboratory test on a single cylinder 4-stroke diesel engine gave the following results while running on full load :*

Area of indicator card	*300 mm²*
Length of diagram	*40 mm*
Spring constant	*1 bar/mm*
Speed of the engine	*400 rpm*
Load on the brake	*370 N*
Spring balance reading	*50 N*
Diameter of brake drum	*1.2 m*
Fuel consumption	*2.8 kg/hr*
Calorific value of fuel	*41800 kJ/kg*
Diameter of cylinder	*160 mm*
Stroke of the piston	*200 mm*

Calculate :

(i) IMEP and IP

(ii) BP

(*iii*) *BMEP*

(*iv*) *BSFC and Brake thermal efficiency*

Solution

(i) *IMEP and IP*

Indicated mean effective pressure is given by

$$\text{IMEP} = \frac{\text{Area of indicator card}}{\text{Length of indicator diagram}} \times \text{Spring No.}$$

$$p_{mi} = \frac{300}{40} \times 1 = 7.5 \text{ bar}$$

$$= \mathbf{750\ kPa}$$

$\therefore$ $\text{IP} = \text{Indicated Power}$

$$= p_{mi} \times \frac{L \times A \times n}{1000}$$

where L = Piston stroke m

$$= \frac{200}{1000} = 0.2 \text{ m}$$

A = Area of cylinder m^2

$$= \frac{\pi D^2}{4} = \frac{\pi}{4} \times \left(\frac{160}{1000}\right)^2 \text{ m}^2$$

n = No. of working cycles/sec

$$= \frac{N}{60 \times 2} = \frac{400}{60 \times 2} = \frac{10}{3}$$

$$\therefore \quad \text{IP} = \frac{7.5 \times 10^5}{1000} \times 0.2 \times \frac{\pi}{4} \times \left(\frac{160}{1000}\right)^2 \times \frac{10}{3}$$

$$= 750 \times \frac{2}{3} \times \frac{\pi}{4} \times \frac{256}{10000}$$

$$= \mathbf{10.57\ kW}$$

(ii) *Brake Power BP*

$$\text{BP} = \frac{2\pi NT}{1000} = \frac{2\pi N(W - S)}{1000}$$

$$= \frac{\pi dN(W - S)}{1000}$$

$$= \pi \times 1.2 \times \frac{400}{60} \times \frac{(370 - 50)}{1000}$$

$$= \frac{\pi \times 8 \times 320}{1000}$$

$$= \mathbf{8.05\ kW}$$

(iii) *Brake mean effective pressure (BMEP)*

On the lines of indicated mean effective pressure brake mean effective pressure is written as

$$p_{bi} = \text{BMEP} = \frac{\text{BP} \times 1000 \times 10^4}{LAn - 10^5} \text{ where } p_{bi} \text{ is in bar.}$$

$$= \frac{8.05 \times 1000}{0.2 \times \frac{\pi}{4} \times \left(\frac{160}{1000}\right)^2 \times \frac{10}{3}}$$

$$= \mathbf{6\ bar\ or\ 600\ kPa}$$

(iv) *Brake specific fuel consumption and brake thermal efficiency*

$$\text{BSFC} = \frac{\text{Wt. of fuel/hour}}{\text{BP}}$$

$$= \frac{2.8}{8.05}$$

$$= \mathbf{0.348\ kg/kWh}$$

Brake thermal efficiency is given by

$$\eta_{bth} = \frac{\text{BP} \times 3600}{\frac{\text{mass of fuel}}{\text{hour}} \times \text{Cal. value}}$$

$$= \frac{3600}{\text{BSFC} \times \text{Cal. val.}}$$

$$= \frac{3600}{0.348 \times 41800}$$

$$= 0.2472$$

$$= \mathbf{24.72\%}$$

Example 23.7 *The following data were recorded during a trial on a 4-stroke diesel engine :*

No. of cylinders	*4*
Diameter of piston	*10 cm*
Stroke	*15 cm*
IMEP	*0.67 MPa*
Speed	*2000 rpm*
No. of explosions	*980 per min.*
Brake torque	*181.5 N-m.*
Fuel consumption	*11.89 kg/hour*
Calorific value of fuel	*41800 kJ/kg*

Relative efficiency on BP basis	*0.5*
Quantity of jacket water	*1020 kg/hr*
Rise in temperature of cooling water	*35°C*
Sp. heat of cooling water	*4.18 kJ/kg/K*

Find :

(i) Mechanical efficiency

(ii) Brake thermal efficiency

(iii) Air standard efficiency

(iv) BSFC

(v) % heat loss to jacket cooling water.

Solution

(i) *Mechanical efficiency*

$$\text{Brake Power BP} = \frac{2\pi NT}{1000}$$

$$= 2\pi \times \frac{2000}{60} \times \frac{181.5}{1000}$$

$$= \mathbf{37.994\ kW}$$

$$\text{Indicated Power} = \text{IP}$$

$$= p_{mi}\frac{LAn}{1000} \times \text{No. of cylinders}$$

$$= \frac{0.67 \times 10^6}{1000} \times \frac{15}{100} \times \frac{\pi}{4} \times \left(\frac{10}{100}\right)^2 \times \frac{980}{60} \times 4$$

$$= \mathbf{51.54\ kW}$$

∴ Mechanical efficiency is given by

$$\eta_m = \frac{\text{BP}}{\text{IP}}$$

$$= \frac{37.994}{51.54}$$

$$= \mathbf{0.7372 \text{ or } 73.72\%}$$

(ii) *Brake thermal efficiency*

$$\eta_{bth} = \frac{\text{BP}}{\frac{\text{mass of fuel}}{\text{sec}} \times \text{Cal. value}}$$

$$= \frac{37.994}{\frac{11.89}{3600} \times 41800}$$

$$= \frac{37.994 \times 3600}{11.89 \times 41800}$$

$$= 0.2752$$
$$= \mathbf{27.52\%}$$

(iii) *Air standard efficiency*

Relative efficiency on BP basis

$$= \frac{\text{Brake thermal efficiency}}{\text{Air standard efficiency}}$$

$$= \frac{\eta_{bth}}{\text{ASE}}$$

$$0.5 = \frac{0.2752}{\text{ASE}}$$

$$\therefore \quad \text{ASE} = \frac{0.2752}{0.5}$$
$$= 0.5504$$
$$= \mathbf{55.04\%}$$

(iv) *Brake specific fuel consumption*

$$\text{BSFC} = \frac{\text{BP}}{\text{Mass of fuel/hour}}$$

$$= \frac{0.2752}{11.89}$$

$$= \mathbf{0.3129\ kg/kWh}$$

(v) *% heat to jacket cooling water*

Heat supplied by fuel/hour

$$= \frac{\text{Mass of fuel}}{\text{hour}} \times \text{Calorific value}$$
$$= 11.89 \times 41800$$
$$= \mathbf{497002\ kJ/hour}$$

Heat to jacket cooling water

$$= m_w \times C_{pw} \times \text{Rise in temperature}$$
$$= 1020 \times 4.18 \times 35$$
$$= \mathbf{149226\ kJ/hr}$$

∴ % heat given to jacket cooling water

$$= \frac{\text{Heat to cooling water}}{\text{Heat supplied by fuel}} \times 100$$

$$= \frac{149226}{497002} \times 100$$

$$= 0.30025 \times 100 = \mathbf{30.025\%}$$

Example 23.8 *In a test on a single cylinder oil engine with 30 cm bore and 45 cm stroke and working on 4-stroke cycle, the following observations were made :*

Duration of trial	*1 hour*
Total fuel consumption	*7.6 kg*
Calorific value of fuel	*45000 kJ/kg*
Total revolutions made	*12000*
Room temperature	*20°C*
MEP	*6 bar*
Net brake load	*150 kg*
Brake drum diameter	*180 cm*
Rope diameter	*3 cm*
Mass of jacket cooling water	*550 kg*
Inlet temperature of jacket cooling water	*15°C*
Exit temperature of jacket cooling water	*60°C*
Temperature of exhaust gases	*300°C*
Total air consumption	*365 kg*
Sp. heat of exhaust gases	*1 kJ/kg/K*

Calculate :

(i) Indicated Power

(ii) Brake Power

(iii) Mechanical Efficiency

(iv) Indicated thermal efficiency

Draw up a heat balance sheet on minute basis.

Solution

(i) *Indicated Power*

Indicated Power is given by

$$\text{IP} = p_{mi}\frac{LAn}{1000} \text{ where } p_{mi} = \text{IMEP}$$

$$= 6 \text{ bar}$$

$$= 6 \times 10^5 \text{ N/m}^2$$

$$L = \text{Piston stroke } m$$

$$= 45 \text{ cm} = 0.45 \text{ m}$$

$$A = \text{Area of cylinder m}^2$$

$$= \frac{\pi}{4}\times\left(\frac{30}{100}\right)^2 = \frac{\pi}{4}\times 0.09$$

$$n = \text{Effective power cycle/sec.}$$

$$= \frac{12000}{3600\times 2} = \frac{60}{36}$$

$$\therefore \quad \text{IP} = \frac{6\times10^5}{1000}\times\frac{0.45}{1}\times\frac{\pi}{4}\times 0.09\times\frac{60}{36}$$

$$= \mathbf{31.8\ kW}$$

(ii) *Brake Power*

Brake Power is given by

$$BP = \frac{2\pi NT}{1000} \text{ where } N = \text{Revolutions per sec}$$

$$= \frac{12000}{3600} = \frac{120}{36}$$

$$T = \text{Torque}$$

$$= \text{Net load} \times \text{Brake arm}$$

$$= 150 \text{ kg} \times \frac{(D+d)}{2}$$

$$= 150 \times 9.81\ N \times \frac{180+3}{2 \times 100}$$

$$\therefore \quad BP = 2\pi \times \frac{120}{36} \times \frac{(150 \times 9.81) \times 183}{1000 \times 100 \times 2}$$

$$= \mathbf{28.1\ kW}$$

(iii) *Mechanical efficiency*

Mechanical efficiency is given by

$$\eta_m = \frac{\text{Brake Power}}{\text{Indicated Power}}$$

$$= \frac{BP}{IP}$$

$$= \frac{28.1}{31.8}$$

$$= 0.883$$

$$= \mathbf{88.3\%}$$

(iv) *Indicated thermal efficiency*

$$\eta_{ith} = \frac{IP}{\frac{\text{mass of fuel}}{\text{sec}} \times \text{Cal. value}}$$

$$= \frac{31.8}{\frac{7.6}{3600} \times 45000}$$

$$= \frac{31.8 \times 4}{7.6 \times 50}$$

$$= 0.355 = \mathbf{35.5\%}$$

Heat balance sheet on minute basis

1. *Heat supplied/min.*

$$= \frac{7.6}{60} \times 45000$$

$$= \mathbf{5700\ kJ/min.} \qquad \mathbf{(100\%)}$$

2. *Heat equivalent of BP*

$$= BP \times 60$$

$$= 28.1 \times 60$$

$$= \mathbf{1686\ kJ/min.} \qquad \mathbf{(29.58\%)}$$

3. *Heat to jacket cooling water/min.*

$$= m_w C_{pw} \Delta T$$

$$= \frac{550}{60} \times 4.187 \times (60 - 15)$$

$$= \mathbf{1725\ kJ/min.} \qquad \mathbf{(30.3\%)}$$

4. *Heat carried away by exhaust gases/min*

$$= \frac{\text{Mass of gases}}{\text{min}} \times \text{Sp. heat of gases} \times (t_{gas} - t_R)$$

where mass of gases/min.

$$= \frac{\text{Mass of air}}{\text{min}} + \frac{\text{Mass of fuel}}{\text{min}}$$

$$= \frac{365}{60} + \frac{7.6}{60} = \frac{372.6}{60}\ \text{kg}$$

$$\text{Sp. heat} = 1\ \text{kJ/kg/K}$$

$$t_g = \text{temperature of exhaust gases}$$

$$= 300°\text{C}$$

$$t_R = \text{room temperature}$$

$$= 20°\text{C}$$

∴ Heat carried away by exhaust gases

$$= \frac{372.6}{60} \times 1 \times (300 - 20)$$

$$= \frac{372.6 \times 280}{60}$$

$$= \mathbf{1739\ kJ/min.} \qquad \mathbf{(30.5\%)}$$

5. *Heat unaccounted for*

$$= 5700 - (1686 + 1725 + 1739)$$

$$= 5700 - 5150$$

$$= \mathbf{550\ kJ/min.} \qquad \mathbf{(9.62\%)}$$

In the table form, the balance sheet is as given below

Heat supply	*Heat expenditure*	*kJ/min*	*Percentage*
Heat supplied/min	1. Heat equivalent of BP	1686	29.58
= 5700 kJ/min	2. Heat to cooling water	1725	30.3
= 100%	3. Heat carried away by exhaust gases	1739	30.5
	4. Heat unaccounted for	550	9.62
	Total	5700	100.00

Example 23.9 *The following observations were taken during a test on a single cylinder low speed four-stroke cycle oil engine having a bore of 30 cm and stroke of 45 cm.*

Ambient temperature	*20°C*
Duration of trial	*1 hour*
Calorific value of fuel	*42000 kJ/kg*
Total fuel consumption	*11.4 kg/hour*
rpm	*300*
IMEP	*6 bar*
Brake drum diameter	*1.8 m*
Brake rope diameter	*2 cm*
Quantity of jacket cooling water	*660 kg/hr*
Temperature of entering cooling water	*20°C*
Temperature of leaving cooling water	*75°C*
Quantity of air consumed	*250 kg/hr*
Sp. heat of exhaust gases	*1 kJ/kg/K*
Exhaust gas temperature	*420°C*

Determine :

(i) Indicated thermal efficiency

(ii) Mechanical efficiency

(iii) Draw up a heat balance sheet on hour basis.

Solution

(i) *Indicated thermal efficiency*

$$= \frac{\text{IP}}{\frac{\text{Mass of fuel}}{\text{sec}} \times \text{Cal. value}}$$

But IP = Indicated Power

$$= p_{mi} \frac{LAn}{1000}$$

where p_{mi} = Indicated mean effective pressure

$= 6 \text{ bar} = 6 \times 10^5 \text{ N/m}^2$

L = Piston stroke m

$= 45 \text{ cm} = 0.45 \text{ m}$

A = Area of cylinder m^2

$$= \frac{\pi}{4} \times \left(\frac{30}{100}\right)^2$$

$$= \frac{\pi}{4} \times 0.09 \text{ m}^2$$

n = Effective power strokes/sec

$$= \frac{N}{2 \times 60} = \frac{300}{2 \times 60}$$

$$= \mathbf{2.5}$$

$$\therefore \quad \text{IP} = \frac{6 \times 10^5}{1000} \times \frac{0.45}{1} \times \frac{\pi}{4} \times 0.09 \times 2.5$$

$$= \mathbf{47.73 \text{ kW}}$$

$$\frac{\text{Mass of fuel}}{\text{sec}} = \frac{11.4}{3600}$$

$\therefore$ Indicated thermal efficiency

$$= \frac{47.73}{11.4/3600 \times 42000}$$

$$= 0.3589$$

$$= \mathbf{35.89\%}$$

(ii) *Mechanical efficiency*

$$\eta_m = \frac{\text{Brake Power}}{\text{Indicated Power}}$$

$$= \frac{\text{BP}}{\text{IP}}$$

We have to calculate the brake power

$$\text{BP} = \frac{2\pi NT}{1000}$$

where N = rps

$$= \frac{300}{60} = 5$$

T = Torque on the shaft

= Net load × Brake arm

$$= 1.5 \times \left(\frac{180+2}{100 \times 2}\right)$$

$$= \mathbf{1.5 \times 0.91 \text{ N-m.}}$$

$$\therefore \quad \text{BP} = 2\pi \times 5 \times 1.5 \times 0.91$$

$$= \mathbf{42.9 \text{ kW}}$$

∴ Mechanical efficiency

$$= \frac{42.9}{47.73}$$

$= \mathbf{0.8988}$ or **89.88%**

(iii) *Heat balance on hour basis*

1. **Heat supplied/hour**

$$= \frac{\text{mass of fuel}}{\text{hour}} \times \text{Cal. value}$$

$= 11.4 \times 42000$

$= \mathbf{478.8 \times 10^3}$ **kJ/hour (100%)**

2. **Heat equivalent of BP (useful work)**

$= \text{BP} \times 3600$

$= 42.9 \times 3600$

$= \mathbf{154.44 \times 10^3}$ **kJ/hour (32.25%)**

3. **Heat equivalent of Friction Power (FP)**

$= \text{FP} \times 3600$

$= (\text{IP} - \text{BP}) \times 3600$

$= (47.73 - 42.9) \times 3600$

$= 4.83 \times 3600$

$= \mathbf{17.388 \times 10^3}$ **kJ/hour (3.63%)**

4. **Heat to jacket water/hour**

$$= \frac{\text{Mass of water}}{\text{hr}} \times \text{Sp. heat} \times \text{Rise in temperature.}$$

$= (660) \times 4.187 \times (75 - 20)$

$= 148.358 \times 10^3$ kJ/hour **(30.98%)**

5. **Heat carried away by exhaust gases/hour**

Mass of exhaust gas per hour

$$= \frac{\text{Mass of air}}{\text{hour}} + \frac{\text{Fuel}}{\text{hour}}$$

$= 250 + 11.4$

$= \mathbf{261.4}$ **kg/hour**

∴ Heat carried away by exhaust gases/hour

$= m_g C_{pg}(t_g - t_R)$

$= 261.4 \times 1 \times (420 - 20)$

$= 261.4 \times 1 \times 400$

$= \mathbf{104.56 \times 10^3}$ **kJ/hour (21.84%)**

6. **Heat unaccounted for (by difference)**

$= 478.8 \times 10^3 - 10^3\,[154.44 + 17.388 + 148.358 + 104.56]$

$= (478 - 424.746) \times 10^3$

$= 63.254 \times 10^3 = \mathbf{63254}$ **kJ/hour (11.3%)**

Example 23.10 *The following readings were obtained during a test on a single cylinder 4-stroke cycle oil engine :*

Cylinder bore	*20*
Stroke length	*35*
IMEP	*700 kN/m²*
Engine speed	*240 rpm*
Fuel used	*3.5 kg/hr*
Calorific value of fuel	*46000 kJ/kg*
Brake torque	*450 N-m*
Mass of jacket cooling water	*5 kg/min*
Rise in temperature of jacket cooling water	*40°C*
Mass of air supplied	*1.35 kg/min*
Temperature of exhaust gases	*340°C*
Room temperature	*15°C*
Mean sp. heat of dry exhaust gases	*1 kJ/kg/K*
Hydrogen in fuel	*13.5% by weight*
C_p of steam in exhaust gases	*2.3 kJ/kg/K*

Calculate :

(i) Mechanical efficiency

(ii) Indicated thermal efficiency

Also draw up a heat balance sheet in kJ/min and as a percentage of the heat supplied to the engine.

Solution

(i) *Mechanical efficiency*

$$\text{Mechanical efficiency} = \frac{BP}{IP}.$$

where BP = Brake Power

$$= \frac{2\pi NT}{1000}$$

$$= 2\pi \times \frac{240}{60} \times \frac{450}{1000}$$

$$= \mathbf{11.304\ kW}$$

and IP = Indicated Power

$$= p_{mi}\,\frac{LAn}{1000}.$$

$$= \frac{700 \times 1000}{1000} \times \frac{35}{100} \times \frac{\pi}{4} \times \left(\frac{20}{100}\right)^2 \times \frac{240}{60 \times 2}$$

$$= 490 \times \frac{\pi}{4} \times 0.04$$

$$= \mathbf{15.4\ kW}$$

∴ Mechanical efficiency is given by

$$\eta_m = \frac{\text{BP}}{\text{IP}}$$

$$= \frac{11.304}{15.4}$$

$$= 0.734$$

$$= \mathbf{73.4\ \%}$$

(ii) *Indicated thermal efficiency*

$$\eta_{ith} = \text{Indicated thermal efficiency}$$

$$= \frac{\text{IP}}{\frac{\text{Mass of fuel}}{\text{sec.}} \times \text{Cal. value}}$$

$$= \frac{15.4}{\frac{3.5}{3600} \times 46000}$$

$$= \frac{15.4 \times 3600}{3.5 \times 46000}$$

$$= 0.3443$$

$$= \mathbf{34.43\%}$$

Calculations for Heat Balance Sheet

1. *Heat supplied/min.*

$$= \frac{\text{Mass of fuel}}{\text{min.}} \times \text{Cal. value}$$

$$= \frac{3.5}{60} \times 46000$$

$$= \mathbf{2683.3\ kJ/min} \qquad \mathbf{(100\%)}$$

2. *Heat equivalent of BP (useful work)*

$$= \text{BP} \times 60$$

$$= 11.304 \times 60$$

$$= \mathbf{678.24\ kJ/min} \qquad \mathbf{(25.28\%)}$$

3. *Heat lost to jacket cooling water/min.*

$$= \frac{\text{Mass of water}}{\text{min.}} \times \text{sp. heat} \times \text{Rise in temperature}$$

$$= m_w C_{pw} \times \Delta T$$

$$= 5 \times 4.187 \times 40$$

$$= \mathbf{837.4\ kJ/min} \qquad \mathbf{(31.2\%)}$$

4. *Heat carried away by exhaust gases/min.*

This item consists of dry exhaust gases and superheated vapours.

Mass of total exhaust gases/min.

$$= \frac{\text{Mass of air}}{\text{min}} + \frac{\text{Fuel}}{\text{min}}$$

$$= 1.35 + \frac{3.5}{60}$$

$$= \mathbf{1.4083\ kg/min}$$

In this, water present = 9 H_2

$$= 9 \times \left(\frac{13.5}{100} \times \frac{3.5}{60}\right)$$

$$= 9 \times 0.007875$$

$$= \mathbf{0.0709\ kg/min}$$

$\therefore$ Mass of dry gas/min $= 1.4083 - 0.0709$

$$= \mathbf{1.3374\ kg}$$

Therefore, heat carried away by dry exhaust gases/min

$$= \frac{\text{Mass of dry gases}}{\text{min}} \times \text{Sp. heat} \times (t_g - t_R)$$

$$= 1.3374 \times 1 \times (340 - 15)$$

$$= \mathbf{434.655\ kJ/min}$$

To find the heat carried away by steam in exhaust gases, generally we take 0.02 bar as the partial pressure of steam vapors in the gases.

$\therefore$ Enthalpy of one kg of steam vapor in the gases

$$= h_g + C_{pv}(t_g - t_{sat}) - h_{fR}$$

where

h_{fr} = sensible heat at room temperature = 62.9 kJ/kg

h_g = enthalpy of saturated steam at 0.02 bar = **2533.5 kJ/kg**

C_{pv} = Sp. heat of vapour = 2.3 kJ/kg/K

t_g = temperature of gases = 340°C

t_{sat} = 17.03°C

$\therefore$ Enthalpy of 1 kg steam in exhaust gases to atmosphere

$$= 2533.5 + 2.3\,(340 - 17.03) - 62.9$$

$$= 2533.5 + 2.3 \times 322.97 - 62.9$$

$$= \mathbf{3213.43\ kJ/kg}$$

Mass of vapor formed/min.

$$= 0.0709\ \text{kg/min}$$

$\therefore$ Heat carried by steam vapor/min

$$= 0.0709 \times 3213.43$$

$$= \mathbf{227.83\ kJ/min}$$

$\therefore$ Total heat carried by exhaust gases/min.

= (Heat dry gas + Heat in vapors)/min.

= 434.655 + 227.83

= **662.485 kJ/min** **(24.69%)**

5. Heat unaccounted for (by difference)

= 2683.3 – (678.24 + 837.4 + 662.485)

= 2683.3 – 2178.125

= **505.175 kJ/min** **(18.83%)**

Heat balance sheet on minute basis

Item	*kJ*	*%*	*Output*	*kJ*	*%*
Heat supplied per min.	2683.3	100	1. Heat equivalent to BP	678.24	25.28
			2. Heat to cooling water	837.4	31.20
			3. Heat in exhaust gases	662.485	24.69
			4. Heat unaccounted (by difference)	505.175	18.83
			Total	2683.3	100.00

Example 23.11 *During a trial of a single cylinder 4-stroke diesel engine the following observations were recorded :*

Bore = 340 mm, Stroke = 440 mm

RPM = 400

Area of indicator diagram	*465 mm*^{2}
Length of the diagram	*60 mm*
Spring constant	*0.6 bar/mm*
Load on hydraulic dynamometer	*950 N*
Dynamometer constant	*7460*
Fuel used	*10.6 kg/hour*
Calorific value of fuel	*49500 kJ/kg*
Cooling water circulated	*25 kg/min*
Rise in temperature of cooling water	*25°C*

The mass analysis of fuel is Carbon = 84%

Hydrogen = 14%

Incombustible = 1%

The volume analysis of exhaust gases is

$CO_2 = 9\%$

$O_2 = 10\%$

$N_2 = 81\%$

Temperature of exhaust gas	*400°C*
Sp. heat of exhaust gases	*1.05 kJ/kg/K*
Ambient temperature	*25°C*
Partial pressure of steam in exhaust gases	*0.03 bar*

Specific heat of superheated steam   *2.1 kJ/kg/K*

Draw up heat balance sheet on minute basis.

Solution

For getting a heat balance sheet, first we will have to calculate the brake power, indicated power and the mass of exhaust gases produced.

First we will find indicated power. For that we will determine indicated mean effective pressure.

$$\therefore \quad \text{IMEP} = p_{mi} = \frac{\text{Area of diagram}}{\text{Length of diagram}} \times \text{Spring No.}$$

$$= \frac{465}{60} \times 0.6$$

$$= \mathbf{4.65\ bar}$$

$$\therefore \quad \text{IP} = p_{mi}\frac{LAn}{1000}$$

$$= \frac{4.65\times10^5}{1000} \times \frac{440}{1000} \times \frac{\pi}{4} \times \left(\frac{340}{1000}\right)^2 \times \frac{400}{60\times2}$$

$$= 465 \times 0.440 \times \frac{\pi}{4} \times 0.1156 \times \frac{20}{6}$$

$$= \mathbf{61.9\ kW}$$

Similarly, BP can be calculated.

$$\text{BP} = \frac{\text{WN}}{\text{Dynamometer constant}}$$

$$= \frac{950\times400}{7460}$$

$$= 50.9\ \text{kW}$$

$$\therefore \quad \text{FP} = \text{Friction power}$$

$$= \text{IP} - \text{BP}$$

$$= 61.9 - 50.9$$

$$= \mathbf{11\ kW}$$

To find the weight of air supplied per kg of fuel, we use the relation

$$W_a = \text{Wt. of air supplied/kg fuel}$$

$$= \frac{NC}{33(\text{CO}_2+\text{CO})}$$

where N = Percentage of nitrogen in gases

$= 81$

C = Percentage of carbon in fuel

$= 84\%$

CO_2 = Percentage of CO_2 in gases

$= 9\%$

CO = Percentage of CO in gases

= Not given i.e. 0%

$$\therefore \quad W_a = \frac{81 \times 84}{33 \times 9}$$

$= \mathbf{22.9\ kg}$

$\therefore$ Weight of gases produced per kg of fuel

$= 22.9 + 1 = \mathbf{23.9\ kg}$

Now we will have the calculations for the heat balance sheet

1. *Heat supplied/min*

= Wt. of fuel/min. × Cal. value

$$= \frac{10.6}{60} \times 49500$$

= **8745 kJ/min** **(100%)**

2. *Heat equivalent of BP (useful work)*

= BP × 60

= 50.9 × 60

= **3054 kJ/min** **(34.92%)**

3. *Heat equivalent of FP*

= FP × 60

= 11 × 60

= **660 kJ/min** **(7.55%)**

4. Heat carried away by cooling water/min.

$$= \frac{\text{Wt. of water}}{\text{min}} \times \text{Sp. heat} \times \text{Rise in temperature}$$

= 25 × 4.187 × 25

= **2612.5 kJ/min** **(29.87%)**

5. Heat carried away by exhaust gases/min

= (Heat carried away by dry gases/min) + (Heat carried away by steam/min)

+ (Wt. of steam/min) × (Heat in 1 kg)

Wt. of steam produced/min

$$= 9H \times \frac{10.6}{60}$$

$$= 9 \times 0.15 \times \frac{10.6}{60}$$

= **0.2385 kg/min**

Partial pressure of steam in the gases

= 0.03 bar

$\therefore$ Heat in steam in exhaust gases/min

$= 0.2385\ [h_g + C_{pv}\ (t_{sup} - t_{sat}) - h_{f_{atm}})$

Here the saturation temperature corresponding to 0.03 bar pressure is 24.1°C and t_{sup} = the temperature of the exhaust gases = 400°C

$h_{f_{atm}}$ = sensible heat at atmospheric temperature

$= 25 \times 4.187 = \mathbf{104.675\ kJ/kg}$

Weight of dry gases/min

$= \left(23.9 \times \frac{10.6}{60}\right) -$ Wt. of steam formed/min

$= 4.2223 - 0.2385$

$= \mathbf{3.9838\ kg/min}$

∴ Heat carried away by dry exhaust gases/min

$= m_g C_{pg} \Delta T$

$= 3.9838 \times 1.05 \times (400 - 25)$

$= \mathbf{1568.63\ kJ/min}$

Heat carried away by steam in exhaust gases/min

$= 0.2385 \times [2545.5 + 2.1\ (400 - 24.1) - 104.675]$

$= 0.2385\ [2440.825 + 2.1 \times 375.9]$

$= 0.2385 \times 3230.215$

$= 770.4$ kJ/min

∴ Total heat carried away by exhaust gases/min.

$= 1568.63 + 770.4$

$= \mathbf{2339.03\ kJ/min}$ **(26.75%)**

6. *Heat unaccounted for (by difference)*

$= 8745 - [3054 + 660 + 2612.5 + 2339.03]$

$= 8745 - 8665.53$

$= \mathbf{79.47\ kJ/min}$ **(0.91%)**

Heat balance table on minute basis

Heat added			*Heat expenditure*		
Item	*kJ*	*%*	*Items*	*kJ*	*%*
Heat supplied by fuel/min	8745	100	1. Heat equivalent of BP	3054	34.92
			2. Heat equivalent of FP	660	7.55
			3. Heat carried away by cooling water	2612.5	29.87
			4. Heat carried away by exhaust gases	2339.03	26.75
			5. Heat unaccounted for (by difference)	79.47	0.91
	8745	100		8745	100

Example 23.12 *During a test on a diesel engine the following observations were made :*

1. *The power developed by the engine used for driving a dc generator* — *Generating output 210 amp 200 volts.*

2. The efficiency of the generator	*82%*
3. The quantity of fuel supplied to the engine	*11.2 kg/hr*
4. Calorific value of fuel	*42600 kJ/kg*
5. The air : fuel ratio	*18 : 1*
6. The exhaust gases were passed through a exhaust gas calorimeter for which the observations were as follows :	
(i) Water circulated	*580 lit/hour*
(ii) Temperature rise of water	*36°C*
(iii) Temperature of exhaust gases at exit of calorimeter	*98°C*
(iv) Ambient temperature	*20°C*
7. Sp. heat of exhaust gases	*1.05 kJ/kg/K*
8. Heat lost to jacket cooling water	*32% of the total heat supplied.*

Draw up a heat balance sheet on minute basis.

Solution

Power output of the generator

$$= \text{Voltage} \times \text{amp.}$$

$$= \frac{200 \times 210}{1000} \text{ kW}$$

$$= 42 \text{ kW}$$

Power input to the generator

$$= \frac{\text{Power output}}{\text{Generator efficiency}}$$

$$= \frac{42}{0.82}$$

$$= \mathbf{51.22 \text{ kW}}$$

This is also the shaft power of the engine i.e., BP of the engine is 51.22 kW.

$$\text{BP} = \mathbf{51.22 \text{ kW}}$$

Heat supplied to the engine/min.

$$= \frac{\text{Wt. of fuel}}{\text{min.}} \times \text{Calorific value}$$

$$= \frac{11.2}{60} \times 42600$$

$$= \mathbf{7952 \text{ kJ/min}} \qquad \mathbf{(100\%)}$$

Again heat equivalent of BP/min

$$= \text{BP} \times 60$$

$$= 51.22 \times 60$$

$$= 3073.2 \text{ kJ/min} \qquad \mathbf{(38.65\%)}$$

Heat carried away by cooling water/min

$$= 0.32 \times 7952$$

$$= \mathbf{2544.64 \text{ kJ/min}} \qquad \mathbf{(32\%)}$$

Heat carried away by exhaust gases/min

$= \text{Wt. of gases/min} \times \text{sp. heat} \times (t_g - t_R)$

Wt. of exhaust gases/min

$= \text{Wt. of air/min} + \text{Wt. of fuel/min}$

$$= \left(\frac{11.2}{60} \times 18\right) + \frac{11.2}{60}$$

$$= \frac{11.2}{60}(18 + 1) = \frac{11.2 \times 19}{60}$$

$= \mathbf{3.55\ kg/min}$

From the observations on exhaust gas calorimeter we can write a heat balance and we get

Heat given by gases = Heat gained by water

$= m_w C_{pw}(t_{w2} - t_{w1})$

$$= \frac{580}{60} \times 4.187 \times 36$$

$= \mathbf{1457.076\ kJ/min}$

In addition to this, the gases leaving the calorimeter are at the temperature of 98°C.

∴ Heat exhausted by gases in atmosphere

$= m_g C_{pg}(t_g - t_R)$

where m_g = 3.55 kg/min.

C_{pg} = 1.05 kJ/kg.K

t_g = 98°C

t_R = room or atmospheric temperature = 20°C

∴ Heat carried by exhaust gases leaving the calorimeter

$= 3.55 \times 1.05\,(98 - 20)$

$= 290.745$ kJ/min

∴ Total heat carried away by exhaust gases/min

$= 1457.076 + 290.745$

$= \mathbf{1747.821\ kJ/min.}$ **(21.98%)**

∴ Heat unaccounted for (by difference)

$= 7952 - [3073.2 + 2544.64 + 1747.821]$

$= 7952 - 7365.661$

$= 586.339$ kJ/min **(7.37%)**

Heat balance sheet on minute basis

Heat input			*Heat output*		
Item	*kJ*	*%*	*Items*	*kJ*	*%*
Heat supplied by fuel/min	7952	100	1. Heat equivalent of BP	3073.2	38.65
			2. Heat carried away by water	2544.64	32.00

Contd.

Contd.

Heat input			Heat output		
Item	*kJ*	*%*	*Items*	*kJ*	*%*
			3. Heat carried away by exhaust gases	1747.821	21.98
			4. Heat unaccounted for (by difference)	586.339	7.37
	7952	100	Total	7952	100

PROBLEMS ON POWER PLANT

Example 23.13 *A Diesel engine power plant was given a test of 45 minutes duration, with results as given below :*

The various performance quantities that these data fix will be calculated. The engine is 8 cylinder, 2-stroke cycle, and has stroke = 45 cm, diameter = 36 cm. Engine runs at 275 rpm and is directly connected to a 750 kW generator. Electrical output was read by watt-hour meter, W_1 = 7562 kWh, W_2 = 8087 kWh. Fuel quantity was measured by direct weighing as 169 kg for the run with CV = 44150 kJ/kg. Water flow is measure by venturimeter as 1370 kg/min. Inlet and outlet temperature of water are 35°C and 46°C. Average fuel temperature is 24°C. Specific gravity = 0.921.

Atmospheric temperature = 18°C. Exhaust gas temperature = 400°C.

Average mean effective pressure of indicator cards take during test is 5.6 bar. Power used for plant pumps = 4 kW. During the test exhaust gas analysis determined that the air : fuel ratio was 26 : 62. Assume generator efficiency = 92.6% and hydrogen content of fuel = 12% by weight sp. heat of gases = 1.07 kJ/kg/K.

Solution

Power Calculations

During 45 minute test, the generator out per 45 minutes is given by watt-meter observations. This is given by :

$$W_2 - W_1 = 8087 - 7562$$
$$= 525 \text{ kW hour.}$$

$$\therefore \text{ Output of the generator} = \frac{525 \times 60}{45}$$
$$= \mathbf{700\ kW.}$$

Efficiency of the generator = 92.6%

∴ Input to the generator = Output of the engine

$$= \frac{700}{0.926}$$
$$= \mathbf{756\ kW}$$
$$= \textbf{Brake Power}$$

Again for indicated power $\text{IP} = p_{mi} \times \dfrac{LAn}{1000} \times \text{No. of cycles}$

where $p_{mi} = 5.6 \text{ bar} = 5.6 \times 10^5 \ \dfrac{\text{N}}{\text{m}^2}$

$$L = \text{Piston stroke } m$$
$$= 45 \text{ cm} = 0.45 \text{ m}$$
$$A = \text{Area m}^2$$
$$= \frac{\pi}{4} \times \left(\frac{36}{100}\right)^2 \text{ m}^2.$$
$$n = \text{No. of power strokes/sec}$$
$$= \frac{275}{60} \text{ (It is 2-stroke cycle)}$$

No. of cycles = 8

$$\therefore \quad \text{IP} = \frac{5.6 \times 10^5}{1000} \times 0.45 \times \frac{\pi}{4} \times \left(\frac{36}{100}\right)^2 \times \frac{275}{60} \times 8$$
$$= \mathbf{941 \text{ kW}}$$

∴ Mechanical efficiency of the engine

$$\eta_m = \frac{\text{BP}}{\text{IP}} = \frac{756}{941}$$
$$= \mathbf{0.8034} \text{ i.e., } \mathbf{80.34\%}$$

Brake thermal efficiency is given by

$$\eta_{bth} = \frac{\text{BP}}{\frac{\text{mass of fuel}}{\text{sec}} \times \text{Cal. value}}$$
$$= \frac{756 \times 60}{\frac{169}{45} \times 44150}$$
$$= \mathbf{0.2736} \text{ or } \mathbf{27.36\%}$$

∴ Indicated thermal efficiency

$$\eta_{ith} = \frac{\eta_{bth}}{\eta_m} = \frac{0.2736}{0.8034}$$
$$= \mathbf{0.3405} \text{ or } \mathbf{34.05\%}$$

For specific fuel consumption we require fuel consumption per hour

$$\therefore \text{ Fuel consumption/hour} = \frac{169}{45} \times 60$$
$$= \mathbf{225.3 \text{ kg/hour}}$$

∴ Brake specific fuel consumption

$$= \mathbf{0.298 \text{ kg/kWh}}$$

Similarly, Indicated specific fuel consumption

$$\text{ISFC} = \frac{\text{Fuel/hour}}{\text{IP}}$$

$$= \frac{225.3}{941}$$

$$= \mathbf{0.2394\ kg/kWh}$$

Overall plant efficiency is given by

$$\eta_P = \frac{\text{Net Output}}{\text{Heat Supplied}}$$

$$= \frac{\text{Generator Output} - \text{Power for pumps}}{\dfrac{\text{mass of fuel}}{\text{sec}} \times \text{Cal. value}}$$

$$= \frac{700 - 4}{\dfrac{169}{45 \times 60} \times 44150}$$

$$= \frac{696 \times 45 \times 60}{169 \times 44150}$$

$$= \mathbf{0.2518} \text{ or } \mathbf{25.18\%}$$

Heat balance calculations

1. *Heat supplied per minute*

$$= \frac{\text{Mass of fuel}}{\text{min}} \times \text{Cal. value}$$

$$= \frac{169}{45} \times 44150$$

$$= \mathbf{165807.7\ kJ/min} \qquad \mathbf{(100\%)}$$

2. *Heat equivalent of useful work*

= Generator output × 60

= 756 × 60

= **45360 kJ/min** **(27.36%)**

3. *Heat to cooling water/min*

= mass of water/min. × sp. heat × Rise in temperature

= 1370 × 4.187 × (46 – 35)

= **63098.1 kJ/min** **(38.05%)**

4. *Heat carried away by exhaust gases.*

= Heat to dry gases + Heat to vapour

Mass of gases/min

$$= \frac{169}{45} \times 26.62 + \frac{169}{45}$$

$$= \mathbf{103.73\ kg/min}$$

Water vapour formed/min.

$= 9H$

$$= 9 \times 0.12 \times \frac{169}{45}$$

$$= \mathbf{4.056 \ kg/min}$$

∴ Weight of dry gases/min

$$= 103.730 - 4.056$$

$$= \mathbf{99.674 \ kg/min}$$

∴ Heat carried away by dry exhaust gases/min

$$= m_g C_{pg} \Delta T$$

$$= 99.674 \times 1.07 \times (400 - 18)$$

$$= \mathbf{40740.75 \ kJ/min} \qquad \mathbf{(24.57\%)}$$

Heat carried away by water vapour/min

$$= 4.056 \,[2545.5 + 2.1\,(400 - 24.1) - 18 \times 4.107]$$

$$= 4.056 \,[2545.5 + 2.1 \times 375.9 - 75.366]$$

$$= \mathbf{13220.63 \ kJ/min} \qquad \mathbf{(7.97\%)}$$

5. *Heat unaccounted for/min (by difference)*

$$= 165807.7 - [45360 + 63098.1 + 40740.75 + 13220.63]$$

$$= \mathbf{3388.22 \ kJ/min} \qquad \mathbf{(2.05\%)}$$

Example 23.14 *The following readings were taken during a 60 min. test at a steady load at 360 rpm. Diesel engine generator unit rated at 650 kW. Initial fuel meter reading 14.043 m³ and final reading is 14.223 m³. Initial watt hour meter reads 86547 kWh and final reading is 87002 kWh. Cal. value of fuel = 44000 kJ/kg.*

Average temperatures are as :

Atmosphere	*4.45°C*
Fuel	*15.6°C*
Exhaust gases	*416.7°C*
Cooling water inlet	*46°C*
Cooling water outlet	*61°C*
Water flow	*660 kg/min*
Sp. gravity of oil	*0.905*
Air : Fuel ratio	*22 : 1*

Calculate a five element heat balance including electrical losses as one item if generator efficiency is 93%.

Solution

Total units delivered in 60 minutes

$$= W_2 - W_1$$

$$= 87002 - 86547$$

$$= 455 \text{ kWh}$$

∴ Output of the generator

$$= \frac{455 \times 60}{60} = \mathbf{455 \ kW}$$

This is 70% of the rated load.

Brake power of the engine

$$\text{BP} = \frac{\text{Generator output}}{\text{Generator efficiency}}$$

$$= \frac{455}{0.93}$$

$= \mathbf{489.25\ kW}$

$\therefore$ Generator loss = 489.25 – 455

$= 34.25$ kW

Heat Balance Calculations

1. *Heat supplied/hour* $= \dfrac{\text{Mass of fuel}}{\text{hour}} \times$ Cal. value

Mass of fuel/hour $= m_f = 14.223 - 14.043\ \text{m}^3/\text{hour}$

$= 0.180\ \text{m}^3/\text{hr}$

$= 0.18 \times$ sp. gravity

$= 0.18 \times 0.905 \times 1000$

$= 0.1629$ kg/hour $\times 1000$

$= 162.9$ kg/hr

$\therefore$ Heat supplied/hour $= 162.9 \times 44000$

$=$ **7167600 kJ/hr. (100%)**

2. *Heat equivalent of useful or shaft power for engine*

$=$ BP $\times 3600$

$= 489.25 \times 3600$

$=$ **1761300 kJ/hr. (24.57%)**

3. *Heat given to cooling water/hour*

$= \dfrac{\text{Mass of water}}{\text{hour}} \times$ sp. heat $\times$ Rise in temp.

$= 660 \times 60 \times 4.187 \times (61 - 46)$

$=$ **2487078 kJ/hour (34.7%)**

4. *Heat carried away exhaust gases/hour*

$= \dfrac{\text{Mass of water}}{\text{hr}} \times$ sp. heat $\times (t_g - t_R)$

where mass of gases/hour $= \dfrac{\text{Mass of fuel}}{\text{hour}} \times$ (Air : Fuel ratio + 1)

$= 162.9 \times [22 + 1]$

$=$ **3746.7 kg/hour**

$\therefore$ Heat carried away by exhaust gases/hour

$= 3746.7 \times 1.07 \times (416.7 - 4.45)$

$=$ **1652697.5 kJ/hour (23.05%)**

5. *Electrical loss/hour*

= Generator input – Generator output

$= 34.25 \text{ kW} \times 3600 \text{ kJ/hour}$

$= \mathbf{123300 \text{ kJ/hour}}$ **(1.72%)**

6. *Heat unaccounted for (by difference)*

$= 7167600 - [1761300 + 2487078 + 1652697.5 + 123300]$

$= \mathbf{1143224.5 \text{ kJ/hour}}$ **(15.96%)**

Example 23.15 *Estimate the fuel storage tank capacity for a diesel plant having 5000 kW installed capacity. Expected plant capacity factor 55%. Fuel contract is to be made for semi-monthly delivery, but allow 100% extra for contingency. Also determine the cpm transfer pump capacity.*

Solution

Installed capacity = 5000 kW

Capacity factor = 55%

∴ Average capacity = 5000 × 0.55

= 2750 kW.

From the chart we get 8.5 kWh per 4.55 kg oil (at 55%) plant capacity factor and 8.0 kWh (at 45%) the lowest one per 4.55 kg fuel.

∴ Required storage capacity required (including 100% extra for contingency) for one month

$$= \frac{2750 \text{ kW}}{\frac{(8.5+8)}{2} \frac{\text{kWh}}{4.55 \text{ kg}}} \, 24 \text{ (day)} \times 30 \text{ (month)}$$

$$= \frac{2750 \times 2 \times 4.55}{16.5} \times 24 \times 30$$

$$= \mathbf{1092000 \text{ kg/month.}}$$

Assuming 905 kg/cu.m as density of oil,

Volumetric capacity of storage tanks for month.

$$= \frac{1092000}{905}$$

$$= \mathbf{1206.63 \text{ m}^3}$$

Generally a tank car capacity is 10,000 gallons i.e., 10,000 × 4.55 kg or 50.276 m^3 and for one month 24 tank cars will be required.

1206.63 cu. m/month

$$= \frac{1206.63}{30 \times 24 \times 60} \text{ cu. m/min.}$$

$$= \mathbf{0.02793 \text{ cu. m/min}}$$

= **Transfer pump capacity**

Example 23.16 *The fuel consumption of a 2000 kW diesel engine power plant is shown in the table given below:*

KW	*400*	*800*	*1200*	*1600*	*2000*	*2400*
Fuel flow kg/hour	*88*	*148*	*224*	*292*	*372*	*472*

And the load on the plant for the week days is given below :

Time	*7 am to 4 pm*	*4 pm to 10 pm*	*10 pm to 7 am*
Average load in kW	*1968*	*576*	*424*

The load on Sunday is 160 kW.

Calculate yearly fuel cost based on oil price at Rs. 400 per barrel of 160 kg capacity.

Solution

The relation between load kW and fuel flow in kg/hour is shown by a graph given. Fig. Ex. 23.16.

For Sunday, the load is 160 kW and the fuel flow is obtained from the graph as 57.5 kg/hour.

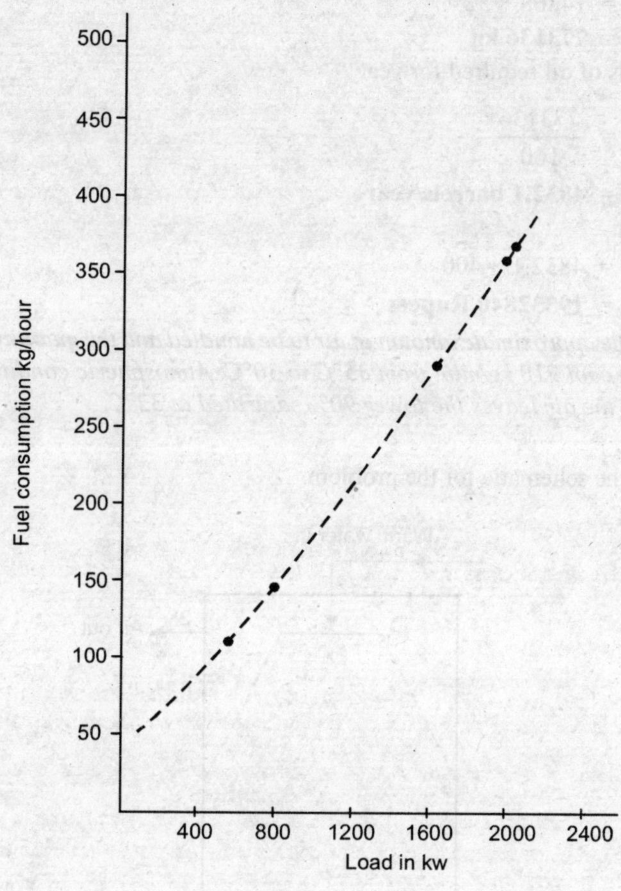

Fig. Ex. 23.16.

∴ Fuel flow on Sunday

$$= 24 \times 57.5$$

$$= 1380 \text{ kg} \qquad \text{(I)}$$

For week days –

7 a.m. to 4 p.m. Load = 1968 kW, Fuel = 362 kg/hour

4 p.m. to 10 p.m. Load = 576 kW, Fuel = 110 kg/hour

10 p.m. to 7 a.m. Load = 424 kW, Fuel = 90 kg/hour

Fuel flow on week day

$$= 24 [362 + 110 + 90]$$

$= 24 \times 562$ kg

$=$ **13488 kg** (II)

$\therefore$ For one week, the fuel flow is

$= \text{I} + \text{II} = 1380 + 13488$

$= 14868$ kg

Assuming 52 weeks in a year, yearly fuel flow is given as

$= 14868 \times 52$

$=$ **773136 kg**

And the number of barrels of oil required for year

$$= \frac{773136}{160}$$

$=$ **4832.1 barrels/year**

$\therefore$ The yearly cost of fuel

$= 4832.1 \times 400$

$=$ **19332840 Rupees**

Example 23.17 *Determine the approximate amount of air to be handled and the quantity of make-up water required by a cooling tower that is to cool 910 kg/min. from 35°C to 30°C. Atmospheric conditions are 34°C dry bulb and 25°C wet bulb. Assume that the air leaves the tower 90% saturated at 32°C.*

Solution

Figure Ex. 23.17 shows the schematic for the problem.

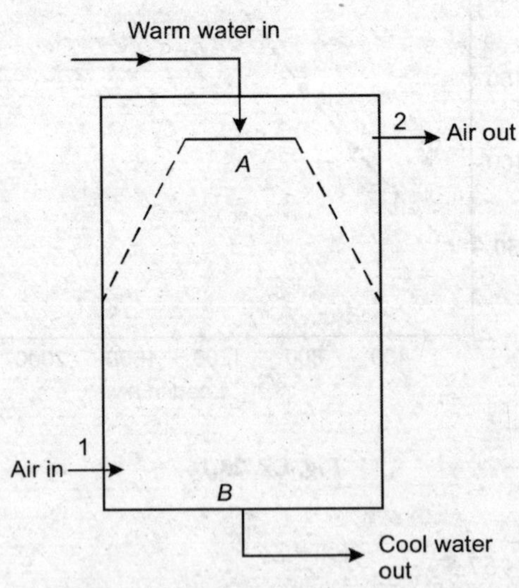

Fig. Ex. 23.17 *Diagrammatic cooling tower.*

The heat balance equation with reference to Figure Ex. 23.17 per kg of dry air is :

$$\left.\begin{array}{l}\text{Enthalpy of entering air/kg}\\ +\text{ Enthalpy of water entering}\end{array}\right] = \left.\begin{array}{l}\text{Enthalpy of air at outlet}\\ +\text{ Enthalpy of water at outlet}\end{array}\right]$$

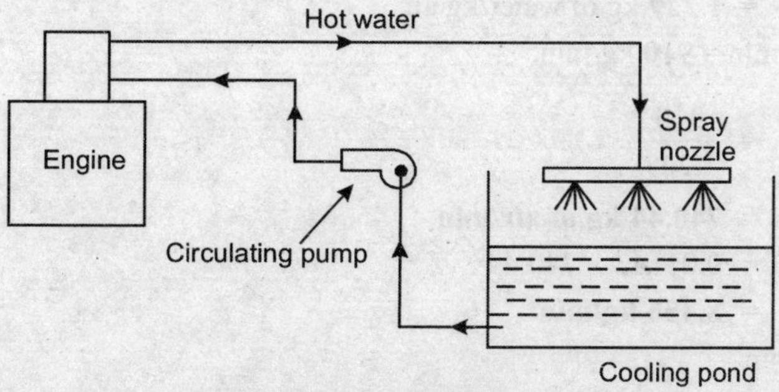

(a) Single circuit cooling system

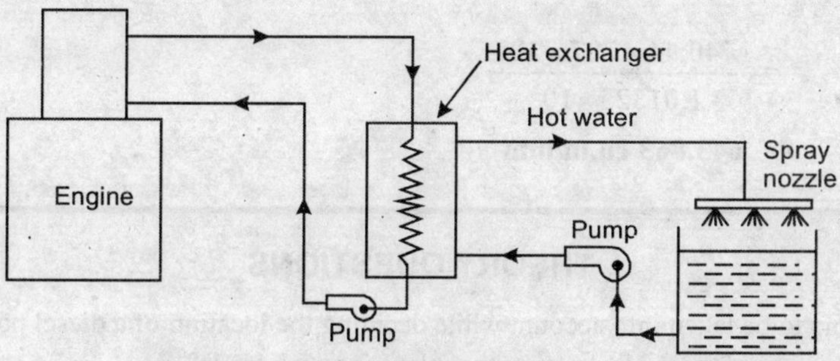

(b) Double Circuit Cooling System

Fig. Ex. 23.17

$$\therefore \quad h_1 + W_{wA}\, h_{fA} = h_2 + W_{wB}\, h_{fB}$$

$$75.187 + W_{wA} \times 146.56 = 102.352 + W_{wB} \times 125.66$$

From chart, $W_{H1} = 0.016$ kg/kg dry air.

$$W_{H2} = 0.02742 \text{ kg/kg dry air.}$$

Make up water $= W_{H2} - W_{H1} = 0.02742 - 0.016 = 0.01142$ kg/kg dry air.

$$\therefore \quad W_{wB} = \frac{75.187 - 102.352 + W_{wA} \times 146.545}{125.61}$$

$$= \mathbf{1.1667\ W_{wA} - 0.2163} \qquad \text{(I)}$$

and
$$W_{wB} = W_{wA} - (W_{H2} - W_{H1})$$

$$= W_{wA} - 0.01142 \qquad \text{(II)}$$

From Eqs. I and II, we get

$$W_{wA} - 0.01142 = 1.16637\, W_{wA} - 0.2163$$

$$\therefore \quad 1.1667\, W_{wA} - W_{wA} = 0.2163 - 0.01142$$

$$0.1667\, W_{wA} = 0.20488$$

$$\therefore \quad W_{wA} = \text{Wt. of water/kg air}$$

$$= \frac{0.20488}{0.1667}$$

$= 1.229$ kg of water/kg air

Actual amount of water/min = 910 kg/min.

$\therefore$ Amount of air/min $= \dfrac{910}{1.229}$

$=$ **740.44 kg of air/min**

$\therefore$ Make up water $= 0.01142 \times 740.44$

$=$ **8.456 kg/min**

Volume of air handled/min

$$= \frac{mRT}{P}$$

$$= \frac{740.44 \times 287 \times 307}{1.01325 \times 10^5}$$

$=$ **643.863 cu.m/min**

THEORY QUESTIONS

23.1. What are the factors to be taken into account while deciding the location of a diesel power plant?

23.2. What are the advantages of a diesel engine power plant for peak loads?

23.3. Why diesel engines are used as standby plants in steam power stations?

23.4. Show with the neat sketch, the water cooling system for diesel engine power plant.

23.5. Explain with a neat sketch or diagram a fuel supply system for a plant with a supply from day tank, two storage tanks, fuel delivery from tank truck. Include all necessary meters, filters, pumps etc.

23.6. Enumerate the different field where use of diesel power plant is mostly required.

23.7. Explain with a neat sketch a diesel engine power plant showing all the systems used.

23.8. What different types of diesel engines that can be used in diesel power plant?

23.9. In what circumstances two-stroke engines are more suitable and economical than four stroke engines?

23.10. What are the different methods of injection used in diesel power plants? Which method is commonly used in large diesel power plants?

23.11. Cooling and cleaning of lubricating oil is necessary. Explain why?

23.12. Draw a neat diagram of lubrication system used for medium capacity power plant.

23.13. What are the different methods used for starting diesel engine? Which method is commonly used and why?

23.14. Why the starting of diesel engine is more difficult than that for petrol engine?

23.15. What is supercharging? Why is it necessary in diesel engine power plants? What are the different methods of supercharging employed?

23.16. What is turbocharging and how is it used in diesel engine?

23.17. Compare the diesel power plant and thermal power plant.

23.18. In what way, the layout is different for small capacity, medium capacity and high capacity plant? Draw the plant layout for all these plants.

PROBLEMS FOR PRACTICE

1. The following particulars refer to a full load test on a two stroke oil engine :

Speed	350 rpm
Load on brake	61.4 kg
MEP	2.69 bar
Cylinder bore	21 cm
Stroke	28 cm
Brake diameter	1 m
Calorific value of oil	44600 kJ/kg
Oil consumption per hour	4.25 kg
Jacket water	497 kg
Jacket water temperature	13°C inlet, 38°C outlet
Air : Fuel ratio	32 : 1
Room temperature	19°C
Exhaust gas temperature	370°C
H_2 in oil	15% by weight
C_p for dry exhaust gases	1.05 kJ/kg/K
C_p for water vapour	2.1 kJ/kg/K

Draw up a heat balance sheet in kJ/hour and also as a percentage of the heat supplied.

2. A marine oil engine works on four stroke cycle and has six single acting cylinders, 62 cm diameter and 99 cm stroke. During a test the following results were obtained :

Mean area of indicator diagram	7.8 sq. cm
Length of diagram	7.5 cm
Spring No.	7
Speed	125 rpm
Brake torque	7800 kg-m
Fuel oil consumption	242 kg/hour
Calorific value of oil	44885 kJ/kg
Jacket cooling water	242 kg/min
Rise of temperature of water	53°C

The exhaust gas from the engine enters a water heater at 332°C and leaves at 165°C. The cooling water flows through the heater at the rate of 78 kg/minute and its temperature is raised by 68°C. The temperature of the engine room is 15°C, the specific heat of the exhaust gas is 1.03 kJ/kg/K.

Calculate the mechanical efficiency, the brake thermal efficiency and draw up a heat account on the basis of 1 kg of fuel. **[Ans:** h_m = 89.5%, h_{bth} = 38.5%]

3. A four cylinder, four stroke diesel engine of 36 cm bore and 41 cm stroke tested at 315 rpm and full load gave the following results :

Brake power	248 kW
Average i.m.e.p.	7.14 bar
Fuel per hour	72.5 kg
Gross calorific value	44000 kJ/kg
Hydrogen content	13%
Air consumption	28 kg/min
Jacket water used	85 kg/min
Temperature rise of jacket water	40°C
Piston cooling oil used	43 kg/min
Sp. heat of piston cooling oil	2.18 kJ/kg/K

Temperature rise of piston oil	24°C
Room temperature	18°C
Exhaust gas temperature	323°C
C_p for dry exhaust gases	1.09 kJ/kg/K
C_p for superheated steam	2 kJ/kg/K

Take partial pressure of steam in gases 0.03 bar.

Draw up a heat balance in kJ/min indicating which items may include friction heat.

4. Cooling water for a 500 kW diesel engine is pumped to a cooling tower at 60°C. It is desired to cool the water to a maximum temperature of 38°C under the atmosphere condition of 32°C dry bulb temperature and 28°C wet bulb temperature. Find the required capacity and efficiency of the cooling tower.
5. A diesel plant has a cooling system employing a cooling tower that loses to the atmosphere approximately 5% of the water circulating. When the two 500 kW engines are operated at full load on an average day the tower cools the water from 54°C to 39°C. What should be the capacity of a water softening plant for the make up be in m^3/min.
6. A cooling tower, needed in conjunction with an IC engine plant, will be required to cool 820 kg of water per minute at 66°C entering temperature. Atmospheric pressure, 1.01 bar, dry bulb temperature 28°C, wet bulb temperature 22°C. Calculate make up (at 16°C).
7. The shop test of an engine whose record is here reported, was made for heat balance. Calculate a four item balance. Dynamometer torque is 654 kg-m. at 450 rpm. A : F ratio = 27 : 1. Atmospheric temperature = 27°C Exhaust gas temperature = 455°C. Inlet and outlet temperature of the cooling water = 55°C and 67°C respectively. Fuel flow per hour = 87 kg/hour. Cooling water flow = 446 kg/min and the sp. heat = 4.187 kJ/kg.K.
8. A cooling tower is needed for a diesel electric power plant. It is required to cool 1000 litres of water at 70°C at inlet to the tower. The dry bulb and wet bulb temperatures of the atmospheric air are 30°C and 24°C respectively. Calculate make-up water at 20°C, air flow required. Assume that air goes out of tower at 65°C and 90% relative humidity.
9. A 500 kW diesel power plant requires cooling water is pumped to the cooling tower at 60°C. It is required to cool the water to a maximum temperature of 35°C under atmospheric conditions of 32°C dry bulb temperature and 28°C wet bulb temperature. Find out the required capacity and efficiency of the cooling tower.
10. A diesel power plant has a cooling system which contains a cooling tower with other elements. 5% of the water circulating loses to the atmosphere. If two 500 kW engines are operated on an average day, the tower cools the water from 60°C to 40°C. What should be the capacity of water softening plant in litres per minute for the make-up.

PART 5

Chapter 24 Gas Turbine Power Plant

24.1 INTRODUCTION

A gas turbine power plant may be defined as one in which the main prime mover is of the rotory type that is turbine and the working medium is a permanant gas. (There is no change of phase for the working substance used in Thermal Power Plant). This gas turbine is one of the oldest prime mover used for power generation. All the advantages meant for the rotory machine namely freedom from vibration, large power production, small size and weights, are attributed to this gas turbine power plant. The major obstacle in the rapid development of this power plant is the high temperature materials required for the combustion chamber and gas turbine unit. Because of this, the efficiency of the plant was also less.

This type of the turbine derives its energy from heat commonly supplied by the combustion of the fuel. As the name implies, the fuel used is the gas or the liquid fuels producing gases and the products of combustion are utilized in turbines to produce power.

All the major processes required in a thermodynamic cycle, namely, compression, combustion and expansion are taking place in the separate components and not in a single cylinder as in case of a diesel engine or petrol engine which are the reciprocating engines.

The industrially successful gas turbine power plant is relatively a newcomer to the power field. Although gas turbines have been the subject of experimental development for decades, only in recent years have there been gas turbines plants of performance and cost that are commercially acceptable. The main obstacle to its use in normal case, is that the net output of the plant is very much less than 30% of the power developed by the turbine itself. The plant uses nearly 70% of the power developed by the turbine. Again, if all the processes are not highly efficient, the losses will easily consume the output of the prime mover, leaving the plant highly inefficient.

It is known that for good efficiency of the thermodynamic cycle for the gas turbine power plant, high temperature in the cycle as far as heat addition process is concerned and the low temperature as far as heat rejection is concerned. The designer and the manufacturer are to face a high temperature problem and for this materials sustaining these highest possible temperatures is the basic problem.

The designer and the manufacturer were confronted with a series of high temperature problems but the commercial development followed when metallurgical and engineering sciences progressed and the result was :

1. Manufacturer would build high-speed turbines for use with gases hot enough to be incandescent (540°C and upto 800–1000°C).

2. The aerodynamic theory of blade action in turbines had developed to the extent that 80% of the theoretically available energy could be converted into shaft or useful work.

3. The high speed compressors of about 80% compression (isentropic) efficiency or better could be built.

Here it is worth to mention that the aeronautical jet engine is a specialized gas turbine plant whose net output is a kinetic propulsion jet of the working medium rather than the shaft work.

The requirement of a high performance aircraft in military and power generation lead to the rapid development of the gas turbine systems. In the field of power generation the use of gas turbine is more recent.

24.2 SELECTION OF SITE FOR GAS TURBINE POWER PLANT

The main factors to be considered while selecting the suitable site for gas turbine power plant/station are :

(a) **Availability of Natural Gas or Liquid Fuel.** For the gas turbine power plant, the fuel (natural gas or liquid fuel) should be available at the site.

(b) **Load Centre.** Usually the gas turbine power plants may be located near the sea-shore where the gaseous fuel can be supplied easily i.e., without any long gas pipelines. The gas turbine power plants generally should be near or at the load centre, so that the transmission cost will be reduced. Long transmission lines mean more transmission losses and also the cost of the lines will increase. This leads to the more cost of the plant project and the efficiency of the plant as a whole will be reduced.

(c) **Access to Site.** The site should be easily accessible by rail or by road.

(d) **Storage of Gas/Fuel.** There should be sufficient storage facilities for gas or liquid fuel. So that the plant should work without any shortage of fuel. At least one week's storage must be there and then the supply of fuel should be at specific interval of period. For these purposes, generally cylindrical, large tanks are preferred.

24.3 APPLICATIONS OF GAS TURBINE POWER PLANT

From application point of view, gas turbine power plants can be classified as :

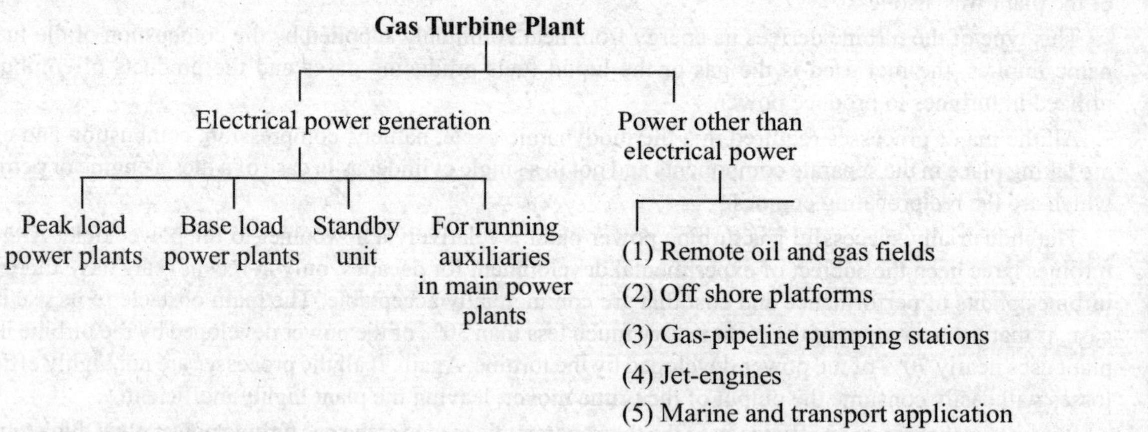

24.3.1 Base-load Power Plant

The major power which is supplied for all the time during the day/month/year is called base-load and the plant which supplies this power is called base-load power plant. If natural gas is available in ample quantities throughout the year, then it will be economical to run the gas turbine power plant as base-load power plant. In case the liquid fuel is used, then the unit will not be economical as the operating cost will be high.

24.3.2 Peak-load Power Plant

If the demand of power is more than the capacity of the base-load power plant, then the gas turbine power plant can be used to supply this load called the peak-load and then the gas turbine power plant will work as a **peak-load**

power plant. The most important requirement for a power plant to work as peak power plant is the **quick starting** of the plant. The gas turbine power plant need not be kept running on no-load and loaded when required.

24.3.3 Stand-by Power Plant

In emergency, this plant can be used to take-up **some important 'load'** when the main power plant is stopped because of certain unforeseen reasons. Generally these gas turbine power plants can be very effectively used as standby units for hydro-power installations.

24.3.4 For Running Auxiliaries

Power is required to start auxiliaries of a steam power plant. Hence, gas turbine plants — generally upto (20–25 MW) are used for starting auxiliaries.

The gas turbine power plant nowadays is universally used as peak load as well as standby unit due to its outstanding operational characteristics.

24.3.5 Mechanical Power

Mechanical power is required to drive certain machines such as compressors used in the following cases :

1. Remote oil and gas fuels
2. Off-shore plant forms
3. Gas pipeline pumping station
4. Jet engines for air-crafts.

24.3.6 Gas Turbine

In industry, a gas turbine plant may be used as standby plant for steam or purchased power. Due to its low weight to power ratio, gas turbine plant is used in marine and transport applications.

24.4 THE SIMPLE GAS TURBINE PLANT

24.4.1 Components

The simple gas turbine plant essentially consists of the following components :

(a) Compressor (b) Combustion chamber (c) Turbine and (d) Generator.

Figure 24.1 shows these components.

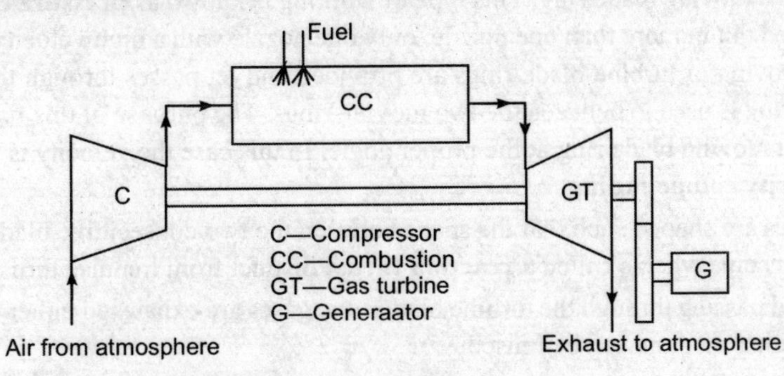

Fig. 24.1 *Simple gas turbine power plant.*

24.4.2 Working of Gas Turbine Power Plant

It has been observed that the gas turbine must receive a continuous supply of gas under a pressure above that of the exhaust region. A gas turbine plant consists of the turbine and a means of supply of the working medium. Modern gas turbine plants are based on a cycle of continuous flow of the medium or working substance. Air is compressed in a machine (compressor of the rotory type) driven by the turbine itself. The compressed air then is admitted to a combustion chamber where fuel is burned in it, thus raising the temperature and increasing the volume of the working gas at constant pressure. The working medium is now ready for the turbine.

A reversible adiabatic or isentropic expansion of a gas is theoretically possible if the gas is made to pass through a properly shaped nozzle from a high pressure region to a low pressure region. Actual expansions are almost reversible, there being but little friction or turbulence generated by nozzle action. The result of this process is a jet of gases containing high grade mechanical (kinetic) energy. Turbine wheels have blades onto which the jet is directed in order that this energy may transformed into a blade thrust and thus to a shaft torque.

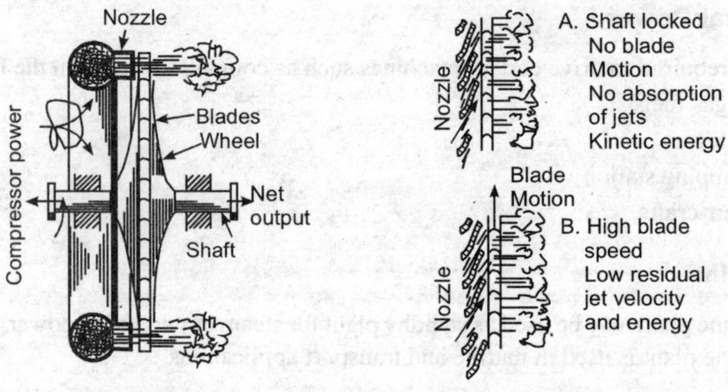

Fig. 24.2 *Nozzle and turbine blade action.*

In the gas turbine a stationary nozzle discharges a jet of gas (usually the products of combustion) against the blades on the periphery of turbine wheel as shown in Fig. 24.2. The jet is thereby deflected and slowed down while the blades receive an impulse force which is transmitted as a mechanical torque to the shaft. If the speed of the jet from the nozzle is high enough, then the expansion may be divided into stages (more than 1) so that each stage will have a nozzle ring and a moving blade ring. This type of working is known as **pressure compounding**. Instead of expansion being carried out in more than one nozzle, only one nozzle with a high velocity may be used and in this case more than one moving or turbine blade rings are provided and jet passes through these moving blade rings. Here one fixed blade ring is used in between the two moving rings. The purpose of this fixed blade ring is to divert the flow on to the next moving blade ring at the proper angle. In this case the velocity is utilised in stages and the method is called **velocity compounding**.

If the moving blades are shaped such that the space between the two consecutive blades forms a nozzle shape, then the blading will receive what is called a **reaction thrust** distinct from impulse thrust.

After expansion and passing through the turbine blades, the gases are exhausted either into atmosphere or to the cooler so that the working medium can be reused.

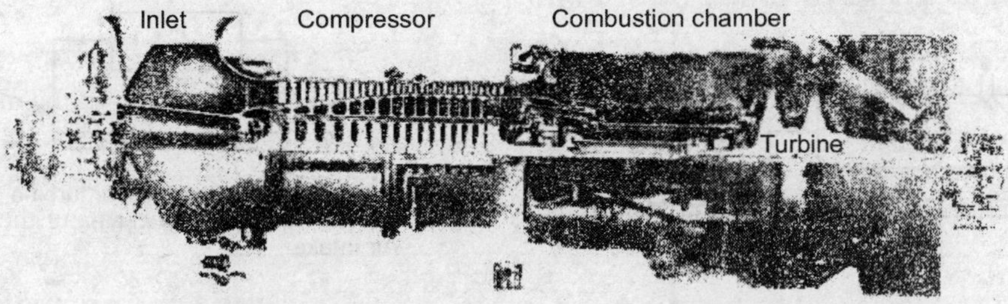

Fig. 24.3 *Package type gas turbine power plant*

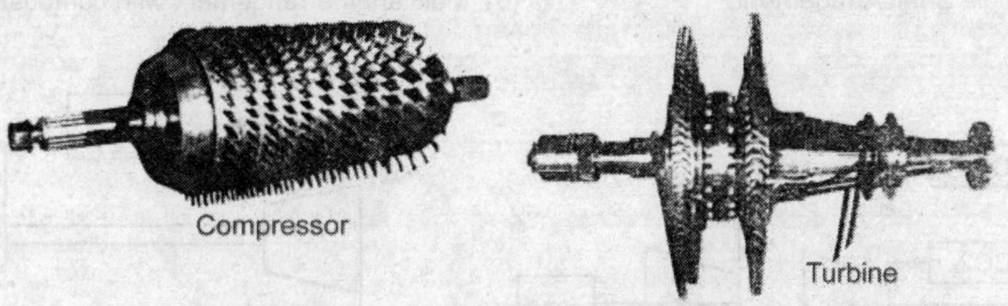

Fig. 24.4

Figures 24.3 and 24.4 show simple gas turbine unit and separate compressor and turbine used in the plant.

24.5 VARIOUS SCHEMATICS OF GAS TURBINE COMPONENTS

The main components of the gas turbine power plant can be arranged in a variety of ways. Some of them are shown in Fig. 24.5 (a) – (d).

(a) **Single Shaft Arrangement.** In this arrangement compressor (C) and gas turbine (GT) and generator (G) are mounted on the same, single shaft. In this arrangement all moving parts — compressor, turbine and generator run at the same speed.

(b) **Multishaft Arrangement.** When the turbine drives the generator only and not compressor, the turbine is called **Free Power Turbine**.

This type of the arrangement has certain advantages over single shaft arrangement. They are :

(i) This allows the turbine driving the compressor and the compressor to operate at its best performance speed and the free power turbine can operate at the desired appropriate speed. This provides the flexibility of control.

(ii) The system can be started more easily as the starting system has to drive the compressor — turbine unit only while the free power turbine unit is at rest.

(iii) It is possible to vary the free power turbine shaft speed over a wide range at constant compressor speed.

(iv) Free power turbine can be accelerated more easily and rapidly from its idling speed because the free power turbine (generator turbine) and compressor turbine coupling is absent.

Only one disadvantage in such cases is that the efficiency of power turbine is reduced at part load.

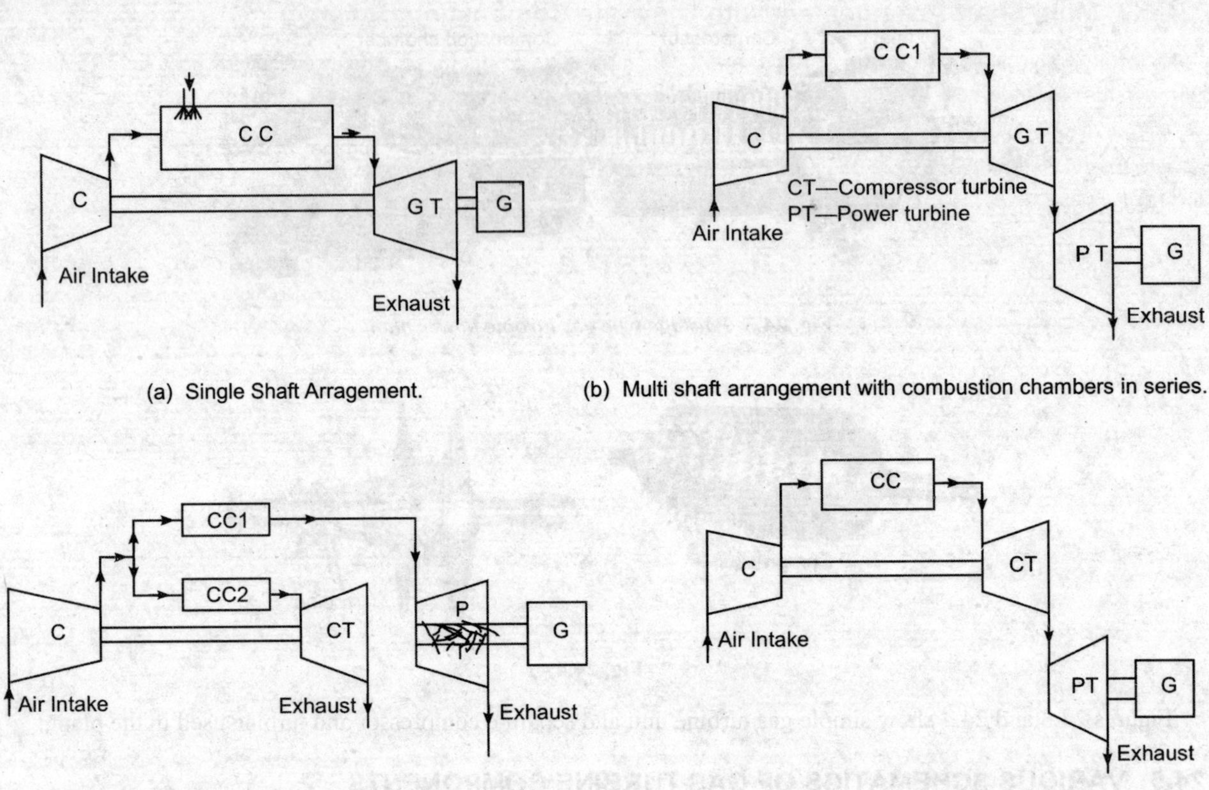

(a) Single Shaft Arragement.

(b) Multi shaft arrangement with combustion chambers in series.

(c) Multi shaft arrangement with combustion chambers in parallel

(d) Multi Shaft with Single Combustion Chamber with PT and Generator.

Fig. 24.5

24.5.1 Free Power Turbine with Combustion Chambers in Series

In this case combustion chamber is provided for every turbine. This type of the arrangement is shown in Fig. 24.5(b).

In this case the control at the various part loads is easer as it is to be done for power turbine only and this can be achieved by temperature variation by controlling the fuel burnt in the combustion chamber for the free power turbine. There is no variation or control is required as far as compressor turbine and its combustion chamber are concerned.

24.5.2 Free Power Turbine with Combustion Chambers in Parallel

In this case, the air delivered by the compressor is divided in two different streams leading to two combustion chambers in parallel as shown in Fig. 24.5(c).

In the previous arrangement, all the air passing through the compressor is passed through the power turbine and hence the power turbine will be large and thus the capital cost will be a problem. This is a disadvantage of the arrangement with combustion chambers in series. To overcome this disadvantage, the combustion chambers in parallel are provided as shown in Fig. 24.5(c). This arrangement reduces the size of the power turbine.

24.5.3 Multi Shaft Arrangement with Only one Combustion Chamber

The arrangement is shown in Fig. 24.5(d) and is the simplest arrangement. This overcomes the disadvantage of a single shaft arrangement. The generator is driven by a separate turbine (Free power turbine).

Other arrangements of compressor, combustion chamber, turbine and generator are shown in Fig. 24.6(a)(b)(c)(d)(e)(f)(g) and (h) below.

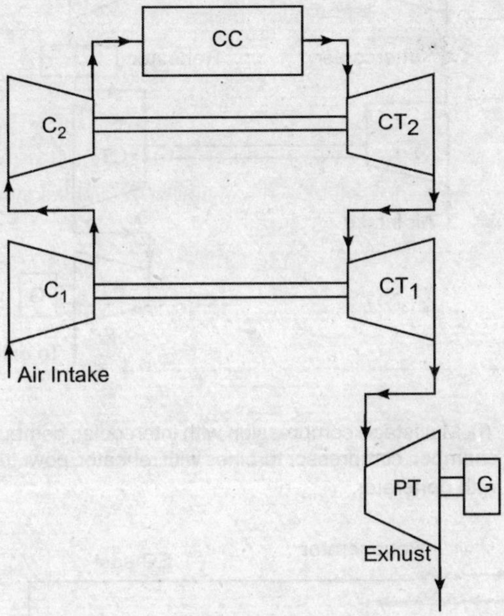

(a) Multistage compression multistage turbines and free power turbine with generetor.

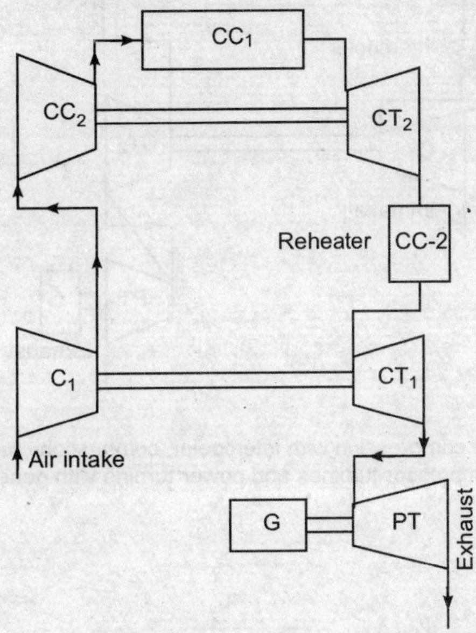

(b) Multistage compression multistage with combustion chamber, reheater, power turbine and generator.

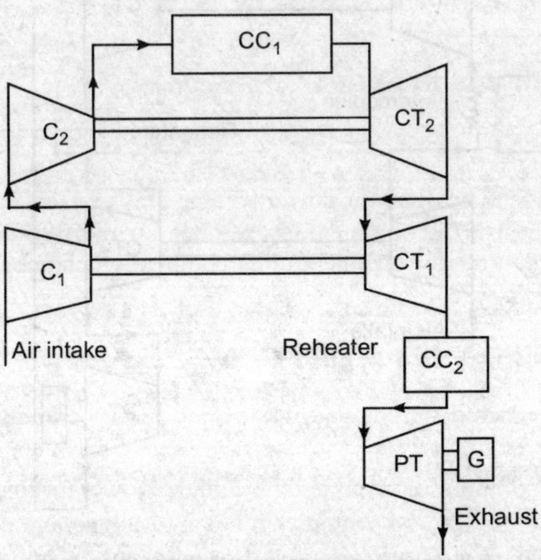

(c) Multistage compression combustion chamber multistage compressor-turbines reheater, power turbine generator.

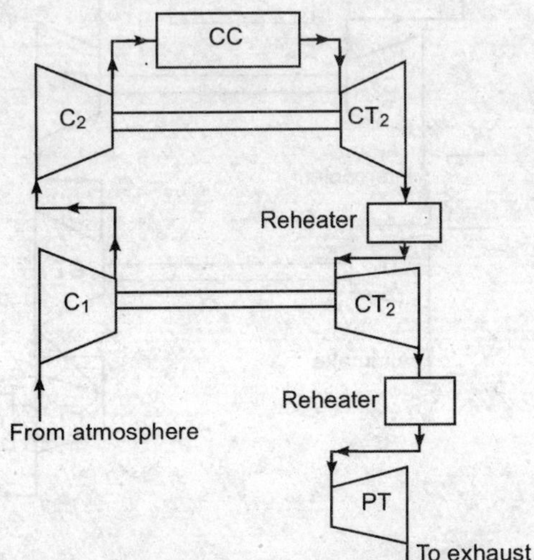

(d) Multistage compression combustion chamber compressor turbines with reheater in between, reheater for PT and generator.

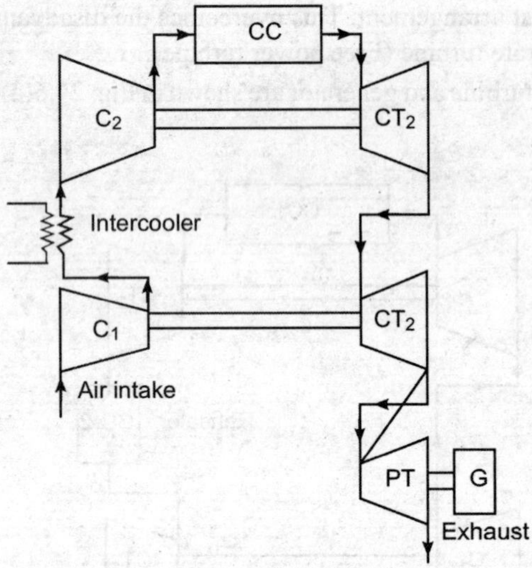

(e) Multistage compression with intercooler, combustion chamber, multistage compressor-turbines and power turbine with generator.

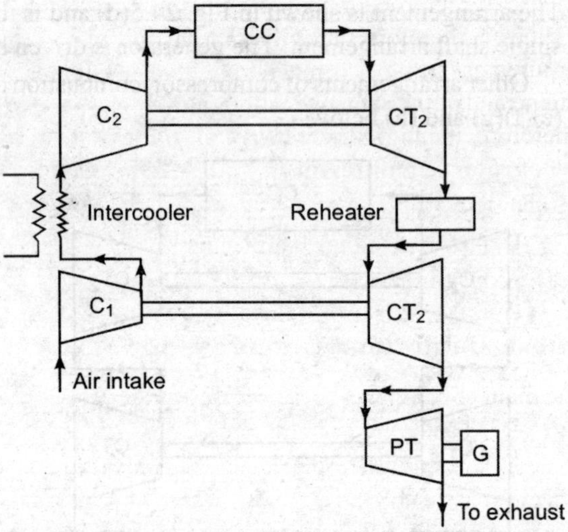

(f) Multistage compression with intercooler, combustion chamber compressor turbines with reheater, powr turbine with generator.

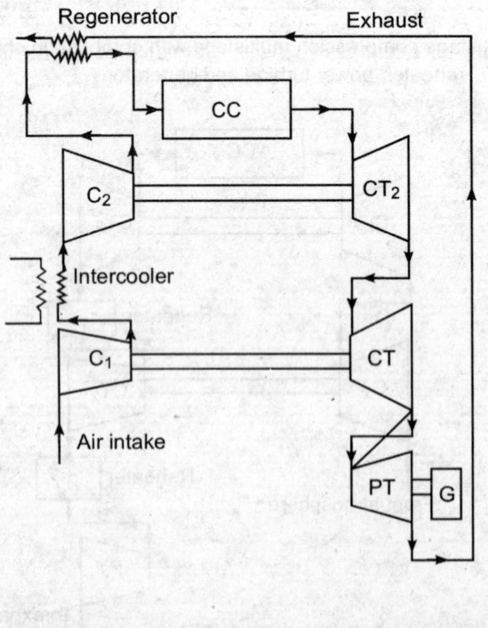

(g) Multistage compression with intercooler regenerative heat exchange, combustion chamber, compressor turbines, power turbine with generator.

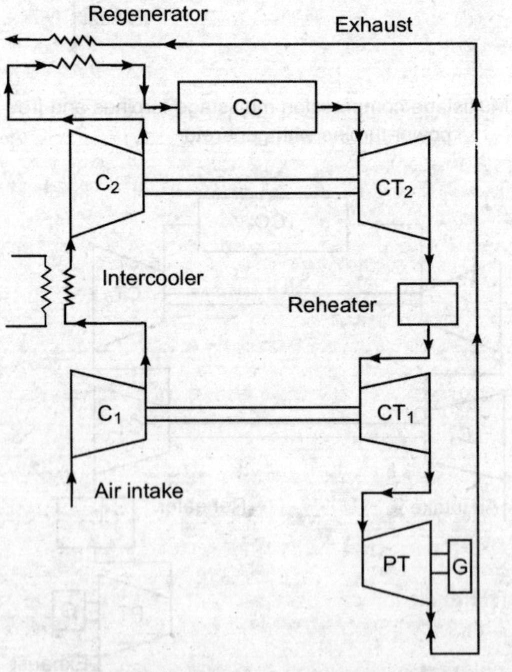

(h) Multistage compression with intercooler, regenerator, combustion chamber, multistage compressor turbines with reheater, power turbine with generator.

24.6 COMPRESSORS IN GAS TURBINE POWER PLANT

In case of reciprocating engines the pressure ratio is high as compared to gas turbines. The range of expansion in engines is therefore more than in gas turbines. Therefore, to produce the required power, quantity of air or gas required gas turbines is much more than in engines. To get the large quantity of air we should have a large diameter engine and at the same we will require high speeds. There is certain limits for the speed for reciprocating engines. Therefore, the high flow rates of air through the turbine and the relatively moderate pressure ratios necessiate the use of **rotory compressors**.

The types of compressors commonly used in the gas turbine plant are :

A. Centrifugal compressors and

B. Axial compressors.

24.6.1 Centrifugal Compressors

The main components of a centrifugal compressors are impeller, diffuser, volute and the driving shaft as shown in Fig. 24.7. The compression is carried out in the following manner :

(i) The fluid enters through the inlet casing with converging nozzle to the impeller.

(ii) The kinetic energy and static pressure rise takes place in the impeller.

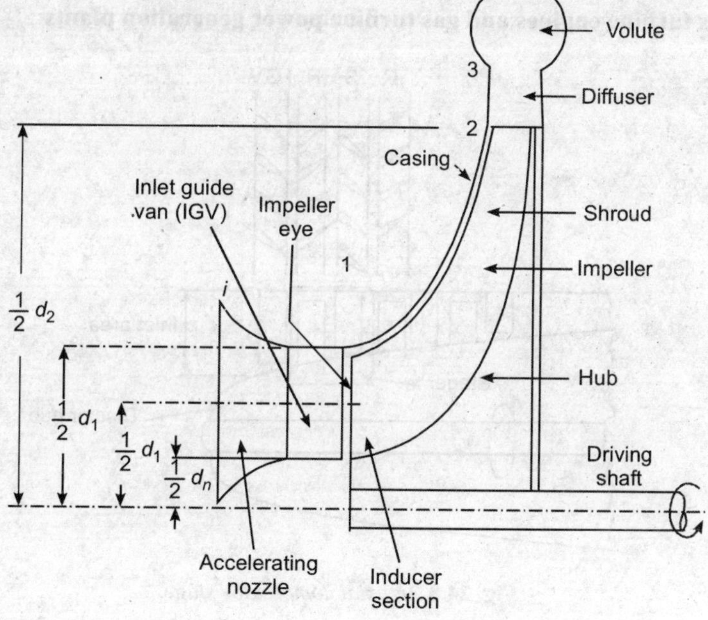

Fig. 24.7 *Elements of a centrifugal compressor stage.*

(iii) High kinetic energy of the fluid is converted into static pressure in the diffuser.

(iv) The fluid is collected in the volute or scroll.

Air enters the eye of the impeller at a very high velocity. The friction in the diffuser will cause some loss in stagnation pressure. Normally compressors are designed such that 50% of the pressure rise takes place in the impeller and the remaining 50% in the diffuser.

As the impellers are high stressed, normally straight vanes are preferred. Curved vanes which are used in hydraulic pumps and fans tend to become straight due to the high centrifugal force. Hence undesirable bending stresses will be set up in the vanes. The choice of radial vanes greatly determines the pressure rise that will be divided equally between the impeller and the diffuser.

In a centrifugal compressor, the energy transfer will be from rotor to fluid.

Centrifugal compressors and fans are turbo-machines employing centrifugal effects to increase the pressure of the fluid. Generally single-stage compressors have the pressure ratio of 4 : 1. For higher pressure ratios, multistage compressors with intercooling are used. The major advantages of these compressors are : (i) smaller length and (ii) wide range of mass-flow rate of air. Also, the disadvantages are :

(a) Large frontal area

(b) Lower maximum efficiency.

24.6.2 Axial Flow Compressors

As centrifugal compressors, axial compressors are turbo machines that increase the pressure of the fluid (air, gas etc). Air or gas flows continuously in the axial direction. Due to lack of knowledge of the aerodynamic behaviour a reversed reaction turbine was used as an axial compressor in early days. In designing the blades for axial compressors, the study of aerodynamics behaviour was very much helpful. Due to this, the efficiency of the axial flow compressors surpass the maximum centrifugal compressor efficiency by about 4 per cent. The efficiency of the axial air compressors is very much sensitive to mass flow rate. Any deviation from the design condition causes the efficiency to drop drastically. Therefore, **the axial flow air compressor is ideal for constant load applications such as in aircraft gas turbine engines and gas turbine power generation plants**.

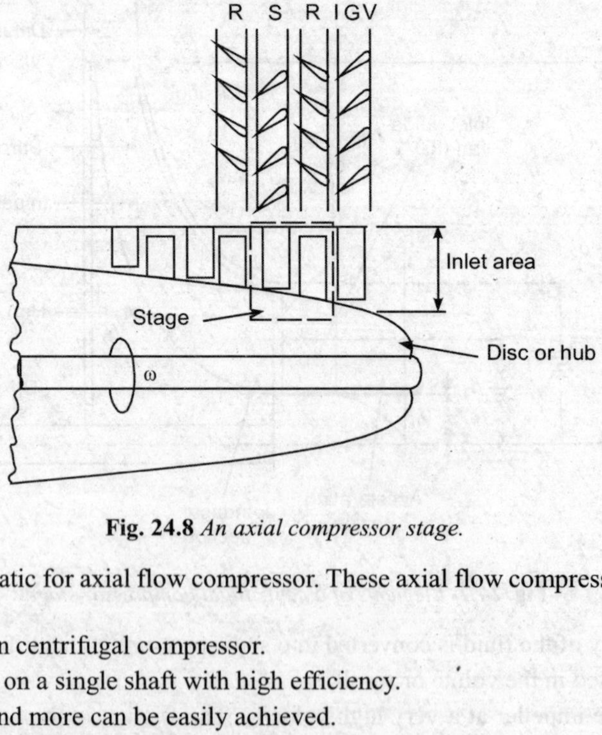

Fig. 24.8 *An axial compressor stage.*

Figure 24.8 shows a schematic for axial flow compressor. These axial flow compressors have the advantages as given below :

(a) Higher efficiency than centrifugal compressor.

(b) Higher pressure ratio on a single shaft with high efficiency.

(c) Pressure ratios of 8 and more can be easily achieved.

(d) It gives high thrust per unit frontal area.

(e) It can handle large amount of air, inspite of small frontal area.

The main disadvantages of axial flow air compressor are :

(a) Large length (b) Construction is complex (c) High cost of manufacture.

24.7 COMBUSTION CHAMBERS IN GAS TURBINE PLANTS

For gas turbines operating on the closed cycle wherein the air is used repeatedly, an air-tight heat exchanger or air-boiler must be used to impart heat to the working fluid. This is not a very efficient method of heat supply. In case of open cycle gas turbines a fresh supply of air is inhaled continuously and direct method of burning fuel in the working stream is always used. It is this type of combustion system which is dealt here. As the combustion is continuous an electric spark is only required for initiating the process, and thereafter the flame will be self sustaining or self piloting. Arrangement of combustion chamber is shown in Fig. 24.9. Primary air is supplied for combustion and secondary air is mixed with primary to reduce the gas temperature suitable for gas turbines.

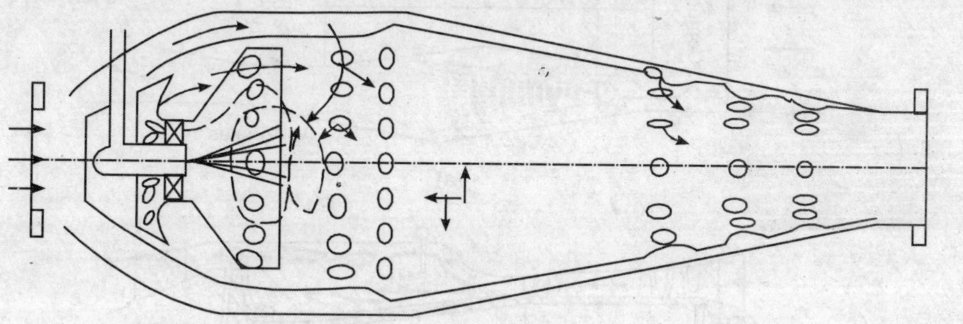

Fig. 24.9 *Combustion chamber*

There are mainly three types of combustion chambers (Geometry wise) :

24.7.1 Cylindrical Combustion Chamber

In case of aircraft practice these chambers are placed around the shaft connecting the compressor and turbine as shown in Fig. 24.10. Each chamber is supplied with separate stream of air from compressor and it is having its own fuel jet fred from a common supply line.

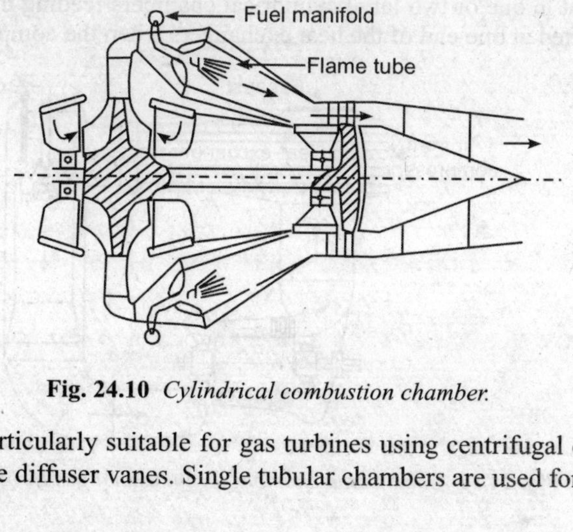

Fig. 24.10 *Cylindrical combustion chamber.*

This type of the layout is particularly suitable for gas turbines using centrifugal compressors because the air streams is already divided by the diffuser vanes. Single tubular chambers are used for small gas turbines.

24.7.2 Annular Combustion Chamber

A single annular combustion chamber surrounding the rotor shaft is suitable with axial compressors. Annular combustion chamber is shown in Fig. 24.11. This type has advantage of low pressure loss due to maximum utilization of space between compressor and turbine. There are some drawbacks. It is difficult to obtain even fuel-air distribution and even temperature distribution. It is difficult to avoid buckling of hot flame tube walls.

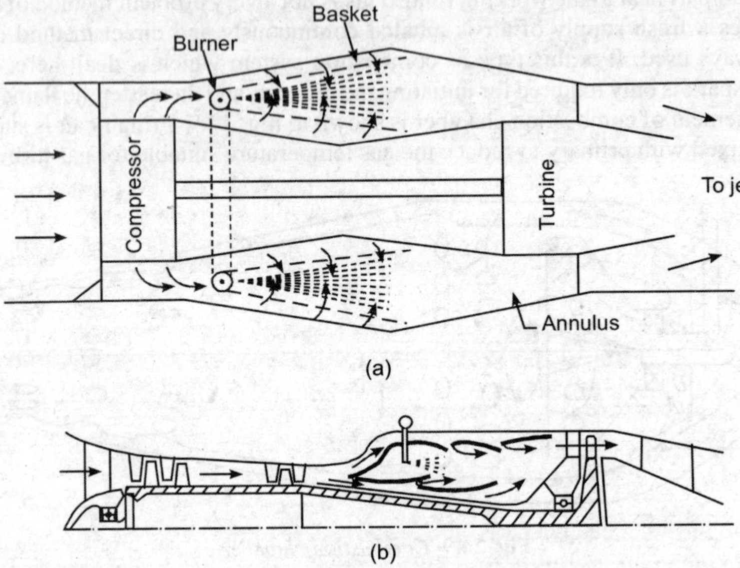

Fig. 24.11 *Annular combustion chamber.*

The combustion of cylindrical and annular combustion chambers is cannular or turbo-annular. This consists of cons or tubular flame tubes uniformly spaced around annular casing. This tube is widely used for large aircraft engines.

24.7.3 Combustion Chamber for Industrial Gas Turbine

The space occupied by the combustion system of an industrial gas turbine is usually one of secondary importance and combustion is carried out in one or two large cylindrical chambers feeding the turbine via a scroll or volute. The chambers are often situated at one end of the heat exchangers when the component is included in the cycle.

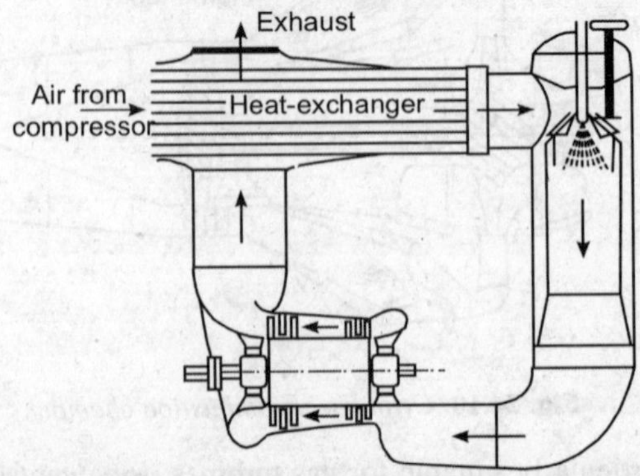

Fig. 24.12 *Combustion chamber for industrial gas turbine.*

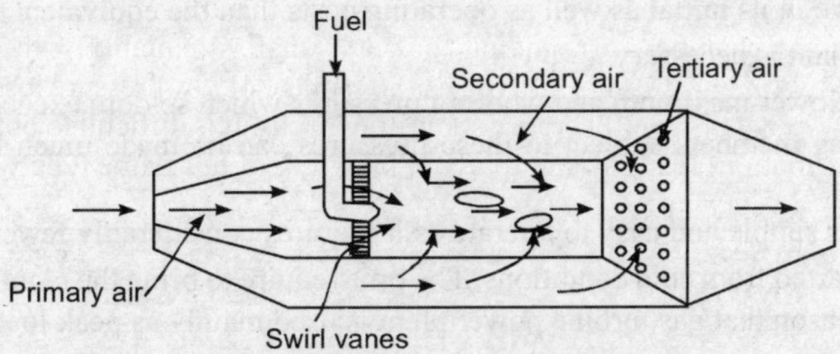

Fig. 24.13 *Downstream type combustion chamber.*

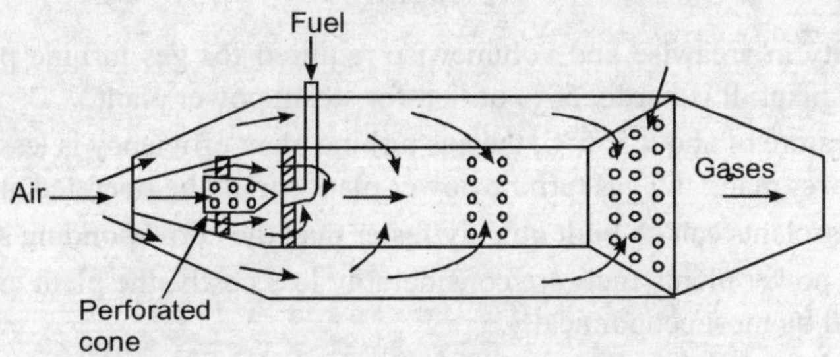

Fig. 24.14 *Upstream.type of combustion chamber.*

Other versions of combustion chambers are shown in Figs. 24.13 and Fig. 24.14.

There are some desirable requirements of the combustion chambers. They are :

1. High efficiency of combustion.
2. Pressure drop should be as small as possible.
3. Easy starting at all atmospheric conditions.
4. Air : Fuel ratio varies from 40 : 1 to 200 : 1, so that the combustion chamber should be able to operate within this range.
5. Carbon deposits should be absent to increase the life of combustion chamber.
6. For aircraft application it should be small and light in weight.
7. Temperature of gases or air should be uniform so as to avoid temperature stresses on the blade. Hence it requires high degree of mixing.
8. Flame should be self-sustaining or self-piloting.
9. Combustion gases going to the turbine should be cooled adequately so as not to affect the blades.
10. It must be capable of handling large amount of air required by hot gases.
11. Combustion chambers must be capable of maintaining the combustion in high velocity of gases as 180 m/s to 60 m/s.

24.8 ADVANTAGES OF GAS TURBINE PLANTS OVER STEAM TURBINE POWER PLANTS

24.8.1 Advantages

Advantages of gas turbine power plants over steam power plant (thermal power plant) are enumerated as follows :

1. It is simpler in design since there are no boiler with their auxiliaries.
2. No feed water is required. Similarly circulating cooling water in condensers—which is a very large quantity is not required.

3. It is less expensive in its initial as well as operating costs than the equivalent steam power plant.

4. No condensing plant is necessary.

5. Due to the much lower maximum and working pressures which seldom exceed 5 to 6 bar, on modern gas turbines, the stress bearing members subject to these pressures can be made much lighter than those of steam turbines.

6. The gas turbine is simple and easy to operate as it requires considerably fewer parts and controls.

7. It can be easily started from cold conditions. The time require to bring the plant to take the load is very very short. (This is the main reason that gas turbine power plant is used mainly as peak load power plant.)

8. Maintenance costs are low.

9. The handling of ash is completely absent in gas turbine power plant. This problem is a major problem in steam turbine power plant.

10. The space capacity in areawise and volumewise required for gas turbine power plant is very small as compared to steam power plant. It is hardly 50% of that for steam power plant.

11. Below the temperature of about 550°C, the gas turbine plant efficiency is less than steam power plant and to compete with steam power plant, the gas turbine power plant should be operated at temperatures above 550°C.

12. Gas turbine power plants can be built quickly/faster than the corresponding steam power plants.

13. If the gas turbine power plants fuels are considerably less costly, the plant can be operated as **Base load** power plant because it will be most economical.

14. The gas turbine power plants uses fewer auxiliaries than the steam power plants. This is the reason that the small size of the gas turbine plant components enables the work tested units, can be transported to the plant site.

15. As the gas turbine plant has fewer components, these components and circuits of the gas turbine plant can be arranged quickly to give the most economic results which is not possible in case of steam power plants.

24.8.2 Disadvantages

1. **Starting Difficulty.** Before starting the turbine which drives the compressor, compressor is to be operated for which driving power from some external source has to be arranged. Once the compressor peaks up sufficient speed, the unit starts and there is no problem.

2. Major work developed by the turbine is required to run/drive the compressor and hence the net available work is less and the efficiency of the gas turbine power plant is less.

24.9 ADVANTAGES OF GAS TURBINE POWER PLANT OVER DIESEL POWER PLANT

24.9.1 Advantages

Advantages of gas turbine power plant over diesel power plant are as follows :

1. Gas turbine unit has less vibrations, particularly absence of cyclic vibrations. This is due to the fact that the gas turbine unit (including compressor and generator) is a rotory machine and hence the perfect balancing of the unit is possible and hence no vibrations.

2. Being rotory, gas turbine unit is very little sliding surfaces and hence less friction. Thus there is less friction power and hence the mechanical efficiency of the turbine plant is high of the order of 90 to 95%. As against this, sliding surfaces in Diesel Power Plant are large and hence large friction power so that the mechanical efficiency of the Diesel Power Plant hardly exceeds 75–80%.

3. Gas turbine power plant has good reliability.

4. In gas turbines, the expansion of gases is upto the atmospheric pressure (complete expansion) and therefore the work developed per kg of gas is larger than that in Diesel Power Plant.

5. Specific weight of gas turbine power plant is very less (0.2 kg/kW) as against 0.34 kg/kW for Diesel Power Plant. Therefore, the space required and the capital cost of the gas turbine plant is considerably less than diesel power plant. Smaller plant dimensions for equivalent output.

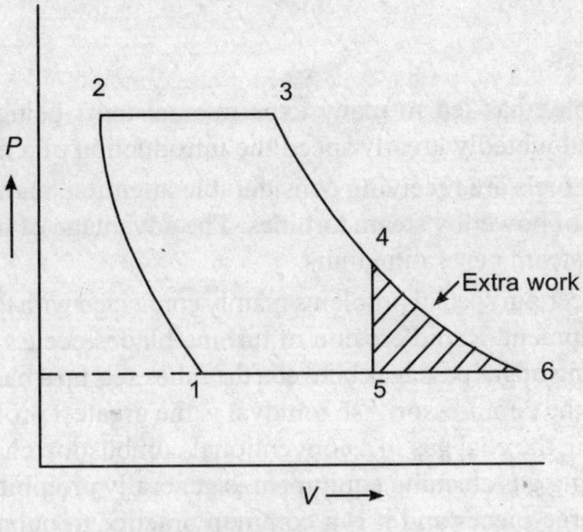

Fig. 24.15

6. Maintenance of gas turbine plants is easy and less costly as compared to diesel power plants.
7. Installation charge are less in case of gas turbine plant than diesel power plant.
8. Being rotory, lubrication is mainly required for bearings and is simpler than that is required for diesel power plant.
9. Ignition problems are more simpler as compared with diesel plant.
10. Smoke is about in the exhaust of gas turbine plant becuase the excess air supplied in gas turbine is nearly 400% (4 times that required theoretically).

24.9.2 Disadvantages

1. Part load efficiency of gas turbine plant is poor or less as compared with diesel power plant.
2. Special metals and alloys are required for gas turbine plant components.
3. Special cooling system for (turbine blade) cooling has to be provided in gas turbine plants.
4. Gas turbines will have short life as compared with diesel plant.

24.10 FUELS FOR GAS TURBINE POWER PLANTS

In general, there are three types of fuels available. They are :

(i) Solid fuels

(ii) Liquid fuels and

(iii) Gaseous fuels.

Gas turbines have been designed to run on all these three types of fuels. In gaseous fuels, natural gas and blast furnace gas are the most common. In liquid fuels cheaper distillate oils and bunker (residual) oils have been used. Development work of many years has striven to design the pulverizers, coal pumps and low pressure loss ash separators that will enable a gas turbine to burn coal.

Thus, an inherent advantage of the gas turbine is its potential ability to burn a very wide range of fuels. In the absence of a very large increse in the operating temperature, it seem unlikely that the overall efficiency of this prime mover will be significantly greater than the diesel engines or steam turbines. Solid, liquid and gaseous fuels have all been utilized to some extent in gas turbines with varying degrees of success where there was an economic advantage in departing from the kerosene used in first gas turbines. In this section, a survey of various types of fuel is made indicating any special influence of particular fuel on the choice of material of costruction.

24.10.1 Solid Fuels

The adaptability of the gas turbine has led to many experimental units being designed for solid fuels and the knowledge thus obtained will undoubtedly greatly speed the introduction of a number of commercial models.

At the present time, two solid fuels are receiving considerable attention, namely coal and peat. These are freely available and used in generation of power by steam turbines. The advantage of using them in the gas turbine would lie in abolition of the expensive steam generating units.

The use of solid fuels presents certain special problems mainly connected with the provision of efficient combustion conditions and ash removal equipment. Rapid erosion of turbine blades occurs due to ash and it blocks passages, causing serious fouling. The fouling of gas passages between the tubes sets up a back pressure which would eventually give rise to surging or stalling in the compressor. Ash removal is the greatest problems. This can be avoided by first gasifying the coal and then issuing the coal gas in a conventional combustion chambers. The size and initial cost of gasification unit and accompanying gas cleaning equipment is generally prohibitive. The pulverized solid fuels can be burnt more efficiently than large pieces and it is a common practice to pulverize even high grade coal. An oil pilot flame is required to maintain ignition of the solid fuels. Two main designs of combustion chambers have been developed for use of coal as a fuel and they are : (i) vortex combustion chamber and (ii) cyclone combustion chamber.

The main difficulty to be overcome in the use of **peat** is to remove the large amount of water present. Peat is formal from dead plants by the action of bacteria and fungi in water logged environments. The natural peat contains approximately 90% water, and 10% fuel by weight. The amount of heat required to drive off this water is greater than the calorific value of the residual fuel. Consequently the greater part of the water must be removed by mechanical means, such as pressing. This method removes about 70% water and the remaining water is unremovable by this technique. It is removed by subjecting the pressed material to the drying action of hot gases issuing from the gas turbine. The peat is then shredded by special machinery before being fluidised for freeding into a refractory combustion chamber. The problem of ash removal is similar to that in the case of coal.

24.10.2 Liquid fuels

The use of oil and oil products as a source of power has increased enormously in the present century. Mineral oil is found in sedimentary rocks. It is believed to originate mainly from the action of bacteria on organic material in stagnant waters, where atmospheric oxygen did not penetrate. The bacterial activity resulted in the formation of the heavy hydrocarbons and the subsequent formation of the lighter paraffins probably occurred as a result of the pressure and temperature arising during the burial of the material by the accumulation of sediments and strata movements. The crude oil gradually drained from the strata where it was formed and tended to collect in particular locations where the over and underlying rocks formed a natural and impervious trap. The gaseous decomposition products were also sometimes prevented from leaking away and remained in the space above the oil pool.

The crude natural oil is a mixture of solids, liquids and gaseous substances. The crude petroleum undergoes special refinery processes and fraction distillation, resulting in production of several materials. The initial distillation separates the crude petroleum into :

(a) Topped crude

(b) Gas oil

(c) Kerosene, otherwise known as paraffin oil

(d) White spirits, solvents etc.

(e) Gasoline (petrol) of various grades of volatility.

The petrol and gas oil fractions provide the main fuels for internal combustion engines. The steam turbines may be operated by burning of heavy fuel oils in the steam generating plant. The gas turbine, however, is capable of producing energy directly from the combustion of a number of these products ranging from kerosene to the low grade 'residual' oils i.e., a fuel containing a proportion of refinery residues. Petrol is, of course, too expensive and too explosive to form a satisfactory fuel.

Ever since the early days of gas turbines, kerosene has been used in aircraft power units. The reasons for this choice are mainly :

(a) Kerosene was widely available at reasonable price.

(b) The danger of fire is appreciably lower than that of the other aviation fuels.

(c) The fluidity, even at the ice point, makes any preheating unnecessary.

(d) Kerosene is satisfactory for the ignition requirements and burns with a steady flame.

In marine or land-based gas turbines, kerosene is too expensive for all, but the low pressure units. At the present time, the various grades of distillate oil are being used in the majority of higher power units, but the developments now being taken place in the use of heavier oils will eventually lead to the adoption of such fuels. This could result in an appreciably lowering of the running costs of a gas turbine.

The main problems to be overcome are concerned with the efficient combustion of low grade fuels and corrosive effect of the products of combustion, particularly when the sulphur and vanadium contents of the fuel are high.

Availability is a further important factor in the selection of a fuel for marine use. The fuel must be readily available at the majority of the ports since the ships cannot be expected to rely entirely on their home port suppliers. The cost of the fuel must also be as low as possible.

One trouble which arises in the use of heavy fuel oils is the deposition of the carbon on the fuel injection tubes and the walls of the flame tube. The origin of this carbon are two-fold, quenching of the burning particles in the latter stages of combustion and the "cracking" of unburnt fuel on the very hot metal parts. The latter will obviously more prevalent at the fuel entry, and will be accentuated by poor atomisation, over rich mixtures and the use of fuels of high aromatic content. The presence of this carbon can lead to carburisation effects in the underlying metals and serious loss of mechanical properties can be caused.

The use of residual oil (cheap fuel) presents some undesirable characteristics. They are :

1. High viscosity and hence require heating before delivery to the atomizers.
2. Overheating may cause polymerization to form sludge or tax.
3. Many form jelly-like substances if come in contact with other oils. This can clog the fuel system.
4. Excessive carbon deposition in the combustion chamber due to high carbon content.
5. Presence of vanadium which is corrosive.
6. Presence of alkali metal such as sodium which forms corrosive sulphates.
7. Large ash formation.

24.10.3 Coal Tar

Coal tar is a very useful fuel for the gas turbine, since it contains no serious impurities. It is unlikely to become an important gas turbine fuel since it is only available in limited quantities. Coal tar is mainly a byproduct of the high temperature carbonisation of coal, although a small amount is obtained by low temperature carbonisation. The crude coal tar is subjected to a distillation process which yields a number of products such as ammoniacal liquor, light oil, middle oil, creosote oil, anthracene oils, soft pitch etc.

A number of fuels have been obtained by blending these various products in different amounts and for convenience these fuels are designated by a number which is the temperature in °C to which the fuel must be preheated for satisfactory atomisation and combustion.

For efficient combustion, good atomisation is essential. The flame temperature is in the region of 1700°C and the flame is very luminous, so that a high rate of heat transfer is experienced. In the gas turbines constructed to use this fuel, it has been found preferable to use a refractory lined combustion chamber so as to avoid the problems which arise with such a luminous flame in the cooling of metal flame tubes.

24.10.4 Gaseous Fuels

The majority of oil fields are rich in sources of **Natural gas** which collects in the rocks above the oil pools. The composition of this gas varies over wide limits, but the percentage of standard saturated hydrocarbons, especially methane, is usually high. The typical average analysis of the gas is :

Methane	80%
Ethane	6%
Propane	3%
Butane	1%
Nitrogen	10%

Since the gas contains a very small percentage of non-combustibles, the calorific values are high.

A similar type of fuel can be obtained from some coal-seams, where the nature of the strata is such that large quantities of methane have been extrapped during the organic decomposition.

Since methane is the main gas produced during the decomposition of organic material, methane-air mixtures are available, and are used for power production, in the large sewage works. There are some recent instances of specially designed gas turbines being installed for power production from these **sludge gases** as they are called.

Blast furnace gas is a valuable by-product of iron-making and since its use in the furnace air preheaters is limited to 20–25% of the amount produced, a considerable proportion is available for power production. The gases issuing from the top of the blast furnace contains a fairly high proportion of carbon monoxide together with a little hydrogen, along with non-inflammable gases such as carbon dioxide and nitrogen. The analysis of the gas varies widely according to the conditions in the furnace.

24.11 MATERIALS FOR GAS TURBINES

The maximum metal temperature in internal combustion engines rarely exceeds 250°C though intermittently the temperatures of the order of 3000°C are encountered. I.C. are water cooled or air cooled. In the steam boilers metal surfaces which come in contact with hot gases are water cooled. The gas turbine blades are exposed to a continuous stream of gas at temperatures of about 1000°C. The blades attain high temperatures. The blades must be cooled to reduce the temperature. The blades are also subjected to a tensile stress of the order of 150–220 MN/m^2. The feasibility of the gas turbine is dependent exclusively on the high temperature supper alloys. The blades are subjected to forces in three directions. Axial forces are caused by the gas flows. The rotor driving force acts along the radial direction and the forces acting normal to the turbine are due to centrifugal forces. Further, the blades are subjected to differential thermal stresses, erosion-corrosion etc. The factors influencing the selection of materials, requirements for high temperature materials, some special materials and blade cooling are to be studied now.

24.11.1 Factors Influencing Selection of Materials

Higher the inlet temperature for turbine, higher will be the thermal efficiency. This conclusion applies to simple as well as regenerative gas turbines. The gas turbine parts—casing, blades, rotor disc and stator blades are in continuous contact with high velocity hot gases. In a low compression plant, the air inlet at the compressor is about 15°C, the temperature at the end of compression is 250°C (523 K), the inner flame-tube temperature is 2000°C, and the temperature of the gases after dilution and mixing with secondary air is about 800 to 1000°C. The temperature of gases while passing through blades is about 500–700°C. The temperature ranges are still higher in modern and advanced plants. Thus the turbine operating conditions are very severe, ranging from regions of low temperature and moderate stress to regions of very high temperature and high stress. Close dimension control must be maintained. The gas turbine problems are—engine life, weight and size, cost and processing methods of materials. Thermodynamically good performance is obtained at high temperatures, but the material availability restricts the performance limit. The designer is not always free to follow the dictates of the thermodynamics and aerodynamic desirabilities. The designer must know that for the given temperatures and a given expected life, what stress can be carried within the assigned criterion of failure. He must also know the type of loading. Static loads lead to consideration of static strength and fluctuating loads require consideration of fatigue and strength.

Following factors must be considered while selecting blade materials :

1. Materials should possess good strength at high temperatures owing to high operating temperatures.
2. Rotation speeds are high (about 30,000 rpm). This leads to centrifugal tensile stresses.
3. It must have high creep strength. Bending stresses due to equivalent impulse load of the gases acting on cantilever blade at a certain distance from fixing should be considered.
4. Corrosive and erosive effects due to high temperature combustion products viz. CO and CO_2 with water.
5. Varying temperature and high structural stability when exposed to varying temperature.

The blade materials should possess the following characteristics :

1. Castability/Forgeability
2. Weldability
3. Machinability
4. No embrittlement during operation.

24.11.2 Requirements for High Temperature Materials

Information about following properties is needed for use in actual design and for comparison of materials in order to make a proper choice.

1. Creep, creep-rupture strength and creep-relaxation characteristics
2. Yield strength
3. Fatigue (endurance) strength
4. Ductility and modulus of elasticity
5. Thermal conductivity and thermal coefficient of expansion
6. Density
7. Hardness
8. Erosion, corrosion, thermal shock, impact and oxidation resistance
9. Effect of temperature cycles and notches
10. Damping capacity
11. Consistency in reproduction of properties
12. Availability and cost
13. Methods, costs of processing.

The most critical parts are blades and rotors. Following table gives an idea of some of the important material properties which must be considered for specific parts.

Part	*Temperature*	*Material properties*
Combustion Liner	100°C	Ductility, strength, corrosion, thermal properties and thermal shock
Turbine buckets	650–850	Fatigue strength, stress rupture, creep, ductility, corrosion, thermal shock
Turbine wheel	300–600	Yield strength, ductility, creep

A metal undergoes elongation with time when subjected to constant stress at constant temperature below the elastic limit. This is known as creep. Only slight recovery is possible after cooling and stress removal. Creep is a matter of great concern owing to danger of rubbing of moving parts and casing. Creep is also important for bolts and other connecting devices. The life of machine depend on its application. Aircraft turbines are designed for 1000–1200 hours, while industrial turbines are designed for 1,00,000 hours.

The materials fails due to the effects of repeated stresses and the failure is due to fatigue. Fatigue failure occurs suddenly and without any warning.

The damping capacity of a material is the amount of internal dissatisfaction per unit volume of the material for one cycle of stress. Damping capacity decreases with higher tensile strength. Blades and rotor discs are subjected to vibrations. Hence, for avoiding fatigue failure, their damping capacity is important. Gas turbine blades are subjected to high stresses at high temperature. Chrome alloys are suitable for blades.

Turbine materials are exposed to highly corrosive and oxidizing combustion gases. Hence, materials used must be corrosion resistant. Corrosion can cause rapid failure and hence it is more important than strength consideration.

24.11.3 Some Typical Materials

1. **Compressors.** The problem of materials is not so acute in compressors. Light weight is desirable in aircraft units. Aluminium and magnesium alloys proved to be satisfactory for compressor casing and accessories.

The impeller of centrifugal compressor is subjected to high centrifugal and thermal stresses. Lighter materials are useful in reducing centrifugal stresses (i.e., aluminium). But these suffer from thermal expansion and allowance is provided for that.

The axial flow compressor blades are now made of titanium alloys. Titanium alloys are low density alloys, strongly corrosion resistant and possesses good strength at higher temperature. Light weight, good creep strength and fatigue resistance are the attractive features of titanium alloys. Iron and steel are found to be economical for use in ships, locomotives and stationary plants where light weight is not essential.

2. **Combustion Chambers.** The gas turbine combustion chamber is generally made of Nimonic-75 alloy. This has an excellent creep resistance. The material used must be corrosion resistant as well as heat resistant. Combustion lines are made from materials such as chromium-nickel, steel and iconel. The inner chamber of the combustion chamber is sometimes created with a refractory material.

3. **Turbine Rotor Discs.** The rotor disc is subjected to high centrifugal and thermal stresses. The outer rim is hottest because it is closer to hot combustion gas and hot blades. The inner region is comparatively at lower temperature. In order to limit the temperature, the cooling air is often directed towards the hot sides of the discs.

The thermal stresses can be reduced by using an alloy of high thermal conductivity. Generally, the discs are forged and iron base alloys are used. The disc hub stresses tend to cause tensile deformation. This can be minimized by using a material of low expansion coefficient. Austentic steels with 12 to 18 percent chromium, 8 to 12 percent nickel and small percentages of tungsten, molybdenum and titanium are used for rotor disc. These days the composite discs are popular. The centre is made up of different materials from the rim. The two are then welded. This type of construction is satisfactory in jet engines where disc cooling is employed.

4. **Turbine and Compressor Blades.** The blades and other parts of turbines and compressors are manufactured by methods like forging, machining, fabrication, powder metallurgy, rolling etc. Forging and precision casting are the most widely used processes of production of turbines' and compressors' blade. The turbine blade materials are Vitallium, Hastelloy B, Haynes alloy 36. Blades are subjected to highest stresses and temperatures. Most satisfactory materials for blades are steel alloys, 8–20 nickel chromium alloys (Nimonic). These have high resistance to oxidation, scaling and deformation and have good creep and fatigue properties. A uniform coating of ceramics like silicon carbine, silicon nitride and aluminium nitride, on the blades of Nimonic alloys provides better mechanical properties. Blades are cooled by compressed air taken by a bleed from the compressor.

5. **Bolts.** A bolt should hold the tightness of tightened joint for a long period of operation. Cracking should not occur in thread roots or under the head. Bolt material should not be brittle and should have strength at elevated temperatures. They should exhibit good stability and corrosion resistance. Inconel X, K-42-B, S-816 and 5590 are used for bolts.

6. **Ceramics.** Ceramics possess very high heat and abrasion resistance. They have a very low conductivity, a high coefficient of thermal expansion. Ceramics are low cost materials. Some ceramics have poor heat shock properties and are brittle. The creep properties are satisfactory and melting points varies from 1200 to 4200°C as against 1250 to 1550°C for steels and heat resistant materials. Experimentation is going on for the use of ceramics since second world war. The attachment of ceramic blades to a metal rotor or even other ceramic material is difficult because of difference in coefficients of expansion. There are two ways to use ceramics — use for complete part such as rotor, stator blades or use as coating material for increasing life of hot parts subjected to corrosion.

The main disadvantage of ceramics is low resistance to mechanical and thermal shock. Hence studies are going on towards the combination of metals and ceramics.

The manufacture of a ceramic body usually involves powdering of the constituents, mixing with plasticizing and bonding agents and forming of desired shape by either pressing or extruding. Drying and curing occur next and then the body is sintered to the final product.

24.12 THERMODYNAMIC CYCLE FOR GAS TURBINE POWER PLANT AND ITS ANALYSIS

The most common type of the gas turbine that is used for electric power generation is a constant pressure gas turbine. Its ideal or theoretical thermodynamic cycle is Joule cycle. This is also called as air standard cycle for gas

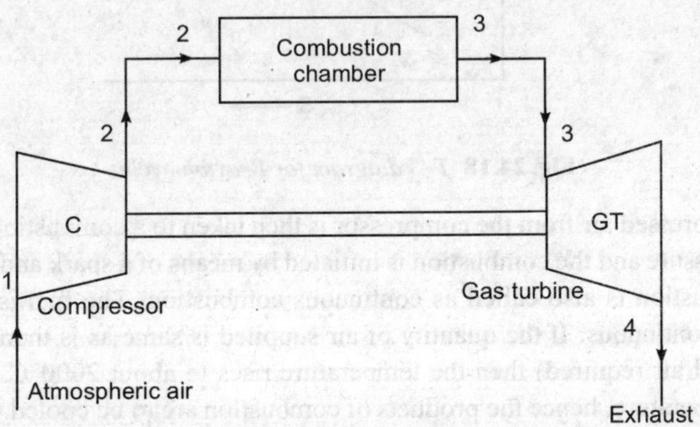

Fig. 24.16 *Schematic for gas turbine*

turbine. The actual cycle is called Brayton cycle for constant pressure gas turbine. The schematic diagram of gas turbine unit is shown in Fig. 24.16.

It is an open cycle, constant pressure gas turbine. The working of this type of gas turbine is as follows :

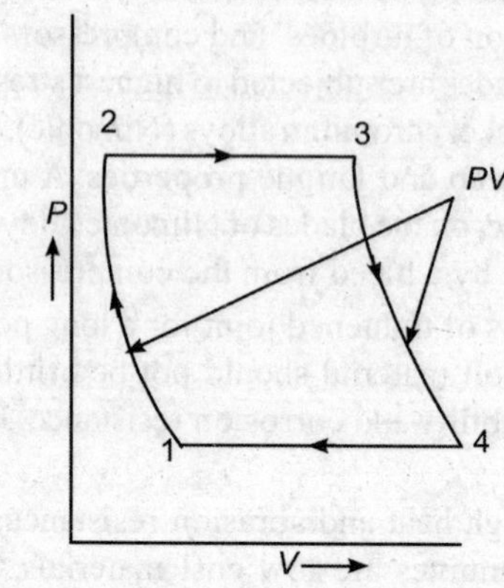

Fig. 24.17 *P–V diagram for Brayton cycle.*

Process 1-2 : Fresh atmospheric air is drawn into the compressor. The compressor used may be either centrifugal type or axial type. The pressure ratio for the compressor is about 4 to 8, and the compression is ideally assumed to be isentropic. This process is shown on *P–V* diagram of Fig. 24.17 and on *T–S* diagram of Fig. 24.18.

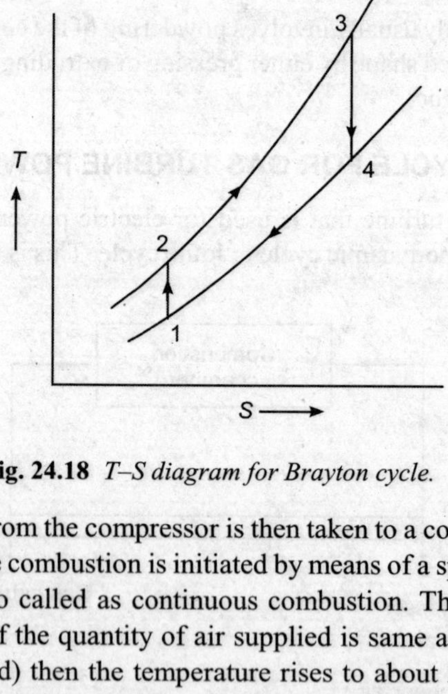

Fig. 24.18 *T–S diagram for Brayton cycle.*

Process 2-3 : The compressed air from the compressor is then taken to a combustion chamber. The combustion takes place at constant pressure and the combustion is initiated by means of a spark and then the burning of the fuel is continuous. This combustion is also called as continuous combustion. The fuel is injected continuously and hence the combustion is continuous. If the quantity of air supplied is same as is there in IC engines (10 to 20% excess over the theoretical air required) then the temperature rises to about 2000°C. The turbine blades cannot withstand such a high temperature, hence the products of combustion are to be cooled to the accepted temperature for the blade materials. This accepted temperature (permissible temperature) is in the range of 750 to 1000°C. The air required for cooling the products of combustion is generally called **secondary air** and the minimum air (including excess air) supplied for combustion of the fuel is called **primary air**. The heat addition process is shown by 2–3 on *P–V* diagram as well as on *T–S* diagram. Figure 24.18 shows the arrangement of cooling the products of combustion.

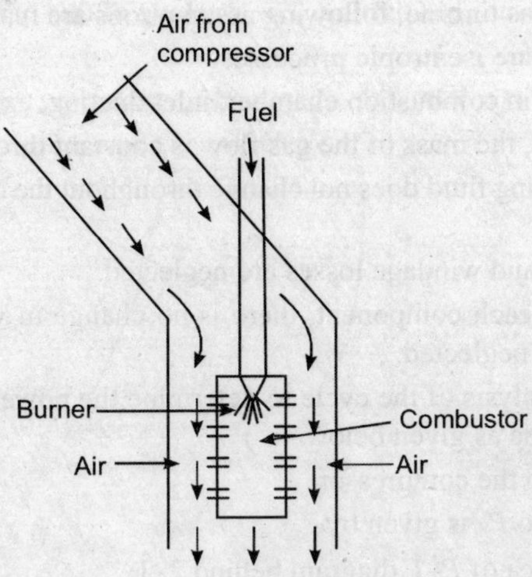

Fig. 24.19 *Cooling of products of combustion.*

Process 3-4 : Products of combustion from the combustion chamber are admitted to the gas turbine where the gases expand isentropically and work is developed. Both pressure and temperature are reduced. This expansion process is shown on *P–V* and *T–S* diagrams of Fig. 24.2 and Fig. 24.3.

Process 4-1 : After expansion the gases are exhausted to atmosphere and the heat is rejected to atmosphere at constant pressure. This process is shown by 4-1 on *P–V* and *T–S* diagram.

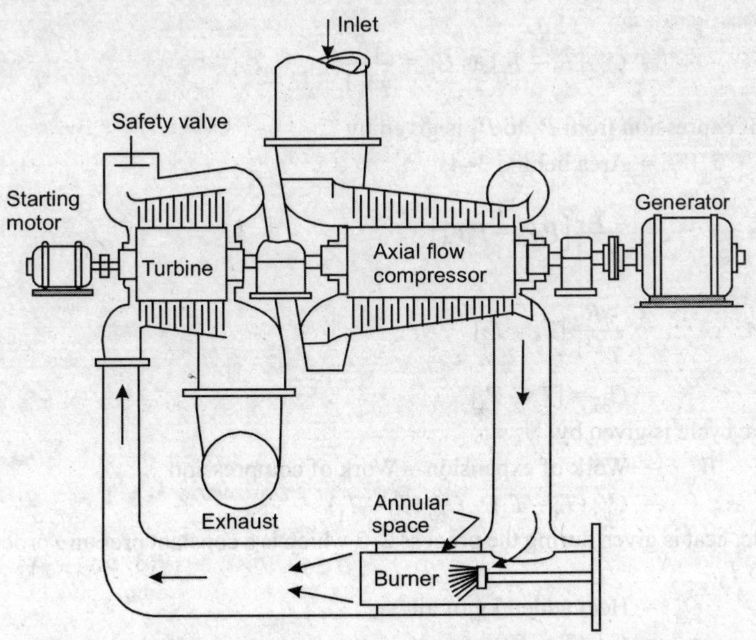

Fig. 24.20 *Simple gas turbine power plant.*

The turbine is coupled to the generator and compressor. The compressor absorbs about 60–70% of the turbine output. The remaining power is used to drive the generator which generates electric power. Figure 24.20 shows a simple gas turbine power plant using axial compressor and generator.

To analyse the ideal cycle for gas turbine, following assumptions are made :

1. Compression and expansion are isentropic processes.

2. There are no pressure losses in combustion chamber, inlet ducting, exhaust ducting, heat exchangers etc.

3. It is a steady flow system i.e., the mass of the gas flow is constant throughout the cycle.

4. The composition of the working fluid does not change throughout the cycle and the working fluid is a perfect gas.

5. Frictional losses at bearings and windage losses are neglected.

6. Between inlet and outlet of each component, there is no change in velocity of gases or change in kinetic energy of gases or fluid flowing is neglected.

With these assumptions, the analysis of the cycle to determine the power developed, thermal efficiency of the Brayton cycle or Joule cycle is done as given below.

Consider 1 kg of air admitted to the compressor.

Work of compression from P_1 to P_2 is given by

$$W_{\text{com}} = \text{Area of } P\text{–}V \text{ diagram behind 2–1.}$$

$$= \text{(Area behind 2–1)}$$

$$= \frac{\gamma}{\gamma-1}\left[P_2V_2 - P_1V_1\right]$$

$$= \frac{\gamma R}{\gamma-1}\left[T_2 - T_1\right] \text{ as } PV = RT$$

$$= C_{pa}\,[T_2 - T_1] \text{ as } C_p = \frac{\gamma R}{\gamma-1}$$

Similarly work of expression from P_3 to P_4 is given by

$$W_t = \text{Area behind 3–4}$$

$$= \frac{\gamma}{\gamma-1}\left[P_3V_3 - P_4V_4\right]$$

$$= \frac{\gamma R}{\gamma-1}[T_3 - T_4]$$

$$= C_{pa} = [T_3 - T_4]$$

∴ Network of the cycle is given by

$$W_{\text{net}} = \text{Work of expansion – Work of compression}$$

$$= C_{pa}\,(T_3 - T_4) - C_{pa}\,(T_2 - T_1) \qquad \text{(I)}$$

Again in the cycle, heat is given during the process 2–3 which is a constant pressure process and the temperature increases from T_2 to T_3.

$$\therefore \quad Q_A = \text{Heat added/kg of air}$$

$$= C_{pa}\,(T_3 - T_2)$$

Similarly, heat rejected during the process 4–1, is given by

$$Q_R = \text{Heat rejected/kg of air}$$

$$= C_{pa}\,(T_4 - T_1).$$

Equation (I) gives

$$\begin{aligned} W_n &= C_{pa}(T_3 - T_4) - C_{pa}(T_2 - T_1) \\ &= C_p[T_3 - T_4 - T_2 + T_1] \\ &= C_{pa}[(T_3 - T_2) - (T_4 - T_1)] \\ &= C_{pa}(T_3 - T_2) - C_{pa}(T_4 - T_1) \\ &= \mathbf{Q_A - Q_R}. \end{aligned} \tag{II}$$

∴ For **ideal cycle** with compression and expansion as isentropic processes,

$$\begin{aligned} \text{Network} &= W_{\text{net}} = \text{Heat added} - \text{Heat rejected} \\ &= Q_A - Q_R \end{aligned}$$

$$\therefore \quad \boxed{W_{\text{net}} = Q_A - Q_R} \tag{III}$$

∴ Thermal efficiency of the cycle or turbine is given by

$$\begin{aligned} \eta_{\text{th}} &= \frac{\text{Net work}}{\text{Heat supplied}} \\ &= \frac{Q_A - Q_R}{Q_A} \\ &= 1 - \frac{Q_R}{Q_A} \\ &= 1 - \frac{C_{pa}(T_4 - T_1)}{C_{pa}(T_3 - T_2)} \\ &= 1 - \frac{T_4 - T_1}{T_3 - T_2} \end{aligned} \tag{IV}$$

This is the expression for thermal efficiency in terms of temperatures at the salient points of the cycle.

Let

$$\begin{aligned} r_p &= \text{Pressure ratio} \\ &= \frac{P_2}{P_1} = \frac{P_3}{P_4}. \end{aligned}$$

For compression process

$$\frac{T_2}{T_1} = \left(\frac{P_2}{P_1}\right)^{\gamma-1/\gamma} = (r_p)^{\frac{\gamma-1}{\gamma}}$$

$$\therefore \quad T_2 = T_1 . r_p^{\gamma-1/\gamma} \tag{V}$$

For expression process

$$\begin{aligned} \frac{T_3}{T_4} &= \left(\frac{P_3}{P_4}\right)^{\gamma-1/\gamma} = \left(\frac{P_2}{P_1}\right)^{\gamma-1/\gamma} \\ &= (r_p)^{\gamma-1/\gamma} \end{aligned}$$

or

$$T_3 = T_4 . r_p^{\gamma-1/\gamma} \tag{VI}$$

Substituting the values of T_2 and T_3 in Eq. IV, we get

$$\eta_{th} = 1 - \frac{T_4 - T_1}{T_4 . r_p^{\gamma-1/\gamma} - T_1 . r_p^{\gamma-1/\gamma}}$$

$$= 1 - \frac{T_4 - T_1}{r_p^{\gamma-1/\gamma}\left[T_4 - T_1\right]}$$

$$= 1 - \frac{1}{r_p^{\gamma-1/\gamma}} \quad \text{(VII)}$$

$$\boxed{\eta_{th} = 1 - \frac{1}{r_p^{\gamma-1/\gamma}}}$$

From Eqs. (V) and (VI), we have

$$\left(r_p\right)^{\frac{\gamma-1}{\gamma}} = \frac{T_2}{T_1} = \frac{T_3}{T_4}$$

$$\therefore \quad \frac{T_2}{T_1} = \frac{T_3}{T_4}$$

From Eq. (VII), we get

$$\eta_{th} = 1 - \frac{1}{T_2/T_1}$$

$$= 1 - \frac{T_1}{T_2} \quad \text{(VIII)}$$

$$= 1 - \frac{1}{T_3/T_4}$$

$$= 1 - \frac{T_4}{T_3}. \quad \text{(IX)}$$

$$\therefore \quad \boxed{\eta_{th} = 1 - \frac{1.}{r_p^{\gamma-1/\gamma}}}$$

$$= \boxed{1 - \frac{T_1}{T_2}}$$

$$= \boxed{1 - \frac{T_4}{T_3}}$$

From Eq. (VII), we observe that the thermal efficiency of the Brayton or Joule cycle is a function of r_p — the pressure ratio.

At $r_p = 1, \eta_{th} = 0.$

As the pressure ratio r_p increases, thermal efficiency increases.

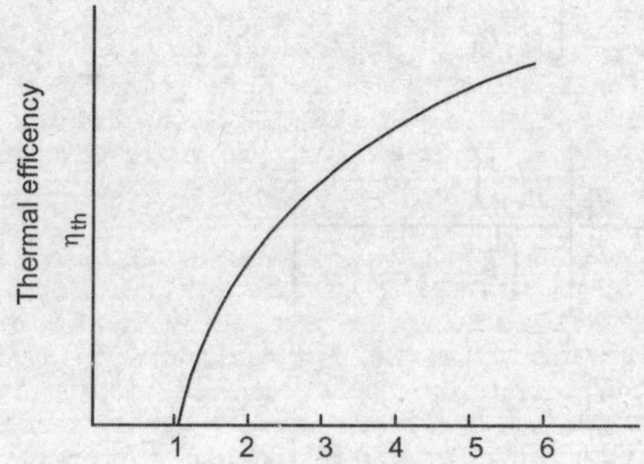

Fig. 24.21 *Pressure ratio vs thermal efficiency.*

Figure 24.21 shows the relation of pressure ratio and the corresponding thermal efficiency.

Though, the thermal efficiency increases with the pressure ratio, the pressure ratio cannot be increased beyond a certain number because the temperature T_2 increases and for the same heat addition in the combustion chamber, temperature T_3 increases. This temperature T_3 is limited by metallurgical considerations. The maximum temperature T_3, the material used for the blades etc. can withstand is called **Ceiling Temperature**. So the ceiling temperature is the limiting value for T_3, T_2 and finally the pressure ratio.

24.13 WORK RATIO

Another term which is useful in gas turbine study is **work ratio**. Work ratio is defined as the ratio of the net work developed by the cycle to the work developed by the turbine.

$$\therefore \quad \text{Work done} = \frac{\text{Net work}}{\text{Turbine work}}$$

$$= \frac{W_T - W_c}{W_T}$$

$$= 1 - \frac{W_c}{W_T}$$

$$= 1 - \frac{C_p(T_2 - T_1)}{C_p(T_3 - T_4)}$$

$$= 1 - \frac{T_1\left(\frac{T_2}{T_1} - 1\right)}{T_3\left(1 - \frac{T_4}{T_3}\right)}$$

$$= 1 - \frac{T_1}{T_3} \cdot \left[\frac{r_p^{\gamma-1/\gamma} - 1}{1 - \frac{1}{r_p^{\gamma-1/\gamma}}} \right]$$

$$= 1 - \frac{T_1}{T_3} \left[\frac{r_p^{\gamma-1}\left(r_p^{\gamma-1/\gamma} - 1\right)}{\left(r_p^{\gamma-1/\gamma} - 1\right)} \right]$$

$$= 1 - \frac{T_1}{T_3} \cdot r_p^{\gamma-1/\gamma} \quad \text{(X)}$$

Work ratio is maximum when

$$\frac{T_1}{T_3} \cdot r_p^{\gamma-1/\gamma} \text{ is minimum}$$

For the same r_p, work ratio is maximum when T_1 is minimum and T_3 is maximum.
Work ratio is denoted by $\mathbf{r_w}$ = **W.R.**

24.14 CONDITION FOR MAXIMUM EFFICIENCY AND NET WORK

We will consider two cases as given below.

I. Some initial temperature and pressure ratio sufficiently large such that the temperature after compression is nearer to the maximum temperature T_3 or T_{max}. The T–S diagram for such a cycle is given in Fig. 24.22. 1–2′–3′–4′–1 is the Brayton cycle.

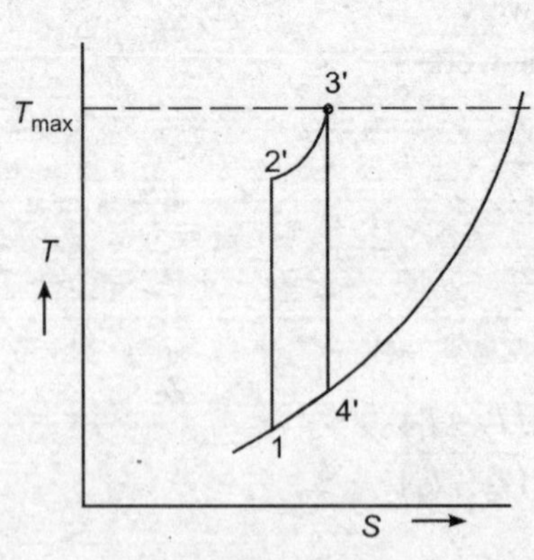

Fig. 24.22 *T–S diagram*

The average temperature of heat addition is $\frac{T_3' + T_2'}{2}$. This will be maximum for maximum efficiency and in that case $T_2' = T_3' = T_{max}$.

Similarly, the average temperature at which the heat is rejected is $\frac{T_4' + T_1'}{2}$ and it should be minimum for maximum efficiency. This will be minimum when $T_1' = T_4'$.

Thus for maximum efficiency, $T_2' = T_{max}$ and $T_4' = T_1$. For this the T–S diagram will be as shown in Fig. 24.23.

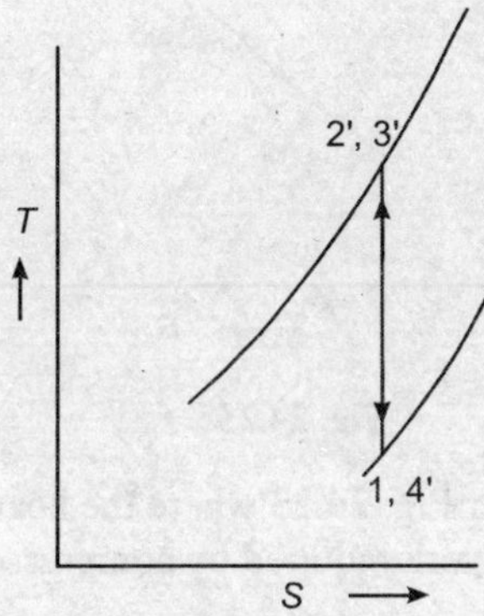

Fig. 24.23 *T–S- diagram*

In such a case, the pressure is the maximum pressure ratio denoted by $(r_p)_{max}$. From Fig. 24.9 we observe that the cycle is given by a single vertical line and work = 0, heat added $Q_A = 0$ and the efficiency is zero $\eta_{th} = 0$.

II. We will consider the second case with the same initial temperature T_1 and the pressure ratio is slightly greater than 1. The cycle is shown in Fig. 24.24 by 1–2′′–3′′–4′′–1.

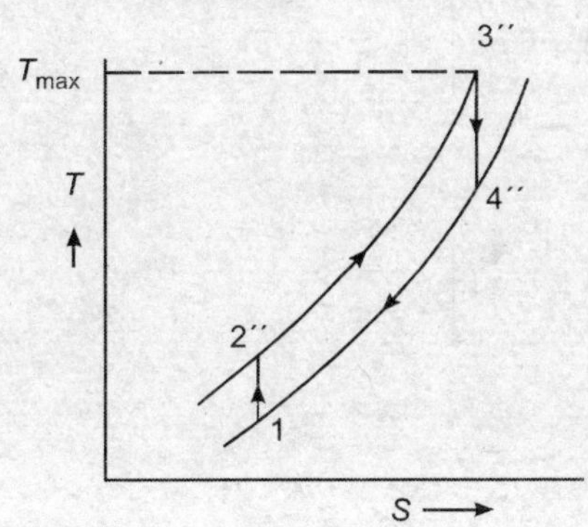

Fig. 24.24

In this case, the heat added is very large as compared to the net work and hence the thermal efficiency is very low. In the limit, the cycle may be considered with $r_p = 1$ so that the cycle diagram will be a single line as shown in Fig. 24.25.

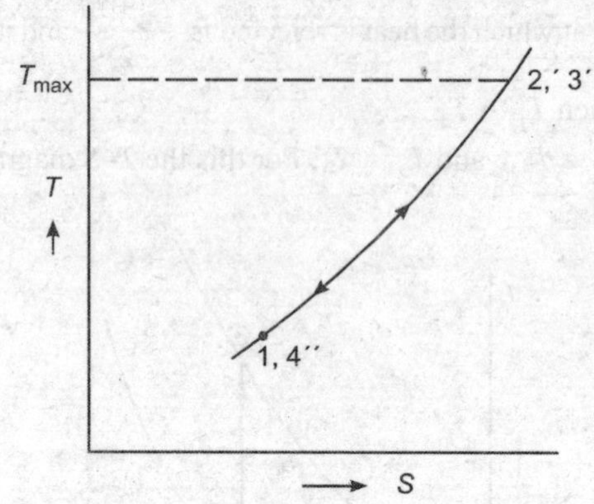

Fig. 24.25

For these two cases considered, we consider Fig. 24.23 where the heat supplied in the combustion chamber is zero and work done by turbine is equal to the work required by compressor or the net work is zero.

$$\therefore \quad W_T - W_c = 0$$

$$W_T = W_c$$

$$C_p(T_{3''} - T_{4''}) = C_p(T_{2''} - T_1)$$

$$\therefore \quad T_3\left(1 - \frac{T_{4''}}{T_{3''}}\right) = T_1\left(\frac{T_{2''}}{T_1} - 1\right)$$

$$T_3\left[1 - \frac{1}{(r_p)^{\gamma-1/\gamma}}\right] = T_1\left[(r_p)^{\gamma-1/\gamma} - 1\right]$$

$$\therefore \quad T_3\left[\frac{r_p^{\gamma-1/\gamma} - 1}{(r_p)^{\gamma-1/\gamma}}\right] = T_1\left[(r_p)^{\gamma-1/\gamma} - 1\right]$$

or
$$\frac{T_3}{T_1} = (r_p)^{\gamma-1/\gamma}$$

$$= (r_p)_{max}^{\gamma-1/\gamma}.$$

or
$$(\mathbf{r_p})_{\mathbf{max}} = \left(\frac{\mathbf{T_3}}{\mathbf{T_1}}\right)^{\gamma/\gamma-1} \qquad \text{(XI)}$$

Thus we have noted two extreme pressure ratios when considered, we get efficiencies and net work as zero. These pressure ratios are $r_p = 1$ and $r_p = (r_p)_{max}$.

Between these two extreme pressure ratios, there lies an optimum value $(r_p)_{opt}$ for which the specific output, the net output is maximum.

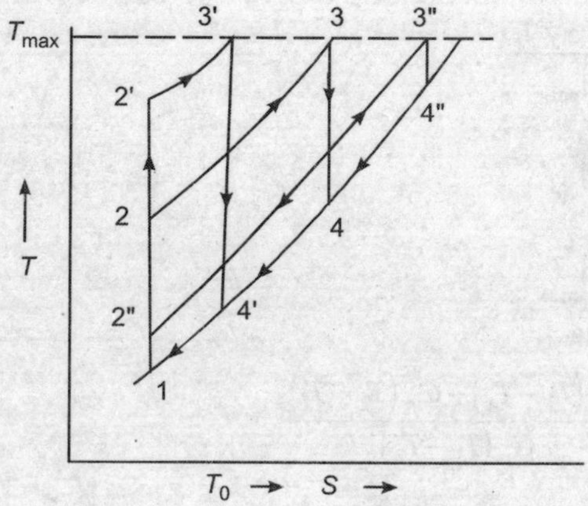

Fig. 24.26 *T–S diagram for optimum* r_p

Figure 24.26 shows a cycle in between the two extremes (1–2–3–4–1) which gives us the maximum net work. Thus,

$$W_{net} = \text{Work of turbine} - \text{Work of compressor}$$

$$= C_p(T_3 - T_4) - C_p(T_2 - T_1)$$

Assuming T_3 and T_1 as constant, and differentiating this equation with respect to (r_p) and equating it to zero, we get the optimum value of the pressure ratio $(r_p)_{opt}$.

$$\therefore \quad \frac{dW_{net}}{d(r_p)} = 0 = \frac{d}{d(r_p)}\left[\left(1 - \frac{1}{r_p^{\gamma-1/\gamma}}\right) - T_1\left\{(r_p)^{\gamma-1/\gamma} - 1\right\}\right]$$

$$= T_3\left[\frac{0 + \frac{\gamma-1}{\gamma}.\, r_p^{\frac{\gamma-1}{\gamma}-1}}{r_p^{2.\frac{\gamma-1}{\gamma}}}\right] - T_1\left[\frac{\gamma-1}{\gamma}.\, r_p^{\frac{\gamma-1}{\gamma}} - 0\right]$$

$$\therefore \quad T_3\left[\frac{\frac{\gamma-1}{\gamma}.\, r_p^{\frac{\gamma-1-\gamma}{\gamma}}}{r_p^{\frac{2(\gamma-1)}{\gamma}}}\right] = T_1\left[\frac{\gamma-1}{\gamma}.\, r_p^{\frac{\gamma-1-\gamma}{\gamma}}\right]$$

$$\frac{T_3}{T_1} = r_p^{\frac{2(\gamma-1)}{\gamma}}$$

$$\frac{\mathbf{T_3}}{\mathbf{T_1}} = (\mathbf{r_p})_{\mathbf{opt}}^{\frac{2(\,-\,)}{\gamma}} \qquad \text{(XII)}$$

We have
$$\frac{T_3}{T_1} = (r_p)_{max}^{\frac{\gamma-1}{\gamma}}$$

or
$$(r_p)_{max}^{\frac{\gamma-1}{\gamma}} = (r_p)_{opt}^{\frac{2(\gamma-1)}{\gamma}}$$

or
$$(r_p)_{max} = (r_p)_{opt}^{2}$$

$$(\mathbf{rp})_{opt} = (\mathbf{r_p})_{max}^{1/2} \quad \text{(XIII)}$$

For maximum efficiency we have

$$\text{Efficiency} = \frac{C_p(T_3 - T_4) - C_p(T_2 - T_1)}{C_p(T_3 - T_2)}$$

Let
$$\frac{T_3}{T_4} = \frac{T_2}{T_1} = x = (r_p)^{\frac{\gamma-1}{\gamma}}$$

and
$$\frac{T_3}{T_1} = y = \text{constant.}$$

$$\therefore \quad \eta_{th} = \frac{T_3\left(1 - \frac{T_4}{T_3}\right) - T_1\left(\frac{T_2}{T_1} - 1\right)}{T_1\left(\frac{T_3}{T_1} - \frac{T_2}{T_1}\right)}$$

$$= \frac{T_1\left\{\frac{T_3}{T_1}\left(1 - \frac{1}{x}\right) - (x-1)\right\}}{T_1(y-x)}$$

$$= \frac{y\left(1 - \frac{1}{x}\right) - (x-1)}{y-x}$$

This is a function of x.

Therefore to get the condition for maximum efficiency, differentiate this relation for thermal efficiency and equate to zero.

$$\therefore \quad \frac{d\eta_{th}}{dx} = 0 = (y-x)\left[\frac{y}{x^2} - 1\right] - \left[y\left(1 - \frac{1}{x}\right) - (x-1)\right]x - \frac{1}{(y-x)^2}$$

$$\therefore \quad (y-x)\left(\frac{y - x^2}{x^2}\right) = \frac{y\left(\frac{x-1}{x}\right) - \left(\frac{x-1}{1}\right)}{(y-x)^2} = \frac{(x-1)\left[\frac{y}{x} - 1\right]}{(y-x)^2}$$

$$= -(x-1)\left(\frac{y-x}{x}\right) \times \frac{1}{(y-x)^2}$$

$$= \frac{-(x-1)}{x(y-x)}$$

$$(y-x)^2(y-x) = -x(x-1).$$

Solving this equation we get

$$y = \frac{T_3}{T_1} = (r_p)_{opt.}^{\frac{2(\gamma-1)}{\gamma}}.$$

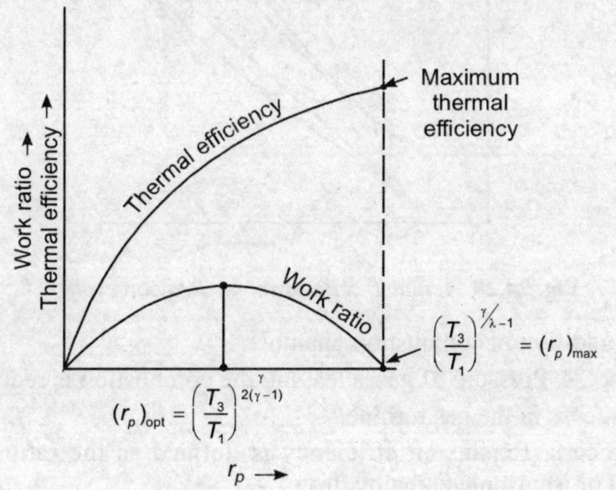

Fig. 24.27 *Relation between r_p, WR, η_{th}*

Figure 24.27 shows relation between pressure ratio and Work ratio and Efficiency.

24.15 ACTUAL BRAYTON OR JOULE CYCLE

We have considered, so far, the ideal working of the gas turbine cycle. In practice, the processes in the cycle are not ideal. Mainly the effect of friction on temperature and pressure is present. Then the cycle of operation becomes practical cycle.

In this actual cycle, the compression and expansion of the working substance are not isentropic and the temperature at the end of process is higher than the theoretical or ideal temperatures. Pressure of the fluid passing through the combustion chamber decreases due to friction. The combustion process in the combustion chamber is again not perfect. With all these irreversibilities, the actual cycle of operation of gas turbine plant is as shown in Fig. 24.28 which is a Temperature Entropy diagram. Thus the actual Brayton cycle is a modified form of theoretical cycle.

With reference to Fig. 24.28, we have

1–2 Isentropic compression process

1–2′ Actual compression process. Entropy increases temperature is more than theoretical. $T_2' > T_2$. Here, one more **term** is to be considered.

Efficiency of compression process or compression efficiency is defined as the ratio of isentropic enthalpy (temperature) rise to the actual enthalpy (temperature rise)

$$\therefore \quad \eta_c = \frac{h_2 - h_1}{h_{2'} - h_1} = \frac{C_p(T_2 - T_1)}{C_p(T_{2'} - T_1)}$$

$$= \frac{T_2 - T_1}{T_{2'} - T_1}$$

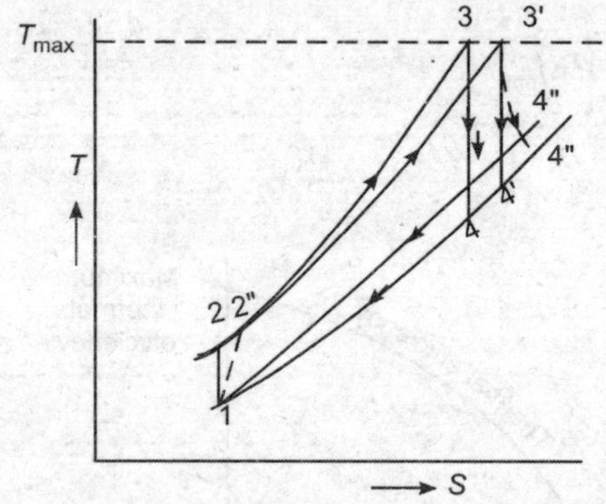

Fig. 24.28 *Actual T–S diagram for Brayton cycle.*

2–3 Constant pressure heat addition in combustion chamber.

2–3′ Actual combustion process. Pressure of gases leaving the combustion is reduced because of friction.

3–4 Isentropic expansion process in the gas turbine.

3′–4′ Actual expansion process. Expansion efficiency is defined as the ratio of actual drop of enthalpy (temperature) to the theoretical or isentropic enthalpy drop.

$$\therefore \quad \eta_t = \frac{h_{3'} - h_4}{h_{3'} - h_{4'}} = \frac{C_p(T_{3'} - T_4)}{C_p(T_{3'} - T_{4'})}$$

$$= \frac{T_{3'} - T_4}{T_{3'} - T_{4'}}$$

4′–1 Theoretical heat rejection process.

In open cycle, gases at 4′ are exhausted into atmosphere.

In closed cycle, the pressure after expansion is slightly greater than the inlet pressure at 1.

In Fig. 24.13, 4′′–1 is the rejection of heat in the cooler and the pressure drops to 1.

Analysis of this actual Brayton cycle is as follows :

$$W_n = \text{Net work/kg}$$

$$= W_T - W_c$$

$$= C_p(T_{3'} - T_{4'}) - C_p(T_{2'} - T_1)$$

$$= C_p(T_3 - T_{4'}) - C_p(T_{2'} - T_1) \qquad \because T_3 = T_{3'}$$

$$\text{But } T_3 - T_{4'} = (T_3 - T_4)\,\eta_t \text{ and } T_{2'} - T_1 = \frac{T_2 - T_1}{\eta_c}$$

$$\therefore \quad W_n = C_p T_3 .\eta_t) \left[1 - \frac{T_4}{T_3}\right] - \frac{C_p T_1}{\eta_c}\left[\frac{T_2}{T_1} - 1\right]$$

$$= C_p T_3 \eta_t \left[1 - \frac{1}{r_p^{\gamma-1/\gamma}}\right] - \frac{C_p - T_1}{\eta_c}\left[r_p^{\gamma-1/\gamma} - 1\right]$$

Considering T_1 and T_3 as constant and also the compression and expansion efficiencies being constant, we can find the condition for maximum net work W_n by differentiating with respect to r_p and equating it to zero.

We get

$$\left(r_p\right)^{\frac{2(\gamma-1)}{\gamma}} = \frac{T_3}{T_1}.\eta_c \eta_t$$

or

$$\mathbf{r_p} = \left(\frac{\mathbf{T_3}}{\mathbf{T_1}}.\boldsymbol{\eta_c \eta_t}\right)^{\gamma/2(\gamma-1)} \qquad \text{(XIV)}$$

This is the pressure ratio at which W_n will be maximum. In ideal cycle, $\eta_t = \eta_c = 100\%$.

$$\therefore \quad \mathbf{r_p} = \left(\frac{\mathbf{T_3}}{\mathbf{T_1}}\right)^{\gamma/2(\gamma-1)} \text{ as before.}$$

For getting the condition for maximum efficiency, we have,

Heat supplied $= C_p(T_{3'} - T_{2'}) = C_p(T_3 - T_{2'})$

$$Q_A \simeq C_p(T_3 - T_2)$$

$$= C_p\left[T_3 - T_1 \cdot r_p^{\frac{\gamma-1}{\gamma}}\right]$$

$$= C_p T_1\left[\frac{T_3}{T_1} - r_p^{\frac{\gamma-1}{\gamma}}\right]$$

$$\therefore \text{Thermal efficiency} = \frac{W_n}{Q_A}$$

$$= \frac{C_p \eta_t T_3\left[1 - \dfrac{1}{r_p^{\frac{\gamma-1}{\gamma}}}\right] - C_p \dfrac{T_1}{\eta_c}\left[r_p^{\frac{\gamma-1}{\gamma}} - 1\right]}{C_p T_1\left[\dfrac{T_3}{T_1} - r_p^{\frac{\gamma-1}{\gamma}}\right]}$$

By differentiating this equation w.r.t. r_p, and equating it zero, we get the condition for maximum efficiency. This condition is given by

$$r_p = \left[\frac{T_3/T_1}{1-\sqrt{\left(\frac{T_3}{T_1}-1\right)\left(\frac{1}{\eta_c\eta_t}-1\right)}}\right]^{\frac{\gamma}{\gamma-1}} \qquad \text{(XV)}$$

and again for ideal cycle, where $\eta_c = \eta_t = 100\%$,

$$\therefore \qquad \mathbf{r_p} = \left(\frac{\mathbf{T_3}}{\mathbf{T_1}}\right)^{\frac{\gamma}{\gamma-1}} \quad \text{same as before.}$$

From these analysis of the actual cycle, we conclude that

1. the thermal efficiency is a function of r_p
2. the thermal efficiency is also a function of η_c, η_t
3. T_3 and also
4. compressor inlet temperature T_1.

24.16 IMPROVEMENTS IN PERFORMANCE OF SIMPLE GAS TURBINE CYCLE

Even the efficiencies of the components like compressor and turbine are improved, the efficiency and the specific work output of the simple gas turbine cycle is quite low. To improve the performance, we should have some modifications in the simple gas turbine plant and for that we can consider the following relation :

Net work/kg = $W_t - W_c$.

To increase this net work, either we should have more turbine work W_t for the same compressor work, or we should decrease the compressor work for the same turbine work. Similarly, to improve the thermal efficiency, heat supplied for the same work has to decrease. Therefore, it is necessary to look towards three main processes :

1. Compression process in compressor
2. Expansion process in gas turbine and
3. Heat supplied in combustion chamber.

24.16.1 Regenerative Gas Turbine Cycle

One of the main reasons for the low efficiency of a simple gas turbine plant is the large amount of heat which is rejected in the turbine exhaust. Due to limitations of maximum turbine inlet temperature and the pressure ratio which may be used with it, the turbine exhaust temperature is always greater than the temperature at the outlet of compressor. So, if this temperature difference is used to increase the temperature of the compressed air before entering the combustion chamber and, thereby, reducing the heat which must be supplied in the combustion chamber for a given turbine inlet temperature, an improvement in efficiency can be attained. This utilization of the heat in turbine exhaust can be effected in a heat exchanger called **regenerator. A regenerator is an heat exchanger where the heat exchange takes place within the system.** No external source of heat is required.

Figure 24.29 shows a schematic diagram of such an arrangement.

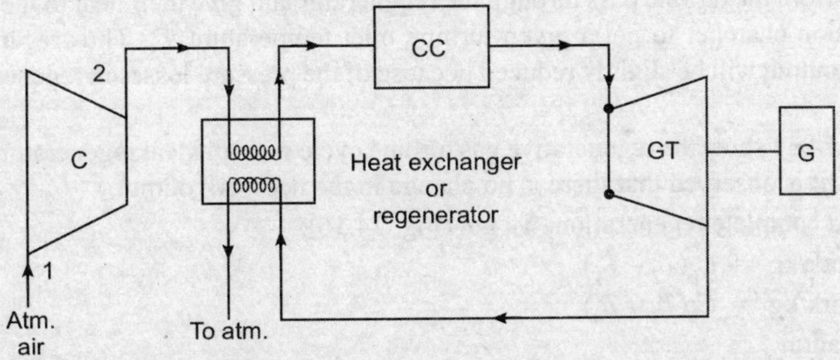

Fig. 24.29 *Regenerative gas turbine cycle.*

Figures 24.30 and 24.31 show the *T–S* diagram for ideal simple regenerative cycle and actual Brayton cycle with 100% regeneration.

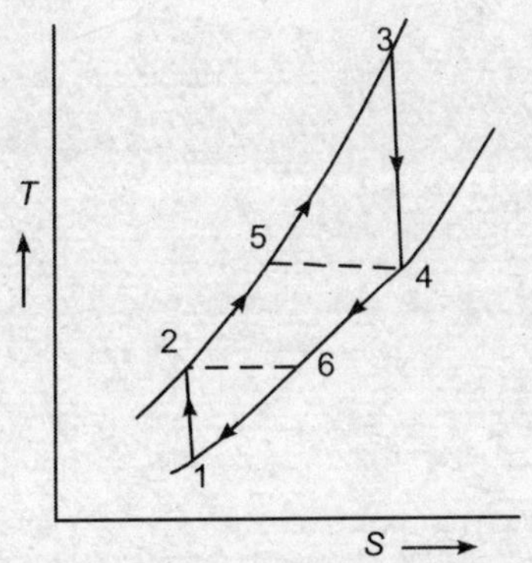

Fig. 24.30 *Theoretical regenerative gas turbine cycle.*

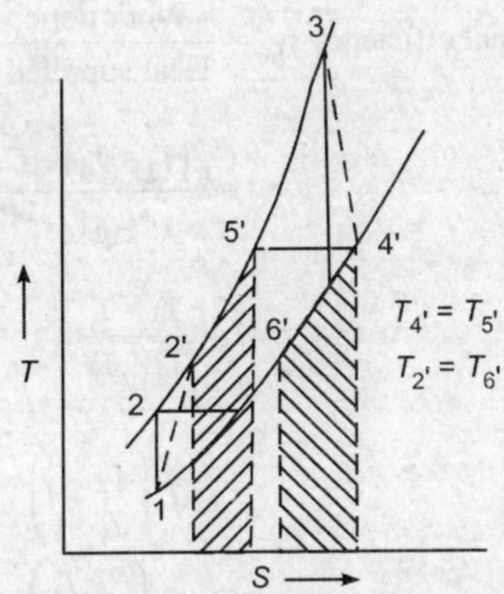

Fig. 24.31 *Actual Brayton cycle with 100% regeneration.*

Figure 24.32 shows the actual Brayton cycle with regenerator of efficiency less than 100%.

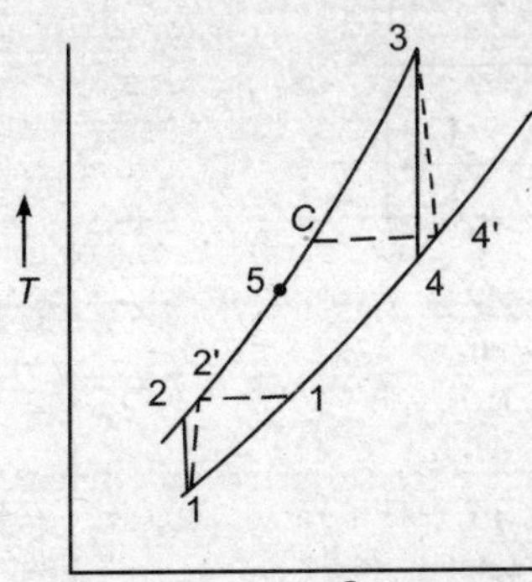

Fig. 24.32 *Actual Brayton cycle with regenerator.*

The exhaust gases from the turbine pass through the regenerator and give their heat to the compressed air before it enters the combustion chamber to get a given turbine inlet temperature T_3. Thus regeneration improves fuel economy. The power output will be slightly reduced because of the pressure losses in regenerator and its associated pipework.

Figures 24.30 and 24.31 show the regenerative gas turbine cycle with an ideal regenerator (100% effectiveness). From these diagrams, it is observed that there is no change in the net work output.

For ideal cycle and complete regeneration, we get (Fig. 24.16).

$$\text{Compressor work/kg} = C_p(T_2 - T_1)$$

$$\text{Turbine work/kg} = C_p(T_3 - T_4)$$

For 100% regeneration

$$T_4 = T_5 \text{ and } T_2 = T_6.$$

∴ The air which is coming out of the regenerator is at $T_5 = T_4$.

$$\therefore \quad \text{Heat supplied/kg} = C_p(T_3 - T_5)$$

$$\therefore \quad \text{Net work done/kg} = C_p(T_3 - T_4) - C_p(T_2 - T_1)$$

and

$$\text{Thermal efficiency } \eta_{th} = \frac{\text{Work done}}{\text{Heat supplied}}$$

$$= \frac{C_p(T_3 - T_4) - C_p(T_2 - T_1)}{C_p(T_3 - T_4)}$$

$$= 1 - \frac{T_2 - T_1}{T_3 - T_4}$$

$$= 1 - \frac{T_1\left(\frac{T_2}{T_1} - 1\right)}{T_3\left(1 - \frac{T_4}{T_3}\right)}$$

$$= 1 - \frac{T_1}{T_3} \cdot \frac{\left(r_p^{\frac{\gamma-1}{\gamma}} - 1\right)}{\left(1 - \frac{1}{r_p^{\frac{\gamma-1}{\gamma}}}\right)}$$

$$= 1 - \frac{T_1}{T_3} \cdot r_p^{\frac{\gamma-1}{\gamma}}.$$

$$\therefore \quad \boldsymbol{\eta_{th} = 1 - \frac{T_1}{T_3} . r_p^{\frac{\gamma-1}{\gamma}}} \quad \text{(XVI)}$$

Again for actual Brayton cycle with 100% regeneration, (Fig. 24.31) we have

Compressor work/kg = $C_p(T_{2'} - T_1)$

Turbine work/kg = $C_p(T_3 - T_{4'})$

$\therefore$ Net work/kg $= C_p(T_3 - T_{4'}) - C_p(T_{2'} - T_1)$

$$= C_p.\eta_t(T_3 - T_4) - C_p\frac{(T_2 - T_1)}{\eta_c}$$

$$= C_p.\eta_t T_3\left[1 - \frac{1}{r_p^{\frac{\gamma-1}{\gamma}}}\right] - C_p\frac{T_1\left(r_p^{\frac{\gamma-1}{\gamma}} - 1\right)}{\eta_c}$$

Heat supplied/kg = $C_p(T_3 - T_{5'}) = C_p(T_3 - T_{4'})$

$$= C_p\eta_t(T_3 - T_4) = C_pT_3\eta_t\left[1 - \frac{1}{r_p^{\frac{\gamma-1}{\gamma}}}\right]$$

$$= C_pT_3\eta_t\left[\frac{r_p^{\frac{\gamma-1}{\gamma}} - 1}{r_p^{\frac{\gamma-1}{\gamma}}}\right]$$

$$\therefore \text{Thermal efficiency} = \frac{\text{Net work}}{\text{Heat supplied}}$$

$$= \frac{C_p\eta_tT_3\left[\frac{(r_p)^{\frac{\gamma-1}{\gamma}} - 1}{r_p^{\frac{\gamma-1}{\gamma}}}\right] - \frac{C_pT_1}{\eta_c}\left[r_p^{\frac{\gamma-1}{\gamma}} - 1\right]}{C_pT_3\eta_t\left[\frac{r_p^{\frac{\gamma-1}{\gamma}} - 1}{r_p^{\frac{\gamma-1}{\gamma}}}\right]}$$

$$= 1 - \frac{T_1}{T_3\eta_t\eta_c}\left[r_p^{\frac{\gamma-1}{\gamma}}\right]$$

$$\boldsymbol{\eta_{th} = 1 - \frac{T_1}{T_3} \times \frac{1}{\eta_t\eta_c} \times r_p^{\frac{\gamma-1}{\gamma}}} \qquad \text{(XVII)}$$

From this relation, it is observed that **the thermal efficiency of a complete regenerative cycle depends upon the turbine and compressor inlet temperatures and the pressure ratio.**

In practice, the regenerator is not 100% efficient because the air from the compressor cannot be heated to the turbine exhaust gas temperature. Thus the effectiveness of the regenerator is defined as the actual temperature rise of air to the ideal temperature rise of air. Refer Fig. 24.18.

Effectiveness can also be defined as the ratio of the actual heat transfer to the maximum possible heat transfer that can be there.

Actual rise of temperature of air = $T_5 - T_{2'}$

Theoretical rise of temperature of air = $T_6 - T_{2'}$

$= T_{4'} - T_{2'}$

$\therefore$ Effectiveness of regenerator = $\dfrac{T_5 - T_{2'}}{T_{4'} - T_{2'}}$

Knowing this effectiveness or efficiency of the regenerator we can find or determine the actual temperature of air leaving the generator and entering the combustion chamber.

$\therefore$ Heat added/kg = $Q_A = C_p(T_3 - T_5)$

$$\therefore \qquad \eta_{th} = \frac{C_p \eta_t T_3 \left(\dfrac{r_p^{\frac{\gamma-1}{\gamma}} - 1}{r_p^{\frac{\gamma-1}{\gamma}}} \right) - \dfrac{C_p \cdot T_1}{\eta_c} \left(r_p^{\frac{\gamma-1}{\gamma}} - 1 \right)}{C_p (T_3 - T_5)} \qquad \text{(XVIII)}$$

Efficiency of regeneration varies from 0.5 to 0.75 in practice.

Generally, counterflow type of regenerator gives greater regeneration efficiency than the parallel flow type regenerator. The main advantage of regeneration is that low specific fuel consumption occurs over a wide range of turbine inlet temperatures and is optimized at high levels of specific power. Thus, it goes without saying that a good regenerative system is of utmost importance.

At present there are three types of regenerative systems usually employed in gas turbine plant. These are :

1. Direct transfer air-to-gas exchanger
2. Periodic flow rotory heat exchanger
3. Liquid metal indirect heat transfer exchanger.

The first system has been widely used over the years.

The second one is used very rarely.

Increased knowledge of liquid metal systems and high level of proven regenerator design technology has made use of liquid metal regenerators in gas turbine plants.

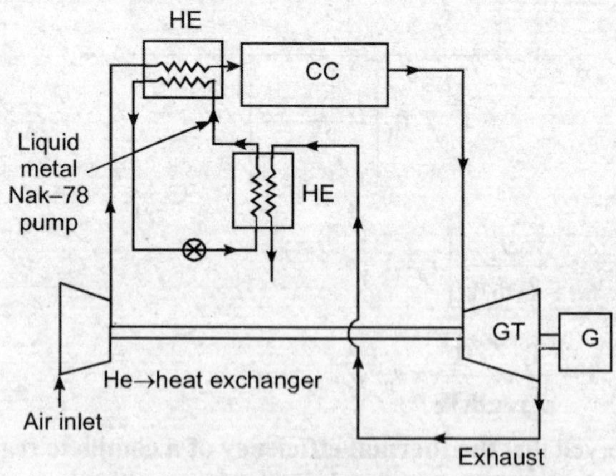

Fig. 24.33 *Liquid metal regenerator circuit.*

A liquid metal system consists of two independent heat exchangers coupled by a liquid metal circuit. A part of the heat energy of the turbine exhaust is passed on to a liquid metal in the turbine heat exchanger which is then pumped to the compressor heat exchanger releasing heat to the compressed air.

The liquid metal system gives significant advantages in terms of installed weight, engine performance and high reliability as compared with direct or rotory type regenerators. The physical separation of hot low pressure and cold high pressure exchangers minimizes structural requirements, air-to-gas carry over losses and seal leakage losses are avoided, and moreover the malfunction of the regenerator does not result in idle time because turbine bypass is avoided in this system. Finned tubes are highly suitable for liquid metal-to-gas heat exchangers.

Figure 24.34 shows the effect of regeneration on efficiency of gas turbine.

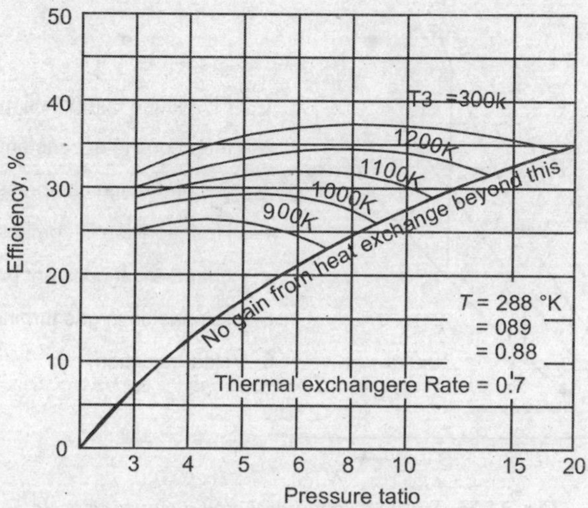

Fig. 24.34 *Efficiency of regenerative gas turbine cycle.*

The requirements of high heat absorbing capacity, high thermal conductivity, low viscosity for high heat transfer coefficients with reduced pressure losses, low vapour pressure and the compatibility with system materials are met by Nak-78 liquid metal alloy with 78 percentage of potassium.

24.16.2 INTERCOOLERS IN GAS TURBINE CYCLE

A regenerator, as discussed in the previous article, does not change the work output of a gas turbine cycle. Therefore, to increase the net work output, there are two methods as :

1. to reduce the work of compression and
2. to increase the turbine work.

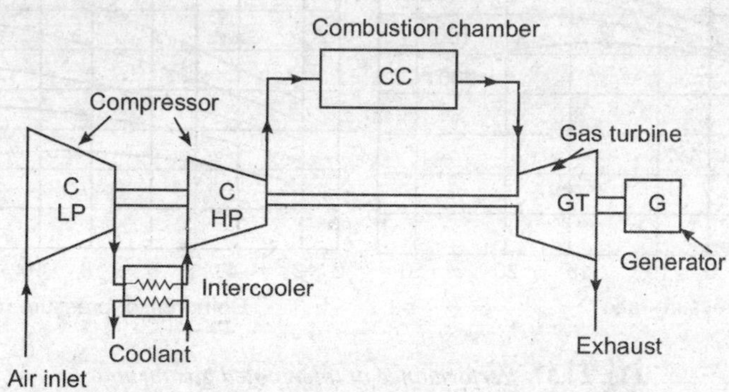

Fig. 24.35 *Multistage compressore with intercooler.*

To reduce the compression work, we use an intercooled cycle for the compressor. In this case we take the advantage of the fact that constant pressure lines on *T–S* diagram converge to the left bottom region of the diagram.

In this cycle, the air is compressed partly and then cooled in the cooler called **intercooler**. This operation is at constant pressure. Air after cooling is admitted to the second stage of the compressor where it is compressed further to the desired pressure. Depending on the pressure ratio more than 1 (one) intercooling stages may be used. Figure 24.35 shows the schematic diagram for the single stage intercooled cycle. Figure 24.36 is the *T–S* diagram for the single stage intercooled cycle.

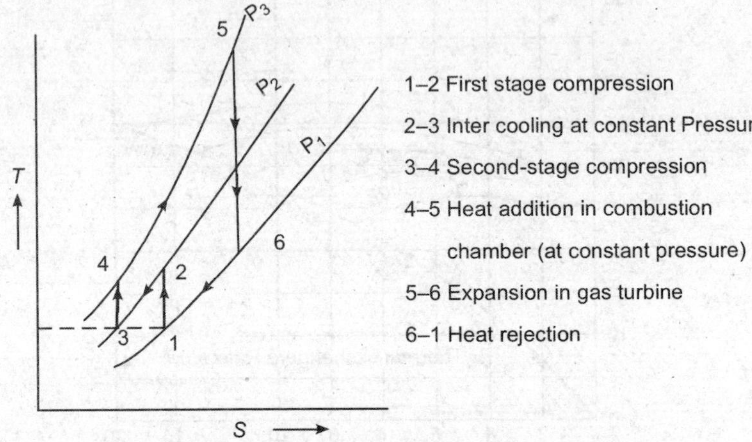

Fig. 24.36 *Single stage intercooled gasturbine cycle.*

For perfect intercooling and minimum work of compression, we have p_2 = Intermediate pressure = $\sqrt{p_1 p_3}$. For perfect intercooling, $T_3 = T_1$.

Intercooling will always increase the net work output of the cycle, and due to lower compressor outlet temperature, there will be more scope for regeneration.

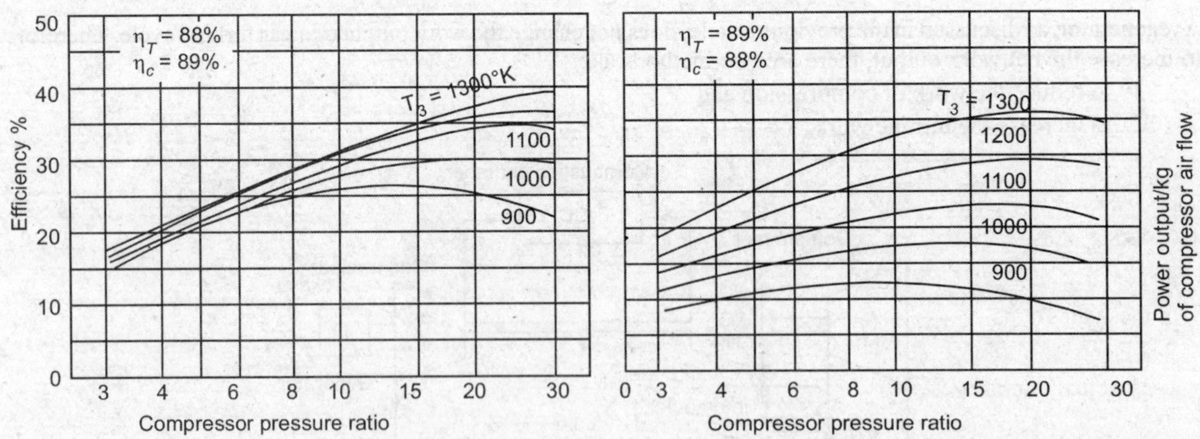

Fig. 24.37 *Performance of intercooled gas turbine.*

Intercooling is useful when the pressure ratio is high and the efficiency of the compressor is high. Figure 24.37 shows the performance of intercooled gas turbine.

24.16.3 GAS TURBINE CYCLE WITH REHEATING

Another method of increasing the specific work output is to use reheating to increase the turbine work. The gain in work output is obtained because of divergence of constant pressure lines on *T–S* diagram with an increase in

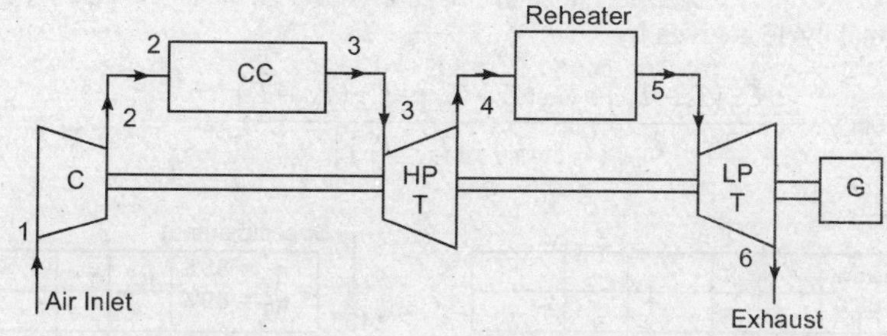

Fig. 24.38 *Schematic diagram of a reheat gas turbine cycle.*

temperature. Therefore, for the same pressure ratio, if the exhaust from one stage is reheated in a separate combustion chamber and expanded in the second stage of the turbine, more output will be obtained than that obtained by expansion in a single stage. Reheating is generally done upto the upper limit of temperature T_3. Usually reheating may be in two stages. With open cycle gas turbine, the limit of reheating stages is the oxygen which is available for combustion.

Figure 24.38 shows a schematic diagram of a reheat gas turbine plant and Figure 24.39 shows the corresponding *T–S* diagram.

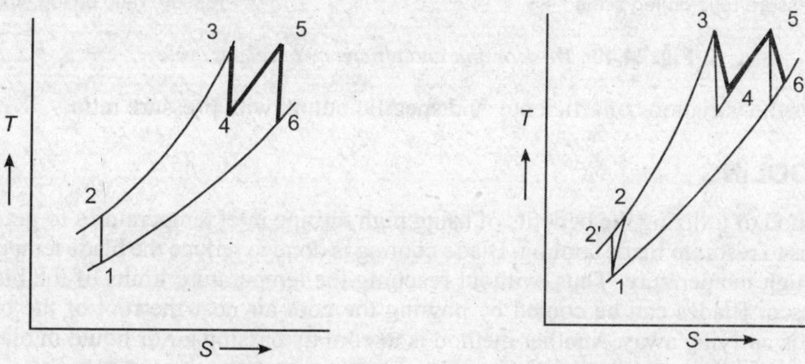

(a) Ideal reheat cycle

(b) Actual reheat cycle

Fig. 24.39 *T–S diagram of reheat cycle.*

In these diagrams :

1–2 Isentropic compression
1–2′ Actual compression
2–3 or 2′–3 Heating in combustion chamber
3–4 Isentropic expansion in HP stage turbine
3–4′ Actual expansion in HP stage turbine
4–5 or 4′–5 Reheating in a reheater
5–6 Isentropic expansion in LP stage turbine
5–6′ Actual expansion in LP stage turbine

6–1 or 6′–1 Heat rejection to sink

If r_{p1} is the pressure ratio in HP turbine and r_{p2} is the pressure ratio in LP stage, then

$$r_p = \text{Total pressure ratio}$$
$$= r_{p1} \times r_{p2} \text{ and } \mathbf{r_{p1} = r_{p2} = (r_p)^{1/2}} \text{ for maximum efficiency}$$

Efficiency of the actual cycle is given by

$$\text{Efficiency} = \frac{C_p(T_3 - T_{4'}) + C_p(T_5 - T_{6'}) - C_p(T_{2'} - T_1)}{C_p\left[(T_3 - T_{2'}) + (T_5 - T_{4'})\right]}$$

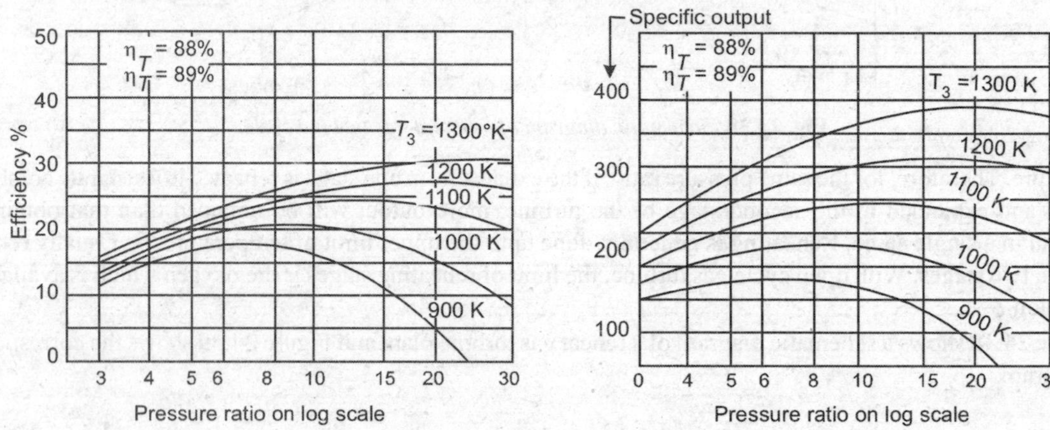

Fig. 24.40 *Work output and efficiency of reheat cycle.*

Figure 24.40 shows the variations of efficiency and specific output with pressure ratio.

24.17 BLADE COOLING

Another possible method of utilizing the benefits of using high turbine inlet temperatures to get more work output at a higher efficiency is to resort to blade cooling. Blade cooling is done to reduce the blade temperature, though the gas will still be at a high temperature. Thus, without reaching the temperature limits of the blades, a higher gas temperature can be used. Blades can be cooled by passing the cool air near the root of the blades so that heat conducted to the root is carrying away. Another method is to cool by passing air or liquid in blade annulus. Blade cooling is normally used for aircraft as turbines.

24.18 CLOSED CYCLE GAS TURBINE PLANTS

Most of the gas turbine power plants in use are open cycle type. The closed cycle gas turbine is a comparatively recent development. It was after years of pioneering work that Escher Wyss of Switzerland first introduced in 1939, a closed cycle version developing 2000 kW (2 MW) at a pressure ratio of 3.8 and turbine inlet temperature of 700°C with a thermal efficiency of over 30%. It took the closed cycle principle long time to establish and it was not until sixties that its use became wide spread all over the world.

The fundamental difference between open and closed cycle gas turbines is in the method of heating the air after compression. In the case of an open cycle gas turbine, the fuel is burned in the air itself to raise it to a high temperature and then products of combustion are passed on to the turbine for expansion and after expansion, the gases are led to the atmosphere. For the next cycle, a fresh supply of air from atmosphere is taken.

In case of closed cycle turbine, on the other hand, the same air or the working fluid is circulated over and over again. The working fluid is heated by burning the fuel in separate supply of air in a combustion chamber and transferring this heat to the working fluid, which passes through tubes fitted in this chamber. Thus the working fluid does not come in contact with the products of combustion.

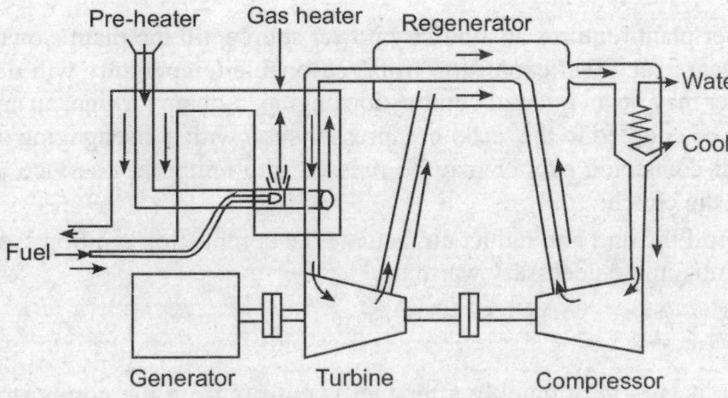

Fig. 24.41 *Closed cycle gas turbine.*

A schematic diagram for the closed cycle gas turbine is shown in Fig. 24.41.

24.19 ADVANTAGES AND DISADVANTAGES OF CLOSED CYCLE GAS TURBINE

Following are the advantages of closed cycle gas turbine :

Advantages

1. Since the pressure of the working fluid is independent of atmospheric pressure, a higher pressure can be used to increase the specific output of the plant. This results in diminution of the sizes of the components both for the machines and the heat transfer apparatus used.

2. Gases other than air, which have more favourable properties, can be used. Helium or Helium-carbon dioxide mixture give hive efficiencies and smaller dimensions for special purposes. The properties of helium at high pressures, such as high heat transfer, low pressure drop, high sound velocity and neutrality towards radioactive materials make it possible to build smaller heat transfer equipment and is highly suitable for nuclear plants.

3. Use of alternative working fluids such as helium etc. gives rise to the possibility of using alternative materials as no oxidation occurs with these inert gases.

4. Existence of constant temperature at all loads result in low thermal stresses.

5. The turbine blades are not fouled by the production.

6. Due to the fact that the working fluids does not come in contact with products of combustion, even very low quality fuel can be used. All types of fuels—solid, liquid, gaseous and nuclear—can be used.

7. The regulation of the closed cycle gas turbine is simpler.

8. In total energy plants direct waste heat utilization at high temperatures for heating purposes, without affecting power cycle efficiency, can be affected because heat rejection is an isobaric process instead of isothermal change of state.

Disadvantages

1. The use of high pressure requires a strong heat exchanger.

2. Since heat transfer is indirect, a part of heat energy is lost in radiation and other losses resulting in poor combustion efficiency.

3. Due to use of air heaters and precooler, the cost and bulk of closed cycle gas turbine plant is much more than that for an open cycle gas turbine plant.

4. A coolant is required for precooler which is a disadvantage as compared to an open cycle plant.

24.20 OPERATION OF GAS TURBINE

24.20.1 Starting

Starting of gas turbine power plant requires an auxiliary power source, till the plant's own compressor inducts air and compresses it to a pressure such that expansion from reasonable temperature will develop enough power to sustain operation. The starter may be an internal combustion engine, a steam turbine, an auxiliary electric motor or another gas turbine. It must be coupled to the turbo-compressor shaft with a disengaging or over running clutch. A main generator or its direct connected exciter may be pressed into temporary service as a motor, in this case, electrical controls supplant the clutch.

Starting procedure is to roll the unit and induct air, actuate the combustion ignition system and inject fuel. The fuel flow is—controlled to obtain the necessary warm up.

24.20.2 Shut Down

Shut down of a gas turbine occurs very quickly after fuel is cut off from the combustor. A rapid shut down is desirable since it minimizes the chilling effect resulting from passing cold air through the hot turbine. Turning gears are generally employed or alternatively device may be operated at reduced speed to ensure symmetrical cooling of rotor. This avoids thermal **kinking** of the shaft.

24.20.3 Plant Layout

In case of a gas turbine power plant, the main building is the turbine house in which major portion of the plant as well as auxiliaries are installed. Usually it is similar in many respects to the steam plant turbine house.

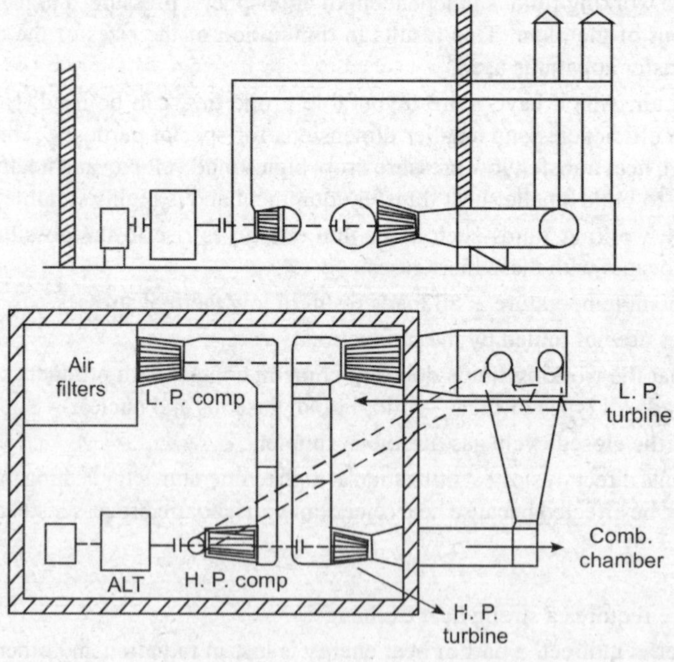

Fig. 24.42 *Gas turbine plant layout.*

The fuel oil storage tanks are arranged outside but adjoining the turbine house and in some installations even heat exchangers are also placed out of doors.

The rotating parts of the plant form a very small part of the total volume of the plant, because the intercoolers, combustion chambers, heat exchangers and interconnecting duct work which have to be arranged and accomodated and also occupy the major portion of the total space.

The layout of a gas turbine plant has a very important effect on the overall performance of the plant, as it is possible to incur a loss to the extent of 20% of the power developed in the interconnecting ducts having a considerable number of sharp bends. Great care is, therefore, taken in the design and layout of the air as well as gas circuits.

A simplified layout to achieve this is shown in Fig. 24.41.

SOLVED EXAMPLES

I. SIMPLE CYCLE

Example 24.1 *In an air standard Brayton cycle with a pressure ratio of 8, the temperature and pressure of the air at inlet to the compressor and 300 K and 1 bar respectively. The maximum allowed temperature in the cycle is 1000 K. Determine the energy added, the net work done per kg of air and the thermal efficiency of the cycle. Give C_p for air as 1.0045 kJ/kg/K.*

Solution

The *T–S* diagram for the given example is shown in Fig. Ex. 24.1

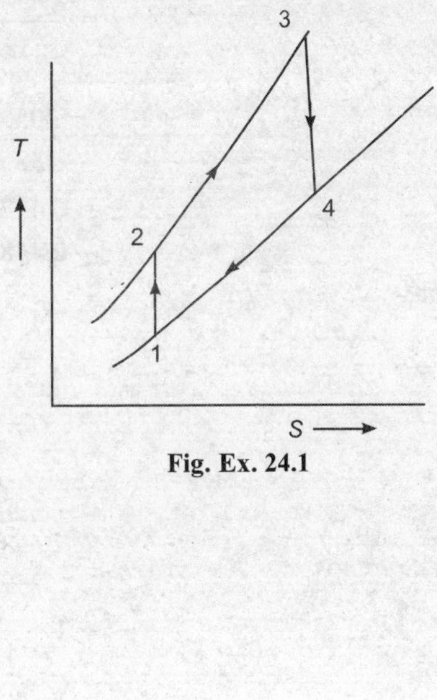

Fig. Ex. 24.1

Given :

$$P_1 = 1 \text{ bar}; T_1 = 300 \text{ K}$$
$$r_p = 8 \; T_3 = 1000 \text{ K}$$
$$C_p = 1.0045 \text{ kJ/kg/K.}$$

First find the temperatures T_2 and T_4

$$\therefore \quad \frac{T_2}{T_1} = \left(\frac{P_2}{P_1}\right)^{\frac{\gamma-1}{\gamma}} = (8)^{\frac{1.4-1}{1.4}} = 8^{0.2857}$$
$$= \mathbf{1.8114}$$
$$\therefore \quad T_2 = 300 \times 1.8114$$
$$= \mathbf{543.42 \text{ K}}$$

Similarly
$$\frac{T_3}{T_4} = \left(\frac{P_3}{P_4}\right)^{\frac{\gamma-1}{\gamma}} = \left(\frac{P_2}{P_1}\right)^{0.2857}$$
$$= \mathbf{1.8114}$$
$$\therefore \quad T_4 = \frac{T_3}{1.8114} = \frac{1000}{1.8114}$$
$$= \mathbf{552.06 \text{ K}}$$

(i) *Energy added/kg*

$$Q_A = C_p (T_3 - T_2)$$
$$= 1.0045 \, (1000 - 543.42)$$
$$= 1.0045 \times 456.58$$
$$= \mathbf{458.635 \text{ K}}$$

(ii) *Work done/kg*

$$W = Q_A - Q_R \text{ for ideal cycle.}$$

where $Q_R = C_p\,(T_4 - T_1)$

$$= W = Q_A - Q_R$$
$$= 458.635 - 253.19$$
$$= \mathbf{205.445}$$
$$= 1.0045\,(552.06 - 300)$$
$$= 1.0045 \times 252.06$$
$$= \mathbf{253.19\ kJ/kg}$$

∴ Work done/kg

(iii) *Thermal efficiency of the cycle*

Thermal efficiency of the cycle is given by

$$\eta_{th} = \frac{\text{Work done}}{\text{Heat supplied}}$$

$$= \frac{Q_A - Q_R}{Q_A} = 1 - \frac{Q_R}{Q_A}$$

$$= \frac{205.445}{458.635}$$

$$= 0.4479$$

$$\simeq \mathbf{0.448} \quad \mathbf{(44.79\%)}$$

or

$$\eta_{th} = \text{A.S.E.} = 1 - \frac{1}{r_p^{\frac{\gamma-1}{1}}}$$

$$= 1 - \frac{1}{8^{0.2587}}$$

$$= 1 - \frac{1}{1.8114}$$

$$= 1 - 0.552$$

$$= 0.448$$

Example 24.2 *The following particulars refer to a gas turbine cycle –*

Air entry at 0.98 bar and 20°C

Compressor discharge pressure = 4.2 bar

Turbine inlet temperature = 850°C

Efficiency of compressor and turbine = 0.8 each

Calculate thermal efficiency of actual and theoretical cycles.

Solution

The cycle is shown on *T–S* diagram in Fig. Ex. 24.2

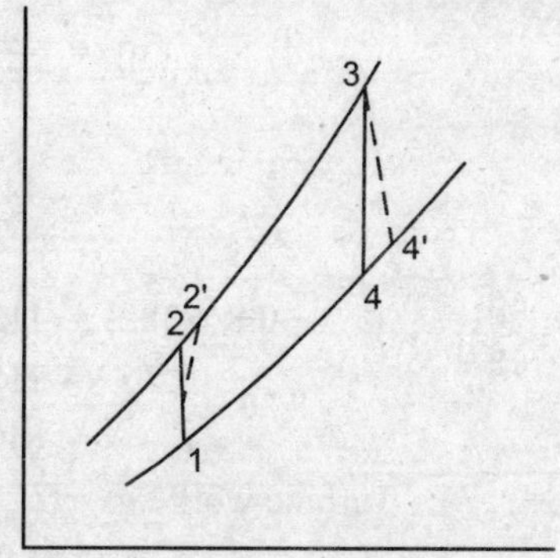

Fig. Ex. 24.2

Given data :

$$P_1 = 0.98 \text{ bar } T_1 = 20 + 273$$
$$= \mathbf{293\ K.}$$
$$P_2 = 4.2 \text{ bar}$$
$$T_3 = 273 + 850 = \mathbf{1123\ K}$$

Compressor efficiency = 0.8

Turbine efficiency = 0.8

Calculations for finding T_2 and T_4.

$$\frac{T_2}{T_1} = \left(\frac{P_2}{P_1}\right)^{\frac{\gamma-1}{\gamma}} = \left(\frac{4.2}{0.98}\right)^{0.2587}$$
$$= \mathbf{1.5155}$$

$\therefore$ $T_2 = \mathbf{444.055\ K}$

Similarly T_4 = Temperature after expansion

$$= \frac{T_3}{1.5155} = \frac{1123}{1.5155}$$
$$= \mathbf{741\ K}$$

$$\therefore \text{ Theoretical efficiency} = 1 - \frac{T_4 - T_1}{T_3 - T_2} \left[\because \eta_{th} = 1 - \frac{Q_R}{Q_A}\right]$$

$$= 1 - \frac{741 - 293}{1123 - 444.055}$$

$$= 1 - \frac{448}{678.945} = 1 - 0.6598$$
$$= \mathbf{0.3402}$$
$$= \mathbf{34.02\%}$$

Actual cycle

$$\text{Compressor efficiency} = 0.8 = \frac{T_2 - T_1}{T_{2'} - T_1}$$

$$= \frac{444.055 - 293}{T_{2'} - 293} = \frac{151.055}{T_{2'} - 293}$$

$$\therefore \quad T_{2'} - 293 = \frac{151.055}{0.8} = 188.82$$

$$\therefore \quad T_{2'} = 293 + 188.82$$
$$= \mathbf{481.82\ K}$$

Similarly turbine efficiency is given by

$$\eta_t = \frac{T_3 - T_{4'}}{T_3 - T_4} = \frac{1123 - T_{4'}}{1123 - 741}$$

$$= \frac{1123 - T_{4'}}{382}$$

$$\therefore \quad 0.8 \times 382 = 1123 - T_{4'}$$

$$T_{4'} = 1123 - 305.6$$

$$= \mathbf{817.4\ K}$$

$$\therefore \quad \text{Turbine work/kg} = C_p (T_3 - T_{4'}) = W_t$$

$$= 1.005\ [1123 - 817.4]$$

$$= 1.005 \times 305.6$$

$$= \mathbf{307.128\ kJ/kg}$$

$$\text{Compressor work } W_c = C_p (T_{2'} - T_1)$$

$$= 1.005\ (481.82 - 293)$$

$$= 1.005 \times 188.82$$

$$= \mathbf{189.76\ kJ/kg}$$

$$\therefore \quad \text{Net work of cycle} = W_n = W_t - W_c$$

$$= 307.128 - 189.76$$

$$= \mathbf{117.368\ kJ/kg}$$

∴ Actual efficiency of cycle

$$= \eta_{act} = \frac{\text{Net work/kg}}{\text{Heat supplied/kg}}$$

$$= \frac{117.368}{Q_A}$$

$$\text{Heat supplied/kg} = C_p (T_3 - T_{2'}) = 1.005\ (1123 - 481.82)$$

$$= 1.005 \times 641.18$$

$$= \mathbf{644.38\ kJ/kg}$$

$$\therefore \quad \eta_{tact} = \frac{117.368}{644.38}$$

$$= 0.1821$$

$$= \mathbf{18.21\%}$$

Example 24.3 *In a gas turbine plant the air pressure and temperature before compression are 1 bar and 28°C. The ratio of maximum pressure to minimum pressure is 3.5 and the temperature of gas before expansion in the turbine is 700°C. If the isentropic efficiencies of the compressor and the turbine are 80% and 85% respectively, calculate*

(a) The efficiency of the cycle

(b) The turbine exhaust temperature

(c) Percentage change in the efficiency of the plant if the regenerator of 0.7 effectiveness is added to the plant.

Solution

Figure Ex. 24.3 shows the cycle on *T–S* diagram

Given :

$$P_1 = 1 \text{ bar}, T_1 = 28 + 273 = 301 \text{ K}$$
$$r_p = 3.5, T_3 = 273 + 700 = \mathbf{923 \text{ K}}$$
$$\eta_c = 80\%, \eta_t = 85\%.$$

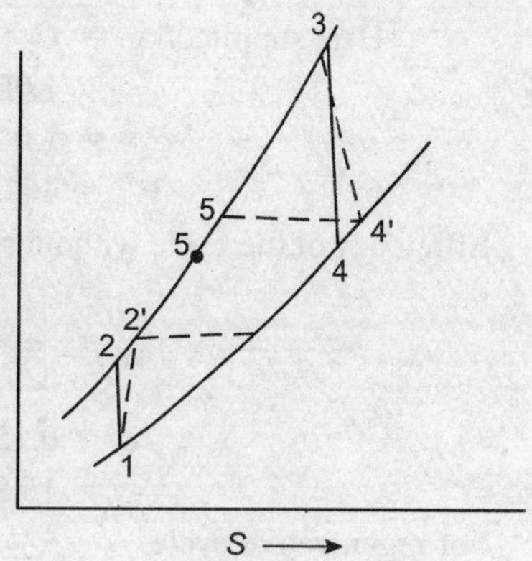

Fig. Ex. 24.3

From *T–S* diagram,

$$\frac{T_2}{T_1} = \left(\frac{P_2}{P_1}\right)^{\frac{\gamma-1}{\gamma}} = (3.5)^{0.2857} = 1.4303$$

$$\therefore \quad T_2 = 1.4303 \times 301 = \mathbf{430.53 \text{ K}}$$

$$\eta_c = \frac{T_2 - T_1}{T_{2'} - T_1} = \frac{430.53 - 301}{T_{2'} - 301}$$

$$\therefore \quad 0.8 = \frac{129.53}{T_{2'} - 301}$$

$$\therefore \quad T_{2'} = 301 + \frac{129.53}{0.8} = 301 + 161.91 = \mathbf{462.91 \text{ K}}$$

Similarly $$T_4 = \frac{923}{1.4303} = \mathbf{645.32 \text{ K}}$$

Similarly, $$\eta_t = \frac{T_3 - T_{4'}}{T_3 - T_4} = \frac{923 - T_{4'}}{923 - 645.32}$$

$$\therefore \quad 0.85 = \frac{923 - T_{4'}}{923 - 645.32}$$

$$923 - T_{4'} = 0.85 \times 277.68 = 236.03$$
$$T_{4'} = 923 - 236.03 = \mathbf{686.97 \text{ K}}$$

(*a*) *Efficiency of the cycle*

$$W_t = \text{Turbine work/kg}$$
$$= C_p (T_3 - T_{4'}) = 1.005 \ (923 - 686.97)$$
$$= 1.005 \times 236.03 = \mathbf{236.74 \text{ K}}$$

$$W_c = \text{Compression work/kg}$$
$$= C_p (T_{2'} - T_1)$$
$$= 1.005 \ [462.91 - 301] = 1.005 \times 161.91 \text{ kJ/kg}$$
$$= \mathbf{162.72 \text{ kJ/kg}}$$

$\therefore$ Net work/kg $= W_t - W_c$

$= 236.74 - 162.72$

$= \mathbf{72.02\ kJ/kg}$

Heat supplied/kg $= C_p\,(T_3 - T_{2'})$

$= 1.005\,(923 - 462.91)$

$= 1.005 \times 460.09$

$= \mathbf{462.39\ kJ/kg}$

Efficiency of the cycle without regeneration

$$\eta_{th} = \frac{W_n}{Q_A} = \frac{72.02}{462.39}$$

$= 0.1558$

$= \mathbf{15.58\%}$

For regenerative cycle,

E = Effectiveness of regenerator

$$= 0.7 = \frac{T_5 - T_{2'}}{T_{4'} - T_{2'}}$$

$$= \frac{T_5 - 462.91}{686.97 - 462.91}$$

$$= \frac{T_5 - 462.91}{224.06}$$

$\therefore$ $0.7 \times 224.06 = T_5 - 462.91 = 156.84$

$\therefore$ T_5 = Temperature after regeneration

$= 462.91 + 156.84$

$= \mathbf{619.75\ K}$

$\therefore$ Heat supplied/kg after regenerative heating

$Q_A = C_p\,(T_3 - T_5)$

$= 1.005\,(923 - 619.75)$

$= 1.005 \times 303.25$

$= \mathbf{304.77\ kJ/kg}$

$\therefore$ Efficiency after regenerative heating

$$= \frac{\text{Work done/kg}}{\text{Heat supplied/kg}}$$

$$= \frac{72.02}{304.77}$$

$= \mathbf{0.2363}$ or $\mathbf{23.63\%}$

(b) The turbine exhaust temperature

$T_{4'} = \mathbf{686.97\ K}$

(c) *Percentage change in efficiency :*

$$\text{Percentage change} = \frac{0.2363 - 0.1558}{0.1558} \times 100$$

$$= \frac{0.0805}{0.1558} \times 100$$

$$= \mathbf{51.67\%}$$

Example 24.4 *The actual minimum and maximum temperatures of a simple gas turbine are T_1 and T_3. The pressure ratio for the cycle is r_p and the isentropic efficiency for the compressor and turbine are η_c and η_t respectively. Neglecting the variation of specific heats and assuming ratio of specific heats as 1.35, prove that the thermal efficiency of the cycle is given by the expression,*

$$\eta = \frac{\eta_c \eta_t T_3 \left[1 - \frac{1}{r_p^{0.259}}\right] - T_1\left(r_p^{0.259} - 1\right)}{\eta_c T_3 - T_1\left(r_p^{0.259} - 1 + \eta_c\right)}$$

Hence estimate the efficiency if $r_p = 6$, $T_1 = 300$ K, $T_3 = 1000$ K, $\eta_c = 80\%$ and $\eta_t = 85\%$.

Solution

Figure Ex. 24.4 shows the *T–S* diagram for the gas turbine cycle.

$$\text{Work of compression} = W_c = C_p\,(T_{2'} - T_1) \text{ kJ/kg}$$

$$\text{Work of turbine} = W_t = C_p\,(T_3 - T_{4'}) \text{ kJ/kg}$$

We have

$$\frac{T_2}{T_1} = \left(\frac{P_2}{P_1}\right)^{\frac{\gamma-1}{\gamma}} = r_p^{\frac{0.35}{1.35}} \left(\frac{P_3}{P_4}\right)^{\frac{0.35}{1.35}} = \frac{T_3}{T_4} = r_p^{0.259}$$

$$= r_p^{0.259}$$

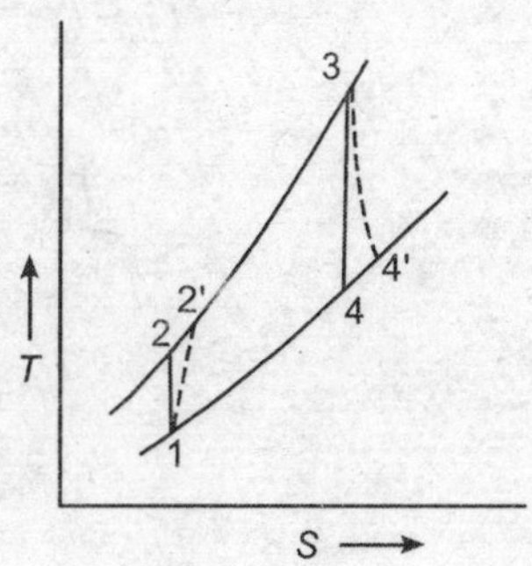

Fig. Ex. 24.4

Again
$$\eta_c = \frac{T_2 - T_1}{T_{2'} - T_1}$$

$$\therefore \quad T_{2'} - T_1 = \frac{T_2 - T_1}{\eta_c}$$

$$= \frac{T_1\left(\frac{T_2}{T_1} - 1\right)}{\eta_c}$$

$$= \frac{T_1}{\eta_c}\left[r_p^{0.259} - 1\right] \qquad \text{(I)}$$

Similarly

$$\eta_t = \frac{T_3 - T_{4'}}{T_3 - T_4} \quad \therefore \; \eta_t \,.\, (T_3 - T_4) = T_3 - T_{4'}$$

$$T_3 - T_{4'} = \eta_t T_3 \left[1 - \frac{1}{r_p^{0.259}}\right] \tag{II}$$

$$\text{Efficiency} = \eta = \frac{\text{Work done/kg}}{\text{Heat supplied/kg}}$$

$$= \frac{C_p(T_3 - T_{4'}) - C_p(T_{2'} - T_1)}{C_p(T_3 - T_{2'})}$$

$$= \frac{(T_3 - T_{4'}) - (T_{2'} - T_1)}{T_3 - T_{2'}}$$

$$= \frac{\eta_t T_3 \left[1 - \frac{1}{r_p^{0.259}}\right] - \frac{T_1}{\eta_c}\left[r_p^{0.259} - 1\right]}{T_3 - T_{2'}}$$

$$= \frac{\eta_c \eta_t T_3 \left[1 - \frac{1}{r_p^{0.259}}\right] - T_1\left[r_p^{0.259} - 1\right]}{\eta_c (T_3 - T_{2'}) = \eta_c T_3 - \eta_c T_{2'}}$$

We know from Eq. (I),

$$T_{2'} - T_1 = \frac{T_1}{\eta_c}\left[r_p^{0.259} - 1\right]$$

$$\therefore \quad T_{2'} = T_1 + \frac{T_1}{\eta_c}\left[r_p^{0.259} - 1\right]$$

$$= T_1 \left[1 + \frac{\left(r_p^{0.259} - 1\right)}{\eta_c}\right]$$

$$\therefore \quad \eta_c T_{2'} = T_1\left[\eta_c + r_p^{0.259} - 1\right]$$

$$\therefore \quad \eta = \frac{\eta_c \eta_t T_3 \left[1 - \frac{1}{r_p^{0.259}}\right] - T_1\left[r_p^{0.259} - 1\right]}{\eta_c T_3 - T_1\left(r_p^{0.259} - 1 + \eta_c\right)}$$

Q.E.D.

When $r_p = 6, r_p^{0.259} = 1.59$

$$= \frac{1}{r_p^{0.259}} = 0.629$$

$$\therefore \quad \eta = \frac{0.8 \times 0.85 \times 1000(1-0.629)-300(1.59-1)}{0.8 \times 1000-300(1.59-1+0.8)}$$

$$= \frac{252.3-177}{800-417}$$

$$= \frac{75.3}{383}$$

$$= \mathbf{0.1966} \text{ or } \mathbf{19.66\%}$$

Example 24.5 *A gas turbine power generating set has two turbines. One turbine drives the compressor and the second turbine drives the generating set. The compressor takes in air at 27°C and compresses it to pressure rotation of 6. Compressed air passes through the combustion chamber where the temperature is raised to 1050 K. Pressure ratio for both turbines is 6.*

Only sufficient hot gas is passed through the turbine driving the compressor. Compressor and turbine efficiencies are 85%. C_{pa} *= 1.004 kJ/kg.K,* $\gamma = 1.4$.

Find

1. *The fraction of hot gases passing through the compressor turbine.*
2. *The power output of the turbine driving the generator.*
3. *Thermal efficiency of the plant.*

Solution The schematic diagram for these two turbines system is shown in Fig. Ex. 24.5. Also *T–S* diagram is shown in Fig. Ex. 24.5(a)

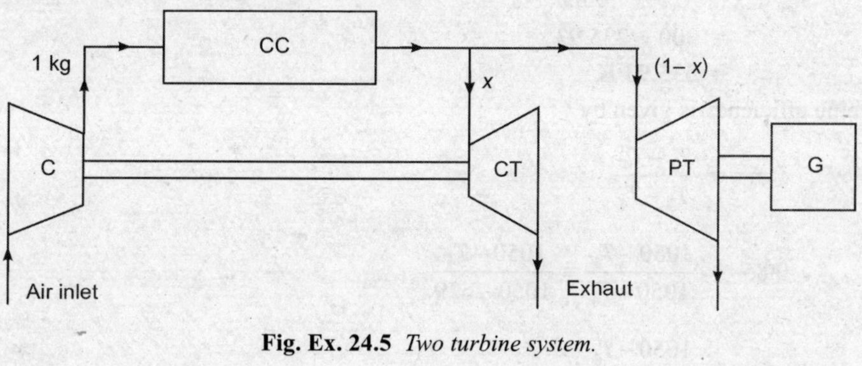

Fig. Ex. 24.5 *Two turbine system.*

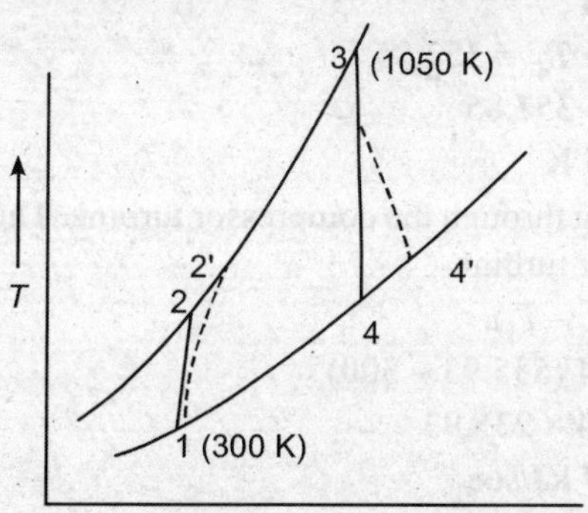

Fig. Ex. 24.5(a)

Given :

$$T_1 = 27 + 273 = 300 \text{ K}$$
$$T_3 = 1050 \text{ K.}$$
$$\text{Pressure ratio} = 6$$
$$\eta_t = \eta_c = 85\%, C_{pa} = 1.004 \text{ kJ/kg.K}$$
$$\gamma = 1.4.$$

First we will find the temperatures at the salient points.

$$\frac{T_2}{T_1} = (6)^{0.2857} = 1.668$$

$$\therefore \quad T_2 = 300 \times 1.668 = \mathbf{500.54\ K}$$

$$T_4 = \frac{T_3}{1.668} = \frac{1050}{1.668} = \mathbf{629\ K}$$

We have compressor efficiency

$$= \eta_c = 0.85$$
$$= \frac{T_2 - T_1}{T_{2'} - T_1} = \frac{500.54 - 300}{T_{2'} - 300}$$

$$\therefore \quad T_{2'} = 300 + \frac{200.54}{0.85}$$
$$= 300 + 235.93$$
$$= \mathbf{535.93\ K}$$

Similarly, turbine efficiency is given by

$$\eta_t = \frac{T_3 - T_{4'}}{T_3 - T_4}$$

or
$$0.85 = \frac{1050 - T_{4'}}{1050 - T_4} = \frac{1050 - T_{4'}}{1050 - 629}$$
$$= \frac{1050 - T_{4'}}{421}$$

$$0.85 \times 421 = \mathbf{1050 - T_{4'}} = 357.85$$
$$\therefore \quad T_{4'} = 1050 - 357.85$$
$$= \mathbf{692.15\ K}$$

Let x be the fraction of gases passing through the compressor turbine. Therefore $(1 - x)$ kg/sec will be the weight of the gases passing through the power turbine.

Work of compressor/kg = $m\ C_p\ (T_{2'} - T_1)$

$$= x\ 1.004\ (535.93 - 300)$$
$$= x\ 1.004 \times 235.93$$
$$= \mathbf{236.87\ kJ/sec}$$

This supplied by the compressor turbine.

∴ Work developed by compressor turbine

$$= x\,C_p\,(T_3 - T_{4'})$$
$$= x \times 1.004\,(1050 - 692.15)$$
$$= x \times 1.004 \times 357.85$$
$$= \mathbf{359.28\ kJ/s}$$

$\therefore \quad 359.28\,x = 236.87$

$\therefore \quad x$ = Mass of gases flowing through compressor turbine out of 1 kg compressed by the compressor

$$= \frac{236.87}{359.28}$$
$$= \mathbf{0.6593\ kg/sec}$$

$\therefore \quad 1 - x = 1 - 0.6593$

= 0.3407 kg of gases are flowing through the power turbine.

$\therefore$ Power developed by the power turbine

$$P = m\,C_p\,(T_3 - T_{4'})$$
$$= 0.347 \times 1.004\,(1050 - 692.15)$$
$$= 0.347 \times 1.004 \times 357.85$$
$$= \mathbf{124.67\ kJ/sec = kW}$$

$\therefore$ Thermal efficiency of the plant

$$\eta_{th} = \frac{\text{Net work}}{\text{Heat supplied}}$$
$$= \frac{124.67}{1.004 \times (T_3 - T_{2'})} = \frac{124.67}{1.004(1050 - 535.93)}$$
$$= \frac{124.67}{1.004 \times 514.07}$$
$$= \mathbf{0.2416}$$
$$= \mathbf{24.16\%}$$

Example 24.6 *An open cycle gas turbine plant consists of a compressor driven by high pressure turbine from which the gas flows directly to a free, low pressure turbine from which the gas flows directly to a heat exchanger. The exhaust gas from the low pressure turbine are used to preheat the air delivered by the compressor before it delivers to the combustion chamber.*

The compressor pressure ratio is 7 and its isentropic efficiency is 0.86. The temperature of air at intake to compressor is 21°C.

The high pressure turbine isentropic efficiency may be assumed to be 0.85 and that for low pressure turbine 0.87. The temperature at entry to high pressure turbine may be taken as 925°C and heat exchanger thermal ratio 0.75. Assume pressure losses in combustion chamber to be 4% of the entry pressure and those in heat exchanger on the hot and cold side to be 6% of the respective entry pressure.

Mechanical efficiency of the compressor and high pressure turbine efficiency is 99%, combustion efficiency is 0.98. Neglect mass of fuel.

Take $C_p = 1.005$ *kJ/kg.K*

Determine :

(a) Air mass flow in kg/sec for 4000 kW power developed at alternate coupling, and

(b) Thermal efficiency of the plant.

Solution The schematic diagram and the T–S diagram are shown in Fig. Ex. 24.6. From diagrams we have

1–2′ Actual compression
2′–6 Heating of air in heat exchanger
6–3 Heating (heat added) in combustion chamber
3–4′ Expansion in HP gas turbine
4′–5′ Expansion in LP gas turbine

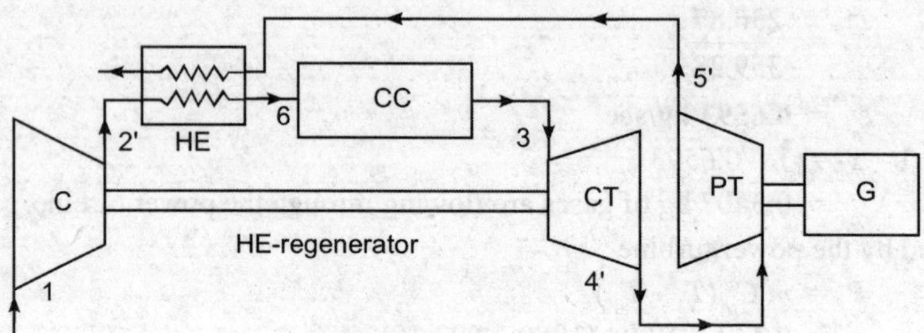

(a) Schematic diagram

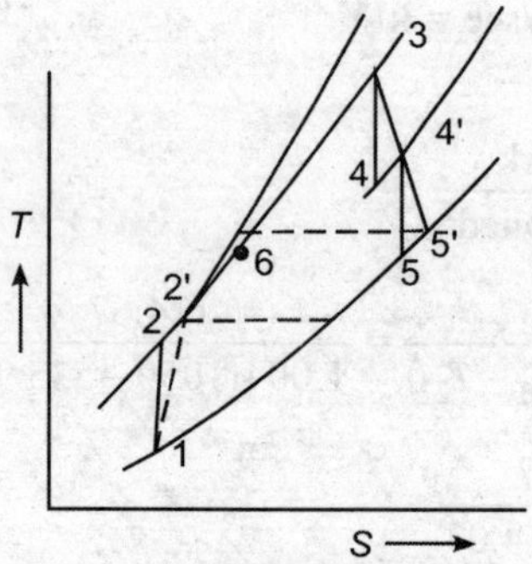

(b) T–S diagram

Fig. Ex. 24.6

We will assume $P_1 = 1$ bar

$$T_1 = 21 + 273 = 294 \text{ K}$$

$$\therefore \quad \frac{P_2}{P_1} = \text{Pressure ratio}$$

$$= 7$$

$$\therefore \quad \frac{T_2}{T_1} = \left(\frac{P_2}{P_1}\right)^{\frac{\gamma-1}{\gamma}} = 7^{\frac{0.4}{1.4}}$$

$$= 1.7436$$

$$\therefore \quad T_2 = 294 \times 1.7436$$

$$= \mathbf{512.61\ K}$$

$$\text{Compressor efficiency} = 0.86 = \frac{T_2 - T_1}{T_{2'} - T_1}$$

$$= \frac{512-294}{T_{2'}-294} = \frac{218}{T_{2'}-294}$$

$$\therefore \quad T_{2'} = \frac{218}{0.86} + 294$$

$$= 253.49 + 294$$

$$= \mathbf{547.49\ K}$$

The pressure drop in heat exchanger is 6%.

$\therefore \quad P_6$ = Pressure of air coming out of heat exchanger

$$= 0.94 \times P_2 = 0.94 \times 7$$

$$= \mathbf{6.58\ bar}$$

Again pressure drop in combustion chamber is 4% of the inlet pressure.

$$\therefore \quad P_3 = 0.96 \times 6.58$$

$$= \mathbf{6.317\ bar}$$

Assuming $\quad C_{pa} = C_{pg}$

$$\text{Work of compressor/kg} = \frac{C_{pa}(T_{2'}-T_1)}{\text{Mech. eff.}}$$

Work done by HP turbine = $C_{pg}(T_3 - T_4') \times$ Mech. eff.

$\therefore$ Work required by compressor = Work given by turbine

$$\frac{C_{pa}(T_{2'}-T_1)}{\text{Mech. eff.}} = C_{pg}(T_3 - T_{4'}) \times \text{Mech. eff.}$$

$$\therefore \quad \frac{(547.49-294)}{0.99} = [(925+273) - T_{4'}]\, 0.99$$

$$\therefore \quad T_{4'} = 1198 - 258.64$$

$$= \mathbf{939.36\ K}$$

$$\text{Turbine efficiency} = \frac{T_3 - T_{4'}}{T_3 - T_4}$$

$$0.85 = \frac{1198-939.36}{1198-T_4} = \frac{258.64}{1198-T_4}$$

$$\therefore \quad T_4 = 1198 - \frac{258.64}{0.85}$$

$$= 1198 - 304.28$$

$$= \mathbf{894.72\ K}$$

$$\therefore \quad \frac{T_3}{T_4} = \frac{1198}{894.72} = \left(\frac{P_3}{P_4}\right)^{0.2857}$$

$$\therefore \quad \frac{P_3}{P_4} = (1.339)^{3.5}$$

$$= 2.778$$

$$\therefore \quad P_4 = \frac{P_3}{2.778} = \frac{6.317}{2.778}$$

$$= \mathbf{2.274\ bar}$$

$$P_7 = 1 \text{ bar}$$

$$= P_5 \times 0.94$$

$$\therefore \quad P_5 = \text{Pressure after expansion in LP turbine}$$

$$= \frac{1}{0.94} = \mathbf{1.064\ bar}$$

$$\therefore \quad \frac{T_{4'}}{T_5} = \left(\frac{P_4}{P_5}\right)^{0.2857} = \left(\frac{2.274}{1.064}\right)^{0.2857}$$

$$= 1.242$$

$$\therefore \quad T_5 = \frac{939.36}{1.242}$$

$$= \mathbf{756.33\ K}$$

LP turbine efficiency is 87%.

$$\therefore \quad 0.87 = \frac{T_{4'} - T_{5'}}{T_{4'} - T_5} = \frac{939.36 - T_{5'}}{939.36 - 756.33}$$

$$939.36 - T_{5'} = 183.03 \times 0.87$$

$$\therefore \quad T_{5'} = 939.36 - 159.24$$

$$= \mathbf{780.12\ K}$$

∴ Work done by LP turbine

$$W_{\text{LP}} = C_{pa}\,(T_{4'} - T_{5'})$$

$$= 1.004\,[939.36 - 780.12]$$

$$= 1.004 \times 159.24$$

$$= \mathbf{159.88\ kJ/kg}$$

$$\text{Net power developed} = 4000 \text{ kW}$$

$$\therefore \quad \text{Mass flow rate/sec} = \frac{4000}{159.88}$$

$$= \mathbf{25.02\ kg/sec}$$

The efficiency of an heat exchanger or effective is given by

$$\varepsilon = 0.75 = \frac{T_6 - T_{2'}}{T_{5'} - T_{2'}}$$

$$= \frac{T_6 - 547.49}{780.12 - 547.49} = \frac{T_6 - 547.49}{232.63}$$

$$\therefore \quad 232.63 \times 0.75 = 174.47 = T_6 - 547.49$$

$$T_6 = 174.47 + 547.49$$
$$= \mathbf{721.96\ K}$$

$\therefore$ Q_A = Heat supplied/kg = $C_p\,(T_3 - T_6)/\eta_{comb}$.

$$= \mathbf{1.004}\ (1198 - 721.96)/0.98$$
$$= \frac{1.004 \times 466.04}{0.98}$$
$$= \mathbf{447.45\ kJ/kg}$$

$\therefore$ Thermal efficiency $= \dfrac{\text{Net work/kg}}{\text{Heat supplied/kg}}$

$$= \frac{159.88}{477.45}$$
$$= 0.3349$$
$$= \mathbf{33.49\%}$$

Example 24.7 *A gas turbine operates on Brayton cycle. The temperature range is 1000 K and 288 K. Find the pressure ratio for maximum power output. Also determine thermal efficiency, work ratio, and specific power output. Take C_p = 1.005 kJ/kg.k γ = 1.4. Expansion and compression are isentropic.*

Solution

The condition for maximum power output or net work is given by

$$\frac{T_3}{T_1} = (r_p)^{2m} = r_p^{\frac{2(\gamma-1)}{\gamma}}$$

$$r_p = \left(\frac{T_3}{T_1}\right)^{\frac{\gamma}{2(\gamma-1)}}$$

$$= \left(\frac{100}{288}\right)^{\frac{1.4}{2\times 0.4}}$$

$$= 3.472^{1.75}$$
$$= \mathbf{8.832}$$

$\therefore$ Pressure ratio for maximum output = **8.832**

We have

$$T_2 = T_1 \times r_p^{\frac{\gamma-1}{\gamma}} = 288 \times (8.832)^{0.2857}$$
$$= 288 \times 1.8633$$
$$= \mathbf{536.64\ K}$$

Similarly, $$T_4 = \frac{T_3}{1.8633} = \frac{1000}{1.8633}$$

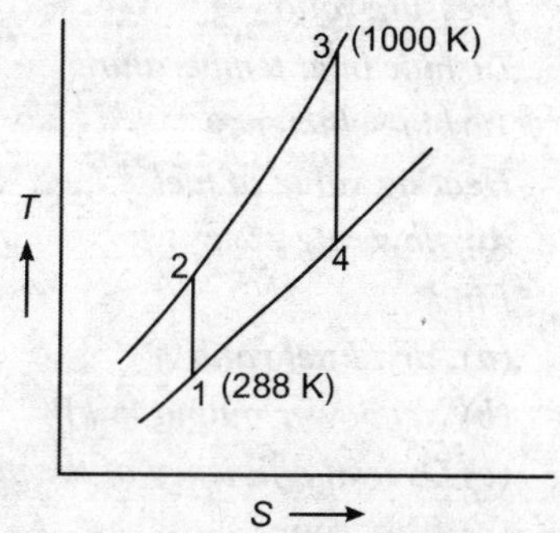

Fig. Ex. 24.7

$$= \mathbf{536.68\ K}$$

$\therefore$ Compressor work $= W_c = C_p\,(T_2 - T_1)$

$$= 1.005\,(536.64 - 288)$$

$$= \mathbf{249.88\ kJ/kg}$$

Also, turbine work/kg $= C_p\,(T_3 - T_4)$

$$W_t = 1.005\,(1000 - 536.68)$$

$$= \mathbf{460.64\ kJ/kg}$$

$\therefore$ Net work/kg (Specific output) $= W_t - W_c$

$$= 460.64 - 249.88$$

$$= \mathbf{210.76\ kJ/kg}$$

Heat supplied/kg $= C_p\,(T_3 - T_2)$

$$= 1.005\,(1000 - 536.64)$$

$$= \mathbf{465.68\ kJ/kg}$$

$\therefore$ Thermal efficiency $= \eta_{th} = \dfrac{\text{Net work}}{\text{Heat supplied}}$

$$= \frac{210.76}{465.68}$$

$$= 0.4526$$

$$= \mathbf{45.26\%}$$

Work ratio is given by $WR = \dfrac{W_t - W_c}{W_t} = \dfrac{\text{Net work}}{\text{Turbine work}}$

$$= \frac{210.76}{460.64}$$

$$= \mathbf{0.4575}$$

Example 24.8 *The following data refers to a gas turbine plant :*

Compressor inlet pressure	*1 bar*
Compressor inlet temperature	*310 K*
Efficiency of compressor	*75%*
Pressure ratio	*4*
Turbine inlet temperature	*840 K*
Turbine efficiency	*83%*
Heating value of fuel	*44000 kJ/kg*
Air flow rate	*75 kg/min*

Find :

(a) Air : Fuel ratio

(b) Net power output in kW

(c) Overall efficiency of the plant

(d) Work ratio.

Assume $C_{pa} = 0.996$, $C_{va} = 0.71$, $C_{pg} = 1.046$, $C_{vg} = 0.774$.

Solution

The *T–S* diagram for the given data is given in Fig. Ex. 24.8

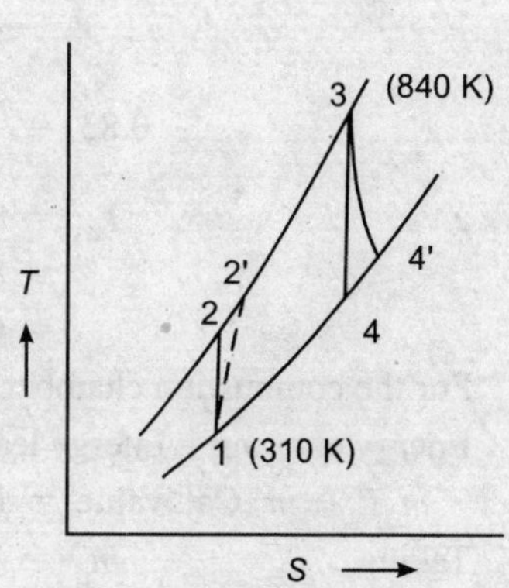

Fig. Ex. 24.8

Adiabatic index for air = $\dfrac{C_{pa}}{C_{va}}$

$$\gamma_a = \frac{0.996}{0.71}$$

$$= \mathbf{1.4}$$

Similarly, adiabatic index for gases

$$= \gamma_g = \frac{C_{pg}}{C_{vg}}$$

$$= \frac{1.046}{0.774}$$

$$= \mathbf{1.3514}$$

From diagram

$$\frac{T_2}{T_1} = \left(\frac{p_2}{p_1}\right)^{\frac{\gamma-1}{\gamma}} = (r_p)^{\frac{0.4}{1.4}} = 4^{0.2857}$$

$$= \mathbf{1.486}$$

$$\therefore \quad T_2 = 310 \times 1.486 = \mathbf{460.66\ K}$$

$$\text{Compressor efficiency} = 0.75 = \frac{T_2 - T_1}{T_{2'} - T_1}$$

$$= \frac{460.66 - 310}{T_{2'} - 310} = \frac{150.66}{T_{2'} - 310}$$

$$\therefore \quad T_{2'} = 310 + \frac{150.66}{0.75} = 310 + 200.88$$

$$= \mathbf{510.88\ K}$$

For expansion of gas in turbine, we have

$$\frac{T_3}{T_4} = \left(\frac{p_3}{p_4}\right)^{\frac{\gamma-1}{\gamma}} = (r_p)^{\frac{1.3514-1}{1.3514}}$$

$$= 4^{0.26} = \mathbf{1.434}$$

$$\therefore \quad T_4 = \frac{T_3}{1.434} = \frac{840}{1.434}$$

$$= \mathbf{585.77\ K}$$

and $$\eta_t = \frac{T_3 - T_{4'}}{T_3 - T_4} = \frac{840 - T_{4'}}{840 - 585.77}$$

$$0.83 = \frac{840 - T_{4'}}{254.23}$$

$$\therefore \quad T_{4'} = 840 - 0.83 \times 254.23$$
$$= 840 - 211$$
$$= \mathbf{629\ K}$$

For the combustion chamber,

Energy entering = Energy leaving

$$m_a T_{2'} + m_p \text{.Cal. value} = (m_a + m_f) T_3$$

Taking $m_a = 1$, and neglecting

Change in kinetic energy, we have

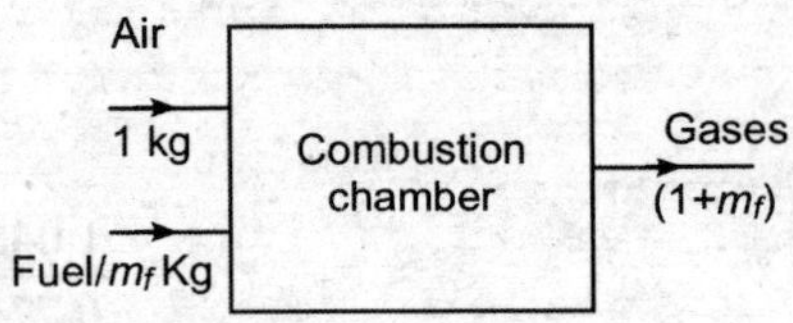

Fig. Ex. 24.9

$$C_{pa}\, T_{2'} + m_f \times \text{Cal. val.} = (1 + m_f) C_{pg}\, T_3$$

$$0.996 \times 510.88 + m_f \times 44000 = (1 + m_f) \times 1.046 \times 840$$

$$508.84 + 44000\, m_f = 878.64 + 878.64\, m_f$$

$$m_f (44000 - 878.64) = 878.64 = 508.84$$

$$m_f \times 43121.36 = 369.80$$

$$m_f = \frac{369.8}{43121.36}$$

$$0.008576 \text{ kg/kg air.}$$

$\therefore$ Air to fuel ratio = A : F ratio

$$= \frac{1}{0.008576}$$

$$= \mathbf{116.6}$$

(b) Net power output

$$\text{Compressor work/sec} = W_c = m_a C_{pa} (T_{2'} - T_1)$$

$$= \frac{75}{60} \times 0.996\,(510.88 - 310)$$

$$= \mathbf{250.1\ kJ/sec\ (kW)}$$

$$\text{Turbine work/sec} = W_t = \left(\frac{75}{60} \times 1.008576\right) \times 1.046\,(840 - 629)$$

$$= \mathbf{278.25\ kW\ (kJ/sec)}$$

$$\therefore \text{ Net power developed} = W_t - W_c$$
$$= 278.25 - 250.1$$
$$= \mathbf{28.15\ kW}$$

(c) Overall efficiency of the plant

Heat supplied/sec = fuel/sec × Cal. value

$$= \frac{75}{60} \times 0.008576 \times 44000$$

$$= 471.68 \text{ kJ/kg}$$

$$\therefore \text{ Overall efficiency of the plant} = \frac{\text{Net work}}{\text{Heat supplied}}$$

$$= \frac{28.15}{471.68}$$

$$= 0.05968$$

$$= 5.968\%$$

(*d*) *Work ratio*

$$\text{Work ratio} = WR = \frac{W_t - W_c}{W_t}$$

$$= \frac{28.15}{278.25}$$

$$= \mathbf{0.1012}$$

$$= \mathbf{10.12\%}$$

II. REGENERATIVE HEATING CYCLE

Example 24.9 *A gas turbine set having a heat exchanger gave the following data at design speed :*

Isentropic efficiency of compressor	*80%*
Isentropic efficiency of turbine	*85%*
Pressure ratio	*5 : 1*
Mass flow of air	*20 kg/hour*
Effectiveness of heat exchanger	*0.8*
Maximum cycle temperature	*725°C*
Ambient pressure and temperature	*1.013 bar, 25°C*

Determine :

(*i*) *Power output*

(*ii*) *Specific fuel consumption*

(*iii*) *Thermal efficiency.*

Take $C_p = 1.05$ *kJ/kg.k,* $\gamma = 1.4$*, heating value of fuel = 43000 kJ/kg*

Solution

The cycle is shown on *T–S* diagram of Fig. Ex. 24.9.

Given :

$$T_1 = 25 + 273 = 298 \text{ K}$$

$$P_1 = 1 \text{ bar}$$

$$\frac{P_2}{P_1} = 5, \; m_a = 20 \text{ kg/hour}$$

$$\eta_c = 0.8, \; \eta_t = 0.85, \; \varepsilon = 0.8$$

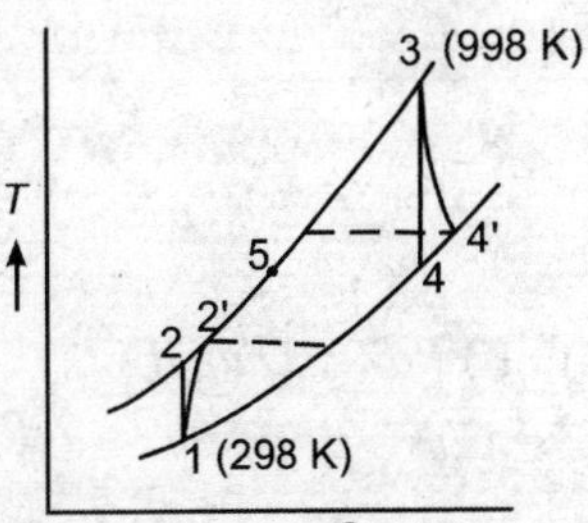

Fig. Ex. 24.9

$$T_3 = 725 + 273 = \mathbf{998\ K}$$

From Fig. Ex. 24.9

$$\frac{T_2}{T_1} = \left(\frac{P_2}{P_1}\right)^{\frac{\gamma-1}{\gamma}} = 5^{0.2857} = 1.5838.$$

$$\therefore \quad T_2 = 298 \times 1.5838 = \mathbf{471.97\ K}$$

But $$\eta_c = 0.8 = \frac{T_2 - T_1}{T_{2'} - T_1} = \frac{471.97 - 298}{T_{2'} - 298}$$

$$\therefore \quad T_{2'} = 298 + \frac{173.97}{0.8} = 298 + 217.46 = 515.46\ \text{K}$$

Similarly for turbine

$$\frac{T_3}{T_4} = \left(\frac{P_3}{P_4}\right)^{\frac{\gamma-1}{\gamma}} = 5^{0.2857} = \mathbf{1.5838}$$

$$\therefore \quad T_4 = \frac{998}{1.5838} = \mathbf{630.13\ K}$$

$$\eta_t = \frac{T_3 - T_{4'}}{T_3 - T_4} \quad \therefore \ 0.85 = \frac{998 - T_{4'}}{998 - 630.13}$$

$$\therefore \quad T_{4'} = 998 - 0.85 \times 367.87 = 998 - 312.69 = \mathbf{685.31\ K}$$

$\therefore$ Compressor work/sec $= mC_{pa}(T_{2'} - T_1)$

$$= \frac{20}{3600} \times 1.05\,(515.46 - 298)$$

$$= \mathbf{1.2685\ (kJ/sec = kW)}$$

Also, turbine work/sec $= mC_{pa}\,(T_3 - T_{4'})$

$$= \frac{20}{3600} \times 1.05\,(998 - 685.31)$$

$$= 312.69 \times 1.05 \times \frac{20}{3600}$$

$$= 1.824\ \text{kJ/sec} = \text{kW}$$

(*i*) *Power Output*

$$P = \text{Power output} = W_t - W_c = 1.824 - 1.2685 = \mathbf{0.5555\ kW}$$

(*ii*) *Specific fuel consumption :*

From diagram, effectiveness of heat exchanger is given by,

$$\varepsilon = \frac{T_5 - T_{2'}}{T_{4'} - T_{2'}} = \frac{T_5 - 515.46}{685.31 - 515.46}$$

$$0.8 = \frac{T_5 - 515.46}{169.85}$$

$$\therefore \quad T_5 = 515.46 + 169.85 \times 0.8$$

$$= 515.46 + 135.88$$

$$= \mathbf{651.34}$$

$$\therefore \quad \text{Heat supplied/hour} = 20 \times 1.05 \times (T_3 - T_5)$$

$$= 20 \times 1.05 \times (998 - 651.34)$$

$$= 7279.86 \text{ kJ/hour}$$

$$= \mathbf{2.0222 \text{ kJ/sec}}$$

$$\therefore \quad \text{Fuel supplied/hour} = \frac{7279.86}{43000}$$

$$= \mathbf{0.1693 \text{ kg/hour}}$$

∴ Specific fuel consumption

$$\text{SFC} = \frac{\text{Fuel/hour}}{\text{Power}}$$

$$= \frac{0.1693}{0.5555}$$

$$= \mathbf{0.3048 \text{ kJ/kW}}$$

(*iii*) *Thermal efficiency*

$$\eta_{th} = \frac{\text{Net work}}{\text{Heat supplied}}$$

$$= \frac{0.5555}{2.0222}$$

$$= 0.2747$$

$$= \mathbf{27.47\%}$$

Example 24.10 *In a gas turbine plant air at 10°C and 1 bar is compressed to 4 bar with compression efficiency of 80%. The air is heated in a regenerator and the combustion chamber till its temperature is raised to 700°C and during the process the pressure falls by 0.14 bar. The air is then expanded in the turbine and then passes to a regenerator which has 75% effectiveness and cause a pressure drop of 0.14 bar. If the isentropic efficiency of the turbine is 85%, determine the thermal efficiency of the plant.*

Solution

The cycle is shown in Fig. Ex. 24.10

$$T_3 = 973 \text{ K}$$

$$P_3 = 4 - 0.14 = 3.86 \text{ bar}$$

Similarly, $P_1 = 1$ bar and $P_4 = 1 + 0.14 = 1.14$ bar

$T_1 = 283$ K

From Fig. Ex. 24.10, $\dfrac{T_2}{T_1} = \left(\dfrac{P_2}{P_1}\right)^{0.2857} = 4^{0.2857}$

$= \mathbf{1.486}$.

$\therefore \quad T_2 = 1.486 \times 283$

$= \mathbf{420.528\ K}$

and $\quad \eta_c = \dfrac{T_2 - T_1}{T_{2'} - T_1} = 0.8.$

$\therefore \quad 0.8 = \dfrac{420.528 - 283}{T_{2'} - 283} = \dfrac{137.528}{T_{2'} - 283}$

$\therefore \quad T_{2'} = 283 + \dfrac{137.528}{0.8}$

$= \mathbf{454.91\ K}$

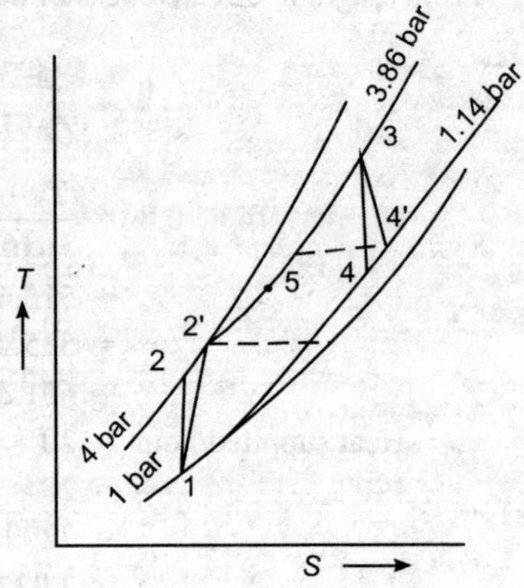

Fig. Ex. 24.10

Similarly, $\dfrac{T_3}{T_4} = \left(\dfrac{P_3}{P_4}\right)^{0.2857} = \left(\dfrac{3.86}{1.14}\right)^{0.2857}$

$= 1.4169$

$T_4 = \dfrac{973}{1.4169} = \mathbf{686.71\ K}$

and

$\eta_t = \dfrac{T_3 - T_{4'}}{T_3 - T_4} = \dfrac{973 - T_{4'}}{973 - 686.71} = 0.85$

$\therefore \quad T_{4'} = 973 - 243.35$

$= \mathbf{729.65\ K}$

Again ε = Effectiveness of regenerator

$= \dfrac{T_5 - T_{2'}}{T_{4'} - T_{2'}}$

$0.75 = \dfrac{T_5 - 454.91}{729.65 - 454.91}$

$T_5 = 0.75 \times 274.74 + 454.91$

$= 206.055 + 454.91$

$= \mathbf{660.965\ K}$

$\therefore$ Heat supplied/kg air $= 1 \times 1.1\ (973 - 660.965)$

$= 1.1 \times 312.035$

$= \mathbf{343.24\ kJ/kg}$

$$\text{Work done/kg} = W_t - W_c = C_{pg}(T_3 - T_{4'}) - C_{pa}(T_{2'} - T_1)$$
$$= 1.1\,(973 - 729.65) - 1.005\,(454.91 - 283)$$
$$= 1.1 \times 240.35 - 1.005 \times 171.91$$
$$= 264.385 - 172.77$$
$$= \mathbf{91.615\ kJ/kg}$$

$$\therefore \quad \text{Thermal efficiency} = \frac{\text{Work done/kg}}{\text{Heat supplied/kg}}$$
$$= \frac{91.615}{343.24}$$
$$= 0.2669$$
$$= \mathbf{26.69\%}$$

III. INTERCOOLING AND REHEATING

Example 24.11 *A gas turbine plant operates between 1 bar and 9 bar with minimum and maximum cycle temperatures of 25°C and 1250°C respectively. The two stage compressor has perfect intercooling and ideal intermediate pressure. Gases are reheated to 1250°C after the expansion in the HP turbine and are expanded in the LP turbine such that output power from each turbine is same. The isentropic efficiencies of compressor and turbine is 0.83. Calculate :*

(i) Efficiency of the cycle

(ii) Power output for a mass flow rate of 5 kg/sec

(iii) Work ratio.

Sketch the plant schematic and the cycle on T–S diagram.

Solution

The schematic diagram for the plant is shown in Fig. Ex. 24.11 and its *T–S* diagram in Fig. Ex. 24.11

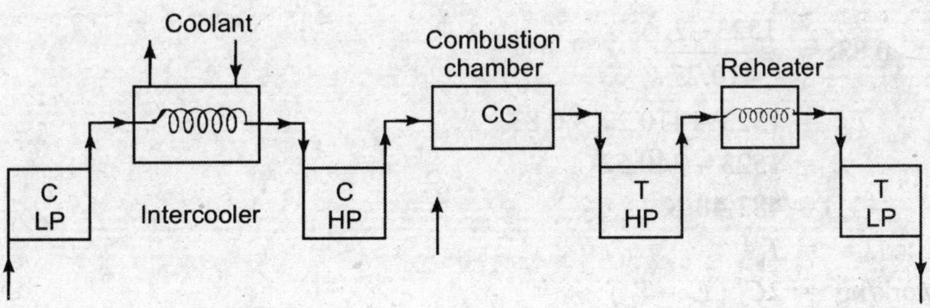

Fig. Ex. 24.11(a) *Schematic of brayton cycle with reheater and intercooler.*

For perfect intercooling, the intermediate pressure P_2 is given by

$$P_2 = \sqrt{P_1 P_4} = \sqrt{1 \times 9} = 3 \text{ bar}$$

$$\therefore \quad \frac{T_2}{T_1} = (3)^{0.2857} = \mathbf{1.3687}$$

$$\therefore \quad T_2 = 298 \times 1.3687 = \mathbf{407.88\ K}$$

But $\eta_c = 0.83 = \frac{T_2 - T_1}{T_{2'} - T_1}$

$= \frac{407.88 - 298}{T_{2'} - 298} = \frac{109.88}{T_{2'} - 298}$

$\therefore \quad T_{2'} = 298 + \frac{109.88}{0.83} = 298 + 132.38$

$= \mathbf{430.38\ K}$

For perfect intercooling and ideal intermediate pressure,

$T_{2'} - T_{4'} = 430.38\ \text{K}$

$\therefore$ Compressor work per kg air $= W_c = 2C_{pa}\,(T_{2'} - T_1)$

$= 2 \times 1.005 \times (430.38 - 298)$

$= \mathbf{266.095\ kJ/kg}$

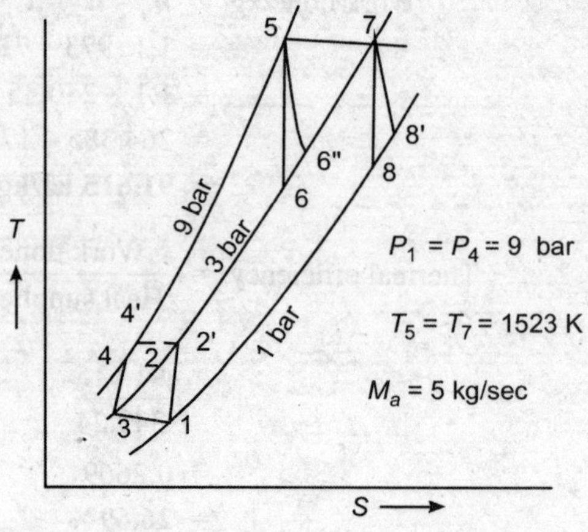

Fig. Ex. 24.11(b) *T-S diagram.*

Work in HP turbine = Work in LP turbine

and $T_5 = T_7$. $\therefore$ Intermediate temperature is also same.

$\therefore$ Reheat pressure is 3 bar only.

$\therefore \quad \frac{T_5}{T_6} = (3)^{0.2857} = 1.3687$

$\therefore \quad T_6 = \frac{1523}{1.3687} = \mathbf{1112.73\ K}$

and $\eta_t = \frac{T_5 - T_{6'}}{T_5 - T_6} = \frac{1523 - T_{6'}}{1523 - 1112.73}$

$0.83 = \frac{1523 - T_{6'}}{410.27}$

$\therefore \quad T_{6'} = 1523 - 410.27 \times 0.83$

$= 1523 - 340.52$

$= \mathbf{482.48\ K}$

$= T_{8'}$

$\therefore$ Turbine work/kg $= 2C_p\,(T_5 - T_{6'})$

$= 2 \times 1.005 \times (1523 - 1182.48)$

$= \mathbf{684.45\ kJ/kg}$

$\therefore$ Net work/sec $= 5\,(W_t - W_c)$

$= 5\,[684.45 - 266.095]$

$= 5 \times 418.355$

$= \mathbf{2091.775\ kJ/s = kW}$

Heat supplied/kg $= C_p\,(T_5 - T_{4'}) + C_p\,(T_7 - T_{6'})$

$= 1.005\,[(1523 - 430.38) + (1523 - 1182.48)]$

$= 1.005 \times (1092.62 + 340.52)$

$$= \mathbf{1440.3057\ kJ/kg}$$

$\therefore$ Heat supplied/sec $= 5 \times 1440.3057$

$$= \mathbf{7201.53\ kJ/sec}$$

(i) Efficiency of the cycle

$$\eta_{th} = \frac{\text{Work}}{\text{Heat supplied}}$$

$$= \frac{2091.775}{7201.53}$$

$$= \mathbf{0.2905}$$

$$= \mathbf{29.05\%}$$

(ii) Power output

Power output = Net work done/sec

$$= \mathbf{2091.775\ kJ/sec = kW}$$

(iii) Work ratio

$$WR = r_W = \frac{W_t - W_c}{W_t}$$

$$= \frac{2091.775}{684.45 \times 5} = \mathbf{0.6112}$$

IV. INTERCOOLING, REHEATING, REGENERATION

Example 24.12 *The following data refers to a gas turbine plant. Two sets of compressors are used.*

The overall pressure ratio is 1 : 4

Intake air pressure is 1 bar

HP turbine drives both the compressors

LP turbine drives the generator

The exhaust from LP turbine plant passes through the heat exchanger. The heat is given to the outlet air from HP compressor. The maximum temperature in the plant is 560°C. The gases are reheated to original temperature after expansion in HP turbine.

Air inlet temperature	*30°C*
η_{comp}	*85%*
η_{exp}	*80%*
Mech. eff. of HP turbine	*95%*
Mech. eff. of LP turbine	*96%*
Combustion efficiency	*96%*
Heat exchanger effectiveness	*75%*

$C_{pa} = 1$ kJ/kg.K, $\gamma_{air} = 1.4$, $C_{pg} = 1.155$ kJ/kg.K, $\gamma_{gas} = 1.33$

Neglect mass of fuel and assume perfect intercooling. Find :

(i) Overall efficiency

(ii) Work ratio

(iii) Mass flow for 3500 kW output.

Solution

The schematic for the plant in question is shown in Fig. Ex. 24.12(a) and the T–S diagram for the same in Fig. Ex. 24.12(b)

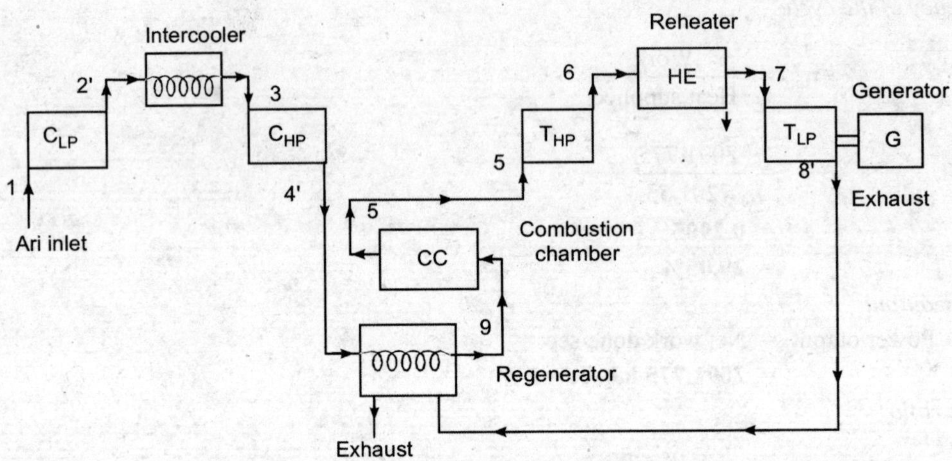

Fig. Ex. 24.12(a) *T–S diagram of Brayton cycle with intercooler, reheater and regenerator.*

In a two stage compressor, the intercooling is perfect and the overall pressure ratio is 4.

$\therefore$ Intermediate pressure $= \sqrt{4} = 2$ bar

$\therefore$ From Fig. Ex. 24.12(b),

$$\frac{T_2}{T_1} = 2^{0.2857} = 1.219$$

$$\therefore \quad T_2 = 1.219 \times 303 = \mathbf{369.36\ K}$$

and

$$\eta_c = 0.85 = \frac{T_2 - T_1}{T_{2'} - T_1}$$

$$= \frac{369.36 - 303}{T_{2'} - 303} = \frac{66.36}{T_{2'} - 303}$$

$$\therefore \quad T_{2'} = \frac{66.36}{0.85} + 303 = 78.07 + 303$$

$$= \mathbf{381.07\ K}$$

$$= T_{4'}$$

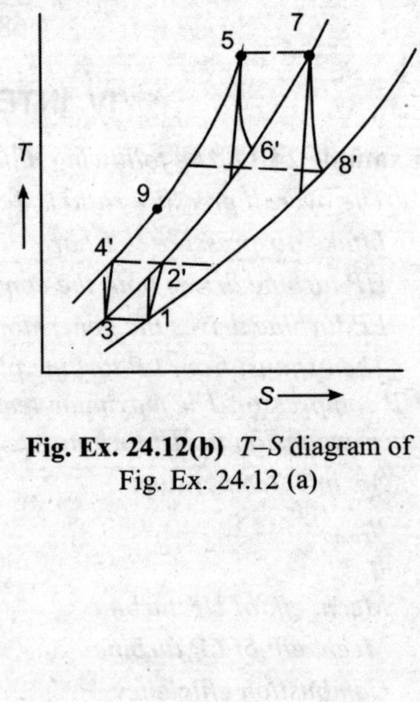

Fig. Ex. 24.12(b) *T–S* diagram of Fig. Ex. 24.12 (a)

$$\text{Compressor work/kg} = 2\, C_{pa}\,(T_{2'} - T_1)$$

$$= 2 \times 1 \times (381.07 - 303)$$

$$= 2 \times 78.07$$

$$= \mathbf{156.14\ kJ/kg}$$

For turbine driving the compressor, we have

Work output of turbine = Work of compressor

$C_{pg}\,(T_5 - T_{6'}) \times$ Mech. eff. = 156.14

$$\therefore \quad 1.155\,(833 - T_{6'}) \times 0.95 = 156.14$$

$$833 - T_{6'} = \frac{156.14}{0.95 \times 1.155}$$

$$= \mathbf{142.3}$$

$$\therefore \quad T_{6'} = 833 - 142.3$$

$$= \mathbf{690.7\ K}$$

But
$$\eta_t = 0.8 = \frac{T_5 - T_6}{T_5 - T_{6'}}$$

$$= \frac{833 - T_6}{833 - 690.7} = \frac{833 - T_6}{142.3}$$

$$T_6 = 833 - 0.8 \times 142.3$$

$$= 833 - 113.84$$

$$= \mathbf{719.16\ K}$$

$$\therefore \quad \frac{T_5}{T_6} = \left(\frac{P_5}{P_6}\right)^{\frac{0.33}{1.33}} = \left(\frac{P_5}{P_6}\right)^{0.2481}$$

$$\frac{833}{719.16} = 1.1583 = \left(\frac{P_5}{P_6}\right)^{0.2481}$$

$$\therefore \quad \frac{P_5}{P_6} = (1.1583)^{1/0.2481} = (1.1583)^{4.031}$$

$$= \mathbf{1.8083}$$

$$\therefore \quad \frac{P_6}{P_7} = \frac{4}{1.8083}$$

$$= \mathbf{2.212}$$

$\therefore$ Pressure ratio for LP turbine

$$\frac{P_6}{P_9} = \frac{4}{1.8083} = \mathbf{2.212}$$

For LP turbine

$$T_7 = 833\text{ K} \therefore \frac{T_7}{T_8} = \left(\frac{P_7}{P_8}\right)^{0.2481} = (2.212)^{0.2481}$$

$$= \mathbf{1.2177}$$

$$\therefore \quad T_8 = \frac{833}{1.2177}$$

$$= \mathbf{684.1\ K}$$

$$\eta_t = 0.8 = \frac{T_7 - T_{8'}}{T_7 - T_8}$$

$$= \frac{833 - T_{8'}}{833 - 684.1}$$

$$\therefore \quad 0.8 \times 148.9 = 119.12 = 833 - T_{8'}$$

$$\therefore \quad T_{8'} = 833 - 119.12$$

$$= \mathbf{713.88\ K}$$

∴ Work done/kg for LP turbine

$$W_{LP} = C_{pg}(T_5 - T_{8'})$$

$$= 1.155\,(833 - 713.88)$$

$$= 1.155 \times 119.12$$

$$= \mathbf{137.584\ kJ/kg}$$

∴ Work delivered to generator

$$= 0.96 \times 137.584$$

$$= \mathbf{132.08\ kJ/kg}$$

Heat Exchanger (Regenerator)

Effectiveness of regenerator is given by

$$\varepsilon = 0.75 = \frac{T_9 - T_{4'}}{T_{8'} - T_{4'}} = \frac{T_9 - 381.07}{713.88 - 381.07}$$

$$= \frac{T_9 - 381.07}{322.81}$$

$$\therefore \quad T_g = 381.07 + 0.75 \times 322.81$$

$$= 381.07 + 242.11$$

$$= \mathbf{623.18\ K}$$

∴ Heat supplied/kg $= C_{pg}(T_5 - T_9) + C_{pg}(T_7 - T_{6'})$

$$= 1.155\,(833 - 623.18) + 1.155\,(833 - 690.7)$$

$$= 1.155 \times 209.82 + 164.36$$

$$= 242.342 + 164.36$$

$$= \mathbf{406.702\ kJ/kg}$$

(*i*) *Overall efficiency*

$$\text{Overall efficiency} = \frac{\text{Net work delivered}}{\text{Heat supplied by fuel}}$$

$$= \frac{132.08}{406.702/0.96}$$

$$= \mathbf{0.3118}$$

$$= \mathbf{31.18\%}$$

(*ii*) *Work ratio*

$$WR = r_w = \frac{\text{Net work}}{\text{Total turbine work}}$$

$$= \frac{132.08}{132.08 + 156.14}$$

$$= \frac{132.08}{288.22}$$

$$= \mathbf{0.4583}$$

(*iii*) *Mass flow*

Power output = **3500 kW = kJ/sec**

Net work delivered = 132.08 kJ/kg

$$\therefore \quad \text{Mass flow rate} = \frac{3500}{132.08}$$

$$= \mathbf{26.5\ kg/sec}$$

Example 24.13 *The following data refers to a gas turbine plant using intercooling, regeneration and reheating arrangement.*

Pressure ratio	*4*
Compressor inlet temperature	*26°C*
Turbine inlet temperature	*600°C*
Compressor efficiency	*0.8*
Turbine efficiency	*0.8*
Regenerator efficiency	*0.8*
Inlet pressure to compressor	*1 bar*

Take C_p = 1.005 kJ/kg.K

Determine :

(*a*) *Cycle thermal efficiency*

(*b*) *Cycle work ratio.*

Solution

Schematic diagram of the plant and corresponding *T–S* diagram are shown in Fig. Ex. 24.13

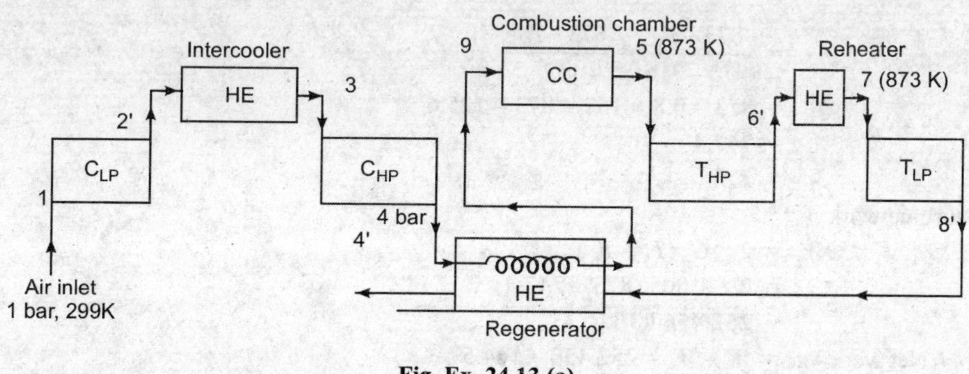

Fig. Ex. 24.13 (a)

Assuming intercooling to be perfect,

$$P_2 = \sqrt{P_1 P_4}$$

$$= \sqrt{1 \times 4} = 2 \text{ bar.}$$

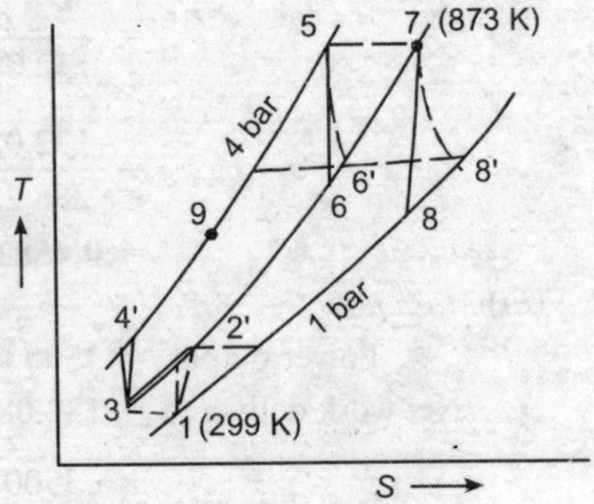

Fig. Ex. 24.13 (b)

$$\therefore \quad \frac{T_2}{T_1} = 2^{0.2857} = 1.219$$

$$\therefore \quad T_2 = 299 \times 1.219 = \mathbf{364.5\ K}$$

and
$$\eta_c = 0.8 = \frac{T_2 - T_1}{T_{2'} - T_1}$$

$$= \frac{364.5 - 299}{T_{2'} - 299}$$

$$\therefore \quad T_{2'} = 299 + \frac{65.5}{0.8} = \mathbf{380.875\ K}$$

$$= T_{4''}$$

$$\therefore \text{ Compressor work/kg} = 2C_p\,(T_{2'} - T_1)$$

$$= 2 \times 1.005\,(380.875 - 299)$$

$$= \mathbf{164.57\ kJ/kg}$$

Now consider turbine

$$T_5 = 600 + 273 = \mathbf{873\ K}$$

For ideal reheating cycle, the intermediate pressure of reheating is again 2 bar.

∴ For HP turbine

$$\frac{T_5}{T_6} = 2^{0.2857} = 1.219$$

$$\therefore \quad T_6 = \frac{873}{1.219} = \mathbf{716\ K}$$

and
$$\eta_t = 0.8 = \frac{T_5 - T_{6'}}{T_5 - T_6}$$

$$= \frac{873 - T_{6'}}{873 - 716} = \frac{873 - T_{6'}}{157}$$

$$\therefore \quad T_{6'} = 873 - 0.8 \times 157 = 873 - 125.6$$

$$= \mathbf{747.4}$$

$$= \boldsymbol{T_{8'}}$$

∴ Total turbine work

$$W_t = 2 \times C_p\,(T_5 - T_{6'})$$

$$= 2 \times 1.005\,(873 - 747.4)$$

$$= \mathbf{252.456\ kJ/kg}$$

$$\therefore \quad \text{Net work/kg} = W_t - W_c = 252.456 - 164.57$$

$$= \mathbf{87.886\ kJ/kg}$$

Again consider regenerator.

$$\text{Regenerator efficiency} = \varepsilon = 0.8 = \frac{T_9 - T_{4'}}{T_{8'} - T_{4'}}$$

$$= \frac{T_9 - 380.875}{747.4 - 380.875}$$

$$= \frac{T_9 - 380.875}{366.525}$$

$$\therefore \quad T_9 = 380.875 + 0.8 \times 366.525$$

$$= 380.875 + 293.22$$

$$= \mathbf{674.095\ K}$$

= Temperature of air at inlet to combustion chamber.

∴ Heat supplied in combustion chamber/kg

$$Q_A = C_p\,(T_5 - T_9) + C_p\,(T_7 - T_{6'})$$

$$= 1.005\,(873 - 674.095) + 1.005\,(873 - 747.4)$$

$$= 1.005 \times 198.905 + 1.005 \times 125.6$$

$$= 199.9 + 126.86$$

$$= \mathbf{326.76\ kJ/kg}$$

(*a*) *Cycle thermal efficiency*

$$\eta_{th} = \frac{\text{Net work/kg}}{\text{Heat supplied/kg}}$$

$$= \frac{87.886}{326.76}$$

$$= 0.269$$

$$= \mathbf{26.9\%}$$

(*b*) *Work ratio*

$$WR = r_w = \frac{\text{Net work}}{\text{Turbine work}}$$

$$= \frac{87.886}{252.456}$$

$$= \mathbf{0.3481}$$

THEORY QUESTIONS

24.1. What is the difference between diesel engine working and gas turbine working?

24.2. (a) What are the different elements of a gas turbine plant? Shown the schematic diagram and explain the working of the plant.

(b) Why a large air to fuel ratio is used in gas turbine plant?

24.3. Enumerate the advantages and disadvantages of a gas turbine plant over Diesel engine power plant.

24.4. Compare the open and closed cycle gas turbine power plants.

24.5. What are the different methods that are practiced to improve the performance of the plant?

24.6. What are the effects of regeneration in a gas turbine on the performance of the plant?

24.7. Explain with neat sketch the working of gas turbine with reheating. In what way, this method improves the performance of the gas turbine plant?

24.8. In what conditions intercooling is beneficial? In what way intercooling improves the performance of the gas turbine plant?

24.9. What is the effect of pressure ratio on Brayton cycle work output and efficiency? Explain.

24.10. What do you understand by maximum pressure ratio $(r_p)_{max}$ and optimum pressure ratio $(r_p)_{opt}$? What is the relation between them?

24.11. Derive the relation for the optimum pressure ratio when compressor and turbine efficiencies are taken into consideration.

24.12. Explain the working of closed and semi-closed cycle gas turbines.

23.13. Briefly describe the different fuels that can be burnt in gas turbine plants.

23.14. Explain in short the materials which are used for different components of the gas turbine power plants.

23.15. Why turbine blade cooling is required? How this cooling is done in practice?

23.16. Discuss the various methods of starting a gas turbine power plant.

23.17. What are the application of gas turbine plants?

23.18. Derive an expression for maximum work in Brayton cycle and prove that –

$$W_{max} = C_p \left[\sqrt{T_{max}} - \sqrt{T_{min}}\right]^2.$$

PROBLEMS FOR PRACTICE

1. The gas turbine power plant consists of compressor and turbine both having isentropic efficiency of 85%. The pressure ratio is 4, and the temperature of air at the inlet to compressor is 15°C. The maximum temperature in the cycle is 700°C. Assume $C_p = 1.1$ and $C_v = 0.786$ kJ/kg.K. Neglecting all the losses, determine the specific output and the overall efficiency of the cycle. **[Ans.** $W = 116.5$ kJ/kg, $\eta_o = 20.35\%$**]**

2. The pressure and temperature of air at inlet to the compressor are 1 bar and 15°C. Pressure ratio is 5 and the compressor efficiency is 85%. Air is heated in the combustion chamber to 550°C. Isentropic efficiency of turbine is 80%. The capacity of the plant is 1500 kW. Estimate the flow rate of air or gases. Assume a pressure drop of 0.1 bar when the air passes through the combustion chamber. Take $C_p = 1.1$ kJ/kg.K and $\gamma = 1.4$. Neglect the mass of fuel. **[Ans.** $W_a = 32.09$ kg/sec**]**

3. Atmospheric air at 1 bar and 10°C enters a compressor. The static pressure at the suction of the compressor is 0.9 bar and compresses it to a total pressure of 6 bar and total (stagnation) temperature of 230°C. Expansion of air in the turbine is upto a total pressure of 1 bar and total temperature of 460°C. The capacity of the plant is 1930 kW. Calculate :

   ^{2}, find mass flow rate of air

   (iv) Temperature of gases at turbine inlet.

   Take $C_p = 1.05$ kJ/kg.K, $C_{pg} = 1.13$ kJ/kg.K, $\gamma = 1.4$ and $R = 300$ N-m/kg-K.

   [Ans. (i) $\eta_c = 86\%$ (ii) $C_1 = 132.8$ m/s (iii) 14.51 kg/s (iv) 1055 K**]**

4. In an oil gas turbine installation air is compressed from a pressure of 1 bar and temperature of 300 K upto a pressure of 5 bar. The oil used has a calorific value of 42500 kJ/kg and combustor efficiency is 95%. Hot gases leave the combustor chamber at 1000 K. The isentropic efficiency of the turbine and compressor are 90% and 85% respectively. Assume a mass flow rate at 1 kg/sec, find :

   (a) Air : Fuel ratio

   (b) Power output of the plant

   (c) Thermal efficiency of the power plant.

   Assume $C_{pa} = 1.005$ kJ/kg.K, $C_{pg} = 1.1$ kJ/kg.K, $\gamma = 1.4$ for both air and gases. $\rho_p = 5$.

   [Ans. (a) 80.71 : 1 (b) 162.36 kW (c) 32.46%**]**

5. In a gas turbine plant, air at 27°C and 0.98 bar is compressed to 6 bar. The temperature of air is increased to

750°C as it passes through the combustion chamber. The isentropic efficiencies of compressor and turbine are 0.8 and 0.85 respectively. Determine the efficiency of the plant. [**Ans.** 20.72%]

6. In a simple gas turbine the inlet air temperature is 300 K and the maximum temperature of the gas is 1000 K, while the pressure range is from 1 bar to 4 bar. Taking C_p = 1.051 kJ/kg.K and C_v = 0.7494 kJ/kg.K, ignoring combustion and ducting losses and assuming that the mass of fuel is small compared with that of air, calculate
   (i) Power per kg air
   (ii) Work ratio and
   (iii) The overall efficiency of the plant, when (a) the compressor and the turbine are cent percent efficient (b) isentropic efficiency of the compressor and the turbine is 80%.

   [**Ans.** (i) (a) 190.5 kJ/kg (b) 83.5 kJ/kg (ii) (a) 0.554 (b) 0.304 (iii) (a) 0.327 (b) 0.153]

7. Determine the power output, efficiency and specific fuel consumption of a gas turbine using 4.5 kg of air per second under following conditions :

Efficiency of the turbine	0.85
Efficiency of the compressor	0.85
Air : Fuel ratio	50
Inlet conditions	1 bar, 27°C
For compression	$C_p = 1.005, \gamma = 1.4$
For combustion and expansion	$C_p = 1.092, \gamma = 1.35$
Lower heat of combustion	42 MJ/kg

   Assume no combustion losses, no heat losses and no heat exchanger losses.

   [**Ans.** Power = 805.5 kW, Efficiency = 21.31% SFC = 0.402 kg/kWh]

8. A gas turbine power plant operates with air entering the compressor at 1 bar and 27°C and a pressure ratio of 5 and the turbine inlet temperature of 1000°C. Calculate the cycle thermal efficiency, the temperature of exhaust gases after heat exchange, assuming the working fluid to be air, with constant specific heat capacity with heat exchanger, making use of 75% of available heat. Assume compressor and turbine efficiencies to be 80 percent. Neglect the pressure drop in the connecting pipes and the mass of fluid added. Find also overall efficiency, without heat exchange. [**Ans.** η_{th} (without heat exchange) = 20.8%, η_{th} with heat exchange = 33.2%, 613.9 K]

9. In a gas turbine plant of power output 2100 kW, a rotory compressor of isentropic efficiency 0.83 is used to compress 30 kg/s air (C_p = 1 kJ/kg.K and γ = 1.4) at compression ratio 5 and total head inlet temperature 295 K, inlet pressure 95 kN/m^2. The air after compression passes through a heat exchanger, in which its temperature is raised to 625 K. The flow rate of fuel of cal. value 45,000 kJ/kg, is 0.25 kg/s. In the turbine, gases (C_p = 1.15 kJ/kg.K, γ = 1.3) expand from 460 to 99 kN/m^2, leave at 682 K. Calculate isentropic efficiency of the turbine. Find combustion efficiency, the thermal ratio of the heat exchanger. Find also the gain in overall efficiency by fitting the heat exchanger. [**Ans.** η_t = 0.875 η_{comb} = 0.9223, Thermal ratio = 0.6817, Gain = 40.8%]

10. In a gas turbine plant, air is compressed from pressure P_1 and temperature T_1 to a pressure ratio r_p. The air is heated to maximum temperature T_{max} and expanded in a two stage turbine with equal pressure drop is stages. The air expanding from HP stage is reheated to the temperature T_{max} and then it expands to initial pressure P_1. Assuming isentropic expansion and compression, show that for maximum output possible,

where

$$s = t^{2/3}$$

$$s = r_p^{\frac{\gamma-1}{\gamma}}$$

$$t = \frac{T_{max}}{T_1}.$$

Chapter 25 Combined Cycle Power Plants

25.1 INTRODUCTION

We know that the heat engine working on Carnot cycle gives us the maximum thermal efficiency than any other type of the heat engine when working between the same temperature limits. This thermal efficiency is given by

$$\eta_{th} = \frac{T_1 - T_2}{T_1}$$

where T_1 is the maximum temperature in the cycle and at which heat is added and T_2 is the minimum temperature in the cycle at which heat is rejected. Generally T_2 is the atmospheric temperature at which the heat is rejected. For a constant atmospheric temperature T_2, efficiency increases with the higher temperature. Therefore, the main objective of a Thermal Power Plant Design Engineer is to consider the higher temperature to the maximum possible limit and similarly the lower temperature to the minimum limit so that the efficiency of the plant will be maximum possible.

Generally, the fluids that are used at present in power plants are gases in gas turbine power plant and steam (water) in thermal power plant.

Therefore, we will consider these two fluids — one is non-considerable or gaseous fluid and the second is water which change phase during the passage through a cycle or a plant.

25.2 WATER VAPOR (STEAM)

Water vapour has physical properties not altogether desirable at either end of the expansion range. It is known that there are vapours which are thermodynamically better suited than steam to one or the other ends of the expansion range. One the low pressure side of expansion, sulphur dioxide, hydrocarbons such as butane, propane, petroleum ether etc. have been found suitable while on the high pressure side of expansion mercury and diphenyl oxide are found attractive. In heat power plants, the temperature difference $T_1 - T_2$ has to be increased in order to get better efficiency. Following are some means to increase the temperature range $(T_1 - T_2)$.

25.2.1 Increase of Degree of Superheat

Comment

(a) More investment in superheaters.

(b) Mean temperature of heat addition is increased for superheating process only.

25.2.2 Increase of Boiler Pressure

(a) High pressure means thicker, heavier and more expensive construction in boilers, piping and turbines.

(b) At high temperature and high pressures the tensile strength of the metal decreases.

25.2.3 Decrease of Exhaust Pressure

(a) Volume of steam at low pressure end of expansion is enormously large and the size is also increased for the turbine.

(b) Condenser required is large and therefore costly.

(c) To get the low exhaust pressure, greater vacuum is to be maintained which is difficult and troublesome (because of the leakage of non-condensable gases such as air).

Critical temperature of water is 374°C and corresponding critical pressure is about 221 bar. If the best available material is used, and considering the best heat transfer coefficients, the maximum temperature of the superheated steam is about 600°C and in practice we are getting 550°C temperature for superheated steam.

At the temperature of atmosphere say 40°C, the saturation pressure is round about 0.075 bar. To maintain such a high vacuum is a main problem because of the leakages of air to the condenser. At the end of expansion to this pressure, another problem is of the large wettness of the steam and the low pressure stages are to be very large due to large volume of steam and the problem of erosion of the turbine blades is severe.

25.3 NON-CONDENSABLE FLUIDS (GAS)

Gases are used in gas turbine power plant and there is a large quantity of air is used as compared to other plants i.e., the air is very high and air to fuel ratio is very large.

The temperature of the exhaust gases of a simple gas turbine, generally, lies between 400 to 500°C and because of the large air to fuel ratio, large quantity of oxygen is present in the exhaust gases. Because of these, the heat that is rejected to atmosphere is also large. In the Brayton cycle, even these exhaust gases are used to preheat air before it enters combustion chamber, very little heat is recovered and the gain in thermal efficiency is also less. There is no effect on the output of the turbine. Also, whatever excess oxygen is present in the exhaust from the turbine, is not utilized and goes into atmosphere without any use.

The fact that the exhaust gases of the gas turbine contain large amount of heat energy, the power engineers made an effort to recover this heat energy and use the same in producing steam which may further be used in steam turbines to produce electrical power. In other words they coupled gas turbine and steam turbine to produce power.

25.4 DESIRABLE PROPERTIES OF WORKING FLUID FOR VAPOUR POWER CYCLE

To get the best thermal efficiency, the working fluid in the vapour power cycle should have the following properties:

(a) **Thermodynamic Properties** (i) The fluid should have a high critical temperature at a relatively low pressure. At the same time it should have a large enthalpy of evaporation at the pressure. Pressure and temperature are limited by metallurgical consideration.

(ii) The saturation pressure at the temperature of heat rejection should be above atmospheric pressure so that the problem of maintaining the vacuum in the condenser will be absent.

(iii) The specific heat of liquid should be small so that the boiling point or saturation temperature is obtained with the less amount heat transfer.

(iv) On *T–S* diagram, the liquid line should be steep so that the expansion line and the liquid line will be close together.

(b) **Physical and Chemical Properties** (i) The fluid should be non-toxic, non-corrosive, not excessively viscous.

(ii) It should be easily available at low cost.

(iii) It should be chemically stable and should not react with the materials of construction at any temperature.

(iv) The freezing point of the fluid should be below the room temperature so that it does not get solidified while flowing through the pipelines.

No working fluid possesses all these characteristics and compromise is to be done while selecting the most suitable working fluid to be used in vapour power cycle.

25.5 BINARY VAPOUR CYCLE

We have seen above that the maximum efficiency of any heat engine is given by

$$\eta_{th} = \frac{T_1 - T_2}{T_1}$$

where T_2 is the temperature at which heat is rejected and is fixed by atmospheric conditions and T_1 is the temperature at which the heat is added and can be increased if efficiency is to be increased. In a steam plant an increase of T_1 may entail the use of corresponding high pressure which is a limiting factor as far as the design factor is concerned.

Mercury is one fluid that can be used in place of steam in the high temperature range of the cycle and increase in T_1 is obtained without any increase in maximum pressure. In this case the heat is rejected at a temperature at which the steam will be generated in water or steam boiler. Thus mercury in condensing rejects heat to water and steam is generated. The device in which this type of the heat exchange is taking place, is called a **Mercury condenser — steam boiler**.

Such a power plant using two vapours is known or called as a **Binary Vapour Plant** and the cycle is called as **Binary Vapour Cycle**.

Both, **mercury vapour power plant** and **steam vapour power plants work on ideal Rankine cycle.**

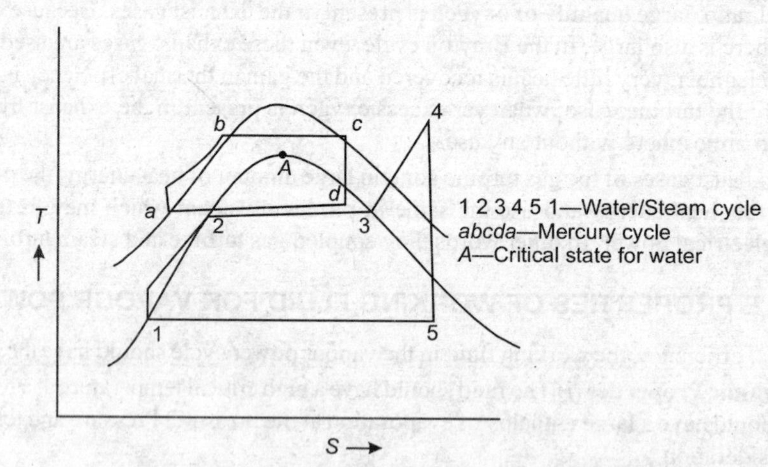

Fig. 25.1 *T–S diagram for mercury-water combined cycle.*

Figure 25.1 shows the *T–S* diagram for this binary vapour plant. In this figure, *A* is the critical state for water and steam and the critical pressure is 221.2 bar and critical temperature is 374.15°C.

1 2 3 4 5 1 is the steam cycle and *a b c d a* is the mercury cycle. Again, in this figure *ab* and *bc* represent the heating and evaporation of liquid mercury plotted on the same temperature scale as the steam. The scale of corresponding pressures is lower for the mercury. The mercury vapour at *c* has a much higher temperature than steam at the same pressure. At *c* the mercury is expanded adiabatically (isentropically) through a separate mercury vapour turbine to

d and then after expansion it is condensed to *a*. While condensing, the latent heat of mercury is given to water for evaporation purposes. Thus the mercury cycle is *a b c d a*.

For steam cycle, 1 2 3 4 5 1, 1–2 represents the heating of the feed water, 2–3 is evaporation of water by means of condensing mercury, 3–4 is superheating of steam by the flue gases. At 4 the steam is expanded adiabatically (isentropically) through a steam turbine to 5 and 5–1 represents the condensing of exhaust steam in the condenser. This completes the steam cycle.

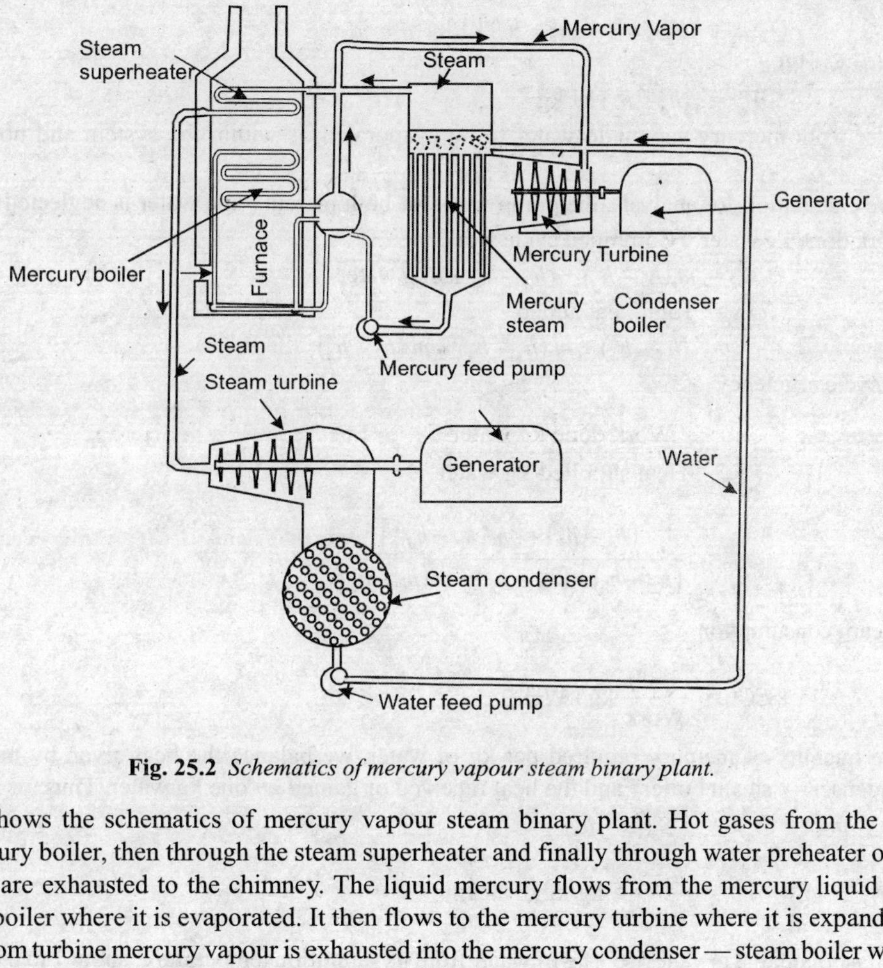

Fig. 25.2 *Schematics of mercury vapour steam binary plant.*

Figure 25.2 shows the schematics of mercury vapour steam binary plant. Hot gases from the furnace pass through the mercury boiler, then through the steam superheater and finally through water preheater or economiser after which they are exhausted to the chimney. The liquid mercury flows from the mercury liquid heater to the mercury vapour boiler where it is evaporated. It then flows to the mercury turbine where it is expanded to the low pressure limit. From turbine mercury vapour is exhausted into the mercury condenser — steam boiler where its latent heat is given up to the hot feed water; in this mercury condenser — steam boiler, mercury vapour condensers while the water is evaporated to steam. After condensing, the mercury liquid returns to the mercury liquid heater and thus the cycle is completed.

With reference to Fig. 25.2, the feed water from the economiser is evaporated to steam in mercury condenser — steam boiler. This saturated vapour is then passed through the steam superheated where it is superheated with the help of hot gases in the mercury boiler. Superheated steam then passes or flows through the steam turbine where it is expanded to the condenser pressure wherein steam is condensed and the condensate is removed by an extraction pump. This pump delivers the condensate to the hot well from where the feed pump delivers the condensate to the economiser and the steam cycle is completed.

To date, the number of mercury – steam boilers is :

Cycle-Analysis for Binary Vapour Cycle : Referring to Fig. 25.1,

Heat added to m kg of mercury required for 1 kg of water.

$$Q_{Am} = (h_c - h_a)\, m \text{ kJ/kg water}$$

Work done in mercury turbine

$$W_m = m\,(h_c - h_d) \text{ kJ/kg water}$$

Heat added to 1 kg of water

$$Q_{AW} = (h_2 - h_1) + (h_4 - h_3) \text{ kJ/kg}$$

Steam turbine work/kg

$$W_s = (h_4 - h_5) \text{ kJ/kg}$$

Heat transfer from mercury vapour to water to be evaporated, is within the system and no external heat is required.

In the above calculations or analysis, the pump work for both mercury and water is neglected.

∴ Total work done/kg water = Combined cycle work

$$W = m\,(h_c - h_d) + (h_4 - h_5) \text{ kJ/kg water}$$

$$Q_A = \text{Total heat added}$$

$$= (h_2 - h_1) + m\,(h_4 - h_3) + m\,(h_c - h_a)$$

Combined cycle efficiency

$$\eta_{th} = \frac{\text{Work done/kg water}}{\text{Heat supplied/kg water}}$$

$$= \frac{(h_4 - h_5) + \left[m(h_c - h_d)\right]}{(h_2 - h_1) + (h_4 - h_3) + m(h_c - h_a)}$$

Specific steam consumption

$$\text{SSC} = \frac{3600}{\text{Work}} \text{ kg/kWH.}$$

To find the quantity of mercury required per kg of water, we balance the heat given by mercury vapour in "mercury condenser — steam boiler" and the heat received or gained by one kg water. Thus

$$m\,(h_c - h_a) = 1 \times (h_3 - h_2)$$

$$\therefore \quad m = \frac{h_3 - h_2}{h_c - h_a} \text{ kg/kg of steam.}$$

Generally it is found that to vaporise 1 kg of water from its saturation temperature, about 7 to 8 kg of mercury are required.

The fluids mentioned below are used in place of mercury in the **Binary Vapour Plants**.

1. Aluminium bromide $AlBr_3$
2. Liquid metals like sodium, potassium
3. Diphenyl ether $(C_6H_5)_2O$
4. Zinc ammonium chloride $Zn(NH_3)_2Cl$

Major disadvantages for mercury as a binary fluid are :

1. Mercury is toxic, highly expensive and difficult to produce.

2. It has low latent heat at exhaust pressure and hence large quantity of mercury (nearly 7 to 8 kg per kg of water) is required to evaporate 1 kg of water.

3. It does not wet the heating surface, so addition of magnesium or titanium is necessary to get better heat transfer.

For water, at sub-critical pressure, but high pressures the saturation temperatures are more than 300°C and the maximum temperatures of superheated steam ranges from 550 to 600°C. Pressures required to get these temperatures, high pressures are needed.

To get nearly 550°C temperature at moderately low pressures, saturation temperature of mercury is 560°C at 12 bar pressure while aluminium bromide has 482°C saturation temperature at the same pressure of 12 bar and water has 187°C saturation temperature. The safe temperature for the best available material for the power plant is about 560°C. Therefore, mercury is the best working fluid in the high temperature range of the binary vapour cycle and the pressure is moderately low.

For mercury, the critical pressure and temperature are nearly 1080 bar and 1460°C respectively.

25.6 COMBINED RANKINE TO RANKINE CYCLES

In the Binary vapour cycle discussed above, mercury plant works on Rankine cycle and steam plant also works on a Rankine cycle. Mercury plant works on the higher temperature side while steam plant works on the lower temperature side. Mercury plants exhaust supplies heat to generate steam in the steam plant. Therefore mercury plant is called **topping cycle plant** while the steam plant is called the bottoming cycle plant.

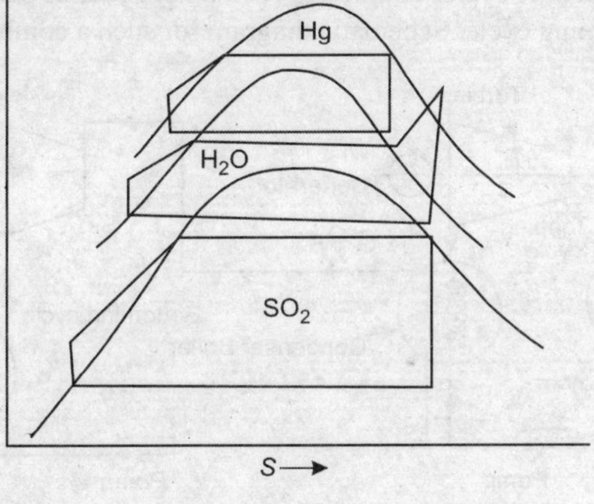

Fig. 25.3 *Mercury steam sulphur dioxide cycle.*

In the similar way, if we consider the third fluid like sulphur dioxide to work in the temperature range below the steam cycle range, the combined cycle is called **tertiary** cycle. In this cycle, the heat rejected by steam in the condenser is utilized in forming or generating sulphur dioxide vapour which will expand in the sulphur dioxide turbine.

Figure 25.3 shows the *T–S* diagram for mercury-steam-sulphur dioxide cycle. Here, sulphur dioxide cycle is the bottoming cycle while mercury cycle is the topping cycle.

In addition to sulphur dioxide, other liquid metals that can be considered for a working fluid in the combined cycle are sodium and potassium. These will be used for topping cycle. For bottoming cycle, fluids such as ammonia, freons etc. may be considered. Figure 25.4 shows the topping cycle using sodium combined with mercury and steam cycles.

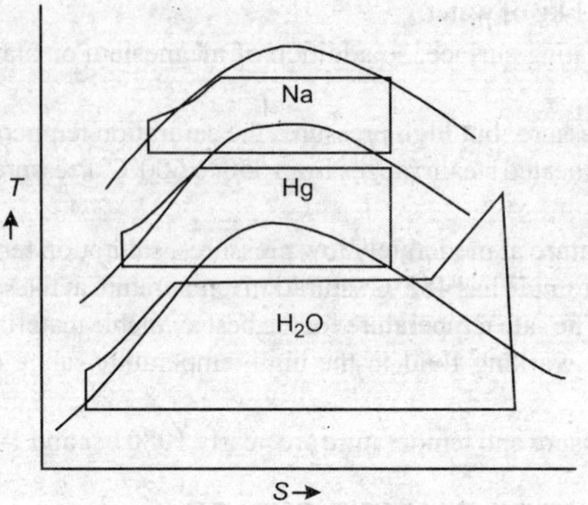

Fig. 25.4 *Sodium mercury steam cycle.*

All these fluids considered above for the combined cycle are, basically, working on the principle of Rankine cycle. These combined cycles may be considered as Rankine-to-Rankine coupled cycles. In such a case, the heat lost by one is absorbed by the other as in a case of binary, mercury steam, cycle.

Let us consider two such Rankine cycles coupled to each other in series such that the heat rejected by a topping cycle is absorbed by the bottoming cycle. Schematic diagram for such a combination is shown in Fig. 25.5.

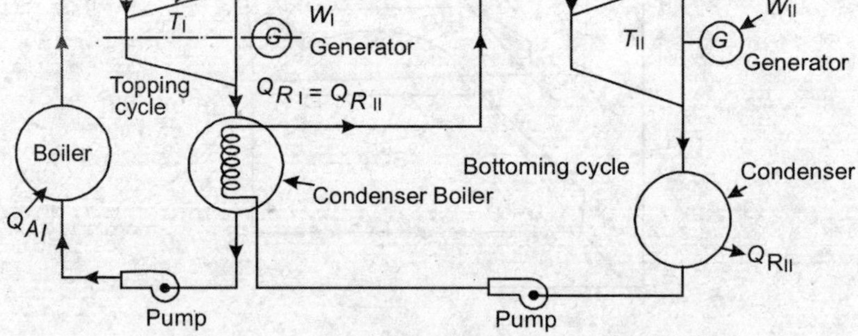

Fig. 25.5 *Two Rankine cycles coupled in series.*

Let η_I and η_{II} be the efficiencies of the topping cycle and bottoming cycle respectively and η be the overall efficiency of the combined cycle.

Then,

$$\eta_I = 1 - \frac{Q_{RI}}{Q_{AI}} \text{ and}$$

$$\eta_{II} = 1 - \frac{Q_{RII}}{Q_{AII}}$$

$$\therefore \quad \frac{Q_{RI}}{Q_{AI}} = 1-\eta_I \text{ or } Q_{RI} = Q_{AI}\,[1-\eta_I]$$

Similarly,

$$Q_{RII} = Q_{AII}\,[1-\eta_{II}]$$

For a combined cycle,

$$\eta = 1-\frac{Q_{RII}}{Q_{AI}}$$

$$= 1-\frac{Q_{AII}(1-\eta_{II})}{Q_{AI}}$$

$$= 1-\frac{Q_{RI}(1-\eta_{II})}{Q_{AI}}$$

$$= 1-\frac{Q_{AI}(1-\eta_I)(1-\eta_{II})}{Q_{AI}}$$

$$= 1-(1-\eta_I)(1-\eta_{II})$$

$$\therefore \quad 1-\eta = (1-\eta_I)(1-\eta_{II})$$

∴ For n number of cycles coupled in series, the overall efficiency will be given by

$$(1-\eta) = (1-\eta_I)(1-\eta_{II})(1-\eta_{III}) \ldots (1-\eta_n) \tag{1}$$

$$= \sum_{i=1}^{i=n}(1-\eta_i)$$

$$\mathbf{1-\eta = \prod_{i=1}^{n}(1-\ _i)}$$

For two cycles coupled in series, we have

$$1-\eta = (1-\eta_I)(1-\eta_{II})$$

$$= 1-\eta_I-\eta_{II}+\eta_I\,\eta_{II}$$

$$\therefore \quad \eta = 1-1+\eta_I+\eta_{II}-\eta_I\,\eta_{II}$$

$$\boldsymbol{\eta = \eta_I+\eta_{II}-\eta_I\,\eta_{II}} \tag{2}$$

From Eq. (1)

$1-\eta$ = **Total loss in the combined cycle**

= **Product of losses in all the cycles in series.**

$$= \sum_{x=1}^{x}(1-\eta_x)$$

From the above discussion we will note that the efficiency of the combined cycle is much more than the efficiency of an individual cycle plant. For example, if the efficiencies of the individual cycles are, say, 0.4 and 0.35, then the efficiency of the combined cycle will be

$$\eta = \eta_I + \eta_{II} - \eta_I \eta_{II}$$
$$= 0.4 + 0.35 - 0.4 \times 0.35$$
$$= 0.75 - 0.140$$
$$= 0.61 \text{ or } \mathbf{61\%}$$

The reason for this is that the work is obtained from the second or bottoming cycle is without consumption of the fresh or additional fuel energy. Thus with combined cycle, the overall efficiency is much more and this is the main reason for working of the combined cycle. Another point to be considered while thinking of combined power plants is that we can utilize the favourable characteristics of the fluids considered in different temperature ranges in the combined cycles.

So far we considered the combined power plants in which different cycles using different fluids in different temperature ranges are used. All these fluids were passing through the Rankine cycle only and hence the coupling of these cycles is Rankine-to-Rankine type.

25.7 COMBINED CYCLE PLANTS

Apart from Rankine-to-Rankine cycle coupling plants, there are few combined cycle plants. Generally, combined cycle plants will consist of topping cycle (high temperature region of h–s diagram) and bottoming cycle working in the low temperature range.

Steam power plant is the power plant which generates and supplies the bulk power (base load) and is combined with other types of the power plants. The combined power plants to be considered are :

1. Steam power plant and gas turbine power plant
2. Steam power plant and MHD power plant
3. Steam power plant and Thermionic power plant
4. Steam power plant and Thermoelectric power plant.

25.8 GAS TURBINE—STEAM TURBINE POWER PLANT

Gas turbine power plant works on an ideal cycle (Air standard cycle) known as Brayton cycle and in this plant, the working substance (gas) does not change phase as in case of steam turbine power plant which works on Rankine cycle and the working substance steam changes phase while passing through the plant.

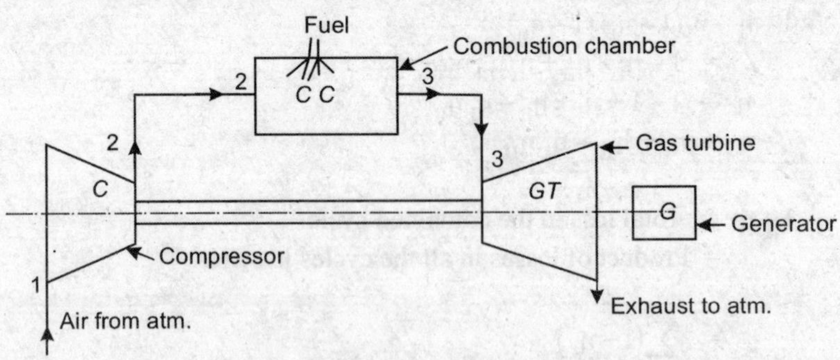

Fig. 25.6 *Open cycle gas turbine plant.*

The schematic diagram for an open cycle gas turbine plant is shown in Fig. 25.6 and the corresponding T–S diagram is shown in Fig. 25.7.

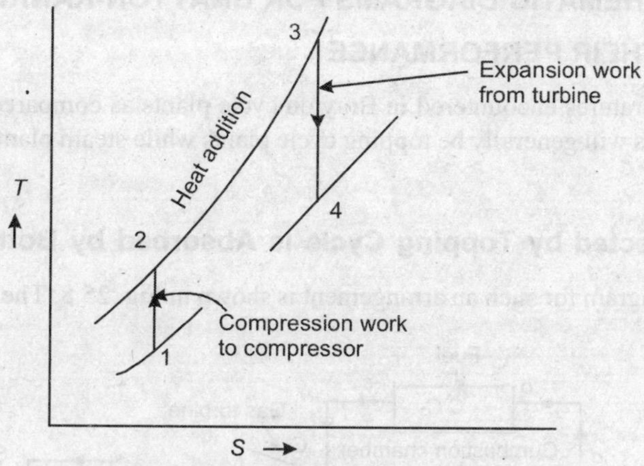

Fig. 25.7 *T–S diagram for gas turbine cycle.*

If r_p = pressure ratio, γ = specific heat ratio.

Then, the Brayton cycle efficiency is given by

$$\eta_{th} = 1 - \frac{1}{r_p^{\frac{\gamma-1}{\gamma}}}$$

The highest temperature of the gases used in the gas turbine is limited by the materials used in the gas turbine unit, and to limit this temperature a high air to fuel ratio is required. Generally, this highest temperature is about 800–1000°C.

If only gas turbine power plant is to be considered, then there are many disadvantages are there compared to the advantages obtained.

The main advantages of gas turbine plant are :

1. Installation cost is low.
2. Installation time required is also less.
3. Quick starting and stopping. This factor is very much required when this is to be used as a peak load and stand-by unit.
4. Quick responses to load changes.

The disadvantages of a gas turbine power plant are as follows :

1. Work input to compressor is very large so that the net work of the cycle and efficiency of the cycle or plant is low.
2. The exhaust gas temperature is high and hence the heat loss is more, the result is low work done.
3. Compressor efficiency and turbine efficiency reduce the net power and hence efficiency of the plant is reduced. Lowering the compressor efficiency, work of compression increases and lowering the turbine efficiency, turbine work is decreased and result is the less net work and hence the efficiency is decreases.
4. Fuel is costly than coal used in steam turbine power plant.

With these disadvantages, the cost of energy is much more in case of gas turbine power plant than that of steam power plant.

For this reason, generally it is advantageous to use gas turbine power plants in conjunction with steam power plant using coal to supply base load.

25.9 DIFFERENT SCHEMATIC DIAGRAMS FOR BRAYTON-RANKINE POWER PLANTS AND THEIR PERFORMANCE

Because of the high temperatures encountered in Brayton cycle plants as compared to the steam Rankine power plants, Brayton cycle plants will generally be topping cycle plants while steam plant will be working as bottoming plants.

25.9.1 All Heat Rejected by Topping Cycle is Absorbed by Bottoming Cycle

The schematic or block diagram for such an arrangement is shown in Fig. 25.8. The two cycles are in series.

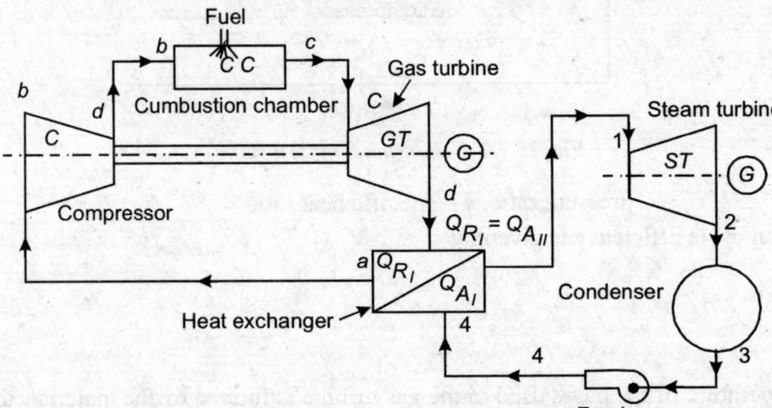

Fig. 25.8 *Brayton–rankine combined cycle plant*

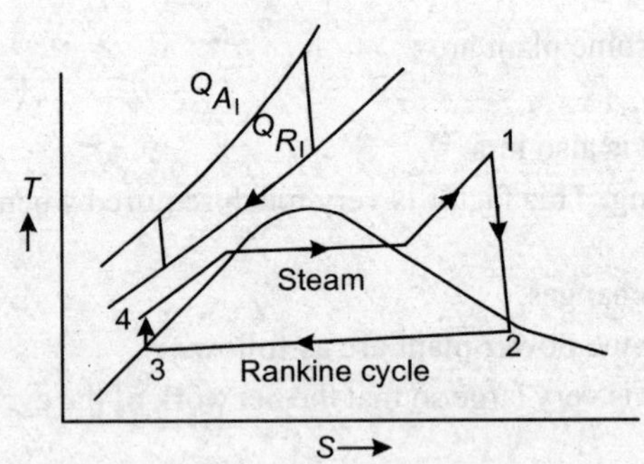

Fig. 25.9 *T–S diagram for Fig. 25.8*

Figure 25.9 shows the *T–S* diagram for the arrangement shown in Fig. 25.8.

In this case, it is assumed that the heat rejected by the gas turbine or Brayton cycle is totally given to the water for evaporation to state one as shown in Fig. 25.8 and Fig. 25.9. In that case, the analysis of this combined cycle is similar to that given in Fig. 25.6. The efficiency of the combined cycle is then given by

$$\eta = \eta_I + \eta_{II} - \eta_I \eta_{II}$$

where η_I and η_{II} are the efficiencies of the Brayton and Rankine cycle plants.

25.9.2 Heat Rejected by Brayton Cycle is Partly Absorbed by Rankine Cycle and Rest is Lost

Figure 25.8 shows the combined plant where all the heat rejected by the topping cycle is absorbed by the bottoming cycle. In practice, generally this does not take place. Instead of that part of the heat rejected by the topping cycle is lost and hence less heat is absorbed by the bottoming cycle. Figure 25.10 shows such an arrangement and Fig. 25.11 is the T–S diagram for the same layout.

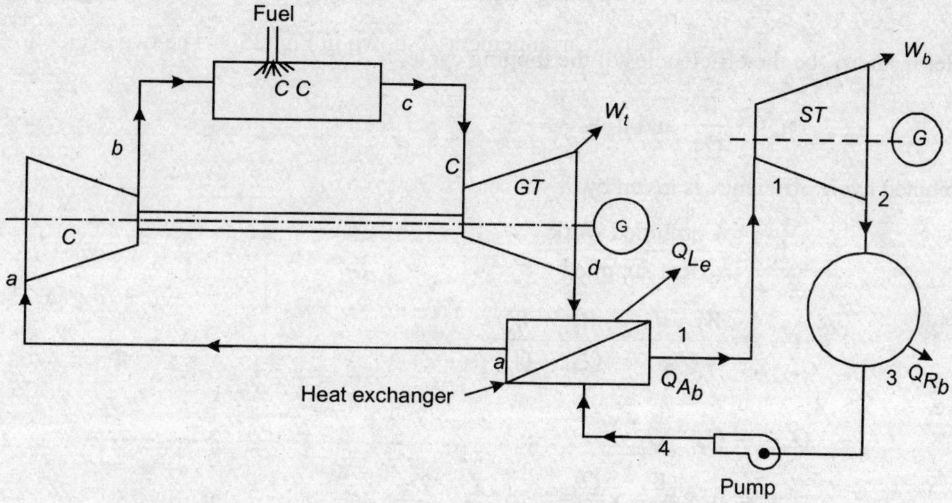

Fig. 25.10 *Schematic for topping and bottoming combined cycle plant with heat loss.*

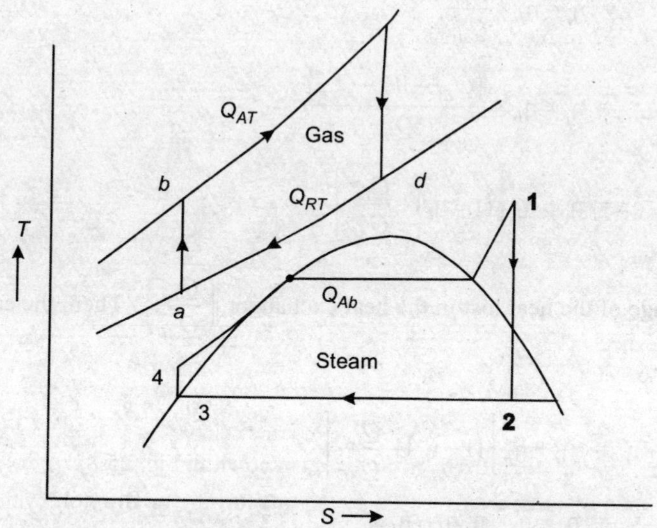

Fig. 25.11 *T–S diagram for arrangement of Fig. 25.10.*

Let,

W_t = Work done of the topping cycle

Q_{At} = Heat added to topping cycle

Q_{Rt} = Heat rejected by topping cycle
W_b = Work done by bottoming cycle
Q_{Ab} = Heat added to bottoming cycle
Q_{Rb} = Heat rejected by bottoming cycle
Q_{Le} = Heat loss in heat exchange

$\therefore \quad Q_{Rt} = Q_{Ab} + Q_{Le}$

and $\quad W$ = Combined cycle work

$$= W_t + W_b.$$

Again, let η_t and η_b be the efficiencies of the topping cycle.

$$\therefore \quad \eta_t = \frac{W_t}{Q_{At}} \text{ and } \eta_b = \frac{W_b}{Q_{Ab}}$$

and combined cycle efficiency is given by

$$\eta = \frac{\text{Combined work}}{\text{Heat supplied}}$$

$$= \frac{W_t + W_b}{Q_{At}} = \frac{W_t}{Q_{At}} + \frac{W_b}{Q_{At}}$$

We have,

$$Q_{Ab} = Q_{Rt} - Q_{Le}$$

$$\therefore \quad \eta = \eta_t + \frac{W_b}{Q_{Ab}} \times \frac{Q_{Ab}}{Q_{At}}$$

$$= \eta_t = \eta_b \times \frac{Q_{Ab}}{Q_{At}}$$

$$= \eta_t + \eta_b \times \frac{(Q_{At} - W_t - Q_{Le})}{Q_{At}}$$

$$= \eta_t + \eta_b \left[(1 - \eta_t) - \frac{Q_{Le}}{Q_{At}} \right]$$

Let x_L be the percentage of the heat lost in the heat exchanger $\left(\frac{Q_{Le}}{Q_{At}} \right)$. Then, the combined cycle efficiency is given by

$$\eta = \eta_t + \eta_b \left[(1 - \eta_t) - \frac{Q_{Le}}{Q_{At}} \right]$$

$$= \eta_t + \eta_b - \eta_t \eta_b - \eta_b x_L$$

$$\boxed{\eta = \eta_t + \eta_b - \eta_t \, \eta_b - \eta_b x_L}$$

From the diagram, it is seen that ratio $\frac{Q_{Ab}}{Q_{Rt}}$ is the boiler efficiency in the bottoming plant i.e.,

$$\eta_B = \text{Boiler efficiency in the bottoming cycle}$$

$$= \frac{Q_{Ab}}{Q_{Rt}}$$

$$= \frac{Q_{Rt} - Q_{Le}}{Q_{Rt}}$$

$$= 1 - \frac{Q_{Le}}{Q_{Rt}}.$$

Then, the combined cycle efficiency or the overall efficiency can be written as

$$\eta = \frac{W_t + W_b}{Q_{At}} = \frac{W_t}{Q_{At}} + \frac{W_b}{Q_{At}}$$

$$= \eta_t + \frac{\eta_b \cdot Q_{Ab}}{Q_{At}}$$

$$= \eta_t + \eta_b \cdot \eta_B \cdot \frac{Q_{Rt}}{Q_{At}}$$

$$= \eta_t + \eta_b \eta_B (1 - \eta_t)$$

$$= \eta_t + \eta_b \eta_B - \eta_t \eta_b \eta_B$$

$$\boxed{\eta = \eta_t + \eta_b \ \eta_B - \eta_t \ \eta_b \ \eta_B}$$

25.9.3 Brayton–Rankine Cycle Plants Operating in Parallel

After considering two plants operating in series, it can then be considered to operate these two cycle plants in parallel. The schematic diagram for such a parallel operation is shown in Fig. 25.12 and corresponding two separate *T–S* diagrams are shown in Fig. 25.13. Note here that there are no topping and bottoming cycle plants as in case of series operations.

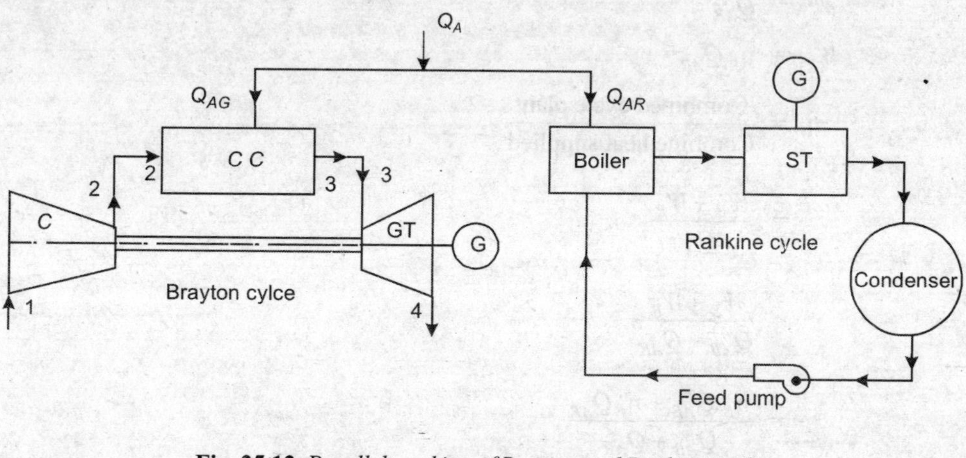

Fig. 25.12 *Parallel working of Brayton and Ranking cycle.*

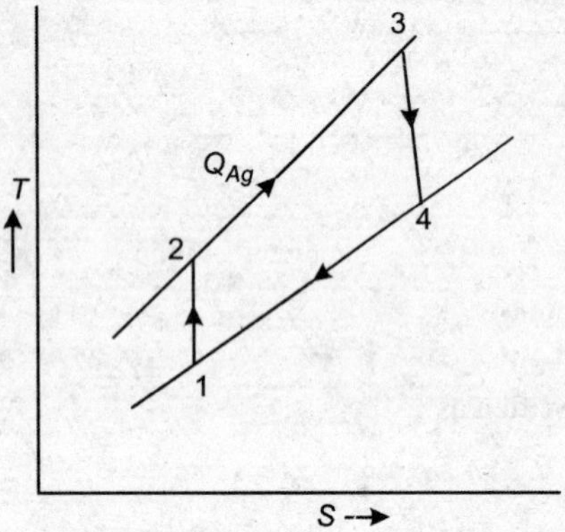

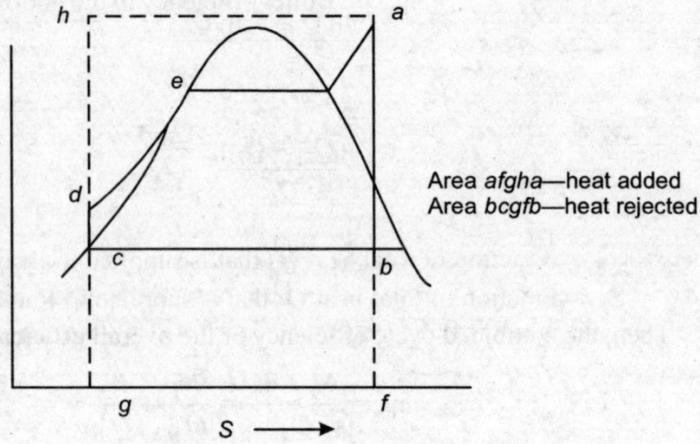

Fig. 25.13 (a) *T–S diagram for Brayton cycle.*

Fig. 25.13 (b) *T–S diagram for Rankine cycle.*

In Fig. 25.12,

$$Q_A = \text{Total heat supplied}$$

$$Q_{AB} = \text{Heat supplied to Brayton cycle gas turbine plant}$$

$$Q_{AR} = \text{Heat supplied to Rankine cycle steam turbine plant}$$

and $$Q_A = Q_{AG} + Q_{AS}$$

Let $$\eta_B = \text{Cycle efficiency of Brayton cycle}$$

$$\eta_R = \text{Cycle efficiency of Rankine cycle}$$

$\therefore$ $$\eta_B = \frac{W_G}{Q_{AB}} \text{ or work done by Brayton cycle plant}$$

$$W_B = \eta_B \times Q_{AB}.$$

Similarly,

$$\eta_R = \frac{W_R}{Q_{AR}} \text{ or work done by Rankine cycle plant}$$

$$W_R = \eta_R Q_{AR}$$

$\therefore$ $$\eta = \frac{\text{Combined cycle plant}}{\text{Combine heat supplied}}$$

$$= \frac{W_B + W_R}{Q_A}$$

$$= \frac{W_B + W_R}{Q_{AB} + Q_{AR}}$$

$$= \frac{\eta_B . Q_{AB} + \eta_R Q_{AR}}{Q_{AB} + Q_{AR}}$$

$$= \frac{\eta_B Q_{AB} + \eta_R Q_{AR}}{Q_A}$$

$$= \eta_B \cdot \frac{Q_{AB}}{Q_A} + \eta_R \cdot \frac{Q_{AR}}{Q_A}$$

$$= \eta_B \cdot x_B + \eta_R \cdot x_R$$

where x_B = Fraction of total heat Q_A that is supplied to Brayton cycle plant

and x_R = Fraction of total heat Q_A that is supplied to Rankine cycle plant

$$= (1 - x_B)$$

$$\therefore \quad \eta = \eta_B x_B + \eta_R (1 - x_B)$$

$$= \eta_B x_B + \eta_R - \eta_R x_B$$

$$= \eta_R + x_B (\eta_B - \eta_R)$$

If we consider $x_B = 1 - x_R$, then we get

$$\eta = \eta_B + x_R (\eta_B - \eta_R).$$

Thus,

$$\boxed{\eta = \eta_R + x_B (\eta_B - \eta_R)}$$

$$\boxed{= \eta_B + x_R (\eta_B - \eta_R)}.$$

It will very logical to note that the efficiency of the combined cycle plant lies between η_B and η_R.

29.9.4 Series Parallel Operation

In this case we will consider the total heat supplied is divided in two ways — one for the Brayton cycle and the other for heating the exhaust of the Brayton cycle. This is called supplementary heating of the exhaust of the Brayton cycle and heating in Rankine cycle. The schematic arrangement for such a case is shown in Fig. 25.14.

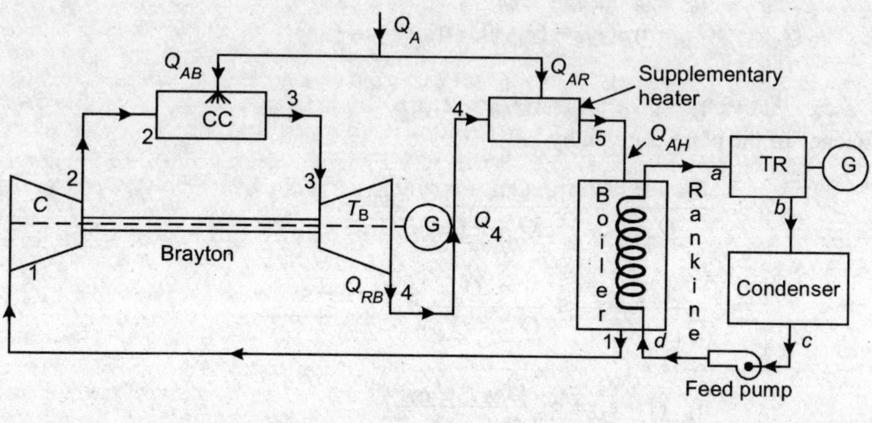

Fig. 25.14 *Schematic of series parallel operation with supplementary heating.*

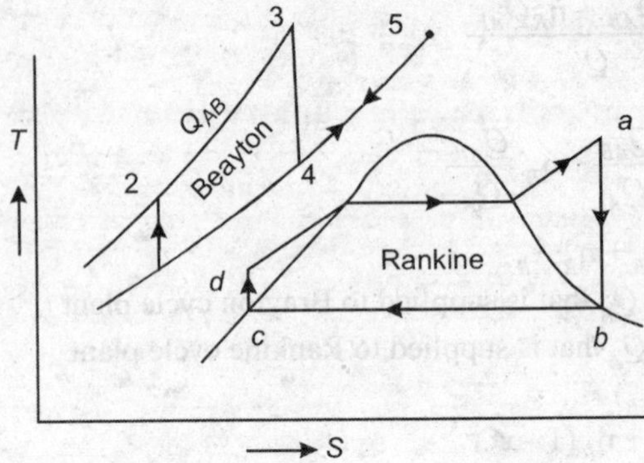

Fig. 25.15 *T–S diagram of arrangement of Fig. 25.14.*

In Fig. 25.14,

$$Q_A = \text{Total heat supplied}$$
$$Q_{AB} = \text{Heat supplied to Brayton cycle plant}$$
$$Q_{AQH} = \text{Heat supplied in the supplementary heater}$$

Let
$$x_1 = \frac{Q_{AB}}{Q_A} \therefore x_2 = \frac{Q_{ASH}}{Q_A} = \frac{Q_A - Q_{AB}}{Q_A}$$
$$= 1 - x_1 \therefore 1 - x_2 = x_1 = \frac{Q_{AB}}{Q_A} = \frac{Q_{AB}}{Q_{AB} + Q_{RBH}}$$
$$\eta_B = \text{Brayton cycle efficiency}$$
$$= \frac{W_B}{Q_{AB}}$$

$\therefore$
$$W_B = \eta_B \cdot Q_{AB} = Q_{AB} - Q_{RB}$$

$\therefore$
$$Q_{RB} = Q_{AB} - \eta_B Q_{AB} = Q_{AB}(1 - \eta_B)$$

Similarly,
$$W_R = \eta_R \times Q_{AR} = \eta_R \times (Q_{RB} + Q_{ASH})$$

$\therefore$ Overall efficiency of the plant is given by

$$\eta = \frac{W_B + W_R}{Q_A} = \frac{\eta_B \cdot Q_{AB} + \eta_R \cdot Q_{AR}}{Q_{AB} + Q_{ASH}}$$
$$= \frac{\eta_B Q_{AB}}{Q_{AB} + Q_{ASH}} + \frac{\eta_R Q_{AR}}{Q_{AB} + Q_{ASH}}$$
$$= \eta_B \cdot (1 - x_2) + \eta_R \frac{Q_{RB} + Q_{ASH}}{Q_{AB} + Q_{ASH}}$$
$$= \eta_B (1 - x_2) + \eta_R [x_2 + (1 - \eta_B)(1 - x_2)]$$
$$\boxed{\eta = \eta_B + \eta_R - \eta_B \, \eta_R - x_2 \, \eta_B \left(1 - \eta_2\right)}$$

Comparing this efficiency with that of the plant employing two coupled cycles in series, we note that the efficiency of the **series parallel** operated plant is less than **series** operated plant.

29.9.5 Series Parallel Plants with Supplementary Heat Supply and Heat Loss Between Two Plants

Schematic block diagram for the arrangement is similar to that in Fig. 25.14 except the supplementary heater as shown in Fig. 25.16.

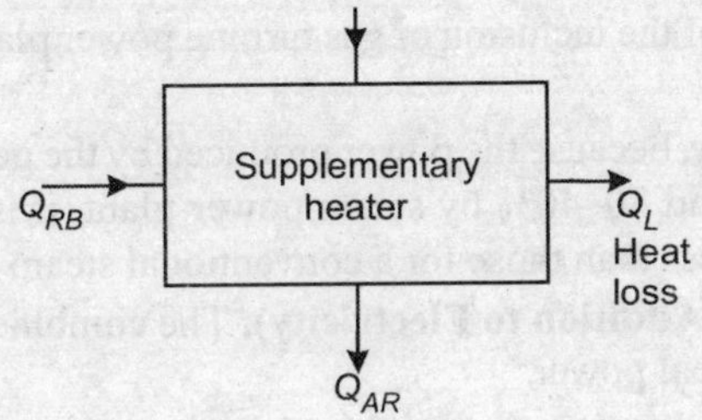

Fig. 25.16 *Supplementary heater with heat loss.*

If $x_L = \dfrac{Q_L}{Q_A}$, then the overall efficiency of the combined plant is given by

$$\eta = \eta_B + \eta_R - \eta_B\,\eta_R - x_L\,\eta_R - \eta_B\,x_2(1-\eta_2)$$

It is clear that the additional heat supply to the exhaust from the Brayton cycle, the power output of the Rankine cycle power plant will increase and efficiency of the Rankine power plant will also increase but the overall efficiency of the combined power plant will decrease. Therefore, in the combined cycle power plants, supplementary heat addition — which is called also as **supplementary firing** — is nowadays becoming less and less attractive.

Instead of supplementary firing, it is profitable to burn the fuel in the combustor or combustion chamber of the Brayton cycle plant (gas turbine power plant) only because this heat supply will take place at an elevated average temperature than that in the Rankine cycle plant (steam power plant).

25.10 ADVANTAGES OF COMBINED CYCLE POWER GENERATION

The Brayton cycle and Rankine cycle combined cycle power plant uses gas turbine plant working on Brayton cycle and steam power plant working on Rankine cycle. Because of the advancement in research and development of gas turbine technology, the efficiency of the Brayton cycle has improved a lot — even upto 30–35% and this development led to the demand of combined cycle power plants employing gas turbine power plant and steam turbine power plants.

The advantages of combined cycle power plants using gas turbines and steam turbines, are as follows :

1. High Total Power. Combined gas turbine power and steam turbine power is available or produced.

2. High Overall Plant Efficiency. Exhaust of the gas turbine power plant is utilized for producing steam, the power obtained from steam turbine power plant is theoretically free because no fuel is required for steam generation. Therefore, overall efficiency of the combined power cycle is much higher exceeding even 50%.

3. Quick Starting of Plants. To start the total combined power plant, gas turbine plant can be started quickly and start supplying power to the consumers or grid. In the meanwhile steam power plant will take the load on the plant. Operating flexibility is present because of such operation.

4. Low Water Requirement. Gas turbine power plants do not require any cooling water. Water is required only for steam power plants. Purpose of water requirement is for boiler feed and for circulating cooling water for condenser working. Therefore, water requirement for the total power generated, is less. This is a distinct advantage.

5. Phased Installation. Gas turbine power plant takes very little time as compared to steam power plant for installation and hence gas turbine power plant is installed first and start producing or generating power while steam power plant is being installed.

6. Low Pollution Level. In the combined cycle plant operation, formation of NO_x and CO_2 is very much lower than that in coal fired boilers in steam power plants.

7. Peak and Base Load. Because of the inclusion of gas turbine power plants, the combined cycle power plants and supply base as well as peak loads.

8. Low Investment Cost. Generally, because the power produced by the gas turbine plant is from 60–70% of the total power supplied by the station, and 30–40% by steam power plant, it is estimated that the investment costs required are approximately 25–35% less than those for a conventional steam power plant.

9. Use for Cogeneration (Heat in Addition to Electricity). The combined cycle power plants can be used for supply of heat while producing electrical power.

In industrial power plants, heat and electricity are produced. Electrical power to the tune of 40% is very common while heating power is required for process heating.

Thus the main attractive features or points of combined power generation are :

(i) Large output
(ii) High cycle efficiency
(iii) Low emission level
(iv) Low investment cost.

10. Power to Ground Area Ratio. Because of the presence of gas turbine plant, the power to ground area is high or much more than that for steam power plant of the same capacity.

11. Manpower Requirement. Because of the simplicity of the gas turbine power plant, the total manpower requirement becomes less as compared to steam power plant of the same capacity.

12. Self-sustaining Features. If, due to any unforeseen fault in the total system, the total station is shut down, then the gas turbine plant can be started with the help of Diesel engine provided for starting the gas turbine plant. Then with the help of power generated by gas turbine plant, steam power plant can be started (auxiliaries of the plant).

13. Floor Area Requirement. Total floor or land area required for the combined cycle power plant is less than that required for steam power plant of the same capacity.

25.11 SOURCE OF GAS FOR GAS TURBINE PLANT

Fuels that can be used for the Brayton cycle or gas turbine power plants are of three types :

(a) Solid fuels
(b) Liquid fuels
(c) Gaseous fuels.

The reserves of the petroleum fuels are limited in the world. Comparatively, the coal reserves are tremendous and therefore it is necessary to use coal as fuel instead of petroleum fuel, in case of stationary plants. Similarly, the supply of natural gas and petroleum products is continuously decreasing. With the present technology available, coal can be gasified and the gases generated can be used in gas turbines. Gasification provides inherent advantages such as heat is generated and gas can be used for a variety of syntheses for chemicals. Also, coal gasification is advantageous in the sense that the gas is purified at high pressures and prior to combustion in the gas turbine power plant. Another advantage of the purified gas is environmental protection.

Therefore, it is required to develop an efficient coal conversion process so that a wide range of coals including high and low sulphur contents can be used for gasification. Again, it is required to develop an efficient coal conversion process so that a wide range of coals can be liquefied. Out of these two possible developments, gasification of coal is more simple than liquefication.

25.11.1 Gas Producer or Gasifier

The plant which generates gas is known as the gas producer plant. When coal is the fuel used for generating gas, the plant is known as **coal gasifier**. These gas producers may be either of the suction type or pressure type. When the works on the gas (gas engine), then the gas producer is of suction type. The essential components of the gas producer are :

(i) the generator
(ii) the scrubber and
(iii) the vaporizer.

When the quantity of gas required is very large as in the case of gas turbines, the pressure type producer or gasifier is used.

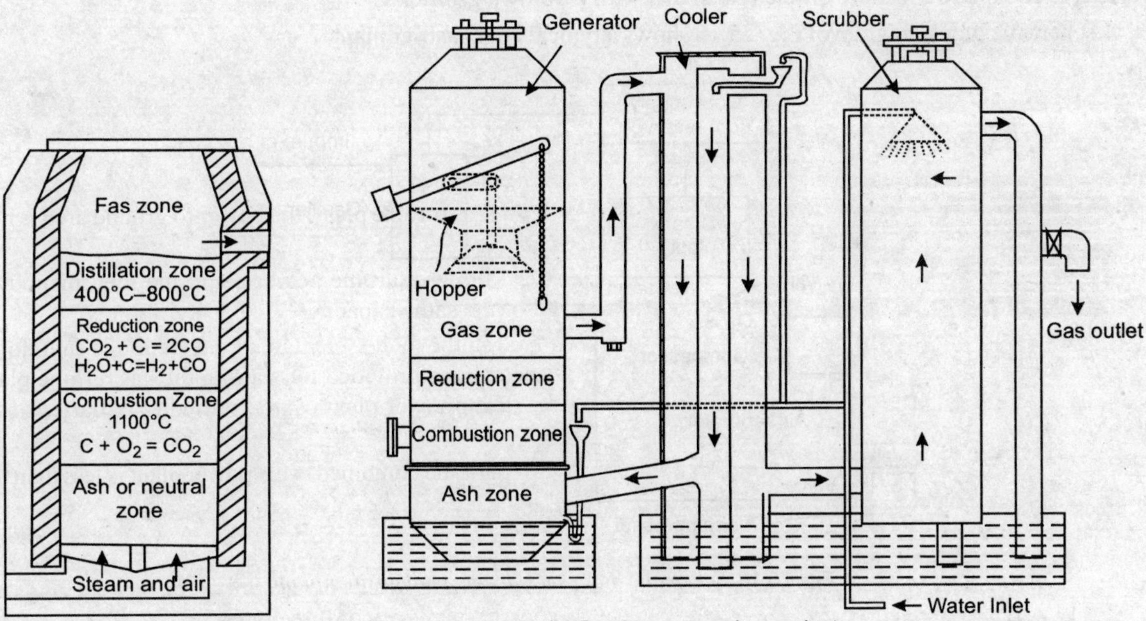

(a) Gas generator showing various zones (b) Suction gas produces plant

Fig. 25.17

There are four zones in the gas generator as shown in Fig. 25.17.

1. Ash zone
2. Combustion zone, where the temperature is near about 1100°C and where the reaction $C + O_2 = CO_2$ takes place.
3. Reduction zone, where the temperature is near about 1000°C and where the reactions $CO_2 + C = 2CO$ and $H_2O + C = CO + H_2$ takes place.

4. Distillation zone.

The gas resulting from the producer plant is a mixture of hydrogen, carbon monoxide, carbon dioxide, nitrogen, hydrocarbons and oxygen also.

The efficiency of the gas producer plant may be as high as 90%.

25.11.2 Coal Gasifier for Gas Turbine Plant

Thus, coal gasification is the first step to convert coal into useful gas turbine fuel. There are many gasification plants as moving bed or Lurgi technology, FBC gasifier and entrained flow gasifier.

For coal gasification, large amount oxygen (O_2) and therefore air is required for operation of coal gasifier. There are mainly two types of gasifiers, one using air and other using O_2 separated from air. It is obvious that the gasifier using O_2 will produce more gas than that produced from gasifier using air. Similarly, O_2 compressor driven gasifier will consume loss power than air compressor because to produce the same quantity of O_2, more air is required to be compressed. But the overall power requirement with oxygen separation, is large as compared to that supplying the same O_2. Therefore, presently, air blown gasifiers appear to offer the preferred choice for coal gasification combine cycles with the better efficiency and lower capital cost. It is further observed that the integrated system with air gasifier is about 3 to 5% more efficient than that with O_2 blown gasifier.

A schematic block diagram of Fig. 25.18 shows a typical coal gasifier plant.

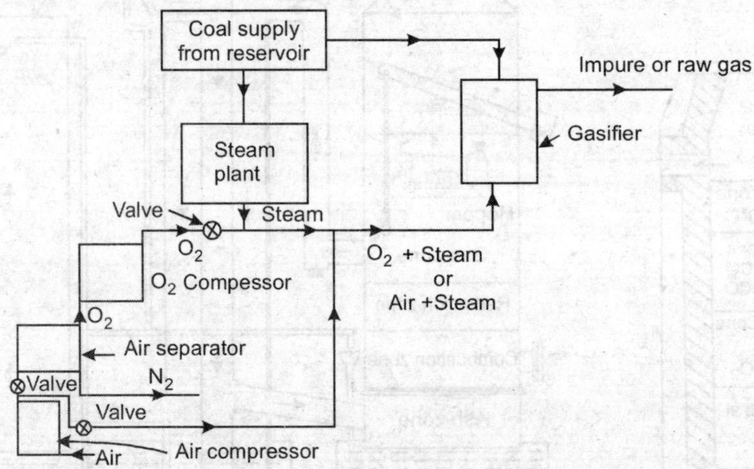

Fig. 25.18 *Schematic diagram showing coal gasifier plant.*

From such a plant (O_2 operated) generally we get the following gases in the mixture as gas

$H_2 = 25–45\%$, $CO_2 = 20–60\%$, $CO_2 = 15–30\%$ and the rest will be hydrocarbon as methane gas CH_4.

A combined cycle power plant using gas fuel from gasifier is shown in Fig. 25.19. Note here that gasifier uses compressed air and steam from boiler and coal from the coal reservoir at the plant site.

In this plant, two stage compressor is used. LP stage of the compressor is driven by a gas turbine which uses gases from the gasifier and the cycle of this section is normal Brayton cycle.

Part of the compressed air from first stage passes, through an intercooler, to the second stage. This air is supplied to the coal gasifier as shown in Fig. 25.19.

Exhaust gases from the gas turbine are admitted to the supplementary combustion chamber which receives gases from gasifier. The gases from the supplementary combustion chamber, then, are used in the boiler which

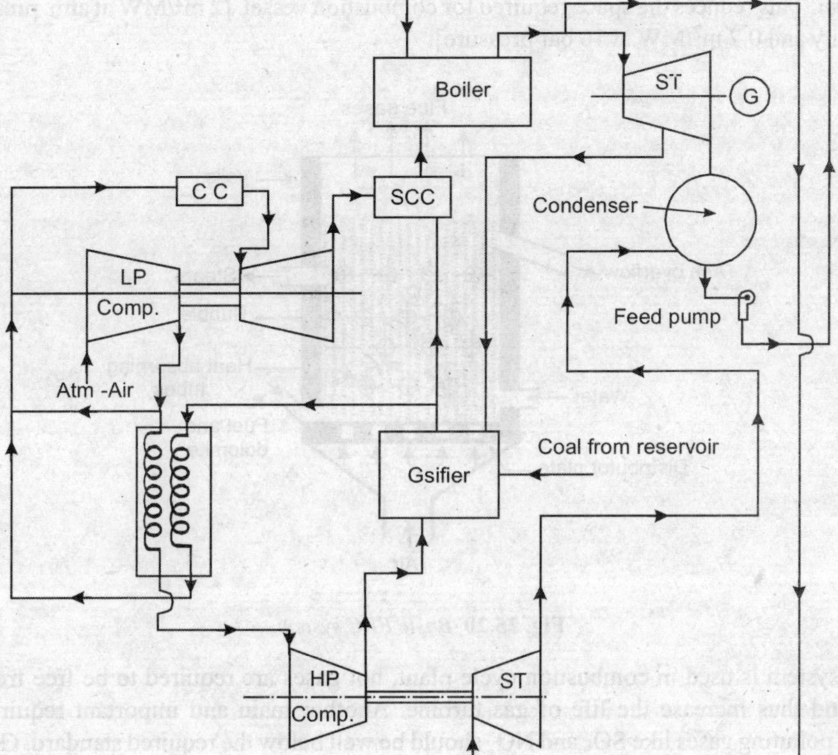

Fig. 25.19 *Combined cycle power plant using gasifier.*

produces sufficient steam to generate power and supply to the coal gasifier as shown in Fig. 25.19. Also the second stage compressor is driven by another steam turbine.

25.11.3 PFBC System Used in Combined Cycle

Fluidised bed may be defined as the bed of solid particles behaving as a fluid. Such type of the bed is used in the combustion system where coal is used. As the reserves of coal are very large in this country, and we are using coal as a fuel in thermal power plants in spite of the fact that coal is a low grade fuel as compared to oil fuel.

Figure 25.20 shows the basic Fluidised Bed combustion system that can be used in boilers used to generate steam to be used in steam turbines. When air is passed through a packed bed of finely divided solid particles (*coal*), it experiences a pressure drop across the bed. At low air velocities, this pressure drop is small and does not disturb the particles. But if the air velocity is increased further, a stage is reached when particles are suspended in the air stream and the pacted bed becomes *fluidised bed*. When air velocity is further increased, the bed becomes turbulent and rapid mixing of particles occurs. In general, the behaviour of this mixture of solid particles and air is like a fluid. Burning of fuel (coal) in such a state is known as *Fluidised bed combustion*.

Generally air is supplied by a centrifugal *blower*. In this case the pressure of air supplied is not much or low pressure is present.

When this FBC system is to be used in a combined cycle, the air supplied to the system will be at high pressure developed in the compressor of the gas turbine unit. Then the system becomes PFBC – *Pressurised Fluidised Bed Combustion system*. This reduces the space required for combustion vessel. [2 m^2/MW at atm. pressure and 2 m/s fluidizing velocity and 0.2 m^2/MW at 10 bar pressure].

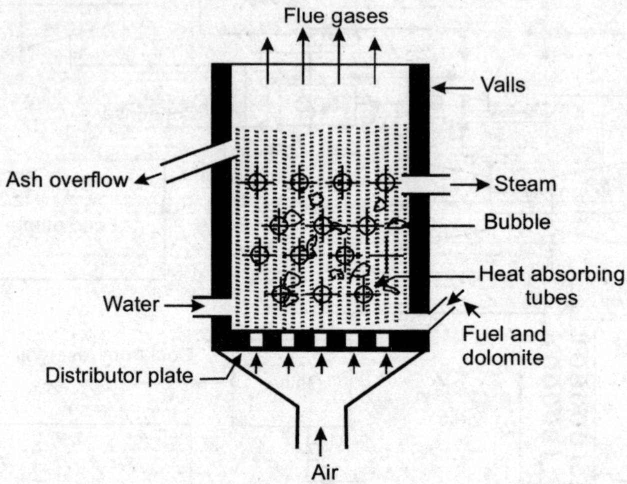

Fig. 25.20 *Basic FBC system.*

When **PFBC** system is used in combustion cycle plant, hot gases are required to be free from particulates to reduce erosion and thus increase the life of gas turbine. Another main and important requirement is that the concentrations of polluting gases like SO_2 and NO_x should be well below the required standard. Generally whatever the advantages are there because of the conventional.

25.11.3.1 Limitations of PFBC for Combined Cycle Power Plants

The *temperature* in PFBC is limited to about 850–1000°C because it is the most favourable or required temperature for the working of gas turbine rotor blades. Also this is the temperature of sulphur retention and is below the ash fusion temperature for most of the coals available. Also for these temperatures NO_x formation is limited to a large extent. Particularly the temperature below 900°C, volatization of all alkalis is avoided.

The second major limitation for the use of PFBC is the *cleaning of gas at high temperatures* to increase the life of gas turbine rotor. At present, the cost of cleaning of gas at high temperatures is very large. Also in the process of cleaning the gas, *heat loss* takes place the effect of which is the *reduction in efficiency*. At present, cyclones to be used to clean-up gas at temperature 800–1100°C is in the developmental stage.

PFBC system for the gas turbine will be accepted or will be very much used even for the low grade coal and other solid fuels like peat, waste wood, sewage sludge and garbage once the economical and efficient methods are developed.

Size of the particulates is also one of the difficult problem. Particulates of the size of 10 mm and above are removed with the help of usual clears. Particulates having 3 to 10 mm can be cleaned either by bagfillers or Electro Static Precipitator (ESP). But again here, the problem of high temperature remains unsolved. If the size of the particulates is reduced below 3 μm, it is observed that the erosion of the turbine blades is not taking place. Then the particulates of the size of 3–10 μm are reduced to the sizes below 3 μm by means of laser bombardment. These particulates of the size less than 3 μm are sometimes called *laser ash* (LASH). Figure 25.22 shows the schematic of the system used for LASH fragmentation method for the removal of particulate. Remember, this fragmentation will be accompanied by shock heating of particulates. Nearly 99% of the energy supplied to LASH can be again

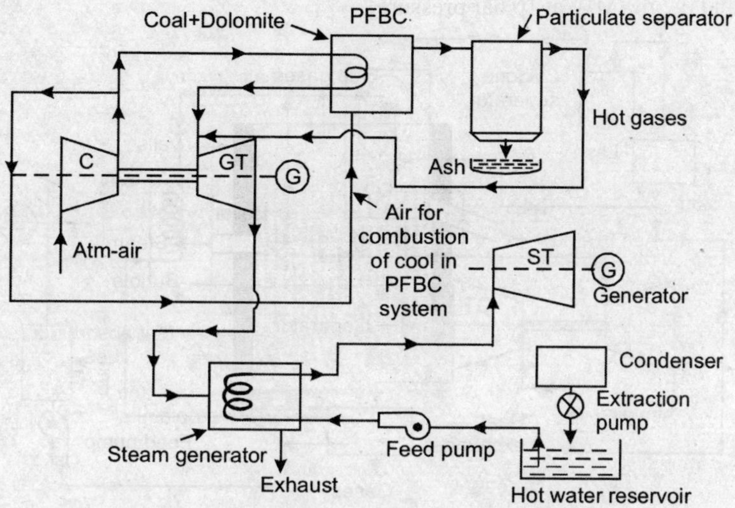

(a) PFBC for hot air and gases

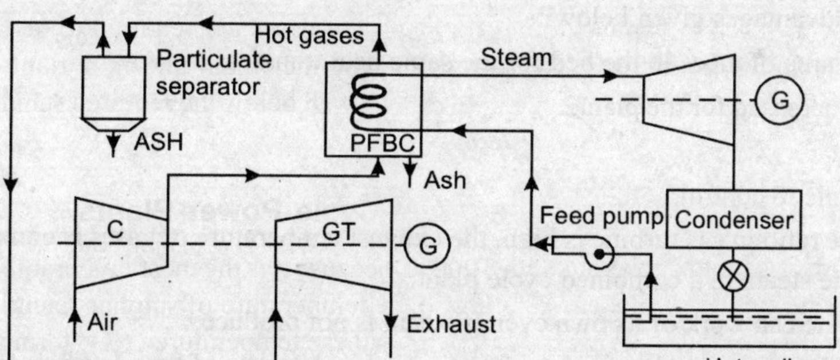

(b) PFBC for hot gases and steam

Fig. 25.21 *Combined cycle power plant with PFBC*

recovered in the form of heat as FBC system in the generation of steam, the same advantages are present when the PFBC system is used in the combined cycle plant.

The exhaust gases from the PFBC are at temperature and therefore can be used in the gas turbine plant, thereby getting additional power.

Thus we observe that PBFC is in the combined cycle power plant, to generate high temperature gases (free from particulate) to be used in gas turbine plant and at the same time steam can be generated at high temperature and pressure to be used in steam power plant. The two different systems are shown in Fig. 25.21.

Figure 25.21(a) shows the generation of steam with the help of hot exhaust gases from gas turbine while hot gases required by gas turbine are supplied by PFBC as shown.

Figure 25.21(b) shows PFBC generating steam as well as hot gases for steam turbine plant and gas turbine plant. The system shown in Fig. 25.21(a) is sometimes known as *Air heating system* while than in Fig. 25.21(b) is known as *steam generating system*.

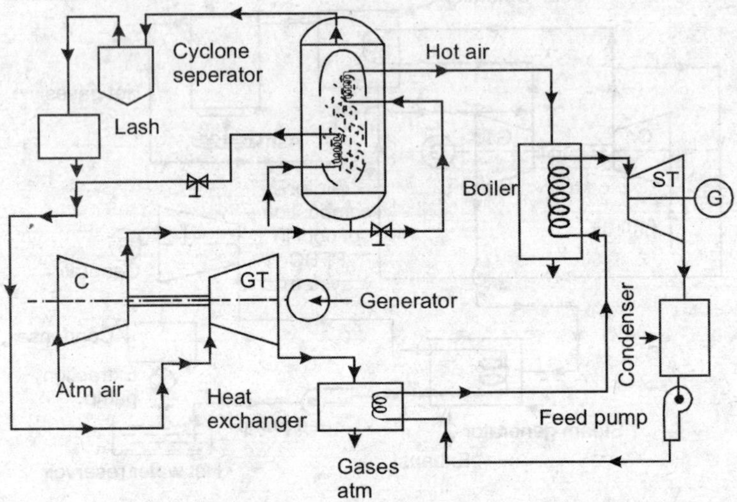

Fig. 25.22 *Combined cycle with lash particulate.*

Steam system has advantages given below :

(a) Required less area of tubes in the bed for the same heat transfer.

(b) Efficiency is increased for the plant.

In air cycle :

(a) System is simple to control.

(b) If the pressure ratio in gas turbine is high, the exhaust temperature of gases is comparatively low and it is not possible to generate steam in a combined cycle plant.

(c) Only gas turbine can work of its own even if steam is not produce.

(d) Because of PFBC, water and steam consumption is low because of the use of air for gas turbine.

25.11.4 PCFBC System Used in Combine Cycle

Another fluidized bed combustor system that can be used in a combined cycle is the pressurised circulating fluidized bed combustion cycle (PCFBC).

Figure 25.23 shows the schematic diagram of PCFB combustor (boiler). The solids in PCFBC are kept in 'fast fluidized' condition.

A major feature of PCFB is that it approaches a condition where solids are fully mixed throughout the combustor, but little mixing of gas in the axial direction. The primary air is supplied to the grid, while secondary air is supplied at some distance from the grid. The staged combustion reduces NO_x. The PCFB units offer all the advantages of PFBC are mentioned above. The combustion efficiency and sulphur retention capability of a PCFB are still better than those in a PFBC. The boiler can be designed to use any of the types of water circulation system i.e., (i) Natural circulation, (ii) Forced circulation, (iii) Once through circulation.

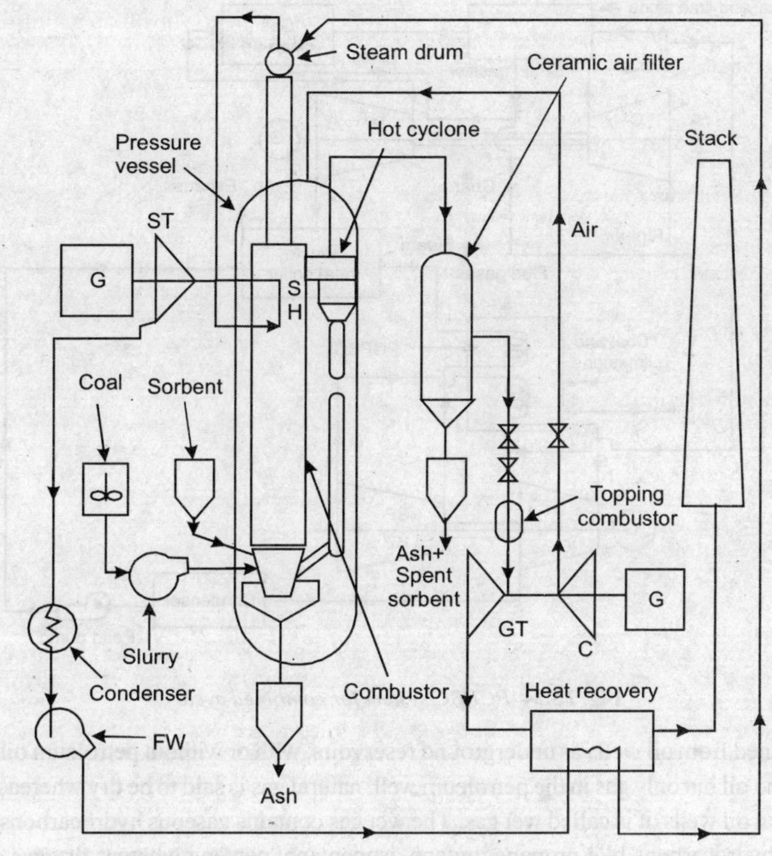

Fig. 25.23 *A pressurized circulating fluidized bed combustor.*

Figure 25.24 shows a combined cycle using coal gasifier with circulating fluidized bed combustor or boiler. Coal gasifier supplies not and clean gases to the gas turbines while the circulating fluidized bed boiler generates steam for steam turbines. Figure 25.24 shows the reheating of steam and thus the cycle used for steam turbines is the reheat cycle.

From the gasifier, char is available while from the gas cyclones or clean-ups 'fines' are obtained and are used in PCFBC in addition to coal and limestone as shown.

Thus when coal is used with coal gasifier, the working of the combined cycle is most efficient and gives higher operating efficiency and at the same time the working of the cycle becomes environmentally benign conversion of coal to electricity.

25.11.5 Natural Gas as Fuel in Gas Turbines Used in Combined Cycle Power Plant

Natural gas is the most important fuel belonging to the gaseous fuels and is found in the vicinity of coal mines or oil reservoirs. Natural gas is a mixture of methane, ethane, propane, butane, pentane, nitrogen and carbon dioxide.

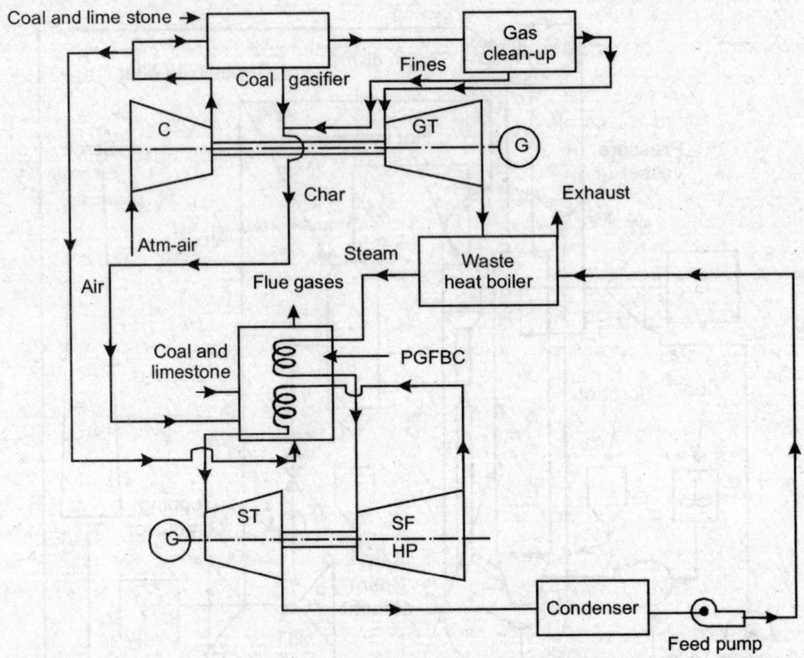

Fig. 25.24 *PCFBC system for combined cycle.*

Natural gas is obtained from oil wells as underground reservoirs, with or without petroleum oil hence it is either dry or wet. When there is no oil but only gas in the petroleum well, natural gas is said to be dry whereas if natural gas occurs along with petroleum in oil wells, it is called wet gas. The wet gas contains gaseous hydrocarbons like methane and its mixtures with higher hydrocarbons like propane, butane, isopentane, pentane whereas dry gas consists of methane, ethane and small amounts of CO, CO_2, H_2S, N_2, H_2 and other inert gases. The calorific value of wet natural gas is higher than that of dry gas because of higher percentage of heavier unsaturated molecules. Dry gas does not form a liquid phase during production conditions whereas a gas is said to be wet if a liquid phase is produced at the surface without retrograde condensation in the reservoir. The wet gas is suitably treated to remove propane, pentane, butane, H_2S (toxic and corrosive), CO_2 (no heating value and corrosive), N_2 (no heating value), heavy hydrocarbons (condenses in the transport system). Such a treated gas is called LPG (Liquified Petroleum Gas).

Composition of Natural Gas. The approximate composition of natural gas is CH_4 = 70–90%, C_2H_6 = 5–10%, C_3H_8 = 1–2%, C_4H_{10} = 0.55–1%, Pentane = 5–1%, H_2 = 3%, CO + CO_2 = Rest. Calorific value varies from 1200–1400 Kcal/m^3 or 33000 to 55000 kJ/m^3. The composition of natural gas varies widely as it depends on the underground conditions, methods used for getting the gas. Figure 25.25 shows the temperature scale of natural gas.

Since the gas contains a very small percentage of non-combustibles, the calorific value is high.

Figure 25.26 shows a schematic diagram showing processing operations of natural gas.

Liquid Natural Gas (LNG). It is mostly used as an industrial fuel and in power generation projects. Liquified natural gas is predominantly methane and it has been reduced to liquid state by cooling to –161°C. It is colourless, odourless, non-toxic and has a specific gravity of about 0.5. On liquification the reduction in the volume is in the ratio 1 : 600. At –161°C and atmospheric pressure the volume of the gas decreases significantly so that it can be liquified.

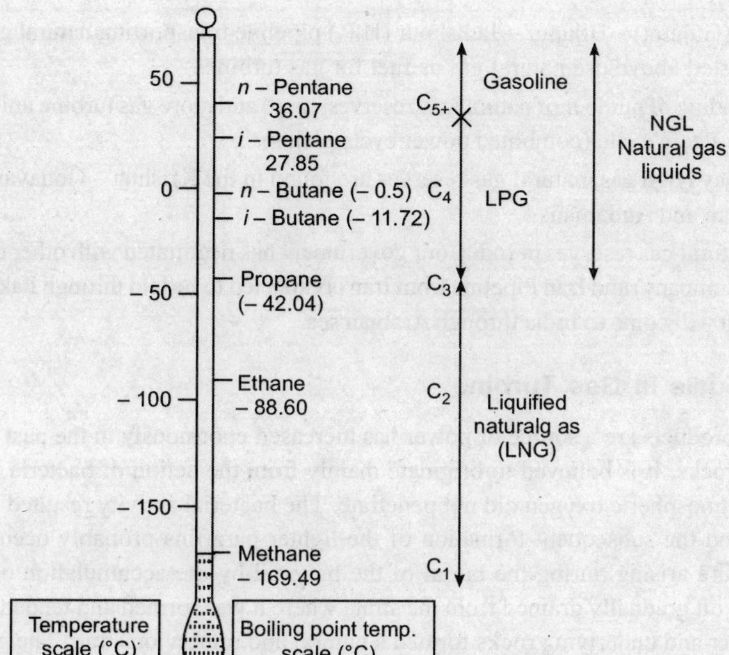

Fig. 25.25 *Temperature scale of natural gas.*

(i) One cubic metre of LNG contains 584 m^3 equivalent of natural gas.

(ii) It is a clean burning fuel and more abundantly available and oil.

(iii) It emits low levels of CO_2, NO_2, SO_2 etc. The gas is produced from the gas well. The raw gas is transported through pipelines for liquifaction. Before liquifaction the impurities like CO_2, sulphur compounds and water are removed along with heavier hydrocarbons leaving the pure natural gas i.e., the natural gas is mainly a methane (CH_4). Then it is cooled to –161°C at atmospheric pressure till the gas becomes liquid and then it is stored in insulated tanks. For using, it has to be heated up so as to regasified it and then transport it through pipelines. Figure 25.26 gives its processing and transportation scheme.

Applications of LNG.

(a) LNG can be used as a fuel where pollution control and efficient heat control are major requirements.

(b) Used in boiler for steam generation and in gas turbines in power plants.

(c) LNG is transported by special tankers and stored in pressure vessels (generally spherical) and can be used when needed particularly during peak load period.

Similar to LNG, natural gas is transported through pipelines.

Large reserves of natural gas under ONGC at Bombay High (off-shore) supply gas to combined gas turbine steam turbine power plant at Uran near Bombay. In addition to this, combined cycle power plants are :

1. at Kawas in Gujarat

2. at Anta in Rajasthan and
3. at Ahraiya in U.P.

There is the Hazira (Gujarat) – Bijapur – Jadishpur (U.P.) pipeline transporting natural gas to be used in power plants. All the plants listed above use natural gas as fuel for gas turbine.

Depending on the finding of number of natural gas reserves, more and more gas turbine units for power generation will be used along with steam cycle (combined power cycle plants).

In addition to Bombay High gas, natural gas reserves are found in the Krishna – Godavari basis, Tripura and to some extent in Rajasthan and Andamans.

In addition to our natural gas reserves in India, our government has negotiated with other oil producing countries like Oman (Oman Oil Company) and Iran Pipeline from Iran is expected to be laid through Pakistan if permitted while the pipeline from Oman will come to India through Arabian sea.

25.11.6 Liquid Fuels in Gas Turbine

The use of oil and oil products are a source of power has increased enormously in the past century. Mineral oil is found in sedimentary rocks. It is believed to originate mainly from the action of bacteria on organic material in stagnant water where atmospheric oxygen did not penetrate. The bacterial activity resulted in the formation of the heavy hydrocarbons and the subsequent formation of the lighter paraffins probably occurred as a result of the pressure and temperature arising during the burial of the material by the accumulation of sediments and strata movements. The crude oil gradually drained from the strata where it was formed and tended to collect in particular locations where the over and underlying rocks formed a natural and impervious trap. The gaseous decomposition products were also sometimes prevented from leaking away and remained in the space above the oil pool.

The crude natural oil is a mixture of solids, liquids and gaseous substances. The crude petroleum undergoes special refinery process and fractional distillation, resulting in production of several materials. The initial distillation separates the crude petroleum into :

1. Topped crude
2. Gas oil
3. Kerosene, otherwise known as paraffin oil
4. White spirits, solvents etc.
5. Gasoline (petrol) of various grades of volatility.

The petrol and gas oil fractions provide the main fuels for internal combustion engines. The steam turbine may be operated by the burning of heavy fuel oils in the steam generating plant. The gas turbine, however, is capable of producing energy directly from the combustion of a number of these products ranging from kerosene to low grade 'residual' oils i.e., a fuel containing a proportion of refinery residues. Petrol is, of course, too expensive and too explosive to form a satisfactory fuel.

Ever since the early days of the gas turbine, kerosene has been used in aircraft power units. The reasons for this choice are mainly –

1. Kerosene is widely available at reasonable price.
2. The danger of fire is appreciably lower than that of the other aviation fuels.
3. The fluidity, even at the ice point, makes any preheating unnecessary.
4. Kerosene is satisfactory for the ignition requirements and burns with a steady flame.

In marine and land based gas turbines, kerosene is too expensive for all, but the low pressure units. At the present times, the various grades of distillate oil are being used in the majority of higher power units, but the developments, now being taking place in the use of heavier oils, will eventually lead to the adoption of such fuels. This could result in an appreciably lowering of the running costs of a gas turbine.

The main problems to be overcome are concerned with the efficient combustion of such a low grade material and the corrosive effect of the products of combustion, particularly when the sulphur and vanadium contents of the fuel are high.

Availability is a further important factor in the selection of a fuel for marine use. The fuel must be readily available at the majority of the ports since ships cannot be expected to rely entirely on their home port supplies. The cost of the fuel must also be as low as possible.

One trouble which arises in the use of the heavy fuel oils is the deposition of carbon on the fuel injection tubes and the walls of the flame tube. The origin of this carbon are two fold, quenching of burning particles in the latter stages of combustion and the **cracking** of unburnt fuel on the very hot metal parts. The latter will obviously more prevalent at the fuel entry and will be accentuated by poor atomization, over-rich mixtures and the use of fuels of high aromatic content. The presence of this carbon can lead to carburisation effects in the underlying metals and serious loss of mechanical properties can be caused.

The use of residual oil (cheap fuel) presents some undesirable characteristics.

1. High viscosity and hence require heating before delivery to the atomizers.
2. Overheating may cause polymerization to form sludge or tar.
3. May form jelly-like substances if come in contact with other oils. This can clog the fuel system.
4. Excessive carbon deposition in the combustion chamber due to high carbon content.
5. Presence of vanadium which is corrosive.
6. Large ash formation.
7. Presence of alkali metal such as sodium which forms corrosive sulphates.

25.11.7 Coal Tar

Coal tar is a very useful fuel for the gas turbine, since it contains no serious impurities. It is unlikely to become an important gas turbine fuel since it is only available in limited quantities. Coal tar is mainly a by product of the high temperature carbonization of coal, although a small amount is obtained by low temperature carbonization. The crude coal tar is subjected to a distillation process which yields a number of products such as ammoniacal liquor, light oil, middle oil, anthracene oil, soft pitch etc.

A number of fuels have been obtained by blending these various products in different amounts and for convenience these fuels are designated by a number which is the temperature in °C to which the fuel must be preheated for satisfactory atomization and combustion.

For efficient combustion, good atomization is essential. The flame temperature is in the region of 1700°C and the flame is very luminous, so that a high rate of heat transfer is experienced. In the gas turbines constructed to use this fuel, it has been found preferable to use a refractory lined combustion chamber so as to avoid the problems which arise with such a luminous flame in the cooling of metal flame tubes.

25.11.8 Combined Cycle Using Brayton Cycle (Topping Cycle) and Rankine Cycle with Organic Fluid (Bottoming Cycle)

In a combined cycle, the cycle operating at the high temperature is called the **topping cycle** and the cycle operating at the lower temperature side is called **bottoming cycle**.

In the present case, normal gas turbine cycle is working at the high temperature portion of the combine cycle and the Rankine cycle using organic fluid is working on the lower temperature side of the combined cycle. The organic

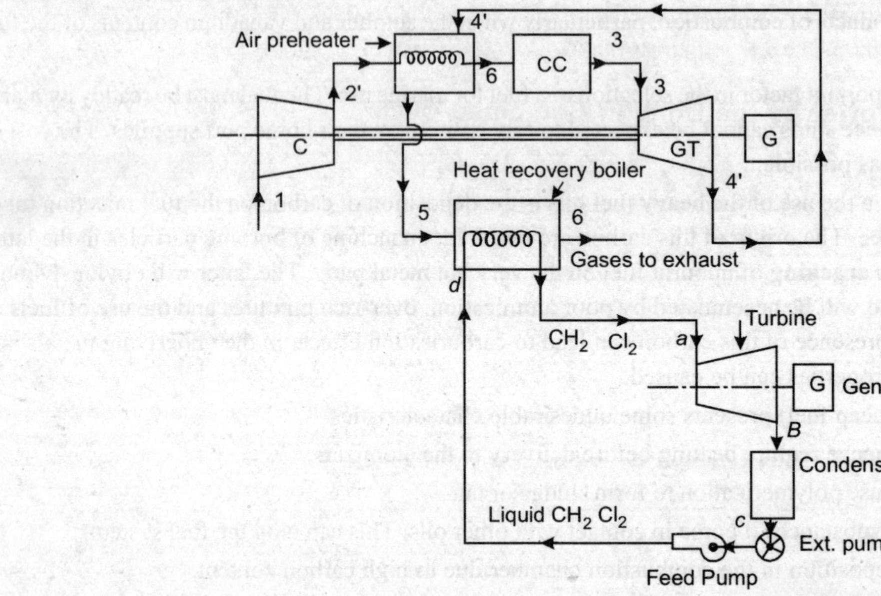

Fig. 25.26 *Combined cycle with open Brayton cycle (Topping) and closed Rankine cycle with organic liquid (Bottoming).*

fluid which is generally used in this cycle is **methylene chloride (CH_2Cl_2)**. Thus, the heat rejected by the Brayton cycle (higher temperature) is absorbed by the Rankine cycle (lower temperature). Here methylene chloride is used in place of water in Rankine cycle. It is found that **Gas-organic fluid cycle** works more efficiently that Gas steam cycle. This is because of the fact that steam requires large latent heat and also high critical pressure. Thus when water is not costly, chemically stable and has high specific heats, it is less efficient. In the combined cycle, everything is in favour of water except efficiency of Rankine cycle. But the combined cycle efficiency is appreciably good.

The schematic diagram for the combined cycle with organic fluid is shown in Fig. 25.27 and the corresponding *T–S* diagram is shown in Fig. 25.27.

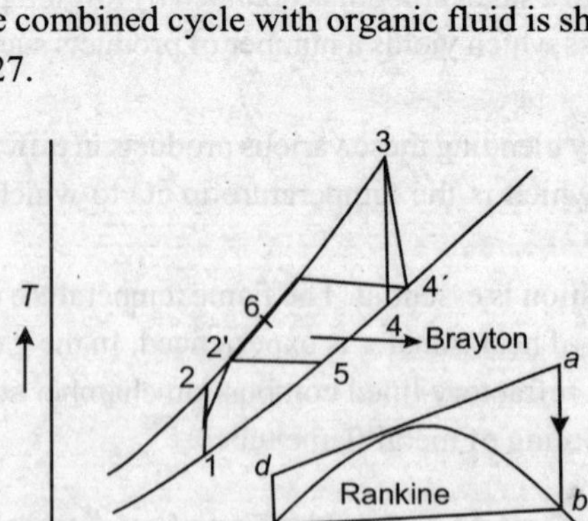

Fig. 25.27 *T–S diagram for plant in Fig. 25.26*

The working of the cycle is simple and the Figs. 25.27 and 25.28 are self-explanatory.

With organic fluids used in the combined cycle, the heat recovery in the heat recovery boiler (HRB) is more as compared to water and also the pressure that can be used is much more so that the work and power obtained with organic fluid is much more that the water/steam for the same gas temperature exhausted by the gas turbine. Thus the thermal efficiency of the combined cycle utilizing organic fluid is also more.

25.11.9 Sharing of Power by the Units in Combined Cycle

After studying the various types of combined cycle power plants, one has to think that in what proportion the units in the combined plants have to share the total power to be delivered to the consumers.

Whatever may be the sharing of the load by both gas turbine and steam turbine units of the combined cycle power plants, the main aim is the maximum efficiency of the plant. The efficiency of the plant will be maximum when, for the given heat input, the heat rejected by the plant is minimum. The heat is rejected to the environment at two points.

1. Exhaust of the gas turbine
2. Heat rejected to the condenser cooling water.

Two ways of sharing the load by gas turbine and steam turbine units may be

(a) Major load being taken by steam turbine unit and less load being taken by gas turbine unit.

(b) Very less load being taken by steam turbine unit and major load being taken by gas turbine unit.

We will consider both of these cases.

(a) When the load shared by steam turbine unit is more, then the steam required will also be more. The steam when condensed in the condenser will reject more heat to cooling water. Similarly, to produce more steam in the heat recovery boiler, more gases from the turbine will be required. But as the load sharing by the gas turbine unit is less, less quantity of gas will be required. Therefore, in the heat recovery boiler, additional fuel is required to be burnt. Therefore, for the particular load on the station, heat rejected to the condenser cooling water and the heat rejected by the gases to the environment is very large. In addition to this, additional fuel burning is required, and the combined effect of all these is the reduction of the thermal efficiency of the combined cycle power plant.

(b) When the load shared by gas turbine unit is large, then the heat reject to the environment is increased but the heat rejected to the condenser cooling water becomes a small quantity as the steam required by the steam turbine unit is small quantity and in that case large quantity of gases are used to produce the required quantity of steam without any additional fuel supply in the heat recovery boiler. Thus the quantity of heat rejected to condenser cooling water is small and there is no additional fuel supply, so that the efficiency of the combined cycle power plant will be more. This efficiency would be maximum when the total heat rejected (gas turbine exhaust and condenser cooling water) to the environment is minimum. At this point, the net heat rate is also minimum. Figure 25.29 shows variation of net heat rate with the percentage of total power developed by the gas turbine unit in the combined cycle power plant. This shows the nature of the curve. From this figure, we note that point *A* expresses the independent steam power plant delivering the total load on the station. Similarly, point *B* represents an independent gas turbine unit delivering the total load. In case of steam power plant, the heat rate (kJ/kWh) drops rapidly with the increase of load sharing by gas turbine units. It is found generally that the gas turbine power shared is approximately 40–50% of the power delivered by the station. The reduction in heat rate (kJ/kWh) occurs because of the recovery of heat from the exhaust gases coming out of the gas turbine units. As the percentage of power of gas turbine unit increases beyond, say 50%, the ratio of heat supplied in the waste heat recovery boiler to the heat supplied in the combustion chamber of gas turbine decreases and therefore, the fuel supplied in the waste heat recovery boiler also decreases. A stage is reached where the waste heat recovery boiler does not require any fuel and the boiler (WHRB) becomes **unfired** boiler. This condition is found to be at 75–80% power developed by the gas turbine unit. In Fig. 25.28, point *A* represent **maximum boiler firing** and point *B* represents **no boiler firing**. In between points *A* and *B*, the heat rate

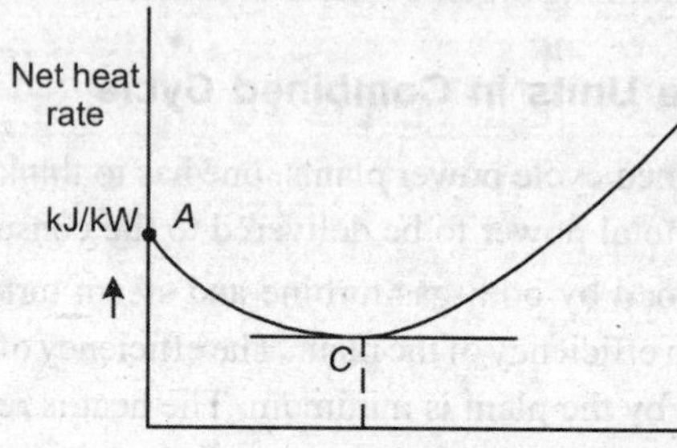

Fig. 25.28 *Net heat rate vs % of power developed by gas turbine.*

passes through a minimum value shown by point *C*. Point *C* thus represents the point of minimum heat rate (kJ/kWh) or a maximum economic running of the combined power plant. Point *C* changes with the change of load on the power plant.

Let T_1 = Air temperature at inlet to the compressor

T_3 = Gas temperature at inlet to the gas turbine

Q = Heat supplied at the gas turbine combustion chamber

W_{gt} = Net power output of the gas turbine

W_{st} = Net power output of the steam turbine

$(\eta_{th})_{gt}$ = Thermal efficiency of gas turbine cycle

$(\eta_{th})_{st}$ = Thermal efficiency of steam turbine

$(\eta_{th})_{comb}$ = Thermal efficiency of the combined cycle

∴ Thermal efficiency of the gas turbine

$$(\eta_{th})_{gt} = \frac{W_{gt}}{Q}$$

and $(\eta_{th})_{comb}$ = Combined cycle thermal efficiency

$$= \frac{\text{Combine power}}{\text{Heat supplied}}$$

$$= \frac{W_{gt} + W_{st}}{Q}$$

$$= \frac{W_{gt}}{Q} + \frac{W_{st}}{Q}$$

$$= (\eta_{th})_{gt} + \frac{\frac{W_{st}}{W_{gt}}}{(\eta_{th})_{gt}}$$

$$= (\eta_{th})_{gt} + (\eta_{th})_{gt} \times \frac{W_{st}}{W_{gt}} \times (\eta_{th})_{gt}$$

$$= (\eta_{th})_{gt} \times \left[1 + \frac{W_{st}}{W_{gt}}\right]$$

25.11.10 Boosting or Repowering of the Existing Steam Power Plant

Power demands are seldom static. Every power plant, whether of public utility or industrial type, may not only expect to serve a variable load, but one whose incidence is in the direction of a growing or adminishing load. Commonly the change is in the direction of a growth of load, this being almost always true for the public service plant. The capacity of an existing power plant may be expanded in a number of ways, viz.

(a) Replace existing plant with a new one having adequate capacity.

(b) Extend the existing plant by purchasing additional equipment similar to that already installed.

(c) Modernize the existing equipment, thereby increasing its output and efficiency.

(d) Purchase power from systems having excess power.

(e) Superpose a high pressure steam plant on the existing plant.

(f) Use of gas turbines to replace the boiler by heat recovery steam generator.

The superposition mentioned above refers to the thermodynamic position of the new equipment, rather than its physical location. Where an existing plant is becoming inadequate chiefly because of load rather than by natural depreciation or obsscence, superposed power may be the economic method of increasing capacity.

Sometimes plants in need of increased capacity have moderate pressure turbines in excellent condition operating at efficiencies not too inferior to new equipment. Such plants may be enlarged by a high pressure, non-condensing addition. In this arrangement, it will be possible to superpose unit which can pass sufficient steam into the original plant needed for the old boilers to be discarded. In this way, not only are the excellent qualities of existing turbines retained, but also it is possible to obtain the benefits of the progress that has occurred in high pressure steam generation. By securing extensions in this way, the new plant will give or exhibit remarkable economies of operation when compared to the original units.

Use of gas turbines which will work as superposed unit to replace the boiler by heat recovery steam generator (HRSG).

There is a definite life for the steam power plant during which its operation will be economical. After this period, the working becomes uneconomical. In this case, the major components of the existing steam power plant can be used except the steam generator.

Existing boiler is to be replaced by the 'Heat Recovery Boiler' by using the exhaust of the gas turbine so that the total operation is the same as plant combined cycle power plant using gas turbine power and the steam power plant. Thus the uneconomical operation of the steam power plant is converted into economical one thus increasing the power output as well as the efficiency. This process of converting old steam power plant into combined power plant by adding a gas turbine unit is called **Boosting or Repowering** of the existing steam plant. In this process of repowering, some parts which are old and economical can be continued to work and the total plant will become economical. In the majority cases, the steam turbines operating at comparatively low pressure and temperatures can still be used in the new set-up. (Life of steam turbine is majority cases is more than the boiler and hence the steam turbines can be used for many year more). The gas turbine unit is going to be modern so that the repowering of the existing steam power plant will be functioning in an economical way.

A simple schematic diagram for combined cycle plant with existing steam turbine and other equipments is shown in Fig. 25.29.

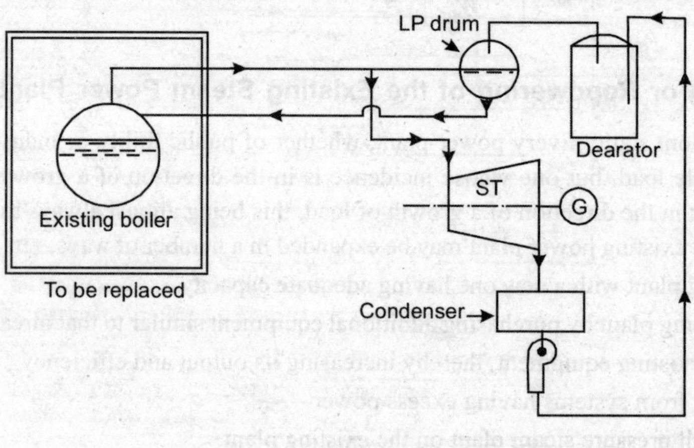

Fig. 25.29 *Existing steam power plant.*

A boiler part of the existing plant is replaced by heat recovery boiler or steam generator and a gas turbine unit is incorporated as shown in Fig. 25.30.

25.12 THERMOELECTRIC POWER GENERATION

Thermoelectric effect or Seebeck effect can be used to produce electrical energy directly from the heat of combustion (hot flue gases from boiler or hot exhaust gases from engines). But in this case the thermal efficiency of the system is very low (1 to 3 percent).

As semiconductor electrodes were found to be the proper and only materials for thermoelectric conversion systems, for thermoelectric generators p and n types of semiconductors are used for a thermoelectric generator. Thus, the heat is supplied to the hot junction and from the cold junction heat is removed, both the junctions being made of copper.

25.12.1 Thermoelectric Power Generator

A schematic diagram of thermoelectric power generator using semiconductors is given in Fig. 25.32. In Fig. 25.32, the two thermocouple materials A and B are joined to the hot end and the other ends are kept cold. An electromotive force EMF is then generated between the cold ends. Then the direct current will flow through the load as shown in Fig. 25.32. The voltage generated and the power output will depend on the given thermocouple and the temperature difference. This power out can be increased by increasing the temperature difference between the hot and cold ends. Generally the cold end will have constant temperature and thus the power output can be varied by varying or changing the hot end temperature.

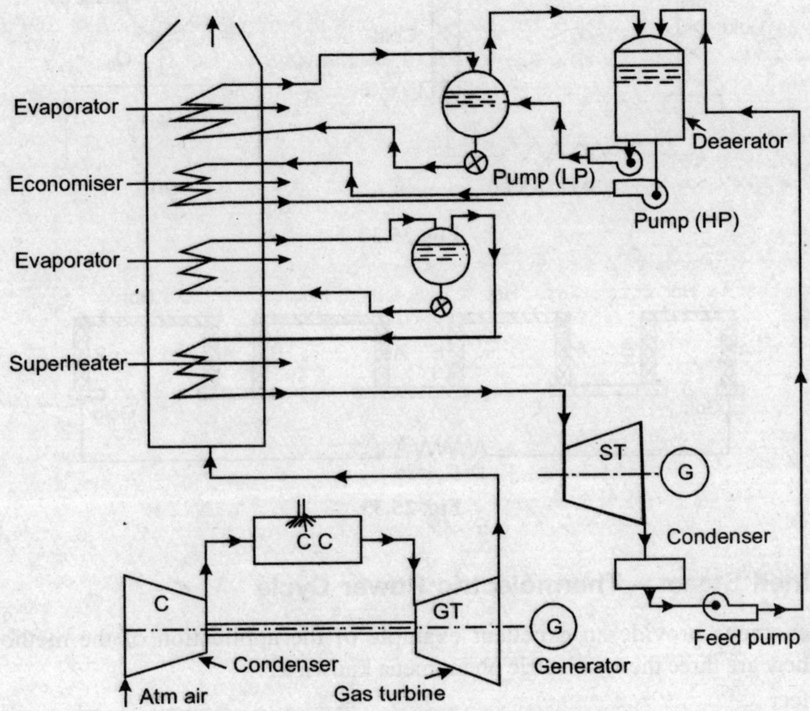

Fig. 25.30 *Repowered combined power plant.*

Generally, power generated for a single thermoelectric convertor or generator is limited and therefore in practice several thermocouples are connected in series to increase both voltage and power as shown in Fig. 25.34.

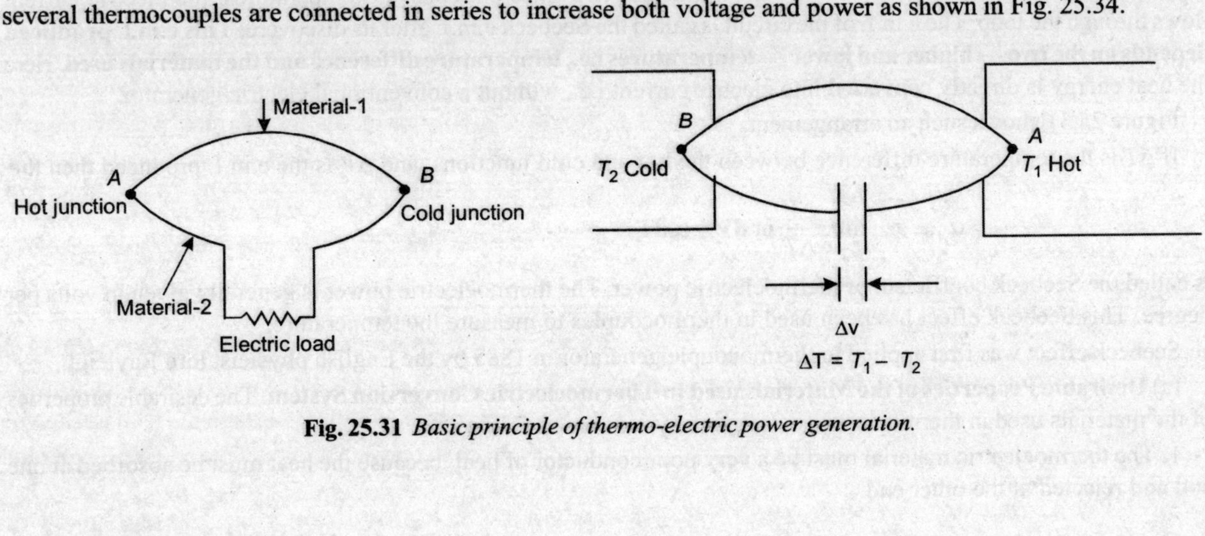

Fig. 25.31 *Basic principle of thermo-electric power generation.*

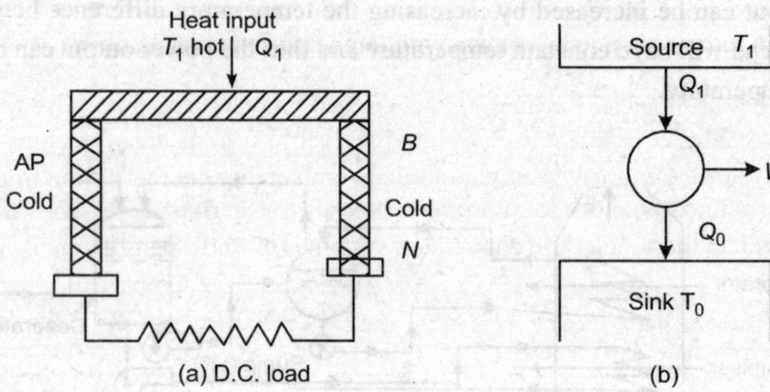

Fig. 25.32

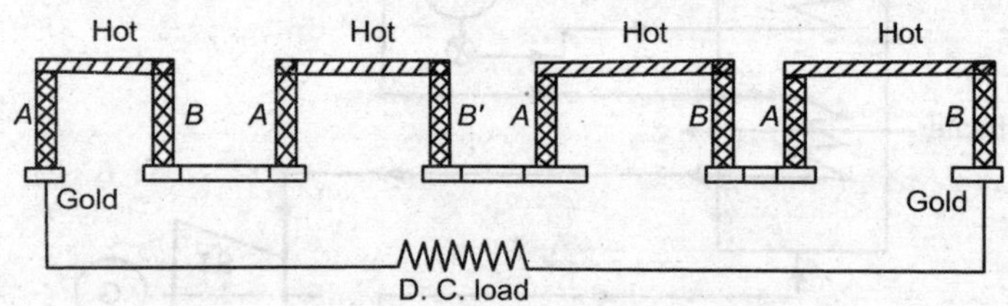

Fig. 25.33

25.12.2 Combined Steam – Thermolectric Power Cycle

Thermoelectric phenomena provide an excellent example of the application of the methods of irreversible thermodynamics. There are three thermoelectric phenomena known as :

(a) Seebeck effect

(b) Peltier effect

(c) Thomson effect.

According to Seebeck, when two dissimilar materials are joined together in the form of a loop to form two junctions, and if these two junctions are maintained at two different constant temperatures, the electric current flows through the loop. The e.m.f. of the circuit is called the Seebeck e.m.f. after its discovere. **This e.m.f. produced depends on the two — higher and lower — temperatures i.e., temperature difference and the materials used.** Here the heat energy is directly converted into electric current i.e., without a conventional electric generator.

Figure 25.31 shows such an arrangement.

If ΔT is the temperature difference between the hot and cold junctions, and ΔV is the e.m.f. produced then the

$$\alpha_{AB} = \lim_{\Delta T \to 0} \frac{\Delta V}{\Delta T} \text{ or } \mathbf{dV = \alpha.dT}$$

is called the Seebeck coefficient or thermoelectric power. The thermoelectric power is generally given in volts per degree. This Seebeck effect has been used in thermocouples to measure the temperature.

Seebeck effect was first applied to thermocouple generator in 1885 by the English physicist lord Rayleigh.

(a) **Desirable Properties of the Materials used in Thermoelectric Conversion System.** The desirable properties of the materials used in thermoelectric conversion system are as follows :

1. The thermoelectric material must be a very poor conductor of heat, because the heat must be absorbed at one end and rejected at the other end.

2. It must be excellent conductor of electricity to minimize the resistance losses (I^2R loss).

3. It should have a high rate of change in voltage with temperature which means that $\frac{dV}{dT}$ must be high.

Metals do not satisfy these properties because metals are very good conductors of heat and insulators are poor conductors. Therefore, the materials that are used for thermoelectric system, are altogether different materials.

In 1929, the Soviet physicist A.F. Loffe pointed out that semiconductors thermoelectrodes had great potential and then it was found that semiconductors were proper materials for thermoelectric conversion system.

The objective of research in thermoelectric conversion system is to find a pair of conductors for which the value of z is defined by

$$(T_h - T_c)_{max} = \frac{1}{2} Z T_c^2$$

is as high as possible.

(b) **Advantages and disadvantages of thermoelectric system :**

I. Advantages

1. Simple and less number of parts are required.
2. The units or systems are much more flexible than conventional systems.
3. It can take overload simply by increasing power output.
4. These units are noiseless and there are no moving parts.
5. The units are static and therefore are more reliable than rotating or reciprocating equipments.
6. Control is easy as it is done early adjusting the heat supply or temperature difference.
7. It is very compact and suitable for low capacity.
8. It can operate in any position.
9. Unit is very much light than conventional system unit.
10. Life may be infinite.
11. There is no leakage (of the working fluid) problem.

II. Disadvantages

As against these advantages, thermoelectric system has some disadvantages also. They are:

1. The main and serious disadvantage of this system is that the thermal efficiency is very low of the order of 1 to 3 percent.
2. It is better for smaller capacity units.

25.12.3 Thermoelectric materials used

Semiconductors have much higher values of Seebeck coefficient (α) and therefore are preferred as compared to metals. The commonly used materials for thermoelectric materials are as follows :

(i) Lead telluride	Pb Te in n and p type forms	Z(°K^{-1})	1.5×10^{-3}
(ii) Bismuth telluride	Bl_2Te_3	Z(°K^{-1})	4×10^{-3}
(iii) Bismuth sulfide	Bi_2S_3	—	—
(iv) Antimony telluride	Sb_2Te_3		
(v) Tin telluride	SnTe	—	—
(vi) Indium arsenide	—	—	
(vii) Germanium telluride	GeTe	—	$Z = 1.5 \times 10^{-3}$
(viii) Cesium sulfide	C_eS	—	$Z = 1.0 \times 10^{-3}$
(ix) Zinc antimonide	Z_nS_b	—	$Z = 1.5 \times 10^{-3}$

where Z is called figure of merit.

Note here that Z is an index or criteria to be used while rating thermoelectric converters.

To achieve higher efficiency, thermoelectric material should have a high value of *Z* and should be able to operate upto very high temperatures.

Let T_H = Temperature of hot source

T_c = Temperature of sink

Z = Figure of merit

Figure of merit depends on the properties of thermoelectric materials used. The high value of figure of merit can be obtained as follows :

(i) by using materials of large Seebeck efficient

(ii) small thermal conductivity

(iii) small electrical resistivity.

∴ Thermoelectric generator efficiency

$$\eta = \left(\frac{T_H - T_C}{T_H}\right)\left[\frac{M-1}{M+\frac{T_H}{T_C}}\right]$$

where $$M = \left[1+\frac{z}{2}(T_H - T_C)\right]^{1/2}.$$

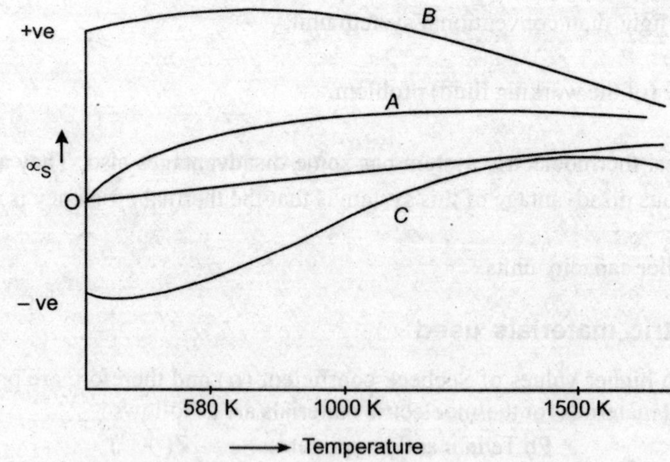

Fig. 25.34 *Variation of Seeback coefficient with temperature.*

Figure 25.34 shows the variation of Seebeck coefficient with temperature for the different materials

(i) Metal

(ii) *P*-type semiconductors

(iii) *N*-type semiconductors

Seebeck coefficient is positive for *P*-type semiconductors and negative for *N*-type semiconductors.

Thermoelectric generator can be introduced as a topping unit to a conventional steam power plant as shown in Fig. 25.35.

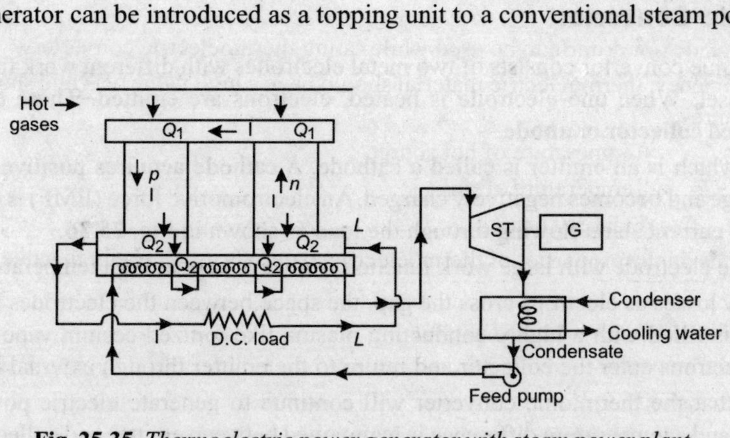

Fig. 25.35 *Thermoelectric power generator with steam power plant.*

The heat from exhaust gases from gas turbines as well as diesel engines can be utilized for thermoelectric power generation. For this purpose even the chimney gases can be used to produce electricity by thermoelectric conversion. Heat Q_2 that is rejected by the system can be utilized to preheat water from condenser and feed pump, then evaporate this water in the evaporator and finally superheated steam coming out of the thermoelectric generator. This steam is used in steam turbines producing power. Steam from the turbine then follows the water circuit and the cycle is repeated.

25.13 COMBINED STEAM AND THERMIONIC POWER GENERATION

Basically power generation by thermionic principles is a non-conventional source of energy, and in this heat energy is directly converted into electrical energy. All metals have free electrons and these electrons are emitted when the metal is heated. Such a metal is called an **emitter**. When the metal is sufficiently heated electrons are emitted from its surface. The energy required to extract electrons from the metal surface is called the **work function** of the metal and depends on the nature of metal and its surface condition.

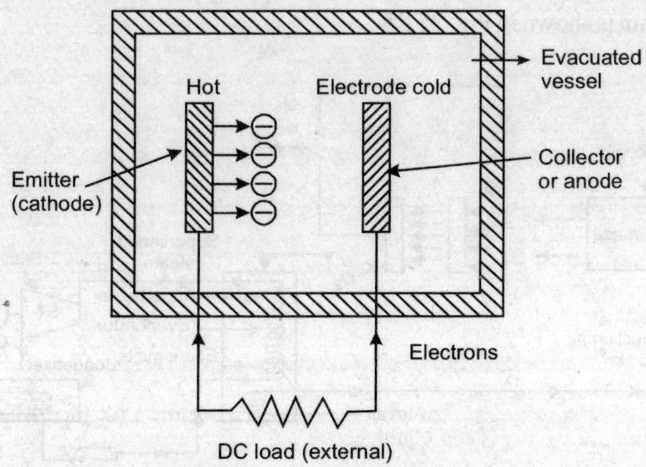

Fig. 25.36 *Basic thermionic generator.*

The electrons emitted from surface of an emitter travel through a small gap between the electrodes and accumulate on a cold metal electrode which is called a **collector**.

25.13.1 Thermionic Convertor

As seen above, thermionic convertor consists of two metal electrodes with different work functions and are sealed into an evacuated vessel. When one electrode is heated, electrons are emitted. These electrons travel to the opposite electrode called **collector** or **anode**.

The hot electrode which is an emitter is called a cathode. A cathode acquires positive charge and the anode acquires negative charge and becomes negatively charged. An electromotive force (EMF) is developed between the electrodes and a direct current starts flowing through the load as shown in Fig. 25.36.

In this converter, the electrode with large work function is always kept at high temperature than the other one.

To minimize energy losses as electrons cross the gap, the space between the electrodes is maintained at a high vacuum or the vessel is filled with a highly conducting plasma like ionized cesium vapour. Electrical power is generated when the electrons enter the collector and return to the emitter through external load.

It is now obvious that the thermionic converter will continue to generate electric power as long as heat is supplied to the emitter and a temperature difference is maintained between emitter and collector. Thermionic power generator can thus be looked upon as a cycle heat engine and its maximum efficiency will be given by Carnot cycle. In normal course, the currents that can be produced by the thermionic converter are very small except in special case of metals at high temperatures. To achieve a substantial electron emission rate and hence a significant current output as well as high efficiency the emitter temperature in a thermionic converter containing **cesium** should be at least 1000°C. The efficiency is then about 10%.

The positively charged cathode tends to pull the electrons back and the electrons already in the gas exert a retarding force on the electrons trying to cross the gap. This produces a space charge barrier. To reduce the space charge barrier and to promote electron emission from the cathode, ionized **cesium vapour** is made to fill the gap. To achieve a higher degree of ionization of the cesium, the emitter temperature must be of the range of 1500–1600°C. With this, the current densities of 20–50 amp/cm^2 have been achieved at voltage 1-2V and thermal efficiency of 10–20% has been achieved.

25.13.2 Thermionic Generator and Steam Power Plant

In this arrangement, thermionic generator cycle is used as a topping unit in the combined thermionic steam power plant. The schematic diagram is shown in Fig. 25.37.

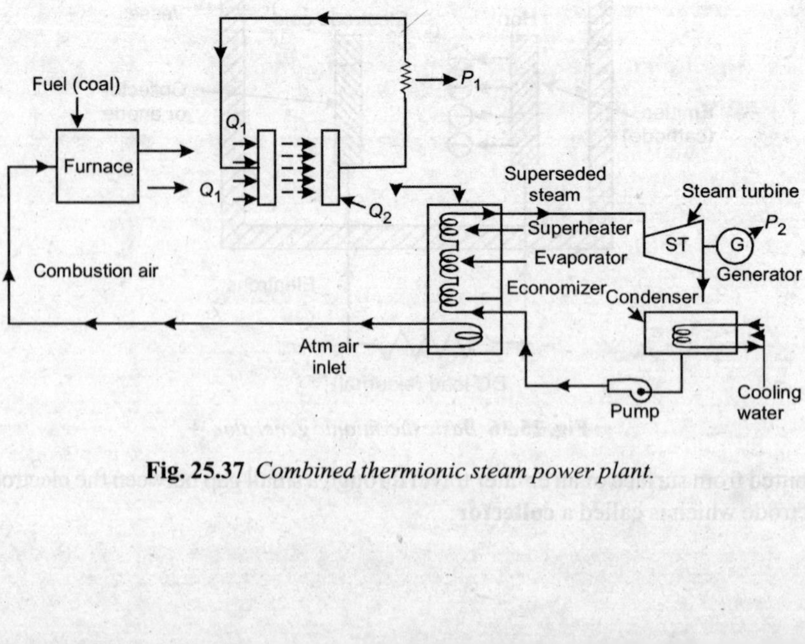

Fig. 25.37 *Combined thermionic steam power plant.*

The working of this unit is self-explanatory. P_1 is the thermionic electrical output and P_2 is the steam plant electrical output.

25.13.3 Other Combined Cycles with Thermionic Generator as Topping Unit

When thermionic generator cycle is to be used as topping cycle, then high temperature source is necessary. If we consider nuclear reactor, then the fuel elements will be the most suitable high temperature heat sources of thermionic generators. Figure 25.38 shows such a schematic for the nuclear reactor and the thermionic generator.

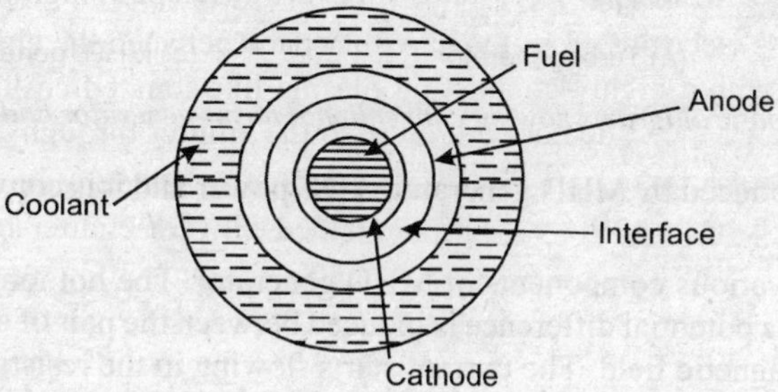

Fig. 25.38 *Thermionic generator and nuclear reactor.*

In this case, the energy released by nuclear reactions is transformed partly into electrical power in addition to the power developed by nuclear power plant (Bottoming steam power plant).

25.14 MAGNETO HYDRO DYNAMIC (MHD) POWER GENERATOR

This is one of the methods of energy conversion directly and may be the most promising method to generate electrical power in future. As the name suggests, MHD deals with the flow of fluids in the presence of magnetic field. In this case, the fluid may be gas at high temperatures or liquids like potassium and sodium. In the MHD generator, the hot ionized gas replaces the copper winding in the conventional electrical generator.

In MHD generator, heat energy of a fluid is directly converted into electrical energy. We can consider a MHD generator as a heat engine the efficiency of which can be increased by increasing the temperature of gas and rejecting the gas at the lowest possible temperature. This type of conversion of heat energy into electrical energy with the help of "Magneto Hydro Dynamic" generator seems to be a promising technique in the years to come.

25.14.1 The Principle of Operation of MHD Power Generation is the Faraday's Law of Electromagnetic Induction

This law states that when a conductor moves in a magnetic field, an electric voltage is induced in the conductor. In this case, conductor need not be a solid conductor but it can be a gas or liquid. MHD generator uses this principle and this is done by using a high pressure and high temperature gases produced by combustion in the combustion chamber, along with strong magnetic field.

If we consider, now, a gas at high pressure and temperature of say 3000°K and expand to the ambient temperature of say 300 K, then according to Carnot cycle, the efficiency will be $\frac{(3000+273)-300}{3273} \frac{3000-300}{3000} = 0.9$ or **90%**.

But it is observed that when the temperature falls to 2000°C or less, the electrical conductivity becomes very very low and then the MHD generator ceases to operate effectively. [With the modern furnaces, we can get the temperatures of more than 2500 K and maximum temperature reachable is 3000 K when the air is preheated in the cycle].

Therefore, for the successful operation of the MHD generator, the outlet temperature of the gases will have to be greater than or equal to 2000°C.

Figure 25.39 shows the two generators — one the conventional turbogenerator and the second a MHD generator.

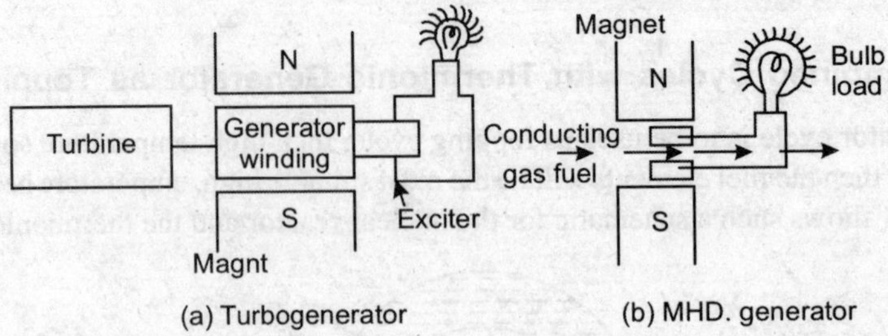

Fig. 25.39 *Schematic diagram showing conventional turbo-generator and MHD generator.*

Note here that the power produced by MHD generator is DC power and it is converted to AC power by using an inverter.

Figure 25.39(b) shows the various components of MHD generator. The hot ionized gas is passed between the poles of a magnet (electro) and a potential difference is induced between the pair of electrodes. These electrodes are placed at right angles of the magnetic field. The current starts flowing in the resistive load connected between the electrodes as shown in Fig. 25.39(b).

As seen before, ionized gas are required at high temperature and pressure and after passing through MHD generator, energy (heat) is added to it and is recirculated in the MHD generator. We observe here that the system is simpler and has large power and temperature handling capacity. It does not have any moving parts and thus have high reliability. These power plants can operate as base load, peak load plants along with a large variations in load without significant loss of efficiency.

MHD system can be classified as

(a) Open cycle system and

(b) Closed cycle system.

The schematic diagrams of these systems is given in Fig. 25.40 and Fig. 25.41. These systems are self explanatory. To increase the electrical conductivity of the gas, the hot gases from the combustion chamber are added (seeded) with a small amount of an ionized alkali metal (cesium or potassium). Generally potassium carbonate is used as seed material.

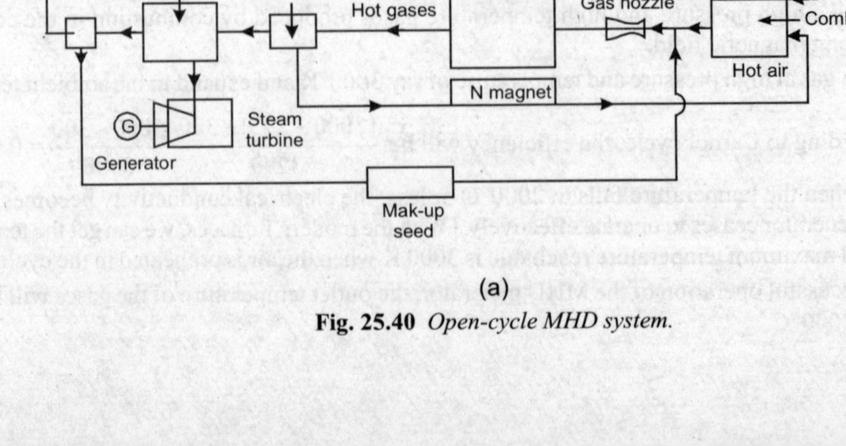

Fig. 25.40 *Open-cycle MHD system.*

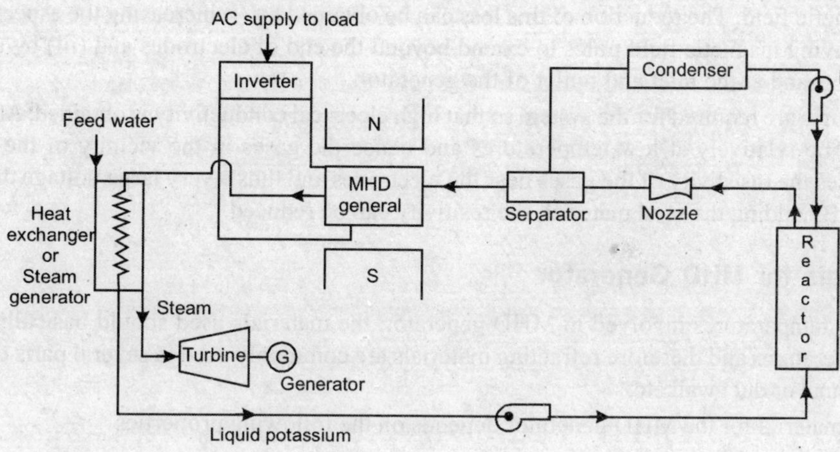

(b)

Fig. 25.41 *Close cycle (Liquid metal)*

A closed cycle system using liquid metal (potassium) as working fluid is shown in Fig. 25.40. Liquid potassium is heated in the breeder reactor and the vapours are passed through convergent–divergent nozzle. Condensed – vapours are separated in the separator, condensed in the condenser and returned to the reactor with the help of the pump.

25.14.2 Advantages of MHD System

Following are the major advantages of the MHD system.

1. This is suitable for peak power generation and emergency service.
2. In this system, fuel can be utilized in a much better manner.
3. As there are no moving parts, it is much more reliable.
4. Power produced by this system is pollution free.
5. The capital cost of MHD plants is comparable to those of conventional steam plants.
6. The overall operational cost is less.
7. The size of the plant is considerably smaller than conventional fuel plants.
8. As soon as it is started it can reach the full power level.
9. The efficiency of the plant (about 50% and more) is comparable to less than 40% for most efficient steam plant.

25.14.2 Disadvantages of MHD System

Following are the disadvantages of the MHD system.

1. This needs very large magnets which is a major expenses.

2. Because of the high velocities of the gases, there will be high friction losses and heat transfer losses also. This loss may be upto 10–12% of the input.

3. When coal is used as a fuel, the short circuiting of the electrodes, because of the molten ash, takes place. Hence oil or natural gas are to be considered as better fuels for this system. This makes the operation more expensive.

4. MHD systems suffer from the reverse flow (short circuits) of electrons through the conducting fluids around the ends of the magnetic field. The reduction of this loss can be obtained by (i) increasing the aspect ratio (L/D) of the generator, (ii) having magnetic field poles to extend beyond the end of electrodes and (iii) by using insulated vans in the fluid ducts and at the inlet and outlet of the generator.

5. High temperatures are required for the system so that high electrical conductivity is obtained. At the same time, the electrodes must be relatively at low temperatures and hence the gases in the vicinity of the electrodes are cooled. This increases the resistivity of the gases near the electrodes and thus a very large voltage drop takes place across the gas film. By adding the seed material, the resitivity can be reduced.

25.14.3 Materials for MHD Generator

Because of the high temperatures involved in MHD generator, the materials used should basically be capable to withstand high temperatures and therefore refracting materials are commonly used in several parts of the generator like electrodes, channel or duct wall etc.

Selection of the material for the MHD generator depends on the following properties

(a) Density of the material
(b) Melting point of the material
(c) Electrical conductivity
(d) Thermal resistance
(e) Erosion resistance
(f) Corrosion resistance
(g) Oxidation resistance etc.

25.14.3.1 *Materials for Electrodes*

To transfer the charge in the better way across the electrodes — plasma interface, the material used for the electrodes should have higher electrical conductivity. Other properties involve should high resistances so that the material used will serve the major purposes for the better working of the MHD generator.

Following are the materials that are commonly used for electrodes.

(i) **Zirconia based Materials.** Good refractory oxide having melting point of 2800°C. To pure zirconia, calcium oxide CaO or yitrium oxide Y_2O_3 or Magnesium oxide MgO are added to eliminate the desruptive phase transition.

(ii) **Chromite based Materials.** Lanthanum chromite – $LaCrO_3$ – is the best suited material for the electrode in MHD generator system. The highly promising material used in MHD electrodes is 20% Mgo doped $LaCrO_3$.

(iii) **Nickel Oxide based Materials.** The most suitable materials for CeO_2 and $ZrO_2 - CeO_2$ solid solution.

(iv) **Aluminate based Materials.** Conducting electrodes material is obtained by adding Fe_3O_4 to $MgAl_2O_4$ (spinel, a high resistivity insulator). The melting point of this material is about 1860°C and cannot be used safety above 1750°C. This material is very much suitable for MHD generators using coal as fuels.

25.14.5.2 *Materials for Channels*

Very commonly used material for channel or duct is ceramics as ceramics offer superior chemical stability towards oxidation and corrosion. Thus corrosion and erosion caused by high velocity plasma will be resisted by ceramics.

25.15 COMBINED CYCLE OF NUCLEAR GAS TURBINE AND STEAM POWER PLANT

The schematic diagram for such combination of Nuclear Gas Turbine Power Plant (NGTPP) and steam power plant using fossile fuel is shown in Fig. 25.42.

In this case, the gas used is Helium and a gas cooled reactor is used so that the gas coming out of the He gas turbine can be and will be used in the conventional steam power plant. As far as He gas is considered, it follows or

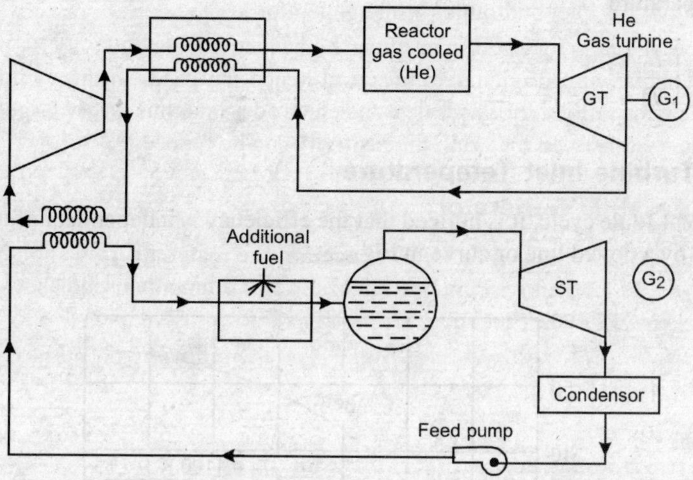

Fig. 25.42 *Combine cycle for nuclear gas turbine and steam power plant.*

undergoes a closed cycle as shown in Fig. 25.42. He gas after compressor passes through the regenerator and then through the nuclear reactor where it is heated and admitted to the gas turbine (He turbine). The exhaust of the turbine passes through a regenerator and preheated He gas from compressor is admitted to the nuclear reactor and the cycle is complete.

In case of steam power cycle, feed water from condenser and feed pump is preheated before being admitted to boiler. In boiler, additional supply of fuel is used to heat the feed water to generate steam. This steam is used in steam turbine to generate power. After expansion in the turbine, the steam is passed through the condenser and then feed pump thus completing the steam cycle.

The steam cycle is a conventional **Rankine cycle.**

The heat generated in the reactor is completely utilised in both cycles and that the heat is rejected only in steam condenser.

The reactor used is high temperature gas cooled reactor.

25.16 PERFORMANCE OF SIMPLE GAS TURBINE AND COMBINED CYCLE

If we take into consideration the compressor and turbine efficiencies, the efficiency of the plant is given by

$$\eta = \frac{T_3 \cdot \dfrac{\eta_t}{R} - \dfrac{T_1}{\eta_c}}{\dfrac{T_3 - T_1}{\left(\dfrac{P_2}{P_1}\right)^{\frac{\gamma-1}{\gamma}}} - \dfrac{T_1}{\eta_c}}$$

From this equation, it is understood that there are main five variables which affect the performance of the gas turbine plant. These variables are

(i) Compressor efficiency η_c

(ii) Turbine efficiency η_t
(iii) Compressor inlet temperature T_1
(iv) Turbine inlet temperature T_3
(v) Pressure ratio $r_p = \left(\frac{P_2}{P_1}\right)$

25.16.1 Effect of Turbine Inlet Temperature

From the equation for ideal Joule cycle, it is noticed that the efficiency is independent of turbine inlet temperature. This efficiency is shown by a dotted line or curve in Fig. 25.43.

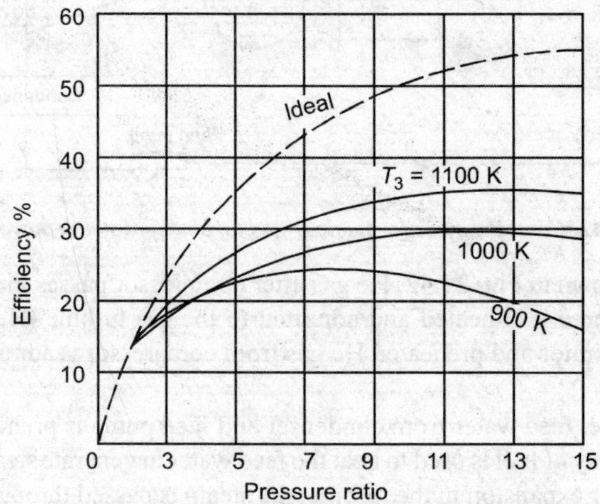

Fig. 25.43 *Effect of turbine inlet temperature.*

For different pressure ratios the efficiency curves for different turbine inlet temperatures can be plotted as shown in Fig. 25.43.

When the turbine inlet temperature T_3 increases, the work output from the turbine increases and this increases the turbine efficiency at a given pressure ratio. Here it will be noticed that to get the higher turbine inlet temperature,

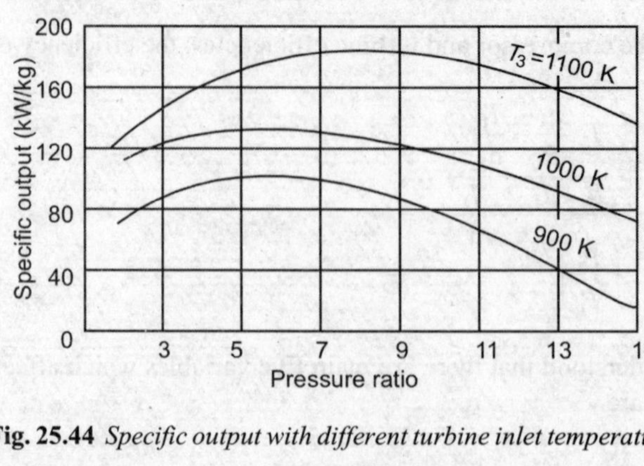

Fig. 25.44 *Specific output with different turbine inlet temperature.*

more heat has to be supplied in the combustion chamber and the effect of this increase in heat supplied, is to decrease the thermal efficiency. But in this case, the rate of increase of turbine output is more than the rate of heat supplied and the thermal efficiency of the plant increases. The maximum temperature of the cycle is limited by metallurgical considerations to approximately 1100 K. The variation of specific output with different turbine inlet temperature is shown in Fig. 25.44.

25.16.2 Effect of Compressor Inlet Temperature

For the given pressure ratio and the turbine inlet temperature (T_3), the work output of the turbine does not change, or in other words, the turbine output is independent of the compressor inlet temperature.

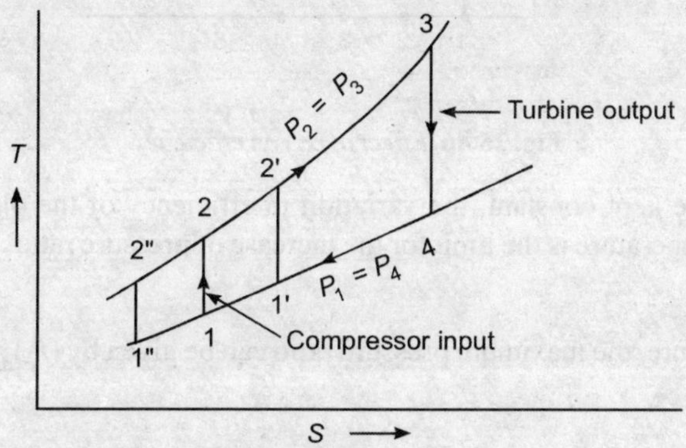

Fig. 25.45 *Effect of compressor inlet temperature.*

If the compressor inlet temperature T_1 is increased to T_1', say, then the corresponding compressor outlet temperature is increased as shown in Fig. 25.45. The effect of this is to reduce the heat supplied which will try to increase the thermal efficiency. But the increase in compressor inlet temperature increases the compressor input work which will reduce the net work output. This is very clear from the *T–S* diagram of Fig. 25.45. In this case the rate of increase of compressor work is greater than the rate of decrease or reduction in heat supply and the net effect is to decrease the thermal efficiency of the simple gas turbine plant. Therefore it is concluded that the increase in compressor inlet temperature decreases the net output of the plant and decreases the thermal efficiency of the plant.

The effect of reduction or decrease in compressor inlet temperature increase the plant output as well as efficiency.

The decrease in compressor inlet temperature is effected at high altitudes where the pressure is lower and the density of air is less than that at sea level conditions. In this case, the mass flow rate is reduced, the effect of which is to decrease the work output of the simple gas turbine plant, but efficiency increases as before.

25.16.3 Effect of Pressure Ratio (r_p)

Efficiency of the simple gas turbine cycle is given by

$$\eta = 1 - \frac{1}{(r_p)^{\frac{\gamma-1}{\gamma}}}$$

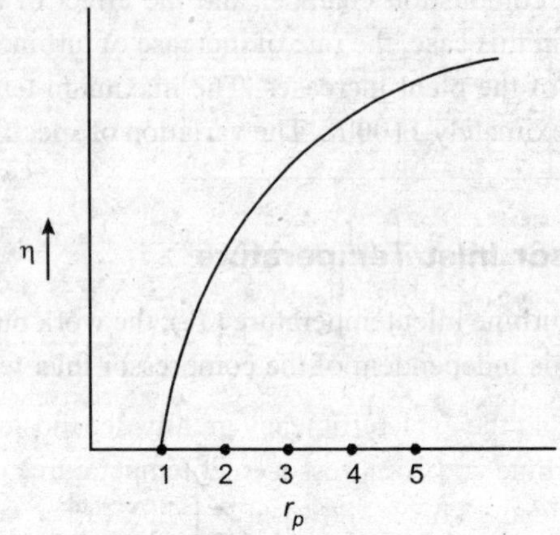

Fig. 25.46 *Effect of* (r_p) *on efficiency.*

If all the other variables are kept constant, the variation in efficiency of the plant is as shown in Fig. 25.46. Maximum gas turbine inlet temperature is the limit for the increase of pressure ratio. If T_1 and T_3 are the compressor inlet and turbine inlet temperature, the maximum pressure ratio can be given by $(r_p)_{max} = \left(\frac{T_3}{T_1}\right)^{\frac{\gamma}{\gamma-1}}$ and the optimum pressure ratio is given by

$$(r_p)_{opt} = \sqrt{(r_p)_{max}} \text{. Theoretically}$$

$$= \left\{\left(\frac{T_3}{T_1}\right).\eta_t\eta_c\right\}^{\frac{\gamma}{2(\gamma-1)}}.$$

25.16.4 Effect of Turbine and Compressor Efficiencies

Efficiency of the cycle with the turbine and compressor efficiencies is given by

$$\eta = \frac{W_t - W_c}{Q_A} = \frac{C_p(T_3 - T_4)\eta_t - C_p\left(\frac{T_2 - T_1}{\eta_c}\right)}{C_p(T_3 - T_{2'})}$$

$$= \frac{C_pT_3\left[1 - \frac{1}{(r_p)^{\frac{\gamma-1}{\gamma}}}\right]\eta_t - C_p\frac{T_1}{\eta_c}\left[(r_p)^{\frac{\gamma-1}{\gamma}} - 1\right]}{C_p(T_3 - T_2')}$$

$$= \frac{T_1\eta_t\left[1-\frac{1}{r_p^{\frac{\gamma-1}{\gamma}}}\right]-\frac{T_1}{\eta_c}\left[r_p^{\frac{\gamma-1}{\gamma}}-1\right]}{T_3-T_1\left[1+\frac{\left(r_p\right)^{\frac{\gamma-1}{\gamma}}-1}{\eta_c}\right]}$$

From this equation, it is evident that the cycle efficiency greatly depends on the efficiencies of the turbine and compressor. For a given value of turbine and compressor inlet temperatures, the efficiency of the simple cycle is linearly proportional to turbine efficiency. This is shown in Fig. 25.47. The effect of compressor efficiency is not

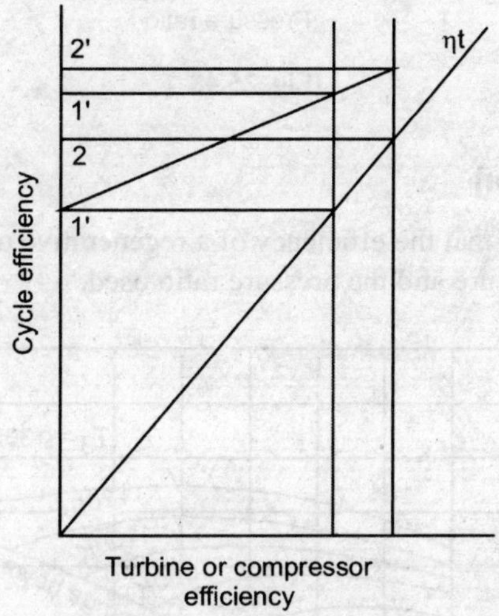

Fig. 25.47

linearly related in that it affects the heat supplied as well as the work output. A decrease in compressor efficiency decreases the heat supplied, but this decrease in the heat supplied is more than offset by the increase in compressor work.

It is observed that a change of 1 percent in the efficiencies of compressor and turbine can result in 3–5% change in the cycle efficiency. Usually the turbine efficiency is higher than that of the compressor and as the turbine is power producing device and produces much more power than compressor consumers, the net output of the plant is affected very much because of the loss of turbine efficiency.

Thus the loss in turbine efficiency reduces the cycle efficiency by a larger amount than would be the case if compressor efficiency is reduced by the same amount. This is also shown in Fig. 25.48.

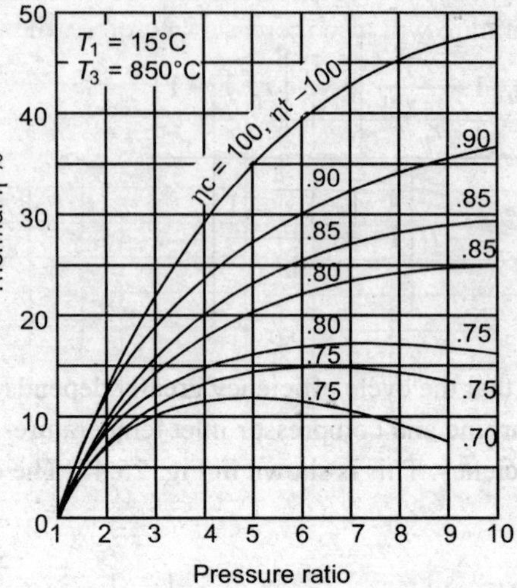

Fig. 25.48

25.16.5 Effect of Regeneration

Without any explanation, it is noticed that the efficiency of a regenerative of a regenerative cycle depends on the turbine and compressor inlet temperature and the pressure ratio used.

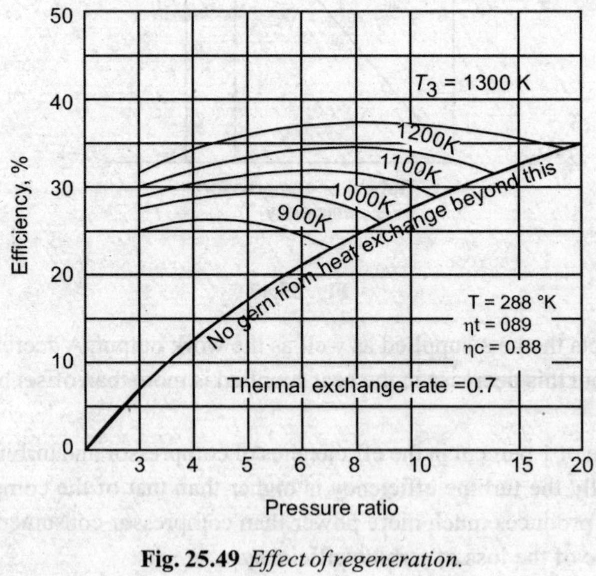

Fig. 25.49 *Effect of regeneration.*

25.16.6 Effect of Intercooling

A regenerator does not change the work output of a gas turbine cycle. Figure 25.50 shows the performance of an intercooled cycle. Intercooling will always increase the net work output of the cycle, and due to lower compressor outlet temperature, there will be more scope for regeneration. However, for the intercooled cycle, the fuel supplied to obtain a given turbine inlet temperature will also increase. Therefore, the thermal efficiency of the intercooled

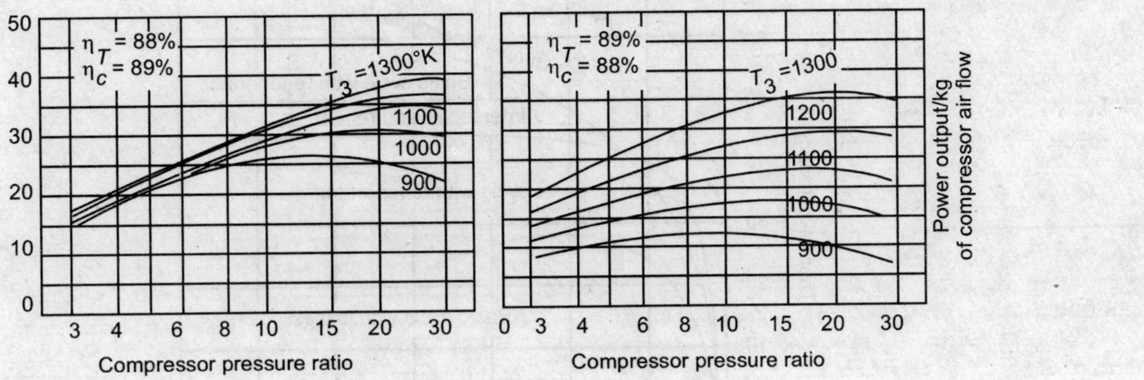

Fig. 25.50 *Effect of intercooling.*

cycle is less than that for a simple cycle. One more reason of loss in efficiency is that heat is supplied at lower temperature.

Intercooling is useful when the pressure ratios are high.

25.16.7 Effect of Reheating

The work output and the efficiency of a reheat cycle is shown in Fig. 25.51. The efficiency of the cycle is given by

$$\eta = \frac{C_p(T_3 - T_4) + C_p(T_5 - T_6) - C_p(T_2 - T_1)}{C_p(T_3 - T_2) + C_p(T_5 - T_4)}$$

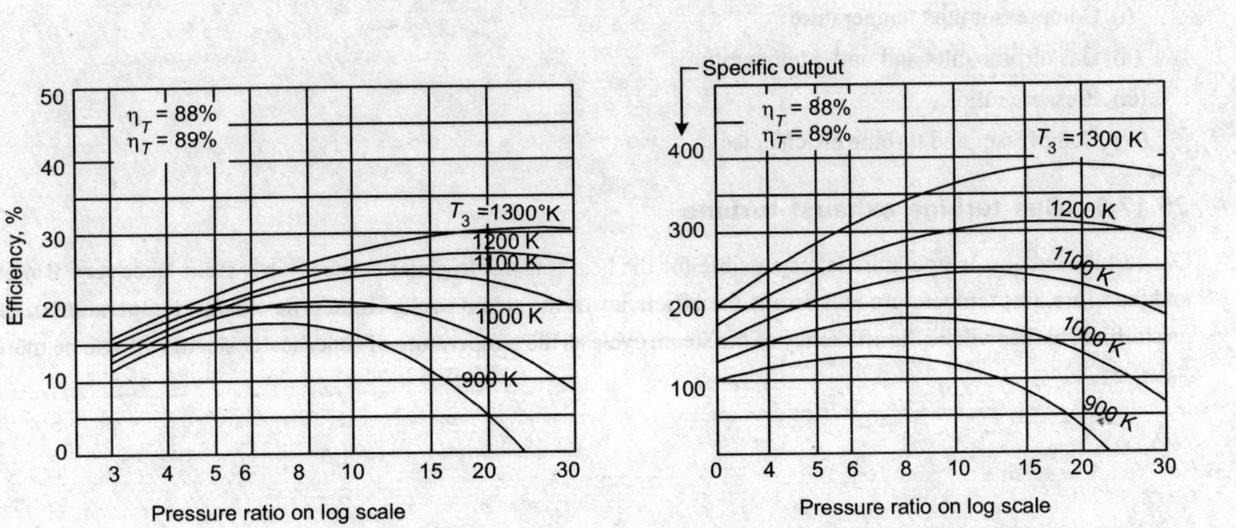

Fig. 25.51 *Effect of reheating.*

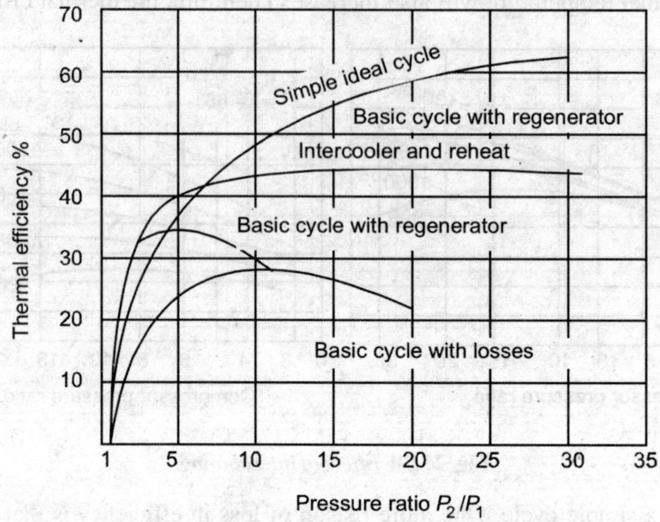

Fig. 25.52 *Performance improvement with various modifications.*

Figure 25.52 shows the improvements in performance with various modifications.

25.17 Performance of Combined Cycle

Different parameters that affect the performance of the gas turbine plant are considered in the previous section. Similarly the parameters which affect the performance of the combined cycle are

(i) Compressor inlet temperature
(ii) Gas turbine inlet and outlet temperatures
(iii) Pressure ratio
(iv) Compressor and turbine efficiencies.

25.17.1 Gas turbine exhaust turbine

Gas turbine exhaust temperature is responsible for the heat transfer to generate steam in a **Heat Recovery Boiler**, and therefore, this temperature will decide the efficiency of the steam cycle. Higher the exhaust temperature of the gas turbine, higher will be the efficiency of the steam cycle as the temperature of superheater steam will also be more.

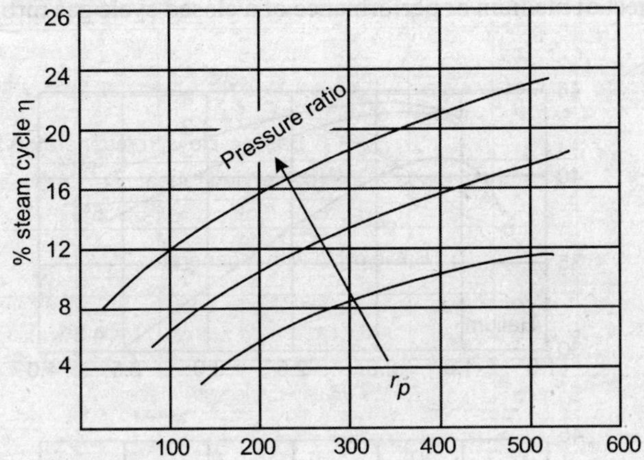

Fig. 25.53 *Effect of exhaust gas turbine temperature.*

25.17.2 Part Load Performance

In case of a combined (gas turbine and steam turbine) cycle operation, two methods can be considered

1. When no supplementary fuel firing is used.
2. When supplementary fuel firing is provided or used.

It is observed in many cases that when no supplementary fuel firing is present, then upto 75% of the station or plant load, the load is controlled by varying the gas turbine inlet temperature. After this load, the supplementary firing of fuel controls the plant load. This is mainly the control of steam plant power generation. When supplementary fuel firing is used, then the load shared by steam plant is instantly increased while that of gas turbine plant remains constant and therefore the percentage of the plant load shared by steam plant increases.

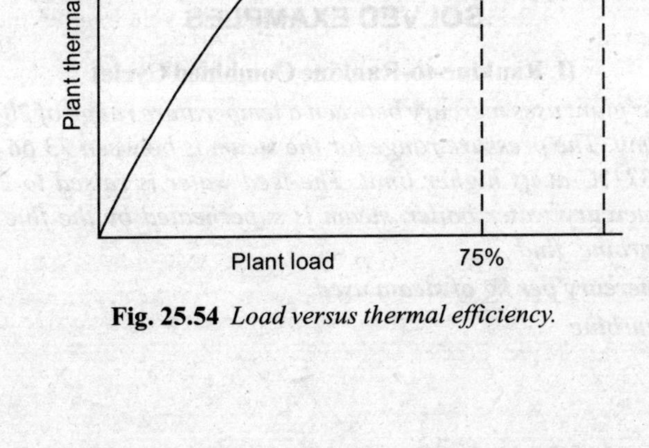

Fig. 25.54 *Load versus thermal efficiency.*

Figure 25.54 shows the variation of the thermal efficiency with load on the combined plant.
Figure 25.55 shows the effect of medium or performance of a closed cycle gas turbine.

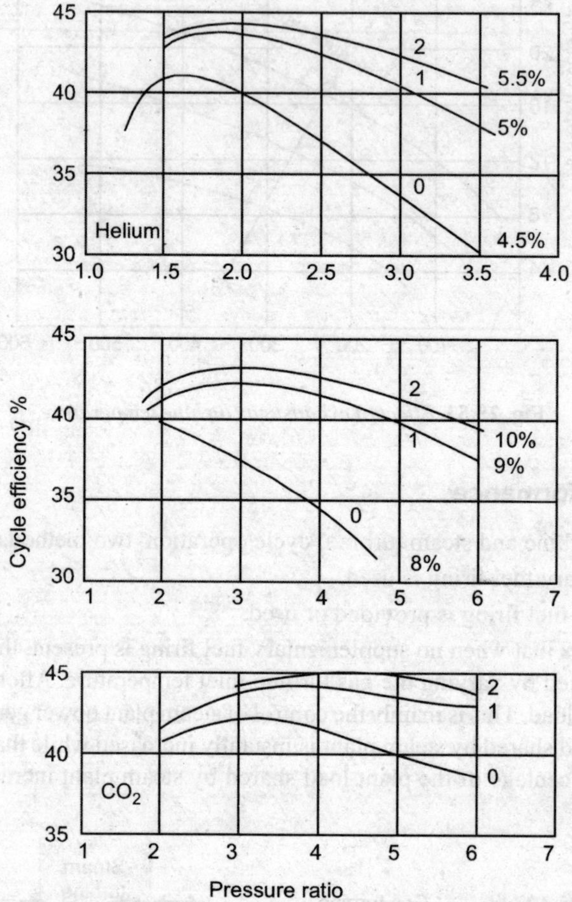

Fig. 25.55 *Actual obtainable effciencies for various gases.*

SOLVED EXAMPLES

(I. Rankine-to-Rankine Combined Cycle)

Example 25.1 *A binary vapour plant uses mercury between a temperature range of 204°C and 537°C, the mercury being just dry at the higher limit. The pressure range for the steam is between 73.66 cm vacuum and 17 bar and the steam is superheated to 371°C at its higher limit. The feed water is raised to 204°C in economiser and is evaporated to dry steam in mercury water boiler, steam is superheated by the flue gases. Assuming adiabatic expansion in Hg and steam turbine, find*

1. *the required weight of mercury per kg of steam used*
2. *work done by mercury turbine*

3. work done by steam turbine

4. the ideal efficiency of the plant

The following are the required values of total enthalpies and entropies of Hg and H_2O.

Mercury

Temperature	*Enthalpy kJ/kg*		*Entropy kJ/kg.K*	
	h_f	h_{fg}	s_f	s_g
204°C	*29*	*300*	*0.044*	*0.3883*
537°C	*75.15*	*288*	*0.08413*	*0.2820*

Steam

Pressure	*Enthalpy kJ/kg*			*Entropy kJ/kg.K*	
	h_f	h_{fg}	h_g	s_f	s_g
73.66 cm Hg	*101*	*2444*	*2545*	*0.354*	*8.576*
17 bar	*872*	*1923*	*2795*	*2.372*	*6.4*

Steam at 17 bar and 371°C, h = 3200 kJ/kg and s = 7.1259.

Solution

Mercury Plant

Refer the *T–S* diagram for Hg Fig. Ex. 25.1

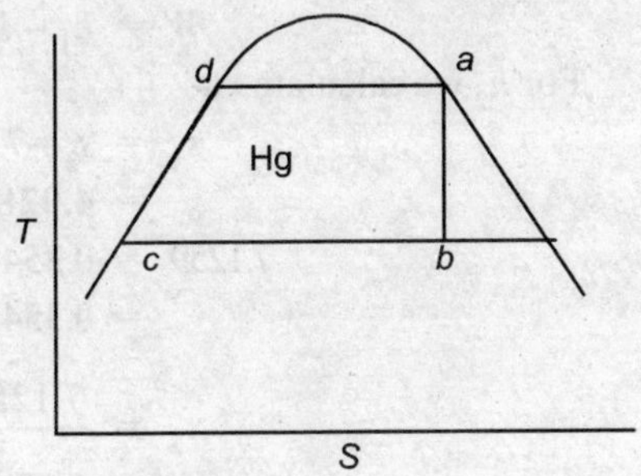

Fig. Ex. 25.1 *T-S diagram for Hg*

$$s_a = 0.282 = s_d$$
$$= s_{fb} + x_b.s_{fg}$$
$$= 0.044 + x_b(0.3883 - 0.044)$$
$$= 0.044 + 0.3443\,x_b$$
$$\therefore \quad x_b = \frac{0.2820 - 0.044}{0.3443}$$
$$= \mathbf{0.6913}$$

1. *Weight of mercury per kg steam*

Let W_m = Weight or mass of mercury required per kg of water or steam.

From mercury – water boiler,

Heat given by mercury = Heat gained by water

$$W_m(x_b h_{fg}) = \text{Latent heat of steam at 17 bar}$$
$$\therefore \quad W_m \times 0.6913 \times 300 = 1923$$
$$\therefore \quad W_m = \frac{1923}{0.6913 \times 300}$$
$$= \mathbf{9.27\ kg/kg\ steam}$$

2. *Work done by mercury turbine*

$$\text{Work done/kg} = h_a - h_b$$
$$= (h_f + h_{fg}) - (h_f + x_b h_{fg})$$
$$= 363.15 - (29 + 0.6913 \times 300)$$
$$= 363.15 - 236.39$$
$$= \mathbf{126.76\ kJ/kg}$$
$$\therefore \quad \text{Total work} = 9.27 \times 126.76$$
$$= \mathbf{1175\ kJ}$$

Heat supplied/kg Hg = $h_a - h_c$

$= 363.15 - 29$

$= \mathbf{334.15\ kJ}$

$= 334.15 \times 9.27$ kJ/kg steam

$= \mathbf{3097.57\ kJ}$

Heat supplied to water in economiser

$= 872 - 101$

$= \mathbf{771\ kJ/kg\ steam}$

Heat to steam for superheating

$= 3200 - 2795$

$= \mathbf{405\ kJ/kg}$

∴ Total heat supplied by fuel/kg of steam

$$Q_A = 3097.57 + 771 + 405$$

$$= \mathbf{4273.57\ kJ/kg\ steam}$$

3. *Steam plant*

Work done/kg steam in steam turbine

$$W = h_1 - h_2$$

For h_2 we calculate x_2.

$$s_1 = s_2 = 7.1259$$

$$= \mathbf{4.028}$$

$$7.1259 = 0.354 + x_2(8.576 - 0.354)$$

$$= 0.354 + x_2 \times 8.222$$

$$\therefore \quad x_2 = \frac{7.1259 - 0.354}{8.222} = \frac{6.7719}{8.222}$$

$$= \mathbf{0.8236}$$

$$\therefore \quad h_2 = h_{f2} + x_2 h_{fg2}$$

$$= 101 + 0.8236 \times 2444$$

$$= \mathbf{2113.88\ kJ/kg}$$

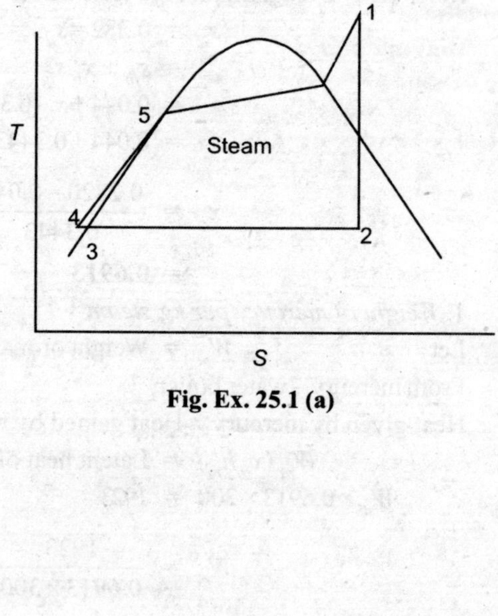

Fig. Ex. 25.1 (a)

∴ Work done/kg steam $= h_1 - h_2$

$= 3200 - 2113.88$

$= \mathbf{1086.12\ kJ/kg}$

4. *Ideal efficiency of plant*

Total work done/kg steam

$= 1175 + 1086.12$

$= \mathbf{2261.12\ kJ}$

Q_A = Heat supplied

$= \mathbf{4273.57\ kJ}$

$$\therefore \text{Ideal efficiency of plant} = \frac{\text{Total work done}}{\text{Heat supplied}}$$

$$= \frac{2261.12}{4273.57}$$

$= \mathbf{0.5291}$

$= \mathbf{52.91\%}$

Example 25.2 *A binary vapour cycle operates on mercury and steam. Saturated mercury vapour at 4.5 bar is supplied to the mercury turbine, from which it exhausts at 0.04 bar. The mercury condenser generates saturated steam at 15 bar which is expanded in a steam turbine to 0.04 bar.*

(a) Find the overall efficiency of the cycle.

(b) If 50,000 kg/hour of steam flow through the steam turbine, what is the flow through the mercury turbine?

(c) Assuming that all processes are reversible, what is the useful work done in the binary vapour cycle for the specified steam flow?

(d) If the steam leaving the mercury condenser is superheated, to a temperature of 300°C in a superheater located in the mercury boiler, and if the internal efficiencies of the mercury and steam turbines are 0.85 and 87% respectively. Calculate the overall efficiency of the cycle. The properties of the saturated mercury are given below:

Pressure bar	Temp.°C	Enthalpy kJ/kg		Entropy kJ/kg K	
		h_f	h_g	s_f	s_g
4.5	450°C	62.93	355.98	0.1352	0.5397
0.04	216.9	29.98	329.85	0.0808	0.6925

Solution

Mercury Plant

Given:

$$P_a = 4.5 \text{ bar}$$
$$P_b = 0.04 \text{ bar}$$
$$h_a = 355.98 \text{ kJ/kg}$$
$$S_a = 0.5397 = S_b$$
$$= 0.0808 + x_b(0.6925 - 0.0808)$$
$$= 0.0808 + 0.6117x_b$$
$$\therefore \quad x_b = \text{Dryness fraction at } b.$$
$$= \frac{0.5397 - 0.0808}{0.6117} = \frac{0.4589}{0.6117}$$
$$= \mathbf{0.7502}$$
$$\therefore \quad h_b = 29.98 + 0.7502 \times (329.85 - 29.98)$$
$$= 29.98 + 224.96$$
$$= \mathbf{254.94 \text{ kJ/kg}}$$

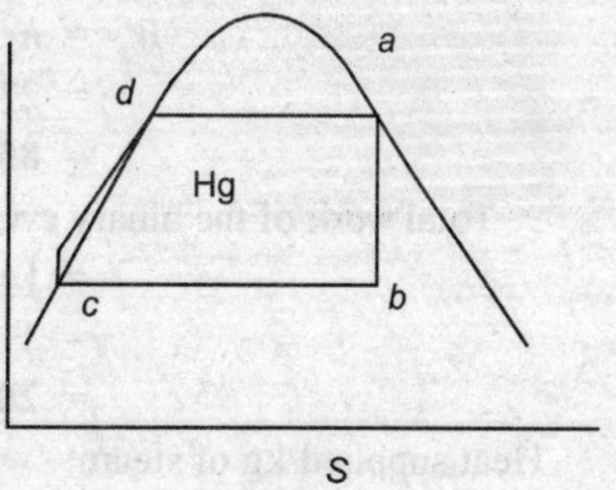

Fig. Ex. 25.2

From mercury condenser steam boiler, heat given by mercury = heat gained by water

$$m_m(h_b - h_c) = (h_1 - h_3)$$
$$m_m(254.94 - 29.98) = 2789.9 - 121.41$$
$$224.96\, m_m = 2668.49$$
$$m_m = \text{Mass of mercury per kg of water}$$
$$= \frac{2668.49}{224.96}$$
$$= \mathbf{11.86 \text{ kg/kg steam}}$$

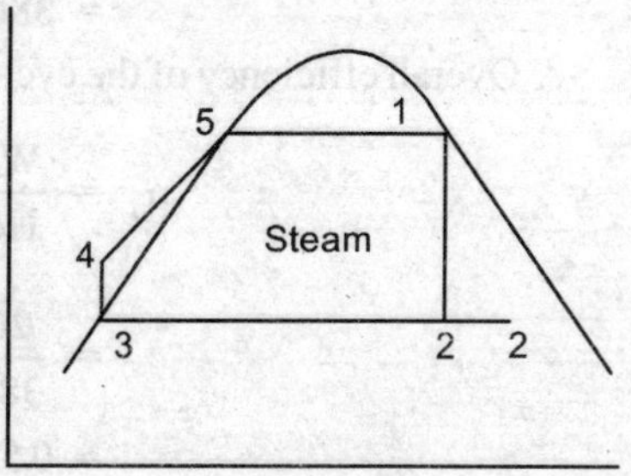

Fig. Ex. 25.2 (a)

Work done in mercury turbine/kg steam

$$= m_m(h_a - h_b)$$
$$= 11.86[355.98 - 254.94]$$
$$= 11.86 \times 101.04$$
$$= \mathbf{1198.5\ kJ} \quad \text{(I)}$$

For steam cycle,

$$h_1 = 2789.9 \text{ kJ/kg}$$
$$s_1 = 6.4406 \text{ kJ/kg.K}$$
$$= s_2 = s_{f2} + x_2 s_{fg2}$$
$$= 0.4225 + x_2 \times 8.0530$$
$$\therefore \quad x_2 = \frac{6.4406 - 0.4225}{8.053}$$
$$= \frac{6.0181}{8.053}$$
$$= \mathbf{0.7473}$$
$$\therefore \quad h_2 = h_{f2} + x_2 h_{fg2}$$
$$= 121.41 + 0.7473 \times 2433.1$$
$$= \mathbf{1939.7\ kJ/kg}$$

$\therefore$ Work done in steam turbine/kg steam

$$W_s = h_1 - h_2$$
$$= 2789.9 - 1939.7$$
$$= \mathbf{850.2\ kJ/kg\ steam} \quad \text{(II)}$$

$\therefore$ Total work of the binary cycle/kg steam

$$= \text{I} + \text{II}$$
$$= 1198.5 + 850.2$$
$$= \mathbf{2048.7\ kJ}$$

Heat supplied/kg of steam

= Heat supplied in mercury boiler/kg steam

$$= m_m[h_a - h_c]$$
$$= 11.86[355.98 - 29.98]$$
$$= 11.86 \times 326$$
$$= \mathbf{3866.36\ kJ/kg\ steam}$$

$\therefore$ Overall efficiency of the cycle

$$\eta_o = \frac{\text{Work of cycle}}{\text{Heat supplied}}$$
$$= \frac{2048.7}{3866.36}$$
$$= 0.5299$$
$$\simeq \mathbf{0.53 \text{ or } 53\%}$$

(b) *Flow through mercury turbine*

Flow of steam through steam turbine

$$m_s = 50{,}000 \text{ kg/hour}$$

Flow of mercury through mercury turbine

$$m_w = 11.86 \times 50{,}000$$
$$= \mathbf{59.3 \times 10^4 \text{ kg/hour}}$$

(c) *Cycle work*

Mercury turbine work/kg steam = 1198.5 kJ

Steam turbine work/kg steam = 850.2 kJ

∴ Total work done/kg steam = **2048.7 kJ**

For the steam flow of 50,000 kg/hour,

$$\text{Cycle work/sec} = \frac{2048.7 \times 50{,}000}{3600}$$
$$= \mathbf{28454.2 \text{ kW}}$$
$$= \mathbf{28.4542 \text{ MW}}$$

(d) *Overall efficiency – Modified cycle*

Mercury cycle remains same.

Steam cycle is modified

Enthalpy of superheated steam/kg

$$h_1 = 3038.9$$

and $$s_1 = 6.9207 = s_2$$
$$= 0.4225 + x_2 \times 8.0530$$

$$\therefore \quad x_2 = \frac{6.9207 - 0.4225}{8.0530}$$
$$= \mathbf{0.8069}$$

$$\therefore \quad h_2 = 121.41 + 0.8069 \times 2433.1$$
$$= \mathbf{2084.75 \text{ kJ/kg}}$$

$$\therefore \quad \text{Work done/kg} = h_1 - h_2$$
$$= 3038.9 - 2084.75$$
$$= \mathbf{954.15 \text{ kJ/kg}}$$

$$\text{Superheat} = h_1 - h_5$$
$$= 3038.9 - 2789.9$$
$$= \mathbf{249.0 \text{ kJ/kg}}$$

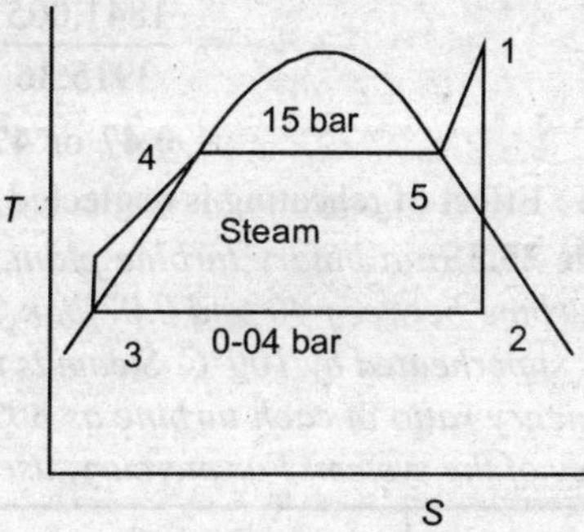

Fig. Ex. 25.2 (b)

Mass of mercury/kg steam will remain same.

∴ Total cycle work/kg steam

$$= 1198.5 + 954.15$$
$$= \mathbf{2152.65 \text{ kJ}}$$

Total heat supplied/kg steam

Q_A = Heat supplied in mercury cycle + Heat supplied for superheating the steam

$$= 3866.36 + 249$$
$$= \mathbf{3915.36 \text{ kJ}}$$

∴ Overall efficiency of the cycle

$$= \frac{2152.65}{3915.36}$$

$$= \mathbf{0.5498 = 54.98\%}$$

This is for 100% efficiencies of the turbines.

Mercury turbine is 0.85 efficient.

∴ Work done by mercury turbine

$$= 1198.5 \times 0.85$$

$$= \mathbf{1018.725\ kJ/kg\ steam}$$

Steam turbine efficiency is 87%

∴ Work done by steam turbine

$$= 945.15 \times 0.87$$

$$= \mathbf{822.28\ kJ}$$

∴ Total cycle work/kg steam

$$= 1018.725 + 822.28$$

$$= \frac{1841.005}{3915.36}\ \text{kJ}$$

$$= \mathbf{0.47}\ \text{or}\ \mathbf{47\%}$$

Note : Effect of reheating is neglected in these calculations.

Example 25.3 *In a binary turbine plant, the mercury turbine plant works between 10 bar and 0.2 bar and the steam turbine between 40 and 0.07 bar. The mercury vapour is supplied in dry and saturated condition but the steam is superheated by 100°C. Steam is tapped at 4 bar and used for feed heating in an open feed heater. Taking the efficiency ratio of each turbine as 80%, find the mass of mercury circulated per kg of steam and the thermal efficiency of the system. For mercury, use the properties given below.*

Pressure	Sat. degree °C	Enthalpy kJ/kg		Entropy kJ/kg.K	
		h_f	h_g	s_f	s_g
10	515.5	72.22	363.0	0.1478	0.5167
0.2	277.3	38.35	336.55	0.0967	0.6385

Solution

Mercury

For adiabatic expansion

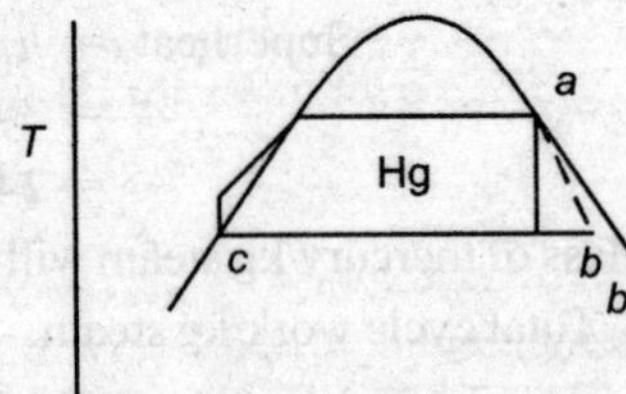

Fig. Ex. 25.3

$$S_a = S_b$$

$$0.5167 = 0.0967 + x_b(0.6385 - 80967)$$

$$\therefore \quad x_b = \frac{0.6385 - 0.1478}{0.5167 - 0.1478} = \mathbf{0.4907}$$

$$= \frac{0.5167 - 0.0967}{0.6385 - 0.0967} = \frac{0.4200}{0.5418}$$

$$= \mathbf{0.7752}$$

Now $\quad h_a = 363\ \text{kJ/kg}$

$$h_b = h_f + x_b\, h_{fg}.$$

$$= 38.35 + 0.7752\,(336.55 - 38.35)$$
$$= 38.35 + 0.7752 \times 298.2$$
$$= \mathbf{269.51\ kJ/kg}$$

Steam

$$h_1 = 3095.1\text{ kJ/kg}, S_1 = 6.587\text{ kJ/kg.K (From Steam Table)}$$
$$= S_4$$
$$\therefore \quad 6.587 = 1.776 + x_4 \times 5.118$$
$$x_4 = \frac{6.587 - 1.776}{5.118}$$
$$= \frac{4.811}{5.118}$$
$$= \mathbf{0.94}$$
$$\therefore \quad h_4 = 604.7 + 0.94 \times 2132.9$$
$$= \mathbf{2609.626\ kJ/kg}$$
$$\therefore \quad m_4\,(h_4 - h_{f4}) = (1 - m_4)\,(h_{f4} - h_{f2})$$
$$m_4 = \mathbf{0.182\ kg}$$

Again,
$$S_1 = S_2$$
$$6.587 = 0.559 + x_2 \times 7.718$$
$$x_2 = \frac{6.587 - 0.559}{7.718} = \frac{6.028}{7.718}$$
$$= \mathbf{0.781}$$
$$\therefore \quad h_2 = 163.4 + 0.781 \times 2409.2$$
$$= \mathbf{2044.98\ kJ/kg}$$

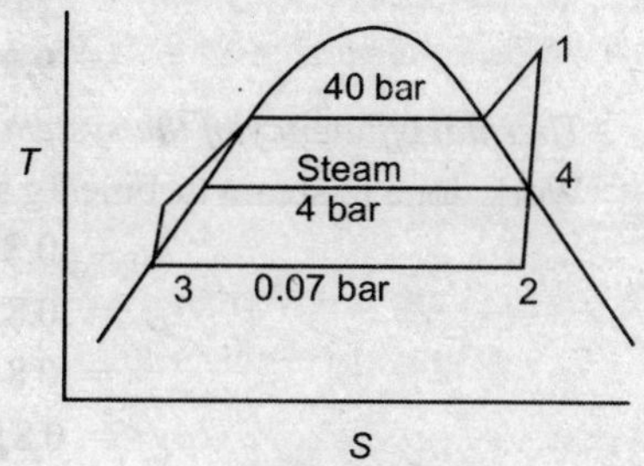

Fig. Ex. 25.3

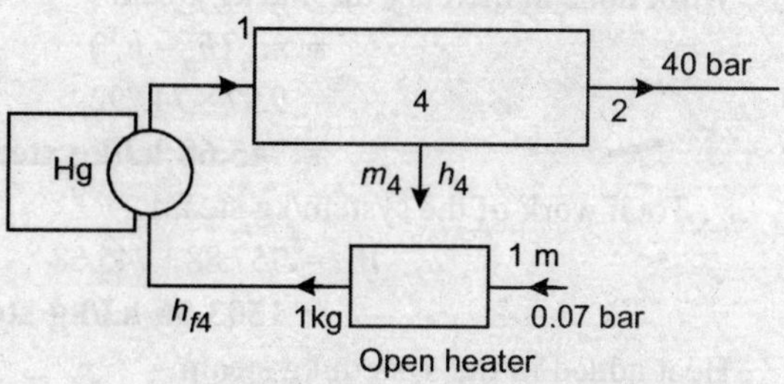

Fig. Ex. 25.3 (a)

Mass of mercury per kg steam

From mercury condenser – steam boiler

Heat given by Mercury = Heat gained by water

$$m_m\,(h_{b'} - h_c) = h_1 - h_{f4}$$
$$\therefore \quad m_m = \text{Mass of mercury}$$
$$= \frac{h_1 - h_{f4}}{h_b{}' - h_c}$$
$$= \frac{3095.1 - 604.7}{h_b{}' - 38.35}$$

$$\text{Actual enthalpy drop} = 0.8 \times (h_a - h_b)$$
$$= 0.8\,[363 - 269.51]$$
$$= \mathbf{74.792\ kJ/kg}$$
$$= h_a - h_{b'} = 363 - h_{b'}$$
$$\therefore \quad h_{b'} = 363 - 74.792$$
$$= \mathbf{288.208\ kJ/kg}$$

Substituting these values in equation for m_m we get

$$m_m = \frac{3095.1 - 604.7}{288.208 - 38.35}$$

$$= \frac{2490.4}{249.858}$$

$$= \mathbf{9.97\ kg/kg\ steam}$$

Thermal efficiency of the system

Work done in steam turbine/kg steam

$$= 0.8\,[(h_1 - h_4) + (1 - m_4)(h_4 - h_2)]$$

$$= 0.8\,[(3095.1 - 2609.626) + (1 - 0.182)(2609.626 - 2044.98)]$$

$$= 0.8\,[485.474 + 0.818 \times 564.646]$$

$$= 0.8\,[485.474 + 461.88]$$

$$= \mathbf{757.88\ kJ/kg\ steam}$$

Work done in mercury turbine/kg steam

$$= m_m\,(h_a - h_{b'})$$

$$= 9.97 \times 74.792$$

$$= \mathbf{745.68\ kJ/kg\ steam}$$

∴ Total work of the system/kg steam

$$W = 757.88 + 745.68$$

$$= \mathbf{1503.56\ kJ/kg\ steam}$$

Heat added in the system/kg steam

$$Q_A = m_m\,(h_a - h_c)$$

$$= 9.97\,(363 - 38.35)$$

$$= 9.97 \times 324.65$$

$$= \mathbf{3236.76\ kJ/kg\ steam}$$

∴ Thermal efficiency of the cycle

$$= \frac{W}{Q_A}$$

$$= \frac{1503.56}{3236.76}$$

$$= \mathbf{0.4645}$$

$$= \mathbf{46.45\%}$$

Example 25.4 *In a binary cycle the pressure in the mercury boiler is 16 bar and in the mercury condenser 0.3 bar. The steam cycle operates between pressure limits of 50 bar and 0.07 bar. Both mercury and steam at entrance to the respective turbines are in dry saturated condition.*

Determine :

(a) Mass of mercury per kg steam

(b) Thermal efficiency of cycle.

Use properties for mercury given below.

16 bar $h_g = 363.4$ $S_g = 0.4962$ $S_f = 0.0994$

0.3 bar $h_f = 40.2$ $h_g = 330.77$ $S_f = 0.0994$ $S_{fg} = 0.52$

Solution

(a) *Mass of Mercury/kg steam*

Mercury T–S diagram

Given :

$$h_a = 363.4 \text{ kJ/kg}$$
$$S_a = 0.4962 \text{ kJ/kg.K}$$

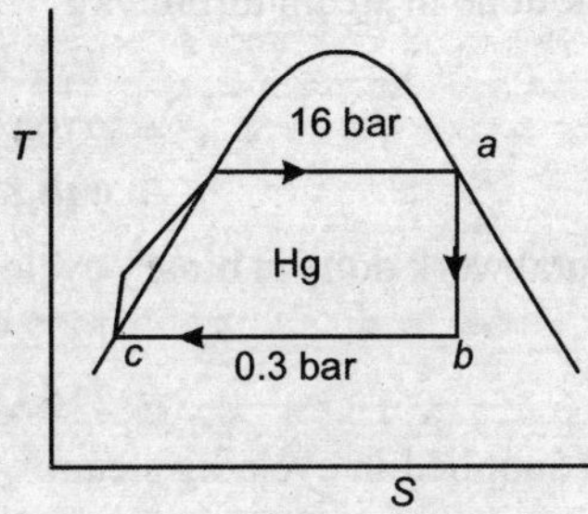

Fig. Ex. 25.4

For expansion in mercury turbine,

$$S_a = S_b$$
$$0.4962 = 0.0994 + 0.52\,x_b$$
$$\therefore \quad x_b = \frac{0.4942 - 0.0994}{0.52} = \frac{0.3948}{0.52}$$
$$= \mathbf{0.7592}$$
$$\therefore \quad h_b = 40.2 + 0.7592 \times (330.77 - 40.2)$$
$$= 40.2 + 0.7592 \times 290.57$$
$$= \mathbf{260.81 \ kJ/kg}$$

$\therefore$ Work done in mercury turbine/kg

$$= h_a - h_b$$
$$= 363.4 - 260.81$$
$$= \mathbf{102.59 \ kJ/kg}$$

Steam T-S diagram

From steam tables

$$h_1 = \mathbf{2794.2 \ kJ/kg}$$
$$S_1 = \mathbf{5.9735 \ kJ/kg.K}$$

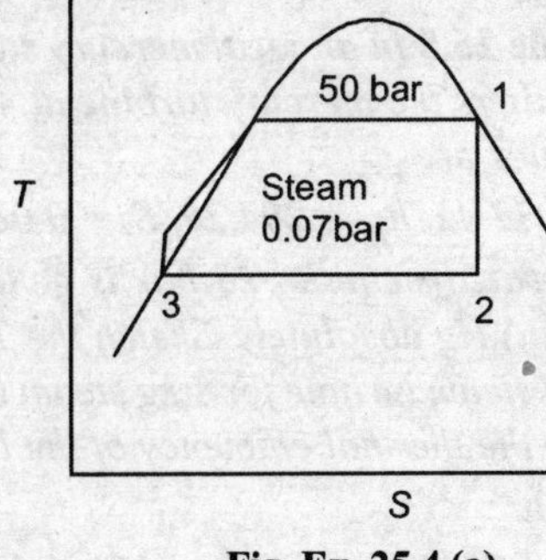

Fig. Ex. 25.4 (a)

For expansion in steam turbine,

$$S_1 = S_2$$
$$5.9735 = 0.5591 + x_2 \times 7.7176$$
$$\therefore \quad x_2 = \frac{5.9735 - 0.5591}{7.7176} = \frac{5.4144}{7.7176}$$
$$= \mathbf{0.7016}$$
$$\therefore \quad h_2 = 163.38 + 0.7016 \times 2409.2$$
$$= \mathbf{1853.591 \ kJ/kg}$$

$\therefore$ From mercury condenser steam boiler, we get

$$[\text{Heat given by mercury}] = [\text{Heat received by water}]$$
$$\therefore \quad (h_b - h_c)\, m_m = 1 \times (h_1 - h_3)$$
$$m_m (260.81 - 40.2) = 2794.2 - 163.38$$
$$m_m = \frac{2630.82}{220.61}$$
$$= \mathbf{11.92 \ kg/kg \ of \ steam}$$
$$= \text{Mass of mercury/kg steam}$$

$\therefore$ Work done in mercury turbine/kg steam

$$= m_m (h_a - h_b)$$
$$= 11.92\,(102.59)$$
$$= \mathbf{1222.90 \text{ kJ/kg steam}}$$

Work done in steam turbine/kg

$$= h_1 - h_2$$
$$= 2794.2 - 1853.591$$
$$= \mathbf{940.609 \text{ kJ/kg}}$$

∴ Total work done in binary cycle

$$W = 1222.9 + 940.609$$
$$= \mathbf{2163.509 \text{ kJ/kg steam}}$$

Heat supplied in cycle/kg steam

$$Q_A = (h_a - h_c)\, m_m$$
$$= 11.92\,(363.4 - 40.2)$$
$$= \mathbf{3892.544 \text{ kJ/kg steam}}$$

∴ Thermal efficiency of binary cycle

$$\eta_{th} = \frac{\text{Work}}{\text{Heat supplied}}$$
$$= \frac{2163.509}{3892.544}$$
$$= \mathbf{0.5558} \text{ or } \mathbf{55.58\%}$$

Example 25.5 *In an ideal mercury steam cycle, saturated mercury vapour (hg = 350.6 kJ/kg and Sg = 0.5342 kJ/kg.K) enters the mercury turbine at 4.22 bar and is exhausted into the condenser – boiler at 0.07 bar where the properties are*

$h_f = 32.42$, $h_{fg} = 294.28$, $S_f = 0.0855$, $S_{fg} = 0.5774$.

Saturated steam at 25 bar is generated in the condenser-boiler and the steam turbine exhausts at 1.513 in (3.85 cm) Hg absolutely. Sketch the T-S diagram and find : (a) the work of mercury turbine per kg of Hg (b) the work of steam turbine for 1 kg steam (c) the gross work of the two turbines per 50,000 kg/hour of steam generated and (d) the thermal efficiency of the binary vapour system.

Solution

The *T–S* diagram for the cycle is shown in Fig. Ex. 25.5.

Mercury Plant

$$h_a = 350.6 \text{ kJ/kg}$$
$$S_a = 0.5342 \text{ kJ/kg.K}$$

For expansion through turbine,

$$S_a = S_b$$
$$0.5342 = 0.0855 + x_b \times 0.5774$$
$$\therefore \quad x_b = \frac{0.5342 - 0.0855}{0.5774} = \frac{0.4487}{0.5774}$$
$$= \mathbf{0.7771}$$
$$\therefore \quad h_b = h_f + x_b + h_{fg}$$

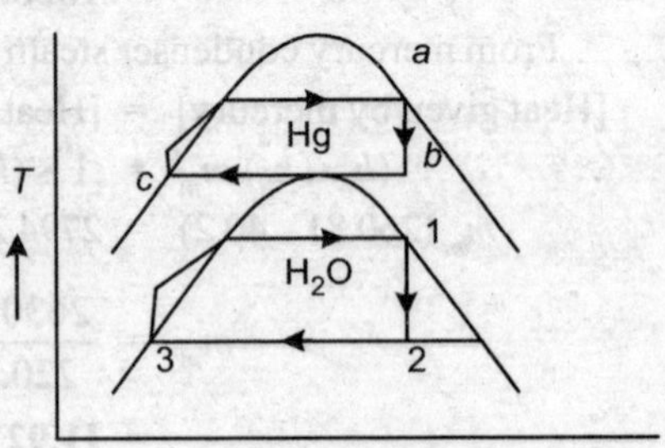

Fig. Ex. 25.5

$$= 32.42 + 0.7771 \times 294.28$$
$$= \mathbf{261.11\ kJ/kg}$$

(a) *The work done in mercury turbine*

∴ Work done/kg Hg

$$W_m = h_a - h_b$$
$$= 350.6 - 261.11$$
$$= \mathbf{89.49\ kJ/kg}$$

(b) *Steam turbine work*

From steam tables,

$h_1 = 2800.9$ kJ/kg, $S_a = 6.2536$ kJ/kg.K at 25 bar and $x = 1$.

For expansion through steam turbine

$$S_a = S_b$$
$$\therefore \quad 6.2536 = S_{f2} + x_2 . S_{fg2}.$$

Exhaust pressure = 3.85 cm Hg absolute.

$$= \frac{3.85}{76} \times 1.01325$$
$$= 0.05133 \text{ bar.}$$
$$\therefore \quad 6.2536 = 1.0990 + x_2 \times 6.4868$$
$$\therefore \quad x_2 = \frac{6.2536 - 1.0990}{6.4868}$$
$$= \mathbf{0.8099}$$
$$\therefore \quad h_2 = 343.31 + 0.8099 \times 2303.8$$
$$= \mathbf{2209.16\ kJ/kg}$$

∴ Work of steam turbine per kg of steam

$$W_s = h_1 - h_2$$
$$= 2800.9 - 2209.16$$
$$= \mathbf{591.74\ kJ/kg}$$

(c) *Gross work*

Assuming steam is generated from condensate, we have

$$m_m (h_b - h_c) = h_1 - h_3$$

or m_m = mass of mercury required per kg steam

$$= \frac{h_1 - h_3}{h_b - h_c}$$
$$= \frac{2800.9 - 343.31}{350.6 - 32.42}$$
$$= \frac{2457.59}{318.18}$$
$$= \mathbf{7.724\ kg/kg\ steam}$$

∴ For steam flow of 50,000 kg/hour.

Total work done in mercury turbine

$$= 89.49 \times 7.724 \times \frac{50,000}{3600} \text{ kW.}$$

$$= \mathbf{9600.3\ kW}$$

Total work done in steam turbine

$$= 591.74 \times \frac{50000}{3600}$$

$$= \mathbf{8218.61\ kJ/kg}$$

∴ Gross work of the cycle

$$W_T = 9600.3 + 8218.61 \text{ kW}$$

$$= \mathbf{17818.91\ kW}$$

(d) *Thermal efficiency of cycle*

Heat added in cycle/kg steam

$$= (h_a - h_c)\, m_m = (350.6 - 32.42)\, 7.724$$

$$= \mathbf{2457.62\ kJ/kg\ steam}$$

$$= \mathbf{34133.61\ kJ/sec}$$

∴ Thermal efficiency of the binary vapour cycle is given by

$$\eta_{th} = \frac{W_T}{Q_A}$$

$$= \frac{17818.91}{34133.61}$$

$$= 0.522$$

$$= \mathbf{52.2\%}$$

(II. Brayton-to-Rankine Combined Cycle)

Example 25.6 *A combined cycle power plant of 190 MW capacity consists of a gas turbine and steam turbine units. The exhaust gases from the gas turbine are supplied to steam boiler at 800°C by supplying extra fuel required in the supplementary combustion chamber. The pressure and temperature of air entering the compressor of a gas turbine are 1 bar and 15°C.*

The pressure ratio of the cycle is limited to 8 and the maximum temperature in the cycle is 800°C. Take the following data for the steam cycle.

Steam is supplied at 60 bar and 500°C. The condenser pressure is 0.05 bar.

Assuming ideal cycle of the unit, find out :

1. Thermal efficiency of the combined cycle

2. Power generated by each unit of the plant

3. Boiler capacity.

Take the following data :

C_{pg} *= 1.1 kJ/kg.K,* γ *= 1.33 for gases CV = 40 MJ/kg.*

Also find out fuel supplied to the plant/hour. The chimney gas temperature is limited to 200°C to avoid acid condensation.

Sketch the schematic of the entire plant and represent the processes on T–S diagram for both the units.

Solution

The schematic diagram for the entire plant is shown in Fig. Ex. 25.6

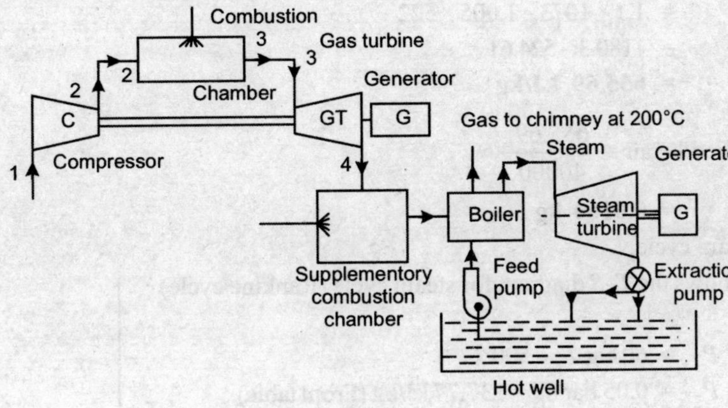

Fig. Ex. 25.6

Gas turbine unit

Given :

$$T_1 = 288\text{ K}$$
$$P_1 = 1\text{ bar}$$
$$r_p = 8$$
$$T_3 = \text{Max. temp.} = 1073\text{ K}$$
$$C_{pa} = 1.005\text{ kJ/kg.K}$$
$$\gamma = 1.4 \text{ for air.}$$

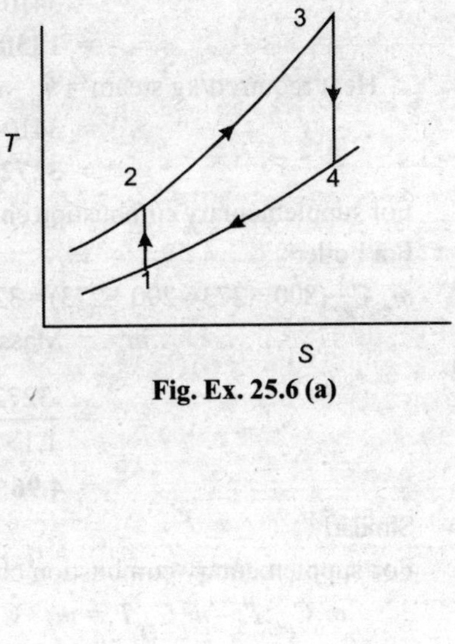

Fig. Ex. 25.6 (a)

$$\therefore \quad \frac{T_2}{T_1} = (8)^{0.2857} = 1.8125$$

$$T_2 = 288 \times 1.8125 = \mathbf{522\ K}$$

Also $\quad \dfrac{T_3}{T_4} = (8)^{\frac{1.33-1}{1.33}} = 1.67523$

$$\therefore \quad T_4 = \frac{1073}{1.67523} = \mathbf{640.51\ K}$$

Compressor work/kg $= C_{pa}(T_2 - T_1)$

$= 1.005\,(522 - 288)$

$= \mathbf{235.17\ kJ/kg}$

Similarly, turbine work/kg $= C_{pg}(T_3 - T_4)$

$= 1.1\,(1073 - 640.51)$

$= \mathbf{475.74\ kJ/kg}$

$\therefore$ Net work of Brayton cycle/kg $= W_t - W_c$

$= 475.74 - 235.17$

$= \mathbf{240.57\ kJ/kg}$

Heat supplied/kg air $Q_A = C_{pg}\,T_3 - C_{pa}\,T_2$

$= 1.1 \times 1073 - 1.005 \times 522$

$= 1180.3 - 524.61$

$= \mathbf{655.69\ kJ/kg}$

$\therefore$ Mass of fuel supplied/kg air $= \dfrac{655.69}{40000}$

$= \mathbf{0.0164\ kg}$

Now we consider steam cycle.

Figure Ex. 25. 6 (b) shows the *T–S* diagram for steam cycle (Rankine cycle)

Given :

$P_1 = 60$ bar $T_1 = 500°C$

$P_2 = 0.05$ bar $h_{f3} = 137.77$ kJ/kg (From table)

From mollier chart, $h_1 = 3410$ kJ/kg

$h_2 = 2080$ kJ/kg

$\therefore$ Steam turbine work/kg steam (neglecting pump work)

$= h_1 - h_2$

$= 3410 - 2080$

$= \mathbf{1330\ kJ/kg}$

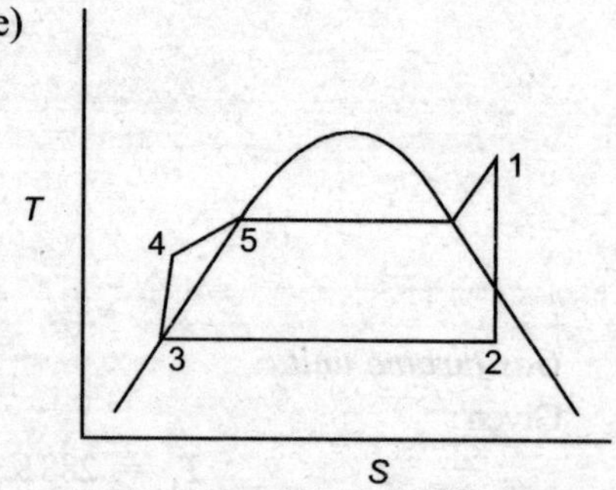

Fig. Ex. 25.6 (b)

$\therefore$ Heat required/kg steam $= h_1 - h_{f3}$

$= 3410 - 137.77$

$= \mathbf{3272.23\ kJ/kg}$

For supplementary combustion chamber (shown in Fig. Ex. 25.6 (c))

For boiler

$m_g.C_{pg}(800 + 273 - 200 + 273) = 3272.23$

$\therefore$ m_g = Mass of gas required/kg steam

$= \dfrac{3272.23}{1.1 \times 600}$

$= \mathbf{4.96\ kg/kg\ steam}$

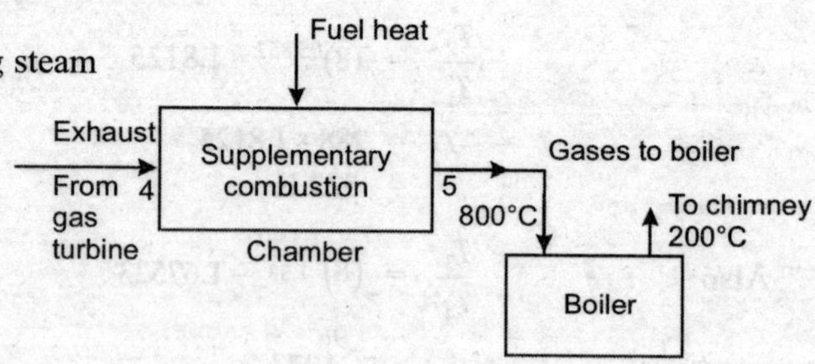

Fig. Ex. 25.6 (c)

Similarly,

For supplementary combustion chamber

$m_g C_{pg}.T_5 - m_g C_{pg} T_4 = m_f \times$ C.V.

$4.96\,[1073 - 640.51] \times 1.1 = m_f \times 40{,}000$

$= \mathbf{2360.14\ kJ/kg\ steam}$

$\therefore$ m_f = mass of fuel/kg steam

$= \dfrac{2360.14}{40000}$

$= \mathbf{0.05918\ kg/kg\ steam}$

1. *Thermal efficiency of combined cycle*

Work done of the combined cycle/kg steam

= (Work done in gas turbine cycle/kg) + (Work done in steam turbine cycle/kg steam)

= (m_g × Work done/kg air) + (Work done/kg steam)

$= 4.96 \times 240.57 + 1330$

$= 1193.23 + 1330$

= **2523.23 kJ/kg steam**

Heat supplied/kg steam

= (Heat supplied in gas turbine cycle per kg steam) + (Heat supplied in supplementary combustion chamber)

$= 655.69 \times 4.96 + 2360.14$

$= 3252.22 + 2360.14$

= **5612.36 kJ/kg steam**

$$\therefore \text{Thermal efficiency of the combined cycle} = \frac{\text{Combined work done/kg steam}}{\text{Total heat supplied/kg steam}}$$

$$= \frac{2523.23}{5612.36}$$

= **0.4496**

= **44.96%**

2. *Power generated by each unit*

Let P_G = Power generated by gas turbine plant

P_S = Power generated by steam turbine plant

and $P_G + P_S = 190 \times 10^3$ kW

$\therefore$ $P_G = 190 \times 10^3 - P_s$

Similarly, Let W_S = mass of steam required/sec

$$= \frac{P_S}{1330}$$

$$W_g = \frac{\text{mass of gases (air)}}{\text{sec}}$$

$P_S = W_S \times 1330$ (1)

$P_G = W_g \times 240.57$

$= 4.96 \times W_S \times 240.57$

$= 1193.23\, W_S$ (2)

But $P_S + P_G = 190 \times 1000$

$= 1330\, W_S + 1193.23\, W_S$

$= 2523.23\, W_S$

$$\therefore \quad W_S = \frac{190 \times 1000}{2523.23}$$

= **75.3 kg/s**

$\therefore$ $P_S = 1330 \times 75.3$

$= 100149$ kW.

$=$ **100.149 MW**

$\simeq$ **100 MW**

$=$ Power generated by steam turbine unit

$\therefore \quad P_G =$ Power generated by gas turbine unit

$= 190 - 100$

$=$ **90 MW**

3. *Boiler capacity*

Steam generated/sec $=$ **75.3 kg/s**

$\therefore$ Boiler capacity $= 75.3 \times 3600$ kg/hr

$= 271080$ kg/hr

$=$ **271.08 T/hour**

4. *Fuel supplied*

Mass of fuel supplied in combustion chamber of gas turbine/kg air = 0.0164 kg/s

$\therefore$ Mass of fuel supplied in combustion chamber of gas turbine/hour $= 0.0164 \times 4.96 \times 75.3 \times 3600$

$=$ **22050 kg/hr**

$=$ **22.050 T/hr**

$\therefore$ Mass of fuel supplied in supplementary combustion chamber/hour $= 0.05918 \times W_S \times 3600$

$= 0.05918 \times 3600 \times 75.3$

$=$ **16042.5 kg/hr.**

$\therefore$ Total fuel supplied/hour $= 16042.5 + 22050$

$=$ **38092.5 kg/hour**

Example 25.7 *In a combined cycle power plant, the air is supplied at the rate of 2000 × 10³ kg/hour at 20°C. The pressure ratio used in gas turbine plant is 7. The suction pressure of the compressor and the exhaust pressure of turbine is 1 bar. The maximum temperature in the gas turbine is limited to 1000°C. The isentropic efficiencies of the gas turbine and compressor are 85% and 80% respectively.*

The following data is given for the steam power plant.

Temperature of exhaust gases from gas turbine is increased to 1200°C by burning fuel before passing into the boiler. The steam is generated at 50 bar and 500°C. The condenser pressure is maintained at 0.1 bar. The temperature of gases entering the chimney is 200°C.

Find the following

(i) Total power generating capacity of the plant

(ii) Overall efficiency of the plant

(iii) Fuel consumption of combined plant

(iv) Air to fuel ratio used for gas turbine plant.

Use the following data :

C.V. of fuel = 45 MJ/kg, $C_{pa} = 1$ *kJ/kg.K,* $\gamma_a = 1.4$, $C_{pg} = 1.1$ $\gamma_g = 1.33$.

Do not neglect the mass of the fuel used.

Draw neat line diagram for the plant components.

Represent the processes on T–S diagram.

Solution

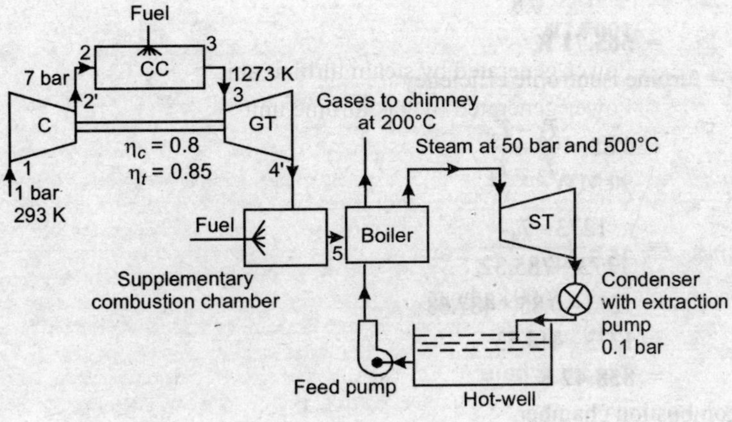

Fig. Ex. 25.7

The schematic diagram of the plant is shown in Fig. Ex. 25.7.

Gas turbine unit

T–S diagram for Brayton cycle for gas turbine is shown in Fig. Ex. 25.7(a)

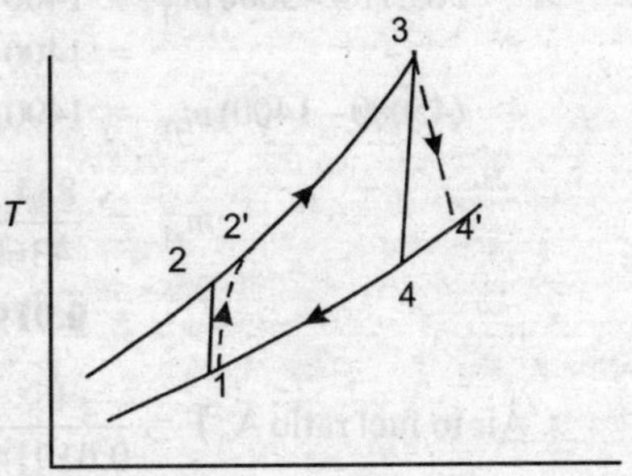

Fig. Ex. 25.7 (a)

Given :

$$P_1 = 1 \text{ bar}, T_1 = 293 \text{ K}$$
$$r_p = 7, T_3 = 1273 \text{ K}$$
$$\eta_c = 0.8\ C_{pa} = 1 \text{ kJ/kg.K}$$
$$\gamma_a = 1.4$$
$$C_{pg} = 1.1 \text{ kJ/kg.K}$$
$$\gamma_g = 1.33.$$

Consider 1 kg of air.

From diagram

$$\frac{T_2}{T_1} = (r_p)^{\frac{\gamma-1}{\gamma}} = (7)^{0.2857}$$
$$= 1.744$$

$\therefore$
$$T_2 = 293 \times 1.744 = \mathbf{510.87\ K}$$

Similarly,

$$\frac{T_3}{T_4} = (r_p)^{\frac{1.33-1}{1.33}} = 7^{0.2481} = \mathbf{1.621}$$

$\therefore$
$$T_4 = \frac{1273}{1.621} = \mathbf{785.32\ K}$$

Now, η_c = Compressor isentropic efficiency

$$= 0.8 = \frac{T_2 - T_1}{T_{2'} - T_1}$$

$$= \frac{510.87 - 293}{T_{2'} - 293} = \frac{217.87}{T_{2'} - 293}$$

$$\therefore \quad T_{2'} = 293 + \frac{217.87}{0.8}$$

$$= \mathbf{565.71\ K}$$

Also, for turbine, η_t = turbine isentropic efficiency

$$= 0.85 = \frac{T_3 - T_{4'}}{T_3 - T_4}$$

$$= \frac{1273 - T_{4'}}{1273 - 785.32}$$

$$\therefore \quad T_{4'} = 1273 - 0.85 \times 487.68$$

$$= 1273 - 414.53$$

$$= \mathbf{858.47\ K}$$

Now, consider the combustion chamber.

Consider 1 kg air. Energy equation gives us,

$$C_{pa} \times T_{2'} + m_f \times \text{CV} = (1 + m_{f1})\, C_{pg} T_3$$

$$1 \times 565.71 + m_{f1} \times 45000 = (1 + m_{f1}) \times 1.1 \times 1273$$

$$565.71 + 45000\, m_{f1} = 1400 \times (1 + m_{f1})$$

$$= 1400 + 1400\, m_{f1}$$

$$(45000 - 1400)\, m_{f1} = 1400 - 565.71$$

$$\therefore \quad m_{f1} = \frac{834.29}{43600}$$

$$= \mathbf{0.01914\ kg/kg\ air}$$

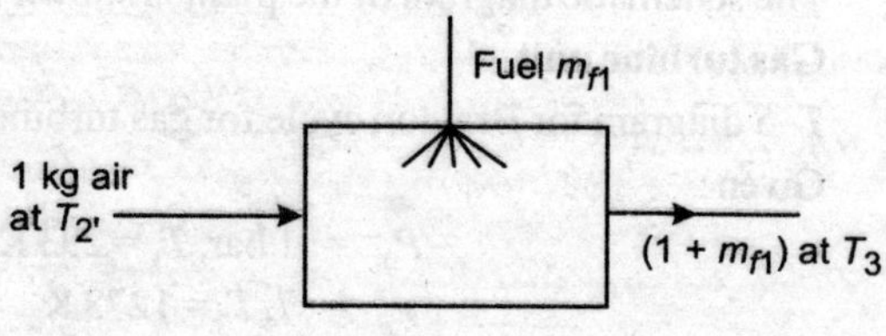

Fig. Ex. 25.7(b)

$\therefore$ Air to fuel ratio A : F $= \dfrac{1}{0.01914}$

$$= \mathbf{52.25}$$

$\therefore$ Total fuel supplied/hour to gas turbine unit $= 2000 \times 10^3 \times 0.01914$

$$= 38280 \text{ kg/hour}$$

$$= \mathbf{10.63\ kg/sec}$$

Net work done/kg air $= W_t - W_c$

$$= m_g C_{pg} (T_3 - T_{4'}) - C_{pa} (T_{2'} - T_1)$$

$$= 1.01914 \times 1.1 (1273 - 858.47) - 1 (565.71 - 293)$$

$$= 1.1 \times 414.53 - 272.71$$

$$= 464.71 - 272.71$$

$$= \mathbf{192\ kJ/kg}$$

$\therefore$ Power produced by gas turbine unit

$$P_1 = P_G = \frac{\text{Work}}{\text{kg air}} \times \text{mass of air/sec}$$

$$= 192 \times \frac{20,00,000}{3600}$$

$= \textbf{106666.7 kJ/s} = \textbf{kW}$

$= \textbf{106.6667 MW}$

Supplementary combustion chamber

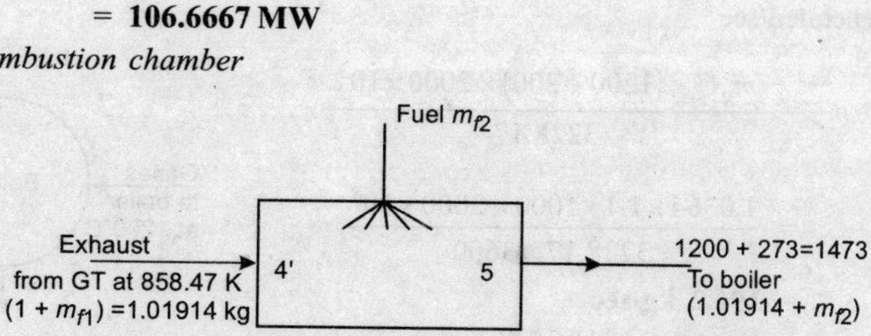

Fig. Ex. 25.7(c)

Energy in + Fuel energy = Energy out

$$m_{g1} C_{pg} T_{4'} + m_{f2} \times \text{C.V.} = (1 + m_{f1} + m_{f2}) C_{pg} T_5$$

$$1.1 \times 1.0914 \times 858.47 + m_{f2} \times 45000 = 1.1 \times (1.0914 + m_{f2}) \times 1473$$

$$1030.63 + 4500\, m_{f2} = 1768.4 + 1620.3\, m_{f2}$$

$$\therefore \quad m_f = \frac{1768.4 - 1030.63}{45000 - 1620.3}$$

$$= \frac{737.77}{43379.7}$$

$$= \textbf{0.017 kg/kg air}$$

Total gases coming to boiler $= 1 + m_{f1} + m_{f2}$

$= 1.0194 + 0.017$

$= \textbf{1.0364 kg/kg air}$

Also total fuel required in supplementary combustion chamber/hour = 0.017 × 2000 × 1000

$= \textbf{34000 kg/hour}$

$\therefore$ Total mass of fuel supplied to the plant = 38280 + 34000

$= \textbf{72280 kg/hour}$

$= \textbf{20.08 kg/sec}$

Steam turbine unit

T–S diagram for steam turbine is shown in Fig. Ex. 25.7 (a)

Given :

$P_1 = 50$ bar

$T_1 = 500 + 273$

$= \textbf{773 K}$

$P_2 = 0.1$ bar

From *h-s* diagram, we get

$h_1 = 3420$ kJ/kg

$h_2 = 2190$ kJ/kg

$h_{f2} = 191.83$ (From table)

$\therefore$ Heat required to generate 1 kg steam $= h_1 - h_{f2}$

$= 3420 - 191.83$

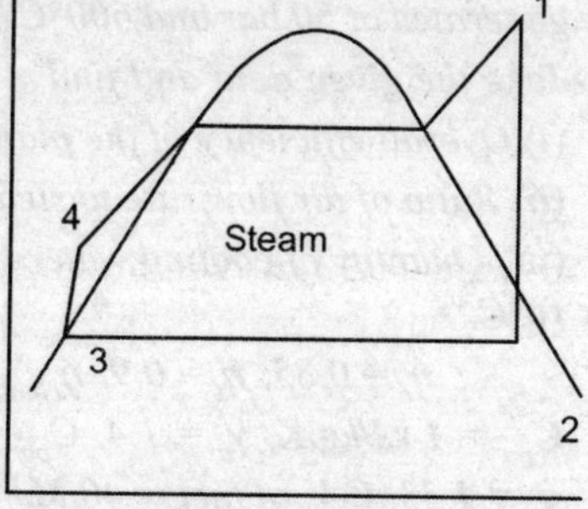

Fig. Ex. 25.7(d)

$= \textbf{3228.17 kJ/kg.}$

∴ Mass of steam generated/sec

$$m_s = \frac{m_g C_{pg}(1200-200)\times 2000\times 10^3}{3228.17}$$

$$= \frac{1.0364\times 1.1\times 1000\times 2000\times 10^3}{3228.17\times 3600}$$

$$= \textbf{196.2 kg/sec}$$

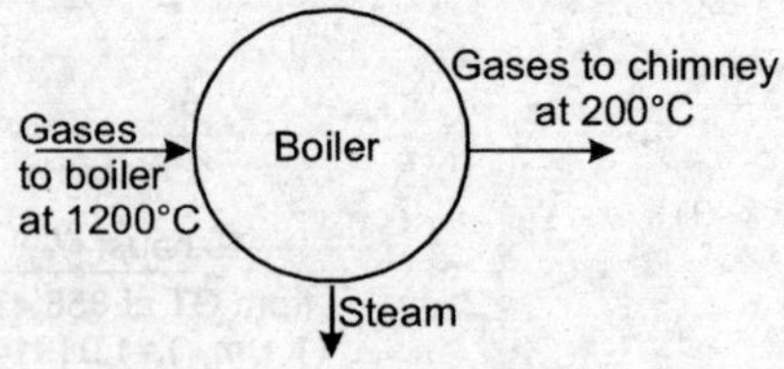

Fig. Ex. 25.7(e)

∴ Work done/sec in steam turbine unit = 196.2 $(h_1 - h_2)$

$$= 196\times(3420-2190)$$

$$= \textbf{241080 kW}$$

$$\mathbf{P_2} = \textbf{241.08 MW}$$

∴ Total power produced by the combined cycle plant

$$P = P_1 + P_2$$

$$= 106.6667 + 241.08$$

$$= \textbf{347.7467 MW}$$

$$\approx \textbf{350 MW}$$

Capacity of the combined power plant

Overall efficiency of the plant

$$\eta_o = \frac{\text{Total work produced}}{\text{Total fuel used} \times \text{Cal. value}}$$

$$= \frac{347747.7\ (\text{kJ/sec})}{20.08\times 45000\ \text{kJ/sec}}$$

$$= 0.3848$$

$$= \textbf{38.48\%}$$

Example 25.8 *A combined cycle power plant is to be designed to develop 100 MW. Out of this 60% power is to be developed in a gas turbine plant and the remaining in a steam power plant. The pressure and temperature of the air entering the compressor are 1 bar and 27°C.*

The maximum pressure and temperature in the gas turbine cycle are limited to 8 bar and 1000°C. Steam is to be generated at 50 bar and 600°C by using supplementary fuel. The condenser pressure is maintained at 0.05 bar.

Take the given data and find

(i) Overall efficiency of the plant

(ii) Ratio of air flow rate to steam flow rate

(iii) Quantity of cooling water required in tonne/hour if the rise in temperature of the cooling water is limited to 10°C.

$\eta_c = 0.85,\ \eta_t = 0.9,\ \eta_{comb} = 0.95$

$C_{pa} = 1$ *kJ/kg.K,* $\gamma_a = 1.4,\ C_{pg} = 1.1$ *kJ/kg.K*

$\gamma_g = 1.33$*, C.V. of fuel = 40 MJ/kg*

Gas temperature going to chimney = 200°C.

Neglect pump work and consider fuel mass.

Solution

Gas turbine plant

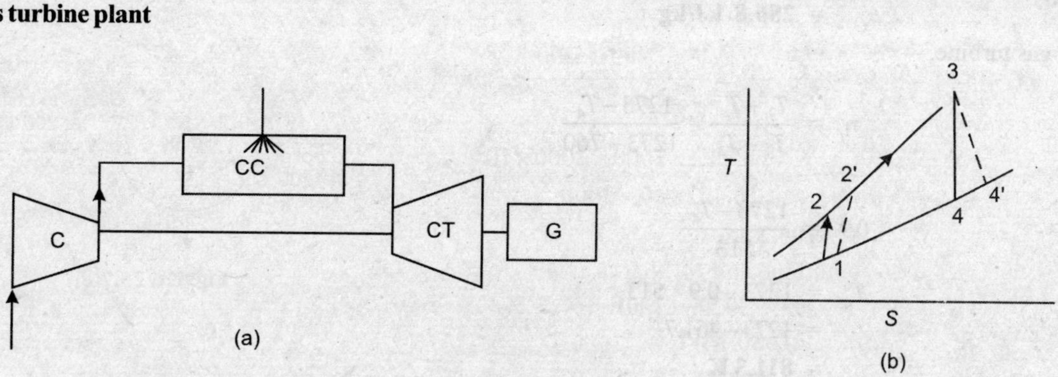

Fig. Ex. 25.8

Schematic diagram for gas turbine plant is shown in Fig. Ex. 25.8 (a) and the corresponding T–S diagram in Fig. Ex. 25.8 (b).

Given :

$$P_1 = 1 \text{ bar } T_1 = 300 \text{ K}, P_2 = 8 \text{ bar } T_3 = 1000 + 273 = 1273 \text{ K}.$$
$$\eta_c = 0.85, \eta_t = 0.9, \eta_{comb} = 0.95$$
$$C_{pa} = 1 \text{ kJ/kg.K } \gamma_a = 1.4, C_{pg} = 1.1 \text{ kJ/kg.K}$$
$$\gamma_g = 1.33 \text{ C.V.} = 40 \text{ MJ/kg}$$

From T–S diagram

$$\frac{T_2}{T_1} = (r_p)^{\frac{\gamma-1}{\gamma}} = 8^{0.2857} = 1.813$$

$$\therefore \quad T_2 = 300 \times 1.813 = \mathbf{543.9\ K}$$

Also,
$$\frac{T_3}{T_4} = (r_p)^{\frac{1.33-1}{1.33}} = 8^{0.2481} = 1.675$$

$$\therefore \quad T_4 = \frac{1273}{1.675} = \mathbf{760\ K}$$

$$\eta_c = \frac{T_2 - T_1}{T_{2'} - T_1} = \frac{543.9 - 300}{T_{2'} - 300}$$

$$\therefore \quad 0.85 = \frac{243.9}{T_{2'} - 300}$$

$$T_{2'} = 300 + \frac{243.9}{0.85} = 300 + 286.8$$

$$= \mathbf{586.8\ K}$$

$\therefore$ Compressor work/kg $= C_{pa}(T_{2'} - T_1)$

$$= 1 \times (586.8 - 300)$$

$$= \mathbf{286.8\ kJ/kg}$$

For gas turbine,

$$\eta_t = \frac{T_3 - T_{4'}}{T_3 - T_4} = \frac{1273 - T_{4'}}{1273 - 760}$$

$$\therefore \quad 0.9 = \frac{1273 - T_{4'}}{513}$$

$$T_{4'} = 1273 - 0.9 \times 513$$

$$= 1273 - 461.7$$

$$= \mathbf{811.3\ K}$$

From combustion chamber we have,

$$C_{pa} T_{2'} + m_{f1} \times \text{C.V.} \times \eta_{\text{comb}} = (1 + m_{f1}) \times C_{pg} T_3$$

$$1 \times 586.8 + (m_{f1} \times 40{,}000) \times 0.95 = (1 + m_{f1}) \times 1.1 \times 1273$$

$$= 1400.3\,(1 + m_{f1})$$

$$\therefore m_{f1}\,[(0.95 \times 40{,}000) - 1400.3] = 1400 - 3 - 586.8$$

$$= 813.5$$

$$m_{f1} = \text{mass of fuel/kg air}$$

$$= \frac{813.5}{36600}$$

$$= 0.0223\ \text{kg/kg air}$$

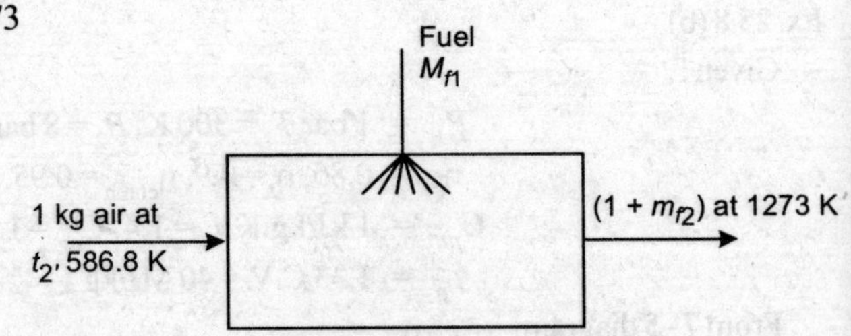

Fig. Ex. 25.8(c)

Net work/kg air $= W_t - W_c$

$$= (1 + m_{f1})\, C_{pg}\,(T_3 - T_{4'}) - C_{pa}\,(T_{2'} - T_1)$$

$$= (1 + 0.0223) \times 1.1\,(1273 - 811.3) - 1\,(586.8 - 300)$$

$$= 519.20 - 286.8$$

$$= \mathbf{232.40\ kJ/kg\ air}$$

Power developed by gas turbine unit is

$$0.6 \times 100 = 60\ \text{MW} = \mathbf{60{,}000\ kW}$$

$$\therefore \text{ Mass flow rate of air } = \frac{60{,}000}{232.40}$$

$$= \mathbf{258.17\ kg/sec}$$

$$\text{Heat supplied/sec} = m_{f1} \times \text{mass of air/sec} \times \text{C.V.}$$

$$= 0.0223 \times 258.17 \times 40000$$

$$= \mathbf{230292.6\ kJ/sec}$$

Now consider the supplementary combustion chamber

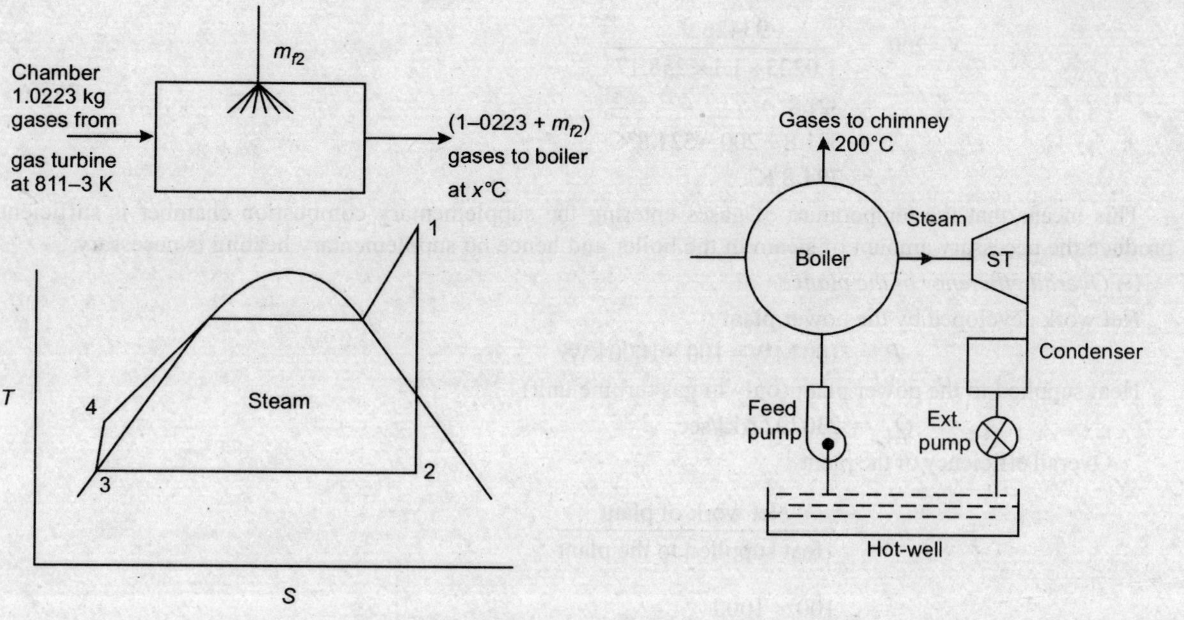

Fig. Ex. 25.8(d)

Given :

$$P_1 = 50 \text{ bar } T_1 = 600 + 273 = 873 \text{ K}$$

$$P_2 = 0.05 \text{ bar.}$$

From chart, $h_1 = 3660$ kJ/kg

$$h_2 = 2195 \text{ kJ/kg}$$

$$h_{f2} = 137.77 \text{ kJ/kg}$$

$\therefore$ Work done/kg steam $= h_1 - h_2$

$$= 3660 - 2195$$

$$= \mathbf{1465 \text{ kJ/kg}}$$

Power produced by steam turbine unit

$$= 0.4 \times 100 = 40 \text{ MW} = 40{,}000 \text{ kW}$$

$\therefore$ Mass of steam required

$$m_s = \frac{40{,}000}{1465}$$

$$= \mathbf{27.3 \text{ kg/sec}}$$

Heat required for producing 1 kg steam

$$= h_1 - h_{f2}$$

$$= 3660 - 137.77$$

$$= \mathbf{3422.23 \text{ kJ/kg}}$$

From supplementary combustion chamber, let X be the temperature of gases entering the boiler.

$\therefore$ Mass of gases/sec $\times C_{pg} (X - 200) = 27.3 \times 3422.23$

$$(258.17 \times 1.0223) \times 1.1\,(X-200) = 93426.9$$

$$X-200 = \frac{93426.9}{1.0223 \times 1.1 \times 258.17}$$

$$= 321.8$$

$$\therefore \quad X = 321.8 + 200 = \mathbf{521.8°C}$$

$$= \mathbf{794.8\ K}$$

This means that the temperature of gases entering the supplementary combustion chamber is sufficient to produce the necessary amount of steam in the boiler and hence no supplementary heating is necessary.

(i) *Overall efficiency of the plant*

Net work developed by the power plant

$$P = 100\ \text{MW} = 100 \times 1000\ \text{kW}.$$

Heat supplied to the power plant (only in gas turbine unit)

$$Q_A = 230292.6\ \text{kJ/sec.}$$

∴ Overall efficiency of the plant

$$h_o = \frac{\text{Net work of plant}}{\text{Heat supplied to the plant}}$$

$$= \frac{100 \times 1000}{230292.6}$$

$$= \mathbf{0.43423}$$

$$= \mathbf{43.423\%}$$

(ii) *Air flow rate to steam flow rate ratio*

m_s = Mass of steam flow = **27.3 kg/sec**

m_a = Mass of air flow = **258.17 kg/sec**

$$\therefore \quad \text{Air to steam ratio} = \frac{A}{S} = \frac{m_a}{m_s}$$

$$= \frac{258.17}{27.3}$$

$$= \mathbf{9.46}$$

(iii) *Cooling water required*

Enthalpy of 1 kg steam entering the condenser is $h_2 = 2195$ kJ/kg.

Enthalpy of 1 kg condensate coming out from condenser is $h_2f = 137.77$

∴ Total heat to be removed from steam by cooling water is

$$= m_s \times (h_2 - h_2f)$$

$$= 27.3\,(2195 - 137.77)$$

$$= 56162.38\ \text{kJ/sec}$$

This heat is given to cooling water and is given by

$$Q = m_w C_{pw} \times \text{Rise in temperature}$$

$$= m_w \times 4.187 \times 10$$

$$\therefore \quad m_w = \text{mass of water/sec}$$

$$= \frac{2057.23 \times 27.3}{4.187 \times 10}$$

$= \mathbf{1341.35\ kg/s}$

$= \mathbf{4828.86\ T/hour}$

Example 25.9 *In a cogeneration plant, the power load is 56 MW and the heating load is 1.163 MW. Steam is generated at 40 bar and 500°C and is expanded isentropically through a turbine to a condenser at 0.06 bar. The heating load is supplied by extracting steam from the turbine at 2 bar, which is condensed, in the heating process, to saturated liquid at 2 bar and then pumped back to the boiler. Compute (a) the steam generation capacity of the boiler in t/hour, (b) the heat input to the boiler in kW, (c) the fuel burning rate of the boiler in t/hour if a coal of calorific value 25 MJ/kg is burned and the boiler efficiency is 88%, (d) the heat rejected to the condenser, (e) the rate of flow of cooling water in the condenser if the temperature rise of water is 6°C. Neglect pump work.*

Solution

T–S diagram for the problem given is shown in Fig. Ex. 25.9.

Given:

P_1 = 40 bar

t_1 = 500°C

P_2 = 2 bar

P_3 = 0.06 bar

From chart

h_1 = 3435 kJ/kg

h_2 = 2680 kJ/kg

h_3 = 2162 kJ/kg

$h_{f2} = h_5 = 504.7$ kJ/kg

$h_{f3} = h_4 = 151.5$ kJ/kg

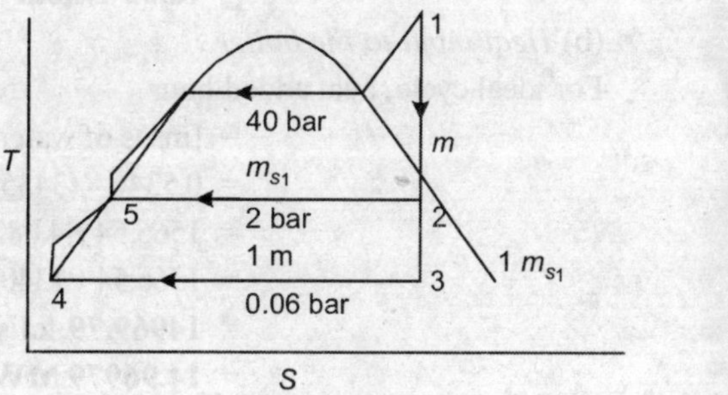

Fig. Ex. 25.9

Consider 1 kg steam admitted to the steam turbine.

Refer *T–S* diagram,

From 1 to 2, steam flow through the turbine is *m* kg and let m_{s1} be the mass of steam extracted from the turbine for heating process.

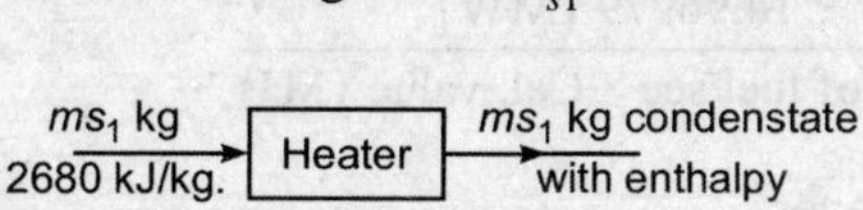

Fig. Ex. 25.9 (a)

$h_S = h_{f_2} = 504.7$ kJ/kg

Heat available for heating/kg = 2680 – 504.7

= **2175.3 kJ/kg**

Total heating load = 1.163 MW

= **1163 kW = kJ/sec**

∴ Steam required for heating purposes

$$m_{s1} = \frac{1163}{2175.3}$$

= **0.5346 kg/sec**

$$\begin{aligned}\text{Power load} &= 5.6\times 10^3\,\text{kW}\\ &= m\,(h_1-h_2)+(1-m_{s1})(h_2-h_3)\\ &= m\,(3435-2680)+(1-0.5346)(2680-2162)\\ &= 755\,m+(m-0.5346)\times 518\\ &= 755\,m+518\,m-276.92\\ &= 1273\,m-276.92\end{aligned}$$

$$\therefore \quad m = \frac{5600+276.92}{1273}=\frac{5876.92}{1273}$$

$$= \mathbf{4.6166\ kg/sec}$$

(a) *Boiler capacity*

$$\begin{aligned}\text{Steam generated/sec} &= m = \mathbf{4.6166\ kg/s}\\ &= 16619.76\ \text{kg/hour}\\ &= \mathbf{16.61976\ T/hour}\\ &\simeq \mathbf{16.62\ T/hour}\end{aligned}$$

(b) *Heat input to the boiler*

For ideal cycle, heat added/hour

$$\begin{aligned}&= [\text{mass of water}\times(h_1-h_{f2})]+(m-m_{s1})\times(h_1-h_{f3})\\ &= 0.5346\times(3435-504.7)+(4.6166-0.5346)\times(3435-151.5)\\ &= 1566.54+4.082\times 3283.5\\ &= 1566.54+13403.25\\ &= \mathbf{14969.79\ kJ/sec = kW}\\ &= \mathbf{14.96979\ MW}\end{aligned}$$

(c) *Fuel consumption*

$$\text{Boiler efficiency} = \frac{\text{Heat supplied}}{\text{Mass of fuel}\times\text{Cal. value}}$$

$$0.88 = \frac{14.96979\ (\text{MW})}{\text{Mass of fuel/sec}\times\text{Cal. value}\ (\text{MJ})}$$

$$= \frac{14.96979}{\frac{m_f}{\text{sec}}\times 25}$$

$$\therefore \quad m_f/\text{sec} = \frac{14.96979}{0.88\times 25}$$

$$\begin{aligned}&= 0.680\ \text{kg/sec}\\ &= \mathbf{2448\ kg/hour}\\ &= \mathbf{2.448\ ton/hour}\end{aligned}$$

(d) *Heat rejected to condenser*

$$\text{Heat rejected to condenser/sec} = \frac{\text{mass of steam}}{\text{sec}}\times(h_3-h_4)$$

$$= (4.6166 - 0.5346)(2162 - 151.5)$$
$$= 4.082 \times 2010.5$$
$$= \mathbf{8206.86\ kJ/sec}$$
$$= \mathbf{8.20686\ MJ/sec}$$
$$\simeq \mathbf{8.21\ MJ/sec}$$

(e) *Cooling water flow through condenser*

Let m_w = mass of cooling water/sec.

Then, heat rejected by steam = heat taken by water

$$\therefore \quad 8206.86 = m_w C_{pw} \times \text{rise in temperature}$$
$$= m_w \times 4.187 \times 6$$
$$\therefore \quad m_w = \frac{8206.86}{4.187 \times 6}$$
$$= 326.68 \text{ kg/sec}$$
$$= \mathbf{0.32668\ m^3/sec}$$

THEORY QUESTIONS

25.1. What are the major advantages of a combined cycle operation of power plants?
25.2. What are the desirable properties of working fluid for vapour power cycle?
23.3. Explain with the help of a neat sketch the working of Binary vapour cycle.
25.4. What are the advantages and disadvantages of mercury as a fluid in binary vapour cycle?
25.5. What do you understand by 'cogeneration'?
25.6. What are topping and bottoming cycles used in combined cycles?
25.7. What is tertiary cycle? Explain, what fluids are used in such cycle.
25.8. What do you understand by 'supplementary firing'? Where it is used? Explain.
25.9. What are the sources of gas for gas turbine cycle plant?
25.10. Explain with a neat sketch the working of combined cycle power plant using gasifier.
25.11. What is Fluidized Bed Combustion system?
25.12. Write a short note on PFBC system and PCFBC system.
25.13. Explain the working of combined cycle with PCFBC system.
25.14. Write in brief about liquid fuels in gas turbines.
25.15. Explain the combine cycle using Brayton cycle and Rankine cycle.
25.16. What is repowering of existing steam power plant? Explain.
25.17. Write short notes on the following
(a) Thermoelectric Power Generation
(b) Thermionic Power Generation
(c) MHD Power Generation
25.18. What are the parameters considered to study the performance of the gas turbine plant and combined power plant? Explain.

Chapter 26 Gas Turbines with Recovery Boiler and Other Systems

26.1 INTRODUCTION

It is experienced nowadays that the cost of the fossil fuels is increasing day by day. In addition to this, availability of the fossil fuels is decreasing. All these facts regarding the fossil fuels lead to thinking and acting to save the fuels as far as possible. The consumption of fuels can be reduced if the heat energy wastage is minimised and also combustion equipment is efficient one. Minimisation of heat wastage can be achieved by utilising all the wastage sources efficiently. Waste heat is typically in the form of unburnt combustible materials, also as sensible and latent heat exhausted through chimneys or stacks. Therefore, it is very much important to minimise the waste heat from all sources. Even then the waste heat energy recovery is of utmost importance.

Recovery and utilisation of waste heat exhausted from industrial processes and combustion equipment leads the improvement of the thermal efficiency of the system. Waste heat can be utilised in heat air, water, to generate steam if sufficient heat is available. It is generally observed that when the temperature of the waste gases exceeds 300°C, steam generation becomes most economical method of heat recovery.

At present, all waste heat recovery systems are of non-storage type but it will be a significant achievement in the waste heat management if the waste heat recovery systems become storage type. In this case, storage of heat may act as an accumulator which will store excess waste heat and will give out when it is required. At present majority of the waste heat energy is exhausted to atmosphere or environment and poses a major problem for ecological concern.

Generally, all the available waste heat energy appears as low temperature heat and the term **utilisation of waste heat** refers to the performance of useful functions before heat being discharged to environment.

26.2 SOURCES OF WASTE HEAT ENERGY AND THEIR GRADES

Waste heat energy can be defined as the heat energy which is not used and is exhausted out as a **waste product**. Sources of waste heat energy can be listed as :

(a) Industries using process heat in the form of steam and water at high temperature.

(b) Heat discharged as exhaust gases from industries.

(c) Dirty gases.

Waste heat may be in the form of :

(i) Sensible heat

(ii) Latent heat.

Hot gases and hot water carry sensible heat and similarly waste latent heat will be exhausted by the industry using process steam.

Grading of such waste heat energy is according to the temperature of waste products. Accordingly, the grades of this waste heat energy are :

(a) Low grade energy

(b) High grade energy.

Waste heat energy is considered **high grade energy** when the temperature of the exhaust or waste substances is 300°C and more and naturally waste heat energy is considered low grade energy when the temperature is below 300°C. Generally, it is not economical to recover low grade energy. The unit of waste heat energy is **quad** which is given as :

$$1 \text{ quad} = 10^{15} \text{ BThU}$$
$$= \mathbf{293 \times 10^{9} \ kWh}$$

In the past, dirty gases were not considered as the source of waste heat energy due to the uneconomical processes of cleaning related to erosion and corrosion problems. But at present, because of the scarcity of fuels and high costs, these gases have become the source of heat energy for industries like cement, iron, paper, glass and ceramic industries.

Flue gases and grain water are another sources of low grade waste heat energy. Low grade heat sources and recovery are commonly found in many industrial plants like aluminium, food processing and chemical industries.

26.3 THERMODYNAMIC CYCLES USED FOR RECOVERY OF HEAT

To use the waste heat energy efficiently, it is necessary to get this energy from the source and then it can be used for power generation. For this purpose, the heat from waste heat energy source is to be used in a thermodynamic cycle so that power is produced. That is to say that waste heat from the source has to be bottomed to the maximum extent and then use this energy for producing power most efficiently. The engines or for that purposes turbines should work efficiently for the temperatures encountered. In majority cases this temperature will not be much higher and hence the cycle efficiency should be optimum as far as possible.

Rankine cycle is one of the thermodynamic cycles which has relatively high thermal efficiency at low temperature and hence Rankine cycle is a unique cycle for the low level temperature heat source. At the same time, the extraction of the heat energy from waste heat source will depend on the temperature of the source and the temperature of the cooling medium or the ambient temperature.

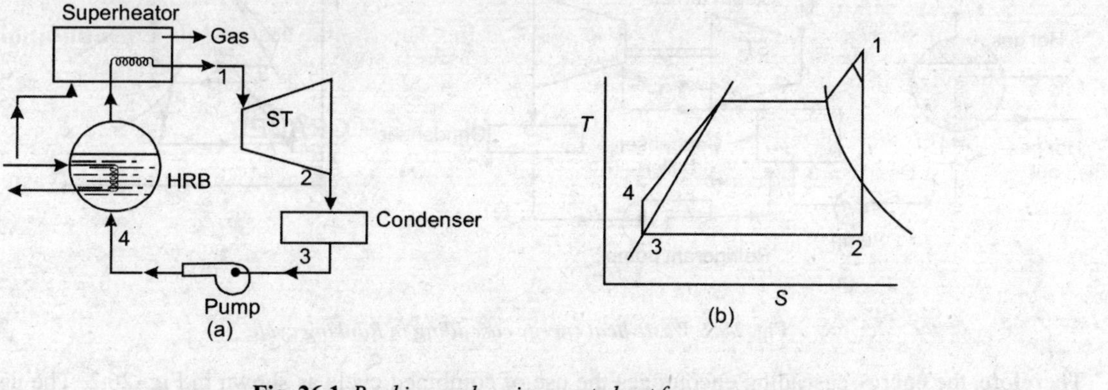

Fig. 26.1 *Rankine cycle using waste heat from gas.*

Consider, now, the internal combustion engine in which chemical energy of the fuel is converted into heat and the mechanical work or power is produced. The efficiency of the engine generally ranges from 30 to 35% and the heat loss from the engine may be considered as :

25–30% in cooling water,

30–35% in exhaust gases.

The heat energy in exhaust gases is exhausted — in normal course — to atmosphere. This energy can be utilised in Rankine cycle as shown in Fig. 26.1, or the waste heat can be used in another system or engine. This second engine cycle is operating on the waste heat from the first engine and the cycle of the second engine is called the bottoming engine or cycle. Thus, in general, any heat engine run by the rejected heat from another engine is called a bottoming cycle.

In addition to Rankine cycle, **stirling cycle** and **absorption cycle** can be considered for using the waste heat in the temperature range of 100–400°C. In absorption refrigeration cycle, only heating and cooling effects are achieved, while stirling engine is yet to be developed for efficient and practical working. Therefore **Rankine bottoming cycle of high efficiency is the only practical cycle.**

Good features of Rankine cycle are :

1. Wide range of temperatures of source and sink.
2. Working fluid can be selected according to the temperatures available.
3. High ratio of cycle work to pump work.

The use of Rankine cycle using waste heat of hot gases is shown in Fig. 26.1. Temperature of the gases in this example is on the higher side of the waste heat energy source, so that we can get moderate pressure and temperature of steam for power generation.

In the lower temperature range, we can use a refrigerant with a low boiling point and small latent heat of vaporisation. For this, the working fluid may be F-12 or R-12 which has lower boiling point than ammonia (NH_3) and also the latent heat of vaporisation is considerably less than that of NH_3. When refrigerant is used as a working fluid in Rankine cycle, sufficient pressure and temperature will be achieved for efficient working of the Rankine cycle. Thus the use of refrigerant as a working fluid in the Rankine cycle makes it possible the conversion of waste heat into work in the lower and medium temperature range.

Again if the high temperature source of waste heat is present, then the principles of energy cascading are receiving greater acceptance as energy costs escalate.

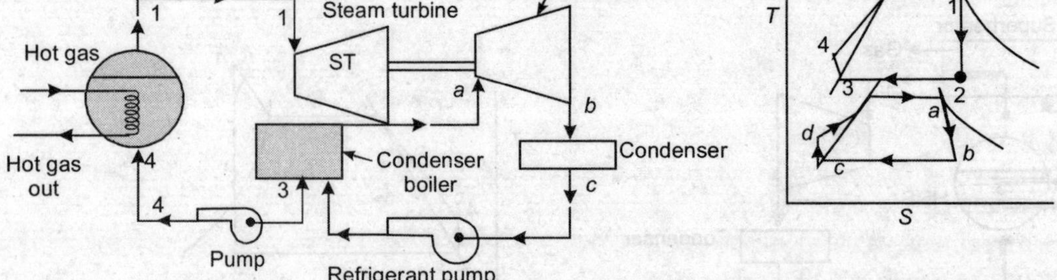

Fig. 26.2 *Waste heat energy cascading in Ranking cycle.*

Therefore, the energy cascading encourages the use of combined cycle as shown in Fig. 26.2. The use of two fluids permits each fluid to operate in a range where it is most effective and maximises the use of thermodynamically available work.

The fluids which are generally used in the bottoming cycle are :

(a) R-85 and (b) R-114.

The desirable properties of the fluids which are commonly used are :

1. It should not be toxic.
2. It should be non-flammable.
3. It should be readily available.
4. It should possess good thermodynamic properties.

Heat which is available from the waste heat source is in the form of sensible heat and latent heat. Generally, air and water are used for recovering sensible heat. In this case air may be used for heating the space in winter air conditioning or this heat air can be used for combustion of the fuels in other systems. The **heat exchangers** used for this purposes are generally **counterflow type**. The optimum heat extraction should be done by increasing area A and overall heat transfer coefficient U in such a way that the cost of extraction is minimum.

The recovery of the waste heat for power generation is through the waste heat recovery boilers where the latent heat is recovered and steam generated is used for power generation from turbines. Depending on the temperatures at which the recovery is effected, the pressure and temperature of steam is decided. The advantages for such recovery of latent heat are :

(a) When latent heat of the working fluid is high, the recovery of latent heat becomes attractive.

(b) The heat transfer coefficient of vaporising sides is very high as compared with air. (Nearly 50–100 times than that for air)

(c) Fluids like refrigerants with lower boiling points can be used to recover the heat from low temperature sources.

(d) For better heat transfer a low temperature difference can be maintained.

26.4 USES OF WASTE HEAT

In addition to the generation of power and process heating, there are many other uses of waste heating. In recent years, certain applications of waste heating are developed for very low temperature waste heat. A few of them are:

(a) Agricultural Use

(i) Spray irrigation

(ii) Soil heating

(iii) Environmental control of animal shelter and green houses are the present day applications where thermal discharges are used without affecting electrical power production. It is observed that the germination of seeds depends on the temperature at the seeding growth. Because of this there is a remarkable growth in the production of soyabean. Similarly the production of tomatoes and many others. Even hot water can be used for soil sterilization and certain diseases are retarded or cured.

(b) Green houses

(c) **Animal shelters.** Feed consumption economy is achieved.

(d) **Aqua cultural uses.** In the controlled environment, the fish species can be cultivated effectively and the growth of the yield can be effected by addition of dissolved oxygen.

(e) **Waste water treatment.** If the municipal and industrial waste water effluents are treated properly using waste heat, then such municipalities and industries can be looked upon as the secondary water sources.

26.5 WASTE HEAT AS SOURCE FOR HEAT PUMP SYSTEMS

If the waste heat is available at the atmospheric temperature and if the system is required to be heated, then its can be possible to use this waste heat source and with the help of compressor, the heat can be delivered to the system as shown in Fig. 26.3. In winter, the sink requires heating and heat pump can be used for this purpose. It is observed

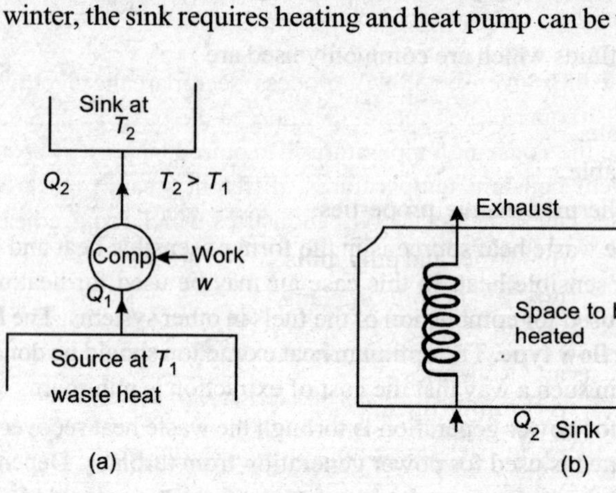

Fig. 26.3 *Heat pump using waste heat.*

that the heat pump will take away the waste heat from the source at low temperature T_1 and will deliver heat energy at the higher temperature as shown in Fig. 26.3.

26.6 POWER GENERATION AND WASTE HEAT

Following are the waste materials which can be used as fuel for generation :

(a) Municipal waste

(b) Industrial waste

(c) Paper waste

(d) Plastic and

(e) Rubber waste.

In old days these materials were dump and destroyed without thinking of its usefulness as fuel for power generation. By using these wastes for power generation, the main problem of dumping these waste can be solved.

There are various applications of this waste heats. They are :

1. Power generation
2. Steam generation for industrial process
3. Supplementary fuels in the boiler
4. Gas production.

The use of waste material as fuel involve the following points :

(a) Cost of preparation involves drying, separating and compacting and requires extensive preparation.

(b) It creates unusual pollution problems because the gases generated are different that SO_2, NO_x, the gases generated create problems of corrosion and health problems.

(c) The ash produced is highly sticky in nature particularly when plastic and rubber are used as fuels. Slag flow and clogging are another problems if proper temperature is not controlled.

Many municipal corporations in the country are thinking and acting in that direction to set-up waste heat power generation.

26.7 COGENERATION PLANTS

26.7.1 Introduction

The requirements of many industries in the power sector and 'process' sector are the electrical power as well as heat for processing purposes. When heat is required then the temperature is the basic requirement and it is found that steam is a very good medium because the constant temperature, if required, supply of heat can be obtained when the steam is condensing. For different constant temperatures, different steam pressures can be used so that constant temperature heating processes can be achieved. The industries which require heat for processing are :

(i) Cement factory	(ii) Paper and pulp mills
(iii) Chemical industries	(iv) Textile mills or plants
(v) Sugar factories	(vi) Rice mills
(vii) Fertilizers plants	(viii) Refineries
(ix) Food processing	(x) Edible oil industries.

In addition to these process heat, the industries do require power to run or drive the various machines, for lighting and other purposes.

In old days, steam was generated separately for power (at the high pressure and temperature) and for processing (at the required temperature pressure) purposes. Having two separate units for process heat and power purposes is wasteful.

The plants which produce or supply heat for the industrial applications, in addition to the main power generation, are called co-generation plants. Also, cogeneration is the combined production of two forms of useful energy from the same fuel. These two forms of energy are thermal and electrical. Generally it is observed that the exhaust from producing one kind of energy is used as input for producing the next or second form of energy.

In cogeneration plants, generally high pressure steam is used for electrical power generation and the exhaust steam is used in manufacturing processes. The efficiency of cogeneration system is as high as 75–85% whereas the efficiency of steam power plant is hardly 35–40%. This is because the exhaust from the power generation system is utilised for further useful purposes.

If, W_e = Power generated

E_a = Available heat energy for industrial or heating applications and

Q_A = Total heat supplied to the plant.

∴ Cogeneration plant efficiency is given by

$$\eta_{co} = \frac{\text{Useful energy}}{\text{Energy supplied}}$$

$$= \frac{W_e + E_a}{Q_A}.$$

26.7.2 Cogeneration Systems

Basically there are two types of cogeneration as :

1. The Topping Cycle
2. Bottoming Cycle

1. **The Topping Cycle.** In such a system, high temperature steam is used for generation of electrical power. The cycle is further carried out in two different ways :

(a) Use of back pressure turbine
(b) Use of extraction turbine.

26.7.3 Use of Back Pressure Turbine

Where steam is required for heating or industrial purposes, such steam is generally of low or comparatively low pressure and may be obtained from a low pressure boiler or through a reducing valve from a high pressure boiler. Where power also is required it is more economical to generate high pressure steam and use it in a turbine which instead of exhausting into a condenser exhaust at the pressure required for the steam to be used for heating or other purposes. The turbine is then called a **Back Pressure Turbine.**

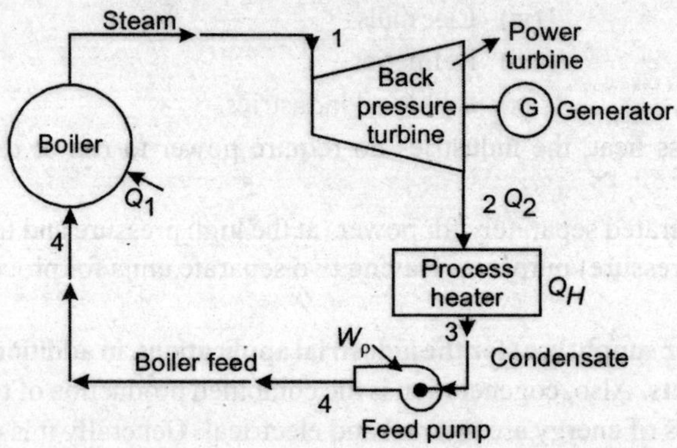

Fig. 26.4 *Back pressure turbine for cogeneration.*

Since the demand of steam for heating or other purposes is liable to be irregular while the supply for power may be more or less constant it follows that a back pressure turbine, which has no condenser, may in many cases become very wasteful. This has led to the introduction of the **reducing turbine**.

The **reducing turbine** or **the bleeder turbine** as it is sometimes called, is practically an ordinary condensing, multistage turbine in which provision is made for leading away some of the steam at an intermediate stage, the steam so take being used for heating or other purposes.

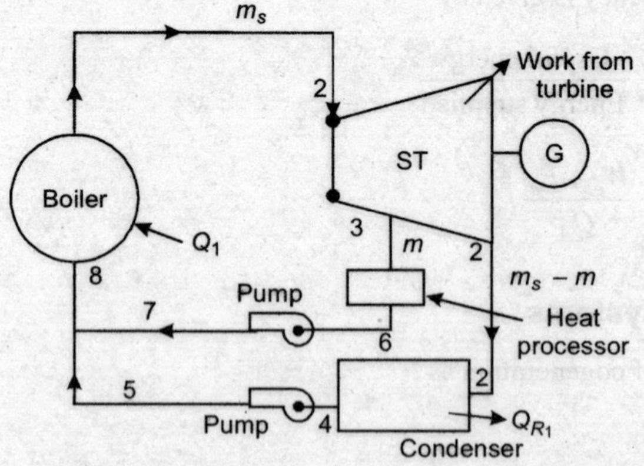

Fig. 26.5 *Cogeneration plant with a reducing turbine or passout turbine.*

In many cases it is possible that the basic need is the process steam and the power is produced as a by product. Such a cycle is often called a by product power cycle. Fig. 26.6 shows the T–S diagram for cogeneration plant with back pressure turbine.

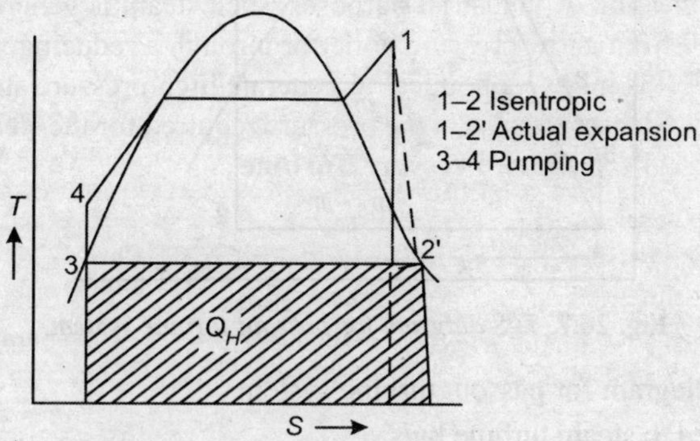

Fig. 26.6 *T-S diagram for back pressure turbine for cogeneration.*

Let, P = Power generated by turbine kW

m_s = Mass of steam kg/hour

$$\therefore \quad P = \frac{m_s(h_1 - h_2)}{3600}$$

or $$3600\,P = m_s(h_1 - h_2)$$

and $$\text{Process heat } Q_H = m_s(h_{2'} - h_3)$$

or $$m_s = \frac{Q_H}{(h_{2'} - h_3)}$$

and $$3600\,P = \frac{Q_H}{h_{2'} - h_3} \times (h_1 - h_{2'})$$

or $$Q_H = \text{Process heat kJ/hour}$$

$$= \frac{3600P \times h_{2'} - h_3}{h_1 - h_{2'}}$$

∴ Cogeneration plant efficiency is given by

$$\boxed{\eta_\omega = \frac{P + W_H}{Q_1}}$$

Generally, cogeneration plant is beneficial if the efficiency of cogeneration given above is greater than that of separate generation (separate electricity power and separate energy for heating).

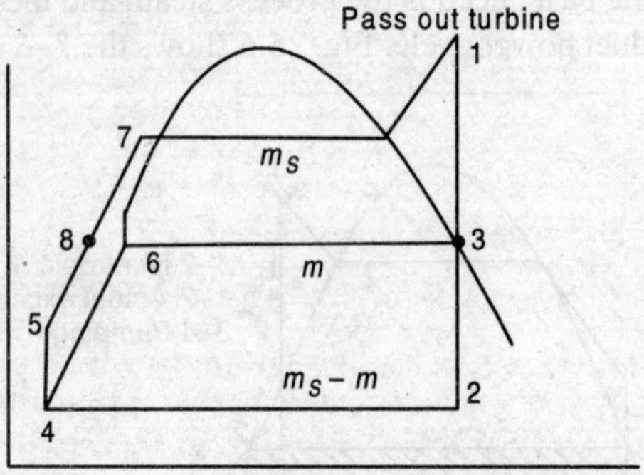

Fig. 26.7 *T–S diagram for passout turbine system.*

Figure 26.7 shows the *T–S* diagram for passout turbine system.

Let m_s = Total steam admitted to steam turbine kg/s

m = Mass of steam bled from turbine kg/s

$\therefore$ Q_1 = Heat given or supplied to boiler

$= m_s (h_1 - h_8)$

where

h_8 = Enthalpy of 1 kg water after mixing of m and $(m_s - m)$

= Enthalpy of water admitted to boiler.

Q_{R1} = Heat rejected in condenser

$= (m_s - m)(h_2 - h_4)$

Q_h = Heat given to proces-heater

$= m(h_3 - h_6)$

W_t = Work produced by turbine

$= m_s(h_1 - h_3) + (m_s - m)(h_3 - h_2)$

W_p = Pump work

$= (m_s - m)(h_5 - h_4) + m(h_7 - h_6)$

$\therefore$ $\boxed{\text{Net power} = W_t - W_p}$

To get state 8, we have

$$(m_s - m) h_5 + mh_7 = [(m_s - m) + m] h_8 = m_s h_8.$$

$\therefore$ Efficiency of this system,

$$\boxed{\eta_\omega = \frac{(\text{Net Power}) + Q_h}{m_s(h_1 - h_8)}}.$$

The main advantage of this cycle is its ability to use wide range of fuels including wood, coal, petroleum, coke, refuse etc. and is used where low quality refuse and byproduct fuels are readily available.

2. **Bottoming Cycle Plant.** In this cycle, first the primary heat at high temperature is directly used for process requirements, and the low temperature or low grade energy – waste heat from process plant is used to generate electrical power.

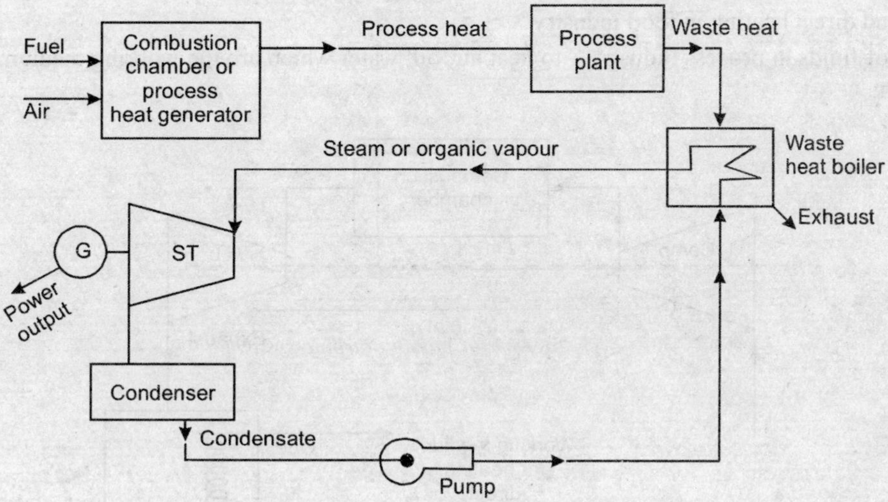

Fig. 26.8 *Cogeneration plant using steam or organic, bottoming cycle.*

The efficiency of this plant cycle is obviously low – 10% low and upto 20% at temperature applications. Given that the waste heat is free, the plant economics may be favourable.

The main applications of bottoming cycles are :

(a) Cement industry (b) Chemical industries

(c) Steel industry and (d) Limestone industry.

Since the temperature of waste heat is low, hydrocarbon fluids are often used as working medium.

Since the bottoming cycle has very low combined efficiency, it is of little thermodynamic or economic interest and only the topping plant, therefore, can provide true savings in primary energy. In addition, most process applications require low temperature and low availability steam which can be conveniently produced in a topping cycle.

Some of the additional arrangements for cogeneration in topping cycle are under :

1. Gas turbine power plant with a heat recovery boiler using gas turbine exhaust to generate steam for further use at the given temperature heating.

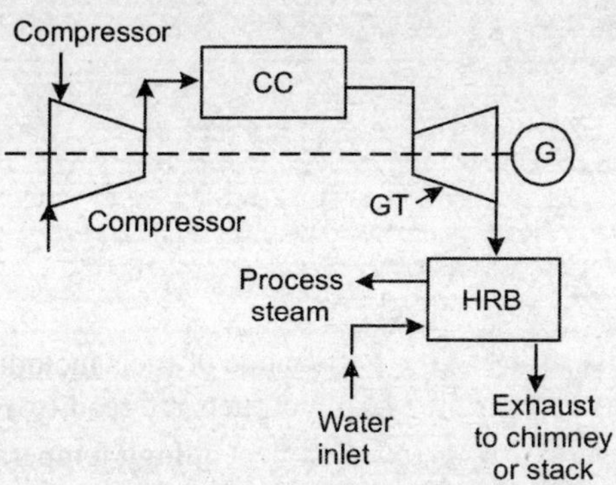

Fig. 26.9 *Cogeneration with gas turbine plant with heat recovery boiler (HRB).*

This arrangement is shown in Fig. 26.9 where steam generated in HRB is used for process industry.

2. Gas turbine cogeneration plant which generates electrical power and the exhaust gases are used for

(a) Drying and direct heating in food industry.

(b) Heating of fluids in process industry – to heat air, oil, water which are the working medium in the process industry or plant.

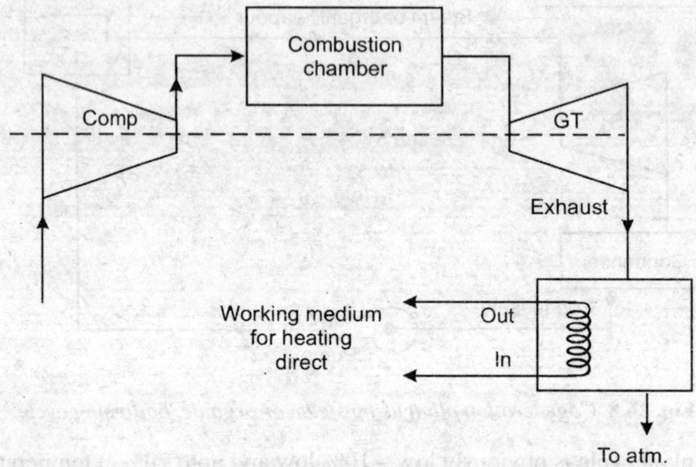

Fig. 26.10 *Cogeneration plant with gas turbine power plant.*

In this case, the exhaust gases from the gas turbine plant are used directly for heating air, water, oil etc. Fig. 26.7 shows such a system.

A simple gas turbine cycle consisting of an air compressor, combustion chamber and the gas turbine is used.

Chapter **27**

Non-conventional Power Generating Plants

27.1 INTRODUCTION

Since ages man has been trying to harness energy to get more power which can be used to do work. Energy appears in many forms but it has ability to produce dynamic and vital effect. Energy is always associated with physical substances but it is not a substance itself. Energy manifests itself by excited state of the matter which gets capacity to do work.

The use of electrical energy is increasing in most industrial countries at a rate between 10 to 15 percent per year i.e., almost doubling every six to seven years. This is due partly to increasing over consumption of primary energy and partly because electrical energy is tending to replace others froms on account of various advantages offered by it.

To support life adequately, a human being requires about 4 kWh per day in the form of food (animate energy). For providing heat, mechanical power and other needs (inimate energy) the total energy requirements are many times more. The standard of living of a nation can be easily correlated with the consumption of electrical energy per capita. The survival of industrial undertakings and our social structure depends primarily upon the low cost, continuous and unlimited supply of electrical energy. Electrical energy is preferred due to the following advantages :

(a) It can be easily transported from one place to another.

(b) Losses in transport are minimum.

(c) It can be easily controlled and regulated to suit requirements.

(d) It can be easily converted into other forms of energy e.g., mechanical, thermal, pneumatic etc.

(e) It can be easily subdivided.

(f) There is overall economy in using this form of energy.

Electrical energy is a manufactured commodity just like other items such as furniture, tools, clothing etc. As in other commodities raw material is processed to give the finished product, in this case also fuels, which are raw forms of energy, are used as raw material out of which energy is released normally in the form of heat which is converted into electrical energy through the medium of mechanical energy by the help of prime movers. So electrical energy is also sold like other commodities at so much per unit quantity.

27.2 CLASSIFICATION OF ENERGY SOURCES AND CORRESPONDING POWER PLANTS

27.2.1 Energy-Basis-Conventional/Non-conventional

Energy

- Conventional
  - 1. Fossile fuels (coal, coke) 2. Vegetation (wood, alcohol)
  - Animal waste (gobar)
  - Water or hydraulic energy
  - Nuclear energy (fission of nuclear fuels like U^{235}, Pu^{239})
- Non-Conventional
  - Solar energy
  - Wind energy
  - Ocean wave energy
  - Ocean thermal energy
  - Tidal energy

27.2.2 Energy-Basis-Renewable/Non-renewable

Energy

Renewable or Regenerative	Non-renewable or Non-regenerative
1. Vegetation	1. Fossile fuels
2. Hydraulic energy	2. Nuclear fuels
3. Solar energy	3. Geothermal energy
4. Wind energy	
5. Tidal energy	
6. Animal waste	
7. Ocean thermal energy	

27.2.3 Energy-Basis-Capital/Celestial Energy

Energy

Capital energy (Existing in or on earth)	Incoming or celestial energy (Reaching earth from outer space)
1. Fossile fuels	1. Solar energy
2. Nuclear fuels	2. Wind energy
3. Geothermal energy	3. Tidal energy
4. Vegetation	
5. Animal	

In this chapter we will describe the following non-conventional energy sources and the power plants –

1. Solar energy power plants
2. Wind energy power plants
3. Ocean wave energy power plants
4. Ocean thermal energy power plants
5. Geothermal or terrestrial energy power plants
6. Tidal energy power plants
7. Fuel cells.

27.3 SOLAR ENERGY (SOLAR POWER PLANTS)

A large energy source is available to us in the form of radiation coming from the sun, and considerable efforts are being made to find ways by which this energy can be utilized effectively. The sun radiates very nearly like a "black" circular disc with a temperature of nearly 6000 K in the wavelength range of 0.3 to 3.0 μm. Approximately one-half of the radiation occurs in the visible-wavelength range and the rest in the infrared upto 3 μm. Part of the solar radiation which is directed towards the earth is absorbed, reflected or refracted away by the atmosphere, and the rest reaches the surface of the earth. In a yearly average the earth absorbs approximately 43% of the radiation coming from the sun (27 percent directly and 16 percent as diffuse sky radiation), 42 percent is reflected or refracted back to the space from clouds and the air, and reflected from the earth's surface, and 15 percent is absorbed in the atmosphere.

The amount of solar radiation which impinges on a unit area of surface normal to the sun's radiation and located outside the atmosphere does not depend on the location on the earth or on the time of the day and is therefore often called the **solar constant**.

Absorption is caused by dust particles and in clean air by ozone and water vapour. On cloudy days the solar radiation drops to very low values (1 percent and less of the radiation on clear days).

At present the conventional energy sources like fossile fuels will be sooner or later –

(a) will be exhausted
(b) will affect the environment (pollution)
(c) will be more expensive and
(d) will have the political or economical control by the nations possessing it.

Solar energy has the advantage that it is readily available. In addition to this, it will not harm the environment. On the other hand,

1. Solar energy flux density is too small for the purpose of power generation and for many other purposes.
2. Solar energy reaching the earth is not easily convertible and for that matter is definitely not 'free'.
3. Solar energy reaching the earth is not constantly available.
4. Solar energy is not available during cloudy weather and at night.
5. For the required quantum, large area is required for the collection of solar energy.
6. Large storage is required so that the solar power plant system can work at night and other cloudy atmosphere.

Solar energy can be made available — by using different devices — at different temperatures as discussed below

27.3.1 Low Temperature Range (Upto 100°C)

(a) Water heating (such as solar water heaters) Domestic
(b) Space heating
(c) Space cooling
(d) Drying certain articles.

} For agricultural products

For these purposes, **flat plate collectors** are used.

27.3.2 Medium Temperature Range (Upto 100 to 200°C)

Applications in this range of temperatures are as follows.

1. Process heating (Industrial process)
2. Cooking
3. Refrigeration and (for preservation food products)
4. Vapour engines and turbines (low boiling temperatures organic vapours)

For these applications, **cylindrical parabolic solar collectors** are used.

27.3.3 High Temperature Range (Temperature T > 200°C)

Applications in this range of temperatures are as below

1. Thermoelectric generator
2. Steam engines and turbines and
3. Sterling engine.

For these applications, **Parabolloid mirror arrays** are used.

27.4 SOLAR COLLECTORS

27.4.1 Flat Plate Collectors

Flat plate collectors are used for temperatures ranging from 90°C to 100°C. Figure 27.1 shows such a Flat Plate Collector.

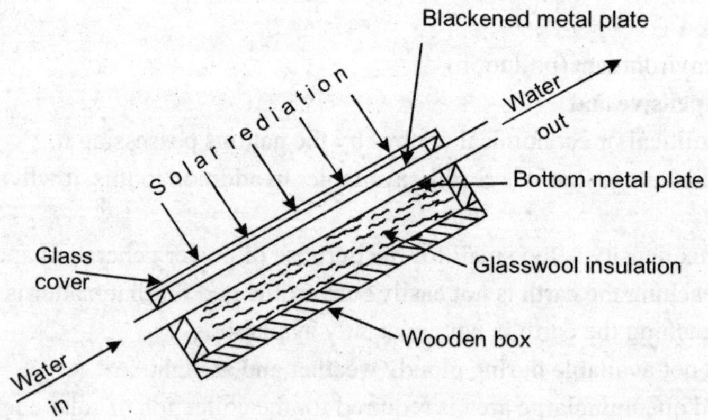

Fig. 27.1 *Flat plate collector.*

The main components of such a collector are —

1. Anabsorber plate (coated black)
2. Transparent glass cover
3. Insulation
4. Copper or identical metal tubes
5. Enclosure–box like structure

Generally absorber plate which is coated black, may be integral with the heat absorbing metallic tubes.

1. *Absorber Plate.* The plate is coated black because it effectively absorbs solar radiation falling on the outer surface (high absorptivity).

2. *Transparent Covers.* These plates may be one or more in number and are made of the material capable of transmitting solar radiation and are placed above the absorbing plate as shown in Fig. 27.1. Thus solar radiation is reaching the absorbing plate while convection, conduction and re-radiation losses are reduced.

3. *Insulation.* This minimises the heat losses.

4. *Box like Enclosure.* This contains the above components and keeps them in tact. The box is insulated at the bottom so that the loss of heat energy is reduced considerably. The collector as a whole system will absorb the sun — (solar) energy — which may be direct or diffused — and transfer it to the water or air or oil flowing through the tubes within the collector.

The flat plate collectors are popularly used for hot water requirements. These flat plate collectors are effective most of the time, and reliable for good many years.

Generally these flat plate collectors do not need tracking to face the sun.

For solar or sun radiation, if P is the intensity of sun radiation to which the body is exposed and if α is the absorption coefficient, ρ is the reflection coefficient and τ is the transmission coefficient then we can write

$$\alpha + \rho + T = 1$$

Again if the losses due to convection and conduction are negligible, then the absorbed solar radiations should be equal to the radiation loss then we can write

$$\alpha P = \varepsilon \delta T^4.$$

or $$\delta T^4 = \frac{\alpha P}{\varepsilon} \text{ with usual notations.}$$

From the above equation it is clear that higher equilibrium temperature will be obtained when $\frac{\alpha}{\varepsilon}$ ratio for flat plate is more. (Ratio of absorption coefficient to emission coefficient of the flat plate is more)

When this ratio is unity, then the collectors are called **Neutral Collectors** and when this ratio is greater than unity the collectors are called **Selective Collectors.**

In tropical regions, if the solar radiation intensity is $P = 800$ W/m^2, then we get maximum equilibrium temperature of 70°C.

(a) A typical version of flat plate collector system is shown in Fig. 27.2 and is self-explanatory.

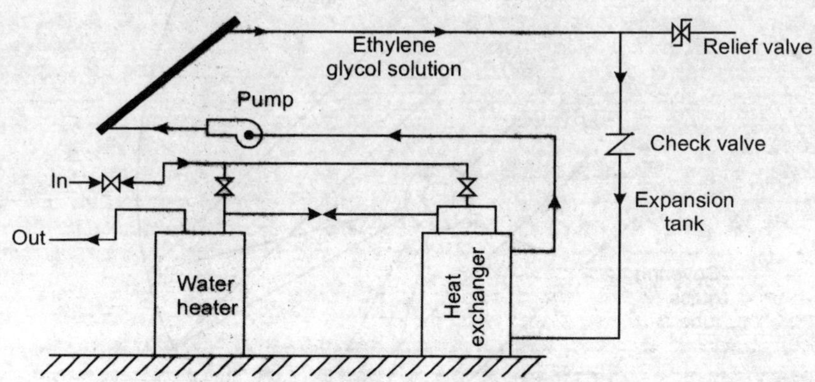

Fig. 27.2 *Low temperature power generation system*

(b) **Low Temperature Power Generation System.** For power generation purposes, fluids like Freon, ammonia are vapourised in the boiler by the water heated in the collector. The high pressure from vapour runs the turbine. The freon is then condensed and circulated in the system. Such systems are used upto 100 kW power generation for agricultural water pumping systems. This is shown in Fig. 27.3.

In this system the working substance may be **butane** vapours.

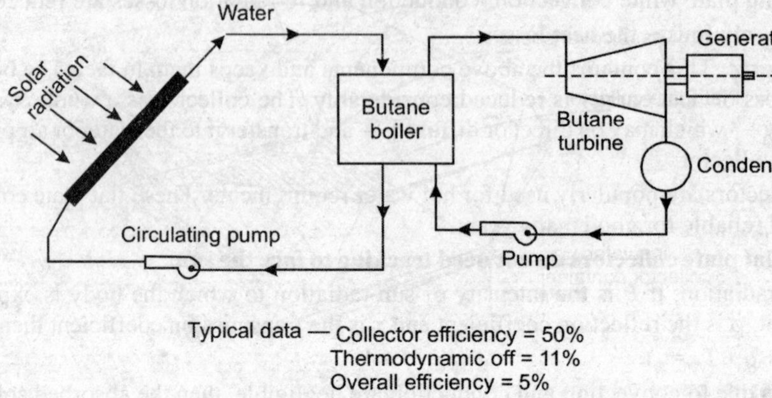

Fig. 27.3 *Low temperature system for power generation.*

27.4.2 Concentrating or Focussing Collectors

A circular or rectangular parabolic mirror can collect the radiation and focus it on to a small area, and for this a mechanism for moving the collector to follow the sun being necessary, such devices are used for metallurgical research where high purity and high temperatures are essential, an example being a 55 m diameter collector giving about 1 MW (th) at Mont Lou is in the Pyrenees. Smaller units having a 20 m diameter reflector can give temperatures of 300°C over an area of about 50 m^2. The collecting efficiency is about 50 percent. On a smaller scale, units about 1 m diameter giving temperatures of about 300°C have been used for cooling in India and other developing countries.

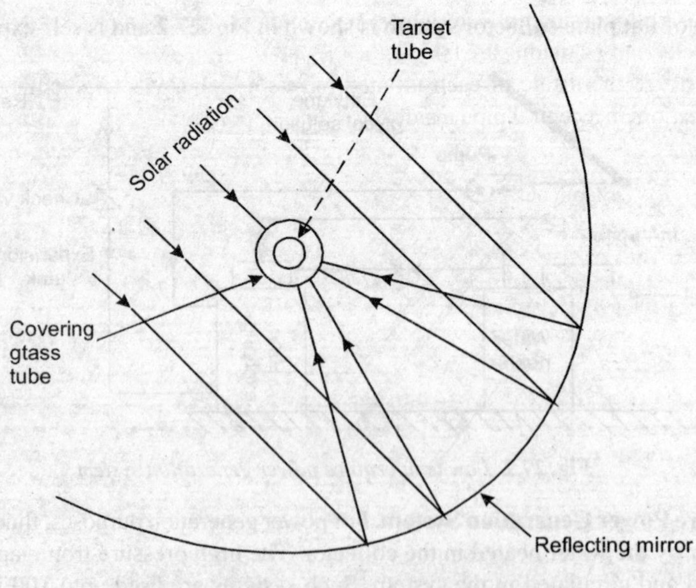

Fig. 27.4 (a) *Focussing solar collector.*

Figure 27.4 shows a focussing solar collector.

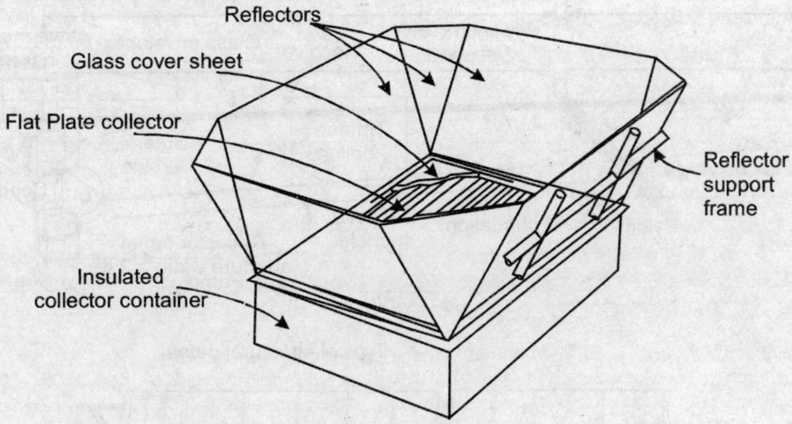

Fig. 27.4 (a) *Concentrating flat plate collector.*

Focussing of sunlight is usually achieved by brightly polished reflectors in the form of cylindrical parabola or paraboloid. In the first type, the energy receiving surface may be the blackened outside of a pipe, placed along the focal line of reflectors. If placed with the axis in east–west direction, their reflectors need not be continuously moved to follow the sun, but must be turned atleast once a week to have the central linear target receiver strip always in focus. They are quite effective upto 150°C.

The second type of focussing collectors, with their reflectors in the shape of parabola of revolution, usually have a small circular plate — coated back at its bottom — for their target. They can easily attain temperatures like 500°C. They must be arranged to continuously track the sun and are limited accordingly to small sizes, usually not much over six feet (upto 2 m) in diameter. Focussing solar collectors are usually expensive and cannot utilize that part of solar radiation scattered by atmosphere.

Relatively inexpensive focussing collectors have been made by Tabor of Israel. Cylindrical bags of transparent plastics 1.5 m in diameter and 12 m long are laid on the ground in an east-west direction and adjusted for the change of seasons. The lower part of the inside of each inflated plastic cylinder has an aluminized surface that reflects and focusses the solar radiation on a central pipe inside the cylinder, producing steam under pressure at 150°C in the pipe.

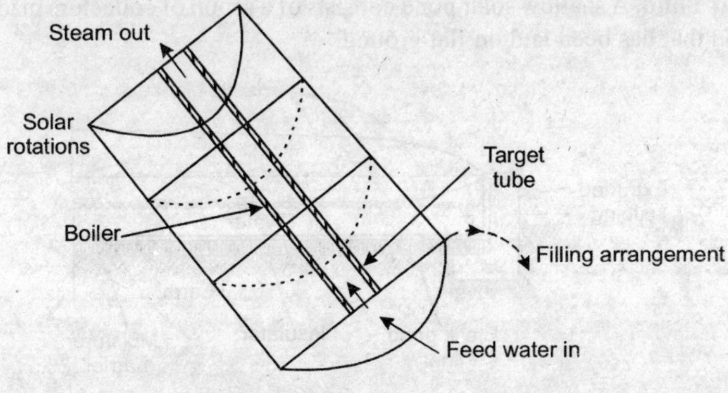

Fig. 27.5 (a) *Boiler at the focal point of a concave solar energy collector.*

Figure 27.5 shows such a concave solar energy collector focussing sun's rays on boiler at focal point.

Figure 27.5 (b) shows the parabolic through system.

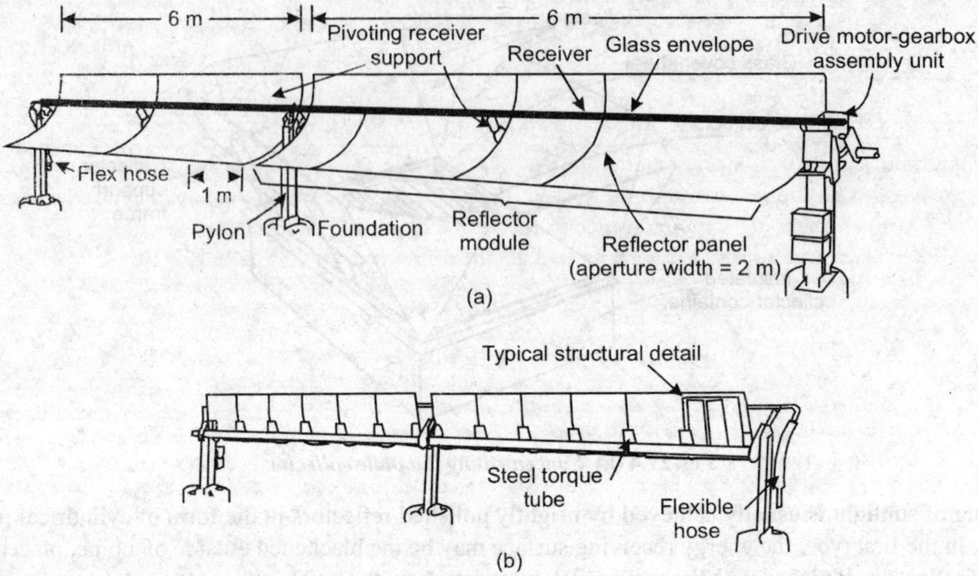

Fig. 27.5 (b)

27.4.2 Solar Ponds

In addition to flat solar collectors and focussing types of solar collectors, there is one least expensive type of solar collector and that is a large **solar pond**.

Basically or primarily these solar ponds were considered for large industrial applications as the cost of the pond decreases considerably with increase in size. Solar ponds are basically of two types :

1. The Shallow pond, and
2. The Salt-gradient pond.

These ponds are equivalent to horizontal flat plate collectors, because they are non-concentrating, accepting both direct and diffuse solar radiations. These ponds are limited to low temperature applications.

1. Shallow Type Solar Pond. A shallow solar pond consists of a group of collectors made of black plastic linears lying on top of insulation that has been laid on flat ground.

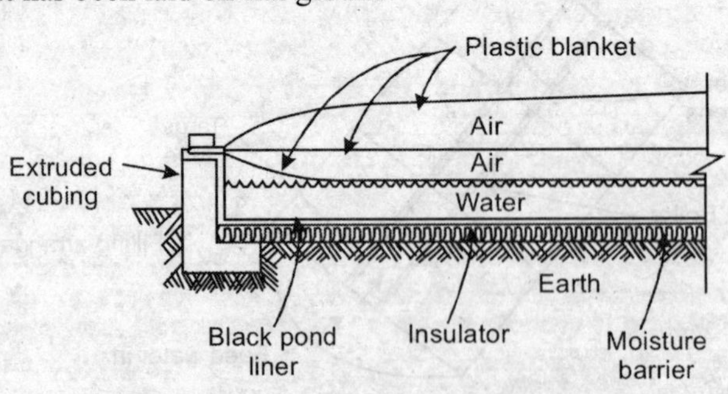

Fig. 27.6 *A shallow solar pond collector.*

Above the water bag is at least one translucent cover sheet, supported by side curbs. When there is adequate sunshine, water is pumped into the collectors from an underground storage tank. The water inside the collectors is heated by solar radiation absorbed by the pond linear and can attain temperatures of upto 60°C. Once heated, the water in the collectors may be pumped to an industrial demand or a hot storage tank for later use. At night or during periods of low insolation, the water in the collectors may be drained back into the underground tank, thereby conserving the heat collected. Mr. Casamajor and Parsons (1979) has design guide for shallow ponds.

Figure 27.6 shows a shallow pond collector design proposed by Dickenson et. al. (1976) for use in a low temperature solar thermal power system. The size of the pond is 4 m wide, 200 m long and 1 m height (deep).

2. Salt-Gradient Ponds. In nature, when an open body of water absorbs solar energy, convection currents are created. As the sun's rays that pass through the surface layers are absorbed in lower layers, this water is heated and rises to the surface, where heat is transferred by convection to ambient air. Once the water cools, the density increases and the surface water moves downwards. This movement of water equalizes the temperature throughout the body of water.

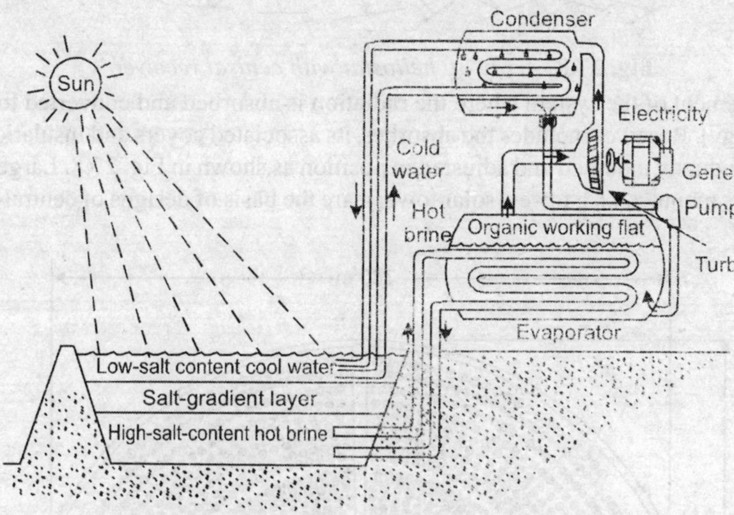

Fig. 27.7 *Electrical power production concept using salt-gradient ponds.*

A salt-gradient solar pond employs a salt concentration gradient to suppress natural convection. Heated water holds more dissolved salt than does cooler water. The salty, heated water is also heavier and thus remains at the bottom of the solar pond. Sunlight penetrating through top layers of the pond is absorbed at the bottom and trapped by the non-convecting gradient layer, which acts as an effective thermal insulator against convection.

In practice, salt-gradient solar ponds consists of three layers, or zones as – (1) a surface convecting zone of low-salinity water, typically 0.2–0.4 m thick; (2) a non-convecting or salinity gradient zone in which salt concentration increases with depth, typically 1.0–1.5 m thick; and (3) a storage zone at the bottom of the pond of uniformly high salt concentration that stores heat and is typically 1–3 m thick.

For utilization of the heat stored at the bottom of the pond, hot brine is drawn from the storage zone (bottom layer) of the pond and pumped through a heat exchanger and back to the bottom of the storage zone. For power production applications where a Rankine cycle is used, condenser cooling water is drawn off the top of the pond and passed through condenser and back to the surface, where it cools.

A typical arrangement for such cycle is shown in Fig. 27.7.

Pioneering work on salt-gradient solar ponds has been done in Israel.

Applications*. The largest solar ponds have been built in Israel. These are in the Dead sea area and are used for

* The physics of the solar ponds by Weinberger. H (1964) Solar Energy 8 (2) 45.

* Solar ponds by Tabor (1981) Solar Energy 27 (3) 181.

power production. These ponds are – (1) 6250 m^2 pond at En Boqek – 150 kW power plant. (2) 40,000 m^2 and 210,000 m^2 together supply energy for 5 MW Rankine cycle peaking plant.

27.4.3 High Temperature Systems

For high temperature systems (temperatures greater than 200°C), paraboloid mirror-arrays are used. For such an assembly of reflecting mirrors, a central receiver system is used. **Heliostats** are the movable mirrors to which sun's rays hit and the rays are reflected to a single point in space (the receiver).

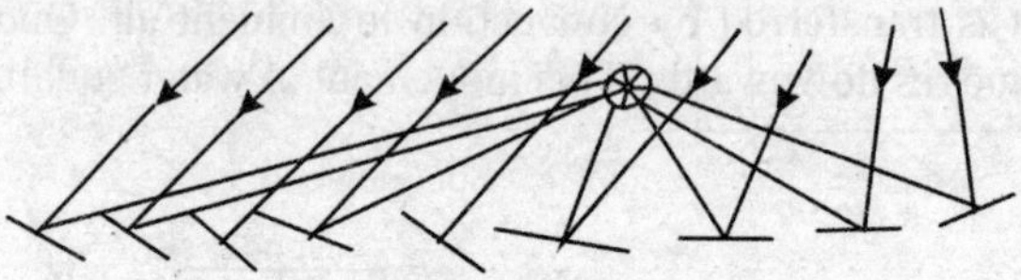

Fig. 27.8 *Array of heliostat with central receiver.*

The receiver is that element of the system where the radiation is absorbed and converted to some other energy form (generally heat energy). Receiver includes the absorber, its associated covers and insulation. The facets of the reflector can also be individually mounted and adjusted in position as shown in Fig. 27.8. Large arrays of heliostats of this type, with receivers mounted on a tower (solar tower), are the basis of designs of central-receiver collectors.

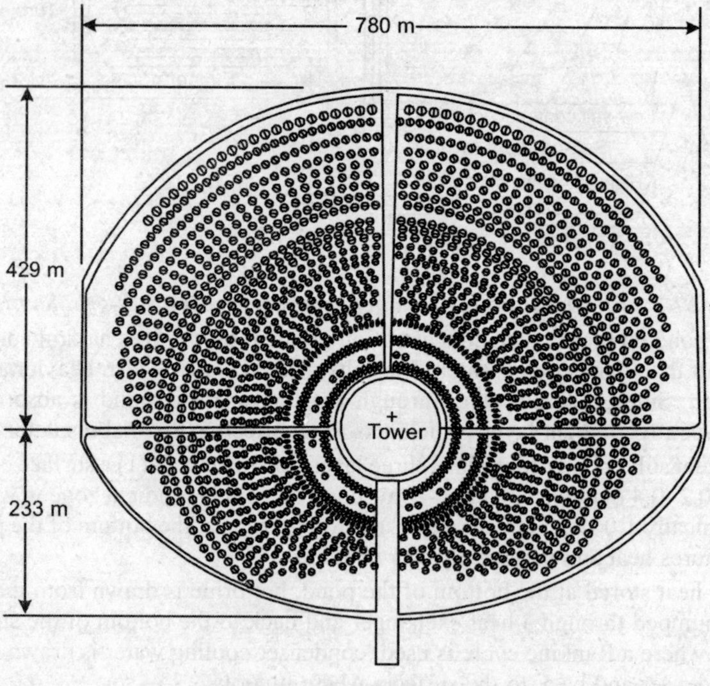

Fig. 27.9 (a) *Power plant with heliostat.*

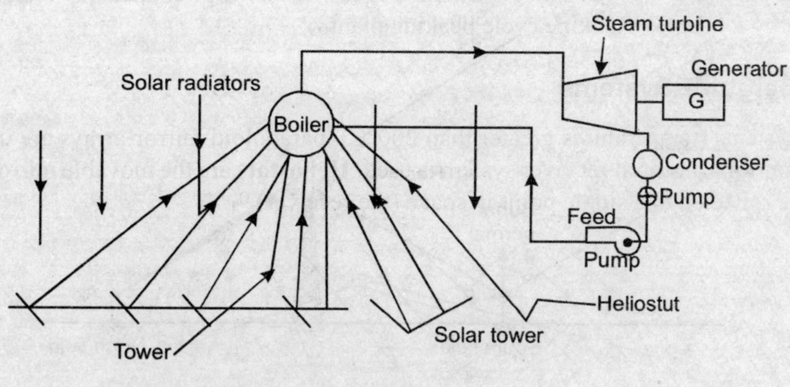

Fig. 27.9 (b) *Solar tower power plant with heliostat.*

Figure 27.9 (b) shows a solar tower power plant.

Figure 27.10 shows the back-side of the heliostat used in a solar tower power plant. Figure 27.11 shows the receiver for the same power plant.

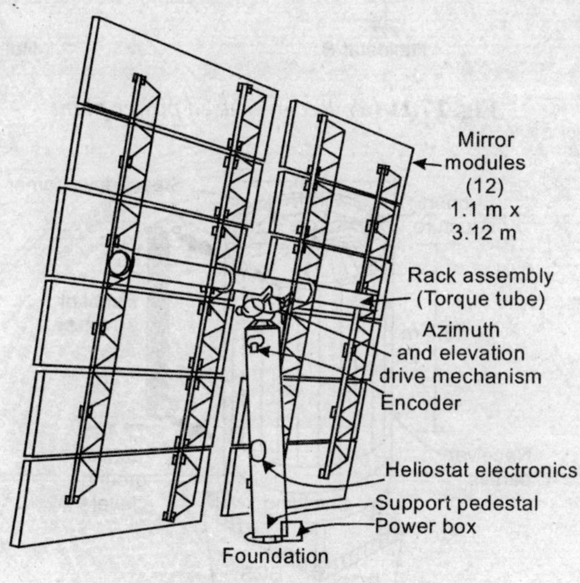

Fig. 27.10 *Back-side of heliostat.*

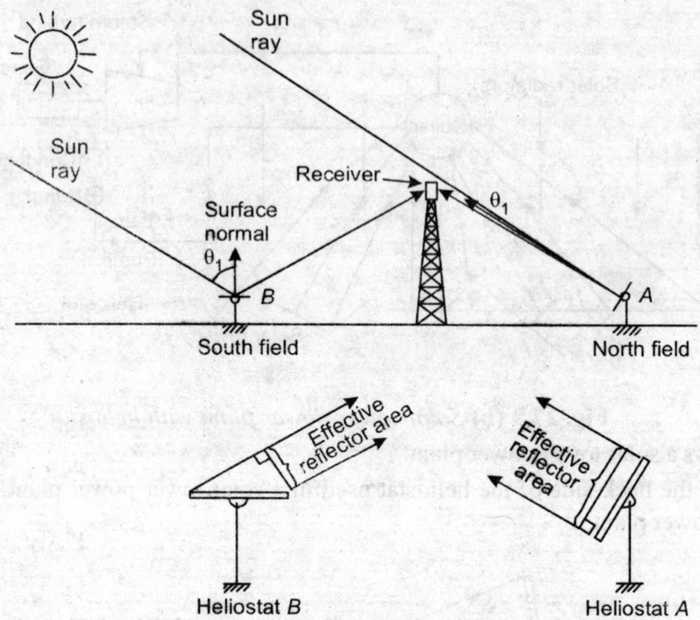

Fig. 27.11 (a) *Receiver used power plant.*

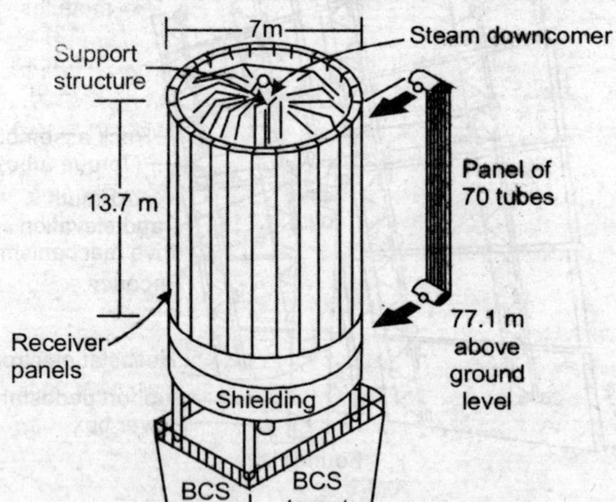

Fig. 27.11 (b) *Central receiver for power plant.*

27.4.4 Photovoltaic Conversion System for Power Generation

High temperatures are required for efficient conversion of heat into work (Carnot cycle), but heat losses in the solar energy collectors become serious at such temperatures. This problem can be avoided by using solar radiation (Photons) directly for production of electricity in a device known as Photovoltaic Cell.

Photovoltaic converters are semiconductor devices that convert part of the incident solar radiation directly into electrical energy. Selinium cells have been used for almost nine decades. Until 1953, an efficiency of 1 was believed to be the maximum attainable. Within a couple of years came the **silicon cells**, and new efficiency of conversion of solar radiation into electricity ranges from 10 to 15%.

Semiconductors are the sensitive elements in a photovoltaic cell. Figure 27.12 shows a photovoltaic cell – selinium cell.

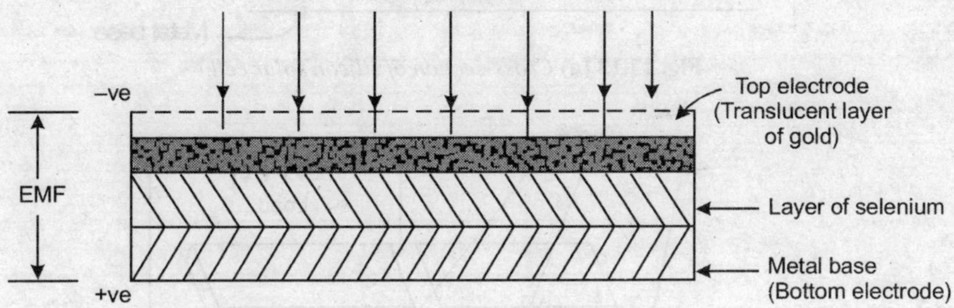

Fig. 27.12 *Selinium cell.*

27.4.4.1 Silicon Solar Cells

Semiconductor has two types of charge carriers, electrons and holes. The electrons carry negative charge while holes carry positive charge. In an intrinsic semiconductor electron and hole concentrations are equal. The conductivity of semiconductors can be modified by introducing impurities. In silicon which is IX group material in the periodic table, phosphorous atmos will make silicon N-type and is rich in electrons. Electrons are in thermal equilibrium and holes are in minority. In thermal equilibrium,

$$\eta_{no} \times p_{no} = n^2{}_1.$$

where η_{no} – electrons of *N* type,

p_{no} = holes in *N* type,

n_1 = Intrinsic carrier concentration.

Silicon is made N-type by doping with phosphorous and P-type by doping it with Boron.

When light is incident on a semiconductor, the photons impart energy to electrons bound in lattice structure. Electrons may get energy from the photon to come out of the bond and make available as free electrons. The vacancy in the bond gives rise to holes. Thus due to the incident radiations electronhole pairs are generated. These normally disappear spontaneously as electrons recombine with holes. The recombination process can be reduced by building into the cells a potential barrier, a thin layer or junction across which a static charge exists. This barrier is created by doping the silicon on one side of the barrier with very small amounts (of the order of one part in 10^6) of boron to form *p*-silicon, which has a deficiency of electrons in its outer shell, and that of the other side with phosphorous to form *n*-silicon, which has an excess of electrons in its outer shell. The barrier inhibits the free migration of electrons, leading to a build-up of electrons in *n*-silicon layer and a deficiency of electrons in the *p*-silicon. If these layers are connected by an external circuit, electrons (i.e. a current) will flow through that circuit. Thus free electrons created by absorption of photons are in excess in the *n*-silicon and flow through the external circuit to the *p*-silicon. Electrical contacts are made by metal bases on the bottom of the cells and by metal grids or meshes on the top layer (which must be largely uncovered to allow penetration of photons). A schematic section of a cell of this type and a schematic of a cell in a circuit are shown in Fig. 27.13.

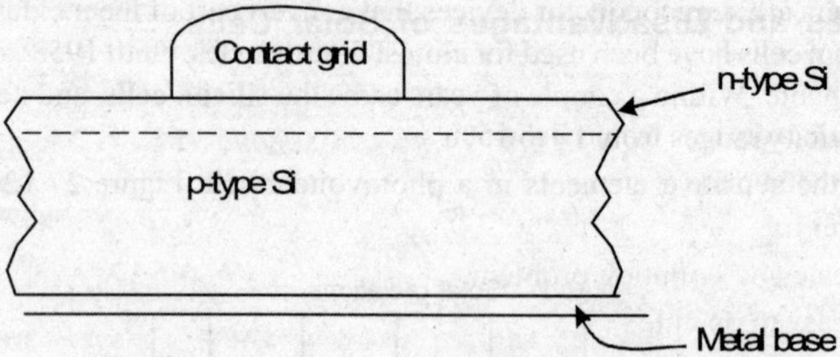

Fig. 27.13 (a) *Cross-section of silicon solar cell.*

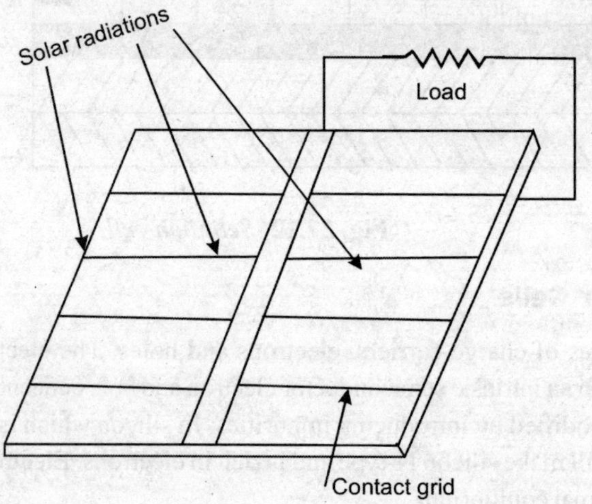

Fig. 27.13 (b) *Schematic of a cell showing up contacts.*

There are many variations on cell material, design and methods of manufacture. Amorphous or polycrystalline silicon (Si), cadmium sulfide (CdS), gallium arsenide (GaAS) and other semiconductors are used for cells.

For silicon-solar cells, high purity silicon must be used in the production of solar batteries. Silicon is purified by zone-refining. Because of its high purity, solar-silicon cells are expensive. Next to oxygen, silicon is the most abundant element on earth. The pure silicon is extracted from sand which is mostly silicon dioxide (SiO_2).

27.4.4.2 Fabrication of Silicon Cells

Procedures for producting the silicon cells are –

(a) The pure silicon is placed in an induction furnace where boron is added to melt. P-type material is produced.

(b) A small seed of single crystal of silicon is dipped into the melt and withdrawn slowly at the rate of 10 cm or less per hour. The rate of growth is adjusted so as to get a single crystal.

(c) Wafers are then sliced from the grown crystal. For this diamond cutting tool is used. To remove any marks, the slices are lapped.

(d) After this the slabs are etched in hydrofluric or nitric acid to get the thin slices of P-type silicon.

(e) The wafers are then sealed in a quartz tube partly filled with phosphorous pentoxide and then the tube is placed in a diffusion furnace wherein the temperature is controlled.

(f) The cells are then etched in a concentrated acid to remove unwanted coatings. Wax or Teflon masking tape is used to protect the surfaces not to be etched.

27.4.4.3 Advantages and Disadvantages of Solar Cells

The advantages are :

(i) They can be made from raw materials which are easily available.
(ii) There are no moving parts.
(iii) They have longer life.
(iv) They do not create any pollution problem.
(v) Solar cells are easy to operate.
(vi) They need little maintenance.
(vii) These cells are highly reliable.
(viii) Their energy source is unlimited.
(ix) They are easy to fabricate.
(x) Solar cells have high power to weight ratio.
(xi) Raw material is easily available in ample quantity.
(xii) Sun tracking is not necessary.

The disadvantages are :

(i) They can be used to generate small amount of electric power.
(ii) The cost is very high.
(iii) The output of the solar cells is not constant. It varies with the time of the day.
(iv) It depends on the sun energy (direct or diffused).

27.5 WIND ENERGY

Wind energy is an indirect form of solar energy since wind is induced chiefly by the uneven heating of the earth by the sun. Wind energy is clean, safe and renewable source and wind power is certainly one of the most attractive solution. Wind power was used in past for several years for propelling ships, driving windwills, pumping water, irrigating fields. Due to the availability of plenty of cheap sources of energy and development of IC engines, wind power was gradually replaced by other energy sources.

Wind can be broadly classified as 'planetary' and 'local planetary' winds which are caused by greater solar heating near the equator than near northern and southern poles. This causes warm tropical air to rise and flow through the upper atmosphere towards the poles and cold air from the poles to flow back to the equator. The direction of such airflow is affected by rotation the earth.

At the same time, the inertia of the cool air moving towards the equator nearer the earth's surface causes it to turn west resulting in the north-east trade winds in the northern hemisphere and the south-east trade winds in the southern hemisphere. Also, because of the earth's axis is inclined to its orbital plane around the sun, seasonal variation in the heat received from the sun results in seasonal variation in the velocity and direction of the wind from the general flow pattern described above.

Local winds are caused by differential heating of land and water. The second mechanism of local wind is caused due to hills and mountain sides.

It has been estimated that about 2 percent of all solar radiation falling on the surface of the earth is converted to the kinetic energy in the atmosphere and that 30 percent of this kinetic energy occurs in the lowest 1000 m of the elevation. This may satisfy the requirement of the half the world. It is also claimed that the wind power is pollution free and its source is free. However, wind energy is erratic and unsteady.

During 1970s and 1980s data about wind energy resources has been collected and various prototype wind turbine generators have been built. Several types have been installed in Canada, Denmark, Netherland, Sweden, UK, USA, German and India.

By 1990 wind energy has become economical to produced electrical energy in Gujarat, Tamil Nadu. The following types of wind turbines are available :

1.	Very Small	0.5 to 1 kW
2.	Small	1 to 15 kW
3.	Medium	15 to 200 kW
4.	Large	250 to 1000 kW
5.	Very large	1000 kW to 6000 kW.

The largest wind generator built in the past is 800 kW in France.

The rotary mechanical energy is used for several applications such as

(a) Pumping water

(b) Grinding flour

(c) Electric energy generators

(d) Operating wood cutting machines (Saw mills).

27.5.1 Advantages and Disadvantages of Wind Energy

Advantages. Following are the advantages of wind energy.

1. It is very clean, cheap and renewable energy source.
2. It is non-polluting energy. Environment is not affected.
3. No fuel is required.
4. It is away from main (conventional) electrical grid.
5. It is very much useful in rural areas, farms, and remote places.
6. It is useful on-shore and off-shore installations.

Disadvantages. Following are the disadvantages of wind energy.

(i) It is not constant but fluctuating energy source and therefore it will required storage.

(ii) Because of its fluctuating nature, energy density is low.

(iii) Quantity wise or capacity wise the plant is very small.

(iv) Because of low capacity, cities and industries requirement cannot be met.

(v) The weight to power ratio of the plant is relatively high.

(vi) As in case of solar forms, the wind forms also need land which is flat, vacant, and free from forests and hills or mountains.

(vii) For continuous supply of energy or power large storage batteries are necessary and diesel generator back-up is required.

(ix) The system is noisy.

(x) Selective places or locations will be available where favourable wind velocities will be available.

(xi) Because of the variabilities of the wind, the design of the turbines and other equipment is complicated.

27.5.2 Site Selection for Wind Mill (Power)

To consider the site for the wind mill, following points or characteristics are considered.

(a) High average wind speed is necessary.

Wind power is given by

$$P_w = \text{Constant} \times V^3_w \text{ Watts/m}^2$$

where

V_w = Wind speed in kmph

Constant = 1.3687×10^{-2}

Minimum speed should be 3.5 to 4.5 m/s to turn the generator.

(b) For a given height of the windmill from the base, there should not be any obstruction to the wind flow within a radius of 3 km or so. Higher altitudes desirable.

(c) We should have an open plain ground or a shore.

(d) Flat plain is required on the top of the hill or mountain.

(e) Land cost should low. Firm foundations are desirable.

(f) Selected site should be nearer to users of the generated electric energy.

The actual power from wind obtained or recovered is about 0.5926 $\left(\frac{16}{27}\right)$. This constant is known as **Betz coefficient** or Gilbert's limit.

27.5.3 Types of Wind Mills

There are various types of wind turbines (wind mills) that are available today. They are

(a) Wind mills with horizontal axis

(b) Wind mills with vertical axis

(a) Wind mills with horizontal axis include

(i) Multiblade type wind mill

(ii) Sail type wind mill and

(iii) Propeller type wind mill.

(b) Wind mills with vertical axis include

(i) Savionius type wind mill

(ii) Darrieus type wind mill.

27.5.3.1 Wind Mills with Horizontal Axis

Various types of the wind mills are shown in Fig. 27.14.

Wind energy conversion devices are commonly known as wind turbines as they convert the energy of the wind stream into energy of rotation. As usual, the component which rotates is called **rotor**. An electric generator is coupled to the turbine to produce electric power. **Aero Generator** is another name is given to the combination of the wind turbine and the generator.

Power produced can be stored in batteries. The electrical energy can be fed to the grid network.

Multiple blade type wind mill generally has 16 to 20 blades. These blades are made from metal sheets. This type of the wind mill run at low speed of 60 to 80 rpm.

Sail type of wind mill has three blades made of three triangular pieces of canvas cloth. This type of wind mill rotates at low speed of 60 to 80 rpm.

Propeller Type of Wind Mill. Propeller type design has a rotor of an aircraft and it may have one/two/three blades mounted on the hub. The rotor shaft is used to rotate the generator through gear box.

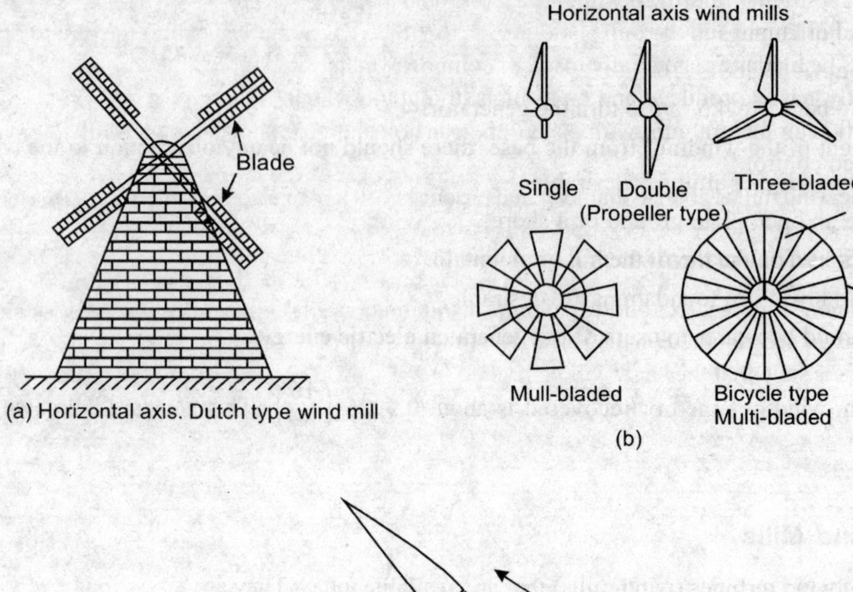

Fig. 27.14 *Types of wind mills.*

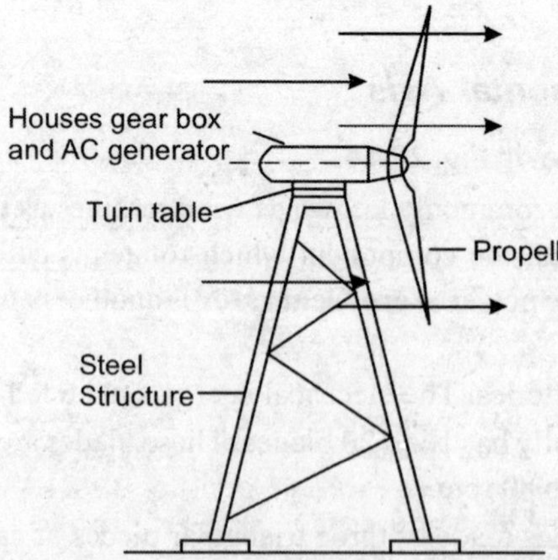

Fig. 27.15 *Propeller type wind mill.*

Figure 27.15 shows a propeller type wind mill. This uses blades of the propeller type having aerofoil cross-section. The number of blades used is restricted to two (2) for economical reason, although the coefficient of performance (efficiency) is slightly improved with three to 4 (four) blades.

The axis of rotation is horizontal and the mill is mounted a top tower. A turn table enables the mill to orient itself in the direction of wind. The blades are generally made of composite material like F.R.P. (Fibre Reinforce Plastic). This material is less costly, easily produced and gives high strength to weight ratio.

The modern wind mill is an integral unit with a step-up gear box and generator together with automatic blade adjusting device for speed control.

A typical propeller type wind mill is 40 m in diameter and produces 120 kW of electrical energy, rotating at 400 rpm.

27.4.3.2 Wind Mills with Vertical Axis

(i) *Savonius type of wind mill.* Figure 27.16 shows savonius wind mill blades. In this mill a hallow circular cylinder is sliced in half and the halves are mounted on the vertical shaft with a gap between them as shown in Fig. 27.16. The arrangement of the halves is such that the shaft will be rotated in the same direction even if the direction of wind is changed.

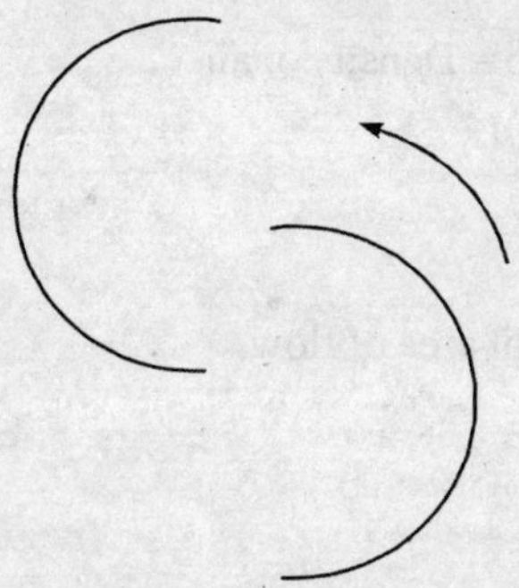

Fig. 27.16 *Savonius type of wind mill.*

Here, the torque is produced by the pressure difference between the two sides of the half facing the wind. To get more power, this needs a large surface area.

This savonius wind mill is self starting but at the same time has low speed and low efficiency.

(ii) *Darrieus type of wind mill.* Amongst all different types of wind mills, Darrieus type of wind mill requires much less surface area.

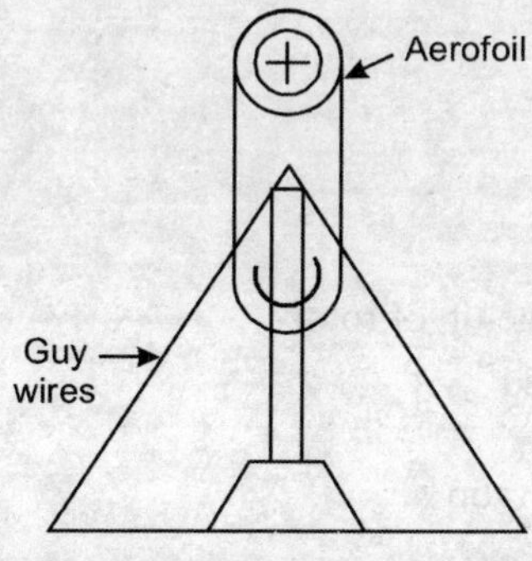

Fig. 27.17 *Darrieus type wind mill.*

Figure 27.17 shows the schematic diagram of Darrieus type of wind mill. It is shaped like an egg-beater and has two or three blades shaped like aerofoils.

Darrieus type of wind mill is not self starting, runs at high speeds, it has high efficiency and its cost is low i.e., the characteristics of this type of the wind mill are

1. It is not self starting.

2. It runs at high speed.
3. It has high efficiency.
4. Its cost is low.

This type of the wind mill is mounted on the vercal shaft and rotate about the vertical axis and because of this, the wind mill runs independently of the direction of the wind. As against this, the horizontal axis machines will have to face the direction of the wind in order to run the generator to generate the power.

The wind stream total power is given by

$$P_t = \frac{mV_a^2}{2}$$

where

m = Mass flow rate kg/s

V_a = Incoming velocity of air m/s

Here $m = \rho AVa$ where ρ = Density of air

$$\therefore \quad P_t = \frac{\rho AV_a^3}{2}$$

or $\frac{P_t}{A}$ = Power per unit area of flow

$$= \frac{eV_a^3}{2}$$

If P = Actual power delivered by the rotor

then C_p = Coefficient of performance

$$= \frac{\text{Actual power delivered}}{\text{Maximum power available}}$$

$$= \frac{P}{P_t}$$

$$= \frac{P}{\frac{\rho AV_a^3}{2}}$$

Also if U = Speed at the tip of rotor

and V_a = Wind speed.

$\therefore \quad \frac{U}{V_a}$ = Tip speed ratio

Generally, the performance of the wind mill rotor is indicated by a plot of variation of the coefficient of performance C_p against the tip speed ratio.

Typical variations of performance for many of the rotors considered are shown in Fig. 27.18.

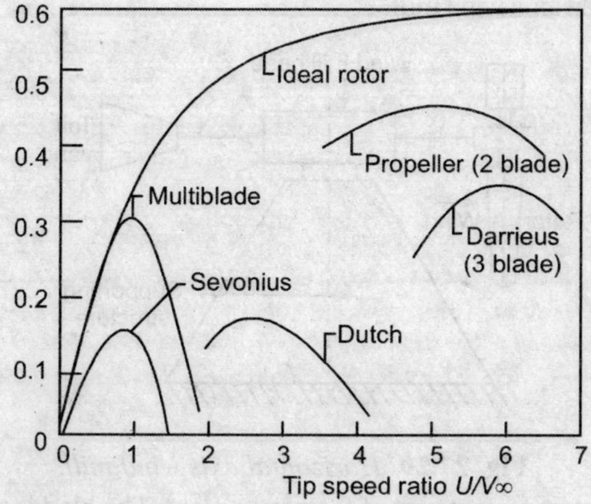

Fig. 27.18

From these curves, it is seen that C_p is the lowest for Savonius and Dutch types whereas the propeller types have the highest value.

Power to weight ratio should be highest.

27.5.4 Elements of Wind Power Plant

Following are the various elements of a wind power plant :

(i) Wind turbine or rotor

(ii) Electric generator

(iii) Supporting structure

(iv) Wind mill head (speed increaser, drive shaft, clutch coupling etc.).

These elements are shown in Fig. 27.19.

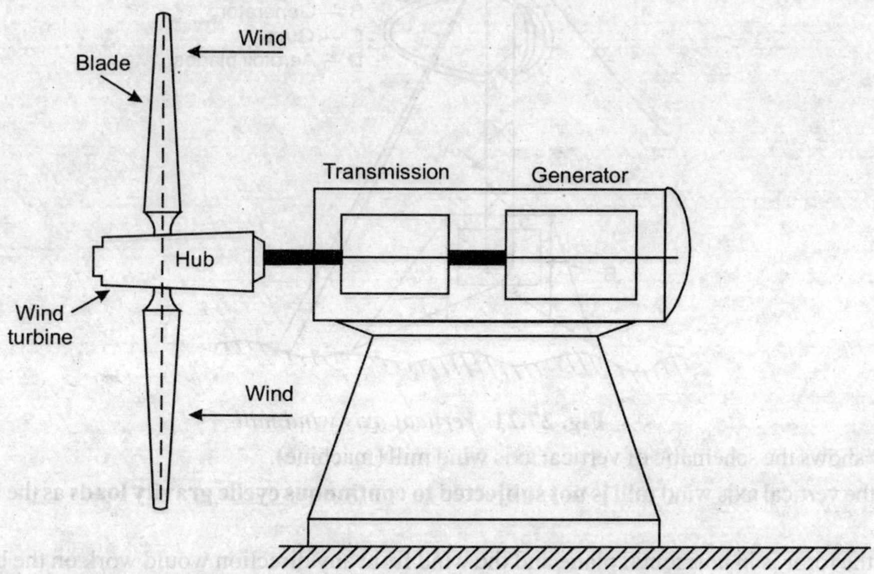

Fig. 27.19 *Elements of wind mill power plant.*

Figure 27.20 shows the horizontal axis wind mill.

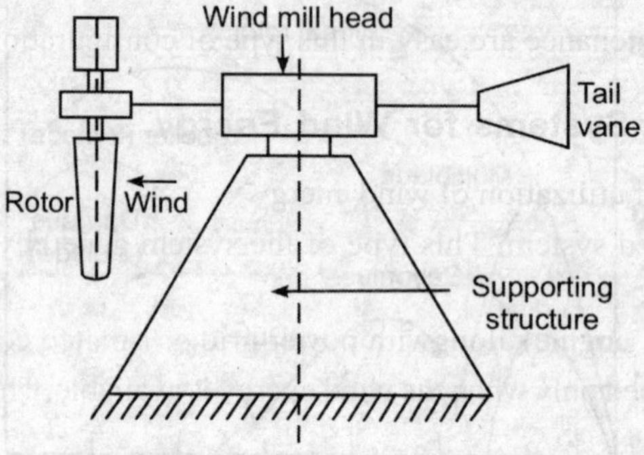

Fig. 27.20 *Horizontal axis wind mill.*

Wind turbine or rotor is the most important part of the power plant. The blades are generally made of composite material like FRP (Fibre Reinforced Plastic). This material is less costly, easily produced and gives high strength to weight ratio.

Following are the functions to be carried out by the wind mill head

(a) It supports the rotor housing and the bearings.

(b) It incorporates the control mechanism — changing the pitch of the blades, tail vane to orient the rotor to face the wind.

The modern wind mill is an integral unit with a step-up gear box and generator together with automatic blade adjusting device for speed control.

A typical propeller type wind mill is 40 m in diameter and produced 120 kW of electrical energy, rotating at 400 rpm.

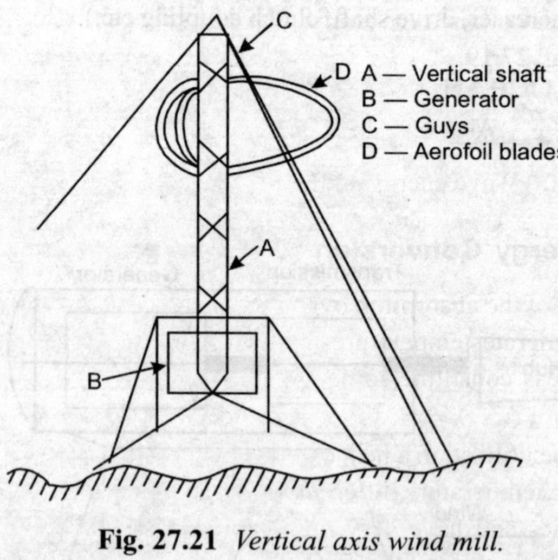

Fig. 27.21 *Vertical axis wind mill.*

Figure 27.21 shows the schematic of vertical axis wind mill (machine).

The rotor of the vertical axis wind **mill is not subjected to continuous cyclic gravity loads** as the blades do not run end over end.

In this case, the rotor is in horizontal plane and the wind from any direction would work on the blades and hence they do not need yawing equipment to turn the rotor.

Heavy components such as gear box, generator etc. can be located at the ground level and thus the machine may need less structural support.

Lastly, the installation and maintenance are easy in this type of configuration.

27.5.5 Types of Utilization Systems for Wind Energy

Basically there are three systems of utilization of wind energy.

(a) Continuous working, isolated system. This type of the system generally require storage and reconversion system.

(b) Fuel-supplement system (IC engine) alongwith power grid or isolated conventional generating units.

(c) Small system which can be used only when the wind energy is available. This is preferred in case of developing countries with large rural areas.

27.5.6 Types of Generating Systems

On the same lines as that for **"Solar Farm"**, wind farms are considered where several identical units are installed and the total power (electrical) produced by the wind farm is fed into the distribution network or can be used to the alone electrical load.

For the electrical generating systems, three factors can be considered as

(i) Type of Electrical Output :

DC

AC with variation of frequency

AC with constant frequency

(ii) Utilisation of Output :

(a) Interconnection with power grid or/and

(b) In conjunction with battery or any other form of storage.

27.6 ENERGY FROM THE OCEAN

There are three forms of ocean energy

(i) Ocean thermal energy (ii) Wave energy and (iii) Tidal energy.

27.6.1 Ocean Thermal Energy Conversion (OTEC) System

Ocean thermal energy is the result of the absorption by the seas and oceans, of the solar radiation which causes like the wind, ocean currents and moderate temperature gradient from the water surface downwards, especially in tropical water. The oceans and seas constitute some 70% of the earth's surface area, so they represent a large storage reservoir of the solar input.

The temperature gradient can be utilized in a heat engine to generate power. This is called "Ocean Temperature Energy Conversion (OTEC)". The temperature difference is small (about 20°C), even in tropics and hence, OTEC systems have very low efficiencies and have very high capital cost. (In tropical regions, the average surface temperature is about 25°C and at the bottom of the sea 1000 m below surface, is about 5°C.)

There are two basic designs for OTEC systems :

(a) Open cycle (Claude cycle)

(b) Closed cycle (Anderson cycle)

27.6.1.1 Open Cycle

In the twenties, the French Engineer Claude made an attempt to build an open cycle 40 kW OTEC plant near Cuba where sea water itself was used as the working fluid which was first degassed and then flash

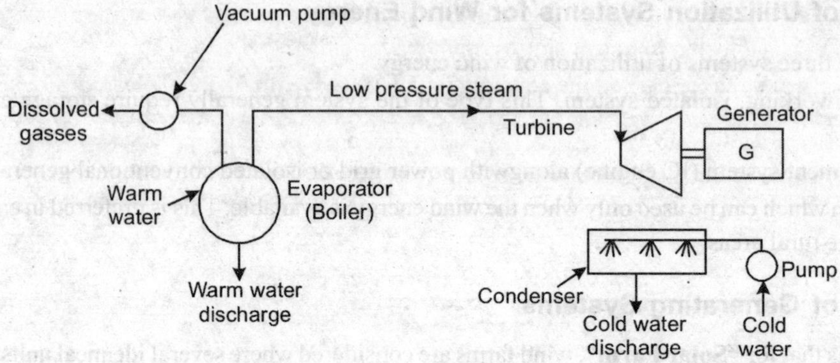

Fig. 27.22 *Schematic of Claude OTEC power plant.*

evaporated in vacuum, so that the turbine could use de-salinated stream at low pressure. A turbine for generating 100 MW is estimated to have a pipe diameter of 20 m.

In the plant warm surface water at about 27°C is admitted into an evaporator in which pressure is maintained at lower pressure below the saturation pressure corresponds to that water temperature. Water entering the evaporator therefore is in superheated state at this pressure and causing it to flash into steam. The low pressure in evaporator in maintained by the vacuum pump that also removes dissolved non-condensable gases from the evaporator.

The steam from evaporator is separated from the water and water is discharged as brine to the ocean. The steam is expanded in the steam turbine. The exhaust steam from the turbine is discharged into condenser. The condensate goes back to ocean. The sea cold water is used to condense the steam. That water is now discharged to the ocean. Such cycle was developed by Claude in 1929.

27.6.1.2 The Closed or Anderson Cycle

There were many practical difficulties in running open cycle plants and was not in favour of working of the plant. All the plants now proposed would operate on a closed Rankine cycle sometimes called Anderson cycle, using low boiling point working fluids like ammonia (NH_3), refrigerant R-12, R-22 and **propane**.

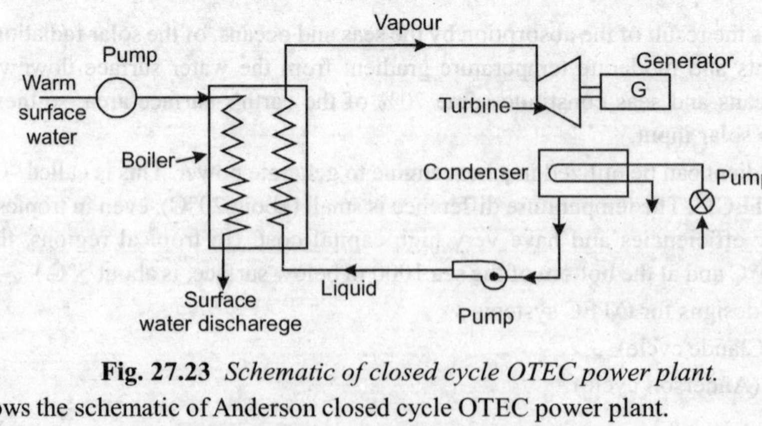

Fig. 27.23 *Schematic of closed cycle OTEC power plant.*

Figure 27.23 shows the schematic of Anderson closed cycle OTEC power plant.

In this cycle propane is used as the working fluid with a 20°C temperature difference between warm surface and cold water. The propane is vaporized in the boiler. The vapour is expanded in the turbine and condensed in surface condenser and sent back to boiler. The ocean cold water is used in the condenser for condensing the propane vapour. The OTEC systems has limitations because very large flow of ocean water is required.

The turbine operates at very low vapour pressures. The turbines required are very large. The cost of the plant is very high. The plant is subjected to ocean storms, high waves etc.

The Indian government has approved the construction of 100 MW off-shore OTEC power plant in Tamil Nadu. It will use open-cycle principle. Technical expertise is from IIT Madras.

27.6.2 Advantages and Disadvantages of OTEC

Following are the advantages of OTEC :

(a) It is clean form of energy conversion.

(b) No payment for the energy required.

(c) It does not require any land area.

(d) It can be a steady source of energy since the temperatures are almost steady.

Disadvantages are as following :

(i) About 30% of the power generated would be used to pump water.

(ii) The system is required to withstand strong convective effect of sea water, hurricanes and presence of debris and fish contribute additional hazard.

(iii) The materials used will have to withstand the highly corrosive atmosphere and working fluid.

(iv) Construction of floating power plants is difficult.

(v) Plant size is limited to about 100 MW due to large size of components.

(vi) Very heavy investment is required.

The US department of energy has built a demonstration unit 50 kW in 1979. It was tested on heat exchangers and biofouling prevention measures.

To get an idea of the gigantic requirement, it is given that for a 150 MW plant,

(i) A flow of 500 m^3/s would be required.

(ii) The area of the heat exchangers will be about 0.5 km^2.

(iii) The length of the cold duct is about 700 m and the diameter would be 25 m.

27.6.3 Wave Energy

Because of the interaction between the winds and the surfaces of the oceans, wave energy is obtained. Like the wind and OTEC, ocean and sea waves are caused indirectly by solar energy. Waves are caused by the wind, which in turn is caused by uneven solar heating and cooling of the earth's crust and the rotation of the earth. Wave energy is more active and concentrated than solar energy. Devices that convert wave energy produce much higher power than solar devices.

Ocean wave energy is an important renewable form of energy and is free. In addition wave power devices do not use up large land area as required in solar and wind devices and are pollution free. Because such devices remove energy from waves, the water is left in placid state and protect coat line from the destructive action of such waves, minimize erosion and even help to create artificial harbours. But it is with some disadvantage like, it requires complicated devices for conversion in useful form of energy. The necessary mechanical strength to withstand the enormous power of stormy sea. Economic factors such as the capital cost, maintenance, repair and replacement and problem of biological growth of marine organism. Also there is scarcity of accessible site of large activity.

In other words, the advantages and disadvantages of the wave energy can be gives as follows.

Advantages

1. It is mostly pollution free.
2. It is a free and renewable energy source.
3. After removal of power, the waves are in placid state.
4. Wave power devices do not require large land areas.
5. The system not only produces electricity but also protects the coast lines from the destructive action of large waves, minimises corrosion, erosion and help create artificial harbour.

Disadvantages

1. Lack of dependability.
2. Relative scarcity of accessible sites of large wave activity.
3. The construction of conversion devices is relatively complicated.
4. The devices have to withstand enormous power of stormy seas.
5. Large capital investment, cost of repair, replacement and maintenance.

27.6.3.1 *Wave Machines*

Several designs of wave machines have been proposed and patented. There are two possible routes to convert wave energy into mechanical energy.

1. Wave machine driving the electrical generator through gear.

2. Wave machine driving air compressor or hydraulic pump to store energy to drive another machine. The stored energy in compressed air or hydro reservoir may be converted into electrical energy for further use.

The total energy of a wave is the sum of its potential and kinetic energies. The potential energy arises from the elevation of the water above the mean level of the wave. The kinetic energy of the wave is that of the liquid between two vertical planes perpendicular to the direct of wave propation.

Wave motion is primarily horizontal but the motion of water is primarily vertical. The latter motion is made use of by floats to get mechanical power.

Following variety of devices have been developed for harvesting of wave energy but these are complicated and fragile in the face of gigantic power of ocean storms.

1. Dolphin-type wave power machine
2. Floatwave machine
3. High-level reservoir wave machine
4. Dam-Atoll wave machine.

1. *Dolphin Type Wave Power Machine.* Figure 27.24 shows the schematic diagram of Dolphin type wave power generator. The design of this type of wave power generator has been done by Tsu Research Laboratory in Japan. The schematic of such a machine shows a Dolphin, a float, a connecting rod and two electrical generators.

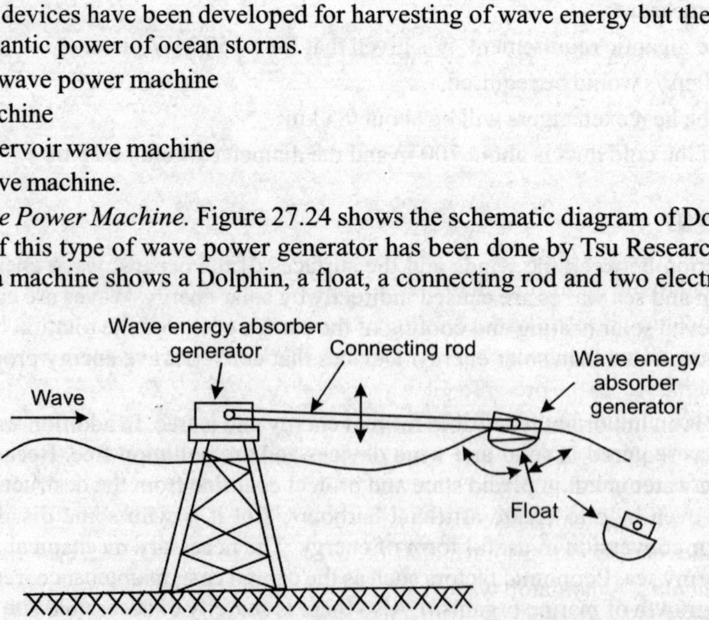

Fig. 27.24 *Schematic of a Dolphin type wave machine.*

The float has two notions. The first is a rolling down motion about its own fulcrum with a connecting rod. The other is nearly vertical or having a motion about the connecting fulcrum. It causes relative revolving movements between the connecting rod and stationary dolphin. All the movements are amplified and converted by gears into continuous rotory motion that drives two electrical generators.

2. *Flot Wave Machine.* The wave motion is primarily horizontal, but the motion of the water is vertical. This latter motion is used by floats, to obtain mechanical power. The concept is that a large float driven up and down by the water within relatively stationary guides. The reciprocating motion is converted into rotory (mechanical) motion and then the generator is driven producing electric power.

Figure 27.25 shows the schematic of float wave machine. The square float moves up and down with the water guided by four vertical manifolds that are the part of a platform. The platform is stabilized within the water by four large under water floatation tanks, so that it is supported by buoyancy forces. The platform is expected to be stationary even in heavy sea. A piston is attached to the float moves up and down inside a cylinder that is attached to the platform. The piston-cylinder arrangement works as a reciprocating compressor. During downward stroke air is taken into cylinder via inlet valve. The upward motion compresses the air and sends it through an outlet valve to four under water floatation tanks via manifold. These floatations serve the dual purpose of buoyancy and air storage.

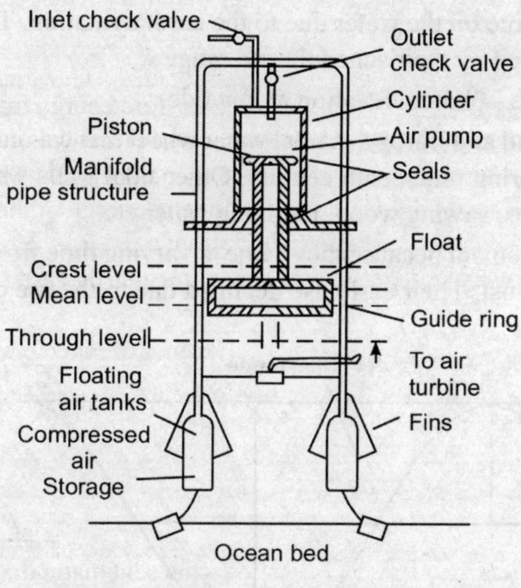

Fig. 27.25 *Schematic of a float wave machine.*

The stored air then used to run air turbine than drives an electrical generator. The electric power is transmitted to the shore by underwater cable.

3. *High Level Reservoir Wave Machine or Hydraulic Accumulator.* A differential area piston is used in a high-level reservoir wave machine where the pressurized water is elevated to a natural reservoir above the wave generator, which has to be near a shoreline or to an artificial water reservoir. The water from the reservoir flows through a turbine coupled to an electric generator and then back to the sea level. It is estimated that to produce 1 MW power a 20 m diameter generator is required.

Hydraulic accumulator wave machines are also used, where the high pressure water is stored in a high pressure accumulator or to a high-level reservoir from which it flows through a water turbine electric generator.

4. *Dam Atoll Wave Machine.* A dam-atoll wave machine is another wave machine used for extracting energy from ocean waves. This machine overcomes the disadvantages of other machines namely complexity and fragility in heavy seas. It is a massive and robust device and strong enough to survive any ocean storm. This machine includes

some of the characteristics of both dams and atolls. The basic principle of operation is based on the observed action of waves as they approach atolls in the ocean (small volcanic islands). The waves wrap themselves around the atolls from all sides, ending in a spiral in the centre, driving a turbine before discharging laterally outward.

A study of the annual wave energy potential for the Indian coast has revealed that around 175–200 MW/m is available. This study was carried out in the laboratory at Ocean Engineering Centre of the IIT Madras. Based on this studies, a pilot plant of 150 kW capacity was build at Vizhinjam near Trivendrum, which was commissioned in October 1991.

27.7 TIDAL ENERGY

The tides are another source of energy from ocean. The tidal energy can be tapped from coastal water by building dams that entrap the water at high tides and release it at low tide back to the sea. Power can then be obtained by turbines from both *in* and *out* flow of water. The power available is very large but only in a few parts of the world.

Tidal energy is a form of hydro energy recurring with every tide. During a lunar day the ocean water rises twice and falls twice. The daily variation in tidal level is mainly due to the changing position of the moon. Thus the tides are rhythmic but not constant and they do not occur on a regular daily schedule. Their occurrence is due to balance forces, mainly the gravitational forces of the moon but also that of the sun, both acting together with that of the earth to balance the centrifugal force on the water due to the earth's rotation. This results in rhythmic rise and fall of water. The tides are characterized by their schedule and range *R*.

R = Water elevation at high tide – Water elevation at low tide.

Tidal mills were used in England and Europe. A tidal water wheel that was constructed under the London Bridge that supplied water to London during nineteenth century. Other tidal mills were common for hundreds of years. They were used for grinding grains, sawing wood, pumping water etc.

Tides though free, were inconvenient because they came at varying time from day to day, having varying ranges (heads) and require large capital cost. Their early use declined due to the use of storm and cheap coal.

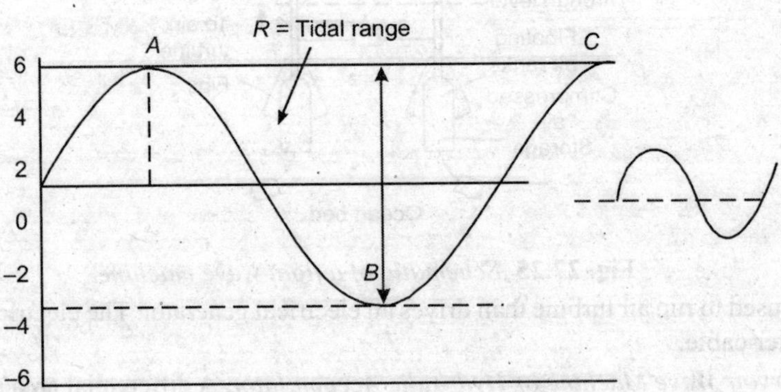

Fig. 27.26 *Tides of sea.*

The rise and fall of wave level follows a sinusoidal curve as shown in Fig. 27.26. With point *A* indicating the high tide point and point *B* indicating the low tide point. The difference between high and low water levels is called the range of the tide.

At times during full or new moon, when the sun, moon and earth are approximately in one line, the gravitational forces of the sun and moon are increased. This high tides are called **spring tides**. On the other hand near the first and third quarters of the moon, when the sun and moon are right angles with respect to the earth, **neap tides** occur.

27.7.1 Selection of Site

Following are the important points for the selection of location of tidal power plant :

1. At the desired location, the tidal range should be adequate throughout the year.
2. The site selected for the tidal power should be free from the wave attack of sea.
3. Tidal pattern should not have an appreciable change at the selected or proposed site.
4. The site at which tidal power plant is to be located should not have excessive sediment load.

27.7.2 Tidal Power Plants

Power obtained from the tides can be determined as

Let

h = Tide range from high to low (in m) and

A = Area of water stored in the basiz (in m^2)

then the energy stored in the full basin is given by

$$E = \rho.g\ A\int_0^h x\ dx$$

$$= \rho\, g\, h^2 A$$

∴ Average power,

$$P_{AV} = \frac{1}{2}\rho g h^2 A / \left(\frac{T}{2}\right)$$

$$= \rho g h^2 \frac{A}{T}$$

where T = Period of Tidal cycle

= 14 hours 44 min. usually taken.

Following are a few locations as sites for tidal power in India

(a) Sever (UK) : 9.8 m, 70 km^2, 8000 MW
(b) Passanaquoddy Bay (North America) : 5.5 m, 262 km^2, 1800 MW
(c) San Jose (South America) : 10.7 m, 777 km^2, 19900 MW
(d) Bhavanagar
(e) Navalakh (Kutch)
(f) Diamond Harbour
(g) Ganga Sagar
(h) Kandla (Gujarat) – 600 MW

27.7.3 Components of Tidal Power Plant

There are three main components of a tidal power plants. They are as follows

1. The dyke to form the basin or pool. This is a barrier between the sea and the basin.
2. Sluice ways from the basin to the sea and vice-versa.
3. The power house, water turbines, electric generators and other auxiliary equipments are the main equipments of power house.

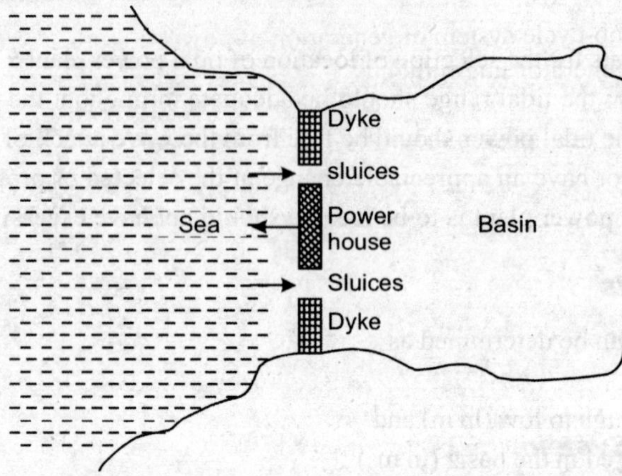

Fig. 27.27 *Simple tidal power plant showing components of the plant.*

27.7.4 Classifications of Tidal Power Plants

The tidal power plants are classified as follows :

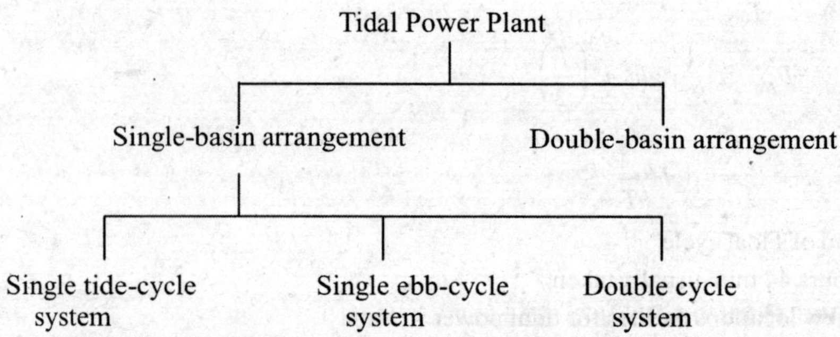

27.7.4.1 Single Basin Tidal System

The simple single basin or pool tidal system has one basin behind the dam. The dam gets filled from the ocean at high tide. During this time of filling the basin, the water turbine is rotated. Generator is connected to the turbine and electrical power is generated. In this case power is generated intermittently.

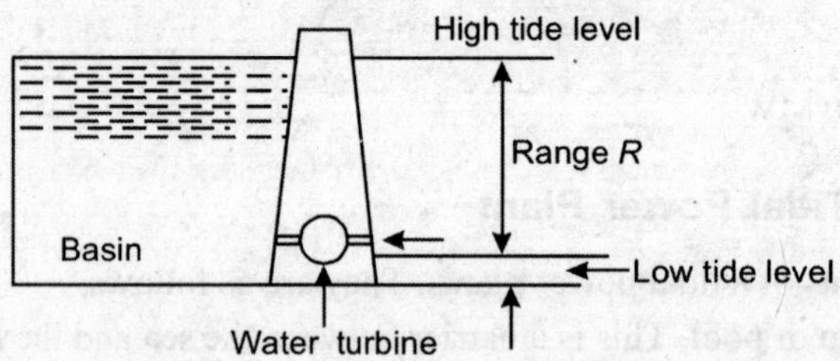

Fig. 27.28 *Single pool or basin arrangement (High tidal system).*

27.7.4.2 Single Basin Ebb-cycle System

When the tide starts receding back to the sea, water from the basin flows back to the sea through another turbine generating power. This is the ebb-cycle system of generation of power. Figure 27.29 shows this system. Here also the power is generated by the generator intermittently.

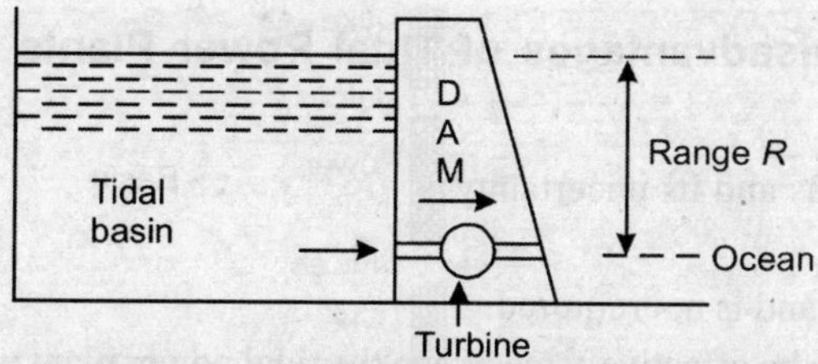

Fig. 27.29 *Ebb-cycle system (Low tidal).*

27.7.4.3 Double Cycle System

In this type of system a single reversible turbine is used so that during both types of flows, the turbine is operated and thus acts in both ways producing power continuously.

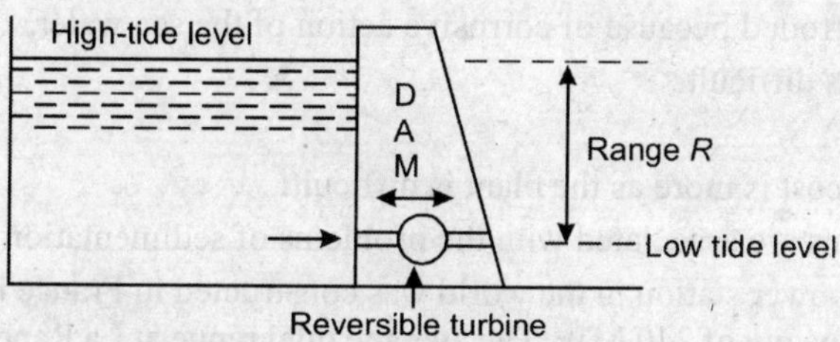

Fig. 27.30 *Single basin double cycle system.*

The flow of water in both direction is used to drive a number of reversible water turbines, each driving an electrical generator.

27.7.4.4 The Two Pool or Double Basin System

The two pool tidal system is one that is much less dependent on tidal fluctuation. But it is more expensive and complex and also requires costly dam construction. Figure 27.31 shows the two pool tidal system.

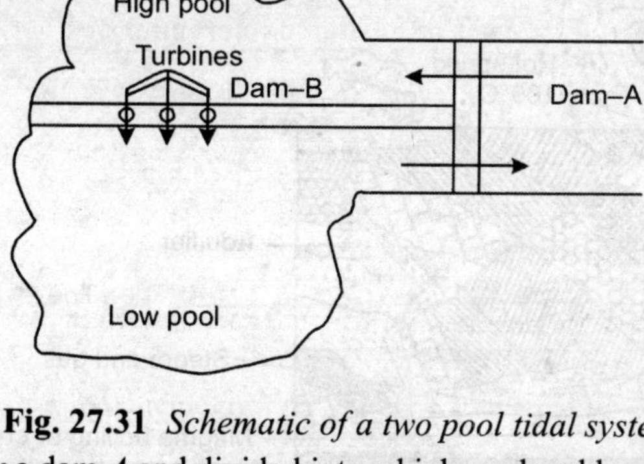

Fig. 27.31 *Schematic of a two pool tidal system.*

An inland basin in enclosed by a dam A and divided into a high pool and low pool by dam B. With proper gate system in dam A, the high pool gets periodically filled at high tide from ocean and the low pool gets periodically emptied at low tide. Water flows from the high to the low pool through the turbines that are situated in dam B. The pools are having large capacities and therefore the fluctuations in the head are minimised—which give more continuous and uniform power.

27.7.5 Advantages and Disadvantages of Tidal Power Plants

Advantages

(i) It is independent of rains and its uncertainty.

(ii) It is unexhaustible.

(iii) Large area of valuable land is not required.

(iv) Peak power demand can be effectively met when the tidal power plant works in combination with thermal or hydroelectric system.

(v) The generation of power with tide is free from pollution.

As against these advantages, following are the disadvantages :

(a) The plant efficiency is affected because the turbines have to work on the wide range of variation of head.

(b) The output is not uniform because of the variation of tidal range.

(c) Machines will be corroded because of corrosive action of the sea water.

(d) Construction in sea is difficult.

(e) It is costly.

(f) Power transmission cost is more as the plant is difficult.

(g) The tidal power plants are associated with the problems of sedimentation and silteration.

The first commercial tidal power station in the world was constructed in France in 1965 across the mouth of La Rance Estuary. It has a high capacity of 240 MW. The average tidal range at La Rance Estuary is 8.4 m and the dam built across the estuary encloses an area of 22 km^2.

27.8 GEOTHERMAL OR TERRESTRIAL ENERGY POWER PLANTS

The earth, having been originally thrown off from the sun according to some theories, still has a molten core. Evidence for this lies in the volcanic action that takes place in many regions of the earth's surface. Geothermal energy is one of the sources of energy available to the human race. This is the thermal energy trapped beneath and within the solid crust of the earth.

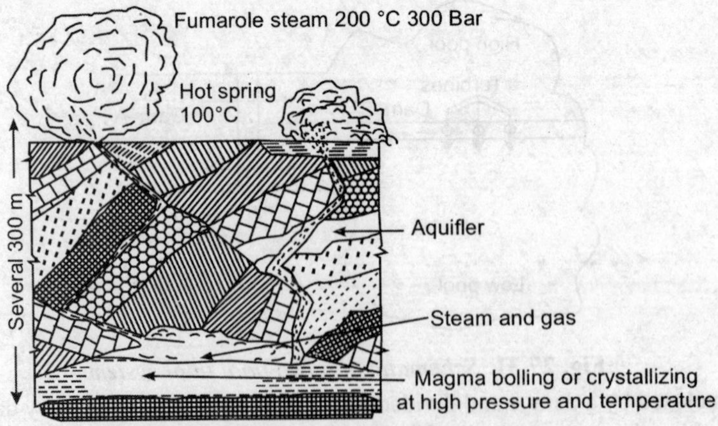

Fig. 27.32 *Typical geothermal deposits.*

Figure 27.32 shows the typical geothermal deposits.

Geothermal energy exists in the form of steam, hot water or hot and molten rock. It is released naturally as geysers, hot-springs and volcanic eruptions. There is tremendous reserve of thermal energy trapped beneath the earth's crust. However, it has been not possible to drill through the earth's crust despite several attempts.

The utilization of geothermal energy is not now new technology, as first geothermal steam well was drilled in Italy in 1904. The capacity of this plant was 400 MW. There are several geothermal power plants in United States, in Mexico, Iceland, New Zealand, Japan and Russia.

In general five geothermal sources are presently known. They are –

1. Hydrothermal convective systems
   (i) Wet steam fields
   (ii) Dry steam fields
   (iii) Hot water fields
2. Geopressure resources
3. Petro-thermal or hot dry rocks (HDR)
4. Magma resources
5. Volcanoes.

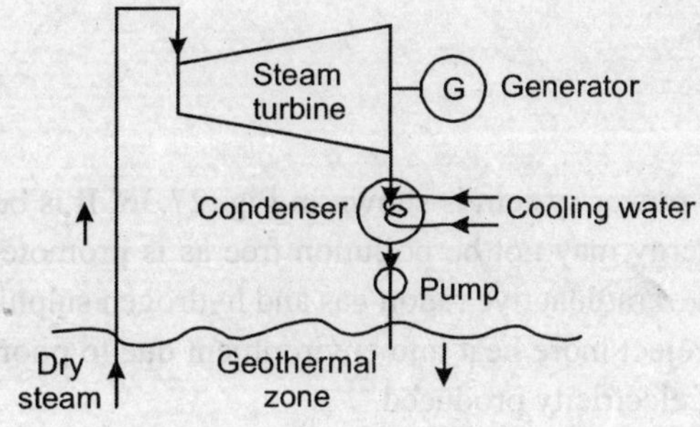

Fig. 27.33 *Dry-steam open system.*

Figure 27.33 shows the dry-steam open system. This is used in Italy and USA.

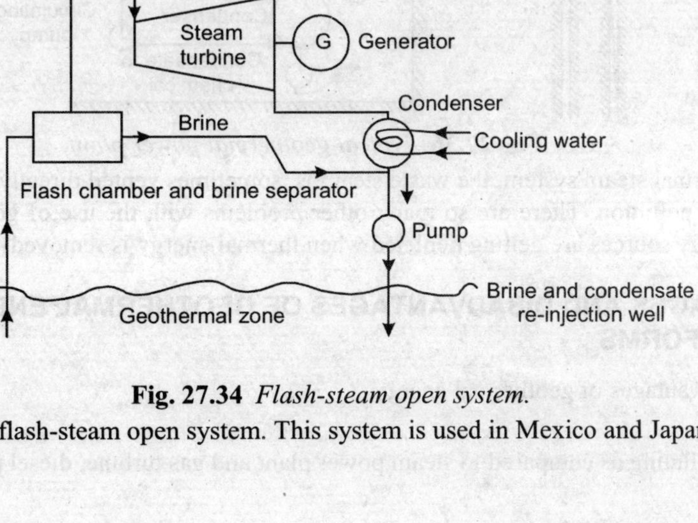

Fig. 27.34 *Flash-steam open system.*

Figure 27.34 shows the flash-steam open system. This system is used in Mexico and Japan.

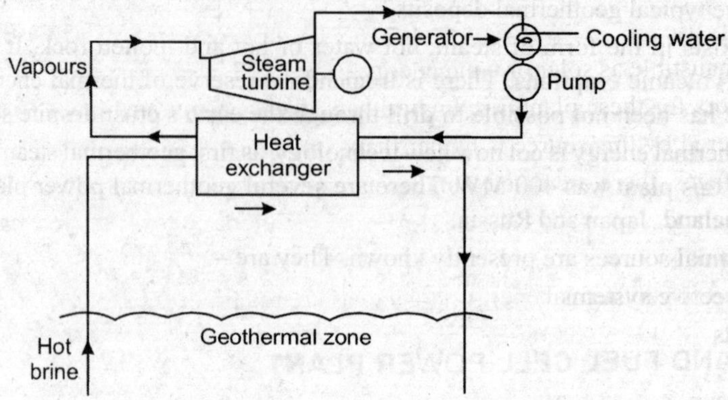

Fig. 27.35 *Hot water closed (binary) system.*

Figure 27.35 shows a hot water closed (Binary) system being developed in Russia and USA.

27.9 APPLICATIONS OF GEOTHERMAL ENERGY

Basically the applications of this energy are

1. Generation of electrical power
2. Space heating of buildings
3. Industrial process heat.

A typical schematic diagram of power system is shown in Fig. 27.36. It is believed that geothermal energy is **pollution free**. But geothermal energy may not be pollution free as is promoted. Air pollution problem may be significant because of the emission of radioactive radon gas and hydrogen sulphide gas. Hydrogen sulphide gas is a poisonous gas. Such plants also reject more heat into environment due to poor steam condition, compared with fossile power plant for each unit of electricity produced.

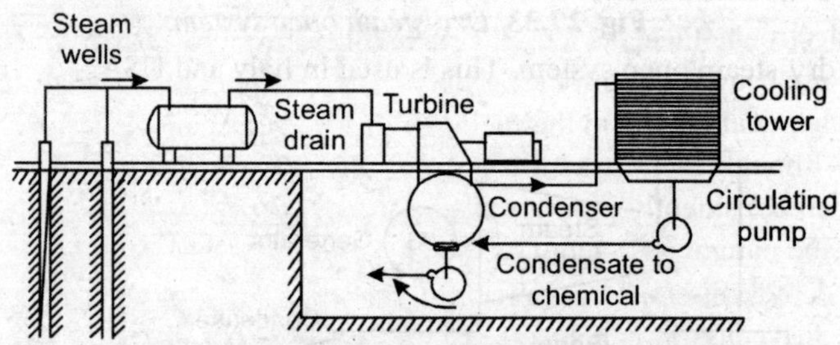

Fig. 27.36 *Typical geothermal power plant.*

In most of geothermal steam system, the waste steam is, sometimes vented directly to the atmosphere with an accompanying noise pollution. There are so many other problems with the use of geothermal energy. Also the packets of such energy sources are getting depleted when thermal energy is removed.

27.10 ADVANTAGES AND DISADVANTAGES OF GEOTHERMAL ENERGY OVER OTHER ENERGY FORMS

Following are the advantages of geothermal energy.

(a) It is cheap.

(b) It is less polluting as compared to steam power plant and gas turbine, diesel power plants.

(c) It is versatile in its use.

(d) It is almost in exhaustible as solar or wind energy.

(e) Annual load factors for these plants are higher than other conventional power plants.

(f) Net energy delivered by this source is higher than conventional power plants.

As against these advantages, there are disadvantages as given below :

(a) Low overall efficiency.

(b) Drilling operation noisy.

(c) To locate the place of geothermal energy, large areas are required.

27.11 FUEL CELLS AND FUEL CELL POWER PLANT

The second law limitations regarding work is not the same for cycles and processes. Gibbs function applies to a process; so if a continuous process could be found in which a 'fuel' is fed to the system and electricity (work) is delivered, then – ΔG would give the amount of net work of a reversible isothermal process. Suppose hydrogen is burned to H_2O (l-liquid); it releases its 'heating value' of about **144452.5 kJ/kg**. Then if this heat is added to a working substance, say, steam or air, the amount that can be converted into work could vary well be within the range of 20-40% — according to the efficiency of the cycle. However, if this combining of H_2 and O_2 could be brought about isothermally at say 25°C, the value of $-\Delta G = W_{max}$ is about 118000 kJ/kg and the efficiency could be 118000/144452.5 = 0.8169 or $\simeq$ 82% on the heating value basis. Observe that for the process to be reversible, the exchange of heat is with the surroundings only, while the system is at the same temperature. As another example, if C_8H_{18} (Octane) – closely analogous to gasoline, could be made to react in an ideal fuel cell at a standard p and T, the ideal work would be about $-\Delta G$ = 462900 kJ/kg (gaseous C_8H_{18}). If the same fuel were used in a certain **ideal cycle** for automotive engines, the ideal work would be only 22330 kJ/kg.

It is the possibility of getting this large percentage of work from a chemical process that has aroused interest in **fuel cells** in recent years, since if this ideal work is approached by an actual chemical reaction (process), the actual work obtained may be two or more times greater than if the same chemicals are used via a heat engine.

Fuel Cells. Fuel cells are electrochemical devices or systems which directly convert hydrogen, or hydrogen-rich fuels into electricity without combustion. This process is much more efficient than traditional thermal power plants, converting up to 80% of the chemical energy in the fuel into electricity as compared to maximum of 30–40% for conventional power plants. Although their structure is somewhat like that of a battery, fuel cells never need recharging or replacing and can considently produce electricity as long as they are supplied with hydrogen and oxygen. Fossil fuels (coal, oil and natural gas), biomass (plant material) or pure hydrogen can be used as the source of fuel. If pure hydrogen is used, the emissions from a fuel cell are only electricity and water. Fuel cells are small and modular in nature and therefore fuel cells power plants can be used to provide electricity in many different applications, from electric vehicles to large, grid-connected utility power plants. First used in the US space programme in the 1950s, fuel cells are the developing technology with a few commercial uses today, but many emerge as a significant source of electricity in the near future.

Fuel Cell Electrochemistry. Fuel cells are quite similar to a battery, with two porous electrodes separated by an electrolyte. Electricity is produced by chemical reaction between hydrogen-based fuel and an oxident—usually oxygen inside the fuel cell. Figure 27.37 is a diagram of the electrochemical reaction which takes place inside a fuel cell. Hydrogen (H_2) flows over the anode (the negative electrode) and splits into positively charged hydrogen ions, and electrons which carry a negative charge. The electrons flow through the anode to the external circuit, performing useful work (this is the electric current generated) while the hydrogen ions pass through the anode and into the electrolyte, moving towards the cathode (the positive electrodes). The electrons eventually return to the cathode which is supplied with oxygen (O_2). At this point the electrons, hydrogen ions and oxygen react to form water (H_2O)

and heat. In stationary fuel cell power plants thus heat can be captured and used for process heat in industries or space heating (co-generation). As long as the fuel cell is supplied with hydrogen and oxygen, this electrical production can continue indefinitely.

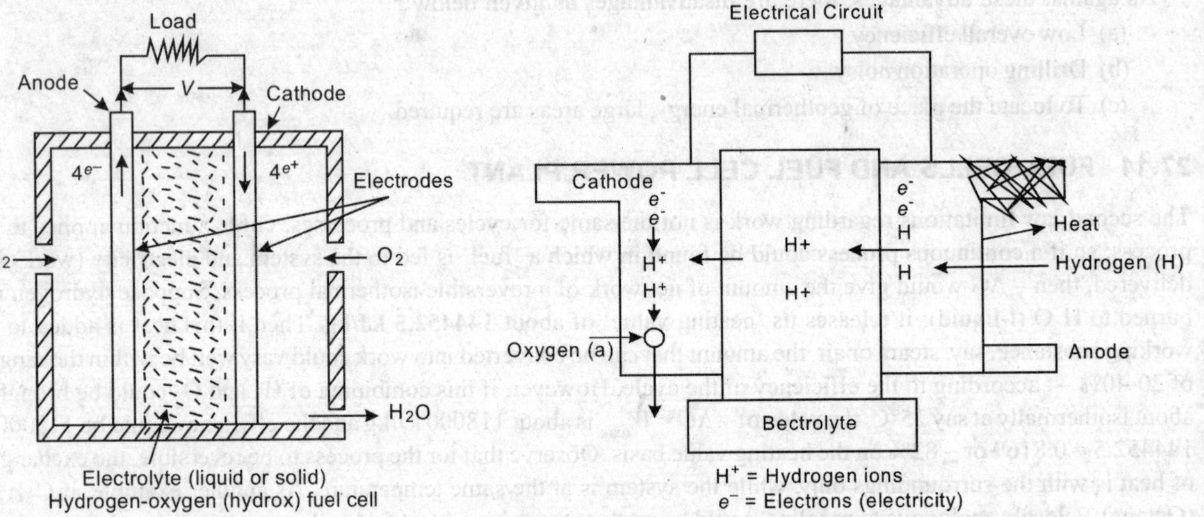

Fig. 27.37 *Different versions of fuel-cell*

Electrolyte. The general design of most fuel cells is similar except for the electrolyte. Several different substances have been used as the electrolyte in fuel cells, each with their own advantages and disadvantages. Main types of fuel cells based on the electrolytes are :

1. alkaline fuel cells
2. solid polymer fuel cells (also known as proton exchange membrane fuel cells)
3. phosphoric acid fuel cells
4. molten carbonate fuel cells and
5. solid oxide fuel cells.

Alkaline and solid polymer fuel cells operate at lower temperature (50–260°C) and are mainly designed for use in transportation applications, while the other three operate at higher temperatures (upto 1000°C) for solid oxide fuel cells and are being developed for use in cogeneration and large central power plants.

Sources of Fuel and Oxidant. Fuel cells operate at maximum efficiency, when operating on pure hydrogen and pure oxygen. Pure oxygen is very expensive, and thus air is used as the source of oxygen in most applications except where the extra cost can be justified, as in the space program. Pure hydrogen is expensive and difficult to transport and store, therefore, like pure oxygen it is used in special cases. Gaseous mixtures of hydrogen (H_2) and carbon dioxide (CO_2) which can be created by the processing of fossil fuels or biomass are used instead of hydrogen in most commercial uses of fuel cells. The most economical sources of the necessary H_2/CO_2 fuel mixture have been found to be gaseous hydrocarbons such as natural gas and propane, light hydrocarbon liquids such as naptha and methanol (from biomass), heavier hydrocarbon liquids such as fuel oil and coal.

Fuel Cell Power Plants. The fuel cell power section contains 'stacks' of one or more fuel cells. The individual fuel cells are small in size and produce between 0.5 and 0.9 volts of electricity. Therefore stacks of large number of fuel cells would be needed for power generation must have a large number of individual fuel cells. This eliminates the problems associated with large conventional power plants which can take up to ten years to build and have a fixed capacity. The fuel cells use air and hydrogen rich fuel and produce water, heat and direct current (DC) electricity.

The power conditioner section of a fuel cell power plant may also have an 'inverter' which converts the DC (direct current) to AC (alternating current).

The hydrogen-oxygen cell is also known as **hydrox**.

The reaction at the anode can be expressed as –

$$2H_2(g) \rightarrow 4H^+ + 4e^-$$

The reaction at the cathode is

$$4e^- + 4H^+ + O_2(g) \rightarrow 2H_2O\,(l)$$

The overall cell reaction is therefore

$$2H_2(g) + O_2(g) \rightarrow 2H_2O\,(l).$$

29.12 BIOMASS

Biomass is organic matter produced by plants on land and in water. This is an organic material which has stored sunlight in the form of chemical energy. The energy obtained from biomass is called **Biomass Energy**. Biomass includes forest crop and residue like wood, wood waste, straw, manure, sugarcane and many other byproducts from a variety of agricultural processes. Unlike coal, oil and natural gas which take millions of years to form, biomass can be considered a renewable energy source because plant life renews and adds to itself every year. It can also be considered a form of solar energy as the latter is used indirectly to grow these plants by photosynthesis.

When burned, the chemical energy is released as heat. Wood had been the chief source of cooking and heating homes and other buildings for thousand of years. Biomass continues to be a major source of energy in much of the developing world even now. Sugarcane is a good example of a biomass crop. The sugar is extracted from the sugarcane by removing the juice, the remainder of the plant called **bagasse** produced heat when burnt.

Biomass Resource. Main sources of biomass energy are

1. Municipal wastes
2. Industrial wastes
3. Agricultural crop residues and energy plantations
4. Animal and human wastes.

Crops that have been used for energy include :

(a) sugarcane
(b) corn
(c) husk
(d) grains
(e) grass
(f) seaweed and many others.

Animal and human wastes are an indirect terrestrial from which methane for combustion and ethylene (used in plastic industry) can be produced while retaining the fertilizer value of the manure.

Seaweed, marine algae can be grown in water.

Biomass Conversion. Bio conversion can take many forms as given below :

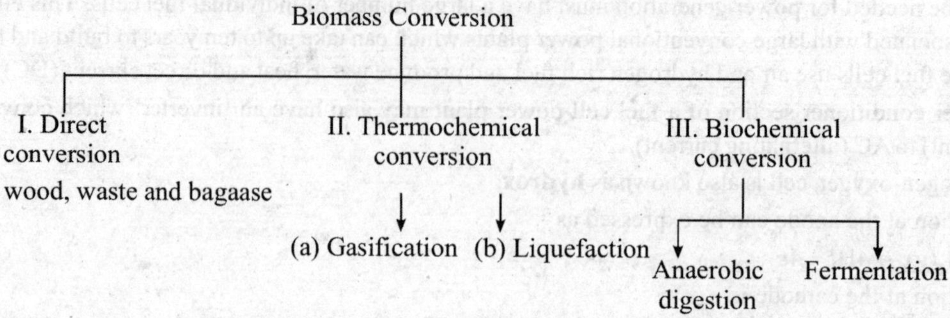

I. Direct Conversion. The simplest and most common method of obtaining energy from biomass is direct combustion. The heat of combustion is used to provide space or process heat, water heating or for producing electricity through the use of a steam turbine. In the developing world, many types of biomass such as dung and agriculture wastes are burned for cooking and heating.

II. Thermochemical Conversion. (a) *Gasification.* In gasification the biomass is heated with limited oxygen to produce low calorific value gas. The same gas can be used as fuel as directly for producing electricity through steam turbine.

(b) *Liquefaction.* The gas produced in gasification can also be used in liquefaction by converting it to methanol (CH_3OH) or ethanol (CH_3CH_2OH). These methanol or ethanol has been blended with gasoline for use in automobiles.

III. Biochemical Conversion. (*a*) *Anaerobic Digestion.* Anaerobic digestion of biomass has been practiced for almost a century, and is very popular in many developing countries such as India and China. Anaerobic digestion involves the microbial digestion of biomass. An anaerobic is a micro-organism that can live and grow without air or oxygen. It gets its oxygen by the decomposition of the matter containing it. The process takes place at low temperature and requires moisture content. The gas generated contents mostly carbon dioxide, methane. This gas is burned directly. The residue can be used as fertilizer or animal feed. This 'biogas' is reasonably clean burning fuel which can be used for cooking, heating or electrical power generation. The residue can be used as fertilizer or animal feed.

(*b*) *Fermentation.* It is the process of breakdown of complex molecules in organic compounds under the influence of a ferment such as yeast, enzymes etc. Fermentation is a wellknown and widely used technology for the conversion of grain and sugar crop into ethanol. Ethanol can be mixed with gasoline to produce **gashol.**

Biomass energy concept are used to cultivate large forest in areas not suitable for food production. The trees can be used for burning in a power plant that would be located in the middle of the forest.

Other scheme to develop aquatic forms growing algae, tropical grasses, floating kelp, water hyacinth (one of the troublesome weed which causes to give shelter to disease carrying organisms, causes floods, disrupts hydroelectric plants). This biomass can be used to produce useful fuel gases.

Thus biomass is the most important renewable energy source for rural and urban application.

THEORY QUESTIONS

27.1 What are the various types of non-conventional energy sources? Explain in brief the working of these energy sources.

PART 6

Power Plant Engineering

Chapter 28 Cost of Electrical Energy and Rates

28.1 INTRODUCTION

The function of a power station is to deliver power at the lowest possible cost per kilowatt-hour (kWh). In engineering thermodynamics, emphasis is often given on the thermal efficiency of a power producing devices like the steam and gas turbines, diesel engines, nuclear power plants. But a power producing device is only a part of the total energy systems which are designed to serve the human needs. The electrical power produced is used for different purposes, such as,

1. Manufacture of different articles using different machines.
2. Lighting of homes, factories, streets, stations, airports and many other items.
3. Communication — via telephone, satellite, TV, radio etc.
4. Transport using railways, conveyors, lifts etc.
5. Heating in furnaces, chemical processes, cooling, hot water etc.

A plant supplying the energy for all these objectives must do so at maximum effectiveness. The effectiveness is measured financially. Thus the consumer must get the services at minimum possible rate and the overall monatory cost of energy production must be minimum. The thermal efficiency of the plant is only one factor of many, that determine the final energy cost. Thus a **most efficienct** plant may not be **most economical** one. The cost of power generation can be minimised by :

1. choosing equipment that will be available for operation during the largest possible percentage of the time in a year
2. reducing the amount invested in the plant
3. simplifying operation by having uniformity of design throughout the station
4. fixing the responsibility of operation in the hands of fewer men.

The energy supply problems may be broadly classified as :

(a) Erection of an entirely new energy system.
(b) Replacement of an existing energy system.
(c) Extension of an existing energy system.

28.2 THE COST OF ELECTRICAL ENERGY

Following are the elements of cost of energy

Elements of cost of energy

1. Fixed cost	2. Operating or running cost variable cost	3. Customer cost	4. Investor's profit
(a) Interest on the capital cost	(a) Cost of fuel	(a) Secondary distribution system	(a) As market conditions allow
(b) Taxes	(b) Cost of maintenance repairs etc.		
(c) Insurance	(c) Cost of oil, grease, water and consumable stores		
(d) Cost of primary system	(d) Wages		
(e) Depreciation	(e) Management		
(f) Management			

28.2.1 Fixed or capital cost

(a) Capital cost of the power plant includes :

(i) Cost of land

(ii) Cost of the building

(iii) Cost of equipment including the cost of transport of equipment to the site and foundation work

(iv) Cost of designing and planning the station

(v) (Architect charges) Cost of right of way.

Cost of land, building and planning may not vary much with the different types of plants, but the cost of equipment is the major factor to be considered carefully. The cost of equipment is usually expressed on the basis of kW installed capacity. Thus, a cost of a non-condensing type steam plant may be Rs. 500/- per kW, whereas the cost of modern steam plant may be of the order of Rs. 1000/- to Rs. 1500/- per kW (as example).

Interest. As an example, consider the case of a 100 MW thermal power station, constructed at the cost of Rs. 5000 per kW, the capital cost including the cost of land, building etc. would be Rs. 500 × 10^6 (Rs. 500 million). When such a heavy sum is invested in a plant, the interest which could have been otherwise obtained by depositing the amount with a banker or by investing it in a profitable business would be a substantial amount. So a plant must pay for the interest also and since the capital investment is fixed, the amount of interest is also fixed and it goes to fixed costs of the plant.

Taxes. A company engaged in generating and selling electric energy (power) is subject to the property tax. Generating equipment, transmission system etc. come under the property of the plant and tax is to be paid to the government as per rules. Besides this, profitable earning of the company is subject to company tax.

Cost of Transmission. Generally transmission system is of two types :

(i) **Primary Transmission Lines (Distribution System).** Under this, following are the items are considered :

1. Cost of construction of towers and lines (High tension lines).
2. Compensation for the right of way.
3. Cost of construction of the receiving stations and substations — its building and the equipment.

Cost of transmission and distribution system or network is also comparable to the capital cost of the plant.

Sometimes it may be of the order of 30% of the total capital cost. In this case plants located near the load centre 1have got a definite advantage. In case of hydroelectric stations, which are usually located away from the load centre, the cost of transmission system is substantial.

The high tension (voltage) transmission lines pass over the lands owned by others and the company has to compensate for the risk by giving certain amount to the land owners and get the **right of way**.

For the receiving stations and substations, equipment is provided. The cost of this comes under the cost of primary transmission system.

(ii) **Secondary Distribution System.** This is considered under the heading customer's cost.

Depreciation or Amortization. Depreciation may be defined as decrease in value of the equipment caused by the wear and tear due to constant use of an equipment. An device or equipment has a span of working life during which it deteriorates or depreciate. The exact rate of depreciation is difficult to determine. Even if the plant is no more functioning and out of regular use, it can fetch some value, which is called its **salvage value**. An amount has to be set aside annually, from the income of the plant, to meet the depreciation due to wear and tear. The capital investment for the plant installation should be recovered during the life span of the equipment, so that it can then be replaced by a new plant. Various methods for providing depreciation are given below :

(i) Straight line method

(ii) Sinking fund method

(iii) Diminishing value method.

(i) **Straight Line Method.** This method provides for setting aside each year an equal proportional part of the cost based on the life span of the property. Thus if an equipment costs Rs. 50,000 initially and if the salvage value or scrap value of the equipment is estimated to be Rs. 5000 only after being use for 15 years, then according to straight line method, the annual charge towards depreciation would be Rs. $\frac{50000-5000}{15}$ = Rs. 3000 only. This is an extremely simple method is easy to apply when it is assumed that the only causes for requirement are wear and tear. This is not necessarily true always.

In general,

Let P = Initial investment to install the plant

S = The salvage value of the plant at the end of plant life

n = Life of the plant-years

r = Annual rate of interest on the invested capital

A = The amount set aside per year to replace the plant at the end of nth year.

Then the amount to be set aside annually is calculated by

$$A = \frac{P-S}{n}$$

This method however does not take into account any interest earned by the depreciation amount. The interest so earned is taken as interest.

This is an extremely simple method and is easy to apply when it is assumed that the only causes for requirement are wear and tear. This is not necessarily true always.

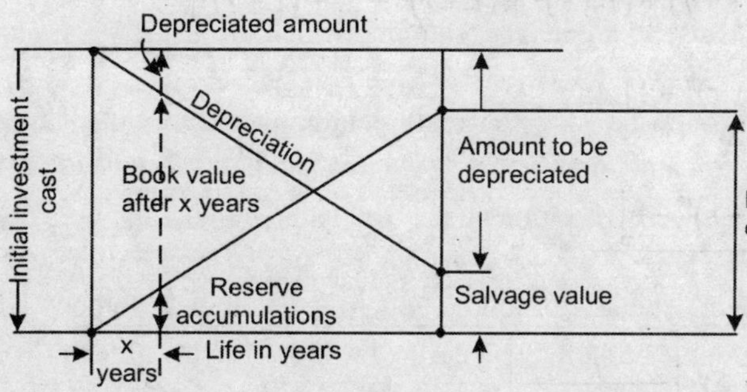

Fig. 28.1 *Depreciation by straight line method.*

Figure 28.1 shows the depreciation by straight line method.

(ii) **Sinking Fund Method.** In this method of calculation of annual depreciation deposits, it is assumed that the occurred depreciation of the property is the amount already accumulated in a **"sinking fund"** that was begun when the equipment was first put into service and the annual deposits of which are such that at compound interest the amount at the end of the expected life of the equipment will be equal to the initial cost.

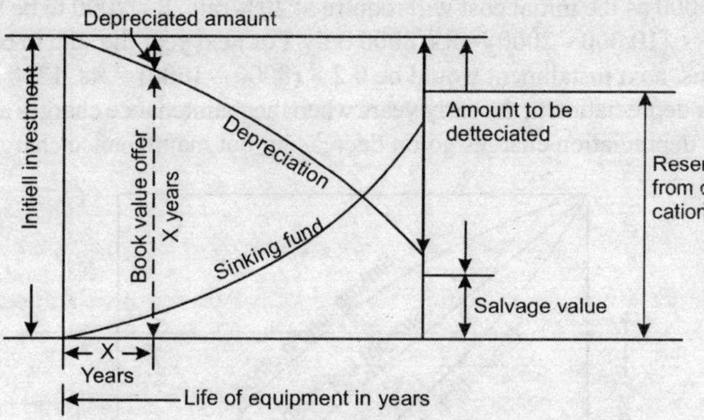

Fig. 28.2 *Sinking fuel method.*

Let A be the amount set aside at the end of each year for n years; then

The amount set aside at the end of the first year = A

Amount at the end of the second year = A + Interest on A

$$= A + Ar$$

$$= A(1 + r)$$

Amount at the end of third year = $A(1 + r)$ + Interest on $A(1 + r)$

$$= A(1 + r) + Ar(1 + r)$$

$$= A(1 + r)^2$$

$\therefore$ Amount at the end of nth year = $A(1 + r)^{n-1}$

$\therefore$ Total amount accumulated in n years is given by the sum of all these amounts accumulated in n years. Let this be y.

$\therefore$

$$y = A + A(1 + r) + A(1 + r)^2 + \ldots + A(1 + r)^{n-1}$$

$$= A[1 + (1 + r) + (1 + r)^2 + \ldots + (1 + r)^{n-1} \quad (1)$$

$$\therefore \quad y(1+r) = A[(1+r) + (1+r)^2 + (1+r)^3 + ... + (1+r)^n] \tag{2}$$

$\therefore$ Eqs. (2) and (1) gives

$$yr = [(1+r)^n - 1]A$$

But

$$y = P - S$$

$$\therefore \quad P - S = \frac{\left[(1+r)^n - 1\right]A}{r}$$

or

$$A = (P-S)^r \left[\frac{r}{(1+r)^n - 1}\right]$$

$$A = (P-S)\left[\frac{r}{(1+r)^n - 1}\right] \tag{3}$$

(iii) **Diminishing Value Method.** In this method, the amount set aside is calculated as a fixed percentage of the balance amount each year. So the installment each year goes on decreasing as the life of the plant increases. Thus an equipment having Rs. 10,000 as the initial cost will require at 20% rate, Rs. 2000 to be set apart during the first year. The balance becomes Rs. (10,000 – 2000) = Rs. 8000 only. For next year the sum to be set aside would be 0.2 × 8000 = Rs. 1600 only. Thus, next installment would be 0.2 × (8000 – 1600) = Rs. 1280 and so on. This method gives the heaviest charges for depreciation in the early years when the maintenance charges are minimum. Thereafter with the passage of time the depreciation charges go on decreasing but maintenance charges increase.

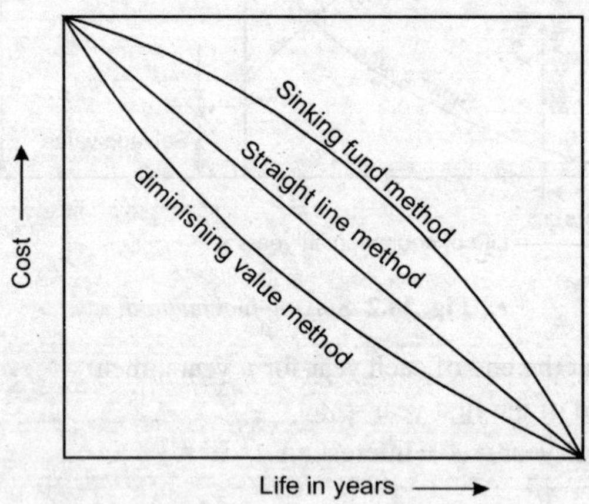

Fig. 28.3 *Representation of depreciation methods.*

Figure 28.3 shows the different methods of depreciation.

Insurance. Costly equipment installed must be insured specially for the fire risks involved. A fixed sum is always paid as premium for insurance. The rate of annual premium may be 2% to 3% but this small percentage becomes substantial amount on account of the heavy capital cost of the plant.

Management. Top management officials drawing substantial amount as salaries and allowances are to be retained by an undertaking irrespective of whether the power is generated or not. Amount paid to such employees also goes under fixed costs.

General Maintenance. Some amount is required for general maintenance, whether or not the plant were operated.

28.2.2 Operating or running costs (Variable costs)

These costs are directly proportional to the energy generated and usually calculated on the basis of annual energy production in kW-hr (kWh). These costs have following main components.

(a) **Cost of Fuel.** They are directly proportional to the amount of energy generated and are about 75 to 85% of the total operating costs.

(b) **Cost of Labour.** These are expenses made to the labour while the plant is running. Payments are made to different crews who are engaged in supervision, inspection, handling of material, boiler and turbine operations, starting and stopping of units etc.

(c) **Cost of Water for**

(i) Boiler fuel

(ii) Condensers; cooling water needed for condensers

(iii) Cooling and house service.

(d) **Cost of Oil, Waste and Consumable Supplies.** These include lubricating oils and greases, cotton waste, chemicals for water treatment plant cleaning agents etc. The cost of these materials is also proportional to the amount of energy generated. These costs increase with the increase of life of the power plant as the efficiency of the power plant decreases with the age.

(e) **Cost of Maintenance.** This means the labour and material cost of the maintenance. These include costs of preventive maintenance, inspecting, cleaning and overhauling of the apparatus on regular schedule to avoid possible breakdown during service.

(f) **Cost of Supervision.** It includes the salary of a station superintendent, chief engineer, chemist etc.

(g) **Cost of Operating Taxes.** They depend on the magnitude of the output rather than size of the investment. They include taxes such as income, excise, sales and employment.

Total operating expenses cannot be expressed as a percentage of the investment since they vary widely with the type of the plant, its geographical location and its capacity.

The range of percentage analysis of the individual items may be indicated as follows :

Item	*Percentage*
Fuel	75 – 85
Labour	5 – 15
Maintenance	5 – 20
Supplies	1 – 5
Supervision	1 – 2
Operating taxes	0 – 10

28.2.3 Customers element or cost

The third general element of cost is proportional to the number of customers. Its components are :

A. **Cost of Secondary Distribution System.**

(a) Depreciation, interest, taxes, insurance on the capital cost of the secondary distribution system.

(b) Line and transformer maintenance and inspection charges.

B. **Labour Cost of Collecting Revenue**

(a) Meter reading

(b) Office (clerical)

(i) Records

(ii) Billing, collecting and accounting.

C. **Cost of Franchise (nominal value)**

D. **Publicity.**

(i) Public relations

(ii) Advertising etc.

28.2.4 The Investor's Profit

The last element which was mentioned as being a part of the cost of electrical energy to the consumer was Investor's profit. As would be expected, this element is also variable owing to variable business conditions prevailing in different localities at different points/times. After interest has been paid and all contingencies present or expected, cared for, the surplus may be declared as dividend.

The private industrialist is entitled to all the profit he can make in the face of business competition.

28.3 TARIFFS FOR ELECTRICAL ENERGY (RATES FOR ELECTRICAL ENERGY)

The total cost of generation of electrical energy is the sum of fixed, operating and consumer costs, and is to be recovered from the consumers to whom the electrical energy generated is supplied. In case of a private company generating electrical energy, another element of cost namely investor's profit is to be included in the total cost to be recovered from the consumers. There are different types of consumers namely domestic, commercial and industrial.

Tariffs or energy rates are different for different types of consumers. Thus whatever may be the type of consumer, the tariffs should recover the total cost of generation of electrical energy. The tariffs must cover the following items:

1. Recovery of capital or fixed cost
2. Recovery of running costs
3. Recovery of consumers cost
4. Recovery of reasonable profit on the invested capital as the power plant is considered a profitable business for both private enterprise and government also.

Tariffs or rates considered for the recovery of the total cost of generation of electrical energy should be such that

1. It should be easier and simpler to understand.
2. It should provide low rates for high consumption (At present, because of the scaresity of power, reverse is the case).
3. It should be uniform over large region and population.
4. It should give benefits to consumers having high load factors.
5. It should provide incentive for using power during off-peak hours thereby the load factor of the power station will increase.
6. It should have a provision of penalty for low power factor.

28.3.1 General tariff (rate) form

The following equation gives the general type of tariff.

$$z = ax + by + C.$$

where

z = Total amount of bill for the period considered

x = Maximum demand in kW

a = Rate per kW of maximum demand

y = Energy consumed in kWH or kWh during the period considered

b = Energy rate per kWh

C = Constant amount charged to the consumer during each billing period. This charge is independent of demand or total energy because a consumer that remains connected the system incurs expenses even if he does not use energy.

28.3.2 Types of tariffs or rates

The various forms of rates used for charging to consumers as per their energy consumed and maximum demand are desired from the general formula as given above. Accordingly, they are :

1. Flat demand rate
2. Straight meter rate
3. Block meter rate
4. Hopkinson demand rate (Two part tariff)
5. Doherty rate (Three part tariff)
6. Wright demand rate

1. **Flat Demand Rate.** This type of charging depends only on the connected load and is independent of the energy consumed. This rate expresses the charge as so many rupees per unit of demand (kW). This system eliminates the use of metering equipments and man power required for the same.

This can be expressed as :

$$z = ax.$$

Under this system of charging, the consumer can theoretically use any amount of energy upto that consumed by all the connected load at 100% use factor continuously at full load.

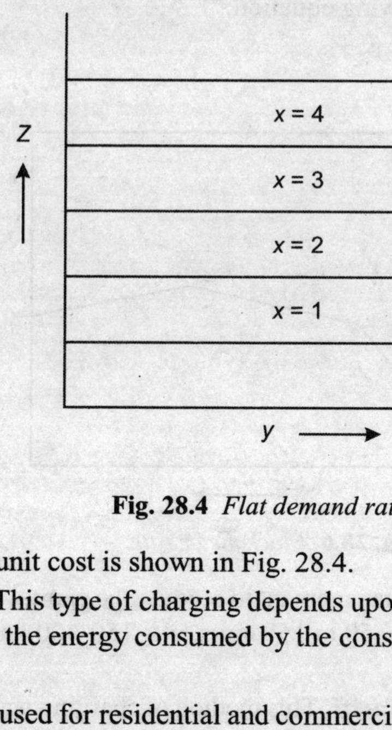

Fig. 28.4 *Flat demand rate.*

The variation of total cost and unit cost is shown in Fig. 28.4.

2. **Straight Line Meter Rate.** This type of charging depends upon the total energy consumed by the consumer. The bill is directly proportional to the energy consumed by the consumer and is represented by :

$z = by$

This type of tariff is sometimes used for residential and commercial consumers. Figure 28.5 shows the variation of bill according to the variation of energy consumed.

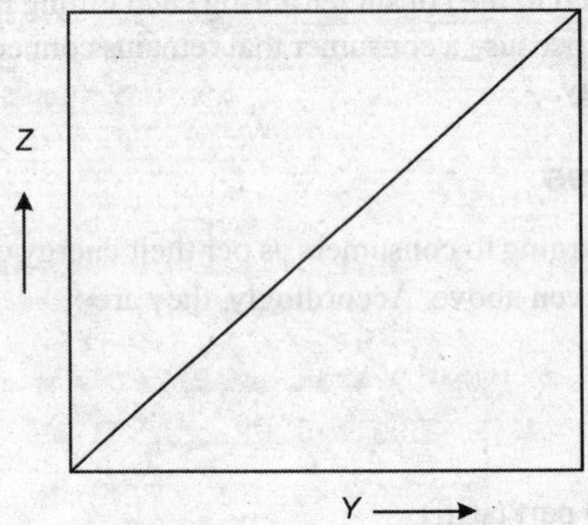

Fig. 28.5 *Straight meter rate.*

Advantage of this system (tariff or rate) is its simplicity.

Disadvantage or the drawback of this system is, the consumer using no energy will not pay any amount although he has incurred some expenses in the power station.

The second drawback is that the rate of energy is fixed and therefore this method of charging does not encourage the consumer to use more power unless the tariff is very low.

3. **Block Meter Rate.** Here the **"block"** means a certain group of units. The straight line meter rate charges the same unit price for all magnitude of energy consumption. The increased generation (and for this purpose consumption) spreads the item of fixed charge over a greater number of units of energy and, therefore, the price of energy should decrease with increase in consumption. To overcome this difficulty, the block meter rate is used. In this method, the charging energy is done according to the following equation :

$$z_1 = b_1y_1 + b_2y_2 + b_3y_3 + ...$$

where $b_3 < b_2 < b_1$ and $y = y_1 + y_2 + y_3 + ...$

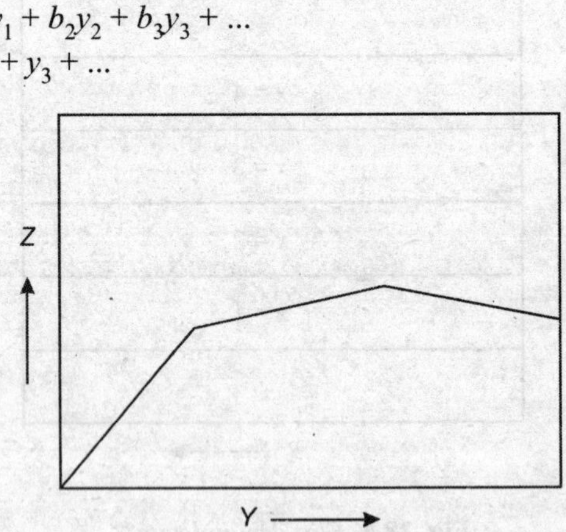

Fig. 28.6 *Block meter rate.*

The level of y_1, y_2, y_3 etc. is decided by the management to recover the capital cost of the plant. Here y_1, y_2, y_3 are the different blocks – say y_1 – 1 to 50 units, y_2 – 50 to 150 units, y_3 – 150-400 units and so on. The variation of bill according to this method is shown in Fig. 28.6.

4. **Hopkinson Demand Rate or Two Part Tariff.** This method of charging was proposed by Dr. John Hopkinson in 1892. This method charges the consumer according to this maximum demand and energy consumption. This can be expressed as :

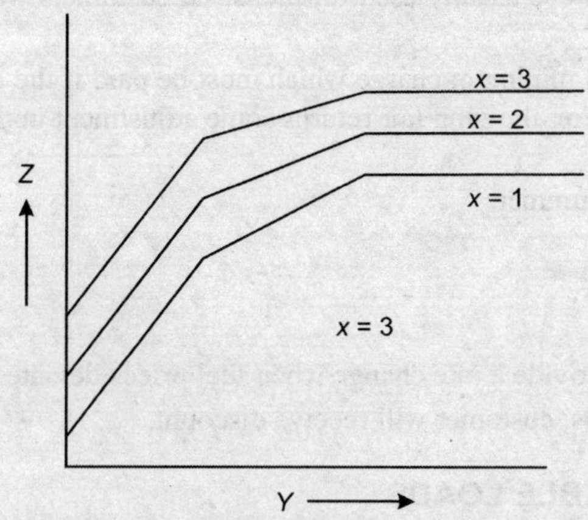

Fig. 28.7 *Hopkinson Rate.*

$$z = a + by$$

This method requires two meters to record the maximum demand and the energy consumption of the consumer.

This method is generally used for industrial customers. The variation z with respect to y taking x as parameter is shown in Fig. 28.7.

5. **Doherty Rate or Three Part Tariff.** This type of charging rate was suggested by Henry L. Doherty at the beginning of the twentieth century. According to this method of charging, the customer has to pay some fixed amount in addition to the charges for maximum demand and energy consumed (Hopkinson method). The fixed amount to be charged (customer charge) depends upon the occasional increase in prices and wages of the workers and so on. This

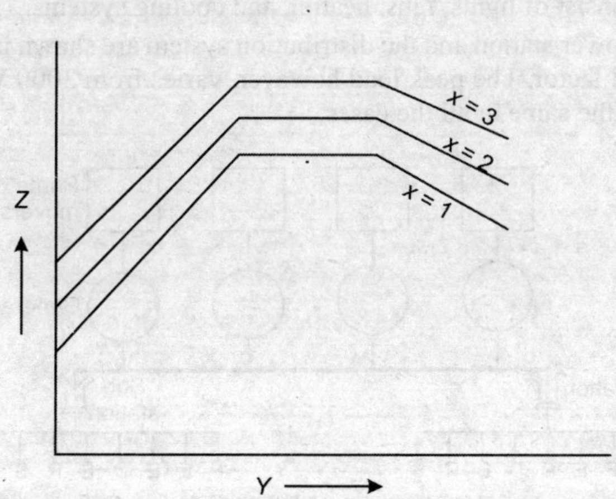

Fig. 28.8 *Doherty Rate.*

is expressed as :

$$z = ax + by + C$$

The variation of z with respect to y taking x as parameter is shown in Fig. 28.8.

6. **Wright Demand Rate.** Mr. Arthur Weight of England introduced this tariff in 1896. This rate intensifies the inducement by lowering both demand and energy charges for a increase in maximum demand so that there is an

improvement in load factor. This rate is usually used for industrial customers who have some measure of control over their maximum demands.

The rate is modified by stating a minimum charge which must be paid if the energy for the billing period falls below the amount by such charge. For allowing fair returns some adjustment in the rate forms are provided. Some of them are :

(a) Higher demand charge in summer
(b) Wage adjustment
(c) Tax adjustment
(d) Power factor adjustment
(e) Fuel price adjustment to provide a rate change when fuel prices deviate from the standard
(f) For prompt payment of bills, customer will receive discount.

28.4 PROBLEMS OF VARIABLE LOADS

The potential customers for a power station may be categorised as :

1. Domestic
2. Industrial
3. Business houses, offices etc.

Each customer has a connected load which is the sum total of the ratings of all the power consuming items. For example, a domestic customer may have 1 m lamps, fans, heaters, air conditioners, radio, television etc. as connected load. Every item has its standard rating, for example, bulbs may be 15 W, 25 W, 40 W, 60 W, 100 W, 200 W or 500 W. Fans generally consumer 65 W to 200 W depending on the size and type e.g., table fans, ceiling fans, exhaust fans etc. Similarly heaters may be of 750 W, 1000 W, 1500 W or even more. An industrial customer generally have motors to drive machinery. Electric power in large quantities is consumed in heating furnaces, rectifiers etc. Business houses, office loads generally consist of lights, fans, heating and cooling systems.

Various types of loads for a power station and the distribution system are shown in Fig. 28.9 and Fig. 28.10. All the loads have got the same load factor. The peak load however, varies from 3000 W to 250 W although average demand spread over 24 hours is the same in all the cases.

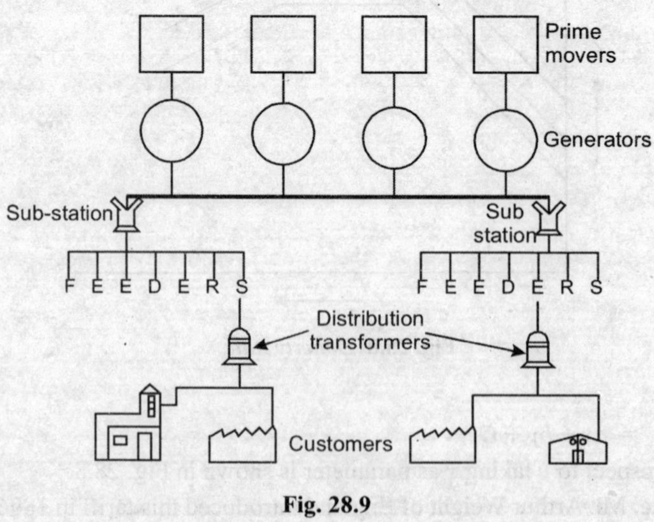

Fig. 28.9

Loads with high value of the ratio of peak load to average load are undesirable.

A power station has to supply power to all customers and customers are generally at liberty to put on any power consuming apparatus at any time. Obviously a customer would normally not put on all the power apparatus at a time and that too for whole day. For example, the power demand of domestic customers is generally high during night hours whereas industrial load is maximum during day time.

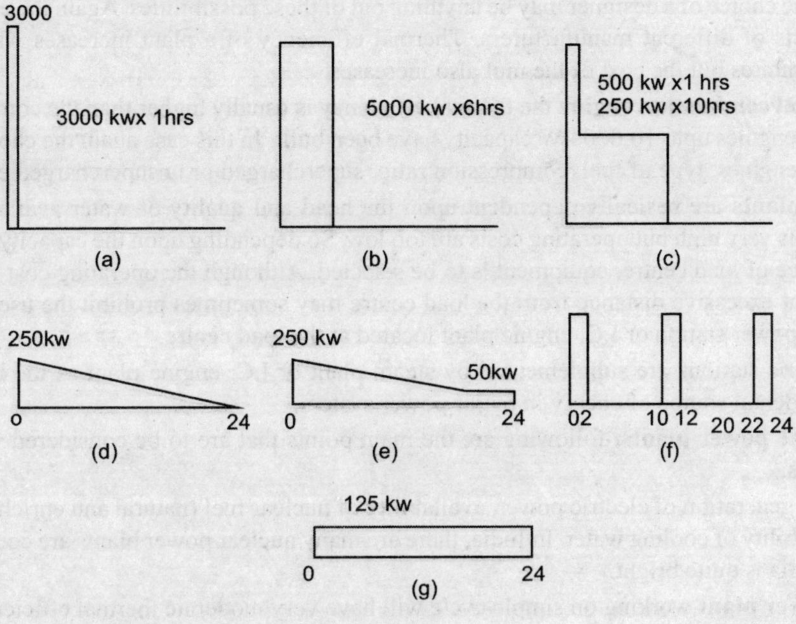

Fig. 28.10

The problem of meeting variable load involves several problems. For example, if a generator is supplying power at 60 MW to a system and if the load suddenly increases to 70 MW, in case of a thermal station, the turbine will require more steam to produce this additional power. To supply extra steam, boiler is to be supplied extra fuel (coal), air and water. Now all this cannot be done instantaneously. Similarly in case of a hydro-electric station, extra discharge of water is require to generate additional power and so on. Governors and other control devices are installed on turbines for this purpose.

To meet the variable loads particularly peak load, following methods are generally used.

1. **Peak Load Plants.** Peak load plants are installed which are operated only during peak load hours. These plants are generally capable of starting quickly so that they can immediately meet the demand. Peak load plants may be diesel engines, gas turbines or even steam plants with quick steam raising boilers. Thermal efficiency of such plants is not an important criteria as these plants operate for limited time only.

2. **Use of Accumulators.** Steam accumulators are used to store steam under pressure which can be used to generate power during peak load periods.

3. **Purchasing Power from Neighbourhood.** Power may be purchased from neighbouring sources of power during peak hours. In India, this practice is common. States purchase power from neighbouring states to meet their demand.

4. **Load Shedding.** In case there is no way out to meet the peak demand, the load shedding is the last alternative. In this case, power supply to a particular feeder is cut off for some time till the load comes down or alternatively load shedding is done by rotation in various areas. With power shortage in most of the states, load shedding in summer has become common feature.

28.5 SELECTION OF EQUIPMENT

Knowing the location and capacity of a plant, the next problem is to select the equipment. For a **thermal power station** the range of pressure may vary from 10 bar to supercritical pressures and the range of temperature may be from 250°C to 650°C. Besides, the cycle used may be simple Rankine cycle, with reheating and regenerative feed heating cycles. So the choice of a designer may be anything out of these possibilities. Again the choice is to be made amongst the products of different manufacturers. Thermal efficiency of a plant increases with the use of high pressure and temperatures but the cost of the unit also increases.

In case of **internal combustion engine** the thermal efficiency is usually higher than the comparable size steam power plant. Diesel engines upto 10,000 kW capacity have been built. In this case again the choice lies between 2-stroke and 4-stroke engines, type of fuel, compression ratio, supercharged or unsupercharged engines.

Hydro-electric plants are basically dependent upon the head and quality of water available. Initial cost of hydro-electric plant is very high but operating costs are too low. So depending upon the capacity, head and quantity of water, and distance of load centre, equipment is to be selected. Although the operating cost of a hydro-electric plant is very low, but excessive distance from the load centre may sometimes prohibit the use of such a plant in favour of a thermal power station or I.C. engine plant located at the load centre.

Some hydroelectric stations are supplemented by steam plant or I.C. engine plant as the hydroelectric plant often is only an incidental source of energy in major power system.

In case of **nuclear power plants,** following are the main points that are to be considered while selecting the nuclear power plants.

Capacity, cost of generation of electric power, availability of nuclear fuel (natural and enriched fuel) and safety measures and availability of cooling water. In India, there are many nuclear power plants are coming. The future of nuclear power in India is quite bright.

Gas turbine power plant working on simple cycle will have very moderate thermal efficiency and the cost of such a plant is also moderate. The cost increases as the simple plant is modified by the inclusion of other equipments such as intercooler, reheater, regenerators and the gain in the thermal efficiency and thereby a reduction in operating cost may justify this additional expense in the first cost. Similarly, the availability of the fuel is also a major consideration because a gas turbine plant (combined with steam plant with heat recovery boiler) at Uran in Maharashtra, is not working at its full capacity for want of fuel. Similarly the availability of gasifiers (from coal which is available in ample quantities) has to taken into consideration while considering the design of the plant for power generation.

28.6 SELECTION OF TYPE OF GENERATION

The following points are usually considered in selecting the type of generation for a power station :

1. Capacity of the plant
2. Probable load factor
3. Cost of fuel and transport facilities
4. Space required
5. Availability of water
6. Degree of reliability
7. Interest and depreciation.

28.7 CHOICE OF SIZE AND NUMBER OF GENERATING UNITS

No hard and fast rules can be given for this but, however, looking at the load curve of the station one can guess for the total generating capacity of a plant must be more than the predicted maximum demand. Obviously, the minimum number of generators can be one but this will not be a wise suggestion. As the load on a power station is never constant, owing to variable demands at the different times of the day, the generator will have to run continuously at variable loads, which will be much less than the rated capacity of the generator for most of the time, without any provision for the maintenance. So a power station which is expected to be reliable in service, must have atleast two generators, whatever the total capacity of the plant may be.

The best way of deciding the size and the number of generating sets in a station is to select the number of sets in such a way so as to fit in the load curve as closely as possible, so that the capacity of the plant may be used efficiently. Extra space capacity is not desired as it increases the capital expenditure. The aim should be to have units of different capacities which will suitably fit in the load curve so that most of the generators when in use can be operated at nearly full load.

28.8 LOAD SUPPLY ARRANGEMENTS

Entire load of the system may be divided into two parts like –

(a) Base load

(b) Peak load

(a) **Base Load Supply.** In interconnected systems with many generating stations of various types, the choice of station to supply the varying load is of considerable economic significance.

Base load is the load which is supplied for most of the time and which remains more or less constant. The main considerations for base load provision are :

(i) high efficiency and

(ii) high availability of the system.

Even a higher capital cost is sometimes favoured if it can ensure resultant given in efficiency, as the cost is spread over a large total energy value.

Thermal Stations. Large coal or oil-fired stations have a capital cost, say, of Rs. 1200 to Rs. 1600 per kW. With an advanced steam cycle they can reach efficiencies upto 40% corresponding to a running cost of about 2 paise/kWh. To attain such results the stations must operate continuously at or near full load, in order to spread the capital charges and to avoid the operating difficulties that arise (especially with boiler plant) with varying loads. Overall generating costs may be in the range 4 paise/kWh for 100 percent load factor to 6.5 paise/kWh for a load factor of 50 percent. Efficient, but expensive modern steam stations should thus be operated to provide the base load; it must, however, not be forgotten that they will eventually be superseded by even more efficient stations and may have to be regulated to two shift working.

Nuclear Stations. The capital cost of a nuclear station is about twice that of a thermal station, but its operating cost is only about one half and in consequence it is even more desirable to operate a nuclear station on a base load of high load factor on economic grounds, reinforced by the technical undesirability of variable load working. Load factors of 90 percent or more have been achieved.

Hydro Stations. Countries with ample water power resources may use them for base load provision, although it is possible that steam generation would be cheaper. Run-of-river stations can usually give firm or base load power according to the minimum flow, and by addition of storage the firm power can be increased upto the limit of average rainfall.

The cost of the civil engineering works associated with the storage for a station to operate at 100 percent load factor is high as the reservoir has to accommodate variations in annual rainfall. The station plant capacity corresponds only to the meanflow so that the cost per unit of rating tends to raise with the load factor. At low load factor the same (or smaller) civil works could be associated with a much higher station power capacity lowering the cost per unit of energy. The cost of a hydro plant is not easy to predict, but once the plant is built, it is settled. In contrast the cost of steam station can be fairly, accurately forecast, but future fuel cost is problematic.

Hydro plant can thus be applied to the base load provision particularly where storage is cheaply obtainable. The plant has the advantage of very high availability and may run for years with only a few very brief shut downs for inspection.

(b) **Peak Load Supply.** It is difficult to secure an economic solution to the problem of peak load provision. The essential requirements are :

(i) Ability to start and take full load within, say, 30 min. and to handle rapidly varying loads.

(ii) Low capital cost in view of the small annual output with the efficiency only a secondary consideration.

(iii) A site near the load centre or accessible transmission facilities to avoid additional transmission costs.

It is advantageous if the peak load plant can in emergency be used as a back-up base load supply. The following are the possible methods of peak load provisions.

Obsolescent Steam Stations. As new stations are built, they supply the increasing base load and replace older base load stations which can be used as two shift, one shift and eventually peak load stations. After 20 years the capital charges on such stations will have been met. However the large stations now coming into this category are less suitable for operation at very low load factors and other means must be sought.

New Steam Stations. One possibility is a steam station designed for peak loads, with a simple thermal cycle forced circulation boilers and rapid start turbines. The economics of building a new peak load station rather than a new base load station must be carefully considered.

Gas Turbine Stations. Gas turbines, particularly of the open cycle type can start within few minutes, facilitate remote or automatic control and because of the small space needed can be located near a load centre or in the premises of a major steam station. Although the size is limited at present to about 200 MW, it is estimated that 15–20 percent of total generating plant in industrial countries will eventually comprise peak load units driven by gas turbines.

Hydro Stations. Its quick starting facility and ability to handle rapidly varying loads makes the hydro station ideal for peak load operation. Such stations are however, often remote from the load centre and the high capital cost and low running cost appear, at first sight, to be the reverse of what is required.

Pumped Storage Stations. If a pumped storage station is used in conjunction with a hydro station the energy cost is low and the efficiency of the storage plant is of minor importance. But in conjunction with a steam or nuclear station the energy cost is of greater significance.

Load Shedding. A final method of dealing with (though not supplying) peak load is by load shedding i.e., by reducing the system voltage and frequency or by switching out certain loads. Voltage drop or frequency reduction causes the load reduction can be achieved but to a limited extent. Otherwise public complaints will be there.

Disconnection of loads is a last resort. Considerable planning in necessary to reduce interruption to essential services like hospitals, water supply and transport systems etc.

Fortuitous Load Stations. Wind, tidal and solar stations are able to generate energy at unpredictable or uncontrollable times. Such stations must therefore, if operating alone or if supplying the major part of the energy network, operate in conjunction with some form of storage. Battery storage is appropriate for small units e.g., a 4 kW wind generator, but pumped water storage must be used with larger installations such as a tidal scheme.

If, however, the stations are feeding into a network large enough to absorb the full output at any time the kilowatt hours generated will result in a corresponding saving in fuel or water consumption at the conventional station. Under such circumstances the fortuitous load stations deliver no firm power and their installation affects no corresponding saving in plant capacity elsewhere on the network. In assuming the economics of such a scheme only the fuel saving can be considered. If storage is added some firm power can be generated but the cost of this may well be higher than that of generating the same power from a conventional station.

Stand-by Stations. In the country where there is a closely interconnected system with numerous generating stations, standby stations are not necessary as part of the main supply system. Individuals large consumers e.g., important manufacturers, hospitals, pumping installations etc. may however consider it desirable to install standby plant to guard against a supply failure. The required outputs do not usually exceed a few hundred kilowatts, and diesel units, which can be started almost instantaneously are usual.

There are circumstances where a load centre is supplied by an overhead transmission line from a distant hydro station, and a standby plant must then form part of the main power system. Steam stations with quick starting forced circulation boilers have been used in such cases.

28.9 LOAD DIVISION BETWEEN SERVICES

In case the load curve has a very high peak value, it is usually supplied from two or more services by interconnection. In that case, total load as shown on load duration curve may be divided into two parts e.g., the base load and the peak load. Base load is supplied by one of the services and the peak load is supplied by the other. In such cases the load is economically apportioned to the various systems in order to achieve the overall economy. In such cases it is not very necessary to interconnect the two systems of the same type. For example, if the base load is supplied by the steam station, then it is not very necessary that the peak load may also be supplied by a steam station. A hydro-electric station can very well be used for supplying the peak load. Similarly, a hydroelectric station can be used for supplying the base load and in that case the peak load can be supplied by steam station or a diesel engine station or any other suitable unit. However, the selection of the services for supplying the base load or peak load is made on that basis of the requirements and the ability of the various services to meet those requirements. For example, as far as the base load is concerned, the requirements of a plant supplying are :

1. The operation cost of the system should be minimum as it has to operate for most of the time.
2. It should be able to supply the load continuously.
3. Maintenance requirement of the plant should be minimum.
4. Capital cost involved should be minimum.
5. Plant should be such that it can easily be located near the load centre.
6. It should require minimum number of operators.
7. Spares etc. should be readily available.

In view of the above requirements, let us consider various types of power plants for their suitability to meet the base load. As far as the hydroelectric stations are concerned, the operating cost in that case is minimum as practically

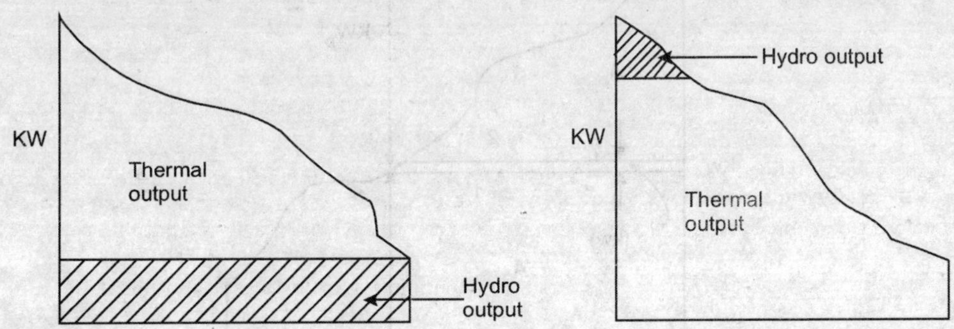

Fig. 28.11

no fuel is required for the purpose of power generation. The problem of procuring the fuel is also eliminated in this case. Also in this case the cost of maintenance is lower than that of other services. However, the major problem in this case is that the initial cost of the plant is very high and sometimes prohibitive. Besides this, hydro-electric plants cannot be necessarily located near the load centre as the same can be located at the site suitable for it. In this case there is more or less dependence on the availability of water which in turn depends on the natural phenomenon of rain.

As far as steam stations are concertained, the capital investment in this case as compared to hydroelectric stations is less but with the modern trend of using higher pressures for the purpose, the cost of such stations has considerably increased. But this increased cost has resulted in lower operating cost so much such that even it may compete with that of hydro-electric stations. Major advantage in this case is that these plants can be easily located near the load centre, as such the cost of transmission lines and the losses occuring can be minimized which also results in economy in operation.

However, in this case the requirement of maintenance is slightly higher.

Diesel power stations have got the major drawback of limited unit generation capacity. So far as their suitability for the base load plants is concerned, these do not find much of favour.

In the case of nuclear power stations, the initial cost is high although the operating cost can be as low as that of a steam station. However, due to limited availability of the fuel for such plants, these are not much favour. Because of the advancement of technology, nuclear fuels can be made available for such stations. But wherever such plants are constructed, these are invariably used for base load as otherwise also such plants suit to constant load conditions as economy in operation in such plants can be achieved only when these are used as base load plants.

As far as the suitability of a plant for supplying the peak load is concerned, the requirements are :

1. Capital cost involved should be minimum.
2. The plant should be capable of being started from cold conditions within minimum time.
3. The plant should be able to withstand peak load for sometime in case of emergency.
4. Operating cost should not be high.

In view of these factors, a hydro-electric station can also be considered for this purpose. A hydroelectric station can be easily started from cold conditions as no warming up period is required in this case. In this case the cost of operation is also low. Major disadvantage in this is that the initial cost is high. However, in this case there is no requirement of fuel, so whenever the output of hydroelectric station is to be utilized, it can be used as a peak station also. Particularly, in case, where the quantity of water available for the purpose of power generation is limited, such stations are used for meeting the peak load in case of a interconnected system.

As far as nuclear power stations are concerned, they do not find their use as peak load plants due to the reasons already mentioned. Diesel engines can also be used for meeting the peak load as they can be easily started from

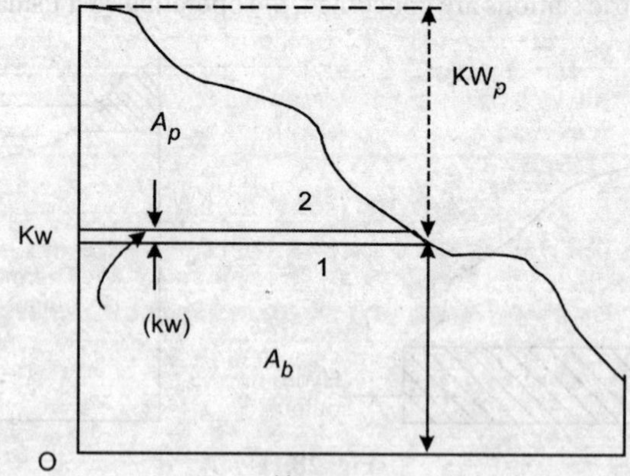

Fig. 28.12

cold conditions and the initial cost in this is also not high. Additional advantage in this case is that the number of auxiliaries required is also limited as a result of which the maintenance work required is also less. Diesel plants have got good overload capacity, minimum being 10% as per Indian Standard Specifications. Thermal efficiency of diesel engines being high, the operating cost is also less as compared to that of a steam power station of equivalent capacity.

Various factors have been considered for the suitability of the various types of plants for base load and peak load. However, this suitability of plant for a particular application also depends on the local conditions. Mostly such cases are decided taking into account the local conditions. However, if the combination of two services is to be used, in that case the next problem is the division of load between the services. All the loads cannot be transferred to one service and also depending upon the operating characteristics of the various services, the load between the services is decided in such a way that the overall economy is achieved. This division of load can be easily manipulated by viewing the load duration curve of the plant. In Fig. 28.12 let the base load be supplied by a plant having the annual cost equation as :

$$(Rs)_1 = A_1 + B_1\text{kW} + C_1\text{kWhr} \qquad (1)$$

For the plant supplying the peak load let the equation be

$$(Rs)_2 = A_2 + B_2\text{kW} + C_2\text{kWhr} \qquad (2)$$

As base load plant is operated for most of the time, normally a plant having $C_1 < C_2$ is used for meeting the base load. Let $B_1 > B_2$.

Load duration curve is shown in Fig. 28.12. Let the load between two services be divided by an arbitrary line drawn on the load duration curve represented by (1). Under these conditions let kW_b be the kW for the base load plant and kW_p be the load for peak load plant. In this case the total operating cost of the combination will be

$$T_1 = A_1 + A_2 + B_1\text{kW}_b + B_2\text{kW}_p + C_1A_b + C_2A_p$$

where A_b is the area of curve for base load plant and A_p that of peak load plant.

If now the base power is extended by the amout of $d(\text{kW}_o)$ to the line (2), the total operating cost of the combination will modify as

$$T_2 = A_1 + A_2 + B_1(\text{kW}_b + d\text{kW}) + B_2(\text{kW}_p + d\text{kW}) + C_1(A_b + \text{kW}h) + C_2(A_p - d\text{kW}h)$$

∴ The change in cost is then

$$T_2 - T_1 = (B_1 - B_2)\, d\text{kW} + (C_1 - C_2)\, d\text{kW}h$$

The optimum condition requirements are that above change must be zero i.e.,

$$h = \frac{B_1 - B_2}{C_2 - C_1}$$

where h represents the hours per years.

Thus division of load between the two services can be achieved and this results in overall economy in operation.

SOLVED PROBLEMS

Example 28.1 *A plant costing Rs. 80000 has a useful life of 12 years. Find the amount which should be saved annually to replace the equipment at the end of than time using straight line method. Assume a salvage value of Rs. 8000.*

Solution

Amount to be saved in 12 years

$$= \text{Rs. } 80000 - 8000$$

$$= \text{Rs. } 72000$$

∴ Amount to be saved annually by straight line method

$$= \text{Rs. } \frac{72000}{12}$$

$$= \textbf{Rs. 6000}$$

Example 28.2 *Power generation equipment in a thermal power plant costs Rs. 15,75,000 and a useful life of 25 years. If the salvage value of the plant be Rs. 75000 and the rate of annual compound interest be 5%, calculate the amount to be saved annually for replacement of equipment after the end of 25 years, by sinking fund method.*

Solution

Annual payment for sinking fund is given by

$$A = (P - S)\left[\frac{r}{(1+r)^{25} - 1}\right]$$

$$= (1575000 - 75000)\left[\frac{0.05}{(1+0.05)^{25} - 1}\right]$$

$$= \frac{1500000 \times 0.05}{(1.05)^{25} - 1} = \frac{15000 \times 5}{3.3863 - 1}$$

$$= \frac{15000}{2.3863} \times 5$$

$$= \textbf{Rs. 31.43}$$

Example 28.3 *Determine the annual cost of a feedwater softener from the following data :*

Cost	*Rs. 80,000*
Salvage value	*Rs. 5%*
Life	*10 years*
Annual repairs and maintenance cost	*Rs. 2,500*
Annual cost of chemicals	*Rs. 5,000*
Labour cost per month	*Rs. 300*
Interest on sinking fund	*5%*

Solution

(a) Operating cost = Cost of repairs and maintenance + Cost of chemicals
= Rs. 2,500 + Rs. 5,000
= Rs. 7,500/-

(b) Labour charges = 300 × 12 = Rs. 3,600

(c) **Depreciation.** Sinking fund method of depreciation is provided so that sufficient amount is provided at the end of useful life of the existing plant to replace it by a new one. In this case,

Total cost of replacement = Capital cost – Salvage value

$$= 80{,}000 - \frac{80{,}000 \times 5}{100}$$

$$= 80{,}000 - 4{,}000$$

$$= \textbf{Rs. 76,000}$$

$$\therefore \quad A = \text{Annual deposit} = (P - S)\left[\frac{r}{(1+r)^{n} - 1}\right]$$

$$= 76{,}000\left[\frac{0.05}{(1+0.05)^{10}-1}\right]$$

$$= \frac{3{,}800}{(1.05)^{10}-1} = \frac{3{,}800}{1.6289-1}$$

$$= \frac{3{,}800}{0.6289}$$

$$= \text{Rs. } 6{,}042.3$$

$\therefore$ Total annual cost $= 7{,}500 + 3{,}600 + 6{,}042.3$

$=$ **Rs. 17142.3**

Example 28.4 *A coal handling plant for a 2 × 120 MW thermal station costs Rs. 40,00,000 and the salvage value is estimated at Rs. 40,000 at the end of useful life of 20 years. What would be the value half-way through its life based on*

(i) straight line depreciation method

(ii) Reducing balance depreciation method.

Solution

(i) *Straight line method*

Total depreciation in 20 years $= 40{,}00{,}000 - 40{,}000$

$= \text{Rs. } 39{,}60{,}000$

Depreciation in 10 years $= \frac{1}{2} \times 39{,}60{,}000$

$=$ **Rs. 19,80,000.**

$\therefore$ Value of plant at the end of 10 years

$= 40{,}00{,}000 - 1980000$

$=$ **Rs. 2,020,000**

(ii) *Reducing balance method*. In the reducing balance method of depreciation, annual depreciation is taken as percentage of the plant value at the beginning of the year.

Let P = Capital outlay required for a certain installation and

p = annual depreciation at percentage.

$\therefore$ Value at the end of 1st year

$= P(1-p)$

Similarly value at the end of 2 years

$= P(1-p)(1-p)$

$= P(1-p)^2$

$\therefore$ Value at the end of n years

$= P(1-p)^n$

$\because$ Value at the end of 20 years is its salvage value.

$\therefore \quad 40{,}000 = 40{,}00{,}000\,(1-p)^{20}$

$1 = 100\,(1-p)^{20}$

$$\therefore \quad (1-p) = \left(\frac{1}{100}\right)^{1/20} = (0.01)^{0.05}$$

$$= 0.7943$$

$\therefore$ Value at the end of 10 years

$$= P(1-p)^{10}$$

$$= 4 \times 10^6 (0.7943)^{10}$$

$$= 4 \times 10^6 \times 0.1$$

$$= \textbf{Rs. } \mathbf{4 \times 10^5}$$

Example 28.5 *A load having a maximum demand of 100 MW may be supplied by one of the following schemes:*

(a) A steam power station capable of supplying the 30% of whole load.

(b) A steam power station in conjunction with a pumped storage scheme capable of supplying 100 × 10⁶ kWh per year with a maximum output of 40 MW.

Calculate the average cost per unit of electrical energy supplied to the load in each of the two cases from the following data :

Capital cost of steam station	*Rs. 1000 per kW installed*
Capital cost of storage station	*Rs. 700 per kW installed*
Operating cost of steam station	*2.5 paise/kWh*
Operating cost of storage station	*0.25 paise/unit*
Efficiency of storage station	*66%*

Interest and depreciation together on the capital cost amount 15% and no spare plant is considered in the schemes.

Solution

(a) Capital cost of station $= 100 \times 1000 \times 1000$

$$= \text{Rs. } 10^8$$

Interest and depreciation on capital cost

$$= 0.15 \times 10^8$$

$$= 15 \times 10^6$$

Interest and depreciation per unit

$$= \frac{15 \times 10^6}{100 \times 1000 \times 8760 \times 0.3}$$

$$= \text{Rs. } 0.0571$$

$$= \textbf{5.71} \text{ paise/kWh.}$$

$\therefore$ Total cost per unit $= 5.71 + 2.5$

$$= \textbf{8.21 paise}$$

(b) Capital cost of steam station

$$= 60 \times 1000 \times 1000$$

$$= \text{Rs. } 6 \times 10^7.$$

Capital cost of pumped storage scheme

$$= 40 \times 1000 \times 700$$

$$= \text{Rs. } 28 \times 10^6$$

∴ Total capital cost of station

$= 60 \times 10^6 + 28 \times 10^6$

$=$ **Rs. 88 × 10⁶**

Interest and depreciation charges per annum

$= 0.15 \times 88 \times 10^6$

$=$ **Rs. 13.2 × 10⁶**

Pumped storage supplies $= 100 \times 10^6$ kWh per year

Efficiency of scheme = 66%.

Therefore, operating cost of storage scheme

$$= \frac{100 \times 10^6 \times 0.25}{0.66 \times 100}$$

= Rs. 3,78,200

Total units supplied $= 100 \times 1000 \times 0.3 \times 8760$

$= 262.8 \times 10^6$ kWh per annum.

∴ kWh supplied by the steam station

$= 262.8 \times 10^6 - 100 \times 10^6$

$=$ **162.8 × 10⁶ kWh**

Operating cost of steam station

$$= \frac{2.5}{100} \times 162.8 \times 10^6$$

= Rs. 40,70,000

∴ Total cost per annum

= 13200000 + 378200 + 4070000

= Rs. 1,76,48,200

$$\text{Hence cost per unit} = \text{Rs. } \frac{17.6482 \times 10^6}{262.8 \times 10^6}$$

= **6.715 paise**

Example 28.6 *Compare the annual cost of supplying a factory load, having maximum demand of 850 kW with 0.29 as the actual energy used by energy obtained from* (a) *a private diesel generating plant and* (b) *a public supply from the following data :*

Cost of plant in case of (a) Rs. 360,000, Cost of fuel Rs. 57 per tonne. Fuel consumption 0.3 kg/kWh generated. Cost of lubricating oil, water and stores 0.15 paise per kWh generated. Wages Rs. 13,500 per annum. Interest and depreciation 12%. Public supply cost Rs. 55 per kW of maximum demand 1.5 paise per kWh.

Solution

(a) **Diesel plant**

Operating charges

$$\text{Fuel} \rightarrow \frac{0.3}{1000} \times 57 = \text{Rs. } \frac{17.1}{1000}$$

$$= \frac{17.1 \times 100}{1000} \text{ paise/kWh}$$

$= \mathbf{1.71\ paise/kWh}$

Cost of lubricating oil, water and stores

$= 0.15$ paise/kWh.

$\therefore$ Total cost $= 1.71 + 0.15$

$= \mathbf{1.86\ paise/kWh}$

kWh generated per annum $= 850 \times 8760 \times 0.29$

$\therefore$ Operating cost per annum

$$= \frac{1.86}{100} \times 850 \times 8760 \times 0.29$$

$= \textbf{Rs. 40100.}$

Fixed charges

$$\text{Interest and depreciation} = \frac{12}{100} \times 360{,}000 = \text{Rs. } 43{,}200$$

Wages $=$ Rs. 13,500

$\therefore$ Total cost of generation per annum = Fixed cost + Operating cost

$= (43200 + 13500) + 40100$

$= \textbf{Rs. 96,800}$

(b) **Public supply**

Fixed cost $=$ Rs. $55 \times 850 = 46750$

$$\text{Operating cost} = \text{Rs. } \frac{1.5}{100} \times 850 \times 8760 \times 0.29$$

$=$ Rs. 32400.

$\therefore$ Total cost per annum $=$ Rs. 46750 + Rs. 32400

$= \textbf{Rs. 79150}$

A public supply will be cheap.

Example 28.7 *The following data relate to a 10 MW thermal station :*

Cost of plant		*Rs. 1200 per kW*
Interest, insurance and taxes		*5% p.a.*
Depreciation		*5%*
Cost of primary distribution		*Rs. 5,00,000*
Interest, insurance, taxes and depreciation		*5%*
Cost of coal including transportation		*Rs. 44 per tonne*
Operating cost		*Rs. 5,00,000*
Plant maintenance cost :		
Fixed	*Rs. 20,000 p.a.*	
Variable	*Rs. 30,000 p.a.*	
Installed plant capacity	*10,000 kW*	
Maximum demand	*9000 kW*	
Annual energy used	*60% of maximum*	
Consumption of coal	*25,500 tonne.*	

Find the cost of power generated per kW per year, cost per kWh generated and the total cost of generation per kWh. Transmission or primary distribution is chargeable to generation.

Solution

Fixed cost

$$\text{Annual interest, insurance and taxes} = \frac{1200 \times 1000 \times 10 \times 5}{100} = \text{Rs. } 600{,}000$$

Annual depreciation charges = Rs. 600,000

Annual fixed maintenance cost = Rs. 20,000

∴ Total annual fixed charges = 6,00,000 + 6,00,000 + 20,000

= **Rs. 12,20,000** (I)

Operating cost. Cost of primary distribution, taxes, insurance and depreciation @ Rs. 5%

= 5,00,000 × 0.05

= **Rs. 25,000**

Cost of coal = 44 × 25,500

= **Rs. 11,21,000**

Cost of maintenance = **Rs. 30,000**

∴ Total annual operating cost = 25000 + 11,21,000 + 30,000

= **Rs. 11,76,000** (II)

∴ Total annual cost = I + II

= 12,20,000 + 11,76,000

= **Rs. 23,96,000** (III)

Annual energy consumption = 9000 × 0.6 × 8760 kWh

∴ Cost of power generated Rs/kW

$$= \frac{\text{Total annual cost (Rs/year)}}{\text{Actual load kW}}$$

$$= \frac{23{,}96{,}000}{9000 \times 0.6}$$

= **Rs. 443.7 per kW**

Cost of energy generated Rs/kWh

$$= \frac{\text{Annual cost}}{\text{Total kWh generated}}$$

$$= \frac{23{,}96{,}000}{9000 \times 0.6 \times 8760}$$

= **Rs. 0.0506**

= **5.06 paise/kWh**

Example 28.8 *The life of a small economiser costing Rs. 5000 is 15 years and the guaranteed annual saving at full load is equal to Rs. 1500. Determine the total sum saved at the end of 15 years, if at the end of each year, after paying 5% on capital, $2\frac{1}{2}$% wear and maintenance (on capital investment) and making an annual contribution to* *the sinking fund (which is allowed to accumulate at 5%), the remainder is reinvested at 5% compound interest.*

Solution

$$5\% \text{ on the capital investment} = \frac{5}{100} \times 5000 = \text{Rs. } 250$$

$$2\frac{1}{2}\% \text{ on the capital investment} = \frac{2.5}{100} \times 5000 = \text{Rs. } 125$$

Total Rs. 375

Annual payment to the sinking fund

$$A = 5000\left[\frac{r}{(1+r)^n - 1}\right]$$

$$= \frac{5000 \times 5}{100}\left[\frac{1}{(1+0.05)^{15} - 1}\right]$$

$$= \textbf{Rs. 231.6}$$

∴ Total annual charges

$$= 375 + 231.6$$

$$= \textbf{Rs. 606.6}$$

Saving on fuel charges =Rs. 1500

∴ Net annual saving $= 1500 - 606.6$

$= \textbf{Rs. 893.4}$

∴ If the saving is reinvested at 5% compound interest on the total savingt in 15 years is given

$$= 893.4 \times \frac{(1.05)^{15} - 1}{0.05}$$

$$= \textbf{Rs. 19270}$$

Out of this saving, a new economiser could be purchased.

Example 28.9 *A power station is offered Rs. P per tonne of ash produced by its boilers and the coal costs. Rs.* $\frac{k}{x+c}$ *per tonne where k and c are constants and x is the ash content of the fuel. If the cost of firing is proportional to the ash control of the fuel, find the best value of x so that the total outlay per tonne of coal incurred by the power company shall be a minimum.*

Solution

Value of ash per tonne ash $= xP$

and cost of fuel $= \dfrac{k}{x+c}$

Let cost of stoking $= A + Bx$

$$\therefore \quad \text{Total cost per tonne} = C = \frac{k}{x+c} + A + Bx - xP$$

$$\therefore \quad \frac{dc}{dx} = 0 = \frac{-k}{(x+c)^2} + B - P$$

$$= -k + (B - P)(x + c)^2$$

$$\therefore \quad (x + c)^2 = \frac{k}{B - P}$$

$$\text{or} \quad x + c = \sqrt{\frac{k}{B - P}}.$$

$$\text{or} \quad x = \sqrt{\frac{k}{B - P}} - C;$$

when $B = P$, then $x = \infty$.

i.e., if we receive as much for the ash as the extra cost of stoking, we can have 100% of ash in the coal.

In case the P is negative, x is a minimum, since we must pay for disposal of the ash.

The increased cost of stoking and ash disposal now offsetting the reduction in the price of coal.

Hence, $x = 0$ when $k = c^2 (B - P)$

Substituting this value in $C = \frac{k}{x+c} + A + Bx - xP$

We get

$$\boxed{C = \frac{k}{c} + A}$$

Ash is advantageous in that it protects the fire-bars.

Example 28.10 *On an isolated mine the fuel supply for the boilers was obtained by drawing upon the hard woods which grew in the vicinity of the mine, and over a period of 15 years this practice exhausted the supply within a radius of 18 km.*

The management then decided to replace the steam plant which would operate on the remaining soft woods, the distribution of which can be taken as the approximately equal to the original distribution of the hard wood.

From the following figures determine the saving effected by this conversion in the 16th year of the life of mine:

Total power (annual) output	*60,00,000 units*
Fuel consump-steam plant	*2.25 kg/unit*
Fuel consumption gas plant	*1.45 kg/unit*

Cost of purchasing, cutting and transporting timber from a radius r km = Rs. 2 (1 + 0.8 r) per tonne. Total cost of supplying, erecting the gas plant on the turn key basis = 6,00,000.

Assume that in the 16th year fuel is drawn from average radius of 1.3 km.

Solution

Steam Plant

Total fuel consumption for supplying 60,00,000 units annually = 60,00,000 × 2.25 kg

Cost of purchasing, cutting and transporting timber from a radius of 18 km

$$= 2\,(1 + 0.8 \times 18) \times \frac{60,00,000 \times 2.25}{1000}$$

= **Rs. 831,600.**

Gas Plant. In this case fuel consumption per annum = $1.45 \times 6 \times 10^6$

= **8700000 kg**

$$\therefore \qquad \text{Cost of fuel} = \frac{87 \times 10^5 \times 2(1+0.8 \times 1.3)}{1000}$$

= **Rs. 35,496**

Hence total annual saving on fuel cost = 8,31,600 – 35496

= Rs. 7,96,104

Cost of gas plant = Rs. 600,000

Hence in the first year = 796104 – 600000

= **Rs. 196,000**

Example 28.11 *Two tariffs are offered :*

(a) Rs. 200 plus 5 paise per unit

(b) A flat rate of 30 paise per unit

At what consumption is the tariff (a) preferable?

Solution

In case of tariff (a) the charges per unit are less but there are some fixed charges. So in lower range this is uneconomical as compared to tariff (b). Let *m* be the number of units at which the charges due to both the tariffs are equal. Then

$$200 + \frac{5}{100} \text{m} = \frac{30}{100} \text{m}$$

$$\therefore \qquad 200 = \frac{25}{100} m$$

$$\therefore \qquad \mathbf{m = 800 \text{ units}}$$

Therefore, at consumption above 800 units the tariff (a) is preferable.

Example 28.12 *A consumer takes a steady load of 200 kW at a power factor 0.85 lagging for 8 hours per day and 315 days per annum. Estimate his annual payment under each of the following tariffs :*

(a) Rs. 80 per annum per kVA plus 10 paise per kWh.

(b) Rs. 80 per annum per kVA plus Rs. 8 per kW plus 2 paise per reactive kVAh.

Solution

$$\text{(a)} \qquad \text{kVA demand} = \frac{200}{0.85} \quad [\text{kW = power factor} \times \text{KVA}]$$

= 235

kWh consumption per annum

= 200 × 8 × 315 (kW × hours/annum)

= **5,04,000**

Hence total charges per annum = 80 × 235 + 0.1 × 5,04,000

= **Rs. 69,200**

(b) Total charges per annum in this case = 80 × 235 + 8 × 200 + 0.02 × 235 × 8 × 315

= **Rs. 32240**

Example 28.13 *Calculate the cost of yearly consumption of 5×10^6 kWh with the maximum demand of 1500 kW on each of the following tariffs :*

(a) a maximum demand charge of Rs. 80 per kW per annum plus 10 paise per kWh.

(b) a sliding scale graduated thus : for first 3000 kWh – 50 paise per kWh; For the next additional 1,000,000 kWh – 40 paise per kWh, for the next additional 2,000,000 kWh – 20 paise per kWh and for any subsequent number of kWh 8 paise per kWh.

Solution

(a) Total annual charges will be

$$= 80 \times 1500 + \frac{10}{100} \times 5 \times 10^6$$

$$= 120{,}000 + 5{,}00{,}000$$

$$= \textbf{Rs. 6,20,000}$$

(b) Total annual charges will be

$$= 0.5 \times 3000 + 0.4 \times 1{,}000{,}000 + 0.2 \times 2000{,}000 + 0.08\,(5 \times 10^6 - 3000 - 1 \times 10^6 - 2 \times 10^6)$$

$$= 1500 + 0.4 \times 10^6 + 0.4 \times 10^6 + 0.08 \times 10^6\,(2 - 0.003)$$

$$= \textbf{Rs. 961,260}$$

Example 28.14 *Compare the cost of supplying energy to a consumer taking 15000 kWh per annum with a maximum demand of 10 kW under the following tariffs :*

(a) 12% of the rateable value plus 6 paise per kWh.

(b) Rs. 10 per room plus 7 paise per kWh.

(c) Rs. 70 per kW of maximum demand plus 3 paise kWh.

The house has eight room and rateable value of Rs. 150 per annum. At what consumption will (a) and (c) be equal?

Solution

(a) Required cost of supply is $= 0.12 \times 150 + 0.06 \times 15000$

$$= 18 + 900 = \textbf{Rs. 918.00}$$

(b) Required cost of supply in this case is $= 10 \times 8 + 0.07 \times 15000$

$$= 80 + 1050$$

$$= \textbf{Rs. 1130}$$

(c) The required cost is $= 70 \times 10 + 0.03 \times 15000$

$$= 700 + 450$$

$$= \textbf{Rs. 1150}$$

Let the consumption under (a) and (c) be equal to m units. Then

$$0.12 \times 150 + 0.06\,m = 70 \times 10 + m \times 0.03$$

$$18 + 0.06\,m = 700 + 0.03\,m$$

$$0.03\,m = 700 - 18 = 682$$

$$\therefore \quad m = \frac{682}{0.03} = 100 \times 227.33$$

$$= \textbf{22733 units}$$

Example 28.15 *An electric supply undertaking with a firm capacity of 5000 kW, has annual maximum demand of 4000 kW and supplies the following categories of load :*

(a) Domestic : 3000 kW maximum demand with average load of 450 kW

(b) Street light : 200 kW maximum demand with 60 kW as average load

(c) Power load : 1800 kW maximum demand with 720 kW as average load.

Calculate the overall rate kWh for each category of load from the following cost data :

Capital cost of equipment Rs. 1,31,40,000

Annual cost of fuel, other consumables, wages repairs and maintenance = Rs. 5,38,740

Annual rate of depreciation and interest at capital cost 10%.

Solution

(a) Power supplied to consumer (domestic) = 60 × 8760 kWh

(b) Power supplied to street light = 60 × 8760 kWh

(c) Power supplied to power load = 720 × 8760 kWh

∴ Total power supplied = 8760 (60 + 450 + 720)

= **1230 × 8760 kWh**

Operating cost per annum = Rs. 5,38,740

$$\therefore \text{ Operating cost per kWh} = \frac{538740}{1230 \times 8760} \times 100$$

= **5 paise per kWh**

Fixed charges

Interest and depreciation = 0.1 × 1,31,40,000

= 1,31,4000

$$\therefore \text{ Fixed charge per kW} = \frac{1314000}{5000}$$

= **Rs. 262.8**

Overall rate per kWh

(a) Fixed charge = 262.8 × 3000

= Rs. 788,400

Operating cost = 0.05 × 450 × 8760

= Rs. 197,100.

∴ Total charges = Rs. 985,500

$$\therefore \text{ Overall cost per unit} = \frac{985500}{450 \times 8760} \times 100$$

= **25 paise/kWh**

(b) Fixed charges = 262.8 × 200 = Rs. 52560

Operating cost = 60 × 8760 × 0.05

= Rs. 26,280

∴ Total charge = **Rs. 78,840**

$$\therefore \text{ Overall cost per unit} = \text{Rs. } \frac{78840}{60 \times 8760} \times 100$$

= **15 paise**

(c) Fixed charge = 262.8 × 1800

= Rs. 473,040

Operating cost = 720 × 8760 × 0.05

= Rs. 315360

∴ Total charges = Rs. 788,400

$$\therefore \text{ Overall charge per unit} = \text{Rs. } \frac{788,400}{720 \times 8760} \times 100$$

= **12.5 paise/kWh.**

Example 28.16 *A contract entered into between a state electricity Board and a mini steel plant is as follows :*

Demand charge :

Rs. 15 per kVA for the first 30 kVA per month

Rs. 12 per kVA for all additional demand for month.

Energy charge :

2.0 paise per kWh for first (100 hr × kVA demand)

1.75 paise per kWh for the next (100 hr × kVA demand)

1.50 paise per kWh for the next (100 hr × kVA demand)

1.00 paise per kWh for all additional energy.

Fuel Adjustment : An additional charge of 0.01 paise/kWh for each Re. 1 increase in the cost of fuel above Rs. 25.00 per tonne of coal shall be applied to the total kWh used.

Minimum Monthly Charge : The minimum bill shall be Rs. 10 per kVA of demand but not less than Rs. 1000 in any month.

The complete rate contract contains details of the method of measuring demand and energy, computing fuel adjustment, prompt payment discount etc.

During the month of March 1976, the maximum demand was 860 kVA and energy consumption of 325000 kWh. The cost of coal applicable for the month is Rs. 37.00 per tonne.

Calculate the bill for the energy supply for the month under reference.

Solution

Demand charges

(i) For first 30 kVA = 30 × 15 = Rs. 450.00

(ii) For the rest (860 – 30) kVA = **830 × 12 = Rs. 9960.00**

Total demand charges = Rs. 10,410.00

Energy charges

First 100 kVA	0.02 × 100 × 860	Rs. 1720.00
Next 100 kVA	0.0175 × 100 × 860	Rs. 1505.00
Next 100 kVA	0.015 × 100 × 860	Rs. 1290.00

Each of these blocks accounts for 860 × 100 = 86000 kWh.

Then 325000 – 3 × 86000

= 67,000 kWh remain for the fourth block.

0.01 × 67000 = Rs. 670.00

Total energy charges = Rs. 5185.00

Fuel adjustment

Rs. 37.00 – Rs. 25.00 = Rs. 12.00 over base price of fuel. This represents 12 times a full one rupee increase in fuel cost, therefore $12 \times 0.01 = 0.12$ paise is the additional charge per kWh.

$$\therefore \text{ Total fuel adjustment charge} = \frac{0.12 \times 325000}{100}$$

$$= \text{Rs. } 390.00$$

$$\therefore \quad \text{Total monthly bill} = \text{Rs. } 1296 + \text{Rs. } 5185 + \text{Rs. } 390$$

$$= \textbf{Rs. 6871.00}$$

But as per contract the minimum bill shall be Rs. 10.00 per kVA. Hence for 860 kVA, minimum bill = 860 × 10

= **Rs. 8600.00.**

THEORY QUESTIONS

28.1. What is meant by power plant economics?

28.2. Explain the terms fixed cost and operating cost of the power station. List the components of each of them.

28.3. Discuss the variation of fixed cost and operating cost with the investments for the power plant. How annual costs are determined?

28.4. What is depreciation? What are the methods to compute the amount of depreciation?

28.5. Discuss the generation of electricity as a production activity. How does electricity differ from an industrial commodity?

28.6. Explain how the cost of electrical energy produced is determined.

28.7. What is the source of power plant income?

28.8. What are the conditions the rate of charge for electrical energy should satisfy?

28.9. From the public standpoint what should be the conditions the 'rates' should meet or satisfy?

28.10. What are the different types of tariffs or rates? Explain each briefly.

28.11. A load having a maximum demand of 50 MW has an average load of 25% of the maximum demand may be supplied in one of the following ways :

(a) A steam power station supplying the whole load.

(b) A steam power station in conjunction with a pumped storage scheme of 20 MW capacity supplying 50×10^6 kWh per year.

Compare the unit cost of two systems on the basis of the following data :

Capital cost of steam station	Rs. 1000 per kW installed
Capital cost of storage plant	Rs. 800 per kW installed
Operating of cost of steam plant	2.5 paise/unit
Operating cost of storage plant	0.24 paise/unit
Efficiency of storage plant	60%
Interest and depreciation	15%

[**Ans:** (a) 9.35 paise/unit (b) 7.84 paise/unit]

28.12. A supply is to be provided to a factory having a maximum demand of 1000 kW and 40% average generation. Compare the cost of providing this load from

(a) Public supply having a tariff of Rs. 70 per kW of maximum demand and 5 paise per kWh consumed.

(b) For a diesel plant costing Rs. 240 per kW having annual cost of transformers, switchgears etc. 1 paise per kWh, annual wages of Rs. 6500 and cost of fuel and oil 2 paise per kWh.

Take annual rate of depreciation as 12%.

[**Ans:** (a) Annual cost Rs. 2,45,200 (b) Annual cost Rs. 1,40,420]

28.13. A plant costs Rs. 10.2×10^5 and it is estimated that after 40 years it will be replaced by a new one and salvage at that instant would be Rs. 2.1×10^5.

Calculate :

1. Annual deposit to be made in order to replace the plant after 40 years.
2. The value of the plant after 15 years on the following basis
   (a) Straight line method
   (b) Sinking fund method at 10% annual compound interest.

28.14. A 500 kW electric lighting plant costs Rs. 1000 per kW installed. Fixed charges are estimated at 14% and operating cost is 13 paise per kWh. The plant averages 200 kW for 5000 hours of the year, 450 kW for 1200 hours, and 80 kW for the remaining period. What is the unit cost of production of electric energy? [**Ans:** 17 paise]

28.15. A plant costing Rs. 65000 has a useful life of 15 years. Find the amount which should be saved annually to replace the equipment at the end of that time

(a) by straight line method

(b) by sinking fund method.

Annual rate of compound interest is 5%.

28.16. A block meter rate is quoted as follows :

First 10 kWh at 10 P per kWh

Next 40 kWh at 6 P per kWh

Next 50 kWh at 4 P per kWh

Next 100 kWh at 3 P per kWh.

Excess over 200 kWh at 2 P per kWh.

Compute the total bills and average unit cost for consumption of 50, 100, 150 and 300 kWh.

[**Ans:** Rs. 3.4, Rs. 5.40, Rs. 6.9, Rs. 10.40]

28.17. A Hopkinson demand rate is quoted as follows :

Demand

First kW of maximum demand at Rs. 5 per kW/month

Next 4 kW of maximum demand at Rs. 4 per kW/month

Excess over 5 kW of maximum demand at Rs. 3/kW/month

Energy

First 50 kWh at 6 P per kWh

Next 50 kWh at 4 P per kWh

Next 200 kWh at 3 P per kWh

Next 400 kWh at 2.5 P per kWh

Excess over 700 kWh at 2 P per kWh.

(a) Compute the monthly bill for a total consumption of 1500 kWh and a maximum demand of 12 kW. Find the unit energy cost.

(b) If the month contains 30 days, what is the lowest possible bill for the given energy consumption?

[**Ans:** (a) Rs. 79, 5.27 P per kWh (b) Rs. 46.25]

28.18. A write demand rate is quoted as under :

For energy purchased monthly, the equivalent of

1. The first 50 hours use of maximum demand at 6 P per kWh
2. The next 50 hours use of maximum demand at 4 P per kWh
3. The next 100 hours use of maximum demand at 3 P per kWh
4. The next 200 hours use of maximum demand at 2 P per kWh

(a) Compute the bill for a monthly energy consumption of 432,000 kWh with a maximum demand of 1600 kW. Also compute the unit energy cost.

(b) Find the lowest possible bill for this energy consumption and the corresponding unit energy cost for a 30 day month. [**Ans:** (a) Rs. 15,040; 3.48 P/kWh (b) 2.11 P/kWh]

Chapter 29 Power Generation and Load Division

29.1 INTRODUCTION

In engineering thermodynamics, emphasis is often given on the thermal efficiency of a power producing devices such as steam and gas turbines, diesel engines and the hydraulic efficiency of a water turbines. But a power producing device is only a part of the total energy systems which are designed to serve the human needs. A plant supplying the energy for all the objectives must do so at maximum effectiveness. The effectiveness is measured financially. Thus the consumer must get the services at minimum possible rate and the overall monetary cost of energy production must be minimum. The thermal efficiency of the plant is only one factor of many, than determine the final energy cost. Thus a most efficient plant may not be most economical thermally.

29.2 ELEMENTS OF ELECTRICAL ENERGY SYSTEM

The manner in which the energy output of a single power plant is subdivided and sent to thousands of individual customers is illustrated in Fig. 29.1. The energy is first sent to **substations** which are located at the ends of the

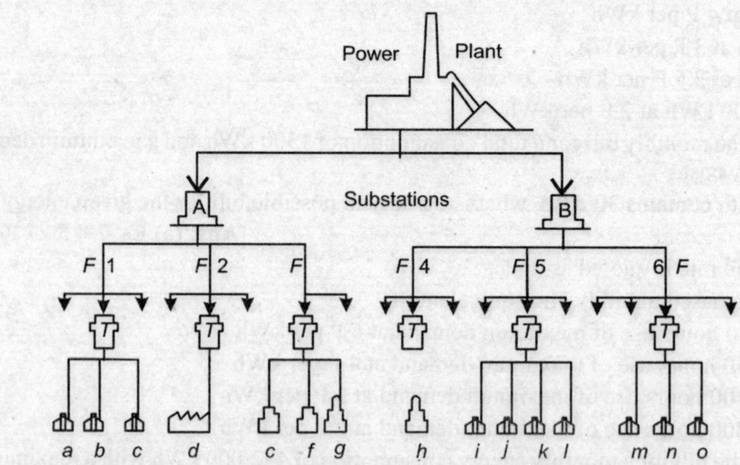

Customers domestic, industrial business etc. *F*=feedes, *T*=distribution transformes

Fig. 29.1 *Elements of an electric power system.*

primary distribution system. **Feeders** which lead out from the substations supply several **distribution transformers** (however the diagram shows only one per feeder). The transformer then is connected by low voltage lines to the system of one or more customers.

29.3 THE VARIABLE DEMAND PROBLEM

Production of electrical energy is like production of any other industrial commodity. The power plant like an industrial unit uses raw material — fuel, water — and processing equipment like boiler, turbine, generator — to produce a salable final product — electrical energy.

But electrical energy differs fundamentally from other commodities in three aspects. Firstly, electrical energy cannot be produced till the instant it is needed and only in quantities exactly equal to its instantaneous demand. No more or no less. Secondly, there is no simple way to store the extra supply of electrical energy and thirdly, the demand for electrical energy is uniform but is continuously variable.

Thus, production of the energy is strictly only when it is required, no storage facilities, and highly fluctuating demand are the main factors of the production and distribution problem of electrical energy.

29.4 LOAD CURVES

A consumer of electrical power uses the power as and when required. As the number of consumers is very large the load on the system is always changing with time and not be constant.

A curve showing the chronological variation in demand for electrical energy with respect to time is known as a **load curve**. Load curve can also be defined as a graphical representation between load in kW and time in hours. It shows variation of load on the power station. When it is plotted for 24 hours of a day, the load curve is called as **Daily Load Curve**, and if the time considered is one year (8760 hours) then it is called **annual load curve**. The monthly load curves are used in establishing rates. The annual load curve is used to determine annual load factor.

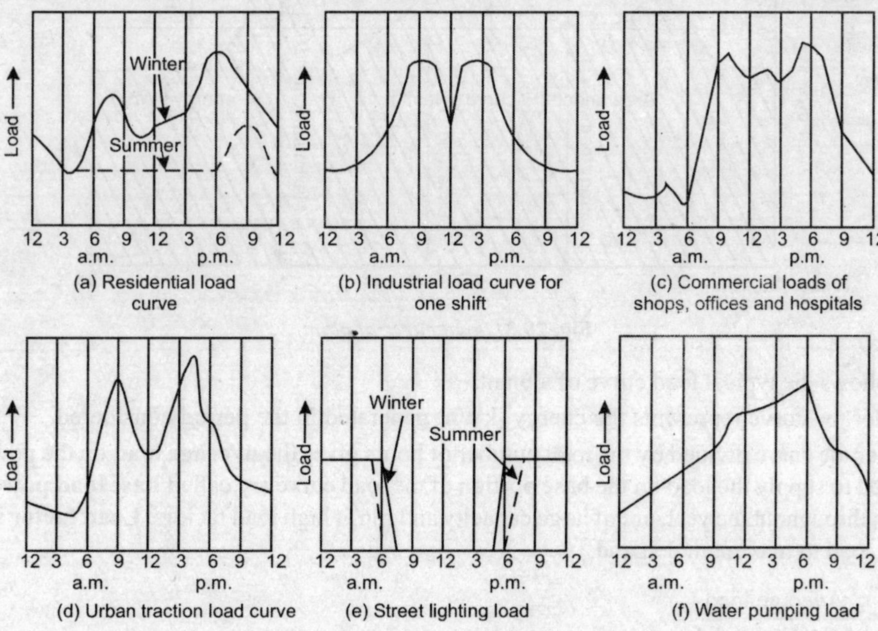

Fig. 29.2 *Load curves of different customers.*

The load curve is useful in predicting the annual requirement of energy and the capacity of the power plant to take the peak load. A power station supplies electrical power to different types of consumers e.g., domestic, commercial, industrial, railways etc. The load curve of each customer is different from the other as per the activities and nature of demand. However, the load curve for the consumers belonging to the same group can be plotted. Figure 29.2 shows a few load curves for different groups of customers.

Figure 29.2(a) shows a typical load curve of a residential community rather than one consumer. In early morning the demand is low which rises upto 9 a.m. when activities like cooking, washing, heating are taking place. The demand then drops down till late in the afternoon. It then steadily rises again till about 10 p.m. and drops down again. The curve in Fig. 29.2(b) shows a load curve for one shift industry. In early hours, demand is only for lighting and continuously running machines like furnaces, refrigerators etc. Then demand rises and at about 8 a.m. industry registers its full demand, which remains constant upto lunch time when some machines are stopped and demand falls. After lunch the demand rises again to its full value which remains upto 4 p.m. The demand drops then gradually and at about 6 p.m. reaches its minimum value.

Students are advised to study and verify the nature of other curves viz. commercial, traction, street lighting and water pumping load curves and varify the variation of demand against the requirements of these consumer groups.

Sometimes a special event may change the local demand drastically and affect the load curve. For example,

1. A cricket or tennis match or broadcasting of special event like independence day.
2. Cloudy day, and rainy season results in increase of lighting demand of domestic consumers, offices, shops.
3. Variations in atmospheric temperatures affect the energy demand.

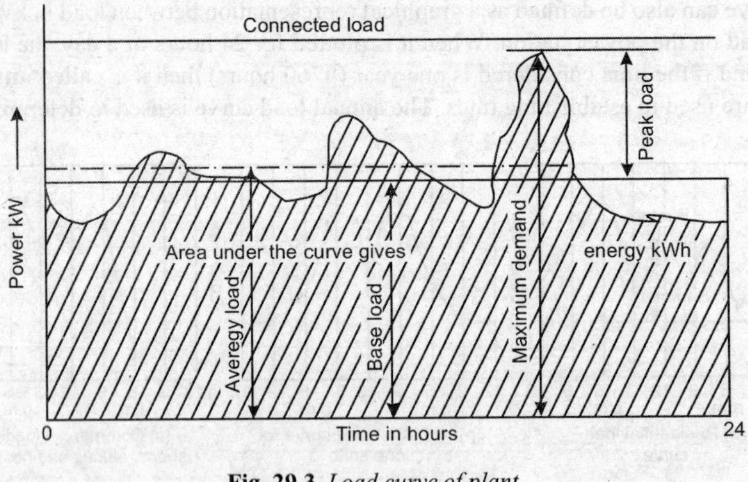

Fig. 29.3 *Load curve of plant.*

Figure 29.3 shows the typical load curve of a plant.

The area under the curve represents the energy (kWh) generated in the period considered.

The area under the curve divided by the total number of hours gives the **average load** on the power station. The power plants used to supply the load on the base portion of the load curve are called **base load power plants**. Base power plants run throughout the year, are of large capacity and run at high load factors. **Load factor** is defined as the ratio of average load to maximum demand.

$$\text{Load factor} = \frac{\text{Average load}}{\text{Minimum load}}$$

The power plant which supply the load on the top portion of the load curve (in excess of base load) are called

Peak Load Power Plants. They run for a short period in the year and work at low load factors. **Peak load** plants should be capable of **quick starting**.

Load factor is less than unity.

29.5 LOAD DURATION CURVE

The variable load problem has injected into the language of power plant a number of terms which need definitions. The basic information is, of course, the operating data of demanded load plotted against the time sequence. This is commonly referred to as load curve.

The load duration curve, useful in financial study is obtained from the same data as the daily load curve. This load duration curve represents rearrangements of all the load elements of chronological load curve in order of descending magnitude.

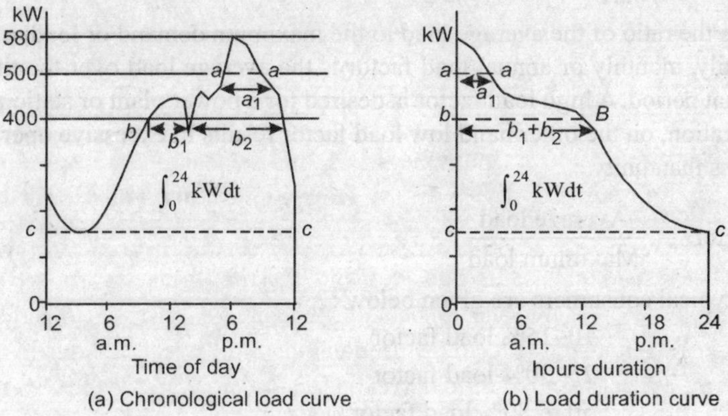

Fig. 29.4 *Load duration curve.*

The ordinates of this curve may extend from 0 to maximum demand in kW or from 0 to 100% maximum demand. The abscissa reaches from 0 hour to 8760 hours or percentage of year. The areas under the load duration curve and a chronological curve are equal. The area of the load curve represents the energy produced is given by

$$E = \int_0^{24} \text{kW}.dt \text{ kWh}$$

Figure 29.4 shows a graphical method of constructing a load duration curve from a chronological load curve. The base of the load duration curve represents the total time span e.g., 24 hours. The peak load on the chronological curve is then plotted at a point at 0 hr on load duration curve. At load ordinate a the intercept a_1 of the load curve is plotted as time ordinate a_1 on the load duration curve as shown. At load ordinate b on the load curve the intercepts b_1 and b_2 are plotted on the load duration curve as $(b_1 + b_2)$ and so on. At minimum load C the intercept covers the entire period of 24 hours. Any point on the load duration curve is a measure of number of hours in a given period during which given load and higher load have prevailed. Thus point B tells us that load of 400 kW and above were demanded for 13 hours during the day.

29.6 TERMS AND DEFINITIONS USED IN POWER PLANT STUDY

Following are some terms and definitions used in power plant practice.

(a) **Connected Load.** It is the rating, in kilowatt, of the installed capacity of a consumer or it is sum of ratings in kW of the equipment installed in the consumer's premises. Thus a consumer having five connections of 100 W lamps, a connection for a radio consuming 70 W and a connection for a heater of 1000 W will have a connected load of $(5 \times 100 + 70 + 1000) = 1570$ kW.

(b) **Maximum Demand.** It is the maximum load which a consumer uses at any time. If all the devices fitted in consumer's house run to their fullest extent simultaneously the maximum demand will be equal to the connected load. But generally the actual maximum demand is less than the connected load, because all the devices never run at full load at the same time. Maximum demand of a power station is the maximum load on the power station in the given period.

(c) **Demand Factor.** It is the ratio of maximum demand to the connected load of a consumer. Thus the demand factor of a consumer having a maximum demand of 1270 W and connected load of 1570 W will be

$$\text{Demand factor} = \frac{\text{Maximum demand}}{\text{Connected load}} = \frac{1270}{1570}$$

$$= \mathbf{0.81}$$

(d) **Load Factor.** It is the ratio of the average load to the maximum demand or load, of a consumer, during the specific period. Thus daily, monthly or annual load factor is the average load over the time specified divided by maximum load during that period. A high load factor is desired for a power plant or station as it generally results in economical power generation, on the other hand low load factor results in excessive operating costs for the plant. Load factor is always less than unity.

$$\text{Load factor} = \frac{\text{Average load}}{\text{Maximum load}}$$

The load factors for typical consumers are given below :

1. **Residential load** 10–15% load factor
2. **Commercial load** 25–30% load factor
3. **Industrial load** 50 to 80% load factor
4. **Municipal load** 25% load factor

(e) **Diversity Factor.** So far it is considered the aspect of energy consumption at the consumer's connection. A power station usually supplied power to several consumers at a time. The nature of demand of various consumers may be different depending on the purpose. Experience shows that the maximum demands of similar types of individual consumers do not occur simultaneously but are spread out over a period of time. There are numerous different types of consumers and the time of maximum demand of each consumer will be different. For example, consumption of electricity for domestic purposes usually has got maximum demand during the night time whereas an industrial consumer may have maximum demand during the day time. A power station is definitely concerned with the maximum demand of individual consumers as the total generating capacity has got to be greater than the maximum demand at any time. Thus, simultaneous maximum demand of all the consumers may be less than the sum of the individual demands.

The time distribution of maximum demands for similar types of consumers is measured by **the diversity factor** where

$$\text{Group diversity factor} = \frac{\text{Sum of individual maximum demands}}{\text{Actual maximum demand of a group}}$$

The group diversity force is always greater than unity. The 'group' here may be the group of consumers erected by a certain transformer. In general, the diversity factor is given by

$$\text{Diversity force} = \frac{\text{Sum of individual maximum demand}}{\text{Simultaneous maximum demand}}$$

The typical diversity factors are given below :

	Residential Lighting	*Commercial Lighting*	*General Power Service*
1. Between consumers	3.4	1.5	1.5
2. Between transformers	1.3	1.3	1.3
3. Between feeders	1.2	1.2	1.2
4. Between substations	1.1	1.1	1.1

The high value of diversity factor is a desirable quality. The high value of diversity factor gives low maximum demand of group and hence high load factor.

(f) **Plant Capacity Factor.** The load and diversity factors do not give any idea of the *reserve capacity* of the plant. The **plant capacity factor** measures the reserve capacity of the plant as well as the degree of utilization of the plant.

$$\text{Plant capacity factor} = \frac{\text{Actual energy produced during a certain period}}{\text{Maximum possible energy that could have produced during same period}}$$

$$= \frac{\text{Average load} \times \text{time}}{\text{Plant capacity} \times \text{time}}$$

$$= \frac{\text{Average load}}{\text{Plant capacity}}$$

Capacity factor shows how close a plant runs to its full rating. The difference between load and capacity factors is an indication of the reserve capacity of the plant.

(g) **Plant Utilization Factor.** It is defined as the ratio of annual energy produced to the maximum possible energy that could been produced during the actual number of hours the plant was in operation.

$$\text{Plant use factor} = \frac{\text{Actual energy produced (generated)}}{\text{Plant capacity} \times \text{Operating hours}}$$

As this factor approaches 1, it indicates the need for additional capacity of the plant. Higher the plant use factor, higher is the operating efficiency of the plant.

(h) **Dump Power.** This term is used in hydroplants and it indicates the power in excess of the load requirements and it can be made available by surplus water.

(i) **Firm Power.** It is the power which is available even under emergency conditions.

(j) **Prime Power.** It is the power in any form — mechanical, hydraulic or thermal that is always available for conversion into electric power.

(k) **Cold Reserve Power.** It is than reserve generating capacity which is not in operation but can be made available for service.

(l) **Hot Reserve Power.** It is that reserve capacity which is in operation but not in service.

(m) **Spinning Reserve.** It is that reserve generating capacity which is connected to the bus bars and is ready to take the load.

29.7 POWER PLANT PERFORMANCE AT VARIABLE LOAD

Performance of generating plants is compared by their average thermal efficiency over a period of time. Average thermal efficiency is the ratio of useful energy output to the total heat input both during the same period of time. This measure of performance varies with uncontrable conditions such as cooling water temperature, shape of load duration curve, total output and quality of fuel. It is not a satisfactory measure of comparison unless all plant performances are corrected to the same controlling conditions. Such corrections are difficult to make and cannot render expected accuracy.

29.8 PERFORMANCE CHARACTERISTICS — INPUT-OUTPUT CURVE

Plant performance can be described most precisely by the input-output curve derived from the testing of the equipment. Figure 29.5 shows such a curve. The equation of the curve may be given by

$$y = a + bx + cx^2 + dx^3 + \dots nx^3 \qquad (29.1)$$

The time consumed in finding the equation of an input-output curve usually cannot be justified and so the curve is always expressed graphically.

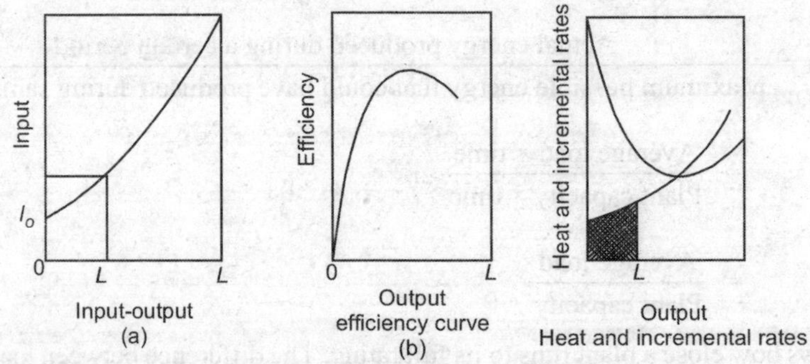

Fig. 29.5 *Input-output curve and corresponding efficiency, heat rate and I-R curves.*

The ordinate of the curve is input while the abscissa is output or load of the plant. At zero load L the positive intercept I_o measures the amount of energy required to keep the apparatus running. This energy is usually dissipated as frictional and heat losses. Any input over the no load input produces certain output. All the additional input does not appear as output owing to losses. Input I is expressed in kJ/hr or MJ/hr and the output L is expressed in MW. The efficiency curve is divided by taking at each load the corresponding input; then

$$\eta = \frac{L}{I} \times 100 \text{ percent}$$

is plotted against L. The efficiency can be expressed in the form of decimal fraction also.

29.8.1 Heat Rate Curve

For variable load analysis, the heat rate *HR* is more useful that thermal efficiency. The heat rate curve is derive by taking at each load the corresponding input.

Then, $\quad \text{HR} = \dfrac{I}{L}$ kJ per kW.

HR is plotted against the corresponding value of L. Figure 29.5(c). The heat rate is inversely proportional to the efficiency. Thus at a load where efficiency is maximum the heat rate happens to be minimum. The reciprocal nature of efficiency and heat rate curves is evident from Fig. 29.5(b) and (c). The heat rate curve can be expressed mathematically also. Thus, if the input-output curve is defined by

$$I = a + bL + cL^2 + dL^3 \text{ etc.}$$

$$HR = \frac{I}{L} = \frac{a}{L} + b + cL + dL^2 \text{ etc.}$$

29.8.2 The Incremental Heat Rate Curve

The incremental heat rate curve is derived from the input-output curve by finding at any load. The incremental rate is defined as

$$IR = \frac{dI}{dL} \text{ kJ/kWh.}$$

IR is plotted against load *L* for the full range of input-output curve. Mathematically, incremental rate is the slope of the input-output curve at the given load. Physically, incremental rate expresses the amount of additional energy required to produce additional output at any given load.

Figure 29.5(a), areas under input-output curve has no significance. However, the areas under the incremental rate curve measure the additional input required to increase output from L_1 to L_2.

$$IR = \frac{dI}{dL} \text{ or } dI = R.dL$$

then $$I_2 - I_1 = \int_{L_1}^{L_2} R.dL.$$

In Fig. 29.5(c) the area under incremental curve from $L = 0$ to $L = L_1$ is equal to the difference $I_1 - I_o$ on input-output curve.

If the input-output curve is defined by

$$I = a + bL + cL^2 + dL^3 + ...$$

then the corresponding incremental rate curve is defined by

$$IR = \frac{dI}{dL} = b + 2cL + 3dL^2 + ...$$

29.8.3 The Incremental Rate and the Heat Rate Relation

Observe Fig. 29.5(c) and note that the incremental rate curve crosses the heat rate curve at the lowest value of heat rate when two curves are plotted on common ordinates. This behaviour can be explained as follows :

Figure 29.6(a) represents a continuous input-output curve. (I_m, L_m) is a point on the curve such that the tangent to the curve at this point passes through the origin.

The equation of this tangent is

$$I_m = a.L_m \text{ where } a \text{ is the slope of the tangent.}$$

But at L_m, the heat rate is

$$HR_m = \frac{I_m}{I_L} = a.$$

Now, if the tangent were considered as input-output curve then corresponding heat rate and incremental rate curve would be horizontal line with $HR = \frac{I_m}{L_m}$ as shown in Fig. 29.6(b). Since the actual input-output curve lies above the tangent at all other loads than L_m, it follows that, the heat rate curve must be higher than I_m/L_m at all other loads. Then heat rate is minimum when $HR_m = \dfrac{I_m}{L_m} = a.$

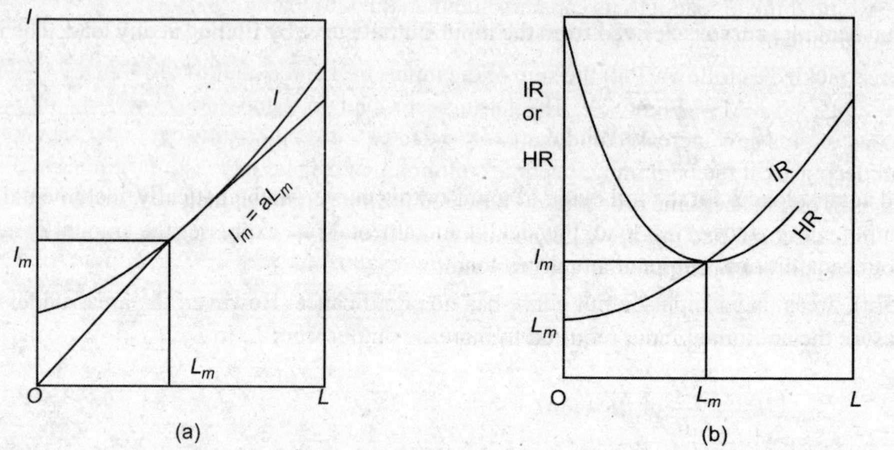

Fig. 29.6 *Curves demonstrating equality of HR and IR at minimum HR for a given input-output curve.*

Now, the slope of the tangent to the actual input-output curve at L_m, is the slope of the curve at L_m and equals the incremental rate at L_m

or, $$a = IR_m = \frac{dI_m}{dL_m} \text{ at } L_m$$

But $$a = HR_m \text{ at } L_m.$$

Therefore $$HR_m = IR_m$$

or the heat rate of a continuous input-output curve is at minimum when it is equal to incremental rate. Since incremental rate curve is a continuously increasing curve, it crosses the heat rate curve at its minimum when $IR_m = HR_m$.

Figure 29.6(b) also shows that for any load between 0 to L_m, the heat rate decreases and for loads above L_m, the heat rate increases.

Equality of heat rate and incremental rate of an input-output curve at minimum heat rate can be demonstrated very briefly by calculus as follows.

At minimum heat rate, the slope of the heat rate curve is zero.

$$\therefore \quad \frac{d(HR)}{dL} = \frac{d(I/L)}{dL} = 0$$

$$= \frac{LdI - I - dL}{L^2}$$

or $$LdI = I.dL$$

or $$\frac{dI}{dL} = \frac{I}{L}$$

or $$IR = HR \text{ where } HR \text{ is minimum.}$$

29.9 LOAD SHARING BETWEEN GENERATORS

For power plants having more than one generating unit, the problem of load sharing is important and should be tackled with care. Improper load division may decrease appreciably the thermal efficiency of input-output curves of two units that are to operate in parallel and supply a common load. Since unit *A* is more efficient throughout its load range, one would think to load it to its capacity and then think of loading *B*. But this would prove wrong as the efficiency would go down.

The problem is tackled as follows. Plot the sum of the inputs to *A* and *B* against the load on *A*. For given constant load on the two units, say 8 MW. Figure 29.7(b) putting zero load on *A* and 8 MW on *B*, the total combined input is $I_c = I_a + I_b$. Load on *A* is now increased and that on *B* is decreased keeping the total load constant i.e., 8 MW. The combined input decreases at the beginning, becomes minimum and again increasing. Minimum combined input I_c gives maximum efficiency. The same procedure is repeated for the combined load say 10 MW and subsequently 12 MW, 14 MW. For each combined input, we get a maximum efficiency. Combining the maximum efficiency points, we get a curve 1–2. This is a optimum load for unit *A* and thus the load on unit *B*.

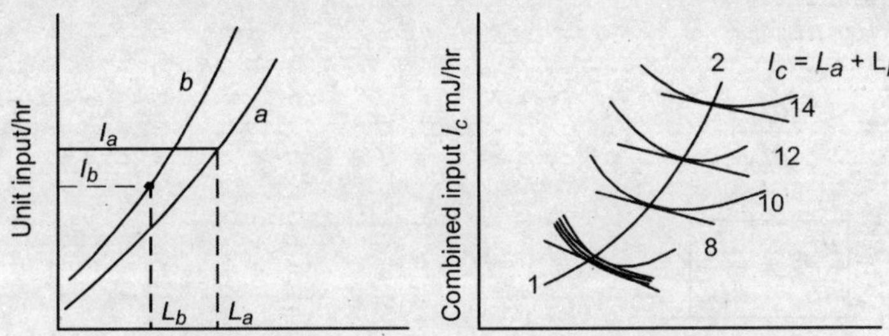

(a) Input-output curves (b)

Fig. 29.7

This method however becomes cumbersome and unwieldy when more than two units run in parallel. A mathematical general solution can be given as follows.

From Fig. 29.7(b), we have at the point of minimum input,

The slope of the curve $\frac{dI_c}{dL_c} = 0$

where $I_c = I_a + I_b \text{ and } L_c = L_a + L_b$

Then,
$$\frac{dI_c}{dL_a} = \frac{dI_a + dI_b}{dL_a} = 0$$

$$= \frac{dI_a}{dL_a} + \frac{dI_b}{dL_a} = 0$$

or
$$\frac{dI_a}{dL_a} = \frac{dI_b}{dL_a}$$

But
$$\frac{dI_b}{dL_a} = -\frac{dI_b}{dL_b} \times \frac{dL_b}{dL_a}$$

But $$\frac{dL_b}{dL_a} = \frac{dL_c - dL_a}{dL_a}$$

$$= \frac{dL_c}{dL_a} - 1.$$

$$\therefore \quad \frac{dI_b}{dL_a} = -\frac{dI_b}{dL_b}\left[\frac{dL_c}{dL_a} - 1\right]$$

$$= \frac{dI_b}{dL_b}[0-1]$$

$$= +\frac{dI_b}{dL_b}$$

or $$\frac{dI_a}{dL_a} = \frac{dI_b}{dL_b}$$

$$\boxed{\frac{dI_a}{dL_a} = \frac{dI_b}{dL_b}}.$$

or for minimum combined input to carry a given combined output, the slopes of the input-output curves for each unit must be equal. But the slopes of these curves are incremental rates. So for most economical load division between a number of units carrying a given combined load, the incremental rates of individual input-output curves must be equal.

This method of load division can be extended to include any number of units.

The performance characteristics discussed so far are typical of steam generators and diesel engines.

SOLVED EXAMPLES

Example 29.1 *(a) Explain the terms fixed cost and operating cost, and average load and load factor.*

(b) In a power station, annual factors are as under :

Load factor = 70%, capacity factor = 60%, and use factor = 65%.

The power plant has a maximum demand of 50,000 kW. Find :

(i) Annual energy production.

(ii) Reserve capacity above peak load.

(iii) Hours per year, the plant is not in use.

Solution

$$\text{Load factor} = \frac{\text{Average load}}{\text{Maximum load}}$$

$$\therefore \quad 0.7 = \frac{\text{Average load}}{50{,}000}$$

$\therefore$ Average load $= 0.7 \times 50{,}000$

$= \mathbf{35000\ kW}$

$\therefore$ Annual energy production = Average load × No. of hours per year

$= 35000 \times 8760$

$= \mathbf{306.6 \times 10^6\ kWh}$

$$\text{Again, Capacity factor} = \frac{\text{Average load} \times 8760}{\text{Capacity} \times 8760}$$

$$\therefore \quad 0.6 = \frac{35000}{\text{Capacity}}$$

$$\text{or} \quad \text{Capacity} = \frac{35000}{0.6} = \mathbf{58333.33\ kW}$$

$\therefore$ Reserve capacity over peak load = Maximum capacity – Peak load

$= 58333.33 - 50{,}000$

$= \mathbf{8333.33\ kW}$

$$\text{Again, use factor} = \frac{\text{Annual energy production}}{\text{Capacity} \times \text{No. of hours of working}}$$

$$\therefore \quad 0.65 = \frac{306.6 \times 10^6}{58333.33 \times x}$$

$\therefore$ x = Number of hours of working the plant per year

$$= \frac{306.6 \times 10^6}{58333.33 \times 0.65}$$

$= 8086.15$ hours

$\therefore$ Number of hours for plant to be closed or non-working/year

$= 8760 - 8086.15$

$= \mathbf{673.85\ hours}$

Example 29.2 (*a*) *For a thermal power plant, use factor is 70% and capacity factor is 60%. How many hours per year the plant will not be in operation?*

(*b*) *Two power plants A and B have incremental fuel costs given by :*

$$\frac{dF_A}{dP_A} = 0.06\, P_A + 30$$

$$\text{and} \quad \frac{dF_B}{dP_B} = 0.07\, P_B + 25$$

where F is the fuel cost in rupees per hour and P is power output in MN. If the total power supplied by both the units is 500 MW, find the economic loading for the two units.

Solution

(a) Let A be the average load for the thermal power plant.

$$\therefore \quad \text{Capacity factor} = \frac{\text{Energy produced/year}}{\text{Maximum energy/year}}$$

$$= \frac{A\times 8760}{C\times 8760} = \frac{A}{C} = 0.6$$

$$\text{Similarly, use factor} = \frac{\text{Energy produced}}{\text{Capacity} \times \text{Time in hours during which plant is in operation}}$$

$$0.7 = \frac{A\times 8760}{C\times t}$$

$$\therefore \quad \frac{0.7\times t}{8760} = \frac{A}{C} = 0.6$$

$$\therefore \quad t = 0.6\times 8760$$

$$= \mathbf{7508.57\ hours}$$

7508.57 hours in year during which the plant was in operation

$\therefore$ 8760 – 7508.57 = 1251.43 = Number of hours during which the plant was not in operation

(b) Let, $P_A + P_B = 500$ MW

$$\therefore \quad P_A = 500 - P_B$$

Then we have,

$$\frac{dF_A}{dP_A} = 0.06P_A + 30$$

$$= 0.06(500 - P_B) + 30$$

$$= 30 + 0.06P_B + 30$$

$$= 0.06P_B + 60$$

$$\therefore \quad F_A = \frac{0.06P_A^2}{2} + 30P_A$$

$$= 0.03P_A^2 + 30P_A$$

$$= 0.03(500 - P_B)^2 + 30(500 - P_B)$$

$$= 0.03\left(25\times 10^4 + P_B^2 - 1000P_B\right) + 15000 - P_B$$

$$= 0.75\times 10^4 + 0.03P_B^2 - 30P_B + 15000 - 30P_B$$

$$= 0.75\times 10^4 + 0.03P_B^2 + 15000 - 60P_B$$

$$= 2.25\times 10^4 + 0.03P_B^2 - 60P_B \qquad \text{(I)}$$

$$\text{Similarly,} \quad \frac{dF_B}{dP_B} = 0.07P_B + 25$$

$$\therefore \quad F_B = \frac{0.07P_B^2}{2} + 25P_B \qquad \text{(II)}$$

$$\therefore \quad \text{Total cost} = \text{I} + \text{II}$$

$$= 2.25\times 10^4 + 0.03P_B^2 - 60P_B + 0.035P_B^2 + 25P_B$$

$$= 2.25\times10^4 + 0.065P_B^2 - 35P_B \quad \text{(III)}$$

For economic loading, this total cost should be minimum.

$$\therefore \quad \frac{dF}{dP_B} = 0$$

$$= \quad P_B - 35$$

$$\therefore \quad 0.130\,P_B = 35$$

$$P_B = \frac{35}{0.13}$$

$$= \mathbf{269.23\ MW}$$

$$\therefore \quad P_A = 500 - 269.23$$

$$= \mathbf{230.77\ MW}$$

∴ For economic loading,

$$P_A = \mathbf{230.77\ MW}$$

$$P_B = \mathbf{269.23\ MW}$$

Example 29.3 (*a*) *What do you understand by heat rate and incremental heat rate?*

(*b*) *What are the elements that contribute to the cost of electricity?*

(*c*) *A power plant of 210 MW installed capacity has the following particulars :*

Capital Cost : Rs. 18000/kW installed capacity.

Interest and depreciation : 12%.

Annual load factor : 60%.

Annual capacity factor : 54%

Annual running charges : 200 × 10⁶ (Rs.)

Energy consumed by power plant auxiliaries = 6%.

Calculate :

(*i*) *Cost of power generation for kWh.*

(*ii*) *Reserve capacity.*

Solution

(c) Total capital cost = 18000 × 210,000 = Rs. 378×10^7

$$\text{Interest and depreciation} = \frac{378 \times 10^7 \times 12}{1 \times 100}$$

$$= 378 \times 10^5 \times 12$$

$$= \mathbf{4536 \times 10^5}$$

Annual running charges = 200×10^6

∴ Total energy cost/year = $(453.6 + 200)\,10^6$

$$= \mathbf{Rs.\ 653.6 \times 10^6}$$

Let, C = Capacity of the plant in kW

= 210,000 kW

$$\text{Now, annual load factor} = 0.6 = \frac{\text{Average load}}{\text{Maximum demand}}$$

$$= \frac{\text{Energy produced/year}}{\text{Maximum demand} \times 8760}$$

$$\text{Also, annual capacity factor} = \frac{\text{Energy produced/year}}{\text{Capacity} \times 8760}$$

$$\therefore \quad 0.54 \times 210{,}000 \times 8760 = 0.6 \times 8760 \times \text{Maximum demand}$$

$$\therefore \quad \text{Maximum demand} = \frac{0.54 \times 210{,}000}{0.6} = 0.9 \times 210{,}000$$

$$= \mathbf{189000\ kW}$$

$$\therefore \text{Annual energy produced} = 0.54 \times 210{,}000 \times 8760$$

$$\therefore \quad \text{Chargeable energy} = 0.94 \times 9933.84 \times 10^5 \text{ kWh}$$

$$= \mathbf{9933.84 \times 10^5\ kWh}$$

(i) Cost of power generation/kWh

$$= \frac{653.6 \times 10^6}{993.384 \times 10^6 \times 0.94}$$

$$= \frac{0.658}{0.94} = 0.7 \text{ Rs.} = \mathbf{70\ paise}$$

(ii) Reserved capacity = Plant installed capacity – Maximum demand

$$= 210{,}000 - 189000 \text{ kW}$$

$$= \mathbf{21000\ kW \text{ or } 21\ MW}$$

Example 29.4 (*a*) *What are base load and peak load power plants? Explain.*

(*b*) *What do you understand by power plant economics? Explain fixed costs and operating costs of a power plant.*

(*c*) *The incremental fuel costs for two generating units A and B of a power plant are given by the following relations :*

$$\frac{dF_A}{dP_A} = 0.06\,P_A + 11.4$$

and

$$\frac{dF_B}{dP_B} = 0.07\,P_B + 10$$

where P is power in MW and F is the fuel cost in Rs/hr.

Determine :

(*i*) *The economic loading of the two units when the total load to be supplied by the power plant is 150 MW.*

(*ii*) *The net increase in fuel cost/hr, if the load is equally shared by each unit.*

Solution

(i) *Economic loading :*

(c) From the data, we get

$$\frac{dF_A}{dP_A} = 0.06\,P_A + 11.4$$

$$\therefore \qquad F_A = \frac{0.06P_A^2}{2} + 11.4\,P_A + \text{constant}$$

Similarly,
$$F_B = \frac{0.07P_B^2}{2} + 10\,P_B + \text{constant}$$

Total power delivered or supplied is 150 MW.

$$\therefore \qquad P_A + P_B = 150$$
$$\therefore \qquad P_B = 150 - P_A$$
$$\therefore \qquad F_B = 0.035\,(150 - P_A)^2 + 10\,(150 - P_A)$$
$$= 0.035\,(22500 - 300\,P_A + P^2{}_A) + 1500 - 10\,P_A$$
$$= 787.5 - 10.5\,P_A + 0.035\,P^2{}_A + 1500 - 10\,P_A$$
$$= 0.035\,P^2{}_A - 20.5\,P_A + 2287.5 \qquad \text{(I)}$$
$$F_A = 0.03\,P^2{}_A + 11.4\,P_A \qquad \text{(II)}$$
$$\therefore \qquad \text{Total cost} = F_A + F_B$$
$$X = 0.065\,P^2{}_A - 9.1\,P_A + 2287.5$$

This will be minimum or economic when

$$\frac{dX}{dP_A} = 0 = 0.065 \times 2\,P_A - 9.1$$
$$= 0.13\,P_A - 9.1$$
$$\therefore \qquad 0.13\,P_A = 9.1$$

or
$$P_A = \frac{9.1}{0.13} = \mathbf{70\ MW}$$
$$\therefore \qquad P_B = 150 - 70 = 80\text{ MW}$$

$\therefore$ Economic loading is

Unit $A \rightarrow P_A = \mathbf{70\ MW}$

Unit $B \rightarrow P_B = \mathbf{80\ MW}$

(ii) *Net increase in cost*

Equal loading means $P_A = 75$ MW and $P_B = 75$ MW

$$F_A = 0.03\,P^2{}_A + 11.4\,P_A$$
$$= 0.03 \times 75^2 + 11.4 \times 75 = 168.75 + 855$$
$$= \mathbf{1023.75\ Rs/h} \qquad \text{(III)}$$
$$F_B = 0.035P_B^2 + 10P_B$$
$$= 0.035 \times 75^2 + 10 \times 75$$
$$= 196.875 + 750$$
$$= \mathbf{946.875\ Rs/h} \qquad \text{(IV)}$$

Total cost for equal loading $= 1023.75 + 946.875$

$$= \mathbf{1970.625\ Rs/h} \qquad \text{(V)}$$

For economic loading,

$$F_A = 0.03\,P^2{}_A + 11.4\,P_A$$
$$= 0.03 \times 70 \times 70 + 11.4 \times 70$$

$$= 147 + 798$$

$$= \mathbf{945\ Rs/h} \tag{VI}$$

$$F_B = 0.035\,P_B^2 + 10\,P_B$$

$$= 0.035 \times 80 \times 80 + 10 \times 80$$

$$= 224 + 800$$

$$= \mathbf{1024\ Rs/h} \tag{VII}$$

∴ Total cost for economic loading

$$= 1024 + 945$$

$$= \mathbf{1969\ Rs/h} \tag{VIII}$$

∴ Net increase in cost $= 1970.625 + 1969$

$$= \mathbf{1.625\ Rs/h}$$

Example 29.5 (*a*) *Define the following* (*any three*) : *load factor, diversity factor, plant use factor, capacity factor.*

(*b*) *An input-output curve of a 10 MW thermal power plant is given by*

$$I = 10^6\,[90 + 60\,L + 2.5\,L^2]\ \text{kJ/h}$$

where I – *Input in kJ/h*

L – *Load on the plant, MW.*

Find :

(*i*) *Load on the plant at which the efficiency of the plant is maximum.*

(*ii*) *Increase in the input-output required to step-up the output of the plant from 5 MW to 7 MW.*

Solution

(b) (i) *Load at maximum efficiency :*

We have, $I = 10^6\,[90 + 60\,L + 2.5\,L^2]$

$$\therefore \quad \frac{I}{L} = 10^6\left[\frac{90}{L} + 60 + 2.5L\right]$$

$$\therefore \quad \eta = \frac{L}{I} = \frac{1}{10^6\left[\dfrac{90}{L} + 60 + 2.5L\right]}$$

For maximum efficiency,

$$\frac{d\eta}{dL} = 0 = \frac{-10^6\left[-\dfrac{90}{L^2} + 2.5\right]}{10^6\left[\dfrac{90}{L} + 60 + 2.5L\right]}$$

$$\therefore \quad -\frac{90}{L^2} + 2.5 = 0$$

$$2.5 = \frac{90}{L^2} \text{ or } L^2 = \frac{90}{2.5} = 36$$

or $L = \mathbf{6\,MW}$

At maximum efficiency, load is 6 MW.

(ii) *Increase in input :*

At 5 MW,
$$\begin{aligned} I &= 10^6\,[90 + 60\,L + 2.5\,L^2] \\ &= 10^6\,[90 + 60 \times 5 + 2.5 \times 25] \\ &= 10^6\,[90 + 300 + 62.5] \\ &= 213.0 \times 10^6\,\text{kJ/h} \\ &= 452.5 \times 10^6\,\text{kJ/h} \end{aligned}$$

At 7 MW,
$$\begin{aligned} I &= 10^6\,[90 + 60\,L + 2.5\,L^2] \\ &= 10^6\,[90 + 60 \times 7 + 2.5 \times 49] \\ &= 10^6\,[90 + 420 + 122.5] \\ &= 10^6 \times 632.5\,\text{kJ/h} \end{aligned}$$

$$\begin{aligned} \therefore \quad \text{Increase in input} &= 10^6\,[I_7 - I_5] \\ &= 10^6\,[632.5 - 452.5] \\ &= \mathbf{10^6 \times 1800\ kJ/h} \end{aligned}$$

Also, alternate method is

$$I = 10^6\,[90 + 60\,L + 2.5\,L^2]$$

or
$$\frac{dI}{dL} = 10^6\,[60 + 5\,L]$$

$$dI = 10^6\,[60 + 5\,L]\,dL$$

$$\therefore \quad \int_5^7 dI = 10^6 \int (60 + 5\,L)\,dL$$

$$\begin{aligned} I_7 - I_5 &= 10^6\,[60\,L + 2.5\,L^2]^7{}_5 \\ &= 10^6\,[60\,(7-5) + 2.5\,(7^2 - 5^2)] \\ &= 10^6\,[60 \times 2 + 2.5\,(49 \times 25)] \\ &= 10^6\,[120 + 2.5 \times 24] \\ &= 10^6\,[120 + 60] = 10^6 \times 180\,\text{kJ/h} \end{aligned}$$

Example 29.6 (*a*) *Discuss the nature of load curves for the following types of consumers :*

(*i*) *Urban residential load and* (*ii*) *Street light.*

(*b*) *An industrial power station has a maximum demand of 70 MW and installed capacity is single generator of 50 MW. Excess energy, if required, is bought from the state grid. The load curve of the plant is defined as follows:*

Time (hr)	*0–6*	*6–8*	*8–12*	*12–14*	*14–18*	*18–22*	*22–24*
Load (MW)	*30*	*46*	*60*	*46*	*66*	*70*	*36*

Determine :

(*i*) *Load factor of power station*

(*ii*) *Load factor of generator* (*station*)

(*iii*) *Energy purchased from grid per day and*

(*iv*) *Plant use factor.*

Solution

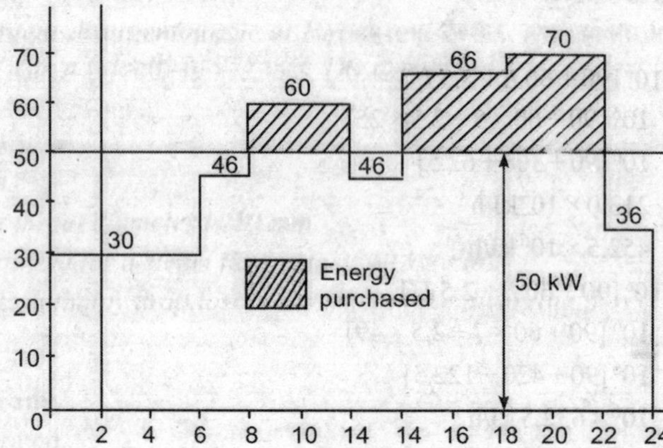

Fig. Ex. 29.6

(i) Assume that the state grid portion for this station is a standby equipment having a capacity of (70 – 50) = 20 kW.

∴ Energy generated (station + grid)

$$= 30 \times 6 + 46 \times 2 + 60 \times 4 + 46 \times 2 + 66 \times 4 + 70 \times 4 + 36 \times 2$$
$$= 180 + 92 + 240 + 92 + 264 + 280 + 72$$
$$= 272 + 332 + 616$$
$$= 1200 \text{ MWH}$$

∴ Average load $= \dfrac{1220}{24} = \mathbf{50.833\ MW}$

∴ Load factor of the power station as a whole

$$= \frac{\text{Average load}}{\text{Peak load}} = \frac{50.833}{70}$$
$$= 0.7262 \text{ or } \mathbf{72.62\%}$$

(ii) Again energy produced by the generator

$$= 30 \times 6 + 46 \times 2 + 50 \times 4 + 46 \times 2 + 50 \times 8 + 36 \times 2$$
$$= 180 + 92 + 200 + 92 + 400 + 72$$
$$= 872 + 164 = 1036 \text{ MWH}$$

∴ Average load on generator $= \dfrac{1036}{24} = 43.17$ MW

∴ Load factor of the generator $= \dfrac{43.17}{50} = \mathbf{0.8633}$ or **86.33%**

(iii) Energy purchased $= 1220 - 1036 = \mathbf{184\ MWH}$

$$= 10 \times 4 + 16 \times 4 + 20 \times 4 = 40 + 64 + 80$$
$$= \mathbf{184\ MWH}$$

(iv) *Plant capacity factor :*

Plant energy generated =1220 MWH

$$\therefore \quad \text{Use factor of plant} = \frac{1220}{70 \times 24}$$

$$= \frac{1220}{1680} = \mathbf{0.7262}$$

$$\text{Generator use factor} = \frac{1036}{50 \times 24} = \mathbf{0.8633}$$

Example 29.7 *An input-output curve of 10 MW capacity power plant is expressed by*

$$I = 10^6 [10 + 8L + 0.04 L^2] \text{ kJ/h}$$

where L is in MW.

Find :

(i) The load at which maximum efficiency is obtained.

(ii) Increase in the input required to increase the plant output from 3 to 5 MW by means of input-output curve and also by incremental rate curve.

Solution

(i)

$$I = 10^6 [10 + 8L + 0.4 L^2] \text{ MJ/hr}$$

$$\therefore \quad \frac{I}{L} = 10^6 \left[\frac{10}{L} + 8 + 0.4 L\right]$$

$$\text{Output} = \text{MW} \times 3600 \text{ MJ/h}$$

$$\therefore \quad \eta \text{ efficiency of the plant} = \frac{\text{Output}}{\text{Input}}$$

$$= \frac{3600 L}{10^6 \left[10 + 8L + 0.4 L^2\right]}$$

Efficiency will be maximum when $\frac{d\eta}{dL} = 0$.

$$\therefore \quad \frac{d\eta}{dL} = 0 = \frac{10^6 \left[10 + 8L + 0.8 L^2\right] \times 3600 - 3600 L \times 10^6 \left[8 + 0.8 L\right]}{\left(10 + 8L + 0.4 L^2\right)^2}$$

$$\therefore \quad 10 + 8L + 0.4 L^2 = 8L + 0.8 L^2$$

$$10 + 0.4 L^2 = 0.8 L^2$$

$$\therefore \quad 10 = 0.4 L^2$$

$$L^2 = \frac{10}{0.4} = \frac{100}{4}$$

$$\therefore \quad L = \sqrt{\frac{100}{4}} = \frac{10}{2} = \mathbf{5\ MW}$$

(ii)

$$I_5 = 10^6 [10 + 8 \times 5 + 0.4 \times 5^2]$$

$$= 10^6\,[10+40+10] = 10^6 \times 60 \text{ MJ/h}$$

$$I_3 = 10^6\,[10+8\times 3+0.4\times 3^2]$$

$$= 10^6\,[10+24+3.6] = 10^6 \times 37.6 \text{ MJ/hr}$$

$\therefore$ Increase in input required $= 10^6\,[60-37.6] = 10^6 \times 22.4$ MJ/h

or $$I = 10^6\,[10+8L+0.4L^2]$$

$$\therefore \quad \frac{dI}{dL} = 10^6\,[8+0.8L]$$

$$\therefore \quad \int_3^5 \frac{dI}{dL} = 10^6[8L+0.4L^2]_3^5$$

$$= 10^6\,[8(5-3)+0.4(25-9)]$$

$$= 10^6\,[16+6.4]$$

$$= \mathbf{22.4 \times 10^6 \text{ MJ/h}}$$

Example 29.8 *The daily load for a power plant is shown by the following equation :*

Load in MW = 350 + 10 t – t²

where t is the time in hours from 0 hours to 24 hours.

Draw the load curve and load duration curve and calculate the load factor.

Solution

$$\text{Load} = 350+10t-t^2$$

For maximum load, $\dfrac{d(\text{Load})}{dt} = 0$

$$\therefore \quad \frac{d}{dt}(350+10t-t^2) = 0$$

$$\therefore \quad (10-2t) = 0$$

$$\therefore \quad t = \frac{10}{2} = \mathbf{5 \text{ hours}}$$

and Maximum load $= 350+10\times 5-5^2$

$$= 400-25$$

$$= \mathbf{375 \text{ MW}}$$

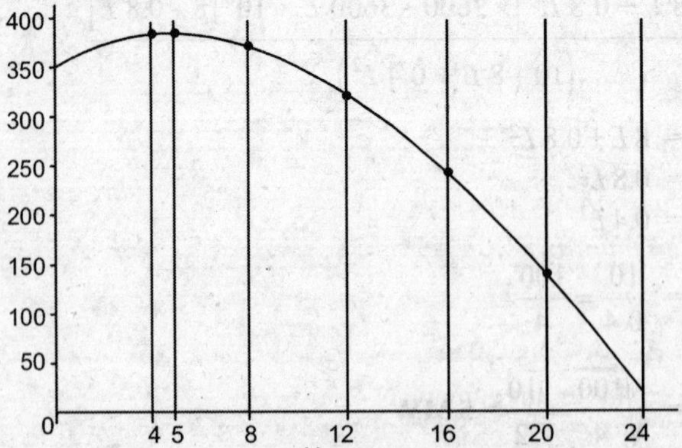

Fig. Ex. 29.8 *Load curve*

Time	Load
$t=0$	350
$t=4$	$350+10\times4-16=374$
$t=5$	$350+50-25=375$
$t=8$	$350+80-64=364$
$t=12$	$350+120-144=326$
$t=16$	$350+160-256=254$
$t=20$	$350+200-400=150$
$t=24$	$350+240-576=14$

Area under the curve is energy, kWh

$$= \int_0^{24}(L-dT) = \frac{10^6}{100}\int_0^{24}\left(350+10\,t-t^2\right)dt$$

$$= 10^3\left[350\,t+\frac{10\,t^2}{2}-\frac{t^3}{3}\right]_0^{24}$$

$$= 10^3\left[350\times24+5\times24\times24-\frac{24\times24\times24}{3}\right]$$

$$= 10^3\,[8400+2880-4608]$$

$$= \mathbf{10^3\times6672\ kWh}$$

$$\therefore\quad \text{Average load} = \frac{10^3\times6672}{24} = 278\times10^3\,\text{kW}$$

$$\therefore\quad \text{Load factor} = \frac{\text{Average load}}{\text{Maximum load}}$$

$$= \frac{278\times10^3}{375\times10^3} = \mathbf{0.7413 = 74.13\%}$$

Example 29.9 *A power plant has the annual factor as follows :*

Load factor = 70%, Capacity factor = 50%

Use factor = 60%, Maximum load = 20 MW.

Find out :

(i) Annual energy production

(ii) Reserve capacity over and above peak load

(iii) Hours during which the plant is not in service per year.

Solution

(i) $$\text{Load factor} = \frac{\text{Average load}}{\text{Maximum demand}} = 0.7$$

$$\therefore\quad \text{Average load} = 20\times0.7 = \mathbf{14\,MW}$$

∴ Annual energy production = Average load × Hours/year

$$= 14 \times 1000 \times 8760$$

$$= \mathbf{122.64 \times 10^6\ kWh}$$

(ii) Capacity factor $= 0.5$

$$= \frac{\text{Energy produced/year}}{\text{Capacity of plant} \times 8760} = \frac{\text{Average load} \times 8760}{\text{Capacity} \times 8760}$$

$$= \frac{14}{C}$$

$$\therefore \quad \text{Capacity of plant} = \frac{14}{0.5} = \mathbf{28\ MW}$$

∴ Reserve capacity over and above peak load = 28 – 20 = **8 MW**

(iii) Use or utilization factor =

$$\frac{\text{Energy produced/year}}{\text{Maximum energy during the actual number of hours the plant was in operation}} = \frac{122.64 \times 10^6}{X}$$

$$= 0.6$$

∴ Maximum energy during the actual number of hours the plant was in operation

$$= \frac{122.64 \times 10^6}{0.6} \text{ kWh} = \mathbf{204.4 \times 10^6\ kWh}$$

∴ Number of hours the plant was in operation/year

$$= \frac{204.4 \times 10^6}{28 \times 10^3} = \mathbf{7300\ hours}$$

∴ Hours during which the plant is not in service/year

$$= 8760 - 7300$$

$$= \mathbf{1460\ hours}$$

Example 29.10 *A 24 hours load curve of a power plant is sinusoidal with maximum and minimum demands of 6000 kW and 2000 kW respectively. If the plant capacity requirement is 7000 kW, find the following :*

(i) Average load on the plant

(ii) Plant load factor

(iii) Plant capacity factor.

Solution

(i) *Average load*

From Fig. Ex 29.10,

$$A \text{ area} = B \text{ area}$$

$$\therefore \quad \text{Average load} = \frac{6000 + 2000}{2}$$

$$= \mathbf{4000\ kW}$$

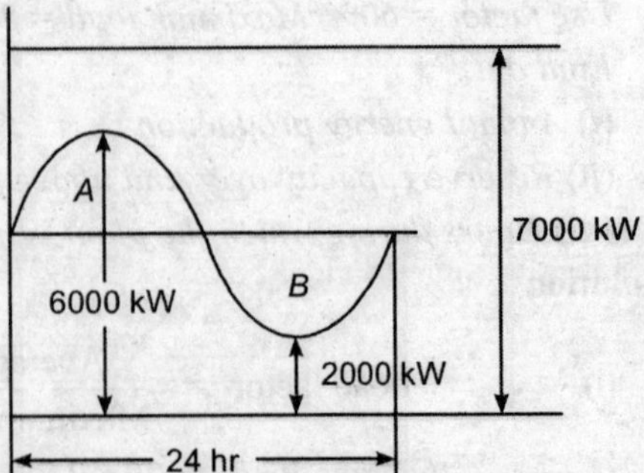

Fig. Ex. 29.10

(ii) *Plant load factor*

$$\text{Load factor} = \frac{\text{Average load}}{\text{Maximum demand}}$$

$$= \frac{4000}{6000} = \frac{2}{3} = \mathbf{0.667}$$

(iii) *Plant capacity factor*

$$\text{Plant capacity factor} = \frac{\text{Energy produced/24 hours}}{\text{Maximum energy that can be produced/24 hours}}$$

$$= \frac{\text{Average load} \times 24}{\text{Capacity of plant} \times 24}$$

$$= \frac{4000 \times 24}{7000 \times 24}$$

$$= \frac{4}{7} = \mathbf{0.5714}$$

Example 29.11 *The estimated cost of two power stations I and II are given as :*

C_I = Rs. (125 kW + 0.0275 kWh)

C_{II} = Rs. (120 kW + 0.03 kWh)

Both the plants supply power to a consumer whose maximum load is 100 MW, and minimum load is 10 MW. Assuming the load duration curve to be a straight line, find the followings for generating cost to be minimum :

(i) installed capacity of each station

(ii) the annual load, capacity and use factor for each station

(iii) average cost of production/kWh for the entire system.

Assume both plants are running in parallel. Reserve capacity of second plant is 20% of its required capacity.

Solution

(i)
$$C_{\text{I}} = 125\,\text{kW} + 0.0275\,\text{kWh}$$
$$C_{\text{II}} = 120\,\text{kW} + 0.03\,\text{kWh}$$

C_{I} is having operational cost less.

∴ It will be a base load station.

∴ $a_1 = 125, b_1 = 0.0275$

∴ $a_2 = 120, b_2 = 0.03$

∴ Number of hours for which second plant is working is given by

$$H = \frac{(a_1 - a_2)}{b_2 - b_1} = \frac{125 - 120}{0.03 - 0.0275}$$

$$= \frac{5}{0.0025} = \mathbf{2000\ hours}$$

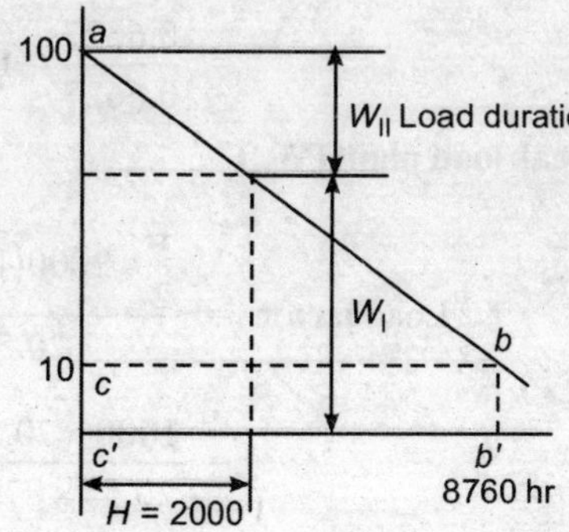

Fig. Ex 29.11

From similar triangles,

$$\frac{100 - W_{\text{I}}}{100 - 10} = \frac{2000}{8076}$$

$$100 - W_{\text{I}} = \frac{90 \times 2000}{8760} = \mathbf{20.55}$$

$$\therefore \quad W_{\text{I}} = 100 - 20.55$$
$$= \mathbf{79.45\ MW}$$

∴ Installed capacity of base-load plant

$$W_{\text{I}} = 79.45\ \text{MW}$$

Installed capacity of peak-load plant

$$W_{\text{II}} = 20.55 + \text{MW}$$
$$= 20.55 + 0.2 \times 20.55$$
$$= \mathbf{24.66\ MW}$$

(ii) *For base load plant (I)*

$$\text{Load factor} = \frac{\text{Average load}}{\text{Maximum load}}$$

$$= \frac{\text{Energy produced/year}}{8760 \times W_{\text{I}}}$$

$$= \frac{\frac{1}{2}(2000+8760)(79.45-10)+10\times 8760}{8760\times 79.45}$$

$$= \frac{373641+87600}{8760\times 79.45} = \frac{461241}{8760\times 79.45}$$

$$= \mathbf{0.6627} \text{ or } \mathbf{66.27\%}$$

Capacity factor = Load factor (as no reserve capacity)

$$= \mathbf{66.27\%}$$

$$\text{Use factor} = \frac{\text{Capacity factor}}{\text{Load factor}}$$

$$= \frac{0.6627}{0.6627} = \mathbf{1}$$

For peak load plant ($\mathbf{W_{II}}$)

$$\text{Load factor} = \frac{\frac{1}{2}\times 2000(100-79.45)}{20.58760}$$

$$= \frac{1000\times 20.55}{20.55\times 8760}$$

$$= \mathbf{0.1142}$$

$$\text{Capacity factor} = \frac{\text{Average load}}{\text{Plant capacity}}$$

$$= \frac{\text{Energy generated}}{8760 \times 24.66} = \frac{\frac{1}{2}\times 20.55\times 2000}{24.66\times 8760}$$

$$= \mathbf{0.09513}$$

$$\text{Use factor} = \frac{\text{Capacity factor}}{\text{Load factor}}$$

$$= \frac{0.09513}{0.1142} = \mathbf{0.833} \text{ or } \mathbf{83.3\%}$$

(iii) *Average cost of production of system :*

Base load plant (W_I)

$$\text{Total units generated} = 461241 \text{ MWh}$$

$$= 10^6 \times 461.241 \text{ kWh}$$

Peak load plant (W_{II})

$$\text{Total units generated} = \frac{1}{2} \times 2000 \times 20.55$$

$$= 20.55 \times 1000 \text{ MW}$$

$$= 20.55 \times 10^6 \text{ kWh}$$

$$\therefore \text{Total units generated} = 10^6 [461.241 + 20.55]$$

$$= \mathbf{10^6 \times 481.791 \text{ kWh}}$$

$$\therefore \quad C_I = 125 \times 79.45 \times 10^3 + 0.0275 \times 461.241 \times 10^6$$

$$= 10^6 [9.93 + 12.68]$$

$$= \mathbf{Rs.\ 10^6 \times 22.61}$$

$$C_{II} = 120 \times 24.66 \times 10^3 + 0.03 \times 20.55 \times 10^6$$

$$= 10^6 [2.9592 + 0.6165]$$

$$= \mathbf{Rs.\ 10^6 \times 3.5757}$$

$$\therefore \quad \text{Total cost} = C_I + C_{II}$$

$$= 10^6 [22.61 + 3.5757]$$

$$= \mathbf{Rs.\ 26.1857 \times 10^6}$$

$$\therefore \quad \text{Average cost} = \frac{\text{Total cost}}{\text{Total units generated}}$$

$$= \frac{10^6 \times 26.1857}{10^6 \times 481.791} = 0.0543 \text{ Rs.} = \mathbf{5.43 \text{ paise}}$$

Example 29.12 *An input-output curve of a power plant is given by the following equation :*

$$I = 18 + 12\,L + 0.5\,L^2 \text{ MJ/h}$$

where L is in MW. Determine the load at which plant efficiency is maximum. Draw the variation of input, heat rate and incremental heat rate with respect to load for the plant. Also find the change in input required when the load changes from 5 to 6 MW and comment on the result.

Solution

$$I = 18 + 12\,L + 0.5\,L^2$$

$$\therefore \quad \eta = \frac{\text{Output}}{\text{Input}} = \frac{L \times 3600 \text{ MJ/h}}{18 + 12\,L + 0.5\,L^2 \text{ MJ/h}}$$

This will be maximum when $\frac{d\eta}{dL}=0$.

$$\therefore \quad 0 = \frac{d\eta}{dL} = \frac{\left(18+12L+0.5L^2\right)\times 3600 - 3600L(12+L)}{\left(18+12L+0.5L^2\right)^2}$$

$$\therefore \quad 0 = 18+12L+0.5L^2-12L-L^2$$

$$= 18-0.5L^2$$

$$\therefore \quad 18 = 0.5L^2$$

$$\therefore \quad L^2 = \frac{18}{0.5}=36$$

$$L = \mathbf{6\,MW}$$

$$I_5 = 18+12\times 5+0.5\times 25=78+12.5=90.5$$

$$I_6 = 18+12\times 6+0.5\times 36=90+18=108$$

$$\therefore \quad I_6-I_5 = 108-90.5=\mathbf{17.5\,MJ/h}$$

For I vs L :

$$I_0 = 18, I_1=30.5, I_2=44, I_3=58.5$$

$$I_4 = 74, I_5=90.5, I_6=108, I_7=126.5, I_8=146.$$

For $\frac{dI}{dL}$ vs L curve :

$L=$	0	1	2	3	4	5	6	7	8
$\frac{dI}{dL}=$	12	13	14	15	16	17	18	19	20

For $\frac{I}{L}$ vs L :

L	0	1	2	3	4	5	6	7	8
$\frac{I}{L}$	∞	30.5	22	19.5	18.5	18.1	8	18.07	18.25

1. The intersection of $\left(\frac{I}{L}\text{ vs }L\right)$ and $\left(\frac{dI}{dL}\text{ vs }L\right)$ gives economical loading of the plant, where input is minimum and efficiency is maximum. At the intersection point, $\frac{dI}{dL}=\frac{I}{L}$.

2. Graph of I vs L shows that the curve is not linear. The intersect on Y-axis shows that some input is required to run the plant at no-load.

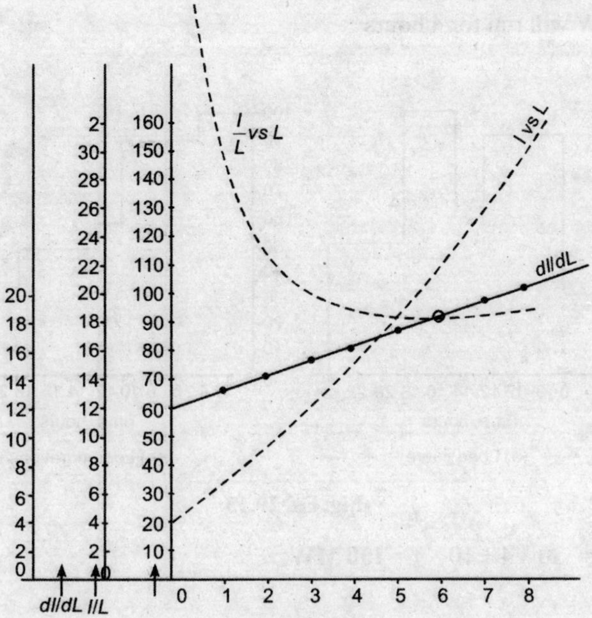

Fig. Ex. 29.12

Example 29.13 *A power station has to supply load as follows :*

Time (hours)	*0–6*	*6–12*	*12–14*	*14–18*	*18–24*
Load (MW)	*30*	*90*	*60*	*100*	*50*

(a) Draw the load curve, (b) Draw the load duration curve, (c) Select suitable generating units to supply the load, (d) Calculate the load factor, (e) Calculate the capacity of the plant and the plant capacity factor.

Solution

The load curve and load duration curve have been drawn in Fig. Ex. 29.13.

$$\text{Energy generated} = 30 \times 6 + 90 \times 6 + 60 \times 2 + 100 \times 4 + 50 \times 6$$

$$= 1540 \text{ MWh} = 1540 \times 10^3 \text{ kWh}$$

$$\text{Average load} = \frac{1540 \times 10^3}{24} \text{ kW}$$

$$\text{Maximum demand} = 100 \times 10^3 \text{ kW}$$

$$\text{Load factor} = \frac{1540 \times 10^3}{24 \times 100 \times 10^3} = \mathbf{0.64}$$

To supply the load, three generating units, each of 30 MW capacity, and one generating unit of 10 MW capacity will be selected. One additional unit will be kept as standby. Its capacity will be equal to that of the largest unit, i.e., 30 MW.

Load duration curve will indicate the operational schedule of different generating units, which will be as follows:

(i) One unit of 30 MW will run for 24 hours

(ii) Second unit of 30 MW will run for 18 hours

(iii) Third unit of 30 MW will run for 10 hours

(iv) Fourth unit of 10 MW will run for 4 hours

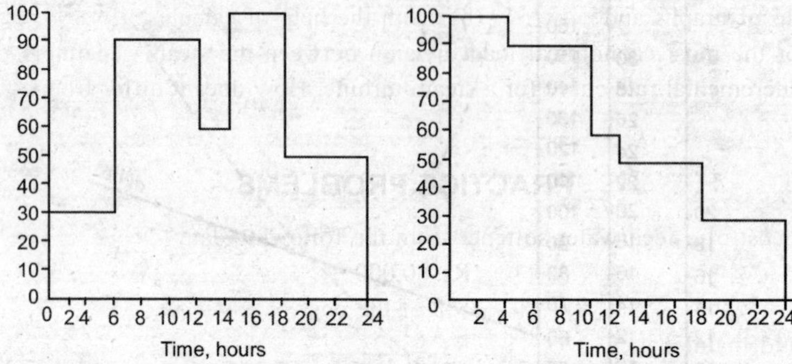

(a) Load curve

(b) Load duration curve

Fig. Ex. 29.13

$$\text{Plant capacity} = 30 \times 4 + 10 \times 1 = \mathbf{130\ MW}$$

$$\text{Capacity factor} = \frac{\text{Energy generated}}{\text{Capacity} \times \text{Operating time}}$$

$$= \frac{1540 \times 10^3 \text{ kWh}}{130 \times 10^3 \times 24 \text{ kWh}} = \mathbf{0.494}$$

THEORY QUESTIONS

29.1. Differentiate between fixed cost and operating cost. List the components of each of them.

29.2. Discuss the variation of fixed cost and operating cost with the investments for the plant. How annual costs are determined?

29.3. What is depreciation? What are the methods to compute the amount of depreciation?

29.4. Discuss the generation of electricity as a production activity. How does electricity differ from an industrial commodity?

29.5. Discuss the variable demand problem of electrical production and distribution.

29.6. What is a load curve? What is its use in the distribution of electrical energy?

29.7. What is a load duration curve? What is its use? How is it constructed?

29.8. Define and explain the following terms :

(i) connected load, (ii) average load,

(iii) demand factor, (iv) load factor

29.9. What is a group diversity factor and a peak diversity factor? How do they help in predicting maximum demand?

29.10. Define a plant capacity factor and a plant use factor. How do they help to find out the reserve capacity of the plant?

29.11. Explain with a diagram :

(a) Input-output curve (b) Heat rate curve (c) Incremental-rate curve

29.12. Explain why incremental-rate curve intersect the heat rate curve at the lowest value of heat rate.

29.13. Prove with the help of performance characteristics of a power plant and/or with the help of calculus that the heat rate of a continuous input-output curve is at a minimum when it is equal to incremental rate.

29.14. Explain :

(a) with the help of graphs and (b) with the help of calculus

the condition for the most economical load division between the steam generators.

29.15. Draw a typical incremental-rate curve for a steam turbine. How does it differ from a similar curve for steam generators?

PRACTICE PROBLEMS

1. Find the annual cost of a feed water softener from the following data :

Cost	Rs. 80,000
Salvage value	5% of the cost
Annual repair and maintenance	Rs. 2500
Annual cost of chemicals	Rs. 300
Interest on sinking fund	5%
Life of plant	10 years

[**Ans.** Rs. 1740]

2. The following data pertains to a power plant of 10 MW installed capacity.

Capital cost	900/kW capacity installed
Interest and depreciation	12%
Annual load factor	60%
Annual capacity factor	50%
Annual running charges	Rs. 15 × 10^6
Energy consumed by power plant auxiliaries	60%

Calculate :

(a) Reserve capacity (b) Cost per kWh. [**Ans.** (a) 25 MW, (b) 4.48 paise/kWh]

3. The following data pertain to a power plant :

Installed capacity = 200 MW; Capital cost = Rs. 4000 × 10^6; Annual cost of fuel, taxes and salaries = Rs. 600 × 10^6; Rate of interest = 5% of the capital; Rate of depreciation = 6% of the capital; Annual load factor = 65%; Capacity factor = 55%; Energy used in running the plant auxiliaries = 4% of total units generated. Determine

(a) the reserve capacity, and (b) the cost of power generation [**Ans.** (a) 31 MW]

4. The incremental fuel costs for units of plant are :

$F_1 = 0.010\,P_1 + 2.2$

$F_2 = 0.010\,P_2 + 1.6$

where P is in MW and F is in Rs. per hour. Assume that both units are operating at all times; that total load varies from 50 MW to 250 MW and the maximum and minimum loads on each unit are to be 120 MW and 25 MW respectively. Find the incremental fuel costs and the allocation of load between the units for minimum cost at the following total load.

(a) 50 MW, (b) 86.7 MW, (c) 180 MW, (d) 200 MW, (e) 215 MW and (f) 250 MW.

[**N.B.** : There are no answers for this example]

5. The incremental fuel costs for two generating units A and B of a power plant are given by the following relations :

$dF_A/dP_A = 0.06\,P_A + 11.4$

$dF_B/dP_B = 0.07\,P_B + 10$

where P is the power in MW and F is the fuel cost in rupees per hour.

(a) Find the economic loading of the two units when the total load to be supplied by the power station is 150 MW.

(b) Find the net increase in fuel cost per hour if the load is equally shared by the two units.

[Ans. (a) P_A = 70 MW, P_B = 80 MW, (b) Rs. 1.63 per hour]

6. A plant costing Rs. 65000 has a useful life of 15 years. Find the amount which should be saved annually to replace the equipment at the end of that time (a) by the straight line method, and (b) by the sinking fund method if the annual rate of compound interest is 5%. Assume that the salvage value of the equipment is Rs. 5000.

[Ans. (a) Rs. 4000, (b) Rs. 2780]

7. A power plant has the following annual factors :
Load factor = 70%; Capacity factor = 50%; Use factor = 60%; Maximum demand = 20 MW.
Estimate (a) annual energy production, (b) reserve capacity over and above the peak load, and (c) the hours per year during which the plant is not in service.
[Ans. (a) 122.8 × 10^6 kWh, (b) 8 MW, (c) 2060 h]

8. The base load power plant load on a 30 MW power plant is 25 MW. The power station supplies load having maximum demand of 10 MW, 8.5 MW, 5 MW and 4.5 MW. The annual load factor is 45%.
Determine : (a) Average load, (b) Energy supplied per year, (c) Diversity factor, (d) Demand factor.

[Ans. (a) 11.25 MW, (b) 98.55 × 10^6 kWh, (c) 1.12, (d) 0.89]

9. The base load power plant and standby power plant share a common load as follows :

Base load station annual output	150 × 10^6 kWh
Base load station capacity	35 MW
Maximum demand on base load plant	30 MW
Standby station capacity	18 MW
Standby station annual output	14 × 10^6 kWh
Maximum demand on standby	15 MW

Determine the following for both the power plants :
(a) Load factor, (b) Capacity or plant factor.

[Ans. For base load (a) 0.57, (b) 0.49; For standby (a) 10.7% (b) 9%]

10. A power generating station has a maximum demand of 10,000 kW and the daily load on the station is as follows :

Time	*kW*	*Time*	*kW*
6 a.m. to 8 a.m.	3,500	5 a.m. to 7 p.m.	8,500
8 a.m. to 12 noon	8,000	7 p.m. to 9 p.m.	10,000
12 noon to 1 p.m.	3,000	9 p.m. to 11 p.m.	4,500
1 p.m. to 5 p.m.	7,500	11 p.m. to 6 a.m.	2,000

(a) Draw the load curve and the load duration curve.
(b) Choose the size and number of generating units.
(c) Draw up the operating schedule of the units.
(d) What reserve plant would be necessary?
(e) Calculate the load factor, plant capacity factor and plant use factor of the station.

[Ans. (b) Three units of 5000, 2500 and 2500 kW capacity, (d) 5000 kW, (e) 55%, 36.66%, 86.5%]

11. A power station has an installed capacity of 210 MW. The cost of the station is Rs. 2 crore per MW. The fixed costs are 14% of the cost of investment. On full load at 100% load factor, the variable costs of the station per year are 1.5 times the fixed costs. Assume that there is no reserve capacity of the plant and that the variable costs are proportional to the energy production. Find the cost of generation per kWh at load factors of 100%, 80%, 60%, 40% and 20%. Plot the curve.

12. Two power stations *A* and *B* supply to steam whose maximum demand is 120 MW and minimum load is 12 MW during the year. The estimated cost of these stations are as follows :
C_A = Rs. (125 kW + 0.0275 kWh), C_B = Rs. (120 kW + 0.03 kWh)

If the load varies as a straight line, find for the minimum cost of generation (i) Installed capacity of each station, (ii) The average cost of generation per kWh for total system.

[**Ans.** (i) 80 MW, 24 MW; (ii) 5.5 paise per kWh]

13. A power station is said to have a use factor of 47% and capacity factor of 40%. For how many hours in a year was the power station not in service? [**Ans.** 1305 h]
14. A central power station has annual load factor, capacity factor and use factor as 60%, 40% and 45% respectively. The maximum demand is 15 MW. Estimate (a) the annual energy production and revenue earned if the cost of energy is Re 1/kWh, (b) the reserve capacity over and above the peak load, and (c) the hours per year the station is not in service. [**Ans.** (a) 78.84×10^6 kWh, Rs. 78.84×10^6, (b) 7.5 MW, (c) 974 K]
15. The input-output curve of a 60 MW power station is given by

$$I = 10^6 [8 + 8L + 0.4L^2] \text{ kJ/hr}$$

where I is the input in kJ/hr and L is load in MW.

(a) Determine the heat input per day to the power station if it works for 20 hours at full load and remaining period at no load.

(b) Also find the saving per kWh of energy produced if the plant works at full load for all 24 hours generating the same amount of energy. [**Ans.** (a) 38592×10^6 kJ/day, (b) 4×10^3 kJ/kWh]

Chapter 30 Electrical Equipment

Smallest particle a chemical element can exist as, and still keep its physical amd chemical characteristics, is an atom. Each atom has a nucleus at centre made up of protons and neutrons. Electrons move around nucleus.

Protons carry positive (+) charges and electrons carry negative (–) charges. Electricity is present when electrons are kept away from protons. They will always try to get together—the electrons moving towards protons. When electrons find a path to move along, current is flowing. Materials through which current flows easily (electrons are easily separated from protons) are called conductors. Materials through which current does not flow easily (protons hold electrons tightly) are non-conductors or insulators.

As electricity depends on presence and flow of electrons, some of the units of measurement are given below :

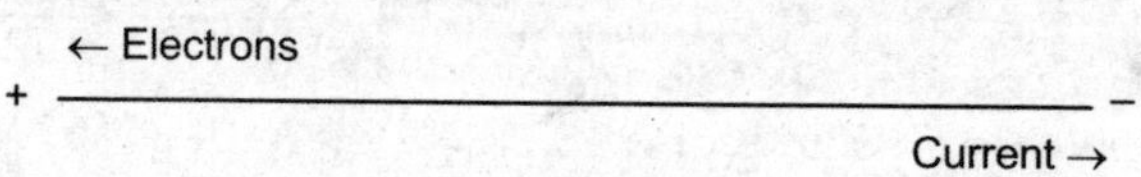

Fig. 30.1 *Current flow is normally measured from* (+) *to* (–)*; electrons go reverse.*

Ampere is the unit of current. One ampere means 6.25 billion electrons flow past a point in a circuit every second.

Volt is the unit of pressure that pushes current through a circuit. One volt will push one ampere through a resistance of one ohm.

Ohm is unit of resistance to current flow. One ohm is quantity of resistance producing a potential drop of 1 volt when the current in it is 1 ampere.

Watt measures power in a circuit. Watts equal current squared times resistance (*PR*). One watt is the power used for 1 ampere to flow through 1 ohm.

Watt hour measures energy or work done in a circuit. Watts multiplied by time in hours gives the power that is flowing.

30.1 GENERATORS

Generators are usually considered along with prime-mover as a combined generating unit with respect to price, weight, dimentions and efficiency. These are generally designed with enclosing revolv-ing fields and operate at high speeds. The voltage of generation is usually high, which in high capacity generators may be of the order of 6 kV or ven more. These generators may be either directly coupled with prime-mover as in case of a diesel engine or low speed turbine driven sets, or they may be coupled through speed reduction gears. Particularly in case of gas and steam turbines, minimum fuel consumption is obtained by operating the turbine at its most efficient speed which may be of the order of 20,000 to 30,000 r.p.m. In all such cases speed reduction gears are invariably used.

The frame of the generator is supported by feet of the sides. Within the frame, sheet steel laminations are built in sections and separated by spacers to form ventilating ducts. Limitations are slotted to receive the armature coils. The generator is ventilated by fans at both ends of the rotor. Rotors are round, with field coils of copper ribbon wound edgewise, placed in slots in the periphery, and fastened by metal wedges. Collector rings are of steel.

As the load from the generator is supplied to a large number of consumers, under certain conditions of operation the load may, therefore, charge suddenly both in magnitude and power factor. For this reason, alternators in conjunction with the exciters and auto-matic voltage regulators are designed in such a way so that there is no undue rise in the voltage at the power station in the event of load being thrown off, assuming transmission lines still under full voltage and therefore drawing a leading current from the station.

30.1.1 Rating of Generators

For a power station some mechanical and electrical features must be specified if a generator is to form a satisfactory link between the turbine and the transmission system.

30.1.2 Mechanical Requirements

1. Speed. The higher the speed the smaller the weight and cost of generator, until increasing stresses necessitate special rotor construction. Normal speed for generation at 50 cycles per second is 3000 revolutions per minute. As steam and gas turbines have high operating speeds, generators coupled to these turbines operate at 3000 r.p.m. with reduction gears, if necessary. However, generators coupled to diesel engines and hydraulic turbines may operate at speeds below 3000 r.p.m.

2. Overspeed. The generator must be designed for the speed attained when maximum load is thrown off, with governor inoperative. With hydraulic turbines this may be 100 percent of rated speed whereas for other units it is much less.

3. Flywheel effect. The minimum inertia necessary for satisfactory speed regulation can only be stated by the manufac-turer; a higher value may be specified to improve transient stability. This is expressed as moment of inertia or kinetic energy acting for a number of seconds, called the inertia constant.

30.2 ELECTRICAL OUTPUT

Power is normally generated at 50 cycles par aacead 3 phase, except where the supply is for a single special purpose, such as traction. The generator and the turbine by which constitutes a generating unit and practical economic consideration dictate that their capabilities are matched. The generator should be capable of meeting the reactive power requirments of the system to which it will be connected and its power rating should be such that when connected to a large power system steady operation of the unit is assured.

30.2.1 Maximum Continuous Rating

The maximum continuous ratings (MCR) of large generators are usually expressed in megawatts (MW) based on nominal values of voltage, current and power factors for which the generator is designed.

The maximum continuous rating may also be expressed in megavolt-amperes (MVA), which is based on the nominal values of the stator and rotor currents. Neither of these should be exceeded since the additional losses would tend to damage their respective insulating materials. Any increase in the MW output above the rated value can only be obtained by reducing the MVAR's, the limiting factor in this case being the stator current. Similarly, the machine can be made to provide additional reactive MVAR's by reducing the MW output, the limiting factor in this instance being the rotor current.

30.2.2 Short Circuit Ratio

The short-circuit ratio is one of the most important characteristics of a generator and in rating the machine strict attention must be given to this parameter. The short-circuit ratio is the ratio between the field current required to produce nominal volts on open circuit and that required to circulate full load current in the stator when short circuited at terminals. This ratio is one of the characteristics of generators which materially influences the stable operation of machines in parallel.

30.2.3 Power Factor

Generator size is determined primarily by MVA rating which is MW rating/power factor. Power factor has further importance, since it determines the ratio of rotor to stator loading. Too low a power factor leads not only to increased kVA rating, but still greater increase in capacity of the rotor and excitation plant. Mostly generators are designed for 0.8 power factor (lagging).

30.2.4 Over Load

When the plant operates at full power for long or unpredictable periods, this becomes continuous rating. But where there are wide seasonal variations of load, the continuous rating of the generators may properly be 10–15 percent below the maximum turbine output, particularly where maximum load coincides with good cooling conditions.

30.2.5 Voltage

Voltage is sometimes dictated by the needs of a local supply; otherwise by minimum combined cost of generators, bus-bars, switchgear and transformers. 11– 14 kV are standard voltages for generators upto 120 MW capacity. 500 MW, 2 pole generator it may be 22–23.5 kV.

30.2.6 Rating of Alternators

Typical rating of a 60 MW alternator is given below :

Rating in MW	60
Rating in MVA	75
Power factor (lagging) at generator terminals	0.8
Voltage, kV (measured at generator terminals)	11.8
R.P.M.	3000
Short Circuit Ratio (at rated MVA and voltage)	0.55

30.3 POWER FACTOR

An ac electrical system carries two types of power:

(1) true power, watts, that pulls the load (mechanical load reflects back into an ac system as resistance).

(2) reactive power, vars, that generates magnetism within, inductive equipment.

The vector sum of these two will give actual volt-amperes, flowing in the circuit.

Power factor is the cosine of the angle between true power and volt-amperes.

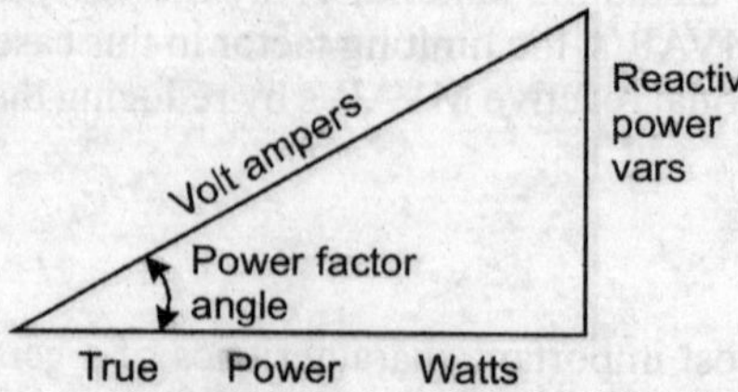

Fig. 30.2 *Power factor*

There are two functions of a synchronous alternator e.g., generating kilowatt (kW) and generating lagging kilovolt amperes (kVAr). The kilowatt output is controlled by adjusting the fuel (steam, gas. or water supply) supply to the turbine. Thus increase of fuel supply increases the kilowatt output. The kilovolt amperes (kVAr) output is controlled by the excitation, thus increase in excitation increases the lagging kVAr output. When an alternator is affording a supply of kW and lagging kVAr, e.g., to a load like induction motor etc., it will operate at a lagging power factor.

The alternator requires magnetisation in order to create its own flux, it can supply only lagging kVAr to the electrical system, if its excitation is increased above the value required to enable it to generate its own flux i.e., over-excitation. If this excitation is reduced below this value the required magnetisation needed to maintain its own voltage will have to be taken from the electrical system i.e., from magnetising or lagging kVAr, supplied by other alternators then operates as under excited and will have a leading, power factor.

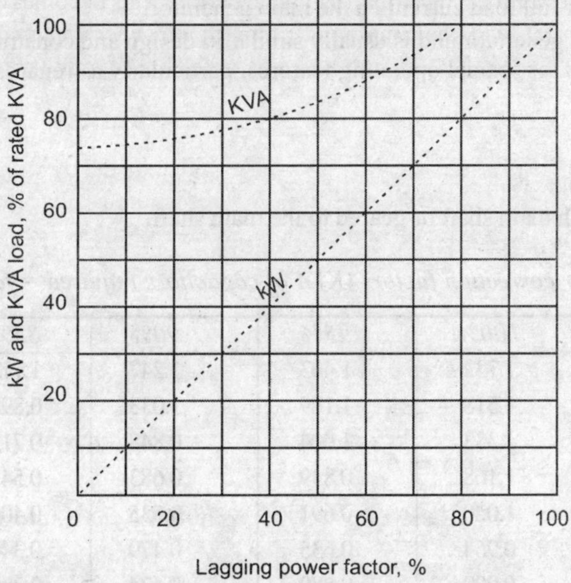

Fig. 30.3 *Effect of power factor on maximum continuous kVA and kW load that can be carried by typical 0·8 p.f. engine driven alternator.*

Power from alternator is transmitted through underground cables and overhead transmission lines. Underground high voltage cables act as capacitors in parallel on the electrical system and may improve the power factor of the load to unity so that it does not require lagging magnetising kVAr to be supplied from the alternators. The necessary magnetisation is supplied by these capacitors as they act as alternators supplying kVAr or in other words they absorb leading kVAr.

While deciding about the kVA rating of the alternator, it is necessary to know probable power factor of transmission or distribution system. Generally 0.8 power factor meets all requirements. However, while designing an alternator for 0.8 power factor, allowance is kept in the stator for 25% more current than would be required for unity power factor.

Low power factor affects the rotor design to a great extent, since the increase in rotor current and hence the energv dissipated in heating the rotor coils is much more than the corresponding increase in the stator current. An alternator when specifically designed for unity power factor load and utilising its full field capacity under that condition must not be loaded in kW to more than 50 percent of its unity power factor rating if used on an 0.8 p.f. load, otherwise the excitation will exceed the safe capacity of the field coils. Further, an alternator designed for 0.8 power factor may show signs of instability if the load has a leading power factor. An alternator is usually designed to remain stable when carrying a specified kVA at zero power factor.

Power Factor Correction. Improving (increasing) system power factor reduces reactive power (vars) and apparent power (volt-amperes) for same amount of true power (watts). Adding capacitors to cancel effect of circuit inductance is one method.

30.4 EXCITATION SYSTEM

The excitation system comprises all that equipment which provides the supply and control of the generator excitation. Basically it consists of the exciter unit, field rheostat, voltage regulator and field suppression equipment.

The excitation system of a generator must have reliability under all conditions of service, simplicity of control and ease of maintenance. The system must respond quickly to changes in excitation current requirements. The continuous voltage and current ratings of the exciter are generally 10 per cent greater than the corresponding excitation values estimated for full load current on the main generator.

An exciter is basically a dc generator and is usually similar in design and construction, its role as a voltage and power factor control unit calls for special operating features, particularly as regards its voltage range and output.

30.4.1 Classification

(A) Based on method of driving

(1) Directly driven through main shaft or geared to the main shaft.

TABLE 30.1 *Capacitor correction factors (KVA of capacitors required = factor kW load)*

Existing Power factor, %	*100%*	*95%*	*90%*	*85%*	*80%*	*75%*
50	1.732	1.403	1.247	1.112	0.982	0.850
55	1.518	1.189	1.033	0.898	0.768	0.636
60	1.333	1.004	0.848	0.713	0.583	0.451
65	1.168	0.839	0.683	0.548	0.418	0.286
70	1.020	0.691	0.535	0.400	0.270	0.138
72	0.964	0.635	0.479	0.344	0.214	0.082
74	0.909	0.580	0.424	0.289	0.159	0.027
75	0.882	0.553	0.397	0.262	0.132	
76	0.855	0.526	0.370	0.235	0.105	
78	0.802	0.472	0.317	0.182	0.052	
80	0.750	0.421	0.265	0.130		
82	0.698	0.369	0.213	0.078		
84	0.646	0.317	0.161			
85	0.620	0.291	0.135			
86	0.594	0.265	0.109			
88	0.540	0.211	0.055			
90	0.485	0.156				
92	0.426	0.097				
94	0.363	0.034				
95	0.329					

(2) Separately driven

(i) motor drive

(ii) auxiliary power drive

(B) Based on method of excitation

(1) Self excited shunt type, D.C. exciter

(2) Main exciter, excited from constant voltage auxiliary exciter (D.C. type)

(3) A.C. exciter

(4) Rectifier.

An exciter may be either directly coupled to the alternator shaft or it may be driven through belt or chain drive.

The advantages of directly coupled excitors are :

(1) The voltage of the system can be maintained independent of auxiliary source of supply and is not affected by faults thereon.

(2) The machine as a unit is complete in itself and independent of outside sources for its operation.

The disadvantages of direct coupled exciters are :

(1) With the high speeds in general use (1500 to 3000 r.p.m.) difficulties such as balancing, coupling up, alignment of shaft and other problems are introduced.

(2) Armature fitting, commutator stresses, brush chatter and ventilation, all require special attention.

(3) Exciter troubles may sometimes necessitate shutting down, or the aim set.

30.5 DC EXCITORS

Direct current exciters are shunt wound machines and compounding can be included to improve response. Figure 30.4 shows the open-circuit characteristic and basic diagram of a self excited compounded shunt wound exciter. *HB*

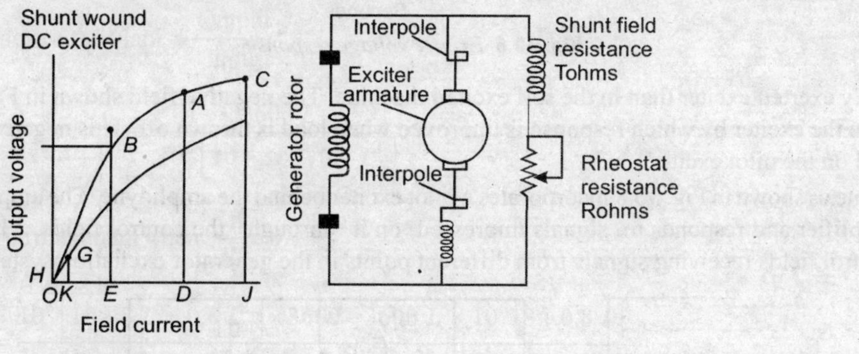

Fig. 30.4 *Shunt wound exciters*

is the unstable voltage region on the open circuit characteristic of the exciter. Since *OE* is the excitation current when voltage *BE* is across the shunt field circuit, *BE*/*OE* is the value of the critical resistance of the circuit. The line of this critical resistance coincides with the slope of the voltage characteristic ; therefore, the voltage is indefinite and can vary freely between the value *H* due to exciter remanent magnetism, and value *B*, the voltage due to field current *OE*. Even small temperature increases in the field-wrinding will contribute to the voltage instability. A method of overcoming this effect is to insert a saturation liner behind each pole piece as shown in Fig. 30.5. Because of reduced magnetic section the liner is saturated much sooner than -the pole body thereby introducing the required non-linearity in the open circuit characteristic.

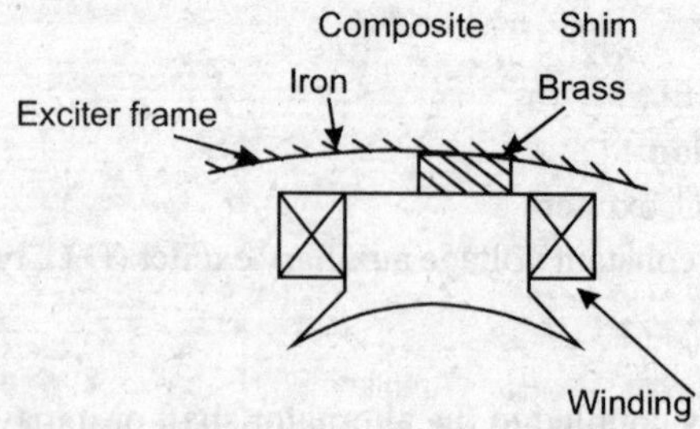

Fig. 30.5 *Shunt wound exciter pole with saturation liner.*

Except for small generators the main exciter is usually separately excited from a pilot exciter and the advantage of this is not only more stable voltage, but also improved voltage response to changes in field current. This is shown in Fig. 30.6 for an actual case in which, when measured within 0.5 sec, the voltage build up is about five times greater

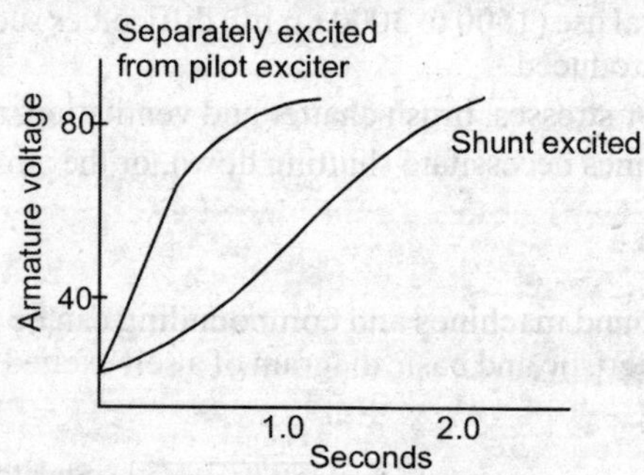

Fig. 30.6 *Exciter voltage response.*

in the separately exerted exciter than in the self excited machine. The negative field shown in Fig. 30.7 provides a negative bias in the exciter by which response is improved when load is thrown off. This negative field constitutes a constant load on the pilot exciter.

An amplidyne as shown in Fig. 30.8 incorporates a pilot exciter behind the amplidyne. The amplidyne is one make of rotating amplifier and responds to signals impressed on it through the control fields. There are usually a number of control fields receiving signals from different points in the generator excitation system.

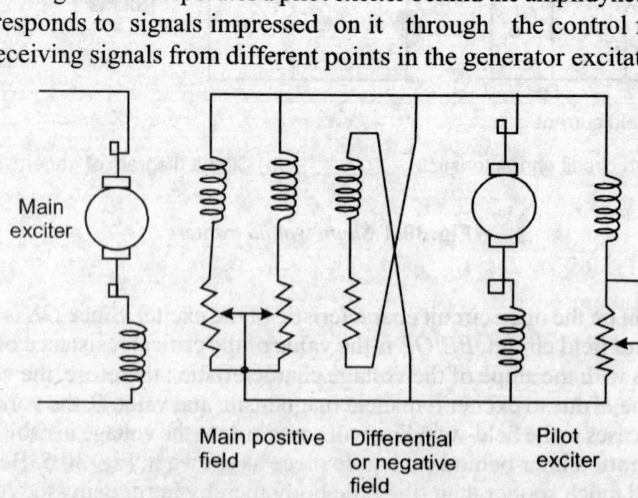

Fig. 30.7 *Arrangement of exciter field winding with pilot exciter.*

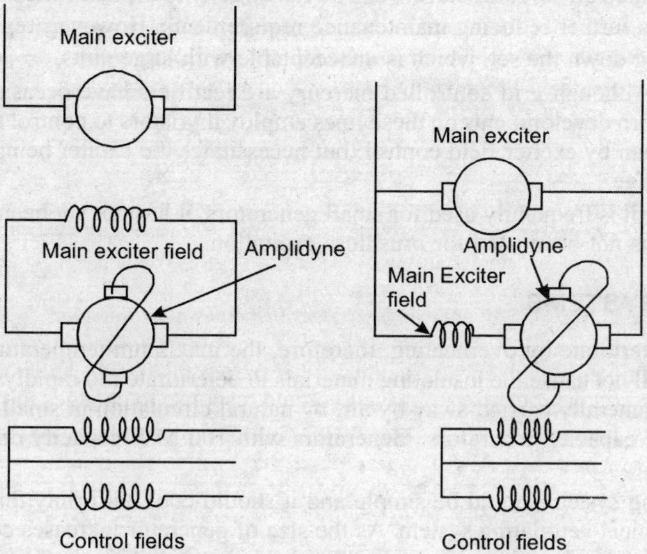

Fig. 30.8 *Amplidyne field control circuits.*

30.6 AC EXCITORS

To avoid commutation difficulties as well as commutator maintenance, ac exciters with silicon or germanium rectifiers have been developed for the largest turbo-generators as well as for smaller machines ; the current from the static floor mounted rectifiers is fed to the rotor of the main machine through conventional slip.

TABLE 30.2 *Some data about exciters*

Generator rating, MW	100	200	300	375	500
Rated power factor	0.80	0.90	0.85	0.85	0.85
Full load excitation, volts	605	520	360	420	555
Full load excitation, amps	1150	3000	3000	3310	3700/5100
Type of exciter	D.C.	D.C.	A.C.	A.C.	A.C.
Speed of exciter, R.P.M.	860	750	3000	300	3000
Type of exciter drive	Gear	Gear	Direct	Direct	Direct

N.B. These figures are typical and there may be variations between generators of different manufacturers.

The ac exciter is mounted on the main generator shaft and may be designed to operate at any economically convenient frequency, e.g. 100–150 Hz or more. The field winding of the exciter is fed from an auxiliary ac exciter, the field of which may, in turn, be provided by a permanent magnet generator.

Rectifier elements must be constructively rated to guard against effects of induced transient currents and voltage due to system short circuits, asynchronous running, pole slip and faulty synchronising as well as due to faults in the excitation system itself.

Instead of the shaft generator a unit transformer could be used to provide the a.c. supply ; although less expensive in first cost such a scheme has the disadvantage of higher sensitivity to system short circuits.

Brushless Excitation. If an a.c. exciter is built with a rotating armature and the rectifier elements, as used with static rectification, are mounted on the rotor current can be fed directly to the main field winding without the use of slip rings and brushes, thus further reducing maintenance requirements. However, replacement of a diode after failure, necessitates shutting down the set, which is unacceptable with large units.

Thyristor Excitation. Although grid controlled mercury are rectifiers have occasionally been used in static rectification schemes, modern developments on these lines employ thyristors to control the main field current; this gives a quicker response than by exciter field control, but necessitates the exciter being designed for continuous operation at its ceiling voltage.

Although thyristor control is frequently used for small generators, it has not yet been fully developed for large turbo-generator sets and has not been used for brushless excitation.

30.7 VENTILATION SYSTEMS

All winding insulations deteriorate by overheating, therefore, the maximum temperature in a generator must be limited to a value which will not cause the insulating materials to deteriorate too rapidly. The heat generated in the stator or rotor winding is generally carried away by air, by natural circulation in small generators and by forced circulation in case of large capacity generators. Generators with 100 MW capacity or more utilise hydrogen as cooling media.

The alternator ventilating system should be simple and it should cool uniformly throughout its entire length. Figure 30.8 illustrates a typical ventilating system. As the size of generator increases cooling becomes more and more difficult. To faciliate matters the air passage is subdivided into parallel air circuits arranged axially, offering a low resistance to the passage of air.

30.7.1 Closed Circuit Cooling

The older system of ventilation in which the cooling air was discharged to the atmosphere after passing through the alternator has been superceded by the closed circuit cooling for all large machines. In this system the same air is circulated continuously through the alternator and is cooled by passing through a heat exchanger.

The advantages of this system are :

(1) Very little or no dirt can collect in the alternator.

(2) Due to limited quantity of oxygen in the circuit fire hazards are reduced. Sometimes even inert gas is also introduced partly, to reduce possibilities of fire.

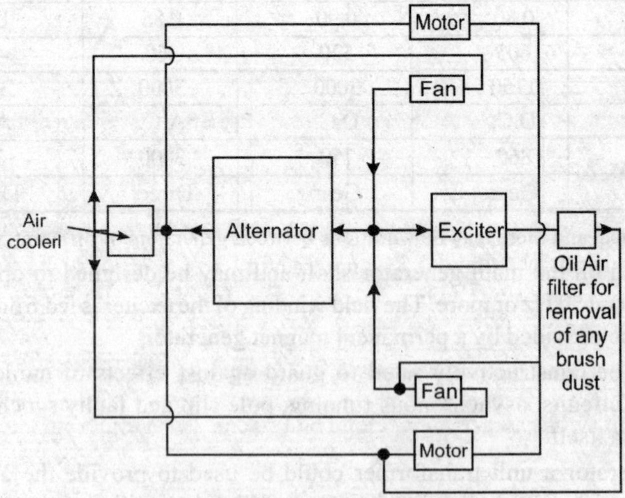

Fig. 30.9 *Alternator ventilating system.*

(3) The system obviates the expense of frequent cleaning of filter.

(4) The noise level in turbine house is reduced.

(5) A very compact system is obtained, for in most cases, it is possible to install cooler and fan in the alternator foundation block itself.

The use of closed circuit system of ventilation opens up the possibilities of application of some other gas as the cooling medium in place of air.

30.7.2 Hydrogen Cooling

With increasing size and length of alternators, cooling by ordinary methods in which air is discharged through the alternator for cooling purposes, becomes inadequate. The use of closed circuit cooling is the only alternative. The use of hydrogen gas as the cooling medium in closed circuit cooling offers the following advantages :

1. The thermal conductivity of hydrogen is almost seven times that of air. Its specific heat is 14.5 times that of air. The overall heat transfer with forced cooling may be as 2 to 3 times that with air.

TABLE 30.3 *Relative characteristics of air and hydrogen*

Propert	*Air*	*Pure hydrogen Pressure kg/cm²*		
		$\frac{1}{30}$ *ata*	*2 ata*	*3 ata*
Density	1	0.07	0.14	0.22
Thermal conductivity	1	6.7	6.7	6.7
Heat transfer coefficient	1	1.55	2.7	3.6

2. The density of hydrogen is $\frac{1}{14}$ that of air. Consequently windage losses are reduced.

3. Due to lower windage losses and better heat transfer in the cooler, less cooling water is required.

4. Hydrogen cooled machines require less space.

5. The fire risk is reduced in the event of electrical break down as oxygen is not present to maintain combustion of the insulation material.

6. Corona appears in hydrogen at about 60 percent of the voltage at which it appears in air of the same pressure, but on the other hand, in hydrogen, corona has no ill effects on fibrous insulation, nitric acid which is formed by corona discharge in air being absent. The importance of this is apparent in view of the desire to increase the voltage on large alternators.

Disadvantages

(1) The cost of cooling is high.

(2) There is every possibility of gas leaking through the casing. For the reason the casing is made explosion proof against any pressure which can be developed by explosive mixture of hydrogen and air.

(3) The effective sealing of the shaft glands at the alternator housing is one of the major problems.

(4) A high degree of hydrogen purity is required to ensure high efficiency and avoid explosions, and this necessitates hydrogen purity, pressure and temperature recorders.

(5) Before admitting hydrogen to the cooling system it is usual to displace the air with carbon dioxide to avoid an explosive mixture in the stator casing. While removing hydrogen from the system, it is displaced by carbon dioxide which in turn is removed by compressed air.

(6) Overall capital and maintenance cost is high.

30.8 TRANSFORMERS

The primary function of a transformer is to transform electrical energy from one alternating voltage to another. To transform large amounts of energy with maximum efficiency many factors must be considered in determining the materials, design and arrangement of primary and secondary coils and the core.

30.8.1 Construction

A transformer consists of two insulated coils of wire linked with a ring of iron. The coils are called high votage and low voltage windings, or primary and secondary windings. The primary winding is connected to the source of energy and the secondary is connected to the load. The high voltage winding is designed for high voltage and has the greater number of turns. The ring is called the core.

Each coil consists of a number of loops of round or rectangular wire. Several strands may be used in parallel but electrically insulated from each other, from the core and from the other coil.

The core consists of thin sheets of high grade silicon steel. The thickness depends somewhat on the frequency at which the transformer is to operate.

30.8.2 Rating

Rating of a transformer consists of the output together with other characteristics, such as voltage, current, frequency, and power factor assigned to it by the manufacturer. It is regarded as a test rating that defines an output, which can be taken from the transformer under prescribed conditions of test and within the limitations of established standards.

The basic items required to specify a transformer are :

single phase or 3 phase)

Frequency (i.e., 50 cycles per second)

Connections (i.e., double-wound or auto-transformer and vector group)

Method of cooling

Rated KVA or MVA

Voltage ratio (at no load)

Tappings (also whether off load or on load)

For example a generator transformer for a 600 MVA unit would be described as a 3 phase 50 c/s 600 MVA transformer with a voltage ratio of 22/430 KV at no load with tappings for + 2% to –16% and with oil cooling arrangement. Vector group Ydl.

30.8.3 Turn Ratio

Turn ratio is the ratio of the number of turns in the high voltage winding to the number in the low voltage winding. At no load the voltage ratio is equal to the turn ratio, but under load the voltage ratio differs from the turn ratio because of the regulation.

30.8.4 Tap Changer

Most large power transformers are provided with tappings by which the ratio (hence voltage) may be varied to suit conditions at different parts of a system. These taps are usually on higher voltage winding to minimise the copper section of tapping devices. The taps may be changed in tree ways :

(1) by internal links accessible only when the transformer is withdrawn from service and oil level is lowered.

(2) by off-load currents switches incapable of interrupting load currents.

(3) by on load tap-changing switches which permit ratio varia-tion when the transformer is in service.

In case of on-load tap changer, the tappings on the windings are brought out through a terminal board to a separate oil filled compartment in which the no-load tap-changer selector is housed. The tap changer is operated by a motor operated driving mechanism by local or remote control and a handle is fitted for manual operation in an emergency.

As tap changing must take place under load, before one tapping is left contact must be made with the next. This could cause a short-circuited number of turns and large currents are prevented by the use of resistors or reactors. Figure 17.10 shows a typical winding diagram connection diagram for ne phase of a transformer with a resistor type tap changer at the neutral point of a star connected winding. In this case one selector switch ($S1$) is on tape 1 and the other ($S2$) on tap (2). The diverter switch ($S3$) is shown connecting tap 1 to the neutral point of the transformer winding and the switching sequence for changeover to tape 2 is as follows :

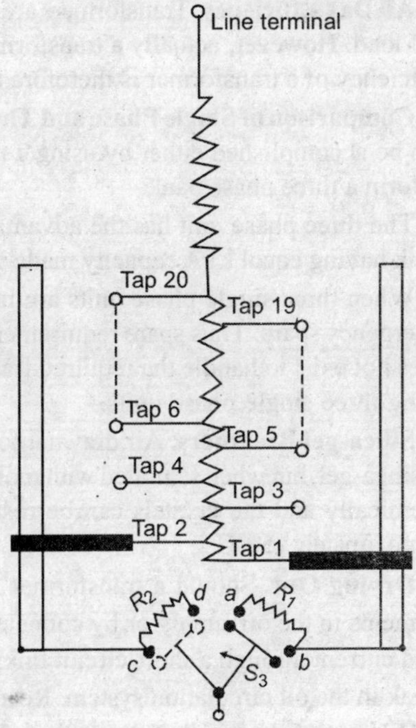

Fig. 30.10 *Tap changer diagram of connections for one phase of transformer.*

1. Contacts *a* and *b* closed as shown ; load current from tap 1 through *b*—this is the running position for tap 1.

2. When the stored energy mechanism operates and the moving contact stem begins its travel from one side of the diverter switch to the other, contact *b* opens; load current from tap 1 through a resistance (R_1) and contact *a*.

3. As the moving contact continues its travel contact *d* closes; both resistances (R_1 and R_2) connected in series across taps 1 and 2 with the load current connected to the mid point of the resistances.

4. Further travel of the moving contact *a* to open ; load current from tap 2 through a resistance (R_2) and contact *d*.

5. Finally, as the moving contact reaches the other side of the divertor switch contact C closes and resistance R_2 is thus shorted out; lead current from tap 2 through contact *a*—this the running position for tap 2.

While on-load tap changers are justifiable for bulk-supply purposes, generally they are unnecessary for generator set-up transformers since the generators can have a large range of stable voltage operation. As the expenditure involved is high for on-load tap changing equipment, for very high-voltage interbus transformers, on-load voltage control is provided more easily by series boosters on the lower voltage side.

Losses on Load Loss. No load loss is the energy consu-med in a transformer that is excited at rated voltage and frequency but which is not supplying load. The current that flows in the primary winding under this condition is the exciting current. The no load loss varies from approximately 1 % of the rated output for small distribution transformers to approximately 0.3% for small power transformers and to approximately 0.25% for very large power transformers. Practically all no load loss is in the core. It includes a small dielectric loss in the insulation and copper loss in the windings.

Load Loss. Load loss is the energy loss incident to carrying load. It includes the losses produced in windings and other metallic parts as a result of the load currents flowing through windings. The voltage required to circulate the rated current through one winding while other is short-circuited is the impedance voltage for the connection at which the test is made. It is usually expressed as a percentage of the current, due to winding resistance, the other 90 degrees out of phase, due to winding reactance.

Total Loss. Total loss is the sum of the no load loss and the load loss. It is the loss that occurs during operation at rated voltage and load. Total loss varies approximately as 3% of the rated output for small distribution transformers to 1% for small power transformers and to 0.5% for every large transformers. Total loss is also a measure of efficiency in transforming power. Thus, transformer efficiency varies from 97% in small distribution transformers to 99% in small power transformers and to 99.5% in very large power transformers.

All Day Efficiency. Transformers are usually designed to operate with minimum losses i.e., maximum efficiency at full load. However, actually a transformer operates at part load and that too varies from time to time. The all day efficiency of a transformer is therefore the ratio of output in kWh to the input in kWh for the day.

Comparison of Single Phase and Three Phase Units for Three Phase Banks. A three phase power transformation can be accomplished either by using a three phase transformer unit, or by interconnecting three single phase unit, to form a three phase bank.

The three phase unit has the advantages of higher efficiency, small size, and less cost when compared with a bank having equal kVA capacity made up of three single phase units.

When three single phase units are used in a bank, it is possible to have a fourth unit at the same location, as emergency spare. Thus spare requirement is reduced to 33% only. Where adequate transportation or rigging facility does not exist to handle the required transformer capacity as a single unit, a definite reason of course develops for using three single phase units.

Silica-gel Breathers. Air drawn into the transformer is dried by passing through a cartridge containing crystals of silica-gel, but when saturated with moisture and no longer effective, they turn pink. The gel, however, is unchanged chemically and the crystals can be restored to their original dryness and blue colour by heating in an oven at approximately 150°C.

Drying Out. Should a transformer require drying out, this is carried out either by placing resistance heater elements in the oil supply or by connecting to a reduced primary voltage with the secondary winding passing full load current through a short-circuit link. The latter method is more convenient on site work as it does not require a break in the oil circulation system. Readings of oil temperature at the top of the coils and the insulation resistance of H.V. and. L.V. windings to earth and to each other are noted. Curves of insulation resistance to time would be plotted ; when these flatten out (after rising as water is driven off) the input power is reduced. When the curve does not rise further, the transformer is dry.

If a transformer is taken out of service for any length of time, it is advisable to have the oil filtered or reconditioned and the bushings should be cleaned and inspected for crazing or cracking. Before being reconnected, the drying out process will have to be carried out again.

30.9 TYPES OF COOLING

30.9.1 Oil Immersed Self Cooled Transformers

In this type of transformers the insulating oil circulates by natural convection within a tank having either smooth sides, corrugated sides, integral tubular sides or detachable radiators. Smooth tanks are used for small distribution transformers upto 50 kVA capacity. Integral tubular type construction is used upto about 3000 kVA.

30.9.2 Oil Immersed, Forced Air Cooled

In this case a fan is added to increase the rate of heat transfer from the cooling surfaces, thereby increasing the permissible transformer output. The transformer may be purchased with fan already installed, or it may be purchased with the option of adding fans latter. 15 to 30 percent higher kVA capacity can be attained in these transformers.

30.9.3 Oil Immersed, Forced oil Cooled

In this case forced circulation of oil increases the heat transfer rate thereby permitting higher kV A ratings. Heat transfer from oil to air is accomplished in external oil to air heat-exchangers. Use of fan to circulate air around heat-exchanger is common in this case.

30.9.4 Oil Immersed, Water Cooled

In this case the cooling water runs through coils of pipe which are in contact with the insulating oil of the transformer. The oil flows around the outside of these pipe coils by natural convection, thereby affecting the desired heat transfer to the cooling water.

30.9.5 Dry Type, Self Cooled

Dry type transformer, available at voltage rating 15 kV and below, contain no oil or other liquid to perform insulating and cooling functions. Air is the medium which surrounds the core and coils, and cooling must be accomplished primarily by air flow inside the transformer. The circulation of air is by natural convection.

30.9.6 Completely Self-Protected Transformer

A power transformer design may include protective devices capable of preventing damage to the unit when it is subjected to electrical conditions that would probably damage conventional transformers. Also, standard switching, metering, and voltage regulating functions may be included within a power transformer assembly. When these protective, switching and metering features are all combined at the factory within a single unit, it may be designated as a CSP power trans-former.

30.9.7 Lighting Protection

Co-ordinated arresters are installed to protect both high and low voltage circuits from lightning and other voltage surges.

30.9.8 Internal Fault Protection

Fusible protective links of high interrupting capacity are connected between the high voltage bushings and the windings, so that the supply circuit can be cleared from internal transformer faults.

30.9.9 Overload Protection

A thermal relay, operates to trip the secondary circuit breaker before damaging temperatures develop in the winding.

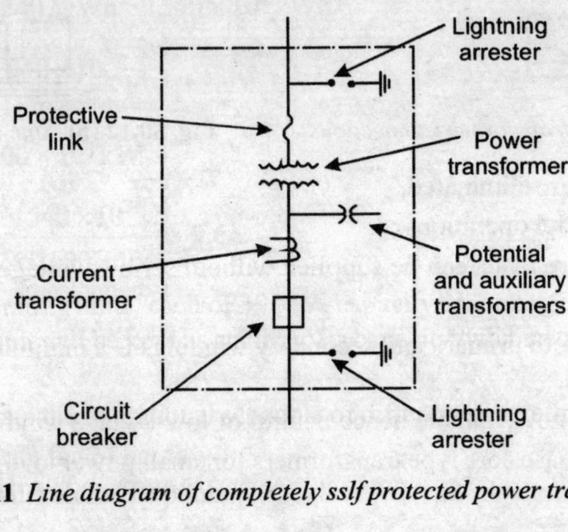

Fig. 30.11 *Line diagram of completely sslf protected power transformer.*

30.9.10 Relaying

Over cuirent relays normally are provided in the low voltage circuit to protect the secondary faults.

30.9.11 Circuit Breaker

Load switching is acomplished by a circuit breaker in the low voltage circuit of the transformer.

30.9.12 Voltage Regulation

Standaid no load taps are provided in the high voltage winding. Tap changing under load equipment for the secondary circuit may be built into the transformer housing.

30.9.13 Distribution and Power Transformers

Transformers upto kVA rating of 500 or less are known as distribution transform-ers. These transformers are used to distribute power directly to points of usage. Transformers with kVA rating of more than 500 are known as power transformers. They are located at power generating stations, at the terminals of power transmission systems, and at the points of large power usage.

30.9.14 Transformer Connections

Connections for the three phase transformations most commonly used are illustrated below. The advantages and disadvantages of these connections are given below :

30.10 DELTA-DELTA

Advantages

1. Most economic connection for large output at low voltage.
2. A three phase bank of three single phase transformers can operate in open-delta at 58% output if one unit fails.

Fig. 30.l2 (a) *Delta-Delta transformer connections.* **Fig. 30.12 (b)** *Star-Star transformer connections.*

3. Third harmonic voltages are eliminated.
4. Easy to phase in for parallel operations.
5. Heavily unbalanced 3-wire loads can be supplied without serious voltage unbalance.

Disadvantages

1. Copper cross section of both primary and secondary windings is a minimum whereas number of turns and insulation per phase is a maximum.
2. Neutral point of winding not available hence neutral of low voltage winding can not be grounded.

Star-Star. Used mainly for 3 phase core type transformers for small power load.

Advantages

1. Copper cross section is maximum.
2. Number of turns per phase is minimum.
3. Most economical for small output at high voltages.
4. Both neutrals available for grounding or for balanced 4 wire supply.
5. Easy to phase in for parallel operations.
6. Can be operated at single phase at 58% output.

Delta-star. Used for step down transformers to supply 4-wire distribution to motor and lighting loads, balanced or unbalanced, also used for stepping up voltage for power transmission.

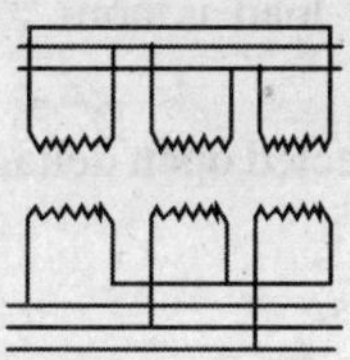

Fig. 30.13 (a) *Delta-star connection.*

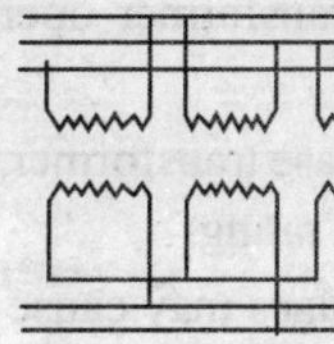

Fig.30.13 (b) *Star delta connections.*

Advantages

(1) No third harmonic.

(2) Secondary neutral available for grounding or for 3 phase, 4-wire supply.

(3) Suitable for unbalanced 4-wire load, resulting unbalanced voltage being relatively small.

(4) Balanced and unbalanced loads may be applied simultaneously.

(5) Transformers of widely different impedances can be used to form a 3 phase bank.

Disadvantages

(1) No primary neutral available for grounding.

(2) A fault in one phase makes 3 phase unit or bank inoperative.

Star-Delta. Star-delta connection is used chiefly for stepping down voltages from high voltage transmission lines.

Advantages

(1) Third harmonic voltages eliminated by delta connected secondary.

(2) Most desirable for stepdown transformers for high voltage transmission.

(3) Primary can be grounded.

(4) Secondary delta connections stablise use the primary neutral.

(5) Transformers of widely different impedances can be used to form a 3 phase bank.

Disadvantages

(1) No secondary neutral available for grounding of fcr 3 phase, 4-wire supply.

(2) A fault on one phase renders a 3 phase unit or bank inoperative.

30.11 OPEN-DELTA

Advantages

(1) A 3 phase shell type transformer can operate in open delta with one damaged phase.

(2) Two units of a transformer bank consisting of three single phase units can be operated in an open delta since a damaged single phase unit can be removed entirely.

Disadvantages

(1) With delta connected, 3 phase shell type transformer a damaged phase must be disconnected and short-circuited on itself to prevent voltage being induced by the good phases.

(2) To operate a 3 phase, core type transformer in open delta, the damaged phase must remain open circuited and yet be capable of withstanding normal voltage induced in it from the other phase windings,

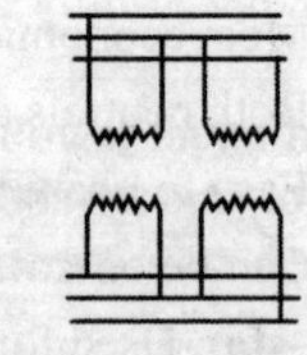

Fig. 30.14 *Open delta connections*

(3) When connected open delta, current in each transformer is 30 degrees out of phase with voltage, and transformer operates at 86.6% power factor if load is non-inductive.

(4) Capacity of a 3-phase transformer, or of a three phase bank, connected open delta, with the damaged phase cut out, is 58% of the bank rating.

(5) Unbalance in voltages may cause burn out of three phase motors.

Earthing of a Power System. It is preferable to earth the neutrals of all industrial power systems. Earthing is necessary because:

(a) it enables fast location and isolation of ground faults.

(b) it minimizes over-voltage due to restriking.

(c) it reduces steady state voltage stress to earth.

(d) it enables use of sensitive protective apparatus.

(e) it eliminates persistent arcing grounds.

Systems should be earthed at the neutral of the supply transformers and generators. If the supply transformers and generators are delta connected, separate earthing transformers may be used. When there are two or more sources of power, the neutrals of 2 or more sources should be earthed to ensure having a neutral earthed in the event one of the sources is out of service.

The high voltage neutral is earthed mainly for the protection of the system while the low voltage neutral is earthed chiefly in order to reduce the possible danger to human life. Where energy is transformed suitable provision should be made to protect the lower voltage system from becoming charged above its normal voltage by leakage or electrostatic induction from the higher voltage.

The system neutral can be earthed by one of the following methods :

(a) Solid Earthing

(b) Resistance Earthing

(c) Reactance Earthing

(d) Resonant Earthing.

Solid Earthing. Solid earthing is generally used on low voltage circuits, 600 volts and less, and on most circuits above 15 kV. In the range of 22 to 15 kV any of the remaining three types may be used. In solid earthing, a direct metallic connection is made from the system neutral to one more earth electrodes cosisting of plates, rods or pipes buried in or driven into the ground, generally in a bed of coke below the permanent water level. Where permanently moist ground cannot be ensured a supply of water should be made available so that periodic watering cannot be done. The value of contact resistance or resistances of the earthing system should be made available so that periodic watering cannot be done. The value of contact resistance or resistance of the earthing system should be as small as possible and this should be checked from time to time.

Nowadays the term 'Solidly earthed' has been replaced by the term 'Effectively earthed'. A sransformer neutral may be 'Solidly earthed' so that there may be no impedance between the neutral and earth. However, the transformer capacity thus solidly earthed may be too small in comparison with the sizes of the system to be effective in stabilising the voltage from phases to ground when the fault occurs.

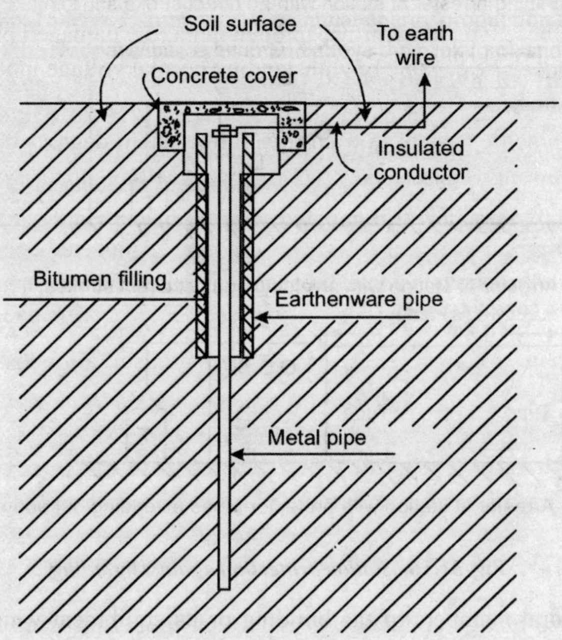

Fig. 30.15. *Solid earthing. Cast iron plates should not be less than 15 mm thick and about 1·5 sq. m in area. Alternatively, 15 cm dia. C.I. pipe not less than 15 mm thick to give equal contact surface area should be used.*

Resistance Earthing. On larger systems resistance earthing is employed to limit the earth fault currents and to lessen the damage when a line to earth fault occurs. The resistance introduced in this case acts as a current limiting device. The resistors used in this case may be metallic, carbon powder or liquid type.

Metallic resistors do not alter with time and little or no maintenance is required. They are, however, inductive and this is a disadvantage with overhead lines exposed to lightning, since travelling waves or impulses are subject to positive reflection and this may unduly stress the insulation of the equipment and cause breakdown. However, the most satisfactory resistor for main earthing is the liquid type and it is inductive and has low surge impedance. These are normally of robust construction and may be moulded outdoors.

There are various time ratings for such registers e.g. 15 seconds, 30 seconds and 1 or 2 minutes.

The merits and demerits of resistance earthing system are :

1. It permits ready relaying of ground faults.
2. It minimizes the hazards of arcing grounds.
3. It has got least inductive influence on neighbouring communications.

However, the system is costlier than the solidly earthed system.

Lightning Protection. The protection of power plant build in ... plant including overhead lines, towers and sub-stations from dangerous over voltages due to lightning i... ...ntial feature indesign these days. The chief features requiring attention when designing a lightning protectio... ...stem are :

(a) to select the points which are vulnerable, to which the conductors should be taken ;

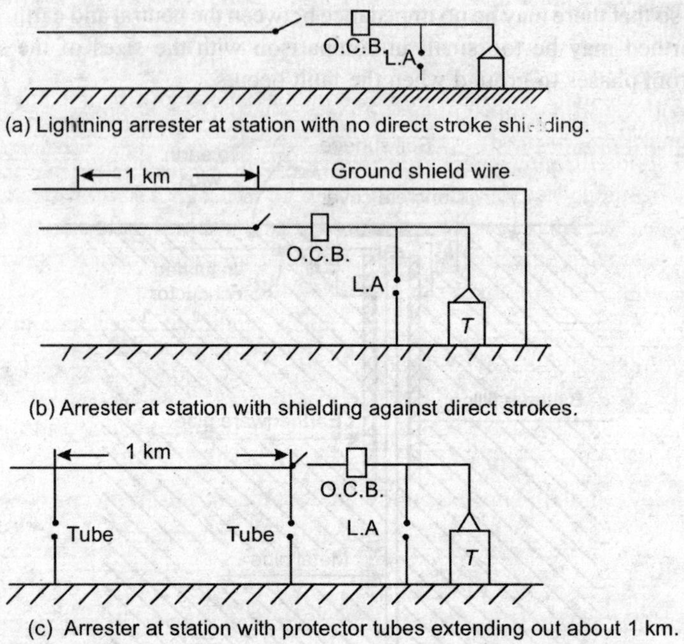

Fig 31.16. *Station protection against lightning.*

(b) to decide such structural members of the building or its equipment which should be connected to the conductor system and which should not be connected ;

(c) to select the most stable routes for the conductor between the elevation rods and the earth plates keeping in mind the metal objects mentioned in (b).

30.11.1 Protection of 'Power Station' and Sub-stations

Power stations are usually housed in high buildings and sub-stations are located generally outdoors or in small buildings. The protection of these structures against lightning involves

(i) interception

(ii) conduction

(iii) dissipation.

For interception an object of sufficient cross-section in good electrical connection with the earth to attract the leader stroke is required. For conduction a path joining this object to earth of such a low impedance (absence of sharp bends and loops) is required so that the discharge will follow it in preference to any other. For dissipation a low resistance connection with the body of the earth is required.

Transmission Lines. Sheilding the lines by means of ground wires affords an effective method of protecting them against direct lightning strokes. Alternatively, the lines can be built without sheilding but provided with auxiliary devices, such as protector tubes, for protection against lightning strokes.

30.12 SWITCHGEAR

Switchgear is a general term covering switching, interrupting, control, metering, protective and regulating devices, also assemblies of those devices with associated interconnections, accessories and supporting structures for use in connection with the generation, transmission, distribution and conversion of electric power. Thus, functions of a switchgear may be summarized as :

(1) To localise the effects of faults by operation of protective equipment and so automatically disconnect faulty plant from the system.

(2) To break efficiently short circuits without giving rise to dangerous condition.

(3) To facilitate redistribution of loads, inspection and maintenance on the system.

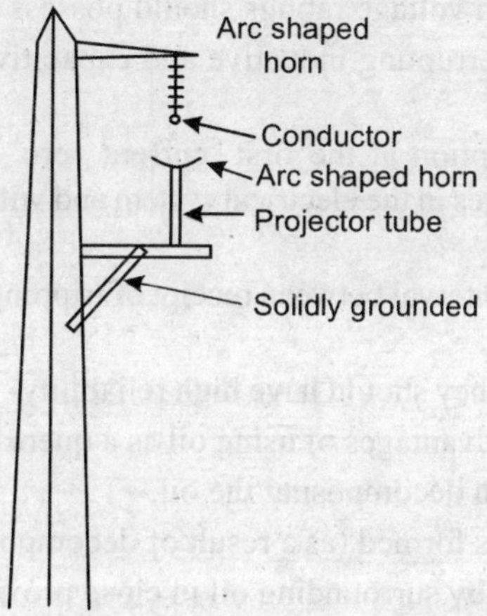

Fig. 30.17 *Protector tube. Here the path of discharge is controlled. The heat of the arc vaporises some of the tube walls, the resulting neutral gas being expelled violently into the arc, thus deionising it to prevent further arc restrisking.*

System Characteristics. In order to make a well co-ordinated application of switchgear to any system, following system characteristics must be known :

(a) Available short circuit current or kVA evaluated at each point in the system where switchgear is to be used, determines the interrupting rating of the circuit breakers or other circuit interrupting devices.

(b) Voltage and frequency of the system

(c) Maximum load requirement of each circuit

(d) Service requirements

(e) Location of equipment, and

(f) Nature of the various loads.

An important part of a switchgear is a circuit breaker which will be considered in details.

Circuit Breakers. Circuit breakers consist essentially of current carrying contacts called electrodes. These are normally engaged but, under predetermined conditions separate out to interrupt the circuit. When the contacts are separated, an arc is struck between them. This arc plays an important part in the interruption process as it provides for the gradual transition from the current carrying to the voltage withstanding the states of the contact, but it is dangerous on account of the energy generated in it in the form of heat which may result in explosive forces. It is, therefore, always desired to incorporate such features in a circuit breaker, so that arc may be extinguished as soon as it is produced.

Classification. Circuit breakers may be classified as follows :

(1) Low voltage circuit breakers.

(2) High voltage circuit breakers.

(a) Oil circuit breakers :

(i) Bulk oil circuit breakers

(ii) Low oil contact circuit breakers.

(b) Oil-less circuit breakers :

(i) Water circuit breakers

(ii) Hard gas circuit breakers

(iii) Air blast circuit breakers.

In general, circuit breakers with high voltage ratings should possess the following characteristics :

(a) They should be capable in interrupting inductive and capacitive circuits and fault currents of all values within their rating.

(b) They should make ths interruption at the first 'current zero' after contact separation, without risk of restriking, without producing overvoltages in the electrical system and without temporary or permanent deterioration of the insulation or interrupting ability.

(c) The opening time i.e. the time interval between receipt of tripping impulse and contact separation should be the minimum mechanically possible,

(d) Electrically and mechanically they should have high reliability.

Oil Circuit Breaker (OGB). The advantages of using oil as a quenching medium are :

(1) The absorption of arc energy in decomposing the oil.

(2) Good cooling property of gases formed (as a result of decomposition of oil).

(3) The cooling surface presented by surrounding oil in close proximity to the arc.

(4) The ability of cool oil of high dielectric strength to flow into the arc space after the current zero.

(5) The action of oil as an insulator.

The disadvantages are:

(1) It is easily inflammable.

(2) It may form an explosive mixture with air.

(3) It requires maintenance.

Plain Break oil circuit Breaker. In this a strong weather tight earthed tank contain oil upto a predetermined level, and an air cushion above the oil level. The fixed and moving contacts remain immersed in oil. It is known as single or double break depending upon the number of contacts provided.

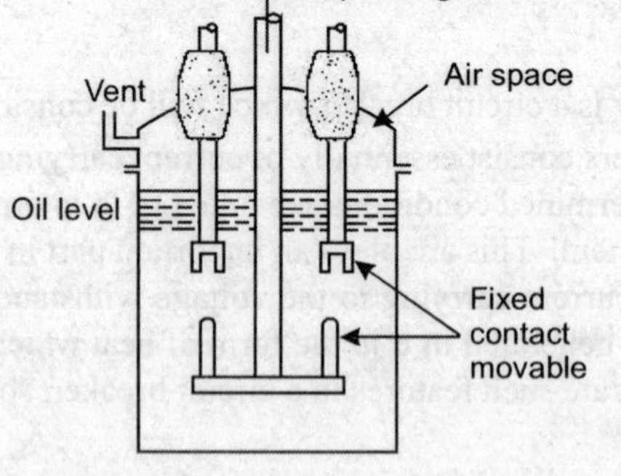

Fig. 30.18 *Double break plain oil circuit breaker.*

An arc is struck when the contacts separate. This vaporizes oil into gas. The sudden formation of the gas by the arc may dissipate heat generated as a result of arc.

However this circuit breaker suffers from the defect of permitting rather long and inconsistent arcing times.

Air Circuit Breakers. In this compressed air (at pressures around 15 kg/cm^2) is used for arc extinction. The advantages of an air-circuit breaker over oil-circuit breaker are :

(1) Elimination of fire hazards due to oil.

(2) High speed operation.

(3) Short and consistent arc duration and hence less burning of contacts.

(4) Suitability for frequent operation, since cooling medium is replaced constantly.

(5) Facility of high speed re-closure.

Limitations

(1) A compressor is constantly required to remain in operation.

(2) Air leakage at the pipe line fittings creates problems.

(3) Sensitivity to restriking voltage.

(4) Current chopping.

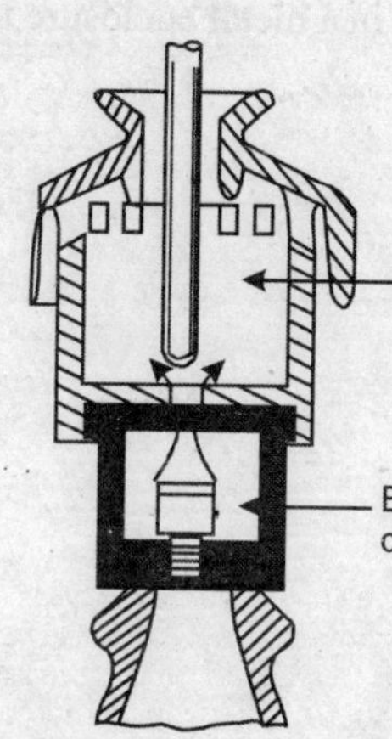

Fig. 30.19 *Air blast circuit breaker.*

A typical air-blast circuit breaker is shown in Fig. 30.19. Flow of air around the moving circuit is responsible for extinction of arc. The breaker is closed by applying pressure at the lower opening, and opened by applying pressure at the upper opening. When the contacts separate, the cold air rushes round the movable contact and blows out the arc.

Water Circuit Breaker. Figure 30.20 shows the principle of the water circuit breaker. The contacts are in water, which is turned into steam by the arc and rushes part the opening to blow out the arc.

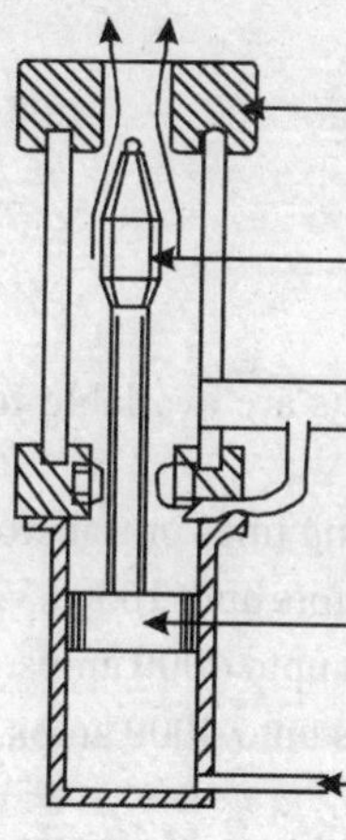

Fig. 30.20 *Water circuit breaker.*

Principles of Switchgear Layout. The principles to be followed in switchgear layout may be summarised as :

(1) The design of the individual units should be such that the risks of failure are reduced to a minimum.

(2) The layout should be such that any section may be isolated without unduly affecting the service.

(3) To provide easy and safe access for maintenance and general routine inspection.

(4) Where necessary reactors should be used to keep the breaking duty within the capacity of the circuit breakers.

(5) Adequate arrangements must be made for fire protection.

Types of Switchgear. Standardized types of switchgear equipment are available to meet a large percentage of the requirements of modern power systems, and leading switchgear manufacturers are in a position to recommend the most advantageous modifications of such standard gear to meet other less frequently encountered situations.

30.12.1 Low Voltage Switchgear

Most industrial power utilised at 400/440 V, 3 phase generally from substations ranging in size form 100 to 1500 kVA. Standard switchgears generally with air circuit breaker equipment are available for this purpose. In this the desired arrangements and number of circuits and circuit breakers of suitable ratings are factory built by manufacturers from standardised sub-assemblies, so that the use provides only suitable foundations, and connects his cables to the proper terminals to complete the installation. These factory assemblies consist of tiers of standard compartments each of which contains a circuit breaker, instruments and control panel. The circuit breakers are built on removable carriages and means are provided to draw the entire breaker out of it compartment for inspection, maintenance, replacement or for isolation of the circuit. Interlocks are provided to prevent withdrawal or insertion unless the breaker is in the open position. Behind the tiers of breakers and instrument and control compartments is a structure carrying busbars, connection bars, terminals and necessary current and potential transformers. The entire structure is housed in a metal enclosure to ensure maximum safety to operators and protection to equipment.

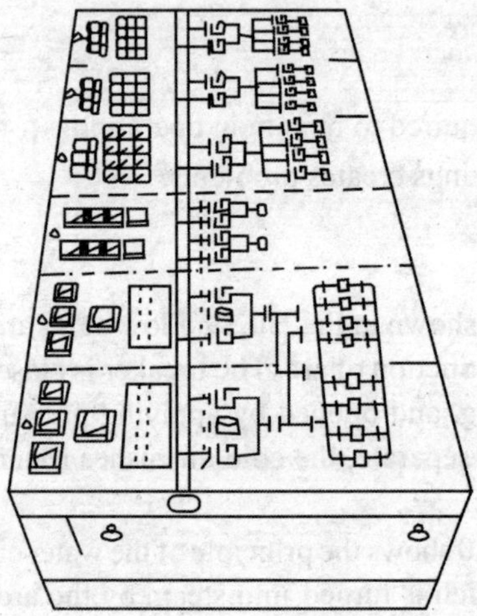

Fig. 30.21 *Switchgear in a power station.*

Standard units are available for the following types of circuits commonly encountered in systems in 220/440 voltage range :

(1) Incoming lines or transformer secondaries upto 4000 amps.

(2) Generators upto 1650 kVA at 220 volts 3000 kVA at 440 Volts.

(3) Busties upto 4000 amps.

(4) Feeders upto 4000 amps.

30.12.2 Medium Voltage Switchgear

Power is distributed to load centres at voltages ranging from 2200 volts to 11000 volts by medium voltage switchgear. The type of equipment now almost universally used in this class is known as Metal Clad Switchgear, which has been now standardized. The circuit interrupting device used in this type of gear is the power circuit breaker.

Manufacturers' listings include standard units for the following types of circuits in commonly used ratings : (1) Generators, (2) Incoming lines or power transformer secondaries, (3) Synchronous or induction motors, (4) Generator neutral grounding, (5) Totalizing metering.

In addition, draw out potential transformer compartments, bus-entrance compartments and instrument brackets are available as standard components of complete equipment. For large size breakers, and where quick removal of breaker units is required, electrically driven elevating mechanisms are provided.

30.12.3 Station Type and High Voltage Switchgear

The transmission of power in large amounts over considerable distances involves voltages rangmg from 34.5 to as high as 400 kV. Switchgears for such systems are specifically designed and manufactured.

For voltages upto 69 kV, the most common type of switchboard in current use is the duplex board. It consists of front and rear panels separated by a passageway entered through doors at the ends. Control switches and instruments are mounted on the front panels and relays and meters on the rear panels. All devices are dead front to give maximum safety to operators. Complete wiring and terminal facilities are provided inside the structure.

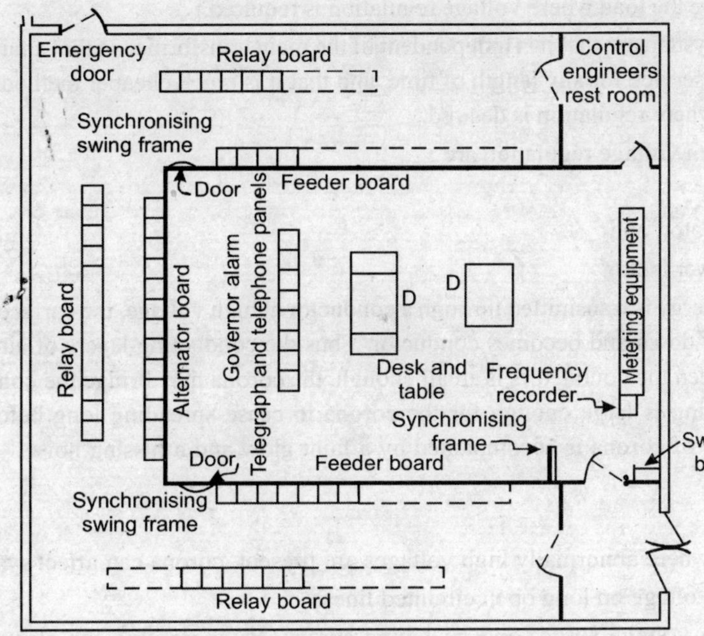

Fig.30.22 *Layout of a control room.*

The most important component of a control and protective relaying switchboard is the very important protective relay. It is the 'eyes' of the power circuit breakers in that it detects the presence and location of abnormal system conditions, such as short circuits or low voltage, and directs the tripping of the correct breakers to remove the faulty part of the system from service quickly and accurately.

Voltage Regulation. It is essential for the satisfactory operation of power system to maintain the supply voltage within narrow limits. There are some legal requirements also in this connection. Following methods are adopted for this porpose :

(a) *Voltage control by Generator Excitation.* In many cases, and in particular in short lines, the voltage at the receiving end is kept within very narrow limits by automatic or hand operated voltage regulators, which act in the field circuit of the alternator exciter. However, this method is unsuitable for long lines, as the voltage at the generator terminals will have to vary too much in order that the voltage at the far end of the line may be constant.

The effect of varying the excitation depends upon the system into which the generator is feeding but in all cases the power output is unaffected as this depends on the fuel supply to the prime mover. In an interconnected system, in which there are two or more alternators, the distribution of load is unchanged by varying the voltage by excitation

; but the reactive kVA can be changed. The sharing load is, of course, determined by the regulation of the governor of the prime-movers.

(b) *Tap Changing Transformers.* The voltage can be varied by having a number of tappings on the secondary winding so that turn-ratio can be changed according to the requirement. This operation may be either manual or automatic depending upon the specific application and requirements.

(c) *Booster Transformer.* Generally it is not convenient to have tap changing gear on the main transformer. Mostly it is desired to increase the voltage at a point in the transmission line at a distance from the main transformers. For this purpose booster transformers, rather than tap changing transformers are preferred. Booster transformers are installed immediately before the load where voltage regulation is required.

The advantages of booster system are that it is independent of the main transformer so that a failure in the former will not throw the latter out of service for any length of time, and that it is much cheaper method when there is no main transformer at the point where regulation is desired.

Some of the other methods of voltage regulation are :

(i) Phase angle control,

(ii) Use of induction regulator, and

(iii) Voltage control by power factor.

Corona. When electrical energy is transmitted through a conductor at high voltage, the surface stress may reach a value at which the air breaks down and becomes conductor. Thus the conducting layers of air form part of the conductor. If the spacing between the conductors is small enough, the corona may bridge the conductor and cause flash over. Generally the spacing is large enough for the corona to cease spreading long before it bridges the conductors. The phenomenons of corona is accompanied by a faint glow and a hissing noise. There is also an energy loss.

Effects of corona.

(1) Under the conditions where abnormally high voltages are present, corona can affect system behaviour.

(2) It can reduce the overvoltage on long open circuited lines.

(3) It attenuates lightning voltage surges and switching surges. By increasing the electrostatic coupling between the shield wire and phase conductors, corona at times of lightning strokes to towers or shield wires reduces the voltage across the supporting string of insulators and thus, in turn, reduces the probability of flash over and improves system performance.

(4) On high voltage lines grounded through a ground fault neutralizer, the inphase current due to corona loss can prevent extinction of the arc during a line to ground fault.

At a given voltage, corona is determined by conductor diameter, line configuration, type of conductor, condition of its surface and weather. Rain is by far the most important aspect of weather in increasing corona.

30.13 TRASMISSION

Generating stations are more often located away from load centres due to various reasons already discussed in earlier chapters. An extensive transmission and distribution system is therefore necessary between the generating plant and the consumers' loads and also a network of inter connection between generating and main load centres, i.e., a grid system.

Normally the transmission and distribution system comprises a network of 3 ph circuits with transforming and/or switching sub-stations at the various junctions. The circuits may be overhead lines or underground cables.

Table 30.3 *Turbo-Alternators Main operating characteristics of large modern Turbo-generator (Practice in various countries)*

Parameter	*Australia*	*France*	*Germany*	*Japan*	*U.K.*	*U.S.S.R.*
Rated voltage (kV)	16.5 for 275 MW and 200 MW 13.8 for 125 MW 11 for lower ratings	20 for 250 MW 15.5 for 125 MW	5.25, 6.3 and 10.5 for 150 MW and below 15 75, 18 5, 21.4 for above 150 MW	13.2 for 75 MW 15 for 125 MW 18 for 156 MW 12.18 for 175 MW 12.6 for 220 MW 24 for 250 MW	19.5/20 for 350 MW 22 for 500 MW	6.3, 10.5, 13.8, 15.75 and 18
Power factor	0.8 to 0.9	0.8 for 125 MW 0.85 for 250 MW	0.7, 0.8	0.85	0.85	0.8 upto 60 MW 0.85 for 100 MW and above
Short circuit Ratio	0.55 for 120 MW 0.60 for 200 MW	0.5 for 125 MW 0.4 for 250 MW	0.47 to 0.63 upto 200 MW 0.42 to 0.8 over 200 MW	0.64	0.4	0.55 to 0.6 when p.f. is 0.8 0.45 to 0.5 when p.f. is 0.85

Main advantages of underground cables are :

1. It does not interfere with amenity.
2. It is less subject to natural hazards such as lightning, birds or human interference.
3. Less chances of pilferage of power which otherwise occurs in case of overhead lines.

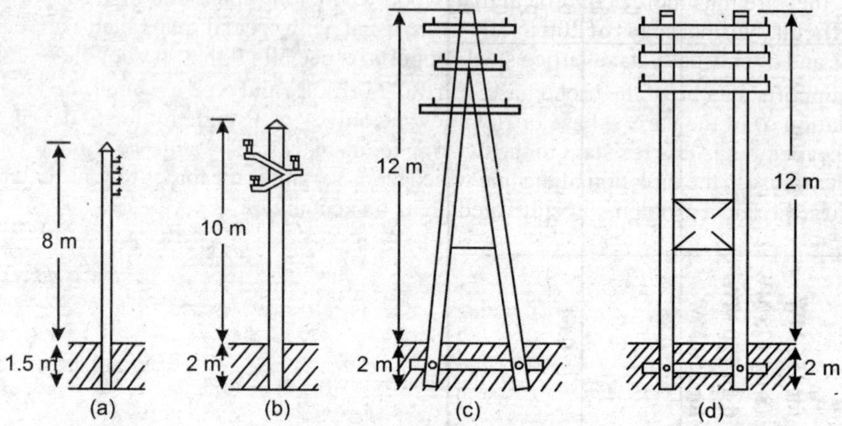

(a) 400 V, 4 phase, 4-wire distributor.
(6) 11 kV, phase feeder.
(c) A pole, double circuit, 33 kV feeder.
(d) H pole double circuit 33 kV feeder.

Fig. 30.23 *Wood pole supports.*

Disadvantages

1. Initial cost is high. 2. Maintenance cost is high.

Over-head Lines. Transmission of electrical power is generally by over-head lines because of the following advantages:

1. Over-head lines are much economical as compared to cables.
2. It is comparatively much easier to transmit at higher voltages by overhead lines.
3. Ease of accessibility for extensions and repairs.

Disadvantages

1. Exposure to sun and rain, smoke, ice etc.
2. Possibility of pilferage of power particularly in rural areas.
3. Interference with communication circuits.

Regulations. There are regulations for securing the safety of public and these have to be observed in the design. All line conductors must comply as regards elongation, breaking load and elasticity with the standard specifications. The lines must be inaccessible except by a ladder or other special appliance. The line conductors must be attached to suitable insulators carried on supports of wood, iron, steel or reinforced concrete. However, supports for a line must be capable of carrying the load due to the conductor and insulators (including ice and coind loads on the conductor), together with the wind load on the support itself. Bending and twisting forces due to the breakage of one or more conductors must be considered.

Wood Poles. These are most economical supports for lines having spans less than about 200 m and are widely used for distribution in rural areas at voltages upto and including 33 kV. For low voltage lines, single pole is generally used although its transverse strength can be increased three or four times by using two poles arranged in A or H formation as shown in Fig. 30.23. The cross arms may be of wood or steel. A zinc or aluminium cap covers the top of the poles to protect the end grain. The life of pole usually ranges from 15 to 25 years.

As wood is an insulator, the pole adds to the insulation strength between the conductors and earth and renders flashover due to lightning less likely. On the other hand, if a pole is struck it may be shattered, causing complete failure of the line. The probability of shattering can be reduced by earthing all metal supports for the insulators either individually at each pole or by connecting to a continuous earth wire.

Lattice Steel Supports. These consist of painted or galvanised steel members which can be transported separately and erected on site—they are thus more economical than wood where long spans and high supports are needed or where transport is difficult. Normal spans of 200 to 500 m are usual, with special spans such as river crossings upto 2 km. Lines of 66 kV and over usually have lattice steel supports, especially if they are double circuit lines.

The lattice steel supports may be of the broad base, Fig. 30.24 (b) or rigid type in which each leg of the support has a separate foundation or of the narrow base or flexible type only with a single foundation, Fig. 30.24 (a). The latter is, of course, cheaper, but is less resistant to the twisting moment caused by a broken conductor 5 on the other hand, it gives some flexibility in the direction of the line which tends to relieve the forces to some extent. The narrow base may also simplify wayleave problems in cultivated areas on account of

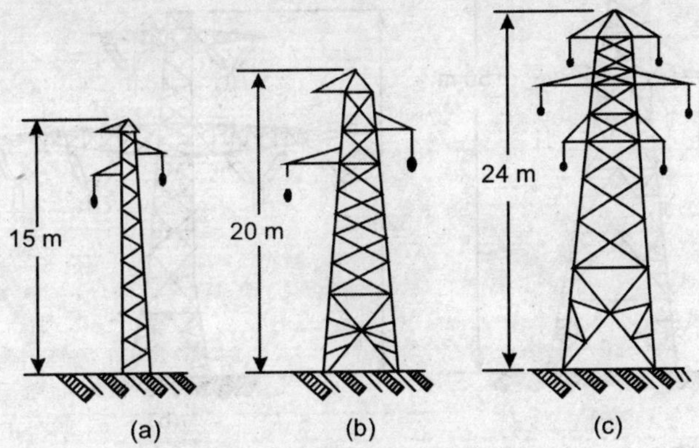

(a) Narrow base, 33 kV, single circuit.
(b) Broad base, 33 kV, single circuit,
(c) 132 kV, double circuit.

Fig. 30.24 *Lattice steel tower types.*

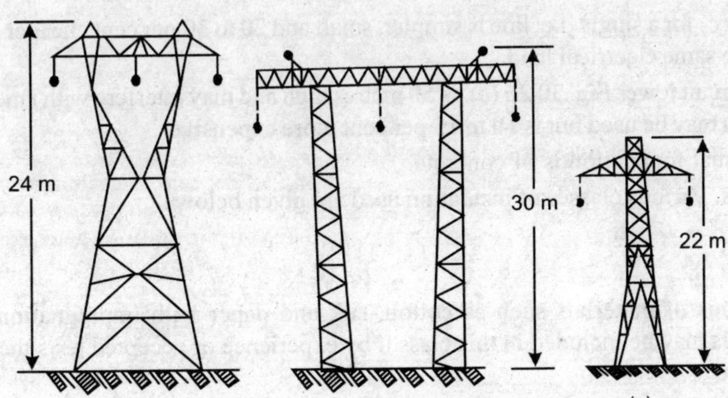

(a) 220 kV, cat's head, single circuit.
(b) 400 kV, portal, single circuit.
(c) 500 kV, direct current, single circuit.

Fig. 30.25 *Lattice steel tower types.*

the smaller area occupied. Lines above 33 or 66 kV cannot, however, be designed with narrow base supports. Double circuit lines Fig. 30.24 (c) are more economical than two single circuit lines although for a given voltage the tower height is greater. There is the possibility of both circuits being affected simultaneously by lightning, but wayleave problems are eased.

The cat's head horizontal configuration, Fig. 30.25 (a), minimizes height and reduces the possibility of conductor clashing, but the line erection is complicated. The portal power Fig. 30.24 (b), also, with horizontal conductor configuration, is longitudinally weak, and even if guyed (as is usual) may give trouble if a conductor breaks.

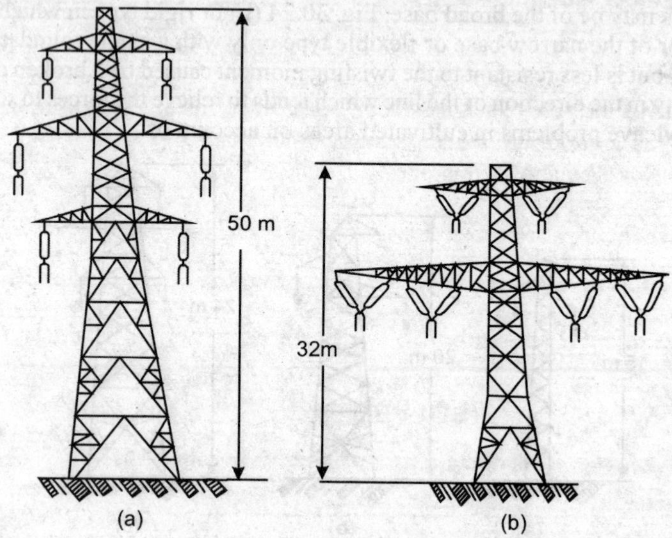

(a) 400 kV, double circuit,
(b) 430 kV, low height, double circuit.

Fig. 30.26 *Lattice steel tower types.*

The tower Fig. 30.25 (c) for a single d.c. line is simpler, small and 20 to 30 per cent cheaper than a 3-phase double circuit tower, to carry the same electrical load.

The 400 kV double circuit tower Fig. 30.26 (a), is 50 metres high and may interfere with amenity. To avoid this the lower type Fig. 30.26 (b) may be used but is 10 to 15 per cent more expensive.

The best and most usual foundation is of concrete.

Insulating Materials. Various classes of insulation used are given below :

30.14 CLASS O

Materials or combinations of materials such as cotton, silk and paper with impregnation. Other materials or combinations of materials may be included in this class if by experience or accepted tests they can be shown to be capable of operation at 90°C.

30.15 CLASS A

Materials or combinations of materials such as cotton, silk and paper when suitably impregnated or coated or when immersed in dielectric liquid such as oil. Other materials or combinations ot materials may be included in this class if by experience or accepted tests they can be shown to be capable of operation at 105°C.

30.16 CLASS B

Materials or combinations of materials such as mica, glass, fibre, asbestos, etc. with suitable bonding substances, for other substitute materials the temperature limit is 130°C in this case.

30.17 CLASS F

Materials or combinations of materials such as mica, glass, fibre, asbestos etc., with suitable bonding substances. The ternperature limit for other substitutes being 155°C in this case.

30.18 CLASS H

Materials or combinations of materials such as silicon, elastomer, mica, glass, fibre, asbestos, etc. with suitable bonding substances such as appropriate silicon resins. Other materials or .combinations of materials may be included in this class if they are capable of operation at 180°C.

30.19 CLASS G

Insulation that consists entirely of mica, porcelain, glassy quartz and similar inorganic materials. Other materials or combinations of materials may be included in this class if they are capable of operation at temperatures over 220°C.

SOLVED EXAMPLES

Example 30.1 *Determine capacitor kVA to bring 1 MW load from 70% to 80% power factor.*

Solution. Factor (from table) is 0.53 5 Capacitor kVA needed = 1000 × 0535 = 535

Example 30.2 *The specific heat of air at normal temperature and pressure (NT.P.) is 0.2375 cal/gm/°C and at temperature of 20°C it rises to 0.242 callgm/°C. Calculate the mass of air required for each kW loss to be dissipated in a generator. Take temperature rise through the generator as T_r°C and ambient temperature as 20°C.*

Solution

1 kW = 240 cal/sec.

Specific heat of air=0.242 cal/gm/°C

Therefore, mas of air required for each kW loss to be dissi-pated in a generator

$$m = \frac{240}{0.242} \times \frac{60}{T_r} \times \frac{1}{1000}$$

$$= 64.9 \text{ kg/min.}$$

Let us consider the case of a 60 MW generator. Taking generator efficiency as 98.2% (excluding exciter losses)

$$\text{Total loss} = \frac{(\text{MW}) \times 1000(100 - \text{eff.})}{100} \text{kW}$$

$$= 1080 \text{ kW}$$

The losses in the generator would be distributed approximately as follows:

Friction	10%	= 108 kW
Windage	30%	= 324 kW
Stator iron	26%	= 280 kW
Stator I^2R	12%	= 130 kW

Rotor I^2R	12%	=	130 kW
Stray loss	10%	=	108 kW
	Total 100%	=	1080 kW

Excluding friction, which is generally taken care of by lubrication system of bearings, and also assigning all losses apart from the rotor I^2R loss, to the stator, we have

Total stator loss = 712 kW

Total rotor loss = 130kW

Quantity of air required for stator ventilation

$$= 712 \times 64.9 = 46208.8 \text{ kg/min}$$

Quantity of air required for rotor ventilation

$$= 130 \times 64.9 = 8437 \text{ kg/min}$$

Therefore total air quantity required

$$= 54{,}645.8 \text{ kg/min} = \text{cu. m/min.}$$

THEORY QUESTIONS

30.1. What is the function of an exciter in an a.c. generator?

30.2. What is the effect of power factor on maximum continuous kVA and kW load that can be carried by typical 0.8 p.f. engine driven alternator ?

30.3. What is the effect of power factor on alternator design ?

30.4. Describe different methods of alternator excitation.

30.5. What is the difference between Thyristor excitation and brushless excitation ?

30.6. What are the advantages of a direct coupled exciter over belt driven exciter ?

30.7. What is the function of alternator ventilating system?

30.8. What are the advantages of hydrogen cooking system for an alternator ?

30.9. Why is transformer considered as vital part of a power plant ?

30.10. Why three phase banks are preferred over three phase units of transformers in big size power stations ?

30.11. Why cooling is necessary in transformers ? Give different methods of cooling generally used in transformers.

30.12. What is the difference between a distribution and a power transformer?

30.13. Why earthing of a power system is necessary ?

30.14. How are power stations protected against lightning ?

30.15. What is the function of a switchgear in a power station ?

30.16. Write short notes on :

(a) Double break plain oil circuit breaker.

(b) Air blast circuit breaker.

(c) Water circuit breaker.

30.17. How is voltage regulated in a power system ?

30.18. Write short note on 'Corona'.

30.19. List few advantages of wood poles over steel poles.

30.20. What is meant by lattice steel supports ?

Chapter 31 Emission and Pollution Control

31.1 INTRODUCTION

The atmosphere which is the largest fraction of biosphere continuously absorbs a wide range of solids, liquids and gases from both man-made and natural sources. These substances travel through air, disperse and react with each other both physically and chemically. The portion of these substances which interacts with the environment to cause toxicity, disease, aesthetic distress, physiological effects or environmental decay, has been labelled by man as a "**pollutant**".

Air pollution is basically the presence of foreign substances in air. Thus it is defined as the presence of one or more contaminants such as dust, fumes, gas, mist, odour, smoke or vapour which is injurious to human, plant or animal life or property. And the day man invented fire air pollution started.

31.2 CLASSIFICATION OF POLLUTANTS

Air pollutants can be classified as follows:

(1) Natural contaminants

e.g., pollen grains, natural fog, bacteria etc.

(2) Aerosols

e.g., dust, smoke, fog and fumes, mists, carbon particles etc.

(3) Gases and vapours

(i) Oxides of Nitrogen (NO_x)

(ii) Oxides of sulphur (SO_x)

(iii) Oxygen compounds O_3, CO, CO_2

(iv) Vapours from various chemical reactions *e.g.* paraffins, olefins, acetylenes, chlorinated hydrocarbons etc.

(v) Radioactive compounds.

In this chapter we will study in detail the air-pollution, causes and remedies, as regards SI and CI engines. The main pollutants contributed by IC engines are CO, unburnt hydrocarbons (UBHC), oxides of nitrogen (NO_x), lead and other particulate emissions. Apart from IC engines, other sources of air pollution are electric power generating stations, industrial and domestic fuel consumption, refuse burning (incinerators), industrial processes etc. also contribute heavily to contamination of environment.

31.3 POLLUTANTS FROM SI ENGINES

The SI engine exhaust gases contain:

(i) Oxides of nitrogen *e.g.* NO, NO_2 etc. collectively called as NO_x

(ii) Carbon Monoxide (CO)

(iii) Organic compounds—unburned or partially burned hydrocarbons (UBHC).

The relative amounts depend on the engine design and operating conditions. Normally their presence is as follows:

(i) NO_x – 500 to 1000 ppm or 20 gm/kg of fuel.

(ii) CO – 1% to 2% or 200 gm/kg of fuel.

(iii) HC – 3000 ppm or 25 gm/kg of fuel.

(1 ml/m³ = 1 ppm)

In gasolene engines other sources of unburned hydrocarbons (UBHC) are:

(a) Piston blow by gases. (Exhaust gases and vapours of fuel which leaks into the crank case are called blow by gases).

(b) Fuel evaporation from carburettor and fuel tank. These are called as non-exhaust sources.

31.3.1 Sources of Pollution from Gasolene Engine

The four possible sources of atmospheric pollution from a petrol engine are as follows.

(i) The fuel tank
(ii) The carburettor

Together they are called as evaporative losses and their contribution is 15 to 20% of HC.

(iii) The crank case blow by—20 to 35% of HC.

(iv) Exhaust pipe—50 to 60% of HC and almost all CO and NO_x

31.3.2 Evaporative Losses

These losses are the direct losses of raw gasolene from the engine fuel system. These emissions amount to 15 to 20% of total HC emission and two main sources are (i) fuel tank, (ii) the carburettor.

(i) Fuel Tank Losses. While the petrol is being filled into the tank these losses occur. The vapours in the fuel tank are released to the atmosphere through breather vent. The tank temperature increases if the vehicle is parked in bright sun light or due to vehicle operation.

Parameters affecting the fuel tank losses are : (a) volatility of fuel (b) location, capacity and design of fuel tank. (c) Amount of fuel in the tank (d) ambient temperature (e) Mode of vehicle operation etc.

(ii) Carburettor Losses. These losses are because of two things (a) float chamber being vented to atmosphere in order to relieve the internal pressure as the carburettor gets heated up and (b) "hot soak" losses which occur when the engine is stopped after the vehicle is run for some time. The loss from the carburettor takes place due to boiling of fuel in it. During hot soak the carburettor temperature rises from 16°C to 48° C above the ambient.

(iii) Crank Case Blow By. The leakage of flue gases past the piston and piston rings from the cylinder to the crank case, is called as blow by. The blow by loss is about 18 to 20% of total HC loss from the engine. This loss may go upto 32% if the piston rings get worn out.

During compression and combustion, the increased cylinder pressure forces some of the gas, in the cylinder, in to crevices viz. volume between piston, rings and cylinder wall. Most of this gas is unburned fuel air mixture and much of the gas escapes into the atmosphere as unburned HC, because these crevices are too narrow for the flame to enter.

The top land clearance and position of the top ring greatly affects the blow by losses, because some of the quenched gas is re-cycled in the combustion chamber. The nearness of this un-burned charge to spark plug, flame speed, local temperature etc. decide the ability to burn, otherwise it goes as UBHC.

(iv) Exhaust Pipe. Mixture of various hydrocarbons is present in petrol fuel. Thus if the combustion is perfect, then exhaust emissions will have CO_2, water vapours and the air that did not take part in combustion process. However because of numerous reasons, the combustion is not complete and hence CO and UBHCs are also present in exhaust gases. The smog (the mixture of smoke and gas) is formed mainly because of presence of HCs in air.

Apart from CO and HCs, exhaust also contains the third main pollutant NO_x. The air in the combustion chamber used for combustion contains 78% of N_2. At low temperature N_2 is not so active but at about 1100 to 1150° C, the reactivity of N_2 with O_2 is very high and various oxides of nitrogen, collectively called as NO_x are formed.

In addition to the pollutants mentioned above, the exhaust also contains organic compounds namely ketones, aldehydes which form smog. To increase the antiknock quality of petrol, it is added with TEL (Tetra- Ethyl-Iead). Thus because of TEL the exhaust emission also contains poisonous lead compounds.

Organic and inorganic compounds of higher molecular weights and lead compounds are exhausted in the form of very small size particles of the order of 0.02 to 0.06 μ.

Table 31.1 shows effect of engine operating conditions on exhaust (car):

Table 31.1

Description	*Engine Operation*			
	Idle	*Acceleration*	*Cruising*	*Deceleration*
A/F ratio	11–12.5	11–13	13–15	11–12.5
Engine speed (rpm)	400–500	3000–4000	2500–3500	3500–500
Exhaust emissions				
(i) CO%	4–5.5	0–5.8	0.9–3.8	2.1–4.8
(ii) NOx ppm.	10–55	1000–3800	980–2800	10–53
(iii) HC ppm	450–1000	55–550	450–280	3850–1900
(iv) Unburned fuel as % of fuel supplied	3.5 to 5.5	1.5–4.5	1.5–4.5	19–59

31.4 DIESEL EMISSION

In diesel engine, emissions can be classified in the same categories as for the gasoline engines. In diesel exhaust concentrations of NO_x are comparable to those from SI engines. Diesel hydrocarbon emissions are significant though exhaust concentrations are lower by about a factor of 5 than typical SI engine levels. Smoke, odour and particulate emission are the other important factors of concern, which cause air pollution. Hydrocarbons in diesel exhaust may also condense to form white smoke during engine starting and warm-up. The particulate emissions are of the order of 0.2 to 0.5% of the fuel mass and diameter of the particles is as small as 0.1 μm. However diesel engines are not a significant source of CO.

Gasolene and diesel contain sulphur. In gasolene it is in small amounts $\leq$ 600 ppm by weight and in diesel fuel $\leq$0.5 p.c.

Sulphur is oxidised to produce SO_2 of which a fraction can be oxidised to SO_3 which combines with water to form a sulphuric acid-aerosol.

Table 31.2 shows approximately the possible variations in concentration of different constituents of diesel exhaust.

Table 31.2. *Range of Concentrations of Different Constituents of Diesel Exhaust*

Sr. No.	*Constituent*	*Range*
1.	HC	a few – 1000 ppm
2.	NO_x	100 – 2000 ppm
3.	CO	0 – 2 P.C.

(i) Engine type and mode of operation are two main factors which influence the diesel exhaust emissions. Emission levels of different engines at full load and rated speed are given in following Table 31.3.

Table 31.3

Sr. No.	*Emission*	*Engine*			
		Two stroke air scavenged	*Four stroke turbo charged*	*Four stroke normally aspirated*	
				Medium speed	*High speed*
1.	CO%	0.34	0.03	0.26	0.14
2.	CO_2%	5.31	7.65	7.18	7.78
3.	UBHC (ppm)	540	250	370	1000
4.	NO_x (ppm)	1240	2550	800	790
5.	Smoke (Hartridge units)	10	8	60	90
6.	Odour (D.I. units)	3.5	3.4	3.3	3.5

Thus with reference to the above table the conclusions can be drawn as under:

1. The two stroke air scavenged engine produces high UBHC and intermediate NO_x emissions. The smoke level remains low.
2. The four stroke medium speed normally aspirated engine has lowest emissions in all categories except for a very high smoke intensity.
3. The four stroke high speed normally aspirated engine has high HC. The odour intensity is also high.
4. The turbocharged four stroke engine is notably low in UBHC and high in NO_x. Generally very low smoke levels are recorded.

(ii) Effect of mode of operation on engine exhaust. The various modes of operation like idle, full load at rated speed and acceleration at full rack and its effect on emission levels in diesel exhaust are shown in Table 31.4.

During idle mode concentration of HC, NO_x emissions are lower than other modes. Emissions at idle are less significant than during any other mode. Smoke and odour are highest during acceleration. Emissions at full load relative to emissions at other modes vary significantly with engine type. Two stroke and 4 stroke turbocharged engines smoke lightly at load, while 4 stroke normally aspirated engines smoke very much at rated full load.

(iii) NO_x in diesel exhaust. The quantity of NO_x varies from few hundred to well over 1000 ppm. The highest local peak temperature and presence of sufficient O_2, causes highest NO_x concentration in diesel exhaust.

A precombustion engine gives rise to less NO_x level than a direct injection (DI) engine. For high fuel-air ratio the additional fuel tends to cool the charge, so the localised peak temperatures are lowered resulting in drop in NO_x level.

The NO_x concentration is also significantly affected by injection system and time. Also the fuel characteristics such as Cetane Number (CN), viscosity, rate of burning etc. all contribute to differences in NO_x levels obtained from different levels.

Table 31.4

Sr. No.	*Engine Exhaust Constituent*	*Con. Values of Exhaust Emission*			
		Idle	*Acceleration*	*Full load*	*Part load*
	2-stroke cycle				
1.	HC (ppm)	250	500	550	350
2.	NO_x (ppm)	180	1200	1250	1.00
3.	Smoke (Hartridge unit)	4	44	10	4
4.	Odour (D.I)	3.6	4.1	3.5	3.0
5.	CO %	0.01	0.24	0.34	0.01
6.	CO_2 %	0.83	5.42	5.29	3.79
	4-stroke cycle normally aspirated				
1.	HC (ppm)	180	330	150	210
2.	NO_x (ppm)	330	920	780	590
3.	Smoke	4	44	10	4
4.	Odour	3.6	4.1	3.5	3.4
5.	CO %	0.02	0.08	0.26	0.04
6.	CO_2 %	2.56	3.40	6.68	5.33

31.5 DIESEL SMOKE

The visible products of combustion is called as smoke. It originates early in the combustion cycle in a localised volume of rich fuel-air (F/A) mixture. If the fuel is burnt in a volume where F/A ratio is greater than 1.5, soot is produced. The amount of soot formed depends upon type of fuel, pressure and local F/A ratio. The soot formed if it finds sufficient O_2, it will burn completely and will become visible. The size of soot particles affects the appearance of smoke. The diesel engine smoke is basically of two types.

(a) Blue white smoke and (b) Black smoke.

(a) Blue white smoke. While engine is started from cold, this smoke is caused by liquid droplets of lubricating oil or fuel oil, due to lower surrounding temperature and the intermediate products of combustion do not burn. If the piston rings are worn out and when the lubricating oil leaks past the rings, it gives rise to this type of smoke.

(b) Black smoke. It is nothing but carbon particles suspended in the exhaust gas. It largely depends upon air-fuel ratio and increases rapidly as load is increased and available air is depleted.

31.6 DIESEL ODOUR

Till date no equipment is developed to measure the intensity of odour. But it is a general practice to have a human panel, who will decide the intensity of odour by comparison—which is an inaccurate method.

Note that aldehydes give a pungent odour for diesel engines, which also causes irritation to nose and eyes.

31.7 CONTROL OF EXHAUST EMISSIONS

The main approaches to minimise exhaust emissions are:

1. Modifications in the engine design and operating variables.
2. Treatment of exhaust gases.

1. Modifications in the Engine Design. The following modifications may help in cleaner exhaust—

(a) Use of lean mixtures and maximum spark retard compatible with good power output and drivability. It is a well known fact that burning leaner mixtures greatly reduces HC and CO emissions refer Fig. 31.1. This reduction will result only if there is good mixture formation and distribution.

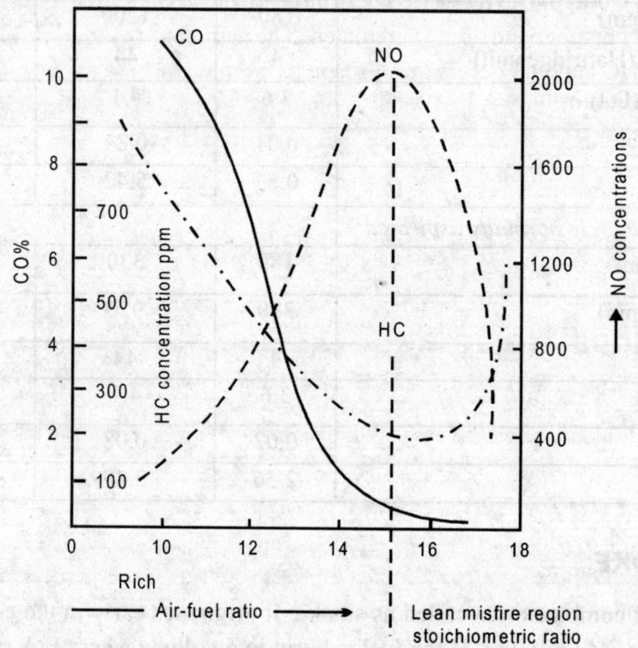

Fig. 31.1. *Effect of A/F ratio.*

One of the methods of producing a uniform mixture and refined fuel metering is the adoption of triple venturi carburettor.

(b) Use of minimum valve over-lap.

(c) Pretreatment of the mixture to improve vaporisation and mixing of fuel with air. To achieve this narrow venturies are used, to produce higher air speeds and better fuel automisation. The exhaust heat can be used to pre-heat the mixture at part loads. Also to minimise the "Crevice volume" between topland and bore and to block the leakage path through the gas into the crank case, a system known as "Sealed ring orifice system" is adopted.

2. Exhaust Treatment Devices. Here the basic technique is to promote oxidation of HC and CO after emission from the engine. Exhaust oxidation devices fall into two categories:

(i) Promotion of after burning of the pollutants by exhaust heat conservation, introduction of supplimentary, air and providing sufficient volume to ensure adequate reaction time.

(ii) Use of catalytic converters.

Catalytic convertors depend on the action of a catalyst containing certain exotic chemicals to convert HC and CO emissions to their oxidised products.

31.8 CONTROL OF HC – EMISSIONS

(i) Blow by Control. Here the blowby vapours are re-circulated back to the intake manifold via air cleaner or inlet of the carburettor. This system is called as positive crank case ventilation (PCV). Thus the blow by gases are consequently re-introduced into the combustion chamber where they are burned along with fresh in coming air and fuel.

(ii) Evaporation Loss Control Device (ELCD). In this case the device collects the evaporative emissions, they are absorbed in an absorbent chamber and are re-circulated. The activated charcoal or foamed polyurethene is used as an adsorbent to hold the HC vapours before they escape to the atmosphere. The two main sources of HC emissions viz. fuel tank and the carburettor bowl, are directly connected to the adsorbent chamber when the engine is turned off. i.e. under hot soak. Apart from it the diurnal cycle loss from the tank and carburettor is also taken care of. Diurnal cycle is the daily cyclic variation in the temperature which causes tank breathing or forcing the gasolene out of the tank.

Adsorbent bed when saturated, is relieved of the vapours, by purging action *i.e.* by purging the air from air cleaner to the intake manifold via the adsorbent bed.

31.9 OTHER METHODS—TO CONTROL EXHAUST EMISSIONS

(a) Petrol Injection. This system provides the flexibility to meet the engine requirements and eliminates the problem of mixture distribution completely and hence the engine will have low levels and exhaust emissions and better fuel economy. Thus HC emissions are controlled by making use of optimum lean A/F ratios at all operating conditions. NO_x level is also controlled as maximum temperature reached in the combustion chamber is low.

(b) Stratified Charge Engine. It operates on very lean air-gasolene mixtures, depending on a localised rich mixture region near the spark plug to initiate combustion. The emission levels of all three pollutants CO, HC and NOx are very much lower with this engine at part throttle due to the leanness of the mixture. Mixture as lean as 30 : 1 can be successfully used.

(c) Proper Maintenance of an Engine. The pollution from a poorly maintained engine is more. Because of mis-fire the entire A/F charge gets exhausted without combustion. If the air cleaner is not periodically cleaned, then its element gets choked and can reduce A/F ratio, generally giving rise to increased emissions of HC and CO. Similar problem is encountered if the choke gets automatically sticked.

(d) Fuel Variation. Methane and propane could burn at leaner ratios than gasoline at a given throttle. Also at leaner mixtures, lower throttle position can be used. Thus methane and propane will produce much lower CO as their A/F mix is lean i.e. it contains more O_2.

Hence if we change from the gasoline to propane as an engine fuel, CO emissions can be substantially reduced with reduced HC and NO_x and if we change propane by methane the CO as well as HC touch zero emission level and only NO_x remains as a significant factor. From point of view of emissions, these fuels are attractive but because of technological progress, we are unable to use these fuels.

The various methods like Catalytic converters, Exhaust gas Recirculation, Passive regeneration, Trap systems are used to control pollution. And Catalytic Converters, EGR systems are discussed in detail.

31.10 CATALYTIC CONVERTER

In order to control the air pollution from IC engines (or automobiles with diesel or gasolene fuel), the necessity arises in after treatment of exhaust gases. Thus it is required to reduce or oxidise HC, CO, NO_x to H_2O, CO_2 and N_2. Figure 31.2 shows the anatomy of a catalytic converter, it starts with a ceramic or a metallic monolithic substrate.

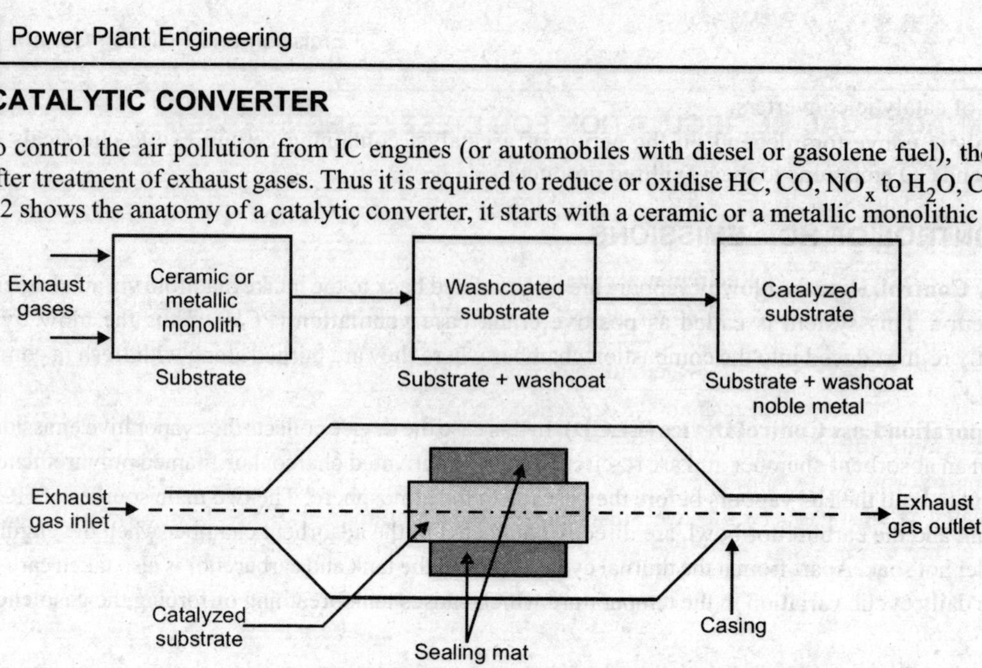

Fig. 31.2. *Diesel catalyst schematic.*

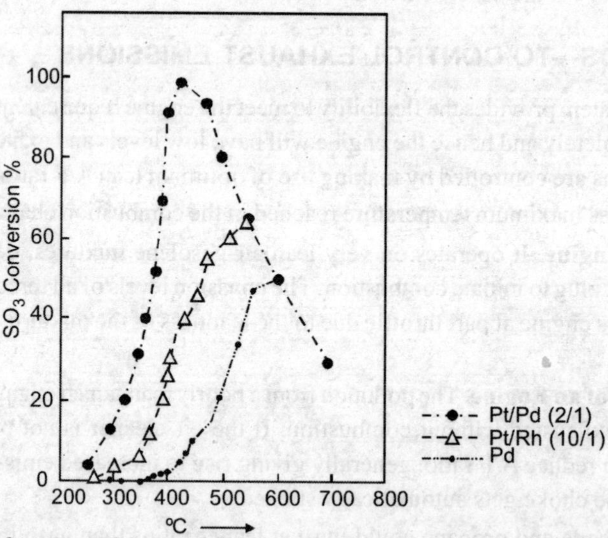

Fig. 31.3. *Performance of three catalyst conversion to SO_3 relative to exhaust temperature.*

This substrate is coated with a wash coat which serves as a primer for the purpose of holding noble metal (Pt, Rh or Pd) and that noble metal molecules dispersed to allow maximum surface exposure to the exhaust gases.

After the washcoat substrate is catalysed, it is canned with a heat-expanding sealing mat between the catalysed substrate and its can. The substrate material is selected to provide adequate mechanical strength, surface area and low flow resistance. Careful consideration has to be given to the type and amount of wash coat used. Alumina (Al-oxide) is very popular because of its surface area and other favourable physical properties. However silica (silicon oxide) oxides are also used as wash coat. The container design, lub-oil additives, fuel sulphur level, application and particulate composition are other important design parameters. The performance of three catalyst viz. Pt, Pd and Rh to convert to SO_3 relative to exhaust temperature is also shown in Fig. 31.3.

31.11 EXHAUST GAS RECIRCULATION FOR DIESELS (EGR SYSTEM)

The most accepted technology for NO_x reduction is exhaust gas recirculation. It is commonly being used in light duty applications but is not accepted in heavy duty diesel engines because of its negative impact on engine wear. EGR is defined by following formula.

$$EGR\% = \frac{\text{Recirculated exhaust}}{\text{Total air charge}} \times 100$$

$$= \frac{\text{Recirculated exhaust}}{\text{Fresh air + Recirculated exhaust}} \times 100$$

EGR is schematically as shown in Fig. 31.4. In this method the peak cylinder combustion temperature is reduced by circulating a portion of exhaust gas thereby reducing the quantity of O_2 required for combustion. Nevertheless the re-circulated exhaust gas serves as heat sink. The exhaust gas from the exhaust manifold goes through an EGR control valve, then a cooler and finally into intake manifold by way of air-to-air intercooler or a bypass system.

With 10% EGR, there is substantial reduction in NOx level. The EGR inter cooler is used in order to reduce the temperature of exhaust gases so that the temperature in the combustion chamber does not reach its peak value and there by reactivity of N_2 with O_2 is reduced.

However the disadvantage of the system is that there is increase in smoke as well as particulates.

Figure 31.5 (a) and (b) shows the effect of EGR on NO_x and smoke level v/s load.

31.12 EFFECT OF AIR POLLUTION ON HUMAN AND PLANT LIFE

The greatest environmental evils is polluted air. The air we breathe has not only life saving but also life damaging properties. Under ideal conditions the air we inhale has a quantitative and qualitative balance that maintains the well being of a human life.

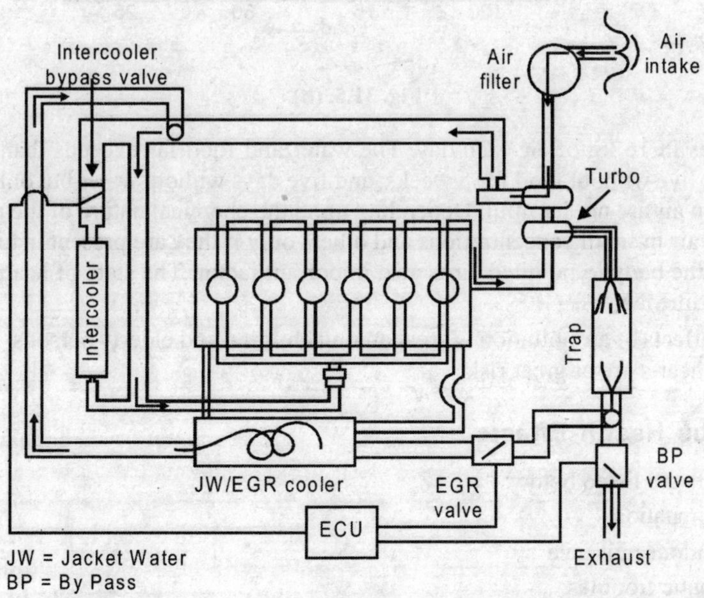

Fig. 31.4. *EGR plus trap systems with Intercooler bypass.*

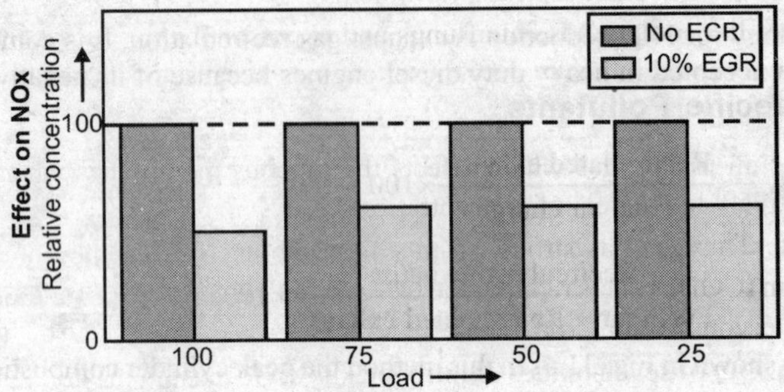

Fig. 31.5. (a)

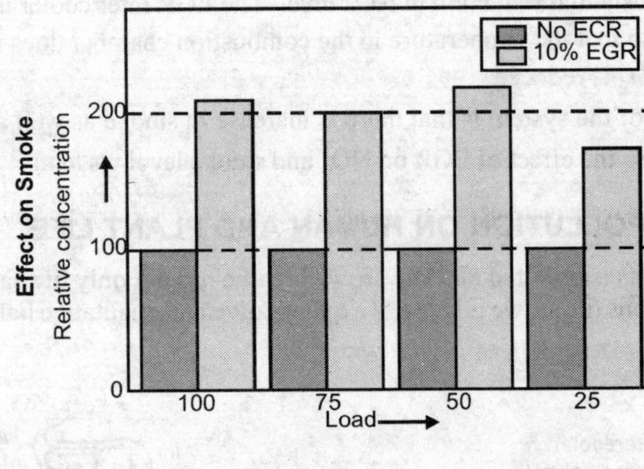

Fig. 31.5. (b)

An average man takes in 16 kg of air each day. The water and food far exceeds than air intake. It has been estimated that a man can live without food for 5 weeks, and five days without water but only for 5 minutes without air. All the constituents in air are not harmful. Depending upon the chemical nature of the pollutants some may be harmful when present in air in small concentrations and others only if they are present in high concentrations. The duration of exposure of the body to polluted air is also important factor. The state of health of the receptor along with his age group are prime factors.

Susceptibility to the effects of air pollution is great among children and elderly persons. Also those with chronic diseases of the lungs or hearts are at great risk.

31.12.1 The Various Health Effects

The various health effects are listed below:

(1) Eye, nose, throat irritation

(2) H_2S, NH_3 cause odour nuisance

(3) Increase of asthmatic troubles

(4) CO combines with haemoglobin and creates problems to those suffering from cardiovascular and pulmonary diseases

(5) Particulate matter, dust etc. create respiratory problems

(6) Lead and its some of the compounds enter lungs and is poisonous.

31.12.2 Effect of Specific Pollutants

(i) **Sulphur di-oxide.** It is an irritant gas which affects the mucous membranes when inhaled. It causes severe bronchospasms at relatively low levels of concentration.

(ii) **Carbon Monoxide.** This gas has strong affinity to combine with haemoglobin of the blood and forms carboxyhaemoglobin (COHb). This reduces the O_2 carrying capacity of haemoglobin to the body tissues. CO affects the central nervous system, may cause heart attacks.

(iii) **Oxides of Nitrogen.** Out of seven oxides of nitrogen known to exist in the ambient air, only two namely NO and NO_2 are thought to affect human health.
Nitrogen dioxide causes occupational disease *e.g.* people working in a manufacturing unit of nitric acid, exposures of farmers to high nitrate fertilisation, electric arc welding and in mining where nitrogen compounds are used as explosives. The eye and nasal irritation is observed after exposure to about 15 ppm of NO_2 and pulmonary discomfort after brief exposure to 25 ppm of NO_2.

(iv) **Hydrogen Sulphide.** It is a foul smelling gas. Exposure to this gas for short periods causes fatigue of the sense of smell.

(v) **Ozone.** This gas has an irritant action in the respiratory treat reaching much deeper into the lungs than the oxides of sulphur.

(vi) **Lead.** The main source of lead in urban atmosphere is the automobile. It causes gastro-intestinal damage, liver and kidney damage, abnormalities in fertility and pregnancy.

More or less, for animals also they cause the same diseases as for human beings.

31.12.3 Pollutants Affecting Plants

They are as under:

(i) SO_2 (ii) O_3

(iii) Nitrogen oxides (NO, NO_2, etc.) (iv) Smog

(v) NH_3 (vi) H_2S

(vii) Herbicides (Weed killers) (viii) Ethylene

The above pollutants interfere with plant growth and the phenomenon of photosynthesis. Smog, dust etc. reduce the amount of light reaching the leaf and also clog the stomata, may reduce CO_2 intake to one extent thus interfere with photosynthesis.

Effects of pollutants on the plants are:

(i) SO_2, causes Interveinal chlorotic bleaching of the leave.

(ii) Ozone suppresses the growth of leaves.

(iii) NO_x suppresses growth and bleaching of leaves etc.

31.13 POLLUTION MEASURING EQUIPMENTS AND CONTEMPORARY POLLUTION NORMS

I. For Petrol Engines (2-stroke and 4-stroke engines) – Idling CO-meter

Keeping the engine running at idle conditions, when the probe is inserted into the exhaust gases i.e. in the silencer, it analyses and displays the constituents *i.e.* it shows the % of CO and HC ppm directly. Note that these are harmful emissions and hence these are to be kept under controlled limits.

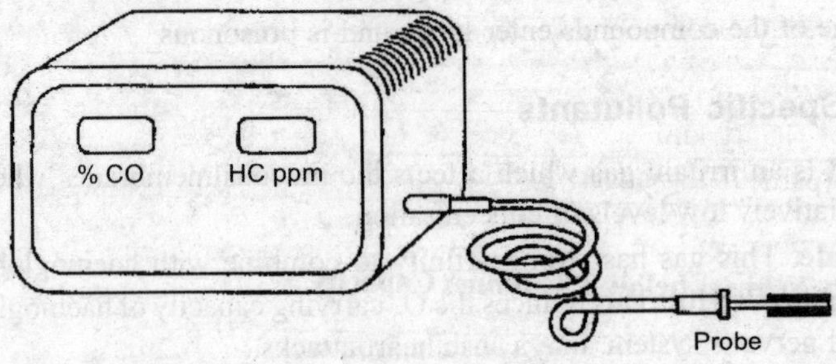

Fig. 31.6. *Exhaust gas analyser unit.*

Pollution Under Control (PUC) contemporary norms are as follows:

"For petrol vehicles vide rule 115 of Central Motor Vehicle (CMV) Rules 1989 emission norms are as follows:

(i) Idling CO emission limit for all four wheeler petrol driven vehicle shall not exceed 3% by volume.

(ii) For 2 and 3 wheeled petrol driven vehicle shall not exceed 4.5% by volume.

II. For Diesel Engines—Hartridge Smoke-meter (Exhaust Gas Analyser)

Figure 31.7 shows the exhaust gas analyser, when the probe is inserted into the exhaust gases, it analyses and displays smoke density directly.

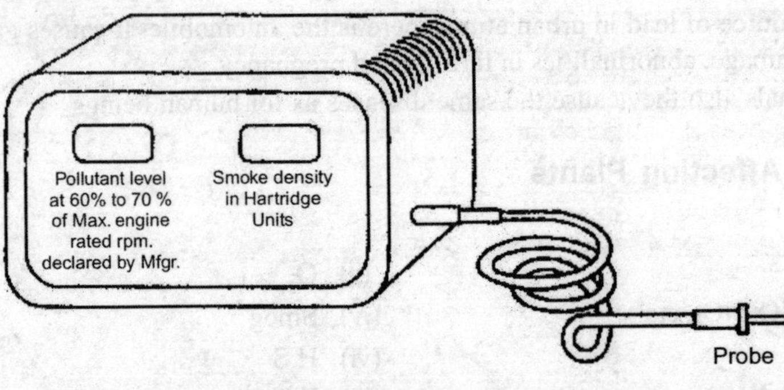

Fig. 31.7. *Exhaust gas analyser unit.*

Pollution Under Control contemporary norms are as follows:

"For diesel vehicles vide rule 115 of CMV Rules 1989 is as follows:

Smoke density on full load at 60% to 70% of Maximum engine rated RPM declared by manufacturer : 75 Hartridge Units OR Free acceleration: 65 Hartridge Units."

31.14 PROPOSED EMISSION NORMS FOR THE YEAR 2000, EURO— I, II, III AND IV

I. Petrol Driven Passenger Car

	India 2000/Euro I	*Euro II*	*Euro III*	*Euro IV*
CO (g/km)	2.72	2.2	1.3	1.0
HC (g/km)	–	–	0.2	0.1
NO_x (g/km)	–		0.15	0.08
HC + NO_x	0.97	0.57		

II. Diesel Passenger Cars

	India 2000	*Euro I*	*Euro II*	*Euro III*
CO (g/km)	2.72 to 6.9	2.2	1.3	1.0
HC (g/km) + NO_x (g/km)	0.97 to 1.7	0.97	0.7	0.56
PM (g/km) (particulate matter)	0.14 to 0.25	0.14 to 0.19	0.08	0.05

III. Diesel Light Duty Vehicles below 3.5 Tonnes Capacity

	India 2000	*Euro I*	*Euro II*
CO (g/km)	2.75 to 6.9	2.72 to 6.9	1.5
HC (g/km) + NO_x (g/km)	0.97 to 1.7	0.97 to 1.7	0.7 to 1.3
PM (g/km)	0.14 to 0.25	0.14 to 0.7	–

IV. Diesel Heavy Duty Vehicles above 3.5 Tonnes Capacity

	India 2000	*Euro I*	*Euro II*	*Euro III*
CO (g/kwh)	4.5	4.5	4.0	2.1
HC (g/kwh)	1.1	1.1	1.1	0.66
NO_x (g/kwh)	8.0	8.0	7.0	5.0
PM (g/kwh)	0.36 – 0.61	0.36 to 0.61	0.15	0.1

Index

R

S

T

U

V

W

Z